Sleep

Physiology, Investigations,
and Medicine

Sleep
Physiology, Investigations, and Medicine

Edited by

Michel Billiard
Hôpital Gui de Chauliac
Montpellier, France

Translated by Angela Kent

Kluwer Academic / Plenum Publishers
New York, Boston, Dordrecht, London, Moscow

Library of Congress Cataloging-in-Publication Data

Sommeil normal et pathologique. English
    Sleep: physiology, investigations, and medicine/edited by Michel Billiard; translated by
Angela Kent.
        p.   ; cm.
    Includes bibliographical references and index.
    ISBN 0-306-47406-9
        1. Sleep disorders.   2. Sleep.   I. Billiard, M. (Michel)   II. Title.
    [DNLM: 1. Sleep Disorders—pathology.   2. Sleep—physiology.   WM 188 D697 2003a]
    RC547.S65713 2003
    616.8′498—dc21

                                                                                    2002043453

Published with the help of the French *Ministère de la Culture—Centre national du livre*

Original French edition: *Le sommeil normal et pathologique: Troubles du sommeil et de l'éveil* © Masson, Paris, 1994, 1998.

ISBN: 0-306-47406-9

©2003 Kluwer Academic/Plenum Publishers, New York
233 Spring Street, New York, New York 10013

http://www.wkap.com

10   9   8   7   6   5   4   3   2   1

This book is dedicated to my wife Annick
and to my children Béatrice and Patrick

Contributors

Joëlle Adrien, INSERM U 288, Faculté de Médecine Pitié Salpêtrière, 91 Boulevard de l'Hôpital, 75013 Paris, France

Michael S. Aldrich (deceased), Sleep Disorders Center, Department of Neurology, UH 8820, Box 0117, 1500 East Medical Center Drive, Ann Arbor, MI 48109-0117, USA

Alain Autret, Clinique Neurologique, Hôpital Bretonneau, 2 Boulevard Tonnelé, 37044 Tours Cedex, France

Michel Averous, Service d'Urologie, Hôpital Lapeyronie, 371 Avenue Doyen Gaston Giraud, 34295 Montpellier Cedex 5, France

Michel Baldy-Moulinier, Service d'Explorations Fonctionnelles et Epileptologie. Hôpital Gui de Chauliac, 80 Avenue Augustin Fliche, 34295 Montpellier Cedex 5, France

Domien Beersma, RUG, Zoological Laboratory. PO Box 14, 9750 AA Haren, The Netherlands

Alain Besset, INSERM, EMI-9930, Hôpital la Colombière, 39 Avenue Charles Flahaut, 34293 Montpellier, Cedex 5 and Service de Neurologie B, Hôpital Gui de Chauliac, 80 Avenue Augustin Fliche, 34295 Montpellier, Cedex 5, France

Michel Billiard, Service de Neurologie B, Hôpital Gui de Chauliac, 80 Avenue Augustin Fliche, 34295 Montpellier Cedex 5, France

Donald L. Bliwise, Sleep Disorders Center, Emory University Medical School, WMB – Suite 6000, Atlanta, GA 30322, USA

Diane Boivin, Center for Study and Treatment of Circadian Rhythms, Douglas Hospital Research Centre, Department of Psychiatry, McGill University, 6875 La Salle Boulevard, suite F-1127, Verdun, Québec, H4H 1R3, Canada

Urbain Calvet, Clinique Saint Jean du Languedoc, 20 Route de Revel, 31077 Toulouse Cedex, France

Bertrand Carlander, Service de Neurologie B, Hôpital Gui de Chauliac, 80 Avenue Augustin Fliche, 34295 Montpellier, Cedex 5, France

Julie Carrier, Sleep Disorders Center, Emory University Medical School, WMB-Suite 6000, Atlanta, GA 30322, USA

Ari Chaouat, Service de Pneumologie, Hôpital de Hautepierre, 67098 Strasbourg Cedex, France

Marie-Josephe Challamel, Service d'Explorations Neurologiques, Centre Hospitalier Lyon Sud, Bâtiment 3B, 69495 Pierre Bénite Cedex, France

Catherine Charpentier, Service de Pneumologie, Hôpital de Hautepierre, 67098 Strasbourg Cedex, France

Bruno Claustrat, Service de Radiopharmacie et de Radioanalyse, Hôpital Neuro-Cardiologique, B.P. Lyon-Monchat 69394, Lyon Cedex 03, France

Giorgio Coccagna, Istituto di Clinica Neurologica, Università degli Studi di Bologna, Via Ugo Foscolo, 7, 40123 Bologna, Italia

Emmanuelle Corruble, Service de Psychiatrie, Centre Hospitalier Universitaire Bicêtre, 78 rue du Général Leclerc, 94275 Le Kremin Bicêtre, France

Louis Crampette, Service ORL et Chirurgie Cervico-Faciale, Hôpital Gui de Chauliac, 80 Avenue Augustin Fliche, 34295 Montpellier Cedex 5, France

Natali Darchia, I. Beritashvili Institute of Physiology, Department of Neurobiology of Sleep-Wakefulness Cycle, 14 Gotua Street, 380060 Tbilisi, Georgia

Yves Dauvilliers, Service de Neurologie B, Hôpital Gui de Chauliac, 80 Avenue Augustin Fliche, 34295 Montpellier Cedex 5, France

Joseph De Koninck, School of Psychology, University of Ottawa, PO Box 450, Ottawa, Ontario, K1N 6X9, Canada

Philippe Ducrotté, Service de Gastro-Entérologie, Hôpital Charles Nicolle, 76031 Rouen Cedex, France

Luigi Ferini-Strambi, Centro del Sonno, IRCCS H San Raffaele, Via Stamira d'Ancona 20, 20127 Milano, Italia

Patricia Franco, Clinique Pédiatrique, Hôpital Universitaire des Enfants Reine Fabiola, Université Libre de Bruxelles, Avenue J.J. Crocq 15, B-1020 Bruxelles, Belgique

Philippe Gajdos, Service de Réanimation Médicale et d'Explorations Fonctionnelles, Hôpital Raymond Poincaré, 92380 Garches, France

Lucile Garma, Fédération des Pathologies du Sommeil. Hôpital Pitié-Salpêtrière, 47-83 Boulevard de l'Hôpital, 75651 Paris Cedex 13, France

Françoise Goldenberg, Laboratoire de Physiologie et d'Explorations Fonctionnelles, Hôpital Henri-Mondor, 51 Avenue du Maréchal de Lattre de Tassigny, 94010 Créteil Cedex, France

Jean Grenier, Centre National de Formation en Santé. Hôpital Montfort, 713, Chemin Montréal, Ottawa, Ontario, K1K 0T2, Canada

José Grosswasser, Clinique Pédiatrique, Hôpital Universitaire des Enfants Reine Fabiola, Université Libre de Bruxelles, Avenue J.J. Crocq 15, B-1020 Bruxelles, Belgique

Christian Guilleminault, Sleep Disorders Clinic, Stanford University Medical Center, 701 Welch Road, Suite 2226, Palo-Alto, CA 94304, USA

Irma Gvilia, I. Beritashvili Institute of Physiology, Department of Neurobiology of Sleep-Wakefulness Cycle, 14 Gotua Street, 380060, Tbilissi, Georgia

Edouard Hirsch, Service de Neurologie, Neuropsychologie et Explorations Foncrionnelles des Epilepsies, Hopitaux Universitaires, 1 Place de l'Hôpital, BP 426, 67091 Strasbourg Cedex, France

André Kahn, Clinique Pédiatrique, Hôpital Universitaire des Enfants Reine Fabiola, Université Libre de Bruxelles, Avenue J.J.Crocq 15, B-1020 Bruxelles, Belgique

Myriam Kerkhofs, Laboratoire de Sommeil, Centre Hospitalier Univesitaire de Charleroi, Hôpital A. Vesale, Rue de Gozée 706, 6110 Montigny-le-Tilleul, Belgique

Romain Kessler, Service de Pneumologie, Hôpital de Hautepierre, 67098 Strasbourg Cedex, France

Martina Kreutzer, Sleep Disorders Clinic, Stanford University Medical Center, 701 Welch Road, Suite 2226, Palo-Alto, CA 94304, USA

Jean Krieger, Service d'Explorations Fonctionnelles du Système Nerveux et de Pathologie du Sommeil, Hôpitaux Universitaires, 1 Place de l'Hôpital, BP 426, 67091 Strasbourg Cedex, France

Odile Lapierre, Centre d'Etude du Sommeil, Département de Psychiatrie, Hôpital du Sacré-Cœur, 5400 Boulevard Gouin Ouest, Montreal, Quebec H4J 1C5, Canada

Damien Leger, Centre de Sommeil, Service de Physiologie, Hotel-Dieu de Paris, 1 Place du Parvis Notre-dame, 75181 Paris Cedex 04, France

Frédéric Lofaso, Service de Réanimation Médicale et Explorations Fonctionnelles. Hôpital Raymond Poincaré, 92380 Garches, France

Elio Lugaresi, Istituto di Clinica Neurologica, Università degli Studi di Bologna, Via Ugo Foscolo 7, 40123 Bologna, Italia

Christian Marescaux, Service de Neurologie, Neuropsychologie et Explorations Fonctionnelles des Epilepsies, Hôpitaux Universitaires, 1 Place de l'Hôpital, BP 426, 67091 Strasbourg Cedex, France

Bruno Maton, Service de Neurologie, Neuropsychologie et Explorations Fonctionnelles des Epilepsies, Hôpitaux Universitaires, 1 Place de l'Hôpital, BP 426, 67091 Strasbourg Cedex, France

Emmanuel Mignot, Stanford University Center for Narcolepsy, 701 Welch Road B, basement, room 145, Palo-Alto CA 94304-5742, USA

Harvey Moldofsky, Sleep Disoders Clinic of the Centre for Sleep and Chronobiology Ltd, 340 College Street, Suite 580, Toronto, Ontario, M5T 3A9, Canada

Pasquale Montagna, Istituto di Clinica Neurologica, Università degli Studi di Bologna, Via Ugo Foscolo 7, 40123 Bologna, Italia

Jaime M. Monti, Departemento de Farmacologia y Terapeutica, Hospital de Clinicas, 2833/602 J. Zudanez, 11300 Montevideo, Uruguay

Jacques Montplaisir, Centre d'Etude du Sommeil, Département de Psychiatrie, Hôpital du Sacré-Cœur, 5400 Boulevard Gouin Ouest, Montreal, Quebec, H4J 1C5, Canada

Charles Morin, Ecole de Psychologie, Université Laval, Sainte-Foy, Québec, G1X 4V4, Canada.

Gassam Mroue, Service de Réanimation Médicale et Explorations Fonctionnelles. Hôpital Raymond Poincaré, 92380 Garches, France

Yvonne Navelet, Explorations Fonctionnelles du Système Nerveux, Centre Hospitalier Universitaire Bicêtre, 78 rue du Général Leclerc, 94275 Le Kremlin Bicêtre Cedex, France

Alain Nicolas, Unité de Sommeil, Unité Clinique de Psychiatrie Biologique, Hôpital du Vinatier, 95 Boulevard Pinel, 69677 Bron Cedex, France

Thomas Penzel, Schlafmedizinisches Labor. Klinik für Innere Medizin, Klinikum der Philipps-Universität, Baldingerstrasse 1, D-35033 Marburg, Deutschland

Jörg H. Peter, Schlafmedizinisches Labor. Klinik für Innere Medizin, Klinikum der Philipps-Universität, Baldingerstrasse 1, D-35033 Marburg, Deutschland

Rosa Peraita Adrados, Unidad de Sueno, Hospital Gregorio Maranon, C/Dr. Esquerdo, 46. 28007, Madrid, Espana

Maria-Antonia Quera-Salva, Unité de Sommeil, Hôpital Raymond Poincaré, 104 Boulevard Raymond Poincaré, 92380 Garches, France

Jean-Claude Raphael, Service de Réanimation Médicale et Explorations Fonctionnelles, Hôpital Raymond Poincarré, 92380 Garches, France.

Elisabeth Rebuffat, Clinique Pédiatrique, Hôpital Universitaire Reine Fabiola, Université Libre de Bruxelles, Avenue J.J.Crocq 15, B-1020 Bruxelles, Belgique

Dominique Samson-Dollfus, Médecine Néonatale et Pédiatrique. Pavillon de la Mère et de l'Enfant. Hôpital Charles Nicolle. 76031 Rouen Cedex, France

Joanthan B. Santo, Center for Study and Treatment of Circadian Rhythms, Douglas Hospital Research Center, Department of Psychiatry, McGuil University, 6875 La Salle Boulevard, suite F-1127, Verdun, Quebec, H4H 1R3, Canada.

François Sellal, Service de Neurologie, Neuropsychologie et Explorations Fonctionnelles des Epilepsies, Hôpitaux Universitaires, 1 Place de l'Hôpital, B.P. 426, 67091 Strasbourg Cedex, France

Cesa Scaglione, Istituto di Clinica Neurologica, Universita degli Studi di Bologna, Via U. Foscolo 7, 40123 Bologna, Italia

Salvatore Smirne, Centro del Sonno, IRCCS H San Raffaele, Via Stamira d'Ancona 20, 20127 Milano, Italia

Martine Sottiaux, Clinique Pédiatrique, Hôpital Universitaire Reine Fabiola, Université Libre de Bruxelles, Avenue J.J.Crocq15, B-1020 Bruxelles, Belgique

Claudine Soubrie, Service de Pharmacologie, Hôpital Pitié-Salpêtrière, 47-83 Boulevard de l'Hôpital, 75651 Paris Cedex 13, France

Mehdi Tafti, HUG, Belle-Idée, Division de Neuropsychiatrie 2, Chemin du Petit-Bel-Air, CH-1225 Chêne-Bourg, Suisse

Anne Thibault-Menard, Service de Neurologie, Neuropsychologie et Explorations Fonctionnelles des Epilepsies, Hôpitaux Universitaires, 1 Place de l'Hôpital, B.P. 426, 67091 Strasbourg Cedex, France

Michel Tiberge, Service d'Explorations Fonctionnelles du Système Nerveux, Hôpital Rangueil, 1 Avenue du Professeur Jean Poulhes, 31403 Toulouse Cedex 4, France

Marie-Françoise Vecchierini, Service Explorations Fonctionnelles, Centre Hospitalier Universitaire Bichat, 46 rue Henri Huchard, 75877 Paris Cedex 18, France

Dominique Warot, Service de Pharmacologie, Hôpital Pitié- Salpêtrière, 47-83 Boulevard de l'Hôpital, 75651 Paris Cedex 13, France

Emmanuel Weitzenblum, Service de Pneumologie, Hôpital de Hautepierre, 67098 Strasbourg, France

Foreword

The question about the function of sleep remains one of the major challenges scientists are faced with. Wherein lies the fascination with sleep? I am convinced that it is the necessity for sleep. No one has failed to experience the overpowering urge to fall asleep after a disturbed night's sleep or after sleep was curtailed or deprived, especially when our daily activities impose restrictions on motor activity. The demand of our body and brain to sleep challenges our understanding of why this is the case, and which are the benefits of a night of profound sleep. Also in animals prolongation of waking consistently increases their attempts to fall asleep. It has been stated that sleep is more necessary to animals than even food! The need for sleep and some insight into the consequences of the preceding daily waking activities on subsequent sleep was wonderfully formulated by Shakespeare in Othello:

Not poppy nor mandragora,
Nor all the drowsy syrups of the world,
Shall ever medicine thee to that sweet sleep
Which thou owed'st yesterday

It is interesting that the most powerful single intervention which invariably influences sleep in a positive and predictable manner is the prolongation of waking. The activities which people or animals engage in during the wakefulness episode are secondary in the magnitude of their effects on sleep. More recently, the proposition by Krueger and Obál that sleep is use-dependent and is initiated at the local neuronal level, perhaps serving to stimulate the use of structures insufficiently activated during wakefulness, renewed the interest in addressing the influence of waking activities on sleep. Their hypothesis is testable, and a new impetus is driving experiments. It has turned out that when brains of human subjects or animals are regionally activated by specific sensory or motor stimulation, the significant changes during sleep are predominant in the delta range, and are related to the stimulated region. Thus, it may be that the EEG is locally enhanced as a consequence of intense wake-time stimulation. Different parts of the brain seem to be able to sleep with differing intensities! The results stemming from the surge of brain imaging studies complement endeavours to illustrate and unravel the regional activities in the sleeping brain and relate them to waking behaviour. Changes in the waking EEG have been shown to correlate with an increase in sleepiness in the course of extended waking and, surprisingly, they correlate with the subsequent changes in sleep. Thus, markers of sleep propensity in the waking EEG are related to the sleep EEG.

The question I am often confronted with: do fish sleep? - or more recently, do flies really sleep? - how can you tell?, reminds us of the uncertainty of defining sleep. It also reflects the increasing awareness of sleep, its evolution and the lack of knowledge on its function. Despite a long tradition of sleep research in animals, it has only been recently recognised that simple models can advance our understanding of the essence of sleep. Recent publications have demonstrated the presence of sleep and sleep regulation in *Drosophila*, and experimental manipulations revealed similarities in response to sleep deprivation and specific drugs such as caffeine in the flies compared to humans and rodents. The inclusion of flies in the repertoire of species considered to sleep is advancing our knowledge on this mysterious third of our lives. After many years of major efforts by the sleep community, sleep is gradually finding its way into the curricula of biology, medicine and psychology. A surge of publications in high-impact journals is addressing issues such as the relationship between sleep, learning and memory, and psychomotor skills, the genetic basis of sleep disorders, the relationship between circadian rhythms and sleep, e.g. the advanced sleep phase syndrome. These problems are attracting the attention of other scientists to our field. A new generation of young scientists is tackling the questions central to the understanding of sleep and its disorders. With the availability and mastery of new and finer methodologies, our knowledge about the physiology of sleep will speedily progress.

Shakespeare was also a master in describing the pains of insomnia:

O sleep, O gentle sleep!
Nature's soft nurse, how have I frightened thee,
That though no more wilt weigh my eyelids down
And steep my senses in forgetfulness?

                                Henry IV

The increasing awareness and understanding of sleep disorders and their therapies has raised the interest of practitioners in topics such as sleep apnea, diagnostics of insomnia, issues related to prescription and use of hypnotics, changes of sleep with aging and the interrelation between the circadian and sleep-wake dependent drive for sleep. The impact of sleep disorders on health and well being is receiving increasing recognition. A new area of sleep research has started with the advances in molecular biology, the neurosciences and biological rhythms. The impetus in sleep research gained by the availability of these methods has been complemented with advances in sleep disorders medicine. The new tools have led to exciting progress in the neurochemical basis and the genetics of sleep and sleep disorders. The investigation of specific pathologies of sleep, is serving to shed light also on basic physiological mechanisms underlying sleep regulation. A typical example is narcolepsy: In narcoleptic patients the homeostatic process of sleep regulation is unaffected, whereas especially through the discovery of the changes in the hypocretin-/orexin system in narcoleptic humans, dogs and mice, the understanding of the role of this system in sleep and wakefulness is advancing. A more specific drug therapy for narcolepsy may ensue, and a new era of hypnotics acting more closely at the physiological level than the ones presently available may ensue. Also the progress in genetics is giving new impetus to enable the understanding of sleep disorders including narcolepsy, restless legs syndrome, and recently sleep-walking!

Michel Billiard has honoured me by asking me to write the foreword to this English edition of the book he first compiled in 1994. It was published first in 1994, and revised in 1998 but was accessible only in French. It has now been updated and translated into English. The present version not only includes an update of many of the original chapters, which were characterised by their large diversity, but several new topics have been included.

The book is aimed towards a readership interested in sleep medicine and sleep disorders, but will also be of interest to scientists and medical doctors from other fields. Readers will obtain information on a diversity of clinically relevant diagnostic and therapeutic aspects of sleep and its disorders.

This is a timely textbook containing interesting and informative chapters which will appeal to sleep specialists and technicians as well as to graduate students.

Irene Tobler, PhD
President of the European Sleep Research Society
Professor of Zoology
University of Zurich
Zurich, Switzerland

Preface

This is the first edition in English of the book *Le Sommeil Normal et Pathologique* published in French by Masson (Paris) in 1994, and re-edited in 1998. More than a simple translation of the 1998 edition, this constitutes a third edition, having benefited widely from the experience of the first two. It was made possible as a result of a grant for translation allocated by the French Ministry of Culture, to whom we would like to express our thanks.

The form of presentation has been maintained as five sections: physiology of sleep, diagnostic procedures, disorders of sleep and wakefulness, medical disorders associated with sleep or worsened during sleep, and sleep as a special circumstance in investigating some medical disorders. However, the chapters are presented independently rather than as sub chapters within the five main sections, as in the previous editions. Many chapters have been entirely rewritten, by their authors or by new authors, and those which were not rewritten have been updated.

Decision trees for insomnia, hypersomnia and circadian rhythm sleep disorders have been added, not as an attempt to comply with current trends but rather to help doctors in their diagnostic approach and in the appropriate use of the various tests and questionnaires available.

The annexes have been enhanced with new questionnaires and their score charts, to provide the clinician with the necessary tools for exploring the principal disorders.

In view of how little time is dedicated to the study of sleep, its physiology and pathology in most medical and paramedical schools, the editor's intention is to provide medical students, doctors and other health professionals, with a tool of reference enabling them to focalise their knowledge.

I am deeply grateful to all the European, North and Latin American authors who agreed to share in this endeavour by sending me their texts; Professor Irene Tobler, President of the European Sleep Research Society, who kindly agreed to write the Foreword to the present volume; Mrs Gail Markham, Foreign Rights Manager at Masson who engaged in the lengthy administrative procedure required for the book to be translated into English; and last but not least, Mrs Angela Kent, translator of the present volume, who has produced a work of very high quality and with whom it has been a true pleasure to collaborate.

And finally to all those associated with the publication of this book, the doctors, researchers, psychologists, nurses, French and foreign students, at the Sleep Clinic at Montpellier Teaching Hospital, without whose help and stimulation this edition would neither have been possible nor brought to a successful conclusion.

Michel Billiard M.D.

Contents

PART 1: PHYSIOLOGY OF SLEEP

1. Normal sleep   3
   C. Guilleminault and M. L. Kreutzer
2. Normal sleep in children   11
   D. Samson-Dollfus
3. Neurobiology of the sleep-wake cycle   31
   J. Adrien
4. The major physiological functions during sleep   45
   Y. Dauvilliers
5. Models of human sleep regulation   61
   D. G. M. Beersma
6. Melatonin and sleep-wake rhythm   71
   B. Claustrat
7. Molecular genetics, circadian rhythms and sleep   83
   Y. Dauvilliers, M. Tafti, and E. Mignot
8. Mental activities during sleep   113
   J. De Koninck

PART 2: DIAGNOSTIC PROCEDURES

9. Polysomnography   127
   A. Besset
10. Ambulatory systems   139
    T. Penzel and J. H. Peter
11. Guidelines for visual sleep analysis   151
    A. Besset
12. Automatic sleep analysis   159
    A. Besset
13. Investigation of vigilance and sleepiness   169
    A. Besset

PART 3: DISORDERS OF SLEEP AND WAKEFULNESS

A) INSOMNIAS

14. Insomnia: Introduction   187
    N. Darchia and I. Gvilia
15. A decision tree approach to the differential diagnosis of insomnia   191
    J. Grenier
16. Transient and short term insomnia   201
    R. Peraita Adrados

17. Primary insomnia   207
     C. M. Morin
18. Insomnia associated with medical disorders   221
     M. Billiard
19. Insomnia associated with psychiatric disorders   227
     L. Garma
20. Insomnia linked to medications   247
     E. Corruble, D. Warot, and Cl. Soubrie
21. Benzodiazepines and new non-benzodiazepine agents   257
     J. M. Monti
22. Other medications used for insomnia   269
     F. Goldenberg
23. Insomnia in children: Clinical aspects and treatment   283
     Y. Navelet
24. Sleep and circadian rhythms in normal aging   297
     J. Carrier and D. Bliwise

B) HYPERSOMNIAS

25. Hypersomnias: Introduction   333
     M. Billiard
26. A decision tree approach to the differential diagnosis of hypersomnia   337
     M. Billiard
27. Insufficient sleep syndrome   341
     M. S. Aldrich
28. Medication and alcohol dependent sleepiness   347
     D. Warot and E. Corruble
29. Obstructive sleep apnoea-hypopnea syndrome and upper airway resistance syndrome   357
     J. Krieger
30. Surgical and prosthetic treatment for sleep apnoea syndrome and upper airway
     resistance syndrome   391
     L. Crampette
31. Narcolepsy   403
     M. Billiard and Y. Dauvilliers
32. Idiopathic hypersomnia   429
     M. Billiard and A. Besset
33. Recurrent hypersomnias   437
     M. Billiard
34. Other hypersomnias   447
     M. Billiard and B. Carlander
35. Hypersomnia in children   457
     M. J. Challamel

C) CIRCADIAN RHYTHM SLEEP DISORDERS

36. Circadian rhythm sleep disorders: Introduction   469
     M. Billiard
37. A decision tree approach to the differential diagnosis of a circadian rhythm sleep
     disorders   471
     M. Billiard
38. Shift work sleep disorder   473
     D. Leger
39. Time zone change (jet lag) syndrome   485
     M. Tiberge
40. Circadian rhythm sleep disorders related to an abnormal escape of the sleep-wake cycle   493
     D. Boivin and J. Santo

D) PARASOMNIAS

41. Parasomnias 513
    M. F. Vecchierini
42. Enuresis 545
    M. Averous

PART 4: MEDICAL DISORDERS ASSOCIATED WITH SLEEP OR WORSENED DURING SLEEP

43. Snoring 555
    L. Crampette
44. Nocturnal hypoxemia in chronic obstructive pulmonary disease 567
    E. Weitzenblum, A. Chaouat, C. Charpentier, R. Kessler, and J. Krieger
45. Central alveolar hypoventilation syndrome 577
    E. Weitzenblum
46. Sleep breathing abnormalities in neuromuscular diseases 581
    M. A. Quera Salva G. Mroue, Ph. Gajdos, J.C Raphael, and F. Lofaso
47. Cardiocirculatory disorders and sleep 589
    G. Coccagna and C. Scaglione
48. Restless legs syndrome in wakefulness and periodic leg movements in sleep 599
    J. Montplaisir, A. Nicolas, and O. Lapierre
49. Abnormal postures and movements during sleep 609
    E. Hirsch, B. Maton, F. Sellal, and C. Marescaux
50. Night epilepsies 617
    M. Baldy-Moulinier
51. Sleep-related headaches 629
    B. Carlander
52. Fatal familial insomnia 635
    E. Lugaresi and P. Montagna
53. Sleep and the gastrointestinal tract 641
    P. Ducrotté
54. Sleep and sudden infant death syndrome 649
    A. Kahn, J. Groswasser, M. Sottiaux, E. Rebuffat, and P. Franco
55. Sleep related painful erections 657
    U. Calvet

PART 5: SLEEP AS A SPECIAL CIRCUMSTANCE IN INVESTIGATING SOME MEDICAL DISORDERS

56. Mood disorders and sleep 665
    M. Kerkhofs
57. Sleep and lesions in the central nervous system 673
    A. Autret
58. Sleep as a tool for investigating epilepsies 689
    M. Baldy-Moulinier
59. Dysautonomias 697
    L. Ferini-Strambi and S. Smirne
60. Fibromyalgia and chronic fatigue syndrome: the role of sleep disturbances 703
    M. Moldofsky

ANNEXES

Questionnaires and scales 717
  1. General questionnaire on sleep 718
     Morning Questionnaire 718
  2. Questionnaires used in the evaluation of insomnia 719

       Sleep Impairment Index   719
       Beliefs And Attitudes About Sleep Scale   722
       The Pittsburgh Sleep Quality Index   725
       The Leeds Sleep Evaluation Questionnaire   730
       The Beck Depression Inventory   732
   3. Scales used in the evaluation of hypersomnia   736
       The Stanford Sleepiness Scale   736
       The Karolinska Sleepiness Scale   738
       The Epworth Sleepiness Scale   739
   4. Questionnaire used in the evaluation of circadian rhythm sleep disorder   740
       The Horne and Östberg Questionnaire   740

Glossary   745

Index   751

PART 1

PHYSIOLOGY OF SLEEP

Chapter 1

Normal sleep

C. Guilleminault and M. L. Kreutzer
Stanford University, Sleep Disorders Clinic, Stanford, USA

Why do we sleep? This simple question is still unanswered, despite the fact that sleep has fascinated man for centuries. Sleep is part of the life cycle. It is a state of consciousness that we experience daily. It is cherished and sought by man, mirrored by the number of plants, amulets, spells, curses, and pills used in relation with its occurrence, independent of civilization or culture. It is a comfortable state, an escape from the daily hardships of life, a mysterious state where "gods may communicate with man". Its disturbance causes grief, brings unhappiness, or announces doom. It is part of our 24-hour life-cycle; it is part of the circadian rhythm.

The quest for the understanding of sleep has taken many faces, and today involves molecular research as well. However, the "sleep gene" remains elusive, despite the fact that many genes involved in the timing of sleep and in some of the events that are part of what constitutes "sleep" have been identified. The "early to bed-early to rise" individual may now blame this personal trait on a long gone ancestor who passed down these genetic characteristics. The search of the 21st century will be to identify the cascade of genes responsible for the occurrence and stereotypic organization of this behaviour. The recognition of genes and gene products involved will allow better understanding and help alleviate the many dysfunctions that have been identified during our time. But this endeavour must begin with well-defined phenotypes and appropriate techniques to recognise and classify abnormalities.

Historically, the definition of human sleep has been based on studies of brain activity as well as on changes that may be seen in vital functions when brain activity is modified. In the early part of the 20th century MacWilliams [9] had already noted that the heartbeat of dogs was markedly different during some segments of sleep, and that these heart beat irregularities reoccurred at regular intervals during the sleep cycle. Conceivably, our nosology might contain a classification of "sleep with irregular heartbeats" instead of "Rapid Eye Movement (REM) sleep" had these observations been pursued further. Understanding the changes in autonomic and vital function associated with the different sleep states is the foundation of Sleep Medicine. Even though the limitations of many sleep characterising definitions arising from the 1960's are recognised today, these definitions are still widely used. They were published in an international manual edited by Drs. Allan Rechtschaffen and Anthony Kales [12]. The rationale behind the perseverance of a classification system that is widely recognised as archaic and poorly adapted to pathological sleep scoring, computerised analysis and automatic scoring, is related to several factors. These include fear of eliminating a system that has guided many advances, absence of consensus for an approved new international system, lack of creativity from many researchers, and finally because the *old* system covers some fundamental states recognised by all. The classic framework for the study of human sleep resulted from the description of REM sleep contrasted by non REM sleep between 1953 and 1955 from the Department of Physiology at the University of Chicago by three individuals: mentor Dr. Nathaniel Kleitman and his two students, Eugene Aserinski and William C Dement.

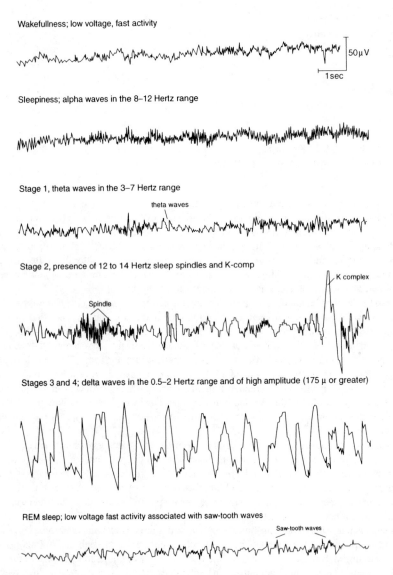

Wakefullness; low voltage, fast activity

Sleepiness; alpha waves in the 8–12 Hertz range

Stage 1, theta waves in the 3–7 Hertz range

theta waves

Stage 2, presence of 12 to 14 Hertz sleep spindles and K-comp

K complex

Spindle

Stages 3 and 4; delta waves in the 0.5–2 Hertz range and of high amplitude (175 μ or greater)

REM sleep; low voltage fast activity associated with saw-tooth waves

Saw-tooth waves

Figure 1.1 Sample electroencephalographic tracings obtained during relaxed wakefulness (eyes open and eyes closed), NREM sleep stages 1 to 4, and REM sleep (from top to bottom)

NORMAL SLEEP

Sleep consists of two different states, Rapid Eye Movement (REM) and Non Rapid Eye Movement (NREM) sleep. REM sleep is defined by the presence of an EEG pattern of low voltage, fast EEG associated with occurrence of rapid eye movements, occurring isolated or in bursts, as well as postural relaxation, i.e. muscle atonia. The muscle atonia is interrupted by bursts of muscle tone leading, at times, to body jerks. These jerks and bursts of eye movements are called *phasic events* occurring on a background of tonic muscle inhibition, called *tonic REM sleep*. Electrophysiologic studies performed on cats have shown that the *phasic events* are associated with bursts of waves simultaneously recorded in the pons, lateral geniculate and occipital lobe [6, 8]. These waves called "ponto-geniculo-occipital (P.G.O.) spikes" or waves separate *phasic* from *tonic*

REM sleep. The term REM sleep has a number of synonyms such as "desynchronised (D) sleep", "dream sleep", "paradoxical sleep" and in infants, "active sleep".

NREM sleep, also called "synchronised sleep", and in infants "quiet sleep" has been subdivided into four sleep stages. Stage 1 is seen at sleep onset and is defined by low voltage mixed frequency (2 to 7 Hertz) waves, in the absence of rapid eye movements and with preserved muscle tone. Vertex sharp waves may be seen, and slow eye movements are often present. Stage 2 is scored when 12 to 14 Hertz sleep spindles and/or K complexes are present against a background activity of relatively low voltage mixed EEG frequencies. Stage 3 is scored when a moderate amount (20 to 50%) of high amplitude (75 microvolts or greater) slow wave (0.5-3.5 Hertz) EEG activity is seen. Stage 4 is defined by the presence of a predominance (greater than 50%) of high amplitude slow wave activity. Stages 3 and 4 combined are often called Slow Wave Sleep (SWS) (fig.1.1).

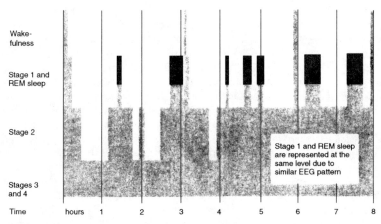

Figure 1.2. Diagram showing the distribution of sleep stages and states in a young, 30 year old subject. Note the predominance of stages 3-4 at the start of the night, whereas the REM sleep state is more abundantly distributed in the second half of the night. This pattern alters with age, showing a lower percentage of stages 3-4 in elderly subjects. This so-called slow wave sleep may be completely absent after the age of 60.

A healthy young adult has the simplest pattern of sleep, regardless of gender. Nocturnal sleep is associated with a pattern of regular reoccurrence of "sleep cycles". Individuals first enter NREM sleep with stage 1 NREM sleep lasting only 1 to 7 minutes. This stage may be interrupted by wakefulness. Stage 2 NREM sleep follows this short transition from wakefulness and continues for 10 to 25 minutes. The gradual appearance of high voltage slow waves signals the development of stage 3 followed by stage 4 lasting between 20 and 40 minutes. A fleeting switch to stage 2 may precede occurrence of REM sleep. The first REM period is of short duration, between 4 and 8 minutes. REM sleep often ends with a brief body movement, and a new sleep cycle begins. NREM and REM sleep continue to alternate throughout the night in a cyclical fashion. As night progresses each sleep cycle differs slightly from the preceding one. The average length of the first sleep cycle is approximately 90 minutes and 100 to 120 minutes from the second to the fourth cycle, which is usually the last one. The last sleep cycle is typically the longest. The organization of NREM and REM sleep also changes with each cycle. There is a predominance of SWS during the first two sleep cycles and REM sleep in the last two cycles. SWS is rare to nonexistent in the last cycle, while REM sleep represents the longest portion (fig. 1.2). These sleep patterns vary with age. The development of sleep is initiated in utero. In the neonatal period, REM sleep generally represents more than 50% of total sleep time. This percentage declines rapidly to around 25% during the prepubertal period and then stabilizes around 20% until old age [13]. SWS is typically of greatest duration during childhood and declines progressively in adulthood with low levels reached around 50 to 55 years of age [3]. There may be a sex difference in this decline, as men reveal a more notable decline [4].

The investigation of sleep, sleep stages and related pathology lead to the development of a standard procedure called *polysomnography,* which is based on the monitoring of sleep states and stages as well as biological variables that vary in relationship with the occurrence of the different sleep states.

POLYSOMNOGRAPHY

Technique

Polysomnography is a term coined to describe the monitoring of many biologic variables during sleep. The value of this monitoring became apparent as the control of vital functions was found to be different during the 3 states of wakefulness, NREM and REM sleep. Most of these functions are regulated by the autonomic nervous system (ANS). Sympathetic and parasympathetic activities are highly correlated to the state of alertness, and the balance between these activities is state dependent. Though this method of recording is different from basic electroencephalographic (EEG) monitoring, it uses the 10-20 international electrode placement system for EEG recording. The EEG electrodes used are C_3, C_4, occipital leads (O_1, O_2), and it is recommended to add a frontal electrode, most commonly F_z. These electrodes refer to the opposite sides of the head (A_1, A_2). Occipital leads may be a bipolar montage: referred to each other. Electrodes recording the electro-oculogram (EOG) are applied to the skin at the outer canthus (or lateral temporal corner) of both eyes. The electromyogram (EMG) of the chin muscles is monitored with two electrodes placed under the chin. The minimum requirements for identification of sleep states and stages are 3 channels of EEG, 2 of EOG, and 1 of EMG. An electrocardiographic (ECG) channel is mandatory. It is usually the derivation of V_2. In a standard montage, leg muscles and respiration are monitored. The electrode placement for leg EMG is on the skin above right and left anterior tibialis muscles. Respiration may be monitored with different methods. The most common montage comprises a nasal cannula/pressure transducer system, a mouth thermistor, a neck microphone, a thoracic and an abdominal band, and a pulse-oximeter, all of which is commercially available. Depending on the goal of the test, CO_2 and other variables can be measured as well. The most common additions are esophageal pressure (Pes) measurement, systematically done in some laboratories, esophageal pH, skin impedance, pulse-transit-time (PTT) and additional EEG leads. The signals obtained from these sensors are amplified by nonspecific amplifiers (i.e. Grass amplifiers) or generic amplifiers of the computerised sleep system on which the signals are monitored and scored. There are many commercially available sleep systems. A minimum of 16 channels must be available and 32 channels are commonly available. However, for economic reasons, many sleep laboratories monitor two beds simultaneously on their 32-channel system. Sleep monitoring must include video monitoring of the sleeping patient, which can be synchronised with the polysomnographic recording. This is especially important for evaluation of nocturnal movement disorders. The frequency at which signals are scanned is an important variable. Usually the minimum is 128 Hertz/channel. However, this will be too low if one wants to perform specific analyses such as Fast Fourier Transformation (FFT) of the ECG, which may allow high and low frequencies of the ECG to indicate sympathetic and parasympathetic activities during sleep. A sampling frequency of 500 Hertz is the minimum required for appropriate and meaningful data collection. Most of the variables collected during sleep are qualitative or semi quantitative at best. For example, thermistors, used to collect airflow, measure only temperature and have a very large margin of error (up to 100%). Nasal cannula/pressure transducer systems are semi-quantitative sensors, but non-linear in the low range. The "flow limitation" is still a visual and subjective interpretation with commercially available systems, despite the fact that this type of airflow monitoring is more sensitive to nasal flow limitation than thermistors. The combined nasal/ oral cannula systems commercially available are not satisfactory, and a mouth thermistor gives better results [11]. End tidal CO_2 measurement as a signal of abnormal breathing results in many more errors than the combined nasal cannula/pressure transducer system and mouth thermistor to date [17].

Polysomnography can be performed using portable equipment. The lower the number of signals monitored, the lower the amount of information that will be collected, and the greater the chance of overlooking pathology. Portable equipment has the advantage of allowing monitoring in hospital beds and at home. The disadvantages are related to loss of signals and the need for repeat studies. There is a correlation between the number of sensors and their invasiveness and the amount of sleep disturbance induced. A one-night polysomnography is not sufficient to thoroughly study sleep. To evaluate the duration of sleep, or effects of drugs on sleep, monitoring should be obtained during several successive nights to lessen the disturbances related to the "first night effect [10] as well as those related to equipment invasiveness.

Scoring of the polysomnogram

Sleep is generally scored in 20 or 30-second segments or *epochs*. The scoring is performed using two international scoring systems: firstly, Rechtschaffen and Kales international manual [12]. The manual gives specific instruction on how to score sleep states and stages. It is based on the 20 or 30 second *epoch*. This scoring system was developed to score the sleep of normal subjects and not of patients. Secondly, the American Sleep Disorders Association Atlas Task Force [16]. This scoring system was developed to complement the international Atlas and to be more helpful when evaluating pathology. It examines smaller epochs (3 seconds and longer) and abnormal events. Both atlases are used simultaneously to study pathology.

Recognising the normal pattern of sleep EEG in the presence of pathology is often difficult and leads to poor inter-scorer agreement in sleep scoring, at times as low as 70%. This observation was an incentive to identify different scoring methods. Furthermore, the realisation that the above sleep scoring system limited the ability to recognise changes occurring during sleep has motivated the development of new scoring definitions and approaches. The American Academy of Sleep Medicine (AASM) and its predecessor the American Sleep Disorders Association (ASDA) [16], indicated that sleep scored visually by analysis of a brain wave pattern lasting 20 or 30 seconds was limiting: a multiplicity of changes in brain activity may occur during that time. The resolution of the human eye to recognise changes in EEG background is insufficient. Computer analysis is superior to the human eye. Despite many drawbacks, spectral analysis of EEG obtained in well defined EEG leads, was one of the first computerised approaches to scoring sleep. However, the eye can recognise more events of short duration. After experimental investigation a task force defined short EEG arousals based on a 3 second analysis of the sleep recording. However, this scoring is just a refinement of the classic sleep stage scoring using a reduced scoring window: It examines smaller epochs (3 seconds and longer) [16] and abnormal events. Both atlases are used simultaneously to study pathology.

Cyclic Alternating Pattern (CAP)

The CAPs scoring system is based on a completely different view of sleep, integrating a more dynamic concept [14]. CAP is defined as a periodic EEG activity of Non-REM sleep. It is characterised by sequences of transient electro-cortical events that are distinct from EEG background and reoccur at up to 1-minute intervals. The concept of CAPs is based on the observation that there are phases of sleep instability contrasting with phases of great stability. This observation is related to neuro-physiological observations made during the studies of sleep in mammals using deep electrodes. Delta, theta, alpha, and beta spindle EEG rhythms are considered uninterrupted tonic bioelectrical cerebral oscillations. They are subdivided into bandwidths and constitute the EEG background that varies according to neurophysiological states (wakefulness, Non REM, and REM sleep). These rhythms though, may be interrupted by periodic activities representing electro-cortical events reoccurring at regular intervals of a few seconds. These interruptions are abrupt shifts of the amplitude or frequency of the EEG.

The periodic activity is characterised by 3 factors: 1) A repetitive EEG element called phase A. 2) An interval with only background EEG that separates the repetitive elements - called phase B. 3) The period or cycle formed by *phase A* and *phase B* which indicates the recurrence rate. CAP is a specific type of periodic activity in which both phase A and phase B range between 2 and 60 seconds. A cycle (called *C*) is formed by a phase A and a following phase B. Phase A is identified by the presence of transient events that include: Delta bursts, vertex sharp transients, polyphasic bursts, K-complexes with or without sleep-spindles, K-alpha, or bursts of alpha EEG arousals.

Changes in EEG amplitude are critical for scoring CAPs. Consistent with the definition, phasic activity initiating a phase A must be 1/3 higher than the background voltage calculated during the 2 seconds before the onset and 2 seconds after the offset of a phase A. Onset and termination of a phase A are established on the basis of an amplitude/ frequency concordance in the majority of the EEG leads: CAP, being a global phenomenon, involves extensive cortical areas. Therefore, phases A should be visible on many EEG leads and bipolar derivations involving frontal, central, parietal and occipital leads are recommended to guarantee a favourable detection of the phenomenon. The monopolar derivations (C3/A2, C4/ A1) are mostly indicated if scoring with a limited number of EEG leads, as indicated by the AASM, though there is a loss of information with the reduction of the number of EEG leads used to define sleep. All EEG events that do not clearly meet phase A characteristics cannot be scored as a phase A. The minimal duration of a phase A or phase B is 2 seconds. If the time between two events is less than 2 seconds, they are combined into one phase. CAPs appear throughout sleep stages 1,2,3 and 4. Phase A morphology has been subclassified into 3 subtypes. These are based on the reciprocal proportion of high voltage low waves (EEG synchrony) and low amplitude fast rhythms (EEG desynchrony) throughout the entire phase A duration.

The advantage of CAPs is to recognise transition from a *stable* state (wake or well-established Non REM sleep) to another state. During this transition a state of *instability* is present. If this state is too long then sleep is perceived as poor and is associated with complaints. CAPs are a better indicator of numerous forms of pathology, such as complaints related to insomnia or periodic movements during sleep. A CAP atlas has been recently published to facilitate scoring sleep with this approach [15].

Computers have been utilised to look at phenomena that are impossible to see with the unassisted eye. Computer-based analyses trying to simulate Rechtschaffen and Kales Atlas [12] have had poor results. Nevertheless, analysis based on approaches more suitable to computers has been helpful. This includes quantitative power spectral analysis of sleep EEG with Fast Fourier Transform (FFT) used to identify specific EEG frequency bands. The EEG during sleep has been described as a stochastic process characterised by spectral properties. In recent years sleep EEG has also been analysed with methods derived from non-linear dynamics. The rationale behind these approaches is numerous: the brain, awake or asleep is a known highly non-linear system, and there are indications that the complex sleep EEG dynamic pattern could be explained by non-linear equations. The Grassberger-Procaccia algorithm [5] for the determination of the correlation dimension to time series of EEG segments recorded during sleep has been used [1, 2]. Additionally, the sleep EEG has been viewed as the result of a low-dimensional chaotic process. However, specific technical factors have to be fulfilled to obtain a reliable estimate of the fractal dimension, particularly stationarity of the time series and a large number of data points. Usage of non-linear dynamics and instability theory has been also implemented for the understanding of spatio-temporal pattern formation. Procedures applying principle component analysis (using the Karhunen-Loewe decomposition technique [7] to the multichannel sleep EEG time series have been tried. This technique showed localized changes of cortical functioning. Kim *et al.* [7] have shown that, secondary to sleep deprivation, there were increased coherence and bilateral hemispheric involvement. This methodology may illuminate changes in local dynamics in brain activity during sleep.

CONCLUSION

Sleep is a complex behaviour. Great effort has been exerted to better understand the cortical functioning of the sleepy brain. Ongoing work will bring a progressively better understanding of normal and pathological sleep.

REFERENCES

1. BABLOYANTZ A., DESTEXHE A. - Low dimensional chaos in an instance of epilepsy. *Proc. Natl. Acad. Sci. USA*, 83, 3513-3517, 1986.
2. BABLOYANTZ A., SALAZAR J.M., NICOLIS G. – Evidence of chaotic dynamics of brain activity during the sleep cycle. *Phys. Lett.* A 111, 152-156, 1985.

3. FEINBERG I., FLOYD T.C. – Systematic trends across the night in human sleep cycles. *Psychophysiology,* 16, 283-291, 1979.
4. FUKUDA N., HONMA H., KOHSAKA M., KOBAYASHI R., SAKAKIBARA S., KOHSAKA S., KOYAMA T. – Gender difference of slow wave sleep in middle aged and elderly subjects. *Psychiatry Clin. Neurosci.,* 53, 151-153, 1999.
5. GRASSBERGER P., PROCACCIA I. – Measuring the strangeness of strange attractors. *Physica D,* 9, 189-208, 1983.
6. JOUVET M. – The role of monoamines and acetylcholine-containing neurons in the regulation of the sleep-wake cycle. *Ergebn. Physiol.,* 64, 166-307, 1972.
7. KIM H., GUILLEMINAULT C., HONG S., KIM D., KIM S., GO H., LEE S. – Pattern analysis of sleep deprived human EEG. *J. Sleep Res.* 10, 193-201, 2001.
8. LAURENT J.P., CESPUGLIO R., JOUVET M. – Délimitation des voies ascendantes de l'activité ponto-géniculo-occipitale chez le chat. *Brain Res.* 65, 29-52, 1974.
9. MACWILLIAMS J.A. – Blood pressure and heart rate action in sleep and dreams: their regulation to hemorrage, angina and sudden death. *BMJ,* 2, 1198-2003, 1923.
10. MOSKO S.S., DICKEL M.J., ASHURST J. – Night-to–night variability in sleep apnea and sleep related periodic leg movements in elderly. *Sleep,* 11, 340-348, 1988.
11. NORMAN R.G., AHMED M.M., WALSLEBEN J.A., RAPOPORT D.M. – Detection of respiratory events during NPSG: Nasal cannula/pressure sensor versus thermistor. *Sleep,* 20, 1175-1184, 1997.
12. RECHTSCHAFFEN A., KALES A. – *A manual of standardized terminology, techniques and scoring system for sleep stages of human subjects.* U.S. Department of Health, Education, and Welfare, Public Health Service-National Institutes of Health, National Institute of Neurological Diseases and Blindness, Neurological Information Network, Bethesda, MD, 1968.
13. ROFFWARG H.P., MUZIO J.N., DEMENT W.C. – Ontogenetic development of the human sleep-dream cycle. *Science,* 152, 604-619, 1966.
14. TERZANO M.G., PARRINO L., SPAGGIARI M.C. – The cyclic alternating pattern sequences in the dynamic organization of sleep. *Electroenceph. Clin. Neurophysiol.* 69, 437-447, 1988.
15. TERZANO M.G., PARRINO L., CHERVIN R., CHOKROVERTY S., GUILLEMINAULT C., HIRSHKOWITZ M., MAHOWALD M., MODOLFSKY H., ROSA A., THOMAS R., WALTERS A. – Atlas, rules and recording techniques for the scoring of cycle alternating pattern (CAP) in human sleep. *Sleep Med.* 2, 2001.
16. THE ATLAS TASK FORCE – EEG arousals: scoring rules and examples: a preliminary report from the Sleep Disorders Atlas Task Force of the American Sleep Disorders Association. *Sleep,* 15, 173-184, 1992.
17. WEESE-MAYER D.E., CORWIN M.J., PEUCKER M.R., DI FIORE J.M., HUFFORD D.R., TINSLEY L.R., NEUMAN M.R., MARTIN R.J., BROOKS L.J., DAVIDSON WARD S.L., LISTER G., WILLINGER M. – Comparison of apnea identified by respiratory inductance plethysmography with that detected by end-tidal CO_2 or thermistor. The CHIME Study Group. *Am. J. Respir. Crit., Care Med.* 162, 471-480, 2000.

Chapter 2

Normal sleep in children

D. Samson-Dollfus
Médecine Néonatale et Pédiatrique. Pavillon de la Mère et de l'Enfant. Hôpital Universitaire Charles Nicolle. Rouen, France

A clear understanding of the child's sleep and its maturation is essential if we are to avoid confusing the physiological phenomena which occur at certain times in life, with disorders of sleep organisation. This applies to the family, the doctor and to the paediatrician. Yet sleep characteristics evolve considerably in the course of maturation: whether in terms of ultradian and circadian sleep-wake rhythms, ultradian sleep stage rhythms, the electroencephalographic (EEG) characteristics of the different stages, sleep-related variables such as respiration or heart rate, changes in temperature, endocrine secretion etc.

In 1965, Monod and Dreyfus-Brisac [75] recalled that sleep in children had been the object of many years study. From the first half of the 20th century, eye movements were reported with phasic facial movements in the sleeping newborn. The different stages of quiet and active sleep were recognised according to the presence of eye, facial and general motility, and the rate and regularity or irregularity of respiration.

Our understanding of sleep maturation was greatly assisted by the advent of EEG recording and the development of polygraphic techniques. The latter were initially simple but became increasingly complex and have proved vital in the accurate analysis of sleep maturation.

Kleitman's team demonstrated the existence of sleep cycles in infants [9, 58, 59] by applying work previously conducted with adults. Most authors refer to the terms quiet sleep (QS) and active sleep (AS) or paradoxical sleep (PS) usually referred to as REM sleep, developed through the use of elaborate polygraphic techniques. The terms regular and irregular sleep have been proposed [108], while stage 1 sleep is also defined in QS and stage 2 in AS [95].

Discussion then followed regarding the terminology for "REM ". The term "seismic sleep" was proposed, in reference to the many spasms occurring during this stage, with background hypotonia. The hypothesis that seismic sleep is a precursor of REM sleep was rejected [1, 52]. This suggested that the two types of sleep were different, with seismic sleep gradually being replaced by REM sleep, as the structures governing REM sleep develop. Seismic sleep was then thought to disappear completely at around 20 days in the full-term infant. This semantic confusion has probably led to the use of the term "active sleep" in newborns, thus avoiding taking sides on the issue.

The value of automatic analysis, quantifying sleep recordings and detecting phasic phenomena has been recognised for many years. Despite the abundant literature, listed in a generalist journal on the quantitative analysis of sleep tracings in infants [95], electronic data is still too technical to be generally applied. Nevertheless, the objective results obtained by these techniques provide a far higher degree of accuracy in evaluating sleep maturation than in visual analysis, provided that the software is appropriate to these very young children. An additional technique used is that of video recording [82]. Polygraphy is advisable and as tracings take from 70 to 90 minutes only, they can be carried out in the clinic to distinguish between normal and pathological conditions. More recent interest in the subject of the ultradian and circadian organisation of full-term newborns, and more specifically, of premature infants, has led clinical researchers to carry out recordings lasting several hours and even for one or two days. However, while it is fairly easy to conduct polysomnography for long tracings (24 to 48 hours) in children and infants, this procedure is too long for premature and newborn infants. The 8 to 11 EEG electrodes and those applied to the chin and eyelids need frequent checking. Testing heart rate is usually straightforward but respiration is more difficult to test. For the newborn, a nasal sensor may have to suffice, although attempts should be made to evaluate abdominal and/or thoracic respiration. This presents problems which are hard to resolve. The improvement in the survival of premature infants of under 27 weeks, is accompanied by a higher risk of intraventricular and peri-ventricular leukomalacic haemorrhage. As a result there are increasing demands for EEG examinations. Besides the specific anomalies (rolandic spikes) a 24 hour study of the organisation of the states of sleep and wakefulness is of undeniable prognostic value [17]. The more premature the infant, the more complex the treatment. EEG technicians no longer have the time to stay at the infants side for several hours. Families tend to be overawed by the medical environment. Thus rather than increase the number of sensors, less invasive means have been explored to evaluate the stages of sleep and wakefulness in newborns. Video, which has been widely used by Prechtl [82] since 1974, and later by his team [83], although greatly improved by digital techniques, is very time-consuming to view. Recordings of motility have proved to be a viable alternative to EEG and video. The simple detection of sleep and wakefulness in newborns and infants began in 1955 with the recording of movements [9, 26, 45], and balisto-cardiographic techniques were perfected [2, 33, 86, 109]. Other parameters were then introduced such as heart rate, respiration rate and their variability [2, 36, 40, 68, 73].

Actimetry, used in adolescents [15], children and young infants [86], and from 1995-96, in full-term newborns [110] is now applied to premature infants. Encouraging results have recently been obtained in premature infants [113]. It is worthwhile supplementing actimetry with other methods which are less disturbing to the newborn and to the nursing staff [2, 33].

NEONATAL PERIOD

More than at any other time of life, the definitions pertaining to sleep and wakefulness during this period are controversial. The neonatal period is very important in mammals. In terms of experimentation, two teams have focused on quiet sleep (QS) and active sleep (AS). In the rat, Jouvet-Mounier *et al.* [53] have demonstrated the relationship between 5-HT receptors and the maturation of both types of sleep. Mirmiran and Van Someren [73], deprived newborn rats of AS and observed a hypersensitivity of the cells of the hippocampus to norepinephrine. The capacities of memorisation were altered in the long term. The surroundings of newborns may affect the duration of wakefulness, quiet sleep (QS) and active sleep (AS), respectively, and have a neuro-psychological effect on the subsequent development of the child. It is clearly impossible to carry out this type of research among human newborns. Nevertheless it is useful to observe the conditions surrounding premature infants in France and abroad: in Northern Europe doctors take steps to reduce noise to a minimum. In France, however, the conditions surrounding newborns in paediatric intensive care units or in neonatology units are quite different: paramedical and medical staff, strident monitoring alarms, and visits by the family generate noise to which these newborns are sensitive, and the more premature they are, the more fragile. Numerous polysomnographic tests conducted for diagnostic purposes have revealed the extent to which various noises can disturb the sleep of premature infants, particularly AS, which they may risk being partly deprived of at a period which is critical for their future [69].

Despite objections, some of which are perfectly valid such as those of Katz-Salamon published in 1997 [34 quatro], long term, comparative, longitudinal studies need to be conducted internationally on the very premature, in regard to premature infants placed in neonatal intensive care units (NICU), looking not only at the early use of nasal continuous positive airway pressure (NCPAP), but also at environmental factors including noise. This would highlight the role of this much neglected factor in the later neuropsychological development of the premature whose short term development was nevertheless satisfactory.

A glossary of neonatal EEG [6] and a recent work [90] have distinguished the characteristics of sleep states in the premature and full-term newborn.

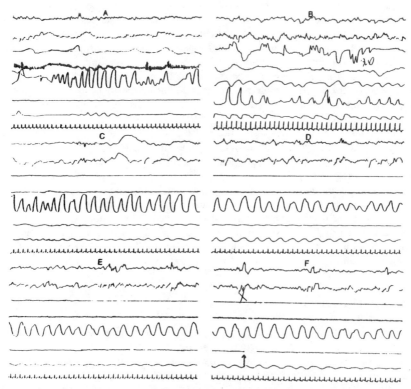

Figure 2.1. Polysomnographic recording of newborn with a gestational age of 41 days, taken at 4 days old. A. During wakefulness: movement artefacts, mean EEG activity showing irregular theta. B. Active sleep: theta is more rhythmical, rapid eye movements, inactive EMG. C. Transitional sleep: active or inactive EMG, irregular respiration, no rapid eye movements. D. Onset of quiet sleep. Tonic EMG activity, regular respiration, slow, almost constant EEG activity. E. Quiet sleep: a tracé alternant pattern appears. F. Slow bursts in response to noise. Each sample tracing comprises from top to bottom: EEG Fz-Cz, EEG Cz-Oz, eye movements, EMG, respiration: nasal, thoracic, abdominal, EKG, timescale 1 s. Scale of amplitude 100µv.

FULL-TERM NEWBORN

Polysomnographic characteristics

The various classifications of sleep and wakefulness states concur in regard to the definitions AS and QS [108] even if there is wide dissent over the definitions of wakefulness and the transitional states [3, 4, 21, 31, 74, 80, 82] as was recently confirmed by Curzi-Descalova [23]

Wakefulness (fig. 2. 1A) As the EEG is very like that of AS, these two stages are distinguished by their polysomnographic parameters. The form of eye movements (EMs) is different. In AS, there are no muscular artefacts on this derivation and EMs are easily recognised [3]. In active wakefulness, there is considerable EMG activity, EMs are numerous and pervaded by muscular artefacts. Even in quiet sleep, EMG shows intermittent and continuous muscular activity [24].

Sleep onset usually occurs in AS or in TS, whether during the daytime or at night. Direct passage to QS, which is slow at first and then alternating, may sometimes occur but is rare.

AS (fig. 2.1B) is marked by rapid eye movements (REM), phasic activity of the limbs, face and body, muscular atonia interrupted by sudden movements, irregular respiration with short pauses and unstable heart rate. EEG shows continuous rhythmic theta activity, of moderate amplitude and which is centrally located [28, 75].

QS (fig. 2.1D, E, F) is characterised by regular respiration, tonic EMG activity and a regular heart rate which is lower than in wakefulness or AS states (fig. 2.1). Although movements rarely occur, they are sometimes observed in the QS of very small infants [45, 112]. These remain occasional but are far more frequent than in the child or adult. EEG shows slow bursts (sometimes in response to noise) which are predominantly anterior, interspersed with theta elements and separated by phases of irregular activity of weak amplitude and variable duration: alternating tracing (fig. 2.1E, 2.1F). Terzano's team has shown that the alternation between bursts of cyclic alternating pattern (very close to the alternating pattern of the newborn) is linked to changes in heart rate [38]. This warrants being verified in the newborn. This type of sleep differs from slow continuous dysrhythmia [2, 27].

Transitional sleep (TS) precedes, follows or replaces AS, corresponding to a stage in which the characteristics of the two stages previously referred to are incomplete [3,4, 23, 43]. TS corresponds to gradual changes in neuro-mediator secretion, as Crochet and Sakai have shown [22]. It is to be considered as a separate stage [23].

AS-QS transitions are more gradual than QS-AS transitions. These always occur in the same way, whatever the term [7, 8, 23]. In the shift from AS-QS the first effect is the disappearance of REM, and the last the discontinuous aspect of QS. In transitions from QS-AS, EEG becomes continuous at the start of TS and REM appears at the end. REM will only appear when the EMG and EEG characteristics of AS are complete [7, 76]. These TS characteristics have a prognostic value: if they fail to occur in the right order, they indicate abnormal neurological development [3, 76, 87].

Sleep-wake architecture

The amount of sleep is 16 hours [21]. In France, 24 hour polysomnographic recordings have shown only 14 hours of sleep. This discrepancy reflects the extent of variability between units in France and abroad. It may also be accounted for by the noise to which babies are exposed in France, as referred to above.

Distribution of QS/AS

Several authors have shown the proportion of AS to be high at birth: approximately 40 to 50% of total sleep time; inter-individual variability is weak in normal newborns [64]. TS probably accounts for 10 to 15% of total sleep time in the newborn.

Circadian organisation

Periods of sleep and wakefulness are equally distributed between day and night during the first two to three weeks. However, sleep does tend to predominate between 11pm and 7am in newborns: continuous phases of uninterrupted sleep are longer during this period than in the daytime [90]. It is thus possible to speak of the onset of circadian organisation in the newborn of less than one week old (fig. 2.2). Highly accurate observations of behaviour [7] have shown the alternation of sleep and wakefulness in newborns to be affected by the presence or absence of the mother.

Ultradian organisation

Conversely, ultradian organisation of the sleep-wake cycle is clearly evident in full-term newborns. Phases of sleep are interrupted by short periods of wakefulness during which the infant cries, suckles at the breast or bottle. There are rare periods of quiet wakefulness. The periodicity of this sleep-wake cycle (fig. 2.2) is roughly three hours [47, 91]. Premature infants who reach full-term age have longer ultradian cycles with longer alternating periods of quiet sleep than in the case of full-term infants [46, 101]. Indeed it is difficult to confirm the differences between infants who were born prematurely but who have reach full-term in age and newborns at full-term.

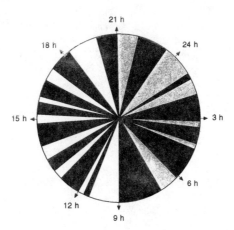

Figure 2.2. Diagram of sleep-wake alternation in the newborn. Little difference is seen to exist between day and night (wakefulness shown in white for daytime and grey for night; sleep in black).

PREMATURE INFANTS

Polysomnographic characteristics

Studies are far more difficult to conduct in this case, than for full-term infants [5, 75, 101]. Polysomnographic recording often poses a problem in the case of premature infants of under 30 weeks. Indeed the head is tiny in size making the choice of a common reference (essential in using digital equipment) difficult: the tip of the nose appears to be the best choice. The ear is too close to the brain, the vertex is at fontanelle level and the chin generates sucking artefacts.

Recording eye movements is often impossible, as the electrodes must be placed too close to the eyes. The operator must thus note these as soon as he/she sees them. For *EMG recording*, the best choice is the chin strap.

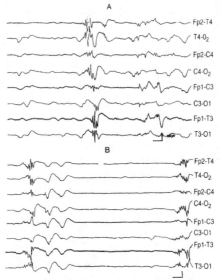

Figure 2.3. Day-old newborn at 29 weeks gestational age. A. Fairly continuous EEG activity probably signifies a state of wakefulness or of active sleep. B. Discontinuous activity corresponds to a state of quiet sleep.

From the age of 27 to 29 weeks from conception, two different types of EEG activity can be discerned [28, 63, 79, 89]. One corresponds to very slow waves overloaded with theta and with continuous rapid frequencies, and the other to more discontinuous elements. This pattern is seen during wakefulness and/or AS (fig. 2.3 A). Polysomnography distinguishes between the two states: strong artefacts and muscular activity during wakefulness, muscular atonia, short phasic movements, occasional starts and above all REM in AS. The ample, continuous tracing of the very premature appears to correspond to wakefulness and/or AS. Polysomnography differentiates between these two states: artefacts and significant muscular activity during wakefulness and jerks with muscular atonia in AS. Periods with eye movement occupy 20 to 50% of AS time. Eye movement density paradoxically appears to be higher in the premature than in the full-term infant [114].

The almost periodic pattern (fig. 2.3 B.) corresponds to QS. Simultaneously, virtually permanent motricity diminishes. This state occupies only 10% of recording time during which short bursts of slow waves occur, interspersed with fast rhythms and rhythmic bursts of theta lasting two to three seconds, separated by periods of minimal amplitude activity of about 20 seconds. The periods between bursts tend to shorten with maturation, assuming the alternating pattern [72] typical of the sleep of full-term infants at about 36 weeks.

Sleep-wake architecture

From 26 to 27 weeks onward, two states are already distinguishable: wakefulness and/or AS, and QS.

At 29 weeks the four states are observed: wakefulness, AS, TS and QS [40]. Before the gestational age of 38 weeks, premature newborns who are stimulated by their environment will remain awake for 25% of the time, whereas when stimulated as little as possible, they will sleep for 95% of the time [40].

No circadian organisation would appear to exist in premature infants, but this is prohibited from developing, in any case, by the stimulation to which the premature infant is subjected: even in infants with no problems, the daily routines undoubtedly disturb their spontaneous sleep-wake rhythm [99, 50]. Moreover materno-foetal interaction [50] is likely to play a role in establishing sleep-wake rhythms (Basic Rest-Activity Cycle); yet these newborns will have been partly deprived of this. During the first year of life of prematurely born children, the evolution of AS and QS depends on environmental conditions far more than on gestational age at birth [5].

The beginnings of ultradian organisation can be seen in the highly premature as from 30 weeks [56], and ultradian rhythms and sleep stages become distinct at 32 weeks [111].

INFANTS UP TO THE AGE OF ONE YEAR

Despite the undeniable variability between individuals, it is possible to establish guidelines to the maturation of sleep-wake states during the first months of life [21].

Polysomnographic characteristics

Before the age of six weeks

Wakefulness. The tracing in wakefulness is usually charged with artefacts of movements and EMG. EEG is barely visible. The rare calm periods show irregular theta rhythms of weak amplitude, which are predominantly rolandic.

Sleep onset. This is practically imperceptible, the baby passing directly from wakefulness to AS or QS.

Quiet sleep. At around the age of six weeks, alternating QS activity disappears. QS rapidly and directly succeeds the wakefulness state when sleep onset does not occur in REM sleep. QS is then characterised by slow, irregular and continuous delta rhythms which clearly predominate in the anterior regions (fig. 2.4.). This pattern persists up to the age of three months.

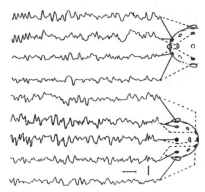

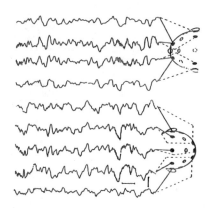

Figure 2.4. 3 weeks: sleep onset NREM (delta with no spindles).   *Figure 2.5.* 2 months: quiet sleep with spindles.

Active sleep. The density of eye movements increases with age, whether the infant was at full-term or premature [60]. There are no other polysomnographic changes.

Six weeks

Wakefulness, sleep onset and arousal: no changes.

Quiet sleep. The first bursts of spindle rhythms appear: short bursts of 12 to 14 c/s lasting about a second, projecting onto the Rolandic regions and the vertex (fig. 2.5). The onset of these spindles represents a prime phase in normal maturation: the onset of stage 2 (QS 2). The maturation of spindles seems to coincide with the rearrangement of the thalamo-cortical structures [107] and with dendrite maturation. As sleep becomes deeper, the slow waves increase in number and amplitude and the spindles are less visible: beginning of stage 3 (QS3).

In defining the frequency and amplitude of waves, Smith *et al*. [105], Salzarulo *et al*. [88] observed the continuous variations in slow activity which appear to be a marker for maturation, in *the second month of life*.

Three months

Wakefulness. Some rhythms become more regular and project onto the centro-occipital regions.

Sleep onset. Onset of theta frequency rhythmicity with diffusion to the central regions.

QS. Distinct presence of deep sleep, rich in delta waves; there is a clear distinction between QS2 with spindles and QS3 which is much richer in delta.

AS. (rarely seen on daytime tracings, must be recorded at night to be seen): the characteristic theta rhythms are irregular and differ in aspect to those of wakefulness. In any case, polysomnography clearly distinguishes the two states.

Arousal: This occurs both in QS and AS. Wakefulness rhythms reappear among numerous artefacts.

Five to six months

Wakefulness. The basic rhythm, of five to six c/s is clearly differentiated in the occipital regions. To observe this the infants eyes must be closed. Crying and artefacts make this more appropriate at the end of the recording, to avoid preventing the onset of sleep.

Sleep onset. Theta frequencies become more rhythmic and diffuse to anterior regions, but their amplitude remains moderate.

QS. Same characteristics as previously.

REM sleep. From five months, REM sleep is clearly defined.

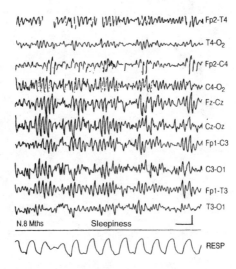

Figure 2.6. Eight months: sleep onset with diffused EEG rhythm hypersynchronisation

Arousal. When this is provoked, beginnings of bursts of theta interspersed with delta which is more ample than on the wakeful tracing. It may occur spontaneously and the wakeful tracing reappears with no artefacts.

From six months to one year.

Wakefulness; Occipital frequencies close to alpha (7 to 8 c/s) begin to appear when the eyes of the infant can be closed.

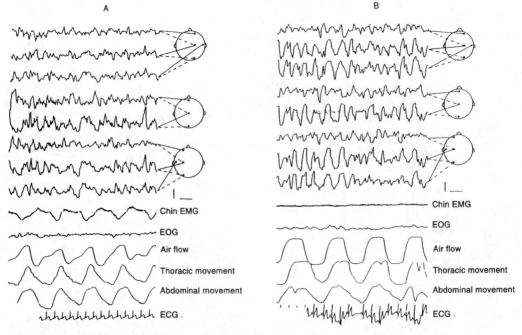

Figure 2.7 A. QS3: delta rhythms are irregular, of moderate amplitude, and spindle rhythms less distinct than in stage 2. B. QS4: delta rhythms become very ample and monomorphic. They predominate in the occipital regions.

Sleep onset. Gibbs first described hypersynchronisation of sleep onset in 1952 [42]. These are ample bursts of diffused delta (fig. 2.6).

NREM sleep. Very deep NREM sleep is now distinguishable in QS3 and QS4 (fig. 2.7 A-B). The delta rhythms are of greater abundance and amplitude than at the same stage in adults [35, 92, 95, 96]. This increase in delta energy correlates to the metabolic activity of the brain [19, 36] and with synaptic density [51].

Night recordings have shown that the transition from stage 2 to stage 3 or 4 may occur with a considerable increase in EEG amplitude, not to be confused with an arousal reaction [97].

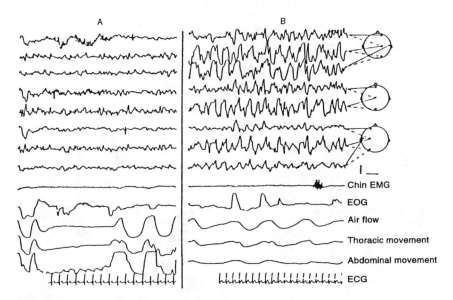

Figure 2.8. 12 months. A. Baby VL: theta activity in REM sleep. B. Baby M.L.: ample delta is seen in a clear state of REM sleep, phasic muscular activity on an inactive background and numerous rapid eye movements.

REM sleep. Weak amplitude theta activity (fig. 2.8 A) noted at birth often gives way to ample delta rhythms which are fairly irregular (fig. 2.8 B), mainly observed in daytime tracings taken to aid diagnosis when the infant, who arrived at the emergency unit the previous night, was probably deprived of REM sleep.

Arousal. This is usually provoked, causing ample hypersynchronous slow waves, similar to those of sleep onset.

Sleep-wake architecture

Daytime sleep decreases in duration [59,80]. This decrease occurs rapidly in the first weeks, becoming more gradual in the months which follow. *Periods of wakefulness* in the first half of the morning and particularly in the second half of the afternoon become longer between the ages of two and six months. Nocturnal sleep becomes slightly shorter.

QS increases in proportion during the first months of life, from an average of 20 to 50% of total time [32, 49]. It is clear that at this age, as at any other age in life, the three stages used in classifying quiet sleep: QS2, QS3 and QS4 are debateable, since the evolution of NREM sleep presents as a continuous variable [88] and maturation is gradual.

The relative amount of REM sleep diminishes in the course of the first year, reaching a level of roughly 35% at six months [49], 25% at eight to twelve months, and 20% from ages one to two years.

Circadian organisation, the presence of which is debateable at birth, develops clearly during the second month. The differences in sleep-wake alternation gradually become more marked between night and day [70, 80]. Sleep phases which are uninterrupted by wakefulness become notably longer in the course of the first six months [21, 46]. Periods of wakefulness become shorter and less frequent at night [29]. Nocturnal sleep really becomes established between the third and fourth months.

During the early weeks, the spontaneous sleep-wake rhythm has a circadian cycle of approximately 25 hours. It appears virtually unaffected by external 24 hour synchronisers, up to the age of 4 to 6 months. Nevertheless the child is sensitive to the mother's behaviour from a very early age [100]. Moreover it is likely that children fed on demand, like those whose parents do not attempt to gradually introduce external synchronisers [37], take longer to acquire sleep/wake rhythms than children subjected to a regular routine.

At about four months, the infant is able to sleep for several consecutive hours during the night, but will still have several periods of sleep during the day at around 11am, 2pm, sometimes 5 or 6pm (2.9A). After this final nap, the end of the afternoon is often a period of agitation during which it is virtually impossible to get the baby to sleep.

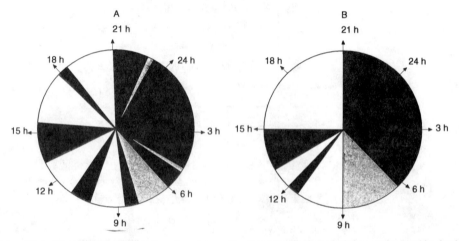

Figure 2.9. A. Diagram of sleep-wake alternation at 4 months. B. At 12 months (wakefulness shown in white for the daytime and grey at night; sleep shown in black). As from 4 months, night time sleep is established, but two or three naps continue to be taken during the day. At 12 months, night time sleep is long and the morning nap becomes shorter.

At the age of one year, a normal child still sleeps for 12 to 14 hours, with practically continuous night time sleep of 10 to 11 hours, and daytime sleep divided into two naps, one at around 11 30am and the other at 1 or 2pm (fig.2.7B). When an EEG is carried out on an infant at a chosen time (just before or just after lunch) a normal baby will often cry but will always fall asleep in less than 20 minutes [94, 115].

A small group of infants studied polysomnographically [97] in their normal surroundings rather than in the laboratory showed no significant first night effect.

A pattern of **ultradian organisation** in QS-REM [70] cycles can be seen once the alternation between nocturnal sleep, wakefulness and napping is established. This alternation lasts for a period of 60 to 70 minutes (fig. 2.10) at about the age of six months [47, 91]. Some studies have reported a much shorter period of duration, of 42 minutes at six months [34]. While REM sleep recurs after a period of approximately 60 minutes during the first months of life, the period for deep NREM sleep is approximately 120 minutes. The regular recurrence can thus be observed during the night of periods of deep NREM sleep with a rhythm of 0.5 c/hr [103], whereas in older children and adults, a reduction in deep NREM sleep is observed during the night, with the absence of ultradian rhythmicity in QS3 and QS4 [41] with distinctly independent cycles of REM and deep NREM sleep.

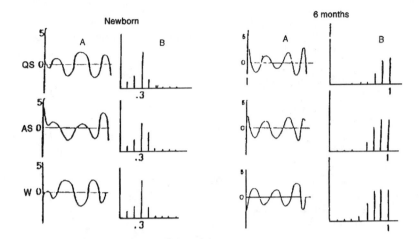

Figure 2.10. Nocturnal ultradian organisation in quiet sleep, active sleep and wakefulness. a. In the newborn, periodicity is 170 to 200 minutes. b. At six months, periodicity is 60 to 70 minutes.

The number of REM sleep episodes is high during the first year, reaching seven or eight and sometimes more. The amount of REM sleep diminishes in proportion to the reduction in the duration of each REM sleep episode at the end of the night, followed by a reduction in the number of periods of REM sleep, whereas the total duration of NREM sleep lengthens. At around the age of three weeks, it is still common for REM sleep onset to occur in AS. From five months onward, REM sleep is clearly established and latency to REM sleep onset lengthens [104]. This is longest between noon and 4pm and shortest between 4 and 8am. Short latencies are usually preceded by a short period of wakefulness while long latencies succeed periods of prolonged wakefulness.

Transitional sleep (TS) known to precede or immediately follow the AS of the newborn [4, 31, 74] is even more evident in the young infant [98]. But this tends to diminish in the course of the first year, giving way to true REM sleep.

As we have seen, **REM sleep onset** in AS, which is practically constant up to *three weeks*, becomes increasingly rare after 5 months. From about 6 weeks to 5 months, EEG shows rhythmic activity of 4 to 5 c/s when the infant closes his eyes. This tends to amplify, before the infant passes into QS [42, 94]. The passage to stage 2 is heralded by the first burst of spindles. Classic hypersynchronisation composed of hypersynchronous theta rhythms, which may be diffused or predominantly central, is constant at 8 months (fig. 2.9.). This period of REM sleep onset may last from two to ten minutes. It then becomes fragmentary before entering stage QS2. The recurrence of hypersynchronisation as the EEG becomes more irregular and less ample is often due to noise causing wakefulness reactions, and thus the passage to a lighter stage of sleep.

Spontaneous arousal still occurs as the infant passes from NREM or REM sleep to wakefulness, characterised by tonic muscular activity, movements and EEG of moderate amplitude.

Evoked arousal, on the contrary, almost always results in slow hypersynchronous bursts from five or six months onward. Toward the age of one year, rhythmic bursts of 7 to 8 c/s are succeeded by delta waves which may be very ample and slow. Arousal responses in NREM sleep differ depending on whether the stimulus (clapping hands, whistle, calling by name) carries an affective connotation.

CHILDREN AGED 1 TO 3 YEARS

EEG characteristics

Wakefulness. The basic theta rhythms, increasingly associated with frequencies of 7 to 8 c/s, are already well localised in the occipital region, especially when the eyes are closed.

Sleep onset. The EEG pattern must be fully understood to avoid interpretation errors: EEG may diminish in amplitude as in the older child and the adult. It is far more common to observe (as in the infant of 7 to 8 months) the hypersynchronisation described by Gibbs and Gibbs [42] *i.e.* slow, ample, diffused bursts, or long regular bursts which are predominantly central. These may be interspersed with sharp waves, described in 1971 by Eeg-Olofson *et al.* [30]. Each of these aspects is normal and should in no way be interpreted as a diagnosis of epilepsy!

NREM sleep. At one year, NREM sleep can be distinguished in all normal children in stages 2, and 3-4.

Stage 2. Delta is of moderate amplitude. K complexes begin to appear towards the 2nd year. At 18 months, vertex sharp waves tend to form a "train of spikes" which can sometimes be very sharp (fig. 2.11). K complexes only appear clearly around the age of three. Spindle rhythms are particularly ample and sharp before the age of three. Frontally located central spindle rhythms of 12 c/s are characteristic of the child [42].

Stages 3 and 4. The number and amplitude of delta waves increases [35], with a high proportion of stage 4 (2 five), higher than that of adults, but lower than that of school age children.

REM sleep. As for the infant, this is rarely recorded during the daytime in normal conditions unless the infant was deprived of sleep the preceding night. In this case the EEG pattern is often ample and slow (fig. 2.6.B). This should not be confused with an arousal reaction.

Arousal. This is usually provoked by an auditory or touch stimulus and gives rise to the hypersynchronous bursts of sleep onset.

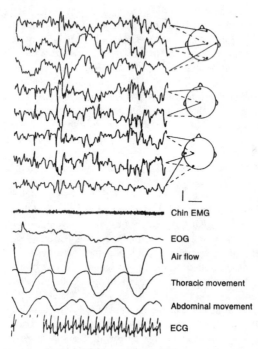

Chin EMG

EOG

Air flow

Thoracic movement

Abdominal movement

ECG

Figure 2.11. Some of the central spikes of this 26 month old child are very sharp, but are normal during sleep.

Sleep-wake architecture and behaviour

Napping. There is a certain degree of consensus between teams and even countries, on the age at which napping ceases [85, 115]. From two naps a day (at 11am and 14pm on average) the child reduces to a single postprandial nap around the age of two years [39] and this disappears completely at ages ranging widely from three to six or seven.

Sleep time and bedtime. When napping ceases, total sleep time continues to diminish, lasting approximately 12 hours. After the age of one nocturnal sleep is established despite considerable inter-individual variations. The duration of sleep may be influenced by lifestyle, particularly as parents behave differently according to their respective work schedules. Bedtime varies from one family to another (between 8 and 10pm).

Many children sleep with the light on.

From 18 months it is fairly common for the child to sleep in the parents' bed. Indeed, according to a questionnaire carried out among 303 families [65], "co-sleeping" occurs in 55% of children aged 2 to 3, at least occasionally. This practice which is more common among poorer families or in cases of problems getting to sleep or of sleep walking, does not lead to any behaviour problems. The reason is often "fear of going to sleep". Others can only get to sleep if there is a night light close to the bed, causing no particular problem. Nevertheless, in the previous history of children aged 10-12 with sleep onset difficulties, night lights were used to encourage sleep when they were younger.

CHILDREN FROM THREE TO SIX YEARS

EEG characteristics

EEG aspects can still be seen during daytime naps. There is no clinical reason to prescribe polysomnographic examinations. All the more so as the lack of staff and resistance from the family are such that night recordings are only rarely taken at this age.

Wakefulness. When the eyes can be closed, alpha frequency bursts appear, still interspersed with numerous theta rhythms. But their occipital localisation prevents them from being mistaken for sleep onset.

Sleep onset. The occipital alpha frequencies disappear. Theta rhythms become increasingly ample and diffused. They may be interspersed with graded sharp wave emissions which are absolutely normal at this age.

NREM sleep. This is observed during napping. It is clearly distinguishable in stage 2, heralded by the first burst of spindles, in stage 3, with an increase of delta waves and even in stage 4, when spindles are hardly visible, as in younger children.

REM sleep. This is hardly ever observed during the daytime.

Arousal reactions. These are similar to those of the younger infant. In the same way, hyperpyretic spasms are fairly frequent. As some doctors still prescribe an EEG, the interpretation must be made with care, bearing in mind that graded sharp waves at sleep onset and arousal are normal.

Sleep-wake architecture and behaviour

Sleep time is further reduced. The morning nap is rare. The child may need to get up earlier because of starting kindergarten, but will continue to take an afternoon nap as most teachers insist upon this.

In the evening, it is easier to get to sleep as the child is tired out by the day's activities. But the fear of night may persist, as well as the desire to stay close to the parents both of whom normally work during the day. Thus many never get to sleep before 10 pm.

CHILDREN FROM SIX TO TWELVE YEARS

EEG characteristics

Sleep onset. This increasingly occurs through the disorganisation of alpha activity, and the onset of irregular theta frequencies of weak amplitude. But ample hypersynchronous bursts are still occasionally seen.

NREM sleep. In stage 2, K complexes appear, with vertex sharp waves becoming increasingly scarce. The most interesting changes take place in stages 3-4. Delta waves increase in amplitude and numbers [35, 55]. There is a higher proportion of stage 4 [20]. The richness of delta waves may be correlated to the very high metabolic levels measured by PET scan [19, 36]. A complement to Borbely's model has been proposed for children [13]. Intense activity in the wakefulness state during the learning stage may encourage the high amplitude of delta during deep sleep [38].

REM sleep. The pattern is probably the same as for adults. As this is never recorded in the daytime descriptions are brief.

Sleep-wake architecture and behaviour

Total sleep time becomes shorter, obviously with a detrimental effect on night time sleep, which is reduced from roughly 10 hours to 8 hours on average. It is likely that children at the beginning of the 20th century slept at least an hour longer than they do today [115]. Permission to stay up and watch television when there is no school the following day leads to later bedtimes. The duration of sleep is the same during school time as at the weekend up to the age of 11. A recent study [77] in Luxembourg on 4255 school children aged 6 to 12, showed that the average time a six year old spends in bed is 11h. 18, while that of a child of 12 is roughly 10h. Girls sleep slightly more than boys. Heavy and light sleepers. At 6, the average bedtime is 8 pm. Children often sleep outside their beds.

From the age of six onwards deep NREM sleep predominates in the first part of the night and REM sleep in the second, as in the case of adults. The number of REM sleep episodes is already at around five at the age of six. Palm *et al.* have shown that the quantity of REM sleep remained stable at 20 to 22%.

The first period of REM sleep latency is roughly 140 minutes at 6 years and is still around 120 minutes at 12 [78] , as though the first episode had been missed.

Arousal reactions diminish up to 11 years, as does the time of intra-sleep wakefulness.

PUBERTY AND ADOLESCENTS OF 12 TO 15 YEARS

Polysomnographic features

The most classic changes affect deep NREM sleep. Indeed the delta rhythms which were so abundant and ample in early childhood begin to diminish, gradually reaching values similar to those of adults.

The "Cyclic Alternating Pattern" (CAP) probably relates to phasic events linked to an arousal. Polysomnography taken at night in six normal children and adolescents showed a relationship to cardiac rate variability. This is accompanied by significant changes between the sympathetic and vagal components of the autonomous system [38].

Arousal reactions increase with age but remain inferior in number than those of the young adult [14].

Sleep-wake architecture and behaviour

Total sleep time during the week decreases with age as adolescents go to bed later and later. From 12 upwards, this duration is shorter during school time than at the weekend. A thousand questionnaires (handed out in an urban area of Belgium [54]) yielded 972 responses which showed that 43% of pre-adolescents complained of sleep disorders: latency to REM sleep onset was over 30 minutes in 14% of these cases. They woke up in the night at least twice a week and 4% were already taking sedatives. Academic problems were more common among poor sleepers. The contributing factors were mainly the low level of education of the parents and family problems. This work was recently supplemented by a survey carried out in Luxembourg [77] on the sleep habits of 4255 children. The study showed that at 12, bedtime was at 9 or 10 pm. There is often a period of excitation just before bedtime: electronic games, loud music etc., which would induce sleep disorders. One child out of two woke up tired [18]. Awakenings during the night were relatively common. Poor sleep quality also appeared to correlate with poor academic achievement, as in the Belgian study.

Studies conducted in Norway [106], Japan [102, 117], Holland [68], France [69] and the USA [81] combine in showing that adolescents whose sleep duration should be roughly 8 hours, sleep around two hours less during the week and slightly more than an hour and a half at the week end. They complain about sleep during school time. Difficulties getting to sleep and frequent awakenings generate poor quality sleep. The effects are those of falling asleep during the daytime, a lack of interest in classwork, low motivation to succeed in studies, and tiredness at school [58, 82 b]. As many go to bed late, the sleep deficit during the week is often high. On the other hand, getting up at set times to go to school is an excellent "Zeitgeber". They recuperate at the weekend, particularly in the morning. The fact that sleep time is less affected by external influences, gives a better idea of sleep structure at this age [14].

OTHER CIRCADIAN AND ULTRADIAN RHYTHMS

We have seen that a child's sleep evolves continuously from the time the premature newborn is viable up to the age of ten or twelve years. This maturation is linked to the evolution of the other circadian and ultradian rhythms [70]. Temperature evolves with sleep-wake alternation. The newborn was seen to have an ultradian thermal rhythm of a period of roughly three hours. Circadian organisation of rhythm and temperature begin to develop in fact from the age of eight to nine months, becoming close to those of the adult from the age of five.

Hormone studies are difficult to carry out on infants and children. The circadian organisation of these rhythms has however been found to establish itself during the very first years of life, particularly as regards the growth hormone, known to be substantially secreted during NREM sleep.

A protocol set up by Carskadon *et al.* to measure sleep-wake phases and melatonin secretion in the adolescent showed the melatonin "offset-phase" to be significantly correlated to age [15].

A Canadian study [61] of the circadian rhythms of young adolescents showed no difference compared to findings for young adults. However the authors stressed that "evening subjects" showed a delayed melatonin secretion with no diminution of sleep consolidation or vigilance.

The rhythm of urinary secretion is also difficult to study, but experience shows that where enuresis is not present, nocturnal secretion diminishes after the second year.

Heart and respiratory rhythms [84] are heavily influenced by the states of wakefulness, AS and QS. Thus they probably participate in the basic rest activity cycle (BRAC) [58], the beginnings of which are perceptible from the first weeks of life. Short pauses in respiration predominate in AS. Sleep deprivation contributes to apnoeas in the infant.

CONCLUSION

1. During the course of childhood, the stages of circadian sleep-wake organisation, ultradian organisation, Basic Rest Activity Cycle or the maturation of electroencephalographic patterns are all important.

The terms used in referring to sleep stages are adapted to this evolution (table 2.1).

The term "active sleep" in fact covers REM sleep and TS. It is often used in preference to "REM sleep", when referring to the full-term newborn or premature infant After the age of six weeks, a distinction must clearly be drawn between the two aspects: "REM sleep" and "transitional sleep".

Due to the particular nature of the EEG for quiet sleep in the newborn (full-term or premature) which evolves from week to week, the sole term "quiet sleep" is employed up to the age of six weeks. From six weeks to two months, the appearance of spindles and continuous delta rhythms marks the distinction between QS2 and QS3.

The terms "sleep onset" and "wakefulness" are used from the age of five months onwards and deep quiet sleep QS4 appears between eight and ten months.

2. An understanding of the circadian and ultradian evolution of the states of sleep and wakefulness enables the paediatrician or general practitioner to explain to anxious parents that:

Intrasleep wakefulness is normal during the first months of life.

Naps are necessary up to the age of three years.

Bedtime rituals are permissible.

3. Finally for the sake of credibility, it is important to allow for the wide variability between individuals and to carry out a detailed interrogation of the habits of the family and the child - often reassuring for the family - before considering that the child may have a real sleep disorder warranting specialised explorations of sleep structure.

Table 2.1. Use of the different terms in function of age in regard to the sleep-wake states of infants and children

	Wakefulness	Sleep onset	Active sleep			Quiet sleep			Arousal
			TS	REM	QS2	QS3	QS4		
NB	Wakefulness			Active sleep		Quiet sleep			
6 weeks	Wakefulness		TS	REM	QS2	QS3			
5 months	Wakefulness	Sleep onset	TS	REM	QS2	QS3			Arousal
8 months	Wakefulness	Sleep onset	TS	REM	QS2	QS3	QS4		Arousal
3 years	Wakefulness	Sleep onset	TS	REM	Light NREM sleep	Deep NREM sleep			Arousal

REFERENCES

1. ADRIEN J. – Ontogenèse du sommeil chez le mammifère. *In: Physiologie du Sommeil*, O. Benoît (éd.), Masson, Paris, 19-29, 1984.
2. ALIHANKA J., VAAHTORANTA K., SAARIKIVI J. – A new method of long-term monitoring of the ballistocardiogram, heart rate and respiration. *Am. J. Physiol.*, 240, 384-392, 1981.
3. ANDERS T.F. – Maturation of sleep patterns in the newborn infant. *Adv. Sleep Res.*, 2, 43-66, 1975.
4. ANDERS T., EMDE R., PARMELEE A. – *A manual of standardized terminology, techniques and criteria for scoring of states of sleep and wakefulness in newborn infants.* UCLA Brain Information Service/BRI Publications Office. Los Angeles, 1971.
5. ANDERS T.F., KEENER M.A., KRAEMER H. – Sleep-wake state organization: neonatal assessment and development in premature infants during the first year of life. II. *Sleep*, 8, 193-206, 1985.
6. ANONYMOUS. Monographic number, EEG in premature and full-term infants: developmental features and glossary, *Neurophysiol. Clin.* 29, 123-219, 1999.
7. ANONYMOUS. – Transitional phases of sleep. *Rev. EEG. Neurophysiol., 3 (special issue):* 1-180, 1973.
8. ARDUINI D., RIZZO G., CAFORIO L., BOCCOLINI M.R., ROMANIC C., MANCUSO S. – Behavioural state transitions in healthy band growth retarded fetuses. *Early hum. Dev.* 19, 155-165, 1989.
9. ASERINSKI E., KLEITMAN N. – A mobility cycle in sleeping infants as manifested by ocular and gross body activity. *J. Applied. Physiol.*, 8, 11-18, 1955.
10. BECKER P.T., THOMAN E.B. – Organization of sleeping and waking states in infants: consistency across contexts. *Physiol. Behavior.*, 31, 405-410, 1983.
11. BES F., SCHULZ H., NAVELET Y., SALZARULO P. – The distribution of slow-wave sleep across the night: a comparison for infants, children and adults. *Sleep*, 14, 5-12, 1991.
12. BONNET M.H., ARAND D.L. – Heart rate variability: sleep stage, time of night, and arousal influences. *Electroencephalogr. Clin. Neurophysiol.*, 102, 390-396, 1997.
13. BORBELY A. – A two-process model of sleep regulation. *Hum. Neurobiol.*, 1, 195-204, 1982.
14. BOSELLI M., PARRINO L., SMERIERI A., TERZANO M.G. – Effect of age on EEG arousals in normal sleep. *Sleep*, 21, 351-357, 1998.
15. CARSKADON M.A. – The second decade, In: *Sleep and its disorders in children.* C. Guilleminault (ed.) Addison-Wesley, Menlo Park: 99-125, 1982.
16. CARSKADON M.A., ACEBO C., RICHARDSON G.S., TATE B.A., SEIFER R. – An approach to studying circadian rhythms of adolescent humans. *J. Biol. Rhythms*, 12, 278-289, 1997.
17. CHAROLLAIS A., MICHEL C., VOISIN F. *et al.* – Sleep and waking states in premature new-borns less than 33 weeks of gestation (in preparation for *J. Biol. Rhythms*).
18. CHOQUET M., TESSON F., STEVENOT A., PREVOST F., ANTHEAUME M. – Les adolescents et leur sommeil: approche épidémiologique. *Neuropsychiatr. de l'enfance*, 36, 399-410, 1988.
19. CHUGANI H.T., PHELPS M.E., MAZZIOTTA J.C. – Positron emission tomography study of human brain functional development. *Ann. Neurol.*, 22, 487-497, 1987.
20. COBLE P.A., KUPFER D.J., REYNOLDS III C.F., HOUCK P. – EEG sleep of healthy children 6 to 12 years of age. *In: Sleep and its disorders in children*, C. Guilleminault (ed.), Raven Press, New York, 29-41, 1987.
21. COONS S. – Development of sleep and wakefulness during the first 6 months of life. *In: Sleep and its disorders in children*, C. Guilleminault (Ed.), Raven Press, New York, 17-27, 1987.
22. CROCHET S., SAKAÏ K. – Effects of microdialysis application of monoamines on the EEG and behavioral states in the cat mesopontine tegmentum. *Europ. J. Neurosc.*, 11, 3738-3752, 1999.

23. CURZI-DASCALOVA L. – Between sleep states transitions in premature babies. *J. Sleep Research.* 10, 153-158, 2001.

24. CURZI-DASCALOVA L., MIRMIRAN M. – *Manual of Methods of Recording and Analysing Sleep-Wakefulnes States in Pre-term and Full-term Infants.* Editions INSERM, Paris, 180p, 1996.

25. DAVENNE D., DUGOVIC C., FRANC B., ADRIEN J. – Ontogeny of slow wave sleep. *In: Slow Wave Sleep: Physiological, Pathophysiological and Functional Aspects*, A. Wauquier, C. Dugovic, M. Radulovacki (eds.), Raven Press, New York, 21-29, 1989.

26 DITTRICHOVA J. – Development of sleep in infancy. *J. Appl. Physiol., 21*, 1143-1146, 1966.

27. DREYFUS-BRISAC C. – The electroencephalogram of the premature infant and full-term new-born. Normal and abnormal development of waking and sleeping patterns. *In: Neurological and electroencephalographic correlative studies in infancy*, P. Kellaway, I. Petersen (eds.), Grune and Stratton, New York, 186-207, 1964.

28. DREYFUS BRISAC C., MONOD N. – The EEG of full-term newborns and premature infants. *In: Handbook of EEG and Clinical Neurophysiology, R.* Lairy (ed.) , *6B. Elsevier*, Amsterdam, 6-23, 1975.

29. EATON-EVANS J., DUGDALE A.E. – Sleep patterns of infants in the first year of life. *Arch. Dis. Child., 63, 6*, 647-649, 1988.

30. EEG-OLOFSONN O., PETERSEN I., SELLDEN U. – The development of the EEG in normal children from the age of 1 to 15 years: paroxysmal activity. *Neuropadiätrie., 4*, 375-404, 1971.

31. ELLINGSON R.J. – Transitional sleep in normal full-term newborns. *Electroencephalogr. Clin. Neurophysiol., 45*, 35P, 1978.

32. EMDE R.N., WALKER S. – Longitudinal study of infant sleep: results of 14 subjects studied at monthly intervals. *Psychophysiology, 13*, 456-461, 1976.

33. ERKINJUTTI M., KERO P., MIKOLA H., HALONEN J.P., SAINIO K. – The SCSB method in sleep scoring in infants. *In: Sleep 1986*, W.P. Koella, H. Schulz, P. Visser (eds.), Gustav Fischer Verlag, Stuttgart, 443-444, 1988.

34. FAGIOLI I., SALZARULO P. – Organisation temporelle dans les 24 heures des cycles de sommeil chez le nourrisson. Rev. EEG. *Neurophysiol., 12*, 344-348, 1982.

35. FEINBERG I. – Effects of maturation and aging on slow-wave sleep in man: implications for neurobiology. *In: Slow wave sleep: physiological, pathophysiological and functional aspects*, A. Wauquier, C. Dugovic, M. Radulovacki (Eds.), Raven Press, New York, 31-48, 1989.

36. FEINBERG I., THODE H.C., CHUGANI H.T., MARCH J.D. – Gamma distribution model describes maturational curves for delta wave amplitude, cortical metabolic rate and synaptic density. *J. Theor. Biol., 142, 2*, 149-161, 1990.

37. FERBER R. – *Solve your child's sleep problems.* First Fireside Book. Simon and Schuster, New York, 1986. Traduit en français par Navelet Y. ESF, Paris, 237p., 1990.

38. FERRI R., PARRINO L., SMERIERI A., TERZANO M.G., ELIA M., MUSUMECI S.A. – Cyclic alternating pattern and spectral analysis of heart rate variability during normal sleep. *J. Sleep Res.,9*, 13-18, 2000.

39. FOSTER J.C., GOODENOUGH F.L., ANDERSON J.E. – The sleep of young children. *J. Genet. Psychol., 35*, 201-232, 1928.

40. GABRIEL M., GROTE B., JONAS M. – Sleep-wake pattern in preterm infants under two different care schedules during four-day polygraphic recording. *Neuropediatrics., 12*, 366-373, 1981.

41. GAILLARD J.M. – Temporal organization of human sleep: general trends of sleep stages and their ultradian cyclic components. *L'Encéphale, 5*, 71-93, 1979.

42. GIBBS F.A., GIBBS E.L. – *Atlas of electroencephalography.* Addison-Wesley Press, Cambridge, Mass., 1952.

43. GROOME L.J., BENANTI J., BENTZ L., SINGH K.P. – Morphology of active sleep-quiet sleep transition in normal human term fetuses. *J. Perinat. Med., 24*, 171-176, 1996.

44. HADDAD G.G., JENG H.J., LAI T.L., MELLINS R.B. – Determination of sleep state in infants using respiratory variability. *Pediatr. Res., 21*, 556-562, 1987.

45. HAKAMADA S., WATANABE K., HARA K., MIYAZAKI S., KUMAGAI T. – Body movements during sleep in full-term newborn infants. *Brain Dev., 4*, 51-55, 1982.

46. HARPER R.M., LEAKE B., MIYAHARA L., HODGMAN J. – Temporal sequencing in sleep and waking states during the first six months of life. *Exp. Neurol., 71*, 294-307, 1981*a*.

47. HARPER R.M., LEAKE B., MIYAHARA L., HOPPEN-BROUWERS T., STERMAN M.B., HODGMAN J. – Development of ultradian periodicity and coalescence at 1 cycle per hour in electroencephalographic activity. *Exp. Neurol., 73*, 127-143, 1981*b*.

48 HARPER R.M., SCHECHTAN V.L., KLUGE K.A. – Machine classification of infant sleep state using cardiorespiratory measures. *Electroencephalogr. Clin. Neurophysiol., 67*, 379-387, 1987.

49. HOPPENBROUWERS T. – Polysomnography in newborns and young infants: sleep architecture. *J. Clin. Neurophysiol., 9*, 32-47, 1992.

50. HOPPENBROUWERS T., UGARTECHEA J.C., COMBS D., HODGMAN J.E., HARPER R.M., STERMAN M.B. – Studies of maternal-fetal interaction during the third trimester of pregnancy. I. Ontogenesis of the basic rest activity cycle. *Exp. Neurol.*, 136-153, 1978.

51. HUTTENLOCHER P.R. – Synaptic density in human frontal cortex – developmental changes and effects of aging. *Brain Res., 163*, 195-205, 1979.

52. JOUVET M. – *Le sommeil et le rêve*. Odile Jacob, Paris, 220p, 1992.
53. JOUVET-MOUNIER D., ASTIC L., LACOTE D. – Ontogenesis of the states of sleep in rat, cat and guinea-pig during the first post-natal month. *Dev. Psychol.*, *2*, 216-239, 1969.
54. KAHN A., VAN de MERCKT C., REBUFFAT E., MOZIN M.J., SOTTIAUX M., BLUM D. – Sleep problems in healthy preadolescents. *Pediatrics* 84, 542-546, 1989.
55. KARACAN I., ANCH M., THORNBY J.I., OKAWA M., WILLIAMS R.L. – Longitudinal sleep patterns during pubertal growth: four-year follow-up. *Pediat. Res.*, *9*, 842-846, 1975.
56. KARCH D., ROTHE R., JURISCH R., HELDT-HILDEBRANDT R., LUBBESMEIER A., LEMBURG P. – Behavioral changes and bioelectric brain maturation of preterm and fullterm newborn infants: a polygraphic study. *Dev. Med. Child. Neurol.*, *24*, 30-47, 1982.
57. KATZ-SALAMON M., FORSSBERG H., LAGERHANS H. – The Stockholm neonatal project: very-low-birth weight infants of the late 20th century in Stockholm. The Stockholm Neonatal project. *Acta pediatrica Suppl.* 419, 1-3, 1997.
58. KLEITMAN N. – *Sleep and wakefulness*. 2nd ed. Chicago, University of Chicago Press. Chicago, 1963.
59. KLEITMAN N., ENGELMANN T.G. – Sleep characteristics of infants. *J. Appl. Physiol.*, *6*, 269-282, 1953.
60. KTONAS P.Y., BES F.W., RIGOARD M.T., WONG C., MALLART R., SALZARULO P. – Developmental changes in the clustering pattern of sleep rapid eye movement activity during the first year of life: a Markov-process approach. *Electroencephalogr. Clin. Neurophysiol.*, *75*, 136-140, 1990.
61. LABERGE L., CARRIER J., LESPÉRANCE P., LAMBERT C., VITARO F., TREMBLAY R.E., MONTPLAISIR J. – Sleep and circadian phase characteristics of adolescent and young adult males in a naturalistic summer time condition. *Chronobiol. Int.* 17, 489-501, 2000.
62. LABERGE L., PETIT D., SIMARD C., VITARO F., TREMBLAY R.E., MONTPLAISIR J. – Development of sleep patterns in early adolescence. *J. Sleep Res.*, 10, 59-67, 2001.
63. LOMBROSO C.T. – Quantified electrographic scales on 10 pre-term healthy newborns followed up to 40-43 weeks of conceptional age by serial polygraphic recordings. *Electroencephalogr. Clin. Neurophysiol.*, *46*, 460-474, 1979.
64. LOMBROSO C.T., MATSUMIYA Y. – Stability in waking-sleep states in neonates as a predictor of long-term neurologic outcome. *Pediatrics*, *76*, 52-63, 1985.
65. MADANSKY D., EDELBROCK C. – Co-sleeping in a community sample of 2 and 3 year-old children. *Pediatrics*, 86, 197-203, 1990.
66. MANTZ J., MUZET A., WINTER A.S. – The characteristics of sleep-wake rhythm in adolescents aged 15-20 years. A survey made at school during ten consecutive days. *Arch. Pediatr.* 7, 256-262, 2000.
67. MARSHALL L., MOLLE M., MICHAELSEN S., FEHM H.L., BORN J. – Slow potential shifts at sleep-wake transitions and shifts between NREM and REM sleep. *Sleep*, 19, 145-151, 1996.
68. MEIJER A.M., HABEKOTHE H.T., VAN den WITTENBOER G.L. – Time in bed, quality of sleep and school functioning of children. *J. Sleep Res.*, 9, 145-153, 2000.
69. MICHEL-ADDE C., SAMSON-DOLLFUS D., MARRET S.- Electroencéphalogrammes des prématurés de moins de 33 semaines enregistrés pour des buts diagnostiques et pronostiques, entre janvier 2000 et juin 2001: répartition des différents stades de veille / sommeil. *Rev. EEG Neurophysiol. (en préparation)*.
70. MINORS D.S., WATERHOUSE J.M. – Rhythms in the infant and the aged. *In: Circadian rhythms and the human.* Wright and son. Bristol. 166-186, 1981.
71. MIRMIRAN M., KOK J.H. – Circadian rhythms in early human development. *Early Hum. Dev.* 26, 121-128, 1991.
72. MIRMIRAN M., KOK J.H., BOER K., WOLF H. – Peri-natal development of human circadian rhythms: role of the fœtal biological clock. *Neurosci. Behav. Rev.*, 16, 371-378, 1992.
73. MIRMIRAN M., VAN SOMEREN E. – The importance of REM sleep for brain maturation. *J. Sleep Res.* 2, 188-192, 1993
74. MONOD N., CURZI-DASCALOVA L. – Les états transitionnels de sommeil chez le nouveau-né à terme. *Rev. EEG Neurophysiol.*, *3*, 87-96, 1973.
75. MONOD N., DREYFUS-BRISAC C. – Les premières étapes de l'organisation du sommeil chez le prématuré et le nouveau-né à terme. *In: Le Sommeil de nuit normal et pathologique: études électroencéphalographiques*, Fischgold (ed.), Masson, Paris, 118-148, 1965.
76. NIJHUIS J.G., VAN de PAS M., JONGSMA H.W. – State transitions in uncomplicated pregnancies after term. *Early Hum. Dev.* 52, 125-132, 1998.
77. NOËL S. – Résultats d'un enquête. *Neurone*, 5, 114-121, 2000.
78. PALM L., PERSSON E., ELMQVIST D., BLENNOW G. – Sleep and wakefulness in normal preadolescent children. *Sleep*, *12*, 299-308, 1989.
79. PARMELEE JR. A.H., SCHULTE F.J., AKIYAMA Y., SCHULTZ M.A., STERN E. – Maturation of EEG activity during sleep in premature infants. *Electroencephalogr. Clin. Neurophysiol.*, *24*, 319-321, 1968.
80. PARMELEE A.H., WENNER W.H., SCHULZ H.R. – Infant sleep patterns: from birth to 16 weeks of age. *Jl. Pediatrics*, *65*, 4, 576-582, 1964.
81. PILCHER J.J., OTT E.S. – The relationships between sleep and measures of health and well-being in college student: a repeated measure approach. *Behav. Med.* 23, 170-178, 1998.

82. PRECHTL H.F.R. – The behavioural states of the newborn infant. *Brain Res.*, *76*, 1304-1311, 1974.

83. PRECHTL H.F.R., NIJHUIS J., MARTIN C.B. – Prenatal development of post-natal behaviour states. In: Koella W.P., Ruther E., Schulz H. (eds.) *Sleep 84.*, Gustave Fischer Verlag, Stuttgart, 97-98, 1985.

84. RICHARDS J.M, ALEXANDER J.R., SHINEBOURNE E.A., DE SWIET M., WILSON A.J., SOUTHALL D.P. – Sequential 22 hour profiles of breathing patterns and heart rate in 110 fullterm infants during their first 6 months of life. *Pediatrics*, *74*, 763-777, 1984.

85. ROSS J.J., AGNEW H.W., WILLIAMS R.L., WEBB W.B. – Sleep patterns in preadolescent children: an EEG-EOG study. *Pediatrics*, *42*, 324-335, 1968.

86. SADEH A., LAVIE P., SCHER A., TIROSH E., EPSTEIN R. – Actigraphic home-monitoring sleep-disturbed and control infants and young children: a new method for pediatric assessment of sleep-wake patterns. *Pediatrics*, *87*, 494-499, 1991.

87. SALZARULO P. – The transitional phases between wakefulness and sleep and between slow sleep and fast sleep in human psychophysiology. *Rev. EEG Neurophysiol., 3 (special issue)*, 3, 69-86, 1973.

88. SALZARULO P., FAGIOLI I., PEIRANO P., BES F., SCHULZ H. – Levels of EEG background activity and sleep states in the first year of life. *In: Phasic Events and Dynamic Organization of Sleep*, M.G. Terzano, P.L. Halasz, A.C. Declerck (eds.), Raven Press, New York, 53-63, 1991.

89. SAMSON-DOLLFUS D. – *L'EEG du prématuré jusqu'à l'âge de 3 mois et du nouveau-né*. Foulon, Paris, 160 p., 1955.

90. SAMSON-DOLLFUS D. – Electroencéphalographie de l'enfant (2ème édition), Masson, Paris, 136p, 2001.

91. SAMSON-DOLLFUS D. – Sleep and sudden infant death syndrome. *In: Sleep in Medical and Neuropsychiatric Disorders*, S. Smirne, M. Franceschi, L. Ferini-Strambi (ed.), Masson, Milan, 65-73, 1988.

92. SAMSON-DOLLFUS D., FORTHOMME J., CAPRON E. – EEG of the human infant during sleep and wakefulness during the first year of life. Normal patterns and their maturational changes; abnormal patterns and their prognostic significance. *In: Neurological and electroencephalographic correlative studies in infancy*, P. Kellaway (ed.), 208-229, 1964.

93. SAMSON-DOLLFUS D., GEFFROY D., NOGUES B., MENESES DA PAULA R. – Le sommeil paradoxal à ondes lentes au cours de la première année de vie. *Rev. EEG. Neurophysiol.*, *16*, 1-7, 1986.

94. SAMSON-DOLLFUS D., NOGUES B., DELAGREE E. – Aspects électroencéphalograpiques et polygraphiques des transitions veille-sommeil chez des nourrissons normaux âgés de 2 à 12 mois. *Rev. EEG. Neurophysiol.*, *11*, 23-27, 1981.

95. SAMSON-DOLLFUS D., NOGUES B., DELAPIERRE G. – Apport de l'analyse automatique des états veille-sommeil au cours de la première année de la vie. *Neurophysiol. Clin.*, *22*, 133-149, 1992.

96. SAMSON-DOLLFUS D., NOGUES B., MENARD J.F., BERTOLDI-LEFEVER I., GEFFROY D. – Delta, Theta, Alpha and Beta power spectrum of sleep electroencephalogram in infants aged two to eleven months. *Sleep.*, *6*, 376-383, 1983.

97. SAMSON-DOLLFUS D., NOGUES B., VERDURE-POUSSIN A., MALLEVILLE F. – Electroencéphalogramme du sommeil de l'enfant normal entre 5 mois et 3 ans. *Rev. EEG. Neurophysiol.*, *7*, 335-345, 1977.

98. SAMSON-DOLLFUS D., POUSSIN A. – Les états transitionnels précédant et suivant les phases de sommeil rapide avec mouvements oculaires chez l'enfant normal de 8 à 30 mois. *Rev. EEG. Neurophysiol.*, *3*, 1, 97-102, 1973.

99. SANDER L.W.., JULIA A., STECHLER G., BURNS P. – Continuous 24-hour interactional monitoring on infants reared in two caretaking environments. *Psychosom. Med.*, *34*, 270-282, 1972.

100. SANDER L.W., STECHLER G., BURNS P., JULIA H. – Early mother-infant interaction and 24-hour patterns of activity and sleep. *J. Am. Acad. Child Psychiatry*, *9*, 103-123, 1970.

101. SCHER M.S., STEPPE D.A., DAHL R.E., ASTHANA S., GUTHRIE R.D. – Comparison of EEG sleep measures in healthy full-term and preterm infants at matched conceptional ages. *Sleep*, *15*, 442-448, 1992.

102. SCHINKODA H., MATSUMOTO K., PARK Y.M., NAGASHIMA H. – Sleep-wake habits of school-age children according to grade. *Psychiatry Clin. Neurosci.* 54, 287-289, 2000.

103. SCHULZ H., BES E., SALZARULO P. – Rhythmicity of slow wave sleep developmental aspects. *In: Slow wave sleep: physiological, pathophysiological and functional aspects*, A. Wauquier, C. Dugovic, M. Radulovacki (eds.), Raven Press, New York, 49-60, 1989.

104. SCHULZ H., SALZARULO P., FAGIOLO I., MASSETANI R. – REM latency: development in the first year of life. *Electroencephalogr. Clin. Neurophysiol.*, *56*, 316-322, 1983.

105. SMITH J.R., KARACAN I., YANG M. – Ontogeny of delta activity during human sleep. *Electroencephalogr. Clin. Neurophysiol.*, *43*, 229-237, 1977.

106. SORENSEN E., URSIN R. – Sleep habiits among adolescents, *Tidsskr. Nor Laegeforen,* 121, 331-333, 2001.

107. STERIADE M., HOBSON J.A. – Neuronal activity during the sleep-waking cycle. *Prog. Neurobiol.*, *6*, 155-376, 1976.

108. THOMAN E.B. – Sleeping and waking states in infants: a functional perspective. *Neurosci. Biobehav. Rev.*, *14*, 93-107, 1990.

109. THOMAN E.B., GLAZIER R.C. – Computer scoring of motility patterns for states of sleep and wakefulness: human infants. *Sleep.*, *10*, 122-129, 1987.

110. VANHULLE C., SAMSON-DOLLFUS D. – Actimétrie: analyse simplifiée des rythmes veille-sommeil chez le nouveau-né. Résultats préliminaires. *Neurophysiol. Clin.*, *26*, 403-416, 1996.

111. VAN SWEDEN B., KOENDERINK M., WINDAU G., VAN DE BOR M., VAN BEL F., VAN DIJK J.G., WAUQUIER A. – Long-term EEG monitoring in the early premature: developmental and chronobiological aspects. *Electroencephalogr. Clin. Neurophysiol.*, *79*, 94-100, 1991.

112. VECCHIERINI-BLINEAU M.-F., NOGUES B., LOWET S. – Étude des mouvements corporels au cours du sommeil chez le nourrisson: comparaison entre nourrissons témoins et nourrissons issus d'une fratrie de mort subite inexpliquée du nourrisson. *Rev. EEG. Neurophysiol.*, *16*, 21 -27, 1986.

113. VOISIN F., CHAROLLAIS A., RADI S. *et al.* – Rest-Activity cycle in premature newborns. *Early human development, (en préparation), 2001.*

114. WATT J.E., STRONGMAN K.T. – The organization and stability of sleep states in fullterm, preterm and small-for-gestational-age infants: a comparative study. *Dev. Psychobiol.*, *18*, 151-162, 1985.

115. WEBB W.B. – *Development of human napping. Sleep and Alertness: Chronobiological, Behavioral, and Medical Aspects of Napping.* Raven Press, New York, 31-51, 1989.

116. WOLFF P.H. – The development of behavioral states and the expression of emotions in early infancy. *In: New proposals for investigation.* The University of Chicago Press, Chicago, 1987.

117. YAMAGUCHI N., KAWAGMURA S., MAEDA Y. – The survey of sleeping time of junior high school students: a study on the sleep questionnaire. *Psychiatry Clin. Neurosci.*, *54*, 290-291, 2000.

Chapter 3

Neurobiology of the sleep-wake cycle

J. Adrien

INSERM U288, Faculté de Médecine Pitié-Salpêtrière, Paris, France

Although all vertebrates demonstrate a basic rest-activity cycle, associations between the polygraphic criteria for sleep and wakefulness are only found in birds and mammals. In the case of the latter, sleep presents as an alternation between two main states, NREM (non-rapid eye movement) sleep and REM (rapid eye movement) sleep, whose respective quantities and rhythms are characteristic for each species.

Qualitatively speaking, states of sleep and wakefulness are traditionally objectified using a group of three physiological variables: the electroencephalogram (EEG) oscillation frequency, muscle tone as shown on the electromyogram (EMG), and eye movements recorded on the electro-oculogram (EOG) (fig. 3.1). Diagrammatically, the nervous system is activated during wakefulness (W) and REM sleep, as shown by the desynchronised EEG pattern and the intense activity of most neurons. Moreover, during REM sleep muscle tone is seen to be absent with bursts of rapid eye movement; NREM sleep, on the other hand, is characterised by a synchronised EEG pattern (spindles and slow waves), corresponding to a phase in which the nervous functions slow down.

Two theories have been put forward to account for the regulatory mechanisms of states of sleep and wakefulness. The first, or **passive** notion, considers sleep simply as a consequence of the cessation of wakefulness. It is the oldest notion, based on a group of experimental arguments conferring a preponderant role to the reticular waking system. The other theory suggests that sleep results from specific mechanisms coming into play. It thus advances the idea of the **active** intervention of certain structures in triggering and maintaining each sleep state.

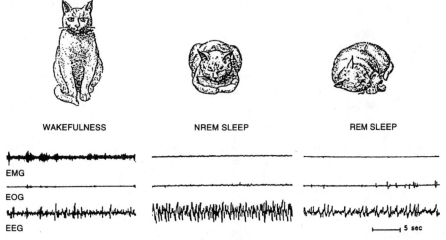

Figure 3.1. The three main states of sleep and wakefulness in the cat: behaviour and polygraphic criteria (from Jouvet, *Pour la Science* 1979).

31

Current evidence suggests that these two notions are not mutually exclusive, but that on the contrary, they describe complementary mechanisms. Sleep-wake regulations are thus thought to result from constant interaction between the systems of sleep and wakefulness, rather like a set of scales in which each side alternately weighs more than the other.

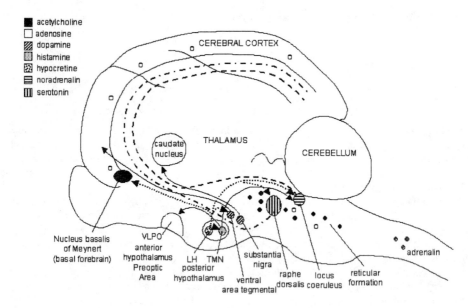

Figure 3.2. Diagram of a sagittal cross section of the brain, showing the main structures involved in regulating the states of sleep and wakefulness

WAKEFULNESS

Wakefulness manifests itself both cortically (EEG activation) and behaviourally (notably as a motor aspect). It is controlled by a complex group of inter-regulated and redundant systems (fig. 3.2). In fact, as the past few decades have shown, no one system is uniquely critical in maintaining wakefulness.

The following structures can be listed, while in no way reflecting their order of importance in regulating wakefulness [23]:

The **reticular formation**, in the midbrain, pontine and medullary regions. This is a system of diffused projection, which plays a preponderant role in cortical wakefulness (EEG desynchronisation) in three main ways (fig. 3.3): one pathway acting as a relay in the non specific thalamic nuclei, one pathway passing through the latero-posterior hypothalamus, and a third, ponto-basalo-cortical pathway which acts as a relay in the Meynert nucleus (see below).

Thus [23, 34]:

At the level of the rostral midbrain reticular formation (cuneate nucleus):

- Neuron activity increases just before cortical arousal and diminishes just before cortical synchronisation.
- Electrical stimulation elicits cortical desynchronisation (cortical wakefulness).
- Electrolytic lesion induces an increase in cortical synchronisation. The same effect is produced transitionally by infusing ibotenic acid in the reticular formation, treatment which locally destroys neuron cell bodies, while leaving the pathway fibres intact.

Meynert's nucleus, a group of neurons which form part of the basal system. It plays a role in the cortical activation of wakefulness [27], but can only do so through its cholinergic afferences which derive from the reticular formation. This nucleus functions in synergy with the posterior hypothalamus and intralaminary thalamus, through cholinergic and GABAergic interactions between the different structures [ref. in 23].

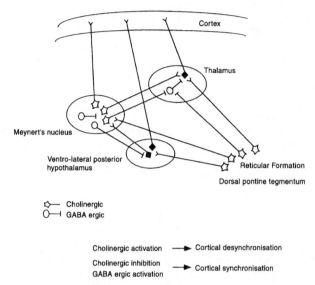

Figure 3.3. Diagram of the different systems involved in the cortical wakefulness desynchronisation system: cholinergic influx is chiefly responsible for inhibiting the functioning of thalamic synchronised rhythm generators, inducing activation of the thalamo-cortical neurons, probably via aspartate or glutamate (from Steriade and McCarley, 1990, © Plenum Press, reproduced with permission).

From a neurochemical point of view, the essential neuromediator of these two waking systems is *acetylcholine*.

Indeed:

- A large number of neurons in the reticular formation as well as in the Meynert's nucleus are cholinergic.
- The release of acetylcholine in the cortex is optimal during wakefulness and minimal during NREM sleep.
- Cortical neurons are activated when acetylcholine is applied to their membrane (micro-iontophoresis technique).
- Nicotine and the anticholinesterasics induce cortical and behavioural wakefulness.
- Muscarinic antagonists (atropine) abolish rapid cortical activity.

Also observed is the intervention of glutamatergic reticulo-basal projections as well as GABAergic basal-cortical pathways which inhibit the intra-cortical interneuron inhibitors (ref. in [23]).

The **locus coeruleus**, with *norepinephrine,* a ubiquitous projection system which innervates the cerebral cortex. It is implied in cortical activation [3, 16], and receives afferences, notably from the posterior hypothalamus and the bulbar adrenergic systems. The coerulean norepinephrine system globally stimulates wakefulness, but also intercedes more specifically in the mechanisms of selective attention and information discrimination.

The **medulla** *epinephrine* system. This represents the main sympathetic excitatory relays which notably regulate sympathetic tone (control of wakefulness autonomic reactions). Moreover, this system is the main source of the excitatory afferences of the locus coeruleus.

The **nigro-striatal** *dopaminergic* system, which issues from the substantia nigra, plays a role in maintaining behavioural wakefulness, whereas the mesolimbic and mesocortical system, issuing from the ventro-medial tegmentum (A10 group) is essential for the processes of cognition and focused attention (ref. in [22]). These dopaminergic systems may be seen as the organizers of the appropriate behaviour for the environment.

The **ventro-lateral posterior hypothalamus.** This is responsible for regulating wakefulness through the intermediary of the *histaminergic* system of which it contains all the cell bodies. It sends projections forward to the anterior hypothalamus, the Meynert basal nucleus, the cerebral cortex (ref. in [23]), and back to the catecholamine systems and the raphe.

- The neurons of the ventro-lateral posterior hypothalamus are selectively active during wakefulness.

-   Electrolytic lesion of this region induces cortical slow waves, and the destruction of cell bodies in this region through ibotenic acid will elicit hypersomnia (transient).

-   Inactivating the neurons of the ventro-lateral posterior hypothalamus by infusing a GABAergic agonist, muscimol, leads to the quasi-disappearance of wakefulness [26].

-   Infusing a histamine synthesis inhibitor in the ventro-lateral posterior hypothalamus will increase sleep. Conversely, histaminergic agonists will accentuate wakefulness. This action takes place via the histaminergic receptors H_1 (activation of thalamo-cortical neurons) and H_2 (activation of basal nuclei neurons) [23].

In the most dorsal part of the **posterior hypothalamus**, *hypocretine (or orexine)* neurons are found, which carry wake-promoting effects, notably in reference to recent studies on narcolepsy (ref. in [23]). These neurons send excitatory projections onto cholinergic (reticular formation and basal forebrain), histaminergic (tubero-mammilary nucleus) and monoaminergic (ventral tegmental area, locus coeruleus, raphe nuclei) systems. Thus, the hypocretinergic network activates all major wake-promoting systems, and is also involved in metabolic control (it regulates food intake and energy expenditure). A functional deficit of this system induces somnolence and is involved in narcolepsy (36).

-   Hypocretin release in the CSF is maximal during the waking period, and intraventricular injection of hypocretin receptor antagonists induces an enhancement of wakefulness and of locomotor activity (see 36).

-   The wake-promoting action of hypocretin is absent in mutant mice that do not express H_1 histaminergic receptors, and its hyperlocomotor effect is blocked by dopaminergic and serotoninergic antagonists (see 36).

From the pharmacological point of view, as a general rule, all the components which stimulate catecholaminergic and histaminergic neurons facilitate wakefulness to the detriment of sleep.

The **raphe system,** with *serotonin*, plays a complex role in regulating wakefulness. Indeed the serotoninergic neurons are active during wakefulness, but paradoxically their inactivation (pharmacological or lesional) leads to insomnia [16].

- During wakefulness, the serotoninergic neurons of the dorsal raphe are active, with a maximal axonal release of 5-HT (9, ref. in [29]).

- In both the cat and the rat, the inhibition of serotonin synthesis (5-HT) by parachlorophenylalanine or PCPA (which blocks the hydroxylase tryptophan enzyme required to transform tryptophan into 5-hydroxytryptophan, 5-HTP) leads to quasi-permanent wakefulness for 2 to 3 days. In this model, sleep is restored by systematically injecting 5-HTP, i.e. by renewing the capacity for serotonin synthesis (11 and ref. in [16]).

Tryptophan $\Rightarrow$ 5-hydroxytryptophan $\Rightarrow$ 5-hydroxytryptamine (5-HT)

Tryptophan hydroxylase

A contradiction exists between the serotoninergic system's high activity level during wakefulness (implying that 5-HT is necessary to maintain wakefulness), and the fact that the depletion of 5-HT leads to insomnia. This paradox may be explained by the fact that serotonin plays a multiple role in controlling states of sleep and wakefulness (fig. 3.4): during wakefulness, through its axonal release, it is thought to "prepare" sleep by favouring the synthesis of hypnogenic substances in certain target structures which are still to be defined [9], thus promoting sleep onset. For the duration of sleep (NREM and REM sleep), serotonin would thus intervene in the "permissive" control of sleep [18].

The serotoninergic system is thus thought to play both a direct and differed role in regulating states of sleep and wakefulness.

Adenosine, like serotonin, is released during wakefulness (especially during prolonged wakefulness, during which it increases sleepiness), and secondarily induces the emergence of sleep [28]. Produced from adenosine triphosphate (ATP) hydrolysis, in function of the output of cellular energy, adenosine will gradually induce the inhibition of the systems of wakefulness, notably through the intermediary of adenosinergic A_1 receptors in the basal telencephalon and cortex. Caffeine, a substance which has long been known to stimulate wakefulness, is, moreover, an antagonist of these receptors.

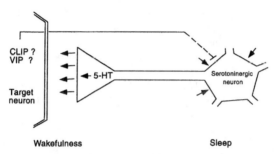

Figure 3.4. Diagram of the double role of serotonin (5-HT) in regulating sleep-wake cycles. – *On the left*: during wakefulness, serotoninergic neurons are active and 5-HT released at the axonal endings. Substances (peptide) are synthesised at the level of the serotoninergic neuron target structures. Among other unknown factors, Corticotrophin -Like Intermediate Lobe Peptide (CLIP) and Vaso-Intestinal Peptide (VIP) are two plausible candidates. – *On the right*: beyond a certain level of concentration, these substances provoke, in return, the inactivation of the serotonin neuron, via the increase in levels of 5-HT in its somatodendritic environment. Conditions are ideal for the onset of sleep (this is the "permissive" role of the serotoninergic system, whose neuronal activity diminishes during NREM sleep, and is totally inhibited during REM sleep).

NREM SLEEP

In the normal course of sleep, NREM sleep occurs before REM sleep.

The **hypothalamus** appears to play a determining role at the stage of sleep onset, through two simultaneous types of action: activation of its anterior part, the preoptic region, which is thought to be a hypnogenic structure [11, 31], and the inhibition by the latter of the ventro-lateral posterior hypothalamus [11, 23, 31] and the midbrain reticular formation, two structures implicated in wakefulness.

-   In cats rendered insomniac by PCPA treatment, the only central site where the infusion of 5-HT or 5-HTP induces sleep is the preoptic region of the anterior hypothalamus [11].
-   Lesion in the preoptic region in the cat will result in quasi-total insomnia which may last several weeks [31].
-   Also found in the preoptic region of the anterior hypothalamus are neurons which are selectively active during NREM sleep [35].
-   Inhibiting the preoptic region by bilaterally infusing this structure with muscimol, will induce insomnia [31].
-   Inhibiting the posterior hypothalamus by the *in situ* infusion of muscimol, or alphafluoromethylhistidine, a histamine synthesis enzyme inhibitor, will trigger hypersomnia in NREM sleep [23, 24].

Serotonin would play a special role in sleep initiation, by facilitating the production of "sleep-inducing" (or wake-reducing) substances in the hypothalamic preoptic area. At the same level, GABAergic neurons would participate in the inhibition of the wake-promoting systems. Finally, adenosine would also inhibit the latter systems, notably after prolonged wakefulness when it induces somnolence. It is not known whether adenosine regulates the process of sleep initiation under normal conditions, but its involvement in the regulation of the states of wakefulness and sleep, if only as an homeostatic factor, is more and more taken into account [28].

This stage of sleep onset is automatically followed by the systems of NREM sleep regulation coming into play. However, the mechanisms involved are still largely unknown.

The anterior hypothalamus, through its GABAergic inhibitory inputs on the wake-promoting systems, participates in the regulation of SWS, but essentially GABAergic neurons of the basal forebrain are involved in this control.

The GABAergic neurons in **Meynert's nucleus** appear to play an important controlling role (fig. 3.3). They appear to inhibit both the cholinergic neurons from the same nucleus and the posterior hypothalamus neurons, which are both known to be involved in cortical wakefulness. The activation of these GABAergic neurons resulting in the simultaneous inhibition of these two waking structures, appears to facilitate the onset of spindles and slow waves in the cortex [8, 34].

From a neurophysiological point of view, the spindles observed on EEG seem to be produced by a thalamic generator (**reticular nucleus**). These would result from the interaction between the GABAergic neurons of this nucleus and the thalamocortical neurons (fig. 3.3).

- Electrical stimulation of the thalamus non specific to a frequency close to that of spindles, will entrain cortical synchronisation.

As for slow delta waves, these are generated by the hyperpolarisation of the thalamocortical neurons, resulting from a reduction of the cholinergic influence of the brain stem [8].

From a pharmacological point of view, the NREM sleep state is known to depend on serotoninergic transmission and the production of certain peptides (blocking of the $5-HT_2$ serotoninergic receptors [12] as well as α-MSH, D_2 prostaglandin, insulin, adenosine (ref. in [28], resulting in an increase in the levels of NREM sleep.

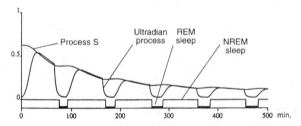

Figure 3.5. Model representing the regulation of NREM sleep in man during the night. Two processes are involved: the first, ultradian controls the periodicity of phases of NREM sleep, and the other, an accumulative process (S) builds up during wakefulness (in proportion to the duration of the latter) and gradually declines during sleep. Episodes of REM sleep appear in between those of NREM sleep, as defined by the ultradian process (after Achermann *et al.*, 1990).

The regulation of NREM sleep in man, during the course of the night has been conceptualised as a model taking three main variables into account [1, 4]: a circadian process, which will not be dealt with here, an ultradian process which shows the periodicity of phases of NREM sleep, and an "accumulative" process (S) which builds up during sleep (fig. 3.5). The S variable is illustrated by the decline in the power of slow waves in the course of the night.

This model has a basically descriptive phenomenological value but does provide a simulation of a certain number of chronobiological or clinical situations. Moreover, it has recently been combined with the more neurophysiological model of REM sleep (see below).

REM SLEEP

The main regulation mechanisms of this state are relatively well known. They can be diagrammatically shown as series of three interlocking groups (fig. 3.6):

- The first group is responsible for directly controlling sleep (onset, maintenance, interruption): this notably involves the ("executive") and authorising ("permissive") systems which order REM sleep - systems composed of relatively well identified neuronal networks, located in the brain stem and spinal medulla [16, 30, 34].

- The second group affects the direct command system by different means: either by neuronal projection onto the executive and permissive structures, or by secreting neuromodulators or peptides whose exact nature is still unknown. This system is thought to be essentially composed of hypothalamo-hypophyseal structures [19].

- Finally, the third group involves more specifically humoral aspects, with a number of endogenous and exogenous substances which "facilitate" wakefulness or sleep [5, 18]. The secretion and/or efficacy of the latter rely on circadian factors which will not be discussed here.

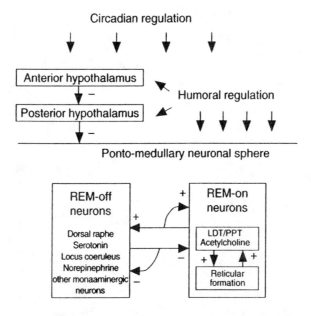

Figure 3.6. Diagrammatic summary of the different hierarchical groups responsible for regulating REM sleep. In the ponto-medullary neuronal sphere, the model is inspired by that of Steriade and McCarley [34], demonstrating the interactions between "REM-off" and "REM-on" neuronal populations. The following stages preside during the course of the REM cycle:

1. The slowing then stopping of REM-off neuronal activity disinhibits the REM-on neuron population.

2. As a result of this disinhibition, the REM-on neurons become increasingly active, a progression due to the excitatory interconnections of the REM-on population.

3. The REM-off neurons become active after being excited by the REM-on neurons. When the REM-off population is sufficiently active, it inhibits the REM-on neurons ending the REM phase.

4. The REM-off population becomes less and less active due to the negative retroactive loop effect of REM-off neurons on themselves, and the cycle reverts to stage 1.

Base command [30, 34]

The production of REM sleep depends directly upon two groups of structures: the neuronal groups, on one hand, responsible for REM executive mechanisms, which are selectively active for the entire duration of REM sleep ("REM sleep-on" neurons) and on the other, by those which control the permissive mechanisms, i.e. those which cease to act for the full duration of REM slep ("REM sleep-off" neurons).

The **executive** REM sleep mechanisms are basically ensured by the *cholinergic* and *cholinoceptive* neurons, located notably in the medullary region in the magnocellular nucleus, and at the pontine level, in the Pedunculopontine Tegmentum (PPT), the Laterodorsal Tegmentum (LDT), the locus coeruleus α (LCα) and the peri-locus coeruleus α (peri-LCα). (Unlike the noradrenergic locus coeruleus proper, the LCα is a cholinergic cell group).

- Bilateral lesion of the dorsomedial pontine tegmentum leads to insomnia in REM sleep [30].
- Infusion of the cholinomimetic, carbachol, in the region of the locus coeruleus α will induce episodes of REM sleep [30] (fig. 3-7).
- The medullary neurons (magnocellular and parvocellular) are the main cholinergic afferents of the peri-LCα and LCα nuclei [30, 34].
- The neurons of the medullary magnocellular nucleus are selectively active during REM sleep [30].

Likewise, REM phasic activity (the prototypes of which are *ponto-geniculo-occipital* activity [PGO] and rapid eye movements), corresponding to intense neuronal activation, originates mainly in cholinergic structures which are also located in the dorsolateral tegmentum of the brain stem (the rostral part of LCα, LDT, PPT and Sakai's X region (ref. in [10]).

- Neurons of the structures making up the pontine generator of phasic activity are selectively released during PGO ("*PGO-on*" neurons) [30, 34].
- The infusion of carbachol in the PPT nucleus induces a burst of REM and PGO activity [30, 34].

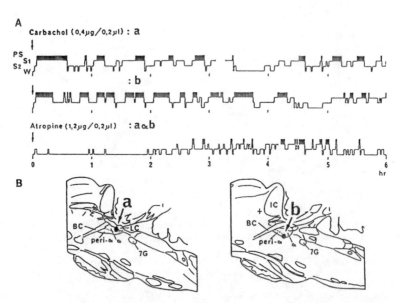

Figure 3.7. Experimental demonstration of the cholinergic nature of the executive mechanisms of REM sleep, in the α locus coeruleus region. In the cat, infusion of the cholinomimetic, carbachol in the locus coeruleus α (a) or the peri-locus coeruleus α (b) (both cholinergic) leads to hypersomnia in REM sleep. The infusion of atropine provokes the reverse effect (Sakai, 1988).

Permissive REM mechanisms depend on *monoaminergic* structures more diffusely disposed in the brain stem: these consist, in particular, of the norepinephrine neurons of the locus coeruleus and the serotoninergic neurons of the raphe (REM-off neurons) [30, 34].

- The neurons of the locus coeruleus proper, like those of the dorsal raphe, show sustained activity during wakefulness, slowing their release during NREM sleep, and ceasing activity throughout REM sleep.

REM sleep production thus corresponds to the activation of all these cholinergic executive systems, and the simultaneous inhibition of all the permissive monoaminergic systems, as shown in the interactive diagram in figure 3.6 [34].

The "permissive" serotoninergic influence notably occurs through the intermediary of the A_1 5-HT receptors located in the pontine tegmentum (ref. in [29]) as well as through the B_1 5-HT receptors [7].

Neurohumoral aspect

A number of mechanisms essentially involving the hypothalamo-hypophysiary axis, influence systems of REM sleep production. Among these, the anterior hypothalamus, particularly the preoptic region, plays a determining role. Indeed, the preoptic region is responsible for inhibiting the ventro-lateral posterior hypothalamus [23, 31]. But the latter not only favours wakefulness, as has already been seen, but also exerts tonic inhibition of the ponto-medullary systems which control REM sleep. The preoptic region thus facilitates sleep by ensuring the disinhibition of the mechanisms which produce REM sleep, via the posterior hypothalamus (fig. 3.8).

- Lesion of the anterior hypothalamus (preoptic region) with ibotenic acid will induce quasi-total long lasting insomnia in cats. In the same way, the inactivation of this area with muscimol (GABAergic agonist) will bring about insomnia [31].
- Infusing muscimol in the posterior hypothalamus will trigger spectacular hypersomnia for several hours [24].

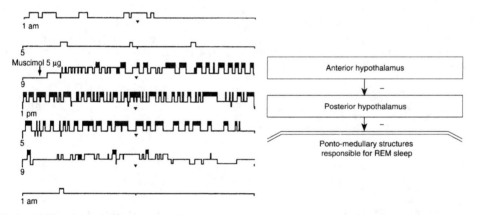

Figure 3.8. Experimental demonstration of the respective influences of the anterior hypothalamus and the posterior hypothalamus on the production of REM sleep. *On the left*, in the cat, the destruction of neurons from the anterior hypothalamus leads to the total abolition of REM sleep (first two lines of the hypnogram). Conversely, the infusion of muscimol in the posterior hypothalamus, thereby inactivating the latter, induces hypersomnia in REM sleep (Sallanon *et al.*, 1989). *On the right*: the hypothalamic influences on REM sleep production may be diagrammatically shown as a series of double inhibitions.

The nature of the neurochemical substrate(s) involved in this control is unknown.

Finally, in terms of neuromodulations, a number of substances are known to be capable of facilitating REM sleep. These include the vaso-intestinal peptide or VIP, which facilitates NREM sleep and REM sleep, Corticotropin-Like Intermediate Peptide (CLIP), the growth hormone and somatostatin, which facilitate REM sleep (6 and ref. in [5]).

Other influences

One of the many influences is the temperature at which the metabolism [20] will modify the balance of the neuronal groups directly controlling REM. These hypothalamic structures which are responsible for thermoregulation thus closely interact with those which control the states of sleep and wakefulness [26]. Moreover both temperature and metabolism are able to exert direct action on the permissive mechanisms of sleep: which explains why, in the pontine cat (submitted to the complete ablation of the anterior brain and thus no longer in control of its thermoregulation) REM sleep disappears completely beyond 35°C, and remains quasi-permanently at a temperature of 24°C [20].

HYPNOGENIC SUBSTANCES?

The study of the mechanisms of sleep regulation has been approached from an entirely different angle i.e. that of research into substances with hypnogenic properties. Two types of strategy have been employed (table 3.1).

In the **first hypothesis**, it is assumed that these possibly hypnogenic substances are gradually synthesised and/or released during sleep. An attempt is made to identify these in sleeping subjects. Thus, the Delta Sleep Inducing Peptide (DSIP) [33] was identified in the blood of a rabbit in the state of NREM sleep. Although the hypnogenic properties of DSIP had been confirmed in certain species, the results are contradictory and remain inconclusive.

The **diachronic hypothesis** is based on observation of the sleep rebound phenomenon which follows sleep deprivation. It postulates that hypnogenic substances accumulate in the brain during wakefulness, and that beyond a certain threshold of accumulation, they will trigger sleep. In line with this hypothesis, attempts have been made to characterise these substances in subjects maintained in prolonged wakefulness.

This approach, whose pioneer was H. Piéron, in 1913, resulted in the identification of the "S Factor", a muramyl-tetrapeptide purified from the CSF of a goat deprived of sleep, and found in human urine (ref. in [21]). This has provided the basis for an immunological theory on sleep, attributing to muramyl-peptides (products of bacteriological origin) or interleukines the role of facilitating deep NREM sleep; however all these factors are also pyrogenic [21].

Table 3.1. Summary of the different research having attempted to define "hypnogenic" substances

Recorded during	From	Nature	Effect on recipient	Authors
Wakefulness	CSF (goat) Urine (human)	"S" factor (close to MDP) Il1, inf α	++ NREM sleep ++ temperature	Krueger Pappenheimer
Wakefulness (total sleep deprivation)	Brain stem extract (rat)	Sleep Promoting Substance (Uridine?)	++NREM sleep ++REM sleep	Nagasaki Inoué
Wakefulness (selective sleep deprivation)	CSF (cat, rat) hypothalamo- hypophysial extracts	? ?	Restores sleep in insomniac recipients	Jouvet Sallanon
NREM sleep	Blood (rabbit)	Delta sleep inducing peptide	+NREM sleep	Adrien Dugovic Monnier Schoenenberger
NREM sleep	Reticular perfusate (cat)	?	+REM sleep (reticular infusion)	Drucker-Colin

Other sleep-inducing substances have been found in the brain stem of rats and cats deprived of sleep [15], in the cerebrospinal fluid [2, 32] or in the hypothalamo-hypophiseal complex [12]. These have not yet been identified.

Finally, a considerable number of hormones or peptides have been studied for their effects on sleep. The results cannot yet be summarised (table 3.2); it is not yet known whether there are any common mechanisms underlying these effects.

Table 3.2. Summary of the effects of the different known peptides and hormones on wakefulness and sleep. (Adrien, 1989)

Substance	Species	Dosage	Effect on sleep	Effect on temperature
Prostaglandin D_2 PGD_2		0.3-2 nmol preopt. reg. 2-20 µg i.cv.	NREM▲	Hypothermia
Vasointestinal polypeptide VIP	Rat	1 pg-100 ng i.cv.	NREM▲ REM▲	
	Cat	100 ng i.cv.	REM▲	
Antiserum	Rat		W▲ REM▼	
Growth hormone GH	Man	5 U	REM▲	
	Cat	50 µg-1 mg i.p.	REM▲	
	Rat	0.1-1 mg i.p.	REM▲	
GH releasing factor GRF	Rat	2 nmol. I.cv.	NREM▲	
	Rabbit	0.01-1 nmol. I.cv.	NREM▲ REM▲	
Somatostatin	Rat	20 µg/day i.cv	REM▲	
Insulin	Rat	200 µUL/day i.cv	NREM▲	
Antiserum	Rat		NREM▼	
Arginin vastocin AVT	Man	20 µg s.c.	REM▲	
	Rabbit	25 µg i.cv	NREM▲	
	Rat	10-50 µg/kg i.p.	TST▲ REM▼	
	Cat	1 pg.i.cv.	NREM▲ REM▼	
Antiserum	cat		NREM▼ ▲REM	
Peptides derived from pro-opio-melanocortin				
Desacetyl-α-melanocyte stimulating hormone des-α-MSH	Rat	1 ng. i.cv.	NREM▲	
Corticotropin-like intermediate lobe peptide CLIP	Rat	10 ng. i.cv	REM▲	
Cholecystokinin octapeptide CCK 8	Rat	10-50 µg/kg	NREM▲	Hypothermia
Adenosine	Rat	1-100 nmol. i.cv.	NREM▲	
Melatonin	Man	1-80 mg p.o	Would mainly act	
	Rat	1-10 mg/kg i.p.	on the circadian	
	Cat		rhythm	
Uridine	Rat	1-10 pmol. I.cv./h	NREM▲ REM▲	
Immunologically active peptides				
Interleukin-1 Il-1	Rabbit	15 µg/kg i.v.	NREM▲ REM▼	Hyperthermia
	Rat	60 U i.cv.	NREM▲ REM▼	Hyperthermia
Interferon α2 INF-α2	Rabbit		NREM▲	Hyperthermia
Tumor necrosis fact. TNF	Rabbit	5 µg i. cv.	NREM▲ REM▼	Hyperthermia
Muramyl dipeptide MDP	Rabbit	1-20 pmol. i.cv.	NREM▲	Hyperthermia
	Cat	10-500 µg/kg i.p.	NREM▲	Hyperthermia
	rat	250-500 µg/kg i.p.	REM▼	

Legend: i.cv. : intracerebroventricularly ▲ increase
         i.p. : intraperitonneally ▼ decrease
         s.c. : subcutaneously
         p.o. : per oral

Whatever the case, none of these substances, whether identified or not, induces sleep in a univocal manner. Rather than "hyponogenics", they are thought to be simple "facilitators" of various stages of sleep and wakefulness.

CONCLUSION

In conclusion, the harmonious alternation between the different states of sleep and wakefulness depends on the complex interactions between multiple factors. Diagrammatically, the latter come under different hierarchical mechanisms, i.e. neurophysiological, humoral, homeostasic and circadian, of which only certain elements are known.

By looking at the two diagrams of regulation developed in figures 3.5 and 3.6, an integrated model of sleep regulation can now be proposed, a model based on experimental physiology, which is applicable to human sleep [25] (fig. 3.9.).

Sleep is one of the rare major physiological functions whose essential role(s) is not entirely understood. The functions of NREM sleep include those of restoring cerebral function, and according to a more recent hypothesis, of maintaining central temperature [15]. In the case of REM sleep, different hypotheses have been proposed concerning its function, from the time this stage of sleep was determined (ref. in [10]). One of the most interesting is that of the genetic programming of behaviour [17]: REM sleep is thus viewed as the state during which there is "blank repetition" of behaviour patterns not only characteristic of the species but also of the individual. Such multiple reprogramming would serve to restrict the influence of the environment on modelling nerve connections [10], and by extrapolation, on behaviour and personality.

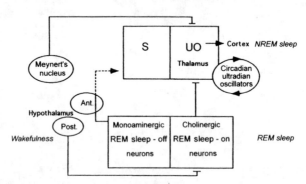

Figure 3.9. Model showing the interaction between wakefulness, NREM sleep and REM sleep regulation systems. During wakefulness, the REM sleep-off system – notably 5HT – induces the synthesis of substances through an accumulative process S. At the same time, cholinergic neurons from Meynert's nucleus (and those of the reticular formation) activate the cortex via the thalamus. During NREM sleep, the S process comes into play, and the thalamo-cortical synchronisation mechanisms develop in accordance with ultradian oscillators (UO). During REM sleep, the cholinergic influence from ponto-medullary structures predominates, notably resulting in thalamo-cortical activation, expressed as EEG desynchronisation. This mathematical modelling, which takes account of the neurophysiological data set out in this chapter, provides a simulation of physiological or pathological situations (after Massaquoi and McCarley, 1992, Achermann *et al.*, 1990).

REFERENCES

1. ACHERMAN P., BEERSMA D.G., BORBELY A.A. – The two-process model: Ultradian dynamics of sleep. *In: Sleep 1990*, J. Horne, (ed.) Pontenagel Press, Bochum, 296-300, 1990.
2. ADRIEN J., DUGOVIC C. – Presence of a paradoxical sleep (PS) factor in the cerebrospinal fluid of PS-deprived rats. *Europ. J. Pharm.*, 100, 223-226, 1984.
3. BERRIDGE C.W., PAGE M.E., VALENTINO R.J., FOOTE S.L. – Effects of locus cœruleus inactivation on electroencephalographic activity in neocortex and hippocampus. *Neurosci.*, 55, 381-393, 1993.
4. BORBELY A.A. – A two process model of sleep regulation. *Human Neurobiology*, 1, 195-204, 1982.
5. BORBELY A., TOBLER I. – Endogenous sleep-promoting substances and sleep regulation. *Physiol. Rev.*, 2, 605-670, 1989.
6. BOURGIN P., AHNAOU A., LAPORTE A.M., HAMON M. and ADRIEN J. REM sleep induction by VIP infused into the oral pontine tegmentum of the rat may involve muscarinic receptors, Neuroscience, 89, 311-312, 1999.

7. BOUTREL B., FRANC B., HEN R., HAMON M., and ADRIEN J. – Key role of 5-HT1B receptors in the regulation of paradoxical sleep as evidenced in 5-HT1B knock-out mice. J. Neuroscience, 19, 3204-3212, 1999.
8. MCCARLEY R.W., MASSAQUOI S.G. – Neurobiological structure of the revisited limit cycle reciprocal interaction model of REM cycle control. *J. Sleep Res.*, *1*, 132-137, 1992.
9. CESPUGLIO R., HONDOUIN F., OULERICH M., EL MANSARI M., JOUVET M. – Axonal and somatodendritic modalities of serotonin release: their involvement in sleep preparation, triggering and maintenance. *J. Sleep Res.*, *1*, 150-156, 1992.
10. DAVENNE D., FREGNAC Y., IMBERT M., ADRIEN J. – Lesion of the PGO pathways in the kitten: II. Impairment of physiological and morphological maturation of the lateral geniculate nucleus. *Brain Res.*, *485*, 267-277, 1989.
11. DENOYER M., SALLANON M., KITAHAMA K., AUBERT C., JOUVET M. – Reversibility of para-chlorophenylalanine-induced insomnia by intrahypothalamic micro-injection of L-5-hydroxytryptophan. *Neuroscience, 28*, 83-94, 1989.
12. DUGOVIC C., ADRIEN J. – Presence of PS-inducing substances in the hypothalamo-hypophyseal complex. *In: Sleep 1986*, W.P. Koella *et al.* (eds), Fisher-Verlag, Stuttgart, 297-299, 1988.
13. DUGOVIC C., WAUQUIER A., LEYSEN J., MARRANNES R., JANSSEN P. – Functional role of 5-HT$_2$ receptors in the regulation of sleep and wakefulness in the rat. *Psychopharmacol.*, *97*, 436-442, 1989.
14. MCGINTY D., SZYMUSIAC R. – Keeping cool: a hypothesis about the mechanisms and functions of slow-wave sleep. *TINS*, *13*, 480-486, 1990.
15. INOUE S., HONDA K., KOMODA Y., UCHIZONO K., KUENO R. – Differential sleep-promoting effects of five sleep substances nocturnally infused in unrestrained rats. *Proc. Natl. Acad. Sci.*, USA, *81*, 6240-6244, 1984.
16. JOUVET M. – The role of monoamine and acetylcholine neurons in the regulation of the sleep-waking cycle. *Ergeb. Physiol.*, *66*, 166-307, 1972.
17. JOUVET M. – Paradoxical sleep as a programming system. *J Sleep Res.*, *7*, (Suppl. 1), 1-5, 1998.
18. JOUVET M. – Hypnogenic indolamine-dependant factors and paradoxical sleep rebound. *In: Sleep 1982*, W.P. Koella (ed.), Karger, Basel, 2-18, 1983.
19. JOUVET M. – The regulation of paradoxical sleep by the hypothalamo-hypophysis. *Arch. Ital. Biol.*, *126*, 259-274, 1988.
20. JOUVET M., BUDA C., SASTRE J.P. – Hypothermia induces a quasi permanent paradoxical sleep state in pontine cats. *In: Living in the cold*, A. Malan, B. Canguilhem (eds), Libbey Eurotext, *193*, 487-497, 1989.
21. KRUEGER JM. AND FANG J.– Cytokines in sleep regulation. In *Sleep and sleep disorders: from molecule to behavior*, O. Hayaishi and S. Inoue (eds), Academic Press, Inc. Tokyo, pp. 261-277, 1997.
22. LE MOAL M., SIMON H. – Mesocorticolimbic dopaminergic network: functional and regulatory roles. *Physiol. Rev.*, *71*, 155-234, 1991.
23. LIN J.S. – Brain structures and mechanisms involved in the control of cortical activation and wakefulness, with emphasis on the posterior hypothalamus and histaminergic neurons. *Sleep Med. Rev, 4*, 471-503, 2000.
24. LIN J.S., SAKAI K., VANNI-MERCIER G., JOUVET M. – A critical role of the posterior hypothalamus in the mechanism of wakefulness determined by micro-injection of muscimol in freely moving cats. *Brain Res.*, *479*, 225-240, 1989.
25. MASSAQUOI S.G., MCCARLEY R.W. – Extension of the Limit Cycle Reciprocal Interaction Model of REM sleep control. An integrated sleep control model. *J. Sleep Res.*, *1*, 138-143, 1992.
26. PARMEGGIANI P.L., ZAMBONI G., PEREZ E., LENZI P. – Hypothalamic temperature during desynchronized sleep. *Exp. Brain Res.*, *54*, 315-320, 1984.
27. PINAULT D., DECHENES M. – Muscarinic inhibition of reticular thalamic cells by basal forebrain neurons. *Neuroreport, 3*, 1101-1104, 1992.
28. PORKKA-HEISKANEN T., STRECKER R.E. and McCARLEY R.W. – Brain site-specificity of extracellular adenosine concentration changes during sleep deprivation and spontaneous sleep: an in vivo microdialysis study. *Neuroscience, 99*, 507-517, 2000.
29. PORTAS C.M., BJORNVATIN B. and URSIN R. – Serotonin and the sleep/wake cycle: special emphasis on microdialysis studies. *Prog. Neurobiol. 60*: 13-35, 2000.
30. SAKAI K. – Executive mechanisms of paradoxical sleep. *Arch. Ital. Biol.*, *126*, 239-257, 1988.
31. SALLANON M., DENOYER M., KITAHAMA K., AUBERT C., GAY N., JOUVET M. – Long-lasting insomnia induced by preoptic neuron lesions and its transient reversal by muscimol injection into the posterior hypothalamus in the cat. *Neuroscience, 32*, 669-683, 1989.
32. SALLANON M., BUDA C., JANIN M., JOUVET M. – Restoration of paradoxical sleep by cerebrospinal fluid transfer in PCPA pretreated insomniac cats. *Brain Res.*, *251*, 137-147, 1992.
33. SCHOENENBERGER G.A., MONNIER M. – Characterization of a delta-electroencephalogram (-sleep)-inducing peptide. *Proc. Natl. Acad. Sci.*, USA, *74*, 1282-1286, 1977.
34. STERIADE M., MCCARLEY R. – REM sleep as a biological rhythm. *In: Brainstem control of wakefulness and sleep*. Steriade and McCarley (eds), Plenum Press, New York, 363-393, 1990.
35. SZYMUSIAK R.S., MCGINTY D.J. – State-dependent neurophysiology of the basal forebrain: relationship to sleep, arousal and thermoregulatory function. *In: The Diencephalon and Sleep*, M. Mancia and G. Marini (eds), Raven Press, New York, 111-124, 1990.
36. TAHERI S., ZEITER J.M., MIGNOT E. – The role of hypocretins (orexins) in sleep regulation and narcolepsy. *Ann. Rev. Neurosci.* 25, 283-313, 2002.

Chapter 4

The major physiological functions during sleep

Y. Dauvilliers
Service de Neurologie B, Hôpital Gui de Chauliac, Montpellier, France

In recent decades we have begun to understand the role of sleep in the vital functions. Three types of sleep and wakefulness are distinguishable in all mammals: wakefulness, NREM sleep and REM sleep. For each of these states, regulation is both global and specific to our different physiological systems, attesting to the existence of a real sleep-related physiology. The major physiological functions, whether ventilatory, cardio-vascular or endocrine, have been explored during sleep to define the changes that occur and determine the main factors involved in regulating them. A disruption of the major vital functions during sleep lies at the origin of several pathological states whose clinical manifestations may be exclusively nocturnal or diurnal or may affect the entire sleep-wake cycle. Research has also focused on a number of other important functions such as the digestive functions, sexual functions and thermoregulation during sleep. Finally, more recently, with the development of new technologies in functional neuro-imagery, it has become possible to determine the levels of global and regional activity of the cerebral metabolism, in different states of sleep and wakefulness, adding to our understanding of cerebral functioning during sleep.

The knowledge emerging from the complex and reciprocal interactions between the various physiological functions, states of sleep and wakefulness and their regulation is the fruit of the multidisciplinary activity of the teams working on this vital and fascinating subject.

THE VENTILATORY FUNCTION

The ventilatory function during sleep has been widely studied over recent years, due notably to the marked prevalence of respiratory disorders during sleep.

Breathing control differs according to the state of sleep and wakefulness, in terms of the type of regulation involved in ventilatory control. Indeed, two types of control have been distinguished, "metabolic" and "behavioural", their degrees of interaction varying in function of the state of sleep and wakefulness [21, 78, 87, 104].

"Metabolic" control is generated by the bulbo-pontine respiratory neurons which are sensitive to the flow of central chemoreceptors (dependent on concentrations of CO_2 and pH) and peripheral chemoreceptors (dependent on O_2 concentrations). These respiratory neurons in turn supply a phasic flow to the spinal motoneurons, on which breathing muscle activity depends *i.e.* the diaphragm, intercostal muscles and the muscles of the upper airways. The main aim of this "metabolic" control is to maintain the plasmatic variations of O_2, CO_2 and pH within physiological limits by varying the ventilatory parameters.

"Behavioural" ventilatory control is generated by a somewhat vast but ill-defined neuron network, mainly involving the frontal cortex and posterior hypothalamus. Indeed, stimuli of diverse neuronal origin are sent to the bulbo-pontine neurons which in turn stimulate the spinal respiratory motoneurons responsible for ventilation or more rarely, directly to the spinal motoneurons. Indeed, it appears that non-respiratory neurons in the reticular formation, whose principle role is to maintain wakefulness also exert a tonic effect on the spinal respiratory motoneurons [104]. This "behavioural" system allows ventilation to adapt automatically to the subject's activity, with the possibility of controlling breathing voluntarily. The neuronal stimuli are not phasic and the neurons are not sensitive to variations in PO_2, PCO_2 and pH.

During wakefulness

Ventilation control is principally under behavioural command which jointly ensures the state of wakefulness. However, metabolic control is also functional even if the hypercapnic or hypoxic trigger thresholds for ventilation are never affected during wakefulness. Thus an impeded metabolic system will not result in serious breathing disorders in wakefulness, due to the presence of behavioural control.

During sleep onset

The shift from wakefulness to sleep is detected on the electroencephalogram by the disappearance of alpha rhythm. This coincides with diminished ventilation related to a drop in current volume. This results in an increase in $PaCO_2$ and a moderate decrease in PaO_2 producing a new ventilation equilibrium mark, with ventilatory response to hypercapnia reduced by 50% at sleep onset compared to wakefulness [87]. The increase in $PaCO_2$ generally precedes the onset of stage 1 by a few seconds [74]. During this transitional phase, breathing rate may increase, diminish or even remain unchanged [26].

The onset of sleep inhibits the behavioural ventilation control which is only present in the state of wakefulness [33]. The inhibition of stimuli affecting neurons with a ventilatory function leads to a reduction of ventilation by diminishing the urge to breathe [28] and by reducing the tonic action of the upper airway respiratory muscles. The decrease in muscular tone, responsible for an increase in upper airway resistance, thus contrasts with the increase in intercostal muscular activity. Diaphragmatic activity remains unchanged in keeping with its weak bundle innervation.

Metabolic control gradually takes command of ventilation, but only becomes effective when sleep stabilises. Ventilation at sleep onset, which is particularly unstable, is usually responsible for periodic breathing characterised by the alternate increase and decrease of breathing amplitude. Periodic breathing is in fact present in 40 – 80% of cases in normal subjects, and more common in adults than in children. "Physiological" apnoea which develops as a result of the gradual reduction in breathing amplitude is sometimes objectified in this type of breathing. Periodic breathing may resemble Cheyne-Stokes breathing when ventilation uptake in post-apnoea (or in minimal post-ventilation) is gradual or in Biot breathing, when ventilation uptake is abrupt and immediately maximal [111].

The physiological regulation of ventilation shows the extent to which ventilation is unstable at sleep onset. The point of equilibrium differs in ventilation during wakefulness and sleep. Reduced ventilation during sleep onset produces the increase in $PaCO_2$ corresponding to the equilibrium point in sleep. As sleep is unstable during this transitional phase, awakening frequently occurs, stimulating ventilation to reduce the (elevated) $PaCO_2$ value to a level compatible with the point of equilibrium during wakefulness. When sleep returns, $PaCO_2$ again becomes insufficient, triggering the inhibition of ventilation and so on [104]. This type of breathing lasts from 10 minutes to an hour, in relation to the time needed for the shift from wakefulness to stable NREM sleep [111]. Breathing is all the more unstable if sleep itself is unstable, as is seen when sleep oscillates between wakefulness, stage 1 and an unstable stage 2. The wider the difference in equilibrium points between wakefulness and sleep, and the greater the ventilatory reflex response to chemical stimuli, the greater the ventilatory irregularity. This transitional phase is the period in which respiratory pathologies frequently occur, particularly that of sleep apnoea syndrome.

The mean periodicity of this breathing is 60 to 90 seconds, breathing which is sometimes marked by the presence of sleep apnoeas lasting from 10 to 40 seconds [111]. Apnoeas are usually of central origin (with no respiratory effort) and more rarely of obstructive origin [51]. The occurrence of these central apnoeas during periodic breathing depends on the type of response to various chemical stimuli. Alternating hyper and hypoventilation leads to changes in CO_2 levels and the resulting hypocapnia may, if its value falls below the "apnoeic" threshold, lead to arrested ventilation [21]. These mechanisms act differently during wakefulness. Hypocapnia and alkalosis secondary to hyperventilation do not lead to apnoeas in healthy subjects. The presence of apnoeas of obstructive origin during periodic breathing also indicates the difference in physiological regulation during sleep and wakefulness. The upper airway muscles are particularly inhibited during sleep compared to the diaphragm; an imbalance between the negative pressure generated by the diaphragm and the tonicity of the upper airway muscles produces upper airway occlusion and, as a result, obstructive apnoeas [20].

During stable NREM sleep (stages 2, 3 and 4)

Ventilation in stable NREM sleep is characterised by marked regularity both in amplitude and rate. Breathing is only dependent on metabolic control; the more stable sleep is, the more effective the ventilatory control. Ventilation is reduced by 13% in stage 2 compared with wakefulness and by 15% in deep NREM sleep [51]. This drop is due to the reduction in tidal volume (from 15 to 20%), breathing rate not producing any notable changes. These changes are different in the animal, where ventilation is reduced to the same degree but relates to a drop in breathing rate, with no notable change to current volume [104].

At least three different mechanisms account for the reduction in ventilation: (1) reduced central control (through the loss of stimuli to neurons with ventilatory functions), (2) reduced ventilatory response to hypercapnic and hypoxic stimuli and (3) increased upper airway resistance [27, 58]. Resistance only increases at the level of the upper airways. Indeed a marked drop is seen in the tonic activity of the upper airway dilatory muscles including the laryngeal muscles, genioglossis, soft palate tensor and the posterior cricoarytenoid muscle [8, 58]. This drop in activity contrasts with the increase in intercostal muscle and diaphragm activity, favouring the notion of a distinct central regulation of these different muscle groups. Thoracic participation in respiration thus increases to the detriment of the abdominal role. The accentuated activity of the intercostal muscles and diaphragm is expressed by a considerable increase in oesophageal pressure variations between wakefulness and NREM sleep, with an increase in upper airway resistance of nearly 250% [41].

The drop in ventilation and metabolic activity results in an increase in alveolar and arterial PCO_2 (+ 2-4%), reduced alveolar and arterial PaO_2 (- 5%) and O_2 saturation (- 2%) [13, 99, 11, 113]. The $O2$ increase seems to be largely due to the accentuation of upper airway resistance [99]. These changes in ventilation parameters occur despite the reduction in the cerebral metabolism during deep NREM sleep, which nevertheless has the effect of lowering O_2 consumption and CO_2 production by about 15% [19, 113].

During REM sleep

REM sleep is characterised by the phasic activation of the motor and autonomous nervous systems in contrast with the atonia of the postural muscles, mainly of the intercostal muscles in regard to respiration. Ventilation decreases in REM sleep compared to wakefulness, but differs only slightly from that of stable NREM sleep, although opinions differ on the subject in regard to variations in the proportion of phasic REM sleep [41, 58]. But breathing is particularly irregular, and variations of tidal volume and respiration rate differ according to REM phasic and/or tonic activity. The activity of the intercostal muscles and upper airway dilators diminishes particularly in REM sleep (reduced by approximately 30% in relation to NREM sleep) contrary to diaphragmatic muscle activity which increases [58]. Thoracic activity thus becomes out of phase with diaphragm activity, and, contrary to NREM sleep, there is an increase in the abdominal role to the detriment of the thoracic. As a result, the increase in upper airway resistance is probably weaker than in NREM sleep, and subject to substantial variations in function of REM sleep phasic and/or tonic activity [41, 75]. Moreover, considerable inter-individual variations have been demonstrated in measurements of upper airway resistance in REM sleep [58].

A reduction is observed in PO_2 with O_2 saturation, in REM sleep, compared with the state of wakefulness, with values relatively close to those of NREM sleep [13, 26]. However, these changes in the level of oxygenation are irregular; an abrupt and intermittent drop of the phasic type is common against a background of tonic diminution. These changes essentially refer to hypoventilation. The hypotonia of intercostal muscles objectified in REM sleep also appears to involve hypoxemia but to a lesser extent. This hypotonia would account for a reduction in thoracic volume, functional respiratory residual capacity and thus a disruption of the ventilation/perfusion relationship [26].

Ventilation is particularly unstable in REM sleep in terms of rate and amplitude with a possibility of apnoeas, even in the normal subject. The apnoeas are often of the central type, more frequent and longer than at other stages of sleep. These respiratory events, although considered as "physiological" when their frequency and duration do not exceed 5 and 10 events per hour and 20 seconds, respectively (in the case of REM sleep only), may still encourage heart rate disorders [35]. Ventilation in REM sleep is quite different to periodic breathing at sleep onset, due to breathing irregularities which abruptly alter breathing rate and amplitude, irregularities which are concomitant with other REM phasic activities such as rapid eye movements…

Homeostasis is different in REM sleep, and ventilation notably escapes classic metabolic control. This irregular ventilation actually persists in animals in conditions of hypoxia, hypercapnia, metabolic alkalosis, as well as after vagotomy or chemoreceptor deafferentation. The awakening reaction set off by hypoxia or hypercapnia is distinctly reduced in REM sleep compared with NREM sleep [87]. But a certain degree of metabolic homeostatic regulation does seem to persist, with REM sleep ventilation remaining correlated with O_2 consumption and CO_2 production, as in NREM sleep [113]. REM sleep ventilation escapes the behavioural control of wakefulness but another behavioural control particular to this stage is necessarily involved. The neurons which generate this ventilatory control are themselves dependent on the bulbo-pontine neuron systems responsible for REM sleep. The same neurons are thought to be responsible for the inactivation of metabolic ventilation control through their inhibiting effect on the hypothalamus [84]. The two main components of REM sleep (phasic and tonic) have a distinct effect on breathing [2, 104]. The spinal respiratory neurons are influenced by the tonic impulses of the bulbo-pontine neurons responsible for REM sleep and phasic impulses; REM phasic activity is at the origin of the saccadic regulation of the motor, autonomic and ventilatory component [2, 104].

No significant difference in ventilation and its general regulation has been demonstrated between the sexes. But episodes of nocturnal desaturation in NREM and REM sleep appear to be more common in men, possibly relating to higher body mass index [51]. On the other hand, ventilation characteristics during sleep do not change to any great extent in relation to age [51]. Breathing pauses tend to be more frequent in elderly subjects however. Sleep apnoea syndrome is present in up to 30% of elderly subjects, although no clear explanation has been found for this age-related increase. Ventilation in the newborn is quite different from that of the adult, notably in regard to the immaturity of the central nervous system, and poses complex, age-specific problems in children, which will not be dealt with here.

Ventilatory regulation thus differs according to states of sleep and wakefulness and other experiments have shown that the regulation of states of sleep and wakefulness is in turn influenced by the structures which regulate breathing. Indeed, ventilation affects the states of sleep and wakefulness via the activation of peripheral and central chemoreceptors; a reduction in the concentration of O_2 in the air gives rise to an increase in NREM sleep and a reduction of REM sleep [66]. Moreover, hypoxia or hypercapnia during sleep leads to an awakening reflex which is vital to enable the effective regulation of ventilatory homeostasis. This awakening reflex necessitates the activation of peripheral chemoreceptors. It is effectively absent in carotid sinus denervation [104]. A ventilation dysfunction during sleep must involve awakening in order to ensure satisfactory ventilatory homeostasis and the survival of the individual.

THE CARDIO-VASCULAR FUNCTION

The gradual reduction of the autonomous nervous system during sleep accounts for the multiple variations in cardio-vascular parameters during this period [62]. The first discoveries were made in animals, because of the potentially harmful and awakening effects of investigation methods [16, 105]. These techniques were later adapted to man and came to focus on the interactions between changes in breathing and cardio-vascular functions. These investigations also enabled research into the periodic rhythmicity of blood pressure and heart and respiration rates during sleep [23].

Variations in heart rate

There is a fairly marked sinus depression from the onset of REM sleep, giving rise to bradycardia with possible sinus pauses which always last under 2.5 s [14, 61]. This bradycardia usually oscillates between 55 and 60/min with little variability and lower values in deep NREM sleep. Bradycardia may however be more marked with pulsations of less than 40/min found in up to 24% of men and 8% of women [14]. A nodal depression may also occur with first and more rarely, second degree auriculo-ventricular blocks. These reductions in sinus and nodal activity are explained both by the activation of parasympathetic tonicity (vagus nerve) and a reduction in sympathetic tonicity during NREM sleep, particularly in deep NREM sleep, which may cause substantial bradycardias [101].

Heart rate is particularly unstable in REM sleep, responding to two types of influence, tonic and phasic [34]. The tonic influence produces bradycardia which may be greater than in deep NREM sleep occurring in relation to sympathetic tone similar to that of calm wakefulness, and marked parasympathetic activity [61, 101]. Phasic influence is accompanied by frequent and sudden increases in heart rate, concomitant with ponto-geniculo-occipital activity and rapid eye movements [61]. These phasic increases account for the wide variations in heart rate in REM sleep. Mean heart rate is higher in REM than in NREM sleep [61]. This marked sympathetic activity in REM sleep may explain the dysrhythmia (notably of the ventricular extrasystole type) and heart conduction (of the auriculo-ventricular block type with sinus arrest which can be very long) in subjects devoid of any heart condition [35]. It is worth noting that heart rate variability in phasic REM sleep is entirely suppressed by destroying the sympathetic and parasympathetic systems [6].

Arterial pressure variations

The earliest data on arterial pressure variations during sleep were obtained in cats, in whom a strong pressure reduction was objectified in NREM sleep and even more so in REM sleep (a difference of up to 25 mmHg compared with wakefulness) [16, 34]. Systemic arterial pressure in man diminishes during NREM sleep, especially in deep NREM sleep (reduction of 10 – 15% compared with wakefulness) related to peripheral vessel vasodilation and bradycardia [1, 69]. The lowering of peripheral vascular resistance is as common in normal as in hypertensive subjects [13]. It should however be noted that in the case of neuropathies affecting the fibres of the autonomous nervous system, regulation of arterial pressure via the vasoregulation of the peripheral arteries ceases to be effective and no drop is observed in arterial pressure during sleep.

In REM sleep, contrary to NREM sleep, there is wide variability in arterial pressure with upsurges of pressure during the phasic component, phasic activity simultaneous with rapid eye movements [23, 34, 57]. In REM sleep, variations in arterial pressure follow those of heart rate and pressure increases simultaneous to phasic activity overlap with tonic reduction in arterial pressure. However, mean arterial pressure in REM sleep does not appear any different to that in calm wakefulness. Contrary to the results obtained in animal experiments where the lowest values of arterial pressure occur in REM sleep, in man mean arterial pressure appears to be higher in REM than NREM sleep [23, 57]. On awakening, arterial pressure increases rapidly whatever the stage of sleep prior to awakening [57]. It is worth noting the marked absence of physiological change in pulmonary arterial pressure during the sleep of healthy subjects.

Another approach to studying the relations between sleep and cardio-vascular functions consists of looking at the effect of arterial pressure variations on the states of sleep and wakefulness [6, 7]. A drug-induced increase of arterial pressure (by intravenous adrenaline) triggers an electrical and behavioural awakening, with a reverse phenomenon observed for a reduction in arterial pressure in the sleeping animal [7]. These effects are independent of the action of the substance on the central nervous system, the mechanical increase in arterial pressure producing the same results as adrenaline; the hypothalamic and midbrain reticular region sensitive to baroreceptor activation are responsible for this. Arterial pressure therefore has a direct influence on states of sleep and wakefulness via (1) the central baroreceptors, which once activated affect the ascending reticular activating system, and (2) the peripheral baroreceptors whose activation facilitates the onset of NREM sleep.

Variations in cardiac output and peripheral circulation

Cardiac output seems to diminish slightly (less than 10%) in NREM sleep, but these findings remain controversial [13, 47]. Subject to confirmation, these minimal changes to cardiac output during sleep may result from a reduction of the pre-charge caused by reduction of the venous return linked to the muscle hypotomia associated with sleep. These possible changes in cardiac output during sleep do not appear to relate to changes in the autonomous nervous system. Moreover, no change has been demonstrated of the ventricular systolic ejection fraction either in NREM or REM sleep [62].

Cerebral blood flow increases slightly in REM sleep phasic activities in man, particularly in the brain stem structures, and appears to diminish slightly in NREM sleep, particularly in precentral regions [92, 107]. Moreover, several animal studies have focused on mesenteric blood flow, both renal and anterior iliac during sleep, showing in addition to weak variations in NREM sleep, substantial variations in REM sleep with an increased flow at mesenteric and renal levels in contrast to a reduced flow in the anterior iliac arteries [61, 89].

The main changes in arterial pressure during NREM and tonic REM sleep are explained by the reduction of overall peripheral vascular resistance, related to a reduction in sympathetic tone [13]. This central sympathetic inhibition is unevenly distributed, the areas most affected being those of the upper and renal mesenteric arteries. Conversely, reflex spinal sympathetic activation occurs during phasic REM sleep leading to vasoconstriction in the arteries to the skeletal muscles [89]. Indeed, the arterial pressure increases which are objectified during REM sleep phasic activities are explained by the transitory increase of peripheral vascular resistance in the skeletal muscles, which in turn relate to sympathetic tone excitation of spinal origin [90].

This results in a reduction of global peripheral resistance, with vasodilation predominating over vasoconstriction during tonic REM sleep, and the contrary effect during phasic events. In REM sleep, variations of global peripheral resistance evolve in the same way as those of arterial pressure.

These changes in the activity of the autonomous system and more specifically of the sympathetic nervous system are regulated by the coming into play of the sino-aortic reflexes to maintain the homeostasis of circulatory variables in NREM and REM sleep [34, 62]. This regulation was chiefly demonstrated by sino-aortic denervation experiments in the cat [61]. Indeed, after sino-aortic deafferentation, arterial pressure proved to be clearly reduced and more unstable in NREM sleep and even more so in REM sleep, linked to the marked reduction in sympathetic activity and thus of global peripheral resistance (including that of skeletal muscle unlike the non denervated state).

As a whole, the different states of sleep and wakefulness clearly influence the central mechanisms involved in regulating arterial pressure. In REM sleep especially, circulatory homeostasis controlled by the superior centres which ensure wakeful behaviour, is suppressed, and homeostasis is only ensured by the sino-aortic reflexes through the intermediary of chemoreceptors and to a lesser extent, baroreceptors [34].

Rhythmicity of cardio-vascular functions

It has been suggested that arterial pressure shows an endogenous circadian rhythmicity but this has not been formally demonstrated [69, 110]. If this rhythmicity does exist, its amplitude is too weak to be detected even in a « constant routine » condition [1]. Nevertheless, Guilleminault *et al.* have objectified a heart rate rhythm in heart transplant patients who were deprived of any autonomous nervous system afferents, with the lowest values in the middle of the night [36].

It has also been suggested that there is a very short (15 – 20 sec) ultradian arterial pressure rhythmicity, but the authenticity of this is still debateable as it is closely interlinked with respiratory irregularities and micro-architecture of sleep [23].

THE ENDOCRINE FUNCTION

The secretory activity of endocrine systems obey a certain rhythmicity, these secretory episodes moreover being a function of their adaptation to the interior and exterior environment. Each hormone in fact possesses a characteristic secretory profile (pulsatile and less frequently, tonic) resulting from the influence of circadian rhythmicity (through the intermediary of the suprachiasmatic nucleus) and/or its relationship with the states of sleep and wakefulness [70, 109]. Three systems have been the focus of particular interest, the hypothalamo-hypophyso-suprarenal system, the renin-angiotensin-aldosterone system and that of melatonin.

Hormones linked to the sleep-wake cycle

Prolactin

The secretory profile of prolactin is characterised by weak diurnal blood concentrations contrasting with high levels at night when sleep is nocturnal. Sleep deprivation reduces this nocturnal secretion and a shift in normal sleeping times leads to a shift in secretory episodes, findings which strongly suggest the dependence of this hormone on the sleep-wake cycle [93]. Nevertheless several studies carried out on transcontinental flights have refined these results by objectivying a certain inertia in the rhythm of secretion of prolactin, testifying also to the influence of a circadian component [102]. It appears that in various sleep pathologies such as narcolepsy, prolactin secretion diminishes distinctly during sleep [38].

Thyrotropin (TSH)

Weak amplitude secretory episodes of TSH are observed throughout the sleep-wake cycle, with important inter and intra-individual variations. A nyctohemeral rhythm exists in these secretions with an increase in blood concentrations before sleep onset, followed by a plateau, then a gradual diminishing during the night and in the morning, in normal subjects. This rhythm is sometimes considered as being of circadian origin but the time and amplitude of the nocturnal peak of secretion depends on the time of sleep; moreover, sleep deprivation will increase nocturnal TSH secretion [86]. Nocturnal variations of TSH are linked to sleep structure: intra-sleep awakenings are accompanied by increased TSH and deep NREM sleep by a reduction in the secretion of this hormone [32]. A disturbance to the nyctohemeral rhythm with the disappearance of the peak of TSH secretion, is encountered in pathology especially in thyroid diseases.

Hormones related to sleep architecture

Growth hormone (GH)

The main secretory peak is objectified at the beginning of the night, associated with the first episode of deep NREM sleep, followed by several peaks of weaker amplitude in the middle and at the end of the night [94]. This secretory activity is abolished at night in cases of sleep deprivation but increases as a consequence during the daytime in the same subjects who continue to be deprived of sleep, with no change to the final level of GH secretion over 24 hours [12]. In transcontinental flights, secretion peaks shift but there appears to be a certain circadian rhythmicity, with the persistence of a GH secretory peak at the moment corresponding to deep NREM sleep before the time shift [108]. The close link between GH secretion and deep NREM sleep has not yet been elucidated and no correlation has been found between the level of delta activity and the quantity of hormones secreted. In children, GH secretion is present from birth and occurs in a tonic pulsatile manner, the tonic secretion gradually diminishing to resemble that of adults from the age of puberty. GH secretion then diminishes with age through the reduced amplitude of secretory episodes. In cases of pathologies like acromegalia and gigantism, nyctohemeral GH secretion is disrupted and tonic secretion increased. In primary sleep pathologies like narcolepsy, the level of GH may collapse [9].

Renin-angiotensin-aldosterone system

Renin action is the limiting stage in the chain of reactions leading to the active hormone, angiotensin II. Renin secretion and thus the activity of the renin-angiotensin-aldosterone system is closely related to sleep architecture as well as to ultradian secretion rhythmicity. In fact, the secretion of this hormone increases in the transition from REM sleep or wakefulness to deep NREM sleep and diminishes as sleep becomes lighter or with awakening 11]. These interactions between sleep and renin secretion persist during sleep deprivation, inversion of the sleep-wake cycle and after a sodium free diet [11]. Likewise, beginning of REM sleep coincides with the descendent phase of renin secretion. The close links between renin secretion and the oscillation between NREM and REM sleep are well established and particularly solid. Other renin secretion stimuli such as diuretic intake do not alter these interactions. Two pathologies responsible for considerable sleep fragmentation i.e. narcolepsy and sleep apnoea syndrome are associated with a flat curve of renin secretion [71].

The secretion profile of aldosterone shows high concentrations at the end of the night and low concentrations at the end of the day [48]. Oscillations in the secretion of aldosterone result from a dual interaction, that of plasma renin activity during sleep and of the adrenocorticotropic axis during periods of wakefulness [19].

Anterior pituitary hormones

Anterior pituitary hormone secretions are usually weak during the transition phases between NREM and REM sleep, independently of any influence of overall sleep on these hormones [30]. There is usually a reduction in the secretory activity of these hormones at sleep onset, notably in the case of luteinising hormone (LH).

Hormones with circadian rhythmicity which is little affected by sleep

Cortisol, ACTH and TSH

The circadian rhythmicity of cortisol and adrenocorticotropin (ACTH) secretion has been known of for some time. Their secretory profiles are characterised by high plasma levels in the morning, which diminish during the daytime with the lowest levels at the end of the evening, followed in turn by an increase after 4 am [109]. The cortisol peak most often occurs at around 8 am and the trough at around 11 pm. The rhythm of these hormone secretions is highly reproducible in the same subject and from one subject to another. This rhythm is not modified by sleep and remains constant after sleep deprivation or the inversion of the sleep-wake cycle. Moreover, after flights involving the crossing of several time zones, these secretions gradually adapt to local time over a period of 10 – 15 days [109].

A disruption of the nyctohemeral rhythm is nevertheless objectified in pathological cases of hypercorticism, notably when this is low at the outset. In primary sleep pathologies like narcolepsy, cortisol secretion rhythmicity remains unchanged [9].

Although cortisol and TSH secretions are only marginally affected by sleep as a whole, they do seem to be linked to deep NREM sleep. Descendent phases of secretion bear a relationship to the onset of deep NREM sleep and ascendant phases to the occurrence of intra-sleep awakenings. These two hormones may hinder the onset of deep NREM sleep. Indeed considerable sleep fragmentation with several long nocturnal awakenings is often associated with an increase of cortisolemia. There seems to be a close temporal relationship between cortisol secretion peaks and EEG arousal beta activity, thus suggesting connections between the mechanisms involved in the process of cerebral activation and those involved in controlling the activity of the hypothalamo-adreno-pituitary axis [18].

A recent study of partial sleep deprivation for 6 days in healthy subjects demonstrated an increase in cortisol levels in the evening and sympathetic nervous system activity, as opposed to glucose tolerance and TSH concentrations which diminished [103]. These results support the hypothesis that sleep affects endocrine and neurovegetative functions, suggesting a possible dysfunctioning in cases of insomnia, particularly in the elderly.

Melatonin

Melatonin is a precious marker of circadian clock activity, this hormone constituting a powerful endogenous synchroniser capable of stabilising or even reinforcing circadian rhythms. The secretory profile of melatonin is closely aligned to transitions between light and darkness, light having a powerful inhibiting, dose-dependent effect on its secretion. The plasma levels which are virtually nonexistent during the daytime are followed by a gradual increase during the night starting before REM sleep onset; the melatonin peak coincides with the thermal minimum towards 3 am [22, 54]. This rhythm has the highest amplitude of the various hormones studied. But it should be added that important inter-individual variations exist, without any considerable intra-individual variations. With long distance flights, levels of melatonin like those of cortisol take from two to three weeks to synchronise with local synchronisers. Melatonin has been recommended to correct functional disorders in cases of jet lag and phase lag syndrome due to its resynchronising power. Disorganisation of the melatonin rhythm in elderly subjects (reduced amplitude and advance phase), especially when insomnia is present, may potentially benefit from the therapeutic effects of this hormone [106].

Testosterone and LH

The secretory profiles of testosterone and LH appear to be independent of the sleep-wake cycle. Testosterone secretion reaches its highest peak early in the morning with a weak amplitude circadian rhythm [24]. The secretory profile of LH is more complex due to numerous interactions with age, sex and the menstrual cycle. The circadian influence does however seem to be weak and sleep does not appear to have any overall influence [43].

Most hormone rhythms in the organism are subjected to several types of regulation, resulting from the influence of the sleep-wake cycle, internal sleep architecture and circadian rhythmicity. However very little attention has been paid to the reciprocity of these interactions with specific regard to the effect of hormone secretions on sleep.

THE DIGESTIVE FUNCTION

Because of the very function of the digestive tube, its activity involves physiological cyclic organisation (absorption, secretion, digestion...). The digestive tube (composed of several intrinsic nervous plexus) is controlled by the autonomous nervous system (ANS) and essentially by the parasympathetic system. The reduced activity of the ANS during sleep suggests variations in the digestive function during the nyctohymera. Indeed, salivary secretion (total and parotidean) and the phenomena of deglutition are particularly reduced during sleep [80]. Moreover the resting pressure of the upper oesophageal sphincter and frequency of peristaltic contractions of the body of the oesophagus and its lower sphincter seem to diminish during sleep, both in NREM and REM sleep [3, 17]. During sleep several factors thus favour the onset of gastro-oesophageal reflux, although no close link has been observed with any particular stage of sleep [25, 31].

The concept of the "migrating motor complex" characterised by the periodic succession of states of motor activity in the small intestine has demonstrated a diminution of the frequency and amplitude of intestinal contractions during sleep [10, 49]. This intestinal motor rhythmicity is interrupted by any intake of food, due to the apparition of a reflex motor response. This motricity was found to have endogenous rhythmicity, differing between wakefulness and sleep, with an intrinsic periodicity of roughly 90 minutes [31, 50]. The duration of these migrating motor complexes appears to be longer and the speed of propagation slower in NREM than in REM sleep [50]. No difference was noted between migrant motor complexes at duodenal or jejunal level, nor between men and women, with nonetheless marked intra-individual variability [10, 100]. But apart from the influence of sleep, there also appears to be circadian variation in the different states of digestive motor activity, mainly at antro-duodenal level [10].

Reduced motor, mechanical and electromyographic activity at colonic level has been reported during sleep, but there appear to be additional regional changes in motor activity during sleep [5]. Reduced motor activity is more marked in the transversal colon, descendant and sigmoid especially in NREM sleep. Note that there is an uptake of colonic motor activities just before awakening, this finding coinciding with the morning sensation of needing to empty the bowels [5]. There is reduced anal sphincter pressure during sleep and a reduction in the amplitude of response of the sphincter external to the rectum with no change to that of the internal sphincter; the sensation of rectal distension is abolished during sleep [79, 112]. These changes are extremely important due to the need to maintain continence during sleep.

There are thus real interactions between sleep and digestive motor physiology, but the origin of variations in digestive activity has not been clearly established. These variations may be produced by sleep itself, by the regulation of the autonomous nervous system during sleep, by a possible circadian component or a variable association of these factors.

THE SEXUAL FUNCTION

The presence of erections in men during sleep has long been established, with a close relationship between erections and REM sleep (association in over 85% of cases). In the rat only 28% of nocturnal erections occur in REM sleep [29, 44]. Erections last longer and are stronger in the morning than at the beginning of the night, although there are considerable inter-individual variations. Moreover puberty and old age are two periods in life during which nocturnal erections exist as frequently in NREM sleep.

For a long time nocturnal erections were used as a measurement to distinguish between psychological and organic impotence, provided the psychological factors involved did not occur during sleep. Indeed, if the methodology is effective, the presence of normal erections in terms of volume and rigidity in a subject complaining of impotence characterises psychological impotence [68]. However it seems that this method of investigation is charged with false positives and false negatives [96].

The physiological mechanisms involved in the occurrence of these erections during sleep remains partially unknown; vascular, muscular (bulbo and erector muscle of penis) and hormonal (androgen sensitivity) mechanisms are probably implicated [39, 45]. Several recent studies have led to defining the cerebral regions involved in triggering erections in REM sleep in animals. Contrary to REM genesis, the brainstem does not appear sufficient to generate erections in REM sleep in the rat and the involvement of higher facilitating structures in the central nervous system (the forebrain, principally) would seem to be necessary [97]. A recent study in rats reported the specific involvement of the lateral preoptic region but not that of the median preoptic region in REM sleep-related erections, with wakeful erections and REM sleep architecture remaining unchanged [98]. In man, REM sleep does not seem to be necessary either, for nocturnal erections to occur [52]. Moreover it appears that REM sleep-related nocturnal erections persist in patients in the vegetative state and are even comparable to those of normal subjects [76].

THERMOREGULATION

Man is considered as homeothermal because of the marked stability of his internal temperature (endotherm, to be more precise), a fact which contrasts with the wide variations possible in his external cutaneous temperature. The type of internal temperature measurement varies according to the experiment: rectal temperature (11 cm from the anal sphincter) and oesophageal temperature are the most commonly used. Internal temperature evolves according to a circadian rhythm with a minimum at about 3 – 4 am (36-36.5°C) and a maximum toward 6 pm (37-37.5°C), the amplitude of the rhythm thus being in the region of one degree [82]. Ultradian temperature modulation has also been reported in relation to the sleep-wake cycle, a reduction in temperature during NREM sleep contrasting with an increase in the passage from NREM to REM sleep [83, 84]. During sleep, body temperature changes in function of the surrounding temperature. A sleeping man represents an efficient thermoregulator. The maintenance of internal temperature is subjected to regulatory mechanisms which come into play when the thermoneutral zone is exceeded. This thermoneutral zone is characterised by a zero balance of exchanges between the body and its surroundings. It differs in wakefulness and sleep, at 28 and 30 – 32°C respectively, in the naked man and also differs according to sleep stage [67]. This zone of thermoneutrality must be distinguished from the zone of thermal comfort (16 to 22°C in classic sleeping conditions) defined by the onset of vasomotor responses, not perceived by the subject, with no notable change to sleep architecture [73]. These vasomotor responses are essential for cutaneous heat transfer to occur and to maintain internal temperature in the most stable manner. Beyond this comfort zone, which features wide interindividual variations, the vasoregulation processes are associated with awakenings experienced by the subject with considerable sleep fragmentation.

The hypothalamus and especially the preoptic region, is a structure of prime importance in thermoregulation, this region being continuously informed of body temperature through the afferences of thermoreceptors in the skin and different organs of the body [53, 95]. In return, the hypothalamus responds by muscular and/or neurohormonal activation, changing cutaneous vasomotricity and respiratory rates to modify the production of metabolic heat. Heat exchanges differ between men and animals, the former, unlike animals, using cutaneous vasomotricity (via variations in the production of sweat) in preference to ventilation. In man, reheating mechanisms are characterised by the vasoconstriction of cutaneous arterioles to limit heat loss through the skin and cooling mechanisms by peripheral vasodilation enabling heat to escape outside the body, with notably, the activation of sudoriparous glands.

The process of regulating internal temperature also depends on states of sleep and wakefulness, the body's reactivity to thermal surroundings being different in deep NREM sleep and in REM sleep [53, 83]. Each sleep state has its own zone of thermoneutrality. The most effective vasoregulation occurs in deep NREM sleep, a period during which reactivity to a change in external temperature is fast (approximately 3 minutes) and effective for slight changes in temperature, testifying to active thermoregulation even if this is less active than in the wakeful state. In NREM sleep, internal temperature drops when the ambient temperature rises and inversely, especially when NREM sleep follows an episode of REM sleep. Indeed, when NREM sleep follows wakefulness, thermoregulation having functioned well during wakefulness, no compensation proves essential in NREM sleep. In REM sleep, the rapidity of triggering and amplitude of sudoral response are reduced. Body temperature follows ambient temperature, with no limits in animals and within certain limits for man [55, 83]. Indeed, thermoregulation in REM sleep exists even if it is largely attenuated in man, again differentiating us from animals which are qualified as poikilothermal (parallel shift of internal and external temperatures) [55, 72]. In the animal, homeothermia is not maintained in REM sleep, thermoregulatory responses are no longer adapted, and there is no effective regulation of vasomotricity. The hypothalamus appears inactivated in REM sleep in animals, a change in its temperature having no effect on body temperature; moreover lesions to the preoptic area considerably reduce the duration of REM sleep [85]. In a situation in which the difference between external temperature and cerebral temperature is too great, the organism reacts in an adapted « homeostatic » manner provoking sleep instability, increasing latency to REM sleep onset and intra-sleep awakenings, reducing total sleep time, especially that spent in REM sleep with preference given to sleep with effective thermoregulation [81]. In situations where internal temperature is moderately raised shortly after REM sleep onset, the organism reacts by increasing the total quantity of sleep, particularly that of deep NREM sleep to reduce central temperature as quickly as possible (favouring sudation) while the duration of REM sleep and latency to REM sleep onset are reduced [40]. If, on the other hand, the temperature is raised too high the organism reacts by reducing deep NREM sleep as well as REM sleep, maintaining light NREM sleep at a constant level and increasing the duration and number of intra-sleep awakenings [83]. Sleep changes in man appear to be more pronounced when exposed to the cold than when exposed to heat (21°C and 37°C respectively in nude subjects) [37]. When the organism is exposed to extreme temperatures for prolonged periods, the mechanisms of thermoregulation prove slightly more effective (amplitude of sudoral response slightly raised) and sleep disturbances do not differ as exposure progresses, in favour of the absence of a short term mechanism of adaptation to sleep regulation and thermoregulation in man [56, 81].

Other experiences have shown that electrical or thermal stimulation of the hypothalamus apart from provoking a metabolic response by modifying body temperature (other than in REM sleep), will also change the states of sleep and wakefulness, confirming the complexity of its role and importance [91]. Thermoregulation thus affects the states of sleep and wakefulness and inversely, the regulation of states of sleep and wakefulness is influenced by the neuro-vegetative structures which control temperature. In REM sleep, thermoregulation is particularly affected compared to the other main physiological functions, because of the hypothalamic sitting of its regulating centre (a region which is particularly inactive in REM sleep) [82]. The onset of REM sleep depends on the hypothalamic and surrounding temperature as well as the need for thermoregulation. The weak effect of thermoregulation in REM sleep is purely of central origin and peripheral thermosensitivity is conserved [82].

There is another thermoregulating mechanism, of the cortical behaviour type, only present in the wakeful state. This modulates the effect of thermal constraints via voluntary motor activity, thus altering the relationship between body posture and the environment.

THE CEREBRAL METABOLISM

Cerebral functioning during sleep in man and the principle functions of sleep still remain a mystery; nevertheless several studies have attempted to provide some of the answers by defining the overall and regional levels of activity of the cerebral metabolism during the various stages of sleep. This has been made possible thanks to the new atraumatic technologies of functional neuro-imagery using PET scan (positron emission tomography), SPECT (single photon emission computed tomography) and MRI (magnetic resonance imaging). These techniques, especially PET which is the most commonly used in this field, provide very accurate measurements of cerebral blood flow, cerebral glucose and O_2 consumption [60, 63]. Coupling is established between neurophysiological data on neuron activity, the energy metabolism (glucose and O_2 consumption) and cerebral blood flow [63].

The first investigations in man consisted in defining the overall variations in the level of cerebral activity during sleep and quantifying them during the different sleep stages. The cerebral glucidic metabolism and use of O_2 vary in man as they do in animals, with a reduction of 25 to 44% in deep NREM sleep compared to calm wakefulness, and with no significant differences for light NREM sleep [46, 60, 64, 88]. On the other hand, cerebral activity is marked in REM sleep, cerebral glucose and O_2 consumption is the same or even higher than in calm wakefulness, but the regional distribution of this activity is different [15, 59, 65].

Regional mapping has been carried out recently of these changes in cerebral activity in function of the various stages of sleep in man. During NREM sleep, PET studies have shown a net reduction in cerebral activity in the thalamic nuclei, brain stem (mesencephalon and tegmentum of pons), basal ganglia, hypothalamus, basal forebrain, the orbito-frontal and anterior cingulate cortical regions, the precuneus, as well as the right medio-temporal regions [15, 46, 64, 88]. These results differ from those obtained in animals in whom the cortical regions appear to be uniformly deactivated in NREM sleep. The techniques of functional neuro-imagery during sleep have been applied to different pathologies, notably to sleepwalking in which a dissociation has been reported during deep NREM sleep between the activation of the thalamo-cingulate pathways and the physiological deactivation of the other thalamo-cortical pathways [4].

In REM sleep, the non homogeneous distribution of cerebral activity has also been objectified with high activation in the brain stem, particularly in the tegmentum pons, the thalamic and limbic regions (amygdala, hippocampus, anterior cingulate cortex) [65]. The least activated regions in REM sleep are the prefrontal cortex (dorsolateral region), the parietal cortex (gyrus supramarginalis), the posterior cingulate cortex and the precuneus [65]. The intense activity of the limbic system found in REM sleep supports several hypotheses notably that which links REM sleep to mnemonic processes.

These regional changes in cerebral activity in REM and NREM sleep provide insight into the principle structures and thereby, in part, into the mechanisms involved in regulating the different states of sleep and wakefulness.

These discoveries combine with those of the past obtained in regard to animals, using other methods of investigation i.e. electrophysiology, pharmacology and biochemistry [42]. The passage from one state of wakefulness and sleep to another appears to correspond to the passage from one state of cerebral equilibrium to another, each with a functional equilibrium in relation to a specific regional metabolism.

CONCLUSION

The regulation of the principle physiological functions differs according to the states of sleep and wakefulness, the passage from one state to another being accompanied by specific variations affecting all the physiological systems. In parallel, the different regulatory mechanisms of our principle physiological systems influence the expression of states of sleep and wakefulness. These close, complex interactions between the structures involved in the various regulatory functions are beginning to be understood, even if numerous questions remain. Better knowledge of these interactions and thus of sleep physiology will provide further insight into sleep-related neuro-vegetative dysfunctioning. The prospect of identifying the pathological states linked to anomalies in the regulation of certain vital functions occuring specifically during sleep adds to the value of conducting research in this area.

REFERENCES

1. AHNVE S., THEORELL T., AKERSTEDT T., FROBERG JE., HALLBERG F. - Circadian variation in cardiovascular parameters during sleep deprivation. *Eur. J. Appl. Physiol.,* 46, 9-19, 1981.
2. ASERINSKY E. - Periodic respiratory pattern occurring in conjunction with eye movements during sleep. *Science.,* 150, 763-766, 1965.
3. AVOTS-AVOTIN A.E., ASHWORTH W.D., STAFFORD B.D., MOORE J.G. - Day and night esophageal motor function. *Am. J. Gastroenterol.,* 85, 683-685, 1990.
4. BASSETTI C., VELLA S., DONATI F., WIELEPP P., WEDER B. - SPECT during sleepwalking. *Lancet.,* 356, 484-485, 2000.
5. BASSOTTI G., BUCANEVE G., BETTI G., MORELLI A. - Sudden awakening from sleep effects of proximal and distal colonic contractile activity in humans. *Eur. J. Gastro. Hepatol.,* 2 , 475-478, 1990.
6. BAUST W., BOHNERT B. - The regulation of heart rate during sleep. *Exp. Brain. Res.,* 7, 169-180, 1969.
7. BAUST W., VIETH J. - The action of blood pressure on the ascending reticular activating system with special reference to adrenaline-induced EEG arousal. *Electroenceph. Clin. Neurophysiol.,* 15, 63-72, 1963.
8. BERGER R.J. - Tonus of extrinsic laryngeal muscles during sleep and dreaming. *Science.,* 134, 840, 1961.
9. BESSET A., BONARDET A., BILLIARD M., DESCOMPS B., CRASTES LE PAULET A.C., PASSOUANT P. - Circadian patterns of growth hormone and cortisol secretions in narcoleptic patients. *Chronobiologia.,* 6, 19-31, 1979.
10. BORTOLOTTI M., ANNESE V., COCCIA G. - Twenty-four hour ambulatory antroduodenal manometry in normal subjects (co-operative study). *Neurogastroenterol. Motil.,* 12, 231-238, 2000.
11. BRANDENBERGER G., FOLLENIUS M., MUZET A., EHRHART J., SCHIEBER J-P. - Ultradian oscillations in plasma renin activity: their relationships to meals and sleep stages. *J. Clin. Endocrinol. Metab.,* 61, 280-284, 1985.
12. BRANDENBERGER G., GRONFIER C., CHAPOTOT F., SIMON C., PIQUARD F. - Effect of sleep deprivation on overall 24h growth hormone secretion. *Lancet.,* 356, 1408, 2000.
13. BRISTOW J.D., HONOUR A.S., PICKERING TG., SLEIGHT P. - Cardio-vascular and respiratory changes during sleep in normal and hypertensive subjects. *Cardiovasc. Res.,* 3, 476-485, 1969.
14. BRODSKY M., WU D., DENES P., KANAKIS C., ROSEN K.M. - Arrhythmias documented by 24-hour continuous electrocardiographic monitoring in 50 male medical students without apparent heart disease. *Am. J. Cardiol.,* 39, 390-395, 1977.
15. BUCHSBAUM M., GILLIN JC., WU JC., HAZLETT E., SICOTTE N., DUPONT RM., BUNNEY WE.- Regional cerebral glucose metabolic rate in human sleep assessed by positron emission tomography. *Life. Sciences.,* 45,1349-1356, 1989.
16. CANDIA O., FAVALE E., GIUSSANI A., ROSSI GF. - Blood pressure during natural sleep and during sleep induced by electrical stimulation of brain stem reticular formation. *Arch. It. Biol.,* 100, 216-233, 1962.
17. CASTIGLIONE F., EMDE C., AMSTRONG D., SCHNEIDER C., BAUERFEIND P., STACHER G., BLUM A.L. - Nocturnal oesophageal motor activity is dependent on sleep stage. *Gut.,* 34, 1653-1659, 1993.
18. CHAPOTOT F., GRONFIER C., JOUNY C., MUZET A., BRANDENBERGER G. - Cortisol secretion is related to electroencephalographic alertness in human subjects during daytime wakefulness. *J. Clin. Endocrinol. Metab.,* 83, 4263-4268, 1998.
19. CHARLOUX A., GRONFIER C., LONSDORFER-WOLF E., PIQUARD F., BRANDENBERGER G. - Aldosterone release during the sleep-wake cycle in humans. *Am. J. Physiol.,* 276, E43-49,1999.
20. CHERNIACK N.S. - Respiratory dysrhythmias during sleep. *New. Engl. J. Med.,* 305, 325-330, 1981.
21. CHERNIACK N.S., LONGOBARDO G.S. - Periodic breathing during sleep. *In: Sleep and Breathing.* NA. Saunders, CE. Sullivan (eds.), 2nd cd., rev. and expanded. Lung Biology in Health and Disease, vol. 71, M. Dekker, New York, 158-190, 1994.
22. CLAUSTRAT B., BRUN J., GARRY P., ROUSSEL B., SASSOLAS G. - A once repeated study of nocturnal plasma melatonin patterns and sleep recording in six normal young men. *J. Pineal. Res.,* 3, 301-310, 1986.
23. COCCAGNA E., MANTOVANI M., BRIGNONI E., MANZINI A., LUGARESI E. - Arterial pressure change during spontaneous sleep in man. *Electroencephal. Clin. Neurophysiol.,* 31, 277-281, 1971.

24. DE LACERDA L., KOWARSKI A., JOHANSON A.J., ATHANASIOU R., MIGEON C.J. - Integrated concentration and circadian variation of plasma testosterone in normal men. *J. Clin. Endocrinol. Metab.*, 37, 366-371, 1973.

25. DENT J., DODDS W.J., HOGAN W.J., TOOULI J. - Factors that influence induction of gastroesophageal reflux in normal human subjects. *Dig. Dis. Sci.* , 33, 270-275, 1988.

26. DOUGLAS N.J. - Control of ventilation during sleep. *Clin. Chest. Med.,* 6, 563-575, 1985.

27. DOUGLAS N.J., WHITE D.P., WEIL I.F., PICKEY C.K., MARTIN R.J., HUDGEL D.W., ZWILLLCH C.W. - Hypoxic ventilatory response decreases during sleep in normal men. *Am. Rev. Resp. Dis.,* 125, 286-289, 1982.

28. FINK B.R. - Influence of cerebral activity in wakefulness on regulation of breathing. *J. Appl. Physiol.,* 16, 15-20, 1961.

29. FISHER C., SCHIAVI R.C., EDWARDS A., DAVIS D.M., REITMAN M., FINE J. - Evaluation of nocturnal penile tumescence in the differential diagnosis of sexual impotence. A quantitative study. *Arch. Gen. Psychiatry.*, 36, 431-437, 1979.

30. FOLLENIUS M., BRANDENBERGER G., SIMON C., SCHLIENGER J.-L. - REM sleep in humans begins during decreased secretory activity of the anterior pituitary. *Sleep.,* 11, 546-555, 1988.

31. GILL R.C., KELLOW J.E., WINGATE D.L. - Gastro-oesophageal reflux and the migrating motor complex. *Gut.*, 28, 929-934, 1987.

32. GOICHOT B., BRANDENBERGER G., SAINI J., WITTERSHEIM G., FOLLENIUS M. - Nocturnal plasma thyrotropin variations are related to slow-wave sleep. *J. Sleep. Res.,* 1, 186-190, 1992.

33. GOTHE B., GOLDMAN MD., CHERNIACK N.S., MANTEY.P. - Effects of progressive hypoxia on breathing during sleep. *Am. Rev. Resp. Dis.,* 126, 97-102, 1982.

34. GUAZZI M., ZANCHETTI A. - Blood pressure and heart rate during natural sleep of the cat and their regulation by carotid sinus and aortic reflexes. *Arch. It. Biol.,* 103, 789-817, 1965.

35. GUILLEMINAULT C., POOL P., MOTTA J., GILLIS A.M. - Sinus arrest during REM sleep in young adults. *N. Engl. J. Med.,* 311, 1006-1010, 1984.

36. GUILLEMINAULT C., WINKLE R., BOUKHABZA D., SILVESTRI R., COBURN S. - 24h holter ECG sleep 82-hour Holter ECG and sleep recordings in heart transplant patients. *In*: W.P. Koella (ed.), *Sleep 82*, Karger, Basel, 346-348, 1983.

37. HASKELL E.H., PALCA J.W., WALKER J.M., BERGER R.J., HELLER H.C. - The effects of high and low ambient temperatures on human sleep stages. *Electroencephalogr. Clin. Neurophysiol.*, 51, 494-501, 1981.

38. HIGUCHI T., TAKAHASHI Y., TAKAHASHI K., NIIMI Y., MIYASITA A. - Twenty-four-hour secretory patterns of growth hormone, prolactin, and cortisol in narcolepsy. *J. Clin. Endocrinol. Metab.*, 49, 197-204, 1979.

39. HIRSHKOWITZ M., MOORE C.A., O'CONNOR S., BELLAMY M., CUNNINGHAM G.R. - Androgen and sleep-related erections. *J. Psychosom. Res.*, 42, 541-546, 1997.

40. HORNE J.A., REID A.J. - Night-time sleep EEG changes following body heating in a warm bath. *Electroencephalogr. Clin. Neurophysiol.*, 60, 154-157, 1985.

41. HUDGEL D.W., MARTIN R.J., JOHNSON B., HILL P. - Mechanics of respiratory system and breathing pattern during sleep in normal humans. *J. Appl. Physiol.,* 56, 133-137, 1984.

42. JOUVET M. - The role of monoamines and acetylcholine-containing neurons in the regulation of the sleep waking cycle. *Ergebn. Physiol.,* 64, 165-305, 1972.

43. KAPEN S., BOYAR R., HELLMAN L., WEITZMAN E.D. - The relationship of luteinizing hormone secretion to sleep in women during the early follicular phase: effects of sleep reversal and a prolonged three-hour sleep-wake schedule. *J. Clin. Endocrinol. Metab.,* 42, 1031-1040, 1976.

44. KARACAN I., GOODENOUGH P.R., SHAPIRO A., STARKER S. - Erection cycle during sleep in relation to dream anxiety. *Arch. Gen. Psychiatry.,* 15, 183-189, 1966.

45. KARACAN I., HIRSHKOWITZ M., SALIS P.J., NARTER E., SAFI M.F. - Penile blood flow and musculovascular events during sleep-related erections of middle-aged men. *J. Urol.,* 138, 177-181, 1987.

46. KENNEDY G., GILLIN JG., MENDELSON W., SUDA S., MIYAOKA M., ITO M., NAKAMURA R.K., STORCH FI., PETTIGREW K., MISHKIN M., SOKOLOFF L. - Local cerebral glucose utilization in non rapid eye movement sleep. *Nature.,* 297, 325-327, 1982.

47. KHATRI L.M., FREIS E.D. - Hemodynamic changes during sleep. *J. Appl. Physiol.,* 22, 867-873, 1967.

48. KRAUTH M.O., SAINI J., FOLLENIUS M., BRANDENBERGER G. - Nocturnal oscillations of plasma aldosterone in relation to sleep stages. *J. Endocrinol. Invest.,* 13, 727-735, 1990.

49. KUMAR D. - Sleep as a modulator of human gastrointestinal motility. *Gastroenterology.,* 107, 1548-1550, 1994

50. KUMAR D., ISIDOWSKI C., WINGATE D.L. - Relationship between enteric migrating tinter complex and the sleep cycle. *Am. J. Physiol.,* 259, 983-990, 1990.

51. KRIEGER J., MAGLASIU N., SFORZA E., KURTZ D. - Breathing during sleep in normal middle-aged subjects. *Sleep.,* 13, 143-154, 1990.

52. LA VIE P. - Penile erections in a patient with nearly total absence of REM: a follow-up study. *Sleep.,* 13, 276-278, 1990.

53. LENZI P., LIBERT J.P., FRANZINI C., CIANCI T., GUIDALOTTI P.L. - Short term thermoregulatory adjustments involving opposite regional temperature changes. *J. Thermal. Biology.,* 11, 151-156,1986.

54. LEWY A.J., WEHR T.A., GOODWIN F.K., NEWSOME D.A., MARKEY S.P. - Light suppresses melatonin secretion in humans. *Science.,* 12, 210, 1267-1269, 1980.

55. LIBERT J.P., CANDAS V., MUZET A., EHRHART J. - Thermoregulatory adjustments to thermal transients during slow wave sleep and REM sleep in man. *J. Physiol.,* 78, 251-257, 1982.

56. LIBERT J.P., DI NISI J., FUKUDA H., MUZET A., EHRHART J., AMOROS C. - Effect of continuous heat exposure on sleep stages in humans. *Sleep.,* 11, 195-209, 1988.

57. LITTLER WA. - Sleep and blood pressure: further observations. *Am. Heart. J.*, 97, 35-37, 1979.

58. LOPES J.M., TABACHNIK E., MULLER N.L., LEXISON H., BRYAN A.C. - Total airway resistance and respiratory muscle activity during sleep. *J. Appl. Physiol.*, 54, 773-777, 1983.

59. MADSEN P.L., HOLM S., FRIBERG L., LASSEN N.A., WILDSCHIODTZ G. - Human regional cerebral blood flow during rapid eye movement sleep. *J. Cereb. Blood. Flow. Metabol.*, 502-507, 1991.

60. MADSEN PL., SCHMIDT J.F., WILDSCHIODTZ G., FRIBERG L., HOLM S., VORSTRUP S., LASSEN N.A. - Cerebral O_2 metabolism and cerebral blood flow in humans during deep and rapid-eye-movement sleep. *J. Appl. Physiol.*, 70, 2597-2601, 1991.

61. MANCIA G., BACCELLI G., ADAMS D.B., ZANCHETTI A. - Vasomotor regulation during sleep in the cat. *Am. J. Physiol.*, 220, 1086-1093, 1971.

62. MANCIA G., ZANCHETTI A. - Cardiovascular regulation during sleep. *In: Physiology in sleep*, J. Orem, CD. Barnes (eds), Academic Press, New York 1-55, 1980.

63. MAQUET P. - Sleep function(s) and cerebral metabolism. *Behav. Brain. Res.*, 1, 69,75-83, 1995.

64. MAQUET P., DEGUELDRE C., DELFIORE G., AERTS J., PETERS J.M., LUXEN A., FRANCK G. - Functional neuroanatomy of human slow wave sleep. *J. Neurosci.*, 15, 17, 2807-2812, 1997.

65. MAQUET P., PETERS J.M., AERTS J., DELFIORE G., DEGUELDRE C., LUXEN A., FRANCK G. - Functional neuroanatomy of human rapid eye movement sleep and dreaming. *Nature.*, 383, 163-166, 1996.

66. McGINTY D. - Physiological equilibrium and the control of sleep states. *In*: D. Mc Ginty, R. Drucker-Colin, A. Morrison and P. Parmeggiani (eds), *Brain mechanisms of sleep*. Raven Press, New York, 361-384, 1985.

67. McPHERSON R.K. - Thermal stress and thermal comfort. *Ergonomics.*, 16, 361-366, 1973.

68. MEISLER A.W., CAREY M.P. - A critical reevaluation of nocturnal penile tumescence monitoring in the diagnosis of erectile dysfunction. *J. Nerv. Ment.*, 178, 78-89, 1990.

69. MILLAR-CRAIG N.W., BISHOP C.N., RAFFTERY E.B. - Circadian variation of blood pressure. *Lancet.*, 1, 795-798, 1978.

70. MULLEN P.E. - Sleep and its interaction with endocrine rhythms. *Br. J. Psychiatry.*, 142, 215-220, 1983.

71. MULLEN P.E., JAMES V.H.T., LIGHTMAN S.L., LINSELL C., PEART W.S. - A relationship between plasma renin activity and the rapid eye movement phase of sleep in man. *J. Clin. Endocrinol. Metab.*, 50, 466-469, 1980.

72. MUZET A., EHRHART J., CANDAS V., LIBERT JP., VOGT JJ. - REM sleep and ambient temperature in man. *Intern. J. Neurosci.*, 18, 117-126, 1983.

73. MUZET A., LIBERT J.P. - Effects of ambient temperature on sleep in man. *In* : Koella WP, Ruther E, Schulz H. (eds), *Sleep 1984*. Gustav Fisher Verlag. Stuttgart-New York, 74-76, 1985.

74. NAIFEH KH., KAMIYA.J. - The nature of respiratory changes associated with sleep onset. *Sleep.*, 4, 49-59, 1981.

75. O'FLAHERTY J.J., SANT'AMBROGIO G., MOGNONI P., SAIBENE F., CAMPORESI E. - Rib cage and abdomen diaphragm contribution to ventilation during sleep in man: *Arch. Physiol.*, 70, 78-80, 1973.

76. OKSENBERG A., ARONS E., SAZBON L., MIZRAHI A., RADWAN H. - Sleep-related erections in vegetative state patients. *Sleep.*, 23, 953-957, 2000.

77. OREM J. - Excitatory drive to the respiratory system in REM sleep. *Sleep.*, 19, S154-S156, 1996.

78. OREM J., MONTPLAISIR J., DEMENT W.C. - Changes in the activity of respiratory neurons during sleep. *Brain Res.*, 82, 309-315, 1974.

79. ORKIN B.A., SOPER N.J., KELLY K.A., DENT J. - Influence of sleep on anal sphincteric pressure in health and after ileal pouch-anal anastomosis. *Dis. Colon. Rectum.*, 35, 137-144, 1992.

80. ORR W.C., JOHNSON L.F., ROBINSON M.G. - Effect of sleep on swallowing, esophageal peristalsis and acid clearance. *Gastroenterology.*, 86, 814-819, 1984.

81. PALCA J.W., WALKER J.M., BERGER R.J. - Thermoregulation, metabolism, and stages of sleep in cold exposed in men. *J. Appl. Physiol.*, 61, 940-947, 1986.

82. PARMEGGIANI P.L. - Regulation of physiological functions during sleep. *Experientia.*, 38, 1405-1408, 1982.

83. PARMEGGIANI P.L. - Interaction between sleep and thermoregulation: an aspect of the control of behavioral states. *Sleep.*, 10, 426-435, 1987.

84. PARMEGGIANI P.L. - Hypothalamic homeothermy across the ultradian sleep cycle. *Arch. Ital. Biol.*,134, 101-107, 1995..

85. PARMEGGIANI P.L., CEVOLANI D., AZZARONI A., FERRARI G. - Thermosensitivity of anterior hypothalamic-preoptic neurons during the waking-sleeping cycle: a study in brain functional states. *Brain. Res.*, 7, 415, 79-89, 1987.

86. PARKER D.C., ROSSMAN L.G., PEKARY A.E., HERSHMAN J.M. - Effect of 64-hour sleep deprivation on the circadian waveform of thyrotropin (TSH) : Further evidence of sleep-related inhibition of TSH release. *J. Clin. Endocrinol. Metab.*, 64, 157-161, 1987.

87. PHILLIPSON E.A. - Control of breathing during sleep. *Am. Rev. Respir. Dis.*, 118, 909-939, 1978.

88. RAMM P., FROST B.J. - Cerebral and local cerebral metabolism in the cat during slow wave and REM sleep. *Brain. Res.*, 365, 112-124, 1986.

89. REIS D.J., MOORHEAD D., WOOTEN G.F. - Redistribution of visceral and cerebral blood flow in the REM phase of sleep. *Neurology.*, 18, 282, 1968.

90. RICHARDSON D.W., HONOUR A.J., GOODMAN A.C. - Changes in arterial pressure during sleep in man. *In*: *Hypertension*, vol. XVI, Neural control of arterial pressure, JE. Woodd (ed.), American heart Association, 62-78, 1968.

91. ROBERTS W.W., ROBINSON T.C.L .- Relaxation and sleep induced by warming of preoptic region and anterior hypothalamus in the cats. *Exp. Neurol.*, 25, 282-294, 1969.

92. SAKAI F., MEYER JS., KARACAN I., YAMAGUCHI F., YAMAMOTO M. - Narcolepsy: regional cerebral

blood flow during sleep and wakefulness. *Neurology.*, 29, 61-67, 1979.

93. SASSIN J-F., FRANTZ A.G., KAPEN S., WEITZMAN E.D. - The nocturnal rise of human prolactin is dependent on sleep. *J. Clin. Endocrinol. Metab.*, 37, 436-440, 1973.

94. SASSIN J-F., PARKER C., MACE JW., GOTLIN R.W., JOHNSON L.C., ROSSMAN L.G. - Human growth hormone release: relation to slow-wave sleep and slow-waking cycles. *Science.*, 165, 513-515, 1969.

95. SATINOFF E. - Neural organization and evolution of thermal regulation in mammals. *Science.*, 201, 16-22, 1978.

96. SCHIAVI R.C, MANDELI J., SCHREINER-ENGEL P., CHAMBERS A. - Aging, sleep disorders, and male sexual function. *Biol. Psychiatry.*, 30, 15-24, 1991.

97. SCHMIDT M.H., SAKAI K., VALATX J.L., JOUVET M. - The effects of spinal or mesencephalic transections on sleep-related erections and ex-copula penile reflexes in the rat. *Sleep.*, 22, 409-418, 1999.

98. SCHMIDT M.H., VALATX J.L., SAKAI K., FORT P., JOUVET M. - Role of the lateral preoptic area in sleep-related erectile mechanisms and sleep generation in the rat. *J. Neurosci.*, 20, 6640-6647, 2000.

99. SIMON P.M., DEMPSEY J.A., LANDRY D.M., SKATRUD JB. - Effect of sleep on respiratory muscle activity during mechanical ventilation. *Am. Rev. Respir. Dis.*, 147, 32-7, 1993.

100. SOFFER E.E., THONGSAWAT S., ELLERBROEK S. - Prolonged ambulatory duodeno-jejunal manometry in humans: normal values and gender effect. *Am. J. Gastroenterol.*, 93, 1318-1323, 1998.

101. SOMERS V.K., DYKEN M.E., MARK A.L., ABBOUD F.M. - Sympathetic-nerve activity during sleep in normal subjects. *N. Engl. J. Med.*, 328, 303-307, 1993.

102. SPIEGEL K., FOLLENIUS M., SIMON C., SAINI J., EHRHART J., BRANDENBERGER G. - Prolactin secretion and sleep. *Sleep.*, 17, 20-27, 1994.

103. SPIEGEL K., LEPROULT R., VAN CAUTER E. - Impact of sleep debt on metabolic and endocrine function. *Lancet.*, 354, 1435-1439, 1999.

104. SULLIVAN CE. - Breathing in sleep. In: *Physiology in Sleep*, J. Orem, CD. Barnes (eds.), Academic Press, New York, 214-272, 1980.

105. TARCHANOFF J. - Quelques observations sur le sommeil normal.. *Arch. It. Biol.*, 21, 318-322, 1894.

106. TOUITOU Y., FEVRE M., LAGOGUEY M., CARAYON A., BOGDAN A., REINBERG A., BECK H., CESSELIN F., TOUITOU C. - Age- and mental health-related circadian rhythms of plasma levels of melatonin, prolactin, luteinizing hormone and follicle-stimulating hormone in man. *J. Endocrinol.*, 91, 467-475, 1981.

107. TOWNSEND R.E., PRINZ P.N., OBRIST W.D. Human cerebral blood flow during sleep and waking. *J. Appl. Physiol.* 35, 620-625, 1973.

108. VAN CAUTER E., KERKHOFS M., CAUFRIEZ A., VAN ONDERBERGEN A., THORNER MO., COPINSCHI G. - A quantitative estimation of growth hormone secretion in normal man: reproductibility and relation to sleep and time of day. *J. Clin. Endocrinol. Metab.*, 74, 1441-1450, 1992.

109. VAN CAUTER E., REFETOFF S. - Multifactorial control of the 24-hour secretory profiles of pituitary hormones. *J. Endocrinol. Invest.*, 8, 381-391, 1985.

110. VEERMAN D.P., IMHOLZ B.P., WIELING W., WESSELING K.H., VAN MONTFRANS G.A. - Circadian profile of systemic hemodynamics. *Hypertension.*, 26, 55-59, 1995.

111. WEBB P.- Periodic breathing during sleep. *J. Appl. Physiol.*, 37, 899-903, 1974.

112. WEBB P., HIESTAND M. - Sleep metabolism and age. *J. Appl. Physiol.*, 38, 257-267, 1974.

113. WHITE D.P., WEIL J.V., ZWILLICH C.W. - Metabolic rate and breathing during sleep. *J. Appl. Physiol.*, 59, 384-391, 1985.

Chapter 5

Models of human sleep regulation[1]

D. G. M. Beersma

RUG, Zoological Laboratory, AA Haren, The Netherlands

INTRODUCTION

One of the reasons to assume that sleep serves important functions to an organism is the fact that timing and duration of sleep are strictly regulated. Specific rebounds occur both upon non-REM sleep deprivation as well as upon REM sleep deprivation, suggesting that the states constitute necessary physiological processes. It is, however, not yet clear what these necessary processes are. The physiological characteristics of non-REM sleep and REM sleep are very different. Apart from the absence or presence of rapid eye movements, there are differences in the electroencephalogram, in thermoregulation, in cardiovascular processes, in breathing. Virtually every physiological process is involved. Each of the processes seems to be crucial for the functioning of the organism, and so each of them may provide the ultimate reason for the rebound responses which occur upon non-REM and REM sleep deprivation.

Although the question why we sleep is still not fully answered, much is known about how we sleep. The available knowledge is consistent with the hypothesis that sleep serves to recover from previous wakefulness or to prepare for proper functioning in the wakeful period to come [27,39,53]. This hypothesis predominantly rests on electrophysiological observations regarding non-REM sleep regulation [8,30,82], which covers 75-80% of total sleep time in healthy adults. The available knowledge about REM sleep regulatory mechanisms [22,66] is not inconsistent with a recovery function of sleep, but this work suggests that the function of REM sleep per se is related to non-REM sleep instead of being related to wakefulness. Similar notions have been formulated in the consolidation hypothesis by Meddis (67). An indirect consequence of this view is that non-REM sleep would fulfil the primary functions of sleep, while REM sleep would serve to keep non-REM sleep going.

The wide variety of data on how we sleep can best be summarised in terms of models of sleep regulation, since models can integrate many observations. Models in turn help to formulate relevant questions for future experimentation.

The present paper reviews currently available models of human sleep regulation, with an emphasis on models concerning electrophysiological processes. For that purpose, the two-process model of sleep regulation [27,38,39] will serve as a starting point. After having presented its concepts and merits, I will discuss alternative views which came up in the last 15 years. Subsequently I will present various new findings and ideas which have modified or will probably soon modify our notions about sleep regulation. Previous reviews in this field were published by Borbély and Achermann [29], and by Borbély [28].

[1] Adapted from *Sleep Medicine Reviews* volume 2, number 1, by D. G. M. Beersma, Models of human sleep regulation, pages 31-43, 1998, by permission of the publisher, W. B. Saunders.

THE TWO-PROCESS MODEL OF SLEEP REGULATION

In the early 1980s ideas on circadian and homeostatic regulation of behaviour were integrated, initially in global terms on the basis of both rat data [25,26] and human data [31]. Subsequently, two proposals for the detailed interaction between the processes in humans were presented at a meeting in Cape Cod in June 1981 [27,38]. Collaborative efforts culminated in a quantitative model of human sleep regulation, which became known as the two-process model of sleep regulation [39]. The model describes the timing of the alternation between human sleep and wakefulness as the result of the interaction of two processes: a homeostatic process keeping track of the instantaneous need for sleep, and a circadian process keeping track of environmental time. Figure 5.1 describes the model's structure.

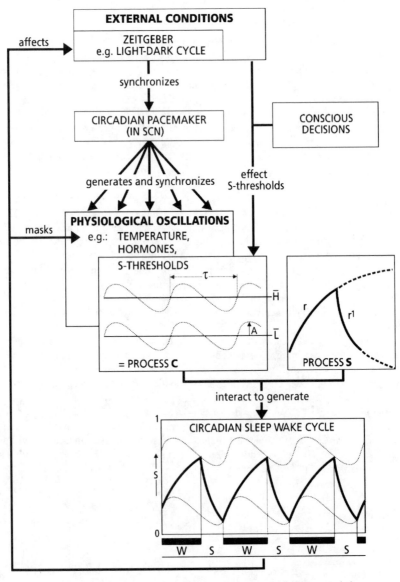

Figure 5.1. Schematic representation of the two-process model of sleep regulation.

The homeostatic process is called process S. This is the regulated variable in the model. It increases during wakefulness in an exponentially saturating way, and it declines exponentially during sleep. Originally it was suggested that the intensity of sleep, measured as the power density of the non-REM sleep EEG, would be proportional to S, because the power density of the non-REM sleep EEG decays in an approximately exponential way upon successive non-REM-REM cycles [30]. Due to the mathematical characteristics of exponential functions, however, it was not possible to distinguish between (A) a proportionality of S and EEG power density, and (B) a proportionality of the rate of change of S and EEG power density [20]. Experimental suppression of non-REM EEG power density [40,42] demonstrated that non-REM sleep intensity (also called slow-wave activity, SWA) must be proportional to the momentary decrement of S.

In the model, the waxing and waning of process S is limited by two thresholds to the S process. One upper threshold determines the transition from the increase during wakefulness to the decrease during sleep. It thereby determines sleep onset and could be called the sleep threshold. The other, lower threshold determines the transition from decreasing towards increasing values of S, i.e. sleep termination. Therefore, this can be seen as the wake threshold. The two thresholds are supposedly under the control of the circadian pacemaker in the suprachiasmatic nuclei (SCN) of the hypothalamus. As a result, the thresholds vary systematically with time of day. The thresholds are presumed to run in parallel and together are called process C.

Apart from the processes S and C, there are many other influences on the alternation of sleep and wakefulness, such as conscious decisions, pain, stress, etc. It is assumed that these factors have an influence on the timing of sleep by modulating the levels of the sleep and wake thresholds, while leaving process S undisturbed.

Simulations demonstrated that with this model a wide variety of phenomena can be explained [39]. It was possible to simulate the characteristics of the sleep-wake patterns of subjects in temporal isolation, in particular upon spontaneous internal desynchronisation [93]. Sleep fragmentation under conditions of continuous bed rest could be explained with the model as well as the duration of daytime sleep of shift workers. Attempts were made to extend the model to include measures of alertness and predict their course under sleep deprivation conditions. For that purpose the distance between S and the sleep threshold was taken as a possible correlate of alertness. There was a reasonable correspondence between simulations and real data. However, later studies demonstrated that several modifications of the two-process model are required to make it a suitable model to describe and predict alertness data [6-8,11-12]. Recently, data collected by Dijk and Czeisler [47-48] revealed that alertness is determined by homeostatic and circadian influences in complex interaction. Even though the current models may be insufficient to explain the complex interaction between homeostatic and circadian processes underlying alertness, there is little doubt that such processes are involved in the regulation of alertness.

The two-process model was not primarily designed to simulate alertness data, but to integrate existing knowledge on sleep regulation. It was used to predict the outcome of new experiments and survived a series of critical tests [40-46,50]. It proved to be compatible with important new data sets, such as SWA data obtained in habitual long and short sleepers [9].

THE INTERACTION OF PROCESS S AND PROCESS C

The concept that sleep timing would be regulated by the interaction of one homeostatic process, S, keeping track of internal needs, and one circadian process, C, keeping track of environmental time was so appealing that it was rapidly accepted by the scientific community. This was not so much a consequence of the qualities of the simulations since a two pacemaker model, developed by Kronauer *et al.* [64] had similar simulation qualities [83]. The attractiveness of the model was predominantly due to its physiological appeal, and its mathematical simplicity. The fact that only a single circadian pacemaker was detected in the mammalian brain [71,81] remained a stumbling block for the two-pacemaker approach. By now there is evidence for the existence of two other pacemakers, one in the mammalian eye [85], and one related to learning temporal patterns of food availability ('Food Entrainable Oscillator'- 75). However, these pacemakers do not seem to be involved in the regulation of the timing of sleep.

Despite the general acceptance of the concept of the two-process model, several publications have proposed alternative kinds of interactions between the two processes. Putilov [77], for instance, suggested to modify the two-process model in such a way that process S, like all processes in the body, is, to some extent, modulated by the circadian pacemaker. By doing so he could demonstrate that process C could be taken to be sinusoidal, which approximates the circadian variation of firing rate in the SCN [60], and the influence of the SCN on core temperature [49,56]. Putilov also demonstrated that some simulations of SWA with the adjusted model were slightly better than with the original two-process model.

Yet, a large range of original simulations were not repeated with the adjusted model, and it remains to be seen whether the benefits of the adjustment outweigh the loss of simplicity of the concept.

Edgar *et al.* [51], on the basis of lesion studies in squirrel monkeys, concluded that one of the functions of the SCN is to consolidate vigilance states, which consolidation occurs in a circadian phase dependent manner. They suggested that sleep timing is the result of the interaction of two opponent processes. SCN dependent processes would actively facilitate the initiation and maintenance of wakefulness, and oppose homeostatic sleep tendency during the subjective day. Along similar lines, Dijk and Czeisler [47-48] suggested that the circadian pacemaker has a function in consolidating wakefulness as well as sleep. This conclusion was based on so-called 'forced desynchrony' studies. In these studies sleep and wakefulness are scheduled to alternate at a period distinctly different from 24 hours. This leads to a desynchronisation between the sleep-wake cycle and the circadian pacemaker. Plotting the values of several variables (alertness data, wakefulness during sleep, total sleep time, sleep latency) as a function of the time since scheduled wake-up and as a function of circadian phase, revealed that the circadian drive for sleep is maximal near habitual wake-up time. Similarly, the circadian drive for wakefulness is maximal near habitual bedtime (this interval of time is sometimes called the 'dead zone' for sleep onset – [65,83]). The data demonstrated that the circadian pacemaker serves to consolidate wakefulness at the end of the day, where alertness would otherwise decrease due to the long duration of prior wakefulness. The pacemaker, likewise, serves to consolidate sleep at the end of the night, where the increase of arousal in response to the long duration of prior sleep would otherwise induce wakefulness. In the two-process model of sleep regulation the circadian drive for wakefulness is modelled in the shape of the wake threshold and the circadian drive for sleep is modelled in the shape of the sleep threshold: high levels of the thresholds represent a low drive for sleep. As a consequence, the data by Dijk and Czeisler [48] suggest that the shape of the wake threshold is qualitatively correct, but that this does not hold for the sleep threshold. In contrast to the shape presented in figure 1, the sleep threshold should increase in the course of habitual wakefulness and sharply decline at the beginning of habitual sleep time, just like Achermann and Borbély [6] proposed on the basis of simulations of alertness data. It must be noted that the shape of the sleep threshold in the original model was not derived from direct experimental results but, in the absence of such data, was simply postulated to run in parallel to the wake threshold. It is obvious that experiments aiming at assessing the shape of the sleep threshold are urgently needed.

Forced desynchrony studies further revealed that SWA is largely independent of circadian phase [48] which is consistent with the assumption of two independent processes underlying human sleep regulation. REM sleep, in contrast, was shown to vary with circadian phase and also with the time since sleep onset. The inhibition of REM sleep at high pressure for non-REM sleep, as proposed by Borbély [27], is consistent with this latter relationship.

Several other models of sleep regulation exist in the literature [36,63,74]. These models simultaneously take into account a wide variety of variables which influence sleep regulation. As a consequence, these models are physiologically and mathematically complex and beyond the scope of the present review.

In the following I will focus on several details of the model and review a series of relevant refinements and alternatives.

Process S

In general, the output of a process can be modulated by modulation of either its intensity or its duration. For non-REM sleep it is clear that it responds to experimental manipulations mainly by changing its intensity [4,20,30,40,42,88]. The two-process model of sleep regulation in its original formulation did not distinguish between REM and non-REM sleep. The non-REM-REM sleep cycle was simply taken as the unit of analysis of sleep intensity. Sleep intensity was defined on the basis of non-REM sleep EEG features only, by calculating the power of the electroencephalogram in the delta frequency range (0.5-4 Hz) by Fourier analysis, or by calculating the amplitude of EEG half-waves with relatively long half-wave durations. It has been shown that the power densities of EEG signals correlate highly with the squares of the half-wave amplitudes [54].

Achermann *et al.* [8] have refined the two-process model, in particular with respect to the details of process S during sleep. For that purpose, they developed a sub-model or module which specifically addressed the time course of SWA in the course of a non-REM-REM sleep cycle. This fine structure was analysed in earlier studies [1,3]. The module receives as inputs information about both the timing of REM sleep as well as the timing of arousals and intermittent wake time. It then produces a SWA profile which almost perfectly mimics the actually observed SWA profile. Apparently, the dynamics of SWA are very well known now. A subtle but important difference with the original formulation of the two-process model is that Achermann had to assume that process S during sleep has two dynamic components. One is the decline which is proportional to SWA, and the other is an increment of S which is presumed to be always present, i.e. both during wakefulness and during sleep. Although the reason to incorporate the latter term in the model was to improve the quality of fit between simulations and data, the term suggests that the activity of cells under all conditions leads to a need for recovery. The recovery occurs during sleep and is superimposed on and outweighs the ongoing activity which creates the need for sleep.

With respect to the module presented by Achermann *et al.*, it is important to note that the timing of the intrusions of wakefulness and of the transitions to and from REM sleep are as yet difficult to predict. For the intrusion of arousals no models are available, and the existing models of the alternation between REM and non-REM sleep are based on the assumption of the existence of an ultradian pacemaker in the human brain [2,66], which remains to be demonstrated. Modelling is hampered by the fact that the functions of REM sleep are not known. Benington and Heller [22] recently revived the discussion regarding the function of REM sleep. They suggested on the basis of various data sets that REM sleep does not serve to recover from some wastage of wakefulness, but that it could serve to recover from the preceding non-REM sleep. (This is a more generalised form of the hypothesis proposed earlier by Wehr [89] that REM sleep serves to counteract the brain cooling in prior non-REM). If this hypothesis survives critical experimentation it could serve as a basis for further modeling. It would then also substantiate the notion that the main function of sleep is fulfilled during the delta waves of non-REM sleep.

Intrusions of wakefulness into sleep as well as intervals of REM sleep have their impact on SWA, e.g., because it takes time before SWA resumes maximum values after interruptions.

Experimental studies using SWA as outcome variables are advised to integrate SWA values over sufficiently long time intervals of accumulated non-REM sleep [18].

The work of Steriade *et al.* [82] has revealed that the delta waves in the EEG originate from cortico-thalamic feedback loops which bring subsets of neurons in mutually synchronous firing and pausing states. The greater the number of neurons participating in a synchronous sample, the larger the amplitude of the delta wave in the EEG. This notion has brought Mourtazaev *et al.* [72] to propose a 'model based' analysis of the sleep EEG. They model the interactions between the neurons with a feedback system which receives a white noise signal as an input, and produces the EEG signal as its output. The feedback strength of the model required to mimic the actually observed EEG is calculated as a function of time during sleep. The strength of the feedback is hypothesised to represent the intensity of the non-REM sleep process. One of the advantages of the method is that the feedback strength is a relative measure, not dependent on the absolute level of the EEG signal. Differences in EEG power density between individuals, which, for instance, may be due to anatomical differences, are automatically set aside with this model-based method of analysis. Another advantage over classical Fourier analysis is that, by its nature, the model-based method discriminates waves in the EEG from transients with other characteristics, such as movement artifacts. An obvious disadvantage is that the results of the analysis depend on the characteristics of the model on which it is based.

The work of Steriade has stimulated other new developments. The electrophysiological findings have triggered research aiming at the biochemical origin of SWA, which may provide clues for the understanding of the fundamental function of sleep itself. Porkka-Heiskanen *et al.* [76] showed that extra-cellular adenosine has a temporal course over the sleep-wake cycle which is similar to process S. Benington and Heller [23] also indicated adenosin as being related to the generation of delta waves. In addition they suggested that the replenishment of astrocytic glycogen represents the ultimate need for sleep. They hypothesised that slow waves are required for the restoration of the glycogen which was depleted during wakefulness. These are highly interesting developments which deserve much attention in the near future.

Merica and Blois [69] also extended on the work by Steriade. They presented a simple model which could explain some of the changes in the composition of the frequency spectrum of the EEG in the course of the non-REM episode [86]. They estimated the electrical consequences when a population of neurons went from the polarised state during waking to the hyperpolarised state during deep SWA. As was shown by Steriade *et al.* [82], the thalamocortical neurons are capable of generating spindle activity at intermediate levels of polarisation. Merica and Blois [69] could show that the temporal distributions of energy of the non-REM sleep EEG signal in the spindle range and in the delta range can be simulated with such a simple model.

Another new development concerning process S during sleep is the work by Werth *et al.* [91,92]. They showed that in the early work by Bos *et al.* [33] on the spatial distribution of electrical activity over the scalp, one important phenomenon had been overlooked: Maximum SWA gradually shifts from occipital to frontal regions in the course of each non-REM episode. Such fine structure of the processes involved in sleep regulation is not yet incorporated in any model. It suggests that the recovery processes occur at specific locations, as was already demonstrated for some aquatic mammals [73] and also for humans [62]. It shows that a non-REM episode is not just containing a bulk of SWA, but that the process develops in a carefully controlled topographical way. Perhaps, this is associated with the dependence of non-REM-REM cycle durations on the brain size of an animal [95]: In larger brains it takes longer to complete a 'wave' of recovery processes from occipital to frontal areas.

Against the background of these new developments in our knowledge of process S during non-REM sleep, it is remarkable that virtually nothing is known about the kind of stimuli to which SWA responds. Few experiments have been performed to test whether different activities during the day lead to differences in SWA during sleep [62,58-59,68]. SWA does seem to increase in response to heat load. Intense physical activity can lead to increased SWA, but such changes are not observed when physical activity is combined with cooling of the brain [57]. There are detectable increases in SWA in response to long sessions of finger tapping]62), but the effects are small. It is as if the wear and tear of the neurons in the brain only depend on their level of operation, which is modulated by brain temperature, and that these cells otherwise operate at a similar average rate, independent from the stimuli they have to deal with. However, it is not sure whether similar statements can be made regarding other types of activity. Aggressive confrontations between rats lead to systematic changes in their subsequent sleep [68]. Obviously, such confrontations entail various changes in emotional, cognitive, and somatic states, such as increased anxiety, alertness, energy metabolism, etc. It is not clear which of these changes is responsible for the changes observed during sleep. In humans, no systematic data on the impact of mental processes on SWA are available.

REM sleep received little attention within the context of the two-process model. One of the reasons is the fact that the non-REM-REM cycle was originally taken as the unit of analysis. However, there are some other reasons contributing to this fact. Concerning the homeostatic regulation of REM sleep the situation is not very clear. No obvious REM sleep intensity measure is available. REM density cannot be a measure of REM sleep intensity because it predominantly responds to changes in non-REM sleep intensity [14,17,32). There are data suggesting that certain frequency characteristics of the REM sleep EEG can serve as a measure of REM sleep intensity [34,35], but this needs independent validation. In contrast to non-REM sleep, it is evident, however, that REM sleep responds to deprivation by a partial rebound REM sleep duration [21,34-35].

Process C

Much less progress than in our knowledge of process S and its underlying physiology has been made with respect to process C. There is not much more than the assumption that process C is driven by the circadian pacemaker. Insight in the physiology and biochemistry of the circadian pacemaker, however, has progressed spectacularly. It is known that the primary circadian pacemaker in mammals is located in the suprachiasmatic nucleus of the hypothalamus [71,81]. The period of the circadian pacemaker can be transplanted by transplanting the SCN [78]. Much is known about the anatomical connections between the pacemaker and other structures in the brain [70,84]. It is clear that the light-dark cycle is the most important stimulus to entrain the circadian pacemaker [16,19,24,61]. Much effort is being given to unravel the pathways which transfer the light-dark information to the SCN [15]. Amir and Stewart [13] noted that phase responses of the circadian pacemaker are not solely dependent on light, but that such phase responses can be conditioned and also occur when the conditioning stimulus is presented alone. This intriguing finding still awaits confirmation by independent research.

There is considerable progress in the knowledge of the circadian pacemaker on the microscopical level. Welsh *et al.* [90] reported that individual SCN cells, cultured in a dish are capable of generating circadian rhythmicity in the production of action potentials. Apparently, single SCN cells in culture are as capable of sustaining circadian rhythmicity as are unicellular algae [79]. This notion was anticipated long ago, and incorporated in a model of the circadian pacemaker by Enright [52]. In this model a series of individual self-sustained ´pacers´ are thought to be coupled to each other on the basis of their integrated output. The coupling can explain many characteristics of circadian pacemakers. Shinohara *et al.* [80] reported the simultaneous presence of independent circadian rhythms in neurotransmitter production in the isolated SCN.

At the molecular level progress in circadian rhythm research is almost overwhelming. It is known now that the circadian oscillations in the pacemaker are due to periodic activation of the transcription of genes that influence their own transcription. It is this autoregulatory feedback mechanism which generates the period of the pacemaking signal [55]. Many of the relevant genes involved in this ticking of the clock are known, and models are being made as to how they interact [37].

Despite all these exciting developments in the field of circadian pacemaker research, it must be concluded that the physiology of process C remains largely obscure. Some data indicate that it is unlikely that the two thresholds run in parallel. Experiments aiming at the shape of the sleep threshold are urgently needed.

CONCLUSIONS

In this review I have tried to discuss a series of modelling approaches to the mechanisms of sleep regulation. Admittedly, the review is limited. There is a clear focus on electrophysiological models, with an emphasis on models describing aspects of the human non-REM sleep EEG. I have tried to emphasise the functional aspects of the models and the functional concepts behind them, more than the mathematical details. The two-process model of sleep regulation was taken as the background for this discussion. The reason for this is that the two-process model conceptually sorts out two major influences on sleep regulation: the circadian pacemaker and homeostatic needs. This approach enables modular modelling: Recognition of the separate influences makes it possible to perform experiments and develop models for each influence separately, which reduces the complexity of the problem [5]. Prospects for a more detailed theoretical understanding of sleep regulation will heavily depend on experimental progress in particular with respect to two questions: what controls the onset of sleep and what controls the timing of REM episodes within sleep.

A remaining issue is whether the modelling approach has consequences for clinical practice. Models of sleep regulation are fundamental to the development of theories about how and why we sleep. The models specify those theories in quantitative detail, which allows critical testing and improvement of the theories. Since the theories about how and why we sleep should form the basis for understanding the nature and the consequences of sleep disorders, there is no doubt that models of sleep regulation are very relevant for diagnosis and treatment.

The clinical problems for which knowledge of sleep regulation could be relevant include all kinds of 'pure' sleep disorders and disorders in which deregulation of sleep is a prominent feature. In this way, models of sleep regulation have, for instance, contributed to the development of therapies for delayed sleep phase syndrome and for jet lag, and large scale investigations are underway to see whether adjustments of the circadian pacemaker by light can reduce the sleep problems of shift workers.

In the field of psychiatric disorders, models have been applied in the study of the pathogenesis and therapy of seasonal and non-seasonal mood disorders [87,94]. In view of these developments it is more than justified to expect increasing impact of models of sleep regulation on clinical practice in the future.

REFERENCES

1. ACHERMANN P. - *Schlafregulation des Menschen: Modelle und Computersimulationen.* Thesis. ADAG Administration & Druck AG, Zürich, 1988.
2. ACHERMANN P., BEERSMA D.G.M., BORBELY A.A. - The two-process model: ultradian dynamics of sleep. In: Horne, JA, (ed) *Sleep 90.* Pontenagel Press, Bochum, 1990, pp 296-300.
3. ACHERMANN P., BORBELY A.A - Dynamics of EEG slow wave activity during physiological sleep and after administration of benzodiazepine hypnotics. *Human Neurobiol.* 6, 203-210, 1987.
4. ACHERMANN P., BORBELY A.A. - Simulation of human sleep: ultradian dynamics of electroencephalographic slow-wave activity. *J. Biol. Rhythms* 5, 141-157, 1990.
5. ACHERMANN P., BORBELY A.A. - Combining different models of sleep regulation. *J. Sleep Res.* 1, 144-147, 1992.
6. ACHERMANN P., BORBELY A.A. - Simulation of daytime vigilance by the additive interaction of a homeostatic and a circadian process. *Biol. Cybern.* 71, 115-121, 1994.
7. ACHERMANN P., DIJK D.J., BRUNNER D.P., BORBELY A.A. - A model of human sleep homeostasis based on EEG slow-wave activity: quantitative comparison of data and simulations. *Brain Res Bull* 31, 97-113, 1993.
8. ACHERMANN P., WERTH E., DIJK D.J., BORBELY A.A. - Time course of sleep inertia after nighttime and daytime sleep episodes. *Arch Ital. Biol.* 134, 109-119, 1995.
9. AESCHBACH D., CAJOCHEN C., LANDOLT H., BORBELY A.A. - Homeostatic sleep regulation in habitual short sleepers and long sleepers. *Am. J. Physiol.* 270, R41-R53, 1996.
10. AKERSTEDT T., FOLKARD S. - Validation of the S and C components of the three-process model of alertness regulation. *Sleep* 18, 1-6, 1995.
11. AKERSTEDT T., FOLKARD S. - Predicting sleep latency from the three-process model of alertness regulation. *Psychophysiol.* 33, 385-389, 1996.
12. AKERSTEDT T., FOLKARD S. - The three-process model of alertness and its extension to performance, sleep latency, and sleep length. *Chronobiol. Int.* 14, 115-123, 1997.
13. AMIR S., STEWART J. - Resetting of the circadian clock by a conditioned stimulus. *Nature* 379, 542-545, 1996
14. ANTONIOLI M., SOLANO L., TORRE A., VIOLANI C., COSTA M., BERTINI M. Independence of REM density from other REM sleep parameters before and after REM deprivation. *Sleep* 4, 221-225, 1981.
15. ARGAMASO S.M., FROEHLICH A.C., McCALL M.A., NEVO E., PROVENCIO I., FOSTER R.G. - Photo pigments and circadian systems of vertebrates. *Biophys-Chem* 56, 3-11, 1995.
16. ASCHOFF J., PÖPPEL E, WEVER R. - Circadiane Periodik des Menschen unter dem Einfluss von Licht-Dunkel-Wechseln unterschiedlicher Periode. *Pflugers Arch.* 306, 58-70, 1969.
17. ASERINSKI E. - Relationship of rapid eye movement density to the prior accumulation of sleep and wakefulness. *Psychophysiology* 10, 545-558, 1973.
18. BEERSMA D.G.M., ACHERMANN P. - Changes of sleep EEG slow-wave activity in response to sleep manipulations: to what extent are they related to changes in REM sleep latency. *J. Sleep Res.* 4, 23-29, 1995.
19. BEERSMA D.G.M., DAAN S. - Strong or weak phase resetting by light pulses in humans. *J. Biol. Rhythms* 8, 340-347, 1993.
20. BEERSMA D.G.M., DAAN S., DIJK D.J. - Sleep intensity and timing: a model for their circadian control. *Lectures on Mathematics in the Life Sciences* 19, 39-62, 1987.
21. BEERSMA D.G.M., DIJK D.J., BLOK C.G.H., EVERHARDUS I. - REM sleep deprivation during 5 hours leads to an immediate REM sleep rebound and to suppression of non-REM sleep intensity. *Electroencephalogr. Clin. Neurophysiol.* 76, 114-122, 1990.
22. BENINGTON J.H., HELLER H.C. - REM-sleep timing is controlled homeostatically by accumulation of REM-sleep propensity in non-REM sleep. *Am. J. Physiol.* 266, R1992-R2000, 1994.
23. BENINGTON J.H., HELLER H.C. - Restoration of brain energy metabolism as the function of sleep. *Prog. Neurobiol.* 45, 347-360, 1995.
24. BOIVIN D.B., DUFFY J.F., KRONAUER R.E., CZEISLER C.A. - Dose-response relationships for resetting of human circadian clock by light. *Nature,* 379, 540-542, 1996.
25. BORBELY A.A. - Sleep: circadian rhythm versus recovery process. In: Koukkou, M, Lehmann, D, Angst, J, (eds), *Functional States of the Brain: Their Determinants.* Elsevier, Amsterdam, 1980, pp 151-161.
26. BORBELY A.A. - Circadian and sleep dependent processes in sleep regulation. In: Aschoff, J, Daan, S, Groos, G (eds). *Vertebrate Circadian Systems: Structure and Physiology.* Springer Verlag, Berlin 1982, pp 237-242.
27. BORBELY A.A. - A two-process model of sleep regulation. *Hum. Neurobiol* 1, 195-204, 1982.
28. BORBELY A.A. - Sleep homeostasis and models of sleep regulation. In: Kryger, MH, Roth, T, Dement, WC (eds) *Principles and Practice of Sleep Medicine,* 2nd edition, Saunders, Philadelphia 1994, pp 309-320.
29. BORBELY A.A., ACHERMANN P. - Concepts and models of sleep regulation: an overview. *J. Sleep Res.* 1, 63-79, 1992.

30. BORBELY A.A., BAUMANN P., BRANDEIS D., STRUACH I., LEHMAN D. - Sleep deprivation. Effect on sleep stages and EEG power density in man. *Electroencephalogr. Clin. Neurophysiol.* 51, 483-495, 1981.

31. BORBELY A.A., TOBLER I., WIRZ-JUSTICE A. - Circadian and sleep dependent processes in sleep regulation: outline of a model and implications for depression, *Sleep Res* 10, 19, 1981.

32. BORBELY A.A., WIRZ-JUSTICE A. - Sleep, sleep deprivation and depression. A hypothesis derived from a model of sleep regulation. *Hum. Neurobiol* 1, 205-210, 1982.

33. BOS K.H.N., Van den HOOFDAKKER R.H., KAPPERS E.J. - An electrode independent function describing the EEG changes during sleep. In: *Sleep 1976*, Koella, WP and Levin, P, eds. Basel: Karger, 1977, pp 470-473.

34. BRUNNER D.P., DIJK D.J., BORBELY A.A. - Repeated partial sleep deprivation progressively changes the EEG during sleep and wakefulness. *Sleep* 16, 100-113, 1993.

35. BRUNNER D.P., DIJK D.J., TOBLER I., BORBELY A.A. - Effect of partial sleep deprivation on sleep stages and EEG power spectra: evidence for non-REM and REM sleep homeostasis. *Electroencephalogr. Clin. Neurophysiol.* 75, 492-499, 1990.

36. CARPENTER G.A., GROSSBERG S. - A neural theory of circadian rhythms: split rhythms, after-effects and motivational interaction. *J. Theor. Biol.* 113, 163-223, 1985.

37. DAAN S., ALBRECHT U., Van der HOST G.T.J., ILLNEROVA H., ROENNEBERG T.,WEHR T.A., SCHWARTZ W.J. - Assembling a clock for all seasons: Are there M and E oscillators in the genes? *J. Biol. Rhythms* 16, 105-116, 2001.

38. DAAN S., BEERSMA D.G.M. - Circadian gating of human sleep and wakefulness. In: *Mathematical Modeling of Circadian Systems,* MC Moore-Ede and CA Czeisler, eds. New York, Raven 1983, pp 129-158.

39. DAAN S., BEERSMA D.G.M., BORBELY A.A. - Timing of human sleep: recovery process gated by a circadian pacemaker. *Am. J. Physiol.* 246, R161-R178, 1984.

40. DIJK D.J., BEERSMA D.G.M. - Effects of SWS deprivation on subsequent EEG power density and spontaneous sleep duration. *Electroencephalogr. Clin. Neurophysiol.* 72, 312-320, 1989.

41. DIJK D.J., BEERSMA D.G.M., DAAN S. - EEG power density during nap sleep: reflection of an hourglass measuring the duration of prior wakefulness. *J. Biol Rhythms* 2, 207-219, 1987.

42. DIJK D.J., BEERSMA D.G.M., DAAN S., BLOEM G.M., Van den HOOFDAKKER R.H. - Quantitative analysis of the effects of slow wave sleep deprivation during the first three hours of sleep on subsequent EEG power density. *Eur. Arch. Neurol. Sci.* 236, 323-328, 1987.

43. DIJK D.J., BEERSMA D.G.M., DAAN S., LEWY A.L. - Bright morning light advances the human circadian system without affecting non-REM sleep homeostasis. *Am. J. Physiol.* 256, R106-R111, 1989.

44. DIJK D.J., BEERSMA D.G.M., DAAN S., Van den HOOFDAKKER R.H. - Effects of seganserin, a 5-HT$_2$ antagonist, and temazepam on human sleep stages and EEG power spectra. *Eur. J. Pharmacol,* 171, 207-218, 1989.

45. DIJK D.J., BEERSMA D.G.M., STIEKEMA M. - Is the midafternoon decline in sleep latency associated with a peak in NREM sleep intensity ? *Chronobiologia* 14, 168-169, 1987.

46. DIJK D.J., BRUNNER D.P., BEERSMA D.G.M., BORBELY A.A. - Electroencephalogram power density and slow wave sleep as a function of prior waking and circadian phase. *Sleep* 13, 430-440, 1990.

47. DIJK D.J., CZEISLER C.A. - Paradoxical timing of the circadian rhythm of sleep propensity serves to consolidate sleep and wakefulness in humans. *Neurosci. Lett.* 166, 63-68, 1994.

48. DIJK D.J., CZEISLER C.A. - Contribution of the circadian pacemaker and the sleep homeostat to sleep propensity, sleep structure, electroencephalographic slow waves, and sleep spindle activity in humans. *J. Neurosci.* 15, 3526-3538, 1995.

49. DIJK D.J., DUFFY J.F., CZEISLER C.A. - Circadian and sleep/wake dependent aspects of subjective alertness and cognitive performance. *J. Sleep Res.* 1, 112-117, 1992.

50. DIJK D.J., VISSCHER C.A., BLOEM G.M., BEERSMA D.G.M., DAAN S. - Reduction of human sleep duration after bright light exposure in the morning. *Neuroscience Letters* 73, 181-186, 1987.

51. EDGAR D.M., DEMENT W.C., FULLER C.A. - Effect of SCN lesions on sleep in squirrel monkeys: evidence for opponent processes in sleep-wake regulation. *J. Neurosci.* 13, 1065-1079, 1993.

52. ENRIGHT J.T. - *The Timing of Sleep and Wakefulness.* New York: Springer, 1980.

53. FEINBERG I. - Changes in sleep cycle pattern with age. *J. Psychiat. Res.* 10, 283-306, 1974.

54. GEERING B.A., ACHERMANN P., EGGIMANN F., BORBELY A.A. - Period-amplitude analysis and power spectral analysis: a comparison based on all-night sleep EEG recordings. *J. Sleep Res.* 2, 121-129, 1993.

55. HASTINGS M.H. - Circadian clockwork: two loops are better than one. *Nat.Rev.Neurosci.* 1, 143-146, 2000.

56. HIDDINGS E.A., BEERSMA D.G.M., Van den HOOFDAKKER R.H. - Endogenous and exogenous components in the circadian variation of core body temperature in humans. *J. Sleep Res.* 6, 156-163, 1997.

57. HORNE J.A., MOORE V.J. - Sleep EEG effects of exercise with and without additional body cooling. *Electroencephalogr. Clin. Neurophysiol.* 60, 33-38, 1985.

58. HORNE J.A., REID A.J. - Night-time sleep EEG changes following body heating in a warm bath. *Electroencephalogr. Clin. Neurophysiol.* 60, 154-157, 1985.

59. HORNE J.A., SHACKELL B.S. - Slow-wave sleep elevations after body heating: proximity to sleep and effects of aspirin. *Sleep* 10, 383-392, 1987.

60. INOUYE S.T., KAWAMURA H. - Persistence of circadian rhythmicity in a mammalian hypothalamic 'island' containing the suprachiasmatic nucleus. *Proc. Natl. Acad. Sci. U.S.A.* 76, 5962-5966, 1979.

61. JEWETT M.E., KRONAUER R.E., CZEISLER C.A. - Light-induced suppression of endogenous circadian amplitude in humans. *Nature* 350, 59-62, 1991.

62. KATTLER H., DIJK D.J., BORBELY A.A. - Effect of unilateral somatosensory stimulation prior to sleep on the sleep EEG in humans. *J. Sleep Res.* 3, 159-164, 1994.

63. KOELLA W.P. - A partial theory of sleep. A novel view of its phenomenology and organization. *Eur. Neurol.*

25, suppl 2, 9-17, 1986.

64. KRONAUER R.E., CZEISLER C.A., PILATO S.F., MOORE-EDE M.C., WEITZMAN E.D. - Mathematical model of the human circadian system with two interacting oscillators. *Am. J. Physiol.* 242, R3-R17, 1982.

65. LAVIE P. - Ultrashort sleep-waking schedule. III. 'Gates' and 'forbidden zones' for sleep. *Electroencephalogr. Clin. Neurophysiol.* 63, 414-425, 1986.

66. MASSAQUOI S.G., McCARLEY R.W. - Extension of the limit cycle reciprocal interaction model of REM cycle control. An integrated sleep control model. *J. Sleep Res.* 1, 138-143, 1992.

67. MEDDIS R. - On the function of sleep. *Anim. Behav.* 23, 676-691, 1975.

68. MEERLO P., PRAGT B.J., DAAN S. - Social stress induces high intensity sleep in rats. *Neurosci. Lett.* 225, 41-44, 1997.

69. MERICA H., BLOIS R. - Relationship between the time courses of power in the frequency bands of human sleep EEG. *Neurophysiol-Clin.* 27, 116-128, 1997.

70. MILLER J.D., MORIN L.P., SCHWARTZ W.J., MOORE R.Y. - New insights into the mammalian circadian clock. *Sleep* 19, 641-667, 1996.

71. MOORE R.Y., EICHLER V.B. - Loss of circadian adrenal corticosterone rhythm following suprachiasmatic lesions in the rat. *Brain Res.* 42, 201-206, 1972.

72. MOURTAZAEV M.S., KEMP B., ZWINDERMAN A.H., KAMPHUISEN H.A.C. - Age and gender affect different characteristics of slow waves in the sleep EEG. *Sleep* 18, 557-564, 1995.

73. MUKHAMETOV L.M. - Sleep in marine mammals. *Exp. Brain Res.*, Suppl 8, 227-238, 1984.

74. NAKAO M., McGINTY D., SZYMUSIAK R., YAMAMOTO M. - A thermoregulatory model of sleep control. *Jpn. J. Physiol.* 45, 291-309, 1995.

75. PHILLIPS D.L., RAUTENBERG W., RASHOTTE M.E., STEPHAN F.K. - Evidence for a separate food-entrainable circadian oscillator in the pigeon. *Physiol. Behav* 53, 1105-1113, 1993.

76. PORKKA-HEISKANEN T., STRECKER R.E., THAKKAR M., BJORKUN A.A., GREEN R.W., McCARLEY R.W. - Adenosine: A mediator of the sleep-inducing effects of prolonged wakefulness. *Science* 276, 1265-1268, 1997.

77. PUTILOV A.A. - Timing of sleep modelling: Circadian modulation of the homeostatic process. *Biol. Rhythm Res.* 26, 1-19, 1995.

78. RALPH M.R., FOSTER R.G., DAVIS F.C., MENAKER M. - Transplanted suprachiasmatic nucleus determines circadian period. *Science* 247, 975-978, 1990.

79. ROENNEBERG T., HASTINGS J.W. - Two photoreceptors control the circadian clock of a unicellular alga. *Naturwissenschaften* 75, 206-207, 1988.

80. SHINOHARA K., HONMA S., KATSUNO Y., ABE H., HONMA K. - Circadian rhythms in the release of vasoactive intestinal polypeptide and arginine-vasopressin in organotypic slice culture of rat suprachiasmatic nucleus. *Neurosci. Lett.* 170, 183-186, 1994.

81. STEPHAN F.K., ZUCKER I. - Circadian rhythms in drinking behavior and locomotor activity of rats are eliminated by hypothalamic lesions. *Proc. Natl. Acad. Sci. U.S.A.* 69, 1583-1586, 1972.

82. STERIADE M., McCORMICK D.A., SEJNOWSKI T.J. - Thalamocortical oscillations in the sleeping and aroused brain. *Science* 262, 679-685, 1993.

83. TROGATZ S.H. - In: Lecture notes in biomathematics, S. Levin (ed), vol 69: *The Mathematical Structure of the Human Sleep-Wake Cycle*. Springer Verlag, Berlin, Heidelberg, New York, London, Paris, Tokyo, 1986.

84. TESSONNEAUD A., COOPER H.M., CALDANI M., LOCATELLI A., VIGUIER-MARTINEZ M.C. - The suprachiasmatic nucleus in the sheep: retinal projections and cytoarchitectural organization. *Cell Tissue Res.* 278, 65-84, 1994.

85. TOSINI G., MENAKER M. - Circadian rhythms in cultured mammalian retina. *Science* 272, 419-421, 1996.

86. UCHIDA S., ATSUMI Y., KOJIMA T. - Dynamic relationship between sleep spindles and delta waves during a NREM period. *Brain Res. Bull.* 33, 351-355, 1994.

87. Van den HOOFDAKKER R.H. - Chronobiological theories of nonseasonal affective disorders and their implications for treatment. *J. Biol. Rhythms* 9, 157-183, 1994.

88. WEBB W.B., AGNEW H.W.Jr. - Stage 4 sleep: influence of time course variables. *Science* 174, 1354-1356, 1971.

89. WEHR T.A. - A brain-warming function for REM sleep. *Neurosci. Biobehav. Rev.* 16, 379-397, 1992.

90. WELSH D.K., LOGOTHETIS D.E., MEISTER M., REPPERT S.M. - Individual neurons dissociated from rat suprachiasmatic nucleus express independently phased circadian firing rhythms. *Neuron* 14, 697-706, 1995.

91. WERTH E., ACHERMANN P., BORBELY A.A. - Brain topography of the human sleep EEG: antero-posterior shifts of spectral power. *Neuroreport* 8, 123-127, 1996.

92. WERTH E., ACHERMANN P., BORBELY A.A. - Fronto-occipital EEG power gradients in human sleep. *J. Sleep Res.* 6, 102-112, 1997.

93. WEVER R. - *The Circadian System of Man*. Berlin: Springer-Verlag, 1979.

94. WIRZ-JUSTICE A. - Biological rhythms in mood disorders. In Psychopharmacology, *Fourth Generation of Progress*, FE Bloom and DJ Kupfer, eds., Raven Press, New York, 1995, pp 999-1017.

95. ZEPELIN H., RECHTSCHAFFEN A. - Mammalian sleep, longevity, and metabolism. *Brain Behav. Ecol.* 10, 425-470, 1974.

Chapter 6

Melatonin and sleep-wake rhythm

B. Claustrat
Service de Radiopharmacie et Radioanalyse, Centre de Médecine Nucléaire, Hôpital Neuro-Cardiologique, Lyon, France

INTRODUCTION

Melatonin is the main hormone secreted by the pineal gland. This is not a recently discovered substance, having been identified in 1958 by Lerner *et al.* [44], and synthesised in the following year. It is a lipophilic compound with an indole structure derived from serotonin using a two-step biochemical process. The role of melatonin, a hormone secreted during the night, is to inform the brain of the alternating periods of day and night, so that the organism is in phase with the light environment. The hormonal message adjusts to variations in the duration of night (scotophase) in the course of the year. Thus in seasonally reproducing animals, modifications in melatonin secretion will prompt mating and the subsequent birth of the litter, at a favourable period of the year. Generally speaking, melatonin participates in regulating rhythmic phenomena, whether daily or annually. The physiological role of melatonin in man is only partially understood and certain data are open to question. Indeed, due to its lack of any known short term toxicity, the substance has been subject to a multitude of studies which are either uncontrolled or involve doses generating blood concentrations far higher than physiological concentrations.

THE MELATONIN NYCTHEMERAL CYCLE

The melatonin plasma profile as a reflection of epiphysial activity

There is no intra-epiphysial storage of melatonin and the substance renews itself at a very fast rate. The plasma profile thus faithfully represents hormone secretion. This tends to occur during the night, persisting for about 10 hours; the maximum levels are reached at about 3 am, while daytime levels are undetectable or very weak in resting subjects. The amplitude of this nycthemeral rhythm is the most marked of the hormone rhythms, even more so than cortisol rhythm.

The plasma profile varies according to the frequency of samples taken. A minimum frequency of one sample per hour is necessary to faithfully evaluate this profile. If samples are taken more frequently (every 20 minutes), secretion episodes are detectable (fig. 6.1) [33]. The episodic or pulsatile character of this secretion, which was contended for some time, has now been ascertained [68]. Moreover, it has not been possible to show a definitive relationship between the peaks and troughs of melatonin secretion and the different stages of sleep [19]. This absence of relationship, possibly resulting from insufficient sampling, warrants reinvestigation.

The melatonin message shows a high degree of heterogeneity between subjects. Conversely, in the same subject, the profile is highly reproducible from one day to the next. In some subjects, episodes of secretion may be very discrete or even exceptionally absent. In addition, melatonin secretion varies with age, the rhythm diminishing gradually with a tendency for phases to become advanced or retarded depending on the author [35, 71]: the rhythm may be completely absent in the elderly [71].

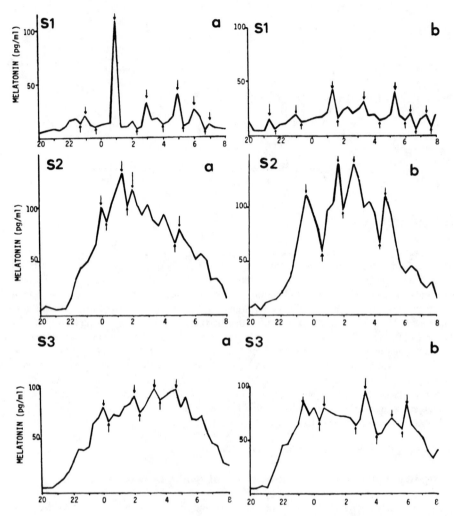

Figure 6.1. Daily profile of plasmatic melatonin in three young volunteers (S1, S2, S3) carried out at a week's interval (*a* and *b*).

The melatonin secretion regulation system

Like many circadian rhythms in mammals (e.g. feeding, sleep/wake, temperature, cortisol or corticosterone behaviour), melatonin rhythm is controlled by an internal clock located in the suprachiasmatic nuclei. In fact, proof of this does not exist in man but by transposing the results observed in non human primates [29]. The polysynaptic regulation pathway uses the central nervous system, particularly the upper cervico-thoracic fluid and sympathetic system: the neurons located in the upper cervical ganglions send axonal extensions with noradrenergic endings into the epiphysis [42, 54]. Noradrenergic predominantly controls epiphysial activity, although most neurotransmission systems are present in the pinealocyte.

Day/night alternation is the main synchroniser of the system: in the presence of light, the nerve influx carried by the retinohypothalamic pathway inhibits noradrenergic transmission. Conversely, the latter is activated as night approaches and melatonin synthesis is stimulated following the increase in N acetyl-transferase synthesis, a key enzyme which catalyses the penultimate stage of melatonin biosynthesis. The hormone is passively released from the pinealocyte in venous blood [59].

A feedback loop probably exists between melatonin and the biological clock, since melatonin receptors have been discovered at suprachiasmatic nuclei level, and melatonin modifies the electrical or metabolic activation of these nuclei [50, 62, 64].

Any disruption of the different levels of the regulatory axis will lead to an alteration in melatonin secretion, a phenomenon which is particularly observed in certain sympathalgias or dysautonomias. In cluster headache for example, phase advance of melatonin secretion has been reported, associated with diminished amplitude [17, 76]. In diabetic dysautonomia, melatonin rhythm diminishes in amplitude [58].

Light has an inhibiting or synchronising effect on melatonin secretion

Light has a double effect on melatonin secretion, either as a synchroniser or as an inhibitor, depending at what time the light is administered. Lewy *et al.* [47] are credited with having demonstrated, in 1980, that melatonin secretion can also be inhibited in man, provided sufficient light is administered over a prolonged period (2,500 lux between 2 and 4 am) (fig. 6.2.a). The inhibiting effect obtained on the plasma profile is dose-dependent [10] – it is perceptible from 300 lux – and varies according to wavelength: green light is the most active whereas red light has no effect [37]. Women, moreover, appear to be more sensitive than men [51]. When light is administered for several consecutive nights, melatonin escapes inhibition and gradually shifts toward the morning (phase delay). With a total absence of light perception – in the blind – melatonin rhythm can free-run, with a period extending slightly over 24 hours [4].

In addition to its inhibiting effect, natural or artificial light exerts a synchronising effect on melatonin secretion: if subjects are exposed to 3,000 lux between 3 and 9 am, phase advance occurs [13]. Conversely, phase delay is observed after evening exposure to bright light (fig.6.2.b and c).

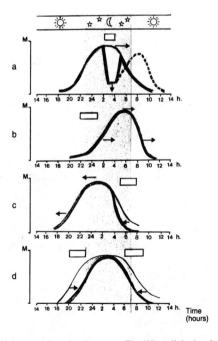

Figure 6.2. Effects of artificial light on melatonin plasma profile. When light is administered once during the night, secretion is inhibited. After repeated doses in the same conditions, secretion gradually shifts to the morning (*a*. phase delay). The plasma profile is delayed, advanced or diminished depending on the time at which light is administered (*b* evening; *c*. morning; *d*. evening + morning).

Can modifications in the natural photoperiod have repercussions on melatonin secretion? The results are somewhat divergent on this question, probably because domestic lighting can exert a masking effect[1] [12]. It is worth noting the results of a Finnish group from Kauppila, who observed a 2 hour lengthening of the melatonin secretion peak in winter, as compared with the summer period [41]. This extension of the melatonin peak appears to be expressed physiologically, as Finnish women present weaker ovarian steroid levels in winter. In our latitudes, seasonal variations of this kind do not occur, due to the possible interference of domestic light (albeit often lower than 150 lux), official time changes in the spring and autumn and milder variations in the natural photoperiod [70].

Melatonin plasma rhythm – a circadian clock marker

A close relationship exists between the melatonin peak and minimum temperature, even in controlled conditions. This phase relation is also observed in "constant routine" conditions: when bright light is administered close to nadir temperature, shifts in melatonin and temperature occur accordingly and equivalently, suggesting that these two rhythms are controlled by the same clock [63].

Unlike temperature rhythm, melatonin rhythm is relatively insensitive to masking effects, apart from that exerted by light and, to a lesser degree, by posture [26]. Thus, sleep does not appear to disrupt melatonin profile. It is for this reason that Lewy *et al.*advocated recording the onset of the melatonon profile under weak light of less than 50 lux [46]. This 'dim light melatonin onset' (DLMO) constitutes a melatonin rhythm phase marker. However, for Shanahan and Czeisler [63], it is the peak rather than the onset of melatonin which is the most reliable phase marker, a view which we share [80]. The study of the melatonin plasma profile provides an excellent diagnostic element for detecting insomnia accompanied by circadian rhythm anomalies (phase advance syndrome, M. Billiard and M.J. Challamel, private paper). Moreover, a non invasive approach exists to evaluate overall melatonin secretion, particularly via saliva doses or fractional hepatic metabolite doses in urine, 6 sulfatoxymelatonin [5, 74]. These protocols which require waking the subject up, unlike taking blood samples, are nevertheless of considerable interest in detecting hypernycthemeral syndrome.

MELATONIN, AN ENDOGENOUS SYNCHRONISER IN MAN

Melatonin secretion occurs as a function of alternations between day and night. Melatonin is considered to be an endogenous synchroniser in man, capable of stabilising or reinforcing circadian rhythms. [7]. This hypothesis normally yields indirect answers only, whether in relation to sleep/wake or cortisol rhythms. For example, the maturation of the melatonin rhythm, which appears at about 3 – 4 months, occurs at the same time as the 24 hour sleep/wake rhythm. The effect of melatonin on temperature rhythm, on the contrary, constitutes a conclusive argument: melatonin secretion reinforces the nocturnal drop in temperature, facilitating the onset of sleep [66]. Moreover, peak melatonin urinary secretion coincides with the nocturnal peak of tiredness and lowest point of vigilance, the phase relation between these different rhythms being maintained over a 64 hour period of sleep deprivation [2].

A relationship also exists between the sleep/wake rhythm and the duration or position of melatonin secretion. Thus Wehr demonstrated in the laboratory that when artificial lighting is

[1] The masking effect results from the influence of an external stimulus on a biological variable, having no relationship to its rhythmic variation. For example, physical activity, at any time of the day, induces an increase in temperature which is superimposed onto the circadian rhythm. Eating, sleeping or changing position "tamper" with the temperature rhythm, which is the witness of the circadian clock. "**Constant routine**" recording suffices to "unmask" the endogenous component of rhythms by eliminating or spreading out over a 24 hour period, the physiological responses to environmental or behavioural stimuli. In "constant routine", subjects stay lying down in weak light but are not allowed to sleep for 24 to 36 hours and are given food (sandwiches) every 3 hours. Experimental conditions are controlled (temperature, noise...). This non physiological situation is the contrary of trained situations in which the subject is submitted to a light/darkness synchroniser and to the masking agents referred to above.

reduced from 16 to 10 hours, sleep duration responds to the lengthening of the melatonin peak [78]. However, melatonin secretion does partially escape this extension of the scotoperiod and sleep does not remain completely synchronous to the latter; on the contrary it shows fragmentation, with arousals in the middle of the night. However, other studies have shown that when melatonin secretion is gradually shifted toward the morning due to the effect of repeated nocturnal lighting, daytime recuperative sleep, synchronous with melatonin secretion, presents a physiological architecture. At the same time, nocturnal vigilance improves in relation to an increase in temperature [11, 15]. Protocols of this kind are currently used among night workers in normal working conditions, with the aim of improving professional performance and the quality of rest.

PHARMACOLOGY OF MELATONIN IN MAN

Kinetic aspects

Melatonin plasma profiles suggest a very fast hormone replacement rate. This is confirmed after intravenous injection of melatonin bolus: plasmatic concentrations present a biexponential decrease with a distribution phase of a very short half-life lasting approximately 2 minutes, followed by a metabolic phase whose half-life lasts about twenty minutes [38, 49]. Melatonin hepatic metabolite – 6 sulfatoxymelatonin – quickly appears in plasma as a consequence of a strong first-pass effect. To offset this very fast metabolic rate, most experimenters administer melatonin at a high oral dose, resulting in circulation rates which are far removed from physiological rates for short term intervals.

Melatonin, a chronobiotic compound

A key idea in understanding the therapeutic potential of melatonin, the administration of this hormone modifies its endogenous secretion according to a phase response curve [2] (PRC) rather than as a classic negative feedback phenomenon: the endogenous rhythm phase is variously modified (phase advance or delay) depending on the time of administering the exogenous stimulus (fig. 6.3). The phase response curve was first described by Lewy *et al.* after modifying the onset of melatonin plasma profile with repeated doses of melatonin (2 capsules dosed at 0.25 mg for 4 days)[45]. A similar phase response curve can be obtained with a single short term dose by perfusion (50 µg in 3 hours) generating physiological rates i.e. of the order of importance of a nocturnal peak (several tens of pg/ml) [80]; when melatonin is administered in the afternoon or evening, phase advance is observed, whereas morning or lunchtime administration induces phase delay (fig. 6.3). The focus of impact of this biochemical signal is probably the circadian clock. A change in shift indication occurs at about 3 pm, at the opposite point of the nocturnal melatonin peak [80]. The effect of melatonin on its own secretion is the opposite of that observed with light. However it is easier to obtain a phase advance with melatonin, whereas light, on the contrary, will more easily provoke phase delay. This chronobiotic property can be exploited to advantage in subjects presenting chronic or transitory rhythm disruption.

Thus, in **the delayed sleep phase syndrome**, administering melatonin will re-establish a time of sleep onset which is more consistent with normal social life [24]. **An advanced sleep phase syndrome** is induced, with no apparent modification of the position of the peak of wakefulness. Entrainment can also be obtained in **the free-running blind,** to eliminate daytime sleepiness [4]; however, in all these cases, the beneficial effect disappears if treatment is suspended and its mechanism appears to be complex: added to the impact on the internal clock, melatonin is likely to produce a direct hypnotic effect, particularly if high doses are used (fig. 6.4). Jan and O'Donnell have confirmed the importance of melatonin in children suffering from neurological disorders, whether congenital or acquired, some of whom presented ocular or cortical blindness [40]. No toxicity was found seven years later.

---

[2] The phase response curve PRC is the graphic representation of the value of a rhythm phase shift induced by a single stimulus, in function of the time the stimulus is applied in relation to the rhythm phase.

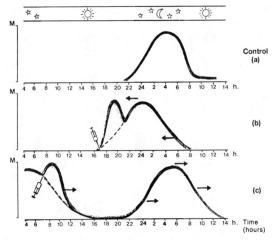

Figure 6.3. Effect of melatonin perfusion on endogenous secretion; control (*a*). The plasma profile presents a phase advance following administration in the afternoon or evening (*b*), a phase delay following administration in the morning or at midday (*c*).

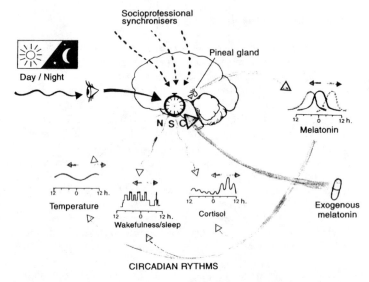

Figure 6.4. Possible point of melatonin impact, during biological rhythm resynchronisation.

The prevention of **jet lag syndrome** after eastbound flights through several time zones requires presynchronisation with melatonin, which must be administered in the late afternoon (to produce phase advance) on the day of departure or several days previously (unnecessary in our opinion), then for the following 4 days in the country of destination, half and hour before bedtime at around 10 – 11 pm [3, 18] (fig. 6.5). The time of melatonin administration for presynchronisation is chosen according to the number of time zones passed through: if the onset of melatonin is considered to occur on average at about 10 pm, the dose of melatonin should be taken at about 10 – n, n representing the number of time zones. If the flight is westbound or if the time difference is close to 12 hours, melatonin will have a less marked effect on jet lag; treatment does not usually start before arrival in the place of destination [60]. An immediate release preparation (0.5 mg capsule) is more effective than a prolonged release preparation. A 5 mg dose will reinforce the hypnotic effect [67]. Finally, combining administration with light may help to improve results.

The advantage of melatonin resynchronisation remains to be established in the case of **night or shift workers [32]**. Indeed there is no consistent disturbance of biological rhythms observed in these populations. In shift workers, the permanent difference between periods of work and recuperation is such that there is no set modification in rhythm phases but rather an attenuation of their amplitude. Night workers show heterogeneous profiles, some subjects spontaneously shifting their melatonin secretion toward the period of morning rest [77]. As was seen earlier, administering artificial light at night favours this shift, with clear benefits to psychomotor performance and restorative sleep; this approach would seem to be preferable to administering melatonin during nocturnal periods of recuperation intended to accelerate readjustment to daytime social habits.

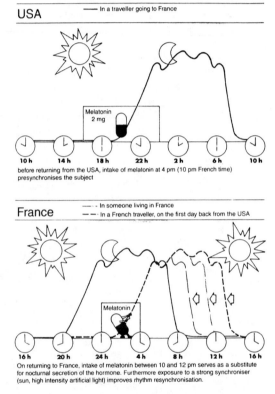

Figure 6.5. Melatonin resynchronisation protocol, for a long distance flight; example of a flight from the eastern coast of the USA to France (6 hour time difference).

The chronobiotic action of melatonin implies exercising caution in administering to patients whose clinical signs may worsen as a result of treatment: thus the failure of melatonin in the treatment of depressives may be due to a reinforcement of the tendency for phase advance in their rhythms, resulting from an inappropriate time of administration, strongly suggesting that these patients should abstain from using melatonin as a hypnotic [16].

Non chronobiotic effects

Melatonin, as a lipophilic compound, crosses the blood brain barrier weakly but significantly. In strong doses, it will interact with most neuromediator systems, producing non specific effects. These central effects have led to its use - often empirical - in psychiatric pathology (depression), or neurological pathology (Parkinson's disease, epilepsy) [1]. Moreover, the fact that secretion occurs nocturnally quickly led to its use as a hypnotic, particularly since the onset of secretion coincides with an evening drop in central temperature and exogenous melatonin lowers temperature in proportion to the dose administered. The results relating to the effects of melatonin on sleep are very heterogeneous in view of the differences in methodology (variable melatonin dosages and times of administration, single or chronic doses ...). Certain effects i.e. the increase in REM sleep latency, can be conceded to be attached to chronobiotic activity, whereas changes in EEG spectral activity result from the hypnotic or non chronobiotic effect of melatonin [28, 39, 73, 79]. However, spectral analysis shows a profile which differs to that of a benzodiazepine, and flumazenil, a benzodiazepine receptor antagonist, does not inhibit its activity [56].

At weak doses, melatonin has a sedative or soporific effect, with an increased sensation of tiredness, slower rate of reaction in performance tests and reduced sleep latency [6, 22]. The drop in vigilance is objectified by an enhancement of θ/α frequency [14]. At high doses, double blind tests have shown a hypnotic effect; altered sleep recording traces vary according to dose or time of administration [79]. After several days' administration, sleep architecture will present modifications (increase in the density of REM and stage II sleep) which are nevertheless compatible with an effect mediated by the circadian clock [25].

These studies, carried out in young volunteer subjects, then proceeded to test the hypnotic effect of melatonin on patients, particularly as its secretion is slightly lower in primary insomniacs [8, 36]. For Ellis *et al.*, melatonin proved to have no effect on a group of patients whose average age was 46, and who presented psychophysiological insomnia [30]. On the contrary, in elderly insomniacs, melatonin administered in the evening for a week increased the quantity of sleep evaluated by actimetry [35]: this group of patients initially presented reduced urinary excretion of melatonin associated with phase delay. The study, involving a control-release formulation, warrants confirmation. It must be stressed that the heterogeneity of the results may be explained by the varied conditions of administering melatonin (time, dose, single or chronic dose, preparation with immediate or prolonged release), of the heterogeneity in the study samples (young, healthy subjects, or elderly insomniacs who possibly presented dementia, submitted to several therapies) who were often low in number, and in protocols (open or double blind). In short, the administration of melatonin appears to reduce latency to sleep and increase overall sleep time, thus improving the quality of sleep perception and reducing daytime sleepiness. Moreover, it permits weaning off benzodiazepine [23]. The effectiveness of treatment bears no relation to the initial levels of endogenous melatonin which can slump in elderly subjects. There is neither intolerance no addiction to the treatment, nor any hangover effect on morning wakening. To summarise, although the effects of melatonin on vigilance and temperature are well established for healthy controls, the results of the few clinical trials in insomniacs do not justify the widespread use witnessed across the Atlantic.

Melatonin exerts a number of endocrine effects whose description is beyond the scope of the present work. Let us however mention that melatonin has been researched as modifying the reproductive function in man, by analogy with the effect observed in certain photoperiodic animal species (hamster). Thus the administration of 250 mg of melatonin for 1 month lowers the level of plasmatic LH in women [57]. An attempt was made to apply this result by developing a contraceptive pill associating a high dose of melatonin (300 mg) with a progestin [72]. The project appears to have been abandoned and the effects of a preparation of this kind on vigilance are questionable. Moreover a weaker dose (3 mg) administered for 6 months has no effect on gonadotropic hormones [65].

Among other possible endocrine effects, it should be noted that prolactin plasma levels are only weakly modified by melatonin, although this parameter should be controlled before setting up chronic treatment.

Action of melatonin at cellular level

The development of an iodised ligand has revealed that the physiological action of melatonin involves membrane receptors coupled with G proteins, the phenomenon of transduction leading to a reduction of intracellular cyclic AMP [55]. In man, these receptors have been detected in the suprachiasmatic nuclei and the cerebral cortex [31, 62]. Moreover, melatonin performs certain activities without involving any prior link to a membrane receptor: it presents antioxidant properties and constitutes a free radical "scavenger" which is stronger than vitamin E, the standard reference for this type of activity [61, 68]. Moreover, melatonin inhibits the formation of adducts generated on DNA sequencing by a chemical carcinogen, Safrol [69]. These protective effects observed *in vitro* suggest possible therapeutic applications in the neurodegenerative processes linked to ageing and have led to studies in oncology. These show, for example, that the use of very high doses of melatonin in conjunction with interleukin-2 in patients presenting a solid tumour at a very advanced stage, will stabilise the neoplastic process, while improving tolerance to treatment and physical conditions [48]. Studies in this field call for confirmation.

CONCLUSION

Melatonin constitutes a reliable biochemical marker for the activities of the circadian clock, useful for exploring sleep disorders accompanied by anomalies in biological rhythm. By determining the melatonin profile in different biological milieus (plasma, urine, saliva) studies can be made of the effect of light on the vigilance levels of night workers.

The therapeutic value of melatonin is real, at least for certain targeted indications (phase delay insomnia). This substance did not initially raise the interest it deserved on the part of pharmaceutical laboratories, as, being a natural substance which is easily synthesised and tested, it is not subject to a government patent, with the result that it has orphan status as medication. Only by presenting it in an original formulation and/or as a therapeutic indication will it be subject to being patented.

Developments in the understanding of the pathophysiology of biological rhythms have revealed the disorganisation of the latter, particularly that of melatonin rhythm, in situations where this was merely suspected hitherto (examples of insomnia in the elderly, migraine, cluster headache) [17, 21, 76]. The use of the resynchronising effect of melatonin is conceivable for these pathologies [20, 34] or to correct the functional problems resulting from jet lag, night or shift work. In such cases, it is still only used as comfort therapy, although the issues at stake in public health and security could alter this situation.

Conversely, and as a possible motivation for development, the progressive implication of melatonin in the rhythmic regulation of the main physiological functions e.g. immunity, may help to extend its therapeutic indications.

The pharmacokinetic characteristics of the exogenous melatonin signal are still to be defined (concentration level and duration). Biopharmacy optimisation may serve to guard against adverse effects, as may a patient by patient appraisal of circadian clock phases. Simple oral administration would seem limited in the future, due to the fact that it causes a heavy overload for too short an interval (some 2 to 3 hours), as opposed to the quick and prolonged release preparations which generate weak levels, imitating the nocturnal profile. Certain pharmaceutical forms i.e. the transmucosal oral patch, respond to this concern. The same result cannot be obtained via transdermal routes since melatonin dissolves badly in the stratum coernum due to its lipophilia, concentrating in the dermis [9]; this results in a weak passage into the blood, slowed down over time resulting in a widely heterogeneous profile between subjects [43]. Finally, the pharmacological approach has led to synthesising numerous analogues (agonist or antagonist) for which an attempt is being made to isolate the specific action among the multiple effects of melatonin [27]. This extremely promising stage, scientifically speaking, has yet to be extended therapeutically.

REFERENCES

1. ANTON-TAY F., DIAZ J.L., FERNANDEZ-GUARDIOLA M. – On the effects of melatonin upon human brain: its possible therapeutic implication. *Life Sci.,* 10, 841-850, 1971.
2. AKERSTEDT T., FRÖBERG J.E., FRIBERG Y., WETTERBERG L. – Melatonin excretion, body temperature and subjective arousal during 64-hours of sleep deprivation. *Psychoneuroendocrinology,* 4, 219-225, 1979.
3. ARENDT J., ALDHOUS M., MARKS V. – Alleviation of jet lag by melatonin: preliminary results of controlled double blind trial. *BMJ,* 292, 1170, 1986.
4. ARENDT J., ALDHOUS M., WRIGHT J. – Synchronisation of a disturbed sleep-wake cycle in a blind man by melatonin treatment. *The Lancet,* i, 772-773, 1988.
5. ARENDT J., BOJKOWSKI C., FRANEY C., WRIGHT J., MARKS V. – Immunoassay of 6-hydroxymelatonin sulfate in human plasma and urine: abolition of the urinary 24-hour rhythm with atenolol. *J. Clin. Endocrinol. Metab.,* 60, 1166-1173, 1985.

6. ARENDT J., BORBELY A.A., FRANEY C., WRIGHT J. – The effets of chronic small doses of melatonin given in the late afternoon on fatigue in man: a preliminary study. *Neurosc. Lett.*, 45, 317-321, 1984.
7. ARMSTRONG S.M. – Melatonin: the internal zeitgeber of mammals? *Pineal Res. Reviews*, 7, 157-202, 1989.
8. ATTENBURROW M.E.J., DOWLING B.A., SHARPLEY A.L., COWEN P.J. – Case-control study of evening melatonin concentration in primary insomnia. *BMJ*, 312, 1263-1264, 1996.
9. BENES L., CLAUSTRAT B., HORRIERE F., GEOFFRIAU M., KONSIL J., PARROTT K.A., DEGRANDE G., MCQUINN R.L., AYRES J.W. – Transmucosal, oral controlled-release, and transdermal drug administration in human subjects: a crossover study with melatonin. *J. Pharm. Sci.*, 86, 1115-1119, 1997.
10. BOJKOWSKI C.J., ALDHOUS M.E., ENGLISH J., FRA-NEY C., POULTON A.L., SKENE D.J., ARENDT J. – Suppression of nocturnal plasma melatonin and 6-sulphatoxymelatonin by bright and dim light in man. *Horm. Metab. Res.*, 19, 437-440, 1987.
11. BOUGRINE S., CABON P., IGNAZI G., GEOFFRIAU M., BRUN J., CLAUSTRAT B. – Phase delay of rhythm of 6-sulfa-toxy-melatonin-excretion by bright light improves sleep and performance. In: Y. Touitou, J. Arendt and P. Pévet (eds.). *Melatonin and the pineal gland – From basic science to clinical application*. Elsevier Science Publishers, 219-223, 1993.
12. BROADWAY J., ARENDT J., FOLKARD S. – Bright light phase shifts the human melatonin rhythm during the Antarctic winter. *Neurosc. Lett.*, 79, 185-189, 1987.
13. BURESOVA M., DVORAKOVA M., ZVOLSKY P., ILLNEROVA H. – Early morning bright light phase advances the human circadian pacemaker within one day. *Neurosc. Lett.*, 121, 47-50, 1991.
14. CAJOCHEN C., KRAUCHI K., VON ARX M.A., MORI D., GRAW P., WIRZ-JUSTICE A. – Daytime melatonin administration enhances sleepiness and theta/alpha activity in the waking EEG. *Neurosc. Lett.*, 207, 209-213, 1996.
15. CAMPBELL S.S., DAWSON D. – Enhancement of nighttime alertness and performance with bright ambient light. *Physiol. and Behav.*, 48, 317-320, 1990.
16. CARMAN J.S., POST R.M., BUSWELL R., GOODWIN F.K. – Negative effects of melatonin on depression. *Am. J. Psychiatry*, 133, 1181-1186, 1976.
17. CHAZOT G., CLAUSTRAT B., BRUN J., JORDAN D., SASSOLAS G., SCHOTT B. – A chronobiological study of melatonin, cortisol growth hormone and prolactin secretion in cluster headache. *Cephalalgia*, 4, 213-220, 1984.
18. CLAUSTRAT B., BRUN J., DAVID M., SASSOLAS G., CHAZOT G. – Melatonin and jet lag: confirmatory result using a simplified protocol. *Biol. Psychiatry*, 32, 705-711, 1992.
19. CLAUSTRAT B., BRUN J., GARRY P., ROUSSEL B., SASSOLAS G. – A once-repeated study of nocturnal plasma melatonin patterns and sleep recordings in six normal young men. *J. Pineal Res.*, 3, 301-310, 1986.
20. CLAUSTRAT B., BRUN J., GEOFFRIAU M., ZAIDAN R., MALLO C., CHAZOT G. – Nocturnal plasma melatonin profile and melatonin kinetics during infusion in status migrainosus. *Cephalalgia*, 17, 511-517, 1997.
21. CLAUSTRAT B., LOISY C., BRUN J., BEORCHIA S., ARNAUD J.L., CHAZOT G. – Nocturnal plasma melatonin levels in migraine: a preliminary report. *Headache*, 29, 241-244, 1989.
22. CRAMER H., RUDOLPH J., CONSBRUCK V., KENDEL K. – On the effects of melatonin on sleep and behaviour on man. In: Costa E., Costa G., Sandler P. (eds.), *Advances in biochemical psychopharmacology*, Raven Press, New York, 11, 187-191, 1974.
23. DAGAN Y., ZISAPEL N., NOF D., LAUDON M., ATSMON J. – Rapid reversal of tolerance to benzodiazepine hypnotics by treatment with oral melatonin: a case report. *Eur. Neuropsychopharmacol.*, 7, 157-160, 1997.
24. DAHLITZ M., ALVAREZ B., VIGNAU J., ENGLISH J., ARENDT J., PARKES J.D. – Delayed sleep phase syndrome response to melatonin. *The Lancet*, 337, 1121-1124, 1991.
25. DAWSON D., ENCEL N. – Melatonin and sleep in humans. *J. Pineal Res.*, 15, 1-12, 1993.
26. DEACON S., ARENDT J. – Posture influences melatonin concentrations in plasma and saliva in humans. *Neurosc. Lett.*, 167, 191-194, 1994.
27. DEPREUX P., LESIEUR D., AIT MANSOUR H., MORGAN P., HOWELL H.E., RENARD P., CAIGNARD D.H., PFEIFFER B., DELAGRANGE P., GUARDIOLA B., YOUS S., DEMARQUE A., ADAM G., ANDRIEUX J. – Synthesis and structure-activity relationships of novel naphthalenic and bioisosteric related amidic derivatives as melatonin receptor ligands. *J. Med. Chem.*, 37, 3231-3239, 1994.
28. DIJK D.J., ROTH C., LANDOLT H.P., WERTH E., AEP-PLI M., ACHERMANN P., BORBELY A.A. – Melatonin effect on daytime sleep in men: suppression of EEG low frequency activity and enhancement of spindle frequency activity. *Neurosc. Lett.*, 201, 13-16, 1995.
29. EDGAR D.M., DEMENT W.C., FULLER C.A. – Effect of SCN lesions on sleep in squirrel monkeys: evidence for opponent processes in sleep-wake regulation. *J. Neurosc.*, 13, 1065-1079, 1993.
30. ELLIS C.M., LEMMENS G., PARKES J.D. – Melatonin and insomnia. *J. Sleep Res.*, 5, 61-65, 1996.
31. FAUTECK J.D., LERCHL A., BERGMANN M., MOLLER M., FRASCHINI F., WITTKOWSKI W., STANKOV B. – The adult human cerebellum is a target of the neuroendocrine system involved in the circadian timing. *Neurosc. Lett.*, 179, 60-64, 1994.
32. FOLKARD S., ARENDT J., CLARK M. – Can melatonin improve shift workers' tolerance of the night shift? Some preliminary findings. *Chronobiol. Int.*, 10, 315-320, 1993.
33. FOLLENIUS M., WEIBEL L., BRANDENBERGER G. – Distinct modes of melatonin secretion in normal men. *J. Pineal Res.*, 18, 135-140, 1995.
34. GARFINKEL D., LAUDON M., ZISAPEL N. – Improvement of sleep quality by controlled-release melatonin in benzodiazepine-treated elderly insomniacs. *Arch. of Gerontology and Geriatrics*, 24, 223-231, 1997.
35. HAIMOV I., LAUDON M., ZISAPEL N., SOUROUJON M., NOF D., SHLITNER A., HERER P., TZISCHINSKY O., LAVIE P. – Sleep disorders and melatonin rhythms in elderly people. *BMJ*, 309, 167, 1994.
36. HAJAK G., RODENBECK A., STAEDT J., BANDELOW B., HUETHER G., RUTHER E. – Nocturnal plasma melatonin levels in patients suffering from chronic primary insomnia. *J. Pineal Res.*, 19, 116-122, 1995.

37. HORNE J.A., DONLON J., ARENDT J. – Green light attenuates melatonin output and sleepiness during sleep deprivation. *Sleep*, 14, 233-240, 1991.
38. IGUCHI H., KATO K.I., IBAYASHI H. – Melatonin serum levels and metabolic clearance rate in patients with liver cirrhosis. *J. Clin. Endocrinol. Metab.*, 54, 1025-1027, 1982.
39. JAMES S.P., SACK D.A., ROSENTHAL N.E., MENDELSON W.B. – Melatonin administration in insomnia. *Neuropsychopharmacology*, 3, 19-23, 1990.
40. JAN J.E., O'DONNELL M.E. – Use of melatonin in the treatment of paediatric sleep disorders. *J. Pineal Res.*, 21, 193-199, 1996.
41. KAUPPILA A., KIVELA A., PAKARINEN A., VAKKURI O. – Inverse seasonal relationship between melatonin and ovarian activity in humans in a region with a strong seasonal contrast in luminosity. *J. Clin. Endocrinol. Metab.*, 65, 823-828, 1987.
42. KLEIN D.C., MOORE R.Y. – Pineal N-acetyltransferase and hydroxyindole-o-methyltrans-ferase: control by the retinohypothalamic tract and the suprachiasmatic nucleus. *Brain Res.*, 174, 245-262, 1979.
43. LEE B.J., PARROTT K.A., AYRES J.W., SACK R.L. – Preliminary evaluation of transdermal delivery of melatonin in human subjects. *Res. Commun. Mol. Pathol. Pharmacol.*, 5, 337-346, 1994.
44. LERNER A.B., CASE J.D., TAKAKASHI Y., LEE T.H., MORI W. – Isolation of melatonin, the pineal gland factor that lightens melanocytes. *J. Am. Chem. Soc.*, 80, 2587, 1958.
45. LEWY A.J., AHMED S., LATHAM JACKSON J.M., SACK R.L. – Melatonin shifts human circadian rhythms according to a phase-response curve. *Chronobiol. Int.*, 9, 380-392, 1992.
46. LEWY A.J., SACK R.L. – The dim light melatonin onset as a marker for circadian phase position. *Chronobiol. Int.*, 6, 93-102, 1989.
47. LEWY A.J., WEHR T.A., GOODWIN F.K., NEWSOME D.A., MARKEY S.P. – Light suppresses melatonin secretion in humans. *Science*, 210, 1267-1269, 1980.
48. LISSONI P., BARNI S., ROVELLI F., BRIVIO F., ARDIZZOIA A., TANCINI G., CONTI A., MAESTRONI G.J.M. – Neuroimmunotherapy of advanced solid neoplasms with single evening subcutaneous injection of low-dose interleukin-2 and melatonin: preliminary results. *Eur. J. Cancer*, 29A, 185-189, 1993.
49. MALLO C., ZAIDAN R., GALY G., VERMEULEN E., BRUN J., CHAZOT G., CLAUSTRAT B. – Pharmacokinetics of melatonin in man after intravenous infusion and bolus injection. *Eur. J. Clin. Pharmacol.*, 38, 297-301, 1990.
50. MCARTHUR A.J., GILLETTE M.U., PROSSER R.A. – Melatonin directly resets the rat suprachiasmatic circadian clock in vitro. *Brain Res.*, 565, 158-161, 1991.
51. MONTELEONE P., ESPOSITO G., LA ROCCA A., MAJ M. – Does bright light suppress nocturnal melatonin secretion more in women than men? *J. Neural Transm.*, 102, 75-80, 1995.
52. MONTI J.M., ALVARVINO F., CARDINALI D.P., SAVIO I., PINTOS A. – Polysomnographic study of the effect of melatonin on sleep in elderly patients with chronic primary insomnia. *Arch. of Gerontology and Geriatrics*, 28, 85-98, 1999.
53. MONTI J.M., CARDINALI D.P. – A critical assessment of the melatonin effect on sleep in humans. *Biol. Signals Recept.*, 9, 328-339, 2000.
54. MOORE R.Y. – The innervation of the mammalian pineal gland. In: Reiter R.J. (ed.), *The pineal and reproduction*. Karger, Basel, 1-29, 1978.
55. MORGAN P.J., BARRETT P., HOWELL H.E., HELLIWELL R. – Melatonin receptors: Localization, molecular pharmacology and physiological significance. *Neurochem. Int.*, 24, 101-146, 1994.
56. NAVE R., HERER P., HAIMOV I., SHLITNER A., LAVIE P. – Hypnotic and hypothermic effects of melatonin on daytime sleep in humans: lack of antagonism by flumazenil. *Neurosc. Lett.*, 214, 123-126, 1996.
57. NORDLUND J.J., LERNER A.B. – The effect of oral melatonin on skin color and on the release of pituitary hormones. *J. Clin. Endocrinol. Metab.*, 45, 768-774, 1977.
58. O'BRIEN I.A.D., LEWIN I.G., O'HARE J.P., ARENDT J., CORRAL R.J.M. – Abnormal circadian rhythm of melatonin in diabetic autonomic neuropathy. *Clin. Endocrinol.*, 24, 359-364, 1986.
59. PARDRIDGE W.M., MIETUS L.J. – Transport of albumin-bound melatonin through the blood-brain barrier. *J. Neurochem.*, 34, 1761-1763, 1980.
60. PETRIE K., CONAGLEN J.V., THOMPSON L., CHAMBERLAIN K. – Effect of melatonin on jet lag after long haul flights. *BMJ*, 298, 705-707, 1989.
61. REITER R.J., MELCHIORRI D., SEWERYNEK E., POEGGELER B., BARLOW-WALDEN L., CHUANG J., ORTIZ G.G., ACUNA-CASTROVIEJO D. – A review of the evidence supporting melatonin's role as an antioxidant. *J. Pineal Res.*, 18, 1-11, 1995.
62. REPPERS S.M., WEAVER D.R., RIVKEES S.A., STOPA E.G. – Putative melatonin receptors in a human biological clock. *Science*, 242, 78-91, 1988.
63. SHANAHAN T.L., CZEISLER C.A. – Light exposure induces equivalent phase shifts of the endogenous circadian rhythms of circulating plasma melatonin and core body temperature in men. *J. Clin. Endocrinol. Metab.*, 73, 227-235, 1991.
64. SHIBATA S., CASSONE V.M., MOORE R.Y. – Effects of melatonin on neuronal activity in the rat suprachiasmatic nucleus in vitro. *Neurosc. Lett.*, 97, 140-144, 1989.
65. SIEGRIST C., BENEDETTI C., ORLANDO A., BELTRAM J.M., TUCHSCHERR L., NOSEDA C.M.J., BRUSCO L.I., CARDINALI D.P. – Lack of changes in serum prolactin, FSH, TSH, and estradiol after melatonin treatment in doses that improve sleep and reduce benzodiazepine consumption in sleep-disturbed, middle-aged and elderly patients. *J. Pineal Res.*, 30, 34-42, 2001
66. STRASSMANN R.J., QUALLS C.R., LISANSKY E.J., PEAKE G.T. – Elevated rectal temperature produced by all-night bright light is reversed by melatonin infusion in men. *J. Appl. Physiol.*, 71, 2178-2182, 1991.
67. SUHNER A., SCHLAGENHAUF P., JOHNSON R., TSCHOPP A., STEFFEN R. – Comparative study to determine the optimal melatonin dosage form for the alleviation of Jet lag. *Chronol. Int.* 15, 655-666, 1998.

68. TAN D.X., CHEN L.D., POEGGELER B., MANCHESTER L.C., REITER R.J. – Melatonin: A potent, endogenous hydroxyl radical scavenger. *Endocr. J.*, 1, 57-60, 1993.

69. TAN, D.X., POEGGELER B., REITER R.J., CHEN L.D., MANCHESTER L.C., BARLOW-WALDEN L.R. – The pineal hormone melatonin inhibits DNA adduct formation induced by the chemical carcinogen safrole in vivo. *Cancer Lett.*, 70, 65-71, 1993.

70. TOUITOU Y., FEVRE M., BOGDAN A., REINBERG A., DE PRINS J., BECK H., TOUITOU C. – Patterns of plasma melatonin with ageing and mental condition: stability of nyctohemeral rhythms and differences in seasonal variations. *Acta Endocr.*, 106, 145-151, 1984.

71. TOUITOU Y., FEVRE M., LAGOGUEY M., CARAYON A., BOGDAN A., REINBERG A., BECK H., CESSELIN F., TOUITOU C. – Age and mental health-related circadian rhythms of plasma levels of melatonin, prolactin, luteinizing hormone and follicle-stimulating hormone in man. *J. Endocr.*, 91, 467-475, 1981.

72. TRINCHARD-LUGAN I., WALDHAUSER F. – The short term secretion pattern of human serum melatonin indicates apulsatile hormone release. *J. Clin. Endocrinol. Metab.*, 69, 663-669, 1989.

73. TZISCHINSKY O., LAVIE P. – Melatonin possesses time-dependent hypnotic effects. *Sleep*, 17, 638-645, 1994.

74. VAKKURI O. – Diurnal rhythm of melatonin in human saliva. *Acta. Physiol. Scan.*, 124, 409-412, 1985.

75. VOORDOUW B.C.G., EUSER R., VERDONK R.E.R., ALBERDA B.T.H., DE JONG F.H., DROGENDIJK A.A.T.C., FAUSER B.C.J.M., COHEN M. – Melatonin and Melatonin-Progestin combination alter pituitary-ovarian function in women can inhibit ovulation. *J. Clin. Endocrinol. Metab.*, 74, 108-117, 1992.

76. WALDENLIND E., GUSTAFSSON S., EKBOM K., WETTERBERG L. – Circadian secretion of cortisol and melatonin in cluster headache during active cluster periods and remission. *J. Neurol. Neurosurg. Psychiatry*, 50, 207-213, 1987.

77. WALDHAUSER F., VIERHAPPER H., PIRICH K. – Abnormal circadian melatonin secretion in night shift workers. *New Engl. J. Med.*, 315, 1614-1615, 1986.

78. WEHR T.A. – The durations of human melatonin secretion and sleep respond to changes in daylength (photoperiod). *J. Clin. Endocrinol. Metab.*, 73, 1276-1280, 1991.

79. WIRZ-JUSTICE A., ARMSTRONG S.M. – Melatonin: nature's soporific? *J. Sleep Res.*, 5, 137-141, 1996.

80. ZAIDAN R., GEOFFRIAU M., BRUN J., TAILLARD J., BUREAU C., CHAZOT G., CLAUSTRAT B. – Melatonin is able to influence its secretion in humans: description of a phase-response curve. *Neuroendocrinology*, 60, 105-112, 1994.

Chapter 7

Molecular genetics, circadian rhythms and sleep

Y. Dauvilliers*, M. Tafti**, and E. Mignot***

*Service de Neurologie B, Hôpital Gui de Chauliac, Montpellier, France; **Unité de Biochimie et Neurophysiologie Clinique, Département de Psychiatrie, Université de Genève, Suisse; *** Narcolepsy Research Centre, Stanford University, Palo Alto, California, USA

INTRODUCTION

Sleep is a vital function present in all mammals and birds, but its phylogenetic origin remains uncertain. In humans, sleep represents over half the first year of life, and roughly a third of adult life. The total suppression of sleep will provoke death in animals [179]. The universal and irrevocable nature of this function, as well as the maintenance (conservation) of NREM sleep – REM sleep organisation in mammals and birds implies the existence of constitutional factors. However, a wide variability in phenotypes has been noted between species, animal strains and individuals of the same species, as well as in the duration of NREM sleep and REM sleep and their distribution throughout the nycthoemeral cycle. These variations may be attributed to environmental factors. Their role in sleep is well known whether in terms of physical effects (light, temperature) or behavioural effects (conditioning, diet). But these factors do not fully account for the variability observed and the physiological differences in "sleep" phenotypes both in animals and humans, suggesting that polymorphic genetic factors are involved in this function. Even for a relatively simple and anatomically localised function such as circadian rhythmicity in the suprachiasmatic nucleus, multiple genes appear to be involved. The situation is probably even more complex for the normal and abnormal regulation of sleep in general. Several sleep disorders (narcolepsy, certain forms of insomnia, sleep apnoea syndrome, restless legs syndrome and periodic limb movement disorder) are known to occur with high frequency in certain families, with a higher rate of concordance in monozygotic than in dizygotic twins, thus suggesting the presence of predisposing genetic factors. The mutations or polymorphisms of certain genes result in pathological phenotypes in animals and humans, whereas others contribute to interindividual differences for the various aspects of sleep.

GENETIC ASPECTS OF NORMAL SLEEP

In animals

Determining genetic factors

The implication of genetic factors is strongly indicated by the many inter-individual variations in the different components of the sleep-wake cycle in laboratory animals placed in a constant environment and receiving the same diet from birth. Indeed significant differences in NREM sleep and REM sleep can be seen within the same species, with far greater inter-strain than intra-strain variability, suggesting that environmental factors play a less important role. Indeed significant sleep variations exist between inbred rodent strains [13, 59, 60, 109, 124, 184, 207, 226]. Furthermore, these differences are highly resistant to prolonged manipulation such as immobilisation, forced activity or sleep deprivation [109, 184, 226, 228, 236].

Gene localization - Methodology

To better localize the different genetic factors, most studies have focused on observing the sleep of pure strains, mainly of mice and rats; mouse models constitute the best animal tool for discovering genes related to sleep [209]. Mice are easy to breed and study, and high density genetic marker maps, like the Whitehead Institute/MIT maps are now available. Moreover there are numerous so-called recombinant or inbred lines which enable the different markers to be precisely localized on the genome [7]. A pure strain is established by successive inbred crossing, and the genome is considered as being at homozygotic state after 60 inbred generations. All the individuals of the same strain thus have the same genetic combination unlike non-inbred or so-called « wild-type » mice. The aim is thus to compare several inbred strains to distinguish the differences for a given phenotype. It is usually necessary to combine several types of approach to localise a gene and determine its function [190].

Genetic linkage analysis studies

To determine the genetic origin of a variation between two strains for a given phenotype, the two relevant strains are crossed using first (F1) and second generation animals (F2, by crossing F1s or crossing a parent and an F1, referred to as a back-cross) to determine the mode of transmission of the studied characteristic. The genetic factor is finally localized on the genome using genetic linkage analysis to detect co-segregation between the studied phenotype and a marker whose position is known on the genome (Fig. 7.1). This linkage analysis is based on the mechanism of meiotic recombination which is responsible for the exchange of material between homologous chromosomes during the formation of gametes. Several hundred markers have to be tested until one is found whose localization is so close that crossing over only occurs rarely between it and the gene

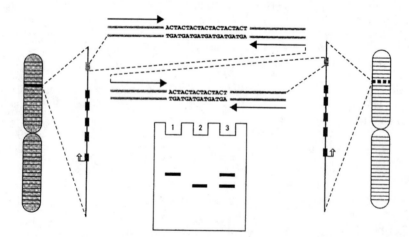

Figure 7.1. Genetic linkage analysis. This method exploits DNA polymorphisms between different individuals or strains that cosegregate with a given phenotype. In principle, DNA polymorphism can be detected by a variety of methods, but at present simple sequence length polymorphism (SSLP) analysis is the most widely used method. Segments carrying a variable number of simple repeats [(ACT)5 or (ACT)7 in the presented example] are amplified by PCR primers complementary to the unique sequences flanking the simple repeats and displayed by polyacrylamide gel electrophoresis. Lane 1, PCR fragment obtained from inbred mouse strain 1; lane 2, PCR fragment obtained from inbred mouse strain 2; lane 3, PCR fragments obtained from an F1 hybrid between inbred mouse strains 1 and 2. Corresponding chromosomes from two different mouse inbred strains are shown. The thin horizontal lines represent SSLPs that differ in the number of repeats between the two chromosomes. The right chromosome is derived from the mouse showing the mutant phenotype, because it is closely linked to the mutated gene. Due to meiotic recombination, none of the other SSLPs exhibits a 100% correlation with the mutant phenotype. (Reprinted from J. Sleep Res., Vol. 8, Suppl. 1, Schibler U., Tafti M. Molecular approaches towards the isolation of sleep-related genes. 3, Copyright (1999), with permission from Journal of Sleep Research).

searched for. The markers used, mainly microsatellites, are selected for their high degree of polymorphism, which increases the possibility of obtaining informative meioses. This method of investigation (genome-wide scan) is the method of choice for discovering new genes involved in sleep [186].

The nature of the phenotype studied can vary widely. The pharmacogenetic approach has been largely used, allowing a particular phenotype to be identified via the specificity of response to a given pharmacological agent. This technique is based on selecting animal strains which are relatively sensitive or resistant to pharmacological agents such as ethanol, benzodiazepines, barbiturates or cholinergic compounds [163, 170]. For example, rats selectively bred for cholinergic compound supersensitivity have been shown to present an increase in REM sleep, a finding which is consistent with the literature on the implication of this transmission system in controlling REM sleep [163].

Candidate gene studies

The study of candidate genes represents another, complementary type of approach, to determine whether a gene which is already known is implicated in sleep regulation. The known physiology of a gene may lead to its suspected implication to account for phenotype variations between two strains. A study of the polymorphism of this gene will look for a relationship between an allele and the phenotype trait studied. By way of example, it has been demonstrated that a strain of rats that were particularly resistant to benzodiazepines and alcohol presented a specific mutation of the alpha-6 subset of the GABA-A receptor [113]. This result is consistent with the idea that alcohol and benzodiazepine sensitivity stem from a common GABAergic mechanism. But other genes and factors also appear to be implicated as this mutation is not found in the other sedative or alcohol sensitive rodent models. The lack of discovery of new genes is a considerable limitation inherent to the strategy of the candidate gene approach.

Genetically modified mice studies

The study of genetically modified mice opens up scope for research in this respect, and appears to have a particularly promising future. This consists of using strains of animals whose genetic heritage has been altered, either by adding supplementary copies of a given gene ("transgenic"), or by deleting the gene of interest ("knock-out") [183, 189]. If the animal is viable, analysis of the obtained phenotype provides information about the normal function of the modified gene. It is then easy to study the sleep of these transgenic mice and draw conclusions about the implication of a given system in sleep regulation.

By way of example, a recent study was made of the sleep of mice deficient in the gene coding for Prion protein (the gene implicated in fatal familial insomnia and Creutzfeldt-Jakob's disease) [216, 217]. These mice present circadian alteration and sleep fragmentation, suggesting the possible role of Prion proteins in regulating normal sleep. Prepro-hypocretin (peptide precursor of orexins A and B) knock-out mice have also been studied; they develop a phenotype that is highly evocative of human or canine narcolepsy [22]. Moreover, transgenic mice for orexin/ataxin 3 responsible for the selective destruction of orexin neurons provide an excellent animal model for narcolepsy [75]. Studies have also looked at the sleep of knock-out mice for TNF alpha receptor [51], 5-HT1B [16], growth hormone and IGF-1[246], and dopamine transporter [240], as well as that of transgenic mice with an overexpression of Prostaglandin D synthase [172] ; all these mice present sleep disorders. These sleep abnormalities, which are constant, irrespective of the strain of transgenic mice studied, probably indicate a non specific effect of the different genetic manipulations and/or the high sensitivity of sleep to diverse alterations in general physiology.

Mutagenesis studies

Mutagenesis is another approach to researching new genes implicated in a function. Its aim is primarily to produce mutants which present abnormalities of sleep or circadian rhythm, secondly to localize these gene mutants, and finally to identify them by means of candidate genes or the technique of positional cloning. Ethylnitrosoureas is the mutagenic agent most commonly used. This technique has already been used successfully in the domain of circadian rhythms, and is notably responsible for the discovery of the clock gene, one of the key genes in circadian rhythmicity [108]. Nevertheless, the feasibility of this approach is limited by the relatively high number of animals which have to be studied to isolate sleep mutations of interest (>3000 and >10,000 respectively to cover the genome and search for dominant or recessive mutations).

Studies using Quantitative Trait Loci or QTL analysis

A number of limitations inherent to each of the techniques described above complicate the process of identifying new genes implicated in sleep physiology. Another strategy was developed to overcome some of these obstacles, consisting of identifying all the loci controlling a given quantitative trait giving rise to a variation between two inbred mice [150, 189]. Sleep physiology and regulation are particularly complex, and probably involve numerous genes and many interactions between these genes and environmental factors. The QTL technique is particularly appropriate when the trait is complex and the number of genes in question is high. It effectively allows numerous loci to be identified, some of which have a major effect and others a minor effect on the different phenotypes related to sleep. This approach usually requires studying either Recombinant Inbred Strains (RI) or crossed mice (F2 or backcross), which are genetically typed and finally the creation of congenic lines, although the last step may not be indispensable [150]. Thus the 2 mouse strains which differ for the given phenotype will be intercrossed, their F1 descendants will then be intercrossed to obtain F2 mice or crossed with one of the parents to obtain a backcross population, or intercrossed to obtain RIs. However, QTL analysis does not localize the gene in question but determines the broad chromosomal region implicated (often between 20 and 30 cM). This region may contain a single gene or a large number. Moreover, the different loci (QTLs) may interact (epistasis), producing a variable QTL effect between genetically different strains [196]. Several subsequent experiments crossing mice are necessary to obtain recombinant animals and thereby reduce the size of the region of interest, and the responsible gene(s) is/are finally determined by means of the candidate gene technique or positional cloning [55]. The discovery of a functional variant of the genome sequence differentiating the strains for the given trait is the final step in this approach, and one which is particularly difficult to establish with certainty.

Several studies using QTL strategy have been conducted successfully for numerous multifactorial traits, from auto-immune diabetes in non-obese mice [218], to drug response in the framework of addiction [29, 41, 139]. QTL candidates for circadian rhythmicity and REM sleep in mice have also been reported recently [82, 142, 196, 207].

Factors implicated in the genesis of circadian rhythms

The alterations in circadian rhythms related to variations in our environment are now well established, temperature and light being able to affect our internal biological clock. But a number of genetic factors have also been seen to be implicated in regulating these rhythms. Basic research on circadian rhythms is one of the most advanced in biology. This is greatly facilitated by the fact that most if not all the circadian rhythms in mammalian organisms are generated or synchronised by a discrete region of the hypothalamus, the suprachiasmatic nucleus [110]. Lesions in these nuclei abolish all rhythmicity and circadian fluctuations are restored by transplanting foetal hypothalamic tissue to lesioned animals [177]. Circadian rhythmicity is a virtually universal property in all unicellular organisms, plants and animals [208]. Transplanting mutant suprachiasmatic tissue to the third ventricle of a normal animal whose suprachiasmatic nuclei was previously lesioned will provoke the expression of an abnormal rhythm in the transplanted animal [177]. This furthers the idea that suprachiasmatic nuclei are necessary and sufficient for the genesis of behavioural activity rhythms. It can thus be demonstrated that the neuronal, metabolic and neurochemical activity of the tissue of these nuclei themselves vary in a circadian fashion even when the tissue is isolated *in vitro* [110]. Moreover, circadian rhythm has been determined in isolated cell cultures taken from the suprachiasmatic nucleus. When the activity of several of these neurons is recorded simultaneously, each cell is seen to present a different circadian rhythmicity whether for the period and/or the phase, with no apparent synchronised activity [130, 238]. Finally, several recent studies have reported finding circadian rhythms in most other tissues in the organism and notably in the liver, and the kidney... [8, 95, 125]. These rhythms also appear to form a hierarchy. Unlike the rhythms entrained by the pacemaker of the suprachiasmatic nucleus, some rhythms of peripheral origin may be entrained by our eating habits [125].

A range of mutations altering circadian rhythmicity have been reported in the *Drosophila* fly, in *Neurospora* fungus, and in the plant *Arabidosis* [30, 42, 43, 76, 116, 145, 192]. Genetic research in this domain has already led to the isolation of several "period" or *per* and "Timeless" or *tim* genes, whose mutations in the drosophile may considerably abolish, shorten, or lengthen circadian rhythmicity in function of the allele type implicated [76, 116, 192, 232, 243].

The *per* gene produces a PER protein that is found in the nuclei of the cells of the eye and the brain with higher levels at night than during the day, indicating circadian rhythmicity. If this gene is destroyed, PER protein is no longer synthesised and the fly becomes arrhythmic; if the gene is merely mutated, the resulting level of production of PER protein explains the differences in length of circadian periods in these animals. The nuclear localization of PER suggested that this protein itself controlled the transcription of other genes, a hypothesis which was confirmed, in addition to PER protein controlling its own *per* gene. The rhythm of production of PER is actually based on the feedback of PER protein on the expression of its own gene; this retroaction provokes oscillation in the production of PER generating the final rhythmicity. Another *Drosophila* mutant strain responsible for disrupting circadian rhythm has led to isolating the *tim* gene. The activity of *per* and *tim* genes is coordinated and mRNA levels always correlate both in amplitude and period (Fig. 7.2).

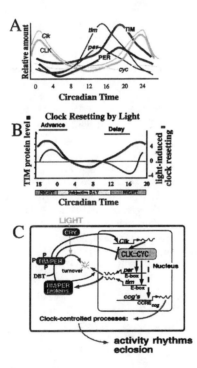

Figure 7.2. Identity and regulation of elements in the *Drosophila* oscillator and their roles in entrainment

A. Temporal regulation of the *per, tim, Clk,* and *cyc* genes and proteins. Care has been taken concerning the relative amplitude of the oscillation and in the timing of peaks

B. How light resets the *Drosophila* clock. Light results in the rapid destruction of TIM whose loss destabilizes PER. If TIM levels are already slowly rising, this rapid loss results in an advance into the next day phase.

C. Elements and control logic in the circadian oscillatory loop of *Drosophila*. Arrows denote positive regulation, and lines terminating in bars denote negative regulation.

(Reprinted from Cell, Vol. 96, Dunlap J.C., Molecular bases for circadian clocks, 281, Copyright (1999), with permission from Elsevier Science)

Moreover, TIM protein is physically associated with PER protein; the ensemble forms the heterodimer required for its translocation to the nucleus. This phenomenon of proteic dimerisation causes two particular domains of PER protein to intervene with TIM, one region called the PAS domain having strong homology with other regulating proteins and a region called PER-CLD [90, 194]. The nuclear localization of the heterodimer is indispensable for regulating the expression of their own genes. In fact, no PER protein circadian rhythmicity has been found in this mutant strain, suggesting that *tim* gene expression is necessary for the development of the rhythmic activity of *per* [73, 191]. In effect, in the absence of protein TIM, protein PER cannot bind to DNA.

Other genes implicated in circadian rhythmicity have also been isolated in *Arabidosis* and in *Neurospora Crassa* ("frequency"or *frq*, "white collar-1 and 2" or *wc-1, wc-2*) [30, 43, 145]. The importance of genes *wc-1* and *wc-2* for photoreception in *Neurospora* suggests that the genes involved in regulating circadian rhythms may have evolved from primitive photoreceptors [30, 103]. Genes *wc-1* and *wc-2* also contain a PAS domain, a fact which favours the importance of the PAS protein family in conceiving the biological clock in the course of evolution [30]. The cyclic activity of the *frq* gene shows the same oscillatory functions as that of the *per* gene. Like the *per* gene in *Drosophila*, the *frq* gene establishes negative feedback loops in *Neurospora* as opposed to genes *wc-1* and *wc-2* which appear to serve as positive regulators in genetic transcription [30].

In mammals, notably in mice, sizeable variations also exist in circadian periods between inbred lines, some are particularly long, others particularly short [178, 232]. To be more precise, mouse strains C57BL and C57BR present significant circadian variations in the Light/Darkness situation, CBA and DBA mice present a less marked sleep-wake nycthoemeral rhythm and finally BALB/c mice have no notable circadian rhythmicity [224, 225, 226]. Numerous genetic factors are likely to be involved to account for these different phenotypes.

Mutations responsible for altering circadian rhythmicity have also been reported in mammals [155, 178, 208, 232]. Two genes which are essential for the production of behavioural rhythmicity, *Clock (Circadian Locomotor Output Cycles Kaput)* and *Wheel*, have indeed been isolated, both resulting from a mutagenesis programme in mice using N-ethyl-N-nitrosurea (ENU) [155, 232]. The mutant *wheel* gene (chromosome 4 of the mouse) exerts a dominant effect and causes complex neurological disruption associating hyperactivity, rotating behaviour and circadian rhythmicity [155]. The mutant *clock* gene (chromosome 5 of the mouse) exerts a semi-dominant effect, responsible only for altering the circadian period which becomes abnormally long [232]. These mutant *clock* mice are capable of following a rhythm entrained by alternating light/darkness but have no endogenous circadian rhythm in temporal isolation. This gene was identified using a combination of techniques associating positional cloning [108] and a transgenic approach [5]. The CLOCK protein has sequence motifs suggesting direct DNA binding properties (the "basic Helix Loop Helix" or bHLH domain), enhancing its implication in regulating the transcription of certain genes [44]. In another rodent species, the golden hamster, a spontaneous semi-dominant mutation of the *tau* gene has been cited as being responsible for the isolated alteration of the circadian period [178]. This mutation in the homozygotic state exclusively induces a shorter circadian period (around 20 h) resistant to variations in alternating light/darkness [178]. This gene codes for a protein belonging to the casein kinase Iε family; the mutation found may be responsible for deactivating the protein via its inability to fix and phosphorylise PER protein [131].

Two other genes affecting circadian rhythm have been localized in *Drosophila*, a homologue of the mouse *clock* gene and a *cyc* gene coding for a CYCLE (or CYC) protein which also belongs to the bHLH family [43, 44, 123]. A homologue of CYC protein has also been found in mammals, BMAL 1 protein. These 2 *clock* and *cyc* genes code for activating proteins in *per* and *tim* gene transcription. A mutation of one of these 2 genes in *Drosophila* will lead to the absence of all circadian rhythmicity. BMAL 1 and CLOCK proteins, like TIM and PER, form a heterodimer which seems to be subjected to a similar process of regulation [123]. More specifically, the PER-TIM protein compound, once translocated to the nucleus then disassembled, strongly inhibits the activity of the CLOCK-BMAL 1 complex, finally influencing the transcriptional activity of the *per* and *tim* genes [123, 193] (Fig. 7.3).

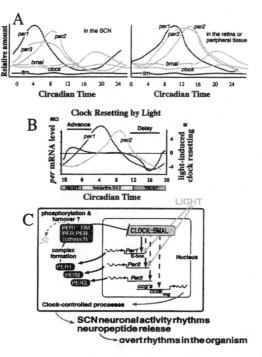

Figure 7.3. Identity and regulation of elements in the mammalian oscillator and their roles in entrainment.

A.   Temporal regulation of the Per1, Per 2, Per 3, tim, Clock, and bmal(mop3) genes.

B.   How light resets the mammalian clock. Light results in the induction of *Per1* and *Per2* but to different extents at different times, so the effect of light on the clock components and therefore on the clock is very much influenced by the time of day.

C.   Elements and control logic in the circadian oscillatory loop of mammals. Arrows denote positive regulation, and lines terminating in bars denote negative regulation. Dashed lines indicate possible regulatory connections.

(Reprinted from Cell, Vol. 96, Dunlap J.C., Molecular bases for circadian clocks. 284, Copyright (1999), with permission from Elsevier Science)

The effect of the environment, and more precisely, the alternation between light and darkness in regulating circadian rhythmicity is well known and its action mechanism is beginning to be understood. It seems in effect that light has a direct influence on the TIM protein via phosphorylation, ubiquitination and finally, degradation. The light-dependent turnover of TIM is nevertheless indirect, relying on the photoreceptive flavoproteins CRYPTOCHROMES (CRY), with light forming an association between proteins CRY and TIM [187]. The transcriptional activity of *per* and *tim* genes is not altered by light, however the formation of the protein complex PER-TIM and its nuclear translocation are clearly affected by light-darkness alternation. But the intermittent presence of light is not essential for the genesis of rhythmic activity. The duration of formation of the PER-TIM complex once in the nucleus appears to be determined by two different endogenous processes: a self-regulation phenomenon and the presence of the DOUBLE-TIME (DBT) protein. This DBT protein binds to PER to destroy it, thus regulating its accumulation both in the cytoplasm and in the nucleus [111]. DBT belongs to the same family as TAU protein, the casein kinase Iε family [105, 131]. The function of these 2 proteins, in *Drosophila* and in mammals appears to be relatively well conserved; in effect, this kinase protein allows PER phosphorylation, the inhibition of its translocation to the nucleus and the reduction of its stability.

After cloning the *clock* gene three homologues of the *Drosophila per* gene were isolated in the mouse and in humans, *mPer 1*, *mPer 2* and *mPer 3* [194, 205, 211]. These *per* genes are expressed in several cerebral regions but significant rhythmic daily fluctuations are only found in the suprachiasmatic nucleus, indicating their implication in generating circadian rhythms [3, 194, 205, 211.] Only *mPer 2* appears to be strongly implicated, knock-out mice developing abnormal circadian rhythmicity ; conversely, *mPer 1* and 3 genes have only added effects [194]. Two homologues of the *cry* gene, *mcry 1* and *mcry 2*, have also been isolated in mammals with demonstrated oscillatory activity. Moreover, these genes appear to have stronger rhythmic activity in mice than in *Drosophila*, with highly altered circadian periodicity in knock-out mice for these genes. Proteins mCRY 1 and 2 are transcription inhibitors of their own *mcry 1 and 2* genes, but also of *mper 1, 2* and *3* genes via protein complex CLOCK-BMAL 1 [193]. A single *mtim* gene has been isolated in humans and in mice but its function, unlike its homologue *tim* in *Drosophila*, is still unknown [43]. Indeed, as knock-out mice for this gene die at the embryonic stage it has not been possible to distinguish any transcriptional rhythmic activity, as light does not appear to exert any effect on protein mTIM and the nuclear translocation of mPER would depend on heterodimerisation between proteins mPER and mCRY and not mTIM [117]. The role of *Drosophila* protein TIM seems to be equivalent to that of CRY in mammals. Finally the two proteins CLOCK and BMAL 1, which are activators of *mPer* transcription and coded respectively by *mclock* et *bmal 1*, also have strong homology with *Drosophila* genes *dclock* and *cycle*. However, the activity of these genes differs between the two species. In *Drosophila*, the expression of the *clock* gene follows a circadian rhythmicity with oscillation regulated by the protein complex PER-TIM-CLOCK inhibiting its own mRNA synthesis. In mice, the transcriptional activity of the *clock* gene is not rhythmical, contrary to the *bmal1* gene whose regulation is thought to be carried out by *mper 2* gene. On the whole, and irrespective of the species studied, *Drosophila* or the mouse, *per* gene is central to the cellular machinery for circadian regulation. It is thus likely that PER and CLOCK proteins work together in producing 24 hour rhythmicity in the suprachiasmatic nuclei.

The expression of genes implicated in circadian regulation also occurs outside the central nervous system, in *Drosophila* as in mammals [8, 125, 199]. However, although these genes function independently of each other, they remain photosensitive in *Drosophila*, contrary to the case for mammals. Regulation which is still only partially understood appears to coordinate and entrain the expression of the different circadian genes. Indeed, in *Drosophila*, a neuropeptide, the pigment dispersing factor (PDF) accumulates at the nerve endings and acts by establishing the circadian phase via coordination of the rhythmic activity of the different neurons [164]. The accumulation of this peptide relies on the presence of other proteins CLOCK and CYC; although no intrinsic PDF rhythmic activity has been determined. Another transcription factor has been isolated, again in *Drosophila*, coded by the *vrille* (*vri*) gene considered to be responsible for the oscillation of PDF neuropeptide production [15]. A similar mechanism has been demonstrated in mammals, the neuropeptide vasopressin appearing to exert the same function as PDF [95]. In mammals, numerous studies have pointed to the presence of soluble factors diffusing from the suprachiasmatic nucleus to other cerebral regions [199], thus entraining sleep/wake rhythms and locomotor activity. These factors are in the process of being identified. One of the factors, TGF alpha, was recently isolated thanks to a yeast secretion tap system [114]. This peptide appears to play an inhibiting role in locomotion by acting on the subparaventricular zone. Other factors stimulating locomotor rhythms are being studied. Lastly, it appears that certain peripheral circadian oscillators depend on food intake, through a hormonal glucocorticoid signal [125].

Factors implicated in the genesis of EEG activity

Several spectral analysis studies of EEG activity during sleep have demonstrated significant variations between different mouse lines for NREM sleep and REM sleep [225, 226]. In EEG, NREM sleep usually associates slow waves characterised by a spectre ranging from 0.5 to 4.5 Hz and sleep spindles ranging from 12 to 15Hz. These sleep spindles are the source of considerable inter-line variations. They are absent in some recessive mouse lines (C57BR). The frequency and amplitude of these spindles may also vary, large amplitudes sometimes being found for frequencies between 10 and 12 Hz, with a dominant character (in the case of BALB/c and CBA mice) [51, 52].

The main EEG activity in REM sleep is theta rhythm whose frequency bands oscillate between 5 and 10 Hz depending on the lines, with frequencies that are particularly low for BALB/c (5Hz) and high for C57BR (8 Hz) (224, 226). Qualitative differences in EEG signals have also been observed with high amplitude points for DBA and BALB/c strains, but not for C57BR [224, 226]. These differences are transmitted genetically and the crosses analysed either using the di-allele method [60] or by simple back-cross segregation analysis [225] suggesting the implication of numerous genes in the expression of each trait [60, 82, 184, 190, 225]. Furthermore, the interactions between these different factors are sometimes complex, and not simply additive, with inbred strain hybrids sometimes presenting marked deviations from the mean for parent strains [60]. A recent EEG spectral analysis study of 6 inbred mouse strains confirmed these variations in the distribution of frequency peaks, notably in the theta band, both in NREM sleep and in REM sleep [58, 59] (Fig. 7.4). Moreover, 80% of inter-strain variability (or heritability) for the theta frequency peak in REM sleep may be attributed to genetic factors. These data strongly suggest the presence of a gene with a major effect, to account for the variability of this trait.

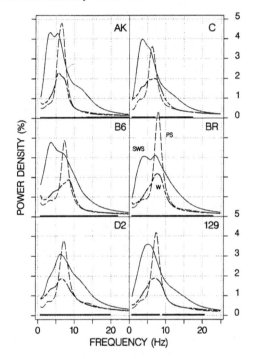

Figure 7.4. Mean spectral profiles for SWS (solid lines), PS (short-dashed lines), and W (long-dashed lines). For 97 consecutive 0.25-Hz bins between 0.6 and 25.1 Hz, mean power density was calculated by averaging over all 4-s epochs scored as SWS, PS, or W in the 24-h recording period. All values were expressed as a percentage of mean total power for all frequencies and behavioral states. Within each panel, a thick horizontal line at the 0% level connects those frequency bins in which power density significantly varied among behavioral states (1-way ANOVA; factor "state" within each strain, $P < 0.01$, $n = 7$/strain).(Reprinted from Am. J. Physiol. Vol. 275, Franken P. *et al.*, Genetic variation in EEG activity during sleep in inbred mice, R1131, Copyright (1998), with permission from the American Physiology Society)

Another team showed that prion gene knock-out mice presented a lower frequency peak in the theta band both in REM sleep and in wakefulness, suggesting the involvement of the prion protein in regulating sleep, particularly in generating hippocampal theta rhythm [91].

Factors implicated in the duration and distribution of sleep in the nycthemer

Sleep time over 24 hours is an individual characteristic that is highly dependent on the environment. Hence many animals of the same strain and of different strains need to be recorded in strictly similar experimental conditions (light, temperature, diet) to verify whether inter-line variability is clearly higher than intra-line variability, and ultimately envisage the presence of an explanatory discriminating genetic factor.

Pharmacogenetics were initially widely applied in this domain; mouse strains LS and SS (long and short sleepers) were initially the most studied models [170]. These mouse strains were created in the 1970s by selecting mice which were the most or least sensitive to the hypnotic effect of ethanol by testing their righting reflex. After 18 generations of selection, the resulting strains had a mean sleep time of 10 minutes (SS) or of two hours (LS) respectively, after similar doses of ethanol. An analysis of the sensitivity of these strains to different pharmacological compounds determined to what extent genetic control of the soporific or anaesthetic effect of these hypnotics was similar or different [34, 49, 50, 140, 170]. Genetic control of the hypnotic effect of the least liposoluble compounds and alcohol consistently showed considerable overlap when objectified; by contrast the effect of the most liposoluble anaesthetics was under different genetic control [34]. Also the different pharmacological effects of alcohol (sedation, hypothermia, toxicity) appeared to be controlled by different genes [50, 170]. QTL analysis of SS X LS hybrids traced intra-strain variations in the differences in alcohol sensitivity [41] to at least seven or eight genes. More recent studies have confirmed this result with a total of 7 QTLs, accounting for 60% of the variation between SS and LS strains [29, 139]. These pharmacogenetic studies are nevertheless considerably restricted: the relationship between these mouse models and the genetic control of sleep is uncertain, no study exists of sleep and circadian rhythms in LS and SS animals. Moreover, the effect of benzodiazepines and alcohol on sleep appears to be indirect and highly dependent on accumulated sleep debt. A better analysis should be made of the physiology of these models, specifically including a study of the circadian and homeostatic regulation of sleep.

Other work has studied and objectified the considerable variations in the time and distribution of sleep over the nyctohemer in several mouse strains. For example, AKR mice sleep 3 hours longer than DBA mice in 24 hours [59]. NREM sleep time and REM sleep time over 24 hours also varies according to the strains studied. C57BL and C57BR strains are characterised by long episodes of REM sleep and short episodes of NREM sleep [59, 225]. At the other end of the spectrum, the BALB/c strain is characterised by episodes of REM sleep of very short duration; DBA mice present intermediate levels for these characteristics [59, 225]. Genetic studies of F1 and F2 mice also indicate the complex nature of the genetic control of these parameters, implicating the presence of several genes. A recent study by QTL on these phenotype differences in sleep/wake behaviour between BALB/c and C57BL mice identified 14 implicated loci with epistatic interactions between them [196]. It should be noted that these loci differ from the localization of genes already known to be involved in circadian regulation, results which underline the complementary value of the QTL method of analysis.

Regulation of NREM sleep clearly differs from that of REM sleep ; the genetic control of these two types of sleep probably reflects this difference. A QTL genetic binding analysis of REM sleep led to identifying several loci to account for variability between different mouse strains in REM sleep time over 24 hours [207]. The loci were different for the duration of diurnal REM sleep (chromosome 7), nocturnal REM sleep (chromosome 5, near the *clock* gene) and for total REM sleep time during 24 hour (several loci are implicated), results which suggest that several genes are implied in the expression and regulation of this sleep stage [207]. Moreover, the genetic factors implicated in the regulation of REM sleep appear to differ for daytime and at night. Another team using the same methodology reported other QTLs for the quantity of diurnal REM sleep among the same mouse strains [222]. To conclude on the subject of REM sleep, a QTL analysis between 2 other mouse strains, C57BL/6 and DBA/2, determined another QTL on chromosome 1 which was significant for the quantity of diurnal REM sleep [206]. Numerous genes are probably implicated in the regulation of this stage of sleep; around 50% of the variance of this trait (REM sleep time) is explained by the presence of at least 6 different genes. It is worth noting that no significant QTL was found between these mouse strains for NREM sleep [207].

One of the regions relevant to REM sleep regulation contains an important candidate gene, the DBP gene coding for Albumin-D binding protein which is a transcription factor expressed with strong circadian rhythmicity [57]. DBP knock-out mice are characterised by a reduction of their circadian activity period and by an overall drop in locomotor activity. The study of their sleep, apart from revealing a total sleep time identical to that of wild-type mice, showed a reduction in circadian amplitude of NREM sleep, as well as alterations in REM sleep [57].

The sleep of other transgenic mice has been studied, each of which showed abnormalities in terms of sleep architecture. Respectively, 5-HT1B receptor, GH-IGF-1 and dopamine transporter (DAT) knock-out mice demonstrate the reduction of NREM sleep, thus testifying to their likely implication in the regulation of this sleep [16, 240, 246]. REM sleep for its part was significantly increased in 5-HT1B-R knock-out mice in contrast to the absence of change in GH-R knock-out mice. Another team studied the sleep of transgenic mice overexpressing PGD2 and reported a significant increase in NREM sleep only after nociceptive stimulation (pinching the tail) with no alterations found for REM sleep [172]. These findings appear to relate to a local inflammation, linked to an increase in the level of cerebral PGD2 released.

Factors implicated in the homeostatic process

After sleep deprivation, both the duration and intensity of sleep increase, seen as a rebound or recuperation of sleep. This rebound is usually proportionate to the sleep debt. The duration of deprivation required depends on the species studied but at least 3 hours of total sleep deprivation are needed in the rodent for a rebound effect to be seen. This rebound in the mouse may either occur immediately after the period of sleep deprivation, or be delayed by several hours (6 to 8), or it may be totally absent [109]. These studies are limited by the inevitable notion of stress involved in depriving the animal of sleep, the rebound may simply be an indication of the intensity of stress during the investigation, rather than the duration of sleep deprivation.

The homeostatic process in sleep has been known of for a good many years, but its neurophysiological mechanisms remain unclear. A recent study involving 6 inbred mouse strains deprived of 6 hours of sleep, in strictly similar circumstances, revealed significant differences in the intensity of the homeostatic rebound of NREM sleep [56]. Later QTL analysis recording 25 recombinant mice (from B6 and D2) identified a significant locus on chromosome 13 with a maximum lod score of 3.57. This locus accounted for 49% of the variance in homeostatic NREM rebound sleep after sleep deprivation [56]. The findings indicate genetic control of this homeostatic process and several candidate genes are likely to be implicated in this area of interest.

Parallel studies of the homeostatic regulation of sleep, have been conducted on transgenic mice. CLOCK knock-out mice in addition to demonstrating important changes in circadian sleep architecture, also presented an alteration in homeostatic sleep regulation with an increased rebound of this type of sleep after sleep deprivation [91, 153]. DBP knock-out mice show a decrease in REM rebound sleep after sleep deprivation with no objectified difference in the rebound of NREM sleep [57].

Alterations in gene expression in sleep and waking states

Several teams have studied the changes in gene expression in the different states of sleep and wakefulness. The genes studied in most detail are the immediate early genes (IEG), generally the proto-oncogene c-fos. The changes occurring in IEG expression are interesting in at least two respects: the expression of c-fos and the other IEG increases with neuron activity [149] and the level of expression of these proteins or their RNA messengers (mRNA) in function of the cerebral region, can be defined thus providing a map of their variations in function of sleep or wakefulness. Moreover, as IEG products are themselves transcription factors, they also alter the expression of numerous other genes.

During the day, *c-fos* expression in the brain correlates directly with the rest/activity cycle. In nocturnal rodents, expression is high at night and low during the day, in most regions of the brain [10, 67, 68]. This effect is reversed when the animals are deprived of sleep during the day, provoking a sleep rebound the following night. Diurnal rodents present the opposite profile of *c-fos* expression, with high levels during the day and low levels at night [161]. Sleep deprivation increases the expression of *c-fos* and the other IEGs whereas sleep recuperation reduces it [10, 67, 68, 157]. Finally, Pompeiano *et al.* [173, 174] examined the levels of the proteins C-FOS and NGFI-A (derived from another IEG) after spontaneous periods of wakefulness and observed a tight correlation between the expression of these genes and the duration of wakefulness.

IEG genes have also been used to dissect the neuro-anatomy of sleep and wakefulness more finely. The locus coeruleus, in particular, appears to play an important role not only because *c-fos* expression changes according to states of sleep and wakefulness in this nucleus, but also because the locus coeruleus appears to control a large part of *c-fos* expression in the entire forebrain in the wakeful state. Unilateral lesions in the locus coeruleus reduce *c-fos* and NGFI-A levels during wakefulness ipsilaterally rather than controlaterally to the lesion [25, 26]. The reduced levels of *c-*

fos and NGFI-A in wakefulness on the lesioned side is comparable to the levels observed during periods of prolonged sleep. Studies have also been conducted during pharmacological REM sleep, induced by injecting cholinergic agonists in the pontine reticular formation. This manipulation activates the transcription of the *c-fos* gene in several nuclei implicated in the regulation of REM sleep [197, 198]. Recently, an exception to this wakefulness/ high c-fos levels correlation was objectified in certain cells in the ventrolateral preoptic region, the neurons expressing high *c-fos* levels during sleep [195]. These neurons probably play a key role in initiating NREM sleep. Despite such correlations, the functional role of IEGs has yet to be established. In fact only two studies suggest a direct *c-fos* role in the regulation of sleep [24, 197]. The former concerns the observation of a reduction in spontaneous sleep and a sleep rebound after deprivation in *c-fos* knock-out mice [197]. In the second study, *c-fos* antisense oligonucleotide injections in the medial preoptic region reduces C-FOS protein levels and increases wakefulness the following day [24].

Other more global approaches than the simple study of IEGs are needed to study gene expression during sleep. Among these, substractive hybridization, PCR differential display ("c-DNA display"), the use of micro-arrays of numerous cDNAs ("DNA chips") or the TaqMan method show promise [26, 189]. The substractive hybridization method has been used on rats deprived of sleep for 24 hours [182]. Four mRNA clones were isolated with lower levels after sleep deprivation and six with higher mRNA levels. An analysis of the structure of one of these clones enabled the primary characteristics of the protein to be deduced, a protein which is close to neurogranin [154].

Several teams have associated molecular biology techniques, sometimes with sleep deprivation, to objectify alterations in the transcriptional activity of several genes usually with selective cerebral tropism of these changes in expression. In fact, variations in mRNA levels over 24 hours of the growth hormone-releasing hormone or GHRH in the hypothalamus [18], of a brain-derived neurotrophic factor or BDNF and its receptor in the hippocampus [17] were demonstrated. Adding the sleep deprivation technique sometimes brought out these variations. For example, mRNA and GHRH levels and those of the adenosine A1 receptor respectively increase at paraventricular and basal telencephalic level after sleep deprivation [9, 220]. Furthermore, mRNA and interleukine 1 beta only increased significantly in the hypothalamus and the brain stem after sleep deprivation [134]. Other experiments have been carried out on the selective deprivation of REM sleep [119, 175, 221]. The increased expression of several mRNA has been objectified, in function of cerebral localization: for instance the level of tyrosine hydroxylase only increases in the locus coeruleus [175]. Finally, corticostatin and hypocretin proteins, strongly implicated in the control of different states of sleep and wakefulness have been identified using the molecular biology approach, demonstrating the importance of this field of investigation [35, 36]. It should be noted that paradoxically, prepro-hypocretin mRNA levels are not modified in the hypothalamus after sleep deprivation (6 hours) [212].

Given that the mammal brain expresses roughly half of the estimated 50,000 genes, it is likely that many genes change their level of expression during the states of sleep and wakefulness [27]. These alterations in gene transcriptional activities reflect a change in neuron functioning, although the role of these genes in sleep regulation is still difficult to ascertain. The sensitisation of these variations by sleep deprivation merely reflects the effects of prolonged wakefulness on the brain. In addition to using sleep deprivation techniques, future studies should systematically take post sleep deprivation into account, considering the marked NREM sleep rebound which may also coincide with alterations in the level of expression of these genes. In any case, all these studies are considerably limited as alterations in the system may occur at a post-transcriptional level, notably in the protein itself.

IN HUMANS

Determining genetic factors

The different human studies also support the notion of a genetic effect on sleep. To determine the impact of these genetic factors on sleep physiology, one interesting approach consists of studying the extent to which the variability of a trait is transmitted for a given aspect of sleep. Twin studies are of interest in this approach, by comparison of the similarities between monozygotic and dizygotic twins. These studies are mainly based on questionnaires on sleep habits (duration, times and quality of nocturnal sleep; frequency of napping) [61, 80, 166]. The results show that for all the variables analysed, the correlation is far higher between monozygotic pairs than between dizygotic pairs. These results remain significant even when the investigated twins no longer live in the same milieu [166] and seem largely independent of the genetic load related to anxiety and depression [106]. In these studies correlations rarely exceed 0.60, thus suggesting that a large part of variance is linked to the environment [80, 166]. Measurements of residual variance between monozygotic twins in fact provides an estimate of the effect of the environmental factors specific to each twin.

Few authors have studied a large number of monozygotic and dizygotic twins polygraphically [127, 128, 129, 235]. These studies have nevertheless confirmed the results obtained by questionnaires. Linkowski *et al.* [127, 128] in studying 26 pairs of twins for 3 consecutive nights, were able to demonstrate significant differences between the correlations obtained with monozygotic and dizygotic twins for all the sleep stages with the exception of REM sleep. Conversely, another team found a higher correlation between monozygotic twins for different REM sleep variables than for dizygotic twins [23].

More recently, one team studying the distribution of the different EEG wakefulness frequency bands in monozygotic and dizygotic twins, reported particularly high heritability of between 80 and 90%, for this trait [227]. Another team demonstrated the existence of a dominant hereditary factor, localized on chromosome 20q, via the exploration of occipital alpha rhythms in wakefulness [233]. Indeed, no alpha waves can be recorded in this area for a certain percentage of the normal population in the, albeit classic, resting-eyes closed state [4, 201]. Certain EEG variations, related to underlying genetic polymorphism, appear to be both of a qualitative and quantitative nature. However, most of these studies take little or no account of recent developments demonstrating that the circadian and homeostatic regulation of sleep occurs independently. Linkowski *et al.* [127, 128, 129] attempted to resolve this question using hormone measurements of cortisol and prolactin in twins. The results indicate the presence of important genetic factors in the regulation and secretion of cortisol but not of prolactin. Another team, after examining the morningness/eveningness of 238 pairs of monozygotic twins, using the Horne and Östberg questionnaire, obtained higher correlations in monozygotic twins [40]. The result thus points to the implication of genetic factors in circadian regulation in humans [40].

Gene localization - Methodology

Two types of approach may be envisaged in humans in the attempt to localize the genes likely to affect sleep. Firstly, genetic binding analysis studies, which try to identify a genetic marker that, through its known localization in the genome close to the gene in question, cosegregates with the phenotype. Studies in man are greatly assisted by the existence of well established genetic maps, enabling chromosomes to be ruled out one by one until a genetic binding is found to the trait in question. The other strategy is that of candidate genes, the known physiology of a gene raising questions as to its possible implication in a particular phenotype.

Circadian factors

The mechanisms involved in the genesis of circadian rhythms are beginning to be elucidated and the genetic factors determined.

The effect of environmental conditions often suffices to explain the variations in many sleep traits found between individuals. Indeed, it is often difficult to define certain parameters like total sleep time over 24 hours, due to marked intra-individual variations. However a distribution of the "sleep time" parameter has been studied in the general population revealing a "normal" mean duration of 7h to 8h with extremes qualified as short sleepers (less than 6 hours) and long sleepers (more than 9 hours). This trait is often characteristic of an individual from a very early age. Furthermore, the existence of families of short and long sleepers is common knowledge, even if this fails to demonstrate the implication of genetic factors any more than environmental ones. But the fact that children are short, medium or long sleepers with parents who have different extreme phenotypes (1 long sleeper parent and 1 short sleeper) argues in favour of the implication of genetic factors. Studies of twins tend to confirm the concept of a genetic effect on sleep, monozygotic twins showing much closer similarities for many sleep traits [80, 166, 235].

As referred to previously, after the cloning of *Clock*, three homologues of the *Drosophila per* gene were isolated in humans and in mice [194, 205, 211]. A human study found an association between a *clock* gene polymorphism and different circadian tendencies attested through questionnaires (morning or evening typology) [100]. Conversely, no polymorphism of *tim* or *hper 1* gene studied proved to be associated with a particular circadian typology [101, 167]. Although the implications of these findings are still unclear, they reinforce the notion that circadian rhythms are generated by similar molecular mechanisms throughout the animal kingdom.

GENETIC ASPECTS OF HUMAN SLEEP DISORDERS

Many sleep disorders such as narcolepsy, certain forms of insomnia, sleep paralysis, hypnagogic hallucinations, sleep apnoeas, restless legs and periodic limb movements, are known to occur with high frequency in particular families [2, 11, 14, 37, 53, 98, 143, 146, 148, 181]. These results all confirm the existence of a set of genes whose function is more specifically attached to sleep. Several studies have been conducted to define these, using either the candidate gene technique, or genetic binding analysis techniques in multiplex families.

Hypersomnias

Narcolepsy

Narcolepsy is one of the most studied sleep disorders on the molecular level. Since its description by Gélineau [62] in 1880, several familial cases of narcolepsy have been reported by various authors [14, 71, 143] suggesting a genetic basis for this disease. Numerous studies have since shown that the risk is 20-40 times greater for the parents of narcoleptic patients [14, 143]. A natural canine model (Doberman and Labrador) of this disease for which transmission is autosomal recessive with complete penetrance, has attributed the origin of the illness to a mutation in the hypocretin (orexin) receptor 2 gene [126]. Moreover, knock-out mice for the precursor ligand for this receptor as well as hypocretin/ataxin 3 transgenic mice present phenotype characteristics of narcolepsy [22, 75]. A mutation on the pre-prohypocretin peptide signal has been found in man in an isolated case, atypical of narcolepsy [169]. Focal hypocretin neuron degeneration has however been demonstrated in human narcolepsy [169, 213] corresponding to the hypocretin/ataxin 3 transgenic mouse, responsible both for the selective destruction of hypocretin neurons and for narcolepsy (Fig. 7.5) [75]. The cause of this neurodegeneration remains a mystery although an auto-immune hypothesis is likely. This subject will be dealt with in a separate chapter (see Chapter 31) however certain points pertaining to the uniquely human aspects of the genetics of narcolepsy warrant closer inspection at this point.

Numerous studies have demonstrated that narcolepsy is associated with the HLA system. The principal predisposing allele is DQB1*0602, an allele which is found in 75-95% cases of narcolepsy with cataplexies (and 25% of Caucasian controls). However this allele is neither necessary, nor sufficient for the disease to develop, especially in the case of narcolepsy without cataplexy. As in other HLA-associated diseases, the associations are complex and other alleles DR and DQ as well as DQB1*0602 have a protective or predisposing effect [144]. These effects are nevertheless far weaker than those of DQB1*0602 and the global contribution of class II HLA system to the total genetic risk of narcolepsy is only partial ($\lambda=3$).

Other non HLA-related genetic factors, which are environmental are also likely to be implicated in this disease. Consistent with this approach, several studies have looked for an association between a polymorphism in certain gene candidates and sporadic narcolepsy. Monoaminergic hypofunctioning in narcolepsy, the reduction of REM sleep by monoamine oxydase A and B (MAO A and B) as well as the association sometimes found between this disease and Norrie's disease (gene localized in Xp11.3/p11.4 close to MAO A and B) have led to the search for an association between a particular MAO A and/or B polymorphism in human narcolepsy. An early study reported an association between VNTR (Variable Number of Tandem Repeats) of MAO A intron 1 and narcolepsy [112], although this was not confirmed by a second team [33]. However, sexual dimorphism in the activity of the gene coding for catechol-O-méthyltransferase (COMT) as well as an effect on the severity of daytime sleepiness were detected, suggesting that COMT is more implicated than MAO-A in favour of a more critical alteration of the dopaminergic/noradrenergic than the serotoninergic pathways [33]. Two other association studies looked for the presence of Single Nucleotide Polymorphism (SNP) which functions in the TNF α gene promoter, with conflicting results, although TNF α may be another susceptibility gene, particularly in association with HLA DRB1*1501 [83, 85]. Finally a recent study reported a possible association between a rare polymorphism of the gene coding for pre-prohypocretin and narcolepsy [63]. However, later studies demonstrated that the loci for pre-prohypocretin, hypocretin receptor-1 and 2 did not contribute to susceptibility to narcolepsy in any significant way [94, 159, 169].

Only one genetic binding analysis study has been published to date for familial forms of human narcolepsy. A Japanese team in effect reported significant genetic binding effect in 4p13-q21 of 8 small multiplex families (lod score de 3.09), favouring the implication of other genes in these families [152].

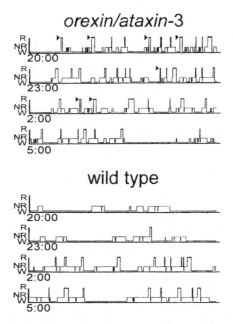

Figure 7.5. Sleep state abnormalities seen in orexin/ataxin-3 Mice. Representative 12hr dark period (20:00 to 08:00) hypnogram for an orexin/ataxin-3 mouse (upper panel) and its transgene-negative littermate (lower panel). The height of the horizontal line above base-line indicates the vigilance state of the mouse at the time. W, wakefulness periods in the hypnogram of the orexin/ataxin-3 mouse. Hypnograms were obtained by simultaneous EEG/EMG recording as described previously (Chemelli *et al.*, 1999). (Reprinted from Neuron, Vol. 30, Hara J. *et al.*, Genetic ablation of orexin neurons in mice results in narcolepsy, hypophagia, and obesity, 349, Copyright (2001), with permission from Elsevier Science)

Sleep apnoea syndrome

This is a complex syndrome in which the upper airways become obstructed during sleep, thus blocking respiration; these repeated breathing arrests prevent the patient from sleeping deeply and the principal complaint of the patient is usually daytime sleepiness. Snoring is one of the first symptoms. Roughly 4 to 5 % of the general population presents a severe form of this syndrome [245] with an increased risk of cardiovascular accidents [70]. Recent studies have also suggested a higher prevalence among black Americans [181]. No twin studies have been carried out on the rate of concordance of sleep apnoea syndrome, but in regard to snoring, two recent studies have shown higher concordance in monozygotic than in dizygotic twins [52]. Elsewhere, numerous familial forms of sleep apnoea syndrome have been reported in the literature [2, 37, 72, 141, 171, 181, 203]. This familial segregation is generally explained by the fact that most of the risk factors involved in the pathophysiology of these sleep apnoeas is largely determined genetically. This is true for instance for obesity, alcoholism and bone and soft tissue anatomic factors which predispose to the obstruction of the upper airways [48, 72, 115, 141]. In certain cases, the implied genetic factor seems to be more particularly central, implying for example an abnormality in ventilation control [2, 48]. A recent study of a large kindred with 11 patients affected both by Charcot-Marie-Tooth 1A disease (duplication of PMP 22 gene) and sleep apnoea syndrome was reported, suggesting a pathophysiological mechanism common to both conditions [37]. Other studies have reported a certain overlap between the genetic predisposition to sleep apnoeas and sudden infant death syndrome [2, 162, 215]. Moreover the marked concordance in chemo-respiratory responses observed in monozygotic twins [102, 215] also appears to underline the importance of central factors in these genetic aspects. Few studies have focused on the implication of possible gene candidates in this condition. A Japanese study nevertheless found higher HLA A2 and B39 in patients suffering from sleep apnoea syndrome [242]. More recently, an American study reported an association between apoliproprotein E ε4 and this disorder independently of age, sex, body mass index and ethnic origin [97].

The multifactorial character of sleep apnoea syndrome probably corresponds to a multitude of genetic factors. The genetic approach to this syndrome is difficult: the most promising line of research would consist of limiting study to rare cases, that are highly genetically determined or that have preselected clinical features.

Kleine-Levin syndrome

The Kleine-Levin syndrome is a rare disorder, characterised by recurrent episodes of hypersomnia, behaviour disorders (megaphagia, hypersexuality), mood and cognitive disorders. A link with the HLA system was initially suggested for 2 subjects [231] and confirmed by a recent study of 30 subjects that found an association with HLA DQB1*0201, suggesting that this allele predisposes to the development of the disorder [32].

Other hypersomnias

Hypersomnia associated with Steinert's myotonic dystrophy is common; an association between this disorder and HLA has also been reported, more precisely with HLA DR6-DQ1 [137].

Insomnias

A recent study reported the familial incidence of the most common sleep disorder, insomnia; 35% of patients report the presence of an immediate or close relative affected by a sleep disorder [11]. These familial forms were all the more frequent if the age of onset of insomnia was before 40 years old and when insomnia was of the sleep onset difficulty type [11]. It is worth noting also that the mother of the propositus was the relative the most often affected [11]. This study confirmed the non negligible family incidence of primary insomnia even if its genetic and/or environmental origin cannot be asserted [11].

Fatal familial insomnia

Fatal familial insomnia is a sleep disorder which has been widely investigated in molecular terms. It is a very rare disease characterised by recalcitrant insomnia, vegetative disorders and mental deterioration [65, 133] rapidly leading to death. The neuropathological lesions are restricted to the atrophy of the dorso-median and anterior nuclei of the thalamus and the inferior olives [28, 133]. The disorder is generally associated with a prion protein gene (PrP) codon 178 mutation sited on chromosome 20 but another association with a mutation on codon 200 has also been described [21]. The same mutations are found in certain forms of dementia of the Creutzfeldt-Jakob type but a second mutation of codon 129 appears to determine fatal insomnia phenotype expression *versus* Creutzfeldt-Jakob [65]. PrP protein is conserved in numerous animal species but its normal function is still unknown. The abnormal form of the prion protein, resistant to protease, is implicated in a group of disorders of the central nervous system (spongiform encephalopathies) where spongiform degeneration is observed with more or less focalised neuron atrophy [89, 176, 237]. These diseases may either appear in a familial context, or in an infectious context, the prion protein being the transmissible agent.

The mechanism by which the prion protein exerts its toxic effects is still unknown. The fact that a simple codon 129 polymorphism not only causes a pathological mutation but can also change the symptomatology to such an extent is also a mystery. The differences in symptomatology are clearly due to a difference in the anatomic distribution of lesions, these being far more diffuse in Creutzfeldt-Jakob disease than in fatal familial insomnia [133, 136]. But the question remains as to how variations if the structure coded by the polymorphism alter the lesional tropism of the pathogenic agent in man [168]. In animals, similar lesions in the thalamus do not lead to lethal insomnia. Moreover, knock-out mice for this gene present abnormalities in sleep continuity and an alteration in homeostatic sleep regulation but with no behaviour disturbance, indicating an important inter-species difference [28, 216, 217]. It is highly likely that thalamic lesions are implicated in the pathophysiology of fatal familial insomnia [28, 132], suggesting that this structure may be implicated in the genesis of other, more common types of insomnia. Insomnia is a very common symptom, affecting at least 10% of the general population. Many insomnias appear to be constitutional and their causes are extremely heterogeneous [79]. Some of these constitutional factors may be genetic implying the abnormal development of the thalamus.

Parasomnias

Sleep walking and night terrors

These parasomnias are considered as arousal disorders and usually occur during NREM sleep (stages III and IV) [104]. These two entities are grouped together because of their frequent association, the similarity of their trigger factors and their very close origins. The prevalence of these parasomnias is close to 10% in children but only rarely warrants consultation; the symptoms usually disappear before adulthood [1, 104]. The familial nature of these parasomnias is recognised by most authors as being in the region of 25% of cases; the exact nature of the mode of transmission is still nevertheless poorly defined [1, 74, 98]. Twin studies have shown marked concordance for sleepwalking with a higher level between monozygotic twins (50%) than for dizygotic twins (under 20% of cases) [6, 88, 93]. Elsewhere, a recent Finnish study on sleepwalking comparing monozygotic and dizygotic twins revealed a strong genetic predisposition both in children and adults and in men as well as women [93]. There seems to be an important overlap in genetic predisposition to sleepwalking, sleep talking and, to a lesser extent, to night terrors [1, 98]. This suggests a close pathophysiological mechanism and a probable similarity in genetic control. A molecular single study has been conducted on sleepwalking to date, which found an association with HLA DQ B1*05 notably for the familial forms, offering a glimpse of a probable genetic predisposition for this parasomnia [122].

REM sleep behaviour disorder

REM sleep behaviour disorder is a relatively rare disorder, more common in elderly subjects and often occurring in a context of neurodegenerative diseases [135, 160]. This parasomnia mainly occurs in the framework of Parkinson's disease, disseminated Lewy body dementia with multi-systematised atrophy, in narcolepsy and in fewer than half the cases in an idiopathic framework [160]. Moreover, the symptomatic character may be secondary [135, 160] A single genetic study has been published to date, reporting an association between this REM sleep parasomnia and HLA-DQ1 or more precisely DQ6 [188].

Bruxism

Bruxism often appears to occur in a family context [77]. This parasomnia is often associated with restless legs syndrome [120]. A study conducted on twins suggests a genetic influence in the incidence of bruxism in view of the higher level of concordance in monozygotic compared to dizygotic twins [88].

Circadian rhythm sleep disorders

Advanced sleep phase syndrome

The existence of mutations responsible for circadian rhythm disorders in the rodent [234] and of polymorphisms in the *clock* gene associated with different circadian typologies in humans [100] suggests that, for the latter, there is genetic determinism in circadian rhythm sleep disorders. The first mutation at the origin of a familial form of advanced sleep phase syndrome was in fact described in man [219]. An advanced sleep phase tendency is often found in the general population (morning type) whereas advanced sleep phase syndrome is a particularly rare disorder. It is characterised by particularly premature sleep onset and abnormally early morning awakening (around 3 – 4 am.) with no associated sleep disorder nor difficulty in awakening. Familial forms are common. A dominant *hper 2* (serine → glycine) gene, the human homologue of the *per* gene in *Drosophila* was found to be responsible for this syndrome in a large kindred with autosomal dominant transmission, and high penetrance [96, 219] (Fig. 7.6).

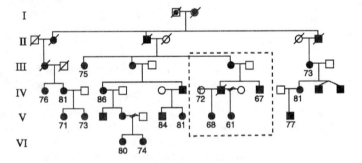

Figure 7.6. Advanced sleep phase syndrome (ASPS) kindred 2174. Horne-Östberg scores are shown below individuals. The dotted line marks a branch (branch 3) where the ASPS phenotype does not cosegregate with the mutation. Circles, women; squares, men; filled circles and squares, affected individuals; empty circles and squares, unaffected individuals. Unknown individuals (not meeting strict criteria for being "affected" or "unaffected") were eliminated from this pedigree for the sake of simplicity. (Reprinted from Science, Vol. 291, Toh K.L. *et al.* An h*Per2* phosphorylation *site* mutation in familial advanced sleep phase syndrome, 1040, Copyright (2001), with permission from Science)

Delayed sleep phase syndrome

Delayed sleep phase syndrome although also rare, is far more common than advanced sleep phase syndrome. Two phenotype forms can be distinguished for this syndrome, the primary form characterised by sleep onset after 2 am. with no associated sleep disorder, and the secondary form associated with difficulties in morning rising, and often of psychopathological origin. The primary form is often familial and there is likely to be an underlying genetic determinism [53]. Two studies have found an association between this condition and HLA DR1 [84, 210]. Moreover, a recent study revealed an association between this delayed sleep phase syndrome and an *hper3* gene polymorphism, another human homologue for the *per* gene, in favour of a genetic predisposition for this disorder [46]. The specific role of *hper 3* in circadian regulation has not been clearly established. PER 3 protein heterodimerizes with PER 1 and 2 and CRY 1 and 2 respectively, the ensemble then entering the nucleus and inhibiting the CLOCK/BMAL 1 transcription factor [117,

241]. A *hper3* phosphorylation disorder may change its function thus altering this complex cellular machinery.

Another team found an association between different polymorphisms of the gene coding for receptor 1a of melatonin (Mel 1a) and the absence of 24 hour sleep-wake rhythm synchronisation (hypernyctohemeral syndrome) with no significant result for the gene coding for receptor 1b (Mel 1b) [44, 45]. These results comply with the selective tropism of the different receptors ; subset 1a of the melatonin receptor being expressed in the suprachiasmatic nucleus (circadian rhythm generator) and subset 1b in the retina. This research on melatonin and its receptors represents a new approach to understanding the mechanisms behind certain sleep disorders and the genetic predisposition of patients suffering from these disorders.

Others

Periodic limb movements and restless legs syndrome

Restless legs syndrome associates discomfort in the legs obliging the patient to get up several times during the night. This syndrome is common (5% of the general population); its prevalence and severity increase with age [47, 147, 148]. It is most often associated with periodic limb movements characterised by rapid jerks in the leg muscles, or more rarely of the arms, at night [147]. Familial forms are common, notably for the idiopathic forms, with in almost half the cases a parent affected to the first degree [121, 148]. The proportion of familial cases appears to vary depending on the geographical origins of the population studied. In Canada, the prevalence and proportion of these families seems to be particularly high; these differences probably relating to a founding effect [121, 148]. The age of onset of the disorder also seems to be younger in the case of familial forms. In up to a third of cases, the disorder is autosomal dominant in transmission [121, 147, 148] with incomplete penetrance and probable anticipation [121, 223]. These variations suggest underlying genetic heterogeneity. A study of monozygotic twins showed a high degree of concordance for restless legs syndrome, 10 of the 12 pairs studied were concordant for the diagnosis although with wide variability in the phenotypes reported both for severity and the age of onset [161].

The aetiology of this complaint is still indeterminate but a dopaminergic mechanism is highly probable. One team focused their research on a possible association between a hydroxylase tyrosin gene polymorphism (a gene which is strongly implicated in the dopaminergic metabolism) and restless legs syndrome, with no significant results [38]. Furthermore, the same team carried out genetic binding analysis in a large kindred of patients (14 patients) affected by restless legs syndrome, to identify the gene(s) implicated [39]. A significant association was distinguished on the long arm of chromosome 12 (maximum lod score of 3.59) [39].

Symptoms non associated with REM sleep

Sleep paralysis and hypnagogic hallucinations are two symptoms of dissociated REM sleep, which often occur in the normal population, aside from narcolepsy of any kind. Isolated sleep paralysis appears to have a high familial incidence, with autosomal dominant transmission with incomplete penetrance and a higher prevalence in black subjects [12, 66, 185]. Twin studies reported higher concordance of this symptom in monozygotic than in dizygotic twins [88]. Unlike narcolepsy, no association was found with HLA DQB1*0602 [31].

Cataplexy without sleepiness is exceptional but rare familial cases have been described with or without associated sleep paralysis [229]. In the case of clearly isolated cataplexy where no other symptom in the narcolepsy tetrade is present, the clinical presentation of this symptom differs from the first episodes of cataplexy in that it is often observed from the first months of life [229]. No study of the association with the HLA system has been reported in these families.

Chromosomal abnormalities

The coexistence of a chromosomal abnormality and a sleep disorder has often led to the localization of the susceptibility gene(s). Unfortunately few sleep studies have been carried out on patients with chromosomal abnormalities and few caryotypes are sought when the sleep disorder is uppermost in the clinical picture. Chromosomal abnormalities are moreover often associated with behavioural effects or produce medical complications which may have side effects on sleep [20]. Despite these limitations, subjects carrying a fragile X chromosome abnormality often have sleep abnormalities and low levels of melatonin [156, 200], whereas subjects affected by Norrie's disease (deletions in the region containing Xp11.3 monoamine oxidase genes) or type C Nieman Pick disease (18q11-q12) may present episodes of cataplexy [112, 165]. The most common case is the observation of sleep apnoeas, these being mainly due to anatomic malformations or to obesity [20, 99]. In certain cases, polygraphic abnormalities nevertheless suggest that central factors in addition to or independent to respiratory factors are implicated, as for example in syndromes like Prader-Willy and Angelman (del (15q)) [99, 204, 230] or Smith-Magendis (del (17p1.2)) [54, 69]. It is also interesting to note that little study has been devoted to sexual chromosome abnormalities [81]. More in-depth studies would be of interest as puberty is associated with a notable reduction in vigilance and an increase in sleep need [20]. Moreover, several disorders like narcolepsy often begin in adolescence and two atypical cases have been reported in the literature of narcolepsy-cataplexy having started at the age of 6 years in a Turner syndrome (XO) [64] conjointly with an abnormally premature puberty.

CONCLUSION

The complexity of sleep-wake cycle regulation, in addition to the many environmental influences also implies a predisposing genetic determinism which is beginning to be understood. Indeed, numerous recent findings demonstrate the role of genes in many physiological and pathophysiological aspects of sleep. A better understanding has been achieved of the regulation of circadian rhythmicity principally as a result of the discovery of new genes over the last decade. The genesis of circadian rhythms can be traced to the cyclical expression of gene activity, in relation to an auto-regulation process that is well conserved between species. Moreover, the differing expression of many other genes according to the various states of sleep and wakefulness, suggests the strong implication of genes at a pre- or post-transcriptional level in the regulation of these states. The identification of most of these genes has enhanced our understanding of sleep physiology and helped to provide better treatment for these disorders. Indeed the complexity of sleep is mirrored by the multitude of disorders, with the result that sleep medicine is a discipline in itself. Most of these disorders are multifactorial and largely genetically determined. Recent progress in molecular genetics and the development of detailed human and animal genomic maps has already led to the identification of the genetic factors responsible for these disorders, and this work continues. In terms of animal research, only a fraction of the genes identified in rodents play a role in a human disorder; nevertheless rodent models will continue to represent an attractive approach to designing well controlled behaviour and genetic studies. In humans, genetic binding analysis studies of a large number of multiplex families affected by a defined disorder or association studies (candidate genes) including a very high number of single case families should be more systematically carried out in the future. Identification must be made of the different genetic factors; these may play a complementary role in the expression of certain insomnias or hypersomnias of probable multifactorial origin. The potential clinical implications of these lines of research are hard to measure as yet, but the discovery of genetic factors with a major effect on certain sleep disorders such as narcolepsy allow new therapeutic strategies to be envisaged.

REFERENCES

1.   ABE K., AMATOMI M., ODA N. - Sleepwalking and recurrent Sleeptalking in children of childhood Sleepwalkers. *Am. J. Psychiatr.*, 141, 800-801, 1984.
2.   ADICKES E.D., BUEHLER B.A., SANGER W.G. - Familial Sleep apnea. *Hum. Genet.*, 73, 39-43, 1986.

3.   ALBRECHT U., SUN Z.S., EICHLE G., LEE C.C. - A differential response of two putative mamalian circadian regulators, mper1 and mper2, to light. *Cell.*, 91, 1055-1064, 1997.
4.   ANOKHIN A., STEINLEIN O., FISCHER C., MAO Y., VOGT P., SCHALT E., VOGEL F. - A genetic study of the human low-voltage electroencephalogram, *Hum. Genet.*, 90, 99, 1992.
5.   ANTOCH M.P., SONG E.J., CHANG A.M., VITATERNA M.H., ZHAO Y., WILSBACHER K.S., SANGORAM A.M., KING D.P., PINTO L.H., TAKAHASHI J.S. - Functional identification of the mouse circadian clock gene by transgenic BAC rescue. *Cell.*, 89, 655-667, 1997.
6.   BAKWIN H. - Sleepwalking in twins. *Lancet.*, 2, 466-467, 1970.
7.   BAILEY D.W. - Definition of inbred strains. *In*: P.L. Altman and D.D. Katz (eds), "Inbred and genetically defined strains of laboratory animals. Part 1. Mouse and rat". FASEB, Bethesda, 4-7, 1979.
8.   BALSALOBRE A., DAMIOLA F., SCHIBLER U. - A serum shock induces circadian gene expression in mammalian tissue culture cells. *Cell.*, 93, 929-937, 1998.
9.   BASHEER R., HALLDNER L., ALANKO L., MC CARLEY R.W., FREDHOLM B.B., PORKKA-HEISKANEN T. - Opposite changes in adenosine A1 and A2A receptor mRNA in the rat following sleep deprivation. *Neuroreport.*, 12, 1577-1580, 2001.
10. BASHEER R., SHERIN J.E., SAPER C.B., MORGAN J.I., MC CARLEY R.W., SHIROMANI P.J. - Effects of sleep on wake-induced c-Fos expression. *J. Neurosci.*, 17, 9746-9750, 1997.
11. BASTIEN C.H., MORIN C.M. - Familial incidence of insomnia. *J. Sleep. Res.*, 9, 49-54, 2000.
12. BELL G.C., DIXIE-BELL D.D., THOMPSON B. - Further studies on the prevalence of isolated sleep paralysis in black subjects. *J. Nat. Med. Ass.*, 78, 7, 649-659, 1986.
13. BENCA R.M., BERGMAN B.M., LEUNG C., NUMMY D., RECHTSCHAFFEN A. - Rat strain differences in response to dark pulse triggering of REM sleep. *Physiol. Behav.*, 49, 83-87, 1991.
14. BILLIARD M., PASQUIE-MAGNETTO V., HECKMAN M., CARLANDER B., BESSET A., ZACHARIEV Z., ELIAOU J.F., MALAFOSSE A. - Family studies in narcolepsy. *Sleep.*, 17, 54-59, 1994.
15. BLAU J., YOUNG M.W. - Cycling vrille expression is required for a functional Drosophila clock. *Cell.*, 99, 661-671, 1999.
16. BOUTREL B., FRANC B., HEN R., HAMON M., ADRIEN J. - Key role of 5-HT1B receptors in the regulation of REM sleep as evidenced in 5-HT1B knock-out mice. *J. Neurosci.*, 19, 3204-3212, 1999.
17. BOVA R., MICHELI M.R., QUAMADRUCCI P., ZUCCONI G.G. - BDNF and trkB mRNAs oscillate in rat brain during the light-dark cycle. *Brain. Res. Mol. Brain. Res.*, 57, 321-324, 1998.
18. BREDOW S., TAISHI P., OBEL F., GUHA-THAKURTA N., KRUEGER J.M. - Hypothalamic growth hormone-releasing hormone mRNA varies across the day in rats. *Neuroreport.*, 7, 2501-2505, 1996.
19. BYRNE E., WHITE O., COOK M. - Familial dystonic choreoathetosis with myokymia: a sleep responsive disorder. *J. Neurol. Neurosurg. Psychiatr.*, 54, 1090, 1991.
20. CARSKADON M.A. - Patterns of sleep and sleepiness in adolescents. *Pediatrician*, 17, 1, 5-12, 1990.
21. CHAPMAN J., ARLAZOROFF A., GOLDFARB L., G. CERVENAKOVA L., NEUFELD M.Y., WEBER E., HERBERT M., BROWN P., GAJDUSEK D.C., KORCZYN A.D. - Fatal Insomnia in a case of familial Creuzfeld-Jakob disease with codon 200(lys) mutation. *Neurology.*, 46, 3, 758-761, 1996.
22. CHEMELLI R.M., WILLIE J.T., SINTON C.M., ELMQUIST J.K., SCAMMELL T., LEE C., RICHARDSON J.A.,WILLIAMS S.C., XIONG Y., KISANUKI Y., FITCH T.E., NAKAZATO M., HAMMER R.E., SAPER C.B., YANAGISAWA M. - Narcolepsy in orexin knockout mice: molecular genetics of sleep regulation. *Cell.*, 98, 437-451, 1999.
23. CHOUVET G., BLOIS R., DEBILLY G., JOUVET M. - The structure of the occurrence of rapid eye movements in REM sleep is similar in homozygotic twins. *C. R. Seances. Acad. Sci. III.*, 296, 1063-1068, 1983.
24. CIRELLI C., POMPEIANO M., ARRIGHI P., TONONI G. - Sleep-waking changes after c-fos antisense injections in the medial preoptic area. *Neuroreport.*, 6, 5, 801-805, 1995.
25. CIRELLI C., POMPEIANO M., TONONI G. - Neuronal gene expression in the waking state: a role for the locus coeruleus. *Science.*, 274, 1211-1215, 1996.
26. CIRELLI C., TONONI G. - Differences in brain gene expression between sleep and waking as revealed by mRNA differential display and cDNA microarray technology. *J. Sleep. Res.*, 8 Suppl 1, 44-52, 1999.
27. CIRELLI C., TONONI G. - Gene expression in the brain across the sleep-waking cycle. *Brain. Res.*, 885, 303-321, 2000.
28. CORTELLI P., GAMBETTI P., MONTAGNA P., LUGARESI E. - Fatal familial insomnia: clinical features and molecular genetics. *J. Sleep. Res.*, 8, Suppl 1, 23-29, 1999.
29. CRABBE J.C., BELKNAP J., BUCK K.J. - Genetic animal models of alcohol and drug abuse. *Science.*, 264, 1715-1723, 1994.
30. CROSTHWAITE S.K., DUNLAP J.C., LOROS J.J. - Neurospora wc-1 and wc-2 transcription, photoresponses, and the origin of circadian rhythmicity. *Science.*, 276, 763-769, 1997.
31. DALHITZ M., PARKES J.D., VAUGHAN R., PAGE G. - The sleep paralysis-excessive daytime sleepiness syndrome. *J. Sleep. Res.*, 1, suppl 1, A52, 1992.
32. DAUVILLIERS Y., MAYER G., LECENDREUX M., NEIDHART E., CHAPPUIS R., BILLIARD M., TAFTI M. - HLA and monoaminergic gene polymorphisms in the Kleine-levin syndrome. *J. Sleep. Res.*, 9, suppl 1, A95, 2000.
33. DAUVILLIERS Y., NEIDHART E., LECENDREUX M., BILLIARD M., TAFTI M. - MAO-A and COMT polymorphisms and gene effects in narcolepsy. *Mol. Psychiatry.*, 6, 367-372, 2001.
34. DE FIEBRE C.M., COLLINS A. - Classical genetic analyses of responses to nicotine and ethanol in crosses derived from long-and short-sleep mice. *J. Pharmacol. Exp. Ther.*, 261, 1, 173-180, 1992.

35. DE LECEA L., CRIADO J.R., PROSPERO-GARCIA O., GAUTVIK K.M., SCHWEITZER P., DANIELSON P.E., DUNLOP C.L., SIGGINS G.R., HENRIKSEN S.J., SUTCLIFFE J.G. - A cortical neuropeptide with neuronal depressant and sleep-modulating properties. *Nature.*, 381, 242-245, 1996.

36. DE LECEA L., KILDUFF T.S., PEYRON C., GAO X., FOYE P.E., DANIELSON P.E., FUKUHARA C., BATTENBERG E.L., GAUTVIK V.T., BARTLETT F.S., FRANKEL W.N., VAN DEN POL A.N., BLOOM F.E., GAUTVIK K.M., SUTCLIFFE J.G. - The hypocretins: hypothalamus-specific peptides with neuroexcitatory activity. *Proc. Natl. Acad. Sci. U S A.*, 95, 322-327, 1998.

37. DEMATTEIS M., PEPIN J.L., JEANMART M., DESCHAUX C., LABARRE-VILA A., LEVY P. - Charcot-Marie-Tooth disease and sleep apnoea syndrome: a family study. *Lancet.*, 357, 267-272, 2001.

38. DESAUTELS A., TURECKI G., MONTPLAISIR J., FTOUHI-PAQUIN N., MICHAUD M., CHOUINARD V.A., ROULEAU G.A. - Dopaminergic neurotransmission and restless legs syndrome: A genetic association analysis. *Neurology.*, 57, 1304-1306, 2001.

39. DESAUTELS A., TURECKI G., MONTPLAISIR J., SEQUEIRA A., VERNER A., ROULEAU G.A. - Identification of a major susceptibility locus for restless legs syndrome on chromosome 12q. *Am. J. Hum. Genet.*, 69, 1266-1270, 2001.

40. DRENNAN M.D., SHELBY J., KRIPKE D.F., KELSOR J., GILLIN J.C. - Morningness/eveningness is heritable. Abstracts Book: *Soc. Neurosci.*, 8, 196, 1992.

41. DUDEK B.C., ABBOTT M.E. - A biometrical genetic analysis of ethanol response in selectively bred long-sleep and short-sleep mice. *Behav. Genet.*, 14, 1-19, 1984.

42. DUNLAP J.C. - The genetic and molecular dissection of a prototypic circadian system. *Ann. Rev. Genet.*, 30, 579-601, 1996.

43. DUNLAP J.C. - Molecular bases for circadian clocks. *Cell.*, 96, 271-290, 1999.

44. EBISAWA T., KAJIMURA N., UCHIYAMA M., KATOH M., SEKIMOTO M., WATANABE T., OZEKI Y., IKEDA M., JODOI T., SUGISHITA M., IWASE T., KAMEI Y., KIM K., SHIBUI K., KUDO Y., YAMADA N., TOYOSHIMA R., OKAWA M., TAKAHASHI K., YAMAUCHI T. - Allelic variants of human melatonin 1a receptor: function and prevalence in subjects with circadian rhythm sleep disorders. *Biochem. Biophys. Res. Commun.*, 262, 832-837, 1999.

45. EBISAWA T., UCHIYAMA M., KAJIMURA N., KAMEI Y., SHIBUI K., KIM K., KUDO Y., IWASE T., SUGISHITA M., JODOI T., IKEDA M., OZEKI Y., WATANABE T., SEKIMOTO M., KATOH M., YAMADA N., TOYOSHIMA R., OKAWA M., TAKAHASHI K., YAMAUCHI T. - Genetic polymorphisms of human melatonin 1b receptor gene in circadian rhythm sleep disorders and controls. *Neurosci. Lett.*, 280, 29-32, 2000.

46. EBISAWA T., UCHIYAMA M., KAJIMURA N., MISHIMA K., KAMEI Y., KATOH M., WATANABE T., SEKIMOTO M., SHIBUI K., KIM K., KUDO Y., OZEKI Y., SUGISHITA M., TOYOSHIMA R., INOUE Y., YAMADA N., NAGASE T., OZAKI N., OHARA O., ISHIDA N., OKAWA M., TAKAHASHI K., YAMAUCHI T. - Association of structural polymorphisms in the human period3 gene with delayed sleep phase syndrome. *EMBO. Rep.*, 2, 342-346, 2001.

47. EKBOM K. - Restless legs syndrome. *Neurology*, 10, 868-873, 1960.

48. EL BAYADI S., MILLMAN R.P., TISHLER P.V., ROSENBERG C., SALISKI W., BOUCHER M.A., REDLINE S. - A family study of sleep apnea. Anatomic and physiologic interactions. *Chest*, 98, 3, 554-555, 1990.

49. ERWIN V.G., KORTE A., JONES B.C. - Central Muscarinic Cholinergic influences on ethanol sensitivity in long-sleep and short-sleep mice. *J. Pharmacol. Exp. Ther.*, 247, 3, 857-862, 1988.

50. ERWIN V.G., JONES B.C., RADCLIFFE R. - Further characterisation of LS x SS recombinant inbred strains of mice activating and hypothermic effects of ethanol. *Alcoholism. Clin. Exp. Res.*, 14, 2, 200-204, 1990.

51. FANG J., WANG Y., KRUEGER J.M. - Mice lacking the TNF 55 kDa receptor fail to sleep more after TNFalpha treatment. *J. Neurosci.*, 17, 5949-5955, 1997.

52. FERINI-STRAMBI L., CALORI G., OLDANI A., DELLA MARCA G., ZUCCONI M., CASTRONOVO V., GALLUS G., SMIRNE S. - Snoring in twins. *Respir. Med.*, 89, 337-360, 1995.

53. FINK R., ANCOLI-ISRAEL S. - Pedigree of one family with delayed sleep phase syndrome. *Sleep Res.*, 26, 713, 1997.

54. FISCHER H., OSWALD H.P., DUBA H.C., DOCZY L., SIMMA B., UTERMANN G., HAAS O.A. - Constitutional interstitial deletion of 17 (p11.2) (Smith Magenis syndrome): A clinically recognizable microdeletion syndrome. Report of two cases and review of the literature. *Klin. Padiatr.*, 205, 3, 162-163, 1993.

55. FLINT J., MOTT R. - Finding the molecular basis of quantitative traits: successes and pitfalls. *Nat. Rev. Genet.*, 2, 437-445, 2001.

56. FRANKEN P., CHOLLET D., TAFTI M. - The homeostatic regulation of sleep need is under genetic control. *J. Neurosci.*, 21, 2610-2621, 2001.

57. FRANKEN P., LOPEZ-MOLINA L., MARCACCI L., SCHIBLER U., TAFTI M. - The transcription factor DBP affects circadian sleep consolidation and rhythmic EEG activity. *J. Neurosci.*, 20, 617-625, 2000.

58. FRANKEN P., MALAFOSSE A., TAFTI M. - Genetic determinants of sleep regulation in inbred mice. *Sleep.* 22, 155-169, 1999.

59. FRANKEN P., MALAFOSSE A., TAFTI M. - Genetic variation in EEG activity during sleep in inbred mice. *Am. J. Physiol.*, 275, 1127-1137, 1998.

60. FRIEDMANN J.K. - A diallelic analysis of the genetic underpinnings of mouse sleep. *Physiol. Behav.*, 12, 169-175, 1974.

61. GEDDA L., BRENCI G. - Twins living apart test : Progress report. *Acta. Genet. Med. Gerontol.*, 32, 17-22, 1983.

62. GELINEAU J. - De la narcolepsie. *Gazette des hôpitaux de Paris*, 53, 626-628 et 54, 635-637, 1880.

63. GENCIK M., DAHMEN N., WIECZOREK S., KASTEN M., BIERBRAUER J., ANGHELESCU I., SZEGEDI A., MENEZES SAECKER A.M., EPPLEN J.T. - A prepro-orexin gene polymorphism is associated with narcolepsy. *Neurology.*, 56, 115-117, 2000.

64. GEORGE C.F., SINGH S.M. - Juvenile onset narcolepsy in an individual with Turner syndrome. A case report. *Sleep.*, 14, 3, 267-269, 1991.

65. GOLDFARB L.G., PETERSEN R.B., TABATON M., BROWN P., LE BLANC A., MONTAGNA P., CORTELLI P., JULIEN J., VITAL C., PENOELBURY W.W., HALTIA M., WILLS P.R.,HAUW J.J., McKEEVER P.E., MONARI L., SCHRANCK B., SWERGOLD G.D., AUTILIO-GAMBETTI L., GAJDUSEK D.C.,LUGARESI E., GAMBETTI P. - Fatal Familial Insomnia and Familial Creutzfeldt-Jakob Disease : A disease phenotype determined by a DNA polymorphism. *Science.*, 258, 806-808, 1992.

66. GOODE G.B. - Sleep paralysis. *Arch. Neurol.*, 6, 3, 228-234, 1962.

67. GRASSI-ZUCCONI G., GIUDITTA A., MANDILE P., CHEN S., VESCIA S., BENTIVOGLIO M. - c-fos spontaneous expression during wakefulness is reversed during sleep in neuronal subsets of the rat cortex. *J. Physiol.*, 88, 91-93, 1994.

68. GRASSI-ZUCCONI G., MENEGAZZI M., CARCERERI DE PRATI A., BASSETTI A., MONTAGNESE P., MANDILE P., COSI C., BENTIVOGLIO M. - c-fos mRNA is spontaneously induced in the rat brain during the activity period of the circadian cycle. *Euro. J. Neurosci.*, 5, 1071-1078, 1993.

69. GREENBERG F., GUZZETTA V., MONTES DE OCA-LUNA R., MAGENIS E., SMITH A.C.M., RICHTER S.F., KONDO I., DOBYNS W.B., PATEL P.I., LUPSKI J.R. - Molecular analysis of the Smith-Magenis Syndrome, a possible continous syndrome associated with del (17) (p11.2). *Am. J. Hum Genet.*, 49, 1207-1218, 1991.

70. GUILLEMINAULT C., ELDRIDGE F.L., SIMMONS F.B. - Sleep apnea syndrome. Can it induce hemodynamic changes? *West. J. Med.*, 123, 7-16, 1975.

71. GUILLEMINAULT C., MIGNOT E., GRUMET F.C. - Familial patterns of narcolepsy. *Lancet.*, 335,1376-1379, 1989.

72. GUILLEMINAULT C., PARTINEN M., HOLLMAN K., POWELL N., STOOHS R. - Familial aggregates in Obstructive Sleep Apnea Syndrome. *Chest.*, 107, 1545, 1995.

73. HALL J. C. - Tripping along the trail to the molecular mechanism of biological clocks. *Trends. Neurosc.*, 18, 230-240, 1995.

74. HALSTROM T. - Night terror in adults through three generations. *Acta. Psychiat. Scand.*, 48, 350-352, 1972.

75. HARA J., BEUCKMANN C.T., NAMBU T., WILLIE J.T., CHEMELLI R.M., SINTON C.M., SUGIYAMA F., YAGAMI K., GOTO K., YANAGISAWA M., SAKURAI T. - Genetic ablation of orexin neurons in mice results in narcolepsy, hypophagia, and obesity. *Neuron.*, 30, 345-354, 2001.

76. HARDIN P.E., HALL J.C., ROSBASH M. - Feedback of the Drosophilia period gene product on circadian cycling of its messenger RNA levels. *Nature.* 343, 6258, 536-540, 1990.

77. HARTMAN E. - Bruxism. *In*: M. Kryeger, T. Roth, W.C. Dement (eds). *Principles and practice of sleep medecine.* WB Saunders, Philadelphia, 385-388, 1989.

78. HASEGAWA Y., MORISHITA M., SUZUMURA. A. - Novel chromosomal abberation in a patient with a unique sleep disorder. *J. Neurol. Neurosurg. Psychiatry.*, 64, 1, 113-116, 1998.

79. HAURI P. - Primary insomnia. *In*: M. Kryeger, T. Roth, W.C. Dement (eds.). *Principles and practice of sleep medicine.* W.B. Saunders, Philadelphia, 442-447, 1989.

80. HEATH A., KENDLER K.S., EAVES L.J., MARTIN N.G. - Evidence for genetic influences on sleep disturbance and sleep patterns in twins. *Sleep.*, 13, 318-335, 1990.

81. HIGURASHI M., KAWAI H., SEGAWA M., IIJIMA K., IKEDA Y., TANAKA F., EGI S., KAMASHITA S., GROWTH S. - Psychologic characteristics, and sleep-wakefulness cycle of children with sex chromosomal abnormalities. *Birth. Defects*, 22, 3,251-275,1986.

82. HOFSTETTER J.R., MAYEDA A.R., POSSIDENTED B., NURNBERGER J.I. - Quantitative trait loci for circadian rhythms of locomotor activity in mice. *Behav. Genet.* 25, 6, 545-546, 1995.

83. HOHJOH H., NAKAYAMA T., OHASHI J., MIYAGAWA T., TANAKA H., AKAZA T., HONDA Y., JUJI T., TOKUNAGA K. - Significant association of a single nucleotide polymorphism in the tumor necrosis factor-alpha gene promoter with human narcolepsy. *Tissue. Antigens.*, 54, 138-145, 1999.

84. HOHJOH H., TAKAHASHI Y., HATTA Y., TANAKA H., AKAZA T., TOKUNAGA K., HONDA Y., JUJI T. - Possible association of human leucocyte antigen DR1 with delayed sleep phase syndrome. *Psychiatry. Clin. Neurosci.*, 53, 527-529, 1999.

85. HOHJOH H., TERADA N., KAWASHIMA M., HONDA Y., TOKUNAGA K. - Significant association of the tumor necrosis factor receptor 2 (TNFR2) gene with human narcolepsy. *Tissue. Antigens.*, 56, 446-448, 2000.

86. HONDA Y., ASAKA A., TANAKA Y., JUJI T. - Discrimination of narcolepsy by using genetic markers and HLA. *Sleep Res.*, 12, 254, 1983.

87. HONRADO G.I., JOHNSON R.S., GOLOMBEK D.A., SPIEGELMAN B.M., PAPAIOANNOU V.E., RALPH M.R. - The circadian system of c-fos deficient mice. *J. Comp. Physiol.*, 178, 4, 563-570, 1996.

88. HORI A., HIROSE G. - Twin studies on parasomnias. *Sleep. Res.*, 24A, 324, 1995.

89. HORWICH A.L., WEISSMAN J. S. - Deadly conformations-protein misfolding in prion disease. *Cell.*, 89, 4, 499-510, 1997.

90. HUANG Z.J., EDERY I., ROSBACH M. - PAS is a dimerisation domain common to Drosophila Perido and several transcription factors. *Nature.*, 364, 259-262, 1993.

91. HUBER R., DEBOER T., TOBLER I. - Prion protein: a role in sleep regulation?. *J Sleep. Res.*, 8, Suppl 1, 30-36, 1999.

92. HUBLIN C., KAPRIO J., PARTINEN M., HEIKKI K., KOSKENVUO M. - Prevalence and genetics of sleepwalking: a population based twin study. *Neurology.*, 48, 1, 177-181, 1997.

93. HUBLIN C., KAPRIO J., PARTINEN M., KOSKENVUO M. - Sleeptalking in twins: epidemiology and psychiatric comorbidity. *Behav. Genet.*, 28, 289-298, 1998.

94. HUNGS M., LIN L., OKUN M., MIGNOT E. - Polymorphisms in the vicinity of the hypocretin/orexin are not associated with human narcolepsy. *Neurology.*,57, 1893-1895, 2001.

95. JIN X., SHEARMAN L.P., WEAVER D.R., ZYLKA M.J., DE VRIES G.J., REPPERT S.M. - A molecular mechanism regulating rhythmic output from the suprachiasmatic circadian clock. *Cell.*, 96, 57-68, 1999.

96. JONES C.R., CAMPBELL S.S., ZONE S.E., COOPER F., DE SANO A., MURPHY P.J., JONES B., CZAJKOWSKI L., PTACEK L.J. - Familial advanced sleep-phase syndrome: A short-period circadian rhythm variant in humans. *Nat. Med.*, 5, 1062-1065, 1999.

97. KADOTANI H., KADOTANI T., YOUNG T., PEPPARD P.E., FINN L., COLRAIN I.M., MURPHY G.M.Jr, MIGNOT E. - Association between apolipoprotein E epsilon4 and sleep-disordered breathing in adults. *JAMA.*, 285, 2888-2890, 2001.

98. KALES A., SOLDATOS C.R., BIXLER E.O., LADDA R.L., CHARNEY D.S., WEBER G., SCHWEITZER P.K. - Hereditary factors in sleepwalking and night terrors. *Br. J. Psychiatr.*, 137,111-118,1980.

99. KAPLAN J., FREDERICKSON P.A., RICHARDSON J.W. - Sleep and breathing in patients with the Prader-Willi syndrome. *Mayo. Clin. Prac.*, 66,1124-1126, 1991.

100. KATZENBERG D., YOUNG T., FINN L., LIN L., KING D.P., TAKAHASHI J.S., MIGNOT E. - A CLOCK polymorphism associated with human diurnal preference. *Sleep.*, 21, 569-576, 1998.

101. KATZENBERG D., YOUNG T., LIN L., FINN L., MIGNOT E. - A human period gene (HPER1) polymorphism is not associated with diurnal preference in normal adults. *Psychiatr. Genet.*, 9, 107-109, 1999.

102. KAWAKAMI Y., YAMAMOTO H., YOSHIKAWA T., SHIDA A. - Chemical and behavioral control of breathing in twins. *Am. Rev. Resp. Dis.* 129, 703, 1982.

103. KAY S.A. - PAS, present future: clues to the origins of circadian clocks. *Science.*, 276, 753-754, 1997.

104. KEEFAUVER S.P., GUILLEMINAULT C. - Sleep Terrors and sleep walking. *In*: M. Kryeger, T. Roth, W.C. Dement (eds). *Principles and practice of sleep medicine*, 2nd ed, W.B. Saunders, Philadelphia, 567-573, 1994.

105. KEESLER G.A., CAMACHO F., GUO Y., VIRSHUP D., MONDADORI C., YAO Z. - Phosphorylation and destabilization of human period I clock protein by human casein kinase I epsilon. *Neuroreport.*, 11, 951-955, 2000.

106. KENDLER K.S., HEATH A.C., MARTIN N.G., EAVES L.J. - Symptoms of anxiety and symptoms of depression: Same genes, different environments? *Arch. Gen. Psychiatry*, 122, 451-457, 1987.

107. KHANNA J.M., KALANT H. - Effects of chronic treatment with ethanol on the development of cross-tolerance to other alcohols and pentobarbital. *J. Pharmacol. Exp. Ther.*, 263, 2, 480-485, 1992.

108. KING D.P., ZHAO Y., SANGORAM A.M., WILSBACHER L D., TANAKA M., ANTOCH M.P., STEEVES T.D.L., VITATERNA M.H., KORNHAUSER J.M., LOWERY P.L., TUREK F.W., TAKAHASHI J.S. - Positional cloning of the mouse circadian *Clock* gene. *Cell.*, 89, 17, 641-653, 1997.

109. KITAHAMA K., VALATX J.L. - Instrumental and pharmacological REM sleep deprivation in mice: strain differences, *Neuropharmacology.*, 19, 529-535, 1980.

110. KLEIN D., MOORE R.Y., REPPERT S.M. - Suprachiasmatic nucleus. *The mind's clock.* Oxford University Press, New York, 467, 1991.

111. KLOSS B., PRICE J.L., SAEZ L., BLAU J., ROTHENFLUH A., WESLEY C.S., YOUNG M.W. - The Drosophila clock gene double-time encodes a protein closely related to human casein kinase Iepsilon. *Cell.*, 94, 97-107, 1998.

112. KOCH H., CRAIG I., DAHLITZ M., DENNEY R., PARKES D. - Analysis of the monoamine oxidase genes and the Norrie disease gene locus in narcolepsy. *Lancet.*, 353, 645, 1999.

113. KORPI E.R., KLEINGOOR C., KETTENMANN H., SEEBURG P.H. - Benzodiazepine-induced motor impairment linked to point mutation in cerebellar GABA A receptor. *Nature.*, 361, 356-359, 1993.

114. KRAMER A., YANG F.C., SNODGRASS P., LI X., SCAMMELL T.E., DAVIS F.C., WEITZ C.J. - Regulation of daily locomotor activity and sleep by hypothalamic EGF receptor signaling. *Science.*, 294, 2511-2515, 2001.

115. KRONHOLM E., AUNOLA S., HYYPPA M.T., KAITSAARI M., KOSKENVUO M., MATTLAR C.E., RANNEMAA T. - Sleep in monozygotic twin pairs discordant for obesity. *J. Appl. Physiol.*, 80, 1, 14-19, 1996.

116. KRONOPKA R.J., BENZER S. - Clock mutants of drosophila melanomaster. *Proc. Natl. Acad Sci.* (USA), 68, 2112-2116, 1971.

117. KUME K., ZYLKA M.J., SRIRAM S., SHEARMAN L.P., WEAVER D.R., JIN X., MAYWOOD E.S., HASTINGS M.H., REPPERT S.M. - mCRY1 and mCRY2 are essential components of the negative limb of the circadian clock feedback loop. *Cell.*, 98, 193-205, 1999.

118. KUSHIDA C.A., GUILLEMINAULT C., MIGNOT E., AHMED O., WON C., O'HARA B., CLERK A.A. - Genetics and craniofacial dysmorphism in family studies of obstructive sleep apnea. *Sleep. Res.*, 25, 275, 1996.

119. KUSHIDA C.A., ZOLTOSKI R.K., GILLIN J.C. - The expression of m1-m3 muscarinic receptor mRNAs in rat brain following REM sleep deprivation. *Neuroreport.*, 6, 1705-1708, 1995.

120. LAVIGNE G., MONTPLAISIR J. - Restless legs syndrome and sleep bruxism: prevalence and association among Canadians. *Sleep.*, 17, 739-743, 1994.

121. LAZZARINI A., WALTERS A.S., HICKEY K., COCCAGNA G., LUGARESI E., EHRENBERG B.L., PICCHIETTI D.L., BRIN M.F., STENROOS E.S., VERRICO T., JOHNSON W.G. - Studies of penetrance and anticipation in five autosomal-dominant restless legs syndrome pedigrees. *Mov. Disord.*, 14, 111-116, 1999.

122. LECENDREUX M., MAYER G., BASSETTI C., NEIDHART E., CHAPPUIS R., TAFTI M. - Genetic susceptibility to sleepwalking: a candidate gene approach. *J. Sleep. Res.*, 9, suppl 1, A227, 2000.

123. LEE C., BAE K., EDERY I. - PER and TIM inhibit the DNA binding activity of a Drosophila CLOCK-CYC/dBMAL1 heterodimer without disrupting formation of the heterodimer: a basis for circadian transcription. *Mol. Cell. Biol.*, 19, 5316-5325, 1999.
124. LEUNG C., BERGMANN B.M., RECHTSCHAFFEN A., BENCA R.M. - Heritability of dark pulse triggering of REM sleep in rats. *Physiol. Behav.*, 52, 127-131, 1994.
125. LE MINH N., DAMIOLA F., TRONCHE F., SCHUTZ G, SCHIBLER U. - Glucocorticoid hormones inhibit food-induced phase-shifting of peripheral circadian oscillators. *EMBO J.*, 20, 7128-7136, 2001.
126. LIN L., FARACO J., LI R., KADOTANI H., ROGERS W., LIN X., QIU X., DE JONG P.J., NISHINO S., MIGNOT E. - The sleep disorder canine narcolepsy is caused by a mutation in the hypocretin (orexin) receptor 2 gene. *Cell.*, 98, 365-376, 1999.
127. LINKOWSKY P., KERKHOFS M., HAUSPIE R., MENDLEWICZ J. - Genetic determinants of EEG sleep : A study in twins living apart. *Electroenceph. Clin. Neurophysiol.*, 79, 114-118,1991.
128. LINKOWSKY P., KERKHOFS M., HAUSPIE R., SUSANNE C., MENDLEWICZ J. - EEG sleep patterns in man: a twin study. *Electroenceph. Clin. Neurophysiol.*, 73, 279-284, 1989.
129. LINKOWSKY P., KERKHOFS M., VAN CAUTER E. - Sleep and biological rhythms in man : a twin study. *Clin. Neuropharmacol.*, 15, Suppl 1A, 42-43, 1992.
130. LIU C., WEAVER D.R., STROGATZ S.H., REPPERT S.M. - Cellular construction of a circadian clock: period determination in the suprachiasmatic nuclei. *Cell.*, 91, 855-860, 1997.
131. LOWREY P.L., SHIMOMURA K., ANTOCH M.P., YAMAZAKI S., ZEMENIDES P.D., RALPH M.R., MENAKER M., TAKAHASHI J.S. - Positional syntenic cloning and functional characterization of the mammalian circadian mutation tau. *Science.*, 288, 483-492, 2000.
132. LUGARESI E., MEOGRI R., MONTAGNA P., BARUZZI A., CORTELLI P., LUGARESI A., TINUPER P., ZUCCONI M., GAMBETTI P. - Fatal familial insomnia and dysautonomia with selective degeneration of thalamic nuclei. *N. Engl. J.Med.*, 315, 997-1003, 1986.
133. LUGARESI E., TOBLER I., GAMBETTI P., MONTAGNA P. - The pathophysiology of fatal familial insomnia. *Brain. Pathol.*, 8 suppl 1, 521-526, 1998.
134. MACKIEWICZ M., SOLLARS P.J., OGILVIE M.D., PACK A.I. - Modulation of IL-1 beta gene expression in the rat CNS during sleep deprivation. *Neuroreport.*, 7, 529-533, 1996.
135. MAHOWALD M.W., SCHENCK C.H. - REM sleep behavior disorder. *In*: M. Kryeger, T. Roth, W.C. Dement (eds). *Principles and practice of sleep medicine*, 2nd ed., W.B. Saunders, Philadelphia, 574-588, 1994.
136. MANETTO V., MEDORI R., CORTELLI P., MONTAGNA P., TINUPER P., BARUZZI A, RANCUREL G., HAUW J.J., VANDERHAEGEN J.J., MAILLEUX P., BUGIANI O., TAGLIAVINI F., BOURAS C., RIZZUTO N., LUGARESI F., GAMBETTI P. - Familial Fatal Insomnia: Clinical and pathological study of five new cases. *Neurology.*, 42, 312-319, 1992.
137. MANNI R., ZUCCA C., MARTINETTI M., OTTOLINI A., LANZI G., TARTARA A. - Hypersomnia in dystrophia myotonica: a neurophysiological and immunogenetic study. *Acta. Neurol. Scand.*, 84, 498-562, 1991.
138. MARINI G., IMERI L., MANCIA M. - Changes in sleepwaking cycle induced by lesions of medialis dorsalis thalamic nuclei in the cat. *Neurosci. Lett.*, 85, 223-227, 1988.
139. MARKEL P.D., FULKER D.W., BENNET B., CORLEY R.P., DEFRIES J.C., ERWIN V.G., JOHNSON T.E. - Quantitive trait loci for ethanol sensitivity in the LS X SS recombinant inbred stains: interval mapping. *Behav. Genet.*, 26, 4, 447-458, 1996.
140. MARLEY R.J., FREUND R.K., WHENER J.M. - Differential response to flurazepam in long-sleep and short-sleep mice. *Pharmacol. Biochem. Behav.*, 31, 453-458, 1988.
141. MATHUR R., DOUGLAS N.J. - Family study in patients with the sleep Apnea-Hypopnea syndrome. *Ann. Intern. Med.*, 122, 174, 1995.
142. MAYEDA A.R., HOFSTETTER J.R., BELKNAP J.R., NURNBERGER JR J.I. - Hypothetical quantitative trait loci (QTL) for circadian period of locomotor activity in CxB recombinant inbred strains of mice. *Behav. Genet.*, 26, 5, 505-511, 1996.
143. MIGNOT E. - Genetic and familial aspects of narcolepsy. *Neurology.*, 50, Suppl 1, S16-S22, 1998.
144. MIGNOT E., LIN L., ROGERS W., HONDA Y., QIU X., LIN X., OKUN M., HOHJOH H., MIKI T., HSU S., LEFFELL M., GRUMET F., FERNANDEZ-VINA M., HONDA M., RISCH N. - Complex HLA-DR and -DQ interactions confer risk of narcolepsy-cataplexy in three ethnic groups. *Am. J. Hum. Genet.*, 68, 686-699, 2001.
145. MILLAR A.J., CARRE I.A., STRAYER C.A., CHAU N.H., KAY S.A. - Circadian Clock mutants in Arabidopsis identified by luciferase imaging. *Science.*, 267, 1161-1166, 1995.
146. MONTAGNA P., COCCAGNA G., CIRIGNOTTA F., LUGARESI E. - Familial restless leg syndrome. *In*: C. Guilleminault E. Lugaresi (eds.). *Sleep Wake Disorders: Natural History. Epidemiology and long term evolution*, Raven Press, New York, 231-235, 1983.
147. MONTPLAISIR J., BOUCHER S., POIRIER G., LAVIGNE G., LAPIERRE O., LESPERANCE P. - Clinical, polysomnographic, and genetic characteristics of restless legs syndrome: a study of 133 patients diagnosed with new standard criteria. *Mov. Disord.*, 12, 61-67, 1997.
148. MONTPLAISIR J., GODBOUT R., BOGHDEN D., POIRIER G. - Familial restless legs with periodic movements in sleep. Electrophysiological, biochemical, and pharmacological study. *Neurology.*, 35, 130-134, 1985.
149. MORGAN J.I., CURRAN T. - Stimulus-transcription coupling in the nervous system: involvement of the inducible proto-oncogenes *fos* and *jun*. *Ann. Rev. Neurosci.*, 14, 421-451, 1991.
150. MOTT R., TALBOT C.J., TURRI M.G., COLLINS A.C., FLINT J. - From the cover: a method for fine mapping quantitative trait loci in outbred animal stocks. *Proc. Natl. Acad. Sci. U S A.*, 97, 12649-12654, 2000.
151. NADEAU J.H. - Modifier genes in mice and humans. *Nat. Rev. Genet.*, 2, 165-174, 2001.

152. NAKAYAMA J., MIURA M., HONDA M., MIKI T., HONDA Y., ARINAMI T. - Linkage of human narcolepsy with HLA association to chromosome 4p13-q21. *Genomics.*, 65, 84-86, 2000.

153. NAYLOR E., BERGMANN B.M., KRAUSKI K., ZEE P.C., TAKAHASHI J.S., VITATERNA M.H., TUREK F.W. - The circadian clock mutation alters sleep homeostasis in the mouse. *J Neurosci.*, 20, 8138-8143, 2000.

154. NEUNNER-JEHLE M., RHYNER T.A., BORBELY A.A. - Sleep deprivation differentially alters the mRNA and protein levels of neurogranin in rat brain. *Brain. Res.*, 685, 1-2, 143-153, 1995.

155. NOLAN P., SOLLARS P.J., BOHNE B.A., EWENS, W.J., PICKARD G.E., BUCAN M. - Heterozygosity mapping of partially congenic lines; mapping of a semi dominant neurological mutation, *Wheels*, on mouse chromosome 4. *Genetics.*, 140, 245-254, 1995.

156. O'HARE J.P., O'BRIEN I.A., ARENDT J., ASTLEY P., RATCLIFFE W., ANDREWS H., WALTERS R., CORRALL R.J. - Does melatonin deficiency cause the enlarged genitalia of the fragile-X syndrome?. *Clin. Endocrinol.*, 24, 3, 327-333, 1986.

157. O'HARA B.F., YOUNG K. A., WATSON F.L., HELLER H.C., KILDUFF T.S. - Immediate early gene expression in brain during sleep deprivation: preliminary observations. *Sleep.*, 16, 1-7, 1993.

158. O'HARA B.F., WATSON F.L., ANDRETIC R., WILER S.W., YOUNG K.A., BITTING L., HELLER H. C., KILDUFF T.S. - Daily variation of CNS gene expression in nocturnal vs. diurnal rodents and in the developing rat brain. *Mol. Brain. Res.*, 48, 73-86, 1997.

159. OLAFSDOTTIR B.R., RYE D.B., SCAMMELL T.E., MATHESON J.K., STEFANSSON K., GULCHER J.R. - Polymorphisms in hypocretin/orexin pathway genes and narcolepsy. *Neurology.*, 57, 1896-1899, 2001.

160. OLSON E.J., BOEVE B.F., SILBER M.H. - Rapid eye movement sleep behaviour disorder: demographic, clinical and laboratory findings in 93 cases. *Brain.*, 123 (Pt 2), 331-339, 2000.

161. ONDO W.G., VUONG K.D., WANG Q. - Restless legs syndrome in monozygotic twins: clinical correlates. *Neurology.*, 55, 1404-1406, 2000.

162. OREN J., KELLY D.R., SHANNON D.C. - Familial occurence of sudden infant death syndrome and apnea of infancy. *Pediatrics.*, 80, 355-358, 1987.

163. OVERSTREET D.H., REZVANI A.H., JANOWSKY D.S. - Increased hypothermic responses to ethanol in rats selectively bred for cholinergic supersensitivity. *Alcohol. Alcoholism.* 25, 1, 59-65, 1990.

164. PARK J.H., HELFRICH-FORSTER C., LEE G, LIU L., ROSBASH M., HALL J.C. - Differential regulation of circadian pacemaker output by separate clock genes in Drosophila. *Proc. Natl. Acad. Sci. U S A.*, 97, 3608-3613, 2000.

165. PARKES JD. - Genetic factors in human sleep disorders with special reference to Norrie disease, Prader-Willi syndrome and Moebius syndrome. *J. Sleep. Res.*, 8, Suppl 1, 14-22, 1999.

166. PARTINEN M., KAPRIO J., KOSKENVUO M., PUTKONEN P., LANGINVAINIO H. - Genetic and Environnmental determination of human sleep. *Sleep.*, 6, 179-185, 1983.

167. PEDRAZZOLI M., LING L., FINN L., KUBIN L., YOUNG T., KATZENBERG D., MIGNOT E. - A Polymorphism in the human *timeless* gene is not associated with diurnal preferences in normal adults. *Sleep. Res. Online.*, 3, 73-76, 2000.

168. PETERSEN R.B., PARCHI P., RICHARDSON S.L, URIG C.B. GAMBETTI P. - Effect of the D178N mutation and the codon 129 polymorphism on the prion protein. *J. Biol. Chem.*, 271, 21, 12661-12668, 1996.

169. PEYRON C., FARACO J., ROGERS W., RIPLEY B., OVEREEM S., CHARNAY Y., NEVSIMALOVA S., ALDRICH M., REYNOLDS D., ALBIN R., LI R., HUNGS M., PEDRAZZOLI M., PADIGARU M., KUCHERLAPATI M., FAN J., MAKI R., LAMMERS G.J., BOURAS C., KUCHERLAPATI R., NISHINO S., MIGNOT E. - A mutation in a case of early onset narcolepsy and a generalized absence of hypocretin peptides in human narcoleptic brains. *Nat. Med.*, 6, 991-997, 2000.

170. PHILLIPS T.J., FELLER D.J., CRABBE J.C. - Selected mouse lines, alcohol and behavior. *Experientia.*, 45, 805-827, 1989.

171. PILLAR G., LAVIE P. - Assessment of the role of inherence in sleep apnea syndrome. *Am. J. Respir. Crit. Care. Med.*, 151, 688-691, 1995

172. PINZAR E., KANAOKA Y., INUI T., EGUCHI N., URADE Y., HAYAISHI O. - Prostaglandin D synthase gene is involved in the regulation of non-rapid eye movement sleep. *Proc. Natl. Acad. Sci. U S A.*, 97, 4903-4907, 2000.

173. POMPEIANO M., CIRELLI C., RONCA-TESTONI S., TONONI G. - NGFI-A expression in the rat brain after sleep deprivation. *Brain. Res.*, 46, 1, 143-153, 1997.

174. POMPEIANO M., CIRELLI C., TONONI G. - Immediate-early genes in spontaneous wakefulness and sleep: expression of c-fos and NGFI-A mRNA and protein. *J. Sleep. Res.*, 3, 80-96, 1994.

175. PORKKA-HEISKANEN T., SMITH S.E., TAIRA T., URBAN J.H., LEVINE J.E., TUREK F.W., STENBERG D. - Noradrenergic activity in rat brain during rapid eye movement sleep deprivation and rebound sleep. *Am. J. Physiol.*, 268, 1456-1463, 1995.

176. PRUSINER S.B. - Molecular biology of prion diseases. *Science.*, 252, 1515-1522, 1991.

177. RALPH M.R., FOSTER R.G., DAVIS F.C., MENAKER M. - Transplanted suprachiasmatic nucleus determines circadian period. *Science.*, 247, 4945, 975-978, 1990.

178. RALPH M.R., MENAKER M. - A mutation of the circadian system in golden Hamsters. *Science.*, 241, 1225-1227, 1988.

179. RECHTSCHAFFEN A., GILLILAND M.A., BERGMANN B.M., WINTER J.B. - Physiological correlates of prolonged sleep deprivation in rats. *Science.*, 221, 182-184, 1983.

180. REDLINE S., TISHLER P.V., HANS M.G., TOSTESON T.D., STROHL K.P., SPRY K. - Racial differences in sleep-disordered breathing in African-Americans and Caucasians. *Am. J. Respir. Crit. Care. Med.*, 155, 186-192, 1997.

181. REDLINE S., TISHLER P. V., TOSTESON T. D. - The familial aggregation of obstructive sleep apnea. *Am. J. Respir. Crit. Care. Med.*, 151, 682-687, 1995.
182. RHYNER T.A., BORBELY A.A., MALLET J. - Molecular cloning of forebrain mRNAs which are modulated by sleep deprivation. *Eur. J. Neurosci.* 2, 1063, 1990.
183. ROEMER K., JOHNSON P.A., FRIEDMANN T. - Knock-in and knock-out. Transgenes, development and disease. *New. Biologist*, 3, 4, 331-335, 1991.
184. ROSENBERG R.S., BERGMANN B.M., SON H.J., ARNASON B.G.W., RECHTSCHAFFEN A. - Strain differences in the sleep of rats. *Sleep.*, 10, 6, 537-541, 1987.
185. ROTH B., BRUHOVA S., BERKOVA L. - Familial sleep Paralysis. *Arch. Suisses Neurol. Neurochir. Psychiatr.*, 102, 321-330, 1968.
186. ROTHENFLUH A., ABODEELY M., PRICE J.L., YOUNG M.W. - Isolation and analysis of six *timeless* alleles that cause short- or long-period circadian rhythms in Drosophila. *Genetics.*, 156, 665-675, 2000.
187. SANCAR A. - Cryptochrome: the second photoactive pigment in the eye and its role in circadian photoreception. *Annu. Rev. Biochem.*, 69, 31-67, 2000.
188. SCHENK C.H., GARCIA-RILL E., SEGALL M., MAHOWALD M.W. - HLA class II genes associated with REM behavior disorder. *Ann. Neurol.* 39, 261-263, 1996.
189. SCHIBLER U., TAFTI M. - Molecular approaches towards the isolation of sleep-related genes. *J. Sleep. Res.*, 8, Suppl 1, 1-10, 1999.
190. SCHWARTZ W. J., ZIMMERMAN P. - Circadian Time keeping in BALB/c and C57BL/6 inbred mouse strains. *J. Neurosci.*, 10, 1, 3685-3694, 1990.
191. SEHGAL A., PRICE J.L., MAN B., YOUNG M.W. - Loss of behavioral rythms and per RNA oscillations in the Drosophila mutant timeless. *Science.*, 263, 1603-1605, 1994.
192. SEHGAL A., OUSLEY A., HUNTER-ENSOR M. - Control of circadian rhythms by a two component clock. *Mol. Cell. Neurosci.*, 7, 165-172, 1996.
193. SHEARMAN L.P., SRIRAM S., WEAVER D.R., MAYWOOD E.S., CHAVES I., ZHENG B., KUME K., LEE C.C., VAN DER HORST G.T., HASTINGS M.H., REPPERT S.M. - Interacting molecular loops in the mammalian circadian clock. *Science.*, 288, 1013-1019, 2000.
194. SHEARMAN L., ZYLKA M.J., WEAVER D.W., KOLAKOWSKI L.F., REPERT S.M. - Two period homologs: circadian expression and photic regulation in the suprachaismatic nuclei. *Neuron.*, 19, 1261-1269, 1997.
195. SHERIN J.E., SHIROMANI P.J., MCCARLEY R.W., SAPER C.B. - Activation of ventrolateral preoptic neurons during sleep. *Science.*, 271, 216-219, 1996.
196. SHIMOMURA K., LOW-ZEDDIES S.S., KING D.P., STEEVES T.D., WHITELEY A., KUSHLA J., ZEMENIDES P.D., LIN A., VITATERNA M.H., CHURCHILL G.A., TAKAHASHI J.S. - Genome-wide epistatic interaction analysis reveals complex genetic determinants of circadian behavior in mice. *Genome. Res.*, 11, 959-980, 2001.
197. SHIROMANI P.J., GRECO M.A., TAKKAR M., MC CARLEY R.W. - c-fos knock out mice have reduced non-REM sleep. *Sleep. Res.*, 26, 42, 1997.
198. SHIROMANI P.J., MALIK M., WINSTON S., MC CARLEY R.W. - Time course of Fos-like immunoreactivity associated with cholinergically induced REM sleep. *J. Neurosci.* 15, 3500-3508, 1995.
199. SILVER R., LE SAUTER J., TRESCO P.A., LEHMAN M.N. - A diffusible coupling signal from the transplanted suprachiasmatic nucleus controlling circadian locomotor rhythms. *Nature.*, 382, 810-813, 1996.
200. STALEY-GANE M.S., HOLLWAY R.J., HAGERMAN M.D. - Temporal sleep characteristics of young fragile X boys. *Am. J. Human. Genet.*, 59, 4, A105, 1996.
201. STEINLEIN O., ANOKHIN A., YPING M., SCHALT E., VOGEL F. - Localization of a gene for low-voltage EEG on 20q and genetic heterogenity. *Genomics.*, 12, 1, 69-73, 1992.
202. STERIADE M. - Basic mechanisms of sleep generation. *Neurology.*, 42, Suppl 6, 9-18, 1992.
203. STROHL K.P., SAUNDERS N.A., FELDMAN N.T., HALLETT M. - Obstructive sleep apnea in family members. *N. Engl. J. Med.*, 229, 18, 969-972, 1978.
204. SUMMERS J.A., LYNCH P.S., HARRIS J.C., BURKE J.C., ALLISON D.B., SANDLER L. - A combined behavioral/pharmacological treatment in sleep-wake schedule disorder in Angelman syndrome. *Dev. Behav. Pediatrics*, 13, 4, 284-287, 1992.
205. SUN Z.S., ALBRECHT U., ZHUCHENKO O., BAILEY J., EICHELE G., LEE C.C. - *RIGUI*, a putative mammalian ortholog of the Drosophila *period* gene. *Cell.*, 90, 1003-1011, 1997.
206. TAFTI M., CHOLLET D., VALATX J.L., FRANKEN P. - Quantitative trait loci approach to the genetics of sleep in recombinant inbred mice. *J. Sleep. Res.*, 8 Suppl 1, 37-43, 1999.
207. TAFTI M., FRANKEN P., KITAHAMA K., MALAFOSSE A., JOUVET M., VALATX J.L. - Localization of candidate genomic regions influencing REM sleep in mice. *Neuroreport.*, 8, 3755-3758, 1997.
208. TAKAHASHI J.S. - Molecular neurobiology and genetics of circadian rhythms in mammals. *Ann. Rev. Neurosci.*, 18, 531-553, 1995.
209. TAKAHASHI J.S., PINTO L.H., VITATERNA M.H. - Forward and reverse genetic approaches to behavior in the mouse. *Science.*, 264, 1724-1733, 1994.
210. TAKAHASHI Y., HOHJOH H., MATSUURA K. - Predisposing factors in delayed sleep phase syndrome. *Psychiatry. Clin. Neurosci.*, 54, 356-358, 2000.
211. TEI H., OKAMURA H., SHIGEYOSHI Y., FUKUHARA C., OZAWA R., HIROSE M. - Circadian oscillation of a mammalian homologue of the *Drosophila period* gene. *Nature.*, 389, 512-516, 1997.
212. TERAO A., PEYRON C., DING J., WURTS S.W., EDGAR D.M., HELLER H.C., KILDUFF T.S. - Prepro-hypocretin (prepro-orexin) expression is unaffected by short-term sleep deprivation in rats and mice. *Sleep.*, 23, 867-874, 2000.

213. THANNICKAL T.C., MOORE R.Y., NIENHUIS R., RAMANATHAN L., GULYANI S., ALDRICH M., CORNFORD M., SIEGEL JM. - Reduced number of hypocretin neurons in human narcolepsy. *Neuron.*, 27, 469-474, 2000.
214. THOMAS D. A., SWAMINATHAN S., BEARDSMORE C. S., MCARDLE E. K., MACFADYEN U. M., GOODENOUGH P. C., CARPENTER R., SIMPSON H. - Comparison of peripheral chemoreceptor responses in monozygotic and dizygotic twin infants. *Am. Rev. Respir. Dis.*, 148, 1605-1609, 1993.
215. TISHLER P.V., REDLINE S., FERRETTE V., HANS M.G., ALTOSE M.D. - The association of sudden unexpected infant death with obstructive sleep apnea. *Am. J. Resp. Crit. Care. Med.*, 153, 6, 1857-1863, 1996.
216. TOBLER I., DEBOER T., FISCHER M. - Sleep and sleep regulation in normal and prion protein-deficient mice. *J. Neurosci.*, 17, 1869-1879, 1997.
217. TOBLER I., GAUS S.E., DEBOER T., ACHERMANN P., FISCHER M., AULICKE T., MOSER M., OESCH B., MCBRIDE P.A. MANSON J.C. - Altered circadian activity rhythms and sleep in mice devoid of prion protein. *Nature.*, 380, 639-642, 1996.
218. TODD J.A., AITMAN T.J., CORNALL R.J., GHOSH S., HALL J.R.S., HEARNE C.M., KNIGHT A.M., LOVE J.M., McALEER M.A., PRINS J.B., RODRIGUES N., LATHROP M., PRESSEY A., DELARATO N.H., PETERSON L.B., WICKER L.S. - Genetic analysis of autoimmune type 1 diabetes mellitus in mice. *Nature.*, 351, 542-547, 1991.
219. TOH K.L., JONES C.R., HE Y., EIDE E.J., HINZ W.A., VIRSHUP D.M., PTACEK L.J., FU Y.H. - An hPer2 phosphorylation site mutation in familial advanced sleep phase syndrome. *Science.*, 291, 1040-1043, 2001.
220. TOPPILA J., ALANKO L., ASIKAINEN M., TOBLER I., STENBERG D., PORKKA-HEISKANEN T. - Sleep deprivation increases somatostatin and growth hormone-releasing hormone messenger RNA in the rat hypothalamus. *J. Sleep. Res.*, 6, 171-178, 1997.
221. TOPPILA J., STENBERG D., ALANKO L., ASIKAINEN M., URBAN J.H., TUREK F.W., PORKKA-HEISKANEN T. - REM sleep deprivation induces galanin gene expression in the rat brain. *Neurosci. Lett.*, 183, 171-174, 1995.
222. TOTH L.A., WILLIAMS R.W. - A quantitative genetic analysis of slow-wave sleep and rapid-eye movement sleep in CXB recombinant inbred mice. *Behav. Genet.*, 29, 329-337, 1999.
223. TRENKWALDER C., COLLADOSO-SEIDEL V., GASSER T., OERTEL W.H. - Clinical symtoms and possible anticipation in a large kindred of familial restless legs syndrome. *Mov. Disord.*, 11, 389-394, 1996.
224. VALATX J.L. - Genetics as a model for studying the Sleep-Waking cycle. *Exp. Brain. Res.*, Suppl 8, 135-145, 1984.
225. VALATX J.L., BUGET R. - Facteurs génétiques dans le déterminisme du cycle veille-sommeil chez la souris. *Brain. Res.*, 69, 315-330, 1974.
226. VALATX J.L., BUGET R., JOUVET M. - Genetic studies of sleep in mice. *Nature.*, 238, 226-227, 1972.
227. VAN BEIJSTERVELDT C.E.M., MOLENAAR P.C.M., DE GEUS E.J.C., BOOMSMA D.I. - Heritability of human brain functioning as assessed by electroencephalography. *Am. J. Hum. Genet.*, 58, 562-573, 1996.
228. VAN TWYVER H., WEBB W.B., DUBE M., ZACKHEIM M. - Effects of environment and strain differences on EEG and behavioral measurement of sleep. *Behav. Biol.*, 9, 105-110, 1973.
229. VELA BUENO A., CAMPOS CASTELLO J.C., BAOS R.J. - Hereditary cataplexy. Is it primary cataplexy? *Waking and Sleeping*, 2, 125-126, 1978.
230. VGONTZAS A.N., KALES A., SEIP J., MASCARI M.J., BIXLER E.O., MYERS D.C., VELA-BUENO A.V., ROGAN, P.K. - Relationship of sleep abnormalities to patient genotypes in Prader-Willi syndrome. *Am. J. Med. Genet.*, 67, 478-482, 1996.
231. VISSCHER F., VAN DER HORST A.R., SMIT L.M. - HLA-DR antigens in Kleine-Levin syndrome. *Ann. Neurol.*, 28, 195, 1990.
232. VITATERNA M.H,. KING D.P., CHANG A.M., KORNHAUSER J.M., LOWREY P.L., MCDONALD J.D., DOVE W.F., PINTO L.H., TUREK F.W., TAKAHASHI J.S. - Mutagenesis and mapping of a mouse gene *Clock*, essential for circadian behavior. *Science.*, 264, 719-725, 1994.
233. VOGEL F. - Brain physiology genetics of the EEG. In: F. Vogel, A.G. Motulsky (eds.). *Human Genetics*. Springer-Verlag, New York, 590-593, 1986.
234. WAGER-SMITH K., KAY S.A. - Circadian rhythm genetics: from flics to mice to humans. *Nat. Genet.*, 26, 23-27, 2000.
235. WEBB W.B., CAMPBELL S.S. - Relationship in sleep characteristics of identical and fraternal twins. *Arch. Gen. Psychiat.*, 40, 1093-1095, 1983.
236. WEBB W.B., FRIEDMANN J.K. - Attempts to modify the sleep patterns of the rats. *Physiol. Behav.*, 6, 459-460, 1971.
237. WEISSMAN C. - A "unified theory" of prion propagation. *Nature.*, 352, 679-682, 1991.
238. WELSH D.K., LOGOTHETIS D.E., MEISTER M., REPPERT S.M. - Individual neurons dissociated from rat suprachiasmatic nucleus express independently phased circadian firing rhythms. *Neuron.*, 14, 697-706, 1995.
239. WINKELMANN J., WETTER T.C., COLLADO-SEIDEL V., GASSER T., DICHGANS M., YASSOURIDIS A., TRENKWALDER C. - Clinical characteristics and frequency of the hereditary restless legs syndrome in a population of 300 patients. *Sleep.*, 23, 597-602, 2000.
240. WISOR J.P., NISHINO S., SORA I., UHL G.H., MIGNOT E., EDGAR D.M. - Dopaminergic role in stimulant-induced wakefulness. *J. Neurosci.*, 21, 1787-1794, 2001.
241. YAGITA K., YAMAGUCHI S., TAMANINI F., VAN DER HORST G.T., HOEIJMAKERS J.H., YASUI A., LOROS J.J., DUNLAP J.C., OKAMURA H. - Dimerization and nuclear entry of mPER proteins in mammalian cells. *Genes. Dev.*, 14, 1353-1363, 2000.
242. YOSHIZAWA T., AKASHIBA T., KURASHINA K., OTSUKA K., HORIE T. - Genetics and obstructive sleep apnea syndrome: a study of human leukocyte antigen (HLA) typing. *Intern. Med.*, 32, 94-97, 1993.

243. YOUNG M.W. - The molecular control of circadian behavioral rhythms and their entrainment in Drosophila. *Annu. Rev. Biochem.*, 67, 135-152, 1998.
244. YOUNG M.W. - Life's 24-hour clock: molecular control of circadian rhythms in animal cells. *Trends. Biochem. Sci.*, 25, 601-606, 2000.
245. YOUNG T., PALTA M., DEMPSEY J., SKATRUD J., WEBER S., BADR S. - The occurrence of sleep-disordered breathing among middle-aged adults. *N. Engl. J. Med.*, 328, 1230-1235, 1993.
246. ZHANG J., OBAL F., FANG J., COLLINS B.J., KRUEGER J.M. - Non-rapid eye movement sleep is suppressed in transgenic mice with a deficiency in the somatotropic system. *Neurosci. Lett.*, 220, 97-100, 1996.

Chapter 8

Mental activities during sleep

J. De Koninck*

School of Psychology, University of Ottawa, Ottawa, Canada

HISTORICAL NOTE

Dreams have played as important a role in the civilisations from which our modern society originates as in primitive societies, not only in political terms but also in daily life [9]. The Egyptians, Greeks and Romans were among those who erected incubation temples where people would spend the night in order to have their dreams interpreted the following morning - dreams which were considered as divine messages [59]. Some have seen dreams as a more natural phenomenon. Hippocrates examined dreams for their expression of certain illnesses. Aristotle searched their origins in the activities of the previous day and insisted that they could influence conscious life. Subsequently, he was probably the first to describe the phenomenon of lucid dreams.

These themes were taken up in the 19th century, notably in France, by Alfred Maury [52] and Hervey de St-Denys [20]. The former had a stimulus applied during sleep to observe how it influenced his dreams. So breathing in eau de Cologne during sleep appeared to him to generate erotic dreams, while a burning matchstick applied near his foot brought on dreams of the sea. A few years later, Hervey de St-Denys [20], in *Les Rêves et les Moyens de les Diriger*, presented a systematic analysis of the methods of controlling dreams and lucid dreams. His work largely influenced Freud, who refers to it, as an indirect source, in *The Interpretation of Dreams* (1900), and is still considered relevant in current discussion on lucid dreams. Finally, while the interest in dreams in the first half of the 20th century was due to the work of psychoanalysts, it was only when the REM (Rapid Eye Movement) sleep phase was identified [3] and associated with the most prominent dreams, that the experimental study of mental activity during sleep began to take off.

PSYCHOPHYSIOLOGY OF MENTAL ACTIVITY DURING SLEEP

When dreaming occurs

The conclusion reached by the work of Kleitman's team [3, 17] was that dream activity occurred almost exclusively during REM sleep. They had thus observed dream recall during REM sleep in 80% of cases as against only 10% in NREM sleep. Their report had such an impact that some researchers and non specialists alike continue to this day to consider REM sleep and dreams as being one and the same. Nevertheless, Foulkes [2] would clearly demonstrate that dream recall could occur in 40% of cases of arousal during NREM sleep. Moreover, he was able to show that, contrary to the hypothesis formulated by Dement and Kleitman [17], dream recall in NREM sleep did not relate to dreams left over from a preceding REM sleep period. Indeed, he obtained dream recall in NREM sleep during the first cycle of night sleep even before a REM sleep period occurs.

* The author wishes to thank Professor Monique Lortie-Lussier for her comments and advice in preparing this text.

Later, Cavallero et al. [12] observed a group of 50 subjects to find that dream recall was indeed more common in REM sleep than in deep NREM sleep (stages 3 and 4) i.e. 86% and 64% respectively. However it is interesting to note that while the accounts of dreams obtained in REM sleep were longer than those in deep NREM sleep, often containing more emotion and characters (other than the dreamer), no significant difference appeared for the ten other dream dimensions studied. Doubtless to say, differences in rates of recall and types of dream obtained in REM and NREM sleep in the various studies can be accounted for by the differences in the methodology of dream collection and defining what a dream is.

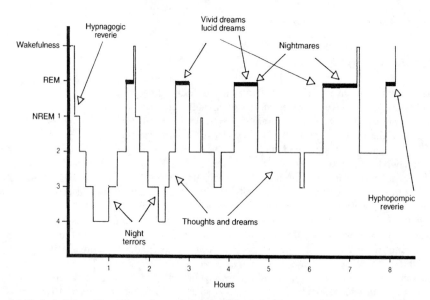

Figure 8.1. Typical picture of the different forms of mental activity during the course of a sleep hypnogram in a young adult.

Despite the permeability of dreams throughout sleep, as demonstrated in recent work, different forms of mental activity are still recognised as being characteristic of the various stages of sleep and periods during a typical night's sleep. Figure 8.1 illustrates their normal distribution on the hypnogram of a young adult. A brief description follows:

1. Hypnagogic dreaming: this consists of sometimes elaborate images occurring at sleep onset. This typically involves imagery with no scenario. This type of dreaming has been the focus of several studies, principally by Vogel's team [60].
2. Hypnopompic dreaming: produced at the time of spontaneous arousal, it is sometimes difficult to distinguish from the imagery produced in REM sleep, as it may also be highly elaborate.
3. Reflexive activity: mainly observed in NREM, particularly in stages 3 and 4. The subject reports thoughts with no hallucinatory content.
4. Classic dreams: these are dreams characterised by vivid visual imagery, one or several scenarios, with a succession of scenes involving characters, interaction and emotions. This type of dream occurs mainly, but not exclusively, in REM sleep.
5. Lucid dreams: these are dreams during which the dreamer is conscious of dreaming while remaining asleep. In the most elaborate form of lucid dreaming, the dreamer controls the way the dream unfolds. Lucid dreams are mainly observed in REM sleep, but may be reported after arousal from stage 2.

6. Nightmares: these dreams of highly anxiogenic content are observed in REM sleep, especially toward the end of the night.
7. Night terrors: these, very brief episodes of panic sometimes accompanied by mental activity, occur in deep NREM sleep (stages 3 and 4) at the beginning of the night.

It is interesting to note that Foulkes [24] obtained dream material (hallucinatory ideation) from normal subjects after periods of relaxation during which they were awake in terms of electrophysical indexes. This permeability of dream imagery between wakefulness and sleep is well known among psychotherapists, who use it as part of the directed waking dream technique [19].

Physiological correlation of dreams

The second notion, which we owe to the work of the Kleitman team, is that of the isomorphism between REM eye movements and the visual imagery of dreams. Several reasons lead to refuting the hypothesis of visually "scanning" the dream image. As was noted above, visual dream content may exist outside REM, with no accompanying eye movements. But even when limited to mental activity in REM sleep, several observations counter this hypothesis. For example, rapid eye movements (REM) occur in those who are blind from birth and who have no visual imagery. Rapid eye movements in REM sleep are diagonal as opposed to vertical and horizontal; they occur in cyclic bursts and regular phases, unlike visual activity in wakefulness. Jouvet [36] noted that the onset of eye movements in REM sleep precedes the onset of PGO activity in the occipital cortex, rendering impossible a causal relationship between exploratory eye movement and visual hallucination. However, while there is no one-to-one relationship between visual imagery and REM, a relationship has been observed between REM density, the level of activity in dream content and its affective character. The more eye movements during a period of REM sleep, the richer the dream content in motor activities and affect [57].

In the mid 70s, Broughton [8] and Bakan [4] put forward the idea that the fanciful nature of dreams could be a manifestation of the reversal of cerebral dominance during REM sleep in favour of the right hemisphere. Broughton's formulation, which was more substantial, suggested the alternating domination of the two hemispheres during Kleitman's "Basic Rest Activity Cycle" (BRAC) [39], such that every 90 minutes, during wakefulness and sleep, this reversal allowed the right hemisphere to become dominant. However interesting this notion may be, Antrobus [1], after making an inventory of recent work, came to the conclusion that the right hemisphere not only fails to dominate in REM sleep, but that the left hemisphere plays a more important role. Pivik [57] suggests for his part that interhemispheric activity is just as complex during REM sleep as in wakefulness. If, in the wakeful state, the contribution of both hemispheres is essential for most functions, the same must be true during REM dreaming. Thus Laberge [46] observed, during the recording of a lucid dream, that hemispheric domination corresponded to the type of dream activity provoked by the dreamer. Finally, Jouvet [35] noted that during the dream it is common to observe "a dissociation between facial recognition of the message-giver and recognition of the semantic content of the message". In his view, the lowered activity of the corpus callosum in REM sleep generates at least a temporary failure of association between the two hemispheres during certain dreams.

DREAM RECALL AND COLLECTION METHODS

For the time being, the study of dreams relies on the account given by the dreamer. The need for introspection thus carries a number of limitations of a methodological nature. Modern research work employs two approaches, either collecting dreams in the laboratory or at home in the subject's natural surroundings. In the former case, subjects are awakened during a polysomnographic recording. While this method may be the most effective way of obtaining the maximum amount of dream content, it is limited in comparison with the dream diary method in natural surroundings. Indeed, in the laboratory, dreams are often lacking in emotion and incorporate the laboratory situation [22]. The dream diary recorded at home, on the other hand, is more likely to contain secondary elaborations extending beyond the original dream story.

Considerable individual differences are also observed in the frequency with which dreams are recalled, both in the laboratory and in natural surroundings. Freud [25], considered that dream content was subjected to systematic censorship. Those who remember very little of their dreams were thus those who most often resorted to the mechanism of repression. This hypothesis, which is difficult to verify experimentally, has little support today. Other factors of a physiological, circumstantial and psychological nature, have been identified to explain the variations in the frequency of spontaneous dream recall [13, 40]. Thus, recall is favoured by sudden arousal in REM sleep, particularly during bursts of REM. Kramer et al. [41] have shown that the successive periods of REM sleep during the course of the night yield the recall of increasingly rich content. In addition, it appears that, during long periods of REM sleep, there is an optimal period for more intense recall [43].

On arousal, there must be no distracting or interfering stimulation. According to the principle of prime and recent events, the elements at the beginning and end of the same dream experience are more accessible to the dreamer, just as the first and last dreams of the night are easier to recall the following morning [58]. The interest and motivation of the subject may ultimately increase the frequency of recall. And finally, there seems to be an increase in dream recall during periods of stress in daily life [5].

DREAM SOURCES

The mechanisms of dream formation remain obscure. Certain neurophysiological models of REM sleep are used as dream formation models [34, 36]. As these are not applicable to mental activity during the rest of sleep, they will not be set out here.

From a phenomenological point of view, it seems that dream content develops from a number of elements related to the previous day's experience, internal (somatic) and external stimulation during sleep, personality, culture, and perhaps even, as Jung has suggested [37], on images of phylogenetic origin.

Research in familiar surroundings using a dream diary has confirmed that day-residue constitutes the main source of contents. However, a delay in incorporating day-residue may be observed when there are changes in physical surroundings. Indeed, Jouvet [35] noted while travelling, that the new environment was not incorporated into his dreams until 6 to 7 days after his arrival in a new country. In his view, dreams may call on two types of memory: "the first devoid of spatial parameters, relates to the most recent events. This memory, responsible for "day-residue", rapidly diminishes (6 to 7 days). The second memory, which is spatial, remains latent in terms of dreaming for 6 to 7 days and is responsible for memories of the environment." According to an inventory study carried out by Nielsen and Powell [54], research tends to suggest that dreams incorporate the events of the previous day, in 65% to 70% of cases, with only 30% to 35% of events going back two days or more. These authors were nonetheless able to observe the 6 day incorporation delay described by Jouvet [35] in two of the three studies undertaken with groups of university students having kept dream diaries.

Laboratory research, where dreams are collected during REM sleep periods, has enabled a more systematic study of the relationship between dream content, the previous day's experiences and the attendant prehypnic environment. Dement and Wolpert [18] presented auditory, skin, and even visual stimulation during sleep, to determine to what extent and in what way dreams were influenced. Subjects have also been placed in stressful situations, by watching films for example [27], or in unfamiliar social or sensory situations, such as isolation [65] or visual field inversion [16a] after which dreams were collected. The conclusion confirms that the previous day's experiences as well as external stimuli tend to be incorporated into dreams but in a disguised form. Incorporation is greater in cases of stress directly experienced and of strong personal significance rather than cases of vicarious stress. Moreover, we were surprised to observe that independent judges were unable to correctly pair dream accounts obtained in REM sleep with descriptions of the corresponding prehypnic thoughts and preoccupations [57bis]. The manifest contents of dreams is thus somewhat removed from the psychological contents of the previous day.

DREAM NORMS

The analysis of the contents of a substantial sample of dreams has provided a quantity of interesting normative data [29]. The established norms involve the relative frequency with which certain elements appear in dreams: characters, their activities, friendly, aggressive or sexual interaction, surrounding objects and emotions felt by the dreamers. These are based on the dreams of 100 male and 100 female American college students, who each recounted five dreams using the dream diary method, kept at home. Table 7.1 presents a summary compiling detailed data from Hall et al. [29]. It transpires, notably, that the predominance of aggressive interaction, anxiety and misfortune give dream contents a negative substance. More recent studies on Canadian populations suggest however, that this negative character is perhaps less pronounced in normal living conditions today [50, 51]. But for the moment, it is impossible to ascertain whether these differences are attributable to cultural differences and/or the evolution of North American society.

Table 8.1. Normative data on the dreams of young American adults*

	Men	Women
Environment (per dream)	1.29	1.31
Interior (%)	44	55
Exterior (%)	47	35
Familiar (%)	31	37
Unfamiliar (%)	19	10
Characters (per dream)	2.4	2.8
Male (%)	53	37
Female (%)	26	40
Animal (%)	6	4
Social interaction		
Aggressive (%)	47	44
Friendly (%)	38	42
Sexual (%)	12	3
Emotions		
Joy (%)	20	20
Sadness (%)	9	13
Anger (%)	16	13
Confusion (%)	22	18
Anxiety (%)	34	37
Outcome		
Success (%)	15	8
Failure (%)	15	10
Fortune		
Fortune (%)	6	6
Misfortune (%)	36	33

*Data obtained in the early 1950s, based on the dreams of 100 men and 100 women aged 18 to 25. Summarised from Hall and Van der Castle [29].

A second aspect of normative work concerns sexual differences, identified not only by Hall et al. [29], but later by other researchers. According to this work, women are more attentive to their dreams and remember them better [2, 10]. Their dreams bring into play more characters of both sexes, whether family, close relatives or children, than those of men, whose unfamiliar, male figures are more numerous. In women's dreams, the surroundings are more often interior than exterior, interactions tend to be friendlier rather than aggressive, emotions pleasant rather than unpleasant [63]. Aggression appears to be a less constant feature, varying depending on the ages of the two sexes [56].

The stability of sexual differences in dreaming is nevertheless a controversial issue. While Hall et al. [30] observed the persistency of these differences in students 30 years later, Kramer et al. [42], for their part, failed to observe the usual differences linked to sex in regard to aggression, emotions, surroundings, misfortune to self and others. In our research on the impact of social roles [50] comparing the dreams of different groups of women, some of the habitual characteristics of female dreams were not found. A comparison of the dreams of three groups of parents of both sexes all with typically male professional qualifications [51] provided no significant differences in emotions, surroundings, or friendly or aggressive interaction.

Comprehensive data is not yet available on the ontogenesis of dreams. The studies carried out on samples from different age groups have yielded some interesting observations [62]. Children's dreams contain more animals than human characters, more aggression, more fanciful content, and magical endings, and are more pleasant than adults' dreams. Their accounts are short and reflect their cognitive functioning level [23]. Adolescents' dreams are longer than children's dreams but not as long as those of adults, and are notable for the amount of friendly interaction between pairs [49]. Finally, in the elderly, themes of vulnerability and loss are found [53]. There is also a reduction in dream recall.

EFFECTS OF DREAMS

Although study has often focused on the source of dreams, with extensive analysis of their content, little attention has been paid to the effects of dreams on wakeful consciousness. For Freud [25], dreams were not related to daily life, their role being to preserve sleep. Recalled dreams had thus failed in their objective. While it is recognised today that a dream which attains wakeful consciousness may have an impact on the latter, the results of the rare studies attempting to evaluate this are often contradictory. Bokert [7] for example, observed that subjects deprived of liquid for 24 hours were less thirsty on awakening in the morning if they had dreamed of quenching their thirst. De Koninck and Koulack [16] presented the same stressful film to subjects before going to bed and upon awakening in the morning. Subjects whose dreams had incorporated elements of the film were the most affected by the second presentation. These two studies suggest that dreams can play a compensatory role in adaptation. Dreams with an adaptive value are those which divert the dreamer's attention from the previous day's preoccupations or which allow a need to be satisfied in dream form. Moreover, other work suggests that incorporating stressful elements helps the dreamer to control tension, if not in the short term, at least on a medium or long term basis. Cartwright [11] observed that a group of women's adaptation to divorce appeared to be facilitated by frequently incorporating the ex-partner in dreams. Wright and Koulack [66] have suggested, by way of reconciling these two notions, that mechanisms of compensation alternate with those of integration through dreams of the previous day's events. Hartmann [33] has suggested, for his part, that stress compensation or integration in dreams is a function of personality type.

CONTROLLING DREAM CONTENT

If dream activity can be said to influence wakeful consciousness, it would be beneficial to be able to manipulate dream contents to increase their adaptive value. It has been clearly demonstrated that suggestion is the most effective means of influencing dream content. De Koninck and Brunette [15] have successfully managed to train subjects not only to incorporate precise objects of phobic connotation into their dreams, but also to change their emotional content and to enhance their pleasant aspect.

Controlling dreams becomes even more specific in the case of lucid dreams. Indeed learning to control the dream while it is taking place, allows several aspects to be modified.

Following in the footsteps of Hervey de St-Denis [20], several authors have systematically attempted to develop the means of controlling dreams [26, 28, 45] and have popularised the phenomenon. Work carried out in the laboratory for several years has allowed certain characteristics to be defined.

Lucid dreams have vivid contents, with characters and surroundings that are very close to reality. Compared to typical REM sleep dreams, they contain more emotion – which tends to be more pleasant. Lucid dreams are not easy to provoke. The factors which favour their production have however been identified, such as becoming aware of the dreaming state. This can be facilitated by false arousals (the dreamer has the impression of awakening but in fact carries on dreaming), by frightening or strange images or activities such as flying, which are impossible to accomplish in the wakeful state. Many authors have proposed even more elaborate recipes such as prehypnic suggestion [26] and mnemonically induced lucid dreams [45] which rely on visualisation, memory and auto-suggestion exercises to be practised on morning awakening. And finally, whether in the laboratory or by using portable devices, some advocate the use of stimulation in REM sleep to force the awareness of dreaming.

Even if lucid dreams appear to be produced essentially in REM sleep, they have also been observed in hypnagogic dreaming at the onset of sleep and during stage 2 [46]. Clearly, one can ask whether during the lucid dream, the dreamer is electrophysiologically "awake". The work of Ogilvie et al. [55] suggests that there is an upsurge of alpha waves. Thus it would appear that periods of "micro-wakefulness" may be responsible for the phenomenon. However, after examining these observations, Laberge [46] challenged the conclusions. He in fact demonstrated that a dreamer could indicate, by voluntary, codified eye movements, the start, end and type of contents provoked in lucid dreaming. He did not note any activation level in lucid dreaming which differed from that of the spontaneous dreaming. But he did observe greater activation of the right hemisphere when the dreamer sang and of the left hemisphere, when he counted. Finally, he noted several signs of activation when a woman experienced orgasm.

Lucid dreams can be said to be of particular interest for the prospects of psychological adaptation. Indeed, they could be beneficial in transforming the contents of nightmares. What is more, if certain types of dream have an adaptive value, the lucid dream technique could be employed to increase the frequency of such dreams. Nevertheless, the medium to long term consequences of the regular and systematic use of lucid dreams are as yet unknown, both in terms of sleep and of dreams.

DREAMS IN PSYCHOPATHOLOGIES

Neuroses and psychoses

Little work has been done on dream content in regard to the various psychopathologies. Neuroses, particularly those of a phobic nature, are known to be accompanied by dreams of anxiogenic content, where the phobic object itself is often represented. In the case of psychoses, the dreams of schizophrenics and depressives are those most often studied. Kramer and Roth [44] report that the following characteristics may be observed in schizophrenics:

1. Dream accounts are short, with little detail.
2. The content of the dream is "primitive" and direct.
3. The content includes few hallucinations as if those experienced in wakefulness resulted from a shift of REM sleep to wakefulness.
4. The content is more violent (open hostility).
5. The dreamer is an observer rather than a participant.
6. The affect of content is neutral.
7. The characters tend to be unfamiliar.

Weinstein, Schwartz and Arkin [61] are nevertheless sceptical as to whether these results can be generalised, as most of the studies are obscured by a number of methodological problems. It is notably difficult to ascertain whether the differences in dream content can be ascribed to cognitive deficiencies during wakefulness or to the schizophrenic state during sleep.

Wolman [64] suggests that signs of an impending attack of psychosis can be discerned in the dreams of preschizoid subjects. He observed extremely violent dreams in several subjects before an attack. The dream could thus serve as a preventive diagnostic tool in schizophrenia.

Accounts of the dreams of depressive patients are also very short. Unlike those of schizophrenics, their dreams almost exclusively contain figures from the dreamer's family. The themes are often masochistic and aggression is as often directed against the dreamer as against other figures [61]. For Kramer and Roth [44], the "manifest" content of their dreams contains few depressive themes, while the "latent" content contains several.

It is commonly observed that the dreams of psychosomatics have high levels of affect and of hostility. Levitan [48] conducted a systematic study of his patients' dreams. He concluded that his patients often experienced the repetition of certain highly traumatic dreams which may themselves have provoked the psychological disorder. A kind of dream "bombardment" would thus generate the psychological disorder.

Night terrors and nightmares

As seen earlier, a distinction must be drawn between night terrors and nightmares; the former, observed in NREM sleep during the first third of the night, are characterised by incomplete arousals with little or no recall of dream content. The content is usually restricted to a single theme only, often a fatal suffocating sensation, which is why it is referred to as "incubus attack". Nightmares, on the other hand, are elaborate, highly anxiogenic dreams, mainly produced in REM sleep during the second part of the night, but also occurring in stage 2 [38]. They are of interest both in terms of content and the personalities of those who regularly experience them.

Belicki and Cuddy [6], after a survey of 1,700 university students, reported that 1.9% of respondents had had at least one nightmare during the preceding week, and 9.6% during the preceding month. In fact only 20.7% had not had any nightmares during the year preceding the survey. There are thus marked individual differences in the frequency of nightmares. Hartmann [32] compared the personality traits between persons who frequently had nightmares, those affected by recurrent traumatic dreams, and those who did not experience nightmares. He concluded that subjects affected by nightmares had sensitive personalities and tended to be artistic or creative. However, using a dichotomy based on the notion of body image boundaries and several aspects of personal relationships, he observed that the boundaries of those who are subject to nightmares are thin, making them vulnerable.

Persistent nightmares are also reported after traumatic events. The most systematically studied are those of war survivors, notably after the holocaust [47].

Several types of treatment are recommended to ease nightmares. Halliday [31] mentions four of these:

1.   Desensitisation or related approaches consisting of desensitising the person in relation to the nightmare, to eliminate the associated fear and the anxiety that induces them.

2.   Psychoanalytical or cathartic treatment, relying on the notion whereby nightmares reflect repressed conflict which is then expressed in dreams.

3.   Altering the typical scenario of nightmares, by prehypnic suggestion.

4.   Using the method applied to lucid dreams, to control the content of nightmares.

The last two methods appear to be the most promising, because of their simplicity and rapidity of effect.

CONCLUSION

A number of relationships have been established between dream content, sex, age and the psychological characteristics of wakeful subjects. Nevertheless, there is no consensus in terms of the meanings attributed to these relationships. The question of where the dream is occurs on the sleep-wake continuum remains open. Dream activity is known to occur during all the stages of sleep. Mental activity similar to that of dreaming is also known to occur during wakefulness. Finally, in lucid dreams, the boundaries between sleep and wakefulness dwindle. The possibility of controlling dream content is attracting increasing interest, both in terms of its implications for the notions of consciousness and unconsciousness during sleep, and of its potential use in psychological adaptation.

REFERENCES

1. ANTROBUS J. – Cortical hemisphere asymmetry and sleep mentation. *Psychological Review,* 94, 359-368, 1987.
2. ARMITAGE R. - Gender differences and the effect of stress on dream recall : A 30-day diary report. *Dreaming,* 2, 137-141, 1992.
3. ASERINSKY E., KLEITMAN N. – Regularly occurring periods of eye motility and concomitant phenomena during sleep. *Science,* 118, 273-274, 1953.
4. BAKAN P. – Dreaming, REM sleep and the right hemisphere : A theoretical integration. *Journal of Altered States of Consciousness,* 3, 285-307, 1978.
5. BELICKI K. – Recalling dreams : An examination of daily variations and individual differences. *In : Sleep and dreams : A source book,* J. Gackenback (ed.), Garland, New York, 1986.
6. BELICKI K., CUDDY M. – Nightmares : Facts, fictions and future directions. *In : Dream images : A call to mental arms,* J. Gackenbach, A.A. Sheikh (eds), Baywood Publishing Company, 5, 99-114, 1991.
7. BOKERT E. - *The effects of thirst and a related verbal stimulus on dream reports.* Unpublished doctoral dissertation, New York University, 1967.
8. BROUGHTON R. – Biorhythmic variations, in consciousness and psychological functions. *Canadian Psychological Review,* 16, 217-240, 1975.
9. CALLOIS R., VON GRUNEBAUM G.E. – Le rêve et les sociétés humaines. Gallimard, Paris, 1967.
10. CARTWRIGHT R.D. – *Nightlife : Explorations in dreaming.* Prentice-Hall, Englewood Cliffs (New Jersey), 1977.
11. CARTWRIGHT R.D. – Dreams that work : the relation of dream incorporation to adaptation to stressful events. *Dreaming,* 1, 3-9, 1992.
12. CAVALLERO C., CICOGNA P., NATALE V., OCCHIONERO M., ZITO A. – Slow wave sleep dreaming. *Sleep,* 15, 562-566, 1992.
13. COHEN D.B. – Toward a theory of dream recall. *Psychological Bulletin,* 81, 138-154, 1974.
14. COHEN D. – *Sleep and Dreaming.* Pergamon Press. New York, 1979.
15. DE KONINCK J., BRUNETTE R. – Presleep suggestion related to a phobic object : Successful manipulation of reported dream affect. *Journal of General Psychology,* 118, 185-201, 1991.
16. DE KONINCK J.M., KOULACK D. – Dream content and adaptation to a stressful situation. *Journal of Abnormal Psychology,* 84, 250-260, 1975.
17. DE KONINCK J., PREVOST F., LORTIE-LUSSIER M. – The effects of inversion of the visual field on REM sleep mentation. *Journal of Sleep Research,* 5, 16-20, 1996.
18. DEMENT W.C., KLEITMAN N. – Cyclic variations in EEG during sleep and their relation to eye movements, body motility and dreaming. *Electroencephalography and Clinical Neurophysiology,* 9, 673-690, 1957.
19. DEMENT W.C., WOLPERT E. – The relation of eye movements, body motility, and external stimuli to dream content. *Journal of Experimental Psychology,* 53, 339-346, 1957.
20. DESOILLE R. – *Théorie et pratique du rêve éveillé et dirigé.* Editions du Mont-Blanc, Genève, 1961.
21. DE SAINT-DENYS H. – *Les rêves et les moyens de les diriger.* Tchou, Paris, 1958. (first edition).
22. FOULKES D. – Dream reports from different stages of sleep. *Journal of Abnormal and Social Psychology,* 65, 14-25, 1962.
23. FOULKES D. – Home and laboratory dreams : Four empirical studies and a conceptual reevaluation. *Sleep,* 2, 233-251, 1979.
24. FOULKES D.- *Children's dreams : Longitudinal studies.* Wiley, New York, 1982.
25. FOULKES D. – *Dreaming : A cognitive psychological analysis.* Hisdale, Lawrence Erlbaum Associates, New Jersey, 1985.
26. FREUD S. – *L'interprétation des rêves.* Presses Universitaires de France, Paris, 1971 (first edition in German, 1900).
27. GARFIELD P. – *Creative dreaming.* Simon and Schuster, New York, 1974
28. GOODENOUGH D.R., WITKIN H.A., KOULACK D., COHEN H. – The effects of stress films on dream affect and on respiration and eye movement during rapid-eye movement sleep. *Psychophysiology,* 15, 313-320, 1975.
29. GREEN C. – *Lucid Dreams.* Hamilton, London, 1968.

30. HALL C., VAN DE CASTLE R.L. – *The content analysis of dreams.* Appleton Century-Crofts, New York, 1966.
31. HALL C., DOMHOFF G.W., BLICK K.A., WEESNER K.E. – The dream of college men and women in 1950 and 1980 : A comparison of dream content and sex differences. *Sleep,* 5, 188-194, 1982.
32. HALLIDAY G. – Direct psychological therapies for nightmares : A review, *Clinical Psychology Review,* 7, 501-523, 1987.
33. HARTMANN E. – *The Nightmare.* Basic Books. New York, 1984.
34. HARTMANN E. – Dreams that work or dreams that poison ? What does dreaming do ? An editorial essay . *Dreaming,* 1, 23-26, 1991.
35. HOBSON A. – *The dreaming brain.* Basic Books, New York, 1988.
36. JOUVET M. – Mémoire et « cerveau dédoublé » au cours du rêve : A propos de 2525 souvenirs de rêve. *L'année du praticien,* 29, 27-32, 1979.
37. JOUVET M. – *le sommeil et le rêve.* Editions Odile Jacob, Paris, 1992.
38. JUNG C.G. – *Essai d'exploration de l'inconscient.* Robert Lafont, Paris, 1964.
39. KELLERMAN H. – *The Nightmare.* Columbia University Press, New York, 1987.
40. KLEITMAN N. – *Sleep and wakefulness.* University of Chicago Press, Chicago, 1963.
41. KOULACK D., GOODENOUGH D.R. – Modèle de rappel des rêves. *Annales médico-psychologiques,* 135, 35-42, 1978.
42. KRAMER M., CZAYA J., ARAND D., ROTH T. – The development of psychological content across the REMP. *Sleep Research,* 3, 121, 1974.
43. KRAMER M., KINNEY L., CHARF M. – Sex differences in dreams. *The Psychiatric Journal of the University of Ottawa,* 8, 1-4, 1983.
44. KRAMER M., McQUARRIE E., BONNET M. – Dream differences as a function of REM period progression. *Sleep Research,* 9, 155, 1979.
45. KRAMER M., ROTH T. – Dreams in psychopathology. *In : Handbook of Dreams,* B.B. Wolman (ed.), Van Nostrand, New York, 1979.
46. LABERGE S. – *Lucid Dreaming.* Ballantine, New York, 1985.
47. LABERGE S. – The psychophysiology of lucid dreaming. *In : Conscious mind, sleeping brain,* J. Gackenback, S. Laberge (eds), Plenum, New York, 135-153, 1988.
48. LAVIE P., KAMINER H. – Dreams that poison sleep : Dreaming in holocaust survivors. *Dreaming,* 1, 11-23, 1991.
49. LEVITAN H.L. – Traumatic events in dreams of psychosomatic patients. *Psychotherapy and Psychosomatics,* 33, 226-232, 1980.
50. LORTIE-LUSSIER M., DE KONINCK J., RENAUD M.F., RINFRET N. – Mothers' and daughters' dreams : reflections of middle adulthood and adolescence. *ASD Newsletter,* 7, 6-7, 1990.
51. LORTIE-LUSSIER M., SCHWAB C., DE KONINCK J. – Working mothers versus homemakers : Do dreams reflect the changing roles of women ? *Sex Roles,* 12, 1009-1021, 1985.
52. LORTIE-LUSSIER M., SIMOND S., RINFRET N., DE KONINCK J. – Beyond gender differences : Family and occupational roles impact on women's and men's dreams. *Sex Roles,* 26, 79-96, 1992
53. MAURY A. – *le sommeil et les rêves.* Paris, 1861.
54. MORGAN K. – *Sleep and aging.* John Hopkins University Press, Baltimore, 1987.
55. NIELSEN T.A., POWELL R.A. – The day-residue and dream-lag effects : A literature review and limited replication of two temporal effects in dream formation. *Dreaming,* 2, 67-77, 1992.
56. OGILVIE R., HUNT H.T., TYSON P.D., LECESCU M.L., JEAKINS D.B. – Lucid dreaming and alpha activity : A preliminary report. *Perceptual and Motor Skills,* 55, 795-808, 1982.
57. PAOLINO A.F., – Dreams : Sex differences in aggressive content. *Journal of Project Technology,* 28, 219-226, 1964.
58. PIVIK T. – Tonic states and phasic events in relation to sleep mentation. *In : The Mind in Sleep,* S.J. Euman, J. Antrobus (eds), Wiley, New York, 214-248, 1991.
59. ROUSSY F., CAMIRAND C., FOULKES D., DE KONINCK J., LOFTIS M., KERR N.H. – Does early-night REM dream content reliably reflect presleep state of mind? *Dreaming,* 6, 2, 1996.
60. TRINDER J., KRAMER M. – Dream recall, *American Journal of Psychiatry,* 128, 296-301, 1971.
61. VAN DE CASTLE R. – *The psychology of dreaming.* General Learning Press, 1971.
62. VOGEL G. – Sleep onset mentation. *In : The Mind in Sleep.* S.J.Ellman, J. Antrobus (eds), Wiley, New York, 125-136, 1991.
63. WEINSTEIN L.N., SCHWARTZ D.G., ARKIN A.M. – Qualitative aspects of sleep mentation. *In : The Mind in Sleep,* S.J. Ellman, J. Antrobus (eds), Wiley, New York, 172-213, 1991.
64. WINGET C., KRAMER M. – *Dimensions of dreams.* University of Florida Press, Gainesville, 1979.
65. WINGET C., KRAMER M., WHITMAN R.M. – Dreams and demography. *Canadian Psychiatry Association Journal,* 17, 203-208, 1972.

66. WOLMAN B.B. – Dreams in Schizophrenia. *In : Handbook of Dreams,* B.B. Wolman (ed.), Van Nostrand Reinhold Company, New York, 1979.
67. WOOD P. – *Dreaming and social isolation.* Unpublished doctoral dissertation. University of North Carolina, 1962.
68. WRIGHT J., KOULACK D. – Dreams and contemporary stress : A disruption-avoidance-adaptation model. *Sleep,* 10, 172-179, 1987.

PART 2

DIAGNOSTIC PROCEDURES

Chapter 9

Polysomnography

A. Besset

INSERM EMI-9930, Hôpital la Colombière, Montpellier and Service de Neurologie B, Hôpital Gui de Chauliac, Montpellier, France

The need to record several different electrograms simultaneously imposes a number of technological and practical constraints which will be clearly set out in this chapter. The principles involved in recording each of the sleep variables will be examined in turn: the electroencelophalogram (EEG), the electromyogram (EMG), followed by the electro-oculogram (EOG). Finally, the different autonomic variables linked to sleep monitoring will be dealt with: cardio-respiratory variables, limb movements measured on the anterior tibialis, and penile plethysmography.

SLEEP VARIABLES

Electroencephalogram (EEG)

EEG activity is the core variable of polysomnographic monitoring. Indeed while it is possible to score part of sleep without using EMG or EOG [14], it is quite impossible to do so without EEG.

Electrode placement

Most laboratories adhere to the manual published by Retschaffen and Kales [20] using the standard 10-20 system terminology developed by Jaspers [11] and recommended by the International Federation of EEG Associations as well as by clinical neurophysiology. This system involves measuring electrode placements using four landmarks: the nasion, inion, and left and right preauricular points. The distances between nasion-inion and left- and right preauricular points must be measured accurately as they are specific to each individual. The electrodes are then placed at intervals of 10 or 20 per cent of the total distance between landmarks (from which the system derives its name).

Once the measurements are taken and the precise location of the electrodes marked, the hair is separated to cleanse the scalp with ether. Electrode cups are then attached to the scalp using small patches of gauze soaked in collodion and dried with compressed air. A conducting gel must then be added through a small hole at the top of the electrode cup. This method does not resist well over a long period, however, so for recordings made overnight or longer, it is preferable to use a conducting paste applied to the inside of the electrode cups before attaching them. In the past, bentonite and salt compounds were used, but these often provoked skin burns, particularly when recordings were taken from subjects over several nights. Modern water-soluble pastes free of sodium chloride which irritates the skin, are preferable, particularly as they improve electrode adhesion to the scalp. It is important to use electrodes of the same metal (silver or stainless steel) as mixing different metals may cause artefacts. The use of teflon lead insulation also considerably improves recording quality.

Derivations

The number of electrodes required to receive the EEG signal depends on the pathology studied. Sleep recordings for epileptic subjects will differ in the number and location of electrodes used compared with recordings for sleep disorders, where only sleep continuity and architecture are studied. In the latter case, a differential understanding of EEG activity in the different regions of the brain is of no value. The manual published by Retschaffen and Kales [20] recommends referential recording with a single EEG lead, either C3 or C4, referenced to an indifferent electrode placed on the ear lobe or contralateral mastoid (A1 or A2). The derivations used are thus C3-A2 or C4-A1.The two proposed derivations are recommended not because of focalised EEG activity but for reasons of safety, maintaining at least one back-up electrode in case of any technical failures with one derivation or the other. Moreover it is better to place the indifferent electrode on the ear lobe rather than on the mastoid which often causes electrocardiogram artefacts, particularly at the end of recording.

There are a number of advantages in using C3-A2 or C4-A1 derivations. On one hand the relatively large interelectrode distance optimises EEG signal amplitudes for sleep analysis, and on the other most sleep grapho-elements, sleep staging criteria (vertex sharp waves, K complexes and spindles) are well visualised in these regions. Moreover, high-voltage NREM slow waves seen maximally in frontal regions yet show clearly on central electrodes. Finally, the relatively large distance from frontal regions minimises the «contamination» of ocular movements in REM sleep on EEG activity. By contrast, the alpha rhythm of relaxed wakefulness is maximal over the occipital poles. This does not pose any particular problem as alpha rhythm is easily centrally characterised in most humans [4] and sleep onset is easy to distinguish particularly as reduced wakefulness is expressed as a spreading of alpha rhythms toward the frontal regions of the brain. Nevertheless, in certain subjects whose alpha rhythm is difficult to assess, a new derivation must be applied over the occipital region: O1-A2 or O2-A1. This derivation can be very useful in recognising sleep onset, particularly in subjects with few slow eye movements.

EEG channel setting (table 9.1)

Gain

The recommended calibration gives a wave deflection of 7.5 to 10 mm for a 50 microvolt signal. This gives the best assessment of the EEG signal in sleep recording. This setting may and indeed should be modified in a number of cases: for children, whose amplitude signal is very high particularly in deep NREM sleep, it may be necessary to use 5 mm amplification for a 50 microvolt signal and in subjects with very low voltage EEG signals, a 14 mm amplification for a 50 microvolt signal.

Time constant

The time constant is 0.3 seconds. It comprises a high-pass 0.5 Hz filter. This time constant must not be changed or shortened in any way, (0.1) even if there is interference from slow potentials due to sweat, as this would have an effect on the EEG signal such as inhibiting the slow waves in deep NREM sleep.

Low-pass filters

These are intended to provide the maximal visualisation of sleep frequency bands, the delta band, spindles as well as the beta band (15 to 18 cs), and minimise high frequency interference, particularly that produced by EMG. Filters of 60 to 75 Hz are recommended. Most modern polygraphs are equipped with 50 Hz rejecter active filters (high-pass low-pass filters) which should obviously be used.

Table 9.1. Different settings for EEG, EMG and EOG channels in polysomnographic monitoring. Time constants are expressed in seconds. The value shown after the dash indicates the corresponding high pass filter.

Variables	Derivation	Gain (50 µv)	Time constant	Low pass filter
standard				
EEG	C3-A2/C4-A1	7.5 -10 mm	0.3 s - 0.5 Hz	60 -75 Hz
EOG	ROC-A1/LOC-A1	7.5 - 10 mm	0.3 s - 0.5 Hz	15 Hz
EMG	Chin	15 - 20 mm	0.0 s - 5 Hz	120 Hz
low voltage subjects				
EEG	C3-A2/C4-A1	14 – 15 mm	0.3 s - 0.5 Hz	60 -75 Hz
high voltage subjects				
EEG	C3-A2/C4-A1	5 mm	0.3 s - 0.5 Hz	60 -75 Hz
subjects with no alpha rhythm				
EEG	O1 – A2/O2 – A1	7.5 -10 mm	0.3 s - 0.5 Hz	60 -75 Hz

Electrooculogram (EOG)

To assess sleep accurately it is essential to recognise eye movement activity during sleep. Indeed not only is the presence of rapid eye movements necessary to carry out a diagnosis of REM sleep but slow eye movements are also important indications of sleep onset. Finally eyelid movements can only occur during wakefulness and are thus a major indication of this state.

Eye movements can be recorded thanks to a potential difference between the cornea (positive) and the retina (negative). Thus electrodes placed next to the eyes can record the differences in potential, due to changes (as the eyeball moves) in the positions of the cornea and the retina relative to the fixed electrode.

Electrode placement

Retchschaffen and Kales [20] recommended placing the electrodes at the right and left outer canthus (right and left eye movements). They further indicated that to detect horizontal and vertical eye movements, electrodes should be offset from the horizontal, at approximately 1 cm above and 1 cm below the horizontal plane.

The skin surface to which the electrodes are applied should be cleaned with ether. It is better to use disposable, self-adhesive electrodes even though they are more expensive. This type of electrode is much more comfortable for the patient and is easy to replace in case of technical problems. This is particularly true for extended recordings. It is important to note that most electrodes come equipped with solidified contact jelly; if this is the case it is wise to remove the jelly and replace it with water soluble conducting paste. The use of collodion must be strictly avoided as technically, it can cause skin retraction resulting in poor contact, but also because it can cause an unpleasant sensation to the face; it is dangerous to use because of the risk of splashing into the eyes – particularly as acetone is used to remove electrodes.

Derivations

It is recommended [20] that the two ocular electrodes be referenced to the same A1 or A2 electrode, either ROC/A1 and LOC/A1, or the contrary. The same reference is recommended as that for the EEG signal. This has the advantage of registering eye movements as out-of-phase pen deflections while EEG activity (K complexes and slow waves, for example) will be seen as in-phase deflections. This eliminates certain artefacts, while having the capacity to distinguish eye movements in certain difficult cases.

These derivations are adequate to analyse human sleep satisfactorily, but if additional information is required on vertical and horizontal eye movements, (dream studies, supra-nuclear paralysis etc.) additional electrodes must be placed supraorbitally and infraorbitally and referenced either to the indifferent electrode or between them with various possibilities depending on the number of tracks available and the type of phenomenon observed. To determine the direction of eye movements, certain direct current (DC) derivations can be used, although this technique is difficult as it generates numerous artefacts due to signal shunting.

EOG channel setting (table 9-1)

EOG channel settings are the same as for EEG although some changes may need to be made: notable for the low-pass filter which must be limited to 15 Hz as there is no rapid component in eye movements and even a nystagmus can be detected with this filter. On the contrary, high frequency interference (EMG) is particularly problematic in observing certain low amplitude eye movements which should thus be completely eliminated. Some laboratories use shorter time constants (0.1s) to minimise contamination from the slow waves in deep NREM sleep on EOG lines. The danger of this type of calibration apart from the fact that it does not completely eliminate contamination from slow sleep waves, is that it makes it impossible to accurately distinguish the slow eye movements which are so useful in determining sleep onset.

Electromyogram (EMG)

The reference manual [20] recommends using the muscles beneath the chin to evaluate muscular activity during sleep. Other muscle groups (intercostal muscles, the anterior tibia, deltoids, masseters etc.) are reserved for monitoring different pathologies (sleep apnoeas, periodic limb movements, bruxism) and must always be studied in addition to chin muscle activity rather than as an alternative.

Electrode placement

The electrodes are placed beneath the chin, overlying the chin muscles. It is advisable to fix at least three electrodes to allow for a back-up electrode during the night. Unlike ocular electrodes, it is advisable to use gauze and collodion rather than self-adhesive electrodes which tend to come unstuck more easily. The skin must be thoroughly cleansed before fixing the electrodes and to make sure they are securely fixed, particularly in the case of bearded patients.

Derivations

Any combination of the three placements can be used. When calibrating at the patient's bedside (see below) the derivation producing the best quality signal should be used.

EMG channel setting

Gain

Gain should be as high as possible. A 14 to 15 mm amplification for a 50 microvolt signal is recommended. If this amplification proves to be too high during wakefulness, it can be reduced and later increased during sleep. Muscle activity during sleep is essentially used to differentiate REM sleep from the other stages. As the EMG signal may be very weak, particularly at the end of the night, it is essential to use high amplification.

Time constant

As the EMG signal is the highest frequency recorded in polysomnography it is necessary to use a short time constant. The reference manual [20] recommends using a time constant of 0.1 second, i.e. a 1.5 Hz high-pass filter. Modern polygraphs have much shorter time constants. It is advisable to use a constant of at least 0.03 seconds, i.e. a 5 Hz high-pass filter.

Low-pass filter

The filter must be set at 120 Hz to monitor the maximum frequencies used in indicating muscular signals.

Calibrating the subject in bed

Preparation must be meticulously checked when the subject is in bed, as any intervention in the room during the night is likely to waken the subject, thus modifying the recording. Each laboratory has its own checking procedure so only the essential steps will be listed here, to ensure that the system operates during the night.

It is essential to make sure that the subject has performed all the necessary bedtime tasks (washing etc.). Once the subject is lying in bed in the most comfortable position, if the electrode wires are not drawn together to form a bundle fixed to a 20-pin plug, they will have to be placed into the electrode box one by one. When the electrodes or 20-pin plug are connected, it is important to check that the electrodes have approximately the same impedance (about 10 Kohms). If one of the electrodes is defective it must be fixed again.

The electrode wires must not hamper the subject's movements. At this stage, physiological calibration should be carried out on the subject.

EOG

Ask the subject to open and shut the eyes, to blink, look straight ahead, look up, down, to the right and to the left. Each movement should be recorded on the tracing, amplification gains are adjusted at this stage.

EMG

Ask the subject to clench the teeth, chew etc. depending on the recording protocol; the amplification gains are adjusted at this stage.

EEG

Ask the subject to relax, with eyes closed for 30 seconds; make sure that the alpha rhythm is adjusted correctly, modifying the amplitude if necessary. Disconnect the 50 Hz rejecter filter and check that tracing continues normally, correcting any faults. Reconnect the 50 Hz rejecter filter and proceed with the other calibrations.

AUTONOMIC VARIABLES

Polysomnography may involve a large number of autonomic variables. As most of these techniques will be dealt with in more detail in other chapters of this book, only a brief reminder of the most common recording techniques for autonomic variables will follow i.e. cardiac and respiratory variables, left and right anterior tibia muscle movements, measurement of penile erection during sleep.

Electrocardiogram

An electrocardiogram (ECG) is taken during sleep to detect variations in heart rate and rhythm. ECG variations may be observed in a wide range of pathologies such as parasomnias, epileptic fits, panic attacks etc., but ECG is most commonly used to assess the severity of respiratory events (pauses, apnoeas) occurring during sleep.

Electrodes may be cupped, fixed with gauze and collodion like those used for sleep recordings, but in the interests of patient comfort, it is preferable to use self-adhesive electrodes which stick easily to skin which has been cleansed with ether.

Placing electrodes

Several electrode positions exist. One of the most common is recording from the right shoulder and the left leg, or the contrary, but the drawback with this derivation is that it requires long leads in addition to those already used if leg muscles are being monitored, for example. It is preferable to use both left and right shoulders, or a derivation using parasternal electrodes.

Table 9.2. Different settings for ECG, respiratory and muscular channels (anterior tibialis). Time constants are expressed in seconds. The value shown after the hyphen indicates the corresponding high pass filter.

Parameter	Derivation	Gain (50 µv)	Time constant	Low pass filter
ECG	R. shoulder/left leg	*	0.1 s – 1.5 Hz	30 Hz
	R. shoulder/left shoulder	*	0.1 s – 1.5 Hz	30 Hz
	Parasternal	*	0.1 s – 1.5 Hz	30 Hz
Respiration	Naso-buccal	*	1 s – 0;15 Hz	15 Hz
	Thoracic	*	1 s – 0;15 Hz	15 Hz
	Abdominal	*	1 s – 0;15 Hz	15 Hz
Leg movements	Right anterior tibialis	*	0.03 s – 0.5 Hz	120 Hz
	Left anterior tibialis	*	0.03 s – 0.5 Hz	120 Hz

ECG channel setting (table 9.2)

ECG sensitivity varies according to the position of the electrodes, which must be adjusted when recording starts and during calibration. A constant of 0.1 sec, i.e. a 1.5 Hz high-pass filter linked with a 15 Hz low-pass filter will give very satisfactory results.

Calibrating the subject

When monitoring starts, the gain must be set according to the position of the electrodes. It is important to note that the speed used during sleep recording (10 or 15 mm/sec) is much slower than for electrocardiography (25 mm/s) and that the subject's movements will modify inter-electrode distances. This means that ECGs carried out in polysomnography can only provide partial elements. It is however necessary to make sure that the P and QRS waves are detectable and that the latter is negative. If these waves prove to be too difficult to detect, the electrode placements must be changed.

Respiratory variables

There are two types of measurement for respiratory parameters: qualitative measurement and semi-qualitative quantitative measurement.

■ **Qualitative measurements**

These measurements qualify and accurately describe the phenomenon but are unable to quantify it.

Thoracic and abdominal straps

Whether mercury strain gauges or graphite rubber, these work on the principle of resistance variations occurring through distortion.

Mercury capillary length gauges [23] consist of a sealed, elastic tube, containing an electric conductor (mercury). The gauge is stretched and reduced by the respiratory movements, which modifies its length and section, and thus increases its resistance. This resistance variation is transformed into an electrical variation through the intermediary of a Wheatstone link operated by a high capacity electric battery. As the current fluctuates in inverse proportion to the length of the gauge, two relatively long gauges (circumference gauges) will give an approximate measurement of volume [15]. The current clinical use of shorter gauges which are easier to use, differentiates between the different types of apnoeas.

Graphite straps comprise a small rubber cylinder made in graphite. Variations in volume created by the movement of the rib cage or the abdomen cause the strap to stretch and modify the section provoking a change in resistance. This resistance variation is transformed into an electrical variation through a Wheatstone link operated by a high capacity electric battery. It is more difficult to obtain a quantitative measure of volume with this type of gauge. But as the amplitude observed is proportionate to stretching, a weak amplitude is theoretically the result of less volume and may indicate a respiratory event. The advantage with this type of monitor is its great stability and high sensitivity.

Placing the straps

Whatever the type of gauge, it is essential to use two straps, one to measure variations in the volume of the rib cage and the other those of the abdomen. The thorax and abdomen normally expand in synchronisation. This mechanism appears on the trace as phased respiratory movement variations recorded for each strap. In the case of particular respiratory events involving either a reduction of diaphragm tone or respiratory muscle tone or the partial or total obstruction of the upper airways, respiration of the paradoxical type may appear. This anomaly is characterised on the recording by dephasing and amplitude changes of the two tracks monitoring the thoracic movements and abdominal movements.

The straps must be carefully attached and fixed with tape to stop them from moving and altering the signal.

Oro-nasal sensors

The use of single straps is insufficient to detect apnoeas during sleep, which are characterised by a blockage of the air passage. A qualitative assessment of the blockage of the air passage can be carried out effectively by detecting the variations in the temperature of exhaled breath which will be warmer than the ambient air. By using sensors which are sensitive to temperature variations [3], thermocouples and thermistors, the passage or non-passage of air can be detected.

The thermocouple consists of a closed circuit composed of two different metallic conductors. A current circulates when the two junctions are kept at different temperatures (Peltier effect). A variation in temperature causes the circuit to generate a variation in electric potential proportionate to the temperature variation. This potential variation is directly monitored by the polygraph.

The thermistor, an electric component in the form of a glass droplet, is a resistor whose value fluctuates in function of temperature variations. The thermistors used are generally of the NTC type (negative temperature coefficient) as their value diminishes when the temperature rises. The warmer exhaled air modifies the ohm resistance which is transformed into an electric variation through a Wheatstone link operated by a high capacity electric battery. This montage ensures high sensitivity and good reliability. A number of precautions must be taken. The thermistors must not be placed under an air conditioning outlet as the cool air could modify thermistor sensitivity; it is also important to check that the temperature of the room is not too high; finally the thermistor must never be in contact with the skin, as this would cause it to heat and remain at body temperature (the temperature to which it is sensitive) and would no longer be able to record changes in temperature.

Placing sensors

The air flow must be monitored at mouth and nostril level. Thermistors can be mounted in series on the same track with two thermistors for the nostrils and one for the mouth.

Unlike straps, thermocouples and thermistors give no indication of volume, as they are sensitive to thermal variations only, and exhaled air, even in small quantities, will not show sufficient thermal variation to proportionately modify the amplitude of the signal.

Respiratory airway setting (table 9.2)

Whether for upper airways or thoraco-abdominal movements, the setting is more or less the same apart from amplitudes.

Gain depends on a large number of factors (position, stretching of the strap etc.) and can only be properly established when calibration is carried out on the subject in bed. As respiratory movements are very slow, a time constant of 1 sec i.e. a 0.15 high-pass filter associated with a 15 Hz low-pass filter, gives very good results.

Calibrating the subject

Respiratory airway amplification is set at this stage.

The straps: ask subjects to breathe normally, inhaling and exhaling deeply. Check that the amplitude is not saturated (plateau signal on the polygraph). Check that the two thoracic and abdominal airways are in phase. It is also important to note the moments of inhaling and exhaling.

Naso-buccal sensors: Check in the same way as for the abdominal and thoracic straps, asking the subject to open his mouth.

▓ Other qualitative measurements

Measurement of respiratory effort

Respiratory effort is generally assessed quantitatively by measuring pleural pressure (see below); it can however be measured qualitatively by using the electrostatically charged mattress [1].

The static charge-sensitive mattress measures variations in the electrostatic field set up by the subject's movements. By using different time constants, it can distinguish respiratory movements from the subject's movements or from the dilation and contraction movements of the heart in the rib cage (balistocardiography). This system not only discriminates between different types of apnoeas but will also indicate respiratory effort and objectify certain events such as partial obstruction or an increase in respiratory resistance which are difficult to isolate without using more invasive quantitative methods.

Nevertheless, this device must operate at a very low speed (< 2 mm/s) to be read on the polygraph, which is incompatible with sleep recordings.

Measurement of tracheal sounds

Tracheal sounds are measured by small microphones placed on the neck overlying the trachea, or on the upper thorax. They will record the presence of any tracheal air flow and provide an indication of snoring and breathing uptake after unspecified respiratory events. They will provide quantitative indications when combined with computerised techniques (spectral analysis) [17].

■ Semi-quantitative and quantitative measurements

These will only be summarised here as the methods (which can be invasive), go beyond the scope of traditional polygraphic recording, which should involve as few constraints as possible.

Impedence plethysmography

Impedence plethysmography uses the thorax as a conductor. By applying a small, high frequency current (100 kHz) across the thorax using a pair of electrodes at low intensity, it is possible to monitor variations in impedance linked to respiratory movements. However, the relation between changes of impedance and variations in the volume of the rib cage are very complex and vary considerably between subjects. By associating a measurement of the volume of air exchanged [25] with a measurement of impedance variations, a quantitative estimate of changes in volume may be obtained. But the fact that the data observed are highly dependent on the subject's position and that these techniques often interfere with other measurements render these methods difficult to use in normal clinical conditions.

Inductance plethysmography

Changes in respiratory volumes (characterised by changes in the cross-sectional area of the rib cage and abdomen) can be measured electronically by determining variations in inductance [23]. Transducers (induction coils) consist of an insulated wire sewn in the shape of a horizontally orientated sinusoid onto a wide elastic band. Two transducers are used, one at the level of the thorax, the other over the abdomen. The respiratory movements stretch and release the transducers producing variations in inductance which are demodulated and then transformed into proportional voltage variations. A calibration process converts these changes into variations of volume, determining the relative contributions of the rib cage and abdomen to the current volume. This technique is accurate but difficult to apply during sleep as, apart from its relatively high cost, it requires long and meticulous calibration and is very sensitive to changes in posture, which may considerably alter the initial volume calibration.

Pneumotachography

This is the standard method for obtaining quantitative information on the volume of airflow. It works on the principle that laminar flow between two points is proportional to the differential pressure between these two points (Poiseuille's law). Several types of pneumotachographs are in use, giving an accurate quantification of airflow during sleep.

Quantitative pneumographs can be calibrated to give linear measurements in relation to a precise range of flow. Measurement is carried out by assessing the differential pressure using either a deflector screen or a calibrated grid. Apart from the fact that these instruments involve wearing an airtight facemask, their weight requires them to be attached in a way which interferes with a good sleep study.

Semi-quantitative pneumotachographs are usable devices which work on the same principle as the preceding ones, but their lightness makes them less awkward to install and disturbs sleep less than conventional pneumotachographs. However they give only a semi-quantitative indication of ventilation.

Measurement of pleural pressure

In order to quantify inspiratory effort, a slightly invasive method must be used such as recording pleural pressure measured by oesophageal pressure. This gives a quantitative assessment of the effect of the respiratory muscles on thoracopulmonary mobility. In the horizontal position, the absolute value of oesophageal pressure is proportionately modified by the weight of the midintestine viscera, depending on the position of the subject; the variation in oesophageal pressure does however remain a good indicator of variations in pleural pressure; it is the only way to analyse ventilatory changes linked to variations in airway resistance during sleep and particularly to distinguish between central and obstructive hypopnoeas. Latex oesophageal balloons remain the standard technique [18] but they are difficult to use at night and it is preferable to use piezoelectric optical fibre or strain gauge diaphragm resistance probes, which often have several sensors providing measurements at various levels [13]. This type of sensor is more costly but far more comfortable and possibly more reliable during sleep.

Detection of flow limitation

While the standard oro-nasal devices such as thermistors or thermocouples (see above) are sufficient to demonstrate clear cases of apnoea, they are quite insufficient to diagnose hypopnoeas and upper airway resistance.

In upper airway resistance events, the increase in breathing effort no longer generates an increase in flow, showing a flat curve characterising air flow limitation. Even if detected flow limitation (DFL) does not fully overlap with upper airway resistance, it can be used as a non intrusive way to identify the latter.

Flow limitation can either be measured by pneumotachograph (see above) or by measuring nasal pressure by means of a nasal cannula pressure transducer system [10]. This device consists of a box containing the pressure transducer supplied by a battery and a disposable nasal cannula with an anti-bacteria filter. The device is linked to the polygraph, either using direct current or via the electrode box, parametered at 30 μvolts per mm, a time constant of 0.05 Hz (3 seconds) and a low pass filter of at least 70 Hz. When ventilation tends to occur through the mouth, there is a partial or total loss of signal, implying the need to add a thermistance or buccal thermo-couple to the system to identify this particular condition.

Nasal pressure cannula is limited in terms of its lowered sensitivity in the case of weak to moderate flow limitation or high overall inspiratory resistance. An additional signal is thus required, such as oesophageal pressure (see above) or pulse transit time.

Pulse transit time (PTT)

PTT is the time (roughly 250 ms) taken by arterial pulsation to cover the distance between the aortic valve and the extremities (usually the fingertip). The rigidity and pressure of the artery wall are the main factors which determine the speed of propagation of the pulse wave, and this largely depends on blood pressure (BP). A rise in BP will increase the rigidity and pressure of the artery wall, thus increasing PTT. As it is difficult to ascertain the precise moment when the aortic valve opens, PTT is measured using a point which is easy to detect: the R wave of the electrocardiogram. This means that the time between the onset of the R wave and the opening of the aortic valve (time of left ventricle isometric contraction) is included in PTT.

BP increases with inspiration, whereas expiration causes it to drop. By measuring PTT it is possible to measure the increase in the difference between inspiration and expiration while assessing the increase in respiratory effort. When this method is validated by oesophageal pressure, it is possible to ascertain whether hypopnoeas are central or obstructive [2].

Finally PTT measurement has been shown to be a highly sensitive marker of sympathetic activation and sleep-related EEG arousals [19].

By providing information on both respiratory events and sleep fragmentation, PTT measurement has proved to be a valuable parameter not only in polygraphy conducted at home, because of its simplicity, but also in those conducted in the laboratory, because of the specific data it provides.

Oxygen and carbon dioxide measurement

If skin electrodes are likely to yield an approximate PaO_2 and $PaCO_2$ in children [4], these estimates are far less accurate in adults, whose skin is thicker; in addition, it is necessary to recalibrate the device frequently (every 4 hours), thus disturbing sleep.

Oxymetry (SaO_2 measurement)

This must be carried out without fail in any polysomnography including a respiratory variable measurement and be coupled with the ECG recording. There are two types of oxymeter: transmitter oxymeters and pulse oxymeters. They are based on the principle of the difference in infrared wavelength absorption for oxyhaemoglobin and reduced haemoglobin. Several models are available on the market and have been subject to review [22].

The sensors can be placed on the ear, the finger or the nose. Both methods are insensitive to skin pigment, venous blood, connective tissue, elements which are known to interfere with the measurement of saturation. The description of SaO_2 response characteristics is problematical because of the wide range of possible variables (minimal, absolute or mean SaO_2 by epoch or stage, the mean minima per epoch etc.). Each laboratory must adapt the different measurements to their diagnostic needs.

Other autonomic parameters

Many other autonomic parameters can be studied using polysomnography: bladder pressure [8], continuous arterial pressure [24] echocardiography [9] etc., but these techniques are often highly sophisticated and go beyond the scope of standard polygraphic monitoring. We will restrict the description here to two widely used techniques: monitoring periodic limb movements and penile plethysmography.

Periodic limb movements

For a wide range of pathologies, it is important to complete the diagnosis with recognition of restless leg during wakefulness and periodic limb movements during sleep.

Cup electrodes may be used, but it is better to use disposable self-adhesive electrodes which are better adapted to this type of examination.

Electrode placement

Electrodes are applied to the anterior tibialis muscles. It is important to place an electrode on both lower limbs, as movements often involve only one leg at a time and can alternate between the limbs during the night [7, 16]. Activity may also be recorded from supplementary muscles such as the soleus, biceps and femoral quadriceps. Finally in rare cases where movement occurs in the upper limbs [6], the arm triceps and biceps must be monitored.

EMG channel setting (table 9.2)

Gain must be adjusted when the subject is calibrated, the other settings being the same as for EMG. But care must be taken as leg muscle recording requires the use of long flex which must be securely fixed with tape to avoid disturbing the subject during sleep.

Calibrating the subject

Ask the subject to imitate the periodic movement, involving:
-    extending the big toe
-    dorsiflecting the foot
-    bending the knees and
-    the hips.

Have the subject perform each stage of the movement in turn, checking that each movement is correctly recorded and making the final adjustment to the gain. If muscular contractions are not correctly recorded, the electrodes must be repositioned.

Penile plethysmography

This consists of measuring variations in penile volume during sleep, qualitatively or quantitatively. The technique involves rubber or mercury-filled strain gauges [12] which, when stretched, will create a variation in resistance transformed into volts using a Wheatstone device. To obtain a semi-quantitative indication, sophisticated devices must be used requiring careful calibration both of the apparatus and the subject.

CONCLUSION

Polysomnographic sleep monitoring demands rigorous and difficult methodology simultaneously using diverse techniques from disciplines as varied as neurophysiology or respiratory physiology. This type of examination provides indispensable information for diagnosing sleep disorders. The value of the recording depends on the accuracy of interpretation, which not only relies on the clinician's knowledge but also on the quality of the preparation and the clinical and technical surveillance of the subject during the night.

REFERENCES

1. ALIHANKA J., VAAHTORANTA K., SAARIKIVI I. – A new method of long-term monitoring of the ballistocardiogram, heart rate and respiration. *Am. J. Physiol., 240*, 384-392, 1981.
2. ARGOD J., PEPIN J.L., SMITH R.P., LEVY P. – Comparison of oesophageal pressure with pulse transit time as a measure of respiratory effort for scoring non apnoeic respiratory events. *Am. J. Respir. Crit. Care Med., 162*, 87-93, 2000.
3. BERENDES J., KIPP J., HANSIOYA P.L. – Technique of polysomnography. *Am. J. EEG Technol., 22*, 63-66, 1982.
4. BRAZIER M.A.B. – The electrical fields at the surface of the head during sleep. *Electroencephalogr. Clin. Neurophysiol., 1*, 195-304, 1949.
5. BURKNI N.K., ALBERT R.K. – Non invasive monitoring of arterial blood gases. A report of the ACCP section on respiratory pathophysiology. *Chest, 83*, 666-670, 1983.
6. COCCAGNA G., LUGARESI E., TASSINARI C.A., AMBROSETO C. – La syndrome delle gambe senza riposo (restless legs). *Omnia Med. Ther.* (Nuova serie), *44*, 619-684, 1966.
7. COLEMAN R.M. – Periodic movements in sleep (nocturnal myoclonus) and restless legs syndrome. *In: Sleeping and waking disorders* (indications and techniques), C. Guilleminault (ed.), Addison-Wesley, Menlo Park, California: 265-295, 1982.
8. GROSSMAN H.B., KOFF S.A., DOKNO A.C. – Cystometry in children. *J. Urol., 117*, 646-648, 1977.
9. GUILLEMINAULT C. – Sommeil et Coeur. *Bulletin veille-sommeil, 5*, 5-9, 1991.
10. HOSSELET J.J., NORMA R.G. AYYAPA I., RAPPOPORT D.M. – Detection of flow limitation with a nasal cannula pressure transducer system. *Am. J. Respir. Crit. Care Med., 12*, 770-775, 1998.
11. JASPER H.H. (Committee chairman) – The ten twenty electrode system of the International Federation. Electroencephalograph. Clin. Neurophysiol., 10, 371-375, 1958.
12. KARAKAN I. – A simple and inexpensive transducer for quantitative measurements of penile erection, during sleep. *Behav. Res. Method Instrum., 1*, 251-252, 1969.
13. LAUNOIS S.H., FEROAH T.R., CAMPBELL W.N., ISSA G., MORRISON D., WITHELAW W.A., ISONO S., REMMERS J.E. – Site of pharyngeal narrowing predicts outcome of surgery for obstructive sleep apnoea. *Am. Rev.Respir. Dis., 147*, 182-189, 1993.
14. LOOMIS A.L., HARVEY E.N., HOBART G.A. – Cerebral states during sleep as studied by human brain potentials. *J. Exp; Psychol., 21*, 127-144, 1937.
15. LOVERIDGE B., PEREZ-PADILLA R., WEST P. – Comparison of the stability of the inductance plethysmography versus mercury strain gauges in measuring ventilation. *Am. Rev. Respir. Dis., 129* (4), A82, 1984.
16. LUGARESI E., TASSINARI C.A., COCCAGNA G., AMBROSETTO C. – Particularités cliniques et polygraphiques du syndrome d'impatiences des membres inférieurs. *Rev. Neurol. Paris, 113*, 545-555, 1965.
17. MESLIER N., RACINEUX J.L. – Use of tracheal sound recordings to monitor airflow during sleep. *In: Sleep Related Disorders and Internal Diseases*, J.H. Peter, T. Podszus, P von Wichert (eds.), Springer, Berlin, 121-124, 1987.
18. MILIC EMILI J., MEAD J. TURNER J.M., GLAUSER E.M. – Improved technique for estimating pleural pressure from oesophageal balloons. *J. Appl. Physiol., 19*, 207-211, 1984.
19. PITSON N., CHLINA S., KNIJN M., Van HERWAADEN M., STRADLING J. – Changes in pulse transit time as markers of arousal from sleep in normal subjects. *Clinical Science, 87*, 269-273, 1994.
20. RECHTSCHAFFEN A., KALES A. (eds.) – *A manual of standardized terminology, techniques and scoring system for sleep stages of human subjects.* Brain Information/Brain Research Institute, Los Angeles, 1968.
21. SACKNER D.J., NIXON A.J., DAVIS B., ATKINS N., SACKNER M.A. – Non invasive measurement of ventilation during exercise using a respiratory inductive plethysmograph. *Am. Rev. Resp. Dis., 122*, 867-871, 1980.
22. SEVERINGHAUS J.W., KELLEHER J.F. – Recent developments in pulse oxymetry. *Anesthesiology, 76*, 1018-1038, 1992.
23. SHAPIRO A., COHEN H.D. – The use of mercury capillary length gauges for the measurement of the volume of thoracic and diaphragmatic components of human respiration: A theoretical analysis and practical method. *Trans NY Acad. Sci. Ser. II, 27*, 634-649, 1965.
24. SMITH N.T., WESSELING K.H., De WITT B. – Evaluation of 2 prototype devices producing non invasive, pulsatile, calibrated blood pressure measurement from a finger, *J. Clin. Monitor., 1*, 17-29, 1985.
25. VICTORIN L., OLSSON T.. – Transthoracic impedance – A tool for the evaluation of air and fluid changes in the lung. *In: Neonatal intensive care*, J.B. Stetson, P. R. Swyer (eds.), Warren H. Green, St Louis, 1976.

Chapter 10

Ambulatory systems

T. Penzel and J.H. Peter

Schlafmedizinisches Labor, Klinik für Innere Medizin, Klinikum der Philipps-Universität, Marburg, Deutschland

THE ROLE OF AMBULATORY SLEEP RECORDING

During the last decade advances in microtechnology have resulted in the development of ambulatory sleep recording systems which fulfil all the demands and are able to record everything that can be recorded in a sleep laboratory. The technical differences have become so small, that the same recording equipment can be used for both sleep laboratory settings and out-patient use. The role of ambulatory sleep recording has thus shifted from the question of what is technically feasible, because nearly anything is possible, to that of the strategies involved. It is necessary to consider the effectiveness of diagnostic procedures in terms of final therapeutic outcome and costs in order to define a new role for ambulatory sleep recording.

Patients complaining and suffering from non-restorative sleep have to be asked a few well thought-out questions first. These questions should clarify sleep habits (bed time, sleep duration, shift work), the use and misuse of sleep disturbing drugs and the presence of other organic or psychiatric disorders (e.g. pain, depression) which could account for non-restorative sleep. These questions can be asked by a family doctor who redefines the complaints in terms of the classical symptoms of insomnia and hypersomnia. If the questions do not clarify the reason for non-restorative sleep and if the answers indicate an intrinsic dyssomnia then the family doctor refers the patient for sleep investigation. Intrinsic dyssomnias, according to the International Classification of Sleep Disorders [18], require sleep investigation and possibly a test of daytime sleepiness using the MSLT or other validated tests. Cardiorespiratory polysomnography and MSLT in a sleep laboratory are the recognised "gold standards" for the diagnosis of most intrinsic sleep disorders according to the review of Chesson *et al.* [34] on indications for polysomnography and related procedures. The gold standard consists of standardised techniques for the recording of sleep EEG, respiration, ECG and limb movements. The recording has to be attended by trained nursing staff. Additional video recording may be important in some cases. The effects of adaptation to a sleep laboratory and reliability in recording and evaluation have been documented for cardiorespiratory polysomnography.

On the basis of this, the question arises as to whether ambulatory sleep recording can be used instead of the gold standard. One argument is cost. Until now the costs of successful high-quality ambulatory sleep recording have not been shown to be lower than costs in a sleep laboratory [28]. There was no clear indication that the management of patients using ambulatory sleep recording is less costly than the use of direct sleep laboratory investigations as the first test.

Screening for sleep disorders could be another argument. There has been some debate over the need to screen for obstructive sleep apnoea due to its high prevalence, increased level of morbidity and mortality, and increased public safety risk [4]. It has been suggested that screening should focus on commercial drivers of dangerous goods, bus drivers, pilots and other limited groups. Even for such specific groups, the cost-effectiveness of screening for sleep apnoea has not been established as yet.

Ambulatory recording of sleep related breathing disorders using a reduced number of parameters is important and useful in many cases with a clear cut risk profile for sleep related breathing disorders, but without reported complaints of excessive sleepiness or non-restorative sleep.

Contributions to the risk profile are obesity, arterial hypertension, nocturnal hypertension, a non-dipping pattern in 24-hour blood pressure recording, predominantly nocturnal arrhythmia polyglobulism, right heart failure, and retrognatia.

Ambulatory sleep recording is very useful for annual treatment control studies in patients with sleep related breathing disorders under ventilation therapy. In addition ambulatory recording of sleep with complete polysomnography is useful in occupational medicine, research studies, space research, and for special questions which cannot be addressed in a fixed sleep laboratory.

INDICATIONS FOR SLEEP RECORDING

The primary reason for sleep recording is to test a suspected sleep disorder. Therefore it is essential that the initial clinical interview be conducted by someone with sound knowledge of sleep medicine. The clinical interview must be complemented by medical, neurological and psychiatric examinations as indicated by the complaints reported by the patient. Standardised questionnaires and sleep diaries can complete the clinical picture and can help in the process of verifying symptoms. Having found non-restorative sleep with either symptoms of hypersomnia or insomnia or both, the question arises as to which patients require a cardiorespiratory sleep recording for differential diagnosis. Indications for polysomnography have been summarised by the ASDA [34] and have been evaluated by the Cochrane Collaboration [35] to structure the findings. The approved indications are diagnosis of sleep related breathing disorders, titration of CPAP pressure in patients with sleep related breathing disorders, follow-up studies in patients where an adjustment of therapy is assumed, narcolepsy, diagnosis of paroxysmal arousals or suspected seizure related sleep disruptions, and periodic limb movements during sleep. In most cases of psychophysiologic insomnia and restless legs syndrome a diagnosis can be reached by a clinical interview. The indications are summarised in 10.1.

Table 10.1. Indications

ICSD term and code	ICD 10 term and code	leading symptoms
Psychophysiologic insomnia 307.42-0	Non-organic insomnia F51-0	insomnia
Sleep state misperception 307.49-1	other non-organic sleep disorder F51.8	insomnia
Idiopathic insomnia 780.52-7	Disorders of initiating and maintaining sleep G47.0	insomnia
Narcolepsy 347	Narcolepsy and cataplexy G47.4	hypersomnia
Recurrent hypersomnia 780.54-2	other sleep disorders (Kleine-Levin-syndrome) G47.8	hypersomnia
Idiopathic hypersomnia 780.54-7	Pathologic increased sleep need G47.1	hypersomnia
Post-traumatic hypersomnia 780.54-8	Pathologic increased sleep need G47.1	hypersomnia
Obstructive sleep apnea 780.53-0	Sleep apnea G47.3	hypersomnia,
	Adipositas permagna with alveolar hypoventilation (Pickwick-syndrome) E66.2	insomnia
Central sleep apnea 780.51-0	Sleep apnea G47.3	hypersomnia,
	Periodic breathing R06.3	insomnia
Central alveolar hypo-ventilation syndrome 780.51-1	Sleep apnea G47.3	hypersomnia, insomnia
Periodic limb movement disorder 780.52-4	other extrapyramidal disorders and movement disorders G25.8	hypersomnia, insomnia
Restless legs syndrome 780.52-5	other extrapyramidal disorders and movement disorders G25.8	hypersomnia, insomnia

Ten out of twelve intrinsic dyssomnias according to ICSD [18] with clear indications for cardiorespiratory polysomnography (PSG) with ICSD and ICD-10 codes are listed. Psychophysiologic insomnia (307.42-0; F51-0) and restless legs syndrome (780.52-5; G25.8) do not usually need a cardiorespiratory polysomnography, like most other symptomatic sleep disorders (extrinsic, circadian sleep disorders, parasomnias and sleep disorders associated with mental, neurologic or other medical disorders).

INDICATIONS FOR AMBULATORY RECORDING

The successful application of ambulatory cardiorespiratory polysomnography requires the experience of a sleep laboratory due to the rough recording environment under home conditions.

Ambulatory cardiorespiratory polysomnography may need more repetitions due to increased electrode or system failure during the recording period [28]. Ambulatory polysomnography has not proved superior to sleep laboratory based polysomnography. In consequence useful ambulatory recording of sleep disorders is limited to problem-oriented recording methods with a reduced number of signals. A limited number of carefully selected signals is sufficient. Clear indications for ambulatory sleep recording are a subset of the above listed indications:

- therapy control studies in patients with sleep related breathing disorders
- patients with no complaints of non-restorative sleep but with clinical findings indicating a high risk of suffering from sleep related breathing disorders
- occupational medicine and research studies

In 1992 the German Sleep Society [26] and in 1994 the American Sleep Disorders Association (ASDA) [33] compiled recommendations for the use of ambulatory sleep recording in sleep related breathing disorders. In both reviews it was stated that the base diagnosis must be done in a sleep laboratory under conditions of cardiorespiratory polysomnography with attending personnel. To exclude sleep related breathing disorders, cardiorespiratory polysomnography is also required, with attending personnel. Regular annual therapy control studies in sleep related breathing disorders can be done with ambulatory recording, unless a change of CPAP pressure or an additional sleep-wake disorder is suspected thus indicating cardiorespiratory polysomnography. The first control study after initiating CPAP therapy should be done in a sleep laboratory. It is good practice to conduct this first study after having applied CPAP for 3 months, and often a change in pressure is required.

THE DEVELOPMENT OF AMBULATORY SLEEP RECORDING DURING THE LAST 30 YEARS

Ambulatory sleep recording started with long-term EEG recording in the mid-70s. The first long-term EEG recording was introduced to track spontaneous rare epileptic seizures in an ambulatory setting [6]. These systems used analogue tape recording and were able to record 4 to 16 channels of EEG if several recorders were coupled [7]. Many long-term recorders were built using the Oxford Medilog 4-24 recorder and later the Oxford Medilog 9200 recorder which both recorded on standard audio cassettes [3, 6, 7, 20]. Most sleep disorders required the recording of other signals beside EEG. In order to cope with these demands, existing EEG systems were expanded for the recording of ECG and respiration first to diagnose patients with sleep apnoea [3] or those with periodic leg movements [20]. With the development of microprocessors it became possible to calculate heart rate and respiratory rate at recording time and store these data in digital memory. One of the first systems of this kind was the Vitalog PMS-8 [21]. The Vitalog was improved over the years and received more memory and more processing power. The Vitalog HMS-5000 could record respiratory effort by means of inductive plethysmography, airflow by means of oro-nasal thermistors, oxygen saturation, heart rate, snoring sounds, EEG alpha- and delta-power, EOG, EMG of the limbs, body movement and body position. Up to 23 different parameters were recorded. For many signals, only the processed parameters were stored in the limited digital memory (512 Kbytes). Data compression allowed a recording duration of slightly more than 24 hours. The Vitalog HMS-5000 was a pioneer system in many aspects. The system also allowed the remote control of a BiPAP ventilator. An algorithm was implemented which imitated the pressure titration based on the measured oxygen saturation. The pressure titration with repetitive ramps of pressure increases could be evaluated to estimate the optimal CPAP pressure for the patient. The goal was to imitate a pressure titration varying between sub-optimal pressure and a pressure higher than required. In that way no automatic pressure titration was implemented, due to the notion that the decision was up to the physician [19]. This option has not been followed in the current development of auto-CPAP equipment.

Today analogue tape recorders have been completely replaced by digital systems which are superior in terms of storage capacity and have fewer or no mechanical parts and consecutive problems. They are also less expensive. The digital systems range from one-channel recorders (e.g. actigraph) to reduced channel sleep recording with 4 to 8 signals (table 10.2) to multi-channel sleep recording with 10 or more signals (table 3). In the following three sections a selection of systems from all three categories is presented. The selection was made on the basis of available publications and validations, and is intended to illustrate the different concepts used in these devices.

AMBULATORY RECORDING OF SLEEP DISORDERS WITH 1-3 CHANNELS

Very simple one to three channel systems cannot be used as sleep recording per se. But such systems can help to verify symptoms specific for sleep disorders.

Activity recording

Activity monitoring can be regarded as a simple recording tool which does not interfere much with patient behaviour. The recording device is very small and records accelerations of the arm in variable time intervals (typical duration: 1 minute). Depending on the time interval the recording may last from one day to many weeks. With published algorithms it is possible to distinguish "sleep" and "wake" [9, 31]. The reliability of this evaluation is limited, so a continuous recording of at least one week is needed [2]. The evaluation of the pattern of movement can give indications for insomnia, PLMS or severe sleep apnoea. Actigraphy is useful in monitoring changes of the sleep-wake schedule in patients with sleep-wake rhythm disorders. Such changes can be found in patients with shift work, narcolepsy or delayed sleep phase syndrome.

Simplified sleep-EEG recording

A one-channel EEG recorder for the on-line calculation of sleep stages has been introduced recently [10]. The Quisi system records one signal derived from three electrodes placed on the forehead (Fp1, Fz – used as reference, Fp2) digitised at 128 Hz. This signal presents a mixture of frontal EEG, EOG and muscle tone. Spectral density is calculated and fed into an artificial neural network for the calculation of sleep stages imitating Rechtschaffen and Kales. The recorder stores the sleep stages and the artificial neural network weight factors, but no raw EEG signal. The recorder is simple to use and after being instructed a patient can mount the system himself before starting the recording. The resulting hypnogram has been validated in only one study so far [10] and the missing raw signal appears to be a severe disadvantage.

ECG recording

A holter ECG can give strong indications for sleep related breathing disorders. Patients with sleep apnoeas have a characteristic cyclical variation of heart rate [14]. In patients with sleep related breathing disorders and cardiac arrhythmia a holter ECG can help to analyse the type and degree of the nocturnal arrhythmia. In these patients a holter ECG should be used in parallel to a cardiorespiratory polysomnography or a reduced sleep recording system.

Blood pressure and autonomous parameter recording

Ambulatory blood pressure recording (ABPM) should not only be performed in patients with hypertension, but also in patients with sleep related breathing disorders as at least 50% have nocturnal hypertension associated with sleep apnoea. The use of intermittent recording systems (e.g. Spacelabs, Dinamap, Accutracer) is of limited value and is still in discussion, because the measurement occurs only every 30 minutes and reflects just one value which may be non representative after an arousal. It may reveal nocturnal hypertension and a non-dipping blood pressure profile. Non-invasive continuous blood pressure recording based on finger-photoplethysmography can give a continuous trace of blood pressure and reflects the apnoea associated blood pressure oscillations very well (e.g. Portapres, Finapres) [17]. This method has been validated in a sleep laboratory and proved to be useful [1] but expensive. A new method to monitor peripheral vasoconstriction is the peripheral arterial tonometry which records volume/pressure changes on the finger. This method monitors autonomous activation and subcortical arousal during sleep [32] and may evolve to an ambulatory arousal recorder.

Table 10.2. Reduced ambulatory recording methods with 4-8 channels

function	biosignals	1	2	3	4	5	6	7
respiration	airflow		X	X	X	X	X	X
	CPAP pressure		X	X	X	X		X
	respiratory effort		X	X		X		X
	snoring sound	X	X	X	X	X	X	X
	SaO2	X	X	X	X	X	X	X
cardiovascular	heart rate	X	X	X	X	X	X	X
sleep	results of EEG analysis							
	results of EOG analysis							
movement	anterior tibialis EMG		X			X		
	body position	X	X	X	X	X	X	X
	acceleration					X		

1. MESAM 4 (Medizintechnik für Arzt und Patient)
2. Polymesam (Medizintechnik für Arzt und Patient)
3. Merlin (Heinen und Löwenstein)
4. Somnocheck (Weinmann)
5. Apnoescreen Pro (Jäger)
6. Sleepdoc Porti II (Fenyves)
7. Sleepdoc Porti III / 4 (Fenyves)

REVIEW OF REDUCED SLEEP RECORDING SYSTEMS WITH 4 – 8 CHANNELS

Some early systems could record up to 8 channels using the analogous audio-tape technique (Oxford Medilog 9200, TEAC Recorder and Brainspy) [3, 7, 20]. These systems were developed for long-term EEG recording, originally aimed at detecting epileptic seizures. With the increasing importance of non-EEG signals in the differential diagnosis of sleep disorders, these systems were enlarged by adding signals for respiration, oxygen saturation and movement recording. As these early systems were limited to 8 channels, a compromise had to be made either by excluding EEG signals or not recording enough respiratory signals. This compromise did not serve the demands of sleep medicine and disappeared as the technology evolved. Dedicated systems for long-term EEG recording and specific sleep disorders remained. Examples are given here.

Sleep related breathing disorder recording

One of the first specialised systems for the early recognition of sleep related breathing disorders is the MESAM4 device [29] which is a successor of the MESAM [14, 24]. This digital device records heart rate, snoring sounds, oxygen saturation and body position. Because sleep is not recorded, the time in bed has to be reported by the patient. By visual analysis the recorded signals allow a clear identification of obstructive sleep apnoea (fig. 10.1) and hypopnoea (fig. 10.2) as has been proven in several validation studies [29]. A differentiation between the different types of sleep related breathing disorders is not possible. The automated evaluation of heart rate and oxygen saturation has only a limited value, because only the number of oxygen desaturations gave a high correlation with polysomnographic findings [36]. The oxygen desaturation index is dependent on the overall blood gas situation of the patient. Therefore it is best to evaluate the recording visually and determine a respiratory disturbance index (RDI) corresponding to a apnoea-hypopnoea index (AHI) in polysomnography. The correlation of visual MESAM4 based RDI and the parallel polysomnography based AHI varied between r=0.92 and 0.96 for three different scorers [24]. The MESAM4 is reliable for the early recognition and therapy control in patients with sleep related breathing disorders.

Snoring

Heart Rate

SaO₂

Body Position

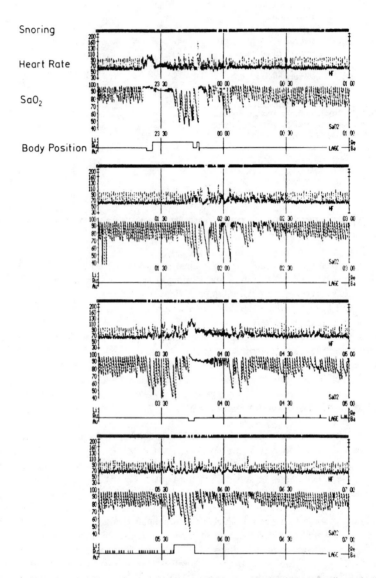

Figure 10.1. The ambulatory recording of a patient with complaints of excessive sleepiness and a diagnosis of obstructive sleep apnoea. The recording was performed with a reduced ambulatory system with the signals snoring sounds, heart rate, oxygen saturation, and body position (MESAM4). Each line of data shows two hours, thus giving a full night's sleep (8 hours) on one page. Apnoea events can be recognised by the repetitive drops of oxygen saturation and the corresponding increases in heart rate. During each block a period with more severe oxygen desaturation accompanying very long apnoeas can be identified. These periods occurred during REM sleep as confirmed by cardiorespiratory polysomnography.

A follow-up development of the MESAM4 has been presented by the Polymesam system [38]. This system also records oronasal airflow, respiratory effort and EMG of the limbs. This addition enables the physician to detect the most important differential diagnoses to sleep apnoea when investigating patients with hypersomnia. Obstructive and central apnoeas can be distinguished and periodic leg movements are recognised.

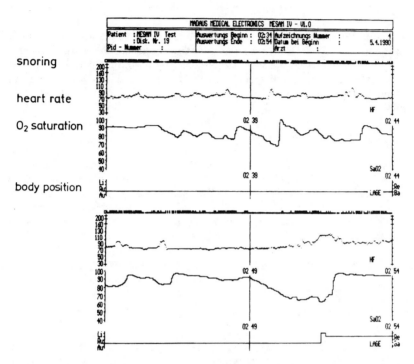

snoring

heart rate

O₂ saturation

body position

Figure 10.2. The ambulatory recording of another patient with a diagnosis of obstructive sleep apnoea is given. In this case, each line of data shows 10 minutes. By the presentation in higher resolution very long hypopnoea events with severe desaturation and increases in heart rate can be identified. The signals were arranged in the same order as in figure 10.1.

The Merlin system consists of a recording box and a junction box for the sensors which is attached to the chest of the body. The Merlin records oro-nasal airflow, ribcage and abdominal movements with piezoelectric sensors and snoring sounds. A body position sensor is part of the junction box on the body. An external oximeter can be connected. The separate recording of airflow, ribcage and abdominal respiratory movements distinguishes apnoeas, hypopnoeas and periodic breathing. This provides a better risk estimate in patients with sleep related breathing disorders and cardiovascular disorders such as congestive heart failure or Cheyne-Stokes breathing. For therapy control studies, the additional recording of CPAP pressure is possible. The signals correspond to a common polysomnographic system (Alice 3) and the system has been validated against this polysomnograph. A comparison of AHI with visual analysis showed a high correlation (0.94, $p<0.001$), and for the AI (0.75, $p<0.001$) and a lower reliability for the HI (0.51, $p<0.001$) [30].

The Sleepdoc Porti II records oxygen saturation, heart rate, body position, snoring and airflow. The Sleepdoc Porti III additionally records respiratory movements and CPAP pressure. This system had been validated against polysomnography with a very high correlation for the apnoea index ($r=0.98$, $p<0.001$). Hypopnoeas were not recognised and therefore sensitivity for the AHI is 71-72% and specificity is 91-100% [37].

The Apnoescreen Pro records the same signals as the Sleepdoc Porti III. Added to this are 7 configurable channels. The unique feature of this system is that the link between the main recording unit and the junction box attached to the patient is wireless using radio telemetry. Oro-nasal airflow can be recorded either by thermistors or by nasal prongs which record changes in air pressure and reflect airflow limitation much better, according to recent studies [22]. This recording unit does not require a computer for reporting. A report generator within the main unit can directly drive a printer. Alternatively the data may be transferred to a polygraph made by the same company for visual evaluation. Validation studies for this new system are not yet available. The validation of a previous version (Apnoescreen I) resulted in a high correlation for apnoea index (r=0.96 p<0.01) and the AHI (r=0.97 p<0.01) [8]. The sensitivity of the apnoea index was 94% and the specificity was either 92% or 100% depending on the criteria [8].

The SOMNOCHECK system consists of just one box attached to the body of the patient and is self applicable by the patient. The system records airflow by thermistors or nasal prongs, snoring sounds, oxygen saturation, heart rate and body position. Optionally it is possible to record CPAP pressure. Data evaluation and reporting is done with computer based software. Several validation studies showed a high sensitivity of 97%, and a high specificity of up to 100% based on visual evaluation [4, 11, 16].

Recording of periodic leg movements

Ambulatory recording of periodic leg movements has been done with analogous recording devices based on the Oxford Medilog system [20]. Many new systems are able to record either EMG or movement of the legs to help in the differential diagnostic process of hypersomnia (see table 10.2 and 10.3). The Polymesam specifically records the anterior tibialis EMG and has a good algorithm for the analysis of periodic leg movements during sleep. No validation studies have been presented for this specific feature.

Table 10.3. Modular ambulatory recording systems which can perform ambulatory polysomnography

function	biosignals	1	2	3	4	5	6	7
maximum duration of recording (hours)		24	12	12	18	V	8	40
maximum number of signals		24	9	12	64	17	18	19
maximum number of electrophysiol. signals		16	8	4	64	16	16	16
respiration	airflow	X	X	X	C	X	X	C
	CPAP pressure	X		X	C	X		C
	respiratory effort	X	X	X	C		X	C
	snoring sound	X	X	X	C	X	X	C
	SaO2	X	X	X		X	X	I
cardiovascular	ECG	X	X	X	C		X	C
	heart rate		X	X			X	I
movement	body position	X	X	X	C	X	X	C
	acceleration		X	C				C

X = being part of the system, either integrated, or to be added as an option
C = can be configured, if the appropriate sensor is connected
I = integrated in the system. The sensor can be connected without configuration requirements
V = variable recording duration, dependent only on the capacity of the PCMCIA storage media (e.g. hard disk, memory card)

1. Handybrain (Schwarzer)
2. Oxford MPA (Oxford Medical Instruments)
3. Compumedics (Resmed Inc.)
4. Vitaport (Temec Inc.)
5. Embla (Flaga)
6. Minisomno (Mallinkrodt)
7. Monet (Medcare Automation)

REVIEW OF AMBULATORY SLEEP RECORDING SYSTEMS (PSG) WITH 10 OR MORE CHANNELS

A review of existing ambulatory polysomnography reveals two root sources. One branch of development stems from electrophysiology with long-term EEG recording and the other stems from reduced ambulatory recording systems created for the diagnosis of sleep related breathing disorders. Both categories of systems have been complemented by the missing amplifiers and sensors in order to meet the requirements of a cardiorespiratory polysomnography as specified by Chesson *et al.* [34].

The latest systems digitise the signals and record them using very small hard disks or memory cards compliant with widely used computer standards (PCMCIA standard). The storage capacity allows many signals (up to 32) to be stored with high sampling rates (200 Hz) sufficient for any demand in sleep research. Because amplifier settings and digital filter characteristics can be changed by microcontrollers in the digital recorders, the input options are much more flexible than analogue technique would allow. Using these techniques, digital data loggers are extremely flexible and can be adapted to any combination of signals. Such systems only need a very small part of analogue signal conditioning, which lowers production costs significantly.

Two of the reduced sleep systems for the early detection of sleep related breathing disorders (Apnoescreen Pro, Sleepdoc Porti 4) described in the previous section have universal digital amplifiers which may be configured to enable the recording of 16 channels low-voltage signals (e.g. EEG).

Table 10.3 shows four ambulatory polysomnography systems which have not been described so far (VITAPORT, EMBLA, MINISOMNO and MONET). These systems provide digital ambulatory cardiorespiratory polysomnography on the basis of embedded computer systems. The recording units are small and can be worn on the body. Signals are digitised at rates of up to 256 Hz and the different channels can be programmed according to the needs. Data are stored on a PCMCIA memory card, or hard disk, or flash-card, or may even be transmitted through a computer network, or a telephone modem connection. The recording unit can be directly connected to a computer for use in a sleep laboratory and for continuous sleep monitoring under the control of attending personnel.

The analysis of the recorded data is made after the recording has ended and can be performed using either visual evaluation or computer based sleep analysis with user adjustable software. The software of the systems differ in various aspects but have the same overall functions for evaluation of sleep, respiration, heart rate, and oxygen saturation. The software packages support user-configurable report generation. These are standard features for any computer based polysomnography systems and not specific to ambulatory sleep recording. No sleep analysis validation studies have been published for any of the systems.

FUTURE DEVELOPMENTS

The technical progress of the last 10 years has overcome the technical limitations which previously dominated ambulatory monitoring. The main indications and concepts for applying the new methods have now been defined and discussed. But questions remain in regard to the implications of mortality and morbidity attached to minor breathing disorders, obstructive snoring, respiratory and subcortical arousal. Long-term outcome studies have to be performed to clarify this and to improve the definitions used for the indications of sleep recording and subsequent therapy in general.

Further technical progress is expected in the area of telemedicine. A few early studies have been conducted using telemedicine for "remote monitoring", during which the sleep data are continuously transmitted through a computer modem to remotely located persons, observing the signals of the sleeping patient. In such a setting, patients may sleep at home and data are transmitted to a co-ordinating sleep centre for recording and monitoring. Feedback can be given using an extra regular phone line if needed [40]. This setting optimises data quality and minimises loss of data due to loose electrodes, if a trained person at the recording site can reattach them. Currently this setting does not seem to be cost effective. But it might be valuable in children at risk of sudden infant death, where it has been used in early investigations.

Modern techniques with the capacity to record many signals in high resolution and to transfer signals to remote locations opens new scope for epidemiological research and to occupational and environmental research with the inclusion of the specific aspects of sleep.

REFERENCES

1.   AARDWEG J.G. - Cyclic changes in breathing pattern. *Thesis*, Amsterdam pp 1-164, 1992.
2.   ACEBO C., SADEH A., SEIFER O., TZISCHINSKY O., WOLFSON A.R., HAFER A., CARSKADON M.A. - Estimating sleep patterns with actigraphy monitoring in children and adolescents: How many nights are necessary for reliable measures? *Sleep 22*, 95-103, 1999.

3. ANCOLI-ISRAEL S. - Evaluating sleep apnea with the portable modified Medilog-Respitrace System. In: Miles LE, Broughton RJ (eds) *Medical Monitoring in the Home and Work Environment*, Raven Press, New York, pp 275-283, 1990.
4. BAUMEL M.J., MAISLIN G., PACK A.I. - Population and occupational screening for obstructive sleep apnea: are we there yet? *Am. J. Respir. Crit. Care Med. 155*, 9-14, 1997.
5. BREDENBRÖKER D., BRANDENBURG U., PENZEL T., PETER J.H., VON WICHERT P. - Somnocheck: Validierung eines neuen ambulanten Meßgerätes zur Erfassung schlafbezogener Atmungsstörungen. *Somnologie 2*, 129-134, 1998.
6. BROUGHTON R.J. - Ambulant home monitoring of sleep and its disorders. In: MH Kryger, T Roth, WC Dement (eds): *Principles and Practice of Sleep Medicine*, WB Saunders Co, Philadelphia, pp 696-701, 1989.
7. BROUGHTON R.J., STAMPI C., DUNHAM W., RIVERS M. - Ambulant monitoring of sleep-wake state, core body temperature and body movement. In: Miles LE, Broughton RJ (eds) *Medical Monitoring in the Home and Work Environment*, Raven Press, New York, pp 139-150, 1990.
8. COHRS S., HERRENDORF G., KINKELBUR J., TITKEMEYER M., HAJAK G. - Validierung eines ambulanten Screening-Gerätes für schlafbezogene Atmungsstörungen: APNOESCREEN 1. *Somnologie 2*, 8-13, 1998.
9. COLE R.J., KRIPKE D.F., GREEN W., MULLANEY D.J., GILLIN J.C. - Automatic sleep/wake identification from wrist activity. *Sleep 15*, 461-469, 1992.
10. EHLERT I., DANKER-HOPFE H., HÖLLER L., VON RICKENBACH P., BAUMGART-SCHMITT R., HERRMANN W.M. - Ein Vergleich zwischen Aufzeichnung und Auswertung von Schlaf-EEG durch QUISI, Version 1.0 und visueller Auswertung durch PSG. Ein einkanaliges, ambulant anwendbares Gerät zur Aufzeichnung und automatischen Auswertung von Schlaf-EEG mittels neuronaler Netzwerk-Techniken. *Somnologie 2*, 104-116, 1998.
11. FICKER J.H., WILPERT J., WIEST G.H., LEHNERT G., HAHN E.G. - Evaluation des "Somnocheck" zur Diagnostik des obstruktiven Schlafapnoe-Syndrom: Automatische versus manuelle Analyse. *Somnologie 2*, 10, 1998.
12. GRAMSBERGEN A., GEISLER H.C., TAEKEMA H., VAN EYKERN L.A. - The activation of back muscles during locomotion in the developing rat. *Dev. Brain Research* 112, 217-228, 1999.
13. GROTE L., PLOCH T., HEITMANN J., KNAACK L., PENZEL T., PETER J.H. - Sleep-related breathing disorder is an independent risk factor for systemic hypertension. *Am. J. Respi.r Crit. Care Med. 160*, 1875-1882, 1999.
14. GUILLEMINAULT C., PENZEL T., STOOHS R., MASTERSON M.B., PETER J.H. - Cyclical variation of heart rate and snoring: an ambulatory device. In: Miles LE, Broughton RJ (eds) *Medical Monitoring in the Home and Work Environment*, Raven Press, New York, pp 265-273, 1990
15. HADDERS-ALGRA M., KLIP-VAN DEN NIEUWENDIJK A.W.J., MARTIJN A., VAN EYKERN L.A. - Assessment of general movements: Towards a better understanding of a sensitive method to evaluate brain function in young infants. *Dev. Med. Child Neurol. 39*, 88-98, 1997.
16. HEIN H., ABDALLAH C., JUGERT C., KUZIEK G., MAGNUSSEN H. - Validierung eines Screening-Gerätes zur Diagnostik und Therapiekontrolle schlafbezogener Atmungsstörungen. *Somnologie 2*, 189-193, 1998.
17. IMHOLZ B.P.M. - Noninvasive finger arterial pressure waveform registration, evaluation of Finapres and Portapres. *Thesis*, Amsterdam pp 1-154, 1991.
18. International Classification of Sleep Disorders – Revised version: Diagnostic and coding manual. Diagnostic classification steering committee, Thorpy MJ, Rochester, *American Sleep Disorders Association*, 1997.
19. JUHÁSZ J., SCHILLEN J., URBIGKEIT A., PLOCH T., PENZEL T., PETER J.H. - Unattended continuous positive airway pressure titration. *Am. J. Respir. Crit. Care Med. 154*, 359-365, 1996.
20. KAYED K., ROBERTS S., DAVIES W.L. - Computer detection and analysis of periodic movements in sleep. *Sleep 13*, 253-261, 1990.
21. MILES LE: A Portable microcomputer for long-term physiological monitoring in the home and work environment. In: Miles LE, Broughton RJ (Eds) *Medical Monitoring in the Home and Work Environment*. Raven Press, New York, pp. 47-58, 1990.
22. NORMAN RG., MUHAMMED M.A., WALSLEBEN J.A., RAPOPORT D.M. - Detection of respiratory events during nPSG: nasal cannula/pressure sensor versus thermistor. *Sleep 20*, 1175-1184, 1997.
23. OHAYON M.M., GUILLEMINAULT C., PRIEST R.G., CAULET M. - Snoring and breathing pauses during sleep: telephone interview Survey of a United Kingdom population sample. *Brit. Med. J. 314*, 860-863, 1997.
24. PENZEL T., AMEND G., MEINZER K., PETER J.H., VON WICHERT P. - MESAM: A heart rate and snoring recorder for detection of obstructive sleep apnea. *Sleep 13*, 175-182, 1990.
25. PENZEL T., BRANDENBURG U. - Diagnostische Verfahren und Standards in der Schlafmedizin. *Internist 37*, 442-453, 1996.
26. PETER J.H., BLANKE J., CASSEL W., CLARENBACH P., ELEK H., FAUST M., FIETZE I., MAHLO H.W., MAYER G., MÜLLER D., PENZEL T., PODSZUS T., RASCHKE F., RÜHLE K.H., SCHÄFER T., SCHLÄFKE M., SCHNEIDER H., SCHOLLE S., STUMPNER J., WIATER A., ZWACKA G. - Empfehlungen zur ambulanten Diagnostik der Schlafapnoe. *Med. Klinik 87* : 310-317, 1992.
27. PETER J.H., PENZEL T. - Portable monitoring of sleep and breathing. In: Saunder N, Sullivan CE (eds.) *Sleep and Breathing*. 2nd edn. Dekker, New York. 379-404, 1994.
28. PORTIER F., PORTMANN A., CZERNICHOW P., VASCAUT L., DEVIN E., BENHAMOU D., CUVELIER A., MUIR J.F. - Evaluation of home versus laboratory polysomnography in the diagnosis of sleep apnea syndrome. *Am. J. Resp. Crit. Care Med. 162*: 814-818, 2000.

29. ROOS M., ALTHAUS W., RHIEL C., PENZEL T., PETER J.H., VON WICHERT P. - Vergleichender Einsatz von MESAM IV und Polysomnographie bei schlafbezogenen Atmungsstörungen (SBAS). *Pneumologie 47*, 112-118, 1993.
30. RÖTTIG J., FIETZE I., WARMUTH R., WITT C., BAUMANN G., MERLIN - A Validation study of a new portable screening device for detection of sleep related breathing disorders. *Am. J. Resp. Crit. Care Med. 153* Suppl.4. A87, 1996.
31. SADEH A., ALSTER J., URBACH D., LAVIE L. - Actigraphically based automatic bedtime sleep-wake scoring. *J. Ambulat. Monit. 2*, 209-216, 1989.
32. SCHNALL R.P., SHLITNER A., SHEFFY J., KEDAR R., LAVIE P. - Periodic, profound peripheral vasoconstriction – a new marker of obstructive sleep apnea. *Sleep 22*, 939-946, 1999.
33. Standards of Practice Committee of the American Sleep Disorders Association: Practice parameters for the use of portable recording in the assessment of obstructive sleep apnea. *Sleep 17*, 372-377, 1994.
34. Standards of Practice Committee of the American Sleep Disorders Association: Practice parameters for the indications for polysomnography and related procedures. *Sleep 20*, 406-422, 1997.
35. Systematic review of the literature regarding the diagnosis of sleep apnea. Summary, Evidence Report/Technology Assessment: Agency for Health Care Policy and Research, Rockville, MD. Number 1, December 1998.
36. STOOHS R., GUILLEMINAULT C. - MESAM4: An ambulatory device for the detection of patients at risk for obstructive sleep apnea syndrome (OSAS). *Chest 101*,1221-1227, 1992.
37. TESCHLER H., HOHEISEL G., SCHUMANN H., WAGNER B., KONIETZKO N. - Validierung des Sleep-Doc-Porti-Systems für die ambulante Schlafapnoediagnostik. *Pneumologie 49*, 496-501, 1995.
38. VERSE T., JUNGE-HÜLSING B., KROKER B., PIRSIG W., ZIMMERMANN E. - First results of a prospective ttudy validating the method of ambulatory polysomnography Using the POLY-MESAM unit. *Sleep and Breathing 2*, 56-64, 1997.
39. WEISS R.K., JAIN A., MUTZ G., STEPHAN E. - Monitoring of physiological key functions of free moving patients via mobile phones. In: Penzel T, Salmons S, Neuman M (eds.) *Biotelemetry XIV, Proc. of 14th Internat. Symp. on Biotelemetry 1997*. Tectum, Marburg: 161-166, 1998.
40. YOUNG T., PALTA M., DEMPSEY J., SKATRUD J., WEBER S., BADR S. - The occurrence of sleep-disordered breathing among middle-aged adults. *New Engl. J. Med. 328*: 1230-1235, 1993.

Chapter 11

Guidelines for visual sleep analysis

A. Besset

INSERM EMI-9930, Hôpital la Colombière, Montpellier and Service de Neurologie B, Hôpital Gui de Chauliac, Montpellier, France

Attempting to describe the methods of visually analysing human sleep at a time when so many sophisticated high performance automatic systems of analysis are being developed may seem more like the work of a palaeontologist than a physiologist. Nevertheless, visual analysis will continue to be an indispensable tool for some time to come, if only to serve as a basis for verifying the analyses produced by computerised methods.

The stages of sleep were first classified by Loomis *et al.* in 1937 [17]. This classification, based solely on electroencephalographic criteria, already distinguished 5 stages of sleep related to the slowing down of electrical activity in the brain. The discovery in 1953 by Aserinsky and Kleitman [2] of sleep with eye movements (REM sleep) and the close tie between this type of sleep and dream recall [9] formed the basis for Dement and Kleitman's classification [10] distinguishing 4 stages of sleep without rapid eye movement (NREM) and one stage of sleep with rapid eye movements (REM). This classification involved combining 2 different parameters: electroencephalogram activity and eye activity. After the discovery by Jouvet and Michel in 1959 on the cat [16], and later by Berger in 1962 [3] on humans, showing muscular atonia accompanying the onset of sleep with eye movements, it became necessary to establish a new classification with three polygraphic parameters: the electroencephalogram (EEG), the electro-oculogram (EOG) and the electromyogram (EMG). This was all the more important as it had been shown that without precise rules, Dement and Kleitman's classification lacked reliability [18]. An international committee was set up and the results of its work were published in 1968, presided over by Allen Rechtschaffen and Antony Kales, under the title "A Manual of Standardised Terminology, Techniques and Scoring System for Sleep Stages of Human Subjects" [19]. The rules laid down in this manual are still applied and used as the standard reference by laboratories throughout the world. They can of course be criticised as lacking flexibility, but their rigidity ensures about 95% reliability between scorers trained in the same laboratory [11]. These criteria are specific to adults but may also be used to score the sleep of older children and adolescents [6, 7]. For newborn infants and young children, different criteria and rules for classification must be applied [1, 14].

PRINCIPLE GUIDELINES FOR SCORING SLEEP STAGES

Several basic notions to consider

A stage diagnosis can only be carried out when several variables are combined for the same epoch (unit of time used as a reference). These variables are the elementary activities and certain grapho-elements (table 11.1) of the three physiological parameters (EEG, EMG, EOG) required to interpret the stages of human sleep.

The notion of epoch

The epoch is the standard 30-sec recording displayed on the screen of the computerized polysomnographic system. In certain cases, a longer period (1 min) may be considered, but this makes the EEG difficult to read and introduces a bias in the results, notably by reducing the number of short stages (less than 1 min). Likewise, for certain reasons, shorter epochs (10 s) may be used, for example when the length of wakefulness must be determined accurately in recordings involving elements like sleep apnoeas or periodic limb movements with 5 to 10 second sleep reactions which would not be scored in 20 to 30 second epochs. However the use of such short epochs amplifies sleep fragmentation and sometimes makes stage 2 difficult to score. Moreover, it is important to emphasise that the shorter the epoch, the greater the number of epochs and the longer the scoring process or the review process after automatic scoring (computer aided analysis). Thus the use of 30 second epochs appears as a good compromise. For recordings for research, the problem could and should be approached differently. Analysis by epoch means respecting the basic guidelines of unity and contiguity.

Table 11.1. Basic activities of the 3 physiological parameters used as scoring criteria for sleep staging.

Sleep parameters	Awake	Stage 1	Stage 2	Stage 3	Stage 4	REM
EEG						
Alpha>50%	+	–	–	–	–	–
Theta>50%	–	+	+	–	–	+
Delta>20%	–	–	–	+	+	–
Delta>50%	–	–	–	–	+	–
Vertex sharp waves	–	(+)	(+)	0	0	–
K complexes	–	–				–
Spindles	–	–	(+)	0	0	
Sawtooth waves	–	–	–	–	–	(+)
EMG						
Tonic activity	0	0	0	0	0	–
Twitches	–	–	–	–	–	(+)
EOG						
Eyelid movement	(+)	–	–	–	–	–
REM (rapid eye movement)	(+)					+
SEM (slow eye movement)	–	(+)	–	–	–	(+)

+ denotes an indispensable criterion for stage diagnosis
- signifies that the absence of the criterion is indispensable for stage diagnosis.
0 denotes that the presence or absence of the criterion will not affect stage diagnosis
(+) signifies that the criterion is not indispensable for diagnosis but that its presence will confirm diagnosis.

Rule of unity

Only one score can be attributed to each epoch. Different portions of an epoch cannot be combined to create a new epoch. During an epoch several elements characteristic of different stages may be present. If this occurs, only the sleep stage corresponding to over 50% of the epoch will be taken into consideration.

Rule of contiguity

This applies to ambiguous epochs where no criterion is present to confirm diagnosis. These epochs can only be scored in function of those immediately preceding or following them (contiguous epochs).

DIAGNOSIS OF DIFFERENT STATES AND STAGES

Sleep states and stages are conventionally scored from 0 to 5. States 0 and 5 are generally referred to as wakefulness and REM sleep but some authors may refer to them by number.

Wakefulness

For most subjects, wakefulness is characterised by EEG rhythmic alpha activity in the range of 8 to 12 cycles per second (c/s). This rhythm is typical of relaxed wakefulness with the eyes closed predominating in the posterior regions. As soon as the eyes open it is replaced by relatively low voltage and more rapid activity. Tonic activity is generally of a high level but is not used as a variable for recognising this state. Eye movement control is voluntary in wakefulness, consisting of rapid movements and eye blinks (fig. 11.1). The latter are pathognomonic of the state of wakefulness and their presence rules out scoring a sleep stage. The presence of eye movements is not indispensable for scoring the state of wakefulness. The onset of sleep (fig. 11.2a) is shown by continuous alpha activity, with the eyes open, which gradually gives way to slower theta activity (5 to 7 cs.). If over 50% of the epoch is composed of alpha and/or beta rhythms, the epoch is scored as wakefulness. This state may be scored independently without taking account of the preceding or following epochs.

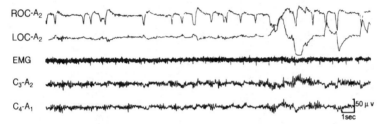

Figure 11.1. Wakefulness; the activity seen on the ROC/A2 lead corresponds to characteristic eyelid movements. The EEG tracings (C3-A2 and C4-A1) show virtually permanent alpha activi⋯
(ROC-A2 = outer canthus of the right eye-A2 lead, LOC-A2 = outer canthus of the left eye-A2 lead)

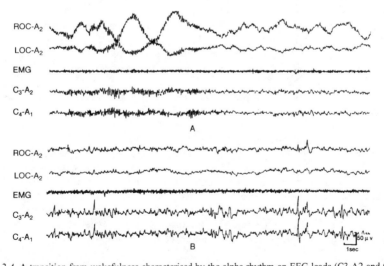

Figure. 11.2 A. A transition from wakefulness characterised by the alpha rhythm on EEG leads (C3-A2 and C4-A1) (left side of the figure) to sleep illustrated by a mixed frequency rhythm (right side of the figure). The movements shown for the eye-movement channels (ROC-A2 and LOC-A2) are slow eye movements. This epoch must be scored stage 1 as over 50% of the epoch is taken up with a mixed frequency rhythm. *B.* Stage 1 tracing characterised by 100% mixed frequency rhythms and the presence of vertex sharp waves.

NREM sleep

This type of sleep comprises stages 1 to 4.

As in the state of wakefulness, the essential variable is EEG activity. Tonic activity is not a significant variable and eye movement activity is only of value in assessing the onset of sleep.

Stage 1

This stage represents the transition from wakefulness to sleep and is commonly accompanied by significant body movements. It is characterised by low-voltage, mixed frequency EEG activity associated with 2 to 7 c/s. rhythms. At the beginning of the night, sharp waves (fig. 11.2b) may occur, with high amplitude (up to 2000 microvolts) known as vertex sharp waves. These particular grapho-elements are specific indicators of stage 1.

Stage 1 is characterised by the complete absence of spindles or K complexes (see stage 2).

At the onset of sleep (fig. 11.2b) particularly during the epochs following wakefulness, stage 1 is accompanied by eye movements lasting several seconds: slow eye movements (SEM). These eye movements reveal eyeball instability: they indicate the loss of control of voluntary movements, signifying a sharp drop in attention.

Muscle tone is fairly high but is still lower than in the state of wakefulness. The transition from wakefulness to sleep is characterised by the slowing down of EEG activity. When mixed frequency EEG activity (2 to 7 cs) takes up over 50% of the epoch, it is scored as stage 1.

This stage cannot always be scored independently. Its interpretation may depend on the preceding or following epochs (see stage 2).

Stage 1 occupies a relatively short period of the night. It lasts for no more than 10% of the total period of sleep.

Stage 2

This stage is characterised by mixed frequency EEG activity of 2 to 7 cs, with K complexes and sleep spindles.

K complexes (fig. 11.3a) are long, slow EEG waves with a well marked negative component immediately followed by a positive component. These complexes last at least half a second. They are often associated with spindles. They occur at a frequency of 1 to 3 per minute [15]. K complexes are best detected in the vertex regions. K complexes can be exteroceptive or interoceptive responses, and thus interpreted as wakefulness responses; nevertheless they can occur in the absence of any detectable stimulus.

Sleep spindles (fig 11.3b) are spindle-shaped waves with a frequency of 12 to 14 c/s lasting at least half a second. The term is used when 6 to 7 consecutive cycles can be counted in half a second.

Spindles are characteristic sleep waves, experienced by most mammals. In humans, they occur at the rate of about 3 to 10 per minute [11]. They are attenuated in insomniacs [5]. They appear in children before the age of 3 months [8], later and at a slower rate for mentally retarded children [21]. Finally, their number and frequency diminish with age [12].

Other particular grapho-elements such as long slow paroxystic waves which do not have the same morphology as K complexes, will often occur during stage 2. If ambiguous epochs occur for over 3 minutes (with neither spindles nor K complexes) and no intervening waking reactions, these epochs will all be scored stage 1. If a waking reaction occurs before 3 minutes, the ambiguous epochs preceding the waking reaction will be scored 2 and those following will be scored 1.

Tonic activity may be present or attenuated; there are no eye movements.

Stage 2 is a stage of light sleep. It accounts for about 50% of overall sleep.

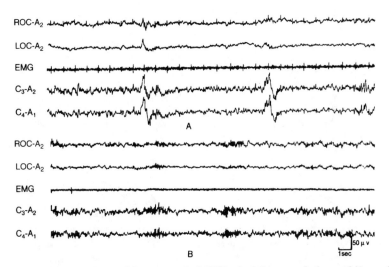

Figure 11.3 A. Stage 2 tracing; EEG activity composed of 100% mixed frequency rhythms and K complexes (well delineated negative sharp wave immediately followed by a positive component). *B.* Stage 2 tracing with background rhythm identical to the preceding one, and with sleep spindles (bursts of activity in the 12 to 14 cps range).

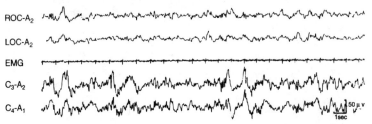

Figure 11.4. Stage 3 tracing; more than 20% of the scoring epoch consists of high voltage slow waves (> 75 μV) of 0.5 to 2 c/s.

Stage 3

Stage 3 (fig. 11.4) is defined by the presence of slow delta waves of 2 c/s or less with a minimum amplitude of 75 microvolts from peak to peak i.e. the difference between the most negative and positive points of the wave.

At least 20% but not more than 50% of the epoch consists of these waves. The word presence refers only to slow delta waves of high voltage with a frequency inferior or equal to 2 c/s, delta waves of lower voltage or higher frequency (2 to 4 c/s) cannot be counted. In practise it is only necessary to carry out an accurate wave by wave count in borderline cases between 15% and 25% or between 40% and 60%. Spindles may be present during this stage.

This stage can be scored independently without referring to the preceding or subsequent epochs.

Stage 4

This stage (fig. 11.5) is characterised by the presence of at least 50% of slow waves in one scoring epoch (see stage 3). As in stage 3 spindles may be present in stage 4. This stage can be scored independently without taking account of the preceding or following epochs. Stage 4 is a stage of deep NREM sleep corresponding to about 10% of overall sleep.

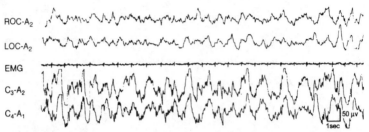

Figure 11.5 Stage 4 tracing; over 50% of the scoring epoch consists of high voltage slow waves (> 75 µV) of 0.5 to 2 c/s.

REM sleep

REM sleep is defined by the combination of low voltage, mixed frequency (2 to 7 c/s) EEG activity, bursts of rapid eye movements and muscular atonia (fig. 11.6a). EEG activity is like that of stage 1 with the virtual absence of K complexes and sleep spindles. But during the course of this REM sleep, no vertex sharp waves are present and EEG alpha activity is sometimes greater than in stage 1; particular waveforms – sawtooth waves may appear in the vertex or frontal regions [22], generally associated with bursts of rapid eye movements (fig. 11.6a). Finally the presence of spindles or K complexes may be considered part of REM sleep in certain conditions relating to the position of spindles at the beginning or the end of the epoch (see basic guidelines).

Rapid eye movements may occur in isolation or in bursts. Eye movements vary in density depending on when they occur: they are rare at the beginning of the night, becoming more profuse in the final episodes of REM sleep.

For an epoch to be scored REM sleep it must be accompanied by skeletal muscle atonia. However, short contractions, or twitches may occur at the extremities (fingertips, nostrils, corners of the mouth etc.) (fig. 11.6b). If these twitches are numerous, they may interfere with sleep analysis. Lastly, in certain disorders such as REM sleep behaviour disorder [20], narcolepsy [13] or under the influence of certain drugs such as clomipramine [4], muscle tone may be maintained or even increased; this type of sleep is thus scored dissociated REM sleep.

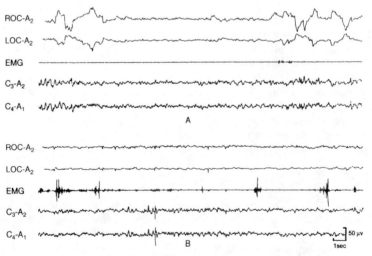

Figure 11.6 A. REM sleep tracing with rapid eye movements on the eye-movement channels (ROC-A2, LOC-A2), muscular atonia seen on the EMG channel and on the EEG channels (C3-A2 and C4-A1) a mixed frequency rhythm with sawtooth waves (left part of figure); *B*. REM sleep tracing with frequent twitches shown on the EMG channel.

Movement time

The term movement time refers to the epochs immediately preceeding and following sleep stages during which EEG and EOG are disturbed for over half the epoch by muscle tension or the subject's postural readjustments. If, despite artifacts, EEG and EOG can be isolated for over half the epoch, it is scored according to the visible EEG and EOG criteria even if accompanied by movement. Nevertheless, if this pattern is preceded or followed by wakefulness, it is scored as an awake pattern. When an epoch is scored "movement time" it is neither attached to wakefulness nor sleep but is interpreted as a separate category.

Table 11.2. Sample sleep stage summary sheet for one night's sleep. Recording time

Recording time	Architecture			
Lights out	Stage 1	length (min)	%(TST)	n° entries
End of night	Stage 2	length (min)	%(TST)	n° entries
Recording time (in minutes)	Stage 3	length (min)	%(TST)	n° entries
	Stage 4	length (min)	%(TST)	n° entries
	Stage 3+4 length (min)	%(TST)	n° entries	
	REM	length (min)	%(TST)	n° entries
General data	**Specific REM data**			
Sleep onset (in minutes)	Number of episodes			
Wake time	Average length of episodes			
Latency to sleep onset (in minutes)	Continuity index			
Number, respective length and times of awakenings	Latency (in minutes)			
Total sleep time (TST in minutes)	REM density			
Latencies	**Cycles**			
(calculated from onset of sleep)	Number			
stage 1	Average length			
stage 2				
stage 3				
stage 4				
REM				

SUMMARISING DATA

After the states and stages of sleep have been diagnosed epoch by epoch, they must be presented in a form providing as accurate an interpretation as possible of sleep continuity and architecture. No consensus format has been arrived at between laboratories on the indispensable data which should appear on sleep data summary sheets; this clearly depends not only on the laboratory but chiefly on the pathology being studied. *Table 11.2* shows a summary of sleep analysis. This table is intended as a guide and items may be added according to the needs of the user.

REFERENCES

1. ANDERS T., EMDE R., PARMELEE A. (eds.) – A Manual of Standardized Terminology, Techniques and Criteria for Scoring of States of Sleep and Wakefulness in Newborn Infants, *Brain Information Service/Brain Research Institute*, UCLA, Los Angeles, 1971.
2. ASERINSKY E., KLEITMAN N. – Regular occurring periods of eye motility and concomitant phenomena during sleep. *Science*, *118*, 273-274, 1974.
3. BERGER R.J. – Tonus of extrinsic laryngeal muscles during sleep and dreaming. *Science*, *134*, 840, 1961.
4. BESSET A. – Effects of antidepressants on human sleep. *Adv. Biosci.*, *21*, 141-148, 1978.
5. BESSET A. – L'analyse du sommeil humain: l'analyse automatique. *In:* M. Billiard, *Le sommeil et ses troubles*, Masson, Paris, 1994.
6. CARSKADON M.A., ORAV E.J., DEMENT W.C. – Evolution of sleep and daytime sleepiness in adolescents. *In: Sleep/Wake Disorders: Natural History, Epidemiology and long term Evolution*, C. Guilleminault, E. Lugaresi (eds.), Raven Press, New York, 201-206, 1983.
7. COBLE P.A., KUPFER D.J., TASKA L.S., KANE J. – EEG. Sleep of normal healthy children. Part 1: Findings using standard measurement methods. *Sleep*, *7*, 289-303, 1984.
8. CROWEL D.H., KAPUNIAI L.E., BOYCHUCK R.B. – Daytime sleep stage organisation, in three month old infants. *Electroencephalogr. Clin. Neurophysiol.*, *53*, 36-47, 1982.

9. DEMENT W.C., KLEITMAN N. – The relation of eye movements during sleep to dream activity: an objective method for the study of dreaming. *J. Exp. Psychol.*, *53*, 339-346, 1957.

10. DEMENT W.C., KLEITMAN N. – Cyclic variations in EEG during sleep and their relation to eye movements, body motility and dreaming. *Electroencephalogr. Clin. Neurophysiol.*, *9*, 673-690, 1957.

11. GAILLARD J.M. – Structure du sommeil humain: analyse traditionnelle, électronique, et pharmacologique. *In: Le sommeil humain - bases expérimentales, physiologiques et physiopathologiques*, O. Benoît, J. Forêt (eds.), Masson, Paris, 47-60, 1992.

12. GUAZZELLI M., FEINBERG I., AMINOFF M., FEIN G., FLOYD T., MAGGINI C. – Sleep spindles in normal elderly: comparison with young adult patterns and relation to nocturnal awakening, cognitive function and brain atrophy. *Electroencephalogr. Clin. Neurophysiol.*, *63*, 526-539, 1986.

13. GUILLEMINAULT C., RAYNAL D., TAKAHASHI S., CARSKADON M., DEMENT W.C. – Evaluation of short term and long term treatment of the narcolepsy syndrome with clomipramine hydrochloride. *Acta Neurol. Scand.*, *54*, 71 - 87, 1976.

14. GUILLEMINAULT C., SOUQUET M. – Sleep states and related pathology. *In: Advances in Perinatal Neurology*, R. Korobkin, C. Guilleminault (eds.), Spectrum, New York, 225-247, 1979.

15. HALASZ P., PAL I., RAJNA P. – K complexes formation of the EEG in sleep. A survey and new examinations. *Acta Physiol. Acad. Sci. Hung.*, *65* (1), 3-35, 1985.

16. JOUVET M., MICHEL M. – Corrélations électromyographiques du sommeil chez le chat décortiqué et mésencéphalique chronique. *C.R. Soc. Biol.*, *153*, Paris, 422-425, 1959.

17. LOOMIS A.L., HARVEY E.N., HOBART G.A. – Cerebral states during sleep as studied by human brain potentials. *J. Exp. Psychol.*, *21*, 127-144, 1937.

18. MONROE L.J. – Inter-rater reliability and the role of experience in scoring EEG sleep. *Psychophysiol.*, *5*, 376-384, 1967.

19. RECHTSCHAFFEN A., KALES A. (eds.) – A manual of standardized terminology, techniques and scoring system for sleep stages of human subjects. *Brain Information Service/Brain Research Institute*, Los Angeles, 1968.

20. SCHENCK C.H., BUNDLIE S.R., ETTINGER M.G., MAHOWALD M.W. – Chronic behavioral disorders of human sleep: a new category of parasomnia. *Sleep*, *9*, 293-308, 1986.

21. SHIBAGAKI M., KIYONO S., WATANABE K. – Spindle evolution in normal and mentally retarded children. A review. *Sleep*, *5*, 47-57, 1982.

22. YASOCHIMA A., HAYASHI H., IJIMA S., SUGITA Y., TESHIMA Y., SHIMIZU T., HISHIKAWA Y. – Potential distribution of vertex sharp wave and saw tooth wave on the scalp. *Electroencephalogr. Clin. Neurophysiol.*, *58*, 73-76, 1984.

Chapter 12

Automatic sleep analysis

A. Besset
INSERM EMI-9930, Hôpital la Colombière, Montpellier and Service de Neurologie B, Hôpital Gui de Chauliac, Montpellier, France

Automatic systems to analyse sleep have been in existence for over twenty years [17, 34]. They were rarely commercialised in the past, being restricted to a limited number of laboratories. There have been such rapid development recently that virtually all the sleep polygraphs designed in the last five years are equipped with automatic sleep stagers, with varying levels of performance. The recent trend in automatic sleep analysis largely stems from the combination of marked technological advances in the field of micro-processing (resulting in the expansion of digital polygraphs) and the demands created by developments in the understanding of sleep medicine (sleep apnoea syndrome, exploration of sleep microstructure etc.). A large number of systems of analysis are available whether automatic, ambulant or stationary. These systems cannot all be described here as this would go beyond the limits of the present article, whose intention is to provide an introduction to the automatic analysis of sleep. Thus we will restrict ourselves to a brief presentation of the most widely used methods of signal analysis employed in the principle systems, and offer some advice on using them, based on our own experience.

CHARACTERISTICS OF AN AUTOMATIC SLEEP STAGING SYSTEM

Despite its widespread use, automatic sleep analysis has not yet replaced visual sleep tracing analysis following the criteria introduced by Rechtschaffen and Kales (R and K) [29]. Indeed, there is still disagreement as to the most appropriate method of electroencephalography (EEG) for studying sleep stages and the parameters to consider in setting up a new classification of sleep states. Most analysis procedures use the standard reference of visual analysis, using the R and K [32] criteria, which, more than thirty years after publication, remains the only consensus classification of sleep stages. It is still rare to find studies comparing the different methods of analysis [16, 19, 26] and much work will need to be done before a new classification emerges, based on the principles of signal analysis. The reference to visual criteria [32] is thus indispensable, which is why the analysis of sleep stages obtained by any automatic system must be accompanied by meticulous visual verification, based on the criteria used by R and K [32].

Most of these systems were developed with the aim of automating sleep tracing interpretation. In addition to sleep stages, they are now required to analyse different physiological variables (muscular, respiratory, cardiac etc.) and to provide indications of EEG signal strength and sleep microstructure (number of spindles, number of K complexes, transient arousals, percentage of occupation of the different frequency bands within each epoch etc.).

Calibrating the system

Collecting data using digital systems can be daunting for someone who is not used to data processing systems (scroll menu). Most systems function with "montages" which are either preset or need to be set up. These montages indicate the different derivations and channel settings to acquire, in addition to a certain number of parameters specific to the system. As with analogue systems, digital systems need to be calibrated with great care as the quality of calibration will determine the quality of recording and thus of analysis. A dual calibration is required: calibration of the system (according to its own norms) and calibration of the parameters explored in the subject (EEG, EMG, EOG, respiratory passageways, oesophageal pressure, pneumotachography etc.). This calibration must take place before recording starts using the customary calibrating techniques when the subject is in bed. The quality of calibration is all the more important as in certain systems, channel settings cannot be altered once the recording starts. In some cases, it is possible to modify some settings on reading, but at the cost of manipulations which are often technically difficult and time consuming.

SIGNAL ANALYSIS

Electroencephalogram (EEG)

To analyse EEG it is necessary to describe its time and frequency characteristics [7]. In the time domain, signals are displayed by their size in function of their evolution in time and space, whereas in the frequency domain, as the time factor is overlooked, signals are displayed in function of the statistical distribution of their frequency and amplitude. Signal analysis is carried out on the basis of the relationship between these two domains; for example, the sampling interval in one domain corresponds to the length of the signal in the other. Hence, the digital interval in the time domain corresponds to the highest level in the frequency domain and vice versa. The closer the technique to the time domain, the closer it remains to the EEG tracing as the clinician sees it; the closer it is to the frequency domain, the further from the EEG tracing. A brief summary follows of the main methods of EEG analysis.

Amplitude analysis [12]

This is the simplest EEG analysis method. It consists of integrating a measure of amplitude for a certain time interval. Even if a good relationship is established between amplitude and sleep stages [3, 20], this procedure is insufficient when used alone, as although it provides a good assessment of amplitude, it gives no information on signal frequency and the wave morphology of which it is composed.

Period analysis [8, 10, 39]

This method consists of counting the number of times the signal crosses the isoelectric line and measuring the time interval between two crossings. Signal frequency is evaluated by estimating the average interval for a given duration. This method is highly sensitive to artefacts, requiring the complete control of signal derivation. The technique is improved by measuring the interval between 2 zero crossings of the first derivative, but the results give an estimate of frequency which is still too inaccurate, particularly in the case of overlapping frequencies. The results of this method are clearly enhanced by prior filtering of the signal.

Phase locked loop [6, 30]

A phase locked loop is a circuit essentially composed of 2 basic elements: the first is an oscillator whose frequency is controlled by voltage, with a phase comparison device which receives the signal from the oscillator, and the second is an external reference on its two input points. The output voltage, proportionate to the phase difference between the inputs, is used to pilot the oscillator. When the loop locks on, the piloted oscillator works at a frequency which is comparable to the frequency of the external control signal. This procedure provides an accurate recognition of the EEG frequency bands utilised in sleep.

Normalised slope descriptors [4, 9]

Also known as Hjorth parameters [31], these introduce the parameters of activity, mobility and complexity.

The filtering function

Digital filtering has now replaced analogue filtering [31]. Digital filters discriminate between the different frequency bands. They are usually formed by associating a high pass element with a low pass element. The utilisation of digital filters imposes the use of anti feedback analogue filters upstream of the sampling function. Digital filters used in association with integrators and comparison devices provide one of the methods which is closest to the time domain.

Spectral analysis [13]

There are two types of method for spectral analysis: parametric methods based on a signal *production model*, and non parametric methods based on fast *Fourier transform* FFT. We will restrict our description to the Fourier transform methods as they are easier to use and are inserted into most sleep polygraph software. The technique relies on 2 basic hypotheses: 1) the normality of signal amplitude distribution and 2) the fixity of the EEG signal, i.e. with no discontinuity, for a period of less than 4 seconds, as the Fourier transform can only be applied to static signals. The EEG signal is considered stationary for short periods (of the order of 1 minute) during which the recording conditions remain constant. Recurrent FFT is conducted for short periods lasting 1 to 10 seconds, for two reasons: 1) to reduce the costs in calculation time and memory space – which was mostly the case several years ago, and 2) chiefly because the power density resulting from FFT is a random variable in terms of its expected value and variance, the latter being reducible by averaging the values obtained on a successive sequence of epochs.

The choice of the length of epoch to which FFT is applied will determine the frequency resolution of the spectre corresponding to the inverse duration of the epoch. Thus for a 2 second epoch, resolution will be 0.5 Hz and for a 4 second epoch, it will be 0.25 Hz. The latter resolution is usually chosen for sleep EEG spectral analysis. Nevertheless, higher resolutions (0.01 Hz) have been used to demonstrate components of less than 1 Hz [1]. Until relatively recently, the number of calculations needed to carry out Fourier transform spectral analysis considerably restricted its use. With the algorithm introduced in 1965 by Cooley and Tukey known as *Fast Fourier Transform* (FFT), the number of basic operations required is only of the order of (N log2 N) i.e. if N = 256 for example, 2,048 instead of 65,536 [28]. Using Fourier transform power spectra which describe signal frequency can be obtained at the same time as amplitude.

Other EEG quantification techniques have been used recently. Mention should be made of one of the most promising: transformation into wavelets enabling the quantification of short life grapho-elements (1 or 2 sec) which are difficult to detect using standard spectral analysis. These include spindles or transient arousals [22], the EEG signal correlation dimension, which decreases during the stages of deep NREM sleep (stages 3 and 4) and increases in REM sleep [1] and principle component analysis of EEG spectral data [25].

Electro-oculogram (EOG)

The detection of eye movements is essential in distinguishing between sleep stages. Slow eye movements can be detected with varying degrees of accuracy using the excess amplitude method [17].

Most systems individually distinguish rapid eye movements only. In order to differentiate between eye movements and NREM sleep slow waves, eye movements are recorded on 2 different channels with a correlation between the two signals providing a method of detecting phase opposition deviations and thus distinguishing them from NREM sleep slow waves. This method is common [11, 34] and generally gives good results. Another method [5] consists of subtracting one EOG from the other. This method has the advantage, apart from its simplicity, of amplifying opposing (EOG) signals, in cases where the signals are not absolutely identical, and attenuating signals which are in phase (slow waves). However, the technique does not eliminate all contamination by slow waves. To be considered as a rapid eye movement (REM) the movement must be sufficiently ample and rapid; so to detect REM, signal slope and amplitude are generally measured (amplitude of the first derivative and amplitude of the primitive). This method suffices to establish a diagnosis of the stage but not to determine the organisation of bursts or the direction of these movements, in particular.

Electromyogram (EMG)

An analysis of EMG frequency provides no useful information for interpreting human sleep in terms of stages. Thus most systems use methods of integrated analysis to determine EMG during sleep. The signal on which we are working is the EMG envelope, which evolves in the same way as the peak to peak amplitude of the EMG signal. This method can be refined by dividing the 30 second epoch into 10 consecutive 3-second segments. The mean value of the envelope is calculated for each segment. A list of 10 values was drawn up with adjustable coefficients, associating the mean amplitude for the epoch, EMG homogeneity and the presence of movements, [5]. The drawback with this type of method is that it is difficult to define a reference for muscle tone for use in evaluating muscular atonia in REM sleep. This can only be done experimentally in each laboratory or *a posteriori* by observing the values obtained by the analyser during irrefutable REM sleep epochs.

AUTOMATIC ANALYSIS OF SLEEP POLYGRAPHY

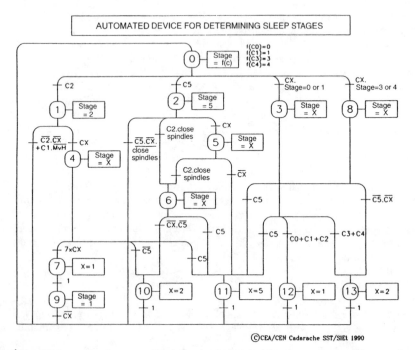

©CEA/CEN Cadarache SST/SIEI 1990

Figure 12.1. Simplified example of an automated device for determining sleep stages. C = condition associated with epoch. X = unidentifiable. Reading of the horizontal bar clearing conditions: "." reads "and", "+" reads "or", a bar over a condition reads "no" or "not". Place a token on the space marked 0. Evaluate the condition, moving the token if necessary. For example, if the condition is evaluated as C1, the token is not moved, the stage is 1 (fC1 = 1). If the condition is C2, the token clears the C2 bar, and the stage is 2.

Classification software

The EEG, EOG and EMG data must then be processed together, to establish a diagnosis by sleep stages, preferably in accordance with the customary criteria [32]. Depending on the method of analysis (methods which are usually close to the frequency domain such as Fourier transform) and the problems they raise (certain overlapping EEG characteristics during sleep), original algorithms have been proposed for classifying sleep stages [18, 27] such as applying algorithms based on fuzzy subset theory [35] characterised by "fuzzy and highly possible" logic [13]. The notion of ambiguous epochs (epochs during which the absence of one or two indicators means that diagnosis depends on knowledge of covert epoch stages) poses a problem which is not completely overcome by most of the decision trees proposed. To compensate for this difficulty, we have set up an automated determining device, which functions in a similar way to the standard GRAFCET, according to the principles of sequential logic [5]. This type of representation (fig. 12.1), traditionally used for sequential automatic devices, implements all the sequential and exceptional rules, by which the classification of an epoch occurs in reference to that of its adjacent epochs.

Most software also offers the possibility of modifying certain acquisition parameters, to improve stage determination. This technique is all the more effective as the system is close to the time domain, and hence to a visual reading. This possibility also allows each laboratory to improve the system's performances - often empirically. Thus the performances of the same system can vary considerably from one laboratory to another. Finally, with a good working knowledge of the system, it is relatively easy to recognise poor interpretations due to artefacts which are undetected by the machine.

Automatic sleep staging

Many systems fail to publish tables of comparison between man (visual analysis) and the machine (automatic analysis). This oversight is often due to the fact that the tables vary in function of a number of adjustment parameters (modified thresholds, inclusion of certain analytical characteristics specific to the laboratory, etc.) which necessarily vary from one scorer to another. Depending on the system, the rate of consistency varies between 65 and 90% in normal subjects and between 65 and 85% in patients affected by sleep disorders. In fact the more sleep is interrupted by movements or anomalies, the greater the risk of artefacts which are difficult for the machine to interpret. In effect, as table 12.1 demonstrates, these scores give a quantitative as opposed to a qualitative perspective of errors. Indeed, for wakefulness to be interpreted as NREM sleep or REM sleep is a substantial error from a diagnostic point of view but often of little importance in terms of the overall quantity of sleep stages (table 12.2). This is another reason why stage diagnosis should be systematically checked visually, and why it is preferable to speak in terms of computer aided analysis rather than automatic analysis.

But to what extent does the system assist interpretation? Several conditions are called for. Signal visualisation and the stage correction or acceptance system must be practical to use, which is not always the case. The data provided by the system must be accessible and useful to the person interpreting the recording. For example, with systems using analysis as the main component in staging, the fact that stage 1 accounts for 22% of variance is of no great benefit for visual diagnosis, because the scorer does not have all the elements enabling him to perceive the error in the system. Systems which use digital filtering with different decision software, indicating the percentages of occupation for each EEG frequency band specific to the stages of human sleep, are a more valuable aid. This is because, in the absence of artefacts, the percentages are correct and the diagnostic error, if there is one, is due to the decision software and is relatively easy to recognise.

Table 12.1. Man-machine comparison for 30-second epochs. The figures underlined indicate the epochs which coincide for man and machine. % refers to the percentage found in relation to the number of epochs in the stage.

		wakefulness	Stage 1	Stage 2	Stage 3	Stage 4	REM sleep	total
V	**Wakefulness**	5599	165	680	36	17	40	6537
I	%	85.6	2.5	10.4	0.5	0.2	0.6	
S								
U	**Stage 1**	24	1801	370	18	10	33	2256
A	%	1	79.8	16.4	0.7	0.4	1.4	
L								
	Stage 2	32	112	17,385	90	24	498	18,141
A	%	0.2	0.6	95.8	0.5	0.1	2.7	
N								
A	**Stage 3**	0	0	13	2661	13	7	2694
L	%	0	0	0.5	98.7	0.5	0.2	
Y								
S	**Stage 4**	0	0	0	3	4123	0	4126
I	%	0	0	0	0.07	99.9	100	
S								
	REM sleep	11	114	654	2	0	6078	6859
	%	0.1	1.8	9.5	0.02	0	88.6	
	Total	5666	2192	19,102	2810	4187	6656	40,613

Table 12.2. Sleep parameters obtained by visual analysis and automatic analysis.

	Visual analysis	Automatic analysis	Difference	%
Time in bed (min)	534.3	534.3	0	0
Total sleep time (min)	448.3	459.8	11.36	2.55
Wakefulness after sleep onset (min)	56.75	56.36	0.39	0.68
Sleep latency (min)	29.2	18.19	11.01	37.7**
REM sleep latency (min)	199.9	86.6	33.3	27.7**
Stage 1 (min)	29.6	28.8	0.84	2.7
Stage 1 (%)	6.61	6.27	0.34	5.14
Stage 2 (min)	238.6	251.3	12.7	5.32
Stage 2 (%)	53.2	54.6	1.4	1
Stage 3 (min)	35.4	36.5	1.1	3.1
Stage 3 (%)	7.90	8.04	0.1	1.77
Stage 4 (min)	54.2	55.09	0.89	1.64
Stage 4 (%)	12.1	11.98	0.12	0.9
REM sleep (min)	90.2	87.5	2.7	2.79
REM sleep (%)	20.01	19.04	0.97	4.84
Continuity index	0.90	0.90	0	0

** = $p < 0.01$

Fine analysis of sleep

This analysis, which incorporates the physiological parameters which cannot be objectified by visual analysis, is undoubtedly the most interesting possibility offered by most automatic sleep analysis systems. One of the most useful measurements is that of slow wave activity, represented by power spectral EEG activity in the 0.5-4.75 Hz delta band, providing a quantification of sleep intensity, evaluation of the duration of voltage band occupation, the quantity of slow waves, number of sleep spindles, number of REM, number of transient arousals etc. It is nevertheless important to note that the utilisation of these parameters in sleep pathology is still restricted to the realms of research (this is likely to change in the near future). Finally, it is not enough to simply analyse the fine variations in sleep; they must be correlated with variations in the other physiological parameters (respiratory, muscular, cardiac etc.). Many systems do not allow for such sophisticated analysis of sleep and those which do require a degree of investment in information technology and mathematics which is beyond the reach of many laboratories.

Automatic analysis of physiological variables

Interpreting sleep stages is the prerequisite to computer aided analysis as it will influence the following step i.e. the analysis of other physiological functions whose variations can only be analysed during sleep and in the 30 minutes of wakefulness tested during calibration. It is therefore important to use systems whose final statistics take account of the modifications made during the various stages. For example, apnoeas recorded during wakefulness must be systematically eliminated, whereas apnoeas which are not recorded by the machine but by the scorer during sleep, must be integrated.

Muscular and respiratory activity

Many systems provide fairly accurate quantification of periodic limb movements or respiratory events occurring in sleep. Analysis of the anterior tibialis is carried out by techniques which have been described elsewhere and that of respiratory functions, by analysing amplitude variations in relation to threshold levels (mean mobility technique) which the clinician can usually determine by himself with good results. As for sleep stages, each respiratory event should be monitored, accepted or corrected; with certain systems this correction is a veritable obstacle course.

Adding new physiological variables to those already established by the system often poses a problem. Indeed while oximetry (measure of SaO_2) is dealt with by practically all the systems, new parameters such as pneumotachometry or the analysis of oesophageal pressure are not included in most systems. Theoretically, acquiring these channels should present no more difficulty than for recording other parameters using IRIG standards. The problems arise in analysing the data from these channels i.e. in accessing the files.

Cardio-respiratory spectral analysis

In the spectral analysis of respiration, the more regular the breathing, the higher and narrower the peak, at virtually fixed frequency. So respiration at around 12 cycles per minute (i.e. a period of $60/12 = 5$ sec.) corresponds to a maximal peak of $1/5$ sec $= 0.20$ Hz.

Heart rate spectral analysis is a recognised method in cardiology used to evaluate the activity of the autonomous nervous system [15]. Two classes of representative spectral components are generally obtained using this technique: 1) the low frequency (LF) component ranging from 0.02 to 0.15 Hz, representing vasomotor activity (0.02 to 0.09 Hz.) and baroreceptor activity (0.09 to 0.15 Hz), and 2) the high frequency (HF) component based on respiratory frequencies ranging from 0.15 to 0.40 Hz.

The high frequency component has been shown to be dependent on the activation of the parasympathetic system, while the low frequency component depends on both the sympathetic and the parasympathetic systems [15].

An interesting means of assessing autonomic activity during sleep using spectral analysis, is to use the LF/BF ratio to define the "sympatho-vagal balance", reflecting both reciprocal and non reciprocal fluctuations of sympathetic and parasympathetic tonicity [29]. Likewise, the low frequency component of blood pressure is considered as a marker for vascular sympathetic modulation whereas the high frequency component reflects the mechanical interaction between respiration and the cardio-vascular system.

Thanks to combined heart rate-blood pressure spectral analysis, *baroreflex sensitivity* can be measured by gains in the relationship between blood pressure and RR heart period. Blood pressure can be continuously and unobtrusively measured using finger inductance photoplethysmography or indirectly by measuring TTP, and respiration using a pressure transducer air flow sensor (cf. polygraphic sleep testing in the same volume).

As respiration plays an important part in the spectral composition of heart rate and blood pressure, it is essential to know the respiration rate value in order to interpret variations in rate and blood pressure.

Data processing

Polygraphic data should be mathematically processed and stored using statistical calculations and data base software. This involves being able to read the data in a format which is accessible to all (ASCII format), not always an easy task, considering the difficulties entailed in accessing the file, transforming binary files into ASCII files (this will depend on the system, as some systems do not cater for this), and above all, processing the ASCII file. It is of interest to note that the raw data produced by an EEG channel for one night's analysis takes up several megabytes – a substantial amount; hence certain systems do not save all the raw data, precluding any later calculations for this data. If an ASCII file can be obtained, it then has to be processed, which requires a certain grasp of information technology, but this is often the price to pay for satisfactory information.

CONCLUSION

The available systems each have their qualities and defects. We have attempted to show that a system must be chosen according to the objectives and technical and financial means of the laboratory. Choosing a system often means choosing an objective, and objectives differ between laboratories. It is also important to remember that after-sales "assistance" is often impossible to obtain. However some systems do provide very effective telephone assistance and telephone assistance via a modem will soon be available to users. Finally, the big laboratories are usually familiar with using IT systems, and they never refuse to offer helpful information and advice. Computer-aided analysis has become a necessary tool; while it provides information which was inaccessible in the past, it requires a different way of thinking, which is not always easy to master.

REFERENCES

1. ACHERMANN P., BORBELY A.A. – Low frequency (1 Hz) oscillations in the human sleep electroencephalogram. *Neuroscience*, 81, 213-222, 1997.
2. ACHERMANN P., HARTMANN R., GUNZIGER A., BORDELY A. – Correlation dimension of the human sleep electroencephalogram: cyclic changes in the course of the night. *Europ. J. Neurosc.*, 6, 497-510, 1994.
3. AGNEW H.W., PARKER J.C., WEBB W.B., WILLIAM R.L. – Amplitude measurement of the sleep electroencephalogram. *Electroenceph. Clin. Neurophysiol.*, 22, 84-86, 1967.
4. ERGLUND K., ELMQVIST D. – Estimation of sleep depth using normalized slope descriptor quantification of the EEG. *In: Sleep 1974*, P. Levin and W.P. Koella (eds.), 141-145, 1975.
5. BESSET A., SPUIG P., DUFAYET J.P., BILLIARD M. – Analyse automatique du sommeil humain selon les critères de Rechtschaffen et Kales. *Neurophysiol. Clin.*, 20, 519, 1990.
6. BROUGHTON R., HEALEY T., MARU J., GREEN D., PAGURECK B. – A phase locked loop device for automatic detection of sleep spindles and stage 2. *Electroenceph. Clin. Neurophysiol.*, 44, 677-680, 1978.
7. BURCH N.R. – Automatic analysis of the EEG: A review and classification of system. *Electroenceph. Clin. Neurophysiol.*, 11, 827-834, 1959.
8. BURCH N.R., NETTELTON W.J., SWEENEY J., EDWARDS R.J. – Period analysis of the electroence-phalogram on a purpose digital computer. *Ann. NY. Acad. Sci.*, 115, 827-843, 1964.
9. CLARENBACH P., KANNO C., KAPP H., CRAMER H. – The time domain methodology in the scoring of human sleep. *In: Sleep 1978*, L. Popoviciu, B. Asgian, G. Badin (eds.), S. Karger, Basel, 587-590, 1980.
10. CHURCH M.W., MARCH J.D., HIBI S., BENSON K., CAVNESS C., FEINBERG I. – Changes in frequency and amplitude of delta activity during sleep. *Electroenceph. Clin. Neurophysiol.*, 39, 1-7, 1975.
11. DEGLER H.E., SMITH J.R., BLACK F.O. – Automatic detection and resolution of synchronous rapid eye movements. *Comp. Biomed. Res.*, 8, 393-404, 1975.
12. DROHOCKI Z. – L'intégrateur de l'électroproduction cérébrale pour l'électroencéphalographie quantitative. *Rev. Neurol.*, 80, 619-624, 1948.
13. DUBOIS D., PRADE H. – Les logiques du flou et du très possible. *La Recherche*, 237, 22, 1309-1315, 1991.
14. DUMERMUTH G., GASSER T., LANGE B. – Aspects of EEG analysis in the frequency domain. *In: CEAN – Computerized EEG analysis*, G. Dolce, H Kunkel (eds.), Gustav Fischer Verlag, Stuttgart, 1975.
15. EUROPEAN SOCIETY OF CARDIOLOGY. Heart rate variability. Standards of measurement. Physiological interpretation and clinical use. *Circulation*, 93, 1043-1065, 1996.
16. FELL J., ROSCHKE J., MANN K., SCHAFFNER C. – Discrimination of sleep stages: a comparison between spectral and non linear EEG measures. *Electroenceph. Clin. Neurophysiol.*, 98, 401-410, 1996.
17. GAILLARD J.M., TISSOT R. – Principles of automatic analysis of sleep record with a hybrid system. *Comp. Biomed. Res.*, 6, 1-13, 1973.
18. GATH I., BAR-ON E. – Computerized method for scoring of polysomnographic sleep recordings. *Comp. Pro. Biomed.*, 11, 217-223, 1980.

19. GEERING B.A., ACHERMAN P., EGGIMANN F., BORBELY A.A. – Period amplitude analysis and power spectral analysis: a comparison based on all night sleep EEG recordings. *J. Sleep Res.*, *2*, 121-129, 1993.

20. GOLDSTEIN L., BURDICK J.A., LAZLO M. – A quantitative analysis of the EEG during sleep in normal subjects. *Acta physiol. Acad. Aci. Hung.*, *37*, 291-300, 1970.

21. GREEN D., PADUREK B., HEALEY T., BROUGHTON R. – Processing components for automatic sleep staging using the EEG. *IEEE Abstracts*, *81*, 202-203, 1977.

22. HACHEMAOUI M., MORVAN C., HERMAN M., NEDELCOUX H., DEBOUZY C., BOURGIN P., ESCOURROU P. – Détection automatique des micro-éveils par l'analyse temps-fréquence en ondelettes. *Neurophysiol. Clin.*, *26*, 433, 1996.

23. HJORTH B. – EEG analysis based on time domain properties. *Electroenceph. Clin. Neurophysiol.*, *29*, 306-310 1970.

24. IITIL T.M., SHAPIRO D.M., FINCK M., KASSEBAUM D. – Digital computer classification of EEG sleep stages. *Electroenceph. Clin. Neurophysiol.*, *27*, 76-83, 1969.

25. JOBERT M., ESCOLA H., POISEAU E., GAILLARD P. – Analysis of sleep using two parameters based on principal component analysis of electromyography spectral data. *Biol. Cybern.*, *71*, 197-207, 1994.

26. KTONAS P.Y., GOSALIA A.P. – Spectral analysis vs period – amplitude analysis of narrow band EEG activity: a comparison based on the delta sleep frequency band. *Sleep*, *4*, 193-206, 1981.

27. KUMAR A. – A real time system for pattern recognition of human sleep stages by fuzzy system analysis. *Pattern Recognition*, *9*, 43-46, 1977.

28 LIFERMANN J. – Théorie et application de la transformée de Fourier rapide. Masson Paris, 1977.

29. PAGANI M., LOMBARDI F., GUZETTI S. , RIMOLDI O., FURLAN R., PIZINELLE P., SANDRONE G., MALFATTO G., DELL'ORTO S., PICCALUGA E. *et al* - Power spectral analysis of heart rate and arterial pressure variability as a measure of sympatho-vagal interaction in man and conscious dog. *Circ. Res., 59, 178-183, 1986.*

30. PIVIK R.T., BYLSMA, F.K., NEVINS R.J. – A new device for automatic sleep spindles analysis: «the spindicator». *Electroenceph clin Neurophysiol.*, *54*, 711-713, 1982.

31. RADIX J.C. – Introduction au filtrage numérique, Eyrolles, Paris, 1970.

32. RECHTSCHAFFEN A., KALES A. (eds.). – A manual of standardized terminology, techniques and scoring system for sleep stages of human subjects. Brain Information Service/Brain Research Institute, Los Angeles, 1968.

33. SHARP F.H., SMITH G.W., SURWILLO W.W. – Period amplitude of the electroencephalogram with recording of interval histograms of EEG half wave durations. *Psychophysiology*, *12*, 471-475, 1975.

34. SMITH J.R., CRONIN M.J., KARACAN I.A. – A multichannel hybrid system for rapid eye movements detection. *Comp. Biomed. Res.*, *4*, 275-290, 1971.

35. ZADEH L. – A fuzzy algorithms. *Inform. Cont.*, *12*, 94-102, 1968.

Chapter 13

Assessing sleepiness

A. Besset

INSERM EMI-9930 Hôpital de la Colombière, Montpellier and Service de Neurologie B, Hôpital Gui de Chauliac, Montpellier, France

Sleepiness generally coincides with the inclination to sleep or "the desire to sleep", recognised as being normal or pathological. Normal sleepiness is mentioned when it relies on processes which may be homeostatic (response to the previous amount of wakefulness) and circadian (time in 24 hours favourable to sleep) and pathological when it is symptomatic of a sleep disorder and when it can occur at any moment in 24 hours. But it is reasonable to question whether there is any real difference in the nature of these two types of sleepiness. Whatever the case, sleepiness which occurs during a period of activity creates repercussions both in terms of safety and the law which must not be overlooked. It is thus involved in numerous transport accidents e.g. trucks, buses, trains, ships and planes and is responsible for serious industrial accidents i.e. chemical or atomic energy plants [92, 83, 32, 48, 62]. Sleepiness also has a marked impact in the educational sphere. The American institute responsible for scheduling school timetables has estimated that in the USA 70 billion dollars is lost in productivity in the workplace due to sleepiness [30]. Finally excessive daytime sleepiness takes its toll on the quality of life of those suffering from the disorder. In narcolepsy for example, a complaint which is essentially characterised by episodes of recurrent daytime sleepiness and a drop in muscle tone in response to emotions (cataplexies), patients experience substantial problems due to their permanent sleepiness, at work, in school, in social relations and driving [15, 16, 38]. In view of the seriousness and pervasiveness of the effects of sleepiness it is clearly important to be able to measure sleepiness rigorously and scientifically.

The ideal way to measure sleepiness would be to take a blood sample of a substance directly responsible for the propensity to sleep. As, unfortunately, this substance has not yet been identified, sleepiness can only be measured indirectly and thus inaccurately. Sleepiness can be evaluated in 3 ways: as a subjective assessment, by studying variations in vigilance in relation to sleepiness and by polygraphic measurements of sleepiness.

SUBJECTIVE ASSESSMENT OF SLEEPINESS

This assessment is essentially introspective. There are two main categories of instruments: those which measure the immediate subjective level of sleepiness at a given moment, and those which measure sleepiness behaviourally, incorporating the level of sleepiness over long periods or measuring the constant level of sleepiness in the course of the various situations in daily life.

Immediate subjective level

This is expressed by means of visual analogue scales and subjective scales.

Visual Analogue Scales (VAS)

VAS are the easiest scales to use. The subject marks the point on a 10cm line corresponding to his level of wakefulness, ranging from the highest wakefulness state (very awake) to the lowest wakefulness state (falling asleep). These scales are highly sensitive to the effects of partial or total sleep deprivation [25].

Subjective scales

These are more sophisticated but also slightly more complicated. The most representative of this group is the Stanford sleepiness scale (SSS) (see annexes). Using a Lickert type scale with eight levels ranging from level 1 "feeling active and vital; alert; wide awake", to level 8 "asleep", the subject has to indicate which one best describes his state. His judgement of sleepiness is based on a purely subjective feeling, with no precise objective criteria, nor any reference to the situation [47]. It is no doubt for these reasons that the scale only provides a good assessment of sleepiness in the normal subject after partial sleep deprivation [45] and is usually unreliable for sleep disorders involving excessive daytime sleepiness such as narcolepsy [74] or sleep apnoea syndrome [78].

On the same principle, the Karolinska sleepiness scale (KSS) (see annexes) measures daytime sleepiness in 9 points based on 5 states ranging from "extremely alert" to "extremely sleepy, fighting sleep". Four intermediary states are noted but not designated in words. This scale is closely linked to the electroencephalographic and oculographic signs of sleep onset [3].

Behavioural assessment of sleepiness

Yawning is a well known sign of the desire to sleep. There is a whole range of behavioural changes due to sleepiness, such as lowering of the eyelids, the brief suspension of consciousness etc. Carskadon [20] thus devised a questionnaire on the behaviour students observed in class according to whether they were wide awake, sleepy or in between. Thirteen types of behaviour were thus listed: yawning, laying the head on the desk, and closing the eyes were the types of behaviour considered as symptomatic of sleepiness, whereas taking a lot of notes or sitting up straight were characteristic of a wide awake state.

Along the same lines, Gillberg *et al* [35] drew up a scale that incorporated subjective sleepiness over long periods: the accumulated time with sleepiness (ATS). This scale is composed of 8 items. The subject must indicate whether he has experienced one of the symptoms: heavy eyelids, sand in the eyes, difficulty concentrating, etc. during the given period and whether these occurred: rarely, 25%, 50% or 75% of the time, most of the time. This scale has proved sensitive to variations in sleepiness and is closely linked to performance level.

But although behavioural observation is interesting the models are often difficult to use in daily life due to problems in determining standard observation conditions, establishing the exact significance of the observed behaviour, and finally in controlling the environment.

Daily situations are more or less "soporific" in terms of the extent to which they generate the desire to sleep. The "soporific" nature of a situation depends on several factors, the main ones being the physical environmental (temperature, noise), the subject's posture (standing, sitting, lying down, eyes closed, eyes open) and his emotional state (worried or relaxed).

Unlike the subjective Stanford sleepiness scale, which questions the subject on his degree of sleepiness, the Epworth sleepiness scale (ESS) [49] (see annexes) asks the subject to rate the probability of dozing from 0 (would *never* doze) to 3 (high chance of dozing) in eight more or less soporific daily situations: 1) sitting and reading, 2) watching TV, 3) sitting, inactive in a public place (e.g. a theatre or a meeting), 4) as a passenger in a car for an hour without a break, 5) lying down to rest in the afternoon when circumstances permit, 6) sitting and talking to someone, 7) sitting quietly after a lunch without alcohol, 8) in a car, while stopped for a few minutes in the traffic. The most soporific situation being number 5, and the least soporific number 6. A score of over 10 is taken to indicate abnormal sleepiness. This scale distinguishes patients with disorders of sleepiness from paired controls [50, 51].

Correlation studies between the Epworth sleepiness scale and the multiple sleep latency test (MSLT, see below) have yielded contradictory results [49, 50, 51, 68, 11, 27]. Sleepiness is a complex phenomenon which takes on various aspects. The Epworth sleepiness scale evaluates a subjective state and behaviour experienced in the recent past, whereas the MSLT measures a physiological tendency to sleep, strongly influenced by the sleep experienced over the preceding 24 hours.

Other questionnaires measure sleepiness behaviour. Among these two multidimensional scales are worth mentioning: the Rotterdam daytime sleepiness scale (RDSS) [91] with 16 items and 3 categories and the Sleep wake activity inventory (SWAI) [75] with 59 items and 6 factors.

The results of subjective scales may be biased in several ways such as the subject's inability to understand the items or the urge to falsify, voluntarily or involuntarily. Finally it is important to take account of individual factors of sleep tendency, the nature of the situation in which sleep tendency is measured and lastly factors relating to the subject and the situation. Sleep tendency measured in a given situation is not necessarily representative of an individual's average sleepiness in everyday life.

STUDY OF VARIATIONS IN VIGILANCE IN RELATION TO SLEEPINESS

Vigilance designates the capacity of the nerve processes (central nervous system) to respond effectively to a stimulus or event. Vigilance is usually measured by psychological performance tests or physiological tests evaluating the brain's capacity to react to standardised stimuli such as evoked potentials or pupillometry. If vigilance is altered by sleepiness it is also altered by "hyperwakefulness" functioning as an inverted U, the optimum being at the top of the U and the minima at the two extremities, so that vigilance (a psychological concept) and sleepiness (a physiological concept) cannot be directly compared.

Psychomotor measure of vigilance

These measurements are carried out on the basis of performance tests related to the subject's capacity to react in everyday life.

Reaction time tests [36]

These measure the rapidity of response to stimuli. They may be divided into simple reaction time and choice reaction time tests.

In simple reaction time tests, the subject has to respond as rapidly and effectively as possible to identical stimuli. In this type of test, the reduction in sleepiness-related performance is characterised by an unusual lengthening of the response time, as well as two kinds of error: errors of omission (lapses): the subject does not respond when the stimulus appears, and errors of instruction (commissions): the subject responds in absence of the stimulus. These tests are classed according to the type of sensorial system implied (visual or auditory). Among the best known, it is worth mentioning the Lisper and Kjellberg auditory reaction time task [57], the Wilkinson visual reaction time task and lastly the Dinges and Powel task [31] or psychomotor vigilance task (PVT) (fig. 13.1), with the added particularity of showing the patient the result of the performance to add a motivating factor. The latter test has proven highly sensitive to "lapses" due to sleepiness [55].

In choice reaction time tests, the subject has to respond as rapidly and effectively as possible to stimuli with clear, specific differences (position, duration, colour etc.) compared to the others. Measurements comprise reaction time, the mean and variability of correct and incorrect responses. Among the most commonly used tests it is worth noting the 4 choice reaction time test [93] in which 4 diode lamps light up one after the other and the subject has to press the button corresponding to each lamp.

It was long thought that to be effective these tests had to last for a long time (roughly an hour) [53, 54, 57, 95, 96]. It was then discovered that with sufficient training, the tests were capable of revealing deficits in performance after only 10 minutes [36, 37, 44]. Other studies with shorter tasks (3 minutes) have shown that performances diminished after 18 hours of continuous wakefulness [65]. In a meta analysis, Pilcher and Huffcutt [69] showed that after short deprivation (<45 hours) performances were affected in short tasks even though more marked effects were seen with long tasks, during long-term sleep deprivation (> 45 hours) performances altered more in short tasks than in long tasks.

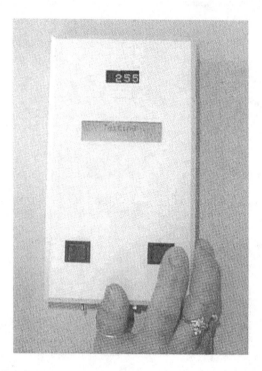

Figure 13.1. Psychomotor vigilance test (PVT): The stimulus appears in the form of time in milliseconds scrolling past in the top window of the device. The subject has to press the lower button as rapidly as possible (right for right-handers, left for left-handers). The score marked up is the reaction time in milliseconds; the interval between stimuli is randomised. The test lasts 10 minutes.

Simultaneous EEG recording has shown that omissions usually corresponded to micro-sleep episodes [21]. Nevertheless it seems that the slowing and variability of responses are a more sensitive sign of lowered performance in relation to sleepiness than the measurement of omissions [89].

Signal detection tests

These tests require subjects to detect the one stimulus (target stimulus) which differs slightly from the other, identical stimuli. These tests originated during the Second World War in devising tasks to test the capacity of radar operators to detect enemy aircraft amongst friendly aircraft. For certain psychologists, the concept of vigilance has become synonymous with the ability to perform well in this type of test. Measurements refer to reaction times, false positives and negatives and the variability of errors and inter stimulus intervals. By applying the "signal detection theory" [86] sensitivity (d'): the subject's ability to discriminate and bias (beta): motivation to reply are distinguished in the same subject.

The most commonly used tests are the Wilkinson auditory test [94] and the continuous performance test (CPT) [76, 73] during which the target stimulus is the letter X among other letters (X type CPT). A variant of the test involves detecting the X target stimulus only when preceded by the letter A (AX type CPT).

Measurement of spontaneous or evoked responses

During the finger tapping test (FTT) [24], the subject lies in a sleeping position in darkness and has to tap a lever as quickly as possible with the index of each hand. The intervals are measured between each tap. The results of the test are correlated with subjective and electroencephalographic sleep latencies. A 20-second stop in response is considered as indicating sleep, and a pause of 60 seconds as indicating the start of a prolonged period of sleep.

In the Oxford sleep resistance test (Osler) [8] (fig. 13.2), the subject sits comfortably in darkness, and has to press a button when a diode lights up, programmed to light up for 1 second every 3 seconds. The button is silent and does not call for pressure to avoid wakening the subject. Sleep onset is defined as overlooking 7 stimuli or 21 seconds without response. The test has proven sensitive to discriminating sleepy subjects in sleep apnoea syndrome.

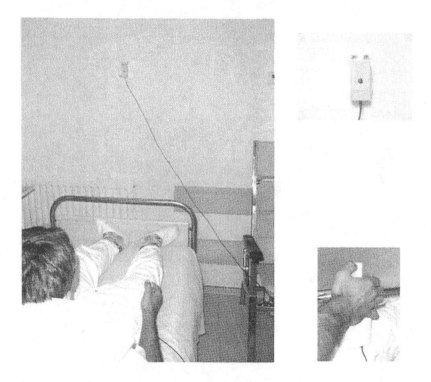

Figure 13.2. Osler test: The subject is half-seated, in darkness. Opposite him at eye-level is a box which emits a red light. The stimulus, the red light, appears for one second every 3 seconds and the subject has to press a handheld button. The subject is considered as having fallen asleep if he keeps his finger on the button, or fails to press the button for 7 consecutive cycles (21 seconds). In addition, the number of consecutive cycles (from 2 to 6) for which the subject keeps his finger on or off the button constitutes as many increasing degrees of sleepiness. The test lasts 40 minutes.

Other attention tests

Other performance tests have been used to evaluate vigilance in relation to sleepiness. These usually consist of attention tests involving the frontal lobe, which is sensitive to sleepiness in normal subjects after sleep deprivation or in subjects with excessive daytime sleepiness. They include letter [25] or symbol cancellation tasks, coding tests, certain tests of manual dexterity like Perdue pegboard (fig 13.3), the Wisconsin card sorting task and the Stroop color-word test (fig. 13.4). (for a more detailed description of these tests see M. Lezack [56]).

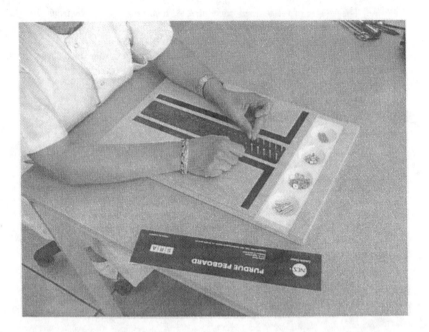

Figure 13.3. Perdue pegboard: The test is carried out using a board with two parallel rows of holes into which the subject has to insert metal pegs, discs and cylinders. The test is in 4 successive parts:

- Dexterous hand: using his dexterous hand, the subject has to insert as many pegs as possible into the row of holes next to his dexterous hand in 30 seconds.
- Non-dexterous hand: using his non-dexterous hand, the subject has to insert as many pegs as possible into the row of holes next to his non-dexterous hand in 30 seconds.
- Both hands simultaneously: using both hands the subject has to insert as many pegs as possible into the two parallel rows of holes in 30 seconds.
- Assembling: the subject has to insert a peg, a disc, a cylinder, and a disc into as many holes as possible in one minute, using both hands.

1	2	3	4
Jaune	Rouge	Vert	Bleu
Vert	Jaune	Bleu	Rouge
Bleu	Vert	Rouge	Jaune
Rouge	Bleu	Jaune	Vert
Vert	Rouge	Bleu	Jaune

1	2	3	4
☐	☐	☐	☐
☐	☐	☐	☐
☐	☐	☐	☐
☐	☐	☐	☐
☐	☐	☐	☐

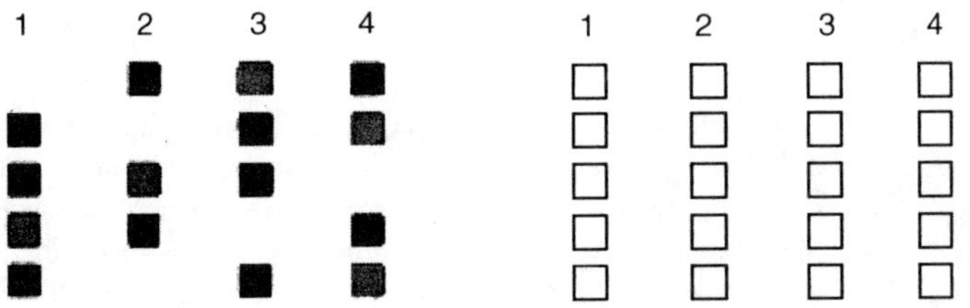

1	2	3	4
Rouge	Vert	Bleu	Jaune
Vert	Bleu	Jaune	Rouge
Bleu	Rouge	Vert	Jaune
Vert	Jaune	Rouge	Bleu
Jaune	Rouge	Bleu	Vert

1	2	3	4
☐	☐	☐	☐
☐	☐	☐	☐
☐	☐	☐	☐
☐	☐	☐	☐
☐	☐	☐	☐

Figure 13.4. Stroop colour-word test (French version). There are 3 parts to the test:

- In the first part the subject has to write the first letter of each word (eg. R for red, B for blue, etc.) in the matching box.
- In the second, the subject has to write the first letter of each colour (B for blue, Y for yellow, etc.) in the matching box.
- In the third, the subject has to write the first letter of the colour in which the word is printed (Y for the word red printed in yellow, R for the word yellow printed in red, etc.) in the matching box.
- Each part of the test lasts a maximum of 90 seconds.

Several problems remain in using psychomotor performances to measure sleepiness. First, it is hard to make deductions about an individual's sleepiness on the basis of a decrement in vigilance, as these alterations in vigilance are the result of numerous factors of which sleepiness is only one. Moreover, the tests themselves are interdependent, requiring the use of a more or less large battery of tests. Finally, the lack of a database generally makes it impossible to compare performances between subjects.

Physiological measurements of vigilance

Several physiological parameters can be measured to assess physiological sleepiness. Among the most important are global motor activity, eye movements, variations in pupil diameter and evoked potentials.

Global motor activity

Developments in new technology, mainly in piezoelectric sensors, have made it possible to measure the accelerations linked to movements, and to miniaturize ambulatory activity monitors referred to as actimetres. These devices, the size and weight of a watch, are worn on the non-dominant wrist, which being less involved in regular activity, favours the recording of the subject's general activity. Actimetres (or actigraphs) continuously record a subject's activity (24 hours a day) mainly to determine the duration and times of the subject's sleep over long periods, usually for several weeks. Several studies have focused on the reliability of such devices in making an effective assessment of sleep and wakefulness based on recording activity, comparing the results of actimetry with those of sleep polygraphy, both in the normal subject and in the subject presenting sleep disorders. The findings clearly show that actimetry overestimates the duration of sleep latency and underestimates total sleep time as well as intra-sleep wakefulness [80, 28]. According to the sleep parameters analysed the rate of correlation varies from 90% (sleep latency, total sleep time) to 70% (intra-sleep wakefulness). Nevertheless the differences observed between polygraphy and actimetry remain acceptable for clinical practice and are far less marked than those observed between polygraphy and data obtained with subjective scales. It should however be noted that the inaccuracy of the actimetric method in measuring sleep latency renders the technique unusable for measuring daytime sleepiness during sleep latency tests and should never be used in place of polysomnographic multiple sleep latency tests (see below). This method (actimetry) is very useful in assessing hypersomnias and insomnias and remains a virtually indispensable tool in studying circadian rhythm sleep disorders [79].

Eye movements

Numerous studies have emphasised the relationship between alterations in oculomotor activity and the decrement of vigilance. The changes basically involve an alteration [70] in voluntary rapid eye movement activity (saccadic), the decrease or disappearance of their spontaneous activity as well as that of eyelid movements [39], the onset of slow eye movements increasing during sleep onset and disappearing when sleep is firmly established [64]. Besides this, slow eye movements correlate to a decrease in spectral power in the alpha band and an increase in the power of theta bands and of delta and theta [89] associated with a drop in performance [88]. Finally these slow eye movements correlate to a subjective impression of sleepiness [3].

Pupillometry

This infrared recording technique of the diameter or surface area of the pupil in response to various stimuli or at rest has been used for some time in measuring alterations in vigilance in order to detect sleepiness [59]. The technique is based on the principle that raised vigilance is associated with a wide pupil diameter (>6 mm for a duration of 10 minutes in total darkness) and one that is stable, whereas sleepiness is synonymous with an unstable, reduced diameter. Pupillometry has been applied to the study of narcolepsy [97] to little effect, showing no differentiation between narcoleptic subjects and those affected by other wakefulness disorders [82] nor even, in certain conditions, normal subjects [67].

Evoked potentials

The amplitude and latency of late components in cerebral evoked potentials (particularly in regard to auditory evoked potentials) respectively diminish and lengthen in sleepy subjects [34, 41]. In addition, a reduced amplitude and lengthened latency can be seen in narcoleptics [16] or in normal subjects after sleep deprivation [71]. The P300 wave appears during a stimulus detection task, 300 ms after the stimulus. Its amplitude is seen to decrease after sleep deprivation [18] in the narcoleptic subject [2]. When evoked potentials are obtained before each MLST session (see below) P300 amplitude is seen to be almost as sensitive as mean latency to sleep onset in differentiating between sleepy narcoleptics and controls [14].

Even more surprising is the contingent negative variation (CVN) in the narcoleptic. It is known that the brain of a subject in the interim between a warning stimulus and a later stimulus requiring a response, generates sustained negativity in the frontal regions referred to as contingent negative variation or the "expectancy wave". This wave will differ in the narcoleptic depending on whether he falls asleep in REM sleep or NREM sleep. CVN is virtually absent before a sleep onset REM period and at wakefulness level before a sleep onset NREM period [13].

These particular physiological measures are of considerable interest in assessing sleepiness but need further development before they can be used in current practice.

POLYGRAPHIC MEASUREMENTS OF SLEEPINESS

These measurements undoubtedly offer the best interpretation and even quantification of the degree of sleepiness.

EEG in standard conditions

The routine EEG is one of the most sensitive indicators of the transitions between sleep and wakefulness. The best known polygraphic signs of sleepiness are the diffusion of alpha waves to the anterior regions, and the occurrence of slow eye movements. These eye movements witness the instability of the ocular globes; they indicate a loss of control over voluntary movements, signifying a drop in the faculties of attention and revealing the state of sleepiness. Such episodes may be seen several times in the course of the day in sleepy subjects, signifying in fact brief recurrent microsleep events.

Digital EEG

Electronic systems of quantitative EEG analysis have recently refined analysis of the different levels of EEG in wakefulness and sleep. Sleepiness increases the power density of alpha and theta EEG frequencies in subjects with eyes open [1, 17, 58]. Conversely, excessive sleepiness decreases alpha band power in subjects with eyes closed in a calm wakefulness state [19, 87]. EEG can reflect sleepiness not only in the calm wakefulness state, but also during performance tests [3, 29, 87]. Monotonous tasks and long lasting tasks were thus shown to be associated with an increase in the power of the EEG alpha band [46]. EEG has thus been used to assess sleepiness in shift workers [88]. Spectral analysis of EEG activity carried out in wakefulness before and after a whole night's sleep has shown that sleep facilitates interhemispheric coupling and intrahemispheric differentiation by increasing intrahemispheric correlation and decreasing interhemispheric correlation. The contrary was found after sleep deprivation [29].

Continuous polygraphic recording

Tests based on repeated recordings (see below) provide an assessment of the degree of sleepiness at a given moment and ultimately an evaluation of circadian variations in sleepiness. However these are not really suitable for investigating certain disorders whose main symptom is the lengthening of nocturnal sleep time, such as idiopathic hypersomnia, recurrent hypersomnia, post-traumatic hypersomnia, infectious hypersomnia or finally hypersomnia related to psychiatric disorders. To assess the quantity of sleep over 24 hours continuous recording must be carried out.

This type of recording, which may be conducted either in laboratory or ambulatory conditions, evaluates the severity of sleepiness over a 24 hour period in the normal subject after sleep deprivation or in a subject suffering from narcolepsy or idiopathic hypersomnia [9, 10]. By this means it is possible to fix the peak times of sleepiness and determine their duration (see chapter 32, fig. 32.3). The advantage of this method is that the subject can be observed in a controlled environment like the laboratory. The main drawbacks apart from the cost, are that the subject's movements are restricted by the cable linking him to the electrode box and the fact that the laboratory situation is not representative of his habitual surroundings.

Ambulatory recording is carried out using portable analogue recorders (tape) or electronic recorders (flash card, microprocessor). The technique allows the subject to be recorded for 24 hours in his usual surroundings, even if wearing electrodes on the scalp and face may alter the subject's habitual situation. This technique has been used to record shift workers [89] and patients affected by sleep disorders [12].

The main drawbacks with the technique are that it is impossible to sort out certain defects during the recording which may result in having to redo the test, and particularly, the lack of objective indications on the subject's behaviour during the recording.

Multiple tests

The alpha attenuation test

This test is based on the principle that when the eyes are closed, alpha activity (8-12 Hz) increases whereas it is attenuated when they open. As in the MSLT (see below), every 2 hours the subject is asked to sit in a normally lit room and to keep the eyes open or closed for alternating 2 minute periods, for 12 minutes. The ratio between the mean power of alpha activity with eyes closed and the mean power of alpha activity with eyes open provides an evaluation of the degree of sleepiness. The higher the ratio, the less sleepy the patient. Correlated with the MSLT, this test provides a satisfactory measurement of sleepiness outside the recording room adapted to MSLT. It has proven to be effective in discriminating between controls and narcoleptic subjects [5]. In any case it is of limited reliability in subjects presenting an abnormally high or low level of alpha, and requires a sophisticated device able to provide Fourier transform spectral analysis of the EEG signal.

The multiple sleep latency test (MSLT)

This test devised by Carskadon and Dement [22] and by Richardson *et al* [74], and considered by many as the gold standard in sleepiness measurement tests, was developed on the basis of the following principle: the sleepier the subject, the faster he falls asleep. The test is based on 20 minute polygraphic recordings (EEG, EOG, EMG) repeated every 2 hours (4 or 5 times a day). The MSLT distinguishes easily between subjects with excessive daytime sleepiness (EDS) and non sleepy subjects (fig. 13.5).

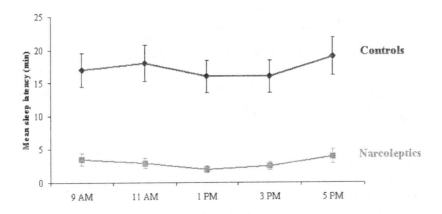

Figure 13.5. Multiple sleep latency test: sleep latencies in normal and narcoleptic subjects (means and standard deviation). Note the severe sleepiness of narcoleptic subjects peaking at 1 and 3 PM.

Even though a protocol has been established [23], a large number of variants to this protocol are practised in many laboratories. As a general rule (table 13.1), the first test begins three hours after the end of the first night of polygraphic recording in the laboratory. If the subject does not fall asleep, the test is stopped after 20 minutes; if the subject falls asleep, the test is allowed to continue (i.e. the subject is left to sleep) for 15 minutes (maximum latency to determine sleep onset REM period) after the first sleep epoch occurs. The global sleepiness index is provided by the mean latency to sleep in 4 or 5 tests. If the most common criterion used to measure sleep latency is the time it takes from turning off the light to the onset of the first scored sleep epoch, usually in stage 1 [23, 74], many authors propose using two consecutive stage 1 epochs [90] or three consecutive stage 1 epochs [40, 63, 74]. One epoch is the suggested criterion in the standard protocol [23], the 3 epoch criterion representing "unequivocal sleep". A study [6] carried out on 100 patients with EDS using all 3 scoring criteria showed that the differences between mean latencies were minimal (6.2 versus 7.2 versus 7.5 minutes). Likewise the use of the median rather than the mean as a global sleepiness index [7] shows only slight differences (0.6 to 5 minutes, mean = 2.3 ± 1.1). Nevertheless the different ways of scoring sleep onset are such that some patients (16%) shift from the category of severe to moderate sleepiness, from moderate to normal sleepiness and from severe to normal sleepiness. The various categories proposed [23, 74, 90] are based on the mean value of latency to sleep onset: > 10 minutes: normal sleepiness, > 5 and < 10: moderate sleepiness, < 5 minutes: severe sleepiness, are recognised more or less unanimously even if there is an arbitrary side to them, being based on a rule of thumb rather than on rigorous validation. So these categories can only be used as a guideline in clinical practice. Moreover the test may give rise to false positives or false negatives. False positives are subjects who are not affected by sleepiness but with a high capacity to fall asleep rapidly: high sleepability with no other evidence of sleepiness (HSNS), showing mean latencies of 5.6 and 7.3 minutes [42]. The false negatives characterise clinically sleepy subjects who do not fall asleep for various reasons.

In the adult, this test shows a high level of test-retest reliability after a period of 4 to 14 months [98]. Although this is the most objective sleepiness test, it remains sensitive to the subject and may be altered by instructions or by subject motivation [4, 43].

Generally speaking, the method consists in trying to place the subject in his/her usual daily conditions (normally dressed) at the time of the test which should be conducted in a non-stimulating environment (lying in a silent, dark, air conditioned room) instructing the patient not to fight against sleep and observing the subject between tests to make sure he does not sleep.

Table 13.1 Principle rules for administering the MLST

Before the test
- No intake of psychoactive drugs for a period of at least 15 days before the test
- In cases of high caffeine intake, a gradual reduction of the dose
- Sleepiness and the quality of night time sleep to be assessed using a sleep diary for 10 to 15 days. Full polysomnography (with respiratory sensors and surface electrodes on the anterior tibialis muscles) the night before the test.

The day of the test
- Full explanation of procedure to be given by the technician
- Dress in street clothes before MSLT begins
- No alcohol or caffeine intake
- The first test to start 1.5 to 3 h after the end of the nocturnal recording
- No smoking 30 min before the test
- Fifteen minutes before the test, the patient must return to his room and stop any excitatory activity
- Go to the toilet if necessary, eat or drink if necessary
- Do not change clothes (i.e. do not put on pyjamas) for the test
- Three minutes before the test, calibration should be used to help the patient relax
- The patient must read the instructions (systematically the same ones)
- The patient should lie in his/her favourite sleeping position. Start the test
- Between tests, the patient should be out of bed and prevented from sleeping

The maintenance of wakefulness test (MWT)

This test is the most common variant of the MLST. It was originally designed to evaluate treatment efficacy in patients with excessive sleepiness [60] and correlates closely with the MLST [81]. It is also based on two or four sessions spaced at 2 hour intervals; but unlike the MLST, the subject is seated in a comfortable position in bed, as opposed to lying down, with low lighting (0.10 to 0.13 lux at the level of the cornea) behind him (7.5 W at 1 metre), at room temperature of 22°C. The patient is asked to remain in a sitting position and to stay awake as long as possible gazing forward and avoiding looking at the light. He must not slap his face or sing to remain awake. Depending on the duration of each session i.e. 20 or 40 minutes and the polygraphic sleep onset criteria: 1) three episodes of stage one or a single epoch of any other stage or 2) ten seconds of microsleep or the first epoch of sleep, several protocols have been tested. The best protocol consists of two 20 minute sessions, stopping the test after 10 seconds of microsleep or the first sleep epoch or after 20 minutes if the subject is not asleep. In these conditions the normative limit is 11 minutes [33]. While the MLST measures the subject's sleepiness, the MWT measures his ability to stay awake. This test should be routine clinical practice to determine a pathological inability to stay awake. When the 12 minute limit is used (5th percentile) the test (four 20 minute sessions) differentiates between narcoleptic patients and controls with 85.5% sensitivity, 96.9% specificity, and a negative predictive value of 44%. Nevertheless, 15% of narcoleptic subjects fall within the norm, indicating that certain narcoleptics do not alter their capacity to stay awake [61].

Index and polygraphic score of sleepiness

As an attempt to provide a less cumbersome test than the multiple sleep latency test, a quantitative evaluation of daytime sleepiness was proposed using a single test [77]. During a single 45 minute polygraphic recording, the "sleepiness index" is calculated. This index corresponds to the percentage of sleep in relation to total recording duration. It is thus possible to calculate the "polygraphic score of sleepiness" evaluated in points according to the latency to sleep onset and the duration of the various stages of sleep on the polygraphic recording. This test has the advantage of taking account of the internal structure of the polygraphic recording, which is not the case with the standard MLST. The test has not been validated to date.

CONCLUSION

Difficulties remain in assessing and quantifying sleepiness, and the multiplicity of possible approaches reflect the extreme complexity involved. The tests, scales and questionnaires explore only one aspect of sleepiness or vigilance, and none of them, however sophisticated, is able to investigate all the aspects. The tools are there to assist the clinician who knows how to distinguish between them and to adapt their potential use and possibilities of interpretation to the diagnosis and treatment of the patient.

REFERENCES

1.   AESCHBACH D., MATTHEWS J.R., POSTOLACHE T.T., JACKSON M.A., GIESEN H.A., WEHR T.A. - Dynamics of the human EEG during prolonged wakefulness: evidence for frequency specific circadian and homeostatic influences. *Neurosci. Lett.*, 239,121-124, 1997.
2.   AGUIRRE R., BROUGHTON R.J. - Complex event-related potentials (P300 and CNV) and MSLT in the assessment of excessive sleepiness in narcolepsy-cataplexy. *Electroenceph. Clin. Neurophysiol.*, 67, 298-316, 1987.
3.   AKERSTEDT T., GILLBERG M., - Subjective and objective sleepiness in the active individual. *Intern. J. Neuroscience,* 52, 29-37, 1990.
4.   ALEXANDER C., BLAGROVE M., HORNE J. - Subject motivation and the Multiple Sleep Latency Test. *Sleep Res., 20,* 403, 1991.
5.   ALLOWAY C.E., OLGIVIE R.D., SHAPIRO C.M. - The alpha attenuation test: assessing excessive daytime sleepiness in narcolepsy-cataplexy. *Sleep,* 4, 258-266, 1997.
6.   BENBADIS S.R., PERRY M., WOLGAMUTH B.R., MENDELSON W.B., DINNER D.S. - The multiple sleep latency test: comparison of sleep onset criteria. *Sleep*, 19, 632-636, 1996.
7.   BENBADIS S.R., PERRY M., WOLGAMUTH B.R., TURNBULL J., MENDELSON W.B. - Mean versus median for the multiple sleep latency test. *Sleep,* 18, 342-345, 1995.
8.   BENNETT L.S., STRADLING J.R., DAVIES R.J.O. - A behavioural test to assess daytime sleepiness in obstructive sleep apnoea. *J. Sleep Res.*, 6, 142-145, 1997.
9.   BILLIARD M., BESSET A. - L'hypersomnie idiopathique. *Le sommeil et ses troubles.* Billiard M., (ed) Masson, Paris, 292-298, 1998
10.   BILLIARD M., QUERA-SALVA M.A., DE KONINCK J., BESSET A., TOUCHON J., CADILHAC J. - Daytime characteristics and their relationships with night sleep in the narcoleptic patient. *Sleep,* 9,167-174, 1986.
11.   BRIONES B., ADAMS N., STRAUSS M., ROSENBERG C., WHALEN C., CARSKADON M., ROEBUCK M., WINTERS M., REDLINE S. - Relation between sleepiness and general health status. *Sleep,* 19, 583-588, 1996.
12.   BROUGHTON R. - Ambulatory sleep wake monitoring in the hypersomnias. *Ambulatory EEG monitoring.* Ebersobe J.S., (ed), Raven Press, New York, 227-298, 1989.
13.   BROUGHTON R., AGUIRRE M. - Differences between REM and NREM sleepiness measured by event-related potentials (P300, CNV), MSLT and subjective estimate in narcolepsy-cataplexy. *Electroenceph. Clin. Neurophysiol.,* 67, 317-326, 1987.
14.   BROUGHTON R., AGUIRRE M., DUNHAM W. - A comparison of multiple and single sleep latency and cerebral evoked potential (P300) measures in the assessment of excessive daytime sleepiness in narcolepsy-cataplexy. *Sleep, 11,* 537-545, 1988.
15.   BROUGHTON R., GHANEM Q. - The impact of compound narcolepsy on the life of the patient. *Narcolepsy.* Guilleminault C., Dement W.C., Passouant P., (eds), Spectrum publications, New York, 659-666, 1976.
16.   BROUGHTON R., LOW R., VALLEY V., DA COSTA B., LIDDIARD S. - Auditory evoked potentials compared to EEG and performance measures of impaired vigilance in narcolepsy-cataplexy. *Sleep Res., 10,* 184, 1981.
17.   CAJOCHEN C., BRUNNER DP., KRAUCHI K., GRAW P., WIRTZ JUSTICE A. - Power density in theta/alpha frequencies of the waking EEG progressively increases during sustained wakefulness. *Sleep,* 18, 890-894, 1995.
18.   CAMPBELL K., CHARBONNEAU S., BEAUDOIN S. - Evoked potentials correlates of total sleep deprivation. *Sleep Res.,9,* 255, 1980.
19.   CANTERO J.L., ATIENZA M., GOMEZ C., SALAS R.M. - Spectral structure and brain mapping of human alpha activities in different arousal states. *Neuropsychobiology,* 39, 110-119, 1999.
20.   CARSKADON M. - Evaluation of excessive daytime sleepiness. *Neurophysiol. Clin.,* 23, 91-100, 1993.
21.   CARSKADON M., DEMENT W.C. - Effects of total sleep loss on sleep tendency. *Percept. Mot. Skills,* 48, 495-506, 1979.
22.   CARSKADON M., DEMENT W.C. - Sleep tendency an objective measure of sleep loss. *Sleep Res.,* 6, 200, 1977.
23.   CARSKADON M., DEMENT W.C., MITLER M.M., ROTH T., WESTBROOK P.R., KEENAN S. - Guidelines for the multiple sleep latency test (MSLT) a standard measure of sleepiness. *Sleep,* 9, 519-524, 1986.
24.   CASAGRANDE M., de GENNARO L., VIOLANI C., BRAIBANTI P., BERTINI M. - A finger tapping task

and a reaction time task as behavioral measures of the transition from wakefulness to sleep: which task interferes less with the sleep onset process ? *Sleep*, 20, 301-312, 1997.

25. CASAGRANDE M., VIOLANI C., CURCIO G., BERTINI M. - Assessing vigilance through a brief pencil and paper letter cancellation task (LCT); effects of one night of sleep deprivation and of time of day. *Ergonomics*, 40, 613-660, 1997.

26. CHERVIN R.D., ALDRICH M.S., PICKETT R., GUILLEMINAULT C. - Comparison of the results of the Epworth sleepiness scale and the multiple sleep latency test. *J. Psychosom. Res.*, 42, 145-155, 1997.

27. CHERVIN R.D., ALDRICH M.S. - The Epworth sleepiness scale may not reflect objective measures of sleepiness or sleep apnea. *Neurology*, 52, 125-131, 1999.

28. COLE R.J., KRIPKE D.F., GRUEN W., MULLANEY F.J., GILLIN J.C. - Automatic sleep wake identification from wrist actigraphy. *Sleep*, 15, 461-469, 1992.

29. CORSI CABRERA M., RAMOS J., ARCE C., GUEVARA M.A., PONCE DE LEON M., LORENZO I. - Changes in the waking EEG as a consequence of sleep and sleep deprivation. *Sleep*, 15, 550-555, 1992

30. DEMENT W.C., GELB M. - Somnolence: its importance in society. *Neurophysiol. Clin.*, 23, 5-14, 1993.

31. DINGES D.F., POWEL J.W. - Microcomputer analyses of performance on a portable simple, visual task RT during sustained operations. *Behav. Res. Method Instrum. Comput.*, 17, 652-655,1985.

32. DINGES D.F. - An overview of sleepiness and accidents. *J. Sleep Res.*, 4 (Suppl. 2), 4-14, 1995.

33. DOGHRAMJI K., MITLER M.M., SANGAL R.B., SHAPIRO C., TAYLOR S., WALSLEBEN J., BLISLE C., ERMAN K.M., HAYDUK R., HOSN R., O'MALLEY E.B., SANGAL J.M., SCHUTTE S.L., YOUAKIM J.M. - A normative study of the maintenance wakefulness test (MWT). *Electroenceph. Clin. Neurophysiol.* 103, 554-562, 1997.

34. FRUHSTORFER H., BERGSTROM R.M. - Human vigilance and auditory evoked response. *Electroencephalogr. Clin. Neurophysiol.*,27, 346-355, 1969.

35. GILLBERG M., KECKLUND G., AKERSTEDT T. - Relations between performance and subjective ratings of sleepiness during a night awake. *Sleep*, 17, 236-241, 1994.

36. GLENVILLE M., BROUGHTON R., WING A.M., WILKISON R.T. - Effects of sleep deprivation on short-duration performance measures compared to the Wilkinson auditory vigilance task. *Sleep*, 1, 169-176, 1978.

37. GODBOUT R., M0NTPLAISIR J. - All day performance variations in normal and narcoleptic subjects. *Sleep*, 9, 200-204, 1986.

38. GOWASMI M., POLLACK C.P. - *Psychosocial aspects of narcolepsy*, Hayworth, Binghampton, New York, 1992.

39. GRAY P.W., ALLEN R.P., THOMAS M.L., THORNE D.R., SING H.C., QUIGG DAVIS H. - Eye blink rates during a simulated driving task before and after sleep deprivation. *Sleep Res.*, 24, 445, 1995.

40. GUILLEMINAULT C., BILLIARD M., MONTPLAISIR J., DEMENT W.-C. - Altered states of consciousness in disorders of daytime sleepiness. *J. Neurol. Sci.*, 26, 377-393, 1975.

41. HAKKINEN V., FRUHSTORFER H. - Correlation between spontaneous activity and auditory evoked responses in the human EEG. *Acta Neurol. Scand.,43:* S 106-S 161, 1967.

42. HARRISON Y, HORN JA. - "High sleepability without sleepiness". The ability to fall asleep rapidly without other signs of sleepiness. *Neurophysiol. Clin.*, 26:15-20, 1996

43. HARTSE K.M., ROTH T., ZORICK F.J. - Daytime sleepiness and daytime wakefulness: the effect of instruction. *Sleep*, 5, S107-S108, 1982.

44. HERSCOVITCH J., BROUGHTON R. - Performance deficit following short term partial sleep deprivation and subsequent recovery oversleeping. *Can. J. Psychol.*, 35, 309-322, 1981a.

45. HERSCOVITCH J., BROUGHTON R. - Sensitivity of the Stanford Sleepiness Scale to the effects of cumulative partial sleep deprivation and subsequent recovery oversleeping. *Sleep*, 4, 83-92, 1981b.

46. HIGUCHI S., WATANUKI S., YASUKOUCHI A. - Effects of reduction in arousal level caused by long lasting task on CNV. *Appl. Hum. Sci.* 16, 29-34, 1997.

47. HODDES E., ZARCONE V., SMYTHE H., PHILIPS R., DEMENT W.C. - Quantification of sleepiness: A new approach. *Psychophysiology*, 10, 431-436, 1973.

48. HORNE J.A., REYNER L.A. - Driver sleepiness. *J. Sleep Res.* 4 (Suppl 2), 23-29, 1995.

49. JOHNS M-W. - A new method for measuring daytime sleepiness: the Epworth Sleepiness Scale. *Sleep, 14,* 540-545, 1991.

50. JOHNS M-W. - Reliability and factor analysis of the Epworth Sleepiness Scale. *Sleep*, 15, 376-381, 1992.

51. JOHNS M-W. - Sleepiness in different situations measured by the Epworth Sleepiness scale. *Sleep, 17,* 703-710, 1994.

52. JOHNSON L.C. - Sleep disturbance and performance. In: A.N. Nicholson (ed), *Sleep wakefulness and circadian rhythm,* AGARD Lecture series n° 105 (NATO Advisory group for aerospace research and development Paris), 1979.

53. JOHNSON L.C., NAITOH P. - *The operational consequences of sleep deprivation and sleep deficit*, AGARD Lecture series n° 193 (NATO Advisory group for aerospace research and development Paris), 1974.

54. JOHNSON L.C., Sleep deprivation: performance tests for partial and selective sleep deprivation in: LE Abt and BF Riess (eds), *Progress in clinical psychology*, vol 3 Grune and Stratton, New York, 28-34, 1968.

55. KRIBBS N.B., DINGES D. - Vigilance decrement and sleepiness. In Harsch J.R., Olgivie R.D. (eds), *Sleep onset mechanisms,* Washington DC: American psychological association, 113-125, 1994.

56. LEZACK M.D. - *Neuropsychological assessment*. University Press, London, Oxford, 1983.

57. LISPER H.O., KJELLBERG A. - Effects of 24-hour sleep deprivation on rate of decrement in a 10-minute auditory reaction time task. *J Exper. Psychol.* 96, 287-290, 1972.

58. LORENZO I., RAMOS J., ARCE C., GUEVARA M.A., CORSI-CABRERA M. - Effects of total sleep deprivation on reaction time and waking EEG activity in man. *Sleep* 18, 346-354, 1995.

59. LOWENSTEIN O., LOWENFIELD I. - Electronic pupillography: a new instrument and some clinical applications. *Arch. Ophtalmol.*, 59, 352-363, 1958

60. MITLER M.M., GUJAVARTY S., BROWMAN C.P. - Maintenance of wakefulness test: a polysomnographic technique for evaluating treatment efficacy in patients with excessive somnolence. *Electrenceph. Clin. Neurophysiol.*, 53, 658-661, 1982.

61. MITLER M.M., WALSLEBEN J., SANGAL B.R., HIRSHKOWITZ M. - Sleep latency on the maintenance wakefulness test (MWT) for 530 patients with narcolepsy while free of psychoactive drugs. *Electroenceph. Clin. Neurophysiol.*, 107, 33-38, 1998.

62. MITLER M.M., CARSKADON M.A., CZEISLER C.A., DEMENT W.C., DINGES D.F., GRAEBER C. – Catastrophes, sleep, and public policy: Consensus report. *Sleep*, 11, 100-109, 1988.

63. MITTLER M.M., - The multiple sleep latency test as an evaluation for excessive daytime somnolence. In: Guilleminault C. (ed) *Sleeping and Waking Disorders: Indication and Techniques.* Menlo Park, CA: Addison Wesley, 145-153, 1982.

64. MORRIS T.L., MITLER J.C., - Electrooculographic and performance indices of fatigue during simulated flight. *Biol. Psychol.*, 42, 343-360, 1996.

65. MULLANEY D.J., KRIPKE D.F., FLECK P.A., JOHNSON LC. - Sleep loss and night affects on sustained continuous performance. *Psychophysiology*, 20, 643-651, 1983

66. National Transportation Safety Board. *Grounding of the US Tankship Exxon Valdez on Blight Reef Prince William Sound Near Valdez, Alaska, March 24, 1989,* NTSB, Washington, PB 91-916401, 1991.

67. NEWMAN J., BROUGHTON R. - Pupillometric assessment of excessive daytime sleepiness in narcolepsy cataplexy. *Sleep*, 14, 121-129, 1991.

68. OLSON L.G., COLE M.F., AMBROGETTI A. - Correlations among Epworth sleepiness scale scores, multiple sleep latency tests and psychological symptoms. *J. Sleep Res.* 7, 248-253, 1998.

69. PILCHER J.J., HUFFCUTT A. - Effect of sleep deprivation on performance: a meta analysis. *Sleep*, 19, 318-326, 1996.

70. PORCU S., FERRARA M., URBANI L., BELLATRECCIA A., CASAGRANDE M. - Smooth poursuit and saccadic eye movements as possible indicators of night time sleepiness. *Physiology & Behavior*, 65, 438-7-443, 1998.

71. PRESSMAN M.R., SPIELMAN A.J., POLLACK C.P., WEITZMAN E.D. - Long latency auditory evoked responses during sleep deprivation and in narcolepsy. *Sleep*, 5, S147-S156, 1982.

72. RADKE R.A. - Sleep disorders: laboratory evaluation in: Pedley T., Daly D. (eds) *Current practice in clinical electroencephalography* 2nd edition, Raven Press, New York, 561-562, 1990.

73. RICCIO C.A., REYNOLDS C.R., LOWE P., MOORE J.J. - The continuous performance test: a window on neural attention? *Archives of Clinical Neuropsychology*, 17: 235-272, 2002.

74. RICHARDSON G.S., CARSKADON M.A., FLAGG W., VAN DEN HOED J., DEMENT W.C., MITLER M.M. - Excessive daytime sleepiness in man: multiple sleep latency measurements in narcoleptic and control subjects. *Electroenceph. Clin. Neurophysiol*, 45, 621-627, 1978.

75. ROSENTHAL L., ROEHRS T., ROTH T. - The sleep wake inventory: a self report measure of daytime sleepiness. *Biol. Psychiatry*, 34, 810-820, 1993.

76. ROSVOLD H., MIRSKY A., SARASON I., BRANSOME E.D.Jr., BECK L.H. - A continuous performance test of brain damage. *Journal of Consulting Psychology*, 20, 343-350, 1954.

77. ROTH B., NEVSIMALOVA S., SONKA K., DOCEKAL P. - An alternative to the multiple sleep latency test for determining sleepiness in narcolepsy and hypersomnia: polygraphic score of sleepiness. *Sleep*, 9, 243-245, 1986.

78. ROTH T., HARTSE K., ZORICK F., CONWAY W. - Multiple naps and evaluation of daytime sleepiness in patient with upper airway sleep apnea. *Sleep*, 3, 425-439,1982.

79. SADEH A., HARI P., KRIPKE D., LAVIE P. - The role of actigraphy in the evaluation of sleep disorders. *Sleep*, 18, 288-302, 1995.

80. SADEH A., SHARKEY K.M., CARSKADON M.A. - Activity-based sleep wake identification: an empirical test of methodological issues. *Sleep*, 17, 201-207, 1994.

81. SANGAL R.B., THOMAS L., MITLER M.M. - Maintenance of Wakefulness Test and Multiple Sleep Latency Test: measurement of different abilities in patients with sleep disorders. *Chest, 101,* 898-902, 1992.

82. SCHMIDT H., FORTIN L. - Electronic pupillography in disorders of arousal. *Sleep and Waking Disorders, Implications and Techniques.* Guilleminault C., (ed) Addison Wesley, Menlo Park, 127-143, 1982

83. SÉCURITÉ ROUTIÈRE. Sécurité sur autoroute. Bilan de l'année 1991. *Bulletin des autoroutes françaises,* 1992, 35.

84. SEIDEL W.F., DEMENT W.C. - The Multiple Sleep Latency Test: Test retest reliability. *Sleep Res.*, 10, 284, 1981.

85. STAMPI C., STONE P., MICHIMORI A. - The Alpha attenuation Test: A new quantitative method for assessing sleepiness and its relationship to the MSLT. *Sleep Res., 22,* 115, 1993.

86. TANNER W.P., SWETS J.A. - A decision making theory of visual detection. *Psychol. Rev.*, 61, 401-409, 1954.

87. TORSVALL L., AKERSTEDT T. - Extreme sleepiness: quantification of EOG and spectral EEG parameters. *Int. J. Neurosci.* 38:435-441, 1988.

88. TORSVALL L., AKERSTEDT T. - Sleepiness on the job: continuously measured EEG changes in train drivers. *Electroencephalogr. Clin. Neurophysiol.* 66, 502-511, 1987.

89. TORSVALL L., AKERSTEDT T., GILLANDER K., KNUTSSON A. - Sleep on the night shift: 24 hour EEG monitoring of spontaneous sleep/wake behavior. *Psychophysiol.*, 26, 352-358, 1989.

90. VAN DEN HOED J, KRAEMER H., GUILLEMINAULT C., ZARCONE V.P., MILES L.E., DEMENT W.C., MITLER M.M. - Disorders of excessive somnolence: polygraphic and clinical data for 100 patients. *Sleep*, 4,

23-37, 1981.

91. VAN KNIPPENBERG F.C., PASSCHIER J., HEYSTECK D., SHACKELTON D., SCHMITZ P., POUBLON R.M.L., VAN DER MECHE F. - The Rotterdam sleepiness scale, a new daytime sleepiness scale. *Psychological reports*, 76, 83-87, 1995.

92. *Wake up America: a national sleep alert.* Executive summary and executive report. Report of the National Commission on sleep Disorder Research, Washington, 1992.

93. WILKINSON R.T., HOUGHTON D. - Portable four choice reaction time test with magnetic tape memory. *Behav. Res. Meth. Instr.Comp.* 7, 441-446, 1975.

94. WILKINSON R.T. - Methods for research on sleep deprivation and sleep function. *Int. Psychiat. Clin.*, 7, 169-176, 1970.

95. WILKINSON R.T. - Sleep deprivation. *The physiology of human survival.* Eldhom O.G., Bacharach A.L., (eds) 339-430, 1965.

96. WILKINSON R.T. - Some factors influencing the effect on environmental stresses upon performance. *Psychological Bulletin*, 72, 260-272, 1969.

97. YOSS R.E., MAYER N.J., OGLE K.N. - The pupillogram and narcolepsy. *Neurology, 19,* 921-928, 1969.

98. ZWYGHUISEN-DOORENBOS A., ROEHRS T., SCHAEFFER M., ROTH T. - Test retest reliability of the MSLT. *Sleep,* 11, 562-565, 1988.

PART 3

DISORDERS OF SLEEP AND WAKEFULNESS

Chapter 14

Insomnia: Introduction

N. Darchia and I. Gvilia
Department of Neurobiology of Sleep-Wakefulness Cycle, Beritashvili Institute of Physiology, Tbilissi, Georgia

Nowadays, a high percentage of the population in modern society suffers from various kinds of sleep disturbances, due to the impact of increased physiological and socio-cultural stress factors. One of the most common sleep complaints, which the human being has suffered from since ancient times, is insomnia. Although its literal meaning is a total lack of sleep, in clinical and practical terms insomnia has come to refer to a difficulty in initiating and/or maintaining sleep, or non-restorative, non-refreshing sleep. In a survey conducted in the general population of different countries, the prevalence of insomnia was estimated to be about one third of the adult population. In particular, an epidemiological study conducted in Los Angeles, USA, found that 32% of the population had a current complaint of insomnia [4]. In a study conducted in Mannheim, Germany, the prevalence of insomnia was 31% [12], and in Upper Bavaria - 29% [29]. In Italy, 13% of people rarely or never slept well and 19% reported complaints of insomnia [6]. An epidemiological survey of the French population, based on DSM-IV criteria for the definition of insomnia, demonstrated that 29% reported at least one sleep problem three times per week for a month, whereas only 9% had two or more sleep problems and were classified as "severe insomniacs" [16]. The 1991 National Sleep Foundation Survey [23] in conjunction with the Gallup Organisation conducted telephone interviews with 1000 Americans and found that 36% of them suffered from some type of insomnia; approximately one in four adults reported occasional insomnia whereas 9% claimed that their sleep difficulty occurred on a chronic basis [3]. It should be noted that all these surveys were based on self-reported assessments of insomnia. An epidemiological study based on polysomnographic recordings is still lacking. These studies have shown that the prevalence of insomnia is multifactorial depending on sex and age, socio-professional category as well as on coexisting physical and/or psychiatric diseases and may vary between countries. This kind of sleep disturbance can affect people of all ages, especially the socioeconomically disadvantaged group. In addition, it has been shown that the complaint of insomnia increases markedly with age [8, 16] and is more frequent in women than in men [18, 29].

In general, about 30-40% of adults indicate some level of insomnia within any given year, and about 10 to 15% indicate that the insomnia is chronic and/or severe [19]. In contrast, the percentage of subjects consulting for insomnia is limited. Only about one in four patients with insomnia has ever spoken to a physician about their sleep problem [23]. Moreover, they do this during a visit intended for another health problem. The Gallup Study in 1995 [27] found that only 5% of insomniacs had visited a doctor specifically to discuss their sleep problem, and only 21% had ever taken medication for sleep.

Insomnia is not only disturbed sleep. It can be serious enough to induce several changes in daytime functioning, leading to a number of adverse consequences for health and safety. Accordingly, insomnia tends to be considered as a sleep difficulty at night with daytime consequences. The consequences of insomnia experienced by patients are not uniform. Insomnia has a decisive influence on health either directly - with respect to the individual's well-being [24] or indirectly - through accidents caused by insufficient sleep [14]. Insomnia can also produce a massive adverse effect on the quality of life, with poor performance at work, memory problems, impaired concentration, moodiness and a lack of promotions. Dramatic differences were found in the quality of insomniacs' lives as compared to those without sleep related problems [20, 26].

In the past few years the impact of insomnia on the quality of life, productivity and safety has been studied and discussed, and its economic cost to society estimated. The assessment of the direct economic cost of insomnia (e.g. $13.9 billion in the USA, 1995) is significant in respect to the evaluation of insomnia as a public health problem. But this is only one aspect and does not reflect the potential consequences of insomnia on a person's well being, safety, family and personal interrelationships and contributions to society i.e. the indirect cost of insomnia. These aspects are certainly the more important; they are difficult to estimate and controversial, as we do not know whether insomnia is the cause or the result of various problems. It seems likely that efforts to reduce the prevalence of insomnia will not only improve the health conditions of patients but will also carry economic benefits.

The DSM-IV classification for sleep disorders [1] determines two main categories of insomnia: primary insomnia and insomnia associated with somatic or psychiatric disorders. Primary insomnia refers to sleep difficulty that appears to be unrelated to any identifiable underlying medical or psychiatric disorder as is the case in the second category of insomnia. The most common psychiatric diagnoses associated with insomnia are anxiety and depression [8, 19, 20]. Bixler *et al.* [4] found that 19.4% of insomniacs had depression and 50% emotional problems. Hohagen *et al.* [13] reported a high co-morbidity of severe insomnia with somatic and psychiatric disorders, especially with depression. It was even hypothesised that the gender difference in the prevalence of insomnia is mainly based on differences in psychiatric co-morbidity [17]. Insomnia may also be associated with a variety of specific sleep disorders, including restless legs syndrome, periodic limb movement disorder, sleep apnoea and circadian rhythm sleep disorders.

In general, sleep in insomniac patients is highly variable. Poor nights may vary with nights of adequate or even good sleep. Insomniacs may perceive their sleep to be insufficient either quantitatively or qualitatively and typically overestimate the extent of their insomnia. Insomnia could also be said to be a subjective phenomenon in the experience of a patient, rather than a simple abnormality of objective sleep measures. There are people who sleep less than 6 hours per night with no perception or complaint of poor sleep, while others complain of habitual difficulty sleeping, although their polygraphic features are normal. The evaluation of insomnia (duration, frequency, onset of the problem, etc.) needs to be carried out carefully in order to understand whether insomnia is serious enough to warrant treatment. To treat insomnia, the diagnostic procedure must be such that the disease is understood. There is considerable debate among sleep specialists regarding the degree to which insomnia is a syndrome with general traits that are always present, and how much it is a symptom of very different underlying dysfunctions that have to be treated differentially [25].

The most common method for insomnia treatment is pharmacological, although nonpharmacological, so-called behavioural approaches, such as relaxation therapy, sleep restriction therapy, and stimulus control therapy are also developed [5, 22]. Pharmacological treatment usually provides rapid symptom relief. Behavioural approaches take more time to improve sleep but the effects continue to persist after training sessions have been completed. Several studies have proven pharmacological, nonpharmacological therapy and combined treatments for insomnia to be successful. Numerous publications worldwide advise physicians how to manage this kind of disturbed sleep. General overviews present the diagnosis and treatment of insomnia, with reviews and summaries of nonpharmacological and pharmacological treatment. The current consensus in the treatment of insomnia suggests a multidimensional approach to the management and appropriate treatment of insomnia: first identify and treat the underlying causes of the patients' complaints, second try hypnotic treatment using modern concepts, and third, use nonpharmacologic treatment techniques.

It should be noted that over the past several decades considerable progress has been made in the evaluation of insomnia including clinical interview, sleep diary, sleep questionnaires, psychological tests, behavioural assessment devices, polysomnography; as well as in the differential diagnosis of insomnia reflected in sleep classifications like the International Classification of Sleep Disorders [15], the Diagnostic and Statistical Manual of Mental Disorders, DSM III-R, DSM-IV [1-2]. Developments have taken place in assessing the consequences of insomnia as well as in both pharmacological and non-pharmacological approaches to treatment. Despite these real advances, many important problems remain in the management, classification, diagnosis and identification of the causes and neural mechanisms of insomnia. Better standard criteria and categories are needed to classify the various sub-types of insomnia. The time has come for more precise characterisation of the cause and consequences of insomnia, and for more accurate diagnosis and treatment. To date no drug has been discovered with unique sleep promoting action without the attendant adverse pharmacological effects (persistent drowsiness during the day, more or less marked impairment of psychomotor performance, decreased capacity for attention and concentration, etc.), minimising the benefits of recovered sleep. This problem clearly calls for further study with a variety of drugs, dose-responses and animal models examined.

Thus, insomnia is a worldwide social and public health problem, the physiological basis and patterns of which remain poorly understood and still need to be recognised. The problem can only be successfully resolved through the experience and know-how of both basic and clinical studies. In this sense, the research based on animal studies, taking account of the clinical features of sleep alteration which are typical for insomnia, will help to refine our knowledge of its basic mechanisms, leading to the development of advanced approaches to the problem.

A few studies have attempted to establish an animal model for insomnia. In some models sleep is impaired nonpharmacologically:

a) by housing the rats under a new light/dark schedule [11]

b) by moving the rats to a novel individual cage [21]

In other models sleep is impaired pharmacologically. Physiological models are preferred to pharmacological ones, since it can always be questioned whether the insomnia and the subsequent restoration of sleep are identical or close to natural insomnia and sleep [28]. No natural models of insomnia have so far been suggested. Recent studies have shown that the main characteristics of the guinea pig's sleep-wakefulness cycle are well-correlated with the objective indices used in clinical practice when diagnosing insomnia. It was found that the guinea pig responded «therapeutically» to both nonpharmacological (long-lasting adaptation to experimental conditions) and pharmacological (benzodiazepine action) interventions [7, 9] Generalising all the data, the authors suggest that the guinea pig's sleep-wakefulness cycle could be considered as a natural model of insomnia. Certain advantages are obvious here - if we have a natural model of insomnia, investigation of biological mechanisms underlying SWC disorders in guinea pigs may have important consequences for clarifying the neurobiological bases of the insomnia problem as a whole, and contribute to establishing safer, more effective treatment.

Thus, the combination of basic and clinical strategies to sleep research should be considered as a positive approach to the problem of insomnia. Finally, there are signs of a breakthrough in our present knowledge of the fundamental brain mechanisms of insomnia, in the not too distant future.

REFERENCES

1. AMERICAN PSYCHIATRIC ASSOCIATION. DIAGNOSTIC AND STATISTICAL MANUAL OF MENTAL DISORDERS - 3rd edition, revised. *APA*, Washington D.C., 1987.
2. AMERICAN PSYCHIATRIC ASSOCIATION. DIAGNOSTIC AND STATISTICAL MANUAL OF MENTAL DISORDERS, 4th edition. *APA*, Washington D.C., 1993.
3. ANCOLI-ISRAËL S., ROTH T - Characteristics of insomnia in the United States: results of the 1991 National Sleep Foundation survey. I, *Sleep,* 22, 347-353, 1999.
4. BIXLER E.O., KALES A., SOLDATOS C.R., KALES J.D., HEALEY S. - Prevalence of sleep disorders: a survey of the Los Angeles metropolitan area. *Am. J. Psychiatry,* 136, 1257-1262, 1979.
5. BOOTZIN R/.R., PERLIS M.L. - Nonpharmacologic treatments of insomnia. *J. Clin. Psychiatry,* 1992, 53, 37-41, 1994
6. CIRIGNOTTA F., MONDINI S., ZUCCONI M., LENZI P.L., LUGARESI E. - Insomnia: an epidemiological survey. *Clin. Neuropharmacol.* 8, 49-54, 1981.
7. DARCHIA N., GVILIA I., ONIANI T.- The sleep-wakefulness cycle of guinea pig as a biosensor for anxiolitics' action. *J. Sleep Res.,* 9, 47, 2000.
8. FORD D., KAMEROW D. - Epidemiologic study of sleep disturbances and psychiatric disorders - an opportunity for prevention? *JAMA,* 262, 1479-1484, 1989.
9. GVILIA I., DARCHIA N., ONIANI T. - The animal model of insomnia. *J. Sleep Res.,* 9, 77, 2000.
10. GVILIA I., DARCHIA N., ONIANI T. - The guinea pig as a natural model of insomnia. *Actas de Fisiologia,* 7, 21, 2001.
11. HALPERIN J.M., MILLER D., IORIO L.C. - Sleep-inducing effects of three hypnotics in a new model of insomnia in rats. *Pharmacol. Biochem. Behav.,* 14, 811-814, 1981.
12. HOHAGEN F., KAPPLER C., SCHRAMM E., RIEMANN D., WEYERER S., BERGER M. - Sleep onset insomnia, sleep maintaining insomnia and insomnia with early morning awakening. Temporal stability of subtypes in a longitudinal study on general practice attenders. *Sleep,* 17, 551-554, 1994.

13. HOHAGEN F., RINK K., KAPPLER C., SCHRAMM E., RIEMANN D., WEYERER S., BERGER M. - Prevalence and treatment of insomnia in general practice: A longitudinal study. *Eur. Arch. Psych. Clin. Neurosci.*, 242, 329-336, 1993.
14. HORNE J., REYNER L. - Sleep-related vehicle accidents. *BMJ,* 310, 565-567, 1995.
15. INTERNATIONA CLASSIFICATION OF SLEEP DISORDERS. DIAGNOSTIC AND CODING MANUAL. DIAGNOSIS CLASSIFICATION STEERING COMMITTEE, Thorpy M.J. Chairman. Rochester, Minnesota: American Sleep Disorders Association, 1990.
16. LEGER D., GUILLEMINAULT C., DREYFUS J.P., DELAHAYE C., PAILLARD M. - Prevalence of insomnia in a survey of 12778 adults in France. *J. Sleep Res.,* 9,35-42, 2000.
17. LINDBERG E., JANSON C., GISLASON T., BJOMSSON E., HETTA J., BOMAN G. - Sleep disturbances in a young adult population: can gender differences be explained by differences in psychological status? *Sleep,* 20,381-387, 1997.
18. LUGARESI E., ZUCCONI M., BIXLER O. - Epidemiology of sleep disorders. *Psychiatric Ann.,* 17, 446-453, 1987.
19. MELLINGER G.D., BALTER M.B., UHLENHUTH E.H.- Insomnia and its treatment, Prevalence and correlates. *Arch. Gen Psychiatry*, 42, 225-232, 1985.
20. MENDELSON W., GARNETT D., LINNOILA M. - Do insomniacs have impaired daytime functioning? *Biol. Psychiatry*, 19, 1261-1263, 1984.
21. MICHAUD J.C., MUYARD J.P., CAPDEVIELLE G., FERRAN E., GIORDANO-ORSINI J.P., VEYRUN J., RONCUCCI R., MOURET J. - Mild insomnia induced by environmental perturbations in the rat: a study of this new model and its possible applications in pharmacological research. *Arch. Int. Pharmacodyn. Ther.*, 259, 93-105, 1982.
22. MORIN C.M., CULBERT J.P., SCHWARTZ S.M., MORIN C.M. Nonpharmacological interventions for insomnia: a meta-analysis treatment efficacy. *Am. J. Psychiatry*, 151, 1172-1180, 1994.
23. NATIONAL SLEEP FOUNDATION. SLEEP IN AMERICA, Gallup Organization, Princeton NJ, 1991.
24. OHAYON M. - Epidemiologic study on insomnia in the general population. *Sleep*, 19,7-15, 1996.
25. REPORT OF AN INTERNATIONAL WORKSHOP, Palm Springs. Sleep and Health: Research and Clinical Perspectives. *Sleep,* 23, 49-88, 1997.
26. ROTH T., ANCOLI-ISRAEL S. - Daytime consequences and correlates of insomnia in the United States: results of the 1991 National Sleep Foundation survey. II, *Sleep*, 22, 354-358, 1999.
27. SLEEP IN AMERICA: 1995. A NATIONAL SURVEY OF U.S. ADULTS. Sleep in America: 1995. A National Survey of U.S. Adults. A report prepared by the Gallup Organization for the National Sleep Foundation. Los Angeles: National Sleep Foundation, 1995.
28. VAN LUIJTELAAR G., COENEN A. - The behavioral pharmacology of sleep. In: *Sleep-Wake research in the Netherlands*, Coenen A. (ed) 93-103, 1994.
29. WEYERER S., DILLING H. Prevalence and treatment of insomnia in the community: results from the upper Bavarian field study. *Sleep*, 14, 392-8, 1991.

Chapter 15

A decision tree approach to the differential diagnosis of insomnia

J. Grenier
Centre National de Formation en Santé, Hôpital Montfort, Ottawa, Canada; School of Psychology, University of Ottawa, Canada

INTRODUCTION

Many health care professionals (general practitioners, psychiatrists, neurologists, psychologists and practicing nurses) are often faced with the responsibility of evaluating and diagnosing a complaint of insomnia. Because these professionals all come from discipline-bound schools of thought and with limited training in sleep medicine, it should be no surprise that the manner in which the assessment and diagnostic process is undertaken can significantly differ from one professional to another, thus increasing the probability of errors in judgment, of errors in diagnosis and ultimately, of errors in the judicious selection of treatment(s). The successful treatment of insomnia requires a comprehensive evaluation and an accurate diagnosis. A common error in the management of insomnia, is to view the sleep complaint as a simple symptomatic disorder. Too often, treatment is initiated prematurely without the benefit of an adequate differential diagnosis and case conceptualisation. Because insomnia is typically associated with multiple aetiologies, a detailed investigation of its contributing and maintaining factors is essential prior to initiating treatment. In this chapter, a decision tree that can be used as a guide to the differential diagnosis of insomnia is provided for clinicians from all medical and mental health professions.

A DECISION TREE APPROACH

The use of this model of reasoning or diagnostic tool can guide and structure the clinical interview. The decision tree is a schematic diagram of the considerations inherent to the differential diagnosis of insomnia. It emphasises specific historical and contextual information in a variety of spheres that is essential in formulating a working diagnosis before initiating treatment: sleep, environmental, pharmacological, medical, psychiatric, psychological, and behavioural. This decision tree does not cover treatment alternatives. However, treatment should be tailored to the working diagnosis, and if response is inadequate, the diagnosis should be reconsidered. Behavioural treatments can be recommended and tailored to help multiple types of insomnia, in addition to other specific treatment modalities.

Three classification systems exist for sleep disorders: The International Classification of Sleep Disorders (ICSD) (1); the International Classification of Diseases (ICD) (2); and the Diagnostic and Statistical Manual of Mental Disorders (DSM-IV) (3). The decision tree uses the nomenclature and diagnostic categories of the International Classification of Sleep Disorders, which is the "reigning" classification for sleep disorders.

Beginning from the sleep complaint, the tree outlines all of the factors potentially responsible for initiating and/or maintaining the insomnia. A plausible diagnosis corresponds to each of the individual levels of questioning. Because insomnia often has multiple causes and contributing factors, the clinician must not abandon the decision-making process once a plausible diagnosis has been identified. Rather, it is essential to employ all levels of questioning, while keeping in mind diagnostic possibilities, before arriving at a more definite diagnosis (single or multiple diagnoses). An element to consider within the context of case conceptualisation, is the fact that with chronic insomnia, the factors responsible for maintaining the chronic nature of the difficulties may be different from the original initiating factors.

In most cases, the differential diagnosis of insomnia can be conducted in the clinician's office via a comprehensive evaluation that should, ideally, involve several assessment procedures: a general clinical interview with an emphasis on a detailed review of the sleep problem as well as on the relative contribution of environmental, psychological, behavioural and medical factors; sleep diary monitoring; and psychological screening. Morin [4] explains in great detail the components of a comprehensive evaluation of insomnia and also provides several questionnaires and forms that may be used within the context of assessing and managing insomnia. Ancillary measures, behavioural assessment devices and nocturnal polysomnography may also be considered at the clinician's discretion. In this respect, the decision tree also guides clinical judgement as to under which diagnostic considerations it becomes necessary to refer a patient to a sleep disorders clinic for a more detailed assessment, often involving nocturnal polysomnography, multiple sleep latency test (MSLT) and other behavioural assessment devices.

DIAGNOSTIC CONSIDERATIONS OF THE DECISION TREE

The first consideration is to distinguish between complaints of insomnia (sleep that is perceived as being difficult to initiate and/or maintain, unsatisfying or nonrestorative) and hypersomnia (excessively deep or prolonged major sleep period and / or excessive daytime sleepiness). Although both conditions can coexist, hypersomnia usually reflects the presence of another sleep pathology, typically narcolepsy, nocturnal respiratory disorders or nocturnal movement disorders. Patients suspected of having hypersomnia or a mixed complaint of insomnia and hypersomnia should be referred to a sleep clinic for further evaluation, usually involving an MSLT and polysomnography.

Once hypersomnia has been ruled out, it becomes important to determine the nature of the insomnia: trouble falling asleep and/or maintaining sleep and/or early morning awakening and/or light sleep and/or nonrestorative sleep. Assessing the impact of the sleep disturbances on daytime functioning and overall quality of life also lends important information as to the clinical significance of the insomnia. Distinguishing between daytime fatigue and somnolence is crucial. Fatigue, irritability, impairment of cognitive and behavioural performance may or may not be related to a sleep problem while true somnolence is usually related to sleep deprivation or disturbances. Individuals with a decreased or short sleep duration and an absence of decreased daytime functioning should not be considered insomniacs. It is also quite common to interview insomniacs and quickly establish that they present with symptoms of anxiety and/or depression. Trouble initiating sleep is usually associated with anxiety while early morning awakening is typically associated with depression. Alternatively, nocturnal respiratory and movement disorders are typically associated with a complaint of excessive daytime sleepiness and sleep-maintenance difficulties rather than sleep-onset or early morning awakening problems. Some medical problems tend to be associated with sleep-maintenance difficulties rather than sleep-onset difficulties. A thorough functional analysis of the variables surrounding sleep should be conducted and include inquiries into elements such as: natural history of the insomnia, onset, duration, identifiable precipitating event (s) (life, behavioural, medical, etc...), pattern of development over time to distinguish between situational (days or weeks) versus chronic (months or years) insomnia, factors that aggravate or ameliorate the problem, bedtime routine, sleep incompatible activities or settings, cognitive and physiological arousal, beliefs and attitudes about sleep, diurnal self-statements concerning sleep, secondary gains, patterns or fluctuations over weeks, months or seasons. Obtaining a thorough sleep history and functional analysis is essential to provide the necessary etiological cues and further guide the diagnostic process.

Did the difficulties start in early childhood and perpetuate in adulthood? If the insomnia is present since early childhood and has been relentless and unvaried through periods of both poor and good emotional adaptation, a diagnosis of "Idiopathic Insomnia" should be considered. Idiopathic insomnia is presumed to be caused by an abnormality in the neurological mechanisms that control sleep-wake system. This is a rare disorder that entails a lifelong inability to obtain adequate sleep and whose diagnosis can be confirmed by a detailed polysomnographic evaluation.

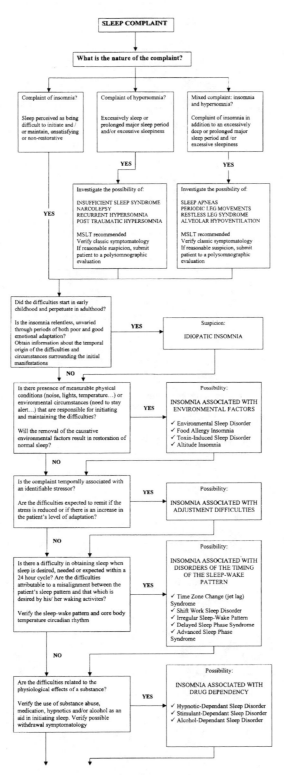

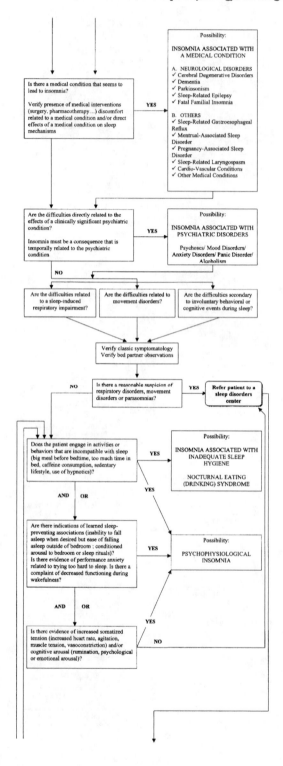

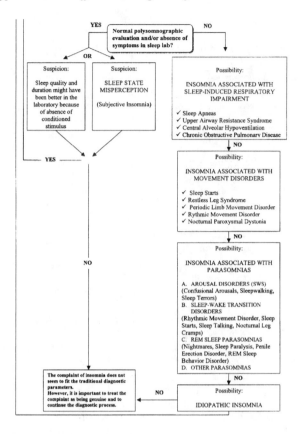

Is there presence of measurable physical conditions (ie., noise, lights, temperature extremes, snoring, movements, mattress comfort, pet) or environmental circumstances (ie., need to stay alert because of a baby) that are responsible for initiating and maintaining the difficulties? Certain environmental factors such as the above-mentioned are known to affect deep or delta sleep. The result is nonrefreshing sleep and various somatic complaints. If the removal of the causative environmental factors result in restoration of normal sleep, then a diagnosis of "Insomnia associated with environmental factors" is to be considered.

Is the complaint temporarily associated with an identifiable stressor? This sleep disturbance is temporarily related to acute stress, interpersonal conflicts, or any other life changes causing emotional, physiological and cognitive arousal. If the patient's sleep is clearly different from his or her usual sleep pattern and the difficulties remit (or are expected to remit) once the stressor is removed or once the patient's level of adaptation is increased, the clinician should consider a diagnosis of "Insomnia associated with adjustment difficulties".

Is there a difficulty in obtaining sleep when sleep is desired, needed or expected within a 24 hour cycle? Are the difficulties attributable to a misalignment between the patient's sleep pattern and that which is desired by his or her waking activities? If this is the case, a diagnosis of "Insomnia associated with disorders of the sleep-wake pattern" may be possible. The main feature in the circadian rhythm disorders is a misalignment between the individual's actual sleep-wake cycle and that which is desired. The patient thus complains of being unable to fall asleep or stay awake when he or she desires to do so within the 24 hour night-day cycle. Rapid time zone change syndrome (jet lag), shift work sleep disorder, irregular sleep wake pattern as well as delayed and advanced sleep phase syndromes are examples of disorders of the sleep-wake pattern. In phase-delay syndrome, the ability to fall asleep and major sleep period is delayed in relation to the desired time. Typically, the patient will report not being able to fall asleep until late at night (ie., 3:00 a.m.) and consequently having difficulty rising at the desired conventional time in the morning. Individuals who work rotating shifts and those who adopt irregular sleep schedules (ie., frequently changing bedtimes and arising times) are more susceptible to develop this disorder. In phase-advance syndrome, the chief complaint is the inability to stay awake in the evening followed by early morning awakening. There are two elements to consider in establishing a differential diagnosis here: the duration of the sleep period and the notion of control over sleep. As for the former, although there is the presence of early morning awakening, the duration of the major sleep period is not shortened. As for the latter, individuals will typically report that although they wish to stay awake beyond for example 8:00 p.m., they are simply unable to stay awake until their desired bedtime.

Are the difficulties related to the physiological effect of a substance? In the affirmative, a diagnosis of "Insomnia associated with drug dependency" should be considered. An important cause of insomnia complaints is the chronic use of hypnotic drugs. The chronic use of hypnotics leads to tolerance and reduced sleep-promoting efficiency. The sleep disturbance may be exacerbated by the medication itself and its chronic use. When medications (especially the barbiturates and benzodiazepines) are stopped abruptly, withdrawal symptoms may include disturbances in sleep architecture as well as rebound insomnia and rebound anxiety. Stimulant use or abuse can obviously lead to insomnia. Alcohol-dependant insomnia is caused by almost nightly use of ethanol as a sleep promoting agent. Drug-dependant insomnia is not limited to a problem of a physical nature but may also be psychological. In fact, some individuals are very reluctant to abandon their nightly tiny dose of a benzodiazepine, which after several years has lost all of its hypnotic effects.

Is there a medical condition that leads to insomnia? "Insomnia associated to a medical condition" is a probable diagnosis when the sleep difficulties develop as a result of several types of medical problems. Some medical problems such as those that involve head trauma and neurological deterioration may directly affect the underlying mechanisms that regulate sleep and sleep architecture. Cardiac disturbances, pulmonary diseases, endocrine and gastrointestinal conditions can also cause sleep-onset or sleep-maintenance insomnia. Almost any source of discomfort or acute or chronic pain may lead to sleep disturbances. Sleep complaints associated with somatic pain are often associated with alpha-delta sleep, that is, frequent intrusions of alpha rhythms (wakefulness) into delta rhythms (stages 3 and 4 or deep sleep) which is subjectively experienced as frequent intrusions of wakefulness into sleep. Finally, insomnia may be directly related to the use of medications prescribed for medical illnesses. Examples of these include some hypertensive medications, bronchodilators and some antidepressants.

Are the difficulties related to the effects of a clinically significant psychiatric condition? Psychiatric patients as well as individuals who suffer from a clinically significant psychological disorder (psychoses, mood disorders, seasonal affective disorder, anxiety disorders, alcoholism) are at risk for experiencing sleep disturbances. Insomnia is the most common sleep complaint in patients with psychiatric conditions. A psychotic decompensation is often associated with significant sleep disruption including severe initial insomnia, severe nocturnal psychomotor agitation and reduced sleep efficiency. In some cases, the day-night cycle may be inverted or delayed. In major depression and dysthymia, sleep abnormalities include disturbances in continuity, diminished delta sleep, early appearance of the first REM period and an altered distribution of REM sleep periods throughout the night. Typically, depressed patients complain of early morning awakening, nonrestorative sleep and excessive daytime sleepiness. In bipolar disorder, hypersomnia is usually part of the depressive phase while insomnia is part of the manic phase. Almost all anxiety disorders, especially generalised anxiety, are associated with difficulties initiating and/or maintaining sleep. Comorbidity between sleep and psychological disturbances is extremely common. However, a patient who suffers from chronic insomnia does not necessarily meet the diagnostic criteria for psychopathology. The clinician must also distinguish between daytime symptoms and psychological distress that are related to the psychiatric condition and those that are the consequence of suffering from chronic sleep difficulties. In order to meet a diagnosis of "Insomnia associated with a psychiatric disorder", the insomnia must be a consequence that is temporally related to the psychiatric condition.

Are the difficulties related to other sleep disorders such as sleep-induced respiratory impairment, nocturnal movement disorders or involuntary behavioural/cognitive events during sleep? Inquiry into the classic symptomatology of these disorders is often overlooked.

Sleep apnoea is a breathing disorder in which breathing intermittently stops during sleep, because of upper airway occlusion (obstructive type), a loss of respiratory effort (central type) or a mixture of both (mixed type). Apneic episodes usually last 10 to 40 seconds and may continue as long as 2 minutes. Apneic episodes are frequently associated with progressive hypoxemia and terminate with brief arousals from sleep. The disordered breathing events are frequently observed by a bed partner. Classic symptoms may include: snoring, witnessed pauses in breathing during sleep, restless sleep, choking, esophageal reflux, cardiac arrhythmias, morning headache, hypertension, intellectual and personality changes. Conditions associated to sleep apnea include: obesity, large neck, nasal obstruction, crowded oropharynx, acromegaly, impotence and hypothyroidism. Patients suffering from sleep apnea are usually unaware of the nocturnal symptoms and most often complain of daytime sleepiness although complaints of sleep maintenance insomnia are also possible.

Restless legs syndrome is a movement disorder characterised by disagreeable leg sensations (calf area) that cause irresistible urges to move the legs. The disagreeable sensation is temporarily relieved with leg motion and return upon cessation of the movements. In severe cases, the disagreeable sensations may be experienced in the feet, thighs and arms. Restless-legs may be experienced during wakefulness but is typically present and most severe at bedtime, thus occasioning sleep-onset difficulties. Many patients with restless-legs also present with periodic limb movement disorder.

Periodic limb movement disorder (PLM) is a movement disorder characterised by episodes of repetitive and stereotyped limb (legs and arms) movements during sleep. Although both legs and arms can be involved, movements typically involve extension of the big toe and flexion of the ankle, knee, and hip. These movements last between 0.5 to 5.0 seconds and typically occur every 20 to 40 seconds. The movements also tend to happen more frequently during the first third of the night. Many patients will not even be aware of these nocturnal movements and it is most often the bed partner that will notice. This condition should not be confused with the normal phenomena of hypnic jerks and twitches that frequently happen at sleep onset. Since some patients do not present any obvious symptoms, many cases are diagnosed incidentally during a polysomnographic evaluation. These limb movements may occasion complete arousals from sleep, but most often, the consequence is limited to unnoticed recurrent micro-arousals, which lead to a complaint of nonrestorative sleep. Patients will complain either of sleep maintenance insomnia or excessive daytime somnolence. Periodic limb movements are also seen more frequently in patients with sleep apnea, restless-legs syndrome and narcolepsy.

Parasomnias are involuntary and sometimes complex behavioural/cognitive events that occur intermittently or episodically during the night. They represent abnormal and excessive activation of the central nervous system often involving autonomic and /or muscular-skeletal activity. Parasomnias may be classified variously, however, the following classification is most seen in the literature: 1) disorders of arousal (confusional arousals, sleep walking, sleep terrors); 2) disorders of the transition between sleeping and waking states (rhythmic movement disorder, sleep starts, sleep talking, nocturnal leg cramps); and 3) dysfunctions occurring primarily in REM sleep (nightmares, sleep paralysis, penile erection disorders, REM sleep behaviour disorder). Other parasomnias are more sustained over time (body rocking, bruxism) or episodic but not necessarily behavioural in nature (nocturnal angina, sleep-related migraines). Many parasomnias are present during childhood but disappear by late adolescence. In some cases, parasomnias may lead to complaints of insomnia or hypersomnia.

If there are no reasons to suspect any of these sleep disorders, the clinician may continue on with the diagnostic process. However, if there is suspicion of one or more of these sleep disorders, it is imperative to refer the patient to a sleep disorders clinic. These sleep disorders must be diagnosed via a more detailed assessment and nocturnal polysomnography. A polysomnographic evaluation will also help confirm the possibility of idiopathic insomnia. In fact, patients should be referred to a sleep disorders clinic only when there is suspicion of respiratory or movement disorders, parasomnias, hypersomnia or idiopathic insomnia. All the other diagnoses discussed in the decision tree can be arrived at in the clinician's office.

Should polysomnographic findings confirm a relatively normal sleep architecture (normal sleep latency, normal number of arousals, normal number of awakenings, normal sleep duration and quality), two possibilities may be considered: 1) "Sleep State Misperception" or 2) sleep quality and duration might have been better in the laboratory because of the absence of conditioned stimulus.

"Sleep State Misperception", also known as "Subjective Insomnia" is a disorder in which a complaint of insomnia occurs but without objective evidence of sleep disturbance. The diagnosis can only be made when there are objective findings confirming that sleep is normal. Patients complain of insomnia with great conviction and typically continue to firmly believe they have insomnia despite objective findings to the contrary. Difficulty initiating or maintaining sleep is most often the chief complaint, however, some individuals insist that they never sleep. This disorder is more common in females and typically starts in middle adulthood. It is a disorder that is not well understood. Patients who receive this diagnosis are not malingering: for reasons that are still unknown and that current technology may be unable to measure, these patients subjectively experience sleep as an almost continual state of wakefulness. It is therefore very important to treat the insomnia complaint of these patients as genuine and to explain the nature of this diagnosis / sleep perception phenomena.

If a diagnosis of "Sleep State Misperception" is unwarranted, then it is possible that the sleep quality and duration might have been better in the laboratory because of the absence of conditioned stimulus. Frequently, patients who suffer from "Psychophysiological Insomnia" (described below) sleep better in the laboratory (or in any setting other than their own bedroom). This phenomena is easily explained by the fact that certain maladaptive sleep hygiene behaviours (ie., watching TV in the bedroom, eating a big meal before bedtime) and/or learned sleep-preventing associations (ie., bed and bedroom are associated with the inability to fall asleep) are suddenly absent. If this is the case, then a careful investigation of sleep hygiene practices and potential learned sleep-preventing associations must be conducted.

Does the patient engage in activities or behaviours that are incompatible with sleep (exercising too late in the day, engaging in intense mental activity at night, meals before bedtime, excessive caffeine consumption, tobacco use at night, excessive time in bed, sedentary lifestyle or use of hypnotics)? "Insomnia associated with inadequate sleep hygiene" is the preferred diagnosis when the difficulties are directly attributable to habits and lifestyle issues that are under the patient's control and that transgress the basic rules of good sleep hygiene. There is no single list of sleep hygiene instructions that is perfect for everyone. This book contains a chapter that features a list of basic sleep hygiene instructions to follow. In fact, sleep hygiene instructions is a common denominator of every insomnia treatment, whether pharmacological or behavioural. Patients should be instructed to follow or implement these sleep hygiene rules with instructions to monitor their progress until the next office visit. Patients should also be aware that the effect of these types of changes in habits produce gradual change rather than dramatic improvement. Although individuals who meet this diagnosis are seen in clinical practice, it is rather uncommon to see patients whose insomnia is directly and solely associated to inadequate sleep hygiene.

Are there indications of learned sleep-preventing associations (inability to fall asleep when desired but ease of falling asleep outside of the bedroom; conditioned arousal to bedroom or sleep rituals)? Is there evidence of performance anxiety related to trying too hard to sleep? Is there a report of decreased functioning during wakefulness? Is there evidence of increased somatised tension (increased heart rate, agitation, muscle tension, vasoconstriction) and /or cognitive arousal (rumination, psychological or emotional arousal)? An affirmative response to these inquiries and an absence of other medical or psychiatric disorders (ie., all other possible causes have been ruled out) that would account for the sleep disturbance, should lead the clinician to consider a diagnosis of "Psychophysiological Insomnia". Psychophysiological insomnia is the classic type of insomnia that is most often seen in the general population. It is also a condition in which there is polysomnographic evidence to corroborate the subjective complaint of increased sleep latency, reduced sleep efficiency and/or increased number and duration of awakenings. It is believed that this type of insomnia emanates from learned sleep-preventing associations, as well as somatic and cognitive arousal. The phenomena of learned sleep-preventing associations stems from the conditioning that takes place between the inability to sleep and factors that are usually conducive to sleep (bed, bedtime, bedtime rituals). Typically, patients will report that they have no problem falling asleep in the living room or in any other setting, but as soon as they are in their own bed, they feel tense and sleepless. This creates a state of apprehension that typically leads to excessive efforts to fall asleep, performance anxiety and excessive time spent awake in bed, which are all incompatible with sleep. The presence of somatic and cognitive arousal is believed to emanate from internal psychological conflicts and anxiety. The anxiety may take on different forms: muscle tension, worries, rumination, rehashing the events of the previous day, excessive worries about the insomnia itself, worries about the loss of control over sleep and its impact on daytime functioning, etc.... This disorder is typically preceded by a more transient form of insomnia: "Insomnia associated with adjustment difficulties". However, when the insomnia persists independently from the original initiating factors, "Psychophysiological Insomnia" is the preferred diagnosis. Hence, after having experienced a few nights of disturbed sleep, usually because of an identifiable stressor, the patient's confidence in the ability to fall asleep or maintain sleep diminishes. The hallmark of this insomnia is that the patient becomes so absorbed with the initial transient sleep loss and the fear of permanently losing control over sleep processes that the insomnia "takes on a life of its own" and becomes chronic in nature. Because the conditioning factors involved in the aetiology are not obvious or intuitive, the patient with psychophysiological insomnia typically reports having no idea why he or she is unable to sleep. The patient will also tend to focus solely on the sleep problem itself and minimise other mental or emotional concerns. Although psychophysiological insomniacs do not necessarily present with any major psychopathology, many present with a history of vulnerability to suffer from anxiety, rigidness, obsessiveness, somatised tension and sometimes depression. Noteworthy, is that the conditioning phenomenon inherent to psychophysiological insomnia is an independent element that may become attached to almost any other type of insomnia presented in this decision tree.

If all hierarchal decisions in this tree have been considered and it is still impossible to arrive at a differential diagnosis, then it is possible that the complaint of insomnia does not fit the traditional diagnostic parameters. This being said, it is important for the clinician to treat the complaint as genuine and to use his or her clinical judgement in continuing the diagnostic process.

CONCLUSION

Given the diverse training populations and the inherent diagnostic uncertainty of insomnia, this decision tree was elaborated in efforts to standardise the diagnostic process so that clinicians from all disciplines may adopt the same implicit reasoning when arriving at a diagnosis of insomnia. The use of this model of reasoning or diagnostic tool should thus reduce reliance on rote memorisation and unsubstantiated inferences based on each interviewer's personal conceptualisation of insomnia. However, one must recognise that a decision tree is not a standalone tool that replaces sound clinical judgment and experience.

The author wishes to thank Dr. Joseph De Koninck for his guidance and contribution in the development of the decision tree.

REFERENCES

1. AMERICAN SLEEP DISORDERS ASSOCIATION (ASDA). - International Classification of Sleep Disorders (ICSD). *Diagnostic and Coding Manual*. Rochester: MN., 1990.
2. WORLD HEALTH ORGANISATION. *International Classification of Diseases*. Geneva. 1992.
3. AMERICAN PSYCHIATRIC ASSOCIATION (APA*). Diagnostic and statistical manual of mental disorders* (4th edition). Washington, DC. 1994.
4. MORIN, C. *Insomnia: Psychological assessment and management*. New York: The Guilford Press, 1993.

Chapter 16

Transient and short term insomnia

R. Peraita Adrados
Unidad de Sueño, Hospital Gregorio Marañon, Madrid, España

The Consensus Development Conference on sleep disorders convened by the National Institute of Mental Health in 1979 [14] subdivided insomnia into transient, short term, and long term or chronic conditions. The first type lasts one to several days, the second, from one to 4 weeks and chronic insomnia for more than 4 weeks. The first and second types will be dealt with in this chapter, i.e. insomnia which is systematically linked to a clearly determined, perfectly identifiable cause, occurring in persons with a history of normal sleep. The International Classification of Sleep Disorders (ICSD) [9] comprises four sections, dyssomnias, parasomnias, sleep disorders associated with medical or psychiatric disorders and "proposed" sleep disorders. The section on dyssomnias itself comprises three subgroups: intrinsic sleep disorders, extrinsic sleep disorders and circadian rhythm sleep disorders. Most transient and short term insomnias are classed in the subgroup of extrinsic disorders, i.e. those whose causes are external to the organism Four of the main ones will be dealt with here: insomnia due to inadequate sleep hygiene, environment-related insomnia, altitude insomnia and adjustment or short term insomnia, to which will be added insomnia due to transient physical stress, pain, coughing, pruritis, not explicitly mentioned in the ICSD, as well as rebound insomnia following the discontinuation of certain hypnotics.

EPIDEMIOLOGY

No epidemiological study has been based exclusively on transient or short term insomnia. The different surveys carried out to date have covered all types of insomnia. Nevertheless, certain studies distinguish between subjects who often or regularly complain of sleep disorders and those who only complain intermittently, providing an indication of the impact of transient and short term insomnia.

The first of these studies was carried out in 1975 on a population of 2,347 residents of the metropolitan area of Houston [11]. According to this, the percentage of respondents who sometimes, often or regularly complained of difficulty falling asleep, staying asleep or of awakening early was on average three to four times higher than the percentage of respondents who complained often or regularly of the same disorders, with, for example, 25.3% of men in the category of those who complained sometimes, often or regularly of trouble getting to sleep, compared with 6% in the category of those who complained often or regularly of the same type of disorder.

A more recent study used a cohort representing the population of the canton of Zurich in Switzerland, numbering 4,547 subjects of both sexes [2]. 591 subjects in this cohort had prospective treatment from the age of 20-21 to the age of 27-28. In this study, the authors distinguished three types of insomnia: continuous (or chronic) insomnia lasting over two weeks, short, intermittent insomnia lasting under two weeks but recurring monthly throughout the year and transient insomnia also lasting under two weeks, but which was non-recurrent. The median age of onset of transient insomnia was 25, that of short, intermittent insomnia was 17 and of chronic insomnia, 22.

During the monitoring period, the incidence of chronic insomnia rose slightly from 8% of men and 12% of women at the age of 21, to 10% of men and 16% of women at the age of 28. The incidence of transient insomnia remained stable (17% of men, 20% of women at the age of 21, 18%

of men and 19% of women at the age of 28) and short, intermittent insomnia fell slightly from 22% of men and 19% of women at the age of 21 to 17% of men and 13% of women at the age of 28. Moreover this study showed that subjects who suffered from transient insomnia scored higher than the control subjects for anxiety, depression, interpersonal sensitivity, obsessive-compulsive disorder and somatisation and that subjects affected by short, intermittent or chronic insomnia had higher scores than those affected by transient insomnia, notably with a marked peak on the depression scale.

A survey carried out on a population of 1,500 residents in the metropolitan area of Madrid [20], men and women of over 18 years, showed that findings on the prevalence of insomnia depended to a large extent on the way in which subjects were questioned. This would account for the wide discrepancies in the literature with prevalence ranging from 1 to 48% of the population depending on the surveys. In this study the authors showed that 25.6% of subjects sometimes complained of sleep onset difficulties and 11.3% of ongoing insomnia.

Finally, a Gallup Institute poll carried out in 1991 [7], based on telephone calls made to 1,950 American men and women, aged 18 and over, showed that 27% of respondents reported transient insomnia and 9%, chronic insomnia. Complaints of insomnia were twice as frequent in women compared to men, probably in part because the former were more prepared to discuss the topic.

On the basis of these four studies, it seems that the percentage of subjects presenting transient or short term insomnia is in the region of 20 to 30% of the population.

INSOMNIA DUE TO POOR SLEEP HYGIENE

This is linked to activities which are non conducive to sleeping well. The ICSD recognises two main categories: activities which lead to increased wakefulness and those which counter good sleep organisation. The first lists the use of substances such as caffeine or alcohol, drugs such as cocaine, excitation provoked by vigorous physical exercise near bedtime, intense mental activity late at night, going out at night or watching television until late. The second lists marked variations in bedtime or time of rising, too much time spent in bed or napping during the day. These factors may also give rise to transient or short term insomnia when they are occasional or to chronic insomnia when they persist for too long. They must always be considered, whatever the type of insomnia, as they are likely to aggravate insomnia brought about by other causes.

No polygraphic studies have been done on this type of insomnia. However when there is distinct subjective improvement in sleep and when a rise in NREM sleep is recorded in polygraphic sleep recordings carried out in sleep disorder units, diagnosis should be oriented toward this type of insomnia.

INSOMNIA LINKED TO ENVIRONMENTAL FACTORS

This includes insomnia caused by sleeping in unfamiliar surroundings and insomnia due to physical factors such as noise, temperature, light or confinement.

Insomnia caused by sleeping in unfamiliar surroundings tends to affect subjects who are used to strict bedtime routines. Travellers, reporters or military personnel do not suffer from this type of insomnia. A perfect illustration is provided by insomnia in hospital, where a number of factors such as discomfort, fear of medical investigations and their results, noise from hospital staff and lighting are likely to result in poor sleep. Another example is that of the subject who spends his first night in the sleep laboratory, characterised by longer sleep latency and repeated nocturnal awakenings (first night effect) [1], with the result that studies tend to include an adaptation period of one or two nights before baseline data are collected.

Among the measurable physical factors likely to lead to insomnia, noise has undoubtedly been the most frequent subject of study. With weak noise stimulation minimal alterations can be detected in polygraphic tracing, such as the appearance of isolated K complexes, short bursts of alpha waves or the momentary disappearance of delta waves. At one stage higher "transient activation phases" are seen in sleep with the return of alpha rhythms for periods of several seconds, which are insufficient to score the epoch as wakefulness [6, 18]. At a still higher level noise alters sleep with the introduction of increased movements, arousals and changes in sleep stages [13, 19]. The amount of time awake and in stage 1 increases, while the amount of NREM sleep and REM sleep may be reduced. Sensitivity to noise varies widely from one individual to another, in terms of intensity as well as significance. Some mothers will wake at the slightest whimper from their child yet sleep through a storm.

Inadequate temperature i.e. too low or too high is also likely to disturb sleep. The amount of time awake and in stage 1 increases while the amount of REM sleep diminishes [7]. The cold is more harmful to sleep than the heat.

As a general rule, a diagnosis of environment-induced sleep disorder depends on the following three criteria: the sleep disorder must be associated with the presence of a physically measurable stimulus or a group of easily identifiable circumstances; the factor(s) responsible affect(s) the subject physically as opposed to psychologically; suppressing such factors will immediately or progressively result in normalised sleep [9].

ALTITUDE INSOMNIA

Altitude insomnia occurs in an acute form in high altitude climbing. It is associated with other symptoms: headaches, fatigue, anorexia and tachycardia. The severity of symptoms varies according to altitude. As a general rule, the number of awakenings increases, the sleep efficiency index falls, NREM disappears and REM sleep diminishes [15, 17]. This sleep alteration depends on periodic respiration [17]. Indeed the hypoxia entrained by high altitude stimulates respiration, provoking hypocapnic alkalosis, leading to apnoea, followed by hypoventilation [3]. Awakening occurs in the transition from apnoea to hyperventilation.

Within a few days a renal compensation mechanism leads to increased urinary excretion of bicarbonates with the gradual correction of alkalosis. Nevertheless hypoxia is not the sole factor at stake: intense cold, trouble finding a comfortable sleeping position, and stress related to the climb are equally likely to play a role.

ADJUSTMENT OR SHORT TERM INSOMNIA

"Adjustment insomnia is a sleep disorder associated with stress, conflict or a change in surroundings causing emotional wakefulness" [9].

This type of insomnia is extremely common. It is the insomnia brought on by family and professional problems; after a separation or while in mourning; before a test, an exam or an election. People do not all react in the same way to emotional stress, which is more likely to affect those who lack self-confidence or who have a low threshold of emotion. To successfully identify this type of insomnia, the stressful circumstances must be determined and insomnia must regress when the circumstances are removed.

The evolution of adjustment insomnia depends on the initial stress level. If this is acute, transitory, accident-related, linked to a reprimand from a superior or passing frustration, sleep disorder is generally short-lived. Where stress is prolonged, however, due to the death of a friend or relation, divorce or financial trouble, adapting to the situation may take longer, and the sleep disorder may persist over several weeks possibly leading to a chronic disorder if therapy is mishandled.

For Morin [12] psychological factors, *i.e.* sleep habits and beliefs and attitudes toward sleep, play a major role in the development of chronic insomnia. During the initial phase of difficulty, subjects who are more vulnerable to insomnia may develop conditioned reactions which are incompatible with sleep. In the long-term, these negative associations produce muscle tension and even more difficulties getting to sleep. Certain subjects thus develop bad sleep habits: irregular sleep-wake times, abuse of the time spent in bed, naps during the day, which become an integral part of the problem in the long term. "The result of this chain reaction is a vicious circle of insomnia with the fear of not sleeping and an increase in emotional distress."

INSOMNIA DUE TO TRANSIENT PHYSICAL STRESS

This type of insomnia is equally common. It is the form of insomnia which accompanies sharp pain, dental or ENT problems, fractured limbs, or a recalcitrant migraine; a pulmonary infection with dyspnoea, coughing; a skin infection associated with pruritis. It may also be the precursory sign of an illness which develops the following day, or several days later, thus explaining the sleep disorder *a posteriori*.

REBOUND INSOMNIA

Hypnotics are indicated for short-term insomnia [5] except in the case of children, and to be used with caution among the elderly.

Rebound insomnia occurs when hypnotic treatment is discontinued. It is characterised by sleep of shorter duration and poorer quality than the subject's basal sleep induced by hypnotic medication. Rebound insomnia can occur in chronic insomniacs who have been taking medication for several months, as well as in subjects suffering from transient insomnia who take the medication for a single night only. Rebound insomnia tends to occur after the discontinuation of short half-life benzodiazepines, brotizolam, or ultra-short midazolam or triazolam [4, 10]. The discontinuation of short- and intermediate-acting half-life products (lorazepam, oxazepam, flunitrazepam, lormetazepam, nitrazepam, temazepam, brotizolam) or long-acting products (flurazepam, diazepam, quazepam) does not usually result in rebound insomnia, due to their slower elimination. The mechanism of this type of insomnia is still largely unknown. Does it result from the separation of an endogenous ligand from its receptor; a modification of benzodiazepine receptors; or again, a disruption in the balance between antagonist neurone systems? Hypnotics which are not structurally linked to benzodiazepines but which have an analogous action mechanism, zolpidem and zopiclone, with short half-lives, have fewer repercussions on withdrawal of treatment and result in fewer tolerance and dependence phenomena [16].

Transient or short term insomnia is thus the most common type of insomnia. It presents no risk, other than when inadequately treated, preparing the way for chronic insomnia.

REFERENCES

1. AGNEW H., WEBB W., WILLIAMS R.L. – The first night effect: an EEG study of sleep. *Psychophysiology*, 7, 263-266, 1966.
2. ANGST J., VOLLRATH M., KOCH R., DOBBLER-MIKOLA A. – The Zurich Study. VII. Insomnia: Symptoms, Classification and Prevalence. *Eur. Arch. Psychiatry Neurol Sci.*, 238, 285-293, 1989.
3. BERSSENBRUGGE A., DEMPSEY J., IBER C., SKATRUD J., WILSON P. – Mechanisms of hypoxia-induced periodic breathing during sleep in humans. *J. Physiol.*, 343, 507-524, 1983.
4. BIXLER E.D., KALES J.D., KALES A., JACOBY J.A., SOLDATOS C.R. – Rebound insomnia and elimination half-life: assesment of individual subject response. *J. Clin. Pharmacol.*, 25, 115-124, 1985.
5. DEMENT W.C. – Normal sleep, disturbed sleep, transient and persistent insomnia. *Acta Psychiatr. Scand. Suppl.*, 332, 41-46, 1986.
6. EHRHART J., MUZET A. – Fréquence et durée des phases d'activation transitoire au cours du sommeil normal ou perturbé chez l'homme. *Arch. Sci. Physiol.*, 28, 213-260, 1964.
7. GALLUP ORGANIZATION – *Sleep in America*, Princeton, New Jersey, 1991.
8. HASKELL E.H., PALCA J.W., WALKER J.M., BERGER R.J., HELLER M.C. – The effects of high and low ambient temperature on human sleep stages. *Electroencephalogr. Clin. Neurophysiol.*, 51, 494-501, 1981.
9. ICSD: The International Classification of Sleep Disorders. – *Diagnostic and Coding Manual.* Diagnostic Classification Steering Committee. MJ. Thorpy Chairman, American Sleep Disorders Association, Rochester, Minnesota, 1990.
10. KALES A., SOLDATOS C.R., BIXLER E.D., KALES J.D. – Rebound insomnia and rebound anxiety: a review. *Pharmacology*, 26, 121-137, 183.
11. KARAKAN I, THORNBY J, WILLIAMS RL. – Sleep disturbance: a community survey. In: *Natural history, epidemiology, and long-term evolution sleep-wake disorders*, C. Guilleminault, E. Lugaresi (eds.), Raven Press, New York, 73-86, 1983.
12. MORIN C.M. – *Vaincre les ennemis du sommeil.* Les Editions de l'Homme. Quebec, 1997.
13. MUZET A. – Les effects du bruit sur le sommeil. *C.R. Soc. Biol.*, 183, 437-422,1989.
14. National Institute of Mental Health, Consensus Development Conference – Drugs and Insomnia. *JAMA*, 251, 2410-2414, 1984.
15. NICHOLSON A.N., SMITH P.A., STONE B.M., BRADWELL A.R., COOTE J.H. – Altitude insomnia: studies during an expedition to the Himalayas. *Sleep*, 11, 534-361, 1988.

16. OSWALD I., ADAM K. – A new look at short-acting hypnotics. In: *Imidazopyridines in sleep disorders.* Sauvanet J.P., Langer S.Z., Morselli P.L. (eds) Raven Press, New York, 253-260, 1988.

17. REITE M., JACKSON D., CAHOON R.L., WEIL J.V. – Sleep physiology at high altitude. *Electroencephalogr. Clin. Neurophysiol.,* 38, 463-471, 1975.

18. SCHIEBER J.P., MUZET A., FERRIERE P.J. – Les phases d'activation transitoire spontanées au cours du sommeil normal chez l'homme. *Arch. Sci. Physiol.,* 25, 443-465, 1971.

19. VALLET M., MOURET J. – Sleep disturbance due to transportation noise: ear plugs or oral drugs. *Experientia,* 40, 429-437, 1984.

20. VELA-BUENO A., DE ICETA M., FERNANDEZ C. – Prevalence of sleep disorders in Madrid, Spain. *Gac. Sanit.,* 13(6), 441-448, 1999.

Chapter 17

Primary insomnia

C. M. Morin
Ecole de Psychologie, Université Laval, Sainte-Foy, Québec, Canada

INTRODUCTION

Insomnia is among the most frequent health complaints in medical practice and the most common of all sleep disorders. It is associated with significant psychological distress, reduced quality of life, and health-care costs. Despite its high prevalence and negative impact, insomnia often goes unrecognised and, consequently, remains untreated. The conceptualisation of insomnia has evolved over the last three decades. Whereas it was initially conceptualised as a symptom of other psychiatric or medical disorders [21] current nosological classifications of sleep disorders make a distinction between the symptom (secondary) and the syndrome (primary) of insomnia [1,2]. Along with these changes in conceptualisation, there have been significant advances in the management of insomnia, from a predominantly symptomatic approach to a more focused behavioral approach on perpetuating factors. This chapter is about the nature and treatment of primary insomnia (fig. 17.1).

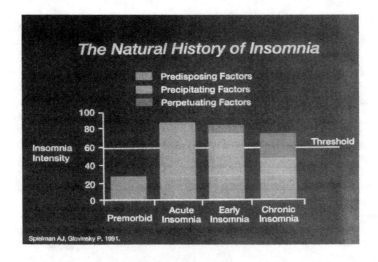

Figure 17.1. The natural history of insomnia

Clinical Presentation and Findings

Insomnia entails a spectrum of complaints reflecting dissatisfaction with the quality, duration, or efficiency of sleep. These complaints may involve problems with falling asleep initially at bedtime, waking up in the middle of the night and having difficulty going back to sleep, or waking up too early in the morning with an inability to return to sleep. These difficulties are not mutually exclusive, as a person may experience mixed problems initiating and maintaining sleep. Insomnia may also involve a complaint of nonrestorative or unrefreshing sleep. In addition, complaints of daytime fatigue, problems with memory and concentration, and mood disturbances (e.g., irritability, dysphoria) are extremely frequent and often the primary concerns prompting patients with insomnia to seek treatment [30].

In addition to standard diagnostic criteria (see table 17.1), several indicators can be used to evaluate the clinical significance of insomnia and the need for treatment. These markers include the severity, frequency, and duration of sleep difficulties and their associated daytime consequences. The time required to fall asleep and the number and duration of awakenings are useful markers of insomnia severity. Although there is no standard definition, sleep-onset insomnia and sleep-maintenance insomnia are respectively defined by a latency to sleep onset and/or time awake after sleep onset greater than 30 minutes, with a corresponding sleep efficiency lower than 85 percent. Likewise, early morning awakening can be operationalised by a complaint of waking up earlier (more than 30 minutes) than desired, with an inability to go back to sleep, and before total sleep time reaches 6.5 hours. These criteria, while arbitrary, are useful to operationalise the definition of insomnia. Total sleep time alone is not a good index to define insomnia because there are individual differences in sleep needs. Some people may function well with as little as 5-6 hours of sleep and would not necessarily complain of insomnia, while others needing 9-10 hours may still complain of inadequate sleep. It is also important to consider the frequency of sleep difficulties in order to distinguish the occasional insomnia that everyone experiences at one time or another in life from the more frequent or recurrent insomnia. The usual cut-off point is more than three nights per week with difficulties initiating and/or maintaining sleep. A distinction is also made between situational/acute insomnia, a condition lasting a few days and often associated with life events or jet lag, short-term/subacute insomnia (lasting between one and four weeks), and chronic insomnia, lasting more than one month. Finally, it is necessary to consider the impact of insomnia on a person's life to judge its clinical significance. As such, a complaint of insomnia must be associated with marked distress or significant impairments of daytime functioning [1-2].

Polysomnographic (PSG) findings in subjectively-defined insomniacs reveal more impairments of sleep continuity parameters (i.e., longer sleep latencies, more time awake after sleep onset, lower sleep efficiency) and reduced total sleep time compared to subjectively-defined good sleepers. Also, insomniacs tend to spend more time in stage 1 sleep, less time in stages 3-4 sleep, and display more frequent stage shifts through the night. Interestingly, sleep disturbances recorded in primary insomniacs are similar to those observed in patients with generalised anxiety disorders or some affective disorders such as dysthymia [18, 46], these similarities may suggest a common underlying thread to these conditions. In addition, there is a significant overlap in the distribution of sleep recordings of subjectively-defined insomniacs and good sleepers such that some insomniacs may show better objective sleep than good sleepers and some good sleepers more sleep impairments than insomniacs. Investigations of the microstructure of sleep reveal that beta activity is increased in primary insomnia relative to good sleepers, both around the sleep onset period and during NREM sleep [22, 27, 43]. These findings are consistent with psychological findings that insomniacs are hypervigilant and ruminative at sleep onset and/or during sleep and with the presumed role of attentional processes and information processing factors [43].

The complaint of fatigue is almost always associated with insomnia. Although some patients may initially report excessive daytime sleepiness, a more in depth investigation usually reveals that insomniacs experience mental and physical fatigue rather than true physiological sleepiness. Findings from the Multiple Sleep Latency Test of primary insomniacs are often comparable to those of good sleepers [55]. Excessive daytime sleepiness is more common among patients with insomnia secondary to another medical (e.g., pain) or sleep disorders (e.g., periodic limb movements, sleep-related breathing disorders). Insomniacs have trouble sleeping at night, in part because of a state of hyperarousal, and this hyperarousal also interferes with their ability to nap during the day.

Neuropsychological evaluations of the daytime performance of primary insomniacs have revealed only mild and fairly selective performance deficits (e.g., attention). Nonetheless, insomniacs tend to have lower expectations and to rate their performance as significantly impaired relative to their own standards (i.e., what they should be able to do) and as more impaired than that of normal controls. Such discrepancies between actual and expected performance and between subjective and objective performance are similar to those observed between objective and subjective measures of sleep. Such discrepancies may reflect a generalised faulty appraisal of sleep and daytime functioning among individuals with insomnia complaints [57]. In addition, performance impairments on neuropsychological testing are more strongly associated with subjective (as measured by daily sleep diaries) than with objective (as measured by PSG) sleep disturbances.

Collectively, these findings suggest that the subjective appraisal/perception of sleep and daytime functioning are partly modulated by psychological and cognitive factors which, in turn, are important determinants of insomnia complaints. These paradoxical findings illustrate the complexity of insomnia and why some individuals with insomnia symptoms do not complain about it, whereas others are dissatisfied with their sleep in the absence of significant evidence of sleep impairment.

Subtypes of Primary Insomnia

The two main nosological classifications of sleep disorders (ICSD and DSM-IV) [1, 2] make an essential distinction between primary (syndrome) and secondary (symptom) insomnia. In secondary insomnia, the sleep disturbances is etiologically linked to the underlying condition, including psychiatric, medical, substance abuse, and other sleep disorders. These conditions are reviewed in another chapter of this volume. In primary insomnia, the sleep disturbance does not occur exclusively in the context of another medical, psychiatric, or substance abuse disorders; it may co-exist with these conditions but it is viewed as an independent disorder. Individuals with primary insomnia often display anxiety and depressive symptomatology, but such clinical findings are not severe enough to reach diagnostic threshold for an anxiety or affective disorder. They are viewed as consequence or co-existing symptoms rather than causes of insomnia. Because of the extensive comorbidity between sleep disturbances and psychopathology [9], the differential diagnosis of insomnia is not always easy to make accurately.

Whereas DSM-IV recognises only one form of primary insomnia, the ICSD distinguishes among three different subtypes: psychophysiological insomnia, sleep-state misperception, and idiopathic insomnia. Psychophysiological insomnia, which is essentially the DSM-IV equivalent of primary insomnia, is the most classic form of insomnia. It is a type of conditioned or learned insomnia that is presumably derived from two sources. The first involves the conditioning of sleep-preventing habits in which repeated pairing of sleeplessness and situational (bed/bedroom), temporal (bedtime), or behavioural (bedtime ritual) stimuli normally associated with sleep leads to conditioned arousal that impairs sleep. The second involves somatised tension believed to result from the internalisation of psychological conflicts and excessive worrying/apprehension about sleep, which are incompatible with sleep [18, 21, 30].

The sleep of individuals with psychophysiological insomnia is very sensitive to minor irritants and daily stressors; it is also characterised by extensive night-to-night variability such that sleep is often unpredictable. Sometimes, their sleep is unexpectedly improved in a novel environment because the conditioned cues that keep them awake at home are not present in that environment. For example, while the sleep of otherwise good sleepers is more disrupted during their first night of recording in the sleep laboratory (i.e., first night effect), insomniacs may actually sleep better during their first night in the laboratory (i.e., the reverse first-night-effect).

Sleep-state misperception, also called subjective insomnia, is a genuine complaint of poor sleep that is not corroborated by objective findings. For example, a patient may perceive very little sleep (e.g., 1-2 hours per night) whereas EEG recordings show normal or near-normal sleep duration and quality. This sleep-state misperception condition is not the result of an underlying psychiatric disorder or of malingering. To some degree, all insomniacs tend to overestimate the time it takes them to fall asleep and to underestimate the time they actually sleep. In sleep-state misperception, however, the subjective complaint of poor sleep is clearly out of proportion with any objective finding. This phenomenon is probably due to several factors including the lack of sensitivity of EEG measures, the influence of cognitive (information processing) variables during the early stages of sleep or, it could also represents the far end of a continuum of individual differences in sleep perception. Interestingly, individuals with subjective insomnia report greater disruption of daily functioning than those with psychophysiological insomnia. Some authors have suggested that sleep-state misperception may be a prodromal phase for more objectively verifiable insomnia [50]. However, this condition is still poorly understood and some have suggested that it should not even be considered a separate diagnostic entity. The main problem with this diagnosis is that most clinicians do not have objective data to confirm or refute the patient's subjective complaint.

Idiopathic (childhood) insomnia is, by definition, of unknown origin. One of the most persistent forms of insomnia, it presents an insidious onset in childhood, unrelated to psychological trauma or medical disorders, and is very persistent throughout the adult life. It does not present the nightly variability observed with other forms of primary insomnia. A mild defect of the basic neurological sleep/wake mechanisms may be a predisposing factor; this hypothesis comes from the observations that patients with this condition often have a history of learning disabilities, attention-deficit hyperactivity, or similar conditions associated with minimal brain dysfunctions. Despite the presence of daytime sequelae (e.g., memory, concentration and motivational difficulties), and the evidence that sleep disturbance is more severe than in psychophysiological insomnia, individuals with idiopathic insomnia often experience less emotional distress than those with the psychophysiological subtype, perhaps due to coping mechanisms they have developed over their lifetime.

These three insomnia subtypes may differ qualitatively from each other and may also require different treatment. For example, patients with sleep-state misperception may require treatments that specifically correct their sleep misperceptions (e.g., cognitive therapy). Although there is still little data to validate these subtypes of insomnia, clinical experience tend to support the existence of various primary insomnia profiles. Additional research is needed, however, to ascertain the number of etiologically distinctive subtypes which warrant clinical attention and also whether they require different treatment.

Epidemiology and Significance

Insomnia is the most common of all sleep disorders. Although prevalence rates vary considerably across surveys due to differences in definition and methodology, estimates from two US epidemiological studies indicate that about one third of the adult population report insomnia, including 10% who complain of persistent and severe insomnia [15, 26]. Other estimates from European countries (France, Germany, and the UK) indicate that about 29% of adults report insomnia symptoms (without insomnia diagnosis), 5.7% are dissatisfied with their sleep (without diagnosis), and about 5.8% meet criteria for an insomnia disorder [40]. Between 7% [26] and 10% [41] of the population use a sleep-promoting drug, most often on a regular basis. Despite similar prevalence rates of insomnia across countries, the use of sleep medications is significantly higher in France and Italy relative to Germany, the UK or the US. Surveys conducted in general medical practice reveal even higher prevalence rates of both insomnia and hypnotic usage. For example, one survey found that more than 30% of medical patients had either moderate or severe insomnia, and almost one quarter of them regularly used prescribed hypnotics [20].

Insomnia complaints are age-related, with an increased prevalence among middle-aged and older adults [26,40-41]. Sleep-onset difficulties are also more common among young adults whereas difficulties maintaining sleep and poor quality sleep occur more frequently among older individuals. Women complain of insomnia more often than men (ratio of 2 to 1), but it is unclear whether such findings reflect reporting biases or actual gender differences. The incidence of insomnia is also slightly higher among first-degree family members (daughter, mother) than in the general population [4]. It is also unclear whether this is suggesting a genetic predisposition or the transmission of poor sleep habits and coping skills through social-learning from parental models.

Whereas earlier epidemiological studies queried respondents only about sleep/insomnia symptoms, recent studies by Ohayon using a computer-based algorithm (Sleep-EVAL) have yielded prevalence estimates of specific insomnia diagnoses. From the estimated 10% of the population who meet criteria for a sleep disorder (of whom 58% have insomnia), about one third would have primary insomnia, another 20% would have insomnia due to a general medical condition, and 10% insomnia related to another mental disorder [39]. These figures are different from those based on patients evaluated at sleep disorders center where more than 40% are diagnosed with a comorbid psychiatric disorder [9].

Chronic insomnia is not a benign problem as it can adversely affect a person's life, causing substantial psychosocial, occupational, health, and economic repercussions. For example, individuals with chronic sleep disturbances experience more psychological distress and impairments of daytime functioning relative to good sleepers; in addition, they take more sick leaves and utilise health care resources more often than good sleepers [26, 52]. Persistent insomnia is also associated with prolonged use of hypnotic medications and with increased risks of major depression [8, 15, 26]. In older adults with cognitive impairments, the presence of sleep disturbances may hasten placement in a nursing home facility [44].

COURSE AND PROGNOSIS

Insomnia can be situational, recurrent, or chronic. The type of insomnia may also change over time, with sleep onset difficulties being more typical of the first episodes of insomnia and sleep maintenance difficulties becoming more common in the more chronic phase of the disorder. The first episode of insomnia may occur at any age, with either a sudden or insidious onset, but individuals with primary insomnia report most frequently that they first experienced sleep difficulties during their 20's or early 30's. That first episode of insomnia was often associated with a stressful life event (e.g., school exam, new job, birth of a child). Subsequently, sleep may have normalised for a few weeks, months, or even years but, eventually, sleep disturbances re-occurred more frequently or for longer periods [30].

For the majority of individuals, sleep difficulties are transient in nature, lasting a few days, and resolving themselves once the initial precipitating event has subsided. For some people, however, perhaps those with a higher predisposition to insomnia, sleep disturbance may persist over time even after the initial cause has disappeared. Insomnia may follow an intermittent course, with repeated brief episodes of sleep difficulties following a close association with the occurrence of stressful events [30] or, it may become a chronic problem. Even in persistent insomnia, there is often extensive night-to-night variability in sleep, with an occasional good night's intertwined with several nights of disrupted sleep. This unpredictability of sleep can be very distressing and it may eventually induce a sense of learned helplessness.

The long-term prognosis of untreated insomnia is not well documented. However, there is some evidence suggesting that insomnia is often a recurrent problem [58] and, when left untreated, it increases the risks for major depression [8, 15]. Even with treatment, some patients will continue to experience intermittent sleep difficulties. For this reason, it is particularly important to teach patients appropriate behavioral and coping skills to manage those residual episodes of insomnia and reduce the risk of chronicity and morbidity associated with insomnia.

AETIOLOGY

The origin of insomnia is often multifactorial and, even in primary insomnia, several psychological, biological, and learning factors have been associated with the onset or maintenance of this condition. A widely accepted etiological theory regarding primary insomnia attributes this condition to a special confluence of endogenous *predisposing characteristics*, sleep-disruptive *precipitating events*, and *perpetuating behaviors or circumstances* [53]. According to this theory, vulnerabilities such as a proneness to worry, repression of negative emotions, physiological hyperarousal, and/or an innate propensity toward light, fragmented sleep may all predispose certain individuals to primary insomnia. Although not everyone with such predisposition will develop insomnia, sleep disturbances are more likely to arise among such individuals when exposed to stressful life events (e.g., bereavement, separation, undergoing a painful medical procedure, etc.). Insomnia may then persist when conditioned environmental cues, poor sleep habits, and dysfunctional cognitions serve to perpetuate sleep disturbance long after the initial precipitating circumstances are resolved. According to this model, psychological (e.g., fear of not sleeping, excessive concerns about the consequences of insomnia) and behavioral factors (e.g., irregular sleep schedules, excessive amounts of time spent in bed) are almost always involved in perpetuating insomnia over time regardless of the nature of the precipitating events. Treatment of chronic insomnia must necessarily address these perpetuating factors.

Clinical observations have provided much support for this etiological model. Many primary insomniacs report an excessive preoccupation with sleep and a heightened arousal as bedtime approaches [30]. Indeed, such patients frankly report that they view bedtime as the worst time of day. A vicious cycle often emerges in which repetitive unsuccessful sleep attempts reinforce the anticipatory anxiety which, in turn, contributes to more insomnia. Through their repetitive association with unsuccessful sleep efforts, the bedroom environment and pre-sleep rituals become cues or stimuli for poor sleep. Moreover, in some cases, formerly benign habits such as watching television, eating, or reading in bed may also reduce the stimulus value of the bed and bedroom for sleep and may further exacerbate the sleep problem. Consequently, it is not unusual for primary insomniacs to report improved sleep in novel settings where conditioned environmental cues are absent and usual pre-sleep rituals are obviated. In addition, many primary insomniacs display poor sleep habits that initially may emerge as a means of coping with sleep disturbances. For example, poor sleep at night may lead to daytime napping or sleeping late on weekends in efforts to catch-up on lost sleep. Alternately such individuals may lie in bed for protracted periods trying to force sleep only to find themselves becoming more and more awake. In addition, other practices such as routinely engaging in physically or mentally stimulating activities shortly before bed or failing to adhere to a regular sleep-wake schedule may emerge due to life style choices or perceived social obligations and also may contribute to the sleep difficulty.

Several cross-sectional studies have corroborated these clinical observations by suggesting that dispositional characteristics, stressful life events, and perpetuating behaviors/circumstances may all play a role in the etiology of primary insomnia. Psychometric studies of insomniacs suggest that these individuals are more likely to show psychological profiles characterised by mild anxiety, depression and a proneness toward worrying and the internalisation of negative affect [11, 21]. In addition, laboratory studies have shown that primary insomniacs evidence less diurnal sleepiness and higher heart rates, core body temperature, and metabolic activity during the night than do age- and gender-matched controls [6, 55]. Collectively, such studies provide indirect support to the assumption that predisposing characteristics (i.e., hyperarousal, worrying, repressive tendencies) may increase the vulnerability of certain individuals to insomnia. Similarly, appropriate precipitating circumstances appear etiologically important inasmuch as approximately 65% of insomniacs report that a stressful life event preceded and led to their insomnia problems [19]. The cross-sectional or retrospective nature of these studies precludes any definite conclusion about a cause-effect relationship between such factors and primary insomnia.

EVALUATION AND DIAGNOSIS

The evaluation of insomnia is based primarily on a detailed clinical assessment of the patient's subjective complaint. The sleep history should cover the type of complaint (initial, middle, late insomnia), its duration (acute vs. chronic), and course (recurrent, persistent); typical sleep schedule, exacerbating and alleviating factors, perceived consequences and functional impairments, and the presence of medical, psychiatric, or environmental contributing factors. A complete history of alcohol and drug use, prescribed and over-the-counter medications, is also essential [34, 51, 53]. Although polysomnography is not indicated for the routine evaluation of insomnia, it is often necessary to rule out other sleep disorders that might contribute to the insomnia complaint (e.g., periodic movements during sleep; sleep apnea) [45]. PSG can also be particularly useful in suspected cases of sleep state misperception. Routine laboratory tests (e.g., complete blood count) may be indicated in some patients, although results from these tests are often negative.

The use of a sleep diary is also an essential component in the evaluation of insomnia. The patient should complete a daily sleep diary for at least one week, preferably two, before initiating and throughout the course of treatment. The diary is useful to document the initial insomnia severity, identify behavioral and scheduling factors that may perpetuate insomnia, and to monitor progress during treatment. Several additional questionnaires provide useful complementary information to evaluate perceived severity and impact of insomnia and to monitor treatment progress. Among those used in our clinic are the Insomnia Severity Index, the Dysfunctional Beliefs and Attitudes about Sleep, the Fatigue Severity Scale, and screening measures of anxiety and depressive symptomatology [30]. A complete psychological evaluation may be necessary for patients with suspected psychiatric disorders.

Primary insomnia is a diagnosis made by exclusion. Thus, the differential diagnosis of primary insomnia requires the ruling out of several other conditions including psychiatric (depression and anxiety), medical (pain), circadian (phase-delay syndrome), or other sleep disorders (restless legs syndrome/periodic limb movements, sleep-breathing disorders). While the essential clinical features of insomnia are often similar for the symptom and the syndrome, in primary insomnia, the sleep disturbance does not occur exclusively during the course of another sleep disorder or mental disorder and is not due to the direct physiologic effects of a substance or a general medical condition (see table 17.1). A diagnosis of secondary insomnia is made when the sleep disturbance is judged to be related temporally and causally to another psychiatric, medical, or sleep disorder [1-2]. Thus, the diagnosis of primary insomnia rests essentially on the subjective complaint of the individual and it is often made by default, i.e., after ruling out all other potential causes.

Table 17.1. Diagnostic Criteria for Primary Insomnia.

A subjective complaint of difficulties initiating or maintaining sleep, or nonrestorative sleep.
Duration of insomnia is longer than 1 month.
The sleep disturbance (or associated daytime fatigue) causes clinically significant distress or impairment in social, occupational or other important areas of functioning
The sleep disturbance does not occur exclusively in the context of another mental or sleep disorder, and is not the direct physiologic effect of a substance or a general medical condition.

In clinical practice, the main differential diagnosis of primary insomnia is with insomnia secondary to anxiety (generalised anxiety disorder; GAD) or depression (dysthymia or major depression). This distinction is not always easy to make because several symptoms (e.g., sleep disturbances, fatigue, concentration problems, irritability) are common to all those conditions. Excessive worrying is the predominant feature of GAD; this characteristic is also present in primary insomnia, but the main focus of worrying is limited to sleep and its potential consequences, whereas worrying in GAD is about multiple sources of preoccupations (e.g., health, finances, family, work). In depression, the predominant clinical feature is sadness and a significant loss of interest for people and activities. In primary insomnia, the interest is present but there is a lack of energy or fatigue, presumably resulting from sleep disturbances, preventing the individual from engaging in potentially pleasurable activities or social interactions. In addition to the nature of the symptoms, the history should also identify relative onset and course of each condition in order to determine whether insomnia is primary or secondary in nature.

NON PHARMACOLOGICAL TREATMENT

In acute insomnia, treatment should focus on the initial triggering factors. For instance, a recent onset of insomnia that is clearly associated with stressful life events (e.g., separation, bereavement, professional burn-out) should be treated by appropriate supportive or problem-solving therapies. Short-term use of hypnotic medications may also be indicated in such circumstances. The most important issue when insomnia is brought to clinical attention is to provide reassurance that temporary sleep disturbances may be normal under such stressful conditions. On the other hand, if insomnia persist even after the initial triggering event has disappeared, treatment should then focus on the underlying perpetuating factors.

Treatment options for primary and chronic insomnia include preventive sleep hygiene education, cognitive-behavioral interventions, and pharmacotherapy. This chapter addresses only the non-pharmacological interventions; drug therapy is discussed in other chapters (see chapters 21 and 22). A summary of non-pharmacological interventions is provided below and in table 17.2; more extensive descriptions are available in other sources [23, 30].

Table 17.2. Cognitive-Behavioural Treatments for Insomnia

Therapy	Description
Stimulus control instructions	A set of instructions designed to reassociate the bed/bedroom with sleep and to re-establish a consistent sleep-wake schedule: Go to bed only when sleepy; get out of bed when unable to sleep; use the bed/bedroom for sleep only (no reading, watching TV, etc); arise at the same time every morning; no napping.
Sleep restriction therapy	A method to curtail time in bed to the actual sleep time, thereby creating mild sleep deprivation, which results in more consolidated and more efficient sleep.
Relaxation training	Methods aimed at reducing somatic tension (e.g., progressive muscle relaxation, autogenic training) or intrusive thoughts (e.g., imagery training, meditation) interfering with sleep.
Cognitive therapy	Psychotherapeutic method aimed at changing faulty beliefs and attitudes about sleep and insomnia (e.g., unrealistic sleep expectations; fear of the consequences of insomnia).
Sleep hygiene education	General guidelines about health practices (e.g., diet, exercise, substance use) and environmental factors (e.g., light, noise, temperature) that may promote or interfere with sleep.

Sleep Hygiene Guidelines

Sleep hygiene education is intended to provide information about lifestyle (diet, exercise, substance use) and environmental factors (light, noise, temperature) that may either interfere with or promote better sleep. It may also include general information about the benefits of maintaining a regular sleep schedule. Sleep hygiene guidelines include:
- avoiding stimulants (e.g., caffeine, nicotine) several hours before bedtime
- avoiding alcohol as a sleep aid as it fragments sleep
- exercising regularly (especially in late-afternoon or early evening)
- allowing at least a 1-hour period to unwind before bedtime
- maintaining a regular sleep schedule
- keeping the bedroom environment quiet, dark, and comfortable [16, 47].

In addition to these guidelines, it is useful to provide some basic information about normal sleep, individual differences in sleep needs, and changes in sleep architecture over the course of the life span. This information is particularly useful to help some patients distinguish clinical insomnia from short-sleep or from normal (age-related) sleep disturbances. Such knowledge can prevent excessive worrying and concerns, which could in itself lead to clinical insomnia.

Inadequate sleep hygiene is rarely the primary cause of insomnia, but it may potentiate sleep difficulties caused by other factors or interfere with treatment progress. Thus, even if patients with insomnia are often well-informed about general sleep hygiene guidelines, it is important to address these factors directly in treatment. On the other hand, while sleep hygiene education may be sufficient in mild insomnia, it is rarely sufficient for more severe and chronic forms of insomnia, which often require more directive and potent behavioral interventions.

Cognitive-Behavioral Therapies

Cognitive-behavioral therapies (CBT) are time-limited, structured, and sleep-focused. Their main objectives are to modify or eliminate factors that perpetuate insomnia, including poor sleep habits, irregular sleep-wake schedules, hyperarousal, and misconceptions about sleep. Another common goal of CBT is to teach patients self-management skills to cope more adaptively with residual sleep disturbances that may persist even after successful treatment. Treatment modalities that have received adequate validation in controlled clinical trials include stimulus control therapy, sleep restriction, relaxation-based interventions, and cognitive therapy. Insight-oriented psychotherapy focusing on predisposing rather than perpetuating factors of insomnia may also be useful in some patients, but there is no evidence that such therapy is effective in the management of insomnia. CBT does not seek to alter the underlying personality structure of the insomnia patient; rather, its main focus is on factors (psychological, behavioral, and cognitive) that perpetuate or exacerbate insomnia.

Sleep restriction. Poor sleepers often increase their time in bed in a misguided effort to provide more opportunity for sleep, a strategy that is more likely to result in fragmented and poor quality sleep. Sleep restriction therapy consists of curtailing the amount of time spent in bed to the actual amount of time asleep [54]. Time in bed is subsequently adjusted based on sleep efficiency (SE; ratio of total sleep/time in bed X 100%) for a given period of time (usually the preceding week). For example, if a person reports sleeping an average of 6 hours per night out of 8 hours spent in bed, the initial prescribed sleep window (i.e., from initial bedtime to final arising time) would be 6 hours. The subsequent allowable time in bed is increased by about 20 minutes for a given week when SE exceeds 85%, decreased by the same amount of time when SE is lower than 80%, and kept stable when SE falls between 80% and 85%. Adjustments are made periodically (weekly) until optimal sleep duration is achieved. Changes to the prescribed sleep window can be made at the beginning of the night (i.e., postponing bedtime), at the end of the sleep period (i.e., advancing arising time), or at both ends. To prevent excessive daytime sleepiness, time in bed should not be reduced to less than 5 hours per night in bed. This procedure leads to rapid improvement of sleep continuity parameters through a mild sleep deprivation and reduction of sleep anticipatory anxiety.

Stimulus control therapy. Chronic insomniacs often become apprehensive around bedtime and associate the bed/bedroom with frustration and arousal. This conditioning process may take place over several weeks or even months, unknowingly to the patient. Stimulus control therapy [7] consists of a set of instructions designed to reassociate temporal (bedtime) and environmental (bed and bedroom) stimuli with rapid sleep onset and to establish a regular circadian sleep-wake rhythm. These instructions are:

- going to bed only when sleepy, not just fatigue but sleepy
- getting out of bed when unable to sleep (e.g., after 20 min), going to another room and returning to bed only when sleep is imminent
- curtailing all sleep-incompatible activities (overt and covert); no eating, TV watching, radio listening, planning or problem solving in bed
- arising at a regular time every morning regardless of the amount of sleep the night before
- avoiding daytime napping

These instructions may appear quite simple on paper but the main challenge is to foster strict compliance with all of them for a few weeks. Also, there may be no need to follow some of those instructions (e.g., getting out of bed when unable to sleep) when time in bed is substantially reduced during the initial compression of time spent in bed; however, as time in bed is gradually increased, these procedures become more relevant.

Relaxation-based interventions. Relaxation is the most commonly used non drug therapy for insomnia. Among the available relaxation-based interventions, some methods (e.g., progressive-muscle relaxation, autogenic training, biofeedback) focus primarily on reducing somatic arousal (e.g., muscle tension), whereas attention-focusing procedures (e.g., imagery training, meditation, thought stopping) target mental arousal in the forms of worries, intrusive thoughts, or a racing mind. Biofeedback is designed to train a patient to control some physiological parameters (e.g., frontalis EMG tension) through visual or auditory feedback. Most of these methods are equally effective for treating insomnia. The most critical issue is to practice diligently and daily the selected method for at least two to four weeks. Professional guidance is often necessary in the initial phase of training.

Also, when stress is a significant contributing factor to insomnia, it may be necessary to implement a more comprehensive stress management program involving relaxation and other components such as time management, problem-solving training, etc.

Cognitive therapy. Cognitive therapy seeks to alter dysfunctional sleep cognitions (e.g., beliefs, attitudes, expectations, attributions). The basic premise of this approach is that appraisal of a given situation (sleeplessness) can trigger negative emotions (fear, anxiety) that are incompatible with sleep. For example, when a person is unable to sleep at night and begins thinking about the possible consequences of sleep loss on the next day's performance, this can set off a spiral reaction and feed into the vicious cycle of insomnia, emotional distress, and more sleep disturbances. Cognitive therapy is designed to identify dysfunctional cognitions and reframe them into more adaptive substitutes in order to short-circuit the self-fulfilling nature of this vicious cycle. Specific treatment targets include unrealistic expectations ("I must get my 8 hours of sleep every night"), faulty causal attributions ("my insomnia is entirely due to a biochemical imbalance"), and amplification of the consequences of insomnia ("Insomnia may have serious consequences on my health") [30]. In addition, patients often perceive themselves as victims of insomnia and as having little resources to cope with sleep difficulties and their daytime consequences. Cognitive therapy can be particularly useful to modify these schemas and teach patients skills to cope with residual sleep disturbances that often persist after treatment. The main therapeutic messages to communicate to patients are as follows:

- Keep realistic expectations
- Do not blame insomnia for all daytime impairments
- Never try to sleep
- Do not give too much importance to sleep
- Do not catastrophise after a poor night's sleep
- Develop some tolerance to the effects of sleep loss

Additional non drug interventions are available for treating insomnia including paradoxical intention, acupuncture, ocular relaxation, and electrosleep therapy. Although potentially useful in practice, those methods have not been evaluated as extensively in controlled studies as the interventions just described. Psychotherapy may also be useful to address predisposing factors to insomnia, but there has been no controlled evaluation of its efficacy.

Combined behavioral and pharmacological approaches. A combined behavioral and drug intervention should theoretically optimise treatment outcome by capitalising on the more immediate and potent effects of drug therapy and the more sustained effects of behavioral therapy. The limited evidence available, however, is not entirely clear as to whether a combined approach has an additive or subtractive effect on long-term outcome. Only a few studies have directly evaluated the combined or differential effects of behavioral and pharmacological therapies for insomnia [17, 25, 28, 32, 49]. Collectively, those studies indicate that drug therapy produces quicker and slightly better results in the acute phase (first week) of treatment, whereas behavioral and drug therapies are equally effective in the short-term interval (4-8 weeks). Combined interventions appear to have a slight advantage over single treatment modality during the initial course of treatment, but it is unclear whether a combined approach produces a better long-term outcome than behavioral therapy alone. For instance, sleep improvements are well sustained after behavioral treatment and those obtained with hypnotic drugs are quickly lost after discontinuation of the medication. The long-term effects of a combined biobehavioral and drug intervention are more equivocal. Some of those patients treated with a combined approach retain their initial sleep improvements over time whereas others return to their baseline values. Thus, despite the intuitive appeal in combining drug and non-drug interventions, it is not entirely clear when, how, and for whom it is indicated to combine behavioral and drug treatments for insomnia. In light of the mediating role of psychological factors in chronic insomnia, behavioral and attitudinal changes appear essential to sustain improvements in sleep patterns. When combining behavioral and drug therapies, patients' attributions of the initial benefits may be critical in determining long-term outcomes. Attribution of therapeutic benefits to the drug alone, without integration of self-management skills, may place a patient at greater risk for insomnia recurrence after the drug is discontinued. Additional research is needed to evaluate the effects of combined treatments for insomnia and to examine potential mechanisms of changes mediating short- and long-term outcomes.

Outcome Evidence

Evidence from clinical trials [33, 35, 37] indicates that non-pharmacological treatments produce reliable changes in several sleep parameters of individuals with chronic insomnia. Between 70% and 80% of patients benefit from treatment, with average effect sizes of 0.88 (sleep-onset latency), 0.65 (time awake after sleep onset), and 0.94 (sleep quality) on the main outcome variables, and smaller effects for the number of awakenings (0.53-0.63) and total sleep time (0.42-0.49). These effect sizes are comparable to those obtained for benzodiazepine-receptor agonists [38]. In terms of absolute changes, treatment reduces subjective sleep-onset latency from an average of 60-65 minutes at baseline to about 35 minutes at post treatment. The duration of awakenings is decreased from an average of 70 minutes at baseline to about 38 minutes following treatment. Total sleep time is increased by a modest 30 minutes, from 6 hours to 6.5 hours after treatment, but perceived sleep quality is enhanced with treatment. These results represent conservative estimates of efficacy because they are based on average treatment effects computed across all non pharmacological treatment modalities. Comparative studies suggest that stimulus control therapy and sleep restriction are the most effective single treatment modalities. In clinical practice, cognitive-behavioral interventions are not incompatible with each other and can be effectively combined. Sleep restriction seems to produce the most rapid changes but the best outcome is obtained from multi-faceted interventions that incorporate behavioral, educational, and cognitive components [33, 35, 37].

Treatment efficacy has been documented primarily by means of prospective daily sleep diaries. The magnitude of improvement is slightly smaller when measured with polygraphic [32] or actigraphic recordings [13], but it is still clearly in the direction of improvements. One particular strength of CBT for insomnia is the durability of sleep improvements over time, which has been documented up to 24 months after treatment completion. Sleep improvements noted during the initial treatment phase are often enhanced at follow-ups, as evidenced by further decreases of sleep latency and time awake after sleep onset and increases in total sleep time. Reduction of hypnotic usage achieved during the course of CBT are also well sustained over time [31]. In summary then, CBT produces reliable and durable improvements of sleep efficiency and continuity in primary insomnia. The majority (70%-80%) of treated patients benefit from CBT, although only a minority become good sleepers. Patients are generally more satisfied with their sleep. In addition, CBT often induces a greater sense of personal control over sleep and a reduced need for hypnotic medications. Despite the evidence documenting the efficacy of CBT, little is known about the specificity of treatment effects and the underlying mechanisms of changes.

INDICATIONS, ADVANTAGES, AND LIMITATIONS

Most treatment studies have focused on primary insomnia in otherwise healthy and medication-free patients. An important question that often arises is whether the findings obtained in those studies generalise to patients with comorbid medical or psychiatric disorders. While the initial treatment of secondary insomnia should focus on the underlying condition, there is essentially no contra-indication to using CBT even in secondary insomnia. Findings from clinical case series [31, 36, 42] suggest that patients with medical and psychiatric conditions can also benefit from CBT for sleep disturbances, even though the outcome with those patients is more modest than with those who suffer from primary insomnia. Recent studies have shown that CBT is effective to treat insomnia associated with chronic pain [10] and with medical conditions in older adults [24]. There is also evidence that CBT can be used successfully with medicated as well as with unmedicated insomniacs [48]. Finally, while some clinicians are pessimistic in their management of chronic hypnotic users, some findings indicate that a supervised, structured, and time-limited withdrawal program, with or without CBT, can facilitate reduction of hypnotic medications by more than 90% among prolonged users [31].

As for all treatment options, CBT presents both advantages and limitations. In general, CBT is well accepted by insomnia patients; nevertheless, this treatment modality is more time consuming and requires more efforts than drug therapy, both for clinicians and patients. As such, compliance is not always optimal. For this reason, it is particularly important to schedule several follow-up visits after the initial evaluation. While only a small percentage of patients achieve complete remission of sleep difficulties, the majority benefit from treatment and, most importantly, CBT produces durable changes in sleep patterns. On the other hand, there is still limited evidence that sleep improvements lead to meaningful changes in daytime functioning. Finally, despite the increasing evidence that CBT is an effective treatment modality for insomnia, it is still under-utilised by health-care practitioners.

Clinical and Practical Issues

CBT is relatively brief in duration with an average consultation time of 5 hours per patient spread over a treatment period of 6-8 weeks. While this treatment duration is considerably shorter than other forms of psychotherapy, and shorter than CBT applied to other conditions (e.g., chronic pain, anxiety, depression), it may still seem unrealistic or not feasible for some practitioners to spend that much time with a single patient. Although not all patients present the same insomnia severity or require the same extent of therapeutic attention, it would be unrealistic to expect to treat everyone in a single session. Time is necessary for patients to integrate self-management skills, and it may be counter-productive to attempt to cover all procedures in a single session and ask patients to return for a follow-up visit only three months later. Until optimal treatment parameters (frequency and duration) are defined [12], the extent of therapeutic involvement should consider several factors including initial insomnia severity, presence of comorbid medical or psychiatric conditions, and patient's motivation and education. For mild forms of insomnia, basic sleep hygiene education may be sufficient, whereas more severe and persistent forms of insomnia will often require more elaborate interventions and more frequent follow-up visits. Some behavioral procedures (e.g., sleep restriction, stimulus control instructions) can be implemented by most clinicians in a limited amount of time, while others (e.g., relaxation, cognitive therapy) may require more time or specialised training. Treatment can also be implemented successfully by nurse practitioners [14]. Group therapy is another cost-effective approach to the clinical management of insomnia [3, 56], as is self-help treatment [29], although this latter approach is appropriate only for individuals who are highly motivated. Therapist guidance is necessary for more complicated cases and, in the presence of significant comorbidy, it is preferable to refer to a behavioral sleep medicine specialist.

CONCLUSION

Significant advances have been made over the last two decades in the conceptualisation, diagnosis, and treatment of insomnia. Despite these progress and increasing recognition that insomnia is a prevalent and costly health problem, only a small proportion of insomnia patients receive treatment. There are several barriers to treating this condition, including the cost and availability of treatment, as well as the time necessary to implement treatment. Additional research is warranted to further our understanding of the natural history of insomnia, its risk factors and pathophysiology, and to continue developing cost-effective therapies to manage this challenging sleep disorder.

REFERENCES

1. AMERICAN PSYCHIATRIC ASSOCIATION. - *Diagnostic and Statistical Manual of Mental Disorders* (DSM-IV). Washington, DC: American Psychiatric Association; 1994.
2. AMERICAN SLEEP DISORDERS ASSOCIATION. - International Classification of Sleep Disorders: *Diagnostic and Coding Manual.* (Revised ed.) Rochester, MN: American Sleep Disorders Association; 1997.
3. BACKHAUS J, HOHAGEN F, VODERHOLZER U, RIEMANN D. - Long-term effectiveness of a short-term cognitive-behavioral group treatment for primary insomnia. *Eur Arch Psychiatry Clin Neurosci.,* 251, 35-41, 2001.
4. BASTIEN CH, MORIN CM. - Familial incidence of insomnia. *J Sleep Res.,* 9, 49-54, 2000.

5. BESSET A, VILLEMIN E, TAFTI M, BILLIARD M. - Homeostatic process and sleep spindles in patients with sleep-maintenance insomnia: effect of partial (21h) sleep deprivation. *Electroenceph Clin Neurophysiol*, 107:122-132, 1998.

6. BONNET MH, ARAND DL. - 24-Hour metabolic rate in insomniacs and matched normal sleepers. *Sleep*, 18, 581-588, 1995.

7. BOOTZIN RR, EPSTEIN D, WOOD JM. - Stimulus control instructions. In: Hauri P. (ed.) *Case Studies in Insomnia*. New York (NY): Plenum Press; 19–28, 1991.

8. BRESLAU N, ROTH T, ROSENTHAL L, ANDRESKI P. - Sleep disturbance and psychiatric disorders: a longitudinal epidemiological study of young adults. *Biol Psychiatry*, 39, 411-418, 1996.

9. BUYSSE DJ, REYNOLDS CF, III, KUPFER DJ, THORPY MJ, BIXLER E, MANFREDI R, KALES A., VGONTZAS A., STEPANSKI E., ROTH T., HAURI P., MESIANO D. - Clinical diagnoses in 216 insomnia patients using the international classification of sleep disorders (ICSD), DSM-IV and ICD-10 categories: a report from the APA/NIMH DSM-IV field trial. *Sleep*, 17, 630-637, 1994.

10. CURRIE SR, WILSON KG, PONTEFRACT AJ, DELAPLANTE L. - Cognitive-behavioral treatment of insomnia secondary to chronic pain. *J Consul Clin Psychol.*, 68, 407-416, 2000.

11. EDINGER JD, STOUT AL, HOELSCHER TJ. - Cluster analysis of insomniacs' MMPI profiles: Relation of subtypes to sleep history and treatment outcome. *Psychosom Med.*, 50, 77-87, 1988.

12. EDINGER JD, WOHLGEMUTH WK, RADTKE RA, MARSH GR. - Dose response effects of behavioral insomnia therapy. *Sleep*, 23, 310, 2000.

13. EDINGER JD, WOHLGEMUTH WK, RADTKE RA, MARSH GR, QUILLIAN E. - Cognitive behavioral therapy for treatment of chronic primary insomnia: a randomized controlled trial. *JAMA*, 285, 1856-1864, 2001.

14. ESPIE CA, INGLIS SJ, TESSIER S, HARVEY L. - The clinical effectiveness of cognitive behaviour therapy for chronic insomnia: implementation and evaluation of a sleep clinic in general medical practice. *Behav Res Ther.*, 39, 45-60, 2001.

15. FORD DE, KAMEROW DB. - Epidemiologic study of sleep disturbances and psychiatric disorders: an opportunity for prevention? *JAMA*, 262, 1479-1484, 1989.

16. HAURI PJ. - Sleep hygiene, relaxation therapy, and cognitive interventions. In: Hauri PJ, (ed). *Case Studies in Insomnia*. New York (NY): Plenum Press; 65–84, 1991.

17. HAURI PJ. - Insomnia: can we mix behavioral therapy with hypnotics when treating insomniacs? *Sleep*, 20, 111-1118, 1997.

18. HAURI PJ, FISHER J. - Persistent psychophysiologic (learned) insomnia. *Sleep*, 9, 38-53, 1986.

19. HEALY ES, KALES A, MONROE LJ, CHAMBERLIN K, SOLDATOS CR. - Onset of insomnia: Role of life-stress events. *Psychosom Med.*, 43, 439-451, 1981.

20. HOHAGEN F, RINK K, KAPPLER C, SCHRAMM E, RIEMANN D, WEYERER S, BERGER M. - Prevalence and treatment of insomnia in general practice: a longitudinal study. *Eur Arch Psychiatry Clin Neurosci.*, 242, 329-336, 1993.

21. KALES A, KALES J D. - *Evaluation and treatment of insomnia*. New York, NY: Oxford University Press, 1984.

22. LAMARCHE CH, OGILVIE RD. - Electrophysiological changes during the sleep onset period of psychophysiological insomniacs, psychiatric insomniacs, and normal sleepers. *Sleep*, 20, 724-733, 1997.

23. LICHSTEIN KL, MORIN CM. (eds) - *Treatment of Late-Life Insomia*. Newberry Park (CA) Sage Publications; 2000.

24. LICHSTEIN KL, WILSON NM, JOHNSON CT. - Psychological treatment of secondary insomnia. *Psychol Aging*, 15, 232-240, 2000.

25. MCCLUSKY HY, MILBY JB, SWITZER PK, WILLIAMS V, WOOTEN V. - Efficacy of behavioral versus triazolam treatment in persistent sleep-onset insomnia. *Am J Psychiatry*, 148, 121-126, 1991.

26. MELLINGER GD, BALTER MB, UHLENHUTH EH. - Insomnia and its treatment: prevalence and correlates. *Arch Gen Psychiatry*, 42, 114-232, 1985.

27. MERICA H, BLOIS R, GAILLARD JM. - Spectral characteristics of sleep EEG in chronic insomnia. *Eur J Neurosci.*, 10, 1826-1834, 1998.

28. MILBY JB, WILLIAMS V, HALL JN, KHUDER S, MCGILL T, WOOTEN V. -Effectiveness of combined triazolam-behavioral therapy for primary insomnia *Am J Psychiatry*, 150, 1259-1260, 1993.

29. MIMEAULT V, MORIN CM. - Self-help treatment for insomnia: bibliotherapy with and without professional guidance. *J Consult Clin Psychol.*, 67, 511-519, 1999.

30. MORIN CM. - *Insomnia: Psychological assessment and management*. New York, NY: Guilford Press; 1993.

31. MORIN CM, BASTIEN C, RADOUCO-THOMAS M, GUAY B, LEBLANC J, BLAIS FC, GAGNÉ A. - Late-life insomnia and chronic use of benzodiazepines: medication tapering with and without behavioral interventions. *Sleep*, 21, Suppl. 99, 1998.

32. MORIN CM, COLECCHI C, STONE J, SOOD R, BRINK D. - Behavioral and pharmacological therapies for late-life insomnia: a randomized clinical trial. *JAMA*, 281, 991-999, 1999.

33. MORIN CM, CULBERT JP, SCHWARTZ SM. - Nonpharmacological interventions for insomnia: a meta-analysis of treatment efficacy. *Am J Psychiatry*, 151, 1172-1180, 1994.

34. MORIN CM, EDINGER JD. - Sleep disorders: evaluation and diagnosis. In: Turner SM, Hersen M, (eds). *Adult Psychopathology and Diagnosis*. 3rd ed. New York (NY): John Wiley and Sons, 483–507, 1993.

35. MORIN CM, HAURI PJ, ESPIE CA, SPIELMAN AJ, BUYSSE DJ, BOOTZIN RR. - Nonpharmacologic treatment of chronic insomnia. *Sleep*, 22, 1134-1156, 1999.

36. MORIN CM, STONE J, MCDONALD K, JONES S. - Psychological management of insomnia: A clinical replication series with 100 patients. *Behav Ther.*, 25, 291-309, 1994.

37. MURTAGH DRR, GREENWOOD KM. - Identifying effective psychological treatments for insomnia: a meta-analysis. *J Consult Clin Psychol.*, 60, 79-89, 1995.
38. NOWELL PD, MAZUMDAR S, BUYSSE DJ, DEW MA, REYNOLDS CF, III, KUPFER DJ. - Benzodiazepines and zolpidem for chronic insomnia: a meta-analysis of treatment efficacy. *JAMA*, 278, 2170-2177, 1997.
39. OHAYON M. - Prevalence of DSM-IV diagnostic crtiteria of insomnia: Distinguishing between insomnia related to mental disorders from sleep disorders. *J Psychiatr Res.*, 31, 333-346, 1997.
40. OHAYON M. - Prevalence, diagnosis, and treatment of chronic insomnia in the general population. In: New developments in the treatment of insomnia: do they really have any impact on the primary health care setting? *Proceedings of the 3rd International WFSRS Congress*, Dresden, Germany, 5-14, 1999.
41. OHAYON M, CAULET M. - Psychotropic medication and insomnia complaints in two epidemiological studies. *Can J Psychiatry*, 41, 457-464, 1996.
42. PERLIS M, ALOIA M, MILLIKAN A, BOEHMLER J, SMITH M, GREENBLATT D, GILES D. - Behavioral treatment of insomnia: a clinical case series study. *J Behav Med.*, 23, 149-161, 2000.
43. PERLIS ML, SMITH MT, ANDREWS PJ, ORFF H, GILES DE. - Beta/Gamma EEG activity in patients with primary and secondary insomnia and good sleeper controls. *Sleep*, 24, 110-117, 2001.
44. POLLAK CP, PERLICK P, LINSNER JP, WENSTON J, HSIEH F. - Sleep problems in the community elderly as predictors of death and nursing home placement. *J Commun Health*, 15, 123-135, 1990.
45. REITE M, BUYSSE D, REYNOLDS C, MENDELSON W. - The use of polysomnography in the evaluation of insomnia. *Sleep*, 18, 58-70, 1995.
46. REYNOLDS CF, TASKA, LS, SEWITCH DE, RESTIFO K, COBLE PA, KUPFER DJ. - Persistent psychophysiologic insomnia: Preliminary research diagnostic criteria and EEG sleep data. *Am J Psychiatry*, 141, 804-805, 1984.
47. RIEDEL BW. - Sleep hygiene. In: Lichstein KL and Morin CM. (eds) *Treatment of Late-Life Insomnia.* Thousand Oaks (CA): Sage Publications; 125–46, 2000.
48. RIEDEL B, LICHSTEIN KL, PETERSON BA, EPPERSON MT, MEANS MK, AGUILLARD RN. - A comparison of the efficacy of stimulus control for medicated and nonmedicated insomniacs. *Behav Modif.*, 22, 3-28, 1998.
49. ROSEN RC, LEWIN DS, GOLDBERG RL, WOOLFOLK RL. - Psychophysiological insomnia: combined effects of pharmacotherapy and relaxation-based treatments. *Sleep Med.*, 1, 279-288, 2000.
50. SALIN-PASCUAL RJ, ROEHRS TA, MERLOTTI LA, ZORICK F, ROTH T. - Long-term study of the sleep of insomnia patients with sleep state misperception and other insomnia patients. *Sleep*, 15, 252-256, 1992.
51. SATEIA MJ, DOGHRAMJI K, HAURI PJ, MORIN CM. - Evaluation of chronic insomnia. An American Academy of Sleep Medicine review. *Sleep*, 23, 243-308, 2000.
52. SIMON G, VONKORFF M. - Prevalence, burden, and treatment of insomnia in primary care. *Am J Psychiatry*, 154, 1417-1423, 1997.
53. SPIELMAN AJ, GLOVINSKY PB. - The varied nature of insomnia. In: Hauri P, (ed). *Case Studies in Insomnia.* New York (NY): Plenum Press; 1-15, 1991.
54. SPIELMAN AJ, SASKIN P, THORPY MJ. - Treatment of chronic insomnia by restriction of time in bed. *Sleep*, 10, 45-56, 1987.
55. STEPANSKI E, ZORICH F, ROEHRS TA, YOUNG D, ROTH T. - Daytime alertness in patients with chronic insomnia compared with asymptomatic control subjects. *Sleep*, 11, 54-60, 1988.
56. VERBEEK I, SCHREUDER K, DECLERCK G. - Evaluation of short-term nonpharmacological treatment of insomnia in a clinical setting. *J Psychosom Res.*, 47, 369-383, 1999.
57. VIGNOLA A, LAMOUREUX C, BASTIEN CH, MORIN CM. - Effects of chronic insomnia and use of benzodiazepines on daytime performance in older adults. *J Gerontol B Psychol Sci Soc Sci.*, 55, P54-P62, 2000.
58. VOLLRATH M., WICKI W., ANGST J. - The Zurich study. VIII: Insomnia: association with depression, anxiety, somatic syndromes, and course of insomnia. *Eur Arch Psychiatry Neurol Sci.*, 239, 113-124, 1989.

Chapter 18

Insomnia associated with medical disorders

M. Billiard
Service de Neurologie B, Hôpital Gui de Chauliac, Montpellier, France

Insomnia is frequently associated with different medical disorders. This may be due to a variety of reasons: repercussions on sleep of the symptoms pertaining to these illnesses, e.g. pain, dyspnoea, nycturia, reduced motility, pruritus; anxiety or even depressive moods induced by the handicap, a failure to understand the course of illness, sometimes the risk to life itself; the adverse effects of certain medications used. This chapter will deal with the sleep disorders encountered in heart, respiratory, gastro-intestinal, renal, rheumatic, endocrine and metabolic conditions. For insomnias related to neurological diseases, the reader should refer to chapter 57 "Sleep and lesions in the CNS" and for insomnia within a framework of fibromyalgia, to chapter 60 "Fibromyalgia and chronic fatigue syndrome: the role of sleep disturbances". As a general rule, the treatment of sleep disturbances associated with medical disorders comes under the treatment of the diseases themselves. Nevertheless it is important to take account of the associated psychological factors, such as anxiety brought on by the illness, stress provoked by a stay in hospital, the possible pain involved in explorations, a scheduled surgical operation, which quite naturally lead to proposing anxiolytic or hypnotic treatment. But there may be drawbacks to prescribing such treatment: it may have a detrimental effect on the immediate course of illness, as in the effect of benzodiazepines on respiratory failure, for example; a risk of tolerance or dependency due to the chronic nature of the underlying illness, not to mention a lack of understanding of the hypnotic effects or non-effects on the sleep of subjects affected by various types of illness.

CARDIOVASCULAR DISORDERS

Congestive heart failure

Heart failure may produce systemic and/or pulmonary venous congestion, presenting signs and symptoms of predominantly right – or left – sided failure, respectively. The condition may develop gradually or may present suddenly as with acute pulmonary oedema. Left-sided failure is exhibited by tachycardia, fatigue, exertional dyspnoea and finally orthopnoea and a cough, producing insomnia. Pulmonary oedema is a sign of acute left-sided failure secondary to pulmonary vein hypertension. Life prognosis is at stake. The subject suffers extreme dyspnoea, cyanosis, tachycardia, major agitation and anxiety which are at variance with sleep. Isolated right-sided failure is rare. It is characterised by systemic venous hypertension exhibited as jugular distention, hepatic stasis, ankle oedema and at an advanced stage, ascites. Cheyne-Stokes respiration is characterised by periods of central apnoeas or hypopnoeas alternating with periods of hyperpnoea in a gradual waxing and waning fashion. It occurs in subjects with congestive heart failure. It is present in sleep and in the most severe cases, may be observed during wakefulness. It is responsible for arousal reactions at peak hyperpnoea periods, sleep fragmentation and excessive daytime sleepiness. In all these cases, the excess of liquid entrains an oedema in the airways, restricting respiratory light and precipitating or exacerbating any obstructive apnoea syndrome; this in turn increases the work of the left ventricle thereby exacerbating the heart condition and sleep disturbance. Finally, the anxiety which is common among patients suffering from heart failure often has a distinctly disturbing effect on sleep. Treatment of congestive heart failure often combines rest, oxygen administration, inotropic and diuretic medication, and sodium restriction. Improvement of the heart function will improve sleep. Oxygen administration is indicated in the presence of Cheyne-Stokes respiration [13]. Nasal CPAP has also been shown to reduce Cheyne-Stokes respiration [2, 10]. In patients suffering from obstructive sleep apnoea syndrome, nasal CPAP may improve left ventricle functioning [31]. Administering a benzodiazepine reduces anxiety but should be prohibited if obstructive sleep apnoea is suspected.

Ischaemic heart disease

Angina pectoris is characterised by pain located below the sternum. This may be vague pain causing very little discomfort or may quickly intensify to become oppressive. It is triggered by physical activity and usually subsides after a few minutes rest. It may occur at night (nocturnal angina). It may be caused by obstructive sleep apnoea syndrome. Prinzmetal's angina, also called variant angina, typically occurs at rest, rather than with effort and is often worse at night. Vaso-spasm is thought to contribute to its pathogenesis.

Myocardial infarction pain is like that of angina pectoris, except that it is usually more severe and is not relieved by nitrate derivatives. Sleep is of very poor quality [4].

Treatment for these disorders relies on short-acting or sustained-release nitrate derivatives, beta-blockers, calcium ion influx inhibitors, and if necessary, on angioplasty or coronary by-pass. It is justifiable to prescribe drugs to relieve anxiety in cases of intense anxiety, but this should be resorted to as a temporary measure. Nasal CPAP is indicated if there is evidence of sleep apnoea syndrome.

Cardiac arrhythmias

Generally speaking cardiac arrhythmias are not themselves a cause of insomnia. It is common, on one hand, for subjects to be unaware of the fact they are suffering from cardiac arrhythmias. And on the other, it is well known today that atrial arrhythmias such as atrial flutter, atrial fibrillation, paroxysmal atrial tachycardia, first or second degree atrioventricular heart blocks [23, 24] are seen in the normal subject during REM sleep. Conversely, cardiac arrhythmias giving rise to a functional symptomatology may be associated with sleep disorders.

CHRONIC OBSTRUCTIVE LUNG DISEASES

Bronchial asthma

Bronchial asthma refers to a reversible obstructive lung disease characterised by increased responsiveness of the airways. The characteristic clinical triad of asthma attacks is paroxysmal wheezing, coughing and shortness of breath. Asthma attacks may occur at any time of the day or night. However nocturnal exacerbation of symptoms during sleep is a frequent finding in asthma patients and there is a high incidence of respiratory arrests and sudden death in adult asthmatics at night [6, 14]. On physical examination during the attack, the patient exhibits varying degrees of respiratory distress. Tachypnea, tachycardia and audible wheezes are frequently present. Chest examination shows a prolonged expiratory phase with relatively high-pitched wheezes through inspiration and most of expiration. Asthma patients are remarkable for their high incidence of sleep disturbances including difficulty in maintaining sleep, early morning awakening and excessive daytime sleepiness [17]. Polysomnographic studies may reveal disruption of sleep continuity and sleep architecture, including frequent awakenings, reduced stage 4 sleep and early morning awakening [19, 22]. Sleep-related apnoeas may be more frequent in asthmatic subjects [15]. Different mechanisms have been suggested such as ventilatory impairment due to supine posture [18], fluctuating upper airway muscle tone in REM sleep [30], gastroesophageal reflux [9], the use of theophylline as a likely cause of insomnia [16, 26], an increase in gastroesophageal reflux episodes [1, 28], prolonged administration of corticosteroids [11], and an increase in the cellular inflammatory response in the broncho-pulmonary tree at night [20].

Bronchodilator therapy and good pulmonary hygiene to improve clearance of secretions are the first steps for most patients. Corticosteroids, often used during exacerbation of asthma, may improve sleep continuity during the episodes.

Chronic obstructive pulmonary disease

The term chronic obstructive pulmonary disease (COPD) refers to generalised airway obstruction, particularly of small airways, associated with varying combinations of chronic bronchitis, asthma and emphysema. Airway obstruction is defined as an increased resistance to airflow during forced expiration. Chronic bronchitis is a condition associated with prolonged exposure to non specific bronchial irritants and accompanied by mucous hypersecretion and certain structural changes in the bronchi. Pulmonary emphysema is enlargement of the air spaces distal to terminal non respiratory bronchioles, accompanied by destructive changes of the alveolar walls. Persistent asthma is defined as asthma which is so persistent that clinically significant chronic airflow obstruction is present most of the time despite antiasthmatic therapy.

Progressive exertional dyspnoea is the most common presenting complaint. Cough, wheezing, recurrent respiratory infections are also present. Physical examination reveals inspiratory and expiratory ronchi and crepitations. Patients with resting hypoxemia and hypercapnia may exhibit cyanosis. Disturbances in sleep include delayed sleep onset, increased wake after sleep onset, frequent stage shifts and frequent arousals [3, 7]. These disturbances may be caused by a variety of factors *i.e.* the use of drugs that have a sleep-reducing effect, increased nocturnal cough, impaired gas exchange leading to hypoxemia and hypercapnia [35]. Treatment is directed at alleviating the conditions that cause symptoms and excessive disability (eg. infection, bronchospasm, bronchial hypersecretion). Avoidance of bronchial irritants is of primary importance. Oxygen administration is indicated in the case of right-sided heart failure. When obstructive sleep apnea syndrome accompanies chronic obstructive pulmonary disease, nasal CPAP is the best treatment. Caution is advised in using benzodiazepines for patients with chronic obstructive pulmonary disease due to the reduction in upper airway muscle tone and blunting of the arousal response to hypercapnia [8]. However, controlled trials with short-acting benzodiazepines suggest that these agents may be safely used in selected patients with mild to moderate chronic obstructive pulmonary disease without daytime hypercapnia.

Idiopathic infiltrative diseases of the lungs

These refer to a spectrum of disorders with different aetiologies, idiopathic pulmonary fibrosis, desquamative interstitial pneumonia, lymphoid interstitial pneumonia and so forth, but similar clinical features and diffuse pathological changes that affect primarily intraalveolar interstitial tissue. Features of interstitial lung diseases include exertional dyspnea of insidious onset, dry cough, anorexia, weight loss, fatigue, weakness and vague chest pain. Tachypnea and laboured breathing are observed and chest examination reveals prominent breath sounds and end-inspiratory crackles at lung bases. Sleep abnormalities consist of repeated awakenings with sleep fragmentation, increased stage 1 and reduced REM sleep [5, 25].

Corticosteroids are indicated in patients without evidence of extensive fibrosis. Supplemental oxygen therapy may be of interest.

GASTROINTESTINAL DISEASES

Gastroesophageal reflux disease

It is characterised by reflux of gastric contents. Heartburn with and without regurgitation of gastric contents into the mouth is the most prominent symptom. Complications of gastroesophageal reflux include oesophagitis, peptic oesophageal striction and oesophageal ulcer.

The burning pain causes difficulty in initiating sleep, frequent awakenings with a bitter taste in the mouth and hoarseness or laryngitis in the morning. Management consists of dietary changes, elevation of the head of the bed, giving an antiacid, use of the H2 antagonists to reduce gastric acidity. Hypnotics should be avoided: the reduction in arousals may lead to less swallowing and slower clearance of acid from the oesophagus.

Peptic ulcer disease

Peptic ulcer is a circumscribed ulceration of the mucous membrane penetrating through the muscularis mucosa and occurring in areas exposed to acid and pepsin. Only about 1/2 of patients present with the characteristic pattern of symptoms. Typical pain is described as burning, gnawing or aching, but distress may also be described as soreness, an empty feeling, or hunger. Pain that wakens the patient at 1 or 2 a.m. is common and highly suggestive of duodenal ulcer.

Treatment is designed to neutralise or decrease gastric acidity. Antacids give symptomatic relief. Histamine H2 receptor blocking agents are generally preferred, specially in the case of nocturnal awakenings caused by pain.

RENAL DISORDERS

Chronic renal failure

Chronic renal failure results from a multitude of pathologic processes that lead to derangement and insufficiency of renal excretory and regulatory function.

Patients affected by mild to moderate renal failure may have only limited symptoms despite high levels of blood urea and creatinine. Nycturia is common at this stage relating to the incapacity of the kidney to concentrate urine during the night. Fatigue, dulled mental activity are often the first signs of uremia. The neuromuscular signs are the next to appear, with muscular jerks, cramps, peripheral neuropathies; gastro-intestinal signs, anorexia nausea. At a more advanced stage other signs appear, skin discoloration, pruritus, hypertension, congestive heart failure. This course was consistently fatal in the past. Nowadays, prolonged survival is ensured thanks to haemodialysis, peritoneal dialysis or kidney transplantation. But the majority of subjects under dialysis report sleep disorders [33]. Between 70 and 90% of patients complain of frequent nocturnal awakenings and roughly 50% of difficulties of sleep onset and excessive daytime sleepiness [34]. Polysomnographic studies have further defined these disorders as a reduction in slow wave sleep, with frequent awakenings [21]. Restless legs and periodic limb movements are almost a standard feature. Sleep apnoea syndrome is common [12, 32, 33].

The treatment of restless legs and periodic limb movements relies on levodopa or dopaminergic agonists and, in the most severe cases, on opiates.

RHEUMATIC DISORDERS

Rheumatoid arthritis and the other forms of inflammatory rheumatism (inflammatory rhizomelic rheumatism in the elderly, chronic juvenile arthritis, psoriasic rheumatism, Gougerot-Sjögren's syndrome, intestinal rheumatism, vascular arthritis, etc.). The common feature of these diseases is an inflammation of the joints causing pain, stiffness and reduced mobility. Sleep disorders include delayed sleep onset but primarily recurrent awakenings and non restorative sleep. Periodic limb movements are more frequent in subjects suffering from rheumatoid arthritis than in controls, adding to the poor quality of sleep.

Treatment for pain and nocturnal stiffness essentially relies on nonsteroidal anti-inflammatory drugs. Hypnotics must be used with caution in view of the chronic nature of these diseases.

ENDOCRINE – METABOLIC DISORDERS

Diabetes may be accompanied by insomnia related to polyuria, peripheral or dysautonomic neuropathy. Diabetics who suffer from sleep disorders are generally those for whom the onset of the disease occurred at an early age [27]. Hyperthyroid subjects including those under substitution therapy may be affected by insomnia. Sleep apnoea syndrome is classic in subjects with hypothyroidism, acromegaly, and diabetic neuropathy with dysautonomia.

CONCLUSION

Insomnia associated with medical disorders is definitely of frequent occurrence, especially in the elderly age group. However, the association between sleep disturbance and medical disorder is probably not straightforward. No epidemiological study has been done on the occurrence of sleep disturbances in any medical disorder. Indeed it might be that among subjects with medical disorders of the same order of severity, some present severe insomnia and others with only a mild one due to various degrees of vulnerability to insomnia. Moreover treatment of insomnia associated with medical disorders is not at all codified as no study has been done in patients with either cardiovascular disorders, or chronic obstructive lung diseases, or gastrointestinal disorders and so forth, of the respective impact of etiological and/or hypnotic treatment(s) on the course of insomnia.

REFERENCES

1. BERQUIST W.E., RACHELEFSKY G.S., KADDEN M., SIEGEL S.C., KATZ R.M., MICKEY M.R., AMENT M.E. – Effect of theophylline on gastroesophageal reflux in normal adults. *J. Allergy Clin. Immunol.* 67, 407-411, 1981.
2. BRADLEY T.D. – Hemodynamic and sympathoinhibitory effects of nasal CPAP in congestive heart failure. *Sleep,* 19, S232-235, 1996
3. BREZINOVA A., CATTERALL J.R., DOUGLAS N.J., CALVERLY P.M.A., FLENLEY D.C. – Night sleep of patients with chronic ventilatory failure and age matched controls: number and duration of the EEG episodes of intervening wakefulness and drowsiness. *Sleep,* 5, 123-130, 1982.
4. BROUGHTON R., BARON R. – Sleep of acute coronary patients in an open ward type intensive care unit. *Sleep Res.* 2, 144, 1973.
5. BYE P.T., ISSA F., BERTHON-JONES M., SULLIVAN C.E. – Studies of oxygenation during sleep in patients with interstitial lung disease. *Am. Rev. Respir. Dis.,* 129, 27- 32, 1984.
6. COCHRANE G.M., CLARK T.J.H. – A survey of asthma mortality in patients between ages 35 and 65 in the greater London hospitals in 1971. *Thorax,* 30, 300-306, 1975
7. FLECHTER E.C., MARTIN R.J., MONLUX R.D. – Disturbed EEG sleep patterns in chronic obstructive pulmonary disease. *Sleep Res.* 11, 186, 1982.
8. GEORGE C.F. – Perspectives on the management of insomnia in patients with chronic respiratory disorders. *Sleep,* 23, Suppl 1, S31-35, 2000.
9. GOODALL R.J.R., EARIS J.E., COOPER D.N., BERNSTEIN A., TEMPLE J.G. – Relationship between asthma and gatroesophageal reflux. *Thorax,* 36, 116-121, 1981.
10. GRANTON G.T., NAUGHTON M.T., BENARD D.C., LIU P.P., GOLDSTEIN R.S., BRADLEY T.D. – CPAP improves respiratory muscle strength in patients with heart failure and central sleep apnea. *Am. J. Respir. Crit. Care Med.* 153, 277-282, 1996.
11. GUILLEMINAULT C., SILVESTRI R. – Aging, drugs and sleep. *Neurobiol. Aging,* 3, 379-386, 1982.
12. HALLETT M., BURDEN S., STEWART D., MAHONY J., FARREL P.C. - Sleep apnea in end-stage renal disease patients on hemodialysis and continuous ambulatory peritoneal dialysis. *ASAIO J.* 41, M435-M441, 1995.
13. HANLY P.J., MILLAR T.W., STELJES D.G., BAERT R., FRAIS M.A., KRYGER M.H. – The effect of oxygen on respiration and sleep in patients with congestive heart failure. *Ann. Intern. Med.* 111, 777- 782, 1989.
14. HETZEL M.R., CLARK T.J.H., BRANTHWAITE M.A. – Asthma: analysis of sudden deaths and ventilatory arrests in hospital. *BMJ,* 1, 808-811, 1977.
15. JANSON C., DE BACKER W., GISLASON T., PLASCHKE P., BJORNSSON E., HETTA J., KRISTBJARNARSON H., VERMEIRE P., BOMAN G. – Increased prevalence of sleep disturbances and daytime sleepiness in subjects with bronchial asthma: a population study of young adults in three European countries. *Eur. Respir. J.,* 9, 2132- 2138, 1996.
16. JANSON C., GISLASON T., ALMQVIST M., BOMAN G. – Theophylline disturbs sleep mainly in caffeine-sensitive persons. *Pulm. Pharmacol.,* 2, 125-129, 1989.
17. JANSON C., GISLASON T., BOMAN G., HETTA J., ROOS B.E. – Sleep disturbances in patients with asthma. *Respir. Med.,* 84, 37- 42, 1990.
18. JONSSON E, MOSSBERG B. – Impairment of ventilatory function by supine posture in asthma. *Eur. J. Respir. Dis.* 65, 496-503, 1984.
19. KALES A., BEALL G.N., BAJOR G.F., JACOBSON A., KALES J.D. – Sleep studies in asthmatic adults: relationship of attacks to sleep stage and time of night. *J. Allergy,* 41, 164-173, 1968.
20. MARTIN R.J., CICUTTO L.C., SMITH H.R., BALLARD R.D., SZEFLER S.J. – Airway inflammation in nocturnal asthma. *Am. Rev. Respir. Dis.* 143, 351-357, 1991.
21. MENDELSON W.B., WADHWA N.K., GREENBERG H.E., GUJAVARTY K., BERGOFSKY E. – Effects of hemodialysis on sleep apnea syndrome in end stage renal disease. *Clin. Nephrol.* 33, 247- 251, 1990.
22. MONTPLAISIR J., WALSH J., MALO J.L. – Nocturnal asthma: features of attacks, sleep and breathing patterns. *Am. Rev. Respir. Dis.* 125, 18-20, 1982.
23. NEVINS D.B. - First and second–degree A-V heart block with rapid eye movement sleep. *Ann. Intern. Med.* 76, 981-983, 1972.

24. OTSUKA K., ICHIMARU Y., YANAGA T., SATO Y. – Studies of arrythmias by 24-hour polygraphic recordings: relationship between atrioventricular block and sleep states. *Am. Heart J.,* 105, 934-940, 1983.

25. PEREZ-PADILLA R., WEST P., LERTZMAN M., KRYGER M.H. - Breathing during sleep in patients with interstitial lung disease. *Am. Rev. Respir. Dis.,* 132, 224-229, 1985.

26. RHIND G.B., CONNAUGHTON J.J., McFIE J., DOUGLAS N.J., FLENLEY D.C. – Sustained release choline theophyllinate in nocturnal asthma. *BMJ,* 291, 1605-1607, 1985.

27. SRIDHAR G.R., MADHU K. – Prevalence of sleep disturbances in diabetes mellitus. *Diabetes Res. Clin. Pract.* 23, 183-186, 1994.

28. STEIN M.R., TOWNER T.G., WEBER R.W., MANSFIELD L.E., JACOBSON K.W., McDONNEL J.T., NELSON H.S. – The effect of thephylline on the lower esophageal sphincter pressure. *Ann. Allergy,* 45, 238-241, 1980.

29. STEPANSKI E., FABER M., ZORICK F., BASNER R., ROTH T. – Sleep disorders in patients on continuous ambulatory peritoneal dialysis. *J. Am. Soc. Nephrol.,* 6, 192-197, 1995.

30. SULLIVAN C.E., ZAMEL N., KOZAR L.F., MURPHY E., PHILLIPSON E.A.– Regulation of airway smooth muscle tone in sleeping dogs. *Am. Rev. Respir. Dis.* 119, 87-99, 1979.

31. TAKAHASHI Y., ORR D., POPKIN J., RUTHERFORD R., LIU P., BRADLEY T.D. – Effect of nasal continuous positive airway pressure on sleep apnea in congestive heart failure. *Am. Rev. Respir. Dis.,* 140, 1578-1584, 1989.

32. WADHWA N.K., MENDELSON W.B. – A comparison of sleep-disordered respiration in ESRD patients receiving hemodialysis and peritoneal dialysis. *Adv. Perit. Dial.* 8, 195-198, 1992.

33. WADHWA N.K., SELIGER M., GREENBERG H.E., BERGOFSKY E., MENDELSON W.B. – Sleep-related respiratory disorders in end-stage renal disease patients on peritoneal dialysis. *Perit. Dial. Int.* 12, 51- 56, 1992.

34. WALKER S., FINE A., KRYGER M.H. – Sleep complaints are common in dialysis unit. *Am. J. Kidney Dis.,* 26, 751-756, 1995.

35. WEITZENBLUM E., CHAOUAT A., CHARPENTIER C., EHRHART M., KESSLER R., SCHINKEWITCH P. – Sleep-related hypoxemia in chronic obstructive pulmonary disease: causes, consequences and treatment. *Respiration,* 64, 187- 193, 1997.

Chapter 19

Insomnias associated with psychiatric disorders

L. Garma
Fédération des Pathologies du Sommeil, Hôpital Pitié-Salpêtrière, Paris, France

Sleep disorders and psychiatric disorders often occur simultaneously. About 35 to 47% of patients mainly suffering from insomnia have an identifiable mental disorder, generally of the nature of anxiety or depression [20, 33, 78]. By comparison, 50 to 80% of patients affected by psychiatric disorders present sleep disorders [62, 66, 91]. A psychiatric evaluation is thus often justifiable in subjects complaining of insomnia.

The associations and interactions between sleep disturbances and psychiatric disorders are complex. Specific alterations to sleep may be caused by the same neurobiological mechanisms as those responsible for psychiatric symptoms. But interaction may also occur between sleep dysfunction and that of the psychiatric pathology.

Table 19.1 Distribution of diagnostic categories of insomnia in 795 patients examined in a sleep disorder centre.

Main diagnosis	Number of patients	%
Primary insomnia	190	24
Insomnia associated with psychiatric disorders		
- mood disorders	150	19
- anxiety disorders		
- generalised anxiety disorder	102	13
- obsessive-compulsive disorder, phobia	32	4
- panic disorder, post-traumatic stress disorder	26	3
- schizophrenia	40	5
- personality disorders	40	5
- other	7	1
Other categories of insomnia		
	208	26

Of the 795 patients classified as chronic insomniacs and examined in our sleep disorder unit [36], 50% presented psychiatric disorders (table 19.1), a frequency compatible with other studies [21, 62, 76].

The category of disorders "associated with mental disorders" was introduced with the publication of the International Classification of Sleep Disorders (ICSD) [47] in 1990. This diagnostic category was later incorporated in the fourth version of the Diagnostic and Statistical Manual of Mental Disorders (DSM-IV) [2] under the category "insomnia associated with a mental disorder". This category groups together subjects who simultaneously present a complete clinical picture of insomnia and a picture of another mental disorder, and for whom insomnia is the dominant complaint. The main symptom, insomnia, is considered to be temporally and aetiologically attributable to a mental disorder and to be severe enough in itself to warrant clinical examination. The accompanying symptoms entirely fulfil the criteria for a specific mental disorder, such as a depressive disorder.

The present chapter will deal successively with insomnia associated with mood disorders, anxiety disorders, schizophrenia, eating disorders and personality disorders.

MOOD DISORDERS

Major depressive disorder

Up to 90% of depressed patients sleep badly and complain of poor sleep. Insomnia is the major complaint. However, hypersomnia replaces insomnia in 10 to 15% of cases [18, 86].

Prevalence

In a study of 7,954 subjects aged 18 and over, Ford and Kamerow [33], showed a major state of depression in 14% of subjects complaining of insomnia and only 1% in non insomniac subjects (dysthymic disorder: 9% versus 2%). Of the 216 patients referred to sleep disorders centres for a complaint of insomnia, 32% received a diagnosis of insomnia (DSM-IV) associated with mood disorder [21]. In a sample of 66 insomniacs, 24 were included in the category of affective disorders according to the DSM-III-R. An epidemiological survey conducted in France of 5,622 people aged 15 and over, representative of the general public, interviewed by telephone, found that for 11.3% of subjects complaining of insomnia, the symptoms were associated with mental disorder, a diagnosis of major depression having been made in 3.3% of these cases [65].

It should be noted that the study by Weyerer and Dilling [91], conducted by psychiatrists on 1,536 subjects aged 15 and over, shows a parallel rise in the frequency of insomnia and depression up to the age of 50-59. After this, the incidence of insomnia continues to rise whereas that of depression (and anxiety disorders) diminishes.

Clinical aspects

Insomnia, a key symptom of major depression, combines - to a varying extent - longer sleep latency, sleep interrupted by frequent nocturnal awakenings and anxious morning awakening (at least 2 hours before the usual time of morning awakening, often 2 to 4 hours after going to sleep). In other words, the sleep of a depressed patient is superficial, fragmented, short and non restorative, leading to a particularly painful feeling of dissatisfaction on morning awakening, with a sensation of tiredness which is often greater than on retiring. It is important to note in passing the well known risk of suicide in the early hours of the morning. While sleepless nights are fairly characteristic of severe depression, insomnia may occur at any time of night, with no particular predominance in the early morning. It is also worth noting the reduction in dream activity which may accompany depression [17]; the proportion of dreams recalled on morning awakening, or after awakenings provoked in the laboratory during REM sleep, is largely inferior in hospitalised depressives compared to healthy controls (20% versus 75%).

Insomnia is followed by alterations in alertness during the day, with difficulty waking up, considerable fatigue and sleepiness. A 24-hour study [49] showed that major depressives do not sleep any longer during the day than healthy controls. But the naps they do take (regardless of the quality of sleep the previous night) can occur at any time of the day as opposed to between 1 p.m. and 3 p.m. as in the case of normal subjects.

Polysomnographic aspects

Macrostructure

Numerous studies have contributed to defining a global description of the objective alterations to sleep in major depression (ref. In [13-14, 17]). One of the best documented biological indexes (table 19.2) presents a fairly consistent group of polysomnographic anomalies - even if none of these are specific when considered individually, and are variously associated depending on the patient.

Table 19.2 Polygraphic characteristics of sleep in mood disorders

Pathology	Sleep latency	Total sleep time	Sleep efficiency index	Duration of deep NREM sleep	REM sleep latency	Eye movement density in the first episode of REM sleep	REM sleep duration
Major depressive disorder	⇑	⇓	⇓	⇓	⇓	⇑	⇑
Manic episode	NS	⇓⇓	NS	NS	⇓ or NS	⇑ or NS	NS

⇑: increased, ⇓: decreased
NS: non significant differences compared with normal controls.

Depressive insomnia associates sleep continuity disorders, decreased deep NREM sleep and anomalies of REM sleep architecture (fig. 19.1).

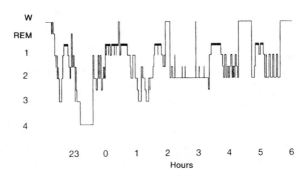

Figure 19.1. Hypnogram of one night's sleep for a 24 year old patient, presenting an episode of major depression. Note the short REM sleep latency. Stage 4 only appears in the second cycle. Nocturnal awakenings are long and morning awakening is premature.

   Sleep continuity alteration is expressed by a reduction in TST, prolonged sleep latency, an increase in the number and duration of nocturnal awakenings (evenly distributed throughout the night), premature morning awakening and a decreased sleep efficiency index (often below 75%).

   Major depressive insomnia is far more severe than primary insomnia, the total duration of the periods of time awake interspersed with periods of sleep being far longer. Considerable variations do however exist in the number of intercurrent awakenings, some patients having far less fragmentary sleep than others while maintaining a substantial level of insomnia.

   The disruption of NREM sleep stages 3 and 4 is characterised by an overall reduction and an abnormal distribution in relation to healthy controls. Stages 3 and 4 are reduced, even absent, in the first cycle, and only reappear in the second or third. Thus the relative quantity of delta waves diminishes between the first and second cycles.

   While the anomalies of REM sleep are not pathognomonic signs of mood disorder, they are nonetheless valuable clues in confirming diagnosis and for differential diagnosis. Insomnia with no depressive characteristics presents no alterations to REM sleep.

   Decreased REM sleep latency, largely corroborated, is the most specific phenomenon of major depressive disorders. With a duration of 100 ± 20 minutes in normal subjects, REM sleep latency in depressed subjects is often 20 to 40 minutes shorter, sometimes lasting only a few minutes. When age and psychological characteristics are monitored, REM sleep latency in major depressives is fairly evenly distributed, varying between 0 and 65-70 minutes, with a peak between 50 and 59 minutes [3]. The premature onset of the first phase of REM sleep is apparent in about two thirds of depressives (this may vary from one night to another in the same patient) and is therefore not systematic.

REM sleep is abnormally distributed throughout the night. In total, in addition to the fact that the first episode of REM sleep begins prematurely, it is also abnormally long and abnormally dense in rapid eye movements [22, 92].

Sleep micro-architecture

Automatic quantification of the spectral components of the EEG signal has confirmed the decreased production of slow wave activity in the first cycle (reduction in the absolute number of delta waves and in their density per minute). Compared with control subjects, depressed patients have more rapid frequencies (particularly in the beta band), less ample delta activity, fewer high amplitude delta waves, as well as greater interhemispheric asymmetry in EEG activity throughout all sleep stages [4-5].

Variation factors

Differences in sleep variables between depressives and non depressives vary according to age, becoming more pronounced in older subjects [52].

Older depressives have a lower sleep efficiency index than healthy subjects of the same age, the differences being more marked than in younger subjects. The data for REM sleep variables is more debateable. Some studies show that older subjects have shorter REM sleep latency than younger subjects, both in the case of normal and depressive subjects [15]. The findings of other authors [40] indicate that age is only significant in depressed patients (fig. 19.2). To enhance the specificity of such measures, Kupfer *et al.* (ref. in [13] suggested only taking decreased REM sleep latency into account where the sum of age and latency was below 90.

For Lauer *et al.* [52] eye movement density in REM sleep does not change with age in normal subjects; as it is high in depressed patients regardless of age, it constitutes a good marker for depression (this will be referred to later). The increase is less marked in younger depressives.

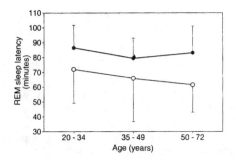

Figure 19.2 Means and standard deviation of REM sleep latency according to age in depressed patients (°) and control subjects (•). In this study the effects of age are only significant in depressed subjects. REM sleep latency is defined as the latency between the onset of stage 2, followed by at least 9 minutes of sleep, and the first episode of REM sleep, including intercurrent awakenings [40].

Male depressives have less deep NREM sleep than female depressives of the same age, the difference being more marked in conditions of automatic delta activity analysis. Female depressives have far more rapid frequencies, in number and amplitude, than male depressives, whereas in control subjects it is the slow frequencies which differ, the number being greater in women than in men. [5, 58].

Degree of severity

Sleep anomalies are evident in the major depressive episodes of 40 to 60% of outpatients and up to 90% of hospitalised patients [2]. Mood disorder severity seems to be the clinical element which is the most closely linked to sleep. The more severe the symptoms [45, 68], the more disturbed the sleep, particularly in the first weeks of illness [31]. A correlation has been found between REM density and the severity of the illness [64]. Within a group of depressives, patients suffering from a more severe form present shorter REM sleep latency, even if there is no linear correlation between the two [14]. Nevertheless some longitudinal studies have shown that REM sleep maintains a stable intra-individual value over time [20], cancelling out any direct link between REM sleep latency and the severity factor.

Relationship between sleep variables and symptoms

In major unipolar depressives, certain signs (insomnia at the end of the night, anhedonia and loss of appetite) are only found in endogenous patients with short REM sleep latency, whereas other signs are present independently of the latency value [38].

Perlis *et al.* [69] conducted a study evaluating the relationship between the clinical symptoms of depression (items on the Hamilton Depression Scale and Beck's inventory) and measured EEG sleep variables. They found the association to be largely independent of the subject's complaint of insomnia and that the variables for deep NREM sleep (number of delta waves in the second cycle) are far more closely correlated to the symptoms of depression than to the length or continuity of sleep.

Progressive aspects

The importance of sleep disorders in depression stems not only from their frequency, but also from their persistence over time. Starting several weeks before the onset of mood disorders, clinical insomnia may be the sign heralding depression in a previously unaffected subject [55]; it is often one of the first symptoms of a relapse and the symptomatic sign of recidivism. It persists in certain subjects after the disappearance of mood disorders.

Rodin *et al.* [72] conducted a study of 196 institutionalised patients with an average age of 72, over a period of three years, and showed a positive relationship between the frequency of depressive moods and the severity of the sleep disorder. Sex, age and state of health were all taken into account. Early morning awakening (the sign the most closely linked to depression in this study) became less frequent as depression lifted.

Clinically speaking, the complaint of insomnia diminishes as the mood improves, but anomalies in polysomnography are not always corrected even after good mood is restored. Decreased levels of deep NREM sleep and shortened REM sleep latency may still persist after complete clinical remission.

For some authors, the persistence of shortened REM sleep latency after clinical recovery, is a predictive factor of relapse, although it lacks specificity. But recent work has shown that a combined analysis of three variables: decreased REM sleep latency, greater density of eye movements in REM sleep and diminished sleep efficiency index, correspond to an abnormal sleep profile characterising episodes of major depression [84]. Subjects whose depression is accompanied by this set of polysomnographic disturbances (compared with depressives whose symptoms are as severe but whose sleep is only slightly altered) responded poorly to psychological therapy (cognitive or inter-relational) and relapsed further [85]. This abnormal hypnic profile may correspond to the expression of a neurobiological substrate in the most severe episodes of endogenous depression, with an increased risk of recidivism. Such anomalies may be signs of severity, identifying patients with major depression requiring anti-depressant treatment.

Other research on the relationship between clinical evolution and sleep alterations, or in other words the characteristics of "state" and "trait", has tended to focus more on deep NREM sleep [20]. A persistent reduction of the level of deep NREM sleep in the first cycle may be a better indication of recidivism than shortened REM sleep latency. The results of combined treatment (pharmacological and psychological) have in fact shown that patients with the lowest level of delta sleep are those who relapse the quickest. Delta sleep may protect against relapse into depression or, more specifically, large quantities of delta activity seem to be necessary to sustain recovery. The components of delta waves have different frequency bands (depending on whether their origin is thalamic or cortical). 2 to 3 Hz waves are more closely linked to the symptomatic depressive episode, whereas slow waves of 0.5 to 1 Hz are those which differ most between patients who do or do not experience a relapse.

To summarise, sleep variables of the "state" type are factors which indicate the presence of a depressive episode, identified and isolated in time, and which are normalised on remission. They are expressed as sleep continuity, REM density or overall delta activity. Variables of the "trait" type, such as a decrease of REM sleep latency and the slowest delta waves (0.5 – 1 Hz), are durable EEG aspects, which persist outside acute attacks, are only slightly modified by psychotherapy, and indicate a constitutional vulnerability to a depressive disorder (or the persistence of a pathological process despite clinical improvement).

Among the follow up studies conducted, mention should be made of the interesting longitudinal study by Cartwright *et al.* [23], carried out on 70 people who developed depression during a divorce. A year later the depressive subjects who had improved clinically maintained short REM sleep latency. The researchers considered short REM sleep latency as a marker associated with positive prognosis. Let us add that depressed subjects reported more bad dreams on arousals provoked in REM sleep than non depressed controls and that patients whose dream content concerned the painful events they were going through, had adjusted better to the changes in their lives one year later.

Familial studies

The sleep anomalies encountered in mood disorders may be markers of genetic vulnerability to this pathology. The presence of decreased REM sleep latency in healthy parents of subjects affected by major depression is abnormally high in relation to that observed in the parents of healthy controls [39]. Research on the sleep of non affected persons (currently or in the past) with parents who suffer from depressive disorders, have shown that one subject in five has less deep NREM sleep and more eye movements in the first cycle of REM sleep (compared with a control group), alterations which may be used to indicate a potential risk of developing an affective disorder at a later date [54].

Manic episode

Insomnia is one of the earliest and most consistent signs of the manic state. A truly symptomatic signal, it is well known and apprehended by the subject's family and friends. In the transition from the depressive to the manic state, the duration of sleep diminishes greatly and mood often shifts abruptly and dramatically. The onset of the manic attack is often accompanied by continuous wakefulness lasting 40 to 50 hours (48 hour sleep-wake cycle). Manic states appear to be triggered in the morning more often than at night. Manic episodes which begin during the night are usually more severe and followed by a virtual absence of sleep the following night. Wehr [89] suggested the hypothesis that insomnia which inaugurates manic shifts may be seen, at least in predisposed subjects, as the final step in the various influences likely to trigger the manic state. Thermoregulatory mechanisms intervene in the clinical (and neuroendocrine) effects of sleep deprivation.

Added to this is the fact that the manic mood shift may be heralded by dreams of death or injury, whereas imminent depression is heralded by the disappearance of dreams [11].

Despite virtually total insomnia, the manic patient finds his sleep adequate and shows no signs of fatigue or daytime sleepiness, but rather an acceleration of all the psychomotor processes. The shorter the night, the more severe the manic symptomatology [32]. Insomnia persists throughout the attack and its recession is one of the best signs of remission.

Polysomnographic studies of these patients (table 19.3) are somewhat rare, being particularly difficult to obtain practically speaking [14]; they illustrate the same types of anomaly as in depressive disorder: prolonged sleep latency, decreased TST and sleep efficiency index. The proportion of NREM sleep stages 3 and 4 does not differ significantly from that found in control subjects. As far as REM sleep is concerned, observations are contradictory, reporting either normal values of latency and percentage of REM sleep, or decreased latency and increased REM density. A recent study carried out in 12 bipolar patients in the manic phase [42] with no treatment, showed a reduction in stages 3 and 4 and a shortening of REM sleep latency, with 3 patients presenting REM sleep latency of less than 20 minutes.

Treatments

Mood disorders and their accompanying sleep disorders require pharmacological treatment, psychological treatment or a combination of the two.

Pharmacological treatments

Patients affected by major depression of average or severe intensity are usually treated by medication. The object of the exercise is to find the right balance between the sedative effects which re-establish sleep continuity, and the stimulant effects which improve depression. Just over half the cases of depression improve regardless of the medication chosen. Pharmacological treatments for mood disorders all have significant effects on sleep, partly because they treat the underlying illness which (usually but not always) tends to normalise sleep parameters, and partly as a result of their direct effects on sleep. Three types of effect are encountered with antidepressants: an immediate anti-awakening effect, a medium-term effect on anxiety and a long-term antidepressant effect.

The selective serotonin reuptake inhibitors (SSRIs), fluoxetine, sertraline, fluvoxamine, citalopram and paroxetine are those most often used today to treat depression. The first three tend to have an awakening effect, while the second two have a more sedative effect. Among the other products used, venlafaxine, a serotonin and norepinephrine inhibitor has an awakening effect. Trazodone and nefazolone on the other hand, which have mixed effects on serotonin receptors, are sedatives. The same is true for mianserin, a tetracyclic antidepressant, and mirtazapine, a norepinephrine and serotonin reuptake inhibitor.

Although the use of tricyclic antidepressants has declined due to their side effects (mainly anticholinergic), they are used in patients who do not respond to other antidepressants. Among the imipraminics, certain substances tend to have a sedative effect (amitriptyline, doxepine, maprotiline, trimipramine), while others tend to have a stimulating effect (imipramine, quinupramine, demexiptiline, desipramine).

The treatment of sleep disorders associated with depression means resorting to antidepressant doses for at least three weeks. A single dose of antidepressant is recommended at bedtime. The disappearance of the sleep complaint is an indicator for the clinician; however it is sometimes difficult to restore sleep to its original level. Poor sleep is a frequent scar of depression. On one hand it is not uncommon to observe the development of insomnia of a psychophysiological mechanism, with a high level of anticipatory anxiety at bedtime, with the patient more or less consciously associating insomnia with the onset of the depressive episode. On the other hand, some antidepressants, if their adverse effects are not taken into consideration, may provoke the onset, persistence and aggravation of insomnia.

Psychological treatments

Psychological therapies have proved their effectiveness in treating insomnia and depression [83]. There are several types of approach: cognitive and behavioural therapy, interpersonal psychotherapy, family therapy. There are several advantages in this type of treatment: the absence of physiological side effects, the possible effectiveness in certain patients who fail to respond to pharmacological treatment, and the prevention of recurrence due to better understanding and/or the avoidance of situations which may provoke a relapse.

Other treatments

Other forms of treatment, such as sleep deprivation or phototherapy may be used. The major indication for light therapy is seasonal affective disorder. However some studies suggest that bright light has a beneficial effect when used as an adjuvant to pharmacological therapies in non-seasonal major depression [12].

An improvement in depression whether or not this is associated with an improvement in sleep has been obtained by various types of sleep manipulation, selective REM sleep deprivation, partial or total sleep deprivation, and possibly, phase advance of the sleep period relative to clock time [41, 57, 96].

Treatment of mania

A number of mood stabilisers are reported to have an acute antimanic effect. However lithium or valproate are the drugs of first choice. Valproate is often preferred for patients with mixed states or rapid cycling disorder.

ANXIETY DISORDERS

Anxiety in its various forms accounts for 45% of cases of insomnia with a psychiatric cause. While anxiety is a symptom in practically all mental disorders, it dominates the clinical picture in the DSM-IV categories of anxiety disorders. In the study carried out on the French population, mentioned earlier, a diagnosis of anxiety disorder was made for 7.7% of subjects complaining of insomnia [65]. It is worth noting, moreover, that insomnia induces anxiety and that 50% of severe insomniacs have high anxiety scores [87].

Anxiety disorders are often but not always associated with disturbances of sleep onset and sleep maintenance.

Anxiety not only causes poor sleep, it also frequently causes an alteration in sleep perception (fear of sleeping or in other words the phenomenological experience of sleep and the opposition between sleep and wakefulness), further aggravating the impression of poor quality sleep. On rising the subject is tense, asthenic, polyalgic, judging sleep to be inadequate and non restorative.

The semiology of the dyssomnias of the anxious subject is seen in laboratory recordings [87]. The differences between the sleep of patients presenting a pathology of anxiety and that of control subjects are minimal and non specific. Globally, insomnia linked to anxiety disorders is far less severe and quantitatively lower than that of mood disorders (table 19.3). It is expressed as sleep continuity disturbances with prolonged sleep latency, short sleep, multiple awakenings, with none of the paradoxical sleep anomalies described in depression.

Table 19.3 Main clinical and polygraphic characteristics of sleep in insomnia associated with a major depressive disorder, insomnia associated with anxiety disorder and in primary insomnia.

Pathology	Insomnia associated with a major depressive disorder	Insomnia associated with an anxiety disorder	Primary insomnia
Clinical aspects	Insomnia far more severe and quantitatively greater, Difficulties getting to sleep, Nocturnal awakenings, Anxious early morning awakening	Insomnia globally less severe than that of major depression, Difficulties getting to sleep, sleep unstable and reduced, but overall preservation of sleep	Isolated insomnia: no psychiatric disorder according to the classifications of mental disorders, somatised stress, acquired mental associations non conducive to sleep
Polysomnographic aspects	Sleep continuity disturbances, Sleep architecture anomalies: - decreased REM sleep latency - increased duration and density of eye movements in the first episode of REM sleep - decreased deep NREM sleep	Sleep continuity disturbances, No anomalies in sleep architecture, No specific anomalies in REM sleep,	

Generalised anxiety disorder (GAD)

Clinical aspects

The DSMIV defines generalised anxiety as anxiety and excessive worry, usually occurring for at least 6 months, concerning a number of events or activities. The person experiences difficulty controlling this preoccupation. The anxiety and worry are associated with three or more of the following six symptoms: agitation or a sensation of being overexcited or at the end of one's tether, proneness to fatigue, difficulty in concentrating, irritability, muscular tension and sleep disturbance (difficulty getting to sleep or maintaining sleep), and anxiety leads to clinically significant suffering or to an alteration in social functioning.

60 to 70% of patients suffering from this permanent state have trouble sleeping. Insomnia, which is usually as marked as the anxiety disorder is severe (impact of the raised level of anxiety) is often persistent. It chiefly entails difficulty getting to sleep, with tension, mentism, rumination and anxious awakenings in the middle of the night [61].

Polysomnographic aspects

Insomnia in the pure form of generalised anxiety [35, 74] is characterised by prolonged sleep latency, sleep instability with a number of nocturnal awakenings and a resulting decrease in total sleep time, and a lower sleep efficiency index. Awakenings are more frequent in the first half of the night, the crucial period for restorative sleep. The temporal organisation of deep NREM sleep and REM sleep is fairly close to the norm. REM sleep latency is normal or prolonged, eye movement density lower than that of non anxious controls.

The objective sleep abnormalities typically seen in anxious subjects are thus quite different from those found in major depression (table 19.4). This distinction is again found in the reactivity of sleep to cholinergic agents, the response of anxious subjects being indistinguishable from that of controls [87].

Table 19.4 Polysomnographic characteristics in major depressive disorder, schizophrenia and anxiety disorder

Pathology	Sleep efficiency index	Duration of deep NREM sleep	REM sleep latency
Major depressive disorder	⇓	⇓	⇓
Schizophrenia	⇓	NS or⇓	NS or⇓
Generalised anxiety disorder	⇓	NS or⇓	NS
Panic disorder	⇓	NS or⇓	NS
Obsessive-compulsive disorder	⇓	NS	NS
Post traumatic stress disorder	⇓	NS or⇓	⇓

⇓ decreased
NS: non significant difference compared to normal controls

Treatments

The main pharmacological treatments are benzodiazepines, azopirones and antidepressants. All the benzodiazepines are probably effective for the treatment of generalised anxiety. Roughly 2/3 of patients feel a moderate or marked improvement after 1 to 2 weeks of treatment. Benzodiazepines are more active in relation to the somatic aspects of the disorder than to the psychological aspects.

The azapirones, mainly buspirone, offer an interesting alternative considering the absence of sedative or myorelaxant effect. But they take longer to come into effect, with an average of 2 to 4 weeks.

Among the antidepressants, the most interesting is venlafaxine, a serotonin and norepinephrine reuptake inhibitor, although it must be born in mind that the product has a somewhat awakening effect.

Psychological treatment relies on cognitive and behavioural approaches which target both cognitive (worry) and behavioural (avoidance) aspects. This type of treatment is administered in the form of a dozen sessions, individually or in a group. The degree of anxiety is reduced by 50% and the benefits are lasting [8]. Treatment acts both on anxiety and on insomnia [63].

Although the benefits of mixing pharmacotherapy and psychological therapy are generally accepted, data is still lacking to confirm this.

Social and specific phobias

Clinical aspects

Phobias are defined as the onset of a fit of anxiety brought about by a specific situation or object, exterior to the subject, presenting no intrinsic danger. Avoidance strategies can extend to major withdrawal (with behaviour which is non conducive to sleep, tending to perpetuate insomnia). Sleep may itself be a phobogenic object. In these sleep phobias, the complaint focuses on difficulties getting to sleep, a disturbing moment due to the loss of control and references.

Polysomnographic aspects

Polysomnographic studies in this diagnostic area are rare and their results, contradictory [25]. For some authors patients who suffer from social phobia do not generally complain of sleep problems (or at least not spontaneously) and their nocturnal recordings do not differ from those of healthy cohorts [80]. In another study the majority of patients complained more than the control subjects of insomnia, difficulties getting to sleep, nocturnal awakenings and the daytime consequences of disturbed sleep [25].

Treatments

The therapeutic strategies available associate pharmacotherapy with psychological intervention.

From the pharmacological point of view 4 therapeutic classes appear to be effective: reversible inhibitors of monoamine oxidase, moclobemide and brofaramine, but these may lead to insomnia in 20% of cases [88]; the selective serotonin reuptake inhibitors, fluvoxamine, sertraline and paroxetine, the first two of which tend to cause awakening; the benzodiazepines, alprazolam and to a lesser degree clonezepam; and the β-blockers (atenolol).

Among the behavioural techniques, the specific effective treatment for agoraphobia is *in vivo* exposure (net improvement in 76% of patients), self-assertion therapies, and cognitive restructuring techniques using Ellis's model of cognitive distortion. These techniques have been validated by controlled testing. It is also worth mentioning the "virtual" immersion techniques (virtual reality, similar to aircraft piloting simulators) used initially before application in the "real" world. The main preoccupation is not the technique itself, but finding the relevant elements to include in virtual worlds so that they induce the emotional charge of the real world. Also noteworthy is the interesting technique of incorporating a phobogenic object into the contents of dream using prehypnic suggestion [29].

Panic disorder (PD)

Clinical aspects

Panic disorder is a form of anxiety characterised by intermittent attacks, at varying intervals, of acute anxiety with neurovegetative signs, interspersed with secondary anticipatory anxiety. Panic attacks, which are unpredictable, usually occur in the daytime, more often in the morning, but may also occur later in the day at bedtime. These attacks sometimes occur during sleep (30 to 58% of patients), or even exclusively during sleep. The patient awakens suddenly, with no recollection of dreaming (patients distinguish clearly between these attacks and the effect of dreaming or nightmares); it is difficult or even impossible to regain sleep due to the intensity of anxiety, neurovegetative symptomatology and hypervigilance. Compared to subjects who only suffer from daytime attacks, patients with night attacks have more severe panic symptoms (as well as more frequent daytime attacks) and complain more of insomnia.

Polysomnographic aspects

The most commonly described polysomnographic anomalies in these patients are moderate, affecting sleep continuity (increased intra-sleep wakefulness, decreased TST and sleep efficiency index) and the decrease of NREM sleep stage 4, with no alterations to REM sleep, unlike depression [7]. REM sleep latency is significantly prolonged when a panic attack occurs during the night. The subjects suffering from PD recorded by Hauri *et al.* [43] slept slightly less well than normal subjects, with no disturbance to sleep architecture or to REM sleep. These patients experienced three times more movements during sleep (without being woken up by this) than normal subjects. Spectral analysis of the EEG signal [73, 75] shows an increase in beta band frequencies with a decrease in the power of theta and delta frequencies (correlated to the level of anticipatory anxiety linked to anxiety attacks).

The main interest of these studies is to show that nocturnal panic attacks (in the few cases recorded) occur during NREM sleep, in the transition from stage 2 to stage 3, i.e. as sleep becomes deeper, distinguishing them radically from nightmares (termed anxiety dreams) which appear during REM sleep, or night terrors which occur in NREM sleep stage 4 [37]. It should be emphasised that because of the frequent coexistence of major depressive states (half the subjects affected by PD with nocturnal attacks, one third of the others) and the sensitivity of PD to antidepressants, it has been suggested that the two categories of disorder share the same aetiology – all the more so since studies have shown decreased REM sleep latency in isolated PD. Nevertheless the polysomnographic - and biological - findings currently available do not indicate any similarity between neurobiological anomalies in major depression and panic disorder [87]. Sleep deprivation, which may be therapeutic in cases of depression, produces no improvement in PD; on the contrary, a third of these patients present an aggravation of anxiety and the onset of panic attacks on the recuperation day. Finally, simulation tests using specific pharmacological agents have not succeeded in objectifying depressive-type modifications to REM sleep. Certain authors have referred to a possible connection between sleep paralysis and nocturnal panic attacks [67].

Treatments

The aim of treatment is to predict the recurrence of attacks by associating medication and cognitive and behavioural therapy.

Several types of product may be prescribed to block daytime panic attacks: selective serotonin uptake inhibitors; benzodiazepines, alprazolam chiefly, at higher doses than for generalised anxiety, clonazepam and diazepam; tricyclic antidepressants (imipramine, clomipramine, desipramine), MAOIs with the known food-related disadvantages, or reversible MAOIs whose effectiveness is not yet confirmed. It is not known whether patients presenting only nocturnal panic attacks respond to the same products as patients suffering from only daytime attacks.

Cognitive and behavioural techniques are effective, acting for example on the catastrophic interpretation of certain bodily sensations. They avoid the occurrence of complications, invalidating phobias, drug addiction or alcoholism.

Obsessive-compulsive disorder (OCD)

Clinical aspects

Very few studies have been carried out on sleep disturbance in obsessive-compulsive disorder. Sleep onset insomnia is a virtually invariable element. Nonetheless, these patients consider that they sleep very well, even when delayed sleep onset and premature morning awakening are clearly evident. Insomnia occurs when the trigger stimuli for anxiety are associated with sleep, or when the illness is so severe that it affects the patient's life as a whole. Obsessions, checking rituals, mental debates and mental suffering may considerably delay bedtime.

Polysomnographic aspects

Some authors have reported decreased REM sleep latency, a result which was refuted by a study carried out on subjects suffering from isolated obsessive-compulsive disorder, who were not depressed and who showed a decrease in sleep efficiency and an increase in intercurrent awakenings in relation to healthy patients, with no change to sleep architecture, particularly to REM sleep [44]. Robinson et al. recorded normal sleep in 13 patients [71]. The (moderate) polysomnographic anomalies found in these patients, should thus not be linked to those of depressive disorder.

Treatments

Treatment for patients suffering from OCD calls for an eclectic combination of several approaches. The association of psychotropic drugs with cognitive and behavioural techniques is the most effective therapeutic strategy, insisting on the former in the case of predominant obsessions and on the latter when rituals predominate.

Numerous placebo-controlled studies have clearly shown clomipramine's effectiveness [26]. Besides clomipramine, the newer non-tricyclic SSRIs are gaining acceptance as effective alternatives for the treatment of OCD in controlled studies. Again the awakening effect of some of these compounds has to be taken in consideration.

Cognitive and behavioural therapy is effective. One of the most effective techniques in the treatment of behavioural and mental rituals is exposure with prevented response. This may be preceded by a stage in which the therapist makes anxiogenic gestures in place of the patient. It is associated with cognitive restructuring techniques centred on the exaggerated importance given to thoughts, excessive responsibility, control, and perfectionism.

Post-traumatic stress disorder (PTSD)

Clinical aspects

The psychological disturbance caused by extreme trauma is common knowledge. Severe insomnia and "traumatic" nightmares in particular are the typical symptoms of a PTSD, which can persist with no improvement for several decades, even after the remission of the other clinical signs [37]. The dreamer relives the traumatic experience through terrifying nightmares, with active bodily movement, finally awakening in a state of terror. While patients suffering from panic disorder deny any knowledge of dreams on awakening during nocturnal attacks, PTSD subjects describe their traumatic dreams with vivid images and intense emotion.

Patients suffering from PTSD complain of poor sleep, without necessarily meaning lighter sleep, as their noise stimulus threshold level is high. They often present sleep misperception. Recordings of their motor activity over several nights with actimetry (which evaluates the number of nocturnal awakenings) have confirmed that many of these subjects overestimate their sleep disturbances [28].

Polysomnographic aspects

Most polysomnographic studies of these patients have indicated a lowering of global efficiency, normal or prolonged REM sleep latency and a decreased duration of REM sleep [51, 56, 95]. However, in a recent study of pure forms of PTSD compared to major depressives and healthy subjects, Mellman *et al.* [60] described an increase in REM density, with no alteration in the duration of this stage, and reported REM sleep latency of less than 60 minutes in 54% of the 24 PTSD patients studied (and less than 20 minutes for 13% of them). Analysis of the spectral components of the EEG signal demonstrates a greater number of beta band frequencies and a reduced production of delta waves [94].

Dreams

Lavie and Kaminer [56] recorded the sleep of two groups of people who had been deported, survivors of the Holocaust and immigrants to Israel, one of which had adjusted well to post-war life and the other which had "adjusted less well", comparing them to controls who had always lived *in situ*. After arousals provoked in the laboratory at each phase of REM sleep, the well-adjusted survivors reported fewer dreams (34%) than subjects who had always lived in normal conditions (80%) or the deportees who had adjusted less well (51%). The decreased recall of dreams (psychogenic amnesia) constituted an adaptive mechanism, for Lavie, protecting against anxiogenic elements.

But the most interesting studies refer to dreams obtained after spontaneous awakenings, as these show that the vast majority of traumatic dreams (in the few cases recorded) are restricted to NREM sleep. Kramer *et al.* [51] made laboratory recordings of Vietnam war veterans suffering from PTSD, with pathological dreams, comparing them to a group of veterans who did not complain of disorders. The first group had more disturbed sleep, with anxiety dreams and traumatic dreams occurring in 84% of cases during NREM sleep.

These traumatic dreams share the distinction of occurring at the beginning of the night, outside REM sleep, between midnight and 3 a.m., i.e. at an earlier stage in the night than typical anxiety dreams, which tend to occur during the last two or three hours of sleep. This time band raises the question of a possible relationship between dreams occurring early and the proximity of preceding deep NREM sleep, the latter possibly being conducive to the expression of dissociative disorders. It is now thought that dissociative processes play a role in PTSD and that repetitive dreams are a manifestation of these disassociations (rupture of the cognitive and mnesic processes). Indeed the flashbacks are so real that some patients live out dissociative experiences linked to sleep.

Treatments

These comprise specific psychological techniques and medications.

The psychological techniques are varied. Exposure techniques consist of two types: virtual exposure whereby the patient is asked to relive the traumatic event including the thoughts and emotions of the moment as many times as necessary until the virtual exposure no longer leads to reactions of distress. The aim of cognitive therapy is to teach the subject to identify and modify erroneous thoughts relating to the trauma. One novel technique is "eye-movement desensitization reprocessing" whereby the patient is instructed to focus on a trauma-related image and its accompanying feelings, sensations, and thoughts, while visually tracking the therapist's fingers as they move back and forth in front of the patient's eyes. After a set of approximately 24 eye movements, cognitive and emotional reactions are discussed with the therapist. [77].

In terms of pharmacology, the SSRIs affect all the symptoms of PTSD including avoidance, numbness and hyperarousal symptoms [34]. MAOIs are less effective. Tricyclic antidepressants appear to be active in less severe cases. Anticonvulsants, carbamazepine and valproate, seem to suggest an interesting line of treatment. As for the benzodiazepines, these may be effective against insomnia and anxiety but do not affect the fact of reviewing the event.

SCHIZOPHRENIA

Clinical aspects

Acute inaugural psychotic attacks and the prolific phases occurring in the history of chronic schizophrenic are, as a general rule, associated with marked sleep disorders. In such periods many schizophrenic patients complain of poor sleep, but insomnia is rarely the main complaint. As in depression, insomnia often heralds the aggravation of symptoms. Some patients recover normal sleep in periods of remission of the illness. Others suffer major, persistent insomnia with multiple awakenings. The processes of withdrawal and apathy, or of general disorganisation are such that alleged sleep disturbances are overestimated in some cases.

Certain acute episodes of schizophrenia are preceded by a period of particularly intense and frequent nightmares. Some authors consider the majority of chronic schizophrenics to be subject to nightmares [36, 46].

Polysomnographic aspects

Studies of sleep in schizophrenics [24, 81] reported sleep continuity disturbance, the frequent presence of decreased REM sleep latency with normal REM density as well as a decrease in deep NREM sleep, particularly of stage 4. Emphasising the importance of the duration of withdrawal from neuroleptics and the need to study patients who had never received treatment, the publications unanimously noted that no variable was specific to clinically defined schizophrenia and that schizophrenics did not all present these anomalies.

A recent study by Lauer *et al.* [55] conducted in a very homogeneous group of paranoid schizophrenics never having received pharmacological treatment, compared with healthy subjects, showed prolonged sleep latency exceeding an hour, increased inter-sleep wakefulness, a decrease in stage 2, but no anomalies in deep NREM sleep or in REM sleep.

Keshavan *et al.* [50] reported a decrease in deep NREM sleep of 14% in schizophrenic patients compared with controls. Kajimura *et al.* [48] described a substantial reduction of delta waves, particularly of high amplitude waves, and considered the decrease in stages 3 and 4 seen in certain schizophrenics as being due to decreased delta wave amplitude.

As regards REM sleep latency, this is shorter (< 40 min) in less than one patient in five in the study by Lauer *et al.* [55]. Seven of the twenty schizophrenic patients studied by Zarcone and Benson [97] showed REM sleep latency of under 20 mins. Taylor *et al.* [82] studied six non narcoleptic schizophrenic patients in detail (out of 36), who presented with a REM sleep latency of less than 10 to 20 min: their first phase of REM sleep was shorter, with fewer eye movements, but with no other differences in sleep. The existence of this short REM sleep latency seems to indicate a more serious case of schizophrenia. According to Poulin et al. [70] the sleep of schizophrenic patients who have never been treated is characterised by longer latency to sleep onset, a shorter total sleep time, awakenings regularly distributed throughout the night, reduced deep NREM sleep and shorter REM sleep latency. Moreover, beta activity increases during episodes of REM sleep in the bifrontal regions and in the anterior region of the right hemisphere.

Circadian aspects

The clinical symptoms of schizophrenics are accompanied by sleep disorders and circadian rhythm sleep disorders [19, 59]. For Wirz-Justice et al. [93] the more disturbed the circadian rhythm, the more severe the cognitive deficiencies. 41 patients treated by haloperidol, stayed in bed for an average of 11 hours at night, with broken sleep (3 hours of awakening) and had an hour's siesta; treated by clozapine, their sleep was less interrupted, their siestas shorter and their results in neuro-psychology tests were better.

Sleep-symptom relationship

Lauer *et al.* [55] described a significant negative correlation between REM sleep latency and the severity of positive symptoms (disorders of thought, conceptual disorganisation). For Keshavan *et al.* [50], conceptual disorganisation is closely and negatively correlated to deep NREM sleep (total quantity of delta waves). Finally Zarcone and Benson found that the length of sleep latency correlated with the severity of thought disorder [97].

Treatments

Treatment for sleep disorders plays an important part in treatment for the schizophrenic. The complaint of poor sleep is moreover one of the signs heralding a relapse [50].

During acute episodes sedative neuroleptics quickly normalise sleep, long before the specific psychotic symptomatology is corrected. The standard sedative antipsychotics alter sleep architecture and their long-term use encourages the risk of insomnia and the occurrence of periodic leg movements in sleep [79]. The atypical, new generation antipsychotics (like clozapine) may be given as a single dose at night; they improve sleep continuity and through their effects on the positive and negative symptoms, allow the patient to be active during the daytime [90].

Good sleep hygiene and cognitive and behavioural techniques contribute to a better quality of sleep.

EATING DISORDERS

Anorexia nervosa

Clinically, the sleep duration of anorexic patients is reduced, discontinuous, with early morning awakening. The last symptom is even more constant in this pathology than in depression, but is unrecognised (or denied) by these patients.

Polysomnographycally, sleep continuity disorders have been reported [16]: decreased TST and increased wakefulness; these disturbances are thought to be linked to the severity of weight loss [53]. It should be noted in passing that studies conducted on anorexics who regain weight, show an improvement in their quality of sleep, later morning awakening as well as increased REM sleep latency. According to Lauer and Krieg [53] nutritional disturbance has a distinct effect on sleep, both in subjective terms and in terms of objective elements in polysomnography, regardless of the psychiatric diagnosis.

Bulimia nervosa

This is characterised by the repeated occurrence of bulimia ("binge eating") involving the absorption, in a limited period of time, of a quantity of food which is largely in excess of normal consumption with the loss of control of eating behaviour during the episode. Added to this binge eating is inappropriate and repeated compensatory behaviour, aimed at preventing weight gain, such as provoked vomiting, the abuse of laxatives, fasting, excessive physical exercise. This disorder does not usually involve any particular sleep disorder.

Bulimia nervosa must be distinguished from nocturnal eating (drinking) syndrome and sleep related eating disorders. Nocturnal eating (or drinking) syndrome is classed as an extrinsic sleep disorder in the ICSD [47]. It associates insomnia, often since childhood, with repeated nocturnal awakening and the inability to go back to sleep without eating or drinking, morning anorexia and complete recall of the nocturnal episodes. A marked preference for sugar is noted, with the absence of any related guilt, this behaviour sometimes being considered by the patient as hypnotic therapy. Polysomnography is normal apart from the awakenings.

Sleep related eating disorder is very different and belongs to the category of parasomnias. This disorder associates confusional awakening of the sleepwalking type and compulsive, involuntary and indiscriminate eating behaviour, followed by partial or total amnesia in regard to the episodes. There is no food intake between the evening meal and bedtime, no feeling of hunger during the night, nor of a disturbed body image. It is thus a form of sleep disorder (sleepwalking) with clinical expression of an eating type.

Treatments

The treatment of anorexia nervosa combines nutritional treatment, individual and family psychotherapy and chemotherapy which is psychotropic, anxiolytic if anxiety is predominant, and antidepressant in the event of depressive symptoms. Sleep disorders quickly disappear when sufficiently balanced diet is re-established.

As indicated above, bulimia nervosa does not involve any particular sleep disorder. The treatment of nocturnal bulimia relies on an association of behavioural techniques and the administration of benzodiazepines. Sleep-related eating behaviour disorder is treated by clonazepam, dopaminergic agents and sometimes opiates. The frequency of associated mood disorders is an indication for the use of fluoxetine or trazodone. Behavioural techniques are ineffective.

PERSONALITY DISORDERS

Clinical aspects

In a sample of 66 insomnia patients examined by psychiatrists, based on questions structured around the DSM-IV [2], 14% corresponded to the category of personality disorders and 39% presented features specific to a maladjusted personality but which were not severe enough to reach the threshold imposed by personality disorder criteria.

Symptoms of morbidity and personality disorder have been assessed, since DSM-III along two distinct diagnostic axes. Ten specific disorders are listed in the DSM-IV [2], including paranoid, schizoid, schizotypal, antisocial, borderline, histrionic, narcissistic, avoidant, dependent and obsessive-compulsive personality disorders. As a case in point the borderline personality is characterised by sustained instability in interpersonal relationships, self-image and affect, with marked impulsiveness which develops in adolescence or early adulthood, hindering social functioning and leading to a chaotic lifestyle.

Polysomnographic aspects

Polysomnographic studies of these patients are few in number and of limited reliability due to the high degree of comorbidity (30 to 80% depending on the authors) with depressive disorders (ref. In [1, 6]).

A single publication deals with subjects presenting a personality disorder of the borderline type (with no mood disorder): these patients sleep less well than control subjects (prolonged sleep latency, increased number and duration of awakenings during the night) and with reduced REM sleep latency (average: 64 ± 9 min, versus 90 ± 23 min in control subjects) [9] or normal REM sleep latency [30]; this decreased REM sleep latency, marking a trait rather than a state, is the expression of a personal predisposition, and/or family vulnerability to depression [10].

Mention must also be made of disturbed circadian sleep rhythm, delayed sleep phase or hypernycthemeral rhythm, due to the particular lifestyle of many of these patients [27].

Treatment

The treatment for sleep disorders associated with personality disorders is as delicate as the treatment of personality disorders themselves, due to the lack of a theoretical basis, uncertain nosology and poor compliance to the treatment proposed.

CONCLUSION

The relationship between sleep disorders and psychiatric illnesses is a complex one, but what can be ascertained is the high frequency of sleep disorders associated with mental disorders.

Whatever the therapy proposed, the long-term treatment of insomnia associated with a mental disorder involves maintaining a prolonged and attentive relationship with the patient. It requires combining the reasonable management of pharmacological treatment whose necessity must be continually reassessed, with psychological approaches which require appropriately trained therapists, a certain directivity and encouragement to make positive existential choices.

REFERENCES

1. AKISKAL H.S., JUDD L.L., GILLIN J.G., LEMMI H. – Subthreshold depressions: clinical and polysomnographic validation of dysthymic, residual and masked forms. *J. Affect. Dis.,* 45, 53-63, 1997.
2. AMERICAN PSYCHIATRIC ASSOCIATION. *Diagnostic and statistical manual of mental disorders* (4th edition). American Psychiatric Association, Wahington D.C., 1994
3. ANSSEAU M., VON FRENCKELL R., FRANCK G., REYNOLDS C.F., KUPFER D.J. – Sommeil et dépression: vers une standardisation de l'utilisation de la latence du sommeil paradoxal en tant que marqueur biologique de dépression majeure. *Rev. EEG Neurophysiol. Clin.,* 17, 411-424, 1987.
4. ARMITAGE R., HOFFMANN R.F., FITCH T., TRIVEDI M., RUSH A.J. – Temporal characteristics of delta activity during NREM sleep in depressed outpatients and healthy adults: group and sex effects. *Sleep,* 23, 607-617, 2000.
5. ARMITAGE R., HUDSON A., TRIVEDI M., RUSH A.J. – Sex differences in the distribution of EEG frequencies during sleep: unipolar depressed outpatients. *J. Affect. Dis.,* 34, 121-129, 1995.
6. ARRIAGA F., CAVAGLIA F., MATOS-PIRES A., LARA A., PAIVA T. – EEG sleep characteristics in dysthymia and major depressive disorder. *Biol. Psychiatry,* 32, 128-131, 1995.

7. ARRIAGA F., PAIVA T., MATOS-PIRES A., CAVAGLIA F., LARA E., BASTOS L. – The sleep of non-depressed patients with panic disorders: a comparison with normal controls. *Acta Psychiatr. Scand.*, 93, 191-194, 1996.

8. BARLOW D.H., ESLER J.L., VITALI B.A. - Psychosocial treatments for panic disorders, phobias, and general anxiety disorder. *In: A guide to treatments that work.* Nathan P.E., Gorman J.M.(eds), Oxford University Press, New York, 288-318, 1997

9. BATTAGLIA M., FERINI-STRAMBI L., SMIRNE S., BERNARDESCHI L., BELLODI S. – Ambulatory polysomnography of never-depressed borderline subjects: a high risk approach to rapid eye movement latency. *Biol. Psychiatry*, 33, 326-334, 1993.

10. BATTAGLIA M., STRAMBI L.F., BERTELLA S., BAJO S., BELLODI L. – First-cycle REM density in never-depressed subjects with borderline personality disorder. *Biol. Psychiatry*, 45, 1056-1058, 1999.

11. BEAUCHEMIN K.M., HAYS P. – Prevailing mood, mood changes and dreams in bipolar disorder. *J. Affect. Dis.* 35, 41-49, 1995.

12. BEAUCHEMIN K.M., HAYS P. – Phototherapy is a useful adjunct in the treatmentof depressed inpatients. *Acta Psychiat. Scand.*, 95, 424-427, 1997.

13. BENCA R.M. – Sleep in psychiatric disorders. *Neurologic Clinics*, 14, 739-764, 1996.

14. BENCA R.M. – Mood disorders. *In: Principles and Practice of Sleep Medicine*, Kryger M.H., Roth T., Dement W.C. (eds), 3rd edition, Saunders, Philadelphia, 1140-1157, 2000.

15. BENCA R.M., OBERMEYER W.H., THISTED R.A., GILLIN J.C. – Sleep and psychiatric disorders: a meta-analysis. *Arch. Gen. Psychiatry*, 49, 651-668, 1992.

16. BENCA R.M., CASPER R.C. – Sleep in eating disorders. *In: Principles and Practice of Sleep Medicine*, Kryger M.H., Roth T., Dement W.C. (eds), 2nd edition, Saunders, Philadelphia, 927-932, 1994.

17. BERGER M., RIEMANN D. – REM sleep in depression – an overview. *J. Sleep Res.*, 2, 211-223, 1993.

18. BILLIARD M., DOLENC L., ALDAZ C., ONDZE B., BESSET A. – Hypersomnia associated with mood disorders: a new perspective. *J. Psychosom. Res.* 38, 41-47, 1994.

19. BOIVIN D.B., MORISET N.J., LAL S. – Abnormal circadian rhythm of sleep propensity in chronic schizophrenia. *Sleep*, 23 (suppl. 2), A363, 2000.

20. BUYSSE D.J., FRANK E., LOWE K.K., CHERRY C.R., KUPFER D.J. – Electroencephalographic sleep correlates of episode and vulnerability to recurrence in depression. *Biol. Psychiatry*, 41, 406-418, 1997.

21. BUYSSE D.J., REYNOLDS C.F., KUPFER D.J., THORPY M.J., BIXLER E., KALES A., MANFREDI R., VGONTAS A., STEPANSKI E., ROTH T., HAURI P., STAPF D. – Effects of diagnosis on treatment recommendations in chronic insomnia. A report from the APA/NIMH DSM-IV field trial. *Sleep*, 20, 542-552, 1997.

22. BUYSSE D.J., TU X.M., HERRY C.R., BEGLEY A.E., KOWALSKI J., KUPFER D.J., FRANK E. – Pretreatment REM slep and subjective sleep quality distinguish depressed psychotherapy remitters and nonremitters. *Biol. Psychiatry*, 45, 205-213, 1999.

23. CARTWRIGHT R.D., KRAVITZ H.M., EASTMAN C.I., WOOD E. – REM latency and recovery from depression: getting over divorce. *Am. J. Psychiatry*, 148, 1530-1535, 1991.

24. CHOUINARD S., POULIN J., STIP E., GODBOUT R. – The sleep of patients with schizophrenia: a meta-analysis. *Sleep*, 24, Suppl, A389-390, 2001.

25. CLARK C.P., GILLIN J.C., GOLSHAN S. – Do differences in sleep architecture exist between depressives with comorbid simple phobia as compared with pure depressives? *J. Affect. Dis.*, 33, 251-255, 1995.

26. CLOMIPRAMINE COLLABORATIVE STUDY GROUP- Clomipramine in the treatment of patients with obsessive-compulsive disorder. *Arch. Gen. Psychiatry*, 48, 730-738, 1991

27. DAGAN Y., SELA H., OMER H., HALLIS D., DAR R. – High prevalence of personality disorders among circadian rhythm sleep disorders (CRSD) patients. *J. Psychosom. Res.*, 41, 357-363, 1996.

28. DAGAN Y., ZINGER Y., LAVIE P. – Actigraphic sleep monitoring in posttraumatic stress disorders (PTSD) patients. *J. Psychosom. Res.* 42, 577-581, 1997.

29. DE KONINCK J., BRUNETTE R. – Presleep suggestion to a phobic object: successful manipulation of reported dream affect. *J. Gen. Psychology*, 118, 185-199, 1991.

30. DE LA FUENTE J.M., BOBES J., VIZUETE C., MENDELWICZ J. – Sleep-EEG in borderline patients without concomitant major depression: a comparison with major depression and normal control subjects. *Psychiatry Res.* 105, 87-95, 2001.

31. DEW M.A., REYNOLDS C.F., BUYSSE D.J., HOUCK P.R., HOCH C.C., MONK T.H., KUPFER D.J. – Electroencephalographic sleep profiles during depression. Effects of episode duration and other clinical and psychosocial factors in older adults. *Arch. Gen. Psychiatry*, 53, 148-156, 1996.

32. FELDMAN-NAIM, TURNER E.H., LEIBENLUFT E. – Diurnal variation in the direction of mood switches in patients with rapid-cycling bipolar disorder. *J. Clin. Psychiatry*, 58, 79-84, 1997.

33. FORD D.E., KAMEROW D.B. – Epidemiological studies of sleep disturbances and psychiatric disturbances. An opportunity to prevention. *JAMA*, 262, 1479-1484, 1989.

34. FRIEDMAN M.J. – Current and future drug treatment for post-traumatic stress disorder. *Psychiatric Annals*, 28, 461-468, 1998.

35. FULLER K.H., WATERS W.F., BINKS P.G., ANDERSON T. – Generalized anxiety and sleep architecture: a polysomnographic investigation. *Sleep*, 20, 370-376, 1997.

36. GARMA L. – *Clinique de l'Insomnie*. Presses Universitaires de France, 1994

37. GARMA L. – Approches critiques de la clinique du rêve et du sommeil. *Confrontations Psychiatriques*, 38, 115-140, 1997.

38. GILES D.E., KUPFER D.J., ROFFWARG H.P., RUSH A.J., BIGGS M.M., ETZL B.A. – Which endogenous depressive symptoms relate to REM latency reduction? *Biol. Psychiatry*, 21, 473-482, 1986.

39. GILES D.E., KUPFER D.J., ROFFWARG H.P., RUSH A.J., BIGGS M.M., ETZL B.A. – Polysomnographic parameters in first-degree relatives of unipolar probands. *Psychiatry Res.,* 27, 127-136, 1989.
40. GILES D.E., ROFFWARG H.P., RUSH A.J., GUZICK D.S. – Age-adjusted threshold values for reduced REM latency in unipolar depression using ROC analysis. *Biol. Psychiatry,* 27, 841-853, 1990.
41. GILLIN J.C., BUSCHSBAUM M., WU J., CLARK C., BUNNEY W. JR. – Sleep deprivation as a model experimental antidepressant treatment: findings from functional brain imaging. *Depress Anxiety,* 14, 37-49, 2001.
42. HANNA M.M., MORENO R.A., TAVARES S.M., TEIXERA V., ALOE F. – Polysomnographic features in manic patients. *Sleep,*23, suppl.2, A353, 2000
43. HAURI P.J., FRIEDMAN M., RAVARIS C.L. – Sleep in patients with spontaneous panic attacks. *Sleep, 12, 323-337, 1989.*
44. HOHAGEN F., LIS S., KRIEGER S., WINKELMANN G., RIEMANN D., FRITSCH-MONTERO R., REY E., ALDENHOFF S., BERGER M. – Sleep EEG of patients with obsessive-compulsive disorder. *Eur. Arch. Psychiatry Clin. Neurosc.,* 243, 273-278, 1994.
45. HUBAIN P., VAN VEEREN C., STANER L., MENDLEWICZ J., LINKOWSKI P. – Neuroendocrine and sleep variables in major depressed inpatients. Role of severity. *Psychiatry Res.* 63, 83-92, 1996.
46. HUBLIN C., KAPRIO J., PARTINEN M., KOSKENVUO M. – Nightmare familial aggregation and associations with psychiatric disorders in a nation wide twin cohort. *Am. J. Med. Gen..* 88, 329-336, 1999.
47. ICSD- *International classification of sleep disorders: diagnostic and coding manual.* Diagnostic Classification Steering Committee, Thorpy M.J., Chairman, Rochester, Minnesota: American Sleep Disorders Association, 1990.
48. KAJIMURA N., KATO M., OKUMA T., SEKIMOTO M., WATANABE T., TAKAHA S. – Relationship between delta activity during all night sleep and negative symptoms in schizophrenia. *Biol. Psychiatry,* 39, 411-418, 1996.
49. KERKHOFS M., LINKOWSKI P., LUCAS F., MENDELWICZ J. – Twenty-four hour patterns of sleep in depression. *Sleep,* 14, 501-506, 1991.
50. KESHAVAN M.S., REYNOLDS C.F., MIEWALD J.M., MONTROSE D.M. – A longitudinal study of EEG sleep in schizophrenia. *Psychiatry Res.,* 59, 203-211, 1996.
51. KRAMER M., KINNEY L. – Sleep patterns in trauma victims with disturbed dreaming. *Psychiat. J. Univ. Ott.,* 13, 12-15, 1988.
52. LAUER C.J., RIEMAN D., WIEGAND M., BERGER M. – From early to late adulthood changes in EEG sleep of depressed patients and healthy volunteers. *Biol. Psychiatry,* 29, 979-993, 1991.
53. LAUER C.J., KRIEG J.C. – Weight gain and all-night EEG-sleep in anorexia nervosa. *Biol. Psychiatry,* 31, 622-625, 1992.
54. LAUER C.J., SCHREIBER W., HOLSBOER F., KRIEG J.C. – In quest of identifying vulnerability markers for psychiatric disorders by all-night polysomnography. *Arch. Gen. Psychiatry,* 52, 145-153, 1995.
55. LAUER C.J., SCHREIBER W., POLLMACHER T., HOLSBOER F., KRIEG J.C. – Sleep in schizophrenia : a polysomnographic study on drug-naive patients. *Neuropsychopharmacol.,* 16, 51-60, 1997.
56. LAVIE P., KAMINER H. – Dreams that poison sleep: dreaming in Holocaust survivors. *Dreaming,* 1, 11-21, 1991
57. LEIBENLUFT E., WEHR T.A. – Is sleep deprivation useful in the treatment of depression? *Am. J. Psychiatry,* 149, 159-168, 1992
58. LISCOMBE M.P., HOFFMANN R.F., TRIVEDI M.H., PARKER M.K., RUSCH A.J., ARMITAGE R. – Quantitative EEG amplitude across REM sleep periods in depression: preliminary report. *J. Psychiatry Neurosci.* 27, 40-46, 2002
59. MARTIN J., JESTE D.V., CALIGUIRI M.P., PATTERSON T., HEATON R., ANCOLI-ISRAEL S. – Actigraphic estimates of circadian rhythms and sleep/wake in older schizophrenia-patients. *Schizoph. Res.,* 49, 77-86, 2001
60. MELLMAN T.A., NOLAN B., HEBDING J., KULICK-BELL R., DOMINGUEZ R. – A polysomnographic comparison of veterans with combat-related PTSD, depressed men and non-ill controls. *Sleep,* 20, 46-51, 1997.
61. MONTI J.M., MONTI D. – Sleep disturbance in generalized anxiety disorder and its treatment. *Sleep Med.,* 3, 263-276, 2000.
62. MORIN C.M., CATESBY WARE J. – Sleep and psychopathology. *Appl. Prevent. Psychol.,* 5, 211-224, 1996
63. MORIN C.M., HAURI P.J., ESPIE C.A., SPIELMAN A.J., BUYSSE D.J., BOOTZIN R.R. – Nonpharmacological treatment of chronic insomnia. *Sleep,* 22,, 1134 -1156, 1999.
64. NOFZINGER E.A., SCHWARTZ R.M., REYNOLDS C.F., THASE M.E., JENNINGS J.R., FRANK E., FASICZKA A.L., GARAMONI G.L;, KUPFER D.J. – Affect intensity and phasic REM sleep in depressed men before and after treatment with cognitive-behavioral therapy. *J. Cons. Clin. Psychiatry,* 62, 83-91, 1994.
65. OHAYON M. – Prevalence of DSM-IV diagnostic criteria of insomnia: distinguishing insomnia related to mental disorders from sleep disorders. *J. Psychiatry Res.* 31, 333-336, 1997.
66. OHAYON M.M., CAULET M., LEMOINE P. – Comorbidity of mental and insomnia disorders in the general population. *Comprehensive Psychiatry,* 39, 185197, 1997.
67. PARADIS C.M., FRIEDMAN S., HATCH M. – Isolated sleep paralysis in African Americans with panic disorder. *Cultural Diversity and Mental Health,* 3, 69-76, 1997
68. .PERLIS M.L., GILES D.E., BUYSSE D.J., THASE M.E., TU X., KUPFER D.J. – Which depressive symptoms are related to which sleep electroencephalographic variables? *Biol. Psychiatry,* 42, 904-913, 1997.
69. PERLIS M.L., GILES D.E., BUYSSE D.J., TU X., KUPFER D.J. – Self-reported sleep disturbances as a prodromal symptom in recurrent depression. *J. Affect. Dis.* 42, 209-212, 1997
70. POULIN J., STIP E., GODBOUT R. – REM sleep EEG beta activity correlates with positive and negative symptoms in drug-naive patients with schzophrenia. *Sleep,* 24, suppl. A390-391, 2001.
71. ROBINSON D., WALSLEBEN J., POLLACK S., LERNER G. – Nocturnal polysomnography in obsessive-compulsive disorder. *Psychiatry Res.* 80, 257-263, 1998.
72. RODIN J., MCAVAY G., TIMKO C. – A longitudinal study of depressed mood and sleep disturbances in elderly adults. *J. Gerontol.,* 43, 45-53, 1988.

73. RÖSCHKE J., KÖGEL P., WAGNER P., GROZINGER M., HEVERS W., SCHLEGAL S. – Electrophysiological evidence for an inverse benzodiazepine receptor agonist in panic disorder. *J. Psychiatr. Res.*18, 849-863, 1999.
74. SALETU-ZYHLARTZ G., SALETU B., ANDERER P., BRANDSTÄTTER N., FREY R., GRUBER G., KLÖSCH G., MANDI M., GRÜNBERGER J., LINZMAYER L. – Nonorganic insomnia in generalized anxiety disorder. 1. Controlled studies on sleep, awakening and daytime vigilance utilizing polysomnography and EEG mapping. *Neuropsychobiology,* 36, 117-129, 1997.
75. SALETU-ZYHLARTZ G.M., ANDERER P.A., BERGER P., GRUBER G., OBERNDOR S., SALETU B. – Nonorganic insomnia in panic disorder. Comparative sleep laboratory studies with normal controls and placebo controlled trials with alprazolam. *Hum. Psychopharm.* 15, 241-254, 2000.
76. SCHRAMM E., HOAGEN F., KAPPLER C., GRASSOHOFF U., BERGER M. – Mental comorbidity of chronic insomnia in general practice attenders using DSM-III-R. *Acta Psychiatr. Scand.,* 91, 10-17, 1995.
77. SHAPIRO F. – *Eye movement desensitization and reprocessing: basic principles, protocols and procedures.* Guilford Press, New York, 1995
78. SKAER T.L., SCLAR D.A., ROBINSON L.M. – Complaints of insomnia in the USA 1995-1998: prevalence, psychiatric comorbidity and pharmacological treatment patterns. *Primary Care Psychiatry,* 7, 145-151, 2001
79. STAEDT J., DEWES D., DANOS P., STOPPE G. – Can chronic neuroleptic treatment promote sleep disturbances in elderly schizophrenic patients. *Int. J. Geriar. Psychiatry* 15 2, 170-176, 2000.
80. STEIN M.B., KROFT C.D.L., WALKER J.R. – Sleep impairment in patients with social phobia. *Psychiatry Res.,* 49, 251-256, 1993.
81. TANDON R., SHIPLEY J.E., TAYLOR S., GREDEN J.F., EISER A., DE QUARDO J., GOODSON J. – Electroencephalographic sleep abnormalities in schizophrenia. *Arch. Gen. Psychiatry,* 49, 185-194, 1992.
82. TAYLOR S.F., TANDON R., SHIPLEY J.E., EISER A., GOODSON J. – Sleep onset REM periods in schizophrenic patients. *Bio. Psychiatry,* 30, 205-209, 1991.
83. THASE M.E., FRIEDMAN E.S., BERMAN S.R., FASICKZA A.L., LIS J.A., HOWLAND R.H., SIMONS A.D. – Is cognitive therapy just a nonspecific intervention for depression? A retrospective comparison of consecutive cohorts treated with cognitive behavior therapy or supportive counselling and pill placebo. *J. Affect. Disord.,* 57, 63-71, 2000.
84. THASE M.E., KUPFER D.J., FASICZKA A.J., BUYSSE D.J., SIMONS A.D., FRANK E. - Identifying an abnormal electroencephalographic sleep profile to characterize major depressive disorder. *Biol. Psychiatry,* 41, 964-973, 1997.
85. THASE M.E., BUYSSE D.J., FRANK E., CHERRY C.R., CORNES C.L., MALLINGER A.G., KUPFER D.J. – Which depressed patients will respond to interpersonal psychotherapy? The role of abnormal EEG sleep profiles. *Am. J. Psychiatry,* 154, 502-509, 1997.
86. THASE M.E., SACHS G.S. – Bipolar depression: pharmacotherapy and related therapeutic strategies. *Biol. Psychiatry,* 48, 558-572, 2000.
87. UHDE T.W. – The anxiety disorders. *In: Principles and Practice of Sleep Medicine.* Kryger M.H., Roth T., Dement W.C. (eds), 3rd edition, Saunders, Philadelphia, 1123-1139, 2000
88. VERSIANI M., NARDI A.E., MUNDIM F.D., ALVES A.B., LIEBOWITZ M.R., AMREIN R. – Pharmacotherapy of social phobia: a controlled study with moclobemide and phenelzine. *Brit. J. of Psychiatry, 161, 353-360, 1992.*
89. WEHR T.A. – Sleep loss: a preventable cause of mania and other excited states. *J. Clin. Psychiatry,* 50 (suppl, 12), 8-16, 1989.
90. WETTER T.C., LAUER C.J., GILLICH G., POLLMACHER T. – The electroencephalographic sleep pattern in schizophrenic patients treated with clozapine or classical antipsychotic drugs. *J. Psychiat. Res.* 30, 411-419, 1996.
91. WEYERER S., DILLING H. – Prevalence and treatment of insomnia in the community: results from the upper Bavarian field study. *Sleep,* 14, 392-398, 1991.
92. WICHNIAK A., RIEMANN D;, KIEMEN A., VODERHOL U., JERNAJCZ W. – Comparison between eye-movement latency and REM-sleep parameters in major depression. *Eur. Arch. Psychiatry Clin. Neurosci.* 250, 48-52, 2000
93. WIRZ-JUSTICE A., CAJOCHEN C., NUSSBAUM P. – A schizophrenic patient with an arythmic circadian rest-activity cycle. *Psychiatry Res.* 73, 83-90, 1997.
94. WOODWARD S.H., ARSENAU N.J., MURRAY C., BLIWISE D.L. – Laboratory sleep correlates of nightmare complaint in PTSD patients. *Biol. Psychiatry,* 48, 11, 1081-1087, 2000.
95. WOODWARD S.H., FRIEDMAN M.J., BLIWISE D.L. – Sleep and depression in combat-related PTSD inpatients. *Biol. Psychiatry,* 39, 182-192, 1996.
96. WU J.C., BUNNEY W.E. – The biological basis of an antidepressant response to sleep deprivation and relapse: review and hypothesis. *Am. J. Psychiatry,* 147, 14-21, 1990
97. ZARCONE V.P., BENSON K.L. – BPRS (Brief Psychiatric Rating Scale) symptom factors and sleep variables in schizophrenia. *Psychiatry Res.,* 66, 111-120, 1997.

Chapter 20

Insomnia linked to medications

E. Corruble*, D. Warot**, and Cl. Soubrie**
* Service de Psychiatrie, Centre Hospitalier Universitaire Bicêtre, Le Kremin Bicêtre, France;
** Service de Pharmacologie, Hôpital Pitié-Salpêtrière, Paris, France

For the clinician, insomnia is a symptom which is difficult to understand as it is a predominantly subjective variable: in clinical practice, it is the patient's complaint which guides the doctor in his diagnostic and therapeutic approach. Insomnia is commonly classified according to different criteria including the time of onset during the night (beginning, middle or end), its course of evolution (occasional, short term or chronic) and the severity of the complaint. The presence of associated signs is a decisive feature in characterising insomnia: it may be isolated or associated with other symptoms. Thus insomnia may evolve in its own right or be part of the framework of a somatic disorder (algic or dyspnoeic, for example) or psychiatric (anxiety disorder, mood disorder, psychose, state of confusion or dementia).

This gives an idea of the number of stones which need to be overturned before ascribing insomnia to medication, all the more so as medication-induced insomnia may occur in a patient presenting a disorder potentially associated with insomnia it is sometimes difficult to distinguish clearly between the symptoms of an illness and the symptoms produced by the medication prescribed to treat the illness.

For the pharmacologist, different aspects of insomnia may be useful to consider. Is the notion of medically-induced insomnia sufficiently soundly based? Is the sleeplessness effect of the molecule an intrinsic and anticipated effect of this substance, i.e. an effect attributable to its pharmacological properties, notably its stimulant properties, or is it an unexpected effect in view of the known pharmacological properties of the drug. At what doses does this sleeplessness effect occur: within the therapeutic dose range, only by overdosing, or regardless of the dose? The present chapter will restrict itself to cases of insomnia attributable to medications when administered at what are considered to be therapeutic doses. Is the sleeplessness effect to be considered as a desired or an undesirable effect? In practice, apart from a few pro-wakening molecules already on the market or being developed for highly restricted indications, insomnia is commonly considered to be an undesirable effect.

INFORMATION SOURCES

The data available on insomnia linked to medications come from different sources: clinical trials, isolated cases and epidemiological studies.

Clinical trials

The research into a sleeplessness effect may be fixed from the outset as the objective of a trial evaluating the tolerance of a molecule, for which (according to its pharmacological properties and animal studies) there is reason to assume that it may lead to sleeplessness. This type of trial may be carried out on healthy volunteers or on patients. Several methods may be used to try to answer the question posed. Subjective methods are the most common, notably sleep questionnaires and analogical visual scales.

It may also be that during a trial intended for an entirely different purpose (pharmacokinetics, for example), cases of insomnia occur in subjects - healthy volunteers or patients - receiving the substance under study. Further studies would then be necessary to define the preliminary data. The trials which provide the most comprehensive answers are comparative trials of a range of doses of the substance under study, against a placebo, demonstrating a progression between the number of subjects with insomnia and the increasing doses. Nevertheless the insomniant effect may not be detected if, for example, it only appears in older subjects, who are often excluded from phases II and III.

Trials can be conducted after the medication is marketed, if cases of insomnia have been reported (see the next paragraph). They often require a substantial number of subjects. In practice, they are rarely conducted.

Spontaneously reported cases

Unless the effect of medication is studied in thousands of subjects with different characteristics before marketing, certain rare effects or those affecting particular groups of subjects not included in trials, may be overlooked.

The best way of rapidly detecting any unexpected effects observed after a new treatment is introduced, is to ask all health personnel to report them to a national or industrial central organisation of the type known in France as the "Psychovigilance system". Another slower and older procedure continues to play a similar role: the system of medical publications. In this case, however, detection is incomplete as the information is selected by doctors and insomnia is a banal symptom, more often attributed to other causes (such as the environment) than to medication.

Epidemiological studies

When confronted with a number of unexpected observations (such as cases of insomnia in young women with urinary infections) three hypotheses are open:
  a) The phenomenon is purely coincidental.
  b) It is a common effect, hitherto undetected. An answer can only be provided by the careful examination of available clinical studies or by conducting new studies.
  c) The effect does exist but is infrequent. Other types of study are then embarked upon:
  - either monitoring a large number of young women treated by the molecule under study (in the present case). By means of a questionnaire, sleep is evaluated, amongst other variables. If possible a large number of women not receiving the "suspected" treatment are monitored in the same way, and the results are then compared; this is known as a cohort study;
  -or a large number of young women with insomnia are found and questioned about their use of medication. The control population consists of women with comparable characteristics but who do not suffer from sleep disorders; this is known as a case control study.

AVAILABLE DATA

In this chapter, the disparity between data will quickly become evident to the reader. For certain classes of medication, it is possible to mention percentage of incidence or the prevalence of the symptom of insomnia. Wherever possible the circumstances are given in which data is collected are given, but not systematically so. A figure only provides real information when accurate information is given on: the exact type of patient (e.g. the presence of a painful pathology being sufficient to explain insomnia), the circumstances of onset, the disappearance of the symptom on discontinuation of treatment, the return of symptoms on reintroduction, the doses, duration of treatment, associated pathologies and corresponding treatment and any changes in habits likely to affect the interpretation of the symptom. Finally, a figure is interesting if it is given in comparison with another, observed in a comparable group of patients receiving a placebo or different medications. In fact, for most of the medications referred to, no figure will be given due to insufficiently accurate data collection conditions.

Amphetamines

Amphetamine is a synthetic amine close to norepinephrine and dopamine, which acts by increasing the release and inhibiting the reuptake of norepinephrine, dopamine and serotonin. Amphetamine and its derivatives are habit-forming. They may induce numerous psychotropic effects including intellectual stimulation, hyperactivity, a reduced feeling of tiredness, insomnia, well-being, euphoria and sometimes anxiety or psychotic decompensation requiring psychiatric treatment. Insomnia is thus an intrinsic effect of amphetamine. For Weiss and Laties [37], studies conducted with soldiers demonstrated the sleeplessness effects of amphetamine for doses of 10, 15 and 20 mg/day.

Insomnia and irritability are commonly observed with methylphenidate and are not always easy to control by changing dosing schedules. The symptoms diminish in certain patients as treatment continues, requiring no reduction in dose [38].

Non amphetamine psychostimulants

Caffeine

Caffeine is a methylxanthine. It has properties which stimulate the central nervous system: intellectual stimulation, hyperesthesia, nervousness, trembling, insomnia or even convulsions. The sleeplessness effect is thus an intrinsic property of this molecule. Karacan *et al.* [21] gave 3 different "doses" (1, 2 or 3 cups, corresponding to 80 to 340 mg) of caffeine, half an hour before bedtime, to 18 healthy volunteers, demonstrating the dose-dependent sleeplessness effect of this molecule (longer sleep latency and decreased sleep duration). Ellinwood and Rockwell [11] reported the sleeplessness effect of caffeine for doses of 500 to 600 mg/day.

Others

Adrafinil and modafinil, have properties which stimulate wakefulness in animals [16] and in man [3]. They are liable to provoke sleep onset disorders and shortened sleep duration. Insomnia, often associated with a feeling of inner tension and irritability, may be alleviated despite the continuation of treatment, with no alterations to the daily dose.

Antidepressants

Imipramine antidepressants

In a double blind study comparing the effects of imipramine and a placebo in depressed patients, Versiani *et al.* [35] reported an incidence of insomnia of 5% using imipramine (n = 164), and of 5% with placebo (n = 162).

In a review of literature on double blind trials comparing the effects of imipramine (200 mg/day on average, 240 patients) and a placebo (227 patients) in patients presenting a major depressive episode, Zerbe [39] reported an incidence of insomnia of 10% with imipramine and 9% with placebo. In a group of 107 patients treated with imipramine (150 mg/day on average) for a major depressive episode, Wagner *et al.* [36] reported an incidence of insomnia of 11% with imipramine compared to 11% with placebo. The percentage rose to 13% with associated treatments. These comparative findings do not come out in favour of imipramine producing a sleeplessness effect.

Monoamine-oxydase inhibitors

Non selective monamine-oxydase inhibitors such as iproniazide and nialamide may induce insomnia, sometimes associated with excitation. In a double blind study comparing the effects of moclobemide and a placebo in depressed patients, Versiani *et al.* [35] reported an incidence of insomnia of 10% with moclobemide (164) compared with 5% with placebo (n = 162).

Selective serotonin and norepinephrine reuptake blockers

According to Blackwell and Simon [4], the sleeplessness effect of fluoxetine occurs in about 10% of cases, requiring the discontinuation of treatment. Lader [22] published a review of efficiency studies of fluoxetine in relation to standard antidepressants. Insomnia is reported in 2 studies out of 10. The reported incidence was 12% in the first study, where the daily dosage varied between 40 and 60 mg/day and 24% in the second study where the average daily dosage was 70 mg/day. Metz and Shader [26] reported an incidence of insomnia of 14% in patients treated with fluoxetine regardless of the dose. In a review of the literature on 3 double blind trials comparing the effects of fluoxetine (20 to 60 mg, 239 patients) and a placebo (227 patients) used to treat major depression, Zerbe [39] reported an incidence of insomnia of 13% with fluoxetine, a stable percentage in function of doses (20, 40 and 60 mg), and 8% with placebo. Rickels *et al* [30] administered fluoxetine at individually fixed doses of between 20 and 80 mg/day to 185 patients presenting an episode of major depression. The frequency of insomnia was 15% in the group receiving one dose per day and 10% receiving two doses per day. These findings suggest that fluoxetine at the doses studied, may induce insomnia.

Blackwell and Simon [4] reported the sleeplessness effect of fluvoxamine. In a group of 110 patients treated with fluvoxamine (150 mg/day on average) for a major episode of depression. Wagner *et al* [36] reported an incidence of insomnia of 12% compared with 11% on placebo. The percentage rose to 13.7% with associated treatment. The findings of the last study do not corroborate with those of Blackwell and Simon [4]. Insomnia is also described with more recent serotonin reuptake blockers such as paroxetin, sertalin and citalopram.

Additional information in regard to paroxetine is provided by a follow-up of 13,741 patients treated with 20 mg/day of paroxetine, the % of patients presenting insomnia or excessive sleepiness is identical i.e. 1.7% [20].

A synthesis of the adverse effects of sertraline on sleep and wakefulness through a metanalysis by Doogan [10] shows that insomnia is more frequent with sertraline than with amitriptyline.

The findings available on how this class of antidepressants interferes with psychomotor performance in the healthy subject show little or no alteration to performance at therapeutic doses, compared to placebo [17, 18].

The above information shows that in the case of recently marketed medication figures can be quoted on the frequency of insomnia; nevertheless they must be interpreted with the same degree of reserve.

Serotonin and norepinephrine reuptake blockers

Like the selective serotonin and norepinephrine reuptake blockers, venlafaxine and milnacipran can induce insomnia; In trials against placebo, percentages of the order of 18% have been reported for patients presenting insomnia under venlafaxine versus 9% in the placebo group [12].

Others

Difficulties falling asleep have been reported particularly at the start of treatment with viloxazine (these findings were confirmed by the electrophysiological study carried out by Brezinova *et al*. [6] which demonstrated an antisleep effect for doses of 200-300 mg/day of viloxazine).

Generally speaking, all antidepressants are susceptible to inducing insomnia. Moreover, manic mood shifts may occur in patients treated with antidepressants, whatever the molecule used. Insomnia may thus appear, even with molecules reputed to be sedative. It recedes with hypomanic or manic symptomatology, after decreasing the dose or by interrupting antidepressant treatment, and with symptomatic treatment.

Hypnotics and anxiolytics

Benzodiazepines and similar drugs

There are three schematic situations in which insomnia is found in subjects treated with benzodiapines (sedative and hypnogenic molecules): the development of tolerance in subjects receiving benzodiazepine prescribed as a hypnotic, *i.e.* the gradual depletion of the hypnogenic effect over time leading to the reappearance of insomnia, a rebound of insomnia *i.e.* the reappearance when treatment is stopped of insomnia whose intensity is increased in relation to its initial level, and the abrupt withdrawal of benzodiazepines where insomnia is often the most apparent symptom [15]. The latency of onset of symptoms depends on the dose and half-life of the benzodiazepines.

Rebound and withdrawal phenomena may be more moderate or rarer with zopiclone and zolpidem than with certain benzodiazepines (although individual data suggests that a small number of patients are liable to present a rebound of insomnia after stopping zolpidem).

Neuroleptic disinhibitors

Difficulties in getting to sleep, or even problems staying asleep have been reported in treatment with carpipramine, sulpiride and amisulpride; insomnia seen in 5% of patients may be linked to uninhibiting or stimulant properties [1]. However, specific studies have found no antisleep effect.

Anti-Parkinsonian drugs

Levodopa induces sleep disorders in 20% of cases, either isolated or associated with other, notably psychiatric symptoms, such as confusion, instability, hyperactivity, anxious agitation or hypomania and episodes of delirium. The contributing factors are the doses administered, length of treatment, the neurological state of the patient and the existence of any previous psychiatric history [7].

Bromocriptine is an ergot derivative with dopaminergic antagonist properties. In the treatment of Parkinson's disease, it can induce states involving various degrees of confusion, agitation, euphoria, delirium or manic attacks. This may give rise to insomnia.

Amantadine, initially developed as an antiviral agent, is used in the treatment of Parkinson's disease for its dopaminomimetic properties. It may induce confusional states, mood disorders of the manic type or episodes of delirium. These disorders are dose-dependent: their incidence appears to increase significantly with doses exceeding 200 mg/day.

Trihexyphenidyle has stimulant properties which can give rise to hyperactivity, insomnia and euphoria. These properties may lead to drug abuse in certain cases [32].

Anticonvulsants

Insomnia has been reported with diphenylhydantoine (sometimes associated with anxiety or agitation) and with carbamazepine in confuso-delirious psychiatric accidents. Phenobarbital and benzodiazepines may induce insomnia within a context of excitation and irritability, particularly in children at the start of treatment [24]. It is difficult to describe the alterations to sleep and wakefulness specific to anticonvulsants like felbamate, tiagabine, gabapentine and lamotrigine to the extent that these are add-on medications to the treatment already used when pathological control is insufficient.

Medication for the cardio-vascular system

In a review of literature, Paykel *et al.* [28] reported cases of insomnia in patients treated with methyldopa and clonidine. These occur in isolation or as is more often the case, in association, with a subconfusional state. Insomnia is also seen when clonidine is discontinued abruptly. Dihydralazine can induce insomnia. Sleep disorders of the insomnia type have been observed in patients receiving β-blockers and calcic inhibitors. Confusional states with insomnia may be encountered in cases of metabolic disturbance with diuretics, particularly in elderly patients.

Theophylline and β-₂ like substances

Ellinwood and Rockwell [11] reported the sleeplessness effect of theophylline, and methylxanthine with psychostimulant properties (increased attention and alertness) close to those of caffeine.

Salbutamol and terbutaline, substances which stimulate the β_2-adrenergic receptors, are used in chest medicine for their bronchodilatory properties and in obstetrics for their tocolytic properties. Their properties have variable stimulating effects on the central nervous system: anxiety, insomnia, concentration difficulties, dysphoria or even delirious episodes in predisposed patients [13]. Inhaling will limit these effects occurring in the treatment of asthma.

Anti-inflammatory drugs

Steroids

Steroids can induce isolated insomnia or insomnia associated with other signs, particularly in the context of mood disorder or an episode of delirium. This type of disorder is all the more frequent with high dosages, recently introduced treatment (first two weeks, essentially) and where there is a predisposition (the female sex, organic damage to the nervous system, previous psychiatric history, depending on the authors) [23, 33].

Non steroid anti-inflammatory drugs

Cases of insomnia, whether isolated or associated with other signs such as irritability, have been reported during treatment with non steroid anti-inflammatory drugs.

Anti-rheumatic drugs

Chloroquine and its derivatives

These may induce neuropsychic disorders. These may be minor events such as tiredness, nightmares, insomnia or more severe disorders such as confuso-delirious states with agitation and behavioural disorders (referred to by [5]).

Gold compounds

Van Riel [34] reported cases of insomnia with gold compounds. These cases are difficult to interpret as they occur in patients presenting a chronic rheumatic pathology which is itself potentially insomniant.

Anti-infectious drugs

Antituberculosis drugs

Isoniazide has a chemical structure which is close to that of certain MAOI. This chemical structure probably accounts for its stimulant properties in certain patients, inducing for example insomnia, maniacal symptomatology or confusion [27].

Antibiotics

Simon and Chermat [31] have described the possible occurrence of cases of insomnia among patients treated with oxylinic acid. An animal study has provided evidence of the excitation effect of this molecule, with doses four times greater if administered intraperitoneally rather than orally. Some authors report the occurrence of insomnia among patients treated with nalidixic acid. Insomnia in this case may be associated with a confusional picture involving agitation and hallucinations. It should be noted that the chemical structure of the molecules in this family is close to those of the amphetamines, whose central stimulant effects are well known; Similarly, Idänpään-Heikkilä and Aranko [19] reported that ciprofloxacine can induce insomnia in 0.5 to 4% of cases.

Antifungal agents

Mood changes of the euphoric type have been observed in association with insomnia in patients treated with imidazole derivatives and griseofulvine [14].

Antiviral drugs

Insomnia has been reported with zidovudine (5%) in a study comparing it with placebo (1%) in patients with AIDS. It must however be emphasised that the percentage of patients with sleepiness is higher in both groups (up to 8.9% of patients) [29]. Aciclovir can also induce insomnia. The multiple association of antiretroviral drugs and lack of systematic study render it impossible to describe the exact nature of sleep disorders induced by the antiretroviral drugs, with any degree of accuracy.

Thyroid hormonotherapy

Insomnia may be a symptom of overdosing thyroid hormones in hypothyroid patients [25].

Retinoids

De Groot and Nater [8] have reported cases of insomnia in patients treated with retinoids.

Alcohol

Alcohol may, in certain conditions have a sleeplessness effect, notably in the early euphoric phase of an acute intoxication at certain doses, during a severe episode of chronic intoxication with dependence, and in withdrawal stages [2].

CONCLUSION

The data on insomnia induced by medication are derived from very different sources. Knowledge of medication-induced insomnia varies from one class of medication to another. Psychotropic medication is subjected to virtually systematic evaluation of any possible effects on sleep and daytime wakefulness, but in the case of the other classes of medication, research is not always conducted into sleep interference. It is sometimes difficult to establish whether insomnia can be ascribed to medication, in view of the subjective nature of the symptom, the range of possible aetiologies and the possible role of associated medication.

Finally, consultation for medically-induced insomnia requires a re-evaluation of the risk benefit relationship of the molecule in question.

REFERENCES

1. ALBERTS J.L., FRANCOIS F., JOSSERAND F. – Etude des effets secondaires rapportés à l'occasion de traitements par Dogmatil®. *Sem. Hôp. Paris,* 61, 19, 1351-1357, 1985.
2. BARRUCAND D., PAILLE F. – Complications neuropsychiatriques de l'alcoolisme . *In: Alcoologie,* Riom Laboratoires (ed.) CERM, 224-237, 1988.
3. BASTUJI H., JOUVET M. – Successful treatment of idiopathic hypersomnia and narcolepsy with modafinil. *Prog. Neuropsychopharmacol. Biol. Psychiatry,* 12, 695-700, 1988.

4. BLACKWELL B., SIMON J.S. – Antidepressant drugs. *In: Meyler's side effects of drugs.* M.N.G. Dukes (ed), Elsevier Science Publishers, 11th ed., 27-70, 1988.
5. BOULANGER J.P., BISSERBE J.C. – Troubles psychiatriques d'origine médicamenteuse, alimentaire ou toxique. *Encycl. Med. Chir.,* Paris, Psychiatrie, 37630 H10, 7, 1983.
6. BREZINOVA V., ADAM K., CHAPMAN K., OSWALD I., THOMSON J. – Viloxazine, sleep and subjective feelings, *Psychopharmacology,* 55, 121-128, 1977.
7. DAMASIO A., LOBO-ANTUNES J., MACEDO C. - Psychiatric aspects in Parkinsonism treated with L-Dopa. *J. Neurol. Neurosurg. Psychiatry,* 24, 502-507, 1971.
8. DE GROOT A.C., NATER J.P. – Dermatological drugs and cosmetics. *In: Meyler's side effects of drugs.* M.N.G. Dukes (ed.), Elsevier Science Publishers, 11th ed., 284-315, 1988.
9. DE RECONDO J., ZIEGLER M. – Les antiparkinsoniens *In: Neuropharmacologie clinique: le médicament en neurologie.* H. Dehen., G. Dordain (eds), Doin, Paris, 275-325, 1989.
10. DOOGAN' D.P. – Toleration and safety of sertraline: experience worldwide. *Int. Clin. Psychopharmacol.* 6(suppl.2), 47-56, 1991.
11. ELLINWOOD E.H., ROCKWELL W.J.K. – Central nervous system stimulants and anorectic agents. *In: Meyler's side effects of drugs.* M.N.G. Dukes (ed), Elsevier Science Publishers, 11th ed., 1-26, 1988.
12. ENTSUAH A.R., RUDOLPH R.L., CHITRA R. – Effectiveness of venlafaxine treatment in a broad spectrum of depressed patients: a meta-analysis. *Psychopharmacol Bull,* 31, 759-766,1995.
13. FELINE A., JOUVENT R. – Manifestations psychosensorielles observées chez des psychotiques soumis à des médications β-mimétiques. *Encéphale,* 3, 149-158, 1977.
14. GRAYBILL J.R., CRAVEN P.C. – Antifungal agents used in systemic mycoses activity and therapeutic use. *Drugs,* 25, 41-62, 1983.
15. HANIN B., MARKS J.- Dépendance aux benzodiazépines et syndrome de sevrage. *Revue de la littérature. Psychiatrie et Psychobiologie,* 3, 347-364, 1988.
16. HERMANT J.F., RAMBERT F.A., DUTEIL J. – Awakening properties of modafinil: effect on nocturnal activity in monkeys *(Macaca mulatta)* after acute and repeated administration. *Psychopharmacology,* 103, 28-32, 1991.
17. HINDMARCH I., BHATTI J.Z. - Psychopharmacological effects of sertraline in normal healthy volunteers. *Eur. J. Clin. Pharmacol.* 35: 221-223, 1988.
18. HINDMARCH I., SHILLINGFORD J., SHILLINGFORD C. - The effects of sertraline in psychomotor performance in elderly volunteers. *J. Clin. Psychiatry,* 51, 34-36, 1990.
19. IDANPAAN-HEIKKILA J.E., ARANKO K. – Miscellaneous antibacterial and antiviral drugs. *In: Side effects of drugs,* Annual 15, M.N.G. Dukes and J.K. Aronson (ed.), Elsevier Science Publishers. 308-326, 1991.
20. IMMAN W. – Paroxetine Drug Safety Research Unit, PEM report United Kingdom, 1993.
21. KARACAN J., THOMBY J.I., ANCH A.M., BOOTH G.H., WILLIAMS R.L., SALIS P.J.- Dose-related sleep disturbances induced by coffee and cafeine. *Clin. Pharmacol. Ther.,* 20, 6, 682-689, 1976.
22. LADER M. – Fluoxetine efficacy vs comparative drugs: an overview. *Br. J. Psychiatry,* 3, S51-S58, 1988.
23. LING M.H., PERRY P.J., MING T.T. - Side effects of corticosteroid therapy. *Arch. Gen. Psychiatry,* 38, 471-477, 1981.
24. LOISEAU P., LAROUSSE C., BOURIN M. – Antiépileptiques. *Sem. Hôp. Paris* 60, 39, 2741-2764, 1984.
25. LOO H. – Les accidents psychiatriques au cours des thérapeutiques hormonales. *Rev. Prat.,* 25, 2411-2413, 1975.
26. METZ A., SHADER R.I. – Adverse interactions encountered when using trazodone to treat insomnia associated with fluoxetine . *Intern. Clin. Pharmacol.,* 5, 191-194, 1990.
27. MIGUERES J., JOVER A., GARBARINI A.- Incidents et accidents neuropsychiatriques de la chimiothérapie antituberculeuse. *Journal des Agrégés,* 3, 111-119, 1970.
28. PAYKEL E.S., FLEMINGER R., WATSON J.P. – Psychiatric side-effects of antihypertensive drugs other than reserpine. *J. Clin. Psychopharmacol.,* 2, 1, 14-39, 1982.
29. RICHMAN D.D., FISCHL M.A., GRIECO M.H., GOTTLIEB M.S., VOLBEROING P.A., LASKIN O.L., LEEDOM J.M., GROOPMAN J.E., MILDVAN D., HIRSH M.S. *et al.* – The toxicity of azidothymidine (AZT) in the treatment of patients with AIDS and AIDS-related complex. A double-blind, placebo-controlled trial. *N. Engl. J. Med.,* 192-197, 1987.
30. RICKELS K., WARD T.S., GAUDIN V. – Comparison of two regimens of fluoxetine in major depression. *J. Clin. Psychiatry,* 46, 3, 38-41, 1985.
31. SIMON P., CHERMAT R. – Insomnie imputable à l'acide oxolinique. *La nouvelle Presse Médicale,* 6, 40, 3754, 1977.
32. SMITH J.M. – Abuse of the antiparkinson drugs: a review of the literature. *J. Clin. Psychiatry,* 41, 351-354, 1980.
33. VANELLE J.M., AUBIN F., MICHEL F. – Les complications psychiatriques de la corticothérapie. *Revue du Praticien,* Paris. 16, 556-558, 1990.
34. VAN RIEL P.L.C.M. – Metals. *In: Meyler's side effects of drugs.* M.N.G. Dukes (ed.), Elsevier Science Publisher. 11th ed., 437-460, 1988.
35. VERSIANI M., OGGERO U., ALTERWAIN P., CAPPONI R., DAJAS F., HEINZE-MARTIN G., MARQUEZ C.A., POLEO M.A., RIVERO-ALMANZOR L.E., ROSSEL L., *et al.* – A double-blind comparative trial of moclobemide versus imipramine and placebo in major depressive episodes. *Br. J. Psychiatry,* 155 (suppl. 6), 72-77, 1989.
36. WAGNER W., CIMANDER K. *et al.* Influence of concomitant psychotropic medication on the efficacy and tolerance of fluvoxamine. *Adv. Pharmacother.* 2, 34-56, 1986.

37. WEISS B., LATIES V.G. – Enhancement of human performance by caffeine and the amphetamines. *Pharmacol. Rev.,* 14, 1, 1-36, 1962.
38. WENDER P.H. – Pharmacotherapy of attention-deficit / hyperactivity disorders in adults. *J. Clin Psychiatry,* 59, 76-79, 1998.
39. ZERBE R. – Safety of fluoxetine . *Br. J. Clin. Pract.,* 40, 7 (suppl. 46), 41-47, 1986.

Chapter 21

Benzodiazepines and new non-benzodiazepine agents

J.M. Monti

Departemento de Farmacologia y Terapeutica, Hospital de Clinicas, Montevideo, Uruguay

The definition of insomnia changes and broadens over time – proof that it is still not completely understood by patients and clinicians. Added to complaints of sleep onset difficulties, and insufficient sleep and recurrent nocturnal awakenings, are those of daytime sleepiness, fatigue, irritability, lack of concentration and inadequate performance. Many chronic insomniacs complain of daytime sleepiness and this may constitute a risk factor, in terms of accidents at home and at work, and the costs to society.

There are three main classifications for insomnia, the World Health Organisation's International Statistical Classification of Diseases and Related Health Problems (ICD-10) [74], the International Classification of Sleep Disorders (ICSD) [19], and the Diagnostic and Statistical Manual of Mental Disorders (DSM-IV) [3]. In medical practice psychologists and psychiatrists tend to use the DSM-IV, whereas sleep specialists prefer the ICSD and general practitioners the ICD-10. However the ICSD seems to be the most adequate to define the therapeutic indications of sleep medication.

The International Classification of Sleep Disorders [19] includes 1) the dyssomnias which comprise, a) intrinsic sleep disorders (they originate or develop within the body or arise from causes within the body); b) extrinsic sleep disorders (they originate or develop from causes outside the body); c) circadian rhythm sleep disorders (they are related to the timing of sleep within the 24-hour day); 2) the parasomnias (they intrude into the sleep process and are not primarily disorders of sleep and wake states per se); 3) sleep disorders associated with medical/psychiatric disorders.

Thus not only is there a lack of agreement between the main sleep classifications of insomnia, but different terms are even used to designate the same entity. Thus in the DSM-IV, primary insomnia includes psychophysiological insomnia, idiopathic insomnia and inadequate sleep hygiene, whereas the ICSD considers the first two entities as intrinsic sleep disorders and the third as an extrinsic disorder.

From the point of view of treatment, two primary types of disorders can be considered: a) the hypersomnias, represented by obstructive sleep apnoea and narcolepsy, and b) the insomnia complaints, which can be due to a variety of psychophysiological, psychiatric, pharmacological, medical or circadian rhythm disorders.

The duration of insomnia has also been considered as an important guide to evaluation and treatment of insomnia [21]. Transient insomnia refers to insomnia related to minor situational stress and lasts no more than a few nights. Short-term insomnia is usually related to stress associated with work and family life and lasts less than about three weeks. Long-term insomnia may be associated with a variety of medical or neuropsychiatric conditions and lasts for months or years.

In the 1990 Cross National Medication Survey [6], conducted in the United States and eight Western European Countries, prevalence rates for mild problems with insomnia varied by country from 23 to 51%. Prevalence for serious problems with insomnia in these same populations varied from 10 to 26%. Serious problems with insomnia appeared at all ages, were slightly more frequent in women than in men (58% to 42%), and became more pronounced with age.

In the study by Hohagen *et al.* [36], in which insomnia was evaluated by means of operational diagnostic criteria (DSM II-R), in 2,512 patients, 18.7% of these had serious problems with insomnia, 12.2% mild insomnia and 15% slight insomnia. The increased prevalence of insomnia with age was more pronounced in the group of patients seriously affected by insomnia and, within this group, women largely predominated. Chronic insomnia is often associated with other disorders. In a follow up study of 7,696 patients affected by sleep and wakeful disorders [15], 80% of which underwent polysomnographic examination, 20 to 25% of diagnosed cases were affected by disorders of initiating and maintaining sleep, or in other words, insomnia. The most common diagnosis was insomnia associated with psychiatric disorders, followed by psychophysiological insomnia, periodic leg movement syndrome and substance dependence.

Buysse *et al.* [14] examined the clinical diagnoses of sleep disorders using the ICSD, DSM IV and ICS 10. Here again, insomnia linked to a psychiatric disorder was the most frequent, followed by psychophysiological insomnia and that linked with inadequate sleep hygiene.

Since insomnia has many causes, the indications for treatment are dependent on the aetiology. Transient and short-term insomnia are treated with a rapidly eliminated hypnotic and sleep hygiene. Chronic insomnia involves different treatment depending on the origin. When related to psychiatric complaints, particularly depression, the antidepressants are useful, mainly those with a sedative effect. In chronic psychophysiological insomnia and generalised anxiety, non drug strategies and sleep-promoting medication are indicated.

In conclusion, effective treatment of insomnia depends upon the duration and the primary cause of the complaint, and a hypnotic could be the drug of choice in many circumstances.

HYPNOTIC DRUGS

The ideal hypnotic

From our present knowledge about sleep and wakefulness in normal subjects, and the architecture of sleep in terms of polysomnographic patterns and states and stages of sleep, the ideal hypnotic should: a) rapidly and reliably induce sleep; b) sustain sleep and prevent repeated awakenings during a period of 7 to 8 hours; c) preserve the architecture of sleep; d) not induce residual effects in the morning following treatment; e) not show a decrease in efficacy with repeated administration; f) not be followed by rebound insomnia after its withdrawal; g) have a wide margin of safety in overdose and not interact with other central nervous system depressants; h) have no potential for abuse following long-term administration of larger-than-usual or therapeutic doses [47].

Several classes of medication have been prescribed as hypnotics over the years. In the 1960s, barbiturates, carbamates, chloral derivatives, piperidinediones and methaqualone were frequently used as hypnotics [31]. Although they were effective in increasing sleep duration in transient and short-term insomnia, tolerance and loss of effectiveness developed during long-term drug administration [39]. In addition, many of these drugs induced severe toxicity in overdose, showed a high liability to physical dependence and interacted with the metabolism of other drugs [26-27].

The benzodiazepine hypnotics

The benzodiazepines (BZDs) were introduced in the1960s, rapidly increased in popularity and have been the leading hypnotics in the field for more than two decades. Recently, 3 compounds that are structurally unrelated to the BZDs (and to each other) have become available for the treatment of insomnia. They are termed zopiclone, zolpidem and zaleplon and belong to the cyclopyrrolone, imidazopyridine and pyrazolopyrimidine class, respectively.

Benzodiazepine derivatives act on the benzodiazepine receptor which constitutes an allosterically active site coupled to the alpha subunit of the GABA-A chloride ionophore complex which exists in 6 isoforms ($\alpha 1$–$\alpha 6$). The GABA-A receptor containing subunit $\alpha 1$ may be of particular importance in sleep mechanisms but of limited importance in those of muscular control [12]. At least three different subtypes of BZD receptors (BZD1, BZD2, and the peripheral BZD site) have been characterised in mammal central nervous systems. Both anxiolytic and sedative effects seem to be mediated by the BZD1 receptor, while myorelaxant and anticonvulsant effects would be related to BZD2 receptor activation [76].

Benzodiazepine derivatives bind simultaneously to the BZD1 and BZD2 receptors. Although the qualitative character of the drug-receptor interaction is similar among different BZDs, they differ considerably in their intrinsic receptor affinity.

Pharmacokinetic aspects

Clinical differences among BZDs are mainly dependent on their pharmocokinetic characteristics. When BZDs are given in a single dose or for a very short period of time, the main determinant of the onset of action is the rate of absorption from the gastrointestinal tract. The duration of action of a single dose is determined by the rate and extent of drug distribution out of the central nervous system. Thus the duration of effect of a single dose of a long elimination half-life drug like flunitrazepam is almost similar, because the former rapidly distributes in peripheral tissue [30].

The half-life becomes significant when BZDs are administered in multiple doses over several weeks or months. During multiple dosage the BZD blood level builds to a steady state in about five half-lives. Under this condition, long half-life drugs accumulate markedly, while short half-life drugs accumulate only minimally. When the BZD has been discontinued, it takes five half-lives to be more than 90% eliminated from the body [2].

Benzodiazepine hypnotics are rapidly absorbed, with peak plasma concentration attained in 60-120 minutes. Plasma protein binding is generally extensive, figures of 90-98% being given for most BZDs. Metabolic processes occur in the liver, and the biotransformation products themselves very often have hypnotic activity (table 21.1). Benzodiazepine inactivation comprises hydroxylation, methylation, oxidation and dealkylation (phase 1), and conjugation to form glucuronides (phase II). Midalozam, triazolam, brotizolam, flunitrazepam and quazepam undergo phase I and phase II metabolism, while temazepam and lormetazepam are mainly metabolised by conjugation with glucuronic acid and excreted by the liver [25, 29].

According to their elimination half-life BZDs can be divided into short-, intermediate or long-acting derivatives (table 21.1).

Table 21.1. Half-life values and active metabolites for various benzodiazepine hypnotics, zopiclone, zolpidem and zaleplon.

Product	Half-life (in hours)	Active metabolite
Short-acting		
Midazolam	1.2 - 2.5	Methylhydroxymidazolam
Triazolam	2.1 – 6.0	Methylhydroxytriazolam
Brotizolam	3.1 – 6.1	Methylhydroxybrotizolam
Zopiclone	3.5 – 6.0	N-oxide derivative
Zolpidem	2.0 – 2.5	
Zaleplon	1.0	
Intermediate action		
Flunitrazepam	9.0 – 31.0	7 – amino derivative
Lormetazepam	9.9	N – desmethyl
Temazepam	10.0 – 20.0	
Long-acting		
Flurazepam	40.0 – 150.0	Hydroxyethylflurazepam
		N-desalkylflurazepam
Quazepam	47.0 –100.0	N-desalkylflurazepam

Effects on sleep variables

Irrespective of their half-life, few differences exist among BZDs in their effectiveness in inducing and maintaining sleep.

The sleep induced by BZD hypnotics is characterised by a shortened sleep-onset latency, decreased number of nocturnal awakenings, reduction in time spent awake which is exclusively related to an increment in stage 2 NREM sleep, a consistent reduction in NREM sleep stages 3 and

4, a mild and dose-dependent REM sleep suppression, and improvement in the subjective quality of sleep [46, 49, 52, 55].

Clinical and polysomnographic studies of triazolam 0.25 - 0.50mg, temazepam 30 mg and flurazepam 30 mg have found improvement in sleep for 4 to 5 weeks [43, 44, 45, 51]. Thus, BZDs are effective agents when administered in terms of a relatively short period of time in chronic insomniacs. Because of their loss of effectiveness after a few weeks, it is best to consider intermittent administration.

Changes in values corresponding to several sleep variables during BZD use clearly seem to be beneficial in patients with chronic insomnia. Accordingly, increased total sleep time and reduced NREM sleep latency, wake time after sleep onset and the number of nocturnal awakenings result in improved subjective estimates of sleep as well as enhanced daytime alertness, performance, mood and sense of well being. The increase of stage 2 sleep and decrease of NREM sleep stages 3 and 4, and REM sleep, are less clearly related to daytime alertness and sleep quality. However, reduction of either sleep stage must be considered *a priori* as a negative change in sleep architecture.

Spectral analysis of sleep in healthy volunteers has shown that benzodiazepine hypnotics reduced low frequency activity (delta and theta bands) and increased spindle activity [11, 20].

In Feinberg's study [23] where EEG profiles were analysed by period-amplitude, triazolam and temazepam reduced the amplitude and incidence of the delta band. Moreover the incidence of sigma waves was significantly increased both by triazolam and temazepam, and that of beta waves by temazepam.

Adverse effects

Commonly reported adverse effects of BZD hypnotics are drowsiness, tiredness, impairment of intellectual function, dizziness, dysarthria, ataxia, depression and anterograde amnesia [27]. Two types of adverse effect warrant particular attention: decreased daytime functioning on many measures of psychomotor performance in patients taking long-acting derivatives, and daytime anxiety in subjects receiving short-acting hypnotics.

Increased daytime sleepiness as quantified by the polysomnographically measured disposition to fall asleep during daytime (multiple sleep latency test), is associated with a rise in accident rate on the road, in the home and at work.

The elderly are particularly susceptible to adverse effects, and this could be related to altered pharmacokinetics as a result of changes in metabolism or because of slowed renal excretion.

Elderly subjects often have concomitant illnesses and take several medications. There is thus a dramatic increase in the incidence of potentially dangerous medicinal interactions, adverse effects and accidents in the home and at work. For this reason, elderly subjects should be prescribed doses equivalent to 1/3 or 1/2 the usual adult dose. Furthermore, adaptations to treatment should be made slowly and reduced to a minimum.

Problems related to long-term use

Rebound insomnia has been shown to develop following withdrawal from short- and intermediate-acting BZD hypnotics [39, 52]. This is characterised by temporary changes from baseline that are opposite to those initially produced by the BZD hypnotic itself. The BZD rebound insomnia includes increased NREM sleep latency, decreased total sleep time, disrupted and fragmented sleep and increased dreaming associated with a REM sleep rebound.

A rebound insomnia has been described for midazolam, triazolam, brotizolam, lormetazepam and flunitrazepam. Rebound insomnia appears to occur more rapidly following administration of short-acting BZDs.

The rebound insomnia induced by triazolam has been thoroughly studied. It was described in poor sleepers and chronic insomniacs who were given 0.25 – 0.5 mg triazolam/night during periods ranging from 4 – 37 nights. The biggest increase in total wake time occurred on the first night of withdrawal. On the second and subsequent nights total wake time tended to return to baseline levels [27].

A withdrawal symptom denoting the presence of psychological and physical dependence follows the abrupt cessation of BZDs administration after treatment with standard therapeutic doses or in high-dose BZD dependent patients. In patients receiving short- or intermediate half-life BZDs, the withdrawal syndrome develops the day after their discontinuation; when BZDs with long half-life are given the withdrawal syndrome shows 3-8 days later. It seems that short-acting BZD hypnotics have a greater potential for causing a withdrawal reaction as compared to the long-acting ones.

The withdrawal syndrome comprises frequent unspecific symptoms including sleep disturbances, anxiety, irritability, headache and nausea; perceptual changes of a qualitative nature such as hypersensitivity to noise, light, smell and touch; perceptual changes of a mainly qualitative nature including kinaesthetic, optical, gustatory, acoustic and olfactory symptoms, and other and less frequent symptoms such as depersonalisation, psychoses and epileptiform crises [67].

Cyclopyrrolone derivatives: Zopiclone

Zopiclone binds to central BZD receptors in a similar manner to the binding of diazepam; like BZDs it does not discriminate between central BZD receptor subtypes [38, 75]. In laboratory animals zopiclone displays sedative-hypnotic, anxiolytic, anticonvulsant and muscle relaxant properties.

Pharmacokinetic aspects

Following a single oral dose of 7.5 mg zopiclone is rapidly absorbed. Peak plasma concentrations are found in 0.5-1.5 hour. The compound undergoes oxidation to the N-oxide active metabolite, and demethylation to inactive N-desmethylzopiclone. The elimination half-lives of zopiclone and its active metabolite range from 3.5 to 6 hours. These values lengthen to 8 hours in cirrhotic and elderly patients.

Effects on sleep variables

Zopiclone is effective in inducing and maintaining sleep after a 7.5 mg dose in chronic insomniacs. Total sleep time increase is related to larger amounts of NREM sleep. No significant change was observed in REM sleep duration or as percentage of total sleep time. In regard to its effects on sleep stages, zopiclone decreased stage 1 and increased stage 2 sleep. Concerning NREM sleep stages 3 and 4, a slight increase of stage 3 or stages 3 and 4 combined was observed in healthy young subjects [9, 70]. However, in adult and old aged insomniac patients zopiclone induced a decrease, no change or an increase in stage 3 and/or stage 4 percentage of total sleep time [42, 56, 59, 60].

On EEG, zopiclone produces spectral changes in the normal patient very close to those caused by benzodiazepine hypnotics, i.e; they tend to suppress low frequency (delta and theta) activity and increase that of spindles.

Adverse effects

Adverse effects were evaluated in a phase 4 study based on 20,513 patients treated with zopiclone [1]. 9.2% of subjects reported at least one adverse effect leading to discontinuation of treatment in 2.8% of cases. A bitter taste was reported by 3.64% of subjects, dry mouth by 1.58%, difficulty getting up in the morning, by 1.30%, drowsiness by 0.52%, nightmares by 0.49%. Administered in the daytime, zopiclone leads impaired memory and psychomotor performance up to 4.5 hours after the dose is administered [34, 40].

Long-term use

Withdrawal of zopiclone, in a restricted number of cases, causes rebound insomnia with characteristics similar to those found after the short half-life BZDs [56, 59]

A withdrawal effect may occur in healthy volunteers and insomniac patients after abrupt zopiclone withdrawal [41, 68, 69]. This must be taken into account.

Imidazopyridine derivatives: Zolpidem

In the primate brain, zolpidem displays very high selectivity for BZD1 receptors. The compound is 73 fold more potent at displacing (3H)-flunitrazepam binding from BZD1 sites than from BZD2 sites [5, 17].

High selectivity for BZD1 sites coupled with high intrinsic activity at the level of the BZD1 site of the GABA-A receptor complex seems to be the determinant for the hypnotic activity of zolpidem.

Zolpidem shares many of the pharmacological actions characteristic of BZDs. Accordingly, it shows anticonvulsant, myorelaxant, sedative, sleep-inducing and anticonflict activity. However, as compared to the BZDs, there are striking quantitative differences. Thus, its sedative effect in laboratory animals occurs at doses 10-20 times lower than those inducing anticonvulsant and myorelaxant effects [4, 18, 76].

Pharmacokinetic aspects

Zolpidem is rapidly absorbed after oral administration, with peak plasma concentrations attained 2 hours after a single therapeutic dose. It is about 92-94% bound to plasma protein and its brain penetration is very rapid. The major metabolic routes in humans include oxidation and hydroxylation, and none of the metabolites are pharmacologically active. The mean elimination is half-life of zolpidem in healthy volunteers is 2.0 – 2.5 hours.

Effects on sleep variables

In the young adult zolpidem 10 and 20 mg reduced sleep onset latency at night and increased total sleep time; NREM sleep stages 3 and 4 were increased after 20 or 30 mg doses [57]. In healthy geriatric non-insomniac subjects, sleep latency was significantly reduced after zolpidem 10 mg. While total wake time and sleep efficiency were improved, no effects wee seen on percentage of NREM sleep stages 3 and 4 [66].

Zolpidem 10 mg did not induce disturbances of respiration during sleep in normal elderly females [61].

Polysomnographic studies in poor sleepers and insomniac patients have shown that zolpidem 10 mg significantly increased sleep duration and diminished time awake after the onset of sleep during 2-4 weeks. NREM sleep latency was also reduced. Zolpidem markedly increased the duration of stage 2 sleep without significantly affecting either NREM sleep stages 3 and 4 or REM sleep [48, 51, 53, 54]. Sleep analysis tends to show that zolpidem only slightly disturbs sleep architecture [48]. No evidence of tolerance was observed in long-term (6-12 months) studies [61, 65]. Brunner *et al.* [13] showed that zolpidem induced a reduction of spectral density in the spindle band in healthy volunteers. Conversely, in quantifying delta power fluctuations from one cycle to another, Patat *et al.* [58] found a significant increase in the latter, while Ferillo *et al.* [24] found no difference between zolpidem nights and placebo nights. In the studies by Declerck *et al.* [16] and Benoit *et al.* [8] carried out among poor sleepers, zolpidem showed no effect on delta activity. Furthermore in the analysis of Feinberg *et al.* [23], zolpidem was shown to diminish mean delta amplitude in healthy subjects, but did not change its incidence within 20 second epochs. Moreover, administration of zolpidem at a daily dose of 10 mg for 15 nights significantly increased power density in the 0.25-1.0 Hz band during the first part of the night in patients with chronic primary insomnia [50].

Adverse effects

Zolpidem did not impair psychomotor performance of patients on the morning after its administration. Adverse effects included dizziness and lightheadedness, somnolence, headache and gastrointestinal upset. To our knowledge no phase 4 study of the adverse effects of the product has so far been published. Zolpidem at doses of 10 and 20 mg had no effect on short-term, retrograde or anterograde memory, the morning after administration [62].

Long-term administration of zolpidem

Rebound effects did not occur on withdrawal of zolpidem [33, 48, 51]. Moreover, withdrawal signs were not apparent after abrupt discontinuation of zolpidem in patients with insomnia treated for up to one year [62, 65].

Pyrazolopyrimidine derivatives: Zaleplon

Zaleplon binds selectively to the brain BZD1 receptor situated on the alpha subunit of the GABA-A-BZD receptor complex [7]. In animal studies, zaleplon has demonstrated sedative, anxiolytic, muscle relaxant, and anticonvulsant activity [64].

Pharmacocinetic aspects

Zaleplon is rapidly and almost completely absorbed following oral administration. However, the compound undergoes extensive presystemic metabolism. As a result, the absolute oral bioavailability of zaleplon is approximately 30%. Peak plasma concentration (Cmax) after single-dose oral zaleplon 10 mg is attained within 1 hour.

Zaleplon is metabolised by aldehyde oxidase to form 5-oxo-zaleplon, and by CYP3A4 to form desethylzaleplon. In man the aldehyde oxidase system predominates. All of zaleplon's metabolites are pharmacologically inactive. The drug is rapidly cleared from the body with elimination half-life (t1/2) of 1 hour.

Effects on sleep variables

Zaleplon (5 and 10 mg) has been administered to normal adults experiencing transient insomnia during the first night in the sleep laboratory. The 10 mg dose was superior to placebo in decreasing latency to persistent sleep.

Adult and elderly outpatients with chronic insomnia were evaluated in double-blind, parallel-group studies that compared the effects of zaleplon 5, 10 and 20 mg with placebo on a subjective measure of time to sleep onset. Zaleplon 10 and 20 mg significantly reduced subjective sleep latency throughout the trial in the adult patients [22].

In a randomised double-blind sleep laboratory study, elderly patients with chronic primary insomnia received zaleplon 5 or 10 mg or placebo for 14 nights. Mean reduction from baseline of latency to persistent sleep was significantly greater with zaleplon 10 mg than with placebo on the first 2 nights of treatment. In other words, a significant difference from placebo was not seen beyond 2 nights [72].

More recently, Walsh *et al.* (2000) re-examined the hypnotic efficacy of zaleplon 10 mg over a period of 35 nights in primary insomniacs. Sleep latency was significantly shortened with zaleplon for all 5 weeks of treatment as assessed by polysomnography and by subjective sleep measures.

Zaleplon did not have a significant effect on total sleep time, the number of nocturnal awakenings, NREM sleep stages 3 and 4 or REM sleep. There was no evidence of tolerance to the sleep promoting effects of zaleplon during the five weeks of administration [73]. Therefore, zaleplon's main clinical usefulness is in the treatment of sleep initiation insomnia.

Adverse effects

No increase in daytime anxiety has been detected after night-time administration of zaleplon. At 5 and 10 mg there was no objective and minimal subjective evidence of rebound insomnia on the first night after discontinuation of treatment with the hypnotic drug. However, rebound has been noted after administration of zaleplon 20 mg [34].

Studies involving the exposure of normal subjects to zaleplon 10 or 20 mg with structured assessment of short-term memory, sedation and psychomotor function, revealed the expected impairment at 1 hour, the time of peak exposure to the drug for both doses. However, no impairment was evident after 4 hours [71].

The abuse potential of zaleplon has been examined in subjects with known histories of sedative drug use. Available evidence tends to indicate that zaleplon has an abuse potential similar to triazolam [63].

The long-term use of hypnotics drugs

In certain patients it may be necessary to prescribe a hypnotic drug for a prolonged period. This is widely recognised by general practitioners whereas many sleep specialists and psychiatrists are reluctant to extend prescriptions beyond several weeks. It has been proven that many patients use hypnotic drugs for long periods. Moreover the use of hypnotics does not necessarily result in a pharmacological escalation and greater flexibility should be applied to the long term use of these drugs.

Table 21.2. Effects of the benzodiazepine hypnotics, zopiclone, zolpidem and zaleplon in controlling insomnia.

	BZDs	Zopiclone	Zolpidem	Zaleplon
Sleep latency	decreased	decreased	decreased	decreased
N° of awakenings	decreased	decreased	decreased	unchanged
Total sleep time	increased	increased	increased	unchanged
stage 2	increased	increased	increased	unchanged
stages 3 and 4	decreased	decreased or unchanged	unchanged	unchanged
REM sleep	slightly or mildly decreased	decreased or unchanged	unchanged	unchanged
Subjective sleep quality	improved	improved	improved	improved
Rebound insomnia	present	present ?	absent	present
Tolerance	present	absent (long-term study)	absent (long-term study)	absent
Dependence	present	?	absent	absent

CONCLUSION

On administration of BZDs, zopiclone or zolpidem, NREM sleep latency is decreased and total sleep time is increased. On the other hand, zaleplon only effectively reduces time to sleep onset. In contrast to BZDs, therapeutic doses of zopiclone, zolpidem or zaleplon do not induce a rebound of insomnia or only among a limited number of subjects, and do not appear to induce phenomena of tolerance or dependence (table 21.2.). A withdrawal reaction is likely to occur following an abrupt discontinuation of benzodiazepines, zopiclone and zaleplon.

Thus in the light of the results available at present, zopiclone and zolpidem appear to present advantages in relation to benzodiazepines and zaleplon, in controlling insomnia.

REFERENCES

1. ALLAIN H., DELAHAYE CH., LE COZ F., BLIN P., DECOMBE R., MARTINET J.P. – Postmarketing surveillance of zopiclone in insomnia: analysis of 20 513 cases. *Sleep*, *14*, 408-413, 1991.
2. AMERICAN PSYCHIATRIC ASSOCIATION - *Benzodiazepine: Dependence, toxicity and abuse*. A Task Force Report. 1st ed., American Psychiatric Association, Washington, DC, 1990.
3. AMERICAN PSYCHIATRIC ASSOCIATION – *Diagnostic and statistical manual of mental disorders*. 4th ed. American Psychiatric Association, Washington D.C., 1994.
4. ARBILLA S., ALLEN J., WICK A., LANGER S.Z. – High affinity (3 H) zolpidem binding in the rat brain: an imidazopyridine with agonist properties at central benzodiazepine receptors. *Eur. J. Pharmacol.*, *130*, 257-263, 1986.
5. ARBILLA S., DEPOORTERE H., GEORGE P., LANGER S.Z. – Pharmacological profile of the imidazopyridine zolpidem at benzodiazepine receptors and electrocorticogram in rats. *N. S. Arch. Pharmacol.*, *330*, 248-251, 1985.
6. BALTER M. – What are the prevalence and characteristic of insomnia? *Insomnia Research Bulletin. WFSRS Founding Congress*, *1*, 1991.
7. BEER B., CLODY D.E., MANGANO R., LEVNER M., MAYER P., BARRETT J.E. – A review of the preclinical development of zaleplon, a novel non-benzodiazepine hypnotic for the treatment of insomnia. *CNS Drug reviews* 3, 207-224, 1997.
8. BENOÎT O., BOUARD G., PAYAN C., BORDERIES P., PRADO J. – Effect of a single dose (10 mg) of zolpidem on visual and spectral analysis of sleep in young poor sleepers. *Psychopharmacology*, *116*, 297-303, 1994.
9. BILLIARD M., BESSET A., DE LUSTRAC C., BRISSAUD L. – Dose-response effects of zopiclone on night sleep and on night-time and day-time functioning. *Sleep*, *10* (suppl., 26-34), 1987.
10. BORBÉLY A.A. – Principles of sleep regulation: Implications for the effect of hypnotics on sleep. *In: The pharmacology of sleep*. A. Kales (ed.). Springer, Berlin, 29-45, 1995.
11. BORBÉLY A.A., MATTMANN P., LOEPFE M., STRAUCH I., LEHMANN D. – Effect of benzodiazepine hypnotics on all-night sleep EEG spectra. *Human Neurobiol.*, *4*, 189-194, 1985.
12. BORMANN J. – Electrophysiology of GABAa and GABAb receptor subtypes. *Trends Neurosci*, *11*, 112-113, 1988.
13. BRUNNER D.P., DIJK D.J., MUNCH M., BORBELY A.A. – Effect of zolpidem on sleep and sleep EEG spectra in healthy young men. *Psychopharmacology*, *104*, 1-5, 1991.
14. BUYSSE D.J., REYNOLDS C.F., KUPFER D.J., THORPY M.J., BIXLER E., MANFREDI R., KALES A., VGONTZAS A., STEPANSKI E., ROTH T., HAURI P., MESIANO D. – Clinical diagnoses in 216 insomnia patients using the international classification of sleep disorders (ICSD), DSM-IV and ICD-10 categories: a report from the APA/NINM field trial. Sleep, 17, 630-637, 1994.
15. COLEMAN R.M. – Diagnosis, treatment and follow-up of about 8 000 sleep/wake disorder patients. *In: Sleep/wake disorders: natural history epidemiology, and long-term evolution*. C. Guilleminault, E. Lugaresi (eds.). Raven Press, New York, 87-97, 1983.
16. DECLERCK A.C., RUWE F., O'HANLON J.F., WAU-QUIER A. – Effects of zolpidem and flunitrazepam on nocturnal sleep of women subjectively complaining of insomnia. *Psychopharmacology*, *106*, 497-501, 1992.
17. DENNIS T., DUBOIS A., BENAVIDES J., SCATTON B. – Distribution of central w1 (benzodiazepine 1) and w2 (benzodiazepine 2) receptor subtypes in the monkey and human brain. An autoradiographic study with (3H) zolpidem and the w1 selective ligand (3H) flunitrazepam. *J. Pharmacol. Exptl. Ther.*, *247*, 309-322, 1988.
18. DEPOORTERE H., ZIVKOVIC B., LLOYD K.G., SANGER D.J., PERRAULT G. – Zolpidem, a novel nonbenzodiazepine hypnotic. I. Neuropharmacological and behavioral effects. *J. Pharmacol. Exptl. Ther.*, *237*, 649-658, 1986.
19. AMERICAN SLEEP DISORDERS ASSOCIATION.– *ICSD - International Classification of Sleep Disorders: diagnostic and coding manual*. American Sleep Disorders Association: Rochester, 1997.
20. DIJK D.J., BEERSMA D.G.M., DAAN S., VAN DEN HOOFDAKKER R.H. – Effects of seganserin, a 5-HT$_2$ antagonist, and temazepam on human sleep stages and EEG power spectra. *Eur. J. Pharmacol.*, *171*, 207-218, 1989.
21. CONSENSUS CONFERENCE. Drugs and insomnia. *JAMA*, *21*, 2410-2414, 1984.
22. ELIE R., DAVIGNON M., EMILIEN G. – Zaleplon decreases sleep latency in outpatients without producing rebound insomnia after 4 weeks of treatment. *J. Sleep Res.* 7 (Suppl. 2): 76, 1998.
23. FEINBERG I., MALONEY T., CAMPBELL I. – Period-amplitude analysis (PAA) of amplitude and incidence of NREM EEG under zolpidem (Z), triazolam (Tr) and temazepam (Te) for delta (0.3-3 Hz), sigma (12-15 Hz) and beta (15-23 Hz) frequencies, *Sleep Res.*, *24A*, 123, 1995.
24. FERRILLO F., BALESTRA V., DE CARLI F., MANFREDI C., PISCHEDDA G.G., ROSADINI G. – Effects of the administration of zolpidem and triazolam on the dynamics of EEG slow waves during sleep. *J. Sleep Res.*, *1* (suppl. 1), 72, 1992.
25. GARZONE P.D., KROBOTH P.D. – Pharmacokinetics of the newer benzodiazepines. *Clin. Pharmacokinet.*, *16*, 337-364, 1989.
26. GILLIN J.C., MENDELSON W.B. – *Neuropharmacology of central nervous system and behavioral disorders*. 1st ed., Academic Press, New York, 1980.
27. GILLIN J.C., BYERLEY W.F. – The diagnosis and management of insomnia. *N. Eng. J. Med.*, *332*, 239-248, 1990.
28. GREENBLATT D.J., DIVOLL M., ABERNETHY D.R., OCHS H.R., SHADER R.I. – Benzodiazepine kinetics: Implication for therapeutics and pharmacogeriatrics. *Drug. Metab. Rev.*, *14*, 251-292, 1983.

29. GREENBLATT D.J., HARMATZ J.S., ENGELHARDT N., SHADER R.I. – Pharmacokinetic determinants of dynamic differences among three benzodiazepine hypnotics. *Arch. Gen. Psychiatry*, *46*, 326-332, 1989.

30. GREENBLATT D.J., MILLER L.G., SHADER R.I. – Neurochemical and pharmacokinetic correlates of the clinical action of benzodiazepine hypnotic drugs. *Amer. J. Med.*, *88* (suppl. 3A), 18S-24S, 1990.

31. HARVEY S.C. – *The pharmacological basis of therapeutics*, 5th ed., MacMillan, New York, 1988.

32. HEDNER J., EMILIEN G., SALINAS E. – Improvement in sleep latency and sleep quality with zaleplon in elderly patients with primary insomnia. *J. Sleep Res.* 7 (Suppl. 2): 115, 1998.

33. HERRMANN W.M., DUBICKI S., WOBER W. – Zolpidem: a four week pilot polysomnographic study in patients with chronic sleep disturbances. *In: Imidazopyridines in sleep disorders*, J.P. Sauvanet, S.Z. Langer, P.L. Morselli (eds.). Raven Press, New York, 261-278, 1988.

34. HEYDORN W.E. – Zaleplon, a review of a novel sedative hypnotic used in the treatment of insomnia. *Exp. Opin. Invest. Drugs* 9, 841-858, 2000.

35. HINDMARCH I. – Immediate and overnight effects of zopiclone 7,5 mg and nitrazepam 5 mg with ethanol, on psychomotor performance and memory in healthy volunteers. *Int. Clin. Psychopharmacol.*, *5* (suppl. 2), 105-143, 1990.

36. HOHAGEN F., RINK K., KAPPLER C., SCHRAMM E., RIEMANN D., WEYERER S., BERGER M. – Prevalence and treatment of insomnia in general practice. *Eur. Arch. Psychiatry Clin. Neurosc.*, *242*, 329-336, 1993.

37. HURST M., NOBLE S. – Zaleplon. *CNS Drugs*, 11, 387-392, 1999.

38. JULOU L., BLANCHARD J.C., DREYFUS J.F. – Pharmacological and clinical studies of cyclopyrrolones: zopiclone and suriclone. *Pharmacol. Biochem. Behav.*, *23*, 653-659, 1985.

39. KALES A., KALES J.D. – *Evaluation and treatment of insomnia*, 1st ed. Oxford University Press, New York, 1984.

40. KUITUNEN T., MATTILA M.J., SEPPALA T. – Actions and interactions of hypnotics on human performance: single doses of zopiclone, triazolam and alcohol. *Int. J. Psychopharmacol.*, *5* (suppl. 2), 115-130, 1990.

41. LADER M., FRECKA G. – Subjective effects during and on discontinuation of zopiclone and of temazepam in healthy subjects. *Pharmacopsychiatry*, *20*, 67-71, 1987.

42. MALEMAK M., SCIMA A., PRICE V. – Effects of zopiclone on the sleep of chronic insomniacs. *Pharmacology*, *27* (suppl. 2), 136-145, 1983.

43. MENDELSON W.B. – *Use and misuse of sleeping pills: A clinical guide*, 1st ed. Plenum Press, New York, 1980.

44. MENDELSON W.B. – Pharmacotherapy of insomnia. *Psychiat. Clin. North. Amer.*, *10*, 555-563, 1987.

45. MITLER M.M., SEIDEL W.F., VAN DEN HOED J., GREENBLATT D.J., DEMENT W.C. – Comparative hypnotic effects of flurazepam, triazolam and placebo. A long-term simultaneous nighttime and daytime study. *J. Clin. Psychopharmacol.*, *4*, 2-13, 1984.

46. MONTI J.M. – Sleep laboratory and clinical studies of the effects of triazolam, flunitrazepam and flurazepam in insomniac patients. *Meth. Find. Exp. Clin. Pharmacol.*, *3*, 303-326, 1981.

47. MONTI J.M. – Concluding remarks. Toward a third generation of hypnotics. *In: Imidazopyridines in sleep disorders.* J.P. Sauvanet, S.Z. Langer, P.L. Morselli (eds.). Raven Press, New York, 365-368, 1988.

48. MONTI J.M. – Effects of zolpidem on sleep in insomniac patients. *Eur. J. Clin. Pharmacol.*, *36*, 461-466, 1989.

49. MONTI J.M., ALTERWAIN P. – Short-term sleep laboratory evaluation of midazolam in chronic insomniacs. Preliminary results. *Arzn. Forsch.*, *37*, 54-57, 1987.

50. MONTI J.M., ALVARINO F., MONTI D. – Conventional and power spectrum analysis of the effects of zolpidem on sleep EEG in patients with chronic primary insomnia. *Sleep*, 23, 1075-1084, 2000

51. MONTI J.M., ATTALI P., MONTI D., ZIPFEL A., DE LA GIGLAIS B., MORSELLI P.L. – Zolpidem and rebound insomnia - a double-blind, controlled polysomnographic study in chronic insomniac patients. *Pharmacopsychiatry*, *27*, 166-175, 1994.

52. MONTI J.M., DEBELLIS J., GRATADOUX P., ALTERWAIN P., ALTIER H., D'ANGELO L. – Sleep laboratory study of the effects of midazolam in insomniac patients. *Eur. J. Clin. Pharmacol.*, *21*, 479-484, 1982.

53. MONTI J.M., MONTI D. – Pharmacological treatment of chronic insomnia. CNS Drugs. *4*, 182-194, 1995.

54. MONTI J.M., MONTI D., ESTEVEZ F., GIUSTI M. Sleep in patients with chronic primary insomnia during long-term zolpidem administration and after its withdrawal. *International Clinical Psychopharmacology*, 11, 255-263, 1996

55. MONTI J.M., TRENCHI H.M., MORALES F., MONTI L. – Flunitrazepam (RO 5-4200) and sleep cycle in insomniac patients. *Psychopharmacologia*, *35*, 371-379, 1974.

56. MOURET J., RUEL D., MAILLARD F., BIANCHI M. – Zopiclone versus triazolam in insomniac geriatric patients, *Int. Clin. Psychopharmacol.*, *5* (suppl. 2), 47-55, 1990.

57. NICHOLSON A.N., PASCOE P.A. – Hypnotic activity of an imidazo-pyridine (zolpidem), *Br. J. Clin. Pharmacol.*, *21*, 205-211, 1986.

58. PATAT A., TROCHERIE S., THEBAULT J.J., ROSEN-ZWEIG P., DUBRUC C., BIANCHETTI G., COURT L.A., MORSELLI P.L. – EEG profile of intravenous zolpidem in healthy volunteers. *Psychopharmacology*, *114*, 138-146, 1994.

59. PECKNOLD J., WILSON R., LE MORVAN P. – Long term efficacy and withdrawal of zopiclone. *Int. Clin. Psychopharmacol.*, *5* (suppl. 2), 57-67, 1990.

60. QUADENS O.P., HOFFMAN G., BUYTAERT G. – Effects of zopiclone as compared to flurazepam in women over 40 years of age. *Pharmacology*, *27* (suppl. 2), 146-155, 1983.

61. RHODES S., PARRY P., HANNING C. – A double blind comparison of the effects of zolpidem and placebo on respiration during sleep in the elderly. *Sleep Res.*, *18*, 69, 1989.

62. ROGER M., DALLOT J.Y., SALMON O., NEVEUX E., GITTON J.P., GERSON M., BROSSEL S., SAUVANET J.P. – Hypnotic effect of zolpidem in geriatric patients: a dose-finding study. *In: Imidazopyridines in sleep disorders*, J.P. Sauvanet, S.Z. Langer, P.L. Morselli (eds.). 1st ed. Raven Press, New York, 279-288, 1988.

63. RUSH C.R., FREY J.M., GRIFFITHS R.R. – Zaleplon and triazolam in humans: acute behavioral effects and abuse potential. *Psychopharmacology,* 145, 39-51, 1999.

64. SANGER D.J., MOREL E., PERRAULT G. – Comparison of the pharmacological profiles of the hypnotic drugs, zaleplon and zolpidem. *Eur. J. Pharmacol.* 313, 35-42, 1996.

65. SAUVANET J.P., MAAREK L., ROGER M., RENAUDIN J., LOUVEL E., OROFIAMMA B. – Open long-term trials with zolpidem in insomnia. *In: Imidazopyridines in sleep disorders*, J.P. Sauvanet, S.Z. Langer, P.L. Morselli (eds.). Raven Press, New York, 339-350, 1988.

66. SCHARF M., VOGEL G., VERNON PEGRAM V., CATESBY WARE J., ROSE V., HOELSCHER V. – Dose-response effects of zolpidem in patients with chronic insomnia. *Sleep Res., 19*, 90, 1990.

67. SCHÖPF J. – Withdrawal phenomena after long-term administration of benzodiazepines. A review of recent investigations. *Pharmacopsychiatria, 16*, 1-8, 1983.

68. SULLIVAN G., MCBRIDE A.J., CLEE W.B. – Zopiclone abuse in South Wales: Three case reports. *Human Psychopharmacol., 10*, 351-352, 1995.

69. THAKORE J., DINAN T.G. – Physical dependence following zopiclone usage: A case report. *Human Psychopharmacol., 7*, 143-145, 1992.

70. TRACHSEL L., DIJK D., BRUNNER D.P., KLENE C., BORBELY A.A. – Effects of zopiclone and midazolam on sleep and EEG spectra in a phase-advanced sleep schedule. *Neuropsychopharmacology, 3*, 11-18, 1990.

71. VERMEEREN A., DANJOU P.E., O'HANLON J.F. – Residual effects of evening and middle-of-the- night administration of zaleplon 10 and 20 mg on memory and actual driving performance. *Human psychopharmacol.* 13, S98-S107, 1998

72. WALSH J.K., FRY J., ERWIN C.W., SCHARF M., ROTH T., VOGEL G.W. – Efficacy and tolerability of 14-day administration of zaleplon 5 mg and 10 mg for the treatment of primary insomnia. *Clin. Drug. Invest.* 16, 347-354, 1998.

73. WALSH J.K., VOGEL G.W., SCHARF M., ERMAN M., ERWIN C.W., SCHWEITZER P.K., MANGANO R.M., ROTH T. A five week polysomnographic assessment of zaleplon 10 mg for the treatment of primary insomnia. *Sleep Medicine* 1, 41-49, 2000

74. WORLD HEALTH ORGANIZATION - *International statistical classification of diseases and related health problems* (ICD-10). 10th ed., World Health Organization, Geneva, 1992.

75. ZIVKOVIC B., ARBILLA S. – Role of w receptor subtypes in the action of hypnotic drugs. *Biol. Psychiat. Intern.*, Congress Series 1968, Excerpta Medica, Amsterdam, 1991.

76. ZIVKOVIC B., PERRAULT G., MOREL E., SANGER D.J. – Comparative pharmacology of zolpidem and other hypnotics and sleep inducers. *In: Imidazopyridines in sleep disorders*, J.P. Sauvanet, S.Z. Langer, P.L. Morselli (eds.). Raven Press, New York, 97-110, 1988.

77. ZIVKOVIC B., PERRAULT G., SANGER D.J. – *Target receptors for anxiolytics and hypnotics: from molecular pharmacology to therapeutics.* Karger, Basel, 1992.

Chapter 22

Other medications used for insomnia

F. Goldenberg
Laboratoire de Sommeil, Hôpital Henti Mondor, Créteil, France

Pharmacological treatment for insomnia is far from being restricted to prescribing hypnotics. Hypnotics are often ineffective in fact (tolerance effect), or even dangerous (in unsuspected cases of sleep apnoea) and may lead to problems of over-consumption and dependence. It must be remembered that since the international consensus conference in Bethesda [11], hypnotics can no longer be prescribed for more than three weeks. A number of other families of psychotropic medication can be used as treatment for insomnia: the foremost is the antidepressants, but others include the centrally active antihistamines, L tryptophan and plant extracts.

Antidepressants

There were a number of major disadvantages in the oldest antidepressants such as cardiovascular problems, anti-cholinergic side effects and the risk of death from overdose. The new antidepressants are far better tolerated. They offer the advantage of not leading to drug dependence or overconsumption. They tend to be prescribed instead of hypnotics whenever insomnia becomes chronic [115].

An antidepressant is the first line treatment for insomnia in a patient with patent depression. Morning insomnia responds well to all types of antidepressant (sedative or stimulant). For insomnia occurring during sleep latency or in the middle of the night, sedative antidepressants appear to be preferable. With the older generation of sedative antidepressants (e.g. amitryptiline), sleep improves from the first night of treatment, whereas 10 to 15 days are necessary before any improvement is seen in depression.

In insomnia which is not depressive, labelled as psychophysiological, we have known for about 20 years that the use of weak doses of sedative antidepressants ranging from a tenth to a third of the standard antidepressant dose, can have a sustained hypnotic effect for several months or even years. The antidepressants used varied from the old tricyclics like amitryptiline or doxepine to a tetracyclic like mianserine; the idea being that a weak dose of antidepressant is preferable to doubling the dose of a hypnotic which was no longer effective [110]. Unfortunately no control studies have proved the efficacy of this attitude in the treatment of chronic insomnia nor of the degree of tolerance. The idea has emerged in recent years that the chronic, so-called psychophysiological insomniac (because the subject currently presents no criteria of major depression or anxiety disorders as described in DSMIV), is very often a subsyndromal depressive or a subject with chronic anxiety but with no current decompensation, or even a personality tending to somatise in the form of the symptom of insomnia. Chronic antidepressant treatment, at a dose which remains to be determined, may provide sustained support for these patients by controlling their anxious tendencies and/or chronic depression and/or somatisation and thereby regulating their sleep. The worsening of poor sleep at certain moments in life would thus be considered as a warning sign indicating that the patient risks slipping into a patent psychiatric pathology.

Table 22.1. Initial effect of antidepressants on sleep

Name	Hypnotic effect (+) Awakening effect (-) No effect (0)	Effect on stage 4 duration	Effect on REM sleep duration
Tricyclic antidepressants:			
Amitriptyline	++	+	-
Clomipramine	-	+ (first 2 hours)	-
Desipramine	-	=	-
Imipramine	-	-	-
Amoxapine	+	+	-
Trimipramine	++	+	=
Doxepine	++	+ or =	-
Tetracyclics:			
Maprotiline	0 (healthy volunteer)	=	-
Mianserin	++	+	-
Serotonin reuptake blockers:			
Fluvoxamine	- -	=	-
Fluoxetine	- -	=	-
Paroxetine	- -	+ or =	-
Sertraline	0	=	-
Citalopram	0	= (depressed)	-
Standard MAOI			
Iproniazide	?	=	- -
MAOI A			
Moclobemide	+ (depressed) 0 or – (healthy volunteer)	=	- (healthy volunteer) + (depressed)
MAOI B			
Toloxatone	0	=	=
Other antidepressants			
Trazodone	++	+	-
Viloxazine	- -	-	-
Venlafaxine	-	=	-
Nefazodone	0	=	= or +
Mirtazapine	+	-	-
Lithium	+ or 0	+ or =	-

Table 22.1 summarises the effects of the different antidepressants on sleep. In fact this refers to the initial effect of the medication only, as very few long term studies have been carried out which monitor depressed subjects under treatment for a year or more. Antidepressants can have a true hypnotic effect (i.e. anti-wakefulness). They generally have a suppressive effect on REM sleep and a beneficial effect on NREM sleep. With the antidepressants which have an awakening rather than a hypnotic effect, NREM sleep may be deficient. Note that polygraphic studies of antidepressants are usually carried out on depressives, sometimes on healthy volunteers, but very rarely on insomniacs or poor sleepers.

The effect of antidepressants on sleep is partly due to the inhibition of serotonin and/or norepinephrine reuptake, and partly to the blocking effects they can have on different types of postsynaptic receptors (muscarinic, alpha 1 noradrenergic, H1 histaminic, 5-HT2 serotoninergic). Antidepressants which block serotonin or norepinephrine reuptake or both, reduce REM sleep and increase REM sleep latency. The antidepressants which moreover block the muscarinic receptors further diminish REM sleep and increase REM sleep latency. Antidepressants which are 5-HT2 antagonists increase deep NREM sleep. Those which are powerful alpha 1 or H1 antagonists increase sleep time and continuity. Finally those which potentialise serotoninergic transmission or noradrenergic transmission in the absence of any antagonist effect on alpha 1 receptors or H1 receptors reduce sleep and its continuity.

The hypnotic or anti-awakening effect

The hypnotic effect of the older antidepressants considered as sedatives, is immediate: an increase in total sleep time, decrease in sleep latency and the number and duration of intercurrent awakenings. In depressed insomniacs, with an equal antidepressant effect, **trimipramine** normalises sleep whereas **imipramine** does not objectively improve sleep, and has even been seen to worsen it due to its effects on variables such as the percentage of wakefulness or of NREM sleep [113].

The newer antidepressants are not immediately hypnotic. The serotonin reuptake inhibitors usually have an awakening effect with the first doses. **Fluoxetine** has a strong awakening effect during sleep [39, 70]. But with regular administration, its effects on reducing sleep efficiency and raising sleep latency are only minor. **Fluvoxamine** has an awakening effect at the start of treatment [46] but effectively re-establishes sleep thereafter. **Paroxetine** has an awakening effect, particularly at doses of 30 to 40 mg [16, 89]. This effect does not always occur [104]. A morning dose induces an increase in sleep latency, while an evening dose has no effect on falling asleep [16]. Paroxetine continues to have an objectively awakening effect after 6 weeks of treatment, although qualitatively, psychophysiological insomniac patients note an improvement in their insomnia [72]. In the healthy volunteer, paroxetine administered for 28 days at a dose of 30 mg has no effect on sleep latency, sleep time, sleep efficiency or the number and duration of awakenings [93]. **Citalopram** (20 mg the first week, 40 mg thereafter) does not alter sleep continuity in major depressives [111].

The effects on REM sleep

Tricyclic antidepressants lead to modifications in sleep from the first night of treatment: delayed onset of first period of REM sleep and diminished percentage of REM sleep [47]. These effects on sleep precede the period of thymic change by about ten days. Decreased REM sleep in the first two nights of treatment with amitriptyline, clomipramine or desipramine correlates positively with the response to treatment of major depressives [26, 48, 86]. **Trazodone** prolongs REM sleep latency but this effect only becomes significant after four weeks of treatment [68]. In another study [62], REM sleep is discretely reduced by trazodone from the outset of treatment (passing from 27.7% to 17%) with a significant rebound on the 5th night of discontinuation. **Mianserin** immediately blocks REM sleep [67] as does **amoxapine** [22], **fluvoxamine** [46], **paroxetine** [73, 89, 105] and **fluoxetine** [39, 73]. Conversely, after regular monotherapy treatment for months or years, fluoxetine considerably prolongs REM sleep latency but does not diminish the percentage of REM sleep [92]. Paroxetine at 30 mg will always block REM sleep on the 28th day, in healthy volunteers: REM sleep latency is increased and the number of REM sleep periods, reduced. The diminished percentage of REM sleep is marginally significant. **Citalopram** increases REM sleep latency and diminishes REM sleep for 5 weeks [111]. During the withdrawal of treatment the following week, the values regain their normal levels after a rebound in the percentage of REM sleep. Maintenance treatment with **imipramine**, monitored every 3 months over a period of 3 years, with no recurrence of depression, continues to have a suppressive effect on REM sleep for 3 years [43]. **Moclobemide** (MAOI A) blocks REM sleep for the first three weeks of treatment [5, 67]. The older generation of MAOI [10] completely suppresses REM sleep but only after one to three weeks. REM sleep suppression persists if treatment is continued for two months. For responding subjects, mood improvement coincides with the (delayed) onset of maximum REM sleep suppression. **Lithium** increases REM sleep latency [9, 10, 49]. It does not alter the duration of the first phase of REM sleep or its density of eye movements [9].

Certain antidepressants such as **trimipramine** with undeniable clinical effects do not have this blocking effect on REM sleep, [104, 106, 113, 116]. REM sleep may even increase with trimipramine [116]. In healthy volunteers trimipramine slightly reduces the duration of REM sleep, while in depressives the duration of REM sleep is unaltered and stage 4 is increased (only at the start of treatment), but REM sleep latency remains higher throughout treatment [17]. **Amineptine** increases the total duration of REM sleep [15], but this increase may correspond to a rebound following the withdrawal of previous treatment. **Nefazodone** increases REM sleep at doses of 100 mg twice a day [101]. In a study by the same team [101] this effect was not found with the placebo control group. Nefazodone did not significantly alter either REM sleep, NREM sleep or sleep efficiency in healthy volunteers [112]. It was conducive to sleep and observed to be anxiolytic in Californian patients taking nefadozone as compared with those receiving other antidepressants (fluoxetine, sertraline, venlafaxine) (paper presented to the Congress of the American Psychiatric Association, San Diego, May 1997). **Toloxatone**, a new generation MAOI, appears to have no effect on REM sleep [4]. **Venlafaxine** which blocks serotonin, norepinephrine and dopamine reuptake, has an arousal effect, delays the onset of REM sleep and reduces its total duration [55]. These effects predominate after a week of treatment in major depressives but are still present after a month of treatment.

Thus the majority of antidepressants increase REM sleep latency and reduce REM sleep. This effect is partly due to their capacity to reduce atonia which is characteristic of REM sleep and which is thought to be controlled by the glycinergic neuron inhibitors of the medullary motoneurons. The anticataplexic effect of some of these (fluoxetine, clomapramine) is evidence of this. Antidepressants have even been involved in the appearance of periodic limb movements or REM sleep behaviour disorder. Antidepressants also increase muscle tone during NREM sleep [31].

The effects on NREM sleep

Amitriptyline, doxepine, amoxapine, mianserin and **trazodone** considerably increase stage 4 from the first night of treatment [22, 62, 68, 74, 110, 114]. **Clomipramine** significantly increases delta sleep in the first two hours of sleep [44]. For these authors, the increase of delta sleep at the beginning of the night is predictive of the clinical response. Indeed the effect of antidepressants seems to be most evident during the first sleep cycle [79]. **Lithium** has been reported to increase stage 4 [35] but this effect has not been corroborated by other studies [9]. The restoration of NREM sleep at the beginning of treatment with **trimipramine** or **quinupramine** merely signals the normalisation of sleep stage distribution from the 30th day [75]. **Fluvoxamine** does not modify the number of delta waves [46]. **Fluoxetine** does not acutely modify NREM sleep [39] but moderately reduces it with regular doses [92]. **Paroxetine** taken at 30 mg in the morning may increase stages 3 and 4 [72]. In healthy volunteers no alteration occurs to NREM sleep on the 28th day with paroxetine. No alteration was seen in NREM sleep among depressed subjects, after 5 weeks of **citalopram** [111].**MAOI**s do not affect NREM sleep [12]. Isocarboxazide and **nialamide** may increase it. **Venlafaxine** does not significantly alter the delta and theta bands of depressed patients, nor stages 3 and 4.

Prescribing antidepressants

In the treatment of ***persistent insomnia with no obvious elements of depression***, **amitriptyline** immediately improves disorders of sleep onset and maintenance, but its anticholinergic effects and the REM sleep rebound on discontinuation may be problematical. Even at weak doses, a persistent increase in NREM sleep is observed. It should be noted, however, that it is the quantity of high voltage slow delta waves which is measured on EEG, which in no way precludes the phenomena of recuperation related to NREM sleep (in the same way as benzodiazepine blockage of slow waves on EEG does not necessarily signify the disappearance of underlying phenomena of recuperation). The increase in delta sleep may account for the action mechanism of certain antidepressants when the deficit of NREM sleep lies at the origin of depression, as in Borbely's theory [7]. **Trimipramine** has been evaluated [27] in primary insomniacs. It improved sleep both subjectively and objectively, with an overall improvement in well-being during the day. The only adverse effect was dryness of the mouth. Discontinuation of trimipramine did not result in an objective rebound of insomnia in the short or long term, for the group as a whole, but some patients' objective total sleep time was reduced after withdrawal, with no subjective worsening of the quality of their sleep. In our centre, paroxetine is the first line treatment for insomnia: at a dose of 10 to 20 mg where there is subsyndromal depression, or a previous history of depression, or current somatisation. At a dose of 20 to 30 mg when there is a history of panic or apyretic tetanus attacks. A higher dose is used if the patient is suffering from patent anxiety disorder. In elderly patients with subsyndromal depression, or in the truly psychophysiological insomniac, a dose of 5 to 10 mg is sufficient. Sleep takes a long time to become regular. A period of at least two months is necessary, and sometimes three or four; so the patient must be warned that treatment may take a long time to become effective, as it acts via a long-term effect on the receptors. No controlled trials have yet been carried out to prove the efficacy of this therapeutic approach. The duration of treatment remains uncertain. It appears that the discontinuation of treatment by chronic insomniacs is followed by a more or less intense recurrence of poor sleep, after a lapse of one or two months. The idea is emerging of antidepressant treatment (the dose still needs to be determined) for a sustained period or even as lifelong treatment in middle aged chronic insomniacs, particularly if there is a history of anxiety or depression. Chronic insomnia is a lifelong illness. Sustained antidepressant treatment reduces the likelihood of a relapse of depression or anxiety [38, 63, 43]. It has also been proven that sustained treatment of chronic insomnia avoids the occurrence of anxious or depressive decompensation [78]. Insomnia generally worsens just before decompensation. It would be of value to conduct a study into the efficacy of antidepressant treatment (at weak doses?) in reducing the likelihood of anxious or depressive decompensation, over 5 or 10 years, in the chronic insomniac.

The limitations of setting up sustained antidepressant treatment are often linked to the side effects, with on one hand, an increase in appetite and a weight gain and on the other a drop in libido with problems of erection or ejaculation. The latter effects would theoretically be minimised by the use of an antidepressant with 5-HT2 blocker properties, such as nefazodone or mirtazapine.

Antidepressants may also be used in the treatment of secondary insomnia with ***sleep apnoea syndrome***, whether this is undetected (antidepressants do not increase apnoeas, unlike the benzodiazepines which can increase the number and duration of apnoeas) or recognised and treated by continuous positive airway pressure, the subject still complaining of poor sleep. Antidepressants not only do not aggravate sleep apnoea syndromes, they may even improve them. **Protriptyline**, a tricyclic antidepressant, has long been the most common antidepressant used in sleep apnoea syndromes. It acts partly by blocking REM sleep, the stage during which apnoeas often last longest, and partly by specifically increasing the tone of the muscles opening the upper airways [6]. It is not considered sufficient, by some, to reduce apnoeas [117]. It is likely to have a positive effect on the frequency and noise of the snoring of non apnoeic snorers [98]. Serotonin reuptake inhibitors such as paroxetine, diminish apnoeas by roughly a third, but not hypopnoeas, during NREM sleep only, in patients affected by obstructive sleep apnoea syndrome (OSAS) [40]. Paroxetine (single 40 mg dose) increases genioglossus activity when inspiration is at a peak, but this effect is insufficient to reduce the frequency of apnoeas in severe cases of OSAS [3]. Antidepressants, which strongly suppress REM sleep, can be used (such as clomipramine) for ***treating insomnia with recurrent*** *dreams and/or panic attacks*. Paroxetine has become the first-line treatment for panic attacks. Treatment is introduced very gradually (5 mg the first week, 10 mg the second, increasing by 5 mg per week to the full dose of a 40 mg taken in two doses of 20 mg, morning and evening). Treatment must be maintained for two years, then reduced by 5 mg in very gradual stages of one to three months. Paroxetine has also been successfully prescribed for treating nocturnal terrors and sleepwalking in adults [52].

In cases of *REM sleep related behaviour disorder* (parasomnia with loss of muscular atonia in REM sleep or aggressive motor activity during REM sleep) clonazepam is the usual treatment. But when this benzodiazepine cannot be prescribed (in cases of apnoea, for example), a tricyclic antidepressant such as desipramine may be used [91].

Antidepressants are often useful in treating *insomnia caused by overconsumption of medication*. Rüther [88] withdraws his chronic insomniacs dependant on benzodiazepine from treatment, by introducing a sedative antidepressant treatment, while gradually diminishing the doses of benzodiazepine. Current practice tends to use specific serotonin reuptake inhibitors or the more recent antidepressants before starting withdrawal. It is advisable to wait 4 to 6 weeks (after reaching the standard antidepressant dose) before starting to reduce the benzodiazepines.

The *interactions between benzodiazepines and antidepressants* are not well known in man. In animals the antidepressive action of serotonin reuptake inhibitors is reduced in association with benzodiazepine [57]. No undesirable interaction appears to exist between benzodiazepines and noradrenergic antidepressants (desipramine, maprotiline).

Antidepressants may be responsible for insomnia or for reinforcing it. Thus tricyclic antidepressants are known to exacerbate periodic leg movements, or even to cause them. It should be noted that specific serotonin reuptake inhibitors do not supposedly increase periodic movements [99]. Our own practice appears to contradict this. Specific serotonin reuptake blockers increase physiological myoclonia during sleep latency, sometimes dramatically. They can also lead to sleep talking and restless dreams, with the persistence of REM sleep.

Overstimulating antidepressants (such as fluoxetine, bupropion and the IMAOs) may induce hypervigilance or anxiety, which are themselves responsible for poor sleep. Fluoxetine is never recommended for insomniacs because of its overstimulating properties in relation to vigilance. Trazodone, a 5-HT2 antagonist, can also be used effectively as a hypnotic to treat antidepressant-related insomnia [71]. A priori, the other 5-HT2 blockers like nefazodone or mirtazapine do not lead to insomnia, even at the start of treatment.

In the case of *eating disorders* during the night, with nocturnal bulimia, fluoxetine is the only antidepressant recommended.

Antihistamines

Most of the antihistamines which bind to the H_1 receptor, cross the blood-meningeal barrier fairly easily and have sedative properties. They are commonly used as hypnotics and sold without prescription in the United States (diphenhydramine, hydroxyzine, doxylamine etc.). Most of these antihistamines have anticholinergic properties [20].

100 mg of **diphenhydramine** reduces REM sleep in healthy volunteers, with no obvious rebound effect after discontinuation [36]. In another polygraphic study [102], diphenhydramine (31 or 62 mg) was administered in association with 8-chlortheophylline. Subjectively, there was a dose-dependent hypnotic effect, but objectively total sleep time, intra-sleep arousals and sleep latency were not significantly altered. Similarly, in the study by Roth *et al.* [84], where diphenhydramine (150 mg taken in three doses: 8 am, 12 pm and 2 pm) is the standard antihistamine, there is no objective hypnotic effect at night but a tendency to fall asleep more easily during the day (MSLT).

Nevertheless, several subjective studies have come out in favour of the hypnotic efficiency of diphenhydramine [59, 81, 108,109,]. On the other hand, a double-blind cross-over study against placebo [8] of the effect of a single dose of 50 or 75 mg of diphenhydramine in ten young, healthy adults, showed no significant difference in the subjective sleep variables, even though subjective evaluation of sleep latency, the number of arousals and soundness of sleep, indicated a slight hypnotic action. Motor activity measured by actimetry even tended to increase during the night, with medication. Neither subjective evaluation in the morning nor tests of performance indicated the presence of any residual effect. There was no insomnia rebound effect the following night. A daytime polygraphic study derived from the MSLT (multiple sleep latency test) [83] with four opportunities to nap for 60 minutes (with advice given to "try to sleep") at 9 am, 11 am, 8 pm and 10 pm after a 50 mg dose of diphenhydramine at 8:30 am, 1 pm and 7 pm showed significantly shorter sleep latency under the drug than with placebo in healthy volunteers. On the other hand, no alterations occurred in sleep stages other than stage 1. Thus diphenhydramine is likely to cause sleepiness but is not really hypnotic. It has few side effects, but care must be taken with older patients as there is a risk of anticholinergic delirium or aggravation of the cognitive state, if taken too frequently or at high doses [12].

The other antihistamines have rarely been studied. **Doxylamine** is subjectively effective in outpatients [80]. It has been judged as effective as zolpidem, compared to placebo, in the treatment of common insomnia [90]. **Hydroxyzine** (25 mg) taken on morning awakening reduces sleep latency on MSLT (mean sleep latency of under 5 minutes in 5 sessions) [97]. **Promethazine** (antihistamine phenothiazine) was evaluated in four healthy subjects [35]. With 25 mg doses, there was no decrease in REM sleep although REM sleep did increase after discontinuation. At a 50 mg dose, promethazine decreased REM sleep with a rebound effect after discontinuation. At doses of 20 and 40 mg in poor sleepers, promethazine prolonged sleep by an hour, respected stage 4 and only reduced REM sleep with the stronger dose [1]. At doses of 100 and 200 mg in healthy volunteers, promethazine reduced REM sleep. It increased stage 4 at 200 mg [82]. Trimeprazine [37] did not alter REM sleep and REM latency at doses of 20 mg (but increased these slightly at 10 mg).

Other substances used as hypnotics

Chloral hydrate has been polygraphically evaluated in the healthy subject [34] at 500 and 1000 mg. Sleep latency was decreased and sleep architecture unaltered. With repeated doses of 500 mg over 28 nights in healthy subjects, followed by 17 nights of discontinued treatment, chloral increased TST, decreased wakefulness and sleep latency, altering neither REM sleep nor NREM sleep [24]. Effects were more pronounced [24] at the outset of treatment. The decrease in wakefulness persisted from the first night after discontinuation, accompanied by an increase in body movement. Chloral lost its effect during the first week in the laboratory [33]. In psychosomatic patients presenting sleep disorders, chloral increased total sleep and deep sleep, reduced the number and duration of intercurrent arousals as well as sleep latency, without altering REM sleep. There was no tolerance effect after 25 days of treatment. Sleep deteriorated again 48 to 72 hours after discontinuation, with no rebound of insomnia [32]. Chloral may be effective in older, non psychotic, slightly demented subjects, to induce and maintain sleep; but there is a rebound of insomnia on discontinuation [54]. Moran *et al.* [66] favour the use of chloral in older subjects. Careful adjustment of the dosage reduces the chances of delirium to a minimum; there is little acquired tolerance or alteration of sleep architecture. One study does however mention confusion and hallucinations in a high percentage of institutionalised elderly patients [41]. Chloral is not recommended for use in elderly subjects taking other medication [66]. Indeed one of the metabolites of chloral shifts warfarine and phenytoine from their usual protein binding sites, resulting in the sudden increase in the unbound fraction of these two medications, with the ensuing risks.

Barbituates are not indicated as hypnotics due to potential dependence and drug abuse and particularly the risk of death from overdose, with a dose of slightly more than double or triple the usual dose.

Neuroleptics should not be administered as hypnotics for patients other than psychotics for whom a neuroleptic sedative would improve both insomnia and the psychiatric symptoms.

Amino acids

L-tryptophan, an essential amino acid which transforms into serotonin, has the reputation of being a "natural" hypnotic agent, at doses of 1 to 5 g. L-trytophane has been shown to reduce sleep latency in the normal subject [103] and this effect correlates in relation to the blood level [19]. It may cause nausea and vomiting. Little is known about its efficacy and tolerance in the long term. Because of recent descriptions of myalgias, eosinophilia and other symptoms following the ingestion of tryptophan (with cases of mortality or invalidity sequellae) great precaution must be exercised, even if these disorders may have been due to an L-trytophan impurity originating in a single production unit [61]. Moreover the concomitant use of tryptophan and antidepressants with a serotonin effect such as fluoxetine [107] or phenelzine [51] is contraindicated. In an unpublished trial, **arginine-L-aspartate** showed objective efficiency in poor sleepers by prolonging sleep time by approximately one hour, without altering sleep architecture.

New products susceptible to modulating sleep

Delta sleep inducing peptide (DSIP) was tested in IV trials on chronic insomniacs for a week [94]. Subjects under 60 years of age showed a net improvement immediately after the end of DISP treatment, lasting throughout the following week. In subjects over 60, whose sleep was twice as disturbed as that of young insomniacs, the immediate effect of DISP was greater but sleep only became normal after a week of treatment. Both groups had objectively normal sleep at the end of the study. In young, healthy subjects, DSIP increased the tendency to sleep and did not disrupt performance during the 8 hours after it had been administered. It has even been observed to increase memory [25, 42].

Melatonin appears to have hypnotic properties in man, particularly in the elderly.

In young, good sleepers, daytime administration of melatonin has shown a dose-dependent reduction of sleep latency, whenever endogenous melatonin levels are low (i.e. when the homeostatic process is low or with a strong circadian process promoting arousal). Let us recall that the physiological secretion of melatonin starts 100 to 120 minutes before entry into sleep (sleep gate). Melatonin has no significant effect on initiating sleep in sleep deprivation experiments, where the homeostatic process is raised. 5 mg of melatonin at 6 pm will slightly reduce sleep latency which is already short, at 11 pm. Taken at 5 pm, at a dose of 0.05, 0.5 or 5 mg, melatonin will (subjectively) reduce sleep latency when sleep is programmed to start at midnight. So it appears that melatonin is susceptible to inducing sleep when the disorder results from ill-matched circadian time and chosen sleep time, as for example in the case of delayed sleep phase. Melatonin is the only effective treatment for this condition [14]. Several studies have focused on REM sleep modifications. Generally speaking, either REM sleep is not affected or it is increased during nocturnal sleep, particularly with high doses of melatonin.

Studies among insomniacs are contradictory. In chronic insomniacs, melatonin significantly improves total sleep time and daytime wakefulness, even though seven subjects out of thirteen showed no improvement in their overall well-being [56]. In elderly insomniacs, 2 mg of sustained release melatonin taken 2 hours before bedtime for three weeks increases sleep efficiency, reduces intercurrent arousal and tends to shorten sleep latency [18]. 2 mg of melatonin, taken 2 hours before going to bed for 7 days will improve sleep installation, when administered in rapid release form,, and an improvement in sleep maintenance, with the sustained release form. This initial treatment is followed by administering 1 mg sustained release melatonin for 2 months, showing an improvement in disorders of sleep installation and maintenance, which persist and involve no problems of tolerance [23]. 3 mg of melatonin in elderly primary insomniacs reduced intercurrent arousal, and increased total sleep time and sleep efficiency for the two weeks of treatment [65].

It has been suggested that the drop in melatonin secretion in elderly subjects is what causes poor sleep. But melatonin secretion does not correlate with objective sleep measures in the elderly volunteer complaining of poor sleep [118]. In fact, the production of melatonin in the elderly insomniac is very variable, varying by a factor of 20. The maximum amplitude and production of melatonin (area on the curve) do not correlate with sleep parameters [28]. Sleep parameters only relate to the circadian parameters of melatonin: time of sleep onset correlated with time of onset of melatonin secretion (DLMO); total sleep time and sleep efficiency correlated with the time melatonin secretion stopped (DLMOff), with a phase difference between bedtime and melatonin rhythm. Elderly insomniacs who have trouble maintaining sleep, and who take 0.5 mg of melatonin with immediate release or with a 30 minute delay before bedtime or 4 hours after bedtime, show no alterations to their sleep latency, sleep time, duration of intercurrent arousal, or sleep efficiency.

Melatonin may even have been responsible for the complete fragmentation of the sleep-wake cycle in a quarter of young healthy volunteers administered with melatonin (5 mg at midday, 8 pm and 4 am if they were awake) exposed to weak lighting for over a week and allowed to sleep and eat at will [60].

Ritanserin is a serotonin agonist acting on the 5-HT$_2$ central receptors. It considerably increases NREM sleep in normal subjects [13, 30]. The effect is dose-dependant, with ritanserin as with seganserin (another 5-HT$_2$ agonist) [29]. However a trial using ritanserin in insomniacs [66] failed to clinically improve sleep (subjective evaluation) in relation to placebo, even though NREM sleep objectively increased.

S-adenosyl-homocytein (SAH) has demonstrated both a subjective and an objective hypnotic effect on poor sleepers, without altering sleep architecture [21]. It reduces anxiety, irritability and bad humour.

Tiagabine, a GABA reuptake inhibitor used in the treatment of partial epilepsies, has been studied in normal elderly subjects to evaluate the acute effect on sleep [58]. Ageing is known to considerably reduce sleep consolidation and the quantity of NREM sleep. Tiagabine improved sleep efficiency, increased NREM sleep and demonstrated a tendency to reduce intercurrent arousal. No trials have been conducted as yet, to evaluate the effect of tiagabine on insomniacs.

Vasopressine administered for 3 months in nasal spray form, increases NREM sleep in elderly subjects with no sleep complaints [75]. The suprachiasmatic nucleus is known to lose part of its vasopressine neurons as an effect of aging. The results of supplementing vasopressine are all the more interesting because, as NREM sleep is not subject to a marked circadian influence, it is highly unlikely that it would have an effect on this type of sleep. These results are still to be confirmed in regard to insomniacs.

Phytotherapy

Plants have always been used to improve sleep. But very few of the plants reputed to be sedative have been evaluated in properly conducted pharmacological trials. In a comparative double blind study against placebo, a commercial preparation containing valerian and an extract of the drug [50], both preparations with valerian had a significant effect on sleep. More recently, a preparation of valerian rich in sesquiterpenes (that of both the sedative components of valerian, which is not cytotoxic) was tested against placebo in 27 subjects with sleep difficulties and daytime sleepiness [53]. Valerian significantly improved sleep. 44% of subjects said they had slept perfectly and 89% that they had slept better with valerian. Using the same extract of valerian without valepotriates on healthy subjects, Balderer and Borbely [2] demonstrated hypnotic activity which was subjective at home but was not objective in the sleep laboratory; No alteration was reported either of stages or of EEG spectra. No adverse effect was reported in these different studies. In a recent polygraphic study [96] conducted with poor sleepers aged around 60, and a parallel control group, an extract of valerian increased NREM sleep, reduced stage 1 and increased the density of spindles. There was no effect on sleep latency either in terms of intra-sleep wakefulness, REM sleep or the subjective quality of sleep. Valerian appears to be a true hypnogenetic which acts by inhibiting transaminase GABA and whose potential pharmacodependence, while still unknown, is certainly far from inexistent [95]. We have tried to objectively evaluate the effect of a sedative extract of passionflower on poor sleepers, using ambulant recordings on cassette. No significant improvement was found but the number of subjects tested was limited. There were no alterations to sleep architecture.

To summarise, the different medications - hypnotic and otherwise - are generally considered by experts as being ineffective in the treatment of chronic insomnia, but by patients as being useful (this is not only a placebo effect). In some cases, where serious chronic insomnia has resisted successive attempts using the whole range of hypnotics, antidepressants, other forms of medication and even psychotherapy and other non medical methods, a single, effective substance is finally found which remains effective for years, as in the various clinical cases reported by Regestein [77]. For one of the subjects described by this author, it was phenobarbital, for others glutethimide or diazepam (to the exclusion of any other BZD), or an opiate ... In theory, the long term use of hypnotics is not supposed to cure chronic insomnia. But some rare cases demand a more flexible approach.

REFERENCES

1. ADAM K., OSWALD I. – The hypnotic effects of an antihistamine: promethazine. *Br. J. Clin. Pharmac.*, *22*, 715-717, 1986.
2. BALDERER G., BORBELY A. – Effect of valerian on human sleep. *Psychopharmacology*, *87*, 406-409, 1985.
3. BERRY R.B., YAMAURA E.M., GILL K., REIST C. – Acute effects of paroxetine on genioglosse activity in obstructive sleep apnea. *Sleep*, *22*, 1087-1092, 1999.
4. BESSET A., BILLIARD M., BRISSAUD L., TOUCHON J. – *Effects of toloxatone on sleep architecture.* 9th European Congress of Sleep Research, Jérusalem, Septembre 1988.
5. BLOIS R., GAILLARD J.M. – Effects of moclobemide on sleep in healthy human subjects. *Acta Psychiatr. Scand.*, *S360*, 73-75, 1990.
6. BONORA M., SAINT-JOHN W.M., BLEDSOE T.A. – Differential elevation by protriptyline and depression by diazepam of upper airway respiratory motor activity. *Am. Rev. Respir. Dis.*, *131*, 41-45, 1985.
7. BORBELY A.A. – The S-deficiency hypothesis of depression and the two-process model of sleep regulation. *Pharmacopsychiat.*, *20*, 23-29, 1987.
8. BORBELY A.A., YOUMBI-BALDERER G. – Effect of diphenhydramine on subjective sleep parameters and on motor activity during bedtime. *Int. J. Clin. Pharmacology Therapy & Toxicology*, *26*, 8, 392-396, 1988.
9. CAMPBELL S.S., GILLIN J.C., KRIPKE D.F., JANOWSKY D.S., RISCH S.C. – Lithium delays circadian phase of temperature and REM sleep in a bipolar depressive: a case report. *Psychiatry Research*, *27*, 23-29, 1989.
10. CHEN C.N. – Sleep, depression and antidepressants. *Br. J. Psychiat.*, *135*, 385-402, 1979.
11. CONSENSUS CONFERENCE. Drugs and insomnia. The use of medications to promote sleep. *JAMA*, *251*, 18, 2410-2414, 1984.
12. DALY M.P. – Sleep disorders in the elderly. *Primary care*, *16*, 2, 475-488, 1989.
13. DECLERCK A.C., WAUQUIER H.M., VAN DER HAMVELTMAN, GELDERS Y. – Increase of slow wave sleep in human volunteers by the serotonin S2 antagonist ritanserin. *Curr. Ther. Res.*, *41*, 427-432, 1987.
14. DUK D.J.., CAJOCHEN C. – Melatonin and the circadian regulation of sleep initiation, consolidation, structure and the sleep EEG. *J. Biol. Rhythms*, *12*, 627-635, 1997.
15. DI PERRI R., MAILLANO F., BRAMANTI P. – The effect of amineptine on the mood and nocturnal sleep of depressed patients. *Prog. Neuro-Psychopharmacol. & Biol. Psychiat.*, *11*, 65-70, 1986.
16. DUNBAR G.C., HUGUES L.S. – A review of the effects of paroxetine, a new selective serotonin reuptake inhibitor, on sleep in healthy volunteers and depressed patients. *Sleep Res.*, *20A*, 130, 1991.
17. FEUILLADE P., PRINGUEY D., BELUGOU J.L., ROBERT P., DARCOURT G. – Trimipramine: acute and lasting effects on sleep in healthy and major depressive subjects. *J. Affective Dis.*, *24*, 135-146, 1992.
18. GARFINKEL D., LAUDON M., NOF D., ZISAPEL N.- Improvement of sleep quality in elderly people by controlled-release melatonin. *Lancet*, 346, 541-546, 1995.
19. GEORGE C.F.P., MILLAR T.W., HANLY P.J., KRYGER M.H. – The effect of L-tryptophan on daytime sleep in normals: correlation with blood levels. *Sleep*, *12*, 345-353, 1989.
20. GILLIN J.C., BYERLEY W.F. – The diagnosis and management of insomnia. *New Engl. J. Med.*, *322*, 239-248, 1990.
21. GOLDENBERG F., BOUR F. – Effects of S-adenosyl-homocysteine on human sleep: polygraphic study. *Sleep Res.*, *16*, 90, 1987.
22. GOLDENBERG F., PLOUIN P., LEYGONIE A. – Ambulatory polygraphic study of amoxapine effect in poor sleepers. *In: Sleep 1988*, J. Horne (ed.), Gustav Fischer Verlag, Stuttgart, New York, 214-216, 1989.
23. HAIMOV I., LAVIE P., LAUDON M., HERER P., VIGDER C., ZISAPEL N. – Melatonin replacement therapy of elderly insomniacs. *Sleep*, 21, 52-68, 1998.
24. HARTMANN E., CRAVENS J. – The effects of long term administration of psychotropic drugs on human sleep. V. The effects of chloral hydrate. *Psychopharmacologia*, *33*, 219-232, 1973.
25. HERMANN-MAURER E.K., ERNST A., SCHNEIDER-HELMERT D., ZIMMERMANN A., SCHOENENBERGER G.A. – Effects of DSIP-therapy on memory in chronic insomnia. *Sleep Res.*, *18*, 53, 1989.
26. HOECHLI D., RIEMANN D., ZULLEY J., BERGER M. – Initial REM sleep suppression by clomipramine: a prognostic tool for treatment response in patients with a major depressive disorder. *Biol. Psychiatry*, *21*, 1217-1220, 1986.
27. HOHAGEN F., MONTERO R.F., WEISS E., LIS S., DRESSING H., RIEMANN D., BERGER M. – Treatment of primary insomnia with trimipramine: an alternative to benzodiazepine hypnotics? *Eur. Arch. Psychiatry Clin. Neurosc.*, *244*, 65-72, 1994.
28. HUGHES R.J., SACK R.L., LEWY A.J. – The role of melatonin and circadian phase in age-related sleep-maintenance insomnia: assessment in a clinical trial of melatonin replacement. *Sleep*, 21, 52-68, 1998.
29. IDZIKOWSKI C., JAMES R., BURTON S.W. – Human SWS is increased dose-dependently by seganserin and ritanserin. *Sleep Res.*, *18*, 55, 1989.
30. IDZIKOWSKI C., MILLS F.J., GLENNARD R. – 5-hydroxytryptamine-2 antagonist increases human slow wave sleep. *Brain Res.*, *378*, 164-168, 1986.
31. JOBERT M., JÄHNIG P., SCHULZ H. – Effect of two antidepressant drugs on REM sleep and EMG activity during sleep. *Neuropsychobiology*, 39, 101-109, 1999.
32. JOVANOVIC U.J. – Die Einwirkung von Chloral-hydrat auf Schlafstörungen bei psychosomatischen Patienten. *Nervenheilkunde*, *11*, 350-358, 1992.
33. KALES A., BIXLER E.O., KALES J.D., SCHARF M.B. – Comparative effectiveness of nine hypnotic drugs: sleep laboratory study. *J. Clin. Pharm.*, *17*, 207-213, 1977.

34. KALES A., KALES J.D., SCHARF M.B., TAN T.L. – Hypnotics and altered sleep-dream patterns. II. All-night EEG studies of chloral hydrate, flurazepam, and methaqualone. *Arch. Gen. Psychiatry, 23*, 219-225, 1970.

35. KALES A., MALSTROM E.J., SCHARF M.B., RUBIN R.T. – Psychophysiological and biochemical changes following use and withdrawal of hypnotics. *In: Sleep, Physiology and Pathology,* A. Kales (ed.), J.B. Lippincott, Philadelphia, 331-343, 1969.

36. KALES A., MALSTROM E.J., TAN T.L. – Drugs and dreaming. *In: Progress in Clinical Psychology,* E. Abt, B.F. Riess (eds.), Grune & Stratton, New York, 154-167, 1969.

37. KAYED K., HANSEN T., GODTLIBSEN O.B. – Controlled clinical investigation of trimeprazine as a sleep-inducer in normal subjects. *Europ. J. Clin. Pharmacol., 11*, 163-167, 1977.

38. KELLER M.B. – The long-term treatment of depression. *J. Clin. Psychiatry, 60* (S17), 41-45, 1999.

39. KERKHOFS M., RIELAERT C., DE MAERTELAER V., LINKOWSKI P., CZARKA M., MENDLEWICZ J. – Fluoxetine in major depression: efficacy, safety and effects on sleep polygraphic variables. *Int. Clin. Psychopharmacol., 5*, 253-260, 1990.

40. KRAICZI H., HEDNER J., EJNELL H., CARLSON J. – Effect of serotonin uptake inhibition on breathing during sleep and daytime symptoms in obstructive sleep apnea. *Sleep, 22*, 61-67, 1999.

41. KRAMER C. – Methaqualone and chloral hydrate: preliminary comparison in geriatric patients. *J. Am. Geriatr. Soc., 15*, 455-461, 1967.

42. KUMAR A., VAN DE WIEL B., BRUINS J., SCHNEIDER-HELMERT D. – Effects of DSIP on human sleep, memory and performance. *Sleep. Res., 15*, 25, 1986.

43. KUPFER D.J., EHLERS C.L., FRANK E., GROCHOCINSKI V.J., McEACHRAN A.B., BUHARI A. Persistent effects of antidepressants: EEG sleep studies in depressed patients during maintenance treatment. *Biol. Psychiatry, 35*, 781-793, 1994.

44. KUPFER D.J., EHLERS C.J., POLLOCK B.J., NATHAN R.S., PEREL J.M. – Clomipramine and EEG sleep in depression. *Psychiatry Res., 30*, 165-180, 1989.

45. KUPFER D.J., FRANK E., PEREL J.M., CORNES C., MALLINGER A.G., THASE M.E., McEACHRAN A.B., GROCHOCINSKI V.J. – Five year outcome for maintenance therapies in recurrent depression. *Arch. Gen. Psychiatry, 49*, 769-773, 1994

46. KUPFER D.J., PEREL J., POLLOCK B.J., NATHAN R.S., GROCHOCINSKI V.J., WILSON M.J., MCEACHRAN A.B. – Fluvoxamine versus desipramine: comparative polysomnographic effects. *Biol. Psychiatry, 29*, 23-40, 1991.

47. KUPFER D.J., SPIKER D.G., COBLE P., MCPARTLAND R.J. – Amitriptyline and EEG sleep in depressed patients. I. Drug effect. *Sleep, 1*, 2, 149-159, 1978.

48. KUPFER D.J., SPIKER D.G., COBLE P.A., NEIL J.F., ULRICH R., SHAW D.H. – Sleep and treatment prediction in endogenous depression. *Am. J. Psychiatry, 138*, 429-434, 1981.

49. KUPFER D.J., WYATT R.J., GREENSPAN K., SCOTT J., SNYDER F. – Lithium carbonate and sleep in affective illness. *Arch. Gen. Psychiatry, 23*, 35-40, 1970.

50. LEATHWOOD P.D., CHAUFFARD F., HECK E., MUNOZ-BOX R. – Aqueous extract of valerian root improves sleep quality in man. *Pharmacol. Biochem. Behav., 17*, 65-71, 1982.

51. LEVY A.B., BUCHER P., VOTOLATO N. – Myoclonus, hyperreflexia and diaphoresis in patients on phenelzine-tryptophan combination treatment. *Can. J. Psychiatry, 30*, 434-436, 1985.

52. LILLYWHITE A.R., WILSON S.J., NUTT D.J. – Successful treatment of night terrors and somnambulism with paroxetine. *Br. J. Psychiatry, 164*, 551-554, 1994.

53. LINDAHL O., LINDWALL L. – Double blind study of a valerian preparation. *Pharmacol. Biochem. Behav., 32*, 1065-1066, 1989.

54. LINNOILA M., VIUKARI M., NUMMINEN A., AUVINEN J. – Efficacy and side effects of chloral hydrate and tryptophane as sleeping aids in psychogeriatric patients. *Int. Pharmaco-psychiat., 15*, 124-128, 1980.

55. LUTHRINGER R., TOUSSAINT M., SCHALTENBRAN N., BAILEY P., DANJOU P.H., HACKETT D., GUICHOUX J.Y., MACHER J.P. – A double-blind, placebo-controlled evaluation of the effects of orally administered venlafaxine on sleep in inpatients with major depression. *Psychopharmacology Bulletin, 32*, 637-646, 1996.

56. MCFARLANE J.G., CLEGHORN J.M., BROWN G.M., STREINER D.L. – The effects of exogenous melatonin on the total sleep time and daytime alertness of chronic insomniacs: a preliminary study. *Biol. Psychiatry, 30*, 371-376, 1991.

57. MARTIN P., PUECH A.J. – Antagonism by benzodiazepines of the effects of serotonin-but not norepinephrine-uptake blockers in the learned-helplessness paradigm in rats. *Biol. Psychiatry, 39*, 882-890, 1996.

58. MATHIAS S., WETTER T.C., STEIGER A., LANCEL M. – The GABA uptake inhibitor tiagabine promotes slow wave sleep in normal elderly subjects. *Neurobiol. Aging, 22*, 247-253, 2001.

59. MEULEMAN J.R., NELSON R.C., CLARK R.L. JR. – Evaluation of temazepan and diphenhydramine as hypnotics in a nursing-home population. *Drugs Intelligence & Clinical Pharmacy, 21*, 716-720, 1987.

60. MIDDLETON B.A., STONE B.M., ARENDT J. – Melatonin and fragmented sleep patterns. *Lancet, 348*, 551-552, 1996.

61. MILBURN D.S., MYERS C.W. – Tryptophan toxicity: a pharmacoepidemiologic review of eosinophilia-myalgia syndrome. *DICP Ann Pharmacother. 25*, 1259-1262, 1991.

62. MONTGOMERY I., OSWALD I., MORGAN K., ADAM K. – Trazodone enhances sleep in subjective quality but not in objective duration. *Br. J. Clin. Pharmac., 16*, 139-144, 1983.

63. MONTGOMERY S.A., REIMITZ P.E., ZIVKOV M. – Mirtazapine versus amitryptiline in the long-term treatment of depression: a double-blind placebo-controlled study. *Int. Clin. Psychopharmacol., 13*, 63-73, 1998

64. MONTI J.M., ALTERWAIN P., MONTI D. – The effects of moclobemide on nocturnal sleep of depressed patients. *J. Affective Disorders, 20*, 201-208, 1990.

65. MONTI J.M., ALVARINO F., CARDINALI D., SAVIO I., PINTOS A. – Polysomnographic study of the effect of melatonin on sleep in elderly patients with chronic primary insomnia. *Archives Gerontology Genetics.* 28, 85-98, 1999.

66. MORAN M.G., THOMPSON II T.L., NIES A.S. – Sleep disorders in the elderly. *Am. J. Psychiatry, 145,* 1369-1378, 1988.

67. MORGAN K., OSWALD I., BORROW S., ADAM K. – Effects of a single dose of mianserin on sleep. *Br. J. Clin. Pharmac., 10,* 525-527, 1980.

68. MOURET J., LEMOINE P., MINUIT M.P., RENARDET M. – Intérêt des études polygraphiques du sommeil dans le suivi des traitements antidépresseurs. Application à la trazodone. *Psychiatr. & Psychobiol., 3,* 29-36, 1988.

69. MOURET J., LEMOINE P., MINUIT M.P., SANCHEZ P., TAILLARD J. – Sleep polygraphic effects of trimipramine in depressed patients. Preliminary report. *Drugs, 38* (suppl. 1), 14-18, 1989.

70. NICHOLSON A.N., PASCOE P.A. – Studies on the modulation of the sleep-wakefulness continuum in man by fluoxetine, a 5-HT uptake inhibitor. *Neuropharmacology, 27,* 597-602, 1988.

71. NIERENBERG A.A., ADLER L.A., PESELOW E., ZORNBERG G., ROSENTHAL M. – Trazodone for anti-depressant associated insomnia. *Am. J. Psychiatry,* 151, 1069-1072, 1994.

72. NOWELL P.D., REYNOLDS C.F., BUYSSE D.J., DEW M.A., KUPFER D.J. – Paroxetine in the treatment of age-related sleep disturbances with prolonged intranasal vasopressin. *J. Clin. Psychiatry,* 60, 89-95, 1999.

73. OSWALD I., ADAM K. – Effects of paroxetine on human sleep. *Br. J. Clin. Pharmac.,* 22, 97-99, 1986.

74. PARRINO L., BARUSI R., SIMEONI S., TERZANO M.G. – Polysomnographic monitoring of insomnia under trazodone treatment. *J. Sleep Res., 1,* S1, 171, 1992.

75. PERRAS B., PANNENBORG H., PIETROWSKY R., BORN J., FEHM H.L. – Beneficial treatment of age-related sleep disturbances with prolonged intranasal vasopressin. *J. Clin. Psychophamacol.,* 19, 28-36, 1999.

76. PRINGUEY D., FEUILLADE P., BELUGOU J.L., TALI-CHET L., DARCOURT G. – Acute response of sleep endogenous depressive pattern to a single-dose test of antidepressant: Biological predictive value. *J. Sleep Res., 1,* S1, 185, 1992.

77. REGESTEIN Q.R. – Specific effects of sedative/hypnotic drugs in the treatment of incapacitating chronic insomnia. *Am. J. Med., 83,* 909-916, 1987.

78. REYNOLDS C.F., BUYSSE D.J., KUPFER D.J. – Treating insomnia in older adults. Taking a long term view. *JAMA,* 281, 1034-1035, 1999.

79. REYNOLDS C.F., KUPFER D.J. – Sleep research in affective illness: State of the art circa 1987 (State-of-the-art review). *Sleep, 10,* 199-215, 1987.

80. RICKELS K., GINSBERG J., MORRIS R.J., NEWMAN H.M., SCHILLER H.M., WEINSTOCK R.M., SCHILLING A.E. – Doxyl-amine succinate in insomniac family practice patients: a double blind study. *Curr. Ther. Res., 35,* 532-540, 1984.

81. RICKELS K., MORRIS R.J., NEWMAN H., ROSENFELD H., SCHILLER H., WEINSTOCK R. – Diphenhydramine in insomniac family practice patients: a double-blind study. *J. Clin. Pharmacol., 23,* 235-242, 1983.

82. RISBERG A.M., RISBERG J., INGVAR D.H. – Effects of promethazine on nocturnal sleep in normal man. *Psychopharmacologia* (Berl.), *43,* 279-284, 1975.

83. ROEHRS T.A., TIETZ E.I., ZORICK F.J., ROTH T. – Day-time sleepiness and antihistamines. *Sleep, 7,* 137-141, 1984.

84. ROTH T., ROEHRS T., KOSHOREK G., SICKLESTEEL J., ZORICK F. – Central effects of antihistamines. *Sleep Res., 15,* 43, 1986.

85. RUIZ-PRIMO E., HARO R., VALENCIA M. – Polysomnographic effects of ritanserin in insomniacs: a crossed double-blind controlled study. *Sleep Res., 18,* 72, 1989.

86. RUSH A.J., GILES D.E., JARRETT R.B., FELDMAN-KOFFLER F., DEBUS J.R., WEISSENBURGER J., ORSULAK P.J., ROFFWARG H.P. – Reduced REM latency predicts response to tricyclic medication in depressed outpatients. *Biol. Psychiatry, 26,* 61-72, 1989.

87. RUSSO R.M., GURURAJ V.J., ALLEN J.E. – The effectiveness of diphenhydramine HCl in pediatric sleep disorders. *J. Clin. Pharmacol., 16,* 284-288, 1976.

88. RUTHER E., LUND L., GRUBER A. – Chronic insomnia in a psychiatric outpatient clinic. *In: Sleep 1984,* W.P. Koella, E. Rüther, H. Schulz (eds.), Gustav Fischer Verlag, Stuttgart, New York, 53-55, 1985.

89. SALETU B., FREY R., KRUPKA M., ANDERER P., GRUNBERGER J., SEE W.R. – Sleep laboratory studies on the single-dose of serotonin reuptake inhibitors paroxetine and fluoxetine on human sleep and awakening qualities. *Sleep, 14,* 439-447, 1991.

90. SCHADECK B., CHELLY M., AMSELLEM D., COHEN A., PERAUDEAU P., SCHECK F. – Efficacité comparative de la doxylamine (15 mg) et du zolpidem (10 mg) dans le traitement de l'insomnie commune. Une étude contrôlée *versus* placebo. *Sem. Hôp. Paris, 13-14,* 428-439, 1996.

91. SCHENK C.H., BUNDLIE S.R., PATTERSON A.L., MAHOWALD M.W. – Rapid eye movement sleep behavior disorder. A treatable parasomnia affecting older adults. *JAMA, 257,* 13, 1786-1789, 1987.

92. SCHENK C.H., MAHOWALD M.W., KIM S.W., O'CONNOR K.A., HURWITZ T.D. – Prominent eye movements during NREM sleep and REM sleep behavior disorder associated with fluoxetine treatment of depression and obsessive-compulsive disorder. *Sleep, 15,* 226-235, 1992.

93. SCHLÖSSER R., RÖSCHKE J., ROSSBACH W., BENKERT O. – Conventional and spectral power analysis of all-night sleep EEG after subchronic treatment with paroxetine in healthy male volunteers. *Eur. Neuropsychopharmacol.,* 8 ; 273-278, 1998

94. SCHNEIDER-HELMERT D. – Efficacy of DSIP to normalise sleep in middle-aged and elderly chronic insomniacs. *Eur. Neurol., 25,* 448-453, 1986.

95. SCHULZ P. – La psychopharmacologie du sommeil chez la personne âgée. *Méd. et Hyg.*, *46*, 3341-3343, 1988.
96. SCHULZ H., STOLZ C., MÜLLER J. – The effect of valerian extract on sleep polygraphy in poor sleepers: a pilot study. *Pharmacopsychiat.*, *27*, 147-151, 1994.
97. SEIDEL W.F., COHEN S., BLIWISE N.G., DEMENT W.C. – Direct measurement of daytime sleepiness after administration of cetirizine and hydroxyzine with a standardized electroencephalographic assessment. *J. Allergy Clin. Immunol.*, *86*, 1029-1033, 1990.
98. SERIES F., MARC I. – Effects of protriptyline on snoring characteristics. *Chest,* 104, 14-18, 1993.
99. SHAFFER J.I., TALLMAN J.B., BOECKER M.R., NEEB M.J., MAHAJAN V.K. – A report on the PLM suppressing properties of serotonin reuptake inhibitors. *Sleep Res.*, *24*, 347, 1995.
100. SHARLEY A.L., WILLIAMSON D.J., ATTENBUR-ROW M.E.J., PEARSON G., SARGENT P., COWEN P.J. – The effects of paroxetine and nefazodone on sleep: a placebo controlled trial. *Psychopharmacology*, *126*, 50-54, 1996.
101. SHARPLEY A.L., WALSH A.E.S., COWEN P.J. – Nefazodone — a novel antidepressant — may increase REM sleep. *Biol. Psychiatry*, *31*, 1070-1073, 1992.
102. SPIEGEL R., ALLEN S.R. – Die Wirkung eines nicht rezeptpflichtigen Schlafmittels auf das Schlafpolygramm gesunder Probanden. *Schweiz. Rundschau Med.* (Praxis), *73*, 169-173, 1984.
103. SPINWEBER C.L., URSIN R., HILBERT R.P., HILDERBRAND R.L. – L-tryptophan: effects on daytime sleep latency and the waking EEG. *Electroenceph. Clin. Neurophysiol.*, *55*, 652-661, 1983.
104. STANER L., KERKHOFS M., DETROUX D., BOUILLON E., DE LA FUENTE J.M., LINKOWSKI P., MENDLEWICZ J. – A double-blind comparison of the effects of paroxetine and amitriptyline on sleep EEG in patients with major depression. *J. Sleep Res.*, *1*, S1, 218, 1992
105. STANER L., KERKHOFS M., DETROUX D., LEYMAN S., LINKOWSKI P., MENDELWICZ J. – Acute, subchronic and withdrawal sleep EEG changes during treatment with paroxetine and amitriptyline: a double-blind randomized trial in major depression. *Sleep*, *18*, 470-477, 1995.
106. STEIGER A., BENKERT O., WOEHRMANN S., STEINSEIFER D., HOLSBOER F. – Effects of trimipramine on sleep EEG, penile tumescence and nocturnal hormonal secretion. A long term study in 3 normal controls. *Neuropsychobiology*, *21*, 71-75, 1989.
107. STEINER W., FONTAINE R. – Toxic reaction following the combined administration of fluoxetine and L-tryptophan: five case reports. *Biol. Psychiatry*, *21*, 1067-1071, 1986.
108. SUNSHINE A., ZIGHELBOIM I., LASKA E. – Hypnotic activity of diphenhydramine, methapyrilene and placebo. *J. Clin. Pharmacol.*, *18*, 425-431, 1978.
109. TEUTSCH G., MAHLER D.L., BROWN C.R., FORREST W.H. JR., JAMES K.E., BROWN B.W. – Hypnotic efficacy of diphenhydramine, methapyrilene and pentobarbital. *Clin. Pharmacol. Ther.*, *17*, 195-201, 1975.
110. TOUCHON J. – Antidepressant drugs and insomnia. *In: Biological Psychiatry 1985*, C. Shagass, R.C. Josiassen, W.H. Bridger, K.J. Weiss, D. Stoff, G.M. Simpson (eds.), *Development in Psychiatry*, vol. 7, Elsevier, 1002-1004, 1985.
111. VAN BEMMEL A.L., VAN DEN HOOFDAKKER R.H., BEERSMA D.G., BOUHUYS A.L. – Changes in sleep polygraphic variables and clinical state in depressed patients during treatment with citalopram. *Psychopharmacology,* 113, 225-230, 1993.
112. VOGEL G., COHEN J., MULLIS D., KENSLER T., KAPLITA S. – Nefazodone and REM sleep. How do antidepressant drugs decrease sleep ? *Sleep,* 21, 70-77, 1998.
113. WARE J.C., BROWN F.W., MOORAD P.J. JR., PITTARD J.T., COBERT B. – Effects on sleep: a double-blind study comparing trimipramine to imipramine in depressed insomniac patients. *Sleep*, *12*, 537-549, 1989.
114. WARE J.C., PITTARD J.T. – Increased deep sleep after trazodone use: a double-blind placebo-controlled study in healthy young adults. *J. Clin. Psychiatry*, *51*, 9 (suppl.), 18-22, 1990.
115. WALSH J.K., SCHWEITZER P.K. – Ten year trends in the pharmacological treatment of insomnia. *Sleep,* 22, 371-375, 1999.
116. WIEGAND M., BERGER M. – Action of trimipramine in sleep and pituitary hormone secretion. *Drugs 38* (suppl. 1), 35-42, 1989.
117. WHYTE K.F., GOULD G.A., AIRLIE M.A.A., SHA-PIRO C.M., DOUGLAS N.J. – Role of protriptyline and acetazomalide in the sleep apnea/hypopnea syndrome. *Sleep*, *11*, 5, 463-472, 1988.
118. YOUNGSTEDT S.D., KRIPKE D.F., ELLIOTT J.A. – Melatonin excretion is not related to sleep in the elderly. *J. Pineal Res.* 24, 142-145, 1998.

Chapter 23

Insomnia in children
Clinical aspects and treatment

Y. Navelet
Explorations Fonctionnelles du Système Nerveux, CHU Bicêtre, Kremlin Bicêtre, France

DEFINITION

Insomnia in children, particularly young children, appears less as a lack of sleep than as an imbalance between the rhythms and sleep needs of the child, and those of the family and social environment. Insomnia is often tolerated over a long period of time before becoming a motive for consultation at the point when encroaches on the life of the parents and the relationship between the child and his family.

In the 1979 Diagnostic Classification of Sleep and Arousal Disorders [8], insomnia in children appears at the beginning of the chapter on "Insomnia in early childhood", a paediatric form of adult psychophysiological insomnia, which it precedes [44]. In the 1997 International Classification of Sleep Disorders [7], it appears under different subheadings of dyssomnia and linked to intrinsic or extrinsic sleep disorders in a manner which may appear artificial to the paediatric reader. Like adult insomnia, it can be linked to problems of sleep latency, frequent nocturnal arousals or arousal at the end of the night [29]. It may be isolated, but is often associated with other sleep disorders in children. Nevertheless as paediatricians and child psychiatrists are familiar with insomnia and the various sleep disorders in children, these have been subject to varied epidemiological studies over the last decade which have helped to define the clinical picture. They cannot all be referred to here. However, whether these studies are American, British, Swedish, Belgian or French, they all stress the age factor. This intervenes both in the frequency of disorders and in their symptomatology and reveals the importance of the maturation of the sleep-wake rhythm of the child in the development and type of disorder.

In 1980, sleep disorders in children represented 1.8% of all requests for consultation, both inpatient and outpatient. This motive for consultation is on the increase today both for general practitioners and paediatricians [42, 62]. This is corroborated by the findings of Anders [5] for whom insomnia in infants under two years of age is the most common complaint of parents in infant consultation. Symptoms often persist for several weeks before the parents decide to consult [76, 77]. Sleep latency difficulties and repeated nocturnal arousals are the most frequent elements in sleep disorder consultation, and these tend to diminish with age. The other associated sleep disorders are, in order of frequency: sleep talking, enuresis, nocturnal terrors, and sleep walking [77]. But true insomnia defined as a lack of sleep, is rare [76]. In a group of 85 children of under 5 seen for the first time in sleep consultation [4], insomnia was the most common complaint (80%) with recurrent nocturnal arousals which are more or less prolonged, occurring mainly before the age of three and the refusal to go to bed tending to occur after the age of two. Insomnia in this age group may be accompanied by parasomnias (20% of cases): headed by circadian rhythm sleep disorders, followed by nocturnal terrors and nightmares, with the exceptional case of sleep talking. In 73% of consulting patients, the disorder had existed for over six months, leading parents to use hypnotic or sedative treatment in 30% of cases, before consulting.

REMINDER OF PHYSIOLOGY

Any study of insomnia in children implies the understanding of the physiology of the sleep-wake rhythm and that of wakefulness during the sleep period or "intra-sleep wakefulness" (see Chapter 2).

EPIDEMIOLOGY

Although the sleep-wake circadian rhythm stabilises at 6 months, 25% of children still wake up during the night between the ages of 6 and 12 months, but are able to get back to sleep without the parents' intervention [5, 74]. The proportion of infants who start to wake up again in the night increases after 9 months, reaching a peak during the second year [33, 73]. This cohort diminishes over the following years. At the age of 3, 21% [59] to 38% [11] of infants still wake up during the night once or several times a week; among these, only 2% continue to wake up after the age of 5 [11].

The problems recede, without disappearing completely, between the ages of 5 and 12 [53]. This period of life, still referred to as the "latency period" is reputed to be relatively free of sleep disorders. In fact these are probably largely unknown. Children of age 3 identified as being "at risk" and longitudinally monitored from birth continue to experience sleep disorders until the age of 6 at least (agitated sleep, frequent arousals, nightmares) as opposed to "controls" of the same age [23, 24]. Kahn *et al.* [51] consider that these sleep disorders remain substantial, with 43% of healthy children aged 8 to 10 suffering from sleep difficulties for over 6 months, 14% of them taking over 30 minutes to get to sleep. However Carskadon [16] found that only 5% of children awoke during the night between ages 10 and 13, with only 1% having difficulties getting back to sleep. In a more recent study; Bruni *et al.* [15] established the Sleep Disturbance Scale for Children, SDSC, using a questionnaire to identify sleep problems (16.6% of which involved insomnia) in 73.4% of normal control children, compared with children in the same age range consulting for sleep disorders. Insomnia relating to a disorganisation of the sleep-wake circadian rhythm, found in handicapped children, was also found in a population of Japanese children aged 6 to 12 attending normal schools [63].

The difficulties reappear in adolescence, as shown by two epidemiological studies [24, 68] conducted among high school children aged 14 to 20, with similar findings. Sleep difficulties appear to be considerable among this age group: 32.6% of young people are dissatisfied with their sleep in terms of the disparity between their expressed needs and actual sleep schedules.

Depending on the surveys and populations studied, varying degrees of sleep latency difficulty are found, with a greater or lesser number of complete or partial arousals, often of long duration, and fewer cases of early morning awakening. Young children are more subject to prolonged, repeated arousals during the night, pre-adolescents and adolescents to sleep latency difficulties or problems getting back to sleep and late bedtimes. As in the case of adults, these problems may only occur intermittently or may occur several times per week or per night. All cases give rise to behaviour problems during the day [52, 67, 95]: irritability, hyperactivity, fatigue on morning awakening and during the day, learning difficulties and academic problems among older children [23], daytime sleepiness and the need to sleep in the afternoon, even a "narcoleptic tendency" in emotional situations [68]. These characteristics recur in the various studies referred to here: Kahn *et al.* [51] observed a 21% school failure rate in children he preferred to call "bad sleepers" than insomniacs. In young children and adolescents, boys and girls are equally affected by insomnia and sleep disorders, while for 14 to 20 year olds, girls complain more often of sleep problems even though overall sleep times are virtually the same for both sexes [68]. Lastly, as has already been noted, these disorders may be associated with other abnormal nocturnal manifestations: nightmares and anxious dreams, different parasomnias with varying degrees of motor behaviour ranging from simple arousal with agitation to violent and dangerous bouts of sleepwalking several times a week [33]. Kahn *et al.* [51] found that 51 % of "bad sleepers" were affected by nocturnal motor behaviour, while the incidence of enuresis was the same as for "good sleepers".

AETIOLOGIES

Insomnia has no single cause in children; it is usually multifactorial, involving predisposition, upbringing, the psychoaffective context and in particular the behaviour of the parents and their own problems, constraints in schedule or imposed by social demands. This chapter will not include treatment of insomnia related to mental handicap, often known to involve sleep disorders [69].

Table 23.1 Aetiologies of insomnia in children

1. Insomnia in infants and young children:
- organic disorder
- perinatal antecedents
- feeding-related
- colic
- parent-child relationship
- sleep ritual, co-sleeping
- daytime behaviour disorder
2. School age insomnia:
- 5 – 12 years, rare, multifactorial
- Adolescent: importance of emotional factors.
3. Special cases:
- Severe insomnia
- Delayed sleep phase insomnia

The environmental factors are multiple, individual tendencies vary widely, with complex interconnections in the formation of insomnia. Cultural factors [55, 79], family or even transgenerational [27] history intervenes in the way in which parents and families perceive sleep and its disorders. Some authors [23, 29, 77] consider that children and adolescents who sleep badly are insomniacs from early infancy. They also tend to have parents who are "bad sleepers" with more distinct same sex parent-child parallelism for boys.

Insomnia in infants and young children

Sleep disorders linked to an organic pathology are less common than those linked to a psychoaffective context, the environment or predisposition.In another study by Kahn *et al.* [50], only 1/5 of children under 5 years of age seen for insomnia in specialised consultation presented an organic cause of the disorder. These included:
- acute or chronic otitis [83]
- atopic dermatosis [60]
- dyssomnia due to sleep fragmentation in acute or chronic upper airway obstructive syndromes of whatever origin, requiring early screening [41]
- arousals due to episodes of oesophageal reflux during sleep [83]
- insomnia due to allergy to cow's milk, detected in 17/146 insomniac children seen in sleep consultations by Kahn *et al.* [48]. This is a major form of insomnia with reduced nocturnal sleep duration and a marked number of intercurrent arousals; insomnia is amended by withdrawing cow's milk from the diet of these infants, who often show a marked reaction to cow's milk antibodies in allergy tests.
- nocturnal sleep disorganisation in epileptics due to fits and intercritical anomalies regardless of the type or stage of intervention. These are alterations to internal sleep architecture, with increasing arousals and micro-awakenings which may be increased by antiepileptic treatment, leading to a certain degree of excessive daytime sleepiness as a result [84].
- encephalopathies, chromosome anomalies, malformations of the nervous system, at the origin of severe disorganisation of the sleep-wake rhythm, electroencephalographic anomalies and sleep behaviour disorders [65]. Some cases of major sensory deprivation, particularly the blind, display insomnia in the form of alterations to the circadian rhythm whose length is prolonged, similar to "freerunning" circadian functioning [64].

- malnutrition leads to retarded growth also accompanied by sleep architecture disorders with a deficit of NREM sleep [78] as well as clinical sleep behaviour disorders when combined with an affective lack [43].

Previous family history and neonatal anoxia are reputed to lead to difficulties in establishing good sleep-wake rhythms due to the combination of an organic disorder and that of the parents' reactive anxiety. Sleep disorders are common among 3 years olds born prematurely, but practically disappear by the age of 5. Various perinatal incidents were found in the previous histories of 1,272 students complaining of sleep difficulties in childhood [25].

Several studies have looked at the possible influence of feeding modes and rhythms i.e. breast as opposed to bottle feeding, on the sleep-wake rhythm. The findings are contradictory. However the infants who were still being breastfed in the second year had trouble establishing a normal sleep-wake rhythm [32]. Their total sleep time over 24 hours is shorter than that of weaned children, with more frequent night awakenings and shorter periods of sleep. The phenomenon is more marked if the child shares a bed with the mother [61]. Using a rubber teat or pacifier to calm a crying infant lowers his/her wakefulness threshold, making the infant more liable to waken than a breast-fed infant of the same age [38]. The habit of automatically giving the bottle to an infant who wakes up during the night can lead to an increase in the volume of urine, discomfort caused by wet diapers, the expectation of a night feed and hormonal and intestinal alterations in response to added calories [33]. Moreover the general tendency to include cereal in the evening feed to make infants sleep, does not appear to be justified: infants from 1 to 4 months fed on a bottle containing rice flour showed no difference in nocturnal sleep to babies of the same age whose feed did not include cereals [57].

"Colic" in babies, consists of paroxysms of loud, prolonged crying, nearly every day, usually toward the end of the day, ending in the agitated onset of sleep. This colic, observed in otherwise healthy babies, does not usually cause the parents concern and represents an extreme form of wakefulness [90]. At 4 months, when colic spontaneously disappears, until the age of 8 months, these children have shorter periods of sleep, more frequent arousals and are difficult to handle, leading to incoherent reactions on the part of the parents. At the age of 3, intrafamilial relationship difficulties persist in the families of these children [71].

Parent-child relationships are difficult to quantify. The parents' sensitivity and powers of observation enable them to help their child establish his own sleep rhythm. It has been noted that parents whose children have sleep problems, tend to present, or previously presented the same type of difficulty more often than other parents. Too lenient an attitude on their part before the age of 2 years, will favour the development of chronic sleep disorders. Their untimely intervention during the spontaneous, even partial awakenings of their baby may prevent him from being able to get back to sleep alone easily, gradually inducing a disorder. Situations which produce anxiety [1, 32]: maternal depression, the mother's unscheduled absence, an illness or accident affecting the baby or a member of the family, a parental history of alcoholism, psychiatric illness, and conjugal problems are encountered more frequently in these families, whereas the mother regularly working outside the home does not correspond with sleep disorders in children [23].

Many children need a sleep ritual to get to sleep, variously associating: a doll, piece of material or soft toy, dummy or thumb sucking, lullabies, stories, a glass of water, rocking, hugs and kisses etc. [91]. This ritual helps separation from the parents and the outside world and reduces anxiety. Each child has his own, more or less complicated ritual. It can be difficult for the parents of a child who wakes up during the night, often several times, to renew the ritual or a part of it each time [33]. The ritual may be totally inappropriate if the child's habitual place of sleep is the living room with the TV on, or if he goes to sleep in the parents' bed after boisterous games. By being firm, the parents can teach the child the right time to go to bed alone, calmly, in his own cot or bed, in a way which is conducive to relaxation and the onset of sleep. Nevertheless, at aged three the young child usually moves from a cot with bars to a bigger open bed, which may lead him to abuse his new found freedom, getting out of bed at bedtime [35] or when waking up at night.

Sharing the bed with other family members, especially the parents (co-sleeping) is a relatively common cultural practice after the first months of life, including in Sweden where 35% to 40% of children aged 2 to 6 and 15% to 19% of children aged 9 sleep in their parents' bed for at least part of the night [53]. In the United States, the practice occurs in 30% of white families and 70% of black American families [56]. It only appears to have a harmful effect on sleep organisation in children from white families, increasing the incidence of sleep problems, particularly those of going to bed and waking during the night, particularly if the habit occurs several times a week.

Daytime behaviour disorders, according to certain authors, are often associated with chronic sleep disorders. They concern children who are often capricious and hard to handle, for whom the mothers have ambivalent feelings, since 20% recognise that the sleep disorders are upsetting to themselves and to family life [32]. This leads to the notion that the child with disturbed sleep enters a vicious circle with his mother in which both show signs of distress. For Guedeney and Kreisler [42], sleep disorders in young children are increasing in frequency and severity. The children concerned are often particularly active. These authors evaluate the child-mother relationship in the course of filmed cónsultations, often finding a lack of positive interaction between mother and child at the root of over-stimulation or incoherent behaviour. Finally, these mothers were found to have experienced serious emotional difficulties, anxiety or depression during pregnancy. For Stoleru *et al.* [87], sleep disorders are more common in children of depressed mothers.

Insomnia in schoolchildren and adolescents

Fewer findings are available for this group than for younger children. Between the ages of 5 and 12, napping stops and night sleeping habits become regular. In principle, a child of this age is able to get to sleep alone, with or without a bedtime ritual, and to stay in bed or in his room alone for the entire night. Bedtime gets later with shorter total sleep time on school days than on holidays. Nevertheless, sleep difficulties lasting over six months in healthy schoolchildren [51], are not significantly linked to any of the usual causes (family or conjugal problems, socio-educational level, depression or parents' psychological difficulties). A multifactorial origin for this poor sleep can however been envisaged, more frequently in a noisy environment [51] or one which is too brightly lit, in children who are emotionally disturbed. The effect on schoolwork of a lack of sleep in preadolescents and adolescents is not simply a state of mind. In the study already referred to by Kahn *et al.* [50], it was more common for "poor sleepers" to repeat their year at school. More recently, Randazzo *et al* [70] showed alterations to the higher cognitive functions (creativity, abstract thinking) in preadolescents and adolescents aged 10 to 14 years, subjected to a single shortened night's sleep from 3 am to 8 am.

For the adolescent, the incidence of sleep disorders mounts in relation to the preceding period, as has been seen. The manifestations of insomnia are the same, with a greater tendency to go to bed and get up later, particularly during holidays and weekends. When this is not checked, it can develop into a "phase delay" resulting in absenteeism at school or work. Dissatisfaction linked to poor sleep increases between the ages of 14 and 19, faster for boys than girls [24]. This sense of dissatisfaction tends to be related to specific sleep disorders for girls, and to tiredness on morning rising in the case of boys. The consumption of psychoactive drugs and irregular sleep habits add to the feeling of tiredness [89]. Emotional factors, stress at school, a noisy environment or cases of parasomnia [58] are the most common aetiological elements related to these difficulties.

The time children and adolescents spend watching television is often evoked by parents and educationalists as a cause of poor sleep [37]. But an inquiry by Weissbluth *et al.* [92] comparing the sleep of schoolchildren in 1980 with that of schoolchildren in 1911 and 1929 showed no difference in the duration of nocturnal sleep and afternoon naps between the three groups, even though children in 1980 spent an average of 2 hours 20 minutes a day watching television.

Insomnia in school age children and adolescents is encouraged by affective stress factors, which are often familial, particularly the mother's depression, a fairly constant factor in all the studies mentioned. Misunderstanding sleep physiology, a rigid upbringing or one which is too lenient or incoherent with irregular hours and habits can also further its development [33]. Some teams have looked at the psychological structure of insomniac children and adolescents and their "temperament", but found no characteristic psychopathological traits [37]. In a retrospective study on a group of adults whose insomnia started in childhood, Hauri and Olmstead [44] found only a number of "slight neurological disorders" (dyslexia, hyperactivity, minimal cerebral dysfunctioning or diffuse EEG anomalies).

Insomnia from "delayed sleep phase" is observed at all ages but is insufficiently documented in normal children. Young children are more inclined to experience "advanced sleep phase" with the need to go to bed early, and early morning rising, which disturbs the rest of the family. A disorganisation of the circadian rhythm is seen in the form of napping at odd or different times of day [33]. "Delayed sleep phase", as we have seen, is more common in the adolescent, whose hours become later, with a preference for night life characteristic of this age group [18].

CLINICAL INVESTIGATION AND EXPLORATIONS

Faced with a child with a sleep disorder, it is important to try to understand the situation before prescribing cumbersome and complex investigations and irrelevant hypnotic treatment.

As we have seen, understanding the problem requires a good knowledge of the physiology of neurobehavioural development. The first thing to eliminate is a medical pathology: otitis, digestive problems, signs of obstruction of the upper airways (snoring, excessive perspiration), disturbing or stereotyped nocturnal motor signs evocative of epilepsy. Anamnesis should give an accurate date of the onset of the disorder, the circumstances at the time, any associated pathology, long term treatment, and any treatment with hypnotics or sedatives. The problem will be easier to discern if a daily sleep diary is kept. This provides a simple means of establishing a day by day account of the hours and events occurring during sleep, whether during napping or in the night. The diary is a good way of interrogating the parents, and will help to determine the occurrence of sleep disorders during the night, at what times and when these occur in relation to the sleep cycles, all at very little cost [36].

Polygraphic sleep recordings in the laboratory should be restricted to difficult cases: severe insomnia or that which is associated with abnormal clinical features. They basically serve to eliminate any neurological pathology (epilepsy) or respiratory pathology (obstruction of the upper airways) or an exceptional case of allergy to cow's milk. The normative polygraphic studies of sleep in children, published in reviews, focus mainly on newborns and infants, rather than on children of school age and adolescents [48] for whom studies tend to concentrate on respiratory anomalies linked to sleep [82]. The current development of ambulant recording techniques which may or may not be coupled with a video recording, will probably redress this situation, with the possibilities for carrying out explorations in the child's own surroundings. An actimeter, worn on the leg for several days is a good means of objectively evaluating nocturnal awakenings [45, 75]. A psychological evaluation is often useful.

TREATMENT OF INSOMNIA IN CHILDREN

Sleep hygiene and the prevention of insomnia in children

Sleep hygiene refers to a number of practices which are necessary for maintaining nocturnal sleep and a good level of daytime alertness [88]. As a general rule, these practices result in a regular cycle of sleep and wakefulness over 24 hours, linked to the underlying circadian rhythms. Environmental factors play a major role in maintaining a normal sleep-wake rhythm.

The days and nights of the young child are punctuated by his needs for food, which he has to learn to control, his need for sleep, which his parents must learn to recognise, by acquiring control of his sphincters, by learning to walk and motor control, by gradually discovering and mastering the external world chiefly represented by the family. The development of all these functions and their inter-relationships, takes place at a stage when the child is fragile and dependent yet very flexible.

A sleep-wake rhythm of good quality develops through a series of pitfalls, which parents ideally face calmly and attentively, helping the child to find the right balance in life to ensure harmonious physical and intellectual development.

Role of the environment

While a "white" noise can help a child get to sleep [86], a room which is too noisy, lights which are too bright or a badly adapted bed can induce or perpetuate sleep disorders. Simple common sense usually suffices to improve these conditions [39], but it may be difficult or even impossible to avoid a noisy environment, or move from an apartment which is too small or overcrowded. Parents' working hours, requiring them to wake children early to undertake the long trip by public transport to get to the nursery or to school, are often a cause of family stress with the consequent repercussions on the child's sleep. Many young parents only see their children late in the day and so tend to overstimulate them in a way which detracts from the balance and relaxation ideal for sleep. Finally children under 3 years of age who are looked after outside the home (nursery or childminder) are usually exposed to different rhythms of activity during the week and at the weekend, which does not always help to resolve the disorder.

Needs and times of sleep and wakefulness, endogenous factors

The parents should be able to recognise their baby's needs for sleep or activity, differentiating these from other needs such as hunger and thirst. The time constraints of the life of the family do not always allow for individual characteristics such as "big sleepers", "short sleepers", "early sleepers" or "late risers" ("night-owls") [34]. Some children always need waking for their bottles while others scream for them. Some go to sleep easily in the midst of noise, bright lights and activity, while others in the same family are fragile sleepers, easily disturbed by the slightest change in their surroundings.

Generations of parents have been raised with a rigid schedule of meals and bedtimes. Habits are more flexible today. Nevertheless, parents may regret being too lenient if their baby has difficulty stabilising because of continually changing bedtimes [34]. Breastfed babies often have greater difficulty acquiring regular sleep habits, especially when demand feeding continues after the first year. After the first 2 or 3 months, the habit of one or more night feeds to help the infant back to sleep creates an unnecessary association between sleep and food, while contributing to the infant's discomfort due to polyuria and modifications in hormonal digestive secretion after this extra feed [33]. The drop in body temperature preceding and conditioning the onset of sleep stabilises in the first six months regardless of environmental factors (clothing, feeding etc.) [6].

Napping stops spontaneously between the ages of 3 and 5, and bedtime gets later, with a loss of half an hour's sleep per year from this age onwards. However some children refuse to sleep in the afternoons from a very early age, while others continue to need a nap up to the age of 6 or 7. Nursery or school staff, like parents, should recognise individual sleep needs and avoid the practice of stopping rest periods too early, in the general interest. A child who has not taken a nap and has not benefited from a rest period during the day, will tend to fall asleep early in the evening, and may then wake up in the middle of the night or too early in the morning, or on the contrary, have trouble winding down in the evening ready for sleep, with the ensuing family battles at bedtime. Regular hours and daily schedules will lead to regular sleeping habits and the role of the parents in acquiring this balance and instilling regularity, is of prime importance.

Once sleep appears, the cycles follow in the same way. Someone who goes to bed late at night will not alter his amount of NREM sleep at the beginning of the night, but reduce the total duration and number of sleep cycles with a reduction of REM sleep in the second part of the night [19]. These late nights, which are common in adolescence, and a chronic lack of sleep may be at the origin of excessive daytime somnolence, which itself gives rise to memory and learning problems [17].

Sleep rituals and separation anxiety

Many of the sleep difficulties encountered in children such as those relating to sleep latency and regaining sleep after a long period of wakefulness during the night, are linked to separation anxiety. The child who goes to bed is afraid of leaving his parents and facing the darkness, he is also afraid of losing control over consciousness and his acts; he is afraid his family will not be there the following day. He tends to prolong going to bed indefinitely. Yet he needs to learn to sleep in his bed and his room alone if he is to become autonomous.

The sleep ritual helps the child to face this moment of solitude. As he gets into bed, he will develop a number of practices to help him to relax and fall asleep, subjecting his parents to this to a greater or lesser extent, particularly the mother. Any excuse serves to delay the moment when the light is turned off and the door closed. The child who cries or screams at bedtime exerts a form of blackmail on those around him, stimulating guilt feelings in the mother and reactivating past fears. She too may apprehend the moment of separation in the evening, being afraid she will not find her baby alive the next morning. She feels guilty about not staying with him and may feel she is abandoning him. A simple sleep ritual, and an affectionate but firm attitude at the point of leaving him are the best ways of helping the child get to sleep.

Whatever the age, it is best to avoid boisterous activities, excitement by older members of the family and frightening television programmes when arriving home at the end of the day. Small children love water play and a warm bath before the evening meal helps them to wind down physically. During the evening meal, where it is best to avoid conflict over food, a quiet game or conversation alone with the child, reading or telling a story before the last hug generally suffice in preparing for bed. It is pointless multiplying the bedtime preparation activities. The simpler the bedtime ritual, the easier it is to repeat during the night if the child is unable to get back to sleep after prolonged wakefulness. Parents avoid the need to stay with the child until he gets back to sleep. It is better to leave the room quietly, turning off the light. The door can be left slightly open with a nightlight nearby so the child does not feel alone, but it is useless to go back to check whether the child is sleeping. Children whose parents leave them alone at the moment of going to sleep wake up less often during the night and for shorter periods of time than those whose parents wait for them to get to sleep, staying by to talk to them [1, 2]. Older children can listen to music or read in bed, turning out the light when they are ready to sleep. The child's fears should in no way be encouraged by looking under the furniture, behind the door or curtains to see whether a ghost or robber is hiding there.

As in the case of very young children [20] the family dynamics are different at preschool age depending on whether the child presents isolated night arousals or problems associated with going to bed [14]. If the child wakes up in the middle of the night, which, as we have seen, is normal and frequent, he generally wakes up only partially without fully regaining consciousness. He may look terrified if he wakes up during an anxious dream or nightmare. Excessive or inappropriate action such as waking the child to talk to him on the pretext of reassuring him, will serve to encourage these repeated, prolonged awakenings so tiring for both the child and the parents. The same calm, reassuring attitude should be adopted as at bedtime. Behaviour therapy aims at reconditioning the child and his parents with increasingly short interventions, separated by longer and longer intervals [25, 27] in order to reassure the child without touching or holding him. The oldest technique which consists of letting the child cry with no intervention other than to check that he is alright, remains the fastest acting [62, 77], even if it is the most testing - for the parents even more than for the child.

Regardless of age, regular bedtimes and rising times with a total sleep time corresponding to the theoretical needs for the age group, are the best guarantees of good quality sleep. Preadolescence and adolescence are times when irregular habits develop, with a tendency to go to bed and get up late, particularly during holidays, while sleep generally becomes lighter [13] even if the need to sleep is still great. This is another period where it is more difficult to get to sleep and go back to sleep after waking up in the night. Intervention is more difficult for the parents and less well tolerated by the child. The adolescent must learn to organise his sleep. If his bedtimes get later and later, he should be advised to get up at the same time, and to avoid stimulants (coffee, alcohol) and psychoactive drugs liable to disorganise sleep [58].

Treating psycho-emotional factors

These are an important aspect of the development of sleep disorders in children of all ages. The disorder may be largely resolved and the atmosphere improved, at least, if the parents are made aware of the problems and assume their responsibilities. The basic elements of sleep physiology and its evolution in relation to age should be explained to them, with advice on simple, caring attitudes [80]. Keeping a sleep diary will help them become aware of the reality of the problem and its frequency. Time is needed, not only to understand what is happening, but to explain and relieve guilt. A "sleep programme" with written instructions and parental support by phone have been shown to be effective in reducing nocturnal awakenings for children of under 5 [81]. By giving the parents the opportunity to discuss their own sleep problems, the family approach can rapidly dispel insomnia [26]. A combined consultation, associating a child psychiatrist and a clinical neurophysiologist can help in understanding the various factors involved in insomnia situating it within the history and psychodynamics of the family [4, 62].

Children who occasionally or regularly take hypnotics are more likely to have mothers who suffer from depression [21, 76]. Mothers often abandon interesting or lucrative professional activities to look after their young children. It is important to stress the value of motherhood and to restore confidence in themselves and their capacities as mothers; they should be advised to set aside time for themselves devoted to motivating activities. They will be all the more relaxed in looking after their children. Fathers have a role to play in resolving sleep problems. For professional reasons or because of a separation they may be absent from the life of an insomniac child, who finds himself alone with his mother. They need assistance in finding and reserving time in the evening or at weekends to devote to their children, even if, *a priori*, their work schedule allows them no time. Their intervention during one of their child's waking episodes will help to alleviate a conflictual relationship or one of renouncement on the part of a fatigued mother, a conflict which frequently continues in the couple's bed.

Treatment for "phase delay" takes the form of chronotherapy or a gradual shifting of the sleep schedule and the strengthening of social synchronisers, to arrive at bedtime and rising hours compatible with social life [3, 34, 10]. In the case of a "deliberate delayed sleep phase", chronotherapy is liable to fail [33, 34]. Severe Delayed Sleep Phase syndrome often associated with "sleepless nights" and suspended school attendance can be brought about by the adolescent suffering from serious psychological and family problems, calling for psychiatric treatment. Listening to the adolescent and giving advice on habits and sleep problems can only be offered as accessories to treatment.

Prescribing melatonin, even after a request for authorisation for temporary use is not currently accepted in France for children and adolescents, despite the fact that this substance has been used successfully in groups of children with retardation and neurological or sensory handicaps of various origins [46, 93].

Should hypnotics be given to children?

Apart from the sleep disorders which form part of a psychopathological picture and the particular case of parasomnias [28], the use of hypnotics should be avoided in general paediatrics.

Depending on the survey, between 4 and 25% of children and adolescents of all ages take hypnotics occasionally or regularly, and 19% have taken them in the past, to which must be added the 13% of children taking drugs for "psychological problems" with a probable effect on sleep [21, 68, 77]. This medication generally comes from the family medicine cabinet, and is usually taken without a medical prescription, by children defined as being "at risk" [21, 68]. It is not unusual to encounter children in consultation, whose sleeping disorders have failed to be calmed by the medication, and have increased on discontinuation of treatment. Apart from the fact that it does not always improve the sleep disorder, even temporarily, this medication is probably to blame for daytime sleepiness and a certain level of fatigue in those who take it. Benzodiazepines and niaprazine are the most often prescribed [30, 72].

Table 23.2. Treatment of insomnia in children

1.	Sleep hygiene and the prevention of insomnia in children:
	- Environment
	- Respect for sleep-wake typology
	- Morning arousal
	- Night arousals
2.	Psychological and family support.
3.	Problem of hypnotics.

Niaprazine (Nopron®) and its English equivalent Trimeprazine Titrate (Vallergan Forte®) which are widely used, have been subject to controlled, double-blind studies using scales of sleep and wakefulness behaviour and sleep diaries. These studies were carried out on hospital inpatients

[30] or in paediatric consultation [12] or child psychology [40] on children presenting clinical sleep disorders. In one of these [85], children were recruited by mailed questionnaire; none of the studies took polygraphic recordings into account to objectively evaluate the sleep organisation of the children submitted for treatment. Work carried out on healthy volunteers [94] aged 18 to 27 showed that niaprazine increased NREM sleep without reducing REM sleep, with no ensuing residual daytime sleepiness, although it must be noted that the dose was inferior (0.75 mg/kg) to those used in the paediatric studies. Although the posology lacked precision, the dose in one study [12] was 1 to 6 mg/kg for 6 to 7 days, with immediate positive results, both on sleep disorders and behaviour [12, 66], but with no sustained improvement after stopping treatment [94] or at longer term [72]. However, as phenothiazines have been cited in incidences of serious illness, or even cot deaths [47, 49], niaprazine may be responsible for the "malaise" encountered in otherwise healthy newborns [83] or older children with neurological symptomatology evoking a convulsive fit [9, 13], with too high a dosage or again, in accidents of abrupt hypotension with bradycardia [54], when given as a premedication dose of 2.7 to 4.2 mg/kg.

CONCLUSION

The understanding and treatment of sleep difficulties in children clearly depends, for a large part, on a good working knowledge of the physiology of the sleep-wake cycle. These difficulties are expressed differently, depending on age and the level of maturation. Emotional disturbance, both personal and familial, often weighs heavily in these cases of sleep disorder, particularly if added to incoherent patterns of upbringing, or disorganisation of the rhythms or schedules of daily life. Good common sense, educating the parents and a simple behavioural approach can largely help to solve these problems, with multidisciplinary support involving both the child and his parents, accompanied, if necessary, by specialist psychotherapy.

Hypnotic treatment should be avoided, particularly in the long term; parents must be taught to recognise the sleep-wake rhythm of their child. They must be given simple but firm rules of life hygiene to strengthen the synchronisers.

Taking the drama out of the situation, calming the family atmosphere and placing value on the affective exchanges between the parents and child will help them to acquire greater reciprocal autonomy, which will in turn help to resolve sleep difficulties.

REFERENCES

1. ADAIR R., BAUCHNER H., PHILIPP B., LEVENSON S., ZUCKERMAN B. – Night waking during infancy: role of parental presence at bedtime. *Pediatrics,* 87, 500-504, 1991.
2. ADAIR R., ZUCKERMAN B., BAUCHNER H., PHILIPP B., LEVENSON S. – Reducing night waking in infancy: a primary care intervention. *Pediatrics,* 89, 585-588, 1992.
3. AGUIRRE BERROCAL A. – Trastornos del ritmo-circadiano-clinica y tratamiento. *In: Trastornos del sueno en la infancia.* R. Peraita Adrados (ed), CEPE, Madrid, 149-165, 1992.
4. AMIN-HANDJANI F., NAVELET Y., FERRARI P. – Sleep disorders in children less than five years old. *J. Sleep Res.,* 5 (suppl.1), 5, 1996.
5. ANDERS J.F., HALPERN L.F., HUA J. – Sleeping through the night: a developmental perspective. *Pediatrics,* 90, 554-560, 1992.
6. ANDERSON E.S., PETERSEN S.A., WAILOO M.P. – Factors influencing the body temperature of 3-4 months old infants at home during the day. *Arch. Dis. Child.,* 65, 1308-1310, 1990.
7. AMERICAN SLEEP DISORDERS ASSOCIATION. *ICSD - International classification of Sleep Disorders, revised: Diagnostic and coding manual.* American Sleep Disorders Association, 1997.
8. ASSOCIATION OF SLEEP DISORDERS CENTERS. - *Diagnostic Classification of Sleep and Arousal Disorders.* First Edition, prepared by the Sleep Disorders Classification Committee, H.P. Roffwarg. Chairman, *Sleep* 2, 1-137, 1979.
9. BARTHEZ M.A., AUTRET A., BILLARD C., SANTINI J.J. – Un bien curieux malaise. *Arch. Fr. Pediatr.,* 44, 622, 1985.
10. BENOIT O. – les troubles du rythme circadien veille-sommeil par échappement pathologique de ce rythme à l'influence des synchroniseurs. *In:* M. Billiard (ed) *Le Sommeil normal et pathologique.* 2ᵉᵐᵉ édition, Masson, Paris, 356-369, 1998.
11. BENOIT O., LACOMBE J., CLODORE M., SALZARULO P. – *Prevalence of sleep disorders in children and developmental aspects of the Sleep-Wake organization.* 5ᵗʰ International Congress of Sleep Research, Copenhagen, 1987 (unpublished data)

12. BESANA R., FIOCCHI A., DE BARTOLOMEIS L., MAGNO F., DONATI C. – Comparison of niaprazine and placebo in pediatric, behaviour and sleep disorders: double-blind clinical trial. *Current therapeutic Research*, 36, 58-66, 1984.

13. BODIOU C., BAVOUX F. – Niaprazine et effets indésirables en pédiatrie. *Thérapie*, 43, 307-311, 1988.

14. BRUNI O., LO RETO F., MIANO S., OTTAVIANO S. – Daytime behavioral correlates of awakenings and bedtime resistance in preschool children. *In: Clinical Neurophysiology at the Beginning of the 21 st Century* (Supplements to Clinical Neurophysiology, Vol. 53). Z. Ambler, S. Nevsimalova, Z. Kadanka, P.M. Rossini (eds), Elsevier Science, Amsterdam, 358-361, 2000.

15. BRUNI O., OTTAVIANO S., GUIDETTI V., ROMOLI M., INNOCENTI M., CORTESI F., GIANNOTTI F. – The sleep disturbance scale for children (SDSC). Construction and validation of an instrument to evaluate sleep disturbance in childhood and adolescence. *J. Sleep Res.*, 251-259, 1996.

16. CARSKADON M.A. – The second decade. *In: Sleeping and Waking Disorders: indications and techniques.* Guilleminault C. (ed), Addison-Wesley, Menlo-Park, California, 99-125, 1982.

17. CARSKADON M.A., DEMENT W.C. – Sleepiness in the normal adolescent. *In: Sleep and its disorders in children.* Guilleminault C. (ed), Raven Press, New York, 53-66, 1987.

18. CARSKADON M.A., VIERA C., ACEBO C. – Association between puberty and delayed sleep preference. *Sleep*, 16, 258-262, 1993.

19. CHALLAMEL M.J., THIRION M. – *Le sommeil , le rêve et l'enfant.* Albin Michel, Paris, 1995.

20. CHAMBRY J., NAVELET Y., FERRARI P. – Insomnia in children. *J. Sleep Res.*, 9, suppl.1, 69, 2000.

21. CHOQUET M., DAVIDSON F. – Les facteurs favorisant l'administration de sédatifs chez les nourrissons et leur signification. *Arch. F. Pediatr.*, 35, 785-792, 1978.

22. CHOQUET M., FACY F., LAURENT F., DAVIDSON F. – Les enfants à risque en âge préscolaire. Mise en évidence par analyse typologique. *Arch. Fr. Pediatr.*, 39, 185-192, 1982.

23. CHOQUET M., LEDOUX S. – La valeur pronostique des indicateurs de risque précoces. Etude longitudinale des enfants à risque à 3 ans. *Arch. Fr. Pediatr.*, 42, 541-546, 1985.

24. CHOQUET M., TESSON F., STEVENOT A., PREVOST E., ANTHEAUME M. – Les adolescents et leur sommeil: approche épidémiologique. *Neuropsychiatrie de l'enfance*, 36, 399-410, 1988.

25. COREN S., SEARLEMAN A. – Birth stress and self-reported sleep difficulty. *Sleep*, 8, 222-226, 1985.

26. DARDENNE P., GUERIN F. – Les insomnies chez le jeune enfant. *Ann. Pediatr.*, (Paris), 33, 705-710, 1986.

27. DE LEERSNYDER H. – *L'enfant et son sommeil.* R. Laffont, Paris, 1998.

28. DE VILLAR R., BASTUJI H., GARDE P., MAUGUIÈRE F., DALERY J., MAILLET J., REVOL O. – Les terreurs nocturnes. Le somnambulisme. *Rev. Prat.* (Paris), 39, 10-14, 1989.

29. DIXON K.N., MONROE L.J., JAKIM S. – Insomniac children. *Sleep*, 4, 313-318, 1981.

30. DOLLFUS S., PETIT M. – Les benzodiazépines chez l'enfant. *In: Efficacité et tolérance des psychotropes chez l'enfant.* Expansion Scientifique Française, Paris, *157-168, 1988.*

31. DURAND V.M., MINDELL J.A. – Behavioral treatment of multiple childhood sleep disorders. *Behav. Modif.*, 14, 37-49, 1990.

32. ELIAS M.F., NICHOLSON N.A., BORA C., JOHNSTON J. – Sleep/Wake patterns of breast-fed infants in the first 2 years of life. *Pediatrics*, 77, 322-329, 1986.

33. FERBER R. – *Solve your child's sleep problems.* Simon and Schuster. New York, 1985. Traduction Française: Navelet Y. *Protégez le sommeil de votre enfant.* ESF, Paris, 1990.

34. FERBER R., - Circadian and schedule disturbances – *In: Sleep and its disorders.* Guilleminault C. (ed), Raven Press. New York, 165-175, 1987.

35. FERBER R. – Sleeplessness in the child. *In Principles and Practice of Sleep Medicine.* Kruger M.H., Roth T., Dement W.C. (eds). W.B. Saunders. Philadelphia, 633-639, 1989.

36. FERBER R. – Sleep schedule dependent causes of insomnia and sleepiness in middle childhood and adolescence. *Pediatrician*, 17, 13-20, 1990.

37. FISHER B.E., RINEHART S. – Stress, arousal, psychopathology and temperament: a multidimensional approach to sleep disturbance in children. *Person. Individ. Diff.*, 11, 431-438, 1990.

38. FRANCO P., SCAILLET S., WERMEMBOL U., VALENTE F., GROSSWASSER J., KAHN A. - The influence of a pacifier on infants' arousals from sleep. *J. Pediatr.* 136, 775-779, 2000.

39. GAILLARD M. – *Le sommeil, ses mécanismes et ses troubles.* Doin, Paris, 1990 – Payot, Lausanne, 1990

40. GALLET J.P. – Etude d'un sédatif à effet hypnotique chez 50 enfants présentant un trouble du sommeil. *Médecine Infantile*, 92, 475-479, 1985.

41. GAULTIER C. – Obstructive sleep apnoea syndrome in infants and children: established facts and unsettled issues. *Thorax*, 50, 1204-1210, 1995.

42. GUEDENEY A., KREISLER L. – Sleep disorders in the first 18 months of life: hypothesis on the role of motherchild emotional exchanges. *Infant Mental Health Journal*, 8, 307-318, 1987.

43. GUILHAUME A., BENOÎT O., GOURMELEN M., RICHARDET J.M. – Relationship between sleep stage IV deficit and reversible HGH deficiency in psychosocial dwarfism. *Pediatr. Res.*, 16, 199-203, 1982.

44. HAURI P.J., OLMSTEAD E. – Childhood onset insomnia. *Sleep*, 3, 59-65, 1980

45. HAURI P.J., WISBEY J. – Wrist actigraphy in insomnia. *Sleep*, 15, 293-301, 1991.

46. JAN E.J., FREEMAN R.D., FAST D.K. – Melatonin treatment of sleep-wake cycle disorders in children and adolescents. *Dev. Med. Child Neurol.,*, 41, 491-500, 1999.

47. KAHN A., BLUM D. – Phenothiazines and sudden infant death syndrome. *Pediatrics*, 70, 75-78, 1982.

48. KAHN A., DAN B., GROSSWASSER J., FRANCO P., SOTTIAUX M. – Normal sleep architecture in infants and children. *J. Clin. Neurophysiol.* 13, 184-197, 1996.

49. KAHN A., HASAERTS D., BLUM D. – Phenothiazine-induced sleep apneas in normal infants. *Pediatrics*, 71, 49-52, 1985.

50. KAHN A., MOZIN M.J., REBUFFAT E., SOTTIAUX M., MULLER M.F. – Milk intolerance in children with persistent sleeplessness: a prospective double-blind cross over evaluation. *Pediatrics*, 84, 595-603, 1989.

51. KAHN A., VAN DE MERCKT C., REBUFFAT E., MOZIN M.J., SOTTIAUX M., BLUM D., HENNART P. – Sleep problems in healthy preadolescents. *Pediatrics*, 84, 542-546, 1989.

52. KAPLAN B.J., McNICOL J., CONTE R.A., MOGHADAM H.K. – Sleep disturbance in preschool-aged hyperactive and non hyperactive children. *Pediatrics*, 80, 839-844, 1987.

53. KLACKENBERG G. – Sleep behaviour studied longitudinally. *Acta Paediatr. Scand.* 71, 501-506, 1982.

54. LOAN W.B., CUTHBERT D. – Adverse cardiovascular response to oral trimeprazine in children. *Br. Med. J.*, 290, 1548-1549, 1985.

55. LOZOFF B. – Culture and family: influences on childhood sleep practices and problems. *In: Principles and Practices of sleep medicine in the child.* R. Ferber, M. Kryger (eds) W.B. Saunders, Philadelphia, 69-73, 1995.

56. LOZOFF B., WOLF A.W., DAVIS N.S. – Sleep problems seen in pediatric practice. *Pediatrics, 75, 477-482, 1985.*

57. MACKNIN M.L., MEDENDORP S.V, MAIER M.C. – Infant sleep and bedtime cereal. *Am. J. Dis. Child*, 143, 1066-1068, 1989.

58. MANNI R., TATTI M.T., MARCHIONI E., CASTELNOVO G., MURELLI R., SARTORI I., GALIMBERTI C.A., TARTARA A. – Poor sleep in adolescents: a study of 869 17-year-old Italian secondary school students. *J. Sleep Res.* 6, 44-49, 1997.

59. MANTZ J., MUZET A. – Le sommeil de l'enfant de trois ans: Enquête en mileu scolaire. *Arch. Fr. Pediatr.*, 48, 19-24, 1991.

60. MONTI J.M., VIGNACE R., MONTI D. – Sleep and night time pruritus in children with atopic dermatis. *Sleep*, 12, 309-314, 1989.

61. MOSKO S., RICHARD C., McKENNA J. – Infant arousals during mother-infant bed sharing: implications for infant sleep and sudden infant death syndrome research. *Pediatrics*, 100, 841-849, 1997.

62. NAVELET Y. – Insomnia in the child and adolescent. *Sleep*, S23-S28, 1996.

63. OISHIBASHI Y., KAKIZAWA T., OTSUKA A., KAWAMA K., SATO Y., SUZUKI M., SASAKI H. – Disturbances of sleep and waking in handicapped children (II): Trend of circadian rhythm disorders in deaf children. *Jpn. J. Psychiat. Neurol.*, 47, 464-465, 1993.

64. OKAWA M., NANAMI T., SHIMIZU T., HISHIKA Y., SASAKI H., NAGAMINE H., TAKANASHI K. – Four congenitally blind children with circadian sleep-wake rhythm disorder. *Sleep*, 10, 101-102, 1987.

65. OKAWA M., SASAKI H. – Sleep disorders in mentally retarded and brain-impaired children. *In: Sleep and its disorders in children.* C. Guilleminault (ed), Raven Press, New York, 269-290, 1987.

66. OTTAVIANO S., GIANNOTTI F., CORTESI F. – The effect of niaprazine on some common sleep disorders in children. *Child's Nerv. Syst.*, 7, 332-335, 1991.

67. PARQUET P.J., BAILLY D. – La prescription des hypnotiques chez l'enfant . *Rev. Ped.*, XXII, 375-388, 1986.

68. PATOIS E., VALATX J.L., ALPEROVITCH A. – Prévalence des troubles du sommeil et de la vigilance chez les lycéens de l'Académie de Lyon. *Rev. Epidém. et Santé Pub.*, 41, 383-388, 1993.

69. QUINE L. – Sleep problems in children with mental handicap. *J. of Ment. Deficiency Res.*, 35, 269-290, 1991.

70. RANDAZZO A.C., MUEHLBACH M.J., SCHWEISTER P.K., WALSH J.K. – Cognitive function following acute sleep restriction in children ages 10-14. *Sleep*, 21, 861-868, 1998.

71. RAUTAVA P., LEHTONEN L., HELENIUS H., SILLANPAA M. – Infantile colic: child and family three years later. *Pediatrics*, 96, 43-47, 1995.

72. RICHMAN N. – A double-blind drug trial of treatment in young children with waking problems. *J. Child Psychol. Psychiat.*, 26, 591-598, 1985.

73. RICHMAN N. – Surveys of sleep disorders in a general population. *In: Sleep and its disorders in children.* C. Guilleminault (ed), Raven Press, New York, 115-127, 1987.

74. RICKERT V.I., JOHNSON C.M. – Reducing nocturnal awakening and crying episodes in infant and young children: a comparison between scheduled awakening and systematic ignoring. *Pediatrics*, 81, 203-212, 1988.

75. SADEH A., LAVIE P., SCHER A., TIROSH E., EPSTEIN R. – Actigraphic home-monitoring sleep disturbed and control infants and young children: a new method for pediatric assessment of sleep-wake patterns. *Pediatrics*, 87, 494-499, 1991.

76. SALZARULO P., CHEVALIER A. – Sleep problems in children and their relationship with early disturbances of the waking-sleeping rhythms. *Sleep*, 6, 47-51, 1983.

77. SALZARULO P., CHEVALIER A., COLVEZ A., BRUNEL M., SENDER C., KASTLER B., ROC M. – Child sleep problems. Parental attitude and recourse: an approach by survey. *Sleep 1978*, S. Karger, Basel 1980, 595-598.

78. SALZARULO P., FAGIOLI I., SALOMON F., RICOUR C., RAIMBAULT G., AMBROSI S., CICCHI I., DUHAMEL J.F., RIGOARD M.T. – Sleep patterns in infants under continuous feeding from birth. *Electroenceph. Clin. Neurophysiol.*, 49, 330-336, 1980.

79. SALZARULO P., GIGANTI F., FICCA G., FAGIOLI I., TOSELLI M. – Gates to awakening in early development. *In: Clinical Neurophysiology at the Beginning of the 21st Century (Supplement to Clinical Neurophysiology Vol. 53)* Z. Ambler, S. Nevsimalova, Z. Kadanka, P.M. Rossini (eds), Elsevier Science, Amsterdam, 352-354, 2000.

80. SCHAEFFER C.E. – Treatment of night wakings in early childhood: maintenance of effects. *Percept. Mot. Skills.* 70, 561-562, 1990.

81. SEYMOUR F.W., BROCK P., DURING M., POOLE G. – Reducing sleep disruptions in young children. Evaluation of therapist-guided and written information approaches: a brief report. *J. Child Psychol. Psychiatry*, 30, 913-918, 1989.

82. SHELDON S.H. – *Evaluating sleep in infants and children.* Lippincott. Raven Publishers, New York. 1996.

83. SHELDON S.H., SPIRE J.P., LEVY H.B. – *Pediatric Sleep Medicine,* W.B. Saunders, Philadelphia, 1992.
84. SHOUSE M.N., MARTINS DA SILVA M., SAMMARITANO M. – Circadian rhythms, sleep and epilepsy. *J. Clin. Neurophysiol.,* 13, 32-50, 1996.
85. SIMONOFF E.A., STORES G. – Controlled trial of trimeprazine tartrate for night waking. *Arch. Dis. Child.,* 62, 253-257, 1987.
86. SPENCER J.A.D., MORAN D.J., LEE A., TALBERT D. – White noise and sleep induction. *Arch. Dis. Child.,* 65, 135-136, 1990.
87. STOLERU S., NOTTELMANN E.D., BELMONT B., RONSAVILLE D. – Sleep problems in children of affectively ill mothers. *J. Child Psychol. Psychiat.* 38, 831-841, 1997.
88. THORPY M., YAGER J. – *The encyclopedia of sleep and sleep disorders.* Facts on File, New York, 1991.
89. TYNJÄLÄ J., KANNAS L., LEVÄLAHTI E. – Perceived tiredness among adolescents and its association with sleep habits and use of psychoactive substances. *J. Sleep Res.* 6, 189-198, 1997.
90. WEISSBLUTH M. – Sleep and the colicky infant. *In: Sleep and its disorders in children.* C. Guilleminault (ed), Raven Press, New York. 129-140, 1987a.
91. WEISSBLUTH M. – *Sleep well.* Unwin Nyman Limited, London, 1987b.
92. WEISSBLUTH W., PONCHER J., GIVEN G., SCHWAB J., MERVIS R., ROSENBERG M. – Sleep duration and television viewing. *J. Pediatr.,* 99, 486-488, 1981.
93. ZHDANOVA I.V., WURTMAN R.J., WAGSTAFF R.J. - Effects of a low dose of melatonin on sleep in children with Angelman syndrome. *J. of Pediat. Endocrinol. and Metabolism,* 12, 57-67, 1999.
94. ZUCCONI M., MONDINI S., GERARDI R., PETRONELLI R., DONATI C., BERTIERI R.S., CIRIGNOTTA F. – Niaprazine: a polysomnographic study of nocturnal sleep and daytime sleepiness in healthy volunteers. *Current Therapeutic Research,* 44, 118-132, 1988.
95. ZUCKERMAN B., STEVENSON J., BAILEY V. – Sleep problems in early childhood: continuities, predictive factors and behavioral correlates. *Pediatrics,* 664-670, 1987.

Chapter 24

Sleep and circadian rhythms in normal aging

J. Carrier and D. Bliwise

Sleep Disorders Center, Emory University Medical School, WMB – Suite 6000, Atlanta, GA 30322, USA

The North American population is aging rapidly. By 2020, more than 40% of the population will be over 40 and more than 25% will be over 55. There is no doubt that aging is associated with an important increase in sleep-wake cycle complaints, which has important individual, social, and economical consequences. Multiple factors including medical problems, side effects of medications, and specific sleep disorders account for this age-dependent increase in sleep difficulties. Notable modifications of the sleep-wake cycle are also observed in "optimal aging," i.e., in people who do not suffer from medical, psychiatric, or specific sleep disorders. These age-dependent changes occur quite early and they may have important repercussions for older individuals, especially when their sleep-wake system faces challenges such as those related to stress, jet lag, and shift work.

DESCRIPTIVE STUDIES OF SLEEP CHARACTERISTICS IN NORMAL AGING

The modification of sleep organisation is a hallmark of the normal aging process (see [6,23,24,251]. Healthy elderly people go to bed and wake up earlier than do younger subjects [67,75,185,295]. Measures of habitual sleep patterns and polysomnographic sleep show a reduction in nighttime sleep duration and an increase in daytime sleep episodes in elderly subjects [51,138,295]. The nocturnal sleep of older people is more fragmented with awakenings than is the sleep of younger people, especially in the second half of the night [104]. Elderly subjects show an increase of about 15-20% in wakefulness during the nighttime sleep episode [138]. One of the most robust age-dependent changes in sleep architecture is the diminution in deep NREM sleep [104]. Elderly subjects also show less spindle activity, lower spindle frequency, and lower spindle amplitude during NREM sleep [105,249,324]. In addition, elderly subjects demonstrate higher percentages of lighter stages of sleep (stages 1 and 2) than do young subjects. The effect of aging on REM sleep is more controversial. Some studies have reported a reduction in REM sleep latency, less REM sleep during the night, and more REM sleep in the first part of the sleep episode whereas other studies have found no effects of aging on these parameters [104,262].

Until now, our understanding of the mechanisms by which sleep deteriorates with age has almost exclusively come from comparisons between young and elderly subjects. However, subjective sleep complaints begin to increase significantly in the middle years of life, and almost all sleep parameters show significant effects of age between the ages of 20 and 60 [29,62,66,88,178,260,295,301,320,322,332]. The results of our study of 110 healthy subjects between the ages of 20 and 60 years indicated that, at home, older subjects woke up earlier, went to bed earlier, and spent less time in bed [66]. In the laboratory, increasing age was associated with less time asleep, more frequent awakenings of longer duration during sleep, less deep NREM sleep and Rapid-Eye Movement sleep (REM), and higher percentages of stage 1 and stage 2 sleep [66].

Quantitative analysis of the sleep EEG across the night is a powerful and sensitive tool for evaluating changes in sleep regulatory processes with advancing age. For example, slow-wave activity (spectral power between 0.75-4.5 Hz during NREM sleep) increases proportionally with the number of hours of wakefulness that precede sleep; it is also an indicator of sleep intensity [38] (see below). Recent studies in both depressed and insomniac populations also suggest that elevated fast frequency activity during NREM sleep might be an indicator of hyperarousal and that it could be related to lower sleep quality [10; 201]. Experimental results on quantitative sleep EEGs point out important modifications between 20 and 60 years of age. The most consistent of these age-dependent changes is a decrease in slow-wave activity, which corroborates that the sleep of middle-aged subjects is less intense than the sleep of younger subjects [62,178]. We recently assessed the effects of age on sleep EEG power spectral density in a group of 100 subjects aged 20y to 60y. The effect of age varied according to frequency. The decrease in power with age was not restricted to slow-wave activity, but also included theta and sigma activity. Increasing age was associated with higher power in the beta range. Across night changes in the sleep EEG have been described with slower frequencies decreasing power across the night and faster frequencies increasing power across the night. With increasing age, the attenuation over the night in power density between 1.25 and 8.00 Hz diminished, and the rise in power between 12.25 and 14.00 Hz across the night decreased. These sleep EEG changes may underlie the aging sleep-wake cycle system's greater difficulty adapting to challenges that ordinarily disrupt sleep (see below).

AGE-DEPENDENT ALTERED MECHANISMS IN SLEEP PROCESSES: THE TWO PROCESS MODEL OF SLEEP-WAKE CYCLE REGULATION

According to contemporary models of sleep-wake cycle regulation, the interaction of homeostatic and circadian processes regulates the sleep-wake cycle [1,37,47,77]. The two-process model has been able to predict the timing and intensity of sleep for a variety of schedules in young subjects (total and partial sleep deprivation, shift-work, constant bed-rest, etc.; reviewed in Borbély and Achermann, 2000 [39]). The homeostatic process represents the accumulation of sleep pressure with increasing time awake and its dissipation during a sleep episode. The circadian process represents the rhythmic variations of sleep and wake propensity over 24 hours. A precise interaction between these two processes ensures optimal quality of both sleep and vigilance. Aging may have a negative impact on the homeostatic and circadian processes individually and in their interaction.

Homeostatic Sleep Process (Process S)

The homeostatic sleep process (process S) represents the sleep debt accumulated during wakefulness. As a result, the homeostatic process increases during waking hours and decreases during sleep. In the two-process model of sleep regulation, the time course of process S stems from the intensity and the dynamic of SWA during NREM sleep. According to this model, the decline in SWA during NREM sleep reflects the dissipation of homeostatic sleep drive. The dissipation of process S during the night can be approximated by an exponential decay with an estimated time constant of 4.2h in young men [77].

Numerous studies have investigated the build-up function of homeostatic sleep drive in young subjects by measuring the effects of various waking intervals on sleep EEGs. For example, investigators varied the duration of prior wakefulness from 2 to 20 hours in a series of experiments by scheduling naps at 10:00, 12:00, 14:00, 16:00, 18:00, 20:00 and 04:00 [83]. In other research, experimenters "displaced" the major sleep episode in order to vary waking duration prior to sleep from 12 to 36 hours [89]. Results have shown an enhancement of both deep NREM sleep and SWA after an extension of prior wakefulness [1]. SWA increases with the number of hours of wakefulness according to a saturating exponential function, which has an estimated time constant of 18.2h in young subjects.

Does sleep homeostatic regulation of SWA change with age?

An alteration of the build-up function of the homeostatic process could explain the decrease of deep NREM sleep/SWA with increasing age. No estimates of age-dependent changes in the build-up function of homeostatic sleep pressure are currently available. Very few studies to date have assessed the effects of manipulations of homeostatic pressure in aging and those that do exist have only compared the effects of two wake intervals, not the build-up function. Two very recent animal studies showed that aged animals exhibited reduced responses following a 12-hour [281] or a 48-hour [200] sleep deprivation. In human studies, elderly individuals have been subject to an acute sleep deprivation or to a sleep fragmentation. These studies have shown that elderly adults respond to sleep deprivation with an increase in deep NREM sleep [30,31,44,69] but some have reported that the rebound tends to be less intense in elderly subjects than in young ones [33,321].

Since the largest decrease in SWA during adult life occurs between age 20 and age 50, we recently investigated the different effects of a 25-hour sleep deprivation in young (20-39y) and middle-aged subjects (40-60y) [119]. Deep NREM sleep and SWA were potentiated in both groups of subjects following sleep deprivation. However, the relative increase of deep NREM sleep and SWA was significantly less pronounced in the middle-aged than in the young (see fig. 24.1). Our results in the middle-aged population strongly suggest that the build-up function of the homeostatic sleep process might be attenuated with increasing age, as do the results of previous studies in the elderly. This would mean that the sleep (SWA) of older people is less sensitive to the accumulation of wakefulness. In the framework of the two-process model, this hypothesis predicts that the time constant of the build-up of the saturation exponential function of Process S will be lower in older subjects than in younger ones (less than 18.2 h). In other words, similar increases in the number of waking hours that precede sleep will produce smaller increases of SWA in older subjects relative to younger subjects. To test this hypothesis, we need to evaluate the effects of varying wake intervals on the sleep of young and older subjects.

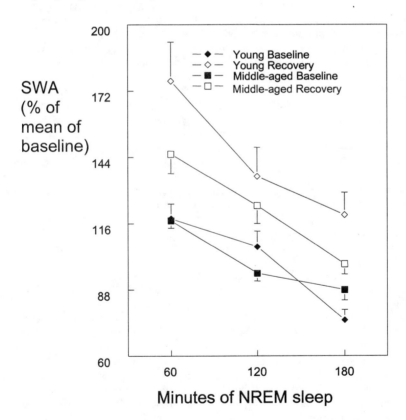

Figure 24.1. Hourly mean SWA (and sem) for the first 180 minutes of NREM sleep for young and middle-aged subjects before and after a 25-hour sleep deprivation. Values are expressed as a percentage of mean SWA during the first 180 minutes of NREM sleep for the baseline sleep episode to make young and middle-aged subjects comparable at baseline (in Gaudreau *et al.*, 2001, *Neurobiology of Aging*).

Does homeostatic regulation of vigilance change with age?

If the sleep of older subjects is shown to be less sensitive to the accumulation of wakefulness, it may or may not have negative consequences on vigilance. If the sleep of older subjects were less sensitive to the accumulation of wakefulness because of a reduced need for sleep, then we would not expect negative effects on vigilance (i.e., less sleep with increasing age would be required to maintain optimal levels of vigilance). On the other hand, if the sleep of older subjects were less sensitive to the accumulation of wakefulness but the need for sleep did not change with increasing age, then we would expect a negative impact of sleep loss on vigilance. This would mean that sleep intensity after longer durations of wakefulness would not be sufficient to maintain optimal levels of wakefulness. This last alternative is more congruent with the increase in sleepiness complaints with increasing age.

Contemporary models of alertness regulation also incorporate the two process model [4,110,153,154,215]. Subjective alertness decreases with the number of hours of wakefulness, and the electroencephalogram (EEG), recorded in a waking state, has recently been proposed as an objective measure of vigilance levels. In addition, recent data suggest that both homeostatic and circadian processes influence waking EEG [2,97]. The relative strengths of these influences vary as a function of waking EEG frequencies. Spectral power in the alpha and theta ranges is particularly sensitive to the accumulation of wakefulness [53]. It has been proposed that in young subjects the wake-dependent modifications observed in both waking and sleep EEGs may represent the same underlying homeostatic process [3]. One recent study of a young population revealed a positive correlation between the increasing rate of theta activity in waking EEG, and the increase in SWA during sleep following a 40-hour sleep deprivation [106].

The majority of the few studies that have compared the effects of an acute sleep deprivation on the vigilance of young and healthy elderly subjects tended to show similar or smaller deteriorations of vigilance in the elderly during an acute sleep deprivation [33,36,44,69,284,311]. To our knowledge, only one study reported that older subjects (40-49 years) exhibited more effects of an acute sleep deprivation on performance than did younger subjects [323]. Altogether, these results may indicate that the vigilance of healthy elderly subjects is less sensitive to the accumulation of wakefulness than is the vigilance of young ones. It is therefore possible that healthy elderly subjects would need less sleep than young subjects for their vigilance to recover from a similar duration of sleep deprivation, as a few studies that show similar or faster recovery rates of vigilance in healthy elderly subjects than in young subjects after recovery sleep following an acute sleep deprivation [36,44,69] have indicated. However, these preliminary experimental results still need further corroboration to reconcile them with the increase in sleepiness complaints with increasing age seen in epidemiologic studies.

We recently investigated the acute effect of a 25-hour sleep deprivation on the subjective alertness and waking EEG of young and middle-aged subjects. We found that middle-aged and young subjects show similar time courses of subjective alertness and spectral power in theta and alpha frequency bands during a 25-hour sleep deprivation in constant behavioural and environmental conditions [94] (See fig. 24.2 for theta power), implying a similar build-up of homeostatic pressure. Thus, the vigilance of young and middle-aged subjects is equally sensitive to the acute accumulation of wakefulness, however the middle-aged subjects had less of an increase in SWA during recovery sleep [119]. These results may signify that there is dissociation in the middle-aged population between the sensitivity of their vigilance on the one hand, and the sensitivity of their sleep to the number of hours of wakefulness on the other. This interpretation is supported by the fact that we found no correlation in the middle-aged subjects between the rate of increase of theta activity in waking EEG and the rate of increase of SWA during recovery sleep following the 25-hour sleep deprivation [94]. In this model, the vigilance system of young and middle-aged subjects would still show the same deterioration with the acute accumulation of wakefulness, yet the sleep of middle-aged subjects would already be less able to increase in intensity following enhanced time awake. This may put the middle-aged population at particular risk to the negative consequences of sleep deprivation.

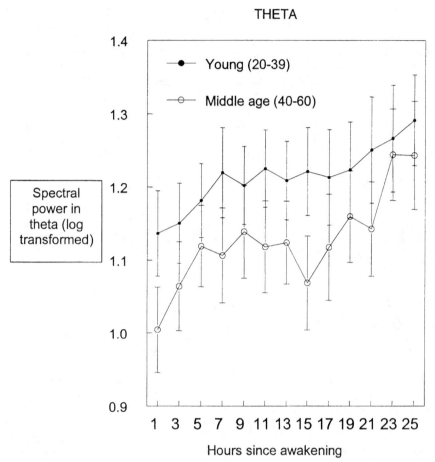

Figure 24.2. Spectral power in theta (4.00-8.00 Hz) frequency band in waking EEGs for young and middle-aged subjects. Waking EEGs with eyes open were recorded every two hours during a 25-hour sleep deprivation (Drapeau *et al.*, 2001 *Sleep*).

Age-Dependent EEG Topographical Differences

The homeostatic process typically has been modelled using the EEG from a single derivation (i.e., central derivation), which is also used for scoring sleep stages. Results from recent studies suggest regional differences in sleep [46] and in the dynamics of the homeostatic component of sleep regulation. A frontal-central dominance for SWA has been observed during baseline sleep [228,229]. Furthermore, it has been reported that the increase in SWA after a sleep deprivation is larger in the frontal than in the occipital region [54,143,276]. The increase in theta and alpha activity in the waking EEG during acute sleep deprivation is also more prominent in anterior than in posterior derivations [55]. These results led to the hypothesis that frontal regions of the brain may have a higher "recovery need" than more posterior areas do.

Very recent results suggest that there are age-dependent topographical differences in the dynamic of low frequencies of the sleep EEG [176,218]. Compared to the young, middle-aged subjects show less SWA particularly in anterior regions of the brain in the first two NREM periods of the nights [218]. There are also topographical differences in the response of SWA to the acute sleep derivation with increasing age [219]. The rebound of SWA is less pronounced in middle-aged subjects than in young subjects, particularly in the most anterior derivations of the brain. Unexpectedly, young and middle-aged subjects also show a difference in SWA rebound in the most posterior derivation of the brain. Again, there is need to evaluate the effects of various wake intervals on the sleep of young and older subjects to estimate if the parameters of the build-up of homeostatic sleep drive differ in anterior and posterior regions. Future research should also determine the functional consequences of topographical differences in SWA rebound in the older population. Executive/pre-frontal neuropsychological functions, attention, memory, and reaction time seem particularly vulnerable to the effects of sleep deprivation in young subjects [133]. Interestingly, the same cognitive functions are also very sensitive to the effects of aging. Even if sleepiness is not necessarily associated to neuropsychological performance in healthy elderly subjects [25], age-dependent sleep-wake modifications may enhance cognitive impairments observed in certain older populations [134,308] or in situations when the sleep-wake cycle is challenged. The interaction between aging, sleep loss, and cognitive functioning must become a priority area of future research.

Circadian Process: Process C

An endogenous pacemaker located in the suprachiasmatic nucleus of the hypothalamus controls the circadian process of sleep regulation. Since it is not possible in humans to measure directly the activity of the endogenous circadian pacemaker (ECP), robust circadian rhythms such as melatonin secretion or rectal temperature are used as markers of its activity. Thus, the period length, phase, and amplitude of the marker rhythm serve as estimates of the corresponding values of the ECP. In healthy young adults, the temperature minimum occurs in the early morning hours (around 6:00h) and temperature peaks in the early evening hours (around 20:00h). Young subjects typically go to bed six hours before and wake up two hours after their temperature minimum. The circadian pacemaker is responsible for the rhythmic variation of wake/sleep propensity throughout the 24-hour day. Subjects can easily initiate and maintain sleep while the biological clock promotes sleep. Subjects have difficulty falling and staying asleep when the biological clock promotes wakefulness. In a normally entrained individual, circadian sleep propensity increases in the evening (when temperature decreases) and reaches its maximum in the early morning hours, at the time of the temperature minimum. Inversely, circadian wake propensity increases during daytime (when temperature increases) and peaks at the time of the temperature maximum (which is a few hours before habitual bedtime) [76,87,90,179,343].

The effects of aging on the circadian timing system: the phase advance hypothesis

It has been suggested that age-dependent changes in the timing and the consolidation of sleep may be linked to age-dependent modifications in the phase of ECP [see reviews [230]]. Thus, the advance in the timing of elderly persons' sleep-wake cycle has been explained in terms of the ECP output signal's earlier clock time phase position. This phase advance would produce an earlier timing of the sleep episode in the evening and an earlier circadian wake signal in the morning.

An individual's tendency to be more of a morning type or an evening type can be measured with standardised questionnaires [142]. These questionnaires assess the times of day that people feel their best in addition to when they prefer to wake-up, to go to bed, to engage in intellectual and physical activity, etc. Older people more often report that they are morning types than do young people. This difference starts during the middle-years of life [66]. Research on the habitual sleep-wake patterns of young and elderly subjects using sleep diaries corroborates this age-dependent tendency toward morningness.

Many circadian markers, such as rectal temperature, do not vary solely as a function of the ECP; they are also influenced by changes in ambient temperature, activity, posture, and waking state. These influences are referred to as "masking effects" because they serve to mask the true status of the ECP. To verify the phase advance hypothesis, the evaluation of circadian phase markers independent of these masking effects is necessary. One approach minimises masking effects by studying subjects under constant behavioural and environmental conditions that are maintained for at least one circadian cycle (the so-called "constant routine protocol") [48,203,204]. Others have proposed extracting the masking influences by mathematical manipulation of the data (demasking algorithm) [63,318]. Studies that controlled for masking effects confirmed that elderly subjects show a phase advance of their temperature circadian rhythm, which is associated with earlier sleep timing, whereas young subjects do not [67,96]. On average, habitual bedtime, habitual waketime, and the minimum of the circadian temperature rhythm all occur one to two hours earlier in elderly subjects than in the young. Recently, we reported that compared to young subjects, people in their forties and fifties exhibit both earlier sleep timing and a phase advance of their circadian temperature rhythm [68]. These age differences are of the same magnitude as those reported in healthy elderly subjects [67,96].

Do we sleep at the same circadian phase as we get older?

Some authors also posit that age-dependent increases in the number and duration of awakenings during sleep are related to a phase angle disturbance. According to this hypothesis, the advance of the ECP signal would be larger than the advance of the sleep-wake cycle. As a result, the increase in the circadian wake signal would occur too early during the sleep episode, leading to more awakenings during sleep in older subjects than in younger ones [57,326]. This interpretation has led to the suggestion that changing the phase position of the circadian rhythms (through such techniques as bright light exposure) might alleviate sleep complaints among elderly subjects who have sleep difficulties (see below) [58]. Studies of the healthy elderly and middle-aged population who do not complain about their sleep do not corroborate this hypothesis. In fact, some authors have reported that young and older subjects sleep at the same circadian phase, whereas others have found that elderly subjects wake up closer to the minimum of their temperature circadian rhythm [67,68,96]. Some have interpreted the observation that elderly subjects awaken closer to the minimum of their temperature circadian rhythm as a reflection of their reduced ability to maintain sleep on the ascending limb of the temperature circadian rhythm when circadian wake propensity increases in the morning (see higher vulnerability to a phase angle misalignment below).

In conclusion, a phase advance of circadian temperature rhythm appears quite early in aging. This phase advance is associated with earlier bedtime and waketime. However, the age-dependent increase in the number and duration of awakenings during sleep does not seem to be related solely to an early phase of the circadian signal during the sleep episode. The mechanisms that underlie the advance of the biological clock have yet to be determined. Some authors have proposed that it is associated with a shorter endogenous period of the circadian pacemaker. However, recent human studies of forced desynchrony and of blind people did not detect an age-dependent modification of the endogenous period [91] for review). Others have suggested that the phase advance may be associated with the phase shifting capacity of the circadian pacemaker in response to the light dark cycle [165]. One study compared the phase shifting effects of 3 consecutive days of bright light (5 hours) exposure before (phase delay) or after (phase advance) the circadian temperature minimum in young and elderly subjects. The phase delays did not differ between the two age groups but the phase advances were attenuated in the elderly. These latest results do not explain the phase advance of the circadian pacemaker in older people [165]. A complete phase response curve and dose response curve is necessary to completely evaluate age-dependent changes in the phase-shifting capacity of the circadian pacemaker in response to the light dark cycle.

The effects of aging on the sleep-wake cycle: inability of the pacemaker to generate a robust circadian signal or problems transducing the signals into rhythms?

To date, the timing of the output signal from the circadian pacemaker has garnered a great deal of attention in the human sleep-wake regulation literature. The relation between sleep-wake regulation and the amplitude of the output from the circadian pacemaker has received much less attention. Yet, in order to function properly the circadian pacemaker should be able to generate a robust output signal. As noted previously, nocturnal sleep becomes shorter, shallower, and more fragmented with advancing age while daytime sleepiness increases. It has been suggested that this age-dependent reduction in amplitude of the sleep-wake cycle is linked to age-dependent attenuation of the output signal from the circadian pacemaker. According to this view, there would be an age-dependent attenuation in the ability of the circadian pacemaker to create the correct internal temporal *milieu* for restful sleep at night and alert wakefulness during the day.

There is evidence for a reduced amplitude of many circadian rhythms with advancing age, including temperature [75,309,326], melatonin [206,270,294,313], cortisol [74,304], and sleep spindle frequency and amplitude [324]. However, some studies strongly suggest that a reduction in the amplitude of circadian rhythms does not necessarily accompany healthy aging. For example, a study in which 48 very old people (>77 y) were compared to 23 young adults showed a similar amplitude of the temperature rhythm in the two groups [212]. Another report compared the amplitude of plasma melatonin profiles during a constant routine between 34 healthy drug free elderly individuals and 98 drug free young men. They detected no significant difference in amplitude between the two age groups [340]. It is possible that these subjects represented the hearty survivors, rather than the "normal aged" and that robust, high amplitude circadian rhythms were part of the cluster of attributes that kept these people active and vital so late into life [212].

The question of why circadian rhythm amplitude should be reduced in advancing age is unknown. The extent to which these age-dependent changes in the amplitude of the circadian oscillator output are intrinsic to the circadian pacemaker itself (or whether they might be better explained by age-dependent changes that are "upstream" or "downstream" from the circadian clock) has yet to be determined and might depend upon the aged population studied [296]. Interestingly, Monk *et al.*. have shown that advancing age, especially in men, may lead to an attenuation of circadian rhythms of subjective alertness and objective performance, even when subjects show robust temperature rhythms [209,213, 214]. These authors concluded that the circadian dysfunction in old age might involve a problem with rhythm transduction rather than rhythm generation. Accordingly, the signal generated by the circadian pacemaker would not be transduced successfully into the downstream rhythms of subjective alertness and objective performance, yet it would be sufficiently strong to generate robust temperature circadian rhythms. Clearly, future work should address age-dependent modification in the efferent pathways from the circadian pacemaker [214].

The effects of aging on the sleep-wake cycle: a higher rigidity to phase shift of the sleep-wake cycle?

There are some indications that older people might be more vulnerable than younger people to phase shifts of the sleep-wake cycle induced by jet lag and shift work [65,207,210]. These studies suggest not only that the output signal from the circadian timing system might be disturbed in advancing age, but also that disturbances might occur in the entertainment mechanisms that allow circadian rhythms to adjust and align appropriately with the external environment. One study evaluated the effects on sleep and temperature rhythms of a 6-hour (h) phase advance of the sleep-wake cycle in healthy elderly subjects. Twenty-five subjects (77-91 y.o.) lived for 15 days in a time-isolation apartment on a routine that the experimenter controlled. The experiment started with five baseline days. The wake time on the 6th night was phase advanced by 6-h, and the routine for the remaining nine days was held constant to the new phase position. Elderly subjects showed an incomplete process of phase adjustment, an important reduction in temperature rhythm amplitude, and a reduction of sleep efficiency that lasted for most of the nine-day recovery period [64, 65,210]. Interestingly, the amount of temperature amplitude reduction was correlated with the amount of sleep disruption in the first few days after the phase shift, which suggests that temperature amplitude reduction may be a predictive circadian variable of sleep disruption in elderly subjects [64]. A comparison of young and middle-aged subjects' ability to recover after a 6-hour phase advance of their sleep-wake cycle yielded similar effects [207]. Middle-aged subjects had larger increases in waking time during the sleep episode and earlier termination of sleep than did the young subjects. As is the case for young subjects [12,164,205], older people show an asymmetry in phase shifting adjustment. Elderly subjects have demonstrated more sleep disruption and greater temperature rhythm amplitude reduction after a 6-hour phase advance of the sleep-wake cycle than after a 6-hour phase delay [208,211]. The mechanisms that underlie this age-dependent decrease in speed of adaptation after a phase shift of the sleep-wake cycle are still unknown. The design of effective preventive and therapeutic strategies for older people who face challenges to their sleep-wake cycle depends greatly on our understanding of the mechanisms that underlie those age-dependent modifications. With an increasing proportion of the older population now facing night work and jet lag, these questions are of more than simply academic interest.

THE EFFECTS OF AGING ON SLEEP: CHALLENGES TO THE SLEEP-WAKE CYCLE

As noted before, one of the most important age-dependent changes in sleep architecture is the decrease in deep NREM sleep and SWA. Importantly, aging is also associated with a reduction in spindle frequency and amplitudes and with concomitant reduction in spectral power in sigma frequencies during NREM sleep. The hyperpolarisation of thalamocortical and cortical neurons is a critical factor for the generation of both delta and sleep spindle oscillations, which is associated with significant changes in neurons' responsiveness to environmental stimuli [99]. Although the functional role of 12-14 Hz sleep spindles and delta waves is still obscure, some authors have suggested that they maintain NREM sleep and protect against disturbing stimuli [288]. If one accepts this assumption, it is possible that with advancing age, NREM sleep of older subjects will become more vulnerable to external and internal disturbing stimuli [91]. Interestingly, recent results have demonstrated that spontaneous awakenings in older subjects are mainly related to a reduction in the consolidation of NREM sleep [85, 272]. The results of new studies strongly suggest that the sleep-wake cycle of healthy older subjects may be particularly vulnerable to situations that involve challenges to their sleep-wake system such as a circadian phase misalignment, stress, caffeine consumption or numerous medical diseases and specific sleep disorders. However, the strength of the association between this augmented vulnerability and an inability to consolidate NREM sleep requires further elucidation.

Lower Tolerance to a Phase Angle Misalignment between Sleep and the Circadian Timing System

Some authors (e.g., [57]) have suggested that the sleep of older subjects might be particularly vulnerable to circadian phases of high wake propensity, which means that it would be more difficult for older people to sleep at the "wrong" circadian phase. This hypothesis might explain in part why subjective sleep problems related to jet lag and shift work increase with age. Forced desynchrony studies (in which sleep episodes were initiated at all circadian phases while homeostatic drive remained constant) have recently corroborated this hypothesis [86]. Compared to baseline sleep, both younger and elderly subjects awoke more often during their sleep episode when they were required to sleep at a circadian phase of high wake propensity. In addition, elderly subjects woke up more often during their sleep than did young subjects at all circadian phases. However, the difference between elderly and young subjects was more prominent when sleep occurred at a time of higher circadian wake propensity, such as during the day [86]. These data support the hypothesis of lower tolerance to a phase angle misalignment in elderly subjects.

In our study of the effects of a 25-hour sleep deprivation in young and middle-aged subjects, recovery sleep was initiated one hour after habitual waketime, which is a time of increasing circadian wake propensity [119]. This experimental situation was similar to what night workers experience when they sleep during the day following their first night shift. In our study, both groups of subjects had more awakenings during their daytime recovery sleep than at their normal sleep times despite the fact that they experienced one night of complete sleep deprivation. Importantly however, middle-aged subjects had even more problems sleeping during the day than the young subjects did. They showed a more substantial increase in awakenings during daytime sleep (see fig. 24.3). Two important new conclusions can be drawn from these results. First, middle-aged subjects show a lower tolerance to an abnormal phase angle, even if homeostatic sleep pressure is increased at sleep onset (sleep deprivation). Second, a lower tolerance to an abnormal phase angle occurs as early as the middle years of life.

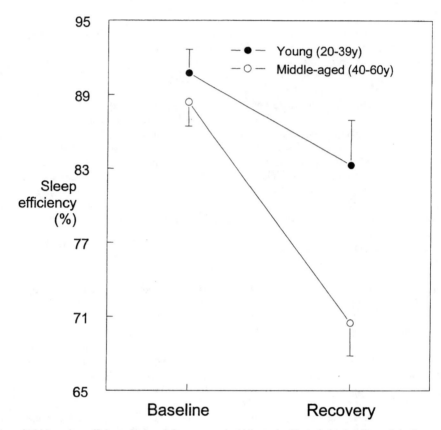

Figure 24.3 Mean sleep efficiency (and sem) for young and middle-aged subjects during baseline and daytime recovery sleep following a 25-hour sleep deprivation (in Gaudreau *et al.*, 2001, *Neurobiology of Aging*).

Age-dependent reduced tolerance to a phase angle misalignment cannot be explained solely by a circadian timing system malfunction. If that were the case, the fact that the sleep of middle-aged subjects is more sensitive to "unfavourable" circadian phases would suggest that the ECP sends a stronger signal as we get older. As noted previously, results from human and animal studies do not support this idea. Studies have shown no change or even reductions in the circadian modulation of many circadian markers with increasing age [91,212,213,340]. In our studies with the middle-aged population, we found no modification in the amplitude of the circadian temperature rhythm [68].

The combined action of the homeostatic and circadian processes is thought to be necessary to maintain consolidated episodes of about 8 hours of sleep during the night and about 16 hours of wakefulness during the day [90]. Throughout the night, increasing circadian sleep propensity counterbalances decreasing homeostatic sleep pressure. Conversely, throughout the day, increasing circadian wake propensity counterbalances increasing homeostatic sleep pressure. A nonlinear interaction of the circadian and the sleep-dependent components of sleep propensity has also been reported in forced desynchrony studies [90]. These studies have shown that the last portion of a sleep episode (when homeostatic sleep pressure is low) is more vulnerable to a circadian phase of high wake propensity than is the beginning of a sleep episode (when homeostatic sleep pressure is high). In our previous study, the increase in awakenings during daytime recovery sleep was more prominent for both groups at the end of the sleep episode, when homeostatic sleep pressure decreased (see fig. 24.4). As the recovery sleep episode progressed, homeostatic sleep propensity decreased and circadian wake propensity increased, leading to more awakenings. Importantly, as also shown in fig. 24.4, the difference between young and middle-aged subjects during daytime recovery sleep was also more prominent at the end of sleep. These results are consistent with the notion of lower sleep homeostatic pressure in middle-aged subjects than in the young. We propose that less sensitivity of sleep to the accumulation of wakefulness (reduction in the strength of homeostatic response of sleep) may explain age-dependent lower tolerance to a phase angle misalignment. As their daytime sleep episode progresses, the smaller homeostatic response of middle-aged subjects' sleep would not be able to "override" as efficiently the increasing circadian wake propensity. This interpretation is supported by data from the forced desynchrony protocol that indicates a strong relation in older people between greater vulnerability to a phase angle misalignment and a reduction of NREM sleep consolidation [85].

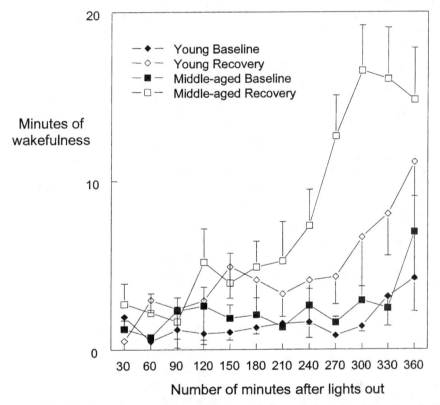

Figure 24.4 Mean number of minutes of wakefulness per 30 minute of the sleep episode (and sem) for young and middle-aged subjects during baseline and daytime recovery sleep following a 25-hour sleep deprivation (in Gaudreau *et al.*, 2001, *Neurobiology of Aging*).

Stress, Cortisol and Age-Dependent Modification of Sleep

The hypothalamic-pituitary-adrenal (HPA) axis controls the secretion of stress hormones. Under normal conditions, the 24-h cortisol profile exhibits a morning maximum, declining levels during the late morning and afternoon, a nocturnal period of low concentration, and an abrupt elevation after the first few hours of sleep. The 24-hour profile of plasma cortisol reflects primarily circadian rhythmicity but it is also closely associated to sleep processes (see [303,305] for reviews). Nevertheless, the precise relationship between the HPA axis and sleep is not yet fully understood.

The initiation of sleep is coupled with a short term inhibition of cortisol secretion, at least when sleep is initiated at habitual bedtime [325]. Periods of wakefulness that interrupt nocturnal sleep are associated with pulses of cortisol secretion [41,103,286]. Decreasing cortisol secretion during nocturnal sleep is associated with deeper sleep, whereas increasing cortisol secretion is associated with lighter sleep [111,125,126]. During late sleep, HPA secretory activity increases to reach a diurnal peak shortly after the final sleep awakening. The implication of the daily fluctuations of ACTH in sleep regulation has been recently outlined in a study showing that a morning rise in ACTH seems to play a critical role in timing the end of nocturnal sleep [40]. Many studies have also demonstrated that exogenous administration of the major mediators of the HPA axis (i.e. CRH, ACTH and cortisol) modifies subsequent sleep architecture [42,141], although the precise effects vary with dose, timing, and route of administration.

The pattern of cortisol secretion changes with age [163,300,301]. In a thorough study of 177 temporal profiles of plasma cortisol in humans aged 18 to 83 years, Van Cauter *et al.* [300] reported significant age-dependent changes in cortisol secretion. Mean cortisol levels and the level of nocturnal nadir increase between the ages of 20 to 80 years. The amplitude of the cortisol circadian rhythm is reduced with increasing age and the timing of the circadian rise of cortisol advances. Importantly, aging is associated with a delay in the onset of the quiescent period. This is thought to reflect a progressive deficit in the negative-feedback control of the HPA axis. Interestingly, similar alterations of the mechanisms of HPA recovery from stimulation are found after sleep loss in normal young subjects. A study performed by Leproult and collaborators showed that a single partial or total sleep deprivation delays the onset of the quiescent period of cortisol secretion the following evening [182]. Total sleep loss was also associated with a 2.4-fold increase in cortisol levels in the evening following the sleep deprivation. Very recent data have corroborated these results using a more chronic sleep restriction (4 hours per night for 6 days) [287]. Sleep debt produced alterations in the 24-h profile of plasma cortisol, including a shorter and delayed quiescent period and higher concentrations of cortisol in the afternoon and early evening. These findings suggest that the mechanism of HPA recovery from stimulation may be affected by sleep.

The fragmentation of sleep that typically occurs in aging may play a role in the elevation of evening cortisol secretion [303]. Alternatively, nocturnal exposure to increased HPA activity may produce sleep fragmentation and impair sleep quality in older people [250]. To date, very few studies have tested these hypotheses. One study demonstrated that poor and good older sleepers reported similar amounts of life stress [115]. However, there was a relationship between life stress and sleep only in the poor sleepers group. Poor sleepers with higher life stress had a longer sleep latency but less early morning awakenings. One recent study has shown that older women and men with higher levels of free cortisol (24-hour urine level) under a mild stress condition (24 hours of indwelling IV catheter) had more impaired sleep, as indicated by lower sleep efficiency, fewer minutes of stages 2,3, and 4 sleep, and more EEG spectral power in the beta frequency range [250]. These last results suggest that the cortisol stress response could contribute to sleep impairment in the elderly. Some authors have proposed that the increased prevalence of insomnia in the older population may be caused by a lower tolerance of sleep to the arousal-producing effects of stress [306]. To our knowledge only one study has tested this hypothesis pharmacologically by comparing the effects of an administration of ovine CRH on the sleep of young and middle-aged subjects [306]. Middle-aged men showed a greater increase in wakefulness and more suppression of deep NREM sleep than did the young subjects, despite similar elevations of ACTH and cortisol. The authors concluded that older subjects might be at higher risk of developing insomnia when faced with equivalent stressors.

The Effects of Caffeine

Caffeine (3,5,7-trimethylxanthin) is certainly the most widely used central nervous system stimulant in North America. It has a half-life of about 3-4 hours and its effects on the sleep-wake cycle mimic some of the effects of a decrease in homeostatic sleep pressure. Caffeine probably affects the sleep-wake cycle via its antagonist action on A1 and A2 adenosine receptors [78]. In the central nervous system, adenosine synthesis is closely associated with neural activity. Interestingly, it has been proposed recently that adenosine is a modulator of the sleepiness associated with the homeostatic sleep pressure induced by prolonged wakefulness [291]. Different lines of evidence support this assertion. Animal studies have shown that cerebral concentrations of adenosine are higher during waking than during sleep and that basal forebrain concentrations of adenosine increase with the amount of wakefulness and decrease with the number of hours of sleep [247]. There is consistent evidence that the systematic application of adenosine agonists over a wide range of doses prolongs deep NREM sleep and enhances SWA [246,248].

Studies have also shown that in young subjects the ingestion of 150 to 400 mg of caffeine in the evening increases sleep latency, decreases sleep consolidation, increases sleep motility, and reduces the amount of deep NREM sleep [34,45,157,161]. Many of these effects are reportedly dose dependent [157]. The administration of 100 mg of caffeine (equivalent to one cup of coffee) just prior to bedtime is sufficient to increase sleep latency and to decrease SWA in young subjects [177].

The majority of information about caffeine's effects on the sleep-wake cycle comes from studies conducted with the young. Caffeine seems to mimic not only some of the effects of a decrease in homeostatic pressure but also some of the effects of age on the sleep-wake cycle. The possible decrease in homeostatic sleep recuperative drive of older individuals may make them less capable of adapting to challenges to the sleep-wake cycle, especially those that affect deep NREM sleep and SWA. To our knowledge, only one study has evaluated the effects of caffeine on sleep in a somewhat older population. Compared to previous reports on the young population, this investigation pointed toward more detrimental effects of caffeine (300mg) on sleep in subjects aged between 50 and 63 years. However, there was no control group in this study [45]. Very recently, preliminary results in rats have also shown stronger effects of caffeine on sleep duration in older than in younger rats [271].

Although the effects of caffeine on habitual sleep in young subjects are well described, we know little about the interaction between sleep deprivation, circadian timing, and caffeine administration. In one animal study, caffeine reduced the elevation of SWA during recovery following sleep deprivation [277]. In another animal study, caffeine slowed the rate of compensatory sleep after sleep deprivation, as measured with the duration of sleep states and sleep continuity [336]. To our knowledge, only one human study has evaluated the effects of caffeine on sleep when bedtime was initiated at an abnormal phase relationship with the circadian signal (i.e. during daytime) and following an acute sleep deprivation [228]. There was no effect of caffeine administration on daytime sleep in this study. However, caffeine was admitted more than eight hours before daytime recovery sleep, which may explain the lack of an effect. More studies are certainly needed to understand how homeostatic and circadian manipulations influence the effects of caffeine on the sleep-wake cycle. No study to date has compared the effects of caffeine or any other stimulant-type medication in young and older subjects under sleep deprivation or when trying to sleep at a time that the biological clock is sending a strong waking signal. A better understanding of how these processes interact may be relevant to older individuals using caffeine as a countermeasure to alleviate sleepiness.

Challenges to Sleep Integrity in Old Age: The Role of Sleep Pathology and Comorbid Diseases

The aforementioned age-dependent alterations in sleep processes occur in the context of diverse and widespread challenges to the integrity of sleep in the elderly population. Some of these arise from the state of sleep itself whereas others represent the influences of co-morbid conditions. Specific sleep disorders such as sleep disordered breathing (SDB) and periodic leg movements/restless legs syndrome (PLMS/RLS) are prevalent in the aged and may be related to age-dependent changes in Process S and Process C. The customary interpretation assumes that the presence of such sleep pathology leads to diminished deep NREM sleep and a dampening of the "amplitude" of sleep/wake, as evidence by increased daytime fatigue and napping. Thus, for example, successful treatment of SDB may result in increased deep NREM sleep and decreased daytime sleepiness. However, causality may not be unidirectional. Both SDB and PLMS are reduced in deep NREM sleep [170,188,317,319]. Particularly for SDB, there is also some suggestion that ventilatory instability is enhanced by rapid cycling of sleep wake state (i.e., sleep fragmentation) [239,240]. Thus, the higher likelihood or vulnerability to such sleep pathology in old age could be interpreted equally as well as at least partially due to diminution of homeostatic drive and weakening of Process S. The scattered reports suggesting that hypnotic medication may improve central sleep apnea, presumably by reducing sleep fragmentation and increasing homeostatic drive [35] are also consistent with this. Curiously, although it may be possible to link sleep pathology to the two-process model of sleep regulation, the role of these specific sleep disorders in insomnia complaints in old age remains uncertain and subject to much debate [7,27,82,156,180,199,226,266,267,338].

In elderly persons, the high prevalence of age-dependent, often overlapping, diseases represents a major challenge to the integrity of sleep. Cross-sectional epidemiologic surveys employing univariate analyses in elderly populations consistently have suggested that such conditions disrupt sleep in the aged. These include: *chronic pain conditions* such as arthritis [17,192,194,232], hip fracture [109,159], fibromyalgia [339], headache [72] and back pain [122,152,159]; *cardiovascular diseases* as manifested by hypertension [122,192,232], myocardial infarct [109,148,192,232], stroke [109,232]; congestive heart failure [159,232], angina [13,194,232]; *respiratory conditions*, such as asthma and bronchitis [17,107,122,166,192,195,232]; and *other systemic diseases* such as diabetes [109,122,194], gastroesophageal reflux [195,254], and duodenal ulcer [159,183,192]. Additionally, *limited mobility* [109,130,192,232,264], *visual impairment* [14]; *menopause* [19]; *lack of regular exercise* [159,232]; *nocturia* [17,159,192,195,202], *alcohol use* [107] and *smoking* [107,232] have all been associated with poorer sleep specifically in the aged population. *Limited social supports* [130], such as widowhood [109], lower socioeconomic status [17,192,232] and *psychiatric conditions* such as anxiety [123,202] and depression [17,71,107,108,109,152,159,192,194,264] often appear to be associated with the greatest risk for insomnia complaints in the elderly [237]. The poor quality of nocturnal sleep associated with the medical conditions noted in these surveys across diverse populations could be construed as a weakened homeostatic drive for sleep (Process S), perhaps facilitated by such diseases.

Unique relationships between Process C and various medical co-morbidities are difficult to appreciate. One might infer effects on the circadian system if sleep latency was uniquely affected by a given medical disease, and indeed a few surveys have reported sufficient data to examine effects of sleep initiation versus sleep maintenance insomnia in old age in relation to selected conditions. Unique associations with sleep latency relative to sleep maintenance insomnia might suggest that a given condition might be associated with relative phase delay in the timing of sleep. Restless legs symptoms might be one such condition [27]. However, most studies report that cardiovascular conditions are equally likely to be associated with both sleep latency and sleep maintenance problems [13,122,232]. Similarly pain [122,123,232], diabetes [122], and depression [107,232] are related to both sets of complaints. Ohayon [237] also reported that problems with initiation and maintenance were both linked to poor physical health. There is some suggestion that particular complaints such as nocturia [192,202] and respiratory symptoms [122,123,232] may be more closely linked to sleep maintenance insomnia, though Middlekoop *et al.* [202] reported anxiety was more closely linked to sleep initiation problems than sleep maintenance in the aged. Even less is known regarding actual phase position of bedtime relative to such conditions. Gislasson *et al.* [123] noted that pain conditions were unrelated to bedtime or arise time and Gale *et al.* [116] note that morningness/eveningness were unrelated to disease. Habte-Gabr *et al.* [130] found that, in elderly men, social activity was associated with later bedtimes, whereas women with poorer overall health, more physical limitations, and more recent hospitalisations appeared phase advanced in their habitual bedtimes. In contrast, Ohayon *et al.* [237] noted that individuals dissatisfied with their sleep generally elected later bedtimes than subjects who were satisfied with their sleep. In summary, despite a large amount of amassed epidemiologic data on sleep in old age, there is little basis to suggest that aged individuals with any type of somatic disease show consistent phase shifts suggestive of alterations in Process C.

As we have shown elsewhere in this chapter, increased sleepiness and napping during the daytime hours in the aged can be viewed as an effect reflecting interactions between homeostatic and circadian processes. Many of the aforementioned studies of sleep and morbidities have shown associations between such conditions and diseases and morning fatigue and/or overt daytime sleepiness [13,14,17,107,122,130,135,192,195,202,232,235,264].

DOES THE SLEEP OF MEN AND WOMEN AGE AT A DIFFERENT RATE?

It is well recognised that there are gender differences in sleep quality in the older population. Compared to older men, older women seek more help for their sleep problems and they report poorer sleep quality, longer sleep latencies, more nighttime awakenings, less frequent napping, and more frequent use of sedative-hypnotic drugs [112,123,158,173,187,198,202,242,327]. The reasons for the preponderance of sleep complaints in elderly women (compared to elderly men) remain poorly understood. In fact, when studied in the laboratory, healthy older women with no sleep complaints usually sleep more deeply (more deep NREM sleep) and wake up less often during the night than do older men [263,320,332], which would lead one to predict fewer sleep difficulties among elderly women on a population-wide basis.

It is not clear when gender differences on polysomnographic sleep emerge over the aging process in healthy subjects. Gender differences in quantitative sleep EEG measures have been reported among subjects as young as 20-29 years of age [84] with young women showing higher slow-wave activity than do young men. In subjects between the ages of 20 and 60 years, we found differences between men and women for a few parameters [66]. Women spent more time in bed than did men, as their sleep diaries indicated. In the laboratory, women showed more deep NREM sleep, fewer awakenings, and shorter REM sleep latency. Importantly, there was no significant interaction between age and gender on any of the sleep parameters, which suggests that the process of aging did not have a different influence on men and women (see fig. 24.5).

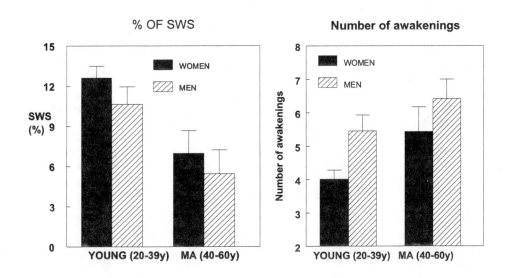

Figure 24.5 Percentage of slow-wave sleep (SWS = deep NREM sleep) and number of awakenings (mean and sem) in young and middle-aged men and women. (data from Carrier *et al.*, 1997, *Journal of Sleep Research*).

We assessed the effects of age and gender on sleep EEG power spectral density in a group of 100 men and women aged 20y to 60y [62].

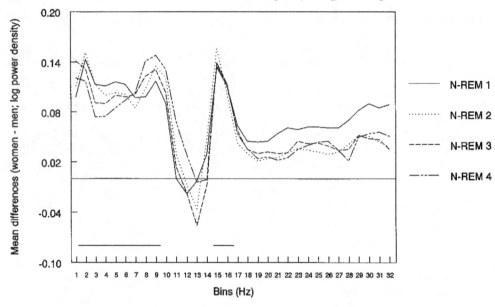

Figure 24.6 Mean power spectral densities of the women relative to the men (men=0) for each frequency bin. The line represents areas where the effect of gender was significant with the regression analyses. (in Carrier *et al.*, 2001, *Psychophysiology*).

Fig. 24.6 illustrates mean power spectral densities of the women relative to the men (men=0) for each frequency bin for the first four NREM periods. There was more elevated activity among women than among men not only in delta frequencies but also in theta, low alpha, and high sigma frequency bins (0.25-9.00 Hz, 14.25-16.00 Hz). These differences were constant across the night. No significant interactions emerged between age and gender, suggesting that the aging process does not differentially influence men and women between 20 and 60 years. We know that sleep and quantitative sleep EEG vary with the level of reproductive hormones across the menstrual cycle [95], during pregnancy [49], and following hormonal replacement therapy in menopausal women [9]. Almost no study to date has carefully controlled for these influences when evaluating whether the sleep-wake cycles of men and women age at a different rate. In our study, although all the pre-menopausal women were studied during the follicular phase of their menstrual cycle, a proportion of them were taking contraceptive pills. In addition, the hormonal status of the peri- and post-menopausal women was unknown. Further research is necessary to understand the influence of hormonal status in gender difference as well as its interaction with the process of aging.

Experimental challenges to the sleep-wake cycle help clarify the mechanisms that underlie gender and age differences. It has been proposed that SWA differences between men and women are linked to gender differences in homeostatic sleep regulation. According to this hypothesis, women would be more sensitive to the accumulation of wakefulness than men. To date, there have been no cross-gender comparisons of estimates of the build-up and dissipation of homeostatic sleep pressure. One recent report showed that young women have a more dramatic increase in SWA following an acute sleep deprivation than do young men [11]. We evaluated gender differences on the effects of a 25-hour sleep deprivation on daytime recovery sleep in middle-aged subjects [61]. Preliminary results suggested that middle-aged men and women do not differ in their response to sleep deprivation during daytime recuperative sleep (see fig. 24.7).

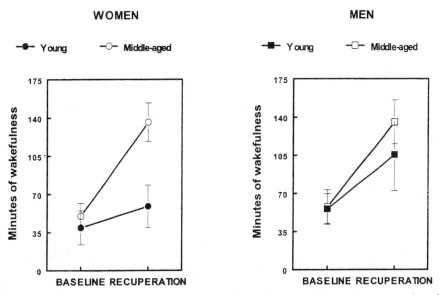

Figure 24.7 Number of minutes of wakefulness (mean and sem) in baseline sleep and in a daytime recovery sleep after a 25-hour sleep deprivation in young (20-39 years) and middle-aged (40-60 years) men and women (unpublished data).

Compared to the young, both middle-aged men and women clearly exhibit greater vulnerability to an abnormal phase angle between sleep and the circadian signal, as measured with a steeper decrease of sleep efficiency during daytime sleep. The observed reduction of SWA following sleep deprivation in the middle-aged subjects suggests that the homeostatic recuperative drive may be attenuated in the middle years of life and that this attenuation is similar in middle-aged men and women (see fig. 24.8).

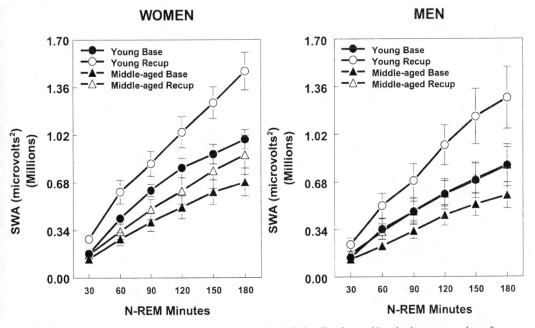

Figure 24.8 Accumulation of SWA (mean power spectral and sem) in baseline sleep and in a daytime recovery sleep after a 25-hour sleep deprivation in young (20-39 years) and middle-aged (40-60 years) men and women (unpublished data).

To evaluate possible gender differences in homeostatic regulation of sleep, a dose response study of wake duration and subsequent recovery sleep is needed in both genders and in different age groups.

Menopause

When we consider sleep quality in menopausal women, we have to take into account factors specific to aging (e.g., sleep vulnerability, increase in sleep disorders, medical conditions, secondary effects of medications), as well as changes associated with hormonal status (night sweats, hot flashes, etc.). Menopause is an important life event for middle-aged women. In the Western world, menopause occurs around age 51 and is characterised by the permanent cessation of menses [197]. Menopause is associated with an increase in gonadotropins (FSH and LH) and a marked diminution in ovarian hormones (progesterone, estrogens). Peri-menopause includes the period prior to menopause when the endocrinological, biologic, and clinical features commence. The phase that marks the transition from the reproductive phase to the non-reproductive state is also known as the climacteric phase [299]. For most women, peri-menopause will last approximately four years [197].

Sleep problems and fatigue are important complaints in menopausal women [15,19,146,147,181,279,289]. Between 47% and 64% of menopausal women complain about their sleep and half of them will suffer from fatigue during daytime [181]. A recent survey by the National Sleep Foundation in the United States revealed that peri- and post-menopausal women are twice as inclined as pre-menopausal women to use sleeping pills, they sleep less than pre-menopausal women, and they report more symptoms of insomnia [231]. Sleep complaints constitute an important cause for medical consultations in menopausal women. In addition, sleep deprivation and fatigue may exacerbate the mood fluctuations that many menopausal women report [18]. The International Classification of Sleep Disorders [149] includes a provisional section about sleep disorders whose classification requires further research, and includes menopausal insomnia.

Despite the lack of doubt that the transitional years throughout menopause constitute a critical period of sleep and vigilance complaints for many middle-aged women, not all women will suffer from disturbed sleep during the menopause transition. In fact, some studies have indicated no effect of menopausal status on sleep. For example, a longitudinal study on subjective sleep quality did not show significant differences between pre and post-menopausal women [196]. Few studies have evaluated the effects of menopausal status on objective sleep parameters. One study reported no significant difference on polysomnographic parameters between pre-, peri-, and post-menopausal women [278]. Another study evaluated objective sleep quality with actigraphy and found more awakenings in peri-menopausal women than in premenopausal women [18]. Obviously, more studies are urgently needed to understand the factors associated with objective sleep quality in menopausal women. Importantly, these studies should measure and control for the severity of vasomotor symptoms such as night sweats and hot flashes since they seem to play a significant role in the deterioration of sleep quality in peri- and menopausal women [278].

75% of American and North European women experience hot flashes during perimenopause. A large Australian study has shown that 13% of pre-menopausal women, 37% of peri-menopausal women, and 62% of post-menopausal women experienced at least one hot flash in the previous two weeks [129]. The highest frequency of hot flashes was reported 3 months or more after the cessation of menses. Women who have hot flashes usually continue to have them for 3 to 5 years. However, for some women, hot flashes continue for a longer interval. Hot flashes show a circadian modulation; they are more prominent at the end of the afternoon/early evening [114,334]. Nevertheless, 44% of peri-menopausal women and 28% of menopausal women report hot flashes during their sleep [231]. In this survey, women experienced hot flashes during sleep three days per week on average [231]. Women who experience hot flashes also report more sleep complaints and show a lower polysomnographic sleep efficiency in the laboratory than do women who do not experience them [146,278,334]. The discomfort of hot flashes does not seem to cause awakenings. In fact, one study has shown that women awaken before the hot flash and not after it [102].

Most studies have shown that both estrogen replacement therapy alone and estrogen/progestogen replacement therapy improve subjective sleep quality and vigilance during the day [101,113,217,241,252,330]. Enhancement of subjective sleep quality is associated with a reduction in vasomotor symptoms and with an improvement in mood [331]. In a longitudinal study, the transition from pre- to post-menopausal status was associated with a significant increase in sleep complaints only in women who chose not to engage in hormone replacement therapy [238]. There are still very few studies on the effects of hormone replacement therapy on polysomnographic sleep. Most studies have indicated only small improvements in polysomnographic sleep on a number of different parameters. Both estrogen alone and estrogen/progestogen replacement therapy may reduce sleep latency [259,275], enhance sleep consolidation [9,217,274,292], and increase REM sleep [9,259,275,292]. A few studies have shown no effect of hormonal replacement therapy on polysomnographic sleep parameters [252]. Differences in selection criteria (sleep complaints, severity of vasomotor symptoms, associated sleep disorders, etc.) and methodology (duration of treatment, evaluation of sleep during estrogen phase treatment or estrogen/progesterone phase treatment) may explain this divergence in results between studies. These parameters need to be controlled in future studies to better identify factors related to the efficacy of hormone replacement therapy to alleviate sleep difficulties in middle-aged women. Long term effects (more than 6 months) of hormonal replacement on sleep should also be addressed. Furthermore, very few studies have compared the effects of different regimens of hormone replacement therapy on sleep. One report suggested that micronised progesterone may work better than medroxyprogesterone to improve the quality of sleep in postmenopausal women who take estrogen [217].

Most studies have shown that the prevalence and severity of SDB increase significantly during and after the menopausal transition [21,28,79,316], yet the possible cause is still a source of debate. While some studies have shown that BMI and neck circumference were positively associated with SDB prevalence and severity [127], recent reports using large number of subjects suggest that the difference between pre- and post-menopausal women is still present after controlling for these two factors [79,316]. However, it remains unclear whether this higher risk is related to age differences or to hormonal changes associated with menopause [337]. Upper airway factors may also be involved. Interestingly, some recent studies have demonstrated that women who take hormonal replacements are at lower risk for developing SDB than are menopausal women who do not [21,160]. Longitudinal studies on sleep apnea during menopausal transition are needed to address the specific role of menopause in the development of SDB and the potential therapeutic effect of hormonal replacement therapy for this disorder [237].

APPROACHES TO ALLEVIATE THE EFFECTS OF AGE ON THE SLEEP-WAKE CYCLE

Traditional Psychopharmacologic Agents in the Treatment of Insomnia in the Elderly

Although the development of agents that may specifically focus on manipulation of Process S, such as Growth Hormone secretagogs, [128,244] or gamma-hydroxybutyrate [302], may hold promise as potential interventions for improving the sleep in aged individuals with insomnia, there can be little doubt that the most frequently used and most adequately studied approach to insomnia in old age remains pharmacotherapy with benzodiazepines or benzodiazepine site-specific agonists. It should be emphasised, however, that recent clinical trials have compared the short to mid-term efficacy of cognitive behavioral therapy for insomnia with traditional benzodiazepines and have found similar efficacy on a number of sleep parameters [224]. Additionally, a recent meta-analysis of studies in each domain yielded similar conclusions [283]. The reader is directed elsewhere for a more thorough discussion of the value of and approach to behavioral interventions for insomnia in old age [98,184].

Epidemiologic Evidence Related to Pharmacologic Treatment for Insomnia

Sleep complaints remain relatively consistent within elderly individuals with as many as 50 to 70 % of elderly individuals showing continued complaints in 2 to 3 year [108,109,117,140,150] or even 10 [167] to 12-year [193] follow-up. Morgan *et al.* [223] and Katz *et al.* [159] even reported figures as high as 83%-84% in a wider age range over 2 years. To some extent, however, estimates of the persistence of the problem may depend upon the insomnia definition employed. For example,

Ford and Kamerow [112] reported persistent insomnia in only about 5% of their over-65 year old population, but their insomnia prevalence in this age group was also somewhat lower than in most other studies of the aged (12.0%). Roberts *et al.* [264] also showed a persistence of insomnia in the elderly in the range of 13%.

Thus, perhaps the initial question in determining the relative importance of treating insomnia in old age is knowing what happens when it is left untreated. In the absence of long-term randomised clinical trials, population based cohort studies following individuals for persistent and incident insomnia over time allows for some appreciation of the broad range of outcomes potentially associated with poor sleep in the aged. Numerous longitudinal studies show a risk of incident (new onset) depression in cases of geriatric insomnia who remain untreated over long periods of time, ranging from one to 12 years [109,112,150,159,186,193,265], although in an African-American population in the United States such relationships could not be detected [108]. Persistent insomnia over time appears to be related to greater use of general medical services as well as psychiatric services [112]. Klink *et al.* [167] reported that persistent insomnia over 10 years was predicted by obstructive respiratory conditions and heart disease. Incident insomnia in the elderly has been associated with poorer overall health, disability, heart disease, stroke, diabetes, ulcer, back pain, hip fracture, and respiratory symptoms [109,150].

Controversy surrounding the possible association between insomnia and mortality has existed for many years. Population-based studies published several decades ago demonstrated that short sleep durations of less than 7 [173] or 6 [333] were associated with higher all-cause mortality rates at 6 or 9 year follow-up, respectively. These findings were not mitigated by age. In another elderly population, Pollak *et al.* [245] found that sleep durations did not predict mortality at 3.5 years after other variables had been controlled, however insomnia (defined as the algebraic product of number of sleep complaints and number of nights with that complaint over a 2-week period) was predictive of mortality. Morgan and Clarke [221] also showed that incident insomnia was associated with mortality in an 8-year follow-up study of a 65 and over population and Kojima *et al.* [169] noted an association between 12-year mortality and short sleep durations. However, most other studies examining either sleep durations or sleep complaints were not able to replicate these findings [5,107,117,152,193,253,269]. Even in the early studies, however, a self-defined insomnia condition was less predictive of mortality than was sleep duration [173]. Recently Kripke *et al.* [171] have replicated these data by continuing to show that self-defined insomnia was not associated with higher rates of mortality, and, in fact, was associated with lower rates of mortality. Reported sleep durations of 6 hours or less continued to demonstrate a small, but nonetheless significant, mortality risk. These findings appeared across age groups [171].

Epidemiologic studies have shed light on a number of interesting aspects of hypnotic medication usage in old age. Ohayon [236] has shown that a wide array of medications see de facto use as sedative/hypnotics, including anxiolytics, anti-depressants, and even neuroleptics., Risk factors for drug use in old age are similar to those risk factors contributing to poor sleep per se, including medical and psychiatric illnesses [81,236,255], visual impairment [14], and possibly urban locale [195]. Elderly women are proportionally higher users than elderly men [26,81,100,123,130,194, 202,220], though, within elderly populations, chronologic age per se may not be [81,100,202]. There may be some suggestion that, in the elderly, use of benzodiazepines specifically are more strongly related to complaints of falling asleep relative to staying asleep [232]. A similar relationship between sleep latency problems and psychotropic use was reported by Habte-Gabr *et al.* [130] in elderly women and by Gislasson *et al.* [123] in elderly men and women.

In those studies that have carefully examined sleeping pill use in relation to medical vs psychiatric comorbidity, longitudinal data suggest that although prevalent (current) use is predicted by both physical and psychiatric diseases, it is the latter that best predicts continued use over periods of 6 years [81]. As many as 5% of the elderly population may fall into this category of continued use [81,236], though some other data suggest that this figure may be as high as 16% over 4.5 years [155], 29% over 8 years [221], 34% over 1 year [222] or 62% over 3 years [109]. Some surveys have noted that, based on specific inquiry, as many as 75% of the elderly report beneficial effects on sleep when taking sedative/hypnotic medications [236], an effect also seen in laboratory-based studies [308]. However many other studies have shown that elderly sedative hypnotic users sleep more poorly than non-users [17,71,100,107,123,130,152,192,195,255]. In fact, Babar *et al.*

[17] reported in a multivariate logistic model that benzodiazepine use was the most powerful predictor of poor sleep when compared to measures of overall physical health, depression and respiratory symptoms. These results might suggest either than the medications lose their efficacy or that the severity of the insomnia in the presence of hypnotics is particularly pernicious and less amenable to modification.

Specific Pharmacologic Treatments

Although examination of selected databases suggests that anti-depressants used in low dosages probably see much use as hypnotics [314], this being particularly true over the decade of the 1990's, there are no randomised clinical trials using low dose anti-depressant therapy to treat primary insomnia in old age. In geriatric major depression, polysomnographically derived data have suggested utility of maintenance doses of a modestly sedating anti-depressant such as nortriptyline may help maintain sleep depth, as indicated by automated counts of delta activity, particularly in the first NREM cycle [261], implying an intensification of Process S. REM sleep measures were also enhanced in this study as well. Poor sleep associated with bereavement-related depression showed a similar pattern of responsiveness to nortriptyline [243]. In these studies, dose was titrated to plasma levels with mean dosages in the 50-100 mg range [52]. In one of the few studies to test anti-depressants in the primary insomnia of old age, Nowell *et al.* [233], in an open label trial, reported that paroxetine at dosages in the 10 to 30 mg range had no effect on sleep stages, though a decrease in alpha band power was correlated with improvement in mood.

Benzodiazepines continue to see much use as hypnotics in the elderly, though reported clinical trials with such medications typically have durations far shorter than their actual use, as indicated in the aforementioned epidemiologic studies. Previous short-term studies of triazolam (0.125 mg for 2 nights) [268], temazepam (7.5 mg for 7 nights) [307], estazolam (1.0 mg for 4 weeks) [310] suggest some improvement in polysomnographic measures of disturbed sleep but without improvement in deep NREM sleep, consistent with known effects of benzodiazepines. A somewhat longer duration (3 month) study with triazolam (0.125 mg) showed similar results [32]. Because of controversy surrounding use of triazolam and possibly its amnestic effects, an independent report from the Institute of Medicine has suggested that further trials be undertaken to examine its efficacy at 0.125 mg in the geriatric population [50].

More recent studies with newer benzodiazepine, site-specific agonists such as zolpidem (5 to 10 mg), zopiclone (7.5 mg) and zaleplon (5 to 10 mg), have suggested that these medications may have a short-term role in treating poor sleep in old age. These medications appear to have favorable adverse effects profiles without rebound insomnia, with some advantage possibly to zaleplon insofar as the latter is concerned [8,136,273]. Total sleep time may be better maintained with zolpidem because of its slightly longer half-life (about 2.5 hrs) relative to zaleplon (about 1 hr) [93, 280]. An open label laboratory study with high dose zolpidem at 20 mg even claimed increases in deep NREM sleep in elderly psychiatric inpatients [174] but these data have yet to be replicated in double blind studies. Zopiclone at 7.5 mg has also been reported to induce an increase in deep NREM sleep in elderly insomniacs [227], thereby putatively altering Process S.

Adverse Effects

It is difficult to ascertain historical trends in hypnotic usage, primarily because the types of medications available continues to expand. A small increase in prevalent usage (primarily benzodiazepines) during the mid-1980s was noted by Dealberto *et al.* [81] in the United States. In the United Kingdom and Hong Kong prevalence rates of prescription sedative/hypnotic medication among the elderly have tended to be lower than in the US, Canada, or other European countries [71,139,192,234,236]. Historical change in such use of medications has considerable importance in examining the controversial findings regarding sedative/hypnotic use and mortality.

The possible association between regular use of hypnotics and mortality remains provocative. Kripke *et al.* [173] originally reported, based on the American Cancer Society Survey 1 (ACS I) of over one million individuals in 1959-1960, that use of hypnotics were associated with higher mortality rates and these relationships were not mitigated by age. More recently Kripke *et al.* [171,

172], using the similarly powered ACS II survey conducted in 1982, reported similar results, even with more stringent controls over covariates. In both women and men the risk for all causes mortality in relation to sleeping pill use was significant in both younger and older (> 70) groups, but, in ACS II, among the older age groups, the highest risk occurred for those using such medications on a nightly basis [172]. The long-standing interpretive problem with these data involves the medications in use at the time of original data collection. At the time of ACS I, most hypnotic medications in use were barbiturates. At the time of ACS II, the most commonly used hypnotic medications were probably anti-depressants and/or benzodiazepines. In a much smaller, population based study in the United Kingdom, Rumble and Morgan [269] were unable to replicate a specific association between hypnotic medication and follow-up mortality during a more contemporary period of benzodiazepine usage, though the ACS I and II findings were replicated by Kojima *et al.* [169] over a 12 year period between the mid-1980's and mid-1990's.

Although mortality associated with hypnotic medication usage would certainly represent an extreme adverse outcome, many other untoward effects of sedative/hypnotics and psychotropics, in general, have been suggested to occur with high frequency in the aged population. Moderate levels of daytime sleepiness appear to be associated with use of hypnotic medication across all age groups [235]. In the elderly, there is considerable evidence suggesting that use of sedative medication may be associated with falls [293], although the definition of sedatives used in that study included not only benzodiazepines but also phenothiazines. In complimentary studies with populations about 100 times as large, Ray *et al.* [257,258], demonstrated risk for hip fracture with anti-psychotics, anti-depressants, and benzodiazepines with long elimination half-lives, but not for benzodiazepines with short elimination half lives. Other work has suggested a higher risk of motor vehicle accidents in elderly drivers who use benzodiazepines with long elimination half-lives and anti-depressants, relative to users of short-acting benzodiazepines [137,256].

In view of these historical data, the widespread use of the newer medications like zolpidem and zaleplon might appear to offer a higher margin of safety for the elderly, particularly given the encouraging adverse effect profiles derived from the aforementioned clinical trials with such medications. Zolpidem, in particular, has been reported to afford a low abuse potential [285], however important new epidemiologic data has suggested that the risk of hip fracture with zolpidem in the aged exceeds that of even benzodiazepines [315]. The risk of hip fracture, even after controlling for comorbid disease and use of other psychoactives, was nearly twice that of a control population who did not use zolpidem. Although specific dosages for zolpidem were not mentioned by Wang *et al.* [315], Golden *et al.* [124] have presented compelling data to suggest that zolpidem is typically prescribed at dosages far exceeding what is currently recommended in geriatrics (5 mg). Taken together with the previously mentioned data on psychotropic medications and falls, these results continue to warn clinicians that all use of sedative/hypnotic medication in the aged must be judicious. If a subject is to take medication, specific warnings about nocturnal ambulation and dose escalation are considered essential.

Finally, a somewhat different perspective on adverse effects (specifically, falls) in relation to sedative/hypnotic use has been provided by Brassington *et al.* [43] who, for the first time, partialed out effects of disturbed sleep from use of sleep medications in accounting for falls in a large, community-dwelling population ages 64-99. In these data, sleep complaints predicted falling better than use of psychotropics in multivariate models. There are several implications of this finding. First, it may be that elderly individuals with poor sleep have particular problems with balance that lead to falls. Secondly, and perhaps more likely, it may well be that people who cannot sleep at night (who are also the most likely users of hypnotic medications) are more likely to arise from bed during the night. In this sense, it may not be medication, but rather the behavioral effects of insomnia, that could account for the reported association between sedative/hypnotic use and falls. Corroboration of these findings from a laboratory perspective comes from the study of Vignola *et al.* [308] who found that elderly chronic insomniacs in general performed more poorly than age-matched good sleepers on a variety of neuropsychological tests; however, there were no differences in the memory or psychomotor performance of those individuals who had taken or not taken benzodiazepines the night prior to testing.

Phototherapy

Bright light exposure in the evening

As noted before, a phase advance of the circadian timing system appears quite early in the aging process. This phase advance produces an earlier timing of high sleep propensity in the evening and an earlier circadian wake signal in the morning. Bright light exposure in the evening has been proposed as a therapeutic intervention to phase delay the signal from the circadian timing system in older subjects [60]. Up to now, very few studies have evaluated the therapeutic potential of evening bright light exposure in older subjects [70]. In one study, elderly subjects who suffered from sleep maintenance insomnia for at least one year were exposed to bright white light (>4000 lux; n=8) in the evening (2 hours) or to dim red light (<50 lux; n=8) for 12 consecutive days [58]. Evening bright light exposure reduced waking time within sleep by an hour while dim red light had no effect on sleep consolidation parameters. Subjects exposed to evening bright light showed a significant delay in the minimum of their temperature circadian rhythm of about 3.1 hours compared to subjects exposed to dim red light. In addition, the minimum of the temperature rhythm occurred later in the sleep episode in the group exposed to bright light, but not in the group exposed to dim red light. Similar results were obtained with a younger population (mean age 53.4 years) suffering from early morning awakenings [175]. In this study, subjects were exposed to four hours of bright light exposure for two consecutive evenings (2,500 lux). Temperature and melatonin circadian rhythms were significantly delayed following the bright light treatment. Wake-up time was also delayed resulting in an increase in total sleep time of more than an hour. Another study showed that evening bright light exposure not only increases sleep quality, but also enhances daytime performance of elderly subjects with sleep maintenance insomnia [229]. In this research, older subjects with sleep maintenance insomnia received an acute bright light treatment in the evening followed by a period of three months of maintenance light treatment [229]. During the maintenance period, subjects exposed themselves twice weekly to either bright light in the evening (Active condition: 21:00-23:00; >4000 lux) or to bright light in the afternoon (Control condition: 15:00-17:00; >4000 lux). During the maintenance period, elderly individuals who received the active evening bright light treatment maintained higher polysomnographic sleep efficiency and better performance compared to those who received bright light treatment in the afternoon. Only one study tested the effects of bright light therapy administered via a visor in elderly subjects [73]. The results suggest that 0.5 hour of bright light (2000 lux) in the evening is modestly sufficient to decrease sleep latency, to increase sleep efficiency and total sleep time, and to enhance daytime vigilance in elderly women who express sleep complaints. Importantly, evening bright light therapy may be beneficial only for elderly subjects with insomnia complaints. Preliminary results in four elderly subjects who did not complain of nighttime insomnia showed that evening bright light therapy initially increased daytime sleepiness and decreased night-time sleep [298].

Bright light exposure during the day

Many reports have demonstrated that elderly subjects are less exposed to environmental light of high intensity than are younger subjects [59,206]. This situation is probably even more severe among nursing home patients. Recent results showed that nursing home patients spend a median of 10.5 minutes of light over 1000 lux and a median of 4 min over 2000 lux [282]. An association between sleep complaints and levels of environmental illumination have been observed in a few studies [206,282] but not in all [121]. In a nursing home, higher light levels predicted fewer nighttime awakenings regardless of dementia severity [282]. Another study showed that, compared to elderly persons with no sleep complaints, older insomniacs are even less exposed to environmental light of high intensity [206]. These observations led to the suggestion that increasing daytime illumination levels may alleviate sleep complaints in elderly subjects. Only a few studies to date have formally tested this hypothesis in healthy elderly subjects. Bright light therapy (6000 lux) for 30 minutes in the morning decreased motor activity during sleep without affecting subjective evaluation of sleep quality in a group of healthy elderly persons [168]. Supplementary exposure to four hours of midday bright light (2500 lux; from 10:00 to 12:00 and from 14:00 to 16:00) for four weeks decreased awake time and increased sleep efficiency in elderly residents of a nursing home with psychological insomnia [206]. Interestingly, the supplementary daytime bright light regimen increased melatonin secretion in the older insomniacs to levels similar to those in the young control group without phase shifting.

Bright light therapy appears to have probable therapeutic efficacy to treat age-dependent modifications to the sleep-wake cycle, especially early morning awakenings and sleep fragmentation. However, many issues need to be addressed. The minimum or optimal duration and intensity of effective light treatment are still unknown [70]. This point is especially important for compliance, as bright light therapy is a time- consuming therapeutic intervention for the patients. Importantly, we should also identify specific populations of people for whom bright light therapy will be the most beneficial. Bright light therapy may in fact have negative adverse effects in certain groups of elderly individuals [120,298]. Finally, further research should address long term efficacy and the possibility of tolerance. To our knowledge, the longest maintenance study to date in elderly people suffering from sleep maintenance insomnia evaluated the effects of three months of bright light therapy [229].

Melatonin administration

Oral exogenous melatonin administration (0.5-5 mg) consistently increases subjective and objective sleepiness at circadian times of low endogenous melatonin levels in young subjects and in older subjects [56,92,131,144,191,290,297,342]. Numerous studies have demonstrated that melatonin levels decrease with age [206,270,294,313]. These results led to the suggestion that melatonin replacement therapy may improve the nighttime sleep of older subjects, particularly in populations suffering from sleep maintenance insomnia [312]. Results from actigraphic recordings and subjective evaluation of sleep have shown that oral administration of melatonin prior to bedtime (0.3-6 mg) in elderly insomniacs decreases sleep latency [151,132] and reduces motility, sleep-wake transitions, and the amount of wakefulness during the sleep episode [118,132,151,335]. These results have been observed both with immediate and controlled-release formulations of melatonin, yet controlled-release melatonin may be more effective to enhance sleep consolidation parameters [132]. Results on the effects of melatonin on polysomnographically measured sleep are mixed. Some studies have shown that melatonin treatment (0.1 mg-3 mg) increases polysomnographic sleep efficiency [216,341] and total sleep time [216] in elderly individuals who present with sleep difficulties. Similar effects of a single dose of melatonin (1mg) were observed in middle-aged subjects who did not complain of sleep disturbances [16]. Conversely, other studies have not reported improvement in polysomnographic sleep consolidation parameters following melatonin administration in older subjects suffering from sleep maintenance insomnia. One study compared the sleep promoting effects of three delivery strategies of 0.5 mg of melatonin in a group of insomniac elders: 1) an immediate-release dose taken 30 minutes before bedtime; 2) a controlled-release dose taken 30 minutes before bedtime; and 3) an immediate release dose taken four hours after bedtime. All three treatments shortened sleep latency compared to a placebo but none improved polysomnographic total sleep time or sleep efficiency [145]. Another study evaluated the effects of 0.5 mg transbucal melatonin in subjects with sleep maintenance insomnia. Results indicated no significant positive effect compared to a placebo on any polysomnographic sleep quality despite a significant increase in nocturnal urinary 6-sulfatoxy melatonin levels in the treatment condition [80].

Factors associated with the efficacy of melatonin replacement therapy to treat age-dependent modifications of the sleep-wake cycle are still unknown and many theoretical questions need be clarified. Recent results do not support the notion that a reduction in melatonin secretion is a general characteristic of healthy aging. Many previous studies on age-dependent decline of melatonin did not control for illumination levels or for medications that suppress melatonin (e.g. aspirin, ibuprofen, B-blockers, etc.). When these factors are controlled for, melatonin concentration does not necessarily differ between healthy older and young subjects [340]. In fact, the decline of melatonin with age seems completed before the age of 30 years [162]. Furthermore, while a few studies have corroborated lower melatonin levels in older subjects complaining of insomnia compared to controls [22,132,206], numerous reports have not observed such a difference [20,189,190]. Indeed, sleep quality seems more related to the phase angle between the timing of sleep and the melatonin circadian rhythm than to the amount of melatonin produced [145]. It is possible that despite similar melatonin levels, the sensitivity to the sleep-promoting effect of endogenous melatonin would be lower in older subjects, particularly in populations suffering from sleep maintenance insomnia. On the other hand, melatonin replacement therapy may also work more efficiently specifically in melatonin deficient older individuals. However, preliminary recent results do not support an association between baseline endogenous melatonin levels and degree of subsequent improvement with melatonin treatment in elderly patients with chronic primary insomnia [216]. The impact of other individual characteristics (e.g., severity of complaints, anxiety, depression, associated sleep disorders, previous history of hypnotic use) on treatment efficacy needs to be addressed. Future studies should also evaluate the influence of timing of administration, dosage, and strategies of administration. One recent study compared three melatonin doses (0.1, 0.3 and 3.0 mg) administered 30-min before bedtime [341]. All doses elevated melatonin levels and significantly increased sleep efficiency in older insomniac subjects. Importantly, the higher dose of melatonin (3 mg) did not enhance the sleep promoting effects of melatonin above levels achieved with the lower doses (0.1 mg, 0.3 mg), but it caused plasma melatonin to remain elevated into the daylight hours, which may have negative side effects. Future research should carefully examine long-term efficacy of melatonin treatment and potential side/residual/withdrawal effects.

CONCLUSION

Many descriptive studies have confirmed important age-dependent modifications to the sleep-wake cycle. However, the mechanisms that underlie many of these changes remain to be fully elucidated. For example, the phase advance hypothesis of the circadian signal has been able to predict age-dependent changes in the timing of the sleep-wake cycle. However, the neurophysiologic mechanisms that spur decreased sleep consolidation, the marked diminution of deep NREM sleep and SWA during baseline sleep, the lower rebound of deep NREM sleep and SWA following an acute sleep deprivation, the higher rigidity to a phase shift of the sleep-wake cycle, and the lower tolerance to a phase angle misalignment between sleep and the circadian signal are still unclear. The design of effective preventive and therapeutic strategies for older people depends greatly on our understanding of these mechanisms. Future research should identify vulnerability factors that put the older population more at risk of suffering from sleep-wake cycle disturbances (e.g. gender, menopause, stress, medical disease, etc.). Importantly, we must evaluate the consequences of age-dependent modifications to the sleep-wake cycle on all aspects of life that we know sleep loss influences, including overall physical health, cognitive functioning, quality of life, psychological adjustment, metabolism, hormonal regulation, and perhaps even survival. This research should lead to the development of new strategies to help normal older individuals adapt to situations that challenge the sleep-wake cycle and to new pharmacological and non-pharmacological treatments for insomnia in older individuals.

REFERENCES

1. ACHERMANN P., DIJK D-J., BRUNNER D.P., BORBELY A. - A model of human sleep homeostasis based on EEG slow-wave activity: quantitative comparison of data and simulations. *Brain. Res. Bull.* 31, 97-113, 1993.
2. AESCHBACH D., MATTHEWS J.R., POSTOLACHE T.T., JACKSON M.A., GIESEN H.A., WEHR T.A. - Dynamics of the human EEG during prolonged wakefulness: evidence for frequency-specific circadian and homeostatic influences. *Neurosci. Lett.*, 239, 121-124, 1997.
3. AESCHBACH D., MATTHEWS J.R., POSTOLACHE T.T., JACKSON M.A., GIESEN H.A., WEHR T.A. - Two circadian rhythms in the human electroencephalogram during wakefulness. *Am. J. Physiol.* 277, R1771-R1779, 1999.
4. AKERSTEDT T., FOLKARD S. - The three-process model of alertness and its extension to performance, sleep latency and sleep length. *Chronobiol. Int.* 14, 115-123, 1997.
5. ALTHUIS M.D., FREDMAN L., LANGENBERG P.W., MAGAZINER J. - The relationship between insomnia and mortality among community-dwelling older women. *J. Am. Geriatr. Soc.* 46 (1998) 1270-1273.
6. ANCOLI-ISRAEL S. - Insomnia in the elderly: a review for the primary care practitioner. *Sleep*, 23, S36-S38, 2000.
7. ANCOLI-ISRAEL S., KRIPKE D.F., KLAUBER M.R., MASON W.J., FELL R., KAPLAN O. - Periodic limb movements in sleep in community-dwelling elderly. *Sleep*, 14, 496-500, 1991.

8. ANCOLI-ISRAEL S., WALSH J.K., MANGANO R.M., FUJIMORI M. - Zaleplon clinical study group. Zaleplon, a novel nonbenzodiazepine hypnotic, effectively treats insomnia in elderly patients without causing rebound effects. *Primary Care Companion to J. Clin. Psychiatry*, 1, 114-120, 1999..

9. ANTONIJEVIC I.A., STALL G.K., STEIGER A. - A Modulation of the sleep electroencephalogram by estrogen replacement in postmenopausal women. *Am. J. Obstet. Gynecol.* 182, 277-282, 2000.

10. ARMITAGE R., HUDSON A., TRIVEDI M., RUSH A.J. - Sex difference in the distribution of EEG frequencies during sleep: unipolar depressed outpatients. *J. Affect. Disord.*, 34, 121-129, 1995.

11. ARMITAGE R., SMITH C, THOMPSON S., HOFFMANN R. - Sex differences in slow-wave activity in response to sleep deprivation. *Sleep Res. Online.* 4, 33-41, 2001.

12. ASCHOFF J., HOFFMAN K., POHL H., WEVER R.A. - Re-entrainment of circadian rhythms after phase-shifts of the zeitgeber. *Chronobiologia*, 2, 23-78, 1975.

13. ASPLUND R. - Sleep and cardiac diseases amongst elderly people. *J. Int. Med.* 236, 65-71, 1994.

14. ASPLUND R. - Sleep, health and visual impairment in the elderly. *Arch. Gerontol. Geriatr.*, 30, 7-15, 2000.

15. ASPLUND R., ABERG H. - Nocturnal micturition, sleep and well-being in women of ages 40-64 years. *Maturitas*, 24, 73-81, 1996.

16. ATTENBURROW M.E.J., COWEN P.J., SHARPLEY A.L. - Low dose melatonin improves sleep in healthy middle-aged subjects. *Psychopharmacology*, 126, 179-181, 1996.

17. BABAR S.I., ENRIGHT P.L., BOYLE P., FOLEY D., SHARP D.S., PETROVICH H., QUAN S.F. - Sleep disturbances and their correlates in elderly Japanese American men residing in Hawaii. *J. Gerontol. A. Biol. Sci. Med. Sci.*, 55A, M406-M411, 2000.

18. BAKER A., SIMPSON S., DAWSON D. - Sleep disruption and mood changes associated with menopause. *J. Psychosom. Res.*, 43, 359-369, 1997.

19. BALLINGER C.B. - Subjective sleep disturbance at the menopause. *J. Psychom. Res.*, 20, 509-513, 1976.

20. BASKETT J.J., WOOD P.C., BROAD J.B., DUNCAN J.R., ENGLISH J., ARENDT J. - Melatonin in older people with age-related sleep maintenance problems: a comparison with age matched normal sleepers. *Sleep*, 24, 418-424, 2001.

21. BIXLER E.O., VGONTZAS A.N., LIN H.M., TEN HAVE T., REIN J., VELA-BUENO A., KALES A. - Prevalence of sleep disordered breathing in women: effects of gender. *Am. J. Respir. Crit. Care Med.*, 163, 608-613, 2001.

22. BLAICHER W., SPECK E., IMHOF M.H., GRUBER D.M., SCHNEEBERGER C., SATOR M.O., HUIBER J.C. – Melatonin in postmenopausal females. *Arch. Gynecol. Obstet.* 263, 116-118, 2000.

23. BLIWISE D. L. - Sleep in normal aging and dementia. *Sleep*, 16, 40-81, 1993.

24. BLIWISE D.L. - Sleep and circadian rhythm disorders in aging and dementia. In Turek FW, Zee PC (eds.), *Regulation of sleep and circadian rhythms,* Marcel Dekker, New York, 487-526, 1999.

25. BLIWISE D.L., CARSKADON M.A., SEIDEL W.F., NEKICH J.C., DEMENT W.C. - MSLT-defined sleepiness and neuropsychological test performance do not correlate in the elderly. *Neurobiol. Aging*, 12, 463-468, 1991.

26. BLIWISE D.L., KING A.C., HARRIS A.C., HASKELL W.L. - Prevalence of self-reported poor sleep in a healthy population aged 50-65. *Soc. Sci. Med.*, 34, 49-55, 1992.

27. BLIWISE D., PETTA D., SEIDEL W., DEMENT W. - Periodic leg movements during sleep in the elderly. *Arch. Gerontol. Geriatr.*, 4, 273-281, 1985.

28. BLOCK A.J., WYNNE J.W., BOYSEN P.G. - Sleep-disordered breathing and nocturnal oxygen desaturation in post-menopausal women. *Am. J. Med.*, 69, 75-79, 1980.

29. BLOIS R., FEINBERG I., GAILLARD J.M., KUPFER D.J., WEBB W.B. - Sleep in normal and pathological aging. *Experientia,* 39, 551-686, 1983.

30. BONNET M. - The effect of sleep fragmentation on sleep and performance in younger and older subjects. *Neurobiol. Aging*, 10, 21-25, 1989.

31. BONNET M.H. - Effect of 64 hours of sleep deprivation upon sleep in geriatric normals and insomniacs. *Neurobiol. Aging*, 7, 89-96, 1986.

32. BONNET M.H., ARAND D.L. - Chronic use of triazolam in patients with periodic leg movements, fragmented sleep and daytime sleepiness, *Aging*, 3, 313-324, 1991.

33. BONNET M.H., ARAND D.L. - Sleep loss in aging. *Clin. Geriatr. Med.*, 5, 405-420, 1989.

34. BONNET M.H., ARAND D.L. - Metabolic rate and the restorative function of sleep. *Physiol. Behav.*, 59, 777-782, 1996.

35. BONNET M.H., DEXTER J.R., ARAND D.L. - The effect of triazolam on arousal and respiration in central sleep apnea patients. *Sleep*, 13, 31-41, 1990.

36. BONNET M.H., ROSA R.R. - Sleep and performance in young adults and older normals and insomniacs during acute sleep loss and recovery. *Biol. Psychol.*, 25,153-172, 1987.

37. BORBELY A.A. - A two-process model of sleep regulation. *Hum. Neurobiol.*, 1, 195-204, 1982.

38. BORBELY A.A., BAUMANN F., BRANDEIS D., STRAUCH I., LEHMANN D. - Sleep deprivation: Effect on sleep stages and EEG power density in man. *Electroencephalogr. Clin. Neurophysiol.* 51, 483-493, 1981.

39. BORBELY A.A., ACHERMANN P. - Sleep homeostasis and models of sleep regulation. In Kryger MH, Roth T, Dement WC (eds.), *Principles and practice of sleep medicine.* W.B. Saunders Company, Philadelphia, 377-390, 2000.

40. BORN J., HANSEN K., MARSHALL L., MOLLE M., FEHM H.L. - Timing the end of nocturnal sleep. *Nature*, 397, 29-30, 1999.

41. BORN J., KERN W., BIEBER K., FEHM-WOLFSDORF G., SCHIEBE M., FEHM H.L. - Night-time plasma cortisol secretion is associated with specific sleep stages. *Biol. Psychiatry*, 21, 1415-1424, 1986.

42. BORN J., SPATH-SCHWALBE E., SCHWAKENHOFER H., KERN W., FEHM H.L. - Influences of corticotropin-releasing hormone, adrenocorticotropin, and cortisol on sleep in normal man. *J. Clin. Endocrinol. Metab.* 68, 904-911, 1989.

43. BRASSINGTON G.S., KING A.C., BLIWISE D.L. - Sleep problems as a risk factor for falls in a sample of community dwelling adults aged 64-99 years. *J. Am. Geriatr. Soc.,* 48, 1234-1240, 2000.

44. BRENDEL D.H., REYNOLDS C.F., JENNINGS J.R., HOCH C.C., MONK T.H., BERMAN T.H., HALL F.T., BUYSSE D.J., KUPFER D.J. - Sleep stage physiology, mood, and vigilance responses to total sleep deprivation in healthy 80-year-olds and 20-year-olds. *Psychophysiology,* 27, 677-685, 1990.

45. BREZINOVA V. - Effects of caffeine on sleep: EEG study in late middle age people. *Br. J Clin. Pharmacol.,* 1, 203-208, 1974.

46. BROUGHTON R., HASAN J. - Quantitative topographic electroencephalographic mapping during drowsiness and sleep onset. *J. Clin. Neurophysiol,* 12, 372-386, 1995.

47. BROUGHTON·R.J. - SCN controlled circadian arousal and the afternoon "nap zone". *Sleep Res. Online,* 1, 166-178, 1998.

48. BROWN E.N., CZEISLER C.A. - A method for quantifying phase position of the deep circadian oscillator and determining a confidence interval. *Sleep Res,* 14, 290. 1985.

49. BRUNNER D.P., MUNCH M., BIEDERMANN K., HUCH R., HUCH A., BORBELY A.A. - Changes in sleep and sleep electroencephalogram during pregnancy. *Sleep,* 17, 576-582, 1994.

50. BUNNEY W.E., JR., AZARNOFF D.L., BROWN B.W., JR., CANCRO R., GIBBONS R.D., GILLIN J.C., HULLETT S., KILLAM K.F., KUPFER D.J., KRYSTAL J.H., STOLLEY P.D., FRENCH G.S., POPE A.M. - Report of the Institute of Medicine Committee on the Efficacy and Safety of halcion. *Arch. Gen. Psychiatry,* 56,349-352, 1999.

51. BUYSSE D.J., BROWMAN B.A., MONK T.H., REYNOLDS C.F., FASCIZKA A.L., KUPFER D.J. - Napping and 24 - hour sleep/wake patterns in healthy elderly and young adults. *J. Am. Geriatr. Soc.,* 40, 779-786, 1992.

52. BUYSSE D.J., REYNOLDS C.F.III., HOCH C.C., HOUCK P.R., KUPFER D.J., MAZUMDAR S., FRANK E. - Longitudinal effects of nortriptyline on EEG sleep and the likelihood of recurrence in elderly depressed patients. *Neuropsychopharmacology,* 14, 243-252, 1996.

53. CAJOCHEN C., BRUNNER D.P., KRAUCHI K., GRAW P., WIRZ-JUSTICE A. - Power density in the theta/alpha frequencies of the waking EEG progressively increases during sustained wakefulness. *Sleep,* 18, 890-894, 1995.

54. CAJOCHEN C., FOY R., DIJK D-J. - Frontal predominance of a relative increase in sleep delta and theta EEG activity after sleep loss in humans. *Sleep. Res. Online,* 2, 65-69, 1999.

55. CAJOCHEN C., KHALSA S.B.R., WYATT J.K., CZEISLER C.A., DIJK D.J. - EEG and ocular correlates of circadian melatonin phase and human performance decrements during sleep loss. *Am. J. Physiol.,* 277, R640-R649, 1999.

56. CAJOCHEN C., KRAÜCHI K., WIRZ-JUSTICE A. - The acute soporific action of daytime melatonin administration: Effects on the EEG during wakefulness and subjective alertness. *J. Biol. Rhythms.,* 12, 636-643, 1997.

57. CAMPBELL S.S., DAWSON D. - Aging young sleep: a test of the phase advance hypothesis of sleep disturbance in the elderly. *J. Sleep Res.,* 1, 205-210, 1992.

58. CAMPBELL S.S., DAWSON D., ANDERSON M.W. - Alleviation of sleep maintenance insomnia with timed exposure to bright light. *J. Am. Geriatr. Soc.,* 41, 829-836, 1993.

59. CAMPBELL S.S., KRIPKE D., GILLIN J.C., HRUBOVCAK J.C. - Exposure to light in healthy elderly subjects and Alzheimer's patients. *Physiol. Behav.,* 42, 141-144, 1988.

60. CAMPBELL S.S., TERMAN M., LEWY A.J., DIJK D.J., EASTMAN C.I., BOULOS Z. – Light treatment for sleep disorders: consensus report. V. Age- related disturbances. *J. Biol. Rhythms,* 10, 151-154, 1995.

61. CARRIER J., GAUDREAU H., LAVOIE H.B., MORETTINI J. - Do middle-aged men and women differ in their ability to recuperate during the day following an acute sleep deprivation? *Sleep,* in press.

62. CARRIER J., LAND S., BUYSSE D.J., KUPFER D.J., MONK T.H. - The effects of age and gender on sleep EEG power spectral density in the "middle" years of life (20y-60y). *Psychophysiology,* 38, 232-242, 2001.

63. CARRIER J., MONK T.H. - Estimating the endogenous circadian temperature rhythm without keeping people awake. *J. Biol. Rhythms,* 12, 266-277, 1997.

64. CARRIER J., MONK T.H., BUYSSE D.J., KUPFER D.J. - Amplitude reduction of the circadian temperature and sleep rhythms in the elderly. *Chronobiol. Int.,* 13, 373-386, 1996.

65. CARRIER J., MONK T.H., BUYSSE D.J., KUPFER D.J. - Inducing a 6-hour phase advance in the elderly: effects on sleep and temperature rhythms. *J. Sleep Res.,* 5, 99-105, 1996.

66. CARRIER J., MONK T.H., BUYSSE D.J., KUPFER D.J. - Sleep and morningness-eveningness in the "middle" years of life (20y-59y). *J. Sleep Res.,* 6, 230-237, 1997.

67. CARRIER J., MONK T.H., REYNOLDS C.F.I., BUYSSE D.J., KUPFER D.J. - Are age differences in sleep due to phase differences in the output of the circadian timing system? *Chronobiol. Int.,* 16, 79-91, 1999.

68. CARRIER J., PAQUET J., MORETTINI J., TOUCHETTE E. - Phase advance of sleep and temperature circadian rhythms in the middle years of life in humans. *Neurosci. Lett.,* 320, 1-4, 2002.

69. CARSKADON M.A., DEMENT W.C. - Sleep loss in elderly volunteers. *Sleep,* 8, 207-221, 1985.

70. CHESSON A.L., LITTNER M., DAVILLA D., ANDERSON W.M., GRIGG-DAMBERGER M., HARTSE K., JOHNSON S., WISE M. - Practice parameters for the use of light therapy in the treatment of sleep disorders. *Sleep,* 22, 641-666, 1999.

71. CHIU H.F.K., LEUNG T., LAM L.C.W., WING Y.K., CHUNG D.W.S., LI S.W., CHI I., LAW W.T., BOEY K.W. - Sleep problems in Chinese elderly in Hong Kong. *Sleep,* 22, 717-726, 1999.

72. COOK N.R., EVANS D.A., FUNKENSTEIN H.H., SCHERR P.A., OSTFELD A.M., TAYLOR J.O., HENNEKENS C.H. - Correlates of headache in a population based cohort of elderly. *Arch. Neurol.,* 46, 1338-1344, 1989.

73. COOKE K.M., KREYDATUS M.A., ATHERTON A., THOMAN E.B. - The effects of evening light exposure on sleep of elderly women expressing sleep complaints. *J. Behav. Med.,* 21, 103-114, 1998.

74. COPINSCHI G., VAN CAUTER E. - Effects of ageing on modulation of hormonal secretions by sleep and circadian rhythmicity. *Horm. Res.,* 43, 20-24, 1995.

75. CZEISLER C.A., DUMONT M., DUFFY J.F., STEINBERG J.D., RICHARDSON G.S., BROWN E.N., SANCHEZ R., RIOS C.D., RONDA J.M. - Association of sleep-wake habits in older people with changes in output of circadian pacemaker. *Lancet,* 340, 933-936, 1992.

76. CZEISLER C.A., WEITZMAN E.D., MOORE-EDE M.C., ZIMMERMAN J.C., KRONAUER R.S. - Human sleep: Its duration and organization depend on its circadian phase. *Science,* 210, 1264-1267, 1980.

77. DAAN S., BEERSMA D.G.M., BORBELY A.A. - Timing of human sleep: Recovery process gated by circadian pacemaker. *Am. J. Physiol.,* 246, R161-R178, 1984.

78. DALY J.W., PADGETT W.L., SHAMIM M.T. - Analogues of caffeine and theophylline: effects of structural alterations on affinity at adenosine receptors. *J. Med. Chem.,* 29, 1305-1308, 1986.

79. DANCEY D.R., HANLY P.J., SOONG C., LEE B., HOFFSTEIN V. - Impact of menopause on the prevalence and severity of sleep apnea. *Chest,* 120, 151-155, 2001.

80. DAWSON D., ROGERS N.L., VAN DEN HEUVEL C.J., KENNAWAY D.J., LUSHINGTON K. - Effect of sustained nocturnal transbucal melatonin administration on sleep and temperature in elderly insomniacs. *J. Biol. Rhythms,* 13, 532-538, 1998.

81. DEALBERTO M.J., SEEMAN T., McAVAY G.J., BERKMAN L. - Factors related to current and subsequent psychotropic drug use in an elderly cohort. *J. Clin. Epidemiol.,* 50, 357-364, 1997.

82. DICKEL M.J., MOSKO S.S. - Morbidity cut-offs for sleep apnea and periodic leg movements in predicting subjective complaints in seniors. *Sleep,* 13, 155-166, 1990.

83. DIJK D.J., BEERSMA D.G.M., DAAN S. - EEG power density during nap sleep: reflection of an hourglass measuring the duration of prior wakefulness. *J. Biol. Rhythms,* 2, 207-219, 1987.

84. DIJK D.J., BEERSMA D.G.M., HOOFDAKKER R.H.- Sex differences in the sleep EEG of young adults: visual scoring and spectral analysis. *Sleep,* 12, 500-507, 1989.

85. DIJK D.J., DUFFY J.F., CZEISLER C.A. - Age-related increase in awakenings: impaired consolidation of NonREM sleep at all circadian phase. *Sleep,* 24, 565-577, 2001.

86. DIJK D.J., DUFFY J.F., RIEL E., SHANAHAN T.L., CZEISLER C.A. - Ageing and the circadian and homeostatic regulation of human sleep during forced desynchrony of rest, melatonin and temperature rhythms. *J. Physiol.,* 516, 611-627, 1999.

87. DIJK D.J., SHANAHAN T.L., DUFFY J.F., RONDA J.M., CZEISLER C.A. - Variation of electroencephalographic activity during non-rapid eye movement and rapid eye movement sleep with phase of circadian melatonin rhythm in humans. *J. Physiol.,* 503, 851-858, 1997.

88. DIJK D.J., BEERSMA D.G.M., HOOFDAKKER R.H. - All night spectral analysis of EEG sleep in young adult and middle-aged male subjects. *Neurobiol. Aging,* 10, 677-682, 1989.

89. DIJK D.J., BRUNNER D.P., BEERSMA D.G.M., BORBELY A.A. - Electroencephalogram power density and slow wave sleep as a function of prior waking and circadian phase. *Sleep,* 13, 430-440, 1990.

90. DIJK D.J., CZEISLER C.A. - Paradoxical timing of the circadian rhythm of sleep propensity serves to consolidate sleep and wakefulness in humans. *Neurosci. Lett.,* 166, 63-68, 1994.

91. DIJK D.J., DUFFY J.F., CZEISLER C.A. - Contribution of circadian physiology and sleep homeostasis to age-related changes in human sleep. *Chronobiol. Int.,* 17, 285-311, 2000.

92. DOLLINS A.B., ZHDANOVA I.V., WURTMAN R.J., LYNCH H.J., DENG M.H. - Effect of inducing nocturnal serum melatonin concentrations in daytime on sleep, mood, body temperature, and performance. *Proc. Natl. Acad. Sci.,* 91, 1824-1828, 1994.

93. DOOLEY M., PLOSKER G.L. - Zaleplon: a review of its use in the treatment of insomnia. *Drugs,* 60, 413-445, 2000.

94. DRAPEAU C., FRENETTE S., LAVOIE H.B., CARRIER J. - Fluctuation of subjective alertness and waking EEG during a 25-hour sleep deprivation in young and middle-aged subjects. *Sleep,* in press.

95. DRIVER H.S., DIJK D.J., WERTH E., BIEDERMANN K., BORBELY A.A. - Sleep and the sleep electroencephalogram across the menstrual cycle in young healthy women. *J. Clin. Endocrinol. Metab.,* 81, 728-735, 1996.

96. DUFFY J.F., DIJK D.J., KLERMAN E.B., CZEISLER C.A. - Later endogenous circadian temperature nadir relative to an earlier wake time in older people, *Am. J. Physiol.* 275, R1478-R1487, 1998.

97. DUMONT M., MACCHI M., CARRIER J., LEFRANCE C., HEBERT M. - Time course of narrow frequency bands in the waking EEG during sleep deprivation. *Neuroreport,* 10, 403-407, 1999.

98. EDINGER J.D., WOHLGEMUTH W.K., RADTKE R.A., MARSH G.R., QUILLIAN R.E. - Cognitive behavioral therapy for treatment of chronic primary insomnia: a randomized controlled trial. *JAMA,* 285, 1856-1864, 2001.

99. ELTON M., WINTER O., HESLENFELD D., LOEWY D., CAMPBELL K., KOK A. - Event-related potentials to tones in the absence and presence of sleep spindles. *J. Sleep Res.,* 6, 78-83, 1997.

100. ENGLERT S., LINDEN M. - Differences in self-reported sleep complaints in elderly persons living in the community who do or do not take sleep medication. *J. Clin. Psychiatry,* 59, 137-144, 1998.

101. ERKKOLA R., HOLMA P., JARVI T., NUMMI S., PUNNONEN R., RAUDASKOSKI T., REHN K., RYYNANEN M., SIPILA P., TUNKELO E. *et al..* - Transdermal oestrogen replacement therapy in a Finnish population. *Maturitas,* 13, 275-281, 1991.

102. ERLIK Y., TATARYN I.V., MELDRUM D.R., LOMAX P., BAJOREK J.G., JUDD H.L. - Association of waking episodes with menopausal hot flushes. *JAMA*, 245, 1741-1744, 1981.

103. FEHM H.L., BIEBER K., BENKOWITSCH R., FEHM-WOLFSDORF G., VOIGT K.H., BORN J. - Relationships between sleep stages and plasma cortisol: a single case study. *Acta Endocrinologica*, 111, 264-270, 1986.

104. FEINBERG I. - Changes in sleep cycle patterns with age. *J. Psychiatr. Res.*, 10, 283-306, 1974.

105. FEINBERG I., KORESKO R.L., HELLER N. - EEG sleep patterns as a function of normal and pathological aging in man. *J. Psychiatr. Res.*, 5, 107-144, 1967.

106. FINELLI L.A., BAUMANN H., BORBELY A.A., ACHERMANN P. - Dual electroencephalogram markers of human sleep homeostasis: correlation between theta activity in waking and slow-wave activity in sleep. *Neuroscience*, 101, 553-529, 2000.

107. FOLEY D.J., MONJAN A.A., BROWN S.L., SIMONSICK E.M., WALLACE R.B., BLAZER D.G. - Sleep complaints among elderly persons: an epidemiologic study of three communities. *Sleep*, 18, 425-432, 1995.

108. FOLEY D.J., MONJAN A.A., IZMIRLIAN G., HAYS J.C., BLAZER D.G. - Incidence and remission of insomnia among elderly adults in a biracial cohort. *Sleep*, 22, S373-S378, 1999.

109. FOLEY D.J., MONJAN A.A., SIMONSICK E.M., WALLACE R.B., BLAZER D.G. - Incidence and remission of insomnia among elderly adults: an epidemiologic study of 6,800 persons over three years. *Sleep*, 22, S366-S372, 1999.

110. FOLKARD S., AKERSTEDT T. - A three-process model of the regulation of alertness-sleepiness. In Broughton RJ, Ogilvie RD (eds.), *Sleep, Arousal, and Performance: a Tribute to Bob Wilkinson*, Birkhauser, Boston, pp. 11-26, 1992.

111. FOLLENIUS M., BRANDENBERGER G., BANDESAPT J.J., LIBERT J.P., EHRHART J. - Nocturnal cortisol release in relation to sleep structure. *Sleep*, 15, 21-27, 1992.

112. FORD D.E., KAMEROW D.B. - Epidemiologic study of sleep disturbances and psychiatric disorders. *JAMA*, 262, 1479-1484, 1989.

113. FRASER D.I., PARSONS A., WHITEHEAD M.I., WORDSWORTH J., STUARD G., PRYSE-DAVIES J. - The optimal dose of oral norethindrone acetate for addition to transdermal estradiol: a multicenter study. *Fertil. Steril.*, 53, 460-468, 1990.

114. FREEDMAN R.R., NORTON D., WOODWARD S., CORNELISSEN G. - Core body temperature and circadian rhythm of hot flashes in menopausal women. *J. Clin. Endocrinol. Metab.*, 80, 2354-2358, 1995.

115. FRIEDMAN L., BROOKS J.O. 3rd., BLIWISE D.L., YESAVAGE J.A., WICKS D.S. - Perceptions of life stress and chronic insomnia in older adults. *Psychol. Aging*, 10, 352-357, 1995.

116. GALE C., MARTYN C. - Larks and owls and health, wealth, and wisdom. *BMJ*, 317, 1675-1677, 1998.

117. GANGULI M., REYNOLDS C.F., GILBY J.E. - Prevalence and persistence of sleep complaints in a rural older community sample: the Movies project. *J. Am. Geriatr. Soc.*, 44, 778-784, 1996.

118. GARFINKEL D., LAUDON M., NOF D., ZISAPEL N. - Improvement of sleep quality in elderly people by controlled-release melatonin. *Lancet*, 346, 541-544, 1995.

119. GAUDREAU H., MORETTINI J., LAVOIE H.B., CARRIER J. - Effects of a 25-h sleep deprivation on daytime sleep in the middle-aged. *Neurobiol. Aging*, 22, 461-468, 2001.

120. GENHART M.J., KELLY K.A., COURSEY R.D., DATILES M., ROSENTHAL L. - Effects of bright light on mood in normal elderly women. *Psychiatry Res.*, 47, 87-97, 1993.

121. GIRARDIN J.L., KRIPKE D.F., ANCOLI-ISRAEL S., KLAUBER M.R., SEPULVEDA R.S., MOWEN M.A., ASSMUS J.D., LANGER R.D. - Circadian sleep, illumination, and activity patterns in women: influences of aging and time reference. *Physiol. Behav.*, 68, 347-352, 2000.

122. GISLASON T., ALMQVIST M. - Somatic diseases and sleep complaints. *Acta Med. Scandinavica*, 221, 475-481, 1987.

123. GISLASON T., REYNISDOTTIR H., KRISTBJARNARSON H., BENEDIKTSDOTTIR B. - Sleep habits and sleep disturbances among the elderly: an epidemiological survey. *J. Int. Med.*, 234, 31-39, 1993.

124. GOLDEN A.G., PRESTON R.A., BARNETT S.D., LLORENTE M. HAMDAN K., SILVERMAN M.A. - Inappropriate medication prescribing in homebound older adults. *J. Am. Geriatr. Soc.*, 47, 948-953, 1999.

125. GRONFIER C., CHAPOTOT F., WEIBEL L., JOUNY C., PIQUARD F., BRANDENBERGER G. - Pulsatile cortisol secretion and EEG delta waves are controlled by two independent but synchronized generators. *Am. J. Physiol.*, 275, E94-E100, 1998.

126. GRONFIER C., LUTHRINGER R., FOLLENIUS M., SCHALTENBRAND N., MACHER J.P., MUZET A., BRANDENBERGER G. - Temporal relationship between pulsative cortisol secretion and electroencephalographic activity during sleep in man. *Electroencephalogr. Clin. Neurophysiol.*, 103, 405-408, 1997.

127. GUILLEMINAULT C., QUERA-SALVA M.A., PARTINEN M., JAMIESON A. - Women and the obstructive sleep apnea syndrome. *Chest*, 92, 104-109, 1988.

128. GULDNER J., SCHIER T., FRIESS E., COLLA M., HOLSBOER F., STEIGER A. - Reduced efficacy of growth hormone-releasing hormone in modulating sleep endocrine activity in the elderly. *Neurobiol. Aging*, 18, 491-495, 1997.

129. GUTHRIE J.R., DENNERSTEIN L., HOPPER J.L., HOPPER J.L., BURGER H.G. - Hot flushes, menstrual status, and hormone levels in a population-based sample of midlife women. *Obstet. Gynecol.*, 88, 437-442, 1996.

130. HABTE-GABR E., WALLACE R.B., COLSHER P.L., HULBERT J.R., WHITE L.R., SMITH I.M. - Sleep patterns in rural elders: demographic, health and psychobehavioral correlates. *J. Clin. Epidemiol.*, 44, 5-13, 1991.

131. HAIMOV I., LAVIE P. - Circadian characteristics of sleep propensity function in healthy elderly: a comparison with young adults. *Sleep*, 20, 294-300, 1997.

132. HAIMOV I., LAVIE P., LAUDON M., HERER P., VIDGER C., ZISAPEL N. - Melatonin   replacement therapy of elderly insomniacs. *Sleep*, 18, 598-603, 1995.

133. HARRISON Y., HORNE J.A., ROTHWELL A. - Pre-frontal neuropsychological effects of sleep deprivation in young adults - a model for healthy aging? *Sleep*, 23, 1067-1069, 2000.

134. HART R., MORIN C.M., BEST A.M. - Neuropsychological performance in elderly insomnia patients. *Aging and Cognition*, 2, 268-278, 1995.

135. HAYS J.C., BLAZER D.G., FOLEY D.J.- Risk of napping: excessive daytime sleepiness and mortality in an older community population. *J. Am. Geriatr. Soc.*, 44, 693-698, 1996.

136. HEDNER J., YAECHE R., EMILIEN G., FARR J., SALINAS E. - Zaleplon shortens subjective sleep latency and improves subjective sleep quality in elderly patients with insomnia. *Int. J. Geriatr. Psychiatry*, 15, 704 – 712, 2000.

137. HEMMELGARN B., SUISSA S., HUANG A., BOIVIN J.F., PINARD G. - Benzodiazepine use and the risk of motor vehicle crash in the elderly. *JAMA*, 278, 27-31, 1997.

138. HOCH C.C., DEW M.A., REYNOLDS C.F. 3rd, MONK T.H., BUYSSE D.J., HOUCK P.R., MACHEN M.A., KUPFER D.J. - A longitudinal study of laboratory- and diary-based sleep measures in healthy "old old" and "young old" volunteers. *Sleep*, 17, 489-96, 1994.

139. HOHAGEN F., KAPPLER C., SCHRAMM E., RINK K., WEYERER S., RIEMANN D., BERGER M. - Prevalence of insomnia in elderly general practice attenders and the current treatment modalities. *Acta Psychiatr. Scand.*, 90, 102-108, 1994.

140. HOHAGEN F., RINK K., KAPPLER C., SCHRAMM E., RIEMANN D., WEYERER S., BERGER M. - Prevalence and treatment of insomnia in general practice: a longitudinal study. *Eur. Arch. Psychiatry Clin. Neurosci.*, 242, 329-336, 1993.

141. HOLSBOER F., VON BARDELEBEN U., STEIGER A. - Effects of intravenous corticotropin-releasing hormone upon sleep-related growth hormone surge and sleep EEG in man. *Neuroendocrinology*, 48, 32-38, 1988.

142. HORNE J.A., ÖSTBERG O. - A self-assessment questionnaire to determine morningness-eveningness in human circadian rhythms. *Int. J. Chronobiol.*, 4, 97-110, 1976.

143. HUBER R., DEBOER T., TOBLER I. - Topography of EEG dynamics after sleep deprivation in mice. *J. Neurophysiol.*, 84, 1888-1893, 2000.

144. HUGHES R.J., BADIA P. - Sleep-promoting and hypothermic effects of daytime melatonin administration in humans. *Sleep*, 20, 124-131, 1997.

145. HUGHES R.J., SACK R.L., LEWY A.J. - The role of melatonin and circadian phase in age-related sleep-maintenance insomnia: Assessment in a clinical trial of melatonin replacement. *Sleep*, 21, 52-68, 1998.

146. HUNTER M. - The south-east England longitudinal study of the climacteric and postmenopause. *Maturitas*, 14, 117-126, 1992.

147. HUNTER M., BATTERSBY R., WHITEHEAD M. - Relationships between psychological symptoms, somatic complaints and menopausal status. *Maturitas*, 8, 217-228, 1986.

148. HYYPPA M.T., KRONHOLM E. - Quality of sleep and chronic illnesses. *J. Clin. Epidemiol.*, 42, 633-638, 1989.

149. ICSD, *International classification of sleep disorders: Diagnostic and coding manual*, American Sleep Disorders Association, Rochester, Minnesota, 1990.

150. JANSON C., LINDBERG E., GISLASON T., ELMASRY A., BOMAN G. - Insomnia in men - a 10-year prospective population based study. *Sleep*, 24, 425-430, 2001.

151. JEAN-LOUIS G., VON GIZYCKI H., ZIZI F. - Melatonin effects on sleep, mood, and cognition in elderly with mild cognitive impairment. *J. Pineal Res.*, 25, 177-183, 1998.

152. JENSEN E., DEHLIN O., HAGBERG B., SAMUELSSON G., SVENSSON T. - Insomnia in an 80-year-old population: Relationship to medical, psychological and social factors. *J. Sleep Res.*, 7, 183-189, 1998.

153. JEWETT M.E., KRONAUER R.E. - Interactive mathematical models of subjective alertness and cognitive throughput in humans. *J. Biol. Rhythms*, 14, 588-597, 1999.

154. JOHNSON M.P., DUFFY J.F., DIJK D.J., RONDA J.M., DYAL C.M., CZEISLER C.A. - Short-term memory, alertness and performance: A reappraisal of their relationship to body temperature. *J. Sleep Res.*, 1, 24-29, 1992.

155. JORM A.F., GRAYSON D., CREASEY H., WAITE L., BROE G.A. - Long-term benzodiazepine use by elderly people living in the community. *Aust. N. Z. J. Public Health*, 24, 7-10, 1999.

156. KALES A., BIXLER E.O., SOLDATOS C.R., VELA-BUENO A., CALDWELL A.B., CADIEUX R.J. - Biopsychobehavioral correlates of insomnia, part 1: role of sleep apnea and nocturnal myoclonus. *Psychosomatics*, 23, 589-600, 1982.

157. KARACAN I., THORNBY J.I., ANCH A.M., BOOTH G.H., WILLIAMS R.L., SALIS P.J. - Dose-related sleep disturbances induced by coffee and caffeine. *Clin. Pharmacol. Ther.*, 20, 682-689, 1976.

158. KARACAN I., THORNBY J.I., WILLIAMS R.L. - Sleep disturbance: a community survey. In Guilleminault C., Lugaresi E. (eds.), *Sleep/Wake Disorders: Natural History, Epidemiology, and Long Term Evolution*. Raven Press, New-York, pp. 37-60, 1993.

159. KATZ D.A., McHOMEY C.A. - Clinical correlates of insomnia in patients with chronic illness. *Arch. Int. Med.*, 158, 1099-1107, 1998.

160. KEEFE D.L., WATSON R., NAFTOLIN F. - Hormone replacement therapy may alleviate sleep apnea in menopausal women. *Menopause: The Journal of The North American Menopause Society*, 6, 196-200, 1999.

161. KELLY T.L., MITLER M.M., BONNET M.H. - Sleep latency measures of caffeine effects during sleep deprivation. *Electroencephalogr. Clin. Neurophysiol.*, 102, 397-400, 1997.

162. KENNAWAY D.J., LUSHINGTON K., DAWSON D., LACK L., VAN DEN HEUVEL C., ROGERS N. - Urinary 6-sulfatoxymelatonin excretion and aging: new results and critical review of the literature. *J. Pineal Res.*, 27, 210-220, 1999.

163. KERN W., DODT C., BORN J., FEHM H.L. - Changes in cortisol and growth hormone secretion during nocturnal sleep in the course of aging. *J. Gerontol. A. Biol. Sci. Med. Sci.*, 51A, M3-M9, 1996.

164. KLEIN K.E., WEGMANN H.M., HUNT B.I. - Desynchronization of body temperature and performance circadian rhythms as a result of out-going and homegoing transmeridian flights. *Aerospace Med.*, 43(2), 119-132, 1972.

165. KLERMAN E.B., DUFFY J.F., DIJK D.J., CZEISLER C.A. - Circadian phase resetting in older people by ocular bright light exposure. *J. Investig. Med.*, 49, 30-40, 2001.

166. KLINK M.E., DODGE R., QUAN S.F. - The relation of sleep complaints to respiratory symptoms in a general population. *Chest*, 105, 151-154, 1994.

167. KLINK M.E., QUAN S.F., KALTERBORN W.T., LEBOWITZ M.D. - Risk factors associated with complaints of insomnia in general adult population: influence of previous complaints of insomnia. *Arch. Intern.Med.*, 152, 1634-1637, 1992.

168. KOHSAKA M., FUKUDA N., KOBAYASHI R., HONMA H., SAKAKIBARA S., KOYAMA E., NAKANO O., MATSUBARA H. - Effects of short duration morning bright light in healthy elderly: II. Sleep and motor activity. *Psychiatry Clin. Neurosci.*, 52, 252-253, 1998.

169. KOJIMA M., WAKAI K., KAWAMURA T., TAMAKOSHI A., AOKI R., LIN Y., NAKAYAMA T., HORIBE H., AOKI N., OHNO Y. - Sleep patterns and total mortality: a 12-year follow-up study in Japan. *J. Epidemiol.*, 10, 87-93, 2000.

170. KRIEGER J., TURLOT J.C., MANGIN P., KURTZ D. - Breathing during sleep in normal young and elderly subjects: hypopneas, apneas, and correlated factors. *Sleep*, 6, 108-120, 1983..

171. KRIPKE D.F., GARFINKEL L., WINGARD D.L., KLAUBER M.R., MARLER M.R. - Mortality associated with sleep duration and insomnia. *Arch. Gen. Psychiatry*, 59, 131-136, 2002.

172. KRIPKE D.F., KLAUBER M.R., WINGARD D.L., FELL R.L., ASSMUS J.D., GARFINKEL L. - Mortality hazard associated with prescription hypnotics. *Biol. Psychiatry*, 43, 687-693, 1998.

173. KRIPKE D.F., SIMONS R.N., GARFINKEL L., HAMMOND E.C. - Short and long sleep and sleeping pills. *Arch. Gen. Psychiatry*, 36, 103-116, 1979.

174. KUMMER J., GUENDEL L., LINDEN J., EICH F.X., ATTALI P., COQUELIN J.P., KYREIN H.J. - Long-term polysomnographic study of the efficacy and safety of zolpidem in elderly psychiatric in-patients with insomnia. *J. Int. Med. Res.*, 21, 171-184, 1993.

175. LACK L., WRIGHT H. - The effect of evening bright light in delaying the circadian rhythms and lengthening the sleep of early morning awakening insomniacs. *Sleep*, 16(5), 436-443, 1993.

176. LANDOLT H.P., BORBELY A.A. - Age-dependent changes in the sleep EEG topography, *Clin. Neurophysiol*, 112, 369-377, 2001.

177. LANDOLT H.P., DIJK D.J., GAUS S.E., BORBELY A.A. - Caffeine reduces low-frequency delta activity in the human sleep EEG. *Neuropsychopharmacology*, 12, 229-238, 1995.

178. LANDOLT H.P., DIJK D.J., ACHERMANN P., BORBELY A.A. - Effect of age on the sleep EEG: slow-wave activity and spindle frequency activity in young and middle-aged men. *Brain Res.*, 738, 205-212, 1996.

179. LAVIE P. - The 24-hour sleep propensity function (SPF): practical and theoretical implications. In Monk TH. (ed.), *Sleep, Sleepiness and Performance*, John Wiley & Sons Ltd., Chichester, pp. 65-93, 1991.

180. LAVIGNE G.J., MONTPLAISIR J.Y. - Restless legs syndrome and sleep bruxism: prevalence and association among Canadians. *Sleep*, 17, 739-743, 1994.

181. LEDESERT B., RINGA V., BREART G. - Menopause and perceived health status among the women of the French GAZEL cohort. *Maturitas*, 20, 113-120, 1995.

182. LEPROULT R., COPINSCHI G., BUXTON O., VAN CAUTER E. - Sleep loss results in an elevation of cortisol levels the next evening. *Sleep*, 20, 865-870, 1997.

183. LEVENSTEIN S., KAPLAN G.A., SMITH M.W. - Psychological predictors of peptic ulcer incidence in the Alameda County study. *J. Clin. Gastroenterol.*, 24, 140-146, 1997.

184. LICHSTEIN K.L., MORIN C.M. (eds.), *Treatment of late-life insomnia*, Sage Publications, Thousand Oaks, 2000.

185. LIEBERMAN H.R., WURTMAN J.J., TEICHER M.H. - Circadian rhythms of activity in healthy young and elderly humans. *Neurobiol. Aging*, 10, 259-265, 1989.

186. LIVINGSTON G., BLIZARD B., MANN A. - Does sleep disturbance predict depression in elderly people? A study in inner London. *Br. J. Gen. Pract.*, 43, 445-448, 1993.

187. LUGARESI E., CIRIGNOTTA F., ZUCCONI M., MONDINI S., LENZI P.L., COCCAGNA G. - Good and poor sleepers: an epidemiological survey of the San Marino population. In Guilleminault C. and Lugaresi E. (eds.), *Sleep/Wake Disorders: Natural History, Epidemiology, and Long Term Evolution*, Raven Press, New-York, pp. 1-12, 1983.

188. LUGARESI E., COCCAGNA G., CERONI G.B., AMBROSETTO C. - Restless legs syndrome and nocturnal myoclonus. In Gastaut,H., Lugaresi,E., Ceroni,G.B., Coccagna,C, (eds.), *The abnormalities of sleep in man: proceedings of the 15th European meeting on electroencephalography*, Bologna 1967, Aulo Gaggi Editore, Bologna, pp. 285-294, 1968.

189. LUSHINGTON K., DAWSON D., KENNAWAY D.J., LACK L. - The relationship between 6-sulphatoxymelatonin and polysomnographic sleep in good sleeping controls and wake maintenance insomniacs. *J. Sleep Res.*, 8, 57-64, 1999.

190. LUSHINGTON K., LACK L., KENNAWAY D.J., ROGERS N., VAN DEN HEUVEL C., DAWSON D. - 6-Sulfatoxymelatonin excretion and self-reported sleep in good sleeping controls. *J. Sleep Res.*, 7, 75-83, 1998.

191. LUSHINGTON K., POLLARD K., LACK L., KENNAWAY D.J., DAWSON D. - Daytime melatonin administration in elderly good and poor sleepers: Effects on core body temperature and sleep latency. *Sleep*, 20, 1135-1144, 1997.

192. MAGGI S., LANGLOIS J.A., MINICUCI N., GRIGOLETTO F., PAVAN M., FOLEY D.J., ENZI G. - Sleep complaints in community-dwelling older persons: prevalence, associated factors, and reported causes. *J. Am. Geriatr. Soc.*, 46, 161-168, 1998.

193. MALLON L., BROMAN J.E., HETTA J. - Relationship between insomnia, depression, and mortality: a 12-year follow-up of older adults in the community. *Int. Psychogeriatr.*, 12, 295-306, 2000.

194. MALLON L., HETTA J. - A survey of sleep habits and sleeping difficulties in an elderly Swedish population. *Ups. J. Med. Sci.*, 102, 185-198, 1997.

195. MANT A., EYLAND E.A. - Sleep patterns and problems in elderly general practice attenders: an Australian survey. *Community Health Studies*, 12, 192-199, 1988.

196. MATTHEWS K.A., WING R.R., KULLER L.H., MEILAHN E.N., KELSEY S.F., COSTELLO E.J., CAGGIULA A.W. - Influences of natural menopause on psychological characteristics and symptoms of middle-aged healthy women. *J. Consult. Clin. Psychol.*, 58, 345-351, 1990.

197. McKINLAY S.M. - The normal menopause transition: an overview. *Maturitas*, 23, 137-145, 1996.

198. MELLINGER G.D., BALTER M.B., UHLENHUTH E.H. - Insomnia and its treatment. Prevalence and correlates. *Arch. Gen. Psychiatry*, 42, 225-232, 1985.

199. MENDELSON W.B. - Are periodic leg movements associated with clinical sleep disturbance? *Sleep*, 19, 219-223, 1996.

200. MENDELSON W.B., BERGMANN B.M. - Age-dependent changes in recovery sleep after 48 hours of sleep deprivation in rats. *Neurobiol. Aging*, 21, 689-693, 2000.

201. MERICA H., BLOIS R., GAILLARD J.M. - Spectral characteristics of sleep EEG in chronic insomnia. *Eur. J. Neurosci.*, 10, 1826-1834, 1998.

202. MIDDELKOOP H.A.M., SMILDE-VAN DEN DOEL D.A., NEVEN A.K., KAMPHUISEN H.A.C., SPRINGER C.P. - Subjective sleep characteristics of 1,485 males and females aged 50 - 93: Effects of sex and age, and factors related to self-evaluated quality of sleep. *J. Gerontol. A. Biol. Sci. Med. Sci.*, 51A, M108-M115, 1996.

203. MILLS J.N., MINORS D.S., WATERHOUSE J.M. - The effects of sleep upon human circadian rhythms. *Chronobiologia*, 5, 14-27, 1978.

204. MINORS D.S., WATERHOUSE J.M. - The use of constant routines in unmasking the endogenous component of human circadian rhythms. *Chronobiol. Int.*, 1, 205-216, 1984.

205. MINORS D.S., WATERHOUSE J.M. - Deriving a "phase response curve" from adjustment to simulated time zone transitions. *J. Biol. Rhythms*, 9, 275-282, 1994.

206. MISHIMA K., OKAWA M., SHIMIZU T., HISHIKAWA Y. - Diminished melatonin secretion in the elderly caused by insufficient environmental illumination. *J. Clin. Endocrinol. Metab.*, 86, 129-134, 2001.

207. MOLINE M.L., POLLAK C.P., MONK T.H., LESTER L.S., WAGNER D.R., ZENDELL S.M., GRAEBER R.C., SALTER C.A., HIRSCH E. - Age-related differences in recovery from simulated jet lag. *Sleep*, 15, 28-40, 1992.

208. MONK T.H., BUYSSE D.J., CARRIER J., KUPFER D.J. - Inducing jet lag in older people: directional asymmetry. *J. Sleep Res.*, 9, 101-116, 2000.

209. MONK T.H., BUYSSE D.J., REYNOLDS C.F., JARETT D., KUPFER D.J. - Rhythmic vs homeostatic influences on mood, activation, and performance in young and old men. *J. Gerontol. B. Psychol. Sci. Soc. Sci.*, 47(4), 221-227, 1992.

210. MONK T.H., BUYSSE D.J., REYNOLDS C.F., KUPFER D.J. - Inducing jet lag in older people: Adjusting to a 6-hour phase advance in routine. *Exp. Gerontol.*, 28, 119-133, 1993.

211. MONK T.H., BUYSSE D.J., REYNOLDS C.F., KUPFER D.J. - Inducing jet lag in an older person: Directional asymmetry. *Exp. Gerontol.*, 30, 137-145, 1995.

212. MONK T.H., BUYSSE D.J., REYNOLDS C.F., KUPFER D.J., HOUCK P.R. - Circadian temperature rhythms of older people. *Exp. Gerontol.*, 30, 455-474, 1995.

213. MONK T.H., BUYSSE D.J., REYNOLDS C.F., KUPFER D.J., HOUCK P.R. - Subjective alertness rhythms in elderly people. *J. Biol. Rhythms*, 11, 268-276, 1996.

214. MONK T.H., KUPFER D.J. - Circadian rhythms in healthy aging: effects downstream from the pacemaker. *Chronobiol. Int.*, 17, 355-368, 2000.

215. MONK T.H., MOLINE M.L., FOOKSON J.E., PEETZ S.M. - Circadian determinants of subjective alertness. *J. Biol. Rhythms.*, 4, 393-404, 1989.

216. MONTI J.M., ALVARINO F., CARDINALI D., SAVIO I., PINTOS A. - Polysomnographic study of the effect of melatonin on sleep in elderly patients with chronic primary insomnia. *Arch. Gerontol. Geriatr.*, 28, 85-98, 1999.

217. MONTPLAISIR J., LORRAIN J., DESNELE R., PETIT D. - Sleep in menopause: differential effects of two forms of hormonal replacement therapy. *Menopause:The Journal of The North American Menopause Society*, 8, 10-16, 2001.

218. MORETTINI J., GAUDREAU H., LAVOIE H.B., CARRIER J. - Slow-wave activity in the antero-posterior axis in young and middle-aged subjects. *Sleep*, 24 (Abstract Suppl) A234, 2001.

219. MORETTINI J., MASSICOTTE-MARQUEZ J., BARBIER S., GAUDREAU H., LAVOIE H.B., CARRIER J. - Topographical differences in SWA rebound after an acute sleep deprivation in the middle years of life, *Sleep*, 2002, in press.

220. MORGAN K. - Sedative-hypnotic drug use and ageing. *Arch. Gerontol. Geriatr.*, 2, 181-199, 1983.

221. MORGAN K., CLARKE D. - Longitudinal trends in late-life insomnia: implications for prescribing. *Age Ageing,* 26, 179-184, 1997.
222. MORGAN K., GILLEARD C.J., REIVE A. - Hypnotic usage in residential homes for the elderly: a prevalence and longitudinal analysis. *Age Ageing,* 11, 229-234, 1982.
223. MORGAN K., HEALEY D.W., HEALEY P.J. - Factors influencing persistent subjective insomnia in old age: a follow-up study of good and poor sleepers aged 65-74. *Age Ageing,* 18, 117-122, 1989.
224. MORIN C.M., COLECCHI C., STONE J., SOOD R., BRINK D. - Behavioral and pharmacological therapies for late-life insomnia: a randomized, controlled trial. *JAMA,* 281, 991-999, 1999.
225. MORIN C.M., MIMEAULT V., GAGNÉ A. - Nonpharmacological treatment of late-life insomnia. *J. Psychom. Res.,* 46, 103-116, 1999.
226. MOSKO S.S., DICKEL M.J., PAUL T., LATOUR T., DHILLON S., GHANIM A., SASSIN J.F. - Sleep apnea and sleep-related periodic leg movements in community resident seniors. *J. Am. Geriatr. Soc.,* 36, 502-508, 1988.
227. MOURET J., RUEL D., MAILLARD F., BIANCHI M. - Zopiclone versus triazolam in insomniac geriatric patients: a specific increase in delta sleep with zopiclone. *Int. Clin.Psychopharmacol.,* 5, 47-55, 1990.
228. MUEHBACH M.J., WALSH J.K. - The effects of caffeine on simulated night-shift work and subsequent daytime sleep. *Sleep,* 18, 22-29, 1995.
229. MURPHY P.J., CAMPBELL SS. - Enhanced performance in elderly subjects following bright light treatment of sleep maintenance insomnia. *J. Sleep Res.,* 5, 165-172, 1996.
230. MYERS B.L., BADIA P. - Changes in circadian rhythms and sleep quality with aging: Mechanisms and interventions. *Neurosci. Biobehav. Rev.,* 19, 553-571, 1995.
231. National Sleep Foundation. *1998 Women and Sleep Poll.* 1998.
232. NEWMAN A.B., ENRIGHT P.L., MANOLIO T.A., HAPONIK E.F., WAHL P.W. - Sleep disturbance, psychosocial correlates, and cardiovascular disease in 5201 older adults: the Cardiovascular Health Study. *J. Am. Geriatr. Soc.,* 45, 1-7, 1997.
233. NOWELL P.D., REYNOLDS C.F.III, BUYSSE D.J., DEW M.A., KUPFER D.J. - Paroxetine in the treatment of primary insomnia: preliminary clinical and electroencephalogram sleep data. *J. Clin. Psychiatry,* 60, 89-95, 1999.
234. OHAYON M. - Epidemiological study on insomnia in the general population. *Sleep,* 19, S7-S15, 1996.
235. OHAYON M., CAULET M., PHILIP P., GUILLEMINAULT C., PRIEST R.G. - How sleep and mental disorders are related to complaints of daytime sleepiness. *Arch. Int. Med.,* 157, 2645-2652, 1997.
236. OHAYON M., CAULET M., PRIEST R.G., GUILLEMINAULT C. - Psychotropic medication consumption patterns in the UK general population. *J. Clin. Epidemiol.,* 51, 273-283, 1998.
237. OHAYON M.M., ZULLEY J., GUILLEMINAULT C., SMIRNE S., PRIEST R.G. - How age and daytime activities are related to insomnia in the general population: consequences for older people. *J. Am. Geriatr. Soc.,* 49, 360-366, 2001.
238. OWENS J.F., MATTHEWS K.A. - Sleep disturbance in healthy middle-aged women. *Maturitas,* 30, 41-50, 1998.
239. PACK A.I., COLA M.F., GOLDSZMIDT A., OGILVIE M.D., GOTTSCHALK A. - Correlation between oscillations in ventilation and frequency content of the electroencephalogram. *J. Appl. Physiol.,* 72, 985-992, 1992.
240. PACK A.I., SILAGE D.A., MILLMAN R.P., KNIGHT H., SHORE E.T., CHUNG D.C.C. - Spectral analysis of ventilation in elderly subjects awake and asleep. *J. Appl. Physiol.,* 64, 1257-1267, 1988.
241. PADWICK M.L., ENDACOTT J., WHITEHEAD M.I. - Efficacy, acceptability, and metabolic effects of transdermal estradiol in the management of postmenopausal women. *Am. J. Obstet. Gynecol.,* 152, 1085-1091, 1985.
242. PARTINEN M., KAPRIO J., KOSKENVUO M., LANGINVAINO H. - Sleeping habits, sleep quality, and use of sleeping pills: a population study of 31,140 adults in Finland. In Guilleminault,C, Lugaresi,E (Eds.), *Sleep/Wake Disorders: Natural History, Epidemiology, and Long-Term Evolution.* Raven Press, New-York, pp. 29-35, 1983.
243. PASTERNAK R.E., REYNOLDS C.F.III, HOUCK P.R., SCHLERNITZAUER M., BUYSSE D.J., HOCH C.C., KUPFER D.J. - Sleep in bereavement-related depression during and after pharmacotherapy with nortriptyline. *J. Geriatr. Psych. Neurol.,* 7, 71-75, 1994.
244. PERRAS B., MARSHALL L., KOHLER G., BORN J., FEHM H.L. - Sleep and endocrine changes after intranasal administration of growth hormone-releasing hormone in young and aged humans. *Psychoneuroendocrinology,* 24, 743-757, 1999.
245. POLLAK C.P., PERLICK D., LINSNER J.P., WENSTON J., HSIEH F. - Sleep problems in the community elderly as predictors of death and nursing home placement. *J. Community Health,* 15, 123-135, 1990.
246. PORKKA-HEISKANEN T. - Adenosine in sleep and wakefulness, *Ann. Med.,* 31, 125-129, 1999.
247. PORKKA-HEISKANEN T., STRECKER R.E., BJORKUM A.A., THAKKAR M., GREENE R.W., McCARLEY R.W. - Adenosine: a mediator of the sleep-inducing effects of prolonged wakefulness. *Science,* 276, 1265-1268, 1997.
248. PORTAS C.M., THAKKAR M., RAINNIE D.G., GREEN D.M., McCARLEY R.W. - Role of adenosine in behavioral state modulation: a microdialysis study in the freely moving cat. *Neuroscience,* 79, 225-235, 1997.
249. PRINCIPE J.C., SMITH J.R. - Sleep spindle characteristics as a function of age. *Sleep,* 5, 73-84, 1982.
250. PRINZ P.A., BAILEY S.L., WOODS D.L. - Sleep impairments in healthy seniors: roles of stress, cortisol, and interleukin beta. *Chronobiol. Int,* 17, 391-404, 2000.
251. PRINZ P.N. - Sleep and sleep disorders in older adults. *J. Clin. Neurophysiol.,* 12, 139-146, 1995.

252. PURDIE D.W., EMPSON J.A., CRICHTON C., MACDONALD L. - Hormone replacement therapy, sleep quality and psychological well being. *Br. J. Obstetr. Gynaecol.*, 102, 735-739, 1995.
253. QURESHI A.I., GILES W.H., CROFT J.B., BLIWISE D.L. - Habitual sleep patterns and risk for stroke and coronary heart disease: a 10-year follow-up from NHANES I. *Neurology,* 48, 904-911, 1997.
254. RAIHA I., IMPIVAARA O., SEPPALA M., KNUTS L.R., SOURANDER L. - Determinants of symptoms suggestive of gastroesophageal reflux disease in the elderly. *Scand. J. Gastroenterol.,* 28, 1011-1014, 1993.
255. RAIHA I., SEPPALA M., IMPIVAARA O., HYYPPA M.T., KNUTS L.R., SOURANDER L. - Chronic illness and subjective quality of sleep in the elderly. *Aging Clin. Exp. Res.,* 6, 91-96, 1994.
256. RAY W.A., FOUGHT R.L., DECKER M.D.- Psychoactive drugs and the risk of injurious motor vehicle crashes in elderly drivers. *Am. J. Epidemiol.,* 136, 873-883, 1992.
257. RAY W.A., GRIFFIN M.R., DOWNEY W. - Benzodiazepines of long and short half-life and the risk of hip fracture. *JAMA,* 262, 3303-3307, 1989.
258. RAY W.A., GRIFFFIN M.R., SCHAFFNER W., BAUGH D.K., MELTON L.J.3rd. - Psychotropic drug use and the risk of hip fracture. *N. Engl. J. Med.,* 316, 363-369, 1987.
259. REGENSTEIN Q.R., SCHIFF I., TULCHINSKY D., RYAN K.J.- Relationship among estrogen-induced psychophysiological changes in hypogonadal women. *Psychosom Med,* 43, 147-155, 1981.
260. REYNER L.A., HORNE J.A. - Gender-and age-related differences in sleep determined by home- recorded sleep logs and actimetry from 400 adults. *Sleep,* 18, 127-134, 1995.
261. REYNOLDS C.F., BUYSSE D.J., BRUNNER D.P., BEGLEY A.E., DEW M.A., HOCH C.C., HALL M., HOUCK P.R., MAZUMDAR S., PEREL J.M., KUPFER D.J. - Maintenance nortriptyline effects on electroencephalographic sleep in elderly patients with recurrent major depression: double-blind, placebo- and plasma-level-controlled evaluation. *Biol. Psychiatry,* 42, 560-567, 1997.
262. REYNOLDS C.F., HOCH C.C., BUYSSE D.J., MONK T.H., HOUCK P.R., KUPFER D.J. - REM sleep in successful, usual, and pathological aging: The Pittsburgh experience 1980-1993. *J. Sleep Res.,* 2, 203-210, 1993.
263. REYNOLDS C.F.III., MONK T.H., HOCH C.C., JENNINGS J.R., BUYSSE D.J., HOUCK P.R., JARRETT D.B., KUPFER D.J. - Electroencephalographic sleep in the healthy "old old": a comparison with the "young old" in visually scored and automated measures. *J. Gerontol. (Medical)*, 46, M39-M46, 1991.
264. ROBERTS R.E., SHEMA S.J., KAPLAN G.A. - Prospective data on sleep complaints and associated risk factors in an older cohort. *Psychosom. Med.,* 61, 188-196, 1999.
265. RODIN J., McAVAY G., TIMKO C. - A longitudinal study of depressed mood and sleep disturbances in elderly adults. *J. Gerontol. B. Psychol. Sci. Soc. Sci.,* 43, P45-P53, 1988.
266. ROEHRS T., CONWAY W., WITTIG R., ZORICK F., SICKLESTEEL J., ROTH T. - Sleep-wake complaints in patients with sleep related respiratory disturbances. *Am. Rev. Respir. Dis.*, 132, 520-523, 1985.
267. ROEHRS T., ZORICK F., SICKLESTEEL J., WITTIG R., ROTH T. - Age-related sleep-wake disorders at a sleep disorders center. *J. Am. Geriatr. Soc.,* 31, 364-370, 1983.
268. ROEHRS T., ZORICK F., WITTIG R., ROTH T. - Efficacy of a reduced triazolam dose in elderly insomniacs. *Neurobiol. Aging*, 6, 293-296, 1985.
269. RUMBLE R., MORGAN K. - Hypnotics, sleep, and mortality in elderly people. *J. Am. Geriatr. Soc.,* 40, 787 – 791, 1992.
270. SACK R.L., LEWY A., ERB D.L., VOLLMER W.M., SINGER C.M. - Human melatonin production decreases with age. *J. Pineal Res.,* 3, 379-388, 1986.
271. SALIN-PASCUAL R., WAGNER D., UPADHYAYA U., SHIROMAN P.J. - Caffeine decreases sleep in middle-aged and old rats but not young rats. *Sleep,* 23, A53, 2000.
272. SALZARULO P., FAGIOLI I., LOMBARDO P., GORI S., GNERI C., CHIARAMONTI R., MURRI L. - Sleep stages preceding spontaneous awakenings in the elderly. *Sleep Res. Online,* 2,73-77, 1999.
273. SCHARF M.B., MAYLEBEN D.W., KAFFEMAN M., KRALL R., OCHS R. - Dose response effects of zolpidem in normal geriatric subjects. *J. Clin. Psych.,* 52, 77-83, 1991.
274. SCHARF M.B., McDANNOLD M.D., STOVER R., ZARETSKY N., BERKOWITZ D.V. - Effects of estrogen replacement therapy on rates of cyclic alternating patterns and hot-flush events during sleep in postmenopausal women: a pilot study. *Clin. Ther.,* 19, 304-311, 1997.
275. SCHIFF I., REGENSTEIN Q., TULCHINSKY D., RYAN K.J. - Effects of estrogens on sleep and psychological state of hypogonadal women. *JAMA,* 242, 2405-2404, 1979.
276. SCHWIERIN B., ACHERMANN P., DEBOER T., OLEKSENKO A., BORBELY A.A. - Regional differences in the dynamics of the cortical EEG in the rat after sleep deprivation. *Clin. Neurophysiol.,* 110, 869-875, 1999.
277. SCHWIERIN B., BORBELY A.A., TOBLER I. - Effects of N^6-cyclopentyladenosine and caffeine on sleep regulation in the rat. *Eur. J. Pharmacol.,* 300, 163-171, 1996.
278. SHAVER J., GIBLIN E., LENTZ M. LEE K. - Sleep patterns and stability in perimenopausal women. *Sleep,* 11, 556-561, 1988.
279. SHAVER J.L., PAULSEN V.M. - Sleep, psychological distress, and somatic symptoms in perimenopausal women. *Fam. Pract. Res. J.,* 13, 373-384, 1993.
280. SHAW S.H., CURSON H., COQUELIN J.P. - A double-blind, comparative study of zolpidem and placebo in the treatment of insomnia in elderly psychiatric in-patients. *J. Int. Med. Res.,* 20, 150-161, 1992.
281. SHIROMANI P.J., LU J., WAGNER D., THAKKAR J., GRECO M.A., BASHEER R., THAKKAR M. - Compensatory sleep response to 12 h wakefulness in young and old rats. *Am. J. Physiol.,* 278, R125-R133, 2000.
282. SHOCHAT T., MARTIN J., MARLER M., ANCOLI-ISRAEL S. - Illumination levels in nursing home patients: effects on sleep and activity rhythms. *J. Sleep Res.,* 9, 373-379, 2000.

283. SMITH M.T., PERLIS M.L., PARK A., SMITH M.S., PENNINGTON J., GILES D.E., BUYSSE D.J. - Comparative meta-analysis of pharmacotherapy and behavior therapy for persistent insomnia. *Am. J. Psychiatry,* 159, 5-11, 2002.
284. SMULDERS F.T., KENEMANS J.L., JONKMAN L.M., KOK A. - The effects of sleep loss on task performance and the electroencephalogram in young and elderly subjects. *Biol. Psychol.,* 45, 217-239, 1997.
285. SOYKA M., BOTTLENDER R., MOLLER H.J. - Epidemiological evidence for a low abuse potential of zolpidem. *Pharmacopsychiatry,* 33, 138-141, 2000.
286. SPATH-SCHWALBE E., GOFFERJE M., KERN W., BORN J., FEHM H.L. - Sleep disruption alters nocturnal ACTH and cortisol secretory patterns. *Biol. Psychiatry,* 29, 575-584, 1991.
287. SPIEGEL K., LEPROULT R., VAN CAUTER E. - Impact of sleep debt on metabolic and endocrine function, *Lancet,* 354, 1435-1439, 1999.
288. STERIADE M., McCORMICK D.A., SEJNOWSKI T.J. - Thalamocortical oscillations in the sleeping and aroused brain. *Science,* 262, 679-685, 1993.
289. STONE A.B., PEARLSTEIN T.B. - Evaluation and treatment of changes in mood, sleep, and sexual functioning associated with menopause. *Obstet. Gynecol. Clin. North Am.,* 21, 391-403, 1994.
290. STONE B.M., TURNER C., MILLS S.L., NICHOLSON A.N. - Hypnotic activity of melatonin, *Sleep,* 23, 663-664, 2000.
291. STRECKER R.E., MORAIRTY S., BASHEER R., THAKKAR M.M., PORKKA-HEISKANEN T., DAUPHIN L.L., RAINNIE D.G., PORTAS C.M., GREENE R.W., McCARLEY R.W. - Adenosinergic modulation of basal forebrain and preoptic/anterior hypothalamic neuronal activity in the control of behavioral state. *Behav. Brain Res.,* 115, 183-204, 2000.
292. THOMSON J., OSWALD I. - Effects of estrogen on the sleep, mood and anxiety of menopausal women. *BMJ,* 2, 1317-1319, 1977.
293. TINETTI M., SPEECHLEY M., GINTER S.F. - Risk factors for falls among elderly persons living in the community. *N. Engl. J. Med.,* 319, 1701-1707, 1988.
294. TOUITOU Y., FEVRE M., BOGDAN A., REINBERG A., DE PRINS J., BECK H., TOUITOU C. - Patterns of plasma melatonin with ageing and mental condition: stability of nycthemeral rhythms and differences in seasonal variations. *Acta Endocrinologica,* 106, 145-151, 1984..
295. TUNE G.S. - Sleep and wakefulness in 509 normal human adults. - *Br. J. Med. Psychol.,* 42, 75-80, 1969.
296. TUREK F.W., PENEV P., ZHANG Y., VAN REETH O., TAKAHASHI J.S., ZEE P. - Alterations in the circadian system in advanced age. In Chadwick DJ, Ackrill K (eds.), *Circadian Clocks and their Adjustment.* John Wiley & Sons, Chichester, pp. 212-232, 1995.
297. TZISCHINSKY O., LAVIE P. - Melatonin possesses time-dependent hypnotic effects, *Sleep,* 17, 638-645, 1994.
298. USUI A., ISHIZUKA Y., MATSUSHITA Y., KUKUZAMA H., KANBA S. - Bright light treatment for night-time insomnia and daytime sleepiness in elderly people: comparison with a short acting hypnotic. *Psychiatry Clin. Neurosci.,* 54, 374-376, 2000.
299. UTIAN W.H. - Menopause - a proposed new functional definition. *Maturitas,* 14,1-2, 1991.
300. VAN CAUTER E., LEPROULT R., KUPFER D.J. - Effects of gender and age on the levels and circadian rhythmicity of plasma cortisol. *J. Clin. Endocrinol. Metab.,* 81, 2468-2473, 1996.
301. VAN CAUTER E., LEPROULT R., PLAT L. - Age-related changes in slow wave sleep and REM sleep and relationship with growth hormone and cortisol levels in healthy men, *JAMA,* 284, 861-868, 2000.
302. VAN CAUTER E., PLAT L., SCHARF M.B., LEPROULT R., CESPEDES S., L'HERMITE-BALERIAUX M., COPINSCHI G. - Simultaneous stimulation of slow-wave sleep and growth hormone secretion by gamma-hydroxybutyrate in normal young men. *J. Clin.Invest.,* 100, 745-753, 1997.
303. VAN CAUTER E., SPIEGEL K. - Circadian and sleep control of hormonal secretions. In Turek FW, Zee PC (e ds.), *Regulation of sleep and circadian rhythms,* Marcel Dekker, New-York, pp. 397-425, 1999.
304. VAN COEVORDEN A., MOCKEL J., LAURENT E., KERKHOFS M., NEVE P., VAN CAUTER E., L'HERMITE-BALERIAUX M., DECOSTER C. - Neuroendocrine rhythms and sleep in aging men. *Am. J. Physiol.,* 260, E651-E661, 1991.
305. VAN REETH O., WEIBEL L., SPIEGEL K., LEPROULT R., DUGOVIC C., MACCARI S. - Interactions between stress and sleep: from basic research to clinical situations. *Sleep Med. Rev,* 4, 201-219, 2000.
306. VGONTZAS A.N., BIXLER E., WITTMAN A.M., ZACHMAN K., LIN H.M., VELA-BUENO A., KALES A.,CHROUSOS G.P. - Middle-aged men show higher sensitivity of sleep to the arousing effects of corticotropin-releasing hormone than young men: clinical implications. *J. Clin. Endocrinol. Metab.,* 86, 1489-1495, 2001.
307. VGONTZAS A.N., KALES A., BIXLER E.O., MYERS D.C. - Temazepam 7.5 mg: Effects on sleep in elderly insomniacs. *Eur. J. Clin. Pharmacol.,* 46, 209-213, 1994.
308. VIGNOLA A., LAMOUREUX C., BASTIEN C.H., MORIN C.M. - Effects of chronic insomnia and use of benzodiazepines on daytime performance in older adults. *J. Gerontol. B. Psychol. Sci. Soc. Sci.,* 55B, P54-P62, 2000.
309. VITIELLO M.V., SMALLWOOD R.G., AVERY D.H., PASCUALY R.A., MARTIN D.C., PRINZ P.A. - Circadian temperature rhythms in young adult and aged men. *Neurobiol. Aging,* 7, 97-100, 1986.
310. VOGEL G.W., MORRIS D. - The effects of estazolam on sleep, performance, and memory: A long-term sleep laboratory study of elderly insomniacs. *J. Clin. Pharmacol.,* 32, 647-651, 1992.
311. VOJTECHOVSKY M., BREZINOVA V., SIMANE Z., HORT V. - An experimental approach to sleep and aging. *Human Development,* 12, 64-72, 1969.
312. WALDHAUSER F., KOVACS J., REITER E. - Age-related changes in melatonin levels in humans and its potential consequences for sleep disorders. *Exp. Gerontol.,* 33, 759-772, 1998.

313. WALDHAUSER F., WEISZENBACHER G., TATZER E., GISINGER B., WALDHAUSER M., SCHEMPER M., FRISCH H. - Alterations in nocturnal serum melatonin levels in humans with growth and aging. *J. Clin. Endocrinol. Metab.,* 66, 648-652, 1988.
314. WALSH J.K., SCHWEITZER P.K. - Ten-year trends in the pharmacological treatment of insomnia. *Sleep,* 22, 371-375, 1999.
315. WANG P.S., BOHN R.L., GLYNN R.J., MOGUN H., AVORN J. - Zolpidem use and hip fractures in older people. *J. Am. Geriatr. Soc.,* 49, 1685-1690, 2001.
316. WARE J.C., McBRAYER R.H., SCOTT J.A. - Influence of sex and age on duration and frequency of sleep apnea event. *Sleep,* 23, 165-170, 2000.
317. WARE J.C., PITTARD J.T., MOORAD P.J., FRANKLIN D. - Reduction of sleep apnea, myoclonus and bruxism in stage 4 sleep. *Sleep Res.,* 11, 181, 1982.
318. WATERHOUSE J., EDWARDS B., MUGARZA J., FLEMMING R., MINORS D., CALBRAITH D., WILLIAMS G., ATKINSON G., REILLY T. - Purification of masked temperature data from humans: some preliminary observations on a comparison of the use of an activity diary, wrist actimetry, and heart rate monitoring. *Chronobiol. Int.,* 16, 461-475, 1999.
319. WEBB P. - Periodic breathing during sleep. *J. Appl. Physiol.,* 37, 899-903, 1974.
320. WEBB W.B. - The measurement and characteristics of sleep in older persons. *Neurobiol. Aging,* 3, 311-319, 1982.
321. WEBB W.B. - A further analysis of age and sleep deprivation effects. *Psychophysiology,* 22 (2), 156-161, 1985..
322. WEBB W.B. - Age-related changes in sleep. *Clin. Geriatr. Med.,* 5, 275-285, 1989.
323. WEBB W.B., LEVY C.M. - Age, sleep deprivation, and performance., *Psychophysiology,* 19, 272-276, 1982.
324. WEI H.G., RIEL E., CZEISLER C.A., DIJK D.J. - Attenuated amplitude of circadian and sleep-dependent modulation of electroencephalographic sleep spindle characteristics in elderly human subjects. *Neurosci Lett,* 260, 29-32, 1999.
325. WEIBEL L., FOLLENIUS M., SPIEGEL K., EHRHART J., BRANDENBERGER G. - Comparative effect of night and daytime sleep on the 24-hour cortisol secretory profile. *Sleep,* 18, 549-556, 1995.
326. WEITZMAN E.D., MOLINE M.L., CZEISLER C.A., ZIMMERMAN J.C. - Chronobiology of aging: temperature, sleep-wake rhythms and entrainment. *Neurobiol. Aging,* 3, 299-309, 1982.
327. WELSTEIN L., DEMENT W.C., REDINGTON D., GUILLEMINAULT C., MITLER M.M. - Insomnia in the San Francisco Bay area: a telephone survey. In Guilleminault C, Lugaresi E (eds.), *Sleep/Wake Disorders: Natural History, Epidemiology, and Long-Term Evolution,* pp. 73-85, 1983.
328. WERTH E., ACHERMANN P., BORBELY A.A. - Brain topography of the human sleep EEG: antero-posterior shifts of spectral power. *Neuroreport,* 8, 123-127, 1996.
329. WERTH E., ACHERMANN P., BORBELY A.A. - Fronto-occipital EEG power gradients in human sleep. *J. Sleep Res.,* 6, 102-112, 1997.
330. WIKLUND I., BERG G., HAMMAR M., KARLBERG J., LINDGREN R., SANDIN K. - Long-term effect of transdermal hormonal therapy on aspects of quality of life in postmenopausal women. *Maturitas,* 14, 225-236, 1992.
331. WIKLUND I., HOLST J., KARLBERG J., MATTSSON L.A., SAMSIOE G., SANDIN K., UVEBRANT M., VON SCHOULTZ B. - A new methodological approach to the evaluation of quality of life in postmenopausal women. *Maturitas,* 14, 211-224, 1992.
332. WILLIAMS R.L., KARACAN I., HURSCH C.J. - *Electroencephalography (EEG) of Human Sleep: Clinical Applications,* John Wiley & Sons, New York, 1974.
333. WINGARD D.L., BERKMAN L.F. - Mortality risk associated with sleeping patterns among adults. *Sleep,* 6, 102-107, 1983.
334. WOODWARD S., FREEDMAN R.R. - The thermoregulatory effects of menopausal hot flashes on sleep. *Sleep,* 17, 497-501, 1994.
335. WURTMAN R.J., ZHADANOVA I. - Improvement of sleep quality by melatonin. *Lancet* 346[8988], 1491, 1995.
336. WURTS S.W., EDGAR D.M. - Caffeine during sleep deprivation: sleep tendency and dynamics of recovery sleep in rats. *Pharmacol. Biochem. Behav.,* 65, 155-162, 2000.
337. YOUNG T. - Menopause, hormonal replacement therapy, and sleep-disordered breathing, *Am. J. Respir. Crit. Care Med.,* 163, 597-601, 2001.
338. YOUNGSTEDT S.D., KRIPKE D.F., KLAUBER M.R., SEPULVEDA R.S., MASON W.J. - Periodic leg movements during sleep and sleep disturbances in elders. *J. Gerontol. A. Biol. Sci. Med. Sci.,* 53, M391-M394, 1998.
339. YUNUS M.B., HOLT G.S., MASI A.T., ALDAG J. - Fibromyalgia syndrome among the elderly: comparison with younger patients. *J. Am. Geriatr. Soc.,* 36, 987-995, 1998.
340. ZEITER J.M., DANIELS J.E., DUFFY J.F., KLERMAN E.B., SHANAHAN T.L., DIJK D.J., CZEISLER C.A. - Do plasma melatonin concentrations decline with age? *Am. J. Med.,* 107, 432-436, 1999.
341. ZHDANOVA I., WURTMAN R.J., REGAN M.M., TAYLOR J.A., SHI J.P., LECLAIR O.U. - Melatonin treatment for age-related insomnia. *J. Clin. Endocrinol. Metab.,* 86, 4727-4730, 2001.
342. ZHDANOVA I.V., WURTMAN R.J., MORABITO C., PIOTROVSKA V.R., LYNCH H.J. - Effects of low oral doses of melatonin given 2-4 hours before habitual bedtime on sleep in normal young subjects. *Sleep,* 19, 423-431, 1996.
343. ZULLEY J., WEVER R., ASCHOFF J. - The dependence of onset and duration of sleep on the circadian rhythm of rectal temperature. *Pflugers Arch.,* 391, 314-318, 1981.

Chapter 25

Hypersomnias
Introduction

M. Billiard
Service de Neurologie B, Hôpital Gui de Chauliac, Montpellier, France

Falling asleep reading a paper or a book, or in conversation with friends, hearing only part of a lecture, falling asleep at the wheel, sleeping 12 hours a night and being unable to wake up in the morning, are all abnormal situations testifying to hypersomnia. But where do the limits lie? Hypersomnia begins at the point when the subject feels that his everyday life is affected, or else does not realise what is happening but suffers from the consequences in the form of car or machine accidents or reduced performance of any kind.

With the exception of a few articles on pathological entities regarded as rare: narcolepsy from 1880 onwards [7] recurrent hypersomnia from 1925 [13], lethargic encephalitis from 1929 [23] and Pickwick's syndrome from 1956 [2], the medical literature devoted to hypersomnia is exceptionally recent. As Lavie points out [15], Kleitman's landmark work *Sleep and Wakefulness*, first published in 1939 [14] contains only 298 references to hypersomnia out of a total of 4,000. Moreover the first sleepiness scale dates from 1973 [10]. It is thus only in the last quarter of the 20th century that sleep specialists began to show an active interest in the different forms of hypersomnia and to attempt to make these known to the medical community.

Unlike insomnia, which is a widespread phenomenon, familiar to patients and doctors alike, and subject to a whole range of therapeutic, pharmacological and non pharmacological treatment, hypersomnias are often poorly understood, overlooked and considered as rare or even exceptional, for which therapeutic means remain limited. In reality, hypersomnias are neither rare nor exceptional, even affecting a significant part of the population. Numerous epidemiological studies stand as proof of this, with prevalence varying according to populations, age, and sex but chiefly in terms of methods of investigation and the framework of questionnaires and interviews (table 25.1). Generally speaking, 4 to 6% of the population is affected by severe hypersomnia, and 15 to 20% by moderate hypersomnia, with women appearing to be slightly more concerned than men.

This is far higher than one would expect. Patients affected by narcolepsy or sleep apnoea syndrome have a far higher rate of accidents than control subjects [5]. In a recent study [18], involving 4002 drivers, 145 of those questioned (3.6%) admitted to regularly feeling sleepy at the wheel. Indeed these subjects had a significantly higher rate of traffic accidents than controls (adjusted odds ratio13.3, confidence interval 4.1 to 43).

Cognitive functions are equally affected, especially memory [22]. Sleep deprivation studies have clearly determined the relationship between sleepiness and memory disorders [3].

Finally, in a population of 3962 subjects of over 65 years, monitored for 4 years, the mortality rate was shown to be 1.73 times higher in subjects who usually had daytime sleep episodes [9] than with controls.

Table 25.1. Prevalence of hypersomnia

Authors	N (gender)	Study population (age range)	Description	Method	Prevalence
Partinen (1982)	2537 (m)	Army draftees (17- 29)	Falling asleep at work	Questionnaire	6.4
Partinen and Rimpelä (1982)	2016 (m, f)	Population sample (15 – 64)	Involuntary sleep attacks	Telephone interview	3.4 (m) 2.5 (f)
Lugaresi et al. (1983)	5713 (m, f)	Population sample (3 – 94)	Sleepiness independent of meal times	Direct interview	8.7
Billiard et al. (1987)	58162 (m)	Army draftees (17 – 22)	Irresistible daily sleep episodes	Questionnaire	4.1
Gislasson and Almqvist (1987)	3201 (m)	Population sample (30 – 69)	Moderate- severe sleepiness	Questionnaire	moderate: 16.7 severe: 5.7
Liljenberg et al. (1988)	3557 (m, f)	Population sample (30 – 65)	Daytime sleepiness, often, very often	Questionnaire	5.2 (m) 5.5 (f)
Ford and Kamerow (1989)	7954 (m, f)	Population sample (18 – 65)	Sleeping too much over 2 weeks or more	Direct structured interview	2.8 (m) 3.5 (f)
Janson et al. (1995)	2394 (m, f)	Population sample (20 – 44)	Daytime drowsiness 3 times per week	Questionnaire	16.0
Hays et al. (1996)	3962 (m, f)	Population sample (65 – 85)	Frequent sleepiness and need to take a nap	Interview	25.2
Hublin et al. (1996)	11354 (m, f)	Population sample (33 – 66)	Daytime sleepiness, daily or almost	Questionnaire	6.7 (m) 11.0 (f)
Enright et al. (1996)	5201 (m, f)	Population sample > 65	Usually sleepy during the day	Questionnaire	17.0 (m) 15.0 (f)
Ohayon et al. (1997)	4952 (m, f)	Population sample (15 – 100)	Daytime sleepiness every day	Telephone interview	moderate: 15.2 severe 4.4 (m) 6.6 (f)

Hypersomnia reveals itself in many ways. It may take the form of more or less irresistible attacks of sleep occurring at various times of the day, more or less permanent drowsiness interfering with daily activities, abnormally prolonged night sleep with major difficulty emerging from sleep in the morning, an irresistible need to sleep in the morning or, alternatively, in the second half of the afternoon, or even periods of more or less continuous sleep lasting several days and recurring at intervals of several months. It may be recognised by the patient or on the contrary overlooked or even denied. It may be provoked by factors which are partially or totally controlled by the subject, and its pathological entities may be isolated, dependent on other illnesses or in response to circadian rhythm or sleep disorders (table 25.2).

Table 25.2. Aetiologies of hypersomnia

Provoked		Spontaneous	
	Primary	Secondary	Circadian rhythm sleep disorders
Insufficient sleep syndrome	OSAS	Associated with:	Delayed sleep phase syndrome
Shift work sleep disorder	Upper airway resistance	mental disorders	
Time zone change (jet lag) syndrome	syndrome	neurological disorders	Advanced sleep phase syndrome
	Narcolepsy	infectious disorders	
Hypnotic dependent sleep disorder	Idiopathic hypersomnia	metabolic disorders	Irregular sleep-wake pattern
	Recurrent hypersomnia	endocrine disorders	Non-24 hour sleep-wake
Toxin induced sleep disorder		Post-traumatic hypersomnia	disorder

Until recently, hypersomnias were difficult to quantify in any way. But over the past 25 years, increasingly accurate questionnaires, subjective scales and objective tests of sleepiness (see chapter 13) have provided general practitioners and specialists with the necessary tools not only to confirm clinically suspected sleepiness, but also to assess its severity and recognise its aetiology.

The mechanisms and causes of hypersomnia are somewhat equivocal. Although simple in cases of induced hypersomnia, they are far more complex in primary and secondary pathologies or within the framework of circadian rhythm sleep disorders. The main pathophysiological advances have been made in hypersomnias such as narcolepsy where an animal model was available. From a pathogenic point of view, the genetic breakthrough in the field of sleep has brought major strides in the development of our understanding.

This used to be limited to amphetamines and their derivatives which have not inconsiderable adverse effects. Moreover, due to incidents of abuse, these medications have gradually been taken off the market, resulting in a virtual therapeutic void. But the last twenty years have seen a remarkable development in quasi-revolutionary technology, including continuous nocturnal ventilation devices and velopharynx or facial surgery, oral devices used for sleep-related respiration disorders; new non-amphetamine wakening medication, modafinil, which is both effective and very well tolerated in primary hypersomnia; and new approaches such as chronotherapy, luxtherapy and melatonin in cases of circadian sleep rhythm disorders.

Thus considerable progress has been made in understanding hypersomnias, in recent decades; their former status as curiosities, has been replaced by that of widespread pathologies open to medical investigation and specific treatment.

REFERENCES

1. BILLIARD M., ALPEROVITCH A., PEROT C., JAMMES A. – Excessive daytime somnolence in young men: prevalence and contributing factors. *Sleep,* 10, 297-305, 1987.
2. BURWELL C.S., ROBIN E.D., WHALEY R.D., BICKELMANN A.G. – Extreme obesity with alveolar hypoventilation: a Pickwickian syndrome. *Am. J. Med.,* 21, 811-818, 1956.
3. DINGES D.F., KRIBBS N.B. - Performing while asleep: effects of experimentally-induced sleepiness. In: Monk T. (ed), *Sleep, Sleepiness and Performance.* John Wiley & Sons. New York, NY. 97-128, 1991.
4. ENRIGHT P.L., NEWMAN A.B., WAHL P.W., MANOLIO T.A., HAPONIK E.F., BOYLE P.J.R.- Prevalence and correlates of snoring and observed apneas in 5201 older adults. *Sleep,* 19, 531-538, 1996.
5. FINDLEY L.J., UNVERZAGT M.E., GUCHU R., FABRIZIO M., BUCKNER J., SURATT P. - Vigilance and automobile accidents in patients with sleep apnea or narcolepsy. *Chest,* 108, 619-624, 1995.
6. FORD D.E., KAMEROW D.B. – Epidemiologic studies of sleep disturbances and psychiatric disorders. *JAMA,* 262, 1479-1484, 1989.
7. GELINEAU J. – De la narcolepsie. *Gaz. des Hôp. Paris.* 55, 626-628, 635-637, 1880.
8. GISLASON T., ALMQVIST M. – Somatic diseases and sleep complaints: an epidemiological study of 3201 Swedish men. *Acta Med Scand.* 221, 475-4 81, 1987.
9. HAYS J.C., BLAZER D.G., FOLEY D.J. – Risk of napping: excessive daytime slepiness and mortality in an older community population. *J. Am. Geriatr. Soc.* 44, 693-698, 1996.
10. HODDES E., ZARCONE V., SMYTHE H., PHILLIPS R., DEMENT W.C. – Quantification of sleepiness: A new approach. *Psychophysiol.,* 10, 431-436, 1973.
11. HUBLIN C., KAPRIO J., PARTINEN M., KOSKENVUO M., HEIKKILA K., KOSKIMIES S., GUILLEMINAULT C. - The prevalence of narcolepsy: an epidemiological study of the Finnish Twin Cohort. *Ann. Neurol.* 35, 709-716, 1994.
12. JANSON C., GISLASON T., DE BACKER W., PLASCHKE P., BJORNSSON E., HETTA J., KRISTBJARNASON H., VERMEIRE P., BOMAN G. – Prevalence of sleep disturbances among young adults in three European countries. *Sleep* 18, 589-597, 1995.
13. KLEINE W. – Periodische Schlafsucht. *Mschr. Psychiat. Neurol.* 57, 285-320, 1925.
14. KLEITMAN N. – *Sleep and Wakefulness.* University of Chicago Press. Chicago. 1939.
15. LAVIE P. - The touch of Morpheus. Pre-20th century accounts of sleepy patients. *Neurology.* 41, 1841-1844, 1991.
16. LILJENBERG B., ALMQVIST M., HETTA J., ROOS B.E., AGREN H. - The prevalence of insomnia: the importance of operationally defined criteria. *Ann. Clin. Res.* 20, 393-398, 1988.
17. LUGARESI E., CIRIGNOTTA F., ZUCCONI M., MONDINI S., LENZI P.L., COCCAGNA G. – Good and poor sleepers: an epidemiological survey of the San Marino population. *In: Sleep-Wake Disorders: Natural History, Epidemiology and Long-term Evolution,* C. Guilleminault and E. Lugaresi (eds), Raven Press. New York, 1-12, 1983.
18. MASA J.F., RUBIO M., FINDLEY L.J. – Habitually sleepy drivers have a high frequency of automobile crashes associated with respiratory disorders during sleep. *Am. J. Respir. Crit. Care Med.* 162, 1407-1412, 2000.

19. OHAYON M.M., CAULET M., PHILIP P., GUILLEMINAULT C., PRIEST R.G. - How sleep and mental disorders are related to complaints of daytime sleepiness. *Arch. Intern. Med.* 157, 2645-2652, 1997.

20. PARTINEN M. – Sleeping habits and sleep disorders in Finnish men, before, during and after military service. *Ann. Med. Milit. Fenn.* 57, (suppl 1), 1-96, 1982.

21. PARTINEN M., RIMPELÄ M. – Sleeping habits and sleep disorders in a population of 2016 Finnish adults. *Yearbook of Health Education Research.* Helsinki, Finland: National Board of Health. 253-260, 1982.

22. ROEHRS T.A., MERRION M., PEDROSI B., STEPANSKI E., ZORICK F., ROTH T. Neuropsychological function in obstructive sleep apnea syndrome (OSAS) compared to chronic obstructive pulmonary disease (COPD). *Sleep,* 18, 382-388, 1995.

23. VON ECONOMO C.- *Encephalitis lethargica.* Urban Schwarzenberg, Wien, 1929.

Chapter 26

A decision tree approach to the differential diagnosis of hypersomnia

M. Billiard

Service de Neurologie B, Hôpital Gui de Chauliac, Montpellier, France

INTRODUCTION

Compared to the differential diagnosis of insomnia, that of hypersomnia is generally simpler, as there is usually a single, non multiple cause, unlike what often occurs in insomnia.

Two points should be emphasised:
- A positive diagnosis of hypersomnia is less evident than in the case of insomnia. Indeed, while some subjects may consult for sleepiness which is more or less debilitating, others may complain of fatigue or a totally different disorder such as snoring or nycturia, while others do not even think to consult, but a drop in performance or repeated accidents will come to the attention of a physician - often the work physician - or else the family might notice the patient falling asleep at idle moments during the day.
- A polysomnographic recording is far more often indicated than in the case of insomnia.

DIAGNOSTIC CONSIDERATIONS OF THE DECISION TREE

The decisional approach should be preceded by an analysis of the type of sleepiness, an evaluation of its severity and a clinical examination seeking signs of any associated pathology.

Analysis of the type of sleepiness

- Abnormal sleep episodes which are more or less irresistible, interspersed by periods of normal wakefulness, often when the subject sits down for a moment, when he drives the car, when he watches a show, or even in totally unpredictable situations.
- More or less permanent sleepiness associated with longer nocturnal sleep time and consistently difficult awakening.
- Recurrent sleepiness, for periods of roughly a week, at intervals of several months.
- Irregular sleepiness

The severity of sleepiness

This is ascertained by clinical interview or in relation to subjective scales i.e., the Stanford sleepiness scale, Epworth sleepiness scale, Karolinska sleepiness scale.

Clinical interview and testing

In search of an associated pathology, whether neurological, psychiatric, infectious, metabolic, endocrine etc.

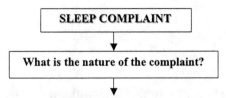

1. **Is the complaint of excessive daytime sleepiness directly related to:**

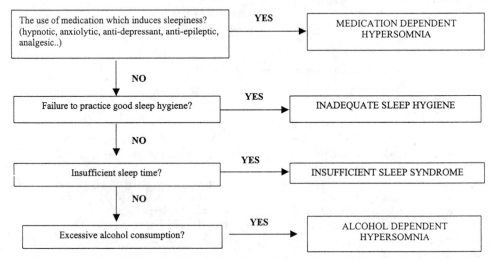

2. **Is the complaint of excessive daytime sleepiness independent of the intake of medication or alcohol, of any behaviour incompatible with sleep, or of any disease (primary hypersomnia):**

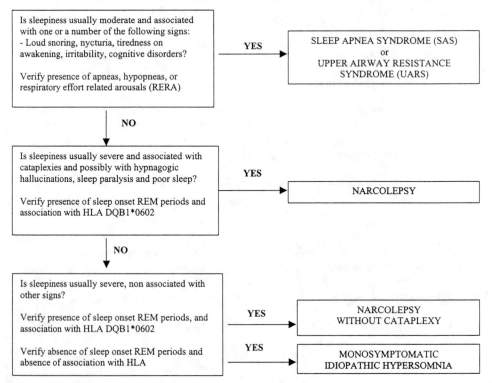

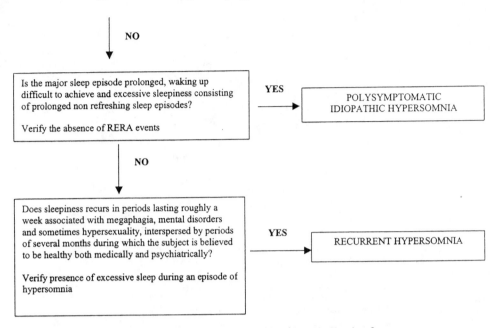

NO

Is the major sleep episode prolonged, waking up difficult to achieve and excessive sleepiness consisting of prolonged non refreshing sleep episodes?

Verify the absence of RERA events

YES → POLYSYMPTOMATIC IDIOPATHIC HYPERSOMNIA

NO

Does sleepiness recurs in periods lasting roughly a week associated with megaphagia, mental disorders and sometimes hypersexuality, interspersed by periods of several months during which the subject is believed to be healthy both medically and psychiatrically?

Verify presence of excessive sleep during an episode of hypersomnia

YES → RECURRENT HYPERSOMNIA

3. **Sleepiness is detected by a doctor or a nurse during hospitalisation for a neurological, psychatric, infectious, metabolic, endocrine, post-traumatic condition.**

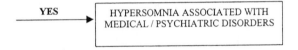

YES → HYPERSOMNIA ASSOCIATED WITH MEDICAL / PSYCHIATRIC DISORDERS

Chapter 27

Insufficient sleep syndrome

M. S. Aldrich *(deceased)*
Department of Neurology, University of Michigan, Ann Arbor, MI, USA

Partial sleep deprivation, or insufficient sleep, has occurred for millennia. Physicians, particularly physicians in training, lose a great deal of sleep. Soldiers suffer partial sleep deprivation during military campaigns; indeed, the military science of sleep logistics involves balancing the cost of man-hours lost to sleep against the beneficial effects of sleep on performance and morale [23]. Insufficient sleep has accompanied pregnancy, child-rearing, and chronic illness throughout humanity's existence.

Insufficient sleep is probably more common in the twentieth century than in earlier times. By some estimates, the average person sleeps about one hour less now than in 1910. A major factor contributing to the reduction in sleep hours is the ready availability of electrical power and artificial light. Electric light bulbs and television contribute to later hours of wakefulness at home and electric power for factories permits greater use of shift work and night work, all of which lead to a reduction in sleep hours. Annual work hours increased by about 10% between 1969 and 1987, and the proportion of the population involved in shift work is probably higher now than ever before [5, 31]. As a result, there is less time for sleep: working adults sleep 7-8 hours/night on average, compared to 8- 9 hours for working adults 40-50 years ago, and feel substantially less rested in the morning [5]. Thus, many otherwise healthy adults accumulate a sleep debt as a result of chronically insufficient amounts of sleep [36].

With the development of clinics devoted specifically to diagnosis of sleep disorders, it became apparent that some of the patients who complain of excessive sleepiness and of repeated episodes of falling asleep during the day have neither narcolepsy nor other pathological sleep disturbances; instead, they are chronically short of sleep. Recognition of such patients led to the formal definition of the insufficient sleep syndrome (ISS) in the diagnostic classification of sleep and arousal disorders published by the Association of Sleep Disorders Centers in 1979 [2].

The current version of this classification, the International Classification of Sleep Disorders [18] includes ISS as an extrinsic sleep disorder - a disorder that originates or develops from causes outside the body - and defines it as a "disorder that occurs in an individual who persistently fails to obtain sufficient nocturnal sleep required to support normally alert wakefulness....The individual engages in voluntary, albeit unintentional, chronic sleep deprivation." [11]. A further requirement for diagnosis is that the person is unaware that increased sleep would alleviate symptoms.

Actually there are two definitions of ISS. One definition requires a complaint of excessive sleepiness in an individual with persistently insufficient sleep to support full alertness [23], and so limits the disorder to patients with a complaint of sleepiness. A broader definition requires only chronically insufficient sleep to support full alertness, and so includes millions of persons who suffer without realising it from sleepiness due to chronic insufficient sleep at night. These persons either assume that afternoon or evening sleepiness is normal or do not recognise it as a problem. The demands of modern life (work, school, family, recreation), the pressure to "get ahead" in one's career, the feeling of "not enough hours in the day" all lead to insufficient sleep.

EPIDEMIOLOGY

In our experience and as reported by others, the ISS accounts for about 2% of all patients presenting to sleep disorders centres and for 6% of patients presenting to sleep disorders centres with complaints of excessive sleepiness [1, 38]. Sixty-three percent of a large series of subjects with ISS presenting to a sleep disorder centre were men, 93% were employed, and the mean age was 40 years with a mean age of onset of symptoms of 27 years [29]. In another series, the average education level was 14 years [12].

The prevalence of ISS using the broader definition that does not require a presenting complaint of sleepiness is much higher. In a recent survey of 220 middle-aged employed men and women without habitual snoring or sleep disordered breathing, 16% of women and 5% of men reported that 2 or more days per week they suffered from uncontrollable sleepiness that interfered with activities [38]. In most of these subjects, sleepiness was probably due to insufficient sleep. Although the prevalence data are not entirely clear, it is likely that insufficient sleep is the leading cause of excessive daytime sleepiness in Western societies.

The psychosocial factors that contribute to the development of ISS include long work hours and demanding family responsibilities. Persons who work two jobs or who are expected to work overtime on a regular basis are at risk for ISS. Shift workers are particularly likely to have insufficient sleep: as they change from day shift to night shift the circadian rhythms cannot adjust quickly enough and sleep-wake patterns are disrupted. Persons who work full-time and raise small children or care for elderly infirm relatives may also develop ISS. In some cases a particular life event such as the birth of a child or change in job responsibilities is associated with the onset of symptoms. Personality factors probably play a role as well. Some persons enjoy the evening hours and may stay up late despite the need to be at work early in the morning.

CLINICAL FEATURES

The principal feature of the insufficient sleep syndrome is daytime sleepiness. In persons with a major sleep period at night, sleepiness is usually most apparent either in the afternoon and early evening, or following meals. Although afternoon drowsiness (especially after a large lunch) is so common that many people believe it is a normal phenomenon, falling asleep during the afternoon, even in boring passive situations, is not "normal" in the sense that fully rested persons do not do so.

Awakening later on weekends or on days off is another common feature of ISS [12, 29]. In one series, patients diagnosed with ISS slept an average of 6.4 hours during the week and 8-8.5 hours on weekends [29].

As with other causes of daytime drowsiness, ISS can be associated with significant decrements in daytime performance, particularly with tasks that require close attention and vigilance [28]. People with the ISS may find it difficult to wake up in the morning, may require alarm clocks to awaken, and may take longer to "get going". Confusion, disorientation, and grogginess upon awakening, a complex of sensations sometimes referred to as sleep inertia or sleep drunkenness, is common [30].

Other symptoms that may accompany ISS include irritability, difficulty with concentration, reduced vigilance, depression, fatigue, restlessness, incoordination, malaise, loss of appetite, gastrointestinal disturbances, painful muscles, visual disturbance, and dry mouth. The apparent increase in shift workers of gastrointestinal disorders, cardiovascular disease, and other medical conditions may be due in part to the effects of chronically insufficient sleep [24].

LABORATORY TESTS

On polysomnographic testing, persons with ISS tend to have high sleep efficiency and short sleep latency, reflecting the increased need for sleep. In one series, the nocturnal sleep latency averaged 7.5 min [39]. People with ISS usually sleep for longer periods in the laboratory than they do at home and total sleep time during polysomnography is generally higher than in patients with sleep apnoea or narcolepsy [8]. Compared to narcoleptics, persons with ISS have less stage REM, less stage 1 sleep, and more stage 3-4 on polysomnography [29].

On the Multiple Sleep Latency Test, subjects with ISS tend to have mean sleep latencies between 5 and 8 minutes, with a dip in alertness in the afternoon corresponding to the usual circadian pattern [29, 34, 39]. Naps during the MSLT have few awakenings and include stage 2 sleep more than 80% of the time, a greater incidence of stage 2 sleep than occurs with narcoleptics [39]. REM sleep occurs in about 10% of naps but rarely with a latency of less than 5 minutes [29, 39]. Some persons with relatively mild sleep debts may show progressively increasing sleep latencies during the MSLT, as each nap reduces the cumulative sleep debt.

COURSE OF ILLNESS

No studies have documented the course of ISS, although some experimental studies suggest that with time there may be adaptation to some of the effects of sleep restriction [18]. Some subjects participating in long term studies of sleep restriction have maintained shorter hours of sleep for months after the experiments ended [24, 31], suggesting that in some cases there may be prolonged behavioural changes in sleep/wake patterns.

CLINICAL VARIANTS

There are some people, probably less than 1% of the population, who apparently do not need much sleep. These persons, referred to as short sleepers, habitually sleep less than 5 hours per day and may sleep as little as 2-3 hours. Short sleepers do not sleep longer on weekends and show no apparent performance decrements or daytime drowsiness.

At the other end of the scale are long sleepers, who require 9-10 hours of sleep each night for full alertness. These patients may develop the insufficient sleep syndrome, but the diagnosis is difficult to make because it can occur even though they sleep 8 hours each night. Diagnostic difficulty can also arise in patients with both insufficient sleep syndrome and another sleep disorder, such as sleep apnoea, because it may be difficult to sort out which problem is the principal cause of daytime sleepiness and other symptoms.

DIFFERENTIAL DIAGNOSIS

Since people with ISS who present to sleep centres usually complain of excessive sleepiness, the differential diagnosis includes narcolepsy, sleep apnoea syndrome, periodic limb movement disorder, and idiopathic hypersomnia. Diagnosis of insufficient sleep is suggested by a history of short sleep at night, recent changes in life style (e.g. a new baby) that have altered sleep patterns, and increased sleep on weekends or holidays. The diagnosis is made more difficult when patients inaccurately report their usual amount of night time sleep. Sleep laboratory studies usually show short sleep latency and long periods of uninterrupted sleep (high sleep efficiency).

The inability to get going in the morning and the occurrence of confusional arousals in association with such nocturnal arousing stimuli as telephone calls may suggest a diagnosis of sleep drunkenness or confusional arousals. These associated problems resolve with increased sleep at night. Later hours of arising on weekends may suggest a diagnosis of delayed sleep phase syndrome, and it is not uncommon for persons with ISS to have a mildly delayed sleep phase, which probably contributes to difficulty attaining full alertness in the morning.

It is sometimes difficult to distinguish the insufficient sleep syndrome from idiopathic hypersomnia. Both syndromes may be associated with high sleep efficiency, difficulty awakening in the morning, and daytime drowsiness. Idiopathic hypersomnia, however, is associated with normal or increased amounts of sleep at night and does not respond to trials of longer nocturnal sleep periods.

If the insufficient sleep syndrome is suspected, a trial of increased sleep at home is indicated. I usually advise patients to try to sleep for at least 8 to 8.5 hours per night, or at least one hour longer than their customary sleep time, whichever is longer, to continue this schedule for at least one month, and to maintain a sleep log. If daytime sleepiness resolves, then the diagnosis is clear. If sleepiness improves only partially or does not improve at all, then sleep testing is indicated to assess for possible narcolepsy, sleep apnoea, or periodic limb movement disorder. However, the clinician must consider the possibility that incomplete compliance with the recommendation for increased sleep accounts for the failure to improve.

PATHOPHYSIOLOGY

The acute effects of sleep deprivation have been investigated by a variety of researchers, indicating in general greater effects on mood, lesser effects on cognitive and vigilance tasks, and little effect on motor tasks [29, 38]. Studies of partial sleep deprivation have demonstrated that during the period of sleep restriction, delta sleep is relatively well preserved and most of the reduced sleep is accounted for by a reduction in stage 1-2 sleep and REM sleep [15] (see [20] for review). Impaired vigilance is evident after just two to four nights with five hours of sleep [7, 37], and chronic reduction of normal sleep hours by 30-60 minutes can lead to increased sleepiness [7]. Webb and Agnew [35] studied 15 subjects who abruptly reduced their nightly sleep from 7.5-8 hours to 5.5 hours for 60 days. On performance tasks, there was a progressive decline in vigilance with no other major alteration in performance. Subjective sleepiness and mood changes were most apparent in the first 2 weeks.

With gradual voluntary sleep restriction over several months, decrements in alertness and performance, along with irritability and daytime sleepiness, are generally mild until sleep is reduced to less than 5 hours per night [15, 21]. Horne and Wilkinson [20] found little change in vigilance or daytime drowsiness following gradual reduction in night time sleep by two hours over several weeks. Persons who sleep longer hours can tolerate more sleep reduction than can shorter sleepers, suggesting that these groups have similar "obligatory sleep" lengths. Horne concludes that these studies support the concept that sleep is composed of "core" sleep and "optional" sleep and that "optional" sleep can be reduced by 1-2 hours in most individuals with little effect on daytime function [18]. The type of performance under study also affects the results: whereas vigilance tasks may be impaired by as little as two nights of reduced sleep, physical tasks may be unaffected despite up to 9 nights of no more than 3 hours of sleep [18, 34].

A model of chronically insufficient sleep is provided by physicians in training. In one study, 27 physicians in their first year of training in the United States slept an average of 5.95 hours per 24 hour day [14]. Physicians in training are more depressed, angry, tense, and fatigued following nights on call with little or no sleep [3, 13] and anaesthesia residents have more difficulty detecting significant changes in critical monitored variables after 24 hours of in-house call than they do in a rested condition [11]. Physicians in training show a progressive increase in anger throughout the first year of training and have a higher than expected incidence of depression [14], but these changes may reflect other aspects of medical training than insufficient sleep.

TREATMENT

Increased sleep time of one hour or more is the best treatment for the insufficient sleep syndrome [6, 27]. The increased sleep can be added on to the night time sleep or it can be taken in the form of several brief naps or one long mid-afternoon "siesta" for patients whose social obligations or work do not permit increased time in bed at night.
Symptoms generally improve within a few days and resolve after a few weeks in patients who comply with these recommendations. Many patients, however, resist the idea that insufficient sleep is the problem and believe that their social responsibilities do not permit them to sleep more hours. Such patients require education about the need for sleep and the consequences of failure to obtain adequate amounts of sleep. I usually point out that they must weigh the costs and benefits of reducing sleep hours below the optimum and that the costs include impaired vigilance and performance as well as an increased risk of motor vehicle and occupational accidents.

Caffeine, which improves vigilance in persons with ISS [19, 22], can be a helpful adjunct for patients who are unwilling or unable to increase their sleep time to optimal amounts. Some patients seek treatment with prescription stimulant medications, particularly if they have received such treatment in the past. In one series, 44% had been treated previously with stimulants, suggesting that they had been misdiagnosed as persons with narcolepsy or idiopathic hypersomnia [29]. Although dextroamphetamine and other stimulants can reduce the performance deficits that occur with acute partial sleep deprivation in military and similar situations, they are not indicated in patients with the insufficient sleep syndrome because of their potential for inducing sleep disruption and tolerance [5].

REFERENCES

1. AMERICAN SLEEP DISORDERS ASSOCIATION. – International classification of Sleep Disorders, Revised: Diagnostic and Coding Manual. American Sleep Disorders Association, Rochester, MN, 1997.
2. ASSOCIATION OF SLEEP DISORDERS CENTERS – Diagnostic Classification of Sleep and Arousal Disorders, First Edition, prepared by the Sleep Disorders Classification Committee, H.P. Roffwarg, Chairman. *Sleep 2*, 1-137,1979.
3. BARTLE E.J., SUN J.H., THOMPSON L., LIGHT A.I., McCOOL C., HEATON S. – The effects of acute sleep deprivation during residency training. *Surgery* 104,311-316, 1988.
4. BLIWISE D.L. – Historical change in the report of daytime fatigue. *Sleep* 19,469-464, 1996.
5. CALDWELL J.A., CALDWELL J.L., CROWLEY J.S. JONES H.D. – Sustaining helicopter pilot performance with Dexedrine during periods of sleep deprivation. *Aviat. Space Environ. Med.* 66, 930-937, 1995.
6. CARSKADON M.A., DEMENT W.C. – Sleep tendency during extension of nocturnal sleep. *Sleep Res.* 8, 147, 1979.
7. CARSKADON M.A., DEMENT W.C. – Cumulative effects of sleep restriction on daytime sleepiness. *Psychophysiology*, 11, 107-113, 1981.
8. CARSKADON M.A., DEMENT W.C. – Nocturnal determinants of daytime sleepiness. *Sleep*, 5, S73-S81, 1982.
9. DEACONSON T.F., O'HAIR D.P., LEVY M.F., LEE M.B., SCHUENEMAN A.L., CODON R.E. – Sleep deprivation and resident performance. *JAMA*, 260, 1721-1727, 1988.
10. DEARY I.J., TAIT R. – Effects of sleep disruption on cognitive performance and mood in medical house officers. *Br.Med.J.*, 295, 1513-1516, 1987.
11. DENISCO R.A., DRUMMOND J.N., GRAVENSTEIN J.S. – The effect of fatigue on the performance of a simulated anesthetic monitoring task. *J. Clin. Monit.*, 3,22-4, 1987.
12. DIPHILLIPO M., FRY J.M., GOLDBERG R. – Characterisation of patients with insufficient sleep syndrome. *Sleep Res.*, 22, 188, 1993.
13. ENGEL W., SEIME R., POWELL V., D'ALESSANDRO R. – Clinical performance of interns after being on call. *South Med. J.*, 80, 761-763, 1987.
14. FORD C.V., WENTZ D.K. – The internship year: a study of sleep, mood states, and psychophysiologic parameters. *South Med. J.*, 77, 1435-1442, 1984.
15. FRIEDMAN J., GLOBUS G., HUNTLEY A.B., MULLANEY D., NAITOH P., JOHNSON L.C. – Performance and mood during and after gradual sleep reduction. *Psychophysiology*, 14, 245-250, 1977.
16. HASLAM D.R. – Sleep loss, recovery sleep, and military performance. *Ergonomics*, 25, 163-178, 1982.
17. HASLAM D.R. – The military performance of soldiers in sustained operations. *Aviat. Space Environ. Med.*, 55, 216-221, 1984.
18. HORNE J. – *Why we sleep.* Oxford University Press, New York, 1988.
19. HORNE J.A., REYNER L.A. – Counteracting driver sleepiness: Effects of napping, caffeine, and placebo. *Psychophysiology*, 33, 306-309, 1996.
20. HORNE J.A., WILKINSON S. – Chronic sleep reduction: daytime vigilance performance and EEG measures of sleepiness, with particular reference to practice effects. *Psychophysiology*, 22, 69-78, 1985.
21. JOHNSON L.C., MACLEOD W.L. – Sleep and awake behaviour during gradual sleep reduction. *Percept. Motor Skills*, 36, 87-97, 1973.
22. LUMLEY M., ROEHRS T., ASKER D., ZORICK F., ROTH T. – Ethanol and caffeine effects on daytime sleepiness/alertness. *Sleep*, 10, 306-312, 1987.
23. NAITOH P., ANGUS R.G. – Napping and human functioning during prolonged work. In: Dinges D.F., Broughton R.J., (eds). *Sleep and Alertness: Chronobiological, Behavioral, and Medical Aspects of Napping.* Raven Press, New York, 221-246, 1989.
24. NAITOH P., KELLY T.L., ENGLUND C. – Health effects of sleep deprivation. *Occup. Med.*, 5, 209-237, 1990.
25. ORTON D.I., GRUZELIER J.H. – Adverse changes in mood and cognitive performance of house officers after night duty. *BMJ*, 298, 21-23, 1989.
26. ROBBINS J., GOTTLIEB F. – Sleep deprivation and cognitive testing in internal medicine house staff. *West J. Med.*, 152, 82-86, 1990.
27. ROEHRS T., TIMMS V., ZWYGHUIZEN-DOORENBOS A., ROTH T. – Sleep extension in sleepy and alert normals. *Sleep*, 12, 449-457, 1989.
28. ROEHRS T., TIMMS V., ZWYGHUIZEN-DOORENBOS A., BUZENSKI R., ROTH T. – Polysomnographic, performance, and personality differences of sleepy and alert normals. *Sleep*, 13, 395-402, 1990.
29. ROEHRS T., ZORICK F., SICKLESTEEL J., WITTIG R., ROTH T. – Excessive daytime sleepiness association with insufficient sleep. *Sleep*, 6, 319-325, 1983.
30. ROTH B., NEVSIMALOVA S., SAGOVA V., PAROUBKOVA D., HORAKOVA A. – Neurological, psychological and polygraphic findings in sleep drunkenness. *Schweiz Arch. Neurol. Neurochir. Psychiatr.*, 129, 209-222, 1981.
31. RUBIN R., ORRIS P., LAU S.L., HRYHORCZUK D.O., FURNER S., LETZ R. – Neurobehavioral effects of the on-call experience in housestaff physicians. *J. Occup. Med.*, 33, 13-18, 1991.

32. SAMKOFF J.S., JACQUES C.H. – A review of studies concerning effects of sleep deprivation and fatigue on residents' performance. *Acad. Med.*, 66, 687-693, 1991.

33. SCHOR J.B. – *The overworked American: the unexpected decline of leisure.* Basic Books, New York, 1991.

34. VAN DEN HOED J., KRAEMER H., GUILLEMINAULT C., ZARCONE V.P., MILES L.E., DEMENT W.C., MITLER M.M. – Disorders of excessive daytime somnolence: Polygraphic and clinical data for 100 patients. *Sleep*, 4, 23-37, 1981.

35. WEBB W.B., AGNEW H.W. Jr. – The effects of a chronic limitation of sleep length. *Psychophysiology*, 11, 265-274, 1974.

36. WEBB W.B., AGNEW H.W. Jr. – The effects on subsequent sleep of an acute restriction of sleep length. *Psychophysiology*, 12, 367-370, 1975.

37. WILKINSON R.T., EDWARDS R.S., HAINES E. – Performance following a night of reduced sleep. *Psychonom. Sci.*, 5, 471-472, 1966.

38. YOUNG T., PALTA M., DEMPSEY J., SKATRUD J., WEBER S., BADR S. – The occurrence of sleep disordered breathing among middle-aged adults. *New Engl. J. Med.*, 328, 1230-1235, 1993

39. ZORICK F., ROEHRS T., KOSHOREK G., SICKLESTEEL J., HARTSE K., WITTIG R., ROTH T. – Patterns of sleepiness in various disorders of excessive daytime somnolence. *Sleep*, 5, 165-174, 1982.

Chapter 28

Medication and alcohol dependent sleepiness

D. Warot* and E. Corruble**

*Service de Pharmacologie, Hôpital Pitié-Salpêtrière, Paris, France;** Service de Psychiatrie, Centre Hospitalier Universitaire Bicêtre, Le Kremlin Bicêtre, France

INTRODUCTION

Sleepiness can be defined as an early stage of sleep disorder. It is a parameter that is difficult to comprehend, owing to its subjective nature.

Sleepiness can also be a difficult symptom for the clinician to discern: it may be discreet and hard to objectify, or fluctuating and transient, or even absent during examination. An aetiological diagnosis of sleepiness sometimes presents problems. The aetiologies of sleepiness may indeed be multiple: induced by insufficient sleep, medication, alcohol; primary; associated with neurological, psychiatric, infectious, metabolic conditions or depending on circadian rhythm sleep disorders. Certain elements must be taken into account in determining whether sleepiness can be attributed to a medication: the search for other possible aetiologies, chronology of the symptom (onset coinciding with the start of treatment with the molecule in question, regression of symptom when treatment is withdrawn) as well as the reference literature available. Associations between medications should be carefully looked at as these often play an important part in inducing sleepiness. Once the aetiological diagnosis has been established, the problem then arises of the importance of the symptom for the patient. Indeed sleepiness is commonly listed as a possible adverse effect of medication. It can sometimes be a desired effect for use in sedation, particularly in anaesthesia (pre-op phase) or in psychiatry (pathologies of anxiety, delirium and/or mania). Whether desired or not, sleepiness can have a considerable impact on the patient's daily life, particularly on occupational and domestic activities, as well as on driving.

The pharmacologist needs to take a number of questions into account:

What is the basis for suspecting drug-related sleepiness? Does sleepiness occur as frequently as with the use of placebo? Is sleepiness an intrinsic, anticipated effect of the molecule, i.e. predictable in terms of its pharmacological properties? Does animal data suggest that the effect might be produced in man? What is the dose range responsible for the effect? Does it occur solely with high, "toxic" doses, potentially therapeutic doses, or is it non dose-related?

Mac Clelland [32] stresses that sleepiness is such a common side effect from medication that its true frequency is probably largely underestimated. It nevertheless remains one of the most frequently reported adverse effects on the nervous system. It concerns not only medication which acts on the central nervous system, but a range of other molecules in the pharmacopoeia.

DATA SOURCES

There is a wide range of data sources in regard to sleepiness linked to medication, including clinical trials, spontaneously reported cases and epidemiological studies.

Clinical trials

Assessed effects in clinical trials

The effects on sleepiness and alertness of a molecule known for its psychotropic properties are systematically evaluated at an early stage – first in healthy volunteers and then in patients. This step may be taken at a later stage in development if a substance produces unexpected psychotropic effects. A range of methods may be used, vigilance tests, on one hand, comprising performance tests and evoked potentials, sleepiness tests, on the other hand, with subjective scales and objective physiological measurements (see chapter 13).

Effects spontaneously reported by subjects during the trial

Cases of sleepiness are also witnessed in subjects receiving a substance in trials carried out for other purposes (e.g. pharmacokinetics). When this is reported, trials must be conducted to research the effects of the molecule on vigilance.

Post-marketing trials

These could be carried out if cases of altered vigilance are reported after a substance has come onto the market. In reality this seldom occurs.

Spontaneously reported cases

The monitoring of suspect drugs (pharmacovigilance) is a system which, among other things, provides a means of gathering and processing information on the adverse effects of drugs, reported after they appear on the market. It is a way of detecting particularly rare adverse effects, unlikely to have been detected during clinical trials, due to too few subjects having been studied. It must be noted that pharmaco-vigilance is limited in terms of common and frequent adverse effects: studies show that these are rarely reported.

The literature sometimes refers to the adverse effects of medication reported as clinical cases. The latter serve to focus attention on such adverse effects and to supplement investigations, if necessary (e.g. clinical trials, pharmacovigilance).

Epidemiological studies

Sleepiness linked to medication has become a public health concern, prompting epidemiological studies. The consequences of sleepiness not only on driving but also on occupational activities and thus on productivity are relatively recent issues. Prospective studies are the most informative but also the most expensive and difficult to carry out. Retrospective studies are less ambitious but nevertheless provide useful information. The main limitation with epidemiological studies is usually the lack of an adequate control group.

AVAILABLE DATA

For certain classes of drugs, the percentage of incidence or prevalence of a symptom of sleepiness is given, for others the information is partial and cannot be represented by figures. Moreover the conditions in which the data are collected are not always stated, and the data lack precision. A figure only really provides information if it includes: the exact nature of the patients, the circumstances of the incident, the doses, length of treatment, associated pathologies, corresponding treatment and method of collection (open, closed questionnaire …). Finally to be of use in evaluating risk/benefits, the figure of incidence or prevalence must be compared to those observed using placebo and the reference substance.

PSYCHOTROPIC DRUGS

Anxiolytics and hypnotics

Benzodiazepines

The benzodiazepines act as sedatives in man. This property is therapeutically useful but can interfere with daytime occupational and domestic activities. Other benzodiazepine properties may reduce performance, such as myorelaxant effects or memory disorders; differences probably exist between the benzodiazepines but satisfactory answers are still lacking on these differences for equi-anxiolytic doses. The data available for benzodiazepines, derived from different methodological approaches, can be summarised as follows [18, 23, 27, 31]:
- Desired sedative effects and residual effects are demonstrated by the same methods;
- The different methods used to quantify sleepiness, as described in the first paragraph, have different degrees of sensitivity (or detection thresholds);
- The results obtained for young healthy volunteers may well be extrapolated and applicable to occasional consumers of benzodiazepine, but most prescriptions involve regular consumers. The effects are likely to differ qualitatively and quantitatively between subjects suffering from different types of anxiety, habitual/occasional insomniacs and other disorders involving associated prescriptions related to co-morbidity. Ideally each category should be studied.

Each substance has a different placebo dose-threshold for reducing performance in tests of alertness; modifications vary in intensity over time, depending on the time the molecule in question was administered.

It is interesting to note that psychomotor performance diminishes after a single 5 mg dose of diazepam, 20 mg of chlordiazepoxide, 15 mg of temazepam, 2 mg (1 mg?) of lorazepam, 10 mg of chlorazepate, 40 mg of clobazam, 10 mg of oxazepam, 5 mg of nitrazepam, 0.25 mg of triazolam… These results must be qualified in the absence (as in most examples) of comparative data of equi-anxiolytic, equi-hypnotic doses and with regard to the diminishing effect over time of the study substances.

Studies which consider the way in which the sedative effect evolves with repeated administration, either in healthy subjects or insomniac/anxious patients, tend to indicate a tolerance effect, although this has not yet been clearly demonstrated.

In the insomniac, using a single dose administered at bedtime, the speed in accomplishing tests the following morning is reduced, this reduction being dose-related. With repeated doses, (4 to 14 nights, depending on the study), the results are consistent or variable depending on the test: either no alteration is observed, or the observed first dose effect is attenuated when the dose is repeated, or the effect after the first dose is accentuated with each administration. So the answer is not unequivocal [23].

In subjects suffering from anxiety, the intensity of the sedative effect of benzodiazepines tends to diminish with repeated administration, but differences occur depending on the test and the substance [29].

Non benzodiazepine hypnotic drugs

Barbiturates are not used in insomnia but many pharmaceutical preparations contain "small" exonerated doses of barbiturates. No evaluation is made of their impact on alertness but their role needs to be systematically taken into account when exploring for morning or daytime sleepiness. Studies available on chloral hydrate fail to provide a clear picture of its efficacy and safety (particularly in the absence of comparative data) and hence of the correct evaluation of the risks of sleepiness.

Hypnotic drugs containing sedative phenothiazines as the sole active principle or in association with benzodiazepines may induce residual sleepiness, which, to our knowledge, has not been comparatively quantified using the methodology applied to benzodiazepines as a whole.

Two recently marketed hypnotics, which belong to chemical classes other than benzodiazepines, but which bind to the same receptors, have been the focus of extensive human clinical and pharmacological studies. At unit doses (7.5 mg for zopiclone, 10 mg for zolpidem), on morning arousal little or no residual sleepiness is observed in insomniacs of under 65 years of age.

Some figures on data derived from clinical studies can be examined, by way of example. In a meta-analysis of 23 clinical studies in which 1,028 insomniacs were treated by zolpidem, Palminteri and Narbonne [39] demonstrated the following effects: zolpidem at 10 mg induces 4.5% of sleepiness (n = 67) and zolpidem at 20 mg induces 9.6% of sleepiness (n = 73) against 3.9% for placebo (n = 76). Zolpidem at 10 mg induces 5.1% of sleepiness (n = 58) against 1.7% with 1 mg of flunitrzepam (n = 57). Zolpidem at 20 mg induces 3.1% of sleepiness (n = 256) against 0.78% with triazolam at 0.5 mg (n = 256). These and other results have led to marketing the unit dose of 10 mg of zolpidem.

As regards zopiclone, reducing the dose from 7.5 mg to 5 mg, would probably provide a better risk/benefit ratio, were this to be demonstrated.

Non benzodiazepine anxiolytic drugs

The prescription of meprobamate (and other carbamates) is based more on experience acquired over time than on controlled, comparative studies. In the absence of quantified data, it is impossible to make a conclusive statement about the present risk/benefit balance in terms of sleepiness.

The sleepiness-inducing effects of buspirone and alpidem appear to be lower at the studied doses compared to benzodiazepines [6, 57]. These results are consistent with those obtained in psychomotor performance studies [52, 57].

In conclusion, it is important to bear in mind that each new drug promises a less harmful effect on performance and often backs this up with the appropriate studies. Can these differences be said to persist, however, at equi-active doses (equi-anxiolytic or equi-hypnotic) or when associations including alcohol are taken into account?

What role, if any, does the consumption of tranquillisers and hypnotics play in the incidence of traffic accidents? What examples are there of retrospective and prospective epidemiological studies:

-In 60% of arrests for driving offences in California, where drivers were suspected of having ingested a substance, benzodiazepines were detected in the blood samples, excluding subjects presenting a positive alcohol count (0.01% - 0.1%) [59].

- Garriot and Latman's study [15] notes that 8.2% of 207 subjects who died in accidents had diazepam in their plasma. No group of controls who had not been involved in accidents was considered for this study.

- Other studies carried out in Finland and the USA later confirmed the link between road accidents and tranquillisers, with or without alcohol [3, 22].

- In a prospective study, in England, Skegg *et al.* [51] demonstrated that the risk of serious road accidents (car, motorbike) was 4.9 times greater in drivers who had been prescribed tranquillisers within the three months preceding the accident.

- Moreover, pedestrians appear to be more frequently subject to accidents under benzodiazepines, although the risk is far lower than for alcohol [23].

Some general remarks on the different epidemiological approaches:
- The presence of a tranquilliser does not always signify a tranquillising effect.
- The consequences of anxiety or untreated insomnia on car handling are not known and a study has yet to be carried out to distinguish between the effects of the drug and the underlying illness in the incidence of traffic accidents. Even though the epidemiological approach provides no final answer, the psychotropes and among these, the benzodiazepines, may contribute to the incidence of fatal traffic accidents.

The accident-prone effect of ethanol has been demonstrated by epidemiology but the added risk of accident in benzodiazepine consumers has not been clearly established (National enquiry 1989-1990: Benzodiazepines and traffic accidents).

Neuroleptic drugs

The sedative property of neuroleptics is usually linked to the blockade of histaminic H_1 receptors (e.g. phenothiazines) and that of the postsynaptic adrenergic α_1 receptors. Clinically speaking however, changes in alertness are not evaluated with the same degree of efficacy for all neuroleptics according to the different indications. The most common co-prescriptions (antidepressants, tranquillisers, correctors of the adverse effects of neuroleptics) are subject to variation, particularly in relation to dose and length of administration. Empirically, three quarters of patients treated by neuroleptic phenothiazines experience dose-dependent sleepiness [34]. Moreover, the sedative effect of neuroleptics varies considerably from one individual to another, in the clinic. Sleepiness, whether a desired or adverse effect, attenuates over time. The tolerance factor, even though it is admitted by all authors, has yet to be demonstrated.

Psychomotor performance is clearly impaired in healthy volunteers taking phenothiazines, whereas butyrophenones may improve certain performances or fail to modify them, as has been shown in the case of benzamides [26]. The analysis of reference data on how neuroleptics interfere with the psychomotor performances of schizophrenics favours the possibility of an improvement of performance in treated schizophrenics – in whom moreover the tolerance factor is found.

Clozapine, risperidone and olanzapine have dose dependent sedative effects, in function of indications. But at therapeutic doses, the sedative effects of risperidone and olanzapine are less marked than in the phenothiazinic neuroleptics [30].

Antidepressants

Tricyclic antidepressants

All the tricyclic antidepressants possess sedative properties to varying degrees depending on the molecules, doses and the subjects to whom they are administered. A number of results are presented here, by way of example. In a review of the literature of 3 double blind trials comparing the effects of imipramine (200 mg/day on average, 240 patients) and placebo (227 patients) in patients presenting a major episode of depression, Zerbe [61] reported 22% of sleepiness using imapramine against 7% using placebo (p<0.05). In a group of 107 patients treated with imipramine (150 mg/day on average) for a major episode of depression, Wagner *et al.* [58] reported 23% of sleepiness against 5.5% using placebo. Lader [28] reports 3 short term double blind studies comparing the effects of amitriptyline and fluoxetine, among which only one study reports 32% of cases of sleepiness using amitriptyline (average dose 159 mg/day). The percentages of sleepiness observed using doxepin increase with the dose: 21% (from 50 to 150 mg/day) to 45% (75 to 200 mg/day).

The sedative property derives from the blocking of histamine H_1 receptors and the postsynaptic adrenergic receptors. This property which lies at the origin of impairment in the performance of tasks linked to driving is found in different antidepressants, more often than not with a dose-effect [21].

IMAOs

In a review based on several studies of the adverse effects of two IMAOs (moclobemide and phenelzine), Amrein *et al.* [1] report 10% of sleepiness using moclobemide (n = 285) against 3% with phenelzine (n = 141) and 8% using placebo (n = 271). Blackwell and Simon [5] report 19% of cases of sleepiness in a controlled trial with patients treated by phenelzine (77 mg/day on average).

In a review of double blind studies comparing moclobemide and imipraminic antidepressants in depressed patients, Amrein *et al.* [1] report 5% of sleepiness using moclobemide (n = 694) against 6% under placebo (N = 271).

At doses between 100 mg and 1 200 mg/day, moclobemide does not impair the psychomotor and cognitive performances of young, healthy volunteers [20], a result which is consistent with the low risk of sleepiness with this IMAO.

Selective serotonin reuptake blockers

As shown by the results of clinical trials versus placebo and from clinical practice, citalopram, paroxetine and sertraline can also induce sleepiness, with high within-patient variability. Lader [28] published a review of short term double blind studies comparing fluoxetine with reference antidepressants. Six studies out of ten reported sleepiness. Levels of sleepiness do not appear to be dose-related in the studied range. In a review of the literature of 3 double blind trials comparing the effects of fluoxetine (20 to 60 mg, 239 patients) with a placebo (227 patients) in treatment for a major episode of depression, Zerbe [61] reported 16% of sleepiness with fluoxetine as against 7% with placebo. Rickels *et al.* [45] administered fluoxetine at individually determined doses ranging from 20 to 80 mg/day, to 185 patients presenting a major episode of depression. Sleepiness occurred in 22.3% of the group receiving a daily dose and in 14.2% in the group receiving the dose twice daily.

In a group of 110 patients treated with fluvoxamine (150 mg/day on average) for a major episode of depression, Wagner *et al.* [58] reported 24.5% of sleepiness against 5.5% with placebo.

Serotonin and norepinephrine reuptake blockers

Like the selective serotonin reuptake blockers, venlafaxine and milnacipran can induce sleepiness. In trials against placebo, percentages of the order of 24% have been reported under venlafaxine versus 10% in the placebo group [12].

Other antidepressant drugs

Trazodone is a sedative by virtue of its H_1 and α_1 antagonist properties. Metz and Shader [36] reported that trazodone can induce sleepiness in 8 to 52% of depressives treated with doses of between 150 and 500 mg/day, the sedative effect being found in healthy volunteers [21, 48].

As mianserin also blocks both types of receptor it also acts as a sedative in patients and healthy volunteers, the effect becoming more marked as the dose increases [5, 53].

The sedative effect of mirtazapine, close to mianserine, appears to be less marked at equi-effective doses in clinical practice. It has not been compared directly with mianserine in clinical trials.

Particular case of the elderly

Sleepiness in the elderly may result in a fall, leading to infirmity (fracture of femur) [40]. One enquiry [45] has shown that the neuroleptics, tricyclic antidepressants and long half-life benzodiazepines were the most commonly cited as increasing the risk of fracturing the neck of femur. Moreover, co-prescriptions are frequently given to the elderly, and the sedative effects of several drugs belonging to different therapeutic classes may accumulate.

MOOD REGULATORS

Lithium

Judd [24] reported several studies demonstrating impaired performance, sometimes accompanied by a sensation of tiredness in healthy volunteers receiving lithium. He confirmed these results in 42 healthy volunteers receiving lithium at therapeutic doses for 14 days. Sleepiness is reported as one of the adverse effects of lithium and a study of cognitive and motor functions has shown impaired performances in patients receiving lithium salts [26].

Carbamazepine

MacPhee *et al.* [34] in a double blind study against placebo in 12 healthy volunteers, using visual analogue scales, showed the sedative effect of carbamazepine at a dose of 10 mg/day between 12 and 6 hours after a single dose.

Daytime sleepiness is reported in patients receiving carbamazepine [8]. Studies quantifying daytime alertness, attention and concentration do not consistently conclude performance impairment, this depending on the patient population [2].

ANTIPARKINSON DRUGS

Ropinirole, a dopaminergic antiparkinsonian agonist, in addition to the sleepiness observed particularly at the onset of treatment, can induce sudden episodes of daytime sleep. These episodes may occur with no prodromes, particularly in the absence of any prior sleepiness. Reducing the dose and withdrawing treatment have usually resulted in the disappearance of these disorders. Similar observations have been reported with pramipexole [14]

ANTIEPILEPTIC DRUGS

Phenobarbital is a sedative with a predictable effect of sleepiness depending on the dose. The same is true for primidone which is partially metabolised as phenobarbital. Progressive dosage helps to control this effect as does taking a single dose at bedtime, wherever possible. The benzodiazepines are sedatives but are rarely prescribed as monotherapy for epilepsy, as this could increase the symptom and should therefore be taken into account in any decision to associate antiepileptics.

Carbamazepine, phenytoin, sodium valproate, progabide and ethosuximide induce sleepiness particularly in the early stages of treatment, although how frequently this occurs, in comparative terms, has not been estimated accurately [8, 10].

A study of psychomotor performance in treated and non treated epileptic patients indicates that the pathology and its treatment can impair psychomotor functions [7].

Sleepiness has been reported, especially at the start of treatment, with topiramate, vigabatrin and fosphenytoin [35].

H$_1$ ANTIHISTAMINES

Sedation is one of the essential properties of this class of molecules [32]. It stems from the molecule's affinity for central or periphery H$_1$ receptors and its capacity to cross the blood brain barrier. The intensity of sedation depends on the molecule, the dose and the subjects [38]. It may be a desired or adverse effect, depending on the case. The results of clinical studies and the evaluation of psychomotor performance show that the more recently commercialised antihistamines such as astemizole, acetirizine and loratadine, taken at therapeutic doses, are less sedative than the classic antihistamines, but it is important to bear in mind that there is wide variability between individuals. Many medications have antihistamine properties, including neuroleptics with a phenothiazine structure, tricyclic antidepressants, mianserine, flunarizine, hydroxyzine and ketotifene.

ANALGESICS

Morphine and opioid central analgesics may induce sleepiness [16] sometimes accompanied by an episode of confusion particularly in elderly subjects [42].

MYORELAXANTS

Baclofene and the benzodiazepines have sedative properties and can induce sleepiness, as can dantrolene. The incidence in relation to comparable efficacy cannot be ascertained owing to the lack of comparative studies available [4].

ANTIMIGRAINE

Sleepiness observed with pizotifene is usually transient and reduced by progressive dosage. Methysergide, a semi-synthetic ergot derivative can also induce sleepiness [4].

CARDIOVASCULAR DRUGS

Clonidine, an α_2 sympathomimetic central action antihypertensor, is a sedative. Paykel *et al.* [41] found that sedation was reported in 24 to 95% of subjects depending on the study and was dose-dependent. Alpha-methyl dopa is also a sedative and will cause sleepiness in 30% of patients; this effect tends to be observed in the early stages of treatment and gradually diminishes. Sleepiness in this case is associated with a feeling of tiredness [47].

Steiner [53] reports that prazosine, a postsynaptic α_1 blocker, can induce sleepiness. For 282 treated patients, he noted 7% of sleepiness in men and 11% in women.

In a review of the adverse neuropsychological effects of the different antihypertensors, Paykel *et al.* [41] cited that 1.2% of patients receiving propranolol complained of sleepiness. Cases of sleepiness have been reported for the other β-blockers. However for Dimsdale *et al.* [12] suggesting the implication of lipophilic β-blockers is greater than that of hydrophilic β-blockers in the incidence of adverse effects, is unfounded.

States of sleepiness and states of confusion in regard to metabolic disorders of the hyponatraemia type, may occur under diuretics, particularly in elderly subjects [42, 43].

ANTICHOLINERGICS

Scopolamine, marketed in the form of a transdermal patch for the prevention of motion sickness, can cause sleepiness. The sedative effect of scopolamine has been clearly evaluated on healthy volunteers, by objective testing (taping, coding) and visual analogue scales [40].

H₂ ANTIHISTAMINES

Nicholson [38] has reported the incidence of sedation in certain conditions, in patients treated by H_2 antihistamines. He notably reports cases of sedation in patients with renal or hepatic insufficiency treated with cimetidine, this effect being absent in healthy volunteers [55].

PROGESTERONE AND PROGESTINS

Sedation is a commonly reported clinical finding for natural progesterone. This has not however been quantified. Progestins can also induce sedation. In a study evaluating a range of micronised progesterone doses, in 24 healthy volunteers, only a dose of 1,200 mg induced the impairment of objective and subjective psychomotor performance. The wide variability of results between individuals should be emphasised [13].

OTHER MEDICATIONS

Sleepiness has been reported with molecules as diverse as griseofulvine, certain imidazole derivatives, gold salts, certain non steroid anti-inflammatory drugs and certain anti-cancer chemotherapies.

ALCOHOL

The effects of alcohol on alertness and psychomotor performance are qualitatively and quantitatively close to those observed with benzodiazepines [18]. The euphoria or state of excitation encountered with alcohol is attributable to a disinhibiting rather than a stimulant effect. A moderate dose of alcohol is likely to reduce REM sleep latency, at bedtime [33] as well as during the daytime [46, 62].

It is necessary to search for possible interactions between alcohol and any medication with sedative properties. In the case of the simultaneous intake of benzodiazepines and alcohol, the data available on evaluating psychomotor performance indicate an accumulation of effects rather than true potentialisation [9, 50]. Moreover, subjects are not always aware of their deficiency, as witnessed by their underestimation of difficulties on self-assessment scales, compared to data collected by objective methods. The inability to correctly assess one's own state is a risk factor in itself. Studies into the interactions between alcohol and antidepressants, alcohol and neuroleptics and alcohol and H_1 antihistamines, demonstrate an increase in sedative effects (for some of these associations at least)

[56]. The mechanism of interaction may be pharmacodynamic and/or pharmacokinetic, depending on the case.

CONCLUSION

It is difficult to diagnose drug-induced sleepiness, owing to the multiple aetiologies of sleepiness, and the fact that several drugs are often associated. Sleepiness induced by therapeutic doses of drugs involves not only central nervous system medications, but many other molecules, notably, placebo. The therapeutic benefits have been widely demonstrated in the case of most sleep-inducing drugs, and the effect may be attenuated by reducing the dose, where possible.

REFERENCES

1. AMREIN R., ALLEN S.R., GUENTERT T.W., HARTMANN D., LORSCHEID T., SCHOERLIN M.P., VRANESIC D. – The pharmacology of reversible monoamine oxidase inhibitors. *Br. J. Psychiatry,* 155, S66-S71, 1989.
2. AMAN M.G., WERRY J.S., PAXTON J.W., TURBOTT S.H., STEWART A.W.- Effects of carbamazepine on psychomotor performance in children as a function of drug concentration, seizure type, and time of medication. *Epilepsia,* 37 (5), 51-60, 1990.
3. ASHTON H. – Drugs and driving. *Adv. Drug Bull.,* 98, 360-363, 1983.
4. BARON J.C. – Les antimigraineux. *In: Neuropharmacologie clinique, le médicament en neurologie,* H. Dehen, G. Dordain (eds.), Doin, France, 219-255, 1989.
5. BLACKWELL B., SIMON J.S. – Antidepressant drugs. *In: Meyler's side effects of drugs,* M.N.G. Dukes,(ed.), Elsevier Science Publishers, 11th edition, 27-70, 1991.
6. BONVALOT T., BOULENGER J.P., ZARIFIAN E. - La buspirone: propriétés pharmacologiques et cliniques du premier représentant d'une famille nouvelle d'anxiolytiques. *Rev. Med. Int.,* IX (1), 97-103, 1988.
7. BRODIE M.J., McPHAIL E., MACPHEE G.J.A., LARKIN J.G., GRAY J.M.B. – Psychomotor impairment and antiepileptic therapy in adult epileptic patients. *Eur. J. Clin. Pharmacol.,* 31, 655-660, 1987.
8. BROGLIN D.- Les antiépileptiques. *In: Neuropharmacologie clinique, le médicament en neurologie.* H. Dehen, G. Dordain (eds). Doin, France, 21-68, 1989.
9. CHAN A.W.K. – Effects of combined alcohol and benzodiazepine: a review. *Drug Alcohol Depend.,* 13, 315-341, 1984.
10. DAVIES-JONES G.A.B. – Anticonvulsants. *In: Meyler's side effects of drugs.* M.N.G. Dukes, (ed.), Elsevier Science Publishers, 11th edition, 120-136, 1988.
11. DIMSDALE J.E., NEWTON R.P., JOIST T. - Neuropsychological side effects of betablockers. *Arch. Intern. Med.,* 149, 514-525, 1989.
12. ENTSUAH A.R., RUDOLPH R.L., CHITRA R. – Effectiveness of venlafaxine treatment in a broad spectrum of depressed patients: a meta-analysis. *Psychopharmacol. Bull.,* 31 (4), 759-766, 1995.
13. FREEMAN E.W., WEINSTOCK I., RICKELS K.L. – A placebo-controlled study of effects of oral progesterone on performance and mood. *Br. J. Clin. Pharmacol.,* 33, 293-298, 1992.
14. FRUCHT S., ROGERS J.D., GREENE P.E., GORDON M.F., FAHN S. - Falling asleep at the wheel: motor vehicle mishaps in persons taking pramipexole and ropinirole. *Neurology,* 52, 1908-1910, 1999.
15. GARRIOT J.C., LATMAN N. – Drug detection in cases of driving under the influence. *J. Forensic. Sci.,* 21, 398-415, 1976.
16. GHODSE A.H., VOHRA A.K. – Opioid analgesics and narcotic antagonists. *In: Side effects of drugs, Annual* 15, M.N.G. Dukes, J.K. Aronson (eds.), Elsevier Science Publishers, 68-84, 1991.
17. HAKKOU F., WAROT D., JAOUEN C., BENSIMON G., SIMON P. – Comparaison des effets du loprazolam et de l'alcool sur les performances psychomotrices et la mémoire chez le sujet sain. *Thérapie,* 43, 51-56, 1988.
18. HINDMARCH I.- Psychomotor function and psychoactive drugs. *Br. J. Clin. Pharmacol.,* 10, 189-209, 1980.
19. HINDMARCH I., KERR J. – Behavioural toxicity of antidepressants with particular reference to moclobemide. *Psychopharmacology,* 106, S49-S55, 1992.
20. HINDMARCH I., SUBHAN Z. – The effects of antidepressants taken with and without alcohol on information processing psychomotor performance and car handling ability. *In: Drug and driving,* J.F. O'Haanlon, J.J. de Gier (eds.) , Taylor and Francis, London, 231-239, 1986.
21. HONKANEN R., ERTAMA L., LINNOILA M., ALHA A., LUKKARI I., KARLSSON M., KIVILUOTO O., PURO M. – Role of drugs in traffic accidents. *Br. Med. J.,* 281, 1309-1312, 1980.
22. IRWIN S.T., PATTERSON C.C., RUTHERFORD W.H. – Association between alcohol consumption and adult pedestrians who sustain injuries in road traffic accidents. *Br. Med. J.,* 286, 522, 1983.
23. JOHNSON L.C., CHERNIK D.A. - Sedative-hypnotics and human performance. *Psychopharmacology,* 76, 101-113, 1982.
24. JUDD L.L. – Effects of lithium on mood, cognition and personality function in normal subjects. *Arch. Gen. Psychiatry,* 36, 860-865, 1979.
25. KING D.J. - The effects of neuroleptics on cognitive and psychomotor function. *Br. J. Psychiatry,* 157, 799-811, 1990.
26. KJELLMAN B.F., KARLBERG B.E., THORELL L.H. – Cognitive and affective functions in patients with affective disorders treated with lithium. *Acta Psychiatr. Scand.,* 62, 32-46, 1980

27. KOELEGA H.S. –Benzodiazepines and vigilance performance: a review. *Psychopharmacology,* 98, 145-156, 1989.
28. LADER M. – Efficacité de la fluoxétine comparée à des médicaments de référence: revue de la littérature. *Br. J. Psychiatry,* 153, (3), S56-S64, 1988.
29. LAPIERRE Y.D. – A critical flicker fusion (CCF) assessment of clobazam and diazepam in anxiety neurosis. *Pharmacopsychiatria,* 15 (suppl 1), 54-56, 1982.
30. LIVINGSTONE M.G., - Risperidone. *Lancet,* 343, 457-460, 1994.
31. LUCKI I., RICKELS K., GELLER A.M. – Chronic use of benzodiazepines and psychomotor and cognitive test performance. *Psychopharmacology,* 88, 426-433, 1986.
32. MAC CLELLAND H.A. – Psychiatric disorders. *In: Textbook of adverse drug reactions.* D.M. Davies (ed.), Oxford medical publications. Second edition, 480-502, 1981.
33. MAC LEAN A.W., CAIRNS J. – Dose–response effects of ethanol on the sleep of young men. *J. Stud. Alcohol,* 43 (5), 434-444, 1982.
34. MAC PHEE G .J.A., GOLDIE C., ROULSTON D., POTTER L., AGNEW E., LAIDLAW J., BRODIE M.J. – Effect of carbamazepine on psychomotor performance in naive subjects. *Eur. J.Clin. Pharmacol.,* 30, 37-42, 1986.
35. MARSON A.G., KADIR Z.A., HUTTON J.L., CHADWICK D.W. - The new antiepileptic drugs: a systematic review of their efficacy and tolerability. *Epilepsia ,* 38, 859-880, 1997.
36. METZ A., SHADER R.I. - Adverse interactions encountered when using trazodone to treat insomnia associated with fluoxetine. *Intern. Clin. Pharmacol.,* 5, 191-194, 1990.
37. NICHOLSON A.N., SCHLOSBERG A., DREYFUS J.F. – *Zopiclone, a third generation of hypnotics.* Karger, Basel, 1983.
38. NICHOLSON A.N., - Central effects of H1 and H2 antihistamines. *Aviat. Space Environ. Med.* 56, 293-298, 1985.
39. PALMINTERI R., NARBONNE G. – Safety profile of zolpidem. *In: Imidazopyridines in sleep disorders.* J.P. Sauvanet, S.Z. Langer, P.L. Morselli (eds), Raven Press, New York, 351-361, 1988.
40. PARROT A.C. - Transdermal scopolamine: Effects upon psychological performance and visual functioning at sea. *Human Psychopharmacology,* 3, 119-125, 1988.
41. PAYKEL E.S., FLEMINGER R., WATSON J.P. – Psychiatric side-effects of antihypertensive drugs other than reserpine. *J. Clin. Psychopharmacol.,* 2 (1), 14-39, 1982.
42. PIETTE F., FERRY M., SOUBRIE C., TEILLET L. - Pathologie iatrogène du sujet âgé, *Encycl. Med. Chir., Thérapeutique,* 25004 A10, 12p., 1992.
43. POROT M., PLENAT M., PEROL J.Y. – Incidences et complications psychiatriques des thérapeutiques. *Encycl. Med.Chir.,* 37 875 A10, 5p., 1977.
44. RAY W.A., GRIFFIN M.R., SCHAFFNER W., BAUGH D.K., MELTON L.J. – Psychotropic drugs use and the risk of hip fracture. *N. Engl. J. Med.,* 316, 363-369, 1987.
45. RICKELS K., WARD T.S., GLAUDIN V. – Comparison of two regimens of fluoxetine in major depression. *J. Clin. Psychiatry,* 46, 38-41, 1985.
46. ROEHRS T., ZWYGHUIZEN-DOORENBOS A., ZWYGHUIZEN H., ROTH T. – Sedating effects of ethanol after a nap. *Alcohol, Drugs and Driving,* 5, 351-356, 1989.
47. ROSEN R.C., KOSTIS J.B. – Biohavioral sequellae associated with adrenergic-inhibiting antihypertensive agents: a critical review. *Health Psychol.,* 4, 579-604, 1985.
48. SAKULSRIPONG M., CURRAN H.V., LADER M. – Does tolerance develop to the sedative and amnesic effects of antidepressants? A comparison of amitriptyline, trazodone and placebo. *Eur. J. Clin. Pharmacol.,* 40, 43-48, 1991.
49. SAUVANET J.P., LANGER S.Z., MORSELLI P.L. – *Imidazopyridines in sleep disorders: A novel experimental and therapeutic approach.* Raven Press, New York, 1988.
50. SELLERS E.M., BUSTO U. – Benzodiazepines and ethanol: Assessment of the effects and consequences of psychotropic drug interactions. *J. Clin. Psychopharmacol.,* 2, 249-262, 1982.
51. SKEGG D.C.G., RICHARDS S.M., DOLL R. – Minor tranquillisers and road accidents. *Br. Med. J.,* 1, 917-919, 1979.
52. SMILEY A., MOSKOWITZ H. - Effects of long-term administration of buspirone and diazepam on driver steering control. *Am. J. Med.,* 80 (suppl. 3B), 22-29, 1986.
53. STEINER J.A. – Antihypertensive drugs. *In: Meyler's side effects of drugs,* M.N.G. Dukes (ed.), Elsevier Science Publishers, 11th edition, 397-415, 1988.
54. STROMBERG C., MATTILA M.J. – Acute comparison of clovoxamine and mianserin, alone and in combination with ethanol, on human psychomotor performance. *Pharmacol. Toxicol.,* 60, 374-379, 1987.
55. THEOFILOPOULOS N., SZABADI E., BRADSHAW C.M. – Comparison of the effects of ranitidine, cimetidine and thioridazine on psychomotor functions in healthy volunteers. *Br. J. Clin. Pharmacol.,* 18, 135-144, 1984.
56. VANDEL B. – Alcool et médicaments psychotropes. *Thérapie,* 36, 269-273, 1981.
57. VANDEL B., PERAULT M.C. – L'alpidem: propriétés pharmacologiques et thérapeutiques. *La lettre du Pharmacologue,* 6, 125-127, 1992.
58. WAGNER W., CIMANDER K., SCHNITKER J., KOCH H.F. - Influence of concomitant psychotropic medication on the efficacy and tolerance of fluvoxamine. *Adv. Pharmacother.,* 2, 34-56, 1986.
59. WHITE J.M., CLARDY D.O., GRAVES M.H., KUO M.C., MAC DONALD B.J., WIERSEMA S.J., FITZPATRIC G. – Testing for sedative-hypnotic drugs in the impaired driver: a survey of 72.000 arrests. *Clin. Toxicol.,* 18, 945-957, 1981.
60. WILLIAMS D., MAC LEAN A., CAIRNS J. – Dose-response effects of ethanol on the sleep of young women. *J. Stud. Alcohol,* 44, 515-523, 1983.
61. ZERBE R. – Safety of fluoxetine. *Br. J. Clin. Pract.,* S46, 41-47, 1986.
62. ZWYGHUIZEN-DOORENBOS A., ROEHRS T., LAMPHERE J., ZORICH F., ROTH T. - Increased daytime sleepiness enhances ethanol sedative effects. *Neuropsychopharmacology,* 1, 279-286.

Chapter 29

Obstructive sleep apnoea-hypopnea syndrome and upper airway resistance syndrome

J. Krieger
Service d'Explorations Fonctionnelles du Système Nerveux et de Pathologie du Sommeil, Hôpitaux Universitaires de Strasbourg, Strasbourg, France

HISTORY

In 1956, [26] Burwell *et al.* used the term "Pickwickian syndrome" to describe a new clinical picture associating obesity, cyanosis, polycythemia, primary alveolar hypoventilation and heart failure. Nearly ten years later, Gastaut *et al.* [64] and Jung and Kühlo [109] demonstrated the presence of repeated apnoeas during the sleep of "Pickwickian" subjects. It appeared rapidly that such apnoeas were seen in patients who did not have the clinical picture described by Burwell *et al.* [26] and the notion of "obstructive sleep apnoea syndrome" (OSAS) was distinguished, particularly through the instigation of Guilleminault *et al.* [85]. This syndrome is now known to be particularly common and plays a decisive role in numerous aspects of the pathology. From the first studies of OSAS, certain authors drew attention to hypopnoeas which they described as differing by a degree of intensity from apnoeas [133]. The mechanism and consequences of these hyponoeas were very easily understood in reference to apnoeas; sleep hypopnoea was later described [72] with clinical manifestations very close to those of sleep apnoea and the notion of obstructive sleep apnoea-hypnopnoea syndrome (OSAHS) was established.

More recently, and again through the impetus of C. Guilleminault, the notion of upper airway resistance syndrome (UARS) [83] appeared, characterised by ventilatory changes which were minor or undetectable (using the usual polygraphic techniques), but where the response of the ventilation system to an increase in upper airway resistance, triggers a cascade of events leading to clinical manifestations very like those of sleep apnoea syndrome.

This upper airway resistance syndrome poses new diagnostic problems in sleep pathology. Chiefly, because of the existence of daytime sleepiness in the absence of detectable ventilatory changes, pathophysiological interest has shifted to the respiratory effort developed in response to an increase in upper airway resistance, and on its consequence - the micro-arousal - whose repetition induces sleep fragmentation.

EPIDEMIOLOGY

The condition is most commonly found in obese men in their fifties. The true frequency of sleep apnoea syndrome (OSAHS) is not known with certainty, as no systematic polysomnographic study has been conducted on a representative sample of the general population. The most convincing study, carried out on 602 subjects (100% snorers and 25% non snorers) randomly selected from among 3,513 responses to questionnaires sent out to 4,284 civil servants in Wisconsin, aged 30 to 60, (excluding pregnant women and any subject with a previous cardio-respiratory history) found an apnoea + hypopnoea index > 5 in 9% of women and 24% of men, > 15 in 4% of women and 9% of men; 2% of men and 4% of women had daytime sleepiness associated with an apnoea + hypopnoea index of > 5. This estimation is certainly minimal, as it concerns an active population and excludes subjects with previous cardio-respiratory history [212].

CLINICAL FEATURES

OSAHS

A diagnosis of OSAHS may be based on the clinical picture, which is often highly evocative, provided the clinician takes the trouble to ask a few questions which unfortunately are not included in the standard "patient's interrogation". But no combination of clinical symptoms can be considered as a reliable test [203].

The clinical picture is dominated by two features: one is diurnal *i.e.* sleepiness, and the other nocturnal *i.e.* snoring.

Daytime sleepiness may vary to a considerable extent. It may simply consist of a tendency to fall asleep in non stimulating situations (reading, repetitive tasks, monotonous driving...). It may be major, occurring in the course of activities, and be compromising to conjugal, social and professional life [43]. Its evaluation requires careful interrogation, defining the precise circumstances and frequency of untimely sleep onsets [160].

The other daytime symptoms are less evocative, but important to recognise as they may be circumstantial to diagnosis. These are headaches which usually occur in the morning, easing off when the subject gets up and into the morning routine, observed in roughly 20% of cases, but whose specificity is subject to debate [192]. Sexual problems in the form of reduced libido are not uncommon [85].

An OSAHS complication can sometimes be the key to the wider clinical picture: heart failure, angina, cardio-respiratory decompensation.

Snoring is often loud, and long-standing. It disturbs the subject's entourage, who sometimes observes that it is repeatedly interrupted by respiratory pauses; when respiration resumes it is accompanied by particularly loud snoring.

Nocturnal polyuria [126] is more common than enuresis, which is usually described [85].

Other nocturnal signs include: sudation, agitation, sleep talking, sleep walking, confused awakenings with automatic behaviour, which add to the disturbed sleep of these patients and their entourage.

UARS

The clinical symptoms of upper airway resistance syndrome overlap widely with those of sleep apnoea syndrome, dominated by daytime sleepiness and snoring. In certain cases snoring may be absent, and because of this, in accordance with the pathophysiology described, it is easy to understand how patients with a high critical pressure but low wakeful pressure threshold can develop sleep fragmentation in response to increased respiratory effort, without snoring.

Two notions worthy of attention are idiopathic hypersomnia on the one hand, and periodic leg movements in sleep, on the other.

A diagnosis of idiopathic hypersomnia should not be made without polygraphic recording which includes a recording of oesophageal pressure; indeed, a study published by the Stanford team reassessing 48 patients diagnosed with idiopathic hypersomnia showed that 15 were in fact subject to UARS [84].

Periodic movements in sleep can also pose a difficult diagnostic problem: the micro-arousal which terminates an episode of increased respiratory effort is often accompanied by motor activation notably affecting the leg muscle flexors. It is thus impossible to distinguish the polysomnographic recording of a patient with periodic leg movements in sleep from that of a patient with UARS. Indeed sleep fragmentation is found, associated with activation of the rear calf muscles, often with a slight increase in ventilation corresponding to the arousal reaction. The only way of detecting the gradual increase in respiratory effort prior to micro-arousal is to record oesophageal pressure.

To this extent it is not legitimate to attribute daytime sleepiness to periodic leg movements in sleep unless prior polysomnography has been conducted with a recording of oesophageal pressure.

Physical examination

Obesity is found to a various extent, depending on the chosen definition of obesity. In our population of 348 patients with an apnoea + hypopnoea index of over 20, 62% had a body mass index exceeding or equal to 30 kg/m². This indicates that although obesity is common, 38% of patients are not obese, and 9% of patients are of normal weight (Krieger, unpublished observations). A particular type of morphology is also worth noting, with a short, thick neck [85].

Arterial hypertension is often observed [85]. Polycythemia is also a classic sign, although this appears to be rare, affecting roughly 10% of patients [77, 114]; on the other hand, the search for the origin of unexplained polycythemia often reveals the presence of OSAHS [27].

The phonation of some patients is particular and may be evocative [62]. Peripheral oedema may be noted, even in the absence of heart or kidney failure [210].

LABORATORY TESTS

Diagnosis is confirmed by polysomnographic recording, which is the only way of obtaining a firm diagnosis [89]. The goal is to look for ventilatory anomalies, to clearly define the conditions in which they occur and, wherever possible, to understand their mechanism in order to adapt treatment accordingly. It is equally indispensable to determine the type and duration of apnoeas, as well as their repercussions, particularly on heart rate, blood oxygenation and sleep organisation. Furthermore, it is essential to evaluate the effects of treatment of whatever kind.

Polysomnography

Objectives

Despite recent developments, there are still too few centres equipped to conduct polysomnographic recordings of quality, in view of the extent of demand. This is why it is fundamental to clarify the target objectives, helping to avoid inadequate recordings which need to be repeated in better conditions. On the other hand, it is certainly useful to use "unravelling" examinations, whose objectives and constraints are very different in terms of sensitivity and specificity. These tests, which are additional to polysomnography, will be dealt with in the section on differential diagnosis.

Polysomnography analyses ventilation and its immediate repercussions, as well as sleep architecture and the behaviour of the sleeper. The indications and methodology of sleep cardio-respiratory polygraphy have been subject to a consensus study approved by the Association of Sleep Disorders Centers [145] and by the management committee of the American Thoracic Society [5]. For an analysis of the techniques used see chapter 9.

Hypopnoeas and apnoeas

Very few problems arise in recognising and characterising apnoeas if the recorded signals are of sufficiently high quality. The initial criteria for describing central, obstructive and mixed apnoeas, such as those proposed by Gastaut *et al.* [64] have been widely adopted: during central apnoeas ventilatory effort is interrupted, whereas it persists during obstructive apnoeas; mixed apnoeas begin as a central apnoea and end as an obstructive apnoea.

During certain apnoeas, repeated expirations are observed with no inspiration, described in terms of a valve phenomenon, expressing the ability of the upper airways to open under the effect of expiratory pressure, and its collapse under the effect of inspiratory depression. The weak volumes of expired gas, of no ventilatory efficacy, have been clearly identified in pneumotachography; using a capnograph or a thermocouple they issue an ample signal (as the expired air is heavily charged with CO_2 and heat) which may mask the apnoea (fig. 29.1).

The definition of hypopnoeas is a more delicate matter, as this basically depends on the type of data obtained. When quantitative data is available on ventilation, hypopnoea is simply defined by a reduction in ventilation to below a certain threshold, usually fixed at 50% of wakeful ventilation, or of stable ventilation prior to the event. In the case of non quantitative data, the definition usually relies on a combination of changes in the thermocouple signal associated with a variation in SaO_2 or sleep structure, these definitions varying from one team to another. An international consensus conference recently attempted to standardise the techniques and criteria used to diagnose respiratory anomalies during sleep, notably emphasising that thermal sensor recording was not adapted to evaluating ventilation [1]. The central or obstructive character of hypopnoeas can only be clearly ascertained by measuring upper airway resistance during the event. The detection of flow limitation is an interesting alternative to measuring upper airway resistance. The notion of inspiratory air flow limitation relies on a variable resistance model of the upper airway passage which acts like a tube with surface adhesion whose resistance increases with the increase of inspiratory effort, incurring flow limitation expressed as a flow curve plateau in the middle of the inspiratory phase. Flow limitation is a sensitive indicator of the increase in upper airway resistance during inspiration [95, 11]. A lengthening of pulse transit time, the transmission time of a pressure wave by the arterial walls between the opening of the aortic valve and the arrival of the pulse wave at the periphery, also constitutes an indirect indicator of the extent of respiratory effort. Indeed the intracardiac pressure at the origin of the pulse wave depends on intrathoracic pressure, and the pulse transit time increases as pressure in the rib cage decreases, with increasing respiratory effort [163]. This information enables a distinction to be made between central and obstructive respiratory events, with a high degree of sensitivity and specificity [7].

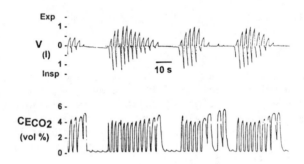

Figure 29.1. Limitations of indirect techniques for scoring ventilation. When ventilation is evaluated by pneumotachography (V, in l), the decrease in ventilation prior to the apnoea appears distinctly, whereas the capnogram (CECO2) indicates an increase in the signal; likewise, when a valve phenomenon occurs i.e. expiration without inspiration, this is clearly identified by pneumotachography, but does not appear with capnography when used alone. Although this cannot be demonstrated, due to technical reasons, recording ventilation by thermocouple is probably prone to the same errors.

By definition, hypopnoeas or apnoeas lasting at least 10 seconds are considered as pathological. Although this definition is widely accepted, its arbitrary nature should not be overlooked. Apnoeas commonly last for 30 or 40 seconds in NREM sleep and more than 60 seconds in REM sleep. The longest apnoeas can last over 3 minutes.

UARS

UARS is defined by the occurrence during sleep of a gradual increase in respiratory effort, with no detectable change in ventilation or SaO_2, ending in a micro-arousal.

The literature has yet to provide accurate data on how many of these events would lead to clinical symptomatology.

A diagnosis of UARS can only be made by recording oesophageal pressure during sleep, this being the only way of visualising the increase in respiratory effort, among the techniques currently available.

Analysing the repercussions of apnoeas

Repercussions on blood oxygenation and heart rate are part of routine polysomnography and will be dealt with in the chapter on the consequences of apnoeas. Many other parameters can be recorded, *i.e.* by gasometry (transcutaneous PCO_2, CO_2 in expired air, which is of particular interest in the case of an alveolar plateau at the end of expiration); from the haemodynamic aspect (systemic arterial pressure, pulmonary arterial pressure). This data is not collected in routine tests, but may be of interest in a particular situation, or in the framework of research.

Sleep analysis

It seems illusory to try to understand sleep-related changes in respiration without accurate information on sleep. In particular, as ventilation control is very different during NREM and REM sleep, the significance of ventilation changes can only be correctly interpreted if the state of sleep in which these occur is known. Likewise, to assess the gravity of an OSAHS, prior information is needed on sleep fragmentation; it is also necessary to recognise REM sleep, this is the state in which apnoeas are usually the most prolonged and hypoxemia, the most severe.

Sleep stage analysis is conducted using the widely accepted standard criteria of Rechtschaffen and Kales [169] which will not be described in detail here. But it is worth noting in passing that these criteria recommend scoring sleep tracings as "epochs" of 20, 30 or 40 seconds, or more. This type of scoring is ill-adapted to describing sleep disturbances during OSAHS where, for a period of 20 seconds, several changes in sleep stage may occur. The problem is partially resolved by introducing the separate scoring of micro-arousals [78] or the finer analysis of sleep stages.

Automatic digital analysis may provide more accurate information than that recommended by Rechtschaffen and Kales [169]. At their present stage of development, these automatic systems of analysis are appropriate for analysing sleep which is relatively undisturbed, but no system has yet proved its worth in analysing sleep as disturbed as that of OSAHS.

Polysomnography demonstrates the presence of strict parallelism between changes in sleep level and changes in ventilation. Sleep onset is accompanied by a reduction in ventilation amplitude which more or less rapidly leads to a pause in respiration of varying duration. When ventilation resumes it is accompanied by a change in EEG indicating lighter sleep or, more often, an arousal (fig. 29.2). The arousal is usually short, lasting several seconds, marked by the reappearance of alpha activity which is often masked by the presence of electromyographic artefacts. The arousals are not usually perceived by patients, who do not habitually complain of poor sleep quality. The ensuing sleep onset is once again associated with apnoea and the cycle repeats itself in a stereotyped pattern throughout the night.

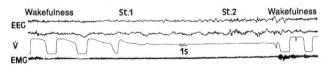

Figure 29.2. Parallel patterns of changes in sleep and wakefulness states and ventilation. Note the occurrence of apnoea after sleep onset, and the arousal concurrent with resumed respiration. Due to the high speed of the recording (15 mm/s) necessary to visualise EEG, the apnoea shown here is short. (V: ventilation, L/min)

Sleep is disorganised, essentially composed of light NREM sleep. When arousals are counted irrespective of their duration, including micro-arousals, they can often number several hundreds in the course of the night. Some patients with less severe OSAHS may attain deep NREM sleep stages 3 – 4; in this case the succession of apnoeas gives way to continuous ventilation (fig. 29.3), often associated with snoring and increased inspiratory oesophageal pressure, revealing the increase in upper airway resistance. These deep NREM sleep periods are usually observed in lateral decubitus.

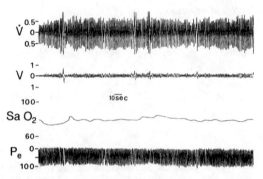

Figure 29.3. Continuous ventilation with increased respiratory effort. During periods of unstable sleep, this patient has obstructive apnoeas; during the period of NREM sleep (stage 3) shown here, recorded in lateral decubitus, there are no apnoeas, but a considerable increase in respiratory effort, witnessed as inspiratory oesophageal pressure (Pe) reaching – 100 cm H_2O. (V: ventilation, L/min; V: ventilatory volume, L; SaO_2:oxygen saturation)

REM sleep is fragmented like NREM sleep [36], with a difference: as apnoeas are usually more prolonged in REM sleep, periods of REM sleep between two arousals are longer. But on the whole, there is usually a reduction in total sleep quality during the night. Moreover it is not unusual for entrance into REM sleep to be associated with a prolonged apnoea. The arousal which ends this apnoea is often succeeded by the return to NREM sleep, "aborting" the REM sleep episode.

This relationship between sleep and ventilation changes is nevertheless liable to a certain degree of variability in time. Changes in ventilation are often less marked at the beginning of the night and restricted to hypopnoeas or apnoeas of relatively short duration. It is only after a certain time that longer apnoeic episodes are observed, probably due to a change in arousal threshold [152] . For this reason, short recordings are liable to give a false impression of the severity of sleep apnoea syndrome.

Recording the patient's behaviour

The patient's general movements are often worth examining to explain a change in ventilation or heart rate. This information can obviously be obtained by direct observation or video and transcribed to the recording. It can also be gathered directly by means of pietzo-electric quartz incorporated into the mattress of the bed used for recording.

Even more important than general movements, is the subject's sleeping position, which is fundamental in analysing variations in the severity of the respiratory disorder. Indeed it is not unusual for repetitive obstructive apnoeas in dorsal decubitus to give way to hypopnoeas associated or otherwise with snoring, or even to normal ventilation in lateral decubitus. Recording sleeping posture is essential for considering positional treatment for OSAHS [28].

Data processing contribution

The quality of the screens now available provide the same degree of accuracy in signal reproduction as in traditional tracing on paper, provided the sample frequency obtained is sufficient.

In addition to signal representation, many systems offer an "automatic" analysis of the respiratory signal, in particular of ventilation and oxymetry, whose validity largely depends on the quality of the signal obtained. These automatic analysis procedures are operational and reliable for apnoeas, but should be used with more caution when it comes to hypopnoeas. It is vital to be able to obtain a raw signal to check the quality of automatic analysis if necessary.

Diagnostic criteria for obstructive sleep apnea-hypopnea syndrome

Apnoea index threshold values (number of apnoeas per hour of sleep) have been proposed to define the diagnosis of obstructive sleep apnoea syndrome. The limit of 5 as initially proposed [85] has been subject to debate [16], particularly in elderly subjects. More recently, a combined index taking account of both apnoeas and hypopnoeas and whose threshold value has been proposed at 10 or at 20 episodes per hour of sleep, has tended to be adopted. The problem is made all the more difficult by the variability in respiratory anomalies from one night to the next, the lower the number of apnoeas, the greater the variability.

In fact, in clinical practice, the arrival of simple effective treatments [198] means that the problem is posed in different terms, all the more as limits with the upper airway resistance syndrome are difficult to fix, hence an emerging tendency to use OSAHS for the clinical spectrum of sleep-disordered breathing, from apnoea, to hypopnoea, to UARS [200]. When unequivocal clinical signs exist, and all other possible aetiologies have been carefully excluded, the identification of sleep fragmentation in relation to respiratory events (hypopnoeas and/or apnoeas and/or respiratory efforts) of sufficiently high number (international recommendations [1] set the threshold to 5 respiratory effort related arousals (RERAs) per hour of sleep) should lead to proposing trial treatment with continuous positive pressure.

On the whole, our view is that the minimum requirements for conducting a polysomnographic examination in good conditions should include an identification of sleep stages (EEG, EOG, EMG), oronasal airflow, respiratory effort and a measurement of SaO_2 and heart rate. These recommendations entirely conform with those of the Association of Sleep Disorders Centers [145] and the American Thoracic Society [5]. The main thing is to insist on the need for well trained medical and paramedical staff. Indeed the most sophisticated electronic device in inexperienced hands may be a source of more errors than no electronic equipment at all.

Other laboratory tests

Few laboratory tests are of any diagnostic value apart from night polysomnography. In return, some are useful in defining the context or orienting treatment.

A multiple sleep latency test (MSLT), or a maintenance of wakefulness test, may be used to objectify daytime sleepiness [164], or to look for it when the subject does not experience it subjectively. In this case the maintenance of wakefulness test (measuring the capacity to remain awake) appears to be better adapted to evaluating functional discomfort than the MSLT (which measures capacity to fall asleep) [179].

The functional respiratory checkup shows normal or will reveal a restrictive syndrome related to obesity. Any associated obstructive syndrome is important to look into because of its prognostic significance (*cf.* associated forms).

ENT examination often reveals a tight congestive canal, restricted by a long soft palate, a hypertrophic pendulous palate, hypertrophic posterior pillar, a large tongue, confirmed by various radiological tests [187]. The pharyngeal canal is restricted particularly at the velopharynx and superior oropharynx (see [188] for a review).

This goes hand in hand with an increase in pharyngeal resistance, and a reduction in the ratio between inspiratory and expiratory flow to 50% of vital capacity, with anomalies in the flow/volume curves [132].

COURSE OF ILLNESS

Natural history

The fact that many apnoeics are long-term asymptomatic snorers has led to the notion of the progressive development of simple snoring toward recognised then complicated OSAHS, leading to the final form known as the Pickwickian syndrome with cardio-respiratory failure [141]. This is a tempting picture but one which fails to account for elderly patients who are long-term snorers and who remain at the simple snoring or OSAHS stage with no complications. Certain factors may favour this progressive course, including alcohol which has been clearly identified [17, 100] as well as overweight, "aging", an associated broncho-pulmonary condition, etc. The development of a local neuropathy related to the vibratory traumatism of snoring may explain the progressive increase in pharyngeal collapsibility [63].

Short term complications

Traffic accidents are roughly 2 to 7 times more common than in a control group which did not show signs of sleepiness [2] so patients should be warned of the risks they incur to themselves and others.

Various heart arrhythms have been described, which is a possible cause of sudden death during sleep, although little is known about the frequency of this, as it is hard to evaluate [71].

Long term complications

Haemodynamics

Hypertension, ischaemic, cardiac or cerebral signs, pulmonary hypertension and chronic pulmonary heart disease are likely to be associated with OSAHS. Their frequency and mechanism are dealt with in the section on pathophysiology (consequences of apnoeas).

Morbidity/mortality

The fundamental question of knowing how untreated OSAHS will develop, is confronted by a major difficulty: no prospective study has been conducted with a random treatment modality. The arguments drawn from retrospective studies are too convincing today to permit a study of this kind, even if it is hard to establish the role played by parasite factors such as android obesity, dyslipemia, smoking, resistance to insulin, etc.).

A retrospective study by He *et al.* [88] carried out on 385 patients (selected from 706 files) shows significantly higher mortality among patients treated by ineffective means compared to patients who had tracheotomy or continuous positive airway pressure. The higher the apnoea index, the higher the mortality rate. Other studies [157] show higher mortality among untreated patients, reporting a high rate of death from vascular causes. Another study [71] showed no difference in mortality between OSAHS and controls, but 2 of the deceased "controls" suffered from chronic obstructive broncho-pneumopathy.

CLINICAL VARIANTS

Associated forms

OSAHS has been described in association with diverse pathologies some of which may have been associated at random. Nevertheless, in certain cases, the treatment of the associated pathology is accompanied by an improvement in sleep apnoeas, suggesting a possible aetiological role.

Anomalies of the upper airways

Almost all the upper airway anomalies in which the calibre is restricted, have been described in association with OSAHS. These mainly include enlarged tonsils, deviation of the nasal wall, mandibular anomalies, supraglottic oedema, tumoral formations. Nevertheless, when it has been possible to redress the upper airway anomaly, there is rarely any significant improvement in sleep apnoea, except in the case of children in whom the removal of the tonsils and/or adenoids or the treatment of cranofacial dysmorphia is generally effective.

Endocrine diseases

Hypothyroidism and acromegaly have often been described in association with OSAHS. The OSAHS appears to be particularly frequent in acromegaly in which 93% of patients with evocative symptoms and 60% of asymptomatic patients had OSAHS [74]. In both cases, treating the endocrine disease can improve sleep apnoeas. The mechanism by which these endocrine diseases induce obstructive apnoeas during sleep may be linked to the morphological changes they induce in the upper airways. However the role of hypophysial hormones in ventilation control has recently been designated [75].

Neuromuscular diseases

Various conditions whose common mechanism may be their interference with the neuromuscular control of ventilation have been described, particularly those of myotonia, cervical cordotomy, poliomyelitis, syringomyelobulbia, olivopontocerebellar degeneration, multiple system atrophy, the diabetic neuropathies [31, 127, 128, 140].

Respiratory diseases

On the basis of studies of patients selected according to daytime sleepiness, OSAHS was more common in the case of chronic obstructive pulmonary disease. A study of a less biased patient group failed to confirm this high frequency. Despite this, there is a frequent association between chronic obstructive pulmonary disease and OSAHS because of the higher frequency of each condition. This association generates severe forms of OSAHS, due to considerable hypoxemia during sleep, daytime hypoventilation and the development of cor pulmonale. This association is often referred to as overlap syndrome [57].

Age-related

In children

The important problem of the relationship between sleep apnoeas and sudden infant death will not be dealt with here.

The true frequency of OSAHS in children is unknown, but appears to be high, since 61% of children listed for adenotonsillectomy unrelated to sleep disorders showed more than 3 desaturations per hour compared to 0 in controls (this frequency drops to 13% six months after the operation) [194]. The condition is seen in both sexes with roughly the same frequency, and the mean age of diagnosis is around 4 years (18 months to 12 years).

The clinical picture in the child may be highly comparable to that of the adult, but certain particularities must be distinguished:
- daytime sleepiness is often replaced by hyperactivity, and behaviour problems which are associated with poor performance at school [25, 194];
- oral breathing is often observed [24];
- height and weight are often retarded; obesity is less common than in the adult [166];
- enuresis is more frequent than in the adult;
- cor pulmonale or systemic hypertension may be a first sign. The first observations of dramatic right cardiac failure during obstruction of the upper airways were made in children;
- tonsil or adenoid hypertrophy often appears as an aetiological factor. Craniofacial morphological anomalies are commonly observed, and the frequency of the syndrome has been emphasised in various dysmorphia (Pierre-Robin, Prader-Willi, Down, Apfert, Pfeiffer, Teacher-Collins, Hurler, de Crouzon etc. [166, 185, 191]).

Diagnosis depends on polygraphic recording which in the young child can be carried out during a nap, provided this includes REM sleep [65].

Tonsillectomy and/or adenoidectomy are often effective in the short term [65]. However, the frequency of respiratory complications which appear at an early post-operative stage in these patients [148], calls for rigorous postoperative surveillance; moreover, the frequency of adult OSAHS sufferers who had tonsillectomies in their childhood, raises questions as to the long-term efficacy of this treatment. In the case of failure, uvulopalatopharyngoplasty (UPP) can be proposed [76]. Continuous positive airway pressure (CPAP) is reserved for very severe cases, or where surgery has failed [80].

The accent has recently been placed on the possibility of a respiratory difficulty expressed only as an increase in respiratory effort (isolated or alternating with periods of hypopnoeas or apnoeas), corrected after adenotonsillectomy.

In the elderly

Sleep-related respiratory anomalies have been described in a high percentage of healthy or randomly selected elderly subjects [92].

But it is not certain that this observation, setting aside clinical symptoms, expresses a condition of greater morbidity or mortality [135, 162]. This apparent better tolerance of apnoeas in elderly subjects may be linked to the absence of hypertension in relation to lower sympathetic reactivity.

When it is symptomatic, OSAHS in the elderly subject shows the same form of clinical expression as in the adult [15].

Sex/pregnancy-related

OSAHS is rare in women before the age of the menopause, but there is no fundamental difference in the clinical picture. Certain particularities have however been reported: morning headaches and fatigue are more frequent, as are difficulties getting to sleep; above all, hypertension is less common [4]. In some cases symptoms may be more discreet giving rise to an underestimation of the true frequency in women [4].

Because it is so rare before the menopause, OSAHS is rare during pregnancy and has not been studied systematically. Observations of foetal hypotrophy [30, 81] have however been reported.

Aetiological

In the vast majority of cases, OSAHS is primitive, i.e. no specific cause can be determined. Commonly observed anomalies in the soft pharyngeal tissue are in fact considered as secondary to the micro-vibratory traumatism of snoring, rather than as aetiological factors. Sometimes an endocrine affection or morphological anomaly of the upper airways may be observed (*cf. supra*), whose causal role is not always easy to demonstrate. More rarely, a toxic factor is responsible (organic solvents [151]).

CENTRAL SLEEP APNOEA SYNDROMES

While the framework for obstructive sleep apnoea syndromes is relatively well defined, that of central sleep apnoeas is far less distinct.

On one hand, the definition of central sleep apnoea syndrome is ambiguous. According to White, the criterion is that 55% of apnoeas are central [208]. However, this criterion may correspond to situations where the initial disturbance is an obstruction of the upper airways, central apnoeas occurring as a response to postapnoeic hyperventilation, bringing into play the hypocapnic apnoea threshold [98] (see pathophysiology below). For this reason, central apnoea syndromes should only be referred to in cases where virtually all the episodes are of central origin.

On the other hand, while the physiological decrease in ventilation during sleep is well tolerated in the normal subject, in cases where there is permanent alveolar hyperventilation – whatever the mechanism - this will lead to a worsening of hyperventilation which may be incompatible with maintaining sleep, creating a conflicting situation between the two functions. This results in ventilation instability with central hypopnoeas and apnoeas, consecutive to sleep instability. These situations, which can be seen in severe chronic obstructive pulmonary disease as well as other conditions of respiratory failure [84], as in muscular or skeletal diseases, are directly linked to broncho-pulmonary pathology. They appear to respond to different "primitive" sleep apnea syndrome mechanisms, with no permanent alveolar hypoventilation. They should not be included in the framework of central sleep apnoea syndromes, just as sleep-related alveolar hypoventilation in chronic obstructive bronchopneumopathies should be excluded from this framework.

These aspects will be dealt with at the same time as alveolar hypoventilation and congenital alveolar hypoventilation with central apnoea, in the section on "Central alveolar hypoventilation syndrome".

Moreover, a number of OSAHS have been described as central using inadequate means of ascertaining the non obstructive nature of the episodes described; all the more so when this involves hypopnoeas, whose characterisation as central or obstructive requires a measurement of upper airway resistance, which rarely features in the routine test.

Furthermore, attempts have been made to distinguish central apnoeas from Cheyne-Stokes dyspnoea [84], without defining the criteria for differentiation, confusing matters even more.

The problem is really complex, as the factors which control inspiratory ventilation activity, upper airway stability and the thoracopulmonary mechanism are interwoven and the information available during sleep does not provide a clear answer. The efficacy of continuous positive airway pressure on Cheyne-Stokes dyspnoea in heart failure is a clear illustration of the way in which these mechanisms are interwoven [22]. The rarity of these symptoms makes it difficult to form homogeneous groups, rendering their study even more complicated.

Clinical features

Guilleminault *et al.* [79] attempted to make insomnia, rather than the hypersomnia, characteristic of obstructive apnoeas, the distinctive feature of central sleep apnoea syndromes. These "central" apnoeas are considered today [84] as being due to an obstruction of the upper airways (*cf.* inframechanisms).

As a result, no specific clinical picture of central sleep apnoea syndrome has emerged. Bradley *et al.* opposed two forms, the more common (13/18) being barely distinguishable from the usual OSAHS picture, the rarest showing cardio-respiratory failure interspersed with acute respiratory failure [20].

In our own experience, the diagnosis is usually made in cases where polygraphic recording is called for after the observation of apnoeas of patients under strict surveillance, due to a cardiac, neurological or metabolic pathology, particularly renal, which is confirmed by a number of published observations. The circumstances vary (papillary necrosis, toxic hypothalamic syndrome, arterial compression of the medulla [150], trichlorethane intoxication [151]; it may sometimes be a case of central apnoeas which persist after tracheostomy for obstructive apnoeas [58].

Mechanisms

A priori, the mechanism of a central apnoea is simple: it is an interruption (or decrease in the case of hypopnoea) of ventilation control. Various lesions of the central nervous system, particularly if they affect the bulbopontine "centres", may conceivably lead to apnoeas. The fact that these are only seen at night may be explained by behavioural ventilation control being inactive during sleep.

The interaction between ventilation regulation and sleep leads to a situation of instability causing ventilation control to oscillate on sleep onset resulting in periodic breathing in healthy subjects. The observation of central apnoeas may from this point onward merely be an expression of the accentuation of this mechanism, with no necessarily pathological significance. One may sometimes ask whether the initial disturbance is not one of sleep rather than ventilation; the demonstration of a drop in the number of central apnoeas under trialozam suggests that sleep instability is at least an aggravating factor [18].

Various situations are known to aggravate this instability, such as altitude, leading to periodic breathing with central apnoeas secondary to altitude hypoxia.

A particular place is occupied by Cheyne-Stokes dyspnoea with its regular waxing and waning of ventilation. It expresses instability in the functioning of gap detection in ventilation regulation [86] and may either result from a slowing in the transmission of information between the receptors and the centres regarding the regulated variable (supposed mechanism of Cheyne-Stokes dyspnoea of cardiac origin), or a gain in control (which may be the mechanism of "neurological" Cheyne-Stokes dyspnoeas by disinhibition). Here again the inactivation of the behavioural system may result in ventilation instability. The question is to draw the borderline between the normal and the pathological and to gather the relevant elements to decide on the need for treatment: hypoxemia severity, sleep fragmentation (in the knowledge that it is not always easy to distinguish between sleep instability and ventilation instability as the primitive factor).

Finally, it has been suggested that central apnoeas are secondary to upper airway obstruction, in the manner of a diving reaction: stimulation of the upper airways leading to a respiratory arrest reflex. This hypothesis is based on an experimental demonstration of apnoeas induced by mechanical stimulation of the upper airways [138] and on the observation that continuous positive airway pressure eliminates central apnoeas in pure central apnoeas [104].

Treatment

Short term success has been reported with acetazolamide [211], a carbonic anhydrase inhibitor, leading to acidosis with a leftward shift in the curve of ventilation response to CO_2; the results remain anecdotic and have not been confirmed in the long term; the possibility of inducing obstructive apnoea by acetazolamide has been reported [205].

One case was improved with oxygenotherapy. In other cases, oxygenotherapy has induced a passage of central apnoea with obstructive apnoea [67].

Apart from the use of mandibulary advance devices, or continuous positive airway pressure [104], central apnoeas of obstructive mechanisms, whose particularity is that of requiring higher pressure than that used for obstructive apnoea syndrome treatment, rely essentially on treatment using techniques of ventilation assistance during sleep.

These use either volumetric intermittent positive airway pressure, or ventilation at differing pressures for inspiration and expiration (BiPAP). The former is a classic technique which has already proven its worth and whose innovation consists of a nasal mask, simplifying its use at home and during sleep [51]. The second is more recent and is promising but its indications and limitations are still being investigated [178].

On the whole, the problem of central apnoea syndromes remains to be clarified [20]. Until this is done, the main thing is to determine when treatment is justified, and whether continuous positive airway pressure or ventilation assistance is effective and tolerated.

DIFFERENTIAL DIAGNOSIS

Snoring

Snoring is, as we have seen, a basic symptom of OSAHS. The question often posed is that of knowing, in the case of a chronic snorer, whether it is "pure" snoring (or simple snoring) or one symptomatic of OSAHS. Neither the clinical aspects nor the dimensions of the upper airways [69] provide an answer to this question. This is probably where sensitive screening tests are the most useful in confirming the absence of apnoeas in a patient with dubious symptomatology. Indeed it appears less and less certain that the distinction between the "pure" and the apnoeic snorer is really important; the notion of "upper airway resistance syndrome" actually places greater emphasis on the increase in respiratory effort during sleep, with or without associated snoring, and on the ensuing sleep fragmentation.

Daytime sleepiness

The conditions responsible for daytime sleepiness are dealt with in other chapters of this work. However it should be remembered that it is not always easy to recognise narcolepsy, when diagnosis is late, and considering that narcolepsy does not preclude snoring. Moreover, periodic leg movements during sleep may be associated with daytime sleepiness and disrupt certain screening tests by producing periodic changes to heart rate reminiscent of the pattern seen in OSAHS.

Polysomnography

Carried out in the right technical conditions, polysomnography always provides answers to the questions posed: are there any abnormalities in ventilation during sleep? What is their nature? How severe are these? The only source of error, apart from technical inadequacies, is the insufficient quality of sleep [41], which is a source of false negatives. For this reason it is vital to dispose of information on sleep quantity and structure.

When sleep quality is insufficient, there should be no hesitation in repeating the recording. The need for sleep recording was discussed [40] in the light of a study showing that an equal number of patients were correctly diagnosed as long as the hypopnoea + apnoea index was over 15, whether sleep parameters were recorded or not. The work concluded that limited means may be adequate when OSAHS is evident. But it is precisely in the other cases that problems arise and accurate information is necessary.

Errors occur when the recording is insufficient:

Insufficient in length: in this respect, a short recording or one without REM sleep may, as we have seen, give a false idea of the severity of apnoeas. This is particularly true for recordings limited to 5 hours in the day, which underestimate sleep destructuration, hypoxemia severity leading to approximately 9% of false negatives compared to polysomnography which lasts all night with a diagnostic criterion of 1A > 5 and/or 1HA > 10. For a restricted recording of 2 hours of night sleep, the percentage of false negatives is around 5% for the same diagnostic criterion [177].

Insufficient data gathered: recordings limited to one or other of the variables mentioned above may give rise to false interpretations. In particular a recording limited to SaO_2 alone risks misinterpreting apnoeas lasting twenty seconds in a young healthy subject. Furthermore, they provide no information on the central or obstructive nature of the disruptions observed; Cheyne-Stokes respiration may thus be mistaken for OSAHS [46]. The problem is particularly important when there is a respiratory and/or cardiac pathology.

Screening problems

The real problem is to evoke this as a possible diagnosis when confronted with any one of the symptoms referred to above, and to indicate recording. The question "Do you snore?" despite its seeming triviality, should be asked as part of the systematic interrogation of all patients. A positive response should raise suspicion of OSAHS; the hypothesis cannot be ruled out by clinical examination alone, due to the possible asymptomatic forms [56]. Thus the need for screening tests to limit the number of patients referred for polygraphic recording.

Flow-volume curves often show a serrated pattern and/or a decrease in the ratio between inspiratory flow and expiratory flow of 50% of vital capacity, but fail to show sufficient sensitivity or specificity to be considered as a satisfactory test [132].

Different techniques have been developed based on recording oxymetry alone; they may well be satisfactory when patients have wakeful hypoxemia, but are insufficient to exclude the diagnosis in patients whose daytime blood gases are normal.

Simultaneous recording of respiratory noises, heart rate and SaO_2 provide a detection of OSAHS on the basis of an A + H index > 10 with sensitivity of 96% when only respiratory noises are scored, 92% when only SaO_2 is scored, 58% when only heart rate is scored and 100% when all three are scored simultaneously, but at the price of poor specificity [193].

PATHOPHYSIOLOGY

Mechanisms of respiratory events

Normal regulation of ventilation creates conditions of instability which are revealed during sleep, and which can be associated with a partial or total obstruction of the upper airways (see Chapter 4).

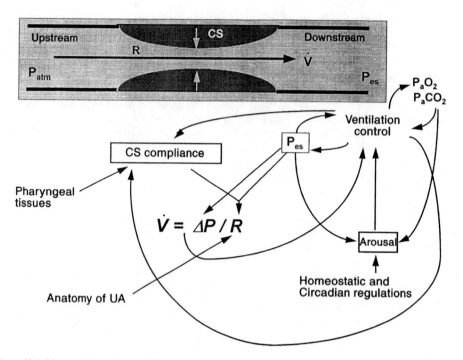

Figure 29.4. Diagram of the relationship between 3 variables (CS compliance, respiratory effort represented by Pes and arousal) responsible for the different syndromes created by partial or complete obstruction of the upper airways (UA) during sleep. The box shows the model of the structure of the upper airways, comprising a collapsible segment (CS) between 2 rigid segments upstream and downstream; the extent of the CS depression regulates the value of R (CS resistance); this depression depends on CS compliance and the value of P_{es} (oesophageal pressure). The flow in the segment (V) is given by Poiseuille's law with DP = atmospheric pressure (P_{atm}) – P_{es}.

Pathophysiology

Figure 29.4 represents a diagram of the structure of the upper airways which can be modelled as a collapsible tube (a soft pipe) corresponding to the pharynx, set between 2 rigid segments corresponding to the nasal fossa upstream and the tracheobronchial tree downstream; this model is referred to as Starling's resistance. According to Poiseuille's law, the flow is proportionate to the pressure differences between the extremities of the tube divided by the resistance of the system. Upstream pressure is equal to atmospheric pressure; downstream pressure, which is variable, is generated by the activity of the inspiratory muscles; it is best evaluated by measuring oesophageal pressure, reflecting pleural pressure. Because of the collapsible nature of the pharyngeal segment, airflow resistance is also variable and will depend on 2 factors: 1) pressure in the system which during inspiration tends to balance out with pleural pressure, thus becoming negative (lower than atmospheric pressure), 2) the capacity of the pharyngeal wall to collapse under this negative pressure; this capacity is measured by the compliance (C = $\Delta V/\Delta P$, the difference of volume produced in an extendible structure by a difference in pressure); it chiefly depends on the tonicity of the pharyngeal tissues and the activity of the dilatory muscles of the pharynx, themselves under the control of the general regulation of muscle tone and ventilation control.

It would thus appear that intrathoracic pressure intervenes in two different ways in the airflow generated during inspiration: on one hand it represents the generating motor force of this flow; on the other, it is also the force which will cause the pharynx to collapse and thus create a resistance to the passage of air. The opposing effects of respiratory effort also explain that after a certain point, any additional increase of inspiratory pressure will not lead to an increase in flow; as a result, although intrathoracic pressure decreases (becomes more negative) during inspiration, the flow curve shows a plateau effect or inspiratory flow limitation (IFL) when the increase of pressure difference (ΔP) is cancelled by an equal increase in resistance (R), cf. Poiseuille's equation (fig. 29.4) [39].

The diagram still requires 2 basic notions to be complete: on one hand, ventilation control is regulated by metabolic needs expressed as arterial PO_2 and PCO_2 values, as well as mechanoreceptors which respond precisely to intrathoracic pressure; the level of vigilance is also an important determinant of the level of ventilation control. On the other hand, as the amplitude of respiratory effort brings the mechanoreceptors into play, it constitutes an arousal stimulus, the efficacy of which will depend on the strength of the stimulus (amplitude of respiratory effort) and the sensitivity of the system (measured by the arousal threshold).

These mechanisms help to explain how an increase in airflow resistance, a physiological phenomenon during sleep, can lead to different situations in function of: 1) the extent of increase in respiratory effort in response to this increased resistance, 2) pharyngeal compliance, and thus capacity to resist the increase in negative pressure thus generated and 3) the arousal threshold, especially in response to an increase in intrathoracic pressure.

Thus,

- if pharyngeal compliance is weak, the increase in respiratory pressure in response to increased resistance will not result in any change to ventilated flow and, if the arousal threshold is high, the system will remain stable; conversely, if the arousal threshold is low, the increase in respiratory effort will result in a micro-arousal, a sudden drop in upper airway resistance and a short increase in flow despite reduced respiratory effort; at the following sleep onset, the increase in upper airway resistance will again be followed by an increase in respiratory effort leading to a new micro-arousal, and so on; the system is destabilised. This situation corresponds to upper airway resistance syndrome.

- if pharyngeal compliance is high, the increase in respiratory effort will lead enhanced resistance such that inspiratory flow is no longer maintained despite (and because of) the increase in respiratory effort; if the decrease in flow is weak, non detectable or lower than the standard definition of hypopnoea, we are again in the framework of upper airway resistance syndrome; if the decrease in flow is greater obstructive hypopnoea occurs.

- if pharyngeal compliance is very high (or the amplitude of respiratory effort very marked), pharyngeal collapse will be complete, resistance is infinite, the flow is zero, and obstructive apnoea occurs (fig. 29.5).

The model will also predict situations which are less often described, where a high arousal threshold would ensure sleep stability, and thus ventilation, at the cost of stable hypoventilation if respiratory effort is low, or at the cost of permanently high respiratory effort if pharyngeal compliance is weak; the latter situation may correspond to that of the "pure" snorer.

This model also adds insight into the importance of the notion of critical occlusion pressure which is precisely the pressure inside the collapsible segment which shows a complete collapse; this pressure best defines pharyngeal behaviour, and depends directly on its compliance; it is estimated at -15 to -8 cm H_2O in normal subjects, -8 to -4 cm H_2O in "pure" snorers, -4 to 0 cm H_2O in the case of obstructive hypopnoea, and 0 to 5 cm H_2O in obstructive apnoea.

This notion explains the way in which the rapid vibrations responsible for snoring are generated: when critical pressure is close to atmospheric pressure, weak inspiratory effort will cause pressure to drop to below critical pressure and thus lead to occlusion; this occlusion will set the collapsible segment in balance with atmospheric pressure, in excess of critical pressure enabling the segment to reopen; but if this inspiratory effort continues it will cause pressure to drop again to below critical pressure, and thus lead to another occlusion which again enables a balance with atmospheric pressure, etc., causing the pharyngeal walls to flutter. Rapid repetition of the open-close cycles causes the quick vibrations of the air column during inspiration, generating snoring. The same pattern applies to ventilation in continuous positive airway pressure when nasal mask pressure approaches critical pressure [68].

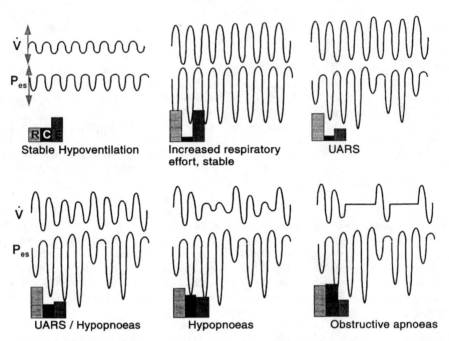

Figure 29.5. Diagram of different syndromes which occur with partial or complete obstruction of the upper airways during sleep, in function of the combination of the 3 variables represented by the height of column. R: respiratory effort, C: compliance of collapsible segment, E: arousal threshold, V: upper airway flow; P_{es} oesophageal pressure; the arrows indicate "normal" levels of V and P_{es}.

Factors of upper airway occlusion

Emphasis has focused on static anatomic factors, on the basis of the morphological anomalies reported in the upper airways (micrognathia, retrognathia, enlargement of the base of the tongue) which are often associated with OSAHS. Apart from characterised anomalies, emphasis has also been placed on the dimensions of the upper airways, especially the nasopharyngeal dimensions, which are narrower than in controls [147]. However all these studies have shown a considerable overlap between patients and controls, with patients both with and without upper airway narrowing [95b, 176]; some studies have found no differences in pharyngeal dimensions between healthy, non apnoeic and apnoeic snorers [165, 174] but greater hypopharyngeal differences in non apnoeic snorers [165]. One interpretation of this surprising result may be that a narrow hypopharynx would cushion negative thoracic pressure, stopping it from mounting to the oropharynx [165]. Inversely, subjects with narrow upper airways do not always develop OSAHS. And it is important to remember that the alterations in soft tissue may be consecutive to the repeated microtraumatism of snoring and the high intrathoracic pressure developed by the inspiratory muscles in the struggle against the obstacle, an interpretation which is supported by the lack of correlation between upper airway dimensions and OSAHS severity [212].

In fact, beyond static dimensions, upper airway occlusion chiefly depends on compliance, i.e. the ability to modify in volume when subjected to pressure gradients. This compliance is greater during sleep in OSAHS than in snorers and controls, illustrating the greater "collapsibility" of the upper airways [102, 103]. The question remains as to whether this greater compliance is due to anatomic changes (linked to fatty deposits, or a functional change in muscle or conjunctive tissue) or to a functional change in the state of activation of the muscles they are composed of. The latter notion suggests a disorder in upper airway muscle control which would become abnormally hypotonic during sleep. Evidence of the inability of patients with OSAHS to increase pharyngeal muscle activity in response to an experimental occlusion of their upper airways during sleep adds considerable weight to this idea [102, 103]. The role of an impairment of the "protective" muscles is also suggested by facilitating upper airway obstruction by oropharyngeal anaesthetic [29]. Lastly, changes in the ultrastructure of UA muscles or the development of a local neuropathy [63] may play a role in this.

In fact the two types of factors, mechanical and functional, appear to intervene to a varying degree depending on the patient. One hypothesis combining mechanical and functional factors would be that upper airway muscle activity is increased during wakefulness to compensate for the reduced calibre, and that at the onset of sleep this compensation no longer functions, leading to the occlusion of the upper airways [208]. It is probably these central factors which account for the failures of surgical treatment for OSAHS.

The role of the timing of this activation has also been stressed; upper airway muscles must indeed already be under pressure when submitted to negative inspiratory pressure; a delay in upper airway muscle activation could also contribute to their collapse [96, 195].

Extrinsic factors also intervene in ventilation control. These include alcohol or benzodiazepine intake [100] which share the capacity to increase muscular hypotonia, on the one hand, and to reduce the arousal threshold on the other, so that the arousal authorising the onset of respiration may be delayed [13]. This effect has not been found for all the benzodiazepines. Another factor of variability is sleeping position: dorsal decubitus aggravates the situation compared to lateral decubitus to the extent that weight on the tongue tends to bring together the anterior and posterior pharyngeal walls.

From the earliest studies on "Pickwickian" syndrome, a decisive role was attributed to obesity in generating OSAHS anomalies [26]. There is still some debate as to whether obesity is a causal factor (as suggested by the cure of certain (rare) OSAHS after weight loss) or simply an aggravating factor, suggested by the presence of OSAHS in patients who are not obese, and the absence of cured sleep apnoea after weight loss in certain cases. It should also be noted that OSAHS was far from being systematically present (41%) in a group of grossly overweight subjects (mean BMI 50 kg/m2), and more common in men than in women (77 vs 7%) [167].

Patterns of occlusion

While causal factors remain a controversial area, the development of the occlusion has been clearly described, thanks to cineradiographic analysis: the onset of obstruction is always oropharyngeal, usually associated with a backward movement of the tongue, followed by the downward aspiration of the soft palate, extending to the hypopharynx (the retrobasilingual pharynx) in 2/3 of cases, and to the pharyngolarynx in 1/3 of cases [159].

However techniques for measuring pressure in stages indicate an obstruction which starts at the velopharynx or the retropalatal region in 56% of subjects, and the retrobasilingual oropharynx in 44% of NREM sleep cases, often more caudal in REM sleep [189]. The locus of obstruction is constant for a given subject, but varies from one patient to another [97] whether nasopharyngeal, oropharyngeal or hypopharyngeal.

Sleep and ventilation instability

All things considered, obstructive sleep apnoea may be interpreted as the incapacity to cope with the increase in upper airway resistance normally associated with sleep, due to the inadequate response of upper airway and/or diaphragmatic muscles; the more narrow and compliant the upper airway, the more likely the occlusion. The physiological consequences of obstructive apnoea tend to maintain patients in a self-perpetuating condition of sleep and ventilation instability.

The intimate mechanisms of these anomalies are unknown. Observation of familial cases [50, 170] as well as evidence of a hereditary transmission or tendencies for the upper airways to narrow during sleep suggests that genetic factors may be implicated. The genetic determinants do not seem to be exclusively linked to anatomic changes [50]. The marked male predominance before the menopause, and the possible induction of obstructive sleep apnoea by introducing male hormones [108] suggests a hormone factor.

Once the upper airways are closed, the response of the ventilation system is utterly inadequate, as the increased activity of the diaphragm will only worsen the situation by generating increasingly negative intrathoracic pressure, aggravating the collapse of the pharynx. This results in gradual asphyxia which can only be relieved by the effect of arousal. This arousal is accompanied by an abrupt increase in upper airway muscle activity and deep inspiration. Arousal has been attributed to hypoxia through the activation of carotid chemoreceptors [197]. The varying duration of apnoeas may be explained by the progressive increase in arousal threshold consecutive to sleep fragmentation. Nevertheless the moderate lengthening of apnoeas during hyperoxia and their termination despite the absence of hypoxemia implies that other mechanisms are at work activated by diaphragmatic fatigue or increased inspiratory effort [66], or by upper airway mechanoreceptors [101]. This hypothesis is supported by the delay in arousal following upper airway occlusion in the healthy subject when upper airways have been anaesthetised [12]. The increase in negative intrathoracic pressure which expresses the increase in respiratory effort, is now considered as the arousal stimulus enabling the apnoea to terminate [66, 112].

Consequences of apnoeas

The consequences of apnoeas during sleep are multiple; we will distinguish those which occur during sleep, in direct relationship with apnoeas, and those which persist into wakefulness.

Immediate (apnoea-related)

Sleep fragmentation

To the extent that the resumption of respiration requires a lightening of sleep or an arousal, the sleep of these patients is seen to be totally disordered, sleep episodes lasting several tens of seconds, interspersed with brief arousals lasting several seconds.

Hypoxemia

During each apnoea, no carbonic gas is rejected, and oxygen is no longer carried to the pulmonary alveoli. The instantaneous evolution of $PaCO_2$ is poorly understood, due to the lack of a measuring device to monitor this instantaneously *in vivo*. The course of blood oxygenation is better known. Each apnoea is accompanied by hypoxia which may be particularly severe, attaining SaO_2 values of 60% or less, corresponding to $PaCO_2$ levels of the order of 20 to 30 mmHg; hypoxemia is not always corrected in periods of ventilation between apnoeas. In fact, apnoeas last from 10 seconds to 3 minutes, usually between 30 and 60 seconds, while periods of interapnoeic ventilation are often limited to 3 to 5 respiratory cycles. Postapnoeic ventilation amplitude, lower in REM sleep than NREM sleep, depends on the extent of hypoxemia during the preceding apnoea, and correlates with the hypoxic ventilation response measured during wakefulness [180].

Apnoeas which are particularly prolonged accompanied by severe hypoxemia may sometimes be responsible for hypoxic convulsions [32]. For apnoeas of equal duration, hypoxemia severity is greater where initial SaO_2 is low and apnoea starts at a lower pulmonary volume [54]. Expiration without inspiration observed during certain apnoeas, especially at their initial phase, may thus contribute to the severity of desaturation during apnoeas, lowing pulmonary volume.

Haemodynamics

The drop in heart rate during apnoeas remains largely unexplained. Its attenuation after administering oxygen and atropine suggests that hypoxemia plays a role by bringing the vagal system into play [214]. It may express a particular reactivity to the stimulation of the upper airways, since in OSAHS a Müller manoeuvre is accompanied by bradycardia of the same order as that associated with apnoeas, while the same manoeuvre is accompanied by an acceleration of heart rate in the control group [6].

Measurements of heart rate have provided contradictory results. A first study showed a reduction in heart rate due to the combination of a reduced heart rate and reduced systolic volume [201]. More recently, the use of different technology showed an increase in systolic volume in inverse relation to heart rate, such that heart rate was maintained in NREM sleep, while in REM sleep systolic volume did not increase, leading to a drop in heart rate. Reduced systolic volume may be a consequence of the reduction in telediastolic left ventricle volume due to the shift to the left of the interventricular septum (demonstrated in echocardiography [190]) itself secondary to right ventricular volume. The intraventricular septum shift appears to be linked to the degree of intrathoracic negative pressure and may be explained by an increase in right ventricular filling [84].

Despite the drop in heart rate, blood pressure increases during apnoeas, even more so when respiration resumes, indicating peripheral vasoconstriction. As a result, the physiological reduction of blood pressure fails during the night; on the contrary blood pressure is slightly higher on morning awakening than at sleep onset [94]. The systemic elevation in blood pressure observed during each apnoea is attributed to hypoxemia. However the role of hypoxemia may be questioned due to the persistent high blood pressure linked to apnoeas after the correction of hypoxemia; on the other hand, an arousal independent of apnoea will induce blood pressure elevation of the same order [173].

An increase in postganglion multiunitary sympathetic muscle activity during sleep may explain the pressure fluctuations by the vasoconstriction it produces. This may be encouraged by sympathetic hyper-reactivity [199] expressed as an increase in pressure response to changes in sleeping position. Local vascular reactivity may also be increased by local factors, such as prostaglandins [119] or endothelin [48].

Cardiac arrhythmias, which are often observed [78] particularly if the heart is ischaemic, have also been attributed to hypoxemia.

Early studies showed an increase in pulmonary artery pressure during apnoeas [34, 44]. More recent studies have also shown an increase in transmural pulmonary artery pressure (PAP) during apnoea, and more so on respiratory reuptake [131]. Alterations in transmural PAP seem to be linked both to SaO_2 level and to the extent of intrathoracic pressure fluctuations [131, 144].

Intrathoracic pressure

The inspiratory thoraco-abdominal muscles increase in strength in response to upper airway occlusion, producing intrathoracic inspiratory pressures of -50 to -100 cm H_2O, whereas ventilation is normally assured at pressures of -5 to -10 cm H_2O.

Intracranial pressure

An increase in intracranial pressure during sleep has also been reported [106]. This may additionally affect cognitive functions.

Endocrine

Growth hormone

A deficiency in growth hormone secretion during OSAHS sleep has been known for some time [134] and may be interpreted as a consequence of the virtual disappearance of deep NREM sleep stages during which peak GH secretion normally occurs. The height-weight deficiency observed in OSAHS in children has been attributed to this reduction in growth hormone secretion; it is also likely to contribute to the development of obesity, common in OSAHS [134].

Hormones regulating hydromineral metabolism

The notion of an increase in nocturnal diuresis and natriuresis [122] in OSAHS gave rise to the discovery of an increase in atrial natriuretic peptide (ANP) in at least 2/3 of patients [119, 123] and the depressed activity of the renin-angiotensin-aldosterone system during sleep [61]. These two mechanisms contribute to increasing diuresis and natriuresis in OSAHS and are thus hardly suspected of contributing to the development of arterial hypertension. On the other hand, the increase in ANP secretion, which augments membrane permeability, may be responsible for the development of the peripheral oedema often associated with OSAHS and a haematocrit, which is reversible from the first night of treatment with continuous positive airway pressure [130].

Long-term consequences

Haemodynamic

Right

Pulmonary hypertension has been found in roughly 20% of patients [207]. Its development seems to depend on permanent hypoxemia during wakefulness. Clinically defined right heart failure was observed in 12% of the 50 cases studied [21].

Left

Hypertension

The haemodynamic consequences of sleep apnoeas undoubtedly constitute the major risk factor to which patients are exposed, both in terms of ischaemic accidents and the development of heart failure.

Hypertension is associated with OSAHS with variable frequency depending on the study, with a mean of around 50% [85]. Conversely, the presence of apnoeas during sleep has been demonstrated in roughly 30% of a random hypertensive population [110, 137].

Frequency appears to be higher in the hypertensive who resist antihypertensive treatment, inciting the search for OSAHS at least in this hypertensive category. However the question of a causal relationship between OSAHS and hypertension is debateable, in view of the common terrain between the two conditions (age, obesity, sex). Analysis of the available epidemiological data has led to contradictory conclusions on this point [93, 121, 139].

Treating sleep apnoeas improves hypertension, but few studies have managed to overcome the problem of simultaneously administered antihypertensins [35]. The only controlled study in this respect showed an improvement in systemic hypertension [147].

The mechanism of systemic hypertension during sleep remains to be elucidated. It does not seem to be linked either to morphometric characteristics (age, BMI), or to sleep architecture, density or duration of apnoeas, or the severity of hypoxemia [53].

Increased catecholamine release during sleep [107], increase in sympathetic tone [90], the repeated pressure jolts during sleep leading to a down-regulation of the baroreceptors, may play a role, as may the failure of defence mechanisms in the form of ANP secretion, a powerful vasodilator [119] or the intervention of local factors modulating sensitivity to catecholamines, such as prostaglandines [117] or endothelins.

These factors remain speculative; nevertheless, in the rat, intermittent hypoxemia applied according to a rhythm close to that observed in OSAHS is susceptible to inducing permanent hypertension [59]. Likewise, an experimental model of obstructive apnoeas in the dog demonstrates the progressive development of permanent hypertension which is reversible after normalisation of ventilation [23].

Cerebral ischaemia

A high percentage of OSAHS was reported in a preliminary study among patients who had been subject to an ischaemic stroke (72% of the 47 patients had an A + H index > 10, 53% an A + H index of > 20). If this result is confirmed, it would imply the need for better surveillance in cases of sleep-related respiratory disorders in ischaemic stroke [111].

Heart failure

Echocardiographic variables of left ventricular functioning have been seen to alter (septum and thicker posterior wall) in OSAHS compared to cohorts; the difference persisted for the non hypertensive OSAHS group and when groups were paired for body mass index [90]. Likewise, an improvement in left ventricle functioning under continuous positive airway pressure [19] suggests that apnoeas have an aggravating effect at least in the development of heart failure. This hypothesis is reinforced by the observation of patients with heart failure, with no cardiopathic cause identified by coronarography and myocardiac biopsy, whose ventricular ejection fraction improves under CPAP [142].

Respiratory

The presiding factors in the development of permanent alveolar hyperventilation remain imprecise [206]. Its frequency, estimated on the basis of the presence of diurnal hypercapnia is of the order of 12%.

Hypercapnia is usually corrected by voluntary hyperventilation, but the quality of correction correlates to maximum expiratory volume per second, suggesting that it is due to faulty ventilation control and mechanical anomalies in ventilation.

Cognitive

Sleep fragmentation diminishes the performances of the normal subject, and is a possible cause of alteration to cognitive functions during OSAHS. The specific alteration of executive functions suggests the prefrontal cortex is functionally or lesionally affected [153]. Hypoxemia also plays a role. Other factors may contribute, such as intracranial hypertension [106]. The reversibility (at least partial) of these disorders by treatment suggests that the alterations are partly functional.

A study showing that hypoxemia rather than sleep macrostructure, distinguishes sleepy from non sleepy patients, raised controversy as to the respective roles of nocturnal hypoxemia and sleep fragmentation in generating daytime sleepiness; several arguments converge in attributing the responsibility to sleep fragmentation, albeit with an independent interactive contribution from hypoxemia [52].

Endocrine

The drop in libido may be explained by the reduction in free and total plasmatic testosterone levels, which is increasingly pronounced according to the severity of sleep hypoxemia. This deficiency appears to correspond to a deficiency of hypothalamic origin as hypophysial stimulation tests prove normal [73].

GH levels measured during wakefulness are also lower, and redressed when apnoeas are eliminated [73].

TREATMENT FOR OSAHS

Continuous positive airway pressure (CPAP)

Implementation

Positive pressure, *i.e.* greater than atmospheric pressure, of 5 to 15 cm H_2O, applied nasally and continuously, during inspiration and expiration, prevents pharyngeal occlusion during inspiration and enables the elimination of apnoeas [198]. Except for rare cases of complete nasal occlusion, the method is always effective (fig. 29.6).

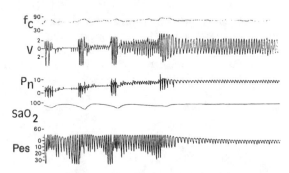

Figure 29.6. Effectiveness of continuous positive airway pressure. Sample of a polygraphic recording of a patient with sleep apnoea syndrome. On the left, pressure applied nasally (P_n, cm H_2O) of 3 cm H_2O is insufficient, with persisting apnoeas authenticated by the interruption of ventilation flow (V, l/s, upward inspiration) of the obstructive type, as shown by the persistence of increasing fluctuations in oesophageal pressure fluctuations (P_{es}, cm H_2O) during apnoeas. Note the repercussions of apnoeas on oxyhaemoglobin saturation (SaO_2 %). Nasal pressure is then increased to 6 cm H_2O. Apnoeas are eliminated, but snoring persists, indicated by the vibrating pattern of the flow curve (V) during inspiration. A minimal additional increase in nasal pressure eliminates snoring and ensures normal ventilation (right of the figure).

Sleep polygraphy is indispensable for setting up CPAP treatment. The criteria for "optimal" treatment remain poorly defined. The eradication of apnoeas and hypopnoeas may not be sufficient. Our view is that the ideal pressure is one which normalises upper airway resistance, and thus minimises inspiratory oesophageal pressure. The eradication of snoring, normalisation of flow curves (disappearance of the inspiratory plateau), and analysis of tracheal sounds seem to represent the right image of this result. Another criterion of effectiveness is sleep restructuration, often with a rebound of NREM sleep and REM sleep [105], and the disappearance of micro-arousals.

These criteria are not universally accepted, as in certain studies, apnoea indexes under treatment may be as high as 20 apnoeas/hour [183], and may result in the non acceptance of treatment. Overnight recording must be conducted, indeed pressure which is effective at the beginning of the night is often less effective at the end of the night, requiring an increase in applied pressure of 1 to 2 cm H_2O [118]. Thus, it would appear to be dangerous to split the night recording into two parts, the first for diagnostic evaluation and the second for adjusting treatment [99] because of the risk of underestimating optimal pressure. Self-operating positive pressure devices, where the pressure delivered is adjusted virtually instantaneously to the patient's ventilation conditions, have not proven their worth clinically, either in terms of efficacy or observation quality; on the other hand, these devices may be a useful aid to ascertaining optimal pressure, in a sleep laboratory or at home [166].

The effective level of pressure may vary in the long term and need to be reassessed; on the other hand it does not appear to be affected by the "moderate" intake of alcohol (60 – 100 mg/dl [14].

Adverse effects are practically nonexistent, limited to discomfort and local irritation from wearing the nasal mask [129, 168, 204]. Watery morning rhinorrhea is usually easy to control by adding a humidifier.

The real limitation to this means of treatment resides in the constraint to the patient. The latter is obliged to sleep with a nasal mask connected to a source of pressure. This constraint, although minimised by adapting the mask to the specific needs of the patient, and by adequate maintenance of the equipment, is not always accepted, and levels of acceptance sometimes do not exceed 50%; nevertheless, several studies have shown that the general level of acceptance and continuation of treatment was over 75 to 80% [115]. The benefits of using lower pressure for expiration than for inspiration [178] have not been confirmed [171].

Effects of treatment

Short term

Apnoeas are rapidly eradicated followed by the disappearance of the symptoms of the illness; in particular, daytime sleepiness disappears very quickly [136]. This improvement in sleepiness is accompanied by an improvement in endogenous evoked responses [175] and reaction time [156]. It results in a decrease in the number of automobile accidents [125, 57].

Long term

Daytime sleepiness continues to improve over the long term, without attaining the same values as the control groups [184]. Daytime arterial blood gas improves, according to the extent to which it was disturbed before treatment; on the other hand, pulmonary pressure is hardly altered [186]. Left ventricular function also improves [22, 120]. From the tenth day of treatment, daytime blood pressure had dropped significantly in a group of hypertensive patients who were also treated with antihypertensins, which remained unchanged [154].

There is an increase in pharyngeal dimensions and a decrease in tongue volume, after 4 to 6 weeks of treatment, with a decrease in the water content of the mucous membranes (evaluated by MRI), compatible with a reduction of oedema [176].

It has been suggested that treatment which is followed regularly will improve OSAHS even after treatment is suspended. This result only applies to the most severe OSAHS, with a modest improvement [38]; it is only in certain cases involving considerable weight loss that treatment was stopped with no recurrence of apnoea [9].

Morbidity/mortality

CPAP is the only treatment which guarantees patients as high a risk of survival as tracheostomy, and is significantly higher than that obtained with other medical means or with uvulopalatopharyngoplasty [88].

Other mechanical means

Mandibular advancement devices

Wearing a dental device to move the mandible forward has positive results when the OSAHS is not too severe. This approach may be a valuable alternative to CPAP [33] in such cases.

Sleeping position

In patients whose OSAHS is position-dependent (reduction of apnoeas by at least 50% in lateral decubitus as opposed to dorsal decubitus), an apnoea index reduced to 5.5 was obtained in 53 to 60% of patients, using postural treatment either voluntarily (recommended change of sleeping position), or with the aid of a posture alarm; the addition of a mandibular advancement device discreetly improves the percentage of good results (73%); in general, the less severe the OSAHS and the greater the effect of sleeping position, the better the results [28].

Electrical stimulation

From 1989, the effects have been described of electrical stimulation applied to the region below the chin [149]. Only isolated observations have as yet been published. Stimulation shortens apnoeas without reducing their frequency. Hypoxemia associated with apnoeas is less severe, and the termination of apnoeas is not accompanied by arousal [91]. Transcutaneous electrical stimulation is however unable to increase the dimensions of the upper airways during wakefulness or to prevent their collapse during sleep [47]; thus the treatment is unlikely to be of clinical use, at least in the transcutaneous stimulation form, an implanted stimulator may provide better results [49, 155].

Surgical treatment

(see Chapter 30).

Other treatments

Drug treatment

The effectiveness of the various drugs has been reported on the basis of non controlled studies [87]. None of these has proved its efficacy in controlled studies, apart from the effect linked to the reduction of REM sleep (protriptyline, fluoxetine [87]). A reduction in REM sleep is also the main effect seen with clonidine; likewise, sleep deterioration seems to be the main improvement factor observed with theophylline.

To the extent that occlusion of the upper airways results from an imbalance of diaphragmatic activity and the muscles of the upper airways, no beneficial effect can be expected from a ventilation stimulant unless it exclusively or preferentially stimulates the upper airway muscles; this is the case of strychnine [172], and possibly of aspirin, but neither of these drugs is clinically useable. The "classic" ventilation stimulants (progesterone, doxapram, almitrine) are not effective [113, 124].

Weight loss

An improvement has often been reported in clinical symptoms after weight loss [158]. Its effectiveness in eradicating apnoeas is less certain, and only appears to affect the morbidly obese [202]. The proportion of patients whose apnoea + hypopnoea index after weight loss was under 10 was 8/23 [158]. Failure occurs among the most obese and those who lose the least weight. Substantial weight loss among the seriously obese (mean BMI of 42.0 to 34.7 kg/m2) is accompanied by a decrease in critical upper airway occlusion pressure and the disappearance of apnoeas when critical pressure drops below − 4 cm H2O [182].

The real problem is that weight loss is difficult to obtain and even more to maintain. Nevertheless it is important to advise patients to lose weight, due to the morbidity related to obesity, and because it worsens the consequences of apnoea.

For this reason, surgery is sometimes proposed to treat obesity. In a recent review of 110 patients, operated using different techniques, (gastroplasty and by-pass) of whom 45 also had permanent alveolar hypoventilation, only 57 were monitored postoperatively, with an average weight loss of 54 kg and BMI reduction of 56 ± 12 to 38 ± 9 kg/m2 (*i.e.* 31% of initial weight) 40 of whom polygraphically: apnoea index had dropped from 64 ± 39 to 26 ± 26.; mortality level was 2.7% with respiratory failure and 0.2% without. Reinterventions were justified by a non specified relapse of obesity [196].

Oxygenotherapy

In acute application, isolated oxygenotherapy increases the mean level of blood oxygenation, at the cost of lengthening the duration of apnoeas, and increasing acidosis. This treatment has failed to show any beneficial effect in the long term. For this reason, oxygenotherapy has no rightful place in OSAHS treatment [60].

On the other hand, in forms associated with chronic obstructive pulmonary disease, oxygenotherapy may be indicated, where this indication relates to broncho-pneumopathy and not to OSAHS itself.

TREATMENT FOR UARS

CPAP treatment has been shown to be effective, and may serve as a diagnostic test in cases where there is doubt, since, as we have seen, no clear indication exists of the number of episodes of increased respiratory effort with micro-arousals needed to produce clinical symptomatology. The efficacy of the other treatments proposed for OSAHS in the treatment of upper airway resistance syndrome remains to be evaluated. It is not certain that improvement of UARS requires any less "effective" means than for improving OSAHS. Indeed in the section on pathophysiology it was seen that the difference between the two states is not necessarily linked to greater pharyngeal compliance, but may simply be related to a different arousal threshold.

Unanswered questions

We have focused on the notion that pathological situations appear when the system is unstable, due to a combination of increased respiratory effort, and decreased arousal threshold, resulting in sleep fragmentation. The question still remains as to the possible consequences of an isolated increase in respiratory effort with no sleep fragmentation.

REFERENCES

1. AASM TASK FORCE. - Sleep-related breathing disorders in adults : Recommendations for syndrome definition and measurement techniques in clinical research. *Sleep* 22, 667-689, 1999.
2. ALDRICH M.S. – Automobile accidents in patients with sleep disorders. *Sleep*, *12*, 487-494, 1989.
3. ALDRICH M.S., CHAUNCEY J.B. – Are morning headaches part of obstructive sleep apnea syndrome? *Arch. Int. Med.*, *150*, 1265-1267, 1990.
4. AMBROGETTI A., OLSON L.G., SAUNDERS N.A. – Differences in the symptoms of men and women with obstructive sleep apnoea. *Aust. N.Z. J. Med.*, *21*, 863-866, 1991.
5. AMERICAN THORACIC SOCIETY – Indications and standards for cardiopulmonary sleep studies. *Am. Rev. Respir. Dis.*, *139*, 559-568, 1989.
6. ANDREAS S., HAJAK G., VONBRESKA B., RUTHER E., KREUZER H. – Changes in heart rate during obstructive sleep apnoea. *Eur. Respir. J.*, *5*, 853-857, 1992.
7. ARGOD J., PEPIN J.L., SMITH R.P., LEVY P. - Comparison of esophageal pressure with pulse transit time as a measure of respiratory effort for scoring obstructive nonapneic respiratory events. *Am. J. Respir. Crit. Care Med.* 162, 87-93, 2000.
8. ATLAS TASK FORCE of the ASDA. – EEG arousals – Scoring rules and examples – A preliminary report from the Sleep Disorders Atlas Task Force of the American Sleep Disorders Association. *Sleep,* 15, 174-184, 1992.
9. AUBERT-TULKENS G., CULÉE C., RODENSTEIN D.O. – Cure of sleep apnea syndrome after long-term nasal continuous positive airway pressure therapy and weight loss. *Sleep*, *12*, 216-222, 1989.
10. AYAPPA I., NORMAN R.G., HOSSELET J.J., GRUENKE R.A., WALSLEBEN J.A., RAPOPORT D.M. - Relative occurrence of flow limitation and snoring during continuous positive airway pressure titration . *Chest* 114, 685-690, 1998.
11. BALLESTER E., BADIA J.R., HERNANDEZ L., FARRE R., NAVAJAS D., MONTSERRAT J.M. – Nasal prongs in the detection of sleep-related disordered breathing in the sleep apnoea/hypopnoea syndrome. *Eur. Respi. J.* 11, 880-883, 1998.
12. BASNER R.C., RINGLER J., GARPESTAD E., SCHWARTZSTEIN R.M., SPARROW D., WEINBERGER S.E., LILLY J., WEISS J.W. – Upper airway anesthesia delays arousal from airway occlusion induced during human NREM sleep. *J. Appl. Physiol.*, *73*, 642-648, 1992.
13. BERRY R.B., BONNET M.H., LIGHT R.W. – Effect of ethanol on the arousal response to airway occlusion during sleep in normal subjects. *Am. Rev. Respir. Dis.*, *145*, 445-452, 1992.
14. BERRY R.B., DESA M.M., LIGHT R.W. – Effect of ethanol on the efficacy of nasal continuous positive airway pressure as a treatment for obstructive sleep apnea. *Chest*, *99*, 339-343, 1991.
15. BERRY D.T.R., PHILLIPS B.A., COOK Y.R., SCHMITT F.A., HONEYCUTT N.A., ARITA A.A., ALLEN, R.S. – Geriatric sleep apnea syndrome – A preliminary description. *J. Gerontol.*, *45*, M169-M174, 1990.

16. BERRY D.T.R., WEBB W.B., BLOCK A.J. – Sleep apnea syndrome: a critical review of the apnea index as a diagnostic criterion. *Chest*, *86*, 529-531, 1984.

17. BLOCK A.J., HELLARD D.W. – Ingestion of either Scotch or Vodka induces equal effects on sleep and breathing of asymptomatic subjects. *Arch. Int. Med.*, *147*, 1145-1152, 1987.

18. BONNET M.H., DEXTER J.R., ARAND D.L. – The effect of triazolam on arousal and respiration in central sleep apnea patients. *Sleep*, *13*, 31-41, 1990.

19. BRADLEY T.D., HOLLOWAY R.M., MCLAUGHLIN P.R., ROSS B.L., WALTERS J., LIU P.P. – Cardiac output response to continuous positive airway pressure in congestive heart failure. *Am. Rev. Respir. Dis.*, *145*, 377-382, 1992.

20. BRADLEY T.D., MCNICHOLAS W.T., RUTHERFORD R., POPKIN J., ZAMEL N., PHILLIPSON E.A. – Clinical and physiologic heterogeneity of the central sleep apnea syndrome. *Am. Rev. Resp. Dis.*, *134*, 217-221, 1986.

21. BRADLEY T.D., RUTHERFORD R., GROSSMAN R.F., LUE F., ZAMEL N., MOLDOFSKY H., PHILLIPSON E.A. – Role of daytime hypoxemia in the pathogenesis of right heart failure in the obstructive sleep apnea syndrome. *Am. Rev. Respir. Dis.*, *131*, 835-39, 1985.

22. BRADLEY T.D., TAKASAKI Y., ORR D., POPKIN J., LIU P., RUTHERFORD R. – Sleep apnea in patients with left ventricular dysfunction – Beneficial effects of nasal CPAP. *In: Sleep and Respiration*, Issa F.G., Suratt P.M., Remmers J.E. (eds.), Wiley-Liss, 363-370, 1990.

23. BROOKS D., HORNER R.L., KIMOFF R.J., KOZAR L.F., RENDER TEIXEIRA C.L., PHILIPSON E.A. – Effect of obstructive sleep apnea versus sleep fragmentation on responses to airway occlusion. *Am. J. Respir. Crit. Care Med.* 155, 1609-1617, 1997.

24. BROUILLETTE R.T., FERNBACH S.K., HUNT C.E. – Obstructive sleep apnea in infants and children. *J. Pediatr.*, *100*, 31-40, 1982.

25. BROWN S., STOOL S. – Behavioral manifestations of sleep apnea in children. *Sleep*, *5*, 200-201, 1982.

26. BURWELL C.S., ROBIN E.D., WHALEY R.D., BICKELMANN A.G. – Extreme obesity with alveolar hypoventilation: a Pickwickian Syndrome. *Am. J. Med.*, *21*, 811-818, 1956.

27. CARLSON J.T., HEDNER J., FAGERBERG B., EJNELL H., MAGNUSSON B., FYHRQUIST F. – Secondary polycythaemia associated with nocturnal apnoea – A relationship not mediated by erythropoietin. *J. Int. Med.*, *231*, 381-387, 1992.

28. CARTWRIGHT R.D., DIAZ F., LLOYD S. – The effects of sleep posture and sleep stage on apnea frequency. *Sleep*, *14*, 351-353, 1991.

29. CHADWICK G.A., CROWLEY P., FITZGERALD M.X., OREGAN R.G., McNICHOLAS W.T. – Obstructive sleep apnea following topical oropharyngeal anesthesia in loud snorers. *Am. Rev. Respir. Dis.*, *143*, 810-813, 1991.

30. CHARBONNEAU M., FALCONE T., COSIO M.G., LEVY R.D. – Obstructive sleep apnea during pregnancy – Therapy and implications for fetal health. *Am. Rev. Respir. Dis.*, *144*, 461-463, 1991.

31. CHOKROVERTY S., SACHDEO R., MASDEU J. – Autonomic dysfunction and sleep apnea in olivopontocerebellar degeneration. *Arch. Neurol.*, *41*, 926-931, 1984.

32. CIRIGNOTTA F., ZUCCONI M., MONDINI S., GERARDI R., LUGARESI E. – Cerebral anoxic attacks in sleep apnea syndrome. *Sleep*, *12*, 400-404, 1989.

33. CLARK G.T. – Mandibular advancement devices and sleep disordered breathing. *Sleep Med. Rev.* 2, 163-174, 1998.

34. COCCAGNA G., MANTOVANI M., BRIGNANI F., PARCHI C., LUGARESI E. – Continuous recording of the pulmonary and systemic arterial pressure during sleep in syndromes of hypersomnia with periodic breathing. *Bull. Physiopathol. Respir.*, *8*, 1159-1172, 1972.

35. COCCAGNA G., MANTOVANI M., BRIGNANI F., PARCHI C., LUGARESI E. – Tracheostomy in hypersomnia with periodic breathing. *Bull. Physiopath. Resp.*, *8*, 1217-1227, 1972.

36. COCCAGNA G., PETRELLA A., BERTI-CERONI G., LUGARESI E., PAZZAGLIA P. – Polygraphic contribution to hypersomnia and respiratory troubles in the pickwickian syndrome. *In: Abnormalities of Sleep in Man*, Gastaut, H., *et al.* (eds.), 215-221. Aulo Gaggi, Bologna, 1968.

37. COLLARD P., PIETERS T., AUBERT G., DELGUSTE P., RODENSTEIN D.O. – Compliance with nasal CPAP in obstructive sleep apnea patients. *Sleep Med. Rev.* 1, 33-44, 1997.

38. COLLOP N.A., BLOCK A.J., HELLARD D. – The effect of nightly nasal CPAP treatment on underlying obstructive sleep apnea and pharyngeal size. *Chest*, *99*, 855-860, 1991.

39. CONDOS R., NORMAN R.G., KRISHNASAMY I., PEDUZZI N., GOLDRING R.M., RAPOPORT D.M. – Flow limitation as a noninvasive assessment of residual upper-airway resistance during continuous positive airway pressure therapy of obstructive sleep apnea. *Am. J. Respir. Crit. Care Med.* 150, 475-480, 1994.

40. DAVIES R.J.O., STRADLING J.R. – The efficacy of nasal continuous positive airway pressure in the treatment of obstructive sleep apnea syndrome is proven. *Am. J. Respir. Crit. Care Med.* 161, 1775-1776, 2000.

41. DEAN R.J., CHAUDHARY B.A. – Negative polysomnogram in patients with obstructive sleep apnea syndrome. *Chest*, *101*, 105-108, 1992.

42. DEMATTEIS M., PEPIN J.L., JEANMART M., DESCHAUX C., LABARRE VILA A., LEVY P. – Charcot-Marie-Tooth disease and sleep apnoea syndrome : a family study. *Lancet* 357, 267-272, 2001.

43. DEMENT W.C., CARSKADON M.A., RICHARDSON G. – Excessive daytime sleepiness in the sleep apnea syndrome. *In: Sleep apnea syndromes*, Guilleminault, C., Dement, W. C. (eds.), 23-46. Alan R Liss, New York, 1978

44. DOLL E., KUHLO W., STEIN H., KEUL J. – Zur Genese des Cor Pulmonale beim Pickwick Syndrom. *Dtsch. Med. Woçhenschr.*, *93*, 2361-2365, 1968.

45. DOUGLAS N.J., THOMAS S., JAN M.A. – Clinical value of polysomnography. *Lancet*, *339*, 347-350, 1992.

46. DOWDELL W.T., JAVAHERI S., MCGINNIS W. – Cheyne-Stokes respiration presenting as sleep apnea syndrome – Clinical and polysomnographic features. *Am. Rev. Respir. Dis.*, *141*, 871-879, 1990.

47. EDMONDS L.C., DANIELS B.K., STANSON A.W., SHEEDY P.F., SHEPARD J.W. – The effects of transcutaneous electrical stimulation during wakefulness and sleep in patients with obstructive sleep apnea. *Am. Rev. Respir. Dis.*, *146*, 1030-1036, 1992.

48. EHLENZ K., HERZOG P., WICHERT P. VON, KAFFAR-NIK H., PETER J.H. – Renal excretion of endothelin in obstructive sleep apnea syndrome. *In: Sleep and Cardiorespiratory Control/Sommeil et Contrôle Cardio-respiratoire*, Gaultier C., Escourrou, P., Curzi-Dascalova, L. (eds), 226. J. Libbey Eurotext/INSERM, Montrouge/Paris, 1991.

49. EISELE D.W., SMITH P.L., ALAM D.S., SCHWARTZ A.R. – Direct hypoglossal nerve stimulation in obstructive sleep apnea. *Arch. Otolaryngol.*, *123*, 57-61, 1997.

50. EL BAYADI S., MILLMAN R.P., TISHLER P.V., ROSENBERG C., SALISKI W., BOUCHER M.A., REDLINE S. – A family study of sleep apnea – Anatomic and physiologic interactions. *Chest*, *98*, 554-559, 1990.

51. ELLIS E.R., MCCAULEY V.B., MELLIS C., SULLIVAN C.E. – Treatment of alveolar hypoventilation in a six year old girl with intermittent positive pressure ventilation through a nose mask. *Am. Rev. Respir. Dis.*, *136*, 188-191, 1981.

52. ENGELMAN H., JOFFE D. – Neuropsychological function in obstructive sleep apnoea. *Sleep Med. Rev.* 3, 59-78, 1999.

53. ESCOURROU P., JIRANI A., NEDELCOUX H., DUROUX P., GAULTIER C. – Systemic hypertension in sleep apnea syndrome – Relationship with sleep architecture and breathing abnormalities. *Chest*, *98*, 1362-1365, 1990.

54. FINDLEY L.J., RIES A.L., TISI G.M., WAGNER P.D. – Hypoxemia during apnea in normal subjects: mechanisms and impact of lung volume. *J. Appl. Physiol.*, *55*, 1777-1783, 1983.

55. FINDLEY L., SMITH C., HOOPER J., DINEEN M., SURATT P.M. – Treatment with nasal CPAP decreases automobile accidents in patients with sleep apnea. *Am. J. Respir. Crit. Care Med.* 161, 857-859, 2000.

56. FLEMONS W.W., McNICHOLAS W.T . – Clinical prediction of the sleep apnoea syndrome. *Sleep Med. Rev.* 1, 19-32, 1997.

57. FLENLEY D.C. – Sleep in chronic obstructive lung disease. *Clin. Chest. Med.*, *6*, 651-661, 1985.

58. FLETCHER E.C. – Recurrence of sleep apnea syndrome following tracheostomy. A shift from obstructive to central apnea. *Chest*, *96*, 205-209, 1989.

59. FLETCHER E.C., LESSKE J., WEI Q., MILLER C.C., UNGER T. – Repetitive, episodic hypoxia causes diurnal elevation of blood pressure in rats. *Hypertension*, *19*, 555-561, 1992.

60. FLETCHER E.C., MUNAFO D.A. – Role of nocturnal oxygen therapy in obstructive sleep apnea – When should it be used? *Chest*, *98*, 1497-1504, 1990.

61. FOLLENIUS M., KRIEGER J., KRAUTH M.O., SFORZA F., BRANDENBERGER G. – Obstructive sleep apnea treatment – Peripheral and central effects on plasma renin activity and aldosterone. *Sleep*, *14*, 211-217, 1991.

62. FOX A.W., MONOSON P.K., MORGAN C.D. – Speech dysfunction of obstructive sleep apnea. A discriminant analysis of its descriptors. *Chest*, *96*, 589-595, 1989.

63. FRIBERG D., ANSVED T., BORG K., CARSSON-NORDLANDER B., LARSSON H., SVANBORG E. – Histological indications of a progressive snorers disease in an upper airway muscle. *Am. J. Respir. Crit. Care Med.* 157, 586-593, 1998.

64. GASTAUT H., TASSINARI C.A., DURON B. – Étude polygraphique des manifestations épisodiques (hypniques et respiratoires) du syndrome de Pickwick. *Rev. Neurol.*, *112*, 568-579, 1965.

65. GAULTIER CL., PRAUD J.P. – Pathologie respiratoire du sommeil de l'enfant. *Rev. Mal. Respir.*, *7*, 475-481, 1990.

66. GLEESON K., ZWILLICH C.W., WHITE D.P. – The influence of increasing ventilatory effort on arousal from sleep. *Am. Rev. Respir. Dis.*, *142*, 295-300, 1990.

67. GOLD A.R., BLEECKER E.R., SMITH, P.L. – A shift from central and mixed sleep apnea to obstructive sleep apnea resulting from low-flow oxygen. *Am. Rev. Respir. Dis.*, *132*, 220-223, 1985.

68. GOLD A.R., SCHWARTZ A.R. – The pharyngeal critical pressure : The whys and hows of using nasal continuous positive airway pressure diagnostically. *Chest*, 110, 1077-1088, 1996.

69. GOLDING-WOOD D.G., BROCKBANK M.J., SWAN-STON A.R., CROFT C.B. – Assessment of chronic snorers. *J. Royal Soc. Med.*, *83*, 363-367, 1990.

70. GOLDMAN J.M., BARNES D.J., POHL D.V. – Obstructive sleep apnoea due to a dermoid cyst of the floor of the mouth. *Thorax*, *45*, 76, 1990.

71. GONZALEZ-ROTHI R.J., FORESMAN G.E., BLOCK A.J. – Do patients with sleep apnea die in their sleep? *Chest*, *94*, 531-538, 1988.

72. GOULD G.A., WHYTE K.F., RHIND G.B., AIRLIE M.A.A., CATTERALL J.R. SHAPIRO C.M., DOUGLAS N.J. – The sleep hypopnea syndrome. *Am. Rev. Respir. Dis.*, 137, 895-898, 1988.

73. GRUNSTEIN R.R., HENDELMAN D.J., LAWRENCE S.J., BLACKWELL C., CATERSON I.D., SULLIVAN C.E. – Neuroendocrine dysfunction in sleep apnea: reversal by continuous positive airway pressure therapy. *J. Clin. Endocrinol. Metab.*, *68*, 352-358, 1989.

74. GRUNSTEIN R.R., HO K.Y., SULLIVAN C.E. – Sleep apnea in acromegaly. *Ann. Int. Med.*, *115*, 527-532, 1991.

75. GRUNSTEIN R.R., KEATING J.M., DONELLY P., COSTAS L.J.V., HO K., SULLIVAN C.E. – Effect of somatostatin analog on sleep apnea, lung function and chemosensitivity in acromegaly. *Eur. Respir. J.*, *1*, 95s, 1988.

76. GUILLEMINAULT C. – Obstructive sleep apnea syndrome and its treatment in children: areas of agreement and controversy. *Pediatr. Pulmonol.*, *3*, 429-436, 1987.

77. GUILLEMINAULT C. – Clinical features and evaluation of obstructive sleep apnea. *In: Principles and Practice in Sleep Medicine*, Kryger M.H., Roth T., Dement W.C. (eds), 552-558. Saunders, Philadelphia, 1989.

78. GUILLEMINAULT C., CONNOLLY S.J., WINKLE R.A. – Cardiac arrhythmia and conduction disturbances during sleep in 400 patients with sleep apnea syndrome. *Am. J. Cardiol.*, *52*, 490-494, 1983.

79. GUILLEMINAULT C., ELDRIDGE F.L., DEMENT W.C. – Insomnia with sleep apnea: a new syndrome. *Science*, *181*, 856-858, 1973.

80. GUILLEMINAULT C., NINO-MURCIA, HELDT G., BALDWIN R., HUTCHINSON D. – Alternative treatment to tracheostomy in obstructive sleep apnea syndrome: nasal continuous positive airway pressure in young children. *Pediatrics*, *78*, 797-802, 1986.

81. GUILLEMINAULT C., SHIOMI T., STOOHS R., SCHNITTGER I. – Echocardiographic studies in adults and children presenting with obstructive sleep apnea or heavy snoring. *In: Sleep and Cardiorespiratory Control/Sommeil et Contrôle Cardio-respiratoire*, Gaultier C., Escourrou P., Curzi-Dascalova L. (eds), 95-103. J. Libbey Eurotext/INSERM, Montrouge/Paris, 1991.

82. GUILLEMINAULT C., STOOHS R., CLERK A., CETEL M., MAISTROS P. – A cause of excessive daytime sleepiness. The upper airway resistance syndrome. *Chest*, 104, 781-787, 1993.

83. GUILLEMINAULT C., STOOHS R., CLERK A., SIMMONS J., LABANOWSKI M. – From obstructive sleep apnea syndrome to upper airway resistance syndrome. – Consistency of daytime sleepiness. *Sleep*, 15, S13-S16, 1992.

84. GUILLEMINAULT C., STOOHS R., QUERASALVA M.A. – Sleep-related obstructive and nonobstructive apneas and neurologic disorders. *Neurology*, *42*, 53-60, 1992.

85. GUILLEMINAULT C., TILKIAN A., DEMENT W.C. – The sleep apnea syndromes. *Ann. Rev. Med.*, *27*, 465-484, 1976.

86. GUYTON A., CROWELL J., MOORE J. – Basic oscillating mechanisms of Cheyne-Stokes breathing. *Am. J. Physiol.*, *187*, 185-200, 1979.

87. HANZEL D.A., PROIA N.G., HUDGEL D.W. – Response of obstructive sleep apnea to fluoxetine and protriptyline. *Chest*, *100*, 416-421, 1991.

88. HE J., KRYGER M.H., ZORICK F.J., CONWAY W., ROTH T. – Mortality and apnea index in obstructive sleep apnea. Experience in 385 male patients. *Chest*, 94, 9-14, 1988.

89. HAPONIK E.F., SMITH P.L., MEYERS D.A., BLEECKER E.R. – Evaluation of sleep disordered breathing. Is polysomnography necessary? *Am. J. Med.*, *77*, 671-677, 1984.

90. HEDNER J., EJNELL H., CAIDAHL K. – Left ventricular hypertrophy independent of hypertension in patients with obstructive sleep apnoea. *J. Hypertension*, *8*, 941-946, 1990.

91. HILLARP B., ROSEN I., WICKSTROM O. – Videoradiography at submental electrical stimulation during apnea in obstructive sleep apnea syndrome – A case report. *Acta Radiologica*, *32*, 256-259, 1991.

92. HOCH C.C., REYNOLDS C.F., MONK T.H., BUYSSE D.J., YEAGER A.L., HOUCK P.R., KUPFER D.J. – Comparison of sleep-disordered breathing among healthy elderly in the seventh, eighth, and ninth decades of life. *Sleep*, *13*, 502-511, 1990.

93. HOFFSTEIN V., CHAN C.K., SLUTSKY A.S. – Sleep apnea and systemic hypertension – A causal association review. *Am. J. Med.*, *91*, 190-196, 1991.

94. HOFFSTEIN V., MATEIKA J. – Evening-to-morning blood pressure variations in snoring patients with and without obstructive sleep apnea. *Chest*, *101*, 379-384, 1992.

95. HOSSELET J.J., NORMAN R.G., AYAPPA I., RAPOPORT D.M. – Detection of flow limitation with a nasal cannula/pressure transducer system. *Am. J. Respir. Crit. Care Med.* 157, 1461-1467, 1998.

95bis. HUDGEL D.W. – Mechanisms of obstructive sleep apnea. *Chest*, 101, 541-549, 1992.

96. HUDGEL D.W., HARASICK T. – Fluctuation in timing of upper airway and chest wall inspiratory muscle activity in obstructive sleep apnea. *J. Appl. Physiol.*, *69*, 443-450, 1990.

97. HUDGEL D.W., ROBERTSON R.D., MARTIN A.J. – Variable site of upper airway narrowing among obstructive sleep apnea patients. *J. Appl. Physiol.*, *61*, 1403-1409, 1986.

98. IBER C., DAVIES S.F., CHAPMAN R.C., MAHOWALD M.M. – A possible mechanism for mixed apnea in obstructive sleep apnea. *Chest*, *89*, 800-805, 1986.

99. IBER C., OBRIEN C., SCHLUTER J., DAVIES S., LEATHERMAN J., MAHOWALD M. – Single night studies in obstructive sleep apnea. *Sleep, 14*, 383-385, 1991.

100. ISSA F.G., SULLIVAN C.E. – Alcohol, snoring and sleep apnea. *J. Neurol. Neurosurg. Psychiatry, 45*, 353-359, 1982.

101. ISSA F.G., SULLIVAN C.E. – Arousal and breathing responses to airway occlusion in healthy sleeping adults. *J. Appl. Physiol., 55*, 1113-1119, 1983.

102. ISSA F.G., SULLIVAN C.E. – Upper airway closing pressures in obstructive sleep apnea. *J. Appl. Physiol., 57*, 520-527, 1984.

103. ISSA F.G., SULLIVAN C.E. – Upper airway closing pressures in snorers. *J. Appl. Physiol., 57*, 528-535, 1984.

104. ISSA F.G., SULLIVAN C.E. – Reversal of central sleep apnea using nasal CPAP. *Chest, 90*, 165-172, 1986.

105. ISSA F.G., SULLIVAN C.E. – The immediate effects of nasal continuous positive airway pressure treatment on sleep pattern in patients with obstructive sleep apnea syndrome. *Electroenceph. Clin. Neurophysiol., 63*, 10-17, 1986.

106. JENNUM P., BORGESEN S.E. – Intracranial pressure and obstructive sleep apnea. *Chest, 95*, 279-283, 1989.

107. JENNUM P., WILDSCHIODTZ G., CHRISTENSEN J., SCHWARTZ T. – Blood pressure, catecholamines, and pancreatic polypeptide in obstructive sleep apnea with and without nasal continuous positive airway pressure (nCPAP) treatment. *Am. J. Hypertension, 2*, 847-852, 1989.

108. JOHNSON M.W., ANCH A.M., REMMERS J.E. – Induction of the obstructive sleep apnea syndrome in a woman by exogenous androgen administration. *Am. Rev. Respir. Dis., 129*, 1023-1025, 1984.

109. JUNG R., KUHLO W. – Neurophysiological studies of abnormal night sleep and the pickwickian syndrome. *In: Sleep Mechanisms*, Albert K., Bally C., Stradlé J.P. (eds.), 140-159. Elsevier, Amsterdam, 1965.

110. KALES A., BIXLER E.O., CADIEUX R.J., SCHNECK D.W., SHAW L.C., LOCKE T.W., VELA-BUENO A., SOLDATOS C.R. – Sleep apnea in a hypertensive population. *Lancet, i*, 1005-1008, 1984.

111. KAPEN S., PARK A., GOLDBERG J., WYNTER J. – The incidence and severity of obstructive sleep apnea in ischemic cerebrovascular disease. *Neurology, 41* (suppl. 1), 125, 1991.

112. KIMOFF R.J., CHEONG T.H., OLHA A.E., CHARBONNEAU M., LEVY R.D., COSIO M.G., GOTTFRIED S.B. – Mechanisms of apnea termination in obstructive sleep apnea. Role of chemoreceptor and mechanoreceptor stimuli. *Am. J. Respir. Crit. Care Med.* 149, 707-714, 1994.

113. KRIEGER J. – Les syndromes d'apnées du sommeil de l'adulte. Sleep apnea syndromes in adults. *Bull. Eur. Physiopathol. Respir., 22*, 147-189, 1986.

114. KRIEGER J. – Regulation of body fluid compartments in obstructive sleep apnea syndrome. *In: Sleep and Cardiorespiratory Control/Sommeil et Contrôle Cardio-respiratoire*, Gaultier C., Escourrou P., Curzi-Dascalova L. (eds.), 123-131. J. Libbey Eurotext/INSERM, Montrouge/Paris, 1991.

115. KRIEGER J. – Long-term compliance with nasal continuous positive airway pressure (CPAP) in obstructive sleep apnea patients and nonapneic snorers. *Sleep, 15*, S42-S46, 1992.

116. KRIEGER J. – Therapeutic use of auto-CPAP. *Sleep Med. Rev.* 3, 159-174, 1999.

117. KRIEGER J., BENZONI D., SFORZA E., SASSARD J. – Urinary excretion of prostanoids during sleep in obstructive sleep apnoea patients. *Clin. Exp. Pharmacol. Physiol., 18*, 551-555, 1991.

118. KRIEGER J., BONIGEN C. – Split-night studies for CPAP titration in obstructive sleep apnea? *J. Sleep Res., 1* (suppl. 1), 121, 1992.

119. KRIEGER J., FOLLENIUS M., SFORZA E., BRANDENBERGER G., PETER J.D. – Effects of treatment with nasal continuous positive airway pressure on atrial natriuretic peptide and arginine vasopressin release during sleep in obstructive sleep apnea. *Clin. Sci., 80*, 443-449, 1991.

120. KRIEGER J., GRUCKER D., SFORZA E., CHAMBRON J., KURTZ D. – Left ventricular ejection fraction in obstructive sleep apnea – Effects of long-term treatment with nasal continuous positive airway pressure. *Chest, 100*, 917-921, 1991.

121. KRIEGER J., IMBS J.L. – Rôle des apnées du sommeil dans l'hypertension artérielle essentielle. *Presse Med., 19*, 1805 – 1809, 1990.

122. KRIEGER J., IMBS J.L., SCHMIDT M., KURTZ D. – Renal function in patients with obstructive sleep apnea. Effects of nasal continuous positive airway pressure. *Arch. Intern. Med., 148*, 1337-1340, 1988.

123. KRIEGER J., LAKS L., WILCOX I., GRUNSTEIN R.R., COSTAS L.J.V., MCDOUGALL J.G., SULLIVAN C.E. – Atrial natriuretic peptide release during sleep in patients with obstructive sleep apnoea before and during treatment with nasal continuous positive airway pressure. *Clin. Sci., 77*, 407-411, 1989.

124. KRIEGER J., MANGIN P., KURTZ D. – Effects of a ventilatory stimulant, almitrine bismesylate, in the sleep apnea syndrome. *Curr. Ther. Res., 32*, 697-705, 1982.

125. KRIEGER J., MESLIER N., LEBRUN T., LEVY P., PHILLIP JOET F., SAILLY J.C., RACINEUX J.L. – Accidents in obstructive sleep apnea patients treated with nasal continuous positive airway pressure. A prospective study. *Chest* 112, 1561-1566, 1997.

126. KRIEGER J., PETIAU C., SFORZA E., DELANOË C., CHAMOUARD V. – Nocturnal polyuria is a symptom of obstructive sleep apnea. *Urol. Int., 50*, 93-97, 1993.

127. KRIEGER A.J., ROSOMOFF H.L. – Sleep-induced apnea. Part.1: A respiratory and autonomic dysfunction following bilateral percutaneous cervical cordotomy. *J. Neurosurg., 39*, 168-180, 1974.

128. KRIEGER A.J., ROSOMOFF H.L. – Sleep-induced apnea. Part.2: Respiratory failure after anterior spinal surgery. *J. Neurosurg.*, *39*, 181-185, 1974.

129. KRIEGER J., SAUTEGEAU A., KURTZ D. – Nasal continuous positive airway pressure for the treatment of obstructive sleep apnea syndrome. Adverse effects, discomfort and suggested solutions. In : *Sleep 1984,* Koella W.P., Ruther E., Schultz H. (eds), Gustave Fischer Verlag, Stuttgart, New York, 58-59, 1985.

130. KRIEGER J., SFORZA E., BARTHELMEBS M., IMBS J.L., KURTZ D. – Overnight decrease in hematocrit after nasal CPAP treatment in patients with OSA. *Chest*, *97*, 729-730, 1990.

131. KRIEGER J., WEITZENBLUM E., REITZER B., KURTZ D. – Pulmonary hemodynamics in obstructive sleep apnea syndrome. *In: Heart and Brain*, Refsum H., Sulg I.A., Rasmussen K. (eds.), 272-279. Springer Verlag, Berlin, 1989.

132. KRIEGER J., WEITZENBLUM E., VANDEVENNE A., STIERLE J.L., KURTZ D. – Flow-volume curve abnormalities and obstructive sleep apnea syndrome. *Chest*, *87*, 163-167, 1985.

133. KURTZ D., KRIEGER J. – Les arrêts respiratoires au cours du sommeil. Faits et hypothèses. *Rev. Neurol.,* *134*, 11-22, 1978.

134. KURTZ D., KRIEGER J., KOWALSKI J., HOFF E., MANGIN P. – Variations nycthémérales des taux de somathormone (GH) plasmatique et syndromes d'apnées du sommeil. Leurs relations avec l'obésité. *Rev. EEG Neurophysiol. Clin.*, *10*, 366-375, 1980.

135. KURTZ D., KRIEGER J., WEITZENBLUM E., BRANDT E. – Respiratory and cardiac impact of sleep apneas in asymptomatic elderly subjects. *In: Sleep and Ageing*, Smirne S., Franceschi M., Ferini-Strambi L. (eds.), 113-121. Masson, Milan, 1991.

136. LAMPHERE J., ROEHRS T., WITTIG R., ZORICK F., CONWAY W.A., ROTH T. – Recovery of alertness after CPAP in apnea. *Chest*, *96*, 1364-1367, 1989.

137. LAVIE P., BEN-YOSEF R., RUBIN A.E. – Prevalence of sleep apnea among patients with essential hypertension. *Am. Heart J.*, *108*, 373-376, 1984.

138. LAWSON E.E., RICHTER D.W., CZYZYK-KRZESKA M.F., BISCHOFF A., RUDESILL R.C. – Respiratory neuronal activity during apnea and other breathing patterns induced by laryngeal stimulation. *J. Appl. Physiol.*, *70*, 2742-2749, 1991.

139. LEVINSON P.D., MILLMAN R.P. – Causes and consequences of blood pressure alterations in obstructive sleep apnea. *Arch. Intern. Med.*, *151*, 455-462, 1991.

140. LEYGÓNIE-GOLDENBERG F., PERRIER M., DUIZABO PH., BOUCHAREINE A., HARF A., BARBIZET J., DEGOS J.F. – Troubles de la vigilance, du sommeil et de la fonction respiratoire dans la maladie de Steinert. *Rev. Neurol.*, *133*, 255-270, 1977.

141. LUGARESI E., MONDINI S., ZUCCONI M., MONTAGNA P., CIRIGNOTTA F. – Staging of heavy snorer's disease. A proposal. *Bull. Eur. Physio-pathol. Resp.*, *19*, 590-594, 1983.

142. MALONE S., LIU P.P., HOLLOWAY R., RUTHERFORD R., XIE A., BRADLEY T.D. – Obstructive sleep apnoea in patients with dilated cardiomyopathy – Effects of continuous positive airway pressure. *Lancet*, *338*, 1480-1484, 1991.

143. MALTAIS F., CARRIER G., CORMIER Y., SERIES F. – Cephalometric measurements in snorers, non-snorers, and patients with sleep apnoea. *Thorax*, *46*, 419-423, 1991.

144. MARRONE O., BELLIA V., PIERI D., SALVAGGIO A., BONSIGNORE G. – Acute effects of oxygen administration on transmural pulmonary artery pressure in obstructive sleep apnea. *Chest*, *101*, 1023-1027, 1992.

145. MARTIN R.J., BLOCK A.J., COHN M.A., CONWAY W.A., HUDGEL D.W., POWLES A.C.P., SANDERS M.H., SMITH P.L. – Indications and standards for cardiopulmonary sleep studies. *Sleep*, *8*, 371-379, 1985.

146. MASA J.F., RUBIO M., FINDLEY L.J., - Habitually sleepy drivers have a high frequency of automobile crashes associated with respiratory disorders during sleep. *Am. J. Respir. Crit. Care Med.* 162, 1407-1412, 2000.

147. MAYER J., BECKER H., BRANDENBURG U., PENZEL T., PETER J.H., VONWICHERT P. – Blood pressure and sleep apnea – Results of long-term nasal continuous positive airway pressure therapy. *Cardiology*, *79*, 84-92, 1991.

148. MCCOLLEY S.A., APRIL M.M., CARROLL J.L., NACLERIO R.M., LOUGHLIN G.M. – Respiratory compromise after adenotonsillectomy in children with obstructive sleep apnea. *Arch. Otolaryngol.*, *118*, 940-943, 1992.

149. MIKI H., HIDA W., CHONAN T., KIKUCHI Y., TAKISHIMA T. – Effects of submental electrical stimulation during sleep on upper airway patency in patients with obstructive sleep apnea. *Am. Rev. Respir. Dis.*, *140*, 1285-1289, 1989.

150. MIYAZAKI M., HASHIMOTO T., SAKURAMA N., YOSHIMOTO T., TAYAMA M., KURODA Y. – Central sleep apnea and arterial compression of the medulla. *Ann. Neurol.*, *29*, 564-565, 1991.

151. MONSTAD P., MELLGREN S.I., SULG I.A. – The clinical significance of sleep apnoea in workers exposed to organic solvents: Implications for the diagnosis of organic solvent encephalopathy. *J. Neurol.*, *239*, 195-198, 1992.

152. MONTSERRAT J.M., KOSMAS E.N., COSIO M.G., KIMOFF R.J. – Mechanism of apnea lengthening across the night in obstructive sleep apnea. *Am. J. Respir. Crit. Care Med.*, *154*, 988-993, 1996.

153. NAEGELE B., PEPIN J.L., LEVY P., BONNET C., PELLAT J., FEUERSTEIN C. – Cognitive executive dysfunction in patients with obstructive sleep apnea syndrome (OSAS) after CPAP treatment. *Sleep.* 21, 392-397, 1998.

154. NAUGHTON M., PIERCE R. – Effects of nasal continuous positive airway pressure on blood pressure and body mass index in obstructive sleep apnoea. *Aust. NZ J. Med.*, *21*, 917-919, 1991.

155. OLIVEN A., ODEH M., SCHNALL R.P. – Improved upper airway patency elicited by electrical stimulation of the hypoglossus nerves. *Respiration*, *63*, 213-216, 1996.

156. OVESEN J., NIELSEN P.W., WILDSCHIODTZ G. – Shortened reaction time during nasal CPAP treatment of obstructive sleep apnea. *Acta Otolaryngol.*, 119-121, 1992.

157. PARTINEN M., JAMIESON A., GUILLEMINAULT C. – Long-term outcome for obstructive sleep apnea syndrome patients. *Chest*, *94*, 1201-1204, 1988.

158. PASQUALI R., COLELLA P., CIRIGNOTTA F., MONDINI S., GERARDI R., BURATTI P., CERONI R., TARTARI F., SCHIAVINA M., MELCHIONDA N., LUGARESI E., BARBARA L. – Treatment of obese patients with obstructive sleep apnea syndrome (OSAS): Effect of weight loss and interference of otorhinolaryngoiatric pathology. *Int. J. Obesity*, *14*, 207-217, 1990.

159. PEPIN J.L., FERRETTI G., VEALE D., ROMAND P., COULOMB M., BRAMBILLA C., LEVY P.A. – Somnofluoroscopy, computed tomography, and cephalometry in the assessment of the airway in obstructive sleep apnoea. *Thorax*, *47*, 150-156, 1992.

160. PETIAU C., DELANOË C., HECHT M.T., CHAMOUARD V., KRIEGER J. – Somnolence et syndrome d'apnées du sommeil. Analyse de 188 questionnaires (102 malades et 86 témoins). *Neurophysiol. Clin./Clin. Neurophysiol.*, *23*, 77-86, 1993.

161. PETROF B.J., KELLY A.M., RUBINSTEIN N.A., PACK A.I. – Effect of hypothyroidism on myosin heavy chain expression in rat pharyngeal dilator muscles. *J. Appl. Physiol.*, *73*, 179-187, 1992.

162. PHILLIPS B.A., BERRY D.T.R., SCHMITT F.A., MAGAN L.K., GERHARDSTEIN D.C., COOK Y.R. – Sleep-disordered breathing in the healthy elderly: Clinically significant? *Chest*, *101*, 345-349, 1992.

163. PITSON D.J., STRADLING J.R. – Value of beat-to-beat blood pressure changes, detected by pulse transit time, in the management of the obstructive sleep apnoea/hypopnoea syndrome. *Eur. Respi. J.* 12, 685-692, 1998.

164. POCETA J.S., TIMMS R.M., JEONG D.U., HO S.L., ERMAN M.K., MITLER M.M. – Maintenance of wakefulness test in obstructive sleep apnea syndrome. *Chest*, *101*, 893-897, 1992.

165. POLO O., TAFTI M., FRAGA J., PORKKA K., DEJEAN Y., BILLIARD M. – Why don't all heavy snorers have obstructive sleep apnea. *Am. Rev. Respir. Dis.*, *143*, 1288-1293, 1991.

166. POTSIC W.P. – Obstructive sleep apnea. *Pediatr. Clin. North Am.*, *36*, 1435-1442, 1989.

167. RAJALA R., PARTINEN M., SANE T., PELKONEN R., HUIKURI K., SEPPÄLÄINEN A.M. – Obstructive sleep apnoea syndrome in morbidly obese patients. *J. Int. Med.*, *230*, 125-129, 1991.

168. RAUSCHER H., POPP W., WANKE T., ZWICK H. – Acceptance of CPAP therapy for sleep apnea. *Chest*, 100, 1019-1023, 1991.

169. RECHTSCHAFFEN A., KALES A. – *A manual of standardized terminology, techniques and scoring system for sleep stages of human subjects.* Brain Information Service. Brain Research Institute, Los Angeles, 1968.

170. REDLINE S., TOSTESON T., TISHLER P.V., CARSKADON M.A., MILLMAN R.P. – Studies in the genetics of obstructive sleep apnea. Familial aggregation of symptoms associated with sleep-related breathing disturbances. *Am. Rev. Respir. Dis.*, *145*, 440-444, 1992.

171. REEVES-HOCHE M.K., HUDGEL D.W., MECK R., WITTEMAN R., ROSS A., ZWILLICH C.W. – Continuous versus bilevel positive airway pressure for obstructive sleep apnea. *Am. J. Respir. Crit. Care Med.* 151, 443-449, 1995.

172. REMMERS J.E., ANCH A.M., DEGROOT W.J., BAKER J.P. JR., SAUERLAND E.K. – Oropharyngeal muscle tone in obstructive sleep apnea before and after strychnine. *Sleep*, *3*, 447-454, 1980.

173. RINGLER J., BASNER R.C., SHANNON R., SCHWARTZSTEIN R., MANNING H., WEINBERGER S.E., WEISS J.W. – Hypoxemia alone does not explain blood pressure elevations after obstructive apneas. *J. Appl. Physiol.*, *69*, 2143-2148, 1990.

174. RODENSTEIN D.O., DOOMS G., THOMAS Y., LIISTRO G., STANESCU D.C., CULEE C., AUBERT-TULKENS G. – Pharyngeal shape and dimensions in healthy subjects, snorers, and patients with obstructive sleep apnoea. *Thorax*, *45*, 722-727, 1990.

175. RUMBACH L., KRIEGER J., KURTZ D. – Auditory event-related potentials in obstructive sleep apnea: effects of treatment with nasal continuous positive airway pressure. *Electroenceph. Clin. Neurophysiol.*, *80*, 454-457, 1991.

176. RYAN C.F., LOWE A.A., LI D., FLEETHAM J.A. – Three-dimensional upper airway computed tomography in obstructive sleep apnea. A prospective study in patients treated by uvulopalatopharyngoplasty. *Am. Rev. Respir. Dis.*, *144*, 428-432, 1991.

177. SANDERS M.H., BLACK J., COSTANTINO J.P., KERN N., STUDNICKI K., COATES J. – Diagnosis of sleep-disordered breathing by half-night polysomnography. *Am. Rev. Respir. Dis.*, *144*, 1256-1261, 1991.

178. SANDERS M.H., KERN N. – Obstructive sleep apnea treated by independently adjusted inspiratory and expiratory positive airway pressures via nasal mask. Physiologic and clinical implications. *Chest*, *98*, 317-324, 1990.

179. SANGAL R.B., THOMAS L., MITLER M.M. – Disorders of excessive sleepiness – Treatment improves ability to stay awake but does not reduce sleepiness. *Chest, 102*, 699-703, 1992.

180. SATOH M., HIDA W., CHONAN T., MIKI H., IWASE N., TAGUCHI O., OKABE S., KIKUCHI Y., TAKISHIMA T. – Role of hypoxic drive in regulation of postapneic ventilation during sleep in patients with obstructive sleep apnea. *Am. Rev. Respir. Dis.*, 143, 481-485, 1991.

181. SCHOENFELD A., OVADIA Y., NERI A., FREEDMAN S. – Obstructive sleep apnea (OSA) – Implications in maternal fetal medicine – A hypothesis. *Med. Hypoth., 30*, 51-54, 1989.

182. SCHWARTZ A.R., GOLD A.R., SCHUBERT N., STRYZAK A., WISE R.A., PERMUTT S., SMITH P.L. – Effect of weight loss on upper airway collapsibility in obstructive sleep apnea. *Am. Rev. Respir. Dis., 144*, 494-498, 1991.

183. SERIES F., CORMIER Y., LAFORGE J. – Influence of continuous positive airway pressure on sleep apnea-related desaturation in sleep apnea patients. *Lung, 170*, 281-290, 1992.

184. SFORZA E., KRIEGER J. – Daytime sleepiness after long-term nasal continuous positive airway pressure (CPAP) treatment in obstructive sleep apnea syndrome. *J. Neurol. Sci., 110*, 21-26, 1992.

185. SFORZA E., KRIEGER J., GEISERT J., KURTZ D. – Sleep and breathing abnormalities in a case of Prader-Willi syndrome. The effects of acute continuous positive airway pressure treatment. *Acta Paediatr. Scand., 80*, 80-85, 1991.

186. SFORZA E., KRIEGER J., WEITZENBLUM E., APPRILL M., LAMPERT E., RATAMAHARO J. – Long-term effects of treatment with nasal continuous positive airway pressure on daytime lung function and pulmonary hemodynamics in patients with obstructive sleep apnea. *Am. Rev. Respir. Dis., 141*, 866-870, 1990.

187. SHELLOCK F.G., SCHATZ C.J., JULIEN P., STEINBERG F., FOO T.K.F., HOPP M.L., WESTBROOK P.R. – Occlusion and narrowing of the pharyngeal airway in obstructive sleep apnea: Evaluation by ultrafast spoiled GRASS MR imaging. *Am. J. Roentgenol., 158*, 1019-1024, 1992.

188. SHEPARD J.W., GEFTER W.B., GUILLEMINAULT C., HOFFMAN E.A., HOFFSTEIN V., HUDGEL D.W., SURATT P.M., WHITE D.P. – Evaluation of the upper airway in patients with obstructive sleep apnea. *Sleep, 14*, 361-371, 1991.

189. SHEPARD J.W., THAWLEY S.E. – Localization of upper airway collapse during sleep in patients with obstructive sleep apnea. *Am. Rev. Respir. Dis., 141*, 1350-1355, 1990

190. SHIOMI T., GUILLEMINAULT C., STOOHS R., SCHNITTGER I. – Leftward shift of the interventricular septum and pulsus paradoxus in obstructive sleep apnea syndrome. *Chest, 100*, 894-902, 1991.

191. STEBBENS V.A., DENNIS J., SAMUELS M.P., CROFT C.B., SOUTHALL D.P. – Sleep related upper airway obstruction in a cohort with Down's syndrome. *Arch. Dis. Child, 66*, 1333-1338, 1991.

192. STEPHAN S., CASSEL W., SCHWARZENBERGER-KESPER F., FETT I., HENN-KOLTER C., PETER J.H. – Psychological problems correlated with sleep apnea. *In: Sleep and Health Risk*, Peter J.H., Penzel T., Podzus T., Wichert P. von (eds.), 167-173. Springer-Verlag, Berlin, 1991.

193. STOOHS R., GUILLEMINAULT C. – MESAM-4 – An ambulatory device for the detection of patients at risk for obstructive sleep apnea syndrome (OSAS). *Chest, 101*, 1221-1227, 1992.

194. STRADLING J.R., THOMAS G., WARLEY A.R.H., WILLIAMS P., FREELAND A. – Effect of adenotonsillectomy on nocturnal hypoxaemia, sleep disturbance, and symptoms in snoring children. *Lancet, 335*, 249-253, 1990.

195. STROHL K.P., HENSLEY M.J., HALLETT M., SAUN-DERS N.A., INGHAM R.H. – Activation of upper airway muscles before onset of inspiration in normal humans. *J. Appl. Physiol., 49*, 638-642, 1980.

196. SUGERMAN H.J., FAIRMAN R.P., SOOD R.K., ENGLE K., WOLFE L., KELLUM J.M. – Long-term effects of gastric surgery for treating respiratory insufficiency of obesity. *Am. J. Clin. Nutr., 55*, S597-S601, 1992.

197. SULLIVAN C.E., ISSA F.G. – Pathophysiological mechanisms in obstructive sleep apnea. *Sleep, 3*, 235-246, 1980.

198. SULLIVAN C.E., ISSA F.G., BERTHON-JONES M., EVES L. – Reversal of obstructive sleep apnoea by continuous positive airway pressure applied through the nares. *Lancet, i*, 862-865, 1981.

199. SVANBORG E., CARLSSON-NORDLANDER B., LARS-SON H., SACHS C., KAIJSER L. – Autonomic nervous system function in patients with primary obstructive sleep apnea syndrome. *Clin. Autonom. Res., 1*, 125-130, 1991.

200. THE REPORT OF AN AMERICAN ACADEMY OF SLEEP MEDICINE TASK FORCE. – Sleep-related breathing disorders in adults : Recommandations for syndrome definition and measurement techniques in clinical research. *Sleep, 22*, 667-689, 1999.

201. TOLLE F.A., JUDY W.V., YU P.L., MARKAND O.N. – Reduced stroke volume related to pleural pressure in obstructive sleep apnea. *J. Appl. Physiol., 55*, 1718-1724, 1983.

202. VICTOR D.W., SARMIENTO C.F., YANTA M., HALVERSON J.D. – Obstructive sleep apnea in the morbidly obese. An indication for gastric bypass. *Arch. Surg., 119*, 970-972, 1984.

203. VINER S., SZALAI J.P., HOFFSTEIN V. – Are history and physical examination a good screening test for sleep apnea? *Ann. Int. Med., 115*, 356-359, 1991.

204. WALDHORN R.E., HERRICK T.W., NGUYEN M.C., ODONNELL A.E., SODERO J., POTOLICCHIO S.J. – Long-term compliance with nasal continuous positive airway pressure therapy of obstructive sleep apnea. *Chest, 97*, 33-38, 1990.

205. WALSH J.K., CORDER J.C., SCHWEITZER P.K., KATSANTONIS G.P. – Obstructive sleep apnea exacerbated by acetazolamide treatment of central sleep apnea: a case report. *Sleep Res.*, *12*, 292 (Abstr), 1983.

206. WEITZENBLUM E., CHAOUAT A., KESSLER R., OSWALD M., APPRILL M., KRIEGER J. – Daytime hypoventilation in obstructive sleep apnoea syndrome. *Sleep Med. Rev.* 3, 79-93, 1999.

207. WEITZENBLUM E., KRIEGER J., APPRILL M., VALLÉE E., ERHART M., RATAMAHARO J., OSWALD M., KURTZ D. – Daytime pulmonary hypertension in patients with obstructive sleep apnea syndrome. *Am. Rev. Respir. Dis.*, *138*, 345-349, 1988.

208. WHITE D.P. – Central sleep apnea. *Med. Clin. N. Amer.*, *69*, 1205-1219, 1985.

209. WHITE D.P., BALLARD R.D. – Pharyngeal muscle activity and upper airway resistance in obstructive apnea patients versus controls. *In: Sleep and Respiration*, Issa F.G., Suratt P.M., Remmers J.E. (eds.), 243-251. Wiley-Liss, 1990.

210. WHYTE K.F., DOUGLAS N.J. – Peripheral edema in the sleep apnea/hypopnea syndrome. *Sleep*, *14*, 354-356, 1991.

211. WHITE D.P., ZWILLICH C.W., PICKETT C.K., DOUGLAS N.J., FLINDLEY L.J., WEIL J.V. – Central sleep apnea. Improvement with acetazolamide therapy. *Arch. Int. Med.*, *142*, 1816-1819, 1982.

212. YOUNG T., PALTA M., DEMPSEY J., SKATRUD J., WEBER S., BADR S. – The occurrence of sleep-disordered breathing among middle-aged adults. *N. Engl. J. Med.*, *328*, 1230-1235, 1993.

213. ZUCCONI M., FERINI-STRAMBI L., PALAZZI S., ORENA C., ZONTA S., SMIRNE S. – Habitual snoring with and without obstructive sleep apnoea – The importance of cephalometric variables. *Thorax*, *47*, 157-161, 1992.

214. ZWILLICH C., DEVLIN T., WHITE D., DOUGLAS N., WEIL J., MARTIN R. – Bradycardia during sleep apnea. Characteristics and mechanism. *J. Clin. Invest.*, *69*, 1286-1292, 1982.

Chapter 30

Surgical and prosthetic treatment for sleep apnoea syndrome and upper airway resistance syndrome

L. Crampette
Service ORL et Chirurgie Cervico-Faciale, Hôpital Gui de Chauliac, Montpellier, France

Obstructive sleep apnoea hypopnea syndrome (OSAHS) and upper airway resistance syndrome (UARS) require treatment in order to prevent cardiovascular morbidity and to cure the excessive daytime sleepiness.

The main methods used are continuous positive airway pressure (CPAP) and surgical treatment. This chapter deals with the principles, basis, methods, results and indications of the different surgical and prosthetic treatments.

PRINCIPLES

Surgical treatment comprises several aims:

1. To short-circuit the obstacle in the Upper Airways (UA): this is the principle of tracheostomy [23], a method which is consistently effective but may cause damage and discomfort.

2. To correct the UA at the point where apnoeic collapse or increase respiratory resistance develop.

The anatomic area in which the obstacle is located during apnoea or narrowing, in the case of increased respiratory resistance, is always that of the oropharynx, either the upper part, known as the velopharynx (VP), or the lower part, referred to as the retrobasilingual oropharynx (RBLO) [10, 11]; it must be emphasised that neither the upper part of the pharynx, *i.e.* the nasopharynx (or epipharynx or cavum), nor the lower part of the pharynx, *i.e.* the hypopharynx (or laryngopharynx) are critical sites: indeed, the nasopharynx retains its shape due to the bony parts of its walls, and the hypopharynx is a purely digestive sector and thus cannot lie at the origin of a sleep-related respiratory disorder; the "hypopharyngeal" obstacles sometimes referred to in the literature are in fact retrobasilingual.

The methods used to correct pharyngeal narrowing may be divided into two groups:

- techniques designed to reduce the volume of pharyngeal soft tissue ("reduction of the content"): uvulopalatopharyngoplasty (UPPP) at VP level, reduction of the base of the tongue at retrobasilingual level.

- techniques aimed at displacing the bone structures where pharyngeal soft tissue is attached, in an attempt to increase pharyngeal light; the common element of these treatments being "enlargement of the form". Some of these techniques are prosthetic *i.e.* in the form of an appliance which advances the mandible in relation to the maxilla. Others come under the heading of surgery, involving various maxillary, mandibular and hyoid bone advancement techniques.

3. Treatment for nasal obstruction may be justified in two ways:

Either nasal dyspermeability is thought to play a role in generating apnoeas or episodes of increased pharyngeal breathing resistance; the nose is clearly not the site of inspiratory collapse during sleep as the walls retain their shape, but nasal obstruction causes the formation of an increased pharyngeal inspiratory depression if the patient maintains nasal breathing; the increase in pharyngeal inspiratory depression facilitates pharyngeal deformation, increasing pharyngeal resistance [16]. On the other hand, when nasal obstruction is incompatible with forced nasal respiration, the subject switches to oral breathing, leading to the deflexion of the base of the tongue and facilitating pharyngeal obstruction [30].

Or, alternatively, the nasal obstruction inhibits nocturnal ventilation: treatment of the nasal obstruction may sometimes be necessary to allow nasal continuous positive airway pressure to be applied.

BASIS FOR TREATMENT

Surgical treatment relies on previously locating the obstacle or narrowing of the pharynx. This seemingly simple approach is in fact difficult. Indeed, the pharyngeal obstacle develops during sleep, in the supine position, and must be conceived of as a dynamic malformation, whereas in fact UA examination is carried out in wakefulness, in a seated position and in a static fashion. The location(s) of the pharyngeal obstacle must therefore be determined on the basis of a body of information, *e.g.* clinical, fibroscopic and imaging-based.

The clinical and fibroscopic tests are like those of the snorer (see chapter 43).

Pharyngeal imaging is of the utmost importance. It comprises computed tomography (CT scan), cephalometric radiograph, and magnetic resonance imaging (MRI).

Pharyngeal computed tomography [2,11,18,43]

The patient lies in a supine position, with the head in a neutral position; he must not swallow; the topogramme profile provides a rough assessment of the length of the palate, the size and position of the tongue, and the size of the soft cervical parts. The standard procedure involves taking a series of cross sections every 8 minutes, with calm respiration; the development of spiral helicoidal scanning has accelerated the rate of images. The study of axial sections provides both "qualitative" and "quantitative" information. From the "qualitative" point of view, it is possible to view the point at which pharyngeal narrowing occurs (VP, retrobasilingual or bifocal?) and to see whether the narrowing is antero-posterior from the posterior displacement of the palate and/or base of the tongue, or transversal due to tonsillar enlargement. From the "quantitative" point of view, a surface calculation gives a cm^2 measurement of the minimal light of the velopharynx and of the retrobasilingual oropharynx. The normal pharynx is characterised by light measuring 2 to 3 cm^2 at the velopharynx level and 3 to 4 cm^2 at the retrobasilingual level. A patent restriction exists where there is less than 1 cm^2 for the velopharynx and 1.5 cm^2 for the hypopharynx (fig. 30.1). Between these values, the interpretation must be made with care and correlated with other methods for investigating the pharynx, as so many parameters can fluctuate in the course of the examination (head position, respiration, pharyngeal muscle contraction) [11, 17].

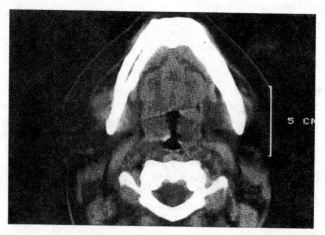

Figure 30.1. CT scan of the axial section of the retrobasilingual pharynx. Note the narrowing, due to the enlargement of the inferior pole of the palatine tonsils.

Cephalometric radiograph [26, 35]

Still referred to as teleradiography, this provides an accurate analysis of maxillo-mandibular bone structure.

The technical principle is simple, consisting of a lateral radiographic view of a scale of 1/1 with a number of measurement landmarks, to determine maxillar and mandibular development in relation to the base of the skull, as well as the position of the hyoid bone in relation to the mandible. The process must be technically perfect [29]. The patient is seated, with a neutral horizontal forward gaze. The head is supported by a cephalostat. The oral occlusion must be natural, the lips relaxed and the radiograph taken during slow expiration. The landmarks (fig. 30.2) are S (center of sella turcica), N (nasion), A (subspinal point, or the lowest point of the premaxilla), B (supramental point, the posterior extremity of the mandibular curve in the section between the alveolar bone and the chin), ANS (anterior nasal spine), PNS (posterior nasal spine), P (tip of the uvula), Go (gonion), Gn (gnathion), H (the anterior, superior extremity of the body of the hyoid bone). Angle SNA, at a norm of 82 ± 2° reflects the development of the mandible in relation to the base of the skull. The distance from PNS – P (normal 37 ± 3mm) corresponds to the length of the soft palate. The posterior air space separates point Go from the posterior pharyngeal wall; the norm is 11 ± 1mm. The distance from the mandibular plane (plane passing through Go and Gn) and point H indicates the position of the hyoid bone in relation to the mandible; the norm is 15.4 ± 3mm.

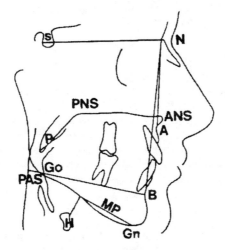

Figure 30.2. Cephalometric bone and soft tissue landmarks (after Riley *et al.*, 1993).

Magnetic Resonance Imaging of the pharynx

Developed by Chabolle *et al.* [5], the subject is in a supine position, providing a very accurate analysis of the hyoid tongue apparatus (position of the hyoid bone, main axis of the body of the tongue, and area of projection of the body of the tongue). The scale is not 1/1, and the measurements of distance, angle and surface area are calculated by expansion coefficient. Projected tongue area is normally between 20 and 25 cm^2. A line is drawn through the genial apophyses, perpendicular to the posterior pharyngeal plane, marking out the supra- and sub-mandibular areas. Normally, the supramandibular area ranges from 17 to 20 cm^2 while the submandibular area is no greater than 5 cm^2.

Indications for the different imaging procedures

The CT scan is the only procedure which takes account of both the VP and the retrobasilingual oropharynx in their transversal and antero-posterior dimensions. We continue to practise its use as a first-line investigation [28]. Conversely if narrowing of the retrobasilingual oropharynx is confirmed, it is impossible to determine the causes using the CT scan. This is where cephalometric radiograph is of interest as it will relate the posterior protrusion of the tongue to an abnormality of the bone structure or an enlarged lingual mass. The respective pros and cons of the CT scan and cephalometric radiograph render these techniques complementary rather than conflicting. Thus we refer to cephalometric radiogaph when the CT scan shows a narrowing of the retrobasilingual oropharynx, and equally when the subject presents a dental occlusion abnormality. Other teams systematically call for cephalometric radiograph. As for MRI, it is of interest only where there is retrobasilingual narrowing with no bone structure abnormality, in the aim of demonstrating the presence of retrobasilingual hypertrophy.

METHODS

Tracheostomy

First proposed in 1969 [23], tracheostomy, by short-circuiting the obstacle to the upper airways, invariably cures OSAHS [17]. It was widely practised in the USA among patients causing particular concern in terms of the severity of their OSAHS and their considerable obesity. CPAP has now, fortunately, replaced tracheostomy for this type of indication, as tracheostomy was never considered as the ideal treatment for OSAHS in view of the local and psychological complications it can lead to [12]. Its status has thus become transitory or palliative.

Transitory tracheostomy

This should be considered in association with "heavy" maxillomandibulolingual surgery where there is a risk of dyspnoea or postoperative haemorrhage which could end in a catastrophe if one recalls the difficulties involved in the intubation of these patients. When there is reluctance to do so, the wise choice would be to prepare for a tracheostomy (cutaneous incision, thyroid isthmectomy, clearing the anterior surface of the trachea, followed by cutaneous suture) so that, if any secondary problems and difficulties of re-intubation arise, in the recovery room or at the patient's bedside, the trachea may be opened very quickly. In the majority of cases where there are no postoperative complications, this precaution results in only a minor scar for the patient to bear.

Palliative tracheostomy

Palliative tracheostomy is now rarely carried out. This indication presupposes both a failure of CPAP treatment and the failure or contraindication of surgical treatment. It is advisable in this case to perform a full tracheostomy, whose principle is based on obtaining perfect continuity between the trachea and cervical skin, to avoid granulations and ostial stenosis, and to adapt an occlusive appliance during the daytime, sealed at the ostium to avoid air leakage or secretion through the orifice [31].

Velopharyngeal surgery

This is dominated by uvulopalatopharyngoplasty (UPPP). In children with obstructive sleep disorders and tonsil enlargement, isolated tonsilectomy is indicated. Laser or radiofrequency treatment applied to the soft palate are appropriate in the case of pure snoring but prove insufficient for OSAHS or UARS.

Uvulopalatopharyngoplasty (UPPP), first described in 1981 [14] is now well codified. It should not be considered as a partial amputation of the soft palate but as an operation with multiple aims: the palate must be shortened, thinned, advanced, and the pharynx enlarged [4, 10, 11, 14, 20, 22, 27, 28, 33]. Local in some cases, most authors prefer general anaesthesia with intubation. Three phases of operation can be described in UPPP surgery (fig.30.3):

- *phase I is tonsillectomy*, scrupulously maintaining the pillars intact.

- phase II is the partial uvulo-velectomy. When excising the soft palate, the main difficulty lies in judging the extent of excision. If this is too limited the result will be partial, or nonexistent, if excessive, it exposes the risk of velar insufficiency. Palpation to ascertain the point of contact between the soft palate and the posterior pharyngeal wall will indicate how much distal soft palate can be removed without a functional risk.

- phase III is pharyngoplasty per se. This involves excising the posterior pillar. The entire posterior pillar, muscle and mucous membrane may be excised [33] or simply the staphylion pharyngeal muscle fibres, avoiding the mucous membrane [27]. The posterior pillar is then moved to one side and lowered to fit into a notch made in the anterior pillar. This operative phase allows for the enlargement of the velopharynx, advancement of the soft palate and suturation of the mucous membrane seams, taking care to avoid leaving raw areas which could become factors for haemorrhaging, retraction and stenosis.

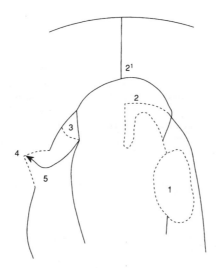

Figure 30.3 Uvulopalatopharyngoplasty
1. Tonsillectomy (or ablation of the mucous membrane in cases of previous history of tonsillectomy.
2. Uvulectomy.
2'. The length of the remaining soft palate, from the posterior edge of the hard palate to the edge of the section should be sufficient to occlude the pharyngonasal isthmus.
3. Upper musculomucous section of the posterior pillar.
4. Cone shaped excision of the anterior pillar.
5. Deflexion and suture of the lateral pharyngeal flap onto the prepared anterior pillar (with suture of the velar edge of section).
1,2. Excision phase ("uvulopalatectomy").
3,4,5. Lateral pharyngoplasty.

Postoperative period. The patient is placed under oral antibiotic treatment. The pain is fairly intense for the first three days [11]. If it is relieved by non sedative type 2 analgesics and viscous *Xylocaine* before meals, it will not prohibit a mixed oral diet. The first 45 days may be marked by disorders, which rarely occur simultaneously, and which are all the easier for the patient to bear if he is prepared for them. These include intermittent false passages of nasal liquid, slight nasal phonation toward the end of the day, taste disorders, pharyngeal paresthesia, a sensation of nasal secretion blocked in the nasal cavities. These transient disorders should be clearly distinguished from real complications. An early complication is haemorrhage, but this is statistically rare. Later complications may take two forms: velar insufficiency resulting from excessive excision, can be corrected by reeducation or even further surgery; the other complication is that of nasopharyngeal stenosis, which requires careful treatment [7]. Such complications are very rare if the technique is carefully adhered to, particularly if lateral pharyngoplasty has been performed correctly. Pharyngoplasty, by enlarging the pharynx avoids relying solely on the shortening of the palate to be effective, preventing velar insufficiency. It also provides a mucous membrane covering of the pharyngo-velar junction, an essential element in preventing velo-pharyngeal stenosis. It must be added, finally, that in our experience of practising UPPP in this way, patients whose OSAHS was not cured, and who had to be given second treatment by ventilation, have never experienced any difficulties in adapting.

Retrobasilingual surgery

This includes on one hand, the operations designed to bring the tongue forward by displacing the mandibular and hyoid bone structures to which it adheres, and on the other, the operations to reduce the size of the tongue.

Tongue advancement surgery

Mandibular osteotomy with genioglossus advancement procedure

The genial apophyses, situated in the concavity of the symphysis menti, constitute the anterior insertions of the tongue. It is possible to advance these without altering dental articulation [37]. Riley *et al* [39] suggest cutting a rectangular window of symphyseal bone, with a long horizontal axis including the genial apophyses. This portion is advanced anteriorly, rotated to allow bony overlap, and immobilized with a titanium screw (fig. 30.3). The advancement obtained is of 13 mm on average. The operation is relatively simple, does not require tracheostomy, bi-maxillary blockage, or orthodontic treatment. From the bio-mechanical point of view it can quite easily be associated with hyoidopexy. Chabolle *et al* [6] propose a variant consisting of advancing the whole of the chin as a block, with osteosynthesis, the advantage of this technique being that it avoids rotating the advanced fragment.

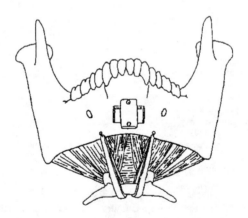

*Figure 30.4.*Genioglossus advancement and hyoidopexy (after Riley *et al.*, 1989).

Hyoid surgery

The hyoid bone is the posterior extremity to which the tongue is fixed. Its position is often too low and too far back in patients suffering from an OSAHS or UARS. Several surgical techniques have been described, almost always coupled with advancement genioplasty (fig. 30.4):
- Hyoidopexy with myotomy consists of moving the hyoid bone closer to the symphyses, by means of a metal wire or facia lata; myotomy consisting of sectioning the sub-hyoid muscles of the stylo-hyoid at the base of the hyoid bone, avoids the need for excess traction [36].

- Expansion hyoidoplasty consists of only advancing the body of the hyoid bone, having freed the 2 lateral parts by sectioning inside the lesser cornu: the lateral hyoid fragments, into which the middle pharyngeal constrictor is inserted, do not accompany the advancement of the body of the hyoid bone, so that increasing the antero-posterior diameter of the retrobasilingual oropharynx does not come at the cost of causing tension to the lateral pharyngeal walls, which could reduce the transversal diameter [32].
- Thyro-hyoïdopexy consists of suspension of the hyoid bone to the superior thyroid cartilage. The hyoid bone is advanced but, unlike the preceding techniques, it is lowered, which would be illogical if carried out in isolation; this operation must thus be coupled with advancement genioplasty [40] (fig. 30.4).

Finally let us mention an affiliated procedure consisting of suspending the base of the tongue: an anchorage screw is inserted into the posterior surface of the mandibular symphyses, a non absorbent thread is passed through the thickness of the tongue from front to back as far as the base of the tongue, then passed from the back to the front, and the two ends of the thread are then attached to the anchorage screw [9].

Mandibular advancement appliances [13, 25, 41]

This is not strictly surgical treatment; however, the principle itself of these appliances consisting in advancing the tongue, their adaptation which often requires referring to a stomatologist or a maxillofacial surgeon, their preparation sometimes with a view to predicting the efficacy of surgical mandibular advancement are all elements which lead to classifying these appliances as part of surgical treatment. These mandibular advancement devices all involve a cast moulded to the upper dental arch, another adapted to the lower arch, the lower cast being held in propulsion in relation to the upper cast either in a fixed manner or by elastic traction. Fixed appliances have the advantage of being simpler; some commercially available models are quite simple in design, made of a thermo-cast material. The disadvantage is that these are uncomfortable, any mandibular movement being rendered impossible, and they are probably even more poorly tolerated in terms of temporo-mandibular articulation. The "elastic" appliances, are more complicated to adapt, impressions need to be taken beforehand and several different specialists referred to, resulting in a higher cost; but they are more comfortable and better tolerated. Their use is limited by the patient's dental condition: they are inconceivable in the case of edentation or if there are too few teeth or if dental condition is poor; there are contraindications in cases of patent dysfunctioning of the temporo-mandibular joint.

Maxillomandibular advancement [3, 24, 37, 38, 39, 45]

This associates mandibular advancement by bilateral saggital split mandibular osteotomy of the Obwegeser-Dalpont type, and maxillary advancement through osteotomy of the Lefort I type (fig. 30.5). An advancement of the order of 10 mm is obtained. Combined advancement of the maxilla avoids the necessity of prior orthodontic treatment, as dental occlusion is preserved. Perioperative intermaxillary blockage is necessary. This may or may not be maintained postoperatively, depending on the team. Heavier surgery is clearly involved, with the risk of sequellae (injury to the inferior dental pedicles, from anaesthetising the region of the jaw) and complications (pseudarthrosis, secondary alterations to dental occlusion). A number of attempts at osteodistraction genesis should be mentioned, a procedure which consists in the gradual mechanical traction of bone segments at an osteotomy site. Ossification occurs in parallel with the distraction; it is the patient who, once the material has been applied, will carry out the distraction. This process is far lighter than standard osteotomy. However, although the technique has been well mastered in terms of mandibular malformations, it proves more delicate when it involves combining mixed maxillomandibular advancement.

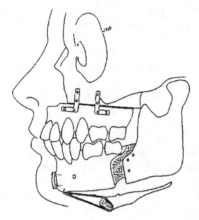

Figure 30.5 Maxillomandibular advancement and hyoidopexy (after Riley *et al.*, 1989).

Lingual reduction surgery

The earliest solution applied consisted of the oral exeresis of a dorsal basilingual segment [1]. Fujita proceeded in a similar way using a CO2 laser endoscope introduced orally [15]. The surgery, although apparently simple, involves a number of risks. Indeed oral introduction precludes the possibility of controlling the lingual pedicles. It is difficult to estimate the volume to remove. For this reason, the current practice tends to favour basilingual reduction performed cervically, which controls lingual pedicles, may remain submucosal and enables associated hyo-epiglottoplasty [8]. A temporary tracheostomy is indicated. One team has experimented a basilingual reduction using radiofrequency [34]. Numerous applications are required (average of 5.5 sessions at 4 week intervals); there are side effects (dysphagia, mucosal ulceration, tongue abscess) the cost is high, as the electrodes are expensive and used once only, and 4 of the 18 patients referred to in the work had not improved their condition.

Treatment for nasal obstruction

Nasal obstruction is treated in function of its cause:
- septoplasty
- reducing the inferior turbinates, by cauterisation or applying radiofrequencies, carried out in simple consultation, or through surgery (partial inferior turbinectomy).
- surgical treatment for nasal valve abnormalities.
- medico-surgical treatment for chronic nasal sinusitis.

INDICATIONS AND RESULTS

The presence of breathing abnormalities, associated with snoring, considerably alters the treatment perspective. The standard treatment used is continuous positive airway pressure. But this treatment is not always well accepted by patients, either from the outset or after a period of time. Moreover, in certain countries like France, the social services only cover the costs of ventilation apparatus in cases with a high respiratory disturbance index, raising the question of surgical treatment for moderate OSAHS and UARS.

When surgical treatment is opted for, the prime aim is obviously to regulate the breathing pattern, rather than to suppress or improve the accompanying snoring. A "complete" recovery is obtained when the postoperative apnoea/hypopnoea index falls below 10, whereas a score of below 20, while not strictly considered as a normalisation of sleep-related breathing, is considered as a good result since patients are protected from the increased risk of mortality linked to OSAHS [19].

The standard reduction by over 50% of the respiratory disturbance index is not a good criterion: if the preoperative index is moderate, a reduction of over 50% is a good result, whereas if the preoperative index is high, at over 40, a reduction of over 50% is still insufficient to be considered a good result. Then there are the problems of objectively verifying surgical treatment using sleep recordings at about sixth months after the operation, and, in the case of a good objective result, the persistence of the efficacy of treatment over time.

Indications to consider:

- the respiratory disturbance index. "Moderate" OSAHS corresponds to an index of below 30, "severe" OSAHS, to an index of over 35

- trial ventilatory treatment for the severe forms

- the location of the sites of pharyngeal, velopharyngeal and/or retrobasilingual obstruction(s), bearing in mind the difficulties in determining these and the need to combine clinical, fiberoptic and imaging approaches: pharyngeal scan, cephalometric radiograph ± pharyngeal MRI

- age and general condition

- body mass index

- cardiovascular and respiratory condition and previous history

- dental condition (good, poor, edentation), dental occlusion, presence of any temporo-mandibular dysfunctioning

- the wishes of the patient

At Montpellier Teaching Hospital we continue to favour taking the therapeutic decision after a multidisciplinary meeting between somnologists involved in sleep-related respiratory disorders diagnosis and treatment, specialists in ENT and in maxillofacial surgery.

For moderate forms of OSAHS, "simple" surgery (UPPP ± nasal surgery) may offer an alternative to ventilation with continuous positive airway pressure, using a nasal mask (CPAP). While the subjective result is almost always excellent, the objective effects of UPPP on OSAHS are less consistent. A 1996 meta-analysis reported a 50% drop in apnea index with a postoperative index of under 10 in 137 of 337 subjects who underwent UPPP [42]. Noticeably nonresponders had a higher baseline apnea index (56.6 ± 30.5) than did responders (31.2 ± 23.3) (p=0.0001).

The general conclusion is that the indications for UPPP in OSAHS need to be very precise: apnoea/hypopnoea index less than 30 to 35; absence of morbid obesity; absence of respiratory failure; unifocal velopharyngeal narrowing; young subject with no contraindications for surgery. In addition to this, a follow-up polysomnographic recording must always be carried out several months after the operation. Indeed if the operation fails, further surgery would only be justified in very rare cases, the usual response being that of secondary adaptation to CPAP.

For severe forms of OSAHS, the indication for CPAP is perfectly clear. However if the patient refuses CPAP treatment, the only solutions are surgical and/or prosthetic treatment. In such cases, UPPP alone would have little chance of curing the patient. It is rather a "multistep" or "multilevel" surgical protocol which should be envisaged. The standard protocol of this type was described by the Stanford School [39]. Phase I comprises UPPP associated with the enlargement of the hypopharynx through genioplasty and hyoidopexy. If there is nasal obstruction, it is treated at a later stage. A polysomnographic evaluation is made six months after the operation. If the patient remains apnoeic, phase II of surgery is decided upon, generally by maxillomandibular advancement osteotomy. The success rate is between 75 and 100% of patients who undergo both phases [3, 38, 39, 45]. Lee *et al* [24] recommend a similar strategy but one involving nasal repermeabilisation surgery, if necessary, as a first phase (phase 0). It should be noted that a number of patients for whom phase I fails, give up on phase II, preferring CPAP treatment: in the work of Lee *et al*. [24], on a population of 35 patients, 11 had failed in phase I (index over 20), but only 3 patients accepted phase II (all 3 were cured).

Some authors have suggested other ideas or variants to the classic treatment, especially in regard to phase I: what should this phase consist of? What is the respective importance of the different surgical acts associated with this phase? Should another stage not be added between phases I and II, inserting a trial period using a mandibular advancement appliance? In some cases, would it not be preferable to short-circuit phase I, passing directly to phase II? Some teams express doubts as to the efficacy of genial transposition in phase I. It is true that this technique is often coupled with UPPP within the framework of phase I, with a success rate of the order of 50 to 60%, so that it is hard to determine how much is attributable to UPPP and how much to genioplasty. Wagner *et al*. [44]. attribute a success rate of only 25% to genioplasty. This team tends to consider phase I as UPPP associated with nasal surgery if necessary, and if this fails, phase II should consist of either maxillomandibular advancement if skeletal abnormalities are shown in cephalometric radiograph, or else basilingual reduction if there are no abnormalities in bone structure and in the presence of an enlarged tongue base shown on pharyngeal MRI.

In the case of failed UPPP ± nasal surgery, other teams tend to suggest adapting the mandibular advancement appliance, possibly followed by maxillomandibular advancement.

Finally, other teams suggest maxillomandibular advancement at the outset for certain patients who are referred with cephalometric abnormalities [3, 21].

The surgical indications for UARS are still unclear.

REFERENCES

1. AFZELIUS L.E., ELMQVIST D., HOUGAARD K., LAURIN S., NILSSON B., RISBERG A.M. – Sleep apnea syndrome; an alternative treatment to tracheostomy. *Laryngoscope*, 91, 285-291, 1981.
2. BOHLMAN M.E., HAPONIK E.F.,SMITH P.L., ALLEN R.P., GOLDMAN S.M. – C.T. demonstration of pharyngeal narrowing in adult obstructive sleep apnea. *Am. J. Radiol.*, 140, 543-548, 1983.
3. BETTEGA G., PEPIN J.L., VEALE D., DESCHAUX C., RAPHAEL B., LEVY P. – Obstructive sleep apnea syndrome. Fifty-one consecutive patients treated with maxillofacial surgery. *Am. J. Respir. Critic. Care*, 162, 641-649, 2000.
4. BLAIR SIMMONS F.,GUILLEMINAULT C., SILVESTRI R. – Snoring and some obstructive sleep apnea syndrome can be cured by oropharyngeal surgery. *Arch. Otolaryngol.*, 109, 503-507, 1983.
5. CHABOLLE F.,SEQUERT C., LACHIVER X., FLEURY B., MARSOT-DUPUCH K., LACAU ST GUILY J., MEYER B., CHOUARD C.H. – Intérêt physiopathologique d'une étude céphalométrique par téléradiographie et IRM dans le syndrome d'apnée du sommeil. Déductions thérapeutiques. *Ann. Otolaryngol. Chir. Cervicofac.* 112, 164- 168, 1995.
6. CHABOLLE F., LACHIVER X., FLEURY B., PANDRAUD L., AZAN L. – Nouvelle technique de transposition génienne dans le traitement chirurgical du syndrome d'apnée du sommeil. *Ann. Otolaryngol. Chir. Cervicofac.*, 112, 164-168, 1995.
7. CHABOLLE F., BIACABE B., SEQUERT C., CABANES J., LACHIVER X., FLEURY B. – Traitement des sténoses vélopharyngées après pharyngectomie pour rhonchopathie chronique. A propos de 13 cas. *Ann. Otolaryngol. Chir. Cervicofac.* 112, 324-329, 1995.
8. CHABOLLE F., WAGNER I., BLUMEN M.B. – Tongue base reduction with hyo-epiglottoplasty: a treatment for severe sleep obstructive apnea. *Laryngoscope*, 109, 1273-1280, 1999.
9. COLEMAN J.A.Jr. – Suspension sutures for the treatment of obstructive sleep apnea and snoring. *Otolaryngol. Clin. North Am.* 32, 277-285, 1999.
10. CRAMPETTE L. – *Syndrome d'apnées au cours du sommeil (SAS) de l'adulte.* Thèse Médecine, Montpellier, March 1986.
11. DEJEAN Y., CHOUARD C.H. – La rhonchopathie chronique. Ronflement et syndrome d'apnée du sommeil. *Rapport de la Société Française d'ORL.* Arnette, Paris, 348p, 1993.
12. DYE J.P. – Living with a tracheostomy for sleep apnea. *N. Engl. J. Med.*, 308, 1167-1168, 1983.
13. FERGUSON K.A., ONO T., LOWE A.A. – A short term control trial of an adjustable oral appliance for the treatment of mild to moderate obstructive sleep apnea. *Thorax*, 52, 362-368, 1997.
14. FUJITA S., CONWAY W., ZORICK F., ROTH T. – Surgical correction of anatomic abnormalities in obstructive sleep apnea syndrome: uvulopalatopharyngoplasty. *Otolaryngol. Head Neck Surg.*, 89, 923-934,1981.
15. FUJITA S., WOODSON T., CLARCK J. – Laser midline glossectomy as a treatment for obstructive sleep apnea. *Laryngoscope*, 101, 805-809, 1991.
16. GLEESON K., ZWILLICH C.W., BENDRICK T.W., WHITE D.P. – Effects of inspiratory nasal loading on pharyngeal resistance. *J. Appl. Physiol.*, 60, 1882-1886, 1986.
17. GUILLEMINAULT C., SIMMONS F.B., MOTTA J., CUMMISKEY J., ROSEKIND M., SCHROEDER J.S., DEMENT W.C. – Obstructive sleep apnea syndrome and tracheostomy. Long term follow up experience. *Arch. Intern. Med.* 141, 985-988, 1981.
18. HAPONIK E.F., SMITH P.L., BOHLMAN M.E., ALLEN R., GOLDMAN S., BLECKER E. – Computerized tomography in obstructive sleep apnea. *Am. Rev. Respir. Dis.* 127, 221-226, 1983.
19. HE J., KRYGER M., ZORICK F., CONWAY W., ROTH T. – Mortality and apnea index in obstructive sleep apnea. Experience in 385 male patients. *Chest*, 94, 9-14, 1988.
20. HERNANDEZ S.F. – Palatopharyngoplasty for the obstructive sleep apnea syndrome.Technique on preliminary report of results in 10 patients. *Am. J. Otolaryngol.*, 3, 229-234,1982.
21. HOCHBAN W., CONRADT R., BRANDEBOURG U., HEITMAN J., PETER J.H. – Surgical maxillofacial treatment of obstructive sleep apnea. *Plat. Reconstr. Surg.* 99, 619-626, 1997.
22. KATSANTONIS G.P., FRIEDMAN W.H., ROSENBLUM B.N., WALSH J.K.- The surgical treatment of snoring: a patient perspective. *Laryngoscope* 100, 138-140, 1990.
23. KUHLO W., DOLL E., FRANC M. – Exfolgreich Behandlung eines Pickwick Syndrome durch eine dauertracheal Kanule. *Dtsch. Med. Wochenschr.* 94, 1286-1290, 1969.
24. LEE N.R., GIVENS C.D., WILSON J., ROBINS R.B.- Staged surgical treatment of obstructive sleep apnea syndrome: a review of 35 patients. *J. Oral Maxillofac. Surg.* 57, 382-385, 1999.

25. MARKLUND M., FRANKLIN K.A., SAHLIN C. – The effect of a mandibular advancement device on apneas and sleep in patients with obstructive sleep apnea. *Chest* 113, 707-713, 1998.
26. MOCHOZUKI T., OKAMOTO M., SANO H., NAGANUMA H. – Cephalometric analysis in patients with obstructive sleep apnea syndrome. *Acta Otolaryngol (Stockh.)*, 524S: 64-72, 1996.
27. MONDAIN M., DEJEAN Y., CRAMPETTE L. – Pharyngoplastie et fonction vélaire dans la chirurgie du ronflement. *J. Fr. ORL*, 43, 429-431, 1994.
28. MONDAIN M., CRAMPETTE L. – Examen ORL et imagerie dans la rhonchopathie. *Cahiers d'ORL*, 22, 413-418, 1997.
29. MOORVES C.F.A., KEENE M.R. – Natural head position: a basic consideration for analysis of cephalometric radiographs. *Am. J. Physiol. Anthropol.*, 16, 213-234, 1958.
30. NIINIMA V., COLE P., SHEPARD R.J. – The switching point from nasal to oral breathing. *Respir. Physiol.*, 42, 61-71, 1980.
31. O'LEARY M.J., FARRELL G. – Myocutaneous fenestration in sleep apnea patients. *Laryngoscope*, 96, 356-359, 1986.
32. PATTON T.J., OGURA J.H., THAWLEY S.E.- Expansion hyoidoplasty. *Otolaryngol. Head Neck Surg.*, 92, 509-519, 1984.
33. PICHE J., GAGNON N.B. – Snoring: physiopathology, surgical treatment and a modified uvulo-palato-pharyngoplasty. *J. Otolaryngol.*, 18, 36-43, 1989.
34. POWELL N.B., RILEY R.W. – Radiofrequency tongue base reduction in sleep disordered breathing: a pilot study. *Otolaryngol. Head Neck Surg.*, 120, 656-664, 1999.
35. RILEY R., GUILLEMINAULT C., HERAN J. – Cephalometric analysis and flow volume loops in obstructive sleep apnea patients. *Sleep*, *6, 303-311, 1983.*
36. RILEY R., GUILLEMINAULT C., POWELL N., DERMAN S. – Mandibular osteotomy and hyoid bone advancement for obstructive sleep apnea; a case report. *Sleep*, 7, 79-82, 1984.
37. RILEY R., POWELL N., GUILLEMINAULT C. – Maxillofacial surgery and obstructive sleep apnoea: a review of 80 patients. *Otolaryngol. Head Neck Surg.*, 101, 353-361, 1989.
38. RILEY R., POWELL N., GUILLEMINAULT C. – Maxillofacial surgery and nasal CPAP. A comparison of treatment for obstructive sleep apnea syndrome. *Chest*, 98, 1421-1425, 1990.
39. RILEY R., POWELL N., GUILLEMINAULT C. – Obstructive sleep apnea syndrome: a review of 306 consecutive treated surgical patients. *Otolaryngol. Head Neck Surg.*, 108, 117-125, 1993.
40. RILEY R., POWELL N., GUILLEMINAULT C. – Obstructive sleep apnea and the hyoid: a revised surgical procedure. *Otolaryngol. Head Neck Surg.*, 111, 717-721, 1994.
41. SCHMIDT-NOWARA W., LOWE A., WIEGAND L., CARTWRIGHT R., PERE-GUERRA F., MENU S. – Oral appliances for the treatment of snoring and obstructive sleep apnea syndrome. *Sleep*, 18, 501-510, 1995.
42. SHER A., SCHECHTMAN K., PICCIRILLO J. – The efficacy of surgical modification of the upper airway in adults with obstructive sleep apnea syndrome. *Sleep*, 19, 156-177, 1996.
43. SURRATT P.M., DEE P., ATKINSON R.L., AMSTRUNG P., WILHOIT S.C. – Fluoroscopic and computed tomographic features of the pharyngeal airway in obstructive sleep apnea. *Am. Rev. Respir. Dis.*, 127, 487-492, 1983.
44. WAGNER I., COIFFIER T., SEQUERT S., LACHIVER X., FLEURY B., CHABOLLE F. – Traitement chirurgical du syndrome d'apnées du sommeil sévère par avancée maxillo-mandibulaire ou par transposition génienne. *ANN. Otolaryngol. Chir. Cervicofac.* 117, 137-146, 2000.
45. WAITE P.D., WOOTEN W., LACHNER J., GUYETTE R.F. – Maxillomandibular advancement surgery in 23 patients with obstructive sleep apnea syndrome. *J. Oral Maxillofac. Surg.* 47, 1256-1261, 1990.

Chapter 31

Narcolepsy

M. Billiard and Y. Dauvilliers
Service de Neurologie B, Hôpital Gui de Chauliac, Montpellier, France

HISTORY

Among the many disorders of daytime sleepiness, narcolepsy was undoubtedly identified the earliest. In 1877, Westphal [177] published the observation of a man affected by sudden attacks of sleep with episodes of motor and language inhibition and in 1880, Gélineau [58] observed the case of a wine barrel merchant who from the age of 36 onwards fell asleep involuntarily during the daytime and experienced falls linked to emotion. Gelineau named this state narcolepsy (from the Greek ναρκωσισ: sleep and λαμβανειν: seize) and the falls, astasia. It was only later that the falls came to be referred to as cataplexy (from the Greek καταπληξισ: fear, stupor) [69]. For several years the identity of the disorder remained uncertain. Some authors [1, 155] considered narcolepsy as a distinct entity whereas for others [101, 178] there was not one but several forms of narcolepsy resulting from several causes *e.g.* traumatic, psychopathological, circulatory and tumoral. It was 1934 when Daniels [38] clearly identified the disorder, as having four distinct symptoms *i.e.* irresistible sleep episodes, cataplexy, hypnagogic hallucinations and sleep paralysis. In 1957, Yoss and Daly [181] referred to this group of signs as the narcoleptic tetrad. The illness was further identified with Vogel's discovery [176] of a direct nocturnal sleep-onset REM period in the narcoleptic subject, which was confirmed in the following years, leading to the notion that cataplexy, hypnagogic hallucinations and sleep paralysis are all partial expressions of REM sleep [147, 154, 171]. In 1973, studies on narcolepsy were given new impetus with the discovery of an animal model of the illness, the narcoleptic dog [92, 124]. A further breakthrough came in the early 1980s with the discovery of the almost exclusive association with HLA-DR2 [18, 87, 99] and toward the end of the 1990s with the discovery that canine narcolepsy is caused by a mutation of the gene coding for hypocretin receptor 2 (Hcrtr2) [33, 103].

EPIDEMIOLOGY

The prevalence of the illness is still not fully known. According to two surveys conducted in California, one in the San Francisco Bay area in 1972 [49] and the other in the Los Angeles agglomeration in 1973 [47], the prevalence of the illness is something like 0.05 to 0.067%. According to a cohort study of 8,000 Finnish twins, 0.026% of the subjects questioned reported irresistible sleep episodes daily with episodes of muscular weakness occurring at least once a week [82], while in a survey recently carried out in France, of 13,058 inhabitants of the Gard, in France, 0.021% of subjects questioned presented complete narcolepsy [141]. The first findings would put the prevalence of narcolepsy in France at a level somewhat lower than that of multiple sclerosis with the number of subjects affected between 25,000 and 38,000. The second, no doubt closer to reality, puts the number of affected subjects in the region of 15,000. Whatever the case, in comparison with other neurological disorders, the illness is not rare (fig. 31.1). Narcolepsy may be more common in Japan [76] and is clearly less common in Israel [100]. There is usually male predominance. In our population of 375 patients, the sex ratio is 2 men to 1 woman. The initial age of onset varies from early childhood to around fifty, with the main peak around the age of 15 and a second peak around the age of 36 [41] (fig. 31.2). The time lapse between the appearance of the first symptom(s) and diagnosis of the illness is typically very long, averaging more than 10 years [3], but has become much shorter in recent years, with better understanding of the characteristics of the illness. The circumstances of onset: high psychological stress, abrupt changes in times of getting up and going to bed, illness, trauma, pregnancy etc. are found in over half the cases, in the days or weeks preceding the first symptoms [13, 142].

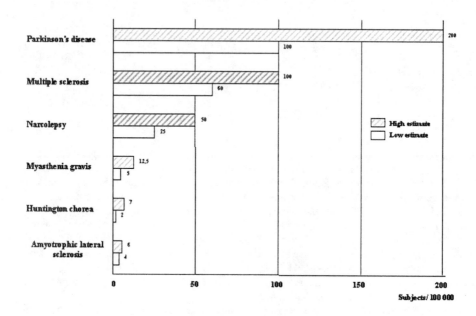

Figure 31.1. Comparative prevalence of various neurological diseases (per 100,000 inhabitants) according to the epidemiological surveys to date. The prevalence of narcolepsy is below that of multiple sclerosis.

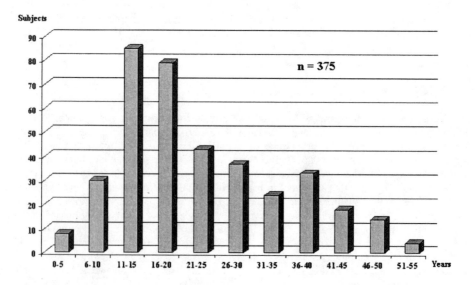

Figure 31.2. Age of onset of the first symptom(s) of the illness in 375 narcoleptic subjects seen consecutively in Montpellier. Note the main peak is from 11 to 15 years and a second peak from 36 to 40.

CLINICAL FEATURES

The two cardinal symptoms are daytime sleepiness and cataplexy. Sleepiness occurs daily but is not permanent, occurring in waves with varying degrees of frequency, depending on the individual, and often at the same time for the same person. It is brought on by passive situations such as watching television or being a passenger in a vehicle. Some subjects manage to overcome sleepiness while others succumb to an irresistible sleep episode. The duration of the sleep episode varies depending on the situation in which the subject finds himself. It always restores normal wakefulness for a period ranging from one to several hours, depending on the subject. This fact is of considerable diagnostic value. It is also useful in evaluating the severity of the illness, in terms of the number of hours during which the subject does not experience sleepiness. Sleepiness may lead to automatism, inappropriate language, arranging objects in unlikely places, driving a vehicle to an unintended destination. Memory disorders have been reported by half the patients [2], but objective tests are usually negative.

Cataplexy is quite distinct from irresistible sleep episodes. It is pathognomonic to the illness. It consists of abruptly releasing muscle tone in reaction to emotional factors, which are usually positive, such as a fit of laughter, receiving a compliment, humour expressed by the subject, the sight of prey for the hunter, perception of a fish biting at the hook for the fisherman, a well-caught ball, and less often negative i.e. anger or stress. It is prompted by poor sleep, and tiredness. All the striated muscles may be affected leading to the progressive slackening of the whole body or of certain muscles only *e.g.* the jaw muscles, producing sudden difficulty in articulating words; the facial muscles producing a grimace, the thigh muscles causing a brief unlocking of the knees. These signs warrant careful investigation, and are not necessarily identified by the subject as pathological. Consciousness is maintained during the episode. A neurological test carried out at the exact moment of the event reveals the deep tendon reflexes to be abolished, with occasional Babinski sign. Cataplexy varies in duration from a fraction of a second to several minutes. Frequency varies from one subject to another, from only a few each year, or even less, to several per day. It must be distinguished from the sensation of physical weakness following a fit of laughter [84].

The other features are deemed accessory to the extent that they are not indispensable for diagnosis. Hallucination, whether hypnagogic (at the onset of sleep) or hypnopompic (on awakening) tends to be auditory, sometimes visual, and somatosensory. Subjects often imagine there is a threatening stranger at the door or in the room. One subject might see a cat walking across the bed and brushing his face with its whiskers, another has the sensation of sleeping on the carapace of a crocodile, feeling its roughness on the skin. These hallucinations are sometimes so terrifying that the subject develops a veritable dread of going to bed and resorts to reassuring strategies such as keeping a weapon within reach or having a dog sleep in the same room. Sleep paralysis is the inability to move the limbs, head or to breathe normally. It is often accompanied by hypnagogic hallucination. It is frightening or even terrifying, and can easily last up to ten minutes. Some subjects establish a code with their partner, getting them to move them to unblock the paralysis. Sleep is usually disturbed. The narcoleptic typically falls asleep as soon as he gets to bed, but his sleep is disturbed by several awakenings. Sleep onset, whether during the day or at night, may be marked by a hypnagogic hallucination. It may also be marked by a disturbing dream. Many narcoleptics report that they begin dreaming at the onset of sleep. Parasomnias such as sleep talking or REM sleep behaviour disorder are common. No relationship has been established between poor quality night time sleep and daytime sleepiness [25].

The order in which the symptoms appear varies from one individual to another. However sleepiness is the first sign, either as an isolated symptom or accompanied by cataplexy in the majority of cases. Cataplexy is rarely the first sign to appear. Hypnagogic hallucinations are present at one stage or another during the course of the illness, in two thirds of subjects, and sleep paralysis in slightly under half [13].

POSITIVE DIAGNOSIS

The diagnosis of narcolepsy is essentially clinical. For Honda [77] it is founded on the association between "recurrent episodes of daytime sleepiness and/or falling asleep, occurring practically every day for a period of at least six months and a clinical confirmation of cataplexy". The International Classification of Sleep Disorders [7], allowing for the fact that cataplexy may sometimes be absent from the clinical picture, uses the same criteria or associates these with a complaint of excessive daytime sleepiness or sudden muscular weakness; the associated features include sleep paralysis, hypnagogic hallucinations, automatisms, disturbed night time sleep; one of the following polygraphic features: latency to sleep onset of less than 10 minutes, REM latency of less than 20 minutes, mean sleep latency of less than 5 minutes in the MSLT, and two or more sleep onset REM periods, a DQB1*0602 association, and the absence of any medical or mental disorders to account for the symptoms, with the possible presence of other sleep disorders, periodic limb movement syndrome or central sleep apnoeas, which may be present, but are not the primary cause of illness.

Polysomnographic recording comprises a night recording followed by a multiple sleep latency test (MSLT) the next day, using current, well-codified practices [32]. Recording at night gives an evaluation of the extent of sleep disturbance [125], and the multiple sleep latency test will confirm diagnosis by scoring an average sleep latency of less than 8 minutes for the 5 test sessions (fig. 31.3) and 2 sleep-onset REM periods (sleep onset REM periods refer to any sleep onset in which REM sleep occurs within a lapse of 0 to 15 minutes after the onset of sleep). It is important to note, however, that not all subjects affected by narcolepsy score a minimum of 2 sleep-onset REM periods during the MSLT. Indeed subjects who suffer from classic forms of narcolepsy may score only one sleep-onset REM period during a test – or even none at all [128, 174] (fig. 31.4).

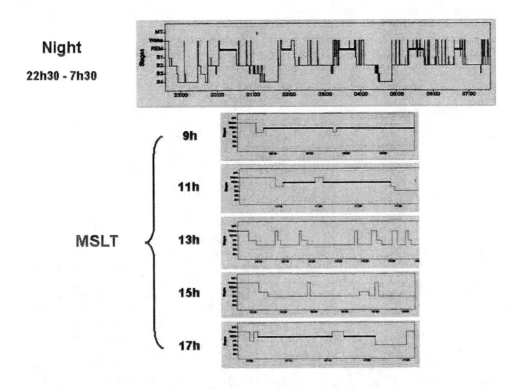

Figure 31.3. Night sleep recording, 22h30 – 7h30, (upper part of the figure) followed by an MSLT, from 9h to 17h (bottom part of the figure). Note, during the night, the normal sleep-onset NREM period (a sleep onset REM period is found in less than half of the cases at night in narcoleptic subjects), the large number of short awakenings, deep NREM sleep (stages 3 and 4) recurring roughly with a 4 hour periodicity; and during the day, latencies to sleep onset between 2 and 5 minutes on the MSLT, and 3 sleep onset REM periods at 9h, 11h and 17h.

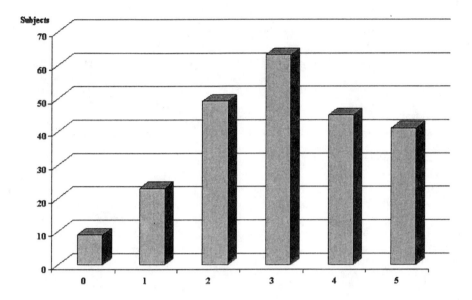

Figure 31.4. Number of sleep-onset REM periods in 225 narcoleptic subjects having undergone MSLT after overnight polysomnography. Note the majority of subjects (61) with 3 sleep onset REM periods (as on figure 31.3) and the number of subjects with a single (23) or no (9) sleep-onset REM period, despite the undoubted presence of daily irresistible sleep episodes and clinical evidence of cataplexy.

HLA typing is not indispensable for diagnosis, but has the advantage of virtually eliminating the diagnosis in uncertain cases if no association with HLA DQB1*0602 is found, at least in the case of Caucasian and Mongolian types (table 31.1). Moreover, testing the children of the patient will determine those at risk of developing narcolepsy.

Table 31.1. HLA-DR B1*1501 and DQB1*0602 typing in 262 unrelated narcoleptic subjects evaluated in Montpellier *

	DR B1*1501+ DQB1*0602+ n=246 (93.8%)	DR B1*1501- DQB1*0602+ n=12 (4.5%)	DR B1*1501- DQB1*0602- n=4 (1.5%)
Caucasians (252)			
France	218	7	4
Maghreb	12		
Europe	8	1	
Middle-East	2		
Blacks (9)			
West-Indies	4	2	
Central Africa		1	
East Africa		1	
West Africa	1		
Asian (1)			
South East Asia	1		

• note the very high proportion of DRB1*1501+, DQB1*0602+ subjects (95.2%) in comparison with the very low proportion of DRB1*1501-, DQB1*0602± subjects (4.7%) in Caucasians, and conversely the almost equal proportion of DRB1*1501+, DQB1*0602+ (55.5%) and DRB1*1501-DQB1*0602+ (44.4%) subjects in Blacks.

COURSE

Narcolepsy is an illness which is more or less debilitating, depending on its degree of severity. Some subjects master their sleepiness fairly well using treatment which provokes wakefulness or by arranging to take a number of naps during the day to fit in with their work schedule, or a combination of the two. But the risks linked to the illness must not be underestimated, risks involving traffic or machine accidents, an occupational handicap necessitating a change of jobs, the need to apply for disability or early retirement grants [3, 4, 28]. Some couples cope with the situation, others separate. Some subjects develop a syndrome of depression. One particular complication is the "status cataplecticus" characterised by subintrant cataplexy confining the subject to bed, occurring either spontaneously or more often after suspending treatment for cataplexy because of side effects [73, 143, 146].

The general course of the illness is hard to systematise. The pattern tends to be for excessive daytime sleepiness and irresistible sleep attacks to persist throughout life even if improvements are commonly observed after retirement, no doubt partly due to a better schedule of sleep and activity. Cataplexy is likely to disappear spontaneously, or once the subject has learnt to control his emotions. Hypnagogic hallucinations and sleep paralysis are more often transient, particularly in the early stages of the illness. Poor sleep does not tend to improve spontaneously. In the majority of subjects, the illness remains stable for several years; in some it gradually improves; in rarer cases it slowly worsens [13, 168].

CLINICAL VARIANTS

Incomplete or atypical forms

In certain subjects, all the elements in the clinical picture correspond to narcolepsy *i.e.* irresistible, restorative sleep episodes, sleep onset REM periods and HLA DQB1*0602 association, but the interrogation fails to reveal cataplexy in the history of the illness. Some of these subjects do eventually develop cataplexy, but most continue along the same course [19, 60]. To avoid introducing a bias in studies on narcolepsy, of whatever type, a descriptive term has been suggested to cover these cases: excessive daytime sleepiness with several sleep onset REM periods [60]. It is far more rare for cataplexy to precede irresistible sleep episodes by several months or years, and when this does occur diagnosis is difficult.

Associated forms

An association between narcolepsy and sleep apnoea syndrome is not uncommon [35, 65]. The two disorders must be treated simultaneously if there is to be any chance of an improvement during wakefulness. An association between narcolepsy and periodic leg movement disorder is even more common [180], but the extent to which abnormal sleep is due to these movements or to the underlying phenomenon is difficult to ascertain.

Symptomatic forms

Observations of so-called symptomatic narcolepsy have been published in relation to brain tumours, usually localised in the posterior hypothalamus and the upper brainstem, multiple sclerosis, encephalitis, cerebral ischaemia, cranial trauma or a degenerative brain disease (see reviews in 9, 22). Generally speaking, a diagnosis of symptomatic narcolepsy requires at the minimum, the presence of irresistible sleep attacks and cataplexy, with an identified time course linked to the neurological process. However sheer coincidence must not be ruled out. As for associations with REM sleep onset and HLA DQB1*0602 haplotype, these do not appear to be indispensable as such markers may be absent in typical non symptomatic narcolepsies.

Childhood forms

(See Ch.35).

DIFFERENTIAL DIAGNOSIS

The first mistake and by far the most common is the hasty conclusion that no pathology is present, and attributing the symptoms to a psychiatric disorder, bladder-bile dysfunction, or epilepsy, because of an inadequate understanding of the illness.

Then there may be some hesitation in regard to other hypersomnias, sleep apnoea syndrome in a subject who snores, idiopathic hypersomnia when the restorative nature of sleep is not clearly marked, hypersomnia linked to depression, but these various pathologies never include cataplexy.

When cataplexies are presenting signs, there may be doubts in regard to syncopes, drop attacks, and functional manifestations.

One can occasionally be duped by a subject who is well informed about the illness, and who hopes to benefit from a diagnosis of narcolepsy as a pretext to stop work or to obtain awakening or stimulant drugs. Such cases are often betrayed by their textbook symptomatology.

Finally, it is important to recognise isolated cataplexy occurring as a separate or hereditary entity, with no sleep onset REM period [57, 6   8 175] and isolated sleep paralysis, which is often familial [37].

PATHOPHYSIOLOGY

Neurophysiology

This remains controversial but the occurrence of irresistible sleep episodes during wakefulness, repeated arousals during sleep, the irruption during wakefulness of muscular atonia typical of REM sleep during cataplexy, hypnagogic hallucinations and sleep paralysis which are more frequent and more unpleasant than in the normal subject, as well as the direct sleep onset REM period, are all indications of disorganised states of sleep and wakefulness.

An early hypothesis was that of a REM sleep disorder [48] but this did not account for reduced sleep latency either in NREM or REM sleep, the frequency of sleep onset NREM periods in certain narcoleptic subjects, or the abnormal distribution of NREM sleep during the night [15]. A second hypothesis is that of a shift in the borders between the states of sleep and wakefulness or the "physiological glue" which holds together the electroencephalographic (EEG), electromyographic (EMG) and electro-oculographic (EOG) elements pertaining to wakefulness, NREM and REM sleep [31]. This would account for cataplexies, where there is a veritable intrusion of wakefulness by REM atonia, hypnagogic hallucinations and dream imagery, with brief episodes of muscular atonia in NREM sleep [125], bursts of eye movements in NREM sleep [64], tympanic phasic activity during sleep onset prior to sleep–onset REM periods [97] and the presence of spindles in REM sleep [44]. A third explanation for the anomalies of sleep and wakefulness is that of a sleep regulation disorder. Kripke [95] had submitted the hypothesis of an alteration of the circadian sleep rhythm after noticing repeated irresistible sleep episodes during the day and numerous arousals at night, although other studies have shown circadian sleep rhythm to persist, in a slightly attenuated form, with the majority of sleep occurring during the night [26, 39]. This line of thinking deserves to be reconsidered today in the light of the two process model of sleep regulation [23, 36]. According to Borbély [23], sleep regulation is subject to two processes, one homeostatic and the other circadian, their reciprocal interaction enabling a harmonious alternation between the states of sleep and wakefulness. An additional ultradian mechanism would account for the alternating episodes of NREM and REM sleep [75]. Using this model, a possible explanation might be an alteration of the homeostatic process. This is suggested by the modified distribution of NREM sleep during the night, and episodes which recur approximately every 4 hours, as opposed to during the first third of sleep as is normally the case [15]. Nevertheless, one study [170] failed to endorse this line of thinking. On the contrary, circadian control in wakefulness appears to be weakened in narcolepsy [39] with ultradian regulation becoming predominant [139], thus accounting for irresistible sleep episodes during the day and nocturnal arousals. These first findings were confirmed by a recent study carried out among narcoleptic subjects and controls subjected to a 4 hour ultradian sleep-wake regimen, showing the disappearance of circadian rhythms in NREM sleep and total sleep in narcoleptics, whereas these were maintained in the controls [52].

Cataplexy is associated with the inhibition of the monosynaptic H-reflex and multisynaptic deep tendon reflexes [66]. But H-reflex activity is fully suppressed only during REM sleep, underlining the relationship between the motor inhibitory compounds of REM sleep and the sudden atonia and areflexia seen during a cataplectic attack. The occurrence of REM sleep muscular atonia during wakefulness remains unexplained but it has been suggested that it may be an exaggerated or disinhibited "stopping and orientating" reaction triggered physiologically by sudden emotion [11].

Sleep paralysis is generally considered as hypnagogic or hypnopompic. Thus it may reflect the combination of a state of partial wakefulness, at sleep onset or on arousal, with motor inhibition occurring at sleep onset or on arousal from REM sleep [70].

Hypnagogic or hypnopompic hallucinations are more vivid in the narcoleptic subject than in the healthy subject, and are often associated with sleep paralysis. According to experiments of provoked arousal carried out on narcoleptics, hypnagogic hallucinations only occur in association with sleep onset REM periods [34, 72, 171].

Neuropharmacology and neurochemistry

Most medication used in narcolepsy facilitates monoaminergic activity – an argument in favour of the hypofunctioning of the monoaminergic systems. The amphetamines used in the treatment of excessive daytime sleepiness and irresistible sleep episodes increase monoaminergic transmission by stimulating the release and blocking the reuptake of monoamines without affecting the cholinergic system. The agents used to counter cataplexy are the tricyclic antidepressants. These products have a complex pharmacological profile comprising the blocking of monoamine reuptake (serotonin, norepinephrine, adrenaline and dopamine) and anticholinergic, antihistamine and anti alpha-adrenergic effects. Conversely, the administration of prazosin, an alpha1-noradrenergic receptor antagonist, will aggravate cataplexy [6]. Also, reduced concentrations of dopamine and its metabolite, homovanilic acid, have been found in the subarachnoid fluid of narcoleptic subjects [126]. Finally, increased dopaminergic D2 receptors have been found *post mortem* in human narcoleptic brains [5], although these results were not confirmed *in vivo* by two PET scan studies [90, 159].

There can be no question of systematically testing all the compounds which act on the different monoaminergic and cholinergic systems in man, measuring the monoamine concentrations and their metabolites in the different regions of the encephalon and analysing the receptors involved. Thus the interest raised by the discovery of a natural animal model for narcolepsy in the dog, first in the Dachshund [92] then in the Poodle [124], followed by other breeds, the Doberman pinscher and Labrador retriever in particular, sharing the partial or full cataplexy seen in man, with direct sleep onset REM period and fragmented sleep. This enabled a number of tests to be carried out to evaluate the effects of different drugs on cataplexy. The first, and most common, is the "food-elicited cataplexy test": 6 to 12 pieces of food (1 cm in diameter) are placed on the ground at 30 cm intervals. The dog is trained to eat the pieces one after the other. A normal dog will eat the pieces in 8 to 15 seconds. A narcoleptic dog presents one or a number of cataplectic attacks at the sight of the food and will thus take longer to absorb all the pieces. A second test, the "game-elicited cataplexy test", can also be carried out: two dogs are brought together in the same room and allowed to play together for 10 minutes. Playing triggers cataplexy attacks which are then counted. Sleepiness can be evaluated by continuous recording for 6 hours during the day.

Cataplexy

A large body of proof exists today in favour of a reciprocal interaction between the cholinergic and monaminergic systems in the regulation of REM sleep and the muscle atonia found in this type of sleep. Use of the canine model has demonstrated similar interaction in the regulation of cataplexy. Administering cholinergic agonists known to increase REM sleep will exaggerate cataplexy whereas cholinergic antagonists will reduce it [45]. However these effects are only obtained with high doses and involve considerable peripheral adverse effects, which is why this therapeutic route has not been pursued in man. These results should be compared to the way in which REM sleep is facilitated by manipulations which increase cholinergic transmission [59, 74]. Moreover drugs which block the reuptake of norepinephrine have a powerful effect on cataplexy [53, 119] but this is not true for drugs blocking the reuptake of dopamine and serotonin. This coincides with the results obtained for cataplexy in narcoleptic subjects i.e. very good for specific norepinephrine reuptake blockers (protriptyline, desipramine, viloxazine), and less good for serotonin reuptake blockers which are only effective at high doses, probably through the weak effect of these products on norepinephrine reuptake. These results are consistent with the fact that compounds which block norepinephrine reuptake have an inhibiting effect on REM sleep. The predominant role of norepinephrine was clarified by the use of agents which act selectively on the alpha- and beta-adrenergic receptors. The alpha-1 adrenergic antagonists (prazosin, phenoxy-benzamine) facilitate cataplexy whereas certain alpha-1 adrenergic agonists (methoxamine, cirazoline) diminish it [111, 112]. Two receptor subsets alpha-1-a and alpha-1-b, were later shown to be susceptible to inducing or blocking cataplexy, a finding which endorses the notion of the privileged role played by alpha-1-b receptors in the control of cataplexy [109, 110, 134].

Excessive daytime sleepiness and irresistible sleep episodes

Dopamine reuptake blockers have no effect on cataplexy but they do have a powerful wake-promoting effect [135, 136]. However these compounds have little effect on REM sleep compared to adrenergic or serotoninergic compounds, their main effect being to reduce total sleep and deep NREM sleep [136]. Abnormal sleepiness and cataplexy are thus differently controlled pharmacologically, the dopaminergic systems acting on the former and the noradrenergic systems on the latter.

In order to define the neuroanatomical basis of the anomalies observed in narcolepsy, measurements were taken of the concentrations of monoamines (norepinephrine, dopamine and serotonin) and their metabolites, 3.4-dihydroxyphenylacetic acid (DOPAC), homovanillic acid (HVA) and 5-hydroxyindoleacetic acid (5-HIAA) in different regions of the encephalon using high performance liquid chromatography. Significant increases in the concentrations of dopamine and its metabolite DOPAC were found in the amygdala. DOPAC was also increased in the parvocellular reticular nucleus, while norepinephrine was increased in the pontis oralis reticular nucleus but reduced in the preoptic hypothalamus [107, 123]. Hence, it is likely that in the canine narcoleptic, specific alterations of the dopaminergic and noradrenergic metabolisms occur in certain regions.

There is also evidence of an increase in the muscarinic M2 cholinergic receptors in the pontine reticular formation [20, 91], the dopaminergic D2 receptors and alpha-1-b adrenergic receptors in the amygdala [24, 109, 134] and in the alpha-2 receptors in the locus coeruleus [56]. The increase in the muscarinic cholinergic receptors in the pontine reticular formation is consistent with the fact that injecting tiny doses of cholinergic agonists into this same structure in narcoleptic dogs will provoke episodes of muscle atonia evoking cataplexy [156 157]. The same results are obtained with animal controls, using far higher doses [156], an argument in support of the hypersensitivity of narcoleptic dogs to cholinergic stimulation of the pontine region. The increase in dopaminergic and adrenergic receptors in the amygdala, indicates the critical role played by this emotion-regulating structure in determining cataplexy. No pathology has yet been identified to account for the altered concentrations of these receptors. However a process of axonal degeneration was demonstrated both in the septum and the amygdala, which appears to peak at the age of 2 months, the age of onset of the disorders in the narcoleptic dog [166, 167].

PATHOGENESIS

The existence of familial cases of narcolepsy points to a genetic basis for the disease but the fact that several cases have been published of monozygotic twins discordant for narcolepsy, underlines the importance of environmental factors.

Human studies

The genetic nature of narcolepsy has been known for some time. Westphal in his first observation indicated that the mother of his patient presented the same symptoms as the latter [177]. As from the 1940s, numerous case histories of patients have been published with varying proportions of index cases presenting a family history of narcolepsy [10, 78, 89, 94, 132, 183]. These findings are questionable to the extent that the clinical criteria used to diagnose narcolepsy varied from one author to another, as did the proportion of index cases presenting a family history of narcolepsy also varied widely, but the existence of familial cases was well and truly demonstrated.

Association studies

In 1984-1985, a major breakthrough was recorded in man, with the demonstration of a 100% association with HLA DR2 [87] which was later confirmed in Europe [18, 99], then in North America [151]. Later, however, non DR2 subjects were reported, suggesting the possible implication of one or a number of other genes [8, 61, 129]. Two possible explanations exist for this HLA association.

The first is that HLA DR2 is implicated in a dysimmunitary process entering the pathophysiology of the disease, as many HLA associated diseases such as insulin-dependent diabetes or multiple sclerosis are autoimmune, although no such process has been demonstrated to date [120]. In the blood, immunoglobulin and complement levels proved normal even at the onset of the disorder [106]. A study of the lymphocyte subpopulations showed no anomalies [106]. The systematic cerebrospinal fluid analysis of 15 patients found two cases of oligoclonal bands after IgG isoelectrofocalisation [55]. But this proportion did not differ from that found in nonimmune conditions. However these results do not rule out the possibility of an autoimmune mechanism, which is either restricted to a limited neuronal network and is thus hard to detect, or else is transitory, appearing at the onset of the condition.

Or HLA DR2 is simply a marker of the disease. This is why further efforts have been made to characterise HLA DR2. First, with the aid of more specific antisera, the exact serological haplotype was shown to be DR15-DQ6 (subtype of HLA DR2-DQl) [80]. The HLA molecules DR and DQ are heterodimers formed by the association of a heavy alpha chain with a molecular weight of 34,000 and a light beta chain with a molecular weight of 28,000. These chains are encoded by specific, polymorphous genes, grouped in several hundred kilobases on chromosome 6, the DQ genes being centromeric in relation to DR genes.

Amplification of the polymorphous exon (second exon) of genes HLA DRB, DQA and DQB using Polymerase Chain Reaction (PCR) then showed that all narcoleptic subjects had the same alleles DRB1*1501, DQA1*0102 and DQB1*0602 [96]. Given the marked linkage imbalance in the binding between DR and DQ in Caucasians and Asians, DRB1*1501 is almost exclusively associated with DQA1*0102 and DQB1*0602, such that the gene which is susceptible to narcolepsy may be one of the genes DRB1*1501, DQA1*0102 or DQB1*0602, or an unknown gene with a linkage imbalance for these alleles. However, in Afro-American narcoleptics the predisposition to narcolepsy is more closely associated with subtype DQ6 than subtype DR15. Indeed only 70% of Afro-American narcoleptics carry DR15, whereas all carry DQ6 [105, 113, 115]. In these conditions the most specific markers for narcolepsy in the various ethnic groups studied to date are DQA1*0102 and DQB1*0602 [114].

Furthermore HLA association studies have been conducted in subjects with narcolepsy without cataplexy and in healthy volunteers. In the former, a DQB1*0602 association was found in only 40.9% of cases, suggesting a direct DQB1*0602 effect on the occurrence of cataplexies [113] and in normal subjects a significant reduction was found in latency to REM sleep when they were DQB1*0602 [121].

The genome contiguous to the HLA DQ region was sequenced to define the genomic region predisposing narcolepsy more closely, but no other genes were found in the region [51].

On the whole, it would seem that HLA genes DQB1 and DQA1 are the genes implied in the predisposition to narcolepsy [114]. But a dominant DQB1*0602 effect does not account for HLA related susceptibility. Indeed, HLA-DQB1*0602 homozygosity doubles or quadruples the risk of developing narcolepsy [148] and the relative risk of narcolepsy varies in heterozygote subjects depending on the allele associated with DQB1*0602 [116]. Indeed nine class 2 HLA alleles appear to affect the predisposition to narcolepsy. The heterozygote combinations carrying DQB1*0301, DQA1*06, DRB1*04, DRB1*08, DRB1*11 and DRB1*12 increase susceptibility whereas those including DQB1*0601, DQB1*0501 and DQA1*01 play a protective role.

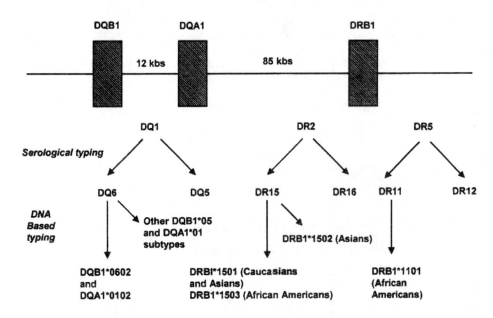

Figure 31.5. HLA DR and HLA DQ alleles most typically observed in narcolepsy. At the DR level, DR2 was split into two subtypes, DR15 and DR16, with the use of serological typing techniques, and DR15 was identified in narcoleptic subjects. At the DQ level, DQ1 was split into two subtypes, DQ5 and DQ6, and DQ6 was identified in all patients. Molecular subtypes of DR15 were further identified at the DRB1 level using DNA sequencing or oligotyping and identified as DRB1*1501 to DRB1*1508. Caucasians and Japanese narcoleptic patients were found to carry the DRB1*1501 gene, whereas most black patients were observed to be DRB1*1503. Similarly molecular subtypes of DQ6 were identified at the DQB1 level, as DQB1*0601 to DQB1*0616, and all patients with narcolepsy were DQB1*0602. (With permission from Dr Emmanuel Mignot).

Nevertheless as indicated above, non DR2 narcoleptic subjects have been identified. This suggests the presence of a second gene or more which are narcolepsy susceptible, located in two completely different parts of the human genome. Gene candidate strategy would appear to be best suited to research on these genes. Various studies have recently been carried out to this end. The association sometimes found between narcolepsy and Norrie's disease (whose gene is localised in Xp11.3/p11.4, close to MAO A and B) has led to the search for a link with a particular MAO A or B polymorphism in narcolepsy. Koch *et al.* [93] found a significant association in terms of the VNTR (Variable Number of Tandem Repeats) of intron I of the monoamine oxidase A gene (MAO A) in 28 subjects affected by narcolepsy, although a later study [169] failed to confirm this association. On the other hand, Dauvilliers *et al.* [42] found a different distribution of the catechol-O-metytransferase (COMT) genotype, the key enzyme in dopaminergic and noradrenergic transmission, in men and women, and an effect of this enzyme on the severity of narcolepsy. Kato *et al.*, [88], examined the presence of a Single Nucleotide Polymorphism (SNP) functional in TNFα promotion but found no significant difference in terms of genotypes or haplotypes.

Familial studies

Parallel to these association studies, familial studies have been pursued. Based on a better definition of narcolepsy, more accurate estimations have been made of the proportion of index cases presenting a family history of narcolepsy, with figures of 4.3% in Japan [77], 6% in the USA [63], 7.6% in France and 9.9% in Quebec [43]. In addition to subjects whose relatives present narcolepsy, the families of narcoleptic subjects are also found to contain subjects who are only affected by recurrent isolated irresistible sleep episodes, possibly corresponding to a milder form of the disorder [17] (fig. 31.6). The morbid risk of this illness has been calculated for subjects who are related: in 236 directly related subjects this was 1.22/0.026 (prevalence of narcolepsy according to Hublin *et al.* [82]) i.e. 46.5 times higher than in the general population [43]. One remarkable finding was that roughly 25% of cases of familial narcolepsy - with narcoleptic subjects and subjects presenting only recurrent isolated irresistible sleep episodes - were negative for DQB1*0602 [117]. These findings support the idea of one or a number of non HLA associated high penetrance genes. These genes were isolated using systematic genome screening methods in multiplex families. A single study has been published so far [131]. It indicates a significant 4p13-q21 link (lod score 3.09) in 8 small, narcoleptic multiplex families, suggesting the implication of other genes including the CLOCK gene and β-1 GABA receptor in these families.

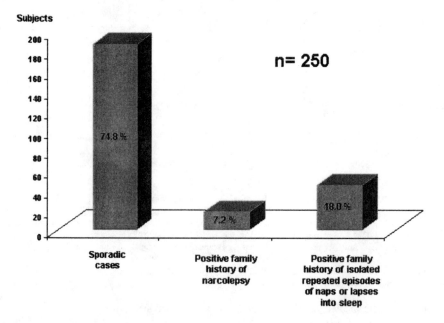

Figure 31.6. Proportion of sporadic cases, cases with a positive family history of narcolepsy and cases with a positive family history of isolated repeated episodes of naps or lapses into sleep in a population of 250 narcoleptic subjects with a second generation family tree.

Twin studies and environmental factors

The cases of 16 pairs of monozygotic twins have been published to date [108] of which 4 or 5 were concordant for narcolepsy. This data points to environmental factors. The nature of these factors is still unknown. Particular circumstances such as severe psychological stress, divorce, mourning, a change of sleep-wake rhythm, an accident, an illness, or pregnancy often precede the onset of irresistible sleep episodes and/or cataplexy, by several weeks or months. Systematic enquiries have been conducted: a scale of the life events of narcoleptics and controls paired by age, sex and social background, demonstrates substantially more events in the year preceding the onset of irresistible sleep episodes or cataplexy in narcoleptics than in the controls [142]; a battery of serological tests carried out on narcoleptics and controls paired by age and sex demonstrates higher levels of antistreptolysines 0 (ASO) and anti-Dnase antibodies (ADB) in narcoleptics than in controls [16, 127]. These are only preliminary studies but they warrant further development, which may lead to preventing the disease.

Animal studies

As previously indicated, a natural canine narcolepsy model was identified in 1973. But it was shown [54] that in certain breeds i.e. Doberman pinschers and Labrador retrievers, the disease was transmitted as an autosomal recessive trait, with full penetrance, referred to as canarc-1, whereas in other breeds such as Poodles and Beagles the disease was not transmitted, implying transmission of a polygenic nature or dependant on environmental factors. Considering the correlation between human narcolepsy and HLA, a correlation with DLA (dog leucocyte antigen) was searched for in the dog, but a genetic link between canarc-1 and the major histocompatibility of the dog was excluded [118]. However a candidate marker cosegregant with narcolepsy in all animals was

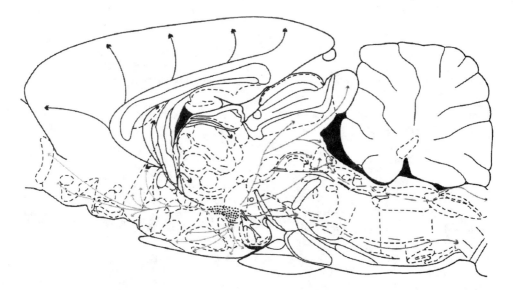

Figure 31.7. Schematic drawing of hypocretin neuron pathways in the rat brain. (Reprinted with permission from J. Neurosci., Vol. 18, Peyron C., Tighe D.K., van den Pol A.N. *et al.*. Neurons containing Hypocretin (Orexin) project to multiple neuronal systems. Page 10013, Copyright 1998 by the Society for Neuroscience).

identified. This marker showed substantial homology with the switch region of the immunoglobulin µ chain gene [122]. But after cloning and sequencing, it turned out that the linkage with canarc-1 was in fact merely a non significant cross reaction with no functional role in the canine immunoglobulin machinery. This work was pursued with the aid of positional cloning strategy used for the first time in the field of sleep, demonstrating a mutation of gene Hcrt2 coding for a G protein-coupled receptor which had close affinity to the hypocretin neuropeptides Hcrt 1 and 2 [103]. These neuropeptides had been identified by Lecea *et al.* [46] and termed hypocretins because of their hypothalamic localisation and their homology to the hormone secretin. These neuropeptides derive from a single protein precursor "preprohypocretin". They are produced by a well defined group of neurons restricted to the lateral hypothalamus and projecting to the olfactory bulb, cerebral cortex, the thalamus, hypothalamus and brainstem, notably the locus coeruleus, raphe nuclei and medullary reticular formation [150] (fig. 31.7). Genome sequencing showed a SINE (Short Interspersed Nucleotide Element) of 226 base pairs inserted upstream of the splice site of the fourth encoded exon in all the narcoleptic Doberman pinschers so that in the transcripts of these animals exon 3 was directly spliced to exon 5, omitting exon 4. This SINE insertion was not found in the Labrador retrievers. But these animals carried a deleted 6th exon (122 base pairs) such that in the transcripts of narcoleptic Labrador retrievers exon 5 was directly spliced to exon 7, omitting exon 6 (fig. 31.8). These mutations are consistent with the autosomal recessive transmission of the disease in these two breeds.

Parallel to this, the endogenous ligands of the two orphan receptors were identified with homologous structure to G protein-coupled receptors expressed in the brain, called orexins A and B due to their stimulating effect on appetite [163]. These were the peptides which had been identified by Lecea *et al.* under the term of hypocretins [46]. But an article [33] published two weeks after that of Lin *et al.* [103] indicated that a null mutation induced by targeted disruption of the mouse orexin gene (orexin knockout mice) resulted in an autosomal recessive phenotype with characteristics remarkably similar to narcolepsy, e.g. brief periods of behavioural arrest during the dark phase, that occurred exclusively in homozygous knockout mice (fig. 31.9).

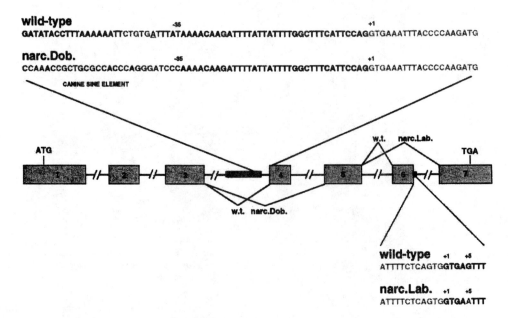

Figure 31.8. Genomic organization of the Canine Hcrtr2 locus, which is encoded by seven exons. For details see text. (Reprinted from Cell, Vol. 98, Lin L., Faraco J., Li R. *et al.*. The sleep disorder canine narcolepsy is caused by a mutation in the *Hypocretin (Orexin) receptor 2* gene, Page 371, Copyright (1999), with permission from Elsevier Science).

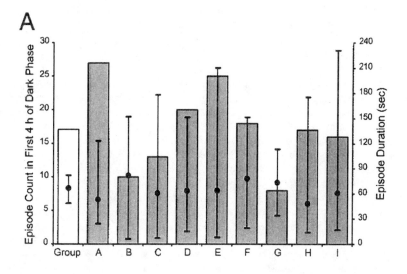

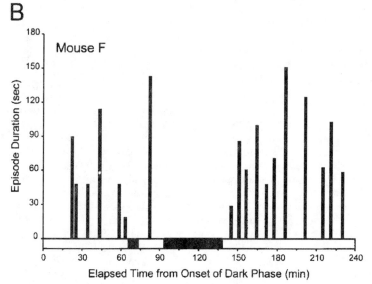

Figure 31.9. Infrared video characterization of narcoleptic episodes in orexin knockout mice.

A. Episode number and duration. Columns represent total number of episodes recorded in the first 4 hr after onset of dark phase. Filled circles represent the mean duration of all recorded episodes. T bars indicate the minimum and maximum duration observed. Data for individual knockout mice are designated A-I. Group is the average count and duration for all mice (A-I), with the T bars indicative of minimum and maximum individual averages. No narcoleptic episodes were observed for any of the wild-type (5) or heterozygote (4) control mice by blinded observers.

B. Typical behaviour during the first 4 hr of the dark phase in knockout mouse. Duration of individual narcoleptic episodes and the time they occurred from dark phase onset are plotted as vertical lines for mouse F. Periods designated as awake and sleeping were judged by gross behavioural observation and are indicated by the white and black bars above the timeline, respectively.

(Reprinted from Cell, Vol. 98, Chemelli R.M., Willie J.T., Sinton C.M. *et al.*, Narcolepsy in *orexin* knockout mice: molecular genetics of sleep regulation, Page 441, Copyright (1999), with permissions from Elsevier Science).

These two consecutive papers imply that hypocretins (orexins) and one of their receptors, receptor 2, are involved in the pathophysiology of canine narcolepsy.

Hypocretin and human narcolepsy

The question which naturally arises following the discovery of the mutation of the gene coding for hypocretin receptor 2 in the narcoleptic dog, is that of the implication of the hypocretin system in human narcolepsy.

With the same aim in mind, Nishino *et al.*, (137-138) failed to detect hypocretin 1 in the cerebrospinal fluid of 32 out of 38 narcoleptics, suggesting a loss of hypocretin peptides in the brains of subjects affected by narcolepsy. On the other hand systematic screening of mutations in preprohypocretin genes OX1R and OX2R was carried out in 74 patients, with or without a family history of narcolepsy, and with or without an HLA-DQB1*0602 association, but who all had cataplexies and sleep onset REM periods. A single mutation has been found to date, a mutation of the gene coding for hypocretin (the substitution of arginine by leucine in the peptide signal) in a narcoleptic who was atypical in terms of the extremely early age of onset of the illness (6 months), severity of sleepiness and cataplexies, and absence of association with HLA DQB1*0602 [149]. Moreover, *in situ* hybridisation of the perifornix region and post-mortem radio-immunological peptide doses introduced in the brains of narcoleptic subjects have shown undetectable levels of pre-prohypocretin in ARN and a targeted loss of hypocretin neurons, with neither gliosis nor signs of inflammation [149]. Thannickal *et al.* [172] also found a global loss of hypocretin neurons, but accompanied this time by residual gliosis in the perifornix region (using different methodology). In view of the association between narcolepsy and HLA, this last observation suggests an auto-immune mediation of human narcolepsy with the highly selective destruction of hypocretin neurons. Conversely, no hypocretin neuron degeneration has been found in the narcoleptic dog [140]. One possible mechanism may be the loss of hypocretin control over the monoaminergic systems. Indeed the projections of hypocretin hypothalamic neurons to the brainstem, particularly to the pedunculopontine nuclei, parabrachialis, dorsal raphe and locus coeruleus, have already been described [150] and these nuclei are known to play an important role in generating REM sleep.

These data correlate with those of two recent studies conducted on dogs. Indeed no trace of hypocretin 1 was found in the brains of 3 dogs affected by sporadic narcolepsy and in the cerebrospinal fluid of two other dogs, whereas it is normally present in the brain and CSF of Dobermans affected by familial narcolepsy, a finding which suggests that animal, sporadic and familial narcolepsies have distinct aetiologies [160]. On the other hand the observation of mice rendered transgenic by selectively destroying the orexin neurons through the use of the poison ataxin-3, has shown them to have a phenotype which is highly similar to that of human narcolepsy, with behavioural arrests, sleep onset REM periods and unstable sleep [67]. An interesting feature of this model in relation to preorexin knockout mice [33] is the post-natal loss of orexin neurons around six weeks, in line with human narcolepsy which often starts in adolescence.

Hence sporadic, canine or human narcolepsy may relate to a deficiency in the production of hypocretin ligands, whereas genetic canine narcolepsy may be caused by a genetically determined loss of function of hypocretin receptors.

Mode of transmission

Two modes of transmission have been proposed for this disease: an autosomal dominant mode with weak penetrance and variable expressivity [10, 94, 183] and a two threshold multifactorial mode with high heritability [77, 89, 132]. The relative frequency of homozygotes and heterozygotes for DR2 and DQl tends to support the autosomal dominant mode of transmission but penetrance is incomplete, because the incidence of narcolepsy in narcoleptic families is low (between 5 and 10%) and because, in families with two or more narcoleptics, the incriminating haplotype is found in both affected and non-affected subjects (fig. 31.10). On the other hand the discordance for narcolepsy noted in 11 recently published pairs of twins [109] supports the multifactorial model, suggesting that environmental factors are involved in triggering the illness.

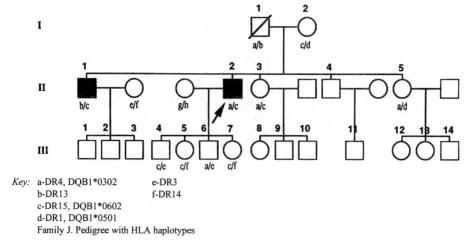

Key: a-DR4, DQB1*0302   e-DR3
      b-DR13   f-DR14
      c-DR15, DQB1*0602
      d-DR1, DQB1*0501
      Family J. Pedigree with HLA haplotypes

Figure 31.10. Multiplex family. The index case and his older brother are both HLA DR15-DQB1*0602. Other subjects (mother, sister, children) also carry this haplotype without being affected by the illness.

The limitations of currently available data, make it difficult to establish a mode of transmission for the disease. Generally speaking, it tends to be multifactorial including genes DQA1*0102 and DQB1*0602 and one or a number of environmental factors, and less frequently multifactorial, including genes other than DQA1*0102 and DQB1*0602 and one or a number of environmental factors, or autosomal dominant.

TREATMENT

This comprises three aspects, the treatment of excessive daytime sleepiness and irresistible sleep episodes; the treatment of cataplexy and other REM sleep-related symptoms, hypnagogic hallucinations and sleep paralysis; and treatment for poor sleep. Each of these forms of treatment comprises two aspects, the main one of which is pharmacological and the second, which is behavioural and must not be overlooked.

Excessive daytime sleepiness and irresistible sleep episodes

Table 31. 2. Practical treatment for excessive daytime sleepiness.

Pharmacological		Behavioural
Modafinil (100 to 400 mg/day)	First treatment	Regular timing of nocturnal sleep Strategically timed naps Patients' support group
Methylphenidate (10 to 60 mg/day) or Mazindol (2 to 8 mg/day) or Pemoline (37.5 to 75 mg/day)	If persistent difficulties	Same
Dextramphetamine (5 to 60 mg/day) or Methamphetamine (20 to 25 mg/day)	If no response	4 - hour schedule of sleep and wakefulness

Pharmacological treatment

Amphetamines and amphetamine derivatives

The first drug to be used to treat sleepiness in narcolepsy was ephedrine, in 1931 [85], followed by the amphetamines, benzedrine [153], d-amphetamine [152], l-amphetamine [147] and methamphetamine [50]. These products release catecholamines and to a lesser extent serotonin in the central nervous system and the periphery and block the reuptake and degradation of its monamines. Their elimination half-life is 8 to 10 hours. Highly effective against sleepiness, these products also carry adverse effects including insomnia, irritability, aggressiveness, high blood pressure, abnormal movements and occasionally amphetamine psychosis. Moreover, they have a high potential for tolerance, addiction and abuse, which has led to their being taken off the market in most countries.

Methylphenidate was introduced in 1959 [182]. Its mode of action is identical but its elimination half-life is distinctly shorter, between 3 and 4 hours, and its adverse effects are far less common and less intense than in the case of the preceding drugs, thus explaining its extensive use over the past forty years or more.

Mazindol is an imidazolidine derivative whose chemical effects are similar to those of amphetamines [144]. Mazindol resembles the tricyclic antidepressants in that it blocks the reuptake of norepinephrine and dopamine. It is effective against sleepiness [165]. The possible adverse effects are of the gastro-intestinal type. Excessive sudation and other sympathomimetic signs may also be encountered.

Selegiline is an l-amphetamine derivative with a specific monoamine oxidase B inhibiting effect. [162]. In a double-blind study against placebo it reduced irresistible sleep episodes and excessive daytime sleepiness in 45% of cases [83]. Its sympathomimetic effects are minor but half the patients experience asymptomatic orthostatic hypotension.

Pemoline is an oxazolidine derivative whose elimination half-life is 16 to 18 hours. It has a less powerful effect. The drawback with this product is a rare but severe form of hepatoxicity, calling for the regular monitoring of liver functioning [79].

Non amphetamine compounds

The first and most ancient of these is caffeine. Its elimination half-life is 3.5 to 5 hours. The adverse effects include palpitations and high blood pressure.

Modafinil is a 2-diphenylmethylsulfinylacetamide. Modafinil's mode of action is not yet fully understood. Although its effects are reduced by prazosin, an alpha 1 adrenergic receptor antagonist, it is neither a direct nor an indirect alpha-1 adrenergic agonist. Two recent studies underlined the role of dopamine. A first study [136] carried out on the narcoleptic dog compared the effects of reuptake blockers of dopamine, norepinephrine, d-amphetamine and modafinil on wakefulness and showed the *in vivo* effect of dopamine and modafinil reuptake blockers on wakefulness to correlate significantly with their *in vivo* affinity for the dopamine transporter (DAT) whereas norepinephrine reuptake blockers had little effect on wakefulness. A second study [179] showed that dopamine transporter gene knockout mice did not respond to the awakening effect of modafinil, methamphetamine and a selective dopamine transporter blocker GBR 12909. This highlights the key role played by the dopamine transporter in the awakening action of amphetamines and modafinil. Modafinil's effect is associated with a reduction in the release of gamma-hydroxybutyric acid (GHB) in the region of the central nervous system, which is likely to play a role in sleep and wakefulness. In a study measuring c-fos expression in the cat [102], the amphetamines and methylphenidate produced diffuse activation of the striatum, the cortex and accumbens nuclei, whereas modafinil led to a relatively specific activation of the anterior hypothalamic sites which may be involved in the mechanisms of sleep and wakefulness. After oral administration the maximal concentration of the product is reached within 2 to 4 hours. The elimination half-life is 10 to 13 hours. The adverse effects are relatively limited and not significant. These include headaches, irritability and insomnia, particularly at the onset of treatment. Modafinil is effective or highly effective in 60 to 70% of subjects [14, 27, 173].

Future trends

Other products appear to be potentially effective and are being studied in narcoleptic dogs. TRH analogues administered intravenously increase wakefulness and reduce sleep [133] and CG-3703 administered orally substantially reduces daytime sleep, acutely and for a period of two weeks [158]. As for hypocretin 1 (orexin A), when administered intravenously to the narcoleptic Doberman, this leads to an extension in the episodes of sleep and of wakefulness over a 24 hour period in dogs recorded by actigraphy, and a substantial reduction of REM sleep, with no modification to NREM sleep, during the 4 hours following the injection of the implanted dog [86].

In man, short term therapeutic perspectives may be envisaged in the form of hypocretin agonists, and long term perspectives in the form of cell transplants.

Behavioural treatment

One long nap at 180° out of phase with the nocturnal mid sleep time has a very favourable effect on daytime performance levels [130]. A further strategy consists in taking short naps (15 to 20 minutes) spread out over the day [12, 130, 161].

Cataplexy and other REM sleep-related symptoms

Pharmacological treatment

Most of the medications used are antidepressants, even though their effect is not oriented to mood improvement.

The first treatment to be used was the tricyclic antidepressants, imipramine [71], protriptyline [164] and clomipramine [145]. These agents act mainly by blocking norepinephrine reuptake. The use of this class of drugs is limited by their adverse effects, e.g. anticholinergic effects, essentially, dryness of the mouth, constipation, difficulty urinating, reduction of libido, impotence, as well as weight increase, trembling and occasionally high blood pressure. In prescribing these medications it is always important to choose the minimum effective dose to avoid or delay the onset of side effects as much as possible.

Over the last ten years, other groups of antidepressants have been tried with success, not because of greater effectiveness but in terms of fewer side effects. These include selective serotonin reuptake blockers: fluoxetine [98], which is the most effective, fluvoxamine, paroxetine, sertraline, a norepinephrine reuptake blocker, viloxazine [62], and the latest on the scene, a norepinephrine and serotonin reuptake blocker, venlafaxine.

Although rarely used, due to their interaction with different medications and foods, the monoamineoxydase blockers are no less effective. Selegiline, the specific monoamine oxidase B blocker already referred to, is of interest in that it acts on cataplexy and to a slightly lesser extent on sleepiness [83] with less interaction with medication or foods.

Table 31.3. Practical treatment for cataplexy

First treatment:	
SSRI group	
Fluoxetine	20 – 60 mg / day
Fluvoxamine	50 – 100 mg / day
Paroxetine	20 – 40 mg / day
Sertraline	50 – 100 mg / day
Norepinephrine reuptake blocker	
Viloxazine	100 – 300 mg / day
Norepinephrine and serotonin reuptake blocker	
Venlafaxine	75 – 350 mg / day
Gamma-hydroxybutyrate	2. 5 mg at bedtime
	+ 1. 25 mg during the night – 5 mg + 2. 5 mg
If persistent difficulties:	
Tricyclics	
Clomipramine	10 – 75 mg/ day
Protriptyline	5 – 60 mg / day
If no response:	
Mazindol	2 – 8 mg / day
Selegiline	20 – 40mg / day

As for mazindol, this is also effective against sleepiness and cataplexy attacks [165].

Finally, gamma-hydroxybutyrate taken in the evening and once at night, reduces cataplexy and the other REM sleep-related symptoms [29]. Its elimination half-life is 1 to 2 hours. Gamma-hydroxybutyrate increases deep NREM sleep, reduces night arousals and consolidates fragmented REM sleep [30]. If the patient wakes up during the night, he may experience some confusion, and memory disturbance. The main problem raised by the use of this product today is its potential for abuse.

Behavioural treatment

This consists of avoiding situations which trigger cataplexy. In the case of certain patients with low tolerance to anti-cataplectic medication, this may be the only effective measure. Some subjects learn to develop a neutral attitude to any emotional situation.

Poor sleep

Pharmacological treatment

Treatment relies on hypnotics, benzodiazepines and non-benzodiazepines, which will not stop night arousals but will delay their occurrence in the course of the night. Our own experience is that these products do not generally bring on tolerance with time in the case of narcoleptics.

The other possibility is gamma-hydroxybutyrate taken at bedtime, with a second dose taken during the night 3 hours before getting up, at the latest.

Behavioural treatment

General measures are indicated: avoiding night sleep deprivation, maintaining regular bedtime and morning rising times, and avoiding strenuous exercise in the evening.

CONCLUSION

Consistent progress has been made in the understanding of narcolepsy. However a number of issues are still challenging. Nothing is known on the environmental factors which may trigger narcolepsy on a special genetic background. The link between hypocretin deficiency and the cholinergic/monoaminergic basis of the disease is still obscure. Sporadic and genetic canine narcolepsies obey different mechanisms and it is certainly timely to complete these data in man. Hypocretin agonists seem promising, but the mode of action of hypocretin-1 administered with success to hypocretin-2 receptor mutated narcoleptic Doberman should be clarified.

REFERENCES

1. ADIE W. – Idiopathic narcolepsy: a disease sui generis; with remarks on the mechanism of sleep. *Brain.*, 49, 257-306, 1926.
2. AGUIRRE M., BROUGHTON R., STUSS D. – Does memory impairment exist in narcolepsy-cataplexy? *J. Clin. Exp. Neuropsychol.* 7,14-24, 1985.
3. ALAILA S.L. – Life effects of narcolepsy: measures of negative impact, social support and psychological well-being. *Loss Grief and Care*, 5, 1-22, 1992.
4. ALDRICH M.S. – Automobile accidents in patients with sleep disorders. *Sleep.*, 12, 487-494,1989.
5. ALDRICH M.S., HOLLINGSWORTH Z., PENNEY J.B. – Dopamine - receptor autoradiography of human narcoleptic brain. *Neurology.*, 42, 410-415, 1992.
6. ALDRICH M.S., ROGERS A.E. – Exacerbation of human cataplexy by prazosin. *Sleep.*, 12, 254-256, 1989.
7. AMERICAN SLEEP DISORDERS ASSOCIATION. ICSD – International classification of sleep disorders, revised: Diagnostic and coding manual. American Sleep Disorders Association, 1997
8. ANDREAS-ZIETZ A., KELLER E., SCHOLZ S., ALBERT E.D., ROTH B., NEVSIMALOVA S., SONKA K., DOCEKAL P., IVASKOVA E., SCHULZ H. – DR2 negative narcolepsy. *Lancet*, 2, 684-685, 1986.
9. AUTRET A., LUCAS B., HENRY-LEBRAS F., de TOFFOL B. – Symptomatic narcolepsies. *Sleep.*, 17, S21-24, 1994.
10. BARAITSER M., PARKES J.D. – Genetic study of narcoleptic syndrome. *J. Med. Genet.*, 15, 254-259, 1978.
11. BASSETTI C., ALDRICH M.S. – Narcolepsy. *Neurologic Clinics*, 14, 545-571, 1996.

12. BILLIARD M. - Competition between the two types of sleep, and the recuperative function of REM sleep versus NREM sleep in narcoleptics. In C. Guilleminault, W.C. Dement, P. Passouant (eds) *Narcolepsy,* Spectrum, New York, 77-96, 1976.
13. BILLIARD M., BESSET A., CADILHAC J. - The clinical and polygraphic development of narcolepsy. In: C. Guilleminault, E. Lugaresi (eds) *Sleep / Wake Disorders: Natural history, epidemiology and long-term evolution,* Raven Press, New-York, 171-185, 1983.
14. BILLIARD M., BESSET A., MONTPLAISIR J., LAFFONT F., GOLDENBERG F., WEILL J.S., LUBIN S. - Modafinil: a double-blind multicentric study. *Sleep* 17, S107-S112, 1994.
15. BILLIARD M., DE KONINCK J., COULOMBE D., TOUZERY A. - Napping behavior in narcoleptic patients: a four hour cycle in slow wave sleep. In: C. Stampi (ed) *Why we nap: Evolution, chronobiology and functions of polyphasic and ultrashort sleep.* Birkhauser, Boston. 245-257, 1992.
16. BILLIARD M., LAABERKI M.F., REYGROBELLET C., SEIGNALET J., BRISSAUD L., BESSET A. - Elevated antibodies to streptococcal antigens in narcoleptic subjects. *Sleep Res.,* 18, 201, 1989.
17. BILLIARD M., PASQUIÉ-MAGNETTO V., HECKMANN M., CARLANDER B., BESSET A., ZACHARIEV Z., ELIAOU J.F., MALAFOSSE A.- Family studies in narcolepsy. *Sleep,* 17, S54-S59, 1994.
18. BILLIARD M., SEIGNALET J. - Extraordinary association between HLA-DR2 and narcolepsy. *The Lancet,* 1, n° 8422, 226-227, 1985.
19. BILLIARD M., SEIGNALET J., BESSET A., CADILHAC J.- Narcolepsy without cataplexy: a clinical variant. *Sleep Res.* 16, 308, 1987.
20. BOEHME R.E., BAKER T.L., MEFFORD I.N., BARCHAS J.D., DEMENT W.C., CIARANELLO R.D. - Narcolepsy: cholinergic receptor changes in animal model. *Life Sciences,* 34, 1825-1828,1984.
21. BOIVIN D.B., LORRAIN D., MONTPLAISIR J. – Effects of bromocriptine on periodic limb movements in human narcolepsy. *Neurology,* 43, 2134-2136, 1993.
22. BONDUELLE M., DEGOS C. - Symptomatic narcolepsies: a critical study. C. Guilleminault, W.C. Dement, P. Passouant (eds) *Narcolepsy,* Spectrum, New-York, 313-332, 1976.
23. BORBELY A.A. - A two process model of sleep regulation. *Human Neurobiol.* 1, 195-204, 1982.
24. BOWERSOX S.S., KILDUFF T.S., FAULL K.F., ZELLER-DE-AMICIS L., DEMENT W.C., CIARANELLO R.D. - Brain dopamine receptor levels elevated in canine narcolepsy. *Brain Res.* 402, 44-48, 1987.
25. BROUGHTON W.A., BROUGHTON R.J. – Psychosocial impact of narcolepsy. *Sleep* 17, S45-S49, 1994.
26. BROUGHTON R., DUNHAM W., NEWMANN J., LUTLEY K., DUCHESNE P., RIVERS M. - Ambulatory 24-hour sleep wake monitoring in narcolepsy-cataplexy compared to matched controls. *Electroenceph. clin. Neurophysiol.* 70, 473-481, 1988.
27. BROUGHTON R, FLEMING J.A.E., GEORGE C.F.P., HILL J.D., KRYGER M.H., MOLDOFSKY H., MONTPLAISIR J., MOREHOUSE R.L., MOSCOVITCH A., MURPHY W.F. - Randomized, double-blind, placebo-controlled trial of modafinil in the treatment of excessive daytime sleepiness in narcolepsy. *Neurology,* 49, 444-451, 1997.
28. BROUGHTON R., GHANEM Q., HISHIKAWA Y., SUGITA Y., NEVSIMALOVA S., ROTH B. - Life effects of narcolepsy in 180 patients from North America, Asia and Europe compared to matched controls. *Can. J. Neurol. Sci.* 8, 299-304, 1981.
29. BROUGHTON R., MAMELAK M. - The treatment of narcolepsy-catapexy with nocturnal gamma-hydroxy-butyrate. *Can. J. Neurol. Sci.* 6, 1-6, 1979.
30. BROUGHTON R., MAMELAK M. - Effects of gamma-hydroxybutyrate on sleep/waking patterns in narcolepsy-cataplexy. *Can. J. Neurol. Sci.* 7, 23-31, 1980.
31. BROUGHTON R., VALLEY V., AGUIRRE M., ROBERTS J., SUWALSKI W., DUNHAM W. - Excessive daytime sleepiness and the pathophysiology of narcolepsy-cataplexy: a laboratory perspective. *Sleep,* 9, 205-215, 1986.
32. CARSKADON M.A., DEMENT W.C., MITLER M.M., ROTH T., WESTBROOK P., KEENAN S. - Guidelines for the multiple sleep latency test (MSLT): a standard measure of sleepiness. *Sleep,* 9, 519-524, 1986.
33. CHEMELLI R.M., WILLIE J.T., SINTON C.M., ELMQUIST J.K., SCAMMEL T., LEE C., RICHARDSON J.A., WILLIAMS S.C., XIONG Y., KISANUKI Y., FITCH T.E., NAKAZATO M., HAMMER R.E., SAPER C.B., YANAGISAWA M. - Narcolepsy in orexin knockout mice: molecular genetics of sleep regulation. *Cell* 98, 437-451, 1999.
34. CHETRIT M., BESSET A., DAMCI D., LELARGE C., BILLIARD M. – Hypnagogic hallucinations associated with sleep onset REM period in narcolepsy-cataplexy. *J. Sleep Res.* 3, 43, 1994.
35. CHOKROVERTY S. – Sleep apnea in narcolepsy. *Sleep,* 9, 250-253, 1986
36. DAAN S., BEERSMA D.G.M., BORBELY A.A.- The timing of human sleep: a recovery process gated by a circadian pacemaker. *Am. J. Physiol.* 246 (Regulatory Integrative Comp. Physiol. 12), R161-R178, 1984.
37. DAHLITZ M., PARKES J.D. - Sleep paralysis. *Lancet,* 341, 406-407, 1993.
38. DANIELS L. - Narcolepsy. *Medicine,* 13, 1-122, 1934.
39. DANTZ B., EDGAR D.M., DEMENT W.C., - Circadian rhythms in narcolepsy: studies on a 90 minute day. *Electroenceph. Clin. Neurophysiol.* 90, 24-35, 1994.
40. DAUVILLIERS Y., BAZIN M., ONDZE B., BERA O., BAZIN M., BESSET A., BILLIARD M. - Severity of narcolepsy among French of different ethnic origins (South of France and Martinique), *Sleep,* 25, 50-55, 2002.
41. DAUVILLIERS Y., MONTPLAISIR J., MOLINARI N., CARLANDER B., ONDZE B., BESSET A., BILLIARD M. - Age of onset of narcolepsy in two large populations of patients in France and Quebec. *Neurology,* 57, 2029-2033, 2001.
42. DAUVILLIERS Y., NEIDHART E., LECENDREUX M., BILLIARD M, TAFTI M. - MAO-A and COMT polymorphisms and gene effects in narcolepsy. *Molecular Psychiatry* 2001, 6, 367-372.

43. DAUVILLIERS Y., PASQUIÉ-MAGNETTO V., CARLANDER B., LESPÉRANCE P., MONTPLAISIR J., BILLIARD M. - Familial and sporadic cases of narcolepsy in France and Quebec (abstract) *J. Sleep Res.,* 7, Supplement 2, 114, 1998.

44. de BARROS-FERREIRA M., LAIRY G.C. - Ambiguous sleep in narcolepsy. In: C. Guilleminault, W.C. Dement, P. Passouant P. (eds) *Narcolepsy,* Spectrum, New-York, 57-75, 1976.

45. DELASHAW J.B., FOUTZ A.S., GUILLEMINAULT CH., DEMENT W.C. - Cholinergic mechanisms and cataplexy in dogs. *Exp. Neurol.* 66,745-757, 1979.

46. de LECEA L., KILDUFF T.S., PEYRON C., GAO X.B., FOYE P.E., DANIELSON P.E.,FUKUHARA C., BATTENBURG E.L.F., GAUTVIK V.T., BARTLETT F.S., *et al..* - The hypocretins: hypothalamic specific peptides with neuroexcitatory activity. *Proc. Natl. Acad. Sci. USA* 95,322-327, 1998.

47. DEMENT W.C., CARSKADON M., LEY R. - The prevalence of narcolepsy II. *Sleep Res.* 2,147, 1973.

48. DEMENT W.C., RECHTSCHAFFEN A., GULEVICH G. - The nature of the narcoleptic sleep attack. *Neurology,* 16, 18-33, 1966.

49. DEMENT W.C., ZARCONE V., VARNER V. *et al.* - The prevalence of narcolepsy. *Sleep Res.* 1, 148, 1972.

50. EATON L.M. Treatment of narcolepsy with desoxyephedrine hydrochloride. *Proc. Staff Meet. Mayo Clin* 18, 262-264, 1943.

51. ELLIS M.C., HETISIMER A.H., RUDDY D.A., HANSEN S.L., KRONMAL G.S., McCLELLAND E., QUINTANA L., DRAYNA D.T., ALDRICH M.S., MIGNOT E. – HLA class II haplotype and sequence analysis support a role for DQ in narcolepsy. *Immunogenetics* 1997, 46, 410-417.

52. ESPA F.,ONZE B., NOBILI L., SCHIAVI G., BILLIARD M., BESSET A. - Lack of sleep/wake circadian rhythms in narcoleptic patients submitted to a 4 hour ultradian regimen. *Sleep* 23, Abstract Supplement 2, A179, 2000.

53. FOUTZ A.S., DELASHAW J.B., GUILLEMINAULT C., DEMENT W.C. - Monoaminergic mechanisms and experimental cataplexy. *Ann. Neurol.* 10, 369-376, 1981.

54. FOUTZ A.S., MITLER M.M., CAVALI-SFORZA L.L., DEMENT W.C. - Genetic factors in canine narcolepsy. *Sleep* 1, 413-422, 1979.

55. FREDRIKSON S., CARLANDER B., BILLIARD M., LINK H. - CSF immune variable in patients with narcolepsy. *Acta Neurol. Scand.,* 81, 253-254, 1990.

56. FRUHSTORFER B., MIGNOT E., BOWERSOX S., NISHINO S., DEMENT W.C., GUILLEMINAULT C. - Canine narcolepsy is associated with an elevated number of alpha-2 receptors in the locus coeruleus. *Brain Res.* 500, 209-214, 1989.

57. GELARDI J.A.M., BROWN J.W. - Hereditary cataplexy. *J. Neurol. Neurosurg. Psychiatry,* 30, 455-457, 1967.

58. GELINEAU J. - De la narcolepsie. *Gaz. des Hôp. (Paris)* 55, 626-628; 635-637, 1880.

59. GILLIN J., SALIN-PASCUAL R., VELASQUEZ-MOCTEZUMA J., SHIROMANI P., ZOLTOSKI R. - Cholinergic receptors subtypes and REM sleep in animals and normal controls. In: Cuello A. (ed.) *Progress in Brain Research.* Elsevier Science Publishers; New York, NY, 379-387, 1993

60. GUILLEMINAULT C. - Narcolepsy syndrome. In: M.H. Kryger, T. Roth, Dement W.C. (eds*). Principles and Practice of Sleep Medicine* (second edition). W.B. Saunders Company, Philadelphia, 547-561, 1994.

61. GUILLEMINAULT C. - HLA-DR2 and narcolepsy: not all narcoleptic cataplectic patients are DR2. *Hum. Immunol.* 17, 1-2, 1986.

62. GUILLEMINAULT C., MANCUSO J., QUERA SALVA M.A., HAYES B., MILTER M., POIRIER G., MONTPLAISIR J. Viloxazine hydrochloride in narcolepsy: a preliminary report. *Sleep* 9, 275-279, 1986.

63. GUILLEMINAULT C., MIGNOT E., GRUMET F.C. - Familial patterns of narcolepsy. *Lancet,* 335, 1376-1379, 1989.

64. GUILLEMINAULT C., REYNAL D., TAKAHASHI S., CARSKADON M. DEMENT W.C. - Evaluation of short-term and long-term treatment of narcolepsy syndrome with clomipramine hydrochloride. *Acta Neurol. Scand.* 54, 71-87.

65. GUILLEMINAULT C., VAN DEN HOED J, MITLER M.M. - Clinical overview of the sleep apnea syndrome. In: C. Guilleminault, W.C.Dement (eds) *Sleep apnea syndromes.* Alain R. Liss, New York, 1-12, 1978.

66. GUILLEMINAULT C., WILSON R.A., DEMENT W.C. - A study on cataplexy. *Arch. Neurol.* 31, 255-261, 1974.

67. HARA J., BEUCKMANN C.T., NAMBU T., WILLIE J.T., CHEMELLI R.M., SINTON C.M., SUGIYAMA F., YAGAMI K, GOTO K., YANAGISAWA M., SAKURAI T. – Genetic ablation of orexin neurons in mice results in narcolepsy, hypophagia, and obesity. *Neuron,* 30, 345-354, 2001.

68. HARTSE K.M., ZORICK F.J., ROTH T., SICKLESTEEL J.M. - Isolated cataplexy: a familial study. *Sleep Res.* 9, 205, 1980..

69. HENNEBERG R. - Über genuine Narkolepsie. *Neurol. Zbl.,* 30, 282-290, 1916

70. HISHIKAWA Y. - Sleep paralysis. In: C. Guilleminault, W.C. Dement, P. Passouant (eds) *Narcolepsy,* Spectrum, New York, 97-125, 1976.

71. HISHIKAWA Y., IDA H., NAKAI K., KANEKO Z. - Treatment of narcolepsy with imipramine (Tofranil) and desmethylimipramine (Pertofran). *J. Neurol. Sci.* 3, 453-461, 1966.

72. HISHIKAWA Y., KOIDA H., YOSHINO K., WAKAMATSU H., SUGITA Y., IIJIMA S., NAN'NO H. - Characteristics of REM sleep accompanied by sleep paralysis and hypnagogic hallucinations in narcoleptic patients. *Waking and Sleeping,* 2, 113-123, 1978.

73. HISHIKAWA Y., SHIMIZU T. – Physiology of REM sleep, cataplexy, and sleep paralysis. *Adv. Neurol.* 67, 245-271, 1995.

74. HOBSON J.A., DATTA S., CALVO J.M., QUATROCHI J. - Acetylcholine as a brain state modulator: triggering and long-term regulation of REM sleep. *Prog. Brain Res.* 98, 389-404, 1993.

75. HOBSON J.A., Mc CARLEY R.W., WYZINSKI P.W. - Sleep cycle oscillations: reciprocal discharge by two brainstem neuronal groups. *Science,* 189, 55-58, 1975.
76. HONDA Y. - Census of narcolepsy, cataplexy and sleep life among teenagers in Fujisawa city. *Sleep Res.* 8, 191, 1979.
77. HONDA Y. - Clinical features of narcolepsy: Japanese experiences. In: Y. Honda, T. Juji, (eds) *HLA in Narcolepsy,* Springer-Verlag, Berlin, 24-57, 1988.
78. HONDA Y., ASAKA A., TANAKA Y., JUJI T. - Discrimination of narcoleptic patients by using genetic markers and HLA. *Sleep Res.* 12, 254, 1983.
79. HONDA Y., HISHIKAWA Y. - A long-term treatment of narcolepsy and excessive daytime sleepiness with pemoline (Betanamin). *Curr. Ther. Res.* 27, 429-441, 1980.
80. HONDA Y., MATSUKI K., JUJI T., INOKO H. - Recent progress in HLA studies and a genetic model for the development of narcolepsy: recent findings of pathophysiology and genetic research. In: S.A. Burton, W.C. Dement, R.K. Ristanovic (eds), *Recent developments in the diagnosis and treatment of narcolepsy: selected proceedings from the third International Symposium on Narcolepsy.* Oak Park III, Matrix communication. 27-29, 1989.
81. HONG S.C., HAYDUK R., LIM J., MIGNOT E. – Clinical and polysomnographic features in DQB1*0602 positive and negative narcolepsy patients: result from the modafinil clinical trial. *Sleep. Med.* 1, 33-39, 2000.
82. HUBLIN C., KAPRIO J., PARTINEN M., KOSKENVUO M., HEIKKILÄ K., KOSMIKIES S., GUILLEMINAULT C. - The prevalence of narcolepsy: An epidemiological study of the finnish twin cohort. *Ann. Neurol.* 35, 709-716, 1994.
83. HUBLIN C., PARTINEN M. HEINONEN EH., PUUKKA P., SALMI T. - Selegiline in the treatment of narcolepsy. *Neurology* 44, 2095-2101, 1994.
84. HUBLIN C., PARTINEN M., KAPRIO J., KOSKENVUO M., GUILLEMINAULT C. - Epidemiology of narcolepsy. *Sleep,* 17, S7-S12, 1994.
85. JANOTA O. - Symptomatische behandlung der pathologischen Schlafsucht, besonders der Narkolepsie. *Med Klin* 27, 278-281, 1931.
86. JOHN J., WU M.F., SIEGEL J.M. - Hypocretin-1 reduces cataplexy and normalizes sleep and waking durations in narcoleptic dogs. *Sleep* 23, *Abstract Supplement* 2, A11, 2000.
87. JUJI T., SATAKE M., HONDA Y., DOI Y. - HLA antigens in Japanese patients with narcolepsy. *Tissue Antigens,* 24, 316-319, 1984.
88. KATO T., HONDA M., KUWATA S. - Novel polymorphism in the promoter region of the tumor necrosis factor alpha gene: No association with narcolepsy. *Am. J. Med. Genet.* 88, 301-304, 1999.
89. KESSLER S. - Genetic factors in narcolepsy. In: C. Guilleminault, W.C. Dement, P. Passouant (eds) *Narcolepsy,* Spectrum. New-York, 285-302, 1976.
90. KHAN N., ANTONINI A., PARKES D., DAHLITZ M.J., MEIER-EWERT K., WEINDL A., LEENDERS K.L. - Striatal dopamine D2 receptors in patients with narcolepsy measured with PET and 11C-raclopride. *Neurology,* 44, 2101-2104, 1994.
91. KILDUFF T.S., BOWERSOX S.S., KAITIN K.I., BAKER T.L., CIARANELLO R.D., DEMENT W.C. - Muscarinic cholinergic receptors and the canine model of narcolepsy. *Sleep,* 9, 102-106, 1986.
92. KNECHT C., OLIVER J., REDDING R., SELCER R., JOHNSON G. - Narcolepsy in a dog and a cat. *J. Amer. Vet. Med. Assoc.* 162, 1052-1053, 1973.
93. KOCH H., CRAIG I., DAHLITZ M., DENNEY R., PARKES D. - Analysis of the monoamine oxidase genes and the Norrie disease gene locus in narcolepsy. *Lancet* 20, 353, 645-646, 1999.
94. KRABBE E., MAGNUSSEN G. - Familial narcolepsy. *Acta Psychiat. Neurol.,* 17, 149-173, 1942.
95. KRIPKE D. - Biological rhythm disturbances might cause narcolepsy. In: C. Guilleminault, Dement W.C., Passouant P. (eds) *Narcolepsy,* Spectrum, New York, 475-483, 1976.
96. KUWATA S., TOKUNAGA K., JIN F., JUJI T., SASAKI T., HONDA Y. - Letter to the editor after Dr. Aldrich's review on narcolepsy. *N. Engl. J. Med.,* 324, 271-272, 1991.
97. LAMSTEIN S.M., SPIELMAN A.J., WEITZMAN E., POLLACK C., ROFFWARG P. - The recording of middle ear activity in narcolepsy. *Sleep Res.* 6, 175, 1977.
98. LANGDON N., SHINDLER J., PARKES J.D., BANDAK S. - Fluoxetine in the treatment of cataplexy. *Sleep* 9, 371-372, 1986.
99. LANGDON N., WELSH K.I., VAN DAM M., VAUGHAN R.W., PARKES J.D. - Genetic markers in narcolepsy. *The Lancet,* 2, n° 8413, 1178-1180, 1984.
100. LAVIE P., PELED R. - Narcolepsy is a rare disease in Israel. *Sleep,* 10, 608-609, 1987.
101. LHERMITTE J. - Les narcolepsies. *Progrès Médical,* 22, 962-975, 1930.
102. LIN J.S., HOU Y., JOUVET M. - Potential brain neuronal targets for amphetamine, methylphenidate and modafinil induced wakefulness, evidenced by *c-fos* immunocytochemistry in the cat. *Nat. Acad. Sci.* 93, 14128-14133, 1996.
103. LIN L., FARACO J., LI R., KADOTANI H., ROGERS W., LIN X., QUI X., de JONG PJ., NISHINO S., MIGNOT E. - The sleep disorder canine narcolepsy is caused by a mutation in the hypocretin (orexin) receptor 2 gene. *Cell* 98, 365-376, 1999.
104. LOWENFELD L. - Uber Narkolepsie. *Munch. Med. Wochenschr.* 49,1041-1045, 1902.
105. MATSUKI K., GRUMET F.C., LIN X., GUILLEMINAULT C., DEMENT W.C., MIGNOT E. - HLA DQB1*0602 rather than HLA DRw15 (DR2) is the disease susceptibility gene in Black narcolepsy. *Lancet,* 339, 1052, 1992.
106. MATSUKI K., HONDA Y., SATAKE M., JUJI T. - HLA in narcolepsy in Japan. In: Y. Honda, T. Juji (eds) *HLA in narcolepsy,* Springer-Verlag, Berlin, 58-75, 1988.

107. MEFFORD I.N., BAKER T.L., BOEHME R., FOUTZ A.S., CIARANELLO R.D., BARCHAS J.D., DEMENT W.C. - Narcolepsy: Biogenic amine deficits in an animal model. *Science,* 220, 629-632, 1983.

108. MIGNOT E. - Genetic and familial aspects of narcolepsy. *Neurology,* 50: S16-S22, 1998.

109. MIGNOT E., BOWERSOX S.S., MADDALUNO J., DEMENT W.C., CIARANELLO R. D. - Evidence for multiple (3 H) - Prazosin binding sites in canine brain membranes. *Brain Res.* 486, 56-66, 1989.

110. MIGNOT E., GUILLEMINAULT C., BOWERSOX S.S., FRUHSTORFER B., NISHINO S., MADDALUNO J., CIARANELLO R.D., DEMENT W.C. - Central alpha-1 adrenoceptor subtypes in narcolepsy-cataplexy: a disorder of REM sleep. *Brain Res.,* 490, 186-191, 1989.

111. MIGNOT E., GUILLEMINAULT C., BOWERSOX S.S, RAPPAPORT A., DEMENT W.C. - Effects of alpha-1-adrenoceptors blockade with prazosin in canine narcolepsy. *Brain Res.* 444, 184-188, 1988.

112. MIGNOT E., GUILLEMINAULT C., BOWERSOX S.S, RAPPAPORT A., DEMENT W.C. - Role of central alpha-1-adrenoceptors in canine narcolepsy. *J. Clin. Invest.* 82, 885-894, 1988.

113. MIGNOT E., HAYDUK R., BLACK J., GRUMET F.C., GUILLEMINAULT C. - HLA Class II studies in 509 narcoleptic patients. *Sleep,* 20, 1012-1020, 1997.

114. MIGNOT E., KIMURA A., LATTERMANN A., LIN X., YASUNUGA S., MUELLER-ECKHARD G., RATTAZI C., LIN L., GUILLEMINAULT C., BRUMET F.C., MAYER G., DEMENT W.C., UNDERHILL P. - Extensive HLA Class II studies in 58 non DRB1*15 (DR2) narcoleptic patients with cataplexy. *Tissue Antigen,* 49, 329-341, 1997.

115. MIGNOT E., LIN X., ARRIGONI J., MACAUBAS C., OLIVE F., HALLMAYER J., UNDERHILL P., GUILLEMINAULT C., DEMENT W.C., GRUMET F.C. - DQB1*0602 and DQA1*0102 (DQ1) are better markers than DR2 for narcolepsy in caucasians and black Americans. *Sleep,* 17, S60-S67, 1994.

116. MIGNOT E., LIN L., ROGERS W., HONDA Y., QIU X., LIN X, OKUN M., HOHJOH H., MIKI T., HSU S.H., LEFFELL M.S., GRUMET F.C., FERNANDEZ-VINA M., HONDA M., RISCH N. – Complex HLA-DR and DQ interactions confer risk of narcolepsy-cataplexy in three ethnic groups. *Am. J. Hum. Genet.* 68, 686-699, 2001

117. MIGNOT E., MEEYAN J., GRUMET F.C., HALLMEYER J., GUILLEMINAULT C., HESLA P.E. - HLA Class II and narcolepsy in thirty-three multiplex families. *Sleep Res.* 25, 303, 1996.

118 MIGNOT E., MOTOYAMA M., WANG C., DEAN R., GAISER C., Mc DEVITT H., DEMENT W.C. - Absence of tight gentic linkage of Canarc-1 and DLA in autosomal recessive canine narcolepsy. *Sleep Res.,* 18, 264, 1989.

119. MIGNOT E., RENAUD A., NISHINO S., ARRIGONI J., GUILLEMINAULT C., DEMENT W.C. - Canine cataplexy is preferentially controlled by adrenergic mechanisms: evidence using monoamine selective uptake inhibitors and release enhancers. *Psychopharmacology* 113, 76-82, 1993.

120. MIGNOT E., TAFTI M., DEMENT W.C., GRUMET F.C. - Narcolepsy and immunity. *Adv. Neuroimmunol.,* 5, 23-37, 1995.

121. MIGNOT E., YOUNG T., LIN L. FINN L., PALTA M. - Reduction of REM sleep latency associated with HLA-DQB1*0602 in normal adults. *Lancet,* 351: 727: 1998.

122. MIGNOT E., WANG C., RATTAZI C., GAISER C., LOVETT M., GUILLEMINAULT C., DEMENT W.C., GRUMET F.C. - Genetic linkage of autosomal recessive canine narcolepsy with an immunoglobulin μ chain switch like segment. *Proc. Natl. Acad. Sci. (USA),* 88, 3475-3478, 1991.

123. MILLER J.D., FAULL K.F., BOWERSOX S.S., DEMENT W.C. - CNS monoamines and their metabolites in canine narcolepsy: a replication study. *Brain Res.* 509, 169-171, 1990.

124. MITLER M.M., BOYSEN B.G., CAMPBELL L., DEMENT W.C. - Narcolepsy cataplexy in a female dog. *Exp. Neurol.* 45, 332-340, 1975.

125. MONTPLAISIR J., BILLIARD M., TAKAHASHI S., BELL I.R., GUILLEMINAULT C, DEMENT W.C. - Twenty-four-hour recording in REM narcoleptics with special reference to nocturnal sleep disruption.. *Biol. Psychiatry,* 13, 73-89, 1978.

126. MONTPLAISIR J., de CHAMPLAIN J., YOUNG S.N., MISSALA K., SOURKES T.L., WALSH J., REMILLARD G. - Narcolepsy and idiopathic hypersomnia: biogenic amines and related compounds in CSF. *Neurology,* 32, 1299-1302, 1982.

127. MONTPLAISIR J., POIRIER G., LAPIERRE O., MONTPLAISIR S. - Streptococcal antibodies in narcolepsy and idiopathic hypersomnia. *Sleep Res.,* 18, 271, 1989.

128. MOSCOVITCH A., PARTINEN M., GUILLEMINAULT C. - The positive diagnosis of narcolepsy: and narcolepsy's borderland. *Neurology,* 43, 55-60, 1993.

129. MUELLER-ECKHARDT G., MEIER-EWERT K., SCHENDEL D.J., REINECKER F.B., MUTHOFF G., MUELLER-ECKHARDT C. - HLA and narcolepsy in a German population. *Tissue Antigens,* 28, 163-169, 1986.

130. MULLINGTON J., BROUGHTON R. - Scheduled naps in the management of daytime sleepiness in narcolepsy. *Sleep* 17, 69-76, 1993.

131. NAKAYAMA J., MIURA M., HONDA M., MIKI T., HONDA Y., ARINAMI T. - Linkage of human narcolepsy with HLA association to chromosome 4p13-q21. *Genomics* 65: 84-86, 2000.

132. NEVSIMALOVA-BRUHOVA S., ROTH B. - Heredofamilial aspects of narcolepsy and hypersomnia. *Schweiz. Arch. Neurol. Neurochir. Psychiat.,* 110, 45-54, 1972.

133. NISHINO S., ARRIGONI J., SHELTON J., KANBAYASHI T., TAFTI M., DEMENT W.C., MIGNOT E. - Effects of thyrotropin-releasing hormone and its analogs on daytime sleepiness and cataplexy in canine narcolepsy. *J. Neurosci.* 17, 6401-6408, 1997.

134. NISHINO S., FRUHSTORFER B., ARRIGONI J., GUILLEMINAULT C., DEMENT W.C., MIGNOT E. - Further characterization of the alpha-1 receptor subtype involved in the control of cataplexy in canine narcolepsy. *J. Pharmacol. Exp. Ther.* 264, 1079-1084, 1993.

135. NISHINO S., MAO J., SAMPATHKUMARAN R., HONDA K., DEMENT W.C., MIGNOT E. - Differential effects of dopaminergic and noradrenergic uptake inhibitors on EEG arousal and cataplexy of narcoleptic canines. *Sleep Res.* 25, 317, 1996.

136. NISHINO S., MAO J., SAMPATHKUMARAN R., SHELTON J., MIGNOT E. - Increased dopaminergic transmission mediates the wake-promoting effects of CNS stimulants. *Sleep Research Online* 17, 436-437, 1994.

137. NISHINO S., RIPLEY B., OVEREEM S., LAMMERS G.L., MIGNOT E. - Hypocretin (orexin) deficiency in human narcolepsy. *The Lancet* 355, 9197, 39-40, 2000.

138. NISHINO S., RIPLEY B., OVEREEM S., NEVSIMALOVA S., LAMMERS G.J., VANKOVA J., OKUN M., ROGERS W., BROOKS S., MIGNOT E. - Low CSF hypocretin (orexin) and altered energy homeostasis in human narcolepsy. *Ann. Neurol.* 50, 381-388, 2001.

139. NOBILI L., BESSET A., FERRILLO F., ROSADINI G., SCHIAVI G., BILLIARD M. - Dynamics of slow wave activity in narcoleptic patients under bed rest condition. *Electroenceph. Clin. Neurophysiol.* 95, 414-425, 1995.

140. OKURA M., RIPLEY B., FUJIFI N., MIGNOT E, NISHINO S. - Hypocretin neurons in narcoleptic and control dobermans *Sleep*, Vol 23, *Abstract Supplement* 2, A299, 2000

141. ONDZE B., LUBIN S., LAVANDIER B., KOHLER F., MAYEUX D., BILLIARD M. - Frequency of narcolepsy in the population of a French " département ". (abstract) *J. Sleep Res.* 7, Supplement 2, 386, 1998.

142. ORELLANA C., VILLEMIN E., TAFTI M., CARLANDER B., BESSET A., BILLIARD M. - Life events in the year preceding the onset of narcolepsy.*Sleep,* 17, S50-S53, 1994.

143. PARKES D. - *Sleep, its disorders.* WB Saunders, Philadelphia, 1985.

144. PARKES J.D., SCHACHTER M. - Mazindol in the treatment of narcolepsy. *Acta Neurol. Scand.* 60, 250-254, 1979.

145 PASSOUANT P., BALDY-MOULINIER M. - Données actuelles sur le traitement de la narcolepsie. Action des imipraminiques. *Concours Med* (Paris) 92, 835-838, 1970.

146. PASSOUANT P., BALDY-MOULINIER M., AUSSILLOUX C. - Etat de mal cataplectique au cours d'une maladie de Gelineau; influence de la clomipramine. *Rev. Neurol.* 123, 56-60", 1970.

147. PASSOUANT P., SCHWAB R.S., CADILHAC J., BALDY-MOULINIER M. - Narcolepsie-cataplexie. Etude du sommeil de nuit et du sommeil de jour. Traitement par une amphétamine lévogyre. *Rev. Neurol.* (Paris). 3, 415-426, 1964.

148. PELIN Z., GUILLEMINAULT C., RISCH N., GRUMET FC., MIGNOT E. – HLA DQB1*0602 homozygosity increases relative risk for narcolepsy but not disease severity in two ethnic groups. US Modafinil in Narcolepsy Multicenter Study Group. *Tissue Antigens.* 51, 96-1200, 1998.

149. PEYRON C., FARACO J., ROGERS W., RIPLEY B., OVEREEM S., CHARNAY Y., NEVSIMALOVA S., ALDRICH M., REYNOLDS D., ALBIN R. *et al.* - A mutation in a case of early onset narcolepsy and a generalized absence of hypocretin peptides in human narcoleptic brains. *Nat Med* 6, 991-997, 2000.

150. PEYRON C., TIGHE D.K., VAN DER POL A.N., de LECEA L., HELLER H.C., SUTCLIFFE J.G., KILDUFF T.S. - Neurons containing hypocretin (orexin) project to multiple neuronal systems. *J Neurosci,* 1, 18, 9996-10015, 1998

151. POIRIER G., MONTPLAISIR J., DECARY F. MOMEGE D., LEBRUN A. - HLA antigens in narcolepsy and idiopathic central nervous hypersomnolence. *Sleep,* 9, 153-158: 1986.

152. PRINZMETAL M., ALLES G.A. - Central nervous system stimulant effects of dextroamphetamine sulfate. *Am. J. Med. Sci.* 200, 665-673, 1940.

153. PRINZMETAL M., BLOOMBERG W. - The use of benzedrine for the treatment of narcolepsy. *JAMA* 105, 2051-2054, 1935.

154. RECHTSCHAFFEN A., WOLPERT E.A., DEMENT W.C., MITCHELL S.A., FISCHER C. - Nocturnal sleep of narcoleptics. *Electroenceph. Clin. Neurophysiol.* 15, 599-609, 1963.

155. REDLICH E. - Epilegomena zur Narkolepsie Frage. *Z. ges. Neurol. Psychiat.* 136, 128-173, 1931.

156. REID M.S., TAFTI M., GEARY J., NISHINO S., SIEGEL J.M., DEMENT W.C., MIGNOT E. - Cholinergic mechanisms in canine narcolepsy: I. Modulation of cataplexy via local drug administration into the pontine reticular formation. *Neuroscience,* 59, 511-522, 1994.

157. REID M.S., TAFTI M., GEARY J., NISHINO S., SIEGEL J.M., DEMENT W.C., MIGNOT E. - Cholinergic mechanisms in canine narcolepsy. II. Acetycholine release in the pontine reticular formation is enhanced during cataplexy. *Neuroscience,* 59, 523-530, 1994.

158. RIEHL J., HONDA K., KWAN M, HONG J., MIGNOT E., NISHINO S. - Chronic oral administration of CG-3703, a thyrotropin releasing hormone analog, increases wake and decreases cataplexy in canine narcolepsy. *Neuropsychopharmacology* 23, 34-45, 2000.

159. RINNE J.O., HUBLIN C., PARTINEN M., RUOTTINEN M., RUOTSALAINEN U., NAGREN K., LEHIKOINEN P., LAIHINEN A. - Positron emission tomography study of human narcolepsy. No increase in striatal dopamine D2 receptors. *Neurology,* 45, 1735-1738, 1995.

160. RIPLEY B., FUJIKI N., OKURA M., MIGNOT E., NISHINO S. – Hypocretin levels in sporadic and familial cases of canine narcolepsy. *Neurobiology of Disease* 8, 525-534, 2001.

161. ROGERS A.E., ALDRICH M. - The effect of regularly scheduled naps on sleep attacks and excessive daytime sleepiness associated with narcolepsy. *Ann. Neurol.* 43, 88-97, 1998.

162. ROSELAAR S.E., LANGDON N., LOCK C.B., JENNER P., PARKES J.D. - Selegiline in narcolepsy. *Sleep* 10, 491-495, 1987.

163. SAKURAI T., AMEMIYA A., ISHII M.,MATSUKAZI I., CHEMELLI R.M., TANAKA H., WILLIAMS S.C., RICHARDSON J.A., KOZLOWSKI G.P., WILSON S. - Orexin and orexin receptors: a family of hypothalamic neuropeptides and G protein-coupled receptors that regulate feeding behavior. *Cell* 92, 573-585, 1998.

164. SCHMIDT M.S., CLARK R.W., HYMAN P.R. - Protriptyline: an effective agent in the treatment of the narcolepsy-cataplexy syndrome and hypersomnia. *Am. J. Psychiatry* 134, 183-185, 1977.
165. SHINDLER J., SCHACHTER M., BRINCAT S., PARKES J.D. - Amphetamine, mazindol, and fencamfamin in narcolepsy. *Br. Med. J.* 290, 1167-1170, 1985.
166. SIEGEL J.M., FAHRINGER H.M., ANDERSON L., NIENHUIS R., GULYANI S., NASSIRI J., MIGNOT E., SWITZER R.C. - Evidence for localized neuronal degeneration in narcolepsy: studies in the narcoleptic dog. *Sleep Res.* 24, 354, 1995.
167. SIEGEL J.M., NIENHUIS R., MIGNOT E., FAHRINGER H.M., JAMGOCHIAN G.M. - Neuronal degeneration in the amygdala of the narcoleptic dog. *Sleep Res.* 25, 370, 1996.
168. SONKA K., TAFTI M., BILLIARD M. - Narcolepsy and ageing. In: S. Smirne, M. Franceschi, L. Ferini-Strambi (eds) *Sleep and ageing,* Masson, Milano, 181-186, 1991.
169 TAFTI M., DAUVILLIERS Y., NEIDHART E., BILLIARD M. - Lack of association between the monoamine oxidase-A (MAO-A) gene and narcolepsy. *Sleep* 23, Abstract Supplement 2, A115, 2000.
170. TAFTI M., RONDOUIN G., BESSET A., BILLIARD M. - Sleep deprivation in narcoleptic subjects: effect on sleep stages and EEG power density. *Electroenceph. Clin. Neurophysiol.* 83, 339-349, 1992.
171. TAKAHASHI Y., JIMBO M. - Polygraphic study of narcoleptic syndrome with special reference to hypnagogic hallucinations and cataplexy. *Folia Psychiat. Neurol. Jap.* S7, 343-347, 1963.
172. THANNICKAL T.C., MOORE R.Y., NIENHUIS R., RAMANATHAN M., KANMAN H. - Reduced number of hypocretin neurons in human narcolepsy. *Neuron* 27, 469-474, 2000.
173. US Modafinil in Narcolepsy Multicenter Study Group. Randomized trial of modafinil for the tretment of pathological sleepiness in narcolepsy. *Ann. Neurol.* 43, 88-97, 1998.
174. VAN DEN HOED J., KRAEMER H., GUILLEMINAULT C., ZARCONE V.P., LAUGHTON E.M., DEMENT W.C., MITLER M.M. - Disorders of excessive daytime somnolence: polygraphic and clinical data for 100 patients. *Sleep,* 4, 23-37, 1981.
175. VELA BUENO A., CAMPOS CASTELLO J.C., BAOS R.J. - Hereditary cataplexy: is it primary cataplexy ? *Waking Sleeping,* 2, 125-126, 1978.
176. VOGEL G. - Studies in the psychophysiology of dreams III. The dream of narcolepsy. *Arch. Gen. Psychiatry,* 3, 421-425, 1960.
177. WESTPHAL C. - Eigentümliche mit Einschlafen verbundene Anfälle. *Arch. Psychiatr. Nervenkr.* 7, 631-635, 1877.
178. WILSON·S. - The narcolepsies. *Brain,* 51, 63-77, 1928.
179. WISOR J.P., NISHINO S., SORA I., UHL G.H., MIGNOT E., EDGAR D.M. – Dopaminergic role in stimulant-induced wakefulness. *J. Neurosci.* 2001, 21, 1787-1794.
180. WITTIG R., ZORICK F., PICCIONE P., SICKLESTEEL J., ROTH T. - Narcolepsy and disturbed nocturnal sleep. *Clin. Electroencephalogr.* 14, 130-134, 1983.
181. YOSS R.E., DALY D.D. - Criteria for the diagnosis of the narcoleptic syndrome. *Proc. Staff. Meet. Mayo Clin.* 32, 320-328, 1957.
182. YOSS R.E., DALY DD. - Treatment of narcolepsy with ritaline. *Neurology,* 9,171-173, 1959.
183. YOSS R.E., DALY D.D. - Hereditary aspects of narcolepsy. *Rans. Am. Neurol. Assoc.,* 85, 239-240, 1960.

Chapter 32

Idiopathic hypersomnia

M. Billiard and A. Besset
Service de Neurologie B, Hôpital Gui de Chauliac, Montpellier, France

Unlike narcolepsy, which has been well documented historically and clearly defined, idiopathic hypersomnia has only recently been identified. Certain limits remain indistinct despite considerable progress in nosology, in recent years.

HISTORY

Idiopathic hypersomnia was for a long time confused with narcolepsy. It was only with the advent of polysomnographic studies that distinctions between the two illnesses began to appear. Dement *et al.* [11] were the first to suggest that subjects affected by excessive diurnal somnolence, but not by attacks of REM sleep, cataplexy, or sleep paralysis, should be considered as having an illness other than narcolepsy. Later, various other terms were proposed to designate this entity: essential narcolepsy [5], NREM narcolepsy [25], hypersomnia [27], the latter authors already giving a complete description of the condition, whereby daytime somnolence was described as being less irresistible than in the case of narcolepsy but usually more durable, with an absence of disturbed nocturnal sleep but a marked lengthening of its duration and great difficulty in wakening, possibly bordering on confusion. Roth was finally the one to provide the clearest description of hypersomnia with sleep drunkenness [32], and was the first to use the term idiopathic hypersomnia [28-29]. Roth described two forms of this hypersomnia: a more typical polysymptomatic form and a monosymptomatic form. In the following years, for simplicity's sake and due to a lack of understanding of illnesses later defined, idiopathic hypersomnia was often misdiagnosed in cases of hypersomnia which did not include apnoea and narcolepsy. In 1972 Guilleminault *et al.* [17] indicated that several cases of hypersomnia formerly considered as idiopathic actually corresponded to upper airway resistance syndromes. Recent publications have begun to clarify the illness [2, 4, 6, 8].

EPIDEMIOLOGY

It is hard to estimate the prevalence of idiopathic hypersomnia in the general public because of the small number of subjects affected and the nosological uncertainties. No epidemiological study has yet been carried out on the illness. The only data available show the ratio of subjects affected by idiopathic hypersomnia / subjects affected by narcolepsy in different published series (table 32.1). As the table indicates, this ratio clearly diminishes over the years, due partly to stricter criteria and partly to an increasingly accurate identification of the illnesses formerly misdiagnosed as idiopathic hypersomnia. The age of onset varies but rarely goes beyond age 30 (fig. 32.1). Unlike narcolepsy, precipitating factors do not seem to play a role.

A heredofamilial pattern is often apparent but this warrants further verification [24].

Table 32.1. Respective numbers of subjects affected by idiopathic hypersomnia and narcolepsy in different series.

Authors	Idiopathic hypersomnia	Narcolepsy	% ratio
Roth (1980)	167	288	57.9
Coleman *et al.* (1982)	150	425	35.2
Baker *et al.* (1986)	74	257	28.7
Aldrich (1996)	42	258	16.2
Billiard *et al.* (unpublished)	53	380	13.9

* Note the declining proportion of subjects diagnosed with idiopathic hypersomnia, due to a better understanding of sleep pathologies with time.

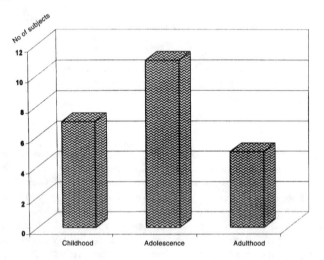

Figure 32.1. Age of onset: before the age of thirty in virtually all cases

CLINICAL FEATURES

The polysymptomatic form is both more complex and easier to distinguish. It is characterised by permanent or more often recurrent daytime sleepiness with sleep episodes which are less irresistible than in the case of narcolepsy but which tend to be of long duration and non-restorative, the subject complaining of never or rarely being fully awake. This sleepiness may lead to automatic acts such as putting objects away in inappropriate places, inadvertently driving a vehicle to somewhere other than the original destination.

Against this background, episodes of sleepiness are less irresistible than in narcolepsy, but are usually protracted or very protracted and do not normally restore alertness. Night time sleep is long-lasting, uninterrupted by waking, or rarely so. Morning arousal is late and difficult, despite repeated alarm bells, telephone alarm calls, or prompting from the parents or partner. It is often marked by "sleep drunkenness" with spatio-temporal disorientation, slow speech and thinking, retrograde and anteretograde memory disorders lasting from a few minutes to an hour or more. There is no history of cataplexy or sleep paralysis.

The monosymptomatic form is far less distinguishable, with night time sleep of normal or slightly longer duration, no morning arousal difficulties, sleep which is more fragmentary than continuous and episodes of restorative or nonrestorative sleep.

Clinical examination is normal but subjects sometimes report migraine headaches, lipothymic tendencies, hot flushes, heavy sweating and Raynaud's syndrome [19].

DIAGNOSTIC PROCEDURES

The diagnosis of idiopathic hypersomnia is essentially based on clinical symptomatology. Polysomnographic recording followed by a multiple sleep latency test is however necessary, not so much to confirm diagnosis as to eliminate other sleep pathologies. Results show a normal sleep pattern with a limited number of arousals and mean sleep latency of slightly under 10 minutes on the multiple sleep latency test (fig. 32.2). There is no sleep-onset REM period. Respiration is normal. However the drawback of this method is that subjects must be woken at 7 to 7 30 am, because of the multiple sleep latency test, thus preventing observations of late spontaneous awakening. Moreover, the multiple sleep latency test is not adapted to observing episodes of prolonged daytime sleep.

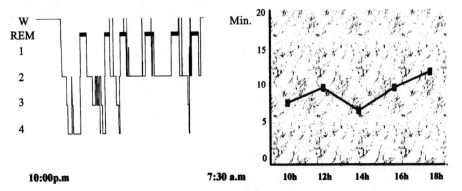

Figure 32.2. Overnight recording and multiple sleep latency test for a 24 year old man suffering from idiopathic hypersomnia (polysymptomatic form). Sleep architecture is normal. The subject had to be woken at 7:30 a.m. for the multiple sleep latency test. The mean sleep latency was 8.8 min.

This explains the interest of a prolonged recording lasting 24 hours or more, subjects being free to turn the light on and off at will, morning and evening, and go to bed or get up whenever they wish. In such conditions, recordings show late morning arousal, one or several naps, and well over 12 hours of sleep within a 24 hour period, at least in the polysymptomatic form (fig. 32.3). A link has been shown between HLA cw2 and HLA DR5 but this has not been confirmed to date [22]. A brain scan should be carried out as a matter of course even if the neurological examination is normal. A personality test is advisable to exclude the aetiology of hypersomnia associated with mental disorder.

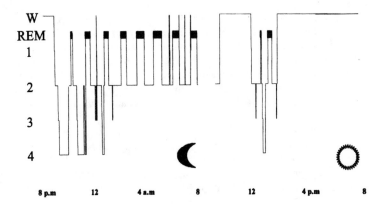

Figure 32.3. Continuous 24 hour recording. The duration of night-sleep is 12 hours 44 minutes and the duration of a single daytime sleep episode is 2 hours 36 minutes, i.e. a total sleep period of 15 hours 20 minutes per 24 hours.

COURSE OF ILLNESS

The illness is usually chronic, does not tend to improve, even if favourable evolution has been seen in some patients. The illness is particularly debilitating socially and professionally [1, 9].

DIFFERENTIAL DIAGNOSIS

Idiopathic hypersomnia is the most "banal" diagnosis among the sleep disorders. Confusion arises not so much with narcolepsy or sleep apnoea syndrome – illnesses which are well known and easy to recognise – as with other sleep disorders, which are less well identified clinically and polysomnographically or which require more detailed investigation.

The first diagnosis to consider is the upper airway resistance syndrome [17]. Excessive daytime somnolence is common. Other symptoms include pronounced snoring, particularly in men, drowsiness on arousal and occasional bruxism. Strikingly, men and women are almost equally affected. On examination, the majority of subjects present no obesity, the face is triangular in shape, with a small chin, the palate is high (ogival), there is a type 2 malocclusion of the mouth with abnormal upper maxillary protrusion in relation to the lower jaw. Polysomnography shows arousal reactions with bursts of alpha rhythm for several seconds followed by a lengthening of the breathing cycle. Final diagnosis calls for polysomnography associated with a recording of oesophageal pressure with pneumotachography if possible. Typically, there is a sequence of breaths characterised by increasing respiratory effort reflected by a pattern of progressively more negative oesophageal pressure, terminated by a sudden change in pressure to a less negative level and an arousal (Respiratory effort related arousal or RERA).

Narcolepsy without cataplexy, also known as atypical narcolepsy or more recently as hypersomnia with sleep-onset REM periods [14] is a clinical form of narcolepsy in which cataplexy has not yet occurred or fails to occur. Positive diagnosis is made when REM sleep onset occurs at least twice during the multiple sleep latency test, associated with the HLA DQB1*0602 allele.

Hypersomnia associated with mental disorder, mood disorder, psychoses, anxiety disorder, is a possible diagnosis requiring psychological evaluation. Polysomnographically, sleep is interrupted by frequent awakenings, REM sleep latency is sometimes shortened, the multiple sleep latency test is non- or only slightly significant despite complaints of excessive sleepiness. In continuous recording, it is striking that these subjects often stay in bed during the day without showing any objective signs of sleepiness (clinophilia) [7].

Chronic fatigue syndrome is characterised by persistent or relapsing drowsiness which is not alleviated by staying in bed. Polysomnographic recordings show a reduction in the sleep efficiency index (time asleep/time in bed) with the intrusion of alpha rhythm in sleep [20].

Consecutive hypersomnia is caused by infection: viral pneumopathy, infectious mononucleosis, Guillain-Barre syndrome or hepatitis, and develops in the weeks or months following the initial phase of illness. Subjects complain of drowsiness and sleepiness and are likely to sleep for excessive periods. Prognosis is favourable but may take months or years [16].

Posttraumatic hypersomnia closely resembles idiopathic hypersomnia. It develops over a period of 6 to 18 months after the trauma [15]. Excessive sleepiness may be the first sign of hydrocephalus.

Daytime somnolence after fragmented sleep, associated with symptoms of nocturnal pain, of the rheumatic type for example, is a common occurrence.

Sleep insufficiency syndrome is characterised by excessive daytime sleepiness, concentration problems and lack of energy. A detailed history of the usual sleep schedule is required for diagnosis [28].

Subwakefulness syndrome is still open to discussion. It consists of normal night time sleep, moderate daytime sleepiness with no attacks of irresistible sleep and polysomnographic findings of intermittent stage 1 episodes during the day [23].

Delayed sleep phase syndrome may create some confusion, in view of the subjects' extreme difficulty in waking in the morning, with morning sleepiness but normal alertness in the afternoon and evening [33].

Finally, at the limit of differential diagnosis, are the long sleepers who are non somnolent during the daytime if they have slept sufficiently at night, as subjects affected by idiopathic hypersomnia may be found at the extreme limit of long sleepers.

PATHOPHYSIOLOGY

Unlike narcolepsy, no natural animal has been reported with idiopathic hypersomnia, thus limiting the possibilities for investigation.

Neurophysiology

The destruction of noradrenergic neurones of the rostral third of the locus coeruleus or of the noradrenergic bundle at isthmus level in cats, entrains harmonious hypersomnia with an increase of NREM and REM sleep, which may be considerable, roughly comparable to idiopathic hypersomnia. This state is accompanied by a reduction in telencephalic norepinephrine with a significant increase of 5-HIAA and tryptophan [26].

Neurochemistry

Current understanding of the neurochemical basis of idiopathic hypersomnia is based on two comparative studies carried out on populations of subjects affected by idiopathic hypersomnia, by narcolepsy and on a control group [12, 21]. Results from the first study show that concentrations of dopamine and indoleacetic acid, a tryptamine metabolite, are significantly reduced in the cerebrospinal fluid of subjects affected by idiopathic hypersomnia as is the case for subjects affected by narcolepsy. Results from the second study, comprising cerebrospinal fluid tests before and after probenecide, show a high accumulation of dopamine, homovanillic acid (HVA) and 3.4-dihydroxyphenylacetic acid (DOPAC) metabolites in both hypersomniac subject groups, suggesting an increased dopamine turn-over in both disorders. But the results of the second study were later reinterpreted using principal component statistical analysis, whereby it was shown that monoamine metabolites DOPAC, MHPG, HVA and 5-HIAA were closely correlated in the CSF of normal subjects, whereas HVA and DOPAC, the dopamine metabolites, were not correlated to the two other metabolites in subjects affected by narcolepsy, and MHPG, the norepinephrine metabolite, was not correlated to the 3 other metabolites in subjects suffering from idiopathic hypersomnia, favouring a dysfunction of the dopaminergic system in narcolepsy and of the noradrenergic system in the case of idiopathic hypersomnia [13].

PATHOGENESIS

The genetic nature of the disorder has been confirmed by several studies. The first [24] does so in a group of 23 participants consisting of 9 subjects (39%) with one or several relatives who were also affected and the second [31] in a population of 167 participants, with 45 subjects (26.9) who also had one or several relatives who were affected. In our own population of 39 subjects selected according to much stricter criteria, 23 subjects (58.9%) reported having 1 to 4 family relatives who were affected by excessive daytime sleepiness non-associated with snoring (fig. 32.4). This data leaves little doubt as to the genetic nature of the disorder. Nevertheless the small number of families available makes it difficult to favour any particular mode of transmission.

Immunogenetic studies have been carried out in an attempt to find a linkage with HLA. Harada *et al.* [18] note no significant differences in the distribution of HLA antigens among 41 subjects affected by essential hypersomnia (combining idiopathic hypersomnia, incomplete forms of narcolepsy and other hypersomnias which are more difficult to classify) and in control subjects. Conversely, Montplaisir and Poirier [22] in a group of 18 subjects affected by idiopathic hypersomnia, found a significant linkage with Cw2 in 22.22% of affected subjects as opposed to 5.7% in control subjects (p<0.05) and with DR5 (in linkage disequilibrium with Cw2) found in 38.88% of affected subjects as opposed to 14.67% of the controls (p<0.05). In our own population all subjects were HLA typed but no significant HLA linkage was found.

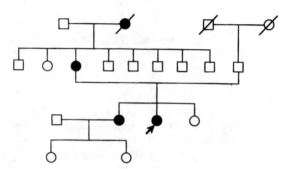

Figure 32.4. Family tree of a multiplex family. Transmission occurs here through the mothers' side.

TREATMENT

As in the case of narcolepsy, pharmacological and behavioural treatment is combined.

Pharmacological treatment for idiopathic hypersomnia relies on the same drugs as for the treatment of daytime sleepiness and irresistible sleep episodes in narcolepsy i.e. amphetamines and derivatives previously and modafinil today, with a mean dose of 300 mg per day. It should be added that amphetamines were often used in the past but were less well tolerated by subjects with idiopathic hypersomnia than in those with narcolepsy.

Unlike the case of narcolepsy, modafinil has not been subjected to systematic investigations in idiopathic hypersomnia and the only results currently available are clinical. In our own experience based on 30 subjects treated with doses of 200 to 400 mg, the results were good to very good in 23 cases, average or mediocre in 3 cases and void in 1 case, with 3 cases showing a fluctuating effect. As a general rule, the results were better for daytime somnolence than for morning awakening. The negative side effects were mainly central, of the headache or irritability type.

Behavioural treatment is more limited in the case of idiopathic hypersomnia. Siestas are discouraged because of their long duration and non restorative nature. The question arises as to whether subjects should be advised to sleep to satiation, but with the social consequences this is likely to produce, some preliminary data suggesting that subjects affected by idiopathic hypersomnia, at least in its polysymptomatic form, may simply be long sleepers who are deprived of enough sleep.

CONCLUSION

Idiopathic hypersomnia is less frequent than narcolepsy. The classic, or polysymptomatic form, is well identified, as opposed to the monosymptomatic form whose frontiers with narcolepsy still need clarification. Pathophysiology is still in infancy. In contrast with narcolepsy characterized by an abnormal propensity to fall asleep, idiopathic hypersomnia, at least in its polysymptomatic form, should be considered as an inability to terminate sleep.

REFERENCES

1. ALDRICH M.S. – Automobile accidents in patients with sleep disorders. *Sleep*, *12*, 487-494, 1989.
2. ALDRICH M.S. – The clinical spectrum of narcolepsy and idiopathic hypersomnia. *Neurology*, *46*, 393-401, 1996.
3. BAKER T.L., GUILLEMINAULT C., NINO-MURCIA G., DEMENT W.C. – Comparative polysomnographic study of narcolepsy and idiopathic central nervous system hypersomnia. *Sleep*, *9*, 323-242, 1986.
4. BASSETTI C., ALDRICH M.S. – Idiopathic hypersomnia. A series of 42 patients. *Brain*, *120*, 1423-1435, 1997.
5. BERTI-CERONI G., COCCAGNA G., GAMBI D., LUGARESI E. – Considerazioni clinico poligrafiche sull narcolessia essenziole a somno lento.*Sist. nerv.*, *19*, 81-89, 1967.
6. BILLIARD M. – Idiopathic hypersomnia. *Neurologic Clinics*, 573-582, 1996.
7. BILLIARD M., DOLENC L., ALDAZ C., ONDZE B., BESSET A. – Hypersomnia associated with mood disorders: a new perspective. *J. of Psychosomatic Research.*, *38* (suppl. 1), 41-47, 1994.
8. BILLIARD M., MERLE C., CARLANDER B., ONDZE B., ALVAREZ D., BESSET A. – Idiopathic hypersomnia. *Psychiatry and Clinical Neurosciences*, *52*, 125-129, 1998.

9. BROUGHTON R., NEVSIMALOVA S., ROTH B. – The socioeconomic effects (including work, education, recreation and accidents) of idiopathic hypersomnia. *Sleep Res.*, 7, 217, 1978.

10. COLEMAN R.M., ROFFWARG H.P., KENNEDY S.J., GUILLEMINAULT C., CINQUE J., COHN M.A., KARACAN I., KUPFER D.J., LEMMI H., MILES L.E., ORR W.C., PHILLIPS E.R., ROTH T., SASSIN J.F., SCHMIDT H.S., WEITZMAN E.D., DEMENT W.C. – Sleep-wake disorders based on a polysomnographic diagnosis. A national cooperative study. *JAMA*, 247, 997-1003, 1982.

11. DEMENT W.C., RECHTSCHAFFEN A., GULEVICH G. – The nature of the narcoleptic sleep attack. *Neurology*, 16, 18-33, 1966.

12. FAULL K.F., GUILLEMINAULT C., BERGER P.A., BARCHAS J.D. – Cerebrospinal fluid monoamine metabolites in narcolepsy and hypersomnia. *Ann. Neurol.*, 13, 258-263, 1983.

13. FAULL K.F., THIEMANN S., KING R.J., GUILLEMI-NAULT C. – Monoamine interactions in narcolepsy and hypersomnia: a preliminary report. *Sleep*, 9, 246-249, 1986.

14. GUILLEMINAULT C. – Narcolepsy syndrome. *In: Principles and Practice of Sleep Medicine*, M.H. Krieger, T. Roth, W.C. Dement (eds.), W.B. Saunders, Philadelphia, 549-561, 1994.

15. GUILLEMINAULT C., FAULL K.F., MILES L., VAN DEN HOED J. – Post-traumatic excessive daytime sleepiness. A review of 20 patients. *Neurology*, 33, 1584-1589, 1980.

16. GUILLEMINAULT C., MONDINI S. – Infectious mononucleosis and excessive daytime sleepiness. A long term follow-up study. *Arch. Intern. Med.*, 146, 1333-1335, 1986.

17. GUILLEMINAULT C., STOOHS R., CLERK. A., CETEL M., MAISTROS P. – A cause of excessive daytime sleepiness: the upper airway resistance syndrome. *Chest*, 104, 781-787, 1993.

18. HARADA S., MATSUKI K., HONDA Y., SOMEYA T. – Disorders of excessive daytime sleepiness without cataplexy, and their relationship with HLA in Japan. *In: HLA in Narcolepsy*, Y. Honda, T. Juji (eds.), Springer-Verlag, Berlin, 172-185, 1988.

19. MATSUNAGA H. – Clinical study on idiopathic CNS hypersomnolence. *Jpn. J. Psychiatr. Neurol.*, 41, 637-644.

20. MOLDOFSKY H. – Non-restorative sleep and symptoms after a febrile illness in patients with fibrositis and chronic fatigue syndrome. *J. Rheumatol*, 16 (suppl. 19), 150-153, 1989.

21. MONTPLAISIR J., DE CHAMPLAIN J., YOUNG S.N., MISSALA K., SOURKES T.L., WALSH J., REMILLARD G. – Narcolepsy and idiopathic hypersomnia: biogenic amines and related compounds in CSF. *Neurology*, 32, 1299-1302, 1982.

22. MONTPLAISIR J., POIRIER G. – HLA in disorders of excessive daytime sleepiness without cataplexy in Canada. *In: HLA in Narcolepsy*, Y. Honda, T. Juji (eds.). Springer-Verlag, Berlin, 186-190, 1988.

23. MOURET J.R., RENAUD B., QUENIN P., MICHEL D., SCHOTT B. – Monoamines et régulations de la vigilance. I. Apport et interprétation biochimique des données polygraphiques. *Rev. Neurol.*, 127, 139-155, 1972.

24. NEVSIMALOVA-BRUHOVA S., ROTH B. – Heredofamilial aspects of narcolepsy and hypersomnia. *Schweiz. Arch. Neurol. Neurochir. Psychiat.*, 110, 45-54, 1972.

25. PASSOUANT P., POPOVICIU L., VELOK G., BALDY-MOULINIER M. – Étude polygraphique des narcolepsies au cours du nycthémère. *Rev. Neurol.*, 118, 431-441, 1968.

26. PETITJEAN F., JOUVET M. – Hypersomnie et augmentation de l'acide 5-hydroxy-indolacétique cérébral par lésion isthmique chez le chat. *Comptes rendus des séances de la Société de Biologie* (Paris), 164, 2288-2293, 1970.

27. RECHTSCHAFFEN A., ROTH B. – Nocturnal sleep of hypersomniacs. *Activ. Nerv.*, (suppl. 11), 229-233, 1969.

28. ROEHRS T., ZORICK F., SICKLESTEEL J., WITTIG R., ROTH T. – Excessive daytime sleepiness associated with insufficient sleep. *Sleep*, 6, 319-325, 1983.

29. ROTH B. – Functional hypersomnia. *In: Narcolepsy*, C. Guilleminault, W.C. Dement, P. Passouant (eds.), Spectrum Publ. New York. 333-349, 1976.

30. ROTH B. – Narcolepsy and hypersomnia. *Schweiz. Arch. Neurol. Psychiatry*, 119, 31-41, 1976.

31. ROTH B. – *Narcolepsy and Hypersomnia*. Karger, Basel, 1980.

32. ROTH B., Nevsimalova S., Rechtschaffen A. – Hypersomnia with "Sleep Drunkenness" *Arch. Gen. Psy.*, 26, 456-462, 1972.

33. WEITZMAN E.D., CZEISLER C.A., COLEMAN R.M., SPIELMAN A.J., ZIMMERMAN J.C., DEMENT W.C. – Delayed sleep phase syndrome: a chronobiological disorder with sleep onset insomnia. *Arch. Gen. Psychiatry*, 38, 737-746, 1981.

Chapter 33

Recurrent hypersomnias

M. Billiard

Service de Neurology B, Hôpital Gui de Chauliac, Montpellier, France

These refer to a group of relatively rare disorders characterised by episodes of more or less continuous sleep, with an average duration of one week, recurring at highly variable intervals of between one and several months or even several years. The classic picture is that of the Kleine-Levin syndrome in which sleep episodes are associated with other symptoms, eating or sexual disorders and psychological disturbance. The origin is still unknown. In addition to the Kleine-Levin syndrome, recurrent hypersomnias symptomatic of organic or psychiatric disorders may be encountered.

THE KLEINE-LEVIN SYNDROME

History

With the exception of Anfimoff's observation, quoted by Kaplinsky and Schumann [1], the first descriptions of the syndrome are attributed to Kleine [32], Lewis [38] and Levin [36]. In 1936, the same Levin [37] collated several case reports, publishing them as examples of "a new syndrome of periodic somnolence and morbid hunger", occurring solely in men, starting after puberty and characterised by attacks of somnolence lasting several days or weeks, associated with excessive appetite, unstable motor control, incoherent remarks and occasionally with hallucinations. In 1942, Critchley and Hoffman [12] introduced the eponymous term Kleine-Levin to describe the syndrome. In 1962, Critchley [11] added eleven cases taken from literature to his initial description. He emphasised four complementary clinical features: the syndrome essentially or even exclusively affects the male sex; onset occurs in adolescence; overeating is of a compulsory nature and is not due to pathological hunger; recovery occurs spontaneously. Billiard [3] has replaced the term periodic hypersomnia with recurrent hypersomnia as cases of periodic attacks are virtually unheard of.

Epidemiology

The Kleine-Levin syndrome is an uncommon disorder, with 25 cases in our own patient group in comparison with 380 narcoleptics. Over the years we have collated 162 cases in literature of subjects suffering from recurrent hypersomnia, megaphagia and psychological disorders. Of these 162 cases, 128 concerned male subjects (79%) and 34 female subjects (21%) i.e. a male/female sex ratio of 4/1. Onset most commonly occurred in adolescence. Later onset does however occur, particularly in women (fig. 33.1). Precipitating factors were indicated in slightly over half the cases observed. These usually consist of an episode of flu or an infection of the upper airways, often occurring during a group trip (school trip, scout camp etc.) and frequently in warm or hot weather. In rarer cases the precipitating factors may consist of an episode of seasickness, extreme drunkenness, a blow to the face, anaesthetic or a first menstrual period for women (table 33.1).

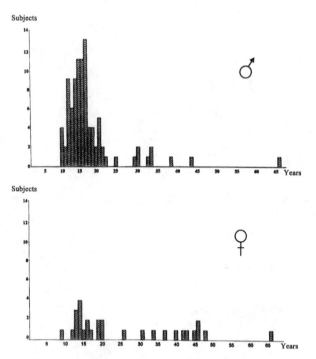

Figure. 33.1. Age of onset of the Kleine-Levin syndrome for 128 male subjects (top) and 34 female subjects (bottom).

Table 33.1. Precipitating factors of first episode in 162 subjects affected by the Kleine-Levin Syndrome.

	Male (128)	Female (34)
Influenza or ENT infection	50 (39%)	9 (26.4 %)
Drunkenness	4 (3.1%)	
Blow to the face	3 (2.3%)	
Emotional shock	3 (2.3%)	
Anaesthetic	2 (1.5%)	
Seasickness	2 (1.5%)	
Exhausting effort	2 (1,5%)	
First menstrual period		3 (2.3%)
Other	3 (2.3%)	2 (1.5%)
No precipitating factors indicated	59 (46%)	20 (58.8%)

Clinical features

The onset of hypersomniac episodes is quick, occurring in a matter of hours or gradually over several days. The subject complains of drowsiness or of sudden sleepiness. He stays in bed continually or gets up only for meals. Sleep may be calm or agitated. Parasomnia may occur. Intense dreaming sometimes occurs. Urinary incontinence is not a feature. Hypersomnia is accompanied by behavioural and cognitive disturbances.

Overeating does not occur with each episode. It is clearly evident in certain subjects. In others it may be less obvious, as in a subject who, in the first episode, ate seven bars of chocolate one after the other but did not indulge in overeating afterwards. It is common for subjects to consume sweet foods such as candies, chocolate and cakes. Above all, as Critchley [11] has emphasised, overeating appears to be of a compulsory nature. Critchley's [11] third case is that of a young boy who does not report feeling abnormally hungry but cannot stop eating the food within his reach. He eats a rotten kidney which his friends considered inedible. Barontini and Zappoli's [2] patient does not seek food outside mealtimes, but as soon as he gets to the table, eats gluttonously even to the point of taking food from other patients' plates while they are eating. This greediness often results in a weight increase by the end of the attack if the episode continues for several weeks.

Hypersexuality in the form of sexual advances, shamelessly expressed sexual fantasies or masturbation in public, is a particularly defining feature but one which occurs in only 1/3 of male cases and more rarely in female cases.

Irritability, with exceptional cases of outright aggression [58], and odd behaviours, are also part of the clinical picture.

Cognitive disturbances, a feeling of unreality, confusion, visual and auditory hallucinations may occur, often resulting in subjects being admitted to psychiatric units.

Examination during the hypersomnia episode is likely to reveal non-specific signs, exaggerated or weak deep tendon reflexes, nystagmus, and more significantly, dysautonomic signs, facial congestion and profuse, foul-smelling perspiration.

After a period which may vary from a number of days to several weeks, the subject emerges from the state within a matter of days or hours. It is not uncommon for the hypersomniac episode to be followed by a phase of manic behaviour with insomnia, lasting one or two days, as if the subject were trying to make up for lost time or conversely of depression, occasionally with suicidal thoughts. The defining element is the recurrence of episodes after a lapse of one to several months, or even of several years, which may lead to academic or professional problems caused by repeated absences.

Between episodes, the alertness and behaviour of subjects is normal even if neurotic traits or slight mental deficiency have been noted in certain subjects.

The attacks usually diminish gradually in frequency and severity, disappearing entirely after several years. Nevertheless cases do exist of prolonged development with attacks continuing to occur after 20 years [7, 8, 11, 23, 35]. One particular form of evolution is that of psychiatric pathology of a psychotic nature [54, 74].

Laboratory tests

Diagnosis of the Kleine-Levin syndrome is essentially clinical and laboratory examinations merely serve to exclude the possibility of secondary recurrent hypersomnia with a recognisable organic or psychiatric cause.

Polysomnography is a useful means of confirming hypersomnia. But this is difficult to obtain as the time lapse between the onset of the episode and the recording must be as short as possible to avoid missing the hypersomniac period.

Biological tests. – According to the various cases reported in scientific literature and with few exceptions, blood count, blood type, plasma electrolytes, urea, creatinine, calcium and phosphorous, SGOT and SGPT are normal. Plasma, bacterial and viral serologies are negative. Electrophoresis and immuno-electrophoresis of plasma proteins are normal.

Electroencephalogram. – The typical pattern shows general slowing of the background activity, often with bursts of irregular high voltage sharp waves [64, 67]. These anomalies often result in the episode being mistakenly diagnosed as an epileptic fit and for antiepileptic treatment to be wrongly prescribed.

Radiological examinations. – Brain scan and magnetic resonance imaging are usually normal. No series is yet available comparing states during and between hypersomniac periods.

Psychological test. – This should always be carried out both during the attack – not an easy task – and when the subject is no longer experiencing an attack, to ensure that there is no personality disorder.

Polysomnography – A multiple sleep latency test has been proposed as a means of diagnosis [52]. The typical pattern shows a shortened mean sleep latency, but the test is difficult to carry out due to the lack of subjects' co-operation. In practice, it is easier and more instructive to carry out a continuous prolonged polygraphic recording for 24 or 48 hours, providing an indication of the total duration of sleep over 24 hours, which is generally high [2, 34,, 50,69] (fig. 33.2) with changes in sleep pattern, reduced stages 3 and 4 of NREM sleep and shortened REM sleep latency [34, 51, 52, 71].

Monoamine assays – A few teams have measured monoamine or metabolites in urine and CSF. Results showed high urine levels of 5-HIAA per 24 hours [51, 70], normal baseline levels of 5-HIAA and HVA in CSF [26, 43, 69], a slightly raised baseline level of noradrenaline [58], and raised levels of 5-HIAA and HVA after probenecide [43, 69]. It is difficult to establish a global interpretation of these results.

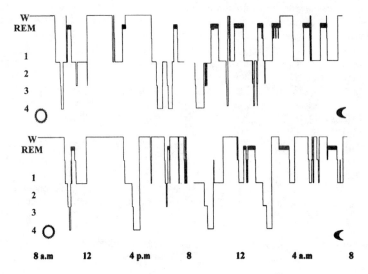

Figure 33.2. – 48 hour polysomnographic recording (from 8 a.m to 8 a.m.) in a young man suffering from the Kleine-Levin syndrome.

Clinical variants

Incomplete forms of the Kleine-Levin syndrome, which do not include eating disorders, are probably as frequent as the complete forms, or more so. There is little to distinguish them from the Kleine-Levin syndrome as overeating may only occur at the outset, and not be repeated later.

Menstruation-linked periodic hypersomnia is less common [5, 19, 23, 31, 39, 40, 47, 56, 59]. It is distinguished by attacks of hypersomnia, which may or may not be associated with other signs of the Kleine-Levin syndrome, and occurs periodically, being linked to the menstruation cycle.

These two variants of the Kleine-Levin syndrome appear to share the same prognosis.

Pathophysiology

This remains unknown, but the combination of hypersomnia, overeating and sexual disinhibition, the intermittent occurrence of the disorder and the absence of clinical abnormalities between attacks would suggest diencephalic, possibly hypothalamic, functional dysregulation.

What evidence exists in favour of this hypothesis?

Several authors have examined the neuroendocrinal functions of subjects affected by the Kleine-Levin syndrome and certain anomalies have been reported. Nevertheless it is necessary to be cautious about accepting and interpreting the data obtained. There are several reasons for this. The first is that the illness is so rare, with a limited number of subjects studied to date and even fewer subjects studied during both hypersomniac and asymptomatic phases. Most of the work so far undertaken by the various teams has focused on single patients only. The second reason is the difficulty in starting investigations from the outset of the illness with the risk that the episode ends by the time tests are carried out. Schematically, two types of examination have been carried out: on the one hand, the measurement of basic hormone levels and levels after stimulation; on the other, the plotting of hormone secretion curves over a 24 hour period or overnight by taking periodic blood samples.

Generally speaking, baseline levels of the main antehypophysary hormones, growth hormone (GH), prolactin (PRL), thyrotropin (TSH), luteinising hormone (LH), follicle-stimulating hormone (FSH), testosterone, cortisol were found to be normal. The same applies to levels obtained after stimulation tests with the following exceptions: a paradoxical GH response to thyrotropin (TRH) [18], attenuated response or no cortisol response to insulin-induced hyperglycaemia [16, 33], absence of TSH to TRH [16] not found by Gadoth *et al.* [18].

The results of studies on hormones secreted at night or over a 24 hour period are summarised in table 33.6. As we can see, the results vary from one author to another, and for the same author [44] from one patient to another, rendering a hypothesis of hypothalamo-hypophysiary dysfunctioning somewhat fragile.

Another argument in favour of a diencephalic attack is the dysautonomic score reported in certain subjects affected by the Kleine-Levin syndrome; a score which was particularly high in the patient described by Hegarty and Merriam [27]. The patient had episodes of face-flushing, heavy perspiration, excessive salivation, high blood pressure and tachycardia lasting from 5 to 15 minutes.

Finally, a certain number of facts point to a relationship between bipolar depression, possibly reflecting a certain form of diencephalo-limbic system dysregulation, and the Kleine-Levin syndrome. Hypersomnia which abruptly gives way to a transitory manic state with hyperactivity and insomnia, is evocative of the sudden mood swings of certain bipolar depressives. Familial anamnesis is often noted in a relative suffering from major recurring or bipolar depression [6, 11, 13, 60]. Certain authors have found a link between the sleep patterns of subjects affected by the Kleine-Levin syndrome and those of subjects affected by mood disorders: reduced REM sleep latency, repeated night-time waking, reduced stages 3 and 4 of NREM sleep [51]. And finally, it is interesting to note that lithium appears to be active in preventing subsequent attacks in certain subjects.

Clearly, the evidence in favour of diencephalic functional dysregulation remains somewhat tenuous. This hypothesis has however been revived by the recent discovery of a hypocretin/orexin system, entirely in the lateral hypothalamus but projecting into different regions of the central nervous system. This system was already known to be involved in several functions, including eating habits, the equilibrium of energy and the neuroendocrine system [48] but we now know that it is at the origin of canine narcolepsy, this disorder being caused by a mutation of the Hcrtr hypocretin receptor 2 gene [41].

Pathogenesis

Although this is not yet known, there is some evidence in favour of an autoimmune origin. On the one hand is the frequency with which a viral or bacterial episode precedes the onset of the first episode (table 33.5). On the other hand is the publication of three anatomo-clinical observations of Kleine Levin syndrome favouring a localised encephalitis of viral origin. The first case in point is that of a 52 year old woman suffering from hypersomnia and megaphagia related to a brain tumour (not confirmed by postmortem examination) [66] in whom the autopsy showed lesions of an inflammatory type (perivascular lymphocytic cuffing) located in the hypothalamus, amygdala and the antero-median grey matter of the temporal lobes. The second is that of a 46 year old man whose first episode occurred at the age of 39 [9]. This man died suddenly nine days after he had chocked on meat while eating and a sausage was removed from his pharynx. The postmortem examination revealed inflammatory lesions of the same type located this time in the median and dorsal intralamminary thalamic nucleï and in the pulvinar. The third case is that of a 9 year old girl who died suddenly from a pulmonary embolism following iliofemoral thrombophlebitis, where the anatomopathological examination showed periventricular inflammatory infiltrates and nodular microglial proliferation localised in the diencephalon and the mesencephalon [15]. In addition, it has recently been suggested for the first time that the allele HLA DQB1*0201 can cause susceptibility to Kleine-Levin syndrome [13]. Indeed the extended haplotype DRB1*0301, DQA1*0501, DQB1*0201 is associated with many autoimmune disorders (Addison's, Grave's, celiac disease and Myasthenia gravis).

Table 33.2. Hormone secretion during sleep or during 24 hours in recurrent hypersomnia or the Kleine-Levin syndrome.

Authors	Hormones	Sleep disorder (n° subjects)	Period of sampling	Periodicity of sample	Mean night or 24h plasma level
Gilligan [19]	hGH	Kleine-Levin (1)	24 h	4 h	=
Kaneda *et al.* [30]	hGH	Recurrent hypersomnia (2)	night	30 min	-
Hishikawa *et al.* [28]	hGH	Recurrent hypersomnia (4)	night	30 min	abnormal
	Cortisol				=
Thomson *et al.* [68]	Cortisol	Kleine-Levin (1)	24 h	1 h	=
	Prolactin				=
	Melatonin				=
Gadoth *et al.* [18]	hGH	Kleine-Levin (1)	night	20 min	↑
	Prolactin				↑
	TSH				=
	FSH				=
Chesson *et al.* [10]	hGH	Kleine-Levin (1)	24 h	20 min	↓
	Cortisol				↑
	Prolactine				↑
	TSH				↑
Mayer *et al.* [44]	hGH	Kleine-Levin (5)	24 h	2 h	↓ (in 2)
	Melatonin				↑ (in 5)
	TSH				=
	Cortisol				=

Treatment

No controlled study has been published to date on treatments for Kleine-Levin syndrome and we are limited to case reports and personal experience. It is logical to use a stimulant like methylphenidate or an awakening drug such as modafinil during the attack of hypersomnia. Nevertheless, the results must be treated with caution to the extent that these treatments are usually started several days after the onset of the attack and thus not long before spontaneous remission, and the methods of evaluation are purely subjective. Neuroleptics are often used in the period before diagnosis is completed and where the symptoms are interpreted as manifestations of a psychotic state. There is no proof that these neuroleptics relieve the symptoms or affect the course of the episode.

Of greater interest is the prophylactic use of mood stabilizer. Positive results with the relief or disappearance of symptoms throughout the period of administration of these products and the recurrence of symptoms when treatment is stopped, have been reported in certain cases with the use of carbamazepine [46, 62, 73], valpromide and lithium carbonate [21, 26, 57] and valproic acid. In other cases, on the contrary, these products failed to prevent the attacks from recurring. These drugs should not be resorted to systematically, but in function of the frequency and severity of symptoms and their socio-professional repercussions. A last line of orientation is that of antiviral drugs of the amantadine type with which significant improvements have been claimed.

In the exceptional cases of periodic hypersomnia punctuated by menstruation, oestroprogestatives which inhibit ovulation seem to be consistently effective [5, 36, 41].

SYMPTOMATIC RECURRENT HYPERSOMNIAS

The Kleine-Levin syndrome cannot fully account for the range of recurrent hypersomnias and the possibility of symptomatic forms of recurrent hypersomnia must thus be considered.

Recurrent hypersomnia of organic origin

These hypersomnias are linked to certain brain tumours or various disturbances of the central nervous system.

The most typical cases are linked to tumours which develop inside the third ventricle. These tumours are pedicled, they do not spread to surrounding structures and consist of colloidal cysts [24, 25, 42], epidermoidal cysts or choroidal papilloma, responsible for paroxystic blockages to the drainage of cerebro-spinal fluid. The onset of sleep is generally preceded by sudden headaches, which may or may not be associated with vomiting and vague sensory disorders. There are no neurological indications of location, at least at the outset. Stasis can be observed at the back of the eyes. Sleep is more or less deep, but reversible. Tumours within the third ventricle are not solely responsible: other intraventricular tumours in the fourth ventricle, for example, may give rise to the same symptomatology, as may intraparenchymatous tumours affecting the walls of the posterior hypothalamus, particularly the pinealoma [17]. Diagnosis in all cases is based on medical imaging techniques.

As for recurrent hypersomnias occurring after encephalitis [22, 45, 61], cranial traumatism [72], cerebral vascular accidents [49, 63, 65], the closer these occur to the event, the more likely they are to be an encephalic effect. Their physiopatholgy is unknown and their treatment non codified.

Recurrent hypersomnia of psychiatric origin

Major recurring depression and bipolar depression. – These disorders are associated with hypersomnia in roughly 20% of cases [53]. Which is why they may present as a form of recurrent hypersomnia and why the aetiology must be established according to previous history and symptomatology. Far more evocative cases do however exist, which are very similar to the Kleine-Levin syndrome, with recurrent hypersomnia and eating and sexual behaviour disorder, corresponding to the typical bipolar depression context [29].

Neurotic recurrent hypersomnias. – The symptomatology is distinctly variable ranging from brief episodes lasting one or two days to episodes lasting several weeks, occurring at long intervals. Subjects say that sleeping is all they are capable of doing, but nevertheless attend to their personal hygiene and meals. Megaphagia may occur. These manifestations are more frequent among women. Neurotic traits are found on examination, with excessive emotionality, attention-seeking, dramatic behaviour indicating a histrionic personality. During the prolonged polygraphic examination, where subjects are free to lie down, sit at a table or in an armchair, they will often tend to stay in bed most of the time, but without manifesting daytime sleeping to any abnormal extent [4, 20].

CONCLUSION

Recurrent hypersomnias are rare disorders often diagnosed with delay. The Kleine-Levin syndrome is the most typical form. Its pathophysiology is still unclear. However recent data suggest an autoimmune aetiology. Moreover the discovery of the hypocretin system may shed further light on the concept.

REFERENCES

1. ANFIMOFF, J. - A. Quoted by Kaplinsky and Schulman, 1935
2. BARONTNI, F. ZAPPOLI R. - A case of Kleine-Levin Syndrome. Clinical and polygraphic study. *Proceedings of the Twentieth European Meeting on Electroencephalography*. Aulo Gaggi, Bologna, 239-245, 1967.
3. BILLIARD M., - Recurring hypersomnias. In: *Sleep 1978*, L. Popoviciu, B. Asgian, G. Badiu (eds), Karger, Basel, 233-239, 1980.
4. BILLIARD M., CADILHAC J. - Les hypersomnies récurrentes. *Rev. Neurol.* (Paris) 144, 249-258, 1980.
5. BILLIARD M., GUILLEMINAULT C., DEMENT W.C. - A menstruation-linked periodic hypersomnia. *Neurology*, 25, 436-443, 1975.
6. BONKALO A. - Hypersomnia: a discussion of psychiatric implication based on three cases. *Br. J. Psychiat.* 114, 69-75 1968.
7. BUCKING P.H., PALMER W.R. - New contribution to the clinical aspects and pathophysiology of the Kleine-Levin syndrome. *Munch. Med. Wochenschr.*, 120, 1571-1572, 1978.
8. BURTON S.A. - Recovery from sleep attacks in Kleine-Levin syndrome may involve a process of gradual phase advance. *Sleep Res.*, 18, 209, 1989.
9. CARPENTER S., YASSA R., OCHS R. - A pathological basis for Kleine-Levin syndrome. *Arch. Neurol.* 39, 25-28, 1982.

10. CHESSON A.L., LEVINE S.N. - Neurendocrine evaluation in Kleine-Levin syndrome: evidence of reduced dopaminergic tone during periods of hypersomnolence. *Sleep,* 14, 226-232, 1991.

11. CRITCHLEY M. - Periodic hypersomnia and megaphagia in adolescent males. Brain, 85, 627-656, 1962.

12. CRITCHLEY M., HOFFMAN H.L. - The syndrome of periodic somnolence and morbid hunger (Kleine-Levin syndrome). *B.M.J.* 1, 137-139, 1942.

13. DAUVILLIERS Y., MAYER G., LECENDREUX M., NEIDHART E., CHAPPUIS R., BILLIARD M., TAFTI M. - Possible association of polymorphisms in the HLA and monoaminergic systems in Kleine-Levin syndrome. *Sleep 23,* Abstract Supplement 2, 2000. A 63-64.

14. DUFFY J.P., DAVISON K. - A female case of Kleine-Levin syndrome. *Brit. J. Psychiat.,* 114, 77-84, 1968.

15. FENZI F., SIMONATI A., CROSATO F., GHERSINI L;, RIZZUTO N. - Clinical features of Kleine-Levin syndrome with lithium. *Can. J. Psychiatry.* 28, 491-493, 1983.

16. FERNANDEZ J.M., LARA I., GILA L., O'NEILL OF TYRONE A., TOVAR J., GIMENO A. - Disturbed hypothalamic-pituitary axis in idiopathic recurring hypersomnia syndrome. *Acta. Neurol. Scand.* 82, 361-363, 1990.

17. GABRIEL P. - *Les pinéalomes* (étude anatomo-clinique) 1 vol. 252p., Maloine, Paris, 1935.

18. GADOTH N., DICKERMAN Z., BECHAR M., LARON Z., LAVIE P. - Episodic hormone secretion during sleep in Kleine-Levin syndrome: evidence for hypothalamus dysfunction. *Brain Dev.* 9, 309-315, 1987.

19. GILLIGAN B.S. - Periodic megaphagia and hypersomnia. An example of the Kleine-Levin syndrome in an adolescent girl. *Proc. Austr. Assoc. Neurol.,* 9, 67-72, 1973.

20. GODENNE G.D. - Report of a case of recurring hysterical pseudostupor. *J. Nerv. Mental Dis.,* 141, 670-677, 1966.

21. GOLDBERG M.A. - The treatment of Kleine-Levin syndrome with lithium. *Can. J. Psychiatry.* 28, 491-493, 1983.

22. GORDON A. - Encéphalite léthargique des centres végétatifs. Syndrome de somnolence périodique avec polyphagie et polydypsie. *Rev. Neurol.* (Paris) 71, 411-416, 1939.

23. GRAN D., BEGEMANN H. - Neue Beobachtungen bei einem Fall von Kleine-Levin -Syndrom. *Munch. Med. Wochenschr.,* 115, 1098-1102, 1973.

24. GROSSIORD A. - *Le kyste colloïde du 3ème ventricule.* Thesis, Paris, 1941.

25. GUILLAIN G., BERTRAND I., PERISSON. - Etude anatomo-clinique d'une tumeur du 3ème ventricule. *Rev. Neurol.* (Paris), 41, 467-473, 1925.

26. HART E.J. - Kleine-Levin syndrome: normal CSF monoamines and response to lithium therapy. *Neurology,* 35, 1395-1396, 1985.

27. HEGARTY A., MERRIAM A.E. - Autonomic events in Kleine-Levin syndrome. *Am. J. Psychiatry,* 147, 951-952, 1990.

28. HISHIKAWA Y., IIJIMA S., TASHIRO T., SUGITA Y., TESHITA Y., MATSUO R., KANEDA H. - Polysomnographic findings and growth hormone secretion in patients with periodic hypersomnia. In: *Sleep 90,* W. P. Koella (ed.), Karger, Basel, 128-133, 1981.

29. JEFFRIES J.J., LEFEBVRE A. - Depression and mania associated with Kleine-Levin-Critchley syndrome. *Can Psychiatr. Assoc. J.* 18, 439-444, 1973.

30. KANEDA H., SUGITA Y., MASAOKA S., IIJIMA S., TANAKA K., WAKAMATSU H., HISHIKAWA Y. - Red blood cell concentration and growth hormone release in periodic hypersomnia. *Waking and Sleeping.* I, 369-374, 1977.

31. KAPLINSKY M.S., SCHULMANN E.D. - Uber die periodische Schlafsucht (I; Mitteilung), *Acta. Med. Scand.,* 85, 107-128, 1935.

32. KLEINE W. - Periodische Schlafsucht. *Mschr. Psychiat. Neurol.,* 57, 285-320, 1925.

33. KOERBER R.K., TORKELSON R., HAVEN G., DONALDSON J., COHEN S.M., CASE M., Increased cerebrospinal fluid 5-hydroxytryptmine and 5-hydroxyindoleacetic acid in Kleine-Levin syndrome. *Neurology,* 34, 1597-1600, 1984.

34. LAVIE P., GADOTH N., GORDON C.R., GOLDHAMMER G., BECHAR M. - Sleep patterns in Kleine-Levin syndrome. *Electroenceph. Clin. Neurophysiol.,* 47, 369-371, 1979.

35. LAVIE P., KLEIN E., GADOTH N., BENTAL E., ZOMER J., BECHAR M., WASJBORT J. - Further observations on sleep abnormalities in Kleine-Levin syndrome: abnormal breathing pattern during sleep. *Electroenceph. Clin. Neurophysiol.,* 52, 98-101, 1981.

36. LEVIN M. - Narcolepsy (Gelineau's syndrome) and other varieties of morbid somnolence. *Arch. Neurol. Psychiatr.,* (Chicago) 22, 1172-1200, 1929.

37. LEVIN M. - Periodic somnolence and morbid hunger: a new syndrome. Brain, 59, 494-504, 1936.

38. LEWIS N.D.C. - The psychoanalytic approach to the problem of children under twelve years of age. *Psychoanalytic Review,* 13, 424-443, 1926.

39. LHERMITTE J., DUBOIS E. - Crises d'hypersomnie prolongée rythmée par les règles chez une jeune fille. *Rev. Neurol.* (Paris) 73, 608-609, 1941.

40. LHERMITTE J., HECAEN H., BINEAU L., Un nouveau cas d'hypersomnie prolongée rythmée par les règles. *Rev. Neurol.,* (Paris) 75, 299, 1943.

41. LIN L., FARACO J., LI R., KADOTANI H., ROGERS W., LIN X., QUI X., DE JONG P.J., NISHINO S., MIGNOT E. - The sleep disorder canine narcolepsy is caused by a mutation in the hypocretin (orexin) receptor 2 gene. *Cell,* 98, 365-376, 1999.

42. LINDON L.C.A. - Cystic tumours of the third ventricle. *Med. J. Aust.,* 2, 122-124, 1938.

43. LIVREA P., PUCA F.M., BARNABA A., DIREDA L. - Abnormal central monoamine metabolism in humans with "true hypersomnia" and "sub-wakefulness", *Eur. Neurol.,* 15, 71-76, 1977.

44. MAYER G., LEONHARD E., KRIEG J., MEIER-EWERT K. - Endocrinological and polysomnographic findings in Kleine-Levin syndrome. No evidence for hypothalamic and circadian dysfunction. *Sleep,* 21, 278-284, 1998.

45. MERRIAM A.E. - Kleine-Levin syndrome following acute viral encephalitis. *Biol. Psychiatry,* 21, 1301-1304, 1986.

46. MUKADDES N.H., KORA M.E., BILGE S. - Carbamezepine for Kleine-Levin syndrome. *J. Am. Acad. Child. Adolesc. Psychiatry,* 38, 791-792, 1999.

47. PAPY J.J., CONTE-DEVOIX B., SORMANI J., PORTO R., GUILLAUME V. - Syndrome d'hypersomnie périodique avec mégaphagie chez une jeune femme, rythmé par le cycle menstruel. *Rev. E.E.G. Neurophysiol. Clin.* 12, 54-61, 1982.

48. PEYRON C., TIGHE D., VAN DEN POL A., DE LECEA L., HELLER H., SUTCLIFFE J., KILDUFF T. - Neurons containing hypocretin (orexin) project to multiple neuronal systems. *J. Neurosci.* 18, 9996-10015, 1998.

49. POHL O., HAIRS G. - Hypomanic psychosis alternating with periodical hypersomnia. *Psychiat. Clin.,* 1, 120-124, 1968.

50. POPOVICIU L., CORFARIU O. - Etude clinique et polygraphique au cours du nycthémère d'un cas de syndrome Kleine-Levin-Critchley. *Rev. Roum. Neurol.,* 9, 221-228, 1972.

51. REYNOLDS C.F., BLACK R.S., COBLE P.A., HOLZER B., KUPFER D.J. - Similarities in EEG sleep findings for Kleine-Levin syndrome and unipolar depression. *Am. J. Psychiatry,* 137, 116-118, 1980.

52. REYNOLDS C.F., KUPFER D.J., CHRISTIANSEN C.L., AUCHENBACH R.C., BRENNER R.P., SEWITCH D.E., TASKA L.S., COBLE P.A. - Multiple sleep latency test findings in Kleine-Levin syndrome. *J. Nerv.Ment. Dis.,* 172, 41-44, 1984.

53. REYNOLDS C.F., SHUBIN R.S., COBLE P.A., KUPFER D.J. - Diagnosis classification of sleep disorders: implications for psychiatric practice. *J. Clin. Psychiatry,* 41, 296-299, 1981.

54. ROBINSON J.T., MCQUILLAN J. - Schizophrenic reaction associated with the Kleine-Levin syndrome. *Medicine,* Baltimore, 96, 377-381, 1951.

55. ROTH B. - Functional hypersomnia. In: *Narcolepsy,* C. Guilleminault, W.C. Dement, P. Passouant (eds.), Spectrum Publ., New York, 333-349, 1976.

56. ROTH B., NEVSIMALOVA S. - The clinical picture of periodic hypersomnia: a study of 38 personally observed cases. In: *Sleep 1980,* W.P. Koella (ed.), Karger, Basel, 120-124, 1981.

57. ROTH B., SMOLIK P., SOUCEK K. - Kleine-Levin syndrome. Lithoprophylaxis. *Ceskolovenska Psychiatrie,* 76, 156-162, 1980.

58. RUSSEL J., GRUNSTEIN R. - Kleine-Levin syndrome: a case report. *Aust. N.Z.J. Psychiatry,* 26, 119-123, 1992.

59. SACHS C., PERSSON H.E., HAGENFELDT K. - Menstuation-related periodic hypersomnia: a case study with successful treatment. *Neurology,* 32, 1376-1379, 1982.

60. SAGRIPANTI P. - Sull'associazione di ipersomnia e bulimia (syndrome di Kleine-Levin). *Il Cervello,* 194, 194-205, 1952.

61. SALTERS M.S., WHITE P.D. - A variant of Kleine-Levin precipitated by both Epstein-Barr and Varicella-Zoster virus infections. *Biol. Psychiatry,* 33, 388-390, 1993.

62. SAVET J.F., ROBERT H., ANGELI C. - Un cas de syndrome de Kleine-Levin stabilisé depuis plus d'un an sous carbamazépine. *Presse Med.* 15, 281, 1986.

63. SEMIONOWA-TIAN-SHANSKAYA V.V. - Periodic pathological sleep. *Zh. Nevropat. Psikhiat.,* 66, 3-9, 1966.

64. SMIRNE S., CASTELOTTI V., PASSERINI D. – EEG study in a case of Kleine-Levin syndrome. *Revista di Neurologia,* 40, 357-365, 1970.

65. TAKAHASHI Y. - Clinical studies of periodic somnolence. Analysis of 28 personal cases. *Psychiat. Neurol. Jap.,* 67, 853-889, 1965.

66. TAKRANI L.B., CRONIN D. - Kleine-Levin syndrome in a female patient. *Can. Psychiatry Assoc. J.,* 21, 315-318, 1976.

67. THACORE V.R., AHMED M., OSWALDI I. - The EEG in a case of periodic hypersomnia. *EEG Clin. Neurophysiol.,* 27, 605-606, 1969.

68. THOMPSON C., OBRECHT R., FRANEY C., ARENDT J., CHECKLEY S.A. - Neuroendocrine rhythms in a patient with the Kleine-Levin syndrome. *Br. J. Psychiatry,* 147, 440-443, 1985.

69. DE VILLARD R., DELIRY J., MOURET J., MAILLET J., CHARRAT A. - Le syndrome de Kleine-Levin: à propos de 4 cas. *Lyon Médical,* 244, 389-394, 1980.

70. VOLLMER R., TOIFL K., KOTHBAUER P., RIEDERER P. - EEG und biochemische Befunde beim Kleine-Levin-Syndrom. *Nervenarzt.,* 52, 211-218, 1981.

71. WILKUS R.J., CHILES J.A. - Electrophysiological changes during episodes of the Kleine-Levin syndrome. *J. Neurol. Neurosurg. and Psychiat.* 38, 1225-1231, 1975.

72. WILL R.G., YOUNG J.P., THOMAS D.J. - Kleine-Levin syndrome: report of two cases with onset of symptoms precipitated by head traumatism. *Br. J. Psychiatry,* 152, 410-412, 1988.

73. WURTHMANN C., HARTUNG H.P., DENGLER W., GERHARDT P. - Kleine-Levin syndrome. The provocation of manic symptoms by an antidepressant and a therapeutic trial of carbamazepine. *Dtsch. Med. Wochenschr.* 114, 1528-1531, 1989.

74. WYSS R., Psychotische Episoden bei einem Fall mit Kleine-Levin-Syndrom. *Schweiz. Arch. Neurol. Psychiatr.,* 1, 01, 203, 1968.

Chapter 34

Other hypersomnias

M. Billiard and B. Carlander
Service de Neurologie B, Hôpital Gui de Chauliac, Montpellier, France

In addition to the excessive sleepiness associated with insufficient sleep, psychotropic medication or alcohol intake, sleep-related breathing disorders, primary hypersomnia, narcolepsy, idiopathic hypersomnia and recurrent hypersomnia, it is important to be able to recognise excessive daytime sleepiness in association with a number of organic and psychological illnesses. Although excessive sleepiness may sometimes be the presenting feature, it is far more likely to be recognised as such during the course of the illness, either when interviewing a member of the family or as a result of observing the patient in hospital.

NEUROLOGICAL CAUSES

Excessive sleepiness may develop in the case of a brain tumour, after a stroke, in degenerative diseases and various other illnesses.

Brain tumour

Clinically, sleepiness is continuous, of variable severity, interspersed with brief arousals whether spontaneous or provoked [90]. Sleepiness may occur in any intracranial hypertension syndrome, but more rarely results from tumours of the diencephalon or peduncular region, with no associated intracranial hypertension [20]. These tumours particularly affect the posterior hypothalamus, glioma, harmatoma; posterior and superior suprasellar craniopharyngioma compressing the floor of the third ventricle; the pineal region, pinealoma, teratoma, affecting the posterior part of the third ventricle; they include tumours which exert a valve action in the third ventricle, colloid cysts in particular, which are responsible for recurrent hypersomnia, and peduncular tumours, even if the latter are usually at the origin of more serious disorders of wakefulness, obnubilation or stupor. A number of cases of narcolepsy symptomatic of brain tumour have been reported, involving sleep episodes, cataplexy and sometimes sleep onset REM periods with associated HLA DR2, DQ1 [1, 3, 12, 49, 62, 71, 82, 87]. In most cases the tumours affect the hypothalamus or midbrain region. Authors have speculated on whether these observations correspond to concomitant manifestations or the effects of the tumours themselves [5, 17]. The predilection of these tumours for the regions of the encephalon controlling wakefulness and sleep is highly evocative. Various questions nevertheless remain unanswered i.e. why these tumours give rise to recurrent sleep episodes rather than to continuous daytime sleepiness? Why cases of narcolepsy symptomatic of tumours are more common than cases of narcolepsy symptomatic of strokes? A particular case is that of daytime sleepiness following preventive cranial radiation, mainly in acutely leukaemic children [53]. The case illustrated in figure 34.1 associates angiomatose lesions of the central grey nuclei, dilation of the third ventricle and possible sequelae resulting from radiotherapy.

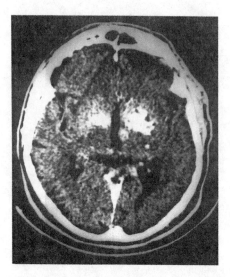

Figure 34.1. A 43 year old man, with encephalic and retinal angiomatosis (Wyburn-Mason syndrome), treated with radiation in 1955, surgery in 1968, then embolisation in 1978; excessive daytime sleepiness appeared several years later; sequelae calcification of the grey nuclei and dilation of the third ventricle on the brain scan.

Stroke

Excessive daytime sleepiness is often a transient state between confusional states, agitation or even coma, marking the initial period of the stroke and a later state of psycho-motor slowing, abulia or akinetic mutism.

Paramedian uni- or bithalamic infarct

The extent of sleepiness varies. The accompanying neurologic characteristics may include vertical paresis, a "skew deviation", paresis of the third cranial pair, dysarthria, and instability in walking. Sleepiness disappears at variable rates. It may give way to a state of apathy, psycho-motor slowing, problems of alertness, and amnesia qualified as "thalamic dementia". Anatomical studies [21] and MRI studies [8, 75] agree in showing that the sites affected in most cases are the inferior part of the dorso-medial nucleus (DM) and in 50 to 75% of cases the antero-medial part of the centro-medial nucleus (CM). These infarcts affect the territories of the paramedian thalamic-subthalamic arteries issuing from the basilar communicating artery. Polysomnographic studies have shown an increase in stage 1 and a reduction in the other stages of NREM sleep, while REM sleep remains unaffected [7].

Paramedian pedunculo-thalamic infarct

Daytime sleepiness may be light, intermittent and easily reversible, but is often deep, approaching obnubilation or stupor. Neurological characteristics include altered ocular motility due to paresis of the third cranial pair, which is partial or total, uni- or bilateral, or due to paresis of the sixth pair [21, 37]. In a series of 19 patients presenting pedunculo-thalamic infarcts, 17 of whom had survived for a long enough period to enable good clinical study, Castaigne *et al*. [21, 37] reported excessive sleepiness in 10 patients. Some patients later developed abnormal movements and memory disorders. The affected vessels included the pedunculo-thalamic arteries and the paramedian peduncular arteries extending from the basilar communicating artery. Schott *et al*. [79] recorded a patient affected by daytime sleepiness associated with akinetic mutism provoked by tegmental thalamic "butterfly wing" softening. Total sleep time over 24 hours was distinctly increased (15 hours 7 minutes) but stages 3 and 4 of NREM sleep and REM sleep were absent.

Tegmental infarct

Those affecting the pontine tegmentum and the reticular formation of this region also result in sleepiness [20, 59]. As at midbrain level, sleepiness may be associated with an alternating state of akinetic mutism [28], but it may be relatively simple [20]. These observations underline the importance of the reticular formation of the pedunculopontine junction. Unlike these infarcts of the upper part of the protuberance, those located at the lower level do not produce sleepiness. Observations of narcolepsy symptomatic of infarcts are less common than those dependent on brain tumours. Two observations have been published, one which refers to a likely tegmental infarct [74], the other linked to an arteriovenous malformation sited in the third ventricle [26].

Degenerative diseases

Excessive daytime sleepiness is common in subjects affected by **Alzheimer type dementia**. This is accounted for in three main ways: sundowning syndrome which affects roughly ¼ of subjects with dementia [14-16]. In this case, night wandering is more or less associated with sleeptalking, cries, dressing and undressing behaviour, breaking objects, incontinence and attempts to escape. Subjects affected by this syndrome are often less violent during the day, partly due to their excessive daytime sleepiness. A second cause relates to psychotropic medication which is sometimes over prescribed. A third cause is associated sleep apnoea syndrome, in which it is still not known whether subjects affected by Alzheimer's are more frequently affected than are normal subjects [22].

Until recently, insomnia was considered to be the characteristic sleep disorder in subjects with **Parkinson's disease**. However recent publications have raised doubts about this issue, showing that a non negligible fraction of patients with Parkinson's present excessive daytime sleepiness [77, 88]. Moreover it is now common to observe sudden irresistible sleep episodes in Parkinsonians taking dopaminergic agonist treatment [38, 60], pramipexole or ropinirole, and for these to disappear when this type of treatment is withdrawn.

Multiple-systems atrophy is a degenerative disease of the central nervous system and of the autonomous nervous system with two distinct clinical and neuropathological forms: striatonigral degeneration (multiple-systems atrophy of the Parkinson type) and sporadic olivopontocerebellar atrophy (multiple-systems atrophy of the cerebellar type). The dominant sleep disorders in this patient group are sleep-related breathing impairments with repeated arousals and hypoxemia resulting in excessive daytime sleepiness [55]. Autret *et al.* [4] have reported observing a patient who was probably affected by striatonigral degeneration presenting recurrent daily sleep episodes, cataplexy and sleep onset REM periods typical of narcolepsy.

Other neurological causes

Excessive daytime sleepiness is not a common symptom in subjects affected by **multiple sclerosis**. However cases associating multiple sclerosis and narcolepsy have been reported [10, 35, 47, 70, 78, 80, 94]. Several possible explanations may account for this association: hypothalamic demyelination plates; the fortuitous presence of two unexceptional conditions; narcolepsy triggered by a non specific inflammatory reaction associated with the multiple sclerosis [5].

Daytime sleepiness is common in **Gayet-Wernick's encephalopathy** no doubt in relation to the petechiae surrounding the third and fourth ventricles and aqueduct of Sylvius.

Normal pressure hydrocephalus is a classic cause of excessive daytime sleepiness. This may in fact relate to psycho-motor slowing.

Arnold-Chiari malformation whether or not this is associated with syringomyelia, is a recognised cause of sleep apnoea syndrome and daytime sleepiness, no doubt via the effect on the neurons involved in controlling the upper airway muscles [6, 32, 36, 72, 83].

Any **neuromuscular disease,** whether motoneuron diseases, motor neuropathy, damaged neuromuscular junctions, or muscular disease, is likely to be accompanied by sleep-related breathing impairment resulting in excessive daytime sleepiness. A case in point is myotonic dystrophy in which daytime sleepiness is common [56, 61, 69, 89]. The origin is not unequivocal. On one hand polysomnographic studies [23, 27, 41,44, 52] have shown that these subjects are frequently affected by sleep apnoea syndrome and that direct sleep onset REM period was often observed [64]. On the other hand, clear anomalies have been shown in the structures involved in controlling sleep and wakefulness i.e. the dorso-medial thalamic nucleus [29], dorsal raphe and central superior nuclei [61].

Head injury

Excessive daytime sleepiness appearing during the year following a head injury may be considered *a priori* as post-traumatic. This typically presents as extended night sleep and episodes of daytime sleep. Sleepiness is usually associated with other characteristics such as headaches, concentration difficulties or memory disorder. Radiography may reveal several possibilities: lesions affecting the hypothalamic region or brainstem, midbrain or pontine tegmentum, typically but rarely, hydrocephalia; or more often than not, the absence of any significant lesions. Polygrpaphically, sleepiness should be objectified by a multiple sleep latency test but is usually is not, pointing to a post-traumatic cerebral syndrome. A particular case after head or brain trauma is that of sleep apnoea syndrome [42]. Post-traumatic narcolepsy has been reported [17, 39, 51, 54]. The more recent publication is remarkable for the number of cases collected i.e. 9, of which 5 had cataplexy; polysomnographic recordings were carried out in all 9 subjects, with at least two latencies to sleep onset REM periods in 8 of the subjects and a single one in the case of the 9th subject; HLA typing was carried out in 5 subjects, 3 of which were HLA DR2 and one HLA DQ1. These were all mild to moderate head injuries involving various sites of impact. Moreover, neurological tests were normal in all these cases and neuroimaging, CT or MRI showed no significant lesions. The findings suggested that either the injury acted as a trigger or that a simple coincidence was at work in subjects who were susceptible to narcolepsy.

PSYCHIATRIC CAUSES

Hypersomnia associated with depression

Hypersomnia is often referred to in the literature on depressive patients [24, 31, 57, 86]. But the objective studies carried out so far [13, 45, 50, 84] have shown that while some subjects do indeed present an increase in total sleep time during the day and night, most spent considerable time in bed, not only at night but during the day (clinophilia) with no significant increase in their total sleep time. Moreover multiple sleep latency tests carried out on subjects with bipolar depression [58] or dysthymia [33] and who complained of excessive daytime sleepiness, showed no reduction in mean latency to sleep onset, indicating that these subjects suffer more from a lack of interest in life, a lack of energy, and withdrawal, than from objective sleepiness.

This disorder corresponds to hypersomnia linked to another DSMIV [2] mental disorder the main criteria of which are the following:
- The main complaint, lasting at least one month, is excessive sleepiness as witnessed by prolonged episodes of sleep or episodes of daytime sleep occurring almost daily.
- Excessive daytime sleepiness causes marked suffering or an impairment of social or professional functioning or in other important domains.
- Hypersomnia is linked to another class I or II disorder (e.g. major depression, dysthymic disorder) but is sufficiently severe in its own right to warrant clinical examination.
- The disturbance is not due to any other sleep disorder (e.g. narcolepsy, breathing related sleep disorder, parasomnia) or to an insufficient amount of sleep.
- The disturbance is not linked to the direct physiological effects of any substance (e.g. the abusive use of any substance or medication) or to the general medical condition.

Seasonal Affective Disorder (SAD) [76]

The seasonal specification applies to modalities in the course of major episodes of depression in bipolar disorder I, bipolar disorder II or major recurrent depression. The main characteristics are the occurrence and remission of episodes of depression at certain periods of the year. In most cases the episodes of depression begin in the autumn or winter and end toward the Spring equinox. Seasonal episodes of depression are often characterised by anergy, hypersomnia, hyperphagia, weight gain and a craving for carbohydrates.

INFECTIOUS CAUSES

Infections caused by Epstein-Barr and other viruses

In the aftermath of infectious mononucleosis, the subject feels intense asthenia inevitably leading to the lengthening of his total sleep time, difficulty awakening and excessive daytime sleepiness evoking idiopathic hypersomnia [43]. Polysomnographically, total sleep time increases and mean latency to sleep onset decreases in the multiple sleep latency test. Hypersomnia tends to go into gradual remission after several months or years. This type of hypersomnia develops following viral pneumopathy, hepatitis B viral infection, and the Guillain-Barré syndrome, probably through the same mechanism.

Encephalitis caused by virus

Disorders of wakefulness and/or consciousness are found in virtually all patients affected by viral encephalitis, with fever in 70 to 90 % of cases, headaches in three quarters of cases and personality changes in over 50% of cases [91]. In the absence of polysomnographic studies it is very difficult to define the border between wakefulness disorders and disorders affecting consciousness. Two nosological contexts are worthy of mention:

The arboviruses represent a heterogeneous group of viruses whose common characteristic is that they are transmitted by a haematophagous arthropod vector. The various arboviruses share the same first symptoms, evoking a fairly severe state of flu with high fever, headache and myalgia. Encephalitic signs then develop, which vary according to the agent responsible. Sleepiness is a fairly characteristic feature of European tick-borne encephalitis. The virus is most commonly found in central and Eastern Europe. In France, the disease is found in Alsace and the Mediterranean region.

First appearing in Europe in 1917, **epidemic encephalitis** or Von Economo-Cruchet's disease [92] affected tens of thousands of subjects in the ten years which followed. Its cause has never been fully identified even if the anatomopathological lesions, and inflamatory sites located mainly in the grey matter of the diencephalon and basal ganglia, strongly suggest a viral infection. Two clinical varieties have been described: the lethargic form, which is the most common, consisting of a febrile flu-like condition, rapidly complicated by sleepiness culminating in a permanent state of sleep, stupor and coma, associated with external ophthalmoplegia and frequent oculogyric crises with nystagmus. The rarer, hyperkinetic form, variously associated major motor instability, ticks, myoclonia, choreic movements, visual hallucinations and insomnia. These two clinical pictures correspond to lesions in different sites: the posterior hypothalamus and midbrain tegmentum in the lethargic form, and the anterior hypothalamus and telencephalon in the hyperkinetic form. Sporadic cases are still exceptionally reported whose description varies slightly from the original. In the four cases reported by Howard and Lees [48], two lethargic-ophthalmoplegic forms and two hyperkinetic forms, the serologic and culture tests showed no infectious agents but oligoclonal IgG bands were present in the cerebrospinal fluid of three patients during the acute phase of the disease.

Human immunodeficiency virus (HIV)

Subjects infected by HIV sometimes complain of excessive daytime sleepiness. A questionnaire on fatigue and sleep disorders was distributed to 14 subjects in stages B and C of the infection (symptomatic), 44 subjects in stage A (asymptomatic or presenting only lymphadenopathy) and 50 HIV seronegative male homosexual subjects [30]. Over half the subjects in stages B and C complained of fatigue as opposed to only 10% of controls, over half also reported frequent episodes of daytime sleepiness against only 12% of controls and 1/3 only felt alert in the morning against the majority of control subjects.

African Trypanosomiasis

African trypanosomiasis (sleeping sickness) is a subacute or chronic parasitic disease caused by the inoculation of a protozoon Trypanosoma brucei transmitted by the tse-tse fly. It is endemic to certain regions of tropical Africa, between the 15th north parallel and the 29th south parallel [34]. It affects roughly 10,000 subjects annually.

The form found in West and Central Africa is due to **Trypanosoma gambiense**, hosted by the Glossina palpalis fly. This produces a slow form of the disease. Several hours after the sting, a trypanoma forms, hardening into a rash around the inoculation. After a phase of incubation varying from several days to several months, the invasion of the blood and tissue constitutes the haemolymphatic phase with successive peaks of fever, fatigue, arthralgia, myalgia and adenopathies. The invasion of the nervous system is characterised by meningoencephalitis with headaches, trembling, dyskenesia, choreoathetosis, mood changes, sleep disorders and epileptic seizures. Patients at first present a disorganised sleep-wake rhythm, with sleep episodes randomly spread over 24 hours, followed by a state of permanent obnubilation leading to coma. The course of the illness lasts from 1 to 3 months.

The East African form is due to **Trypanosoma rhodesiense**, transmitted by the Glossina fuscipes fly. The course of the illness is much more rapid, leading to death within a few months, before any sleep disorders have had time to appear.

Sleep alterations in the West and Central African form have been characterised using polygraphic sleep recordings at night [11], in the afternoon [81] and over 24 hours [18-19]. These tests have revealed alterations in circadian sleep rhythm with sleep/wake episodes of shorter and shorter duration occurring randomly day and night and sleep onset REM periods (fig. 34.2). At the advanced stage the different stages of NREM sleep are no longer distinguishable from one another.

Diagnosis of the illness relies on the subject having travelled to the endemic zone and the presence of an intense inflammatory syndrome in the blood and cerebrospinal fluid (major increase of IgM). It is confirmed, with difficulty, by evidence of trypanosoma in the blood, cerebrospinal fluid or ganglia, by a centrifugate search, or by guinea pig inoculation. It is possible to obtain a specific cerebrospinal fluid immunodiagnosis.

From the pathophysiological point of view, a disrupted expression of the c-fos gene has been demonstrated in the suprachiasmatic nucleus [9] and prostaglandine (PG) D_2 whose hypnotic effects are known [46], is increased in LCR in the advanced stage of the illness [67].

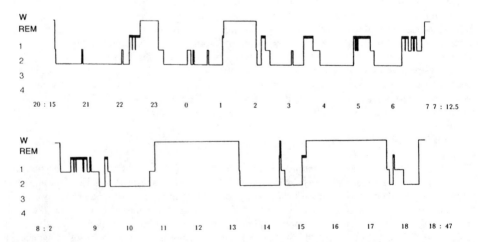

Figure 34.2. Polysomnography (day and night) of a 26 year old Cameroonian suffering from African trypanosomiasis: absence of deep NREM sleep; 295 minutes of daytime sleep, with a sleep onset REM period in the first of three episodes.

METABOLIC AND ENDOCRINAL CAUSES

Post-prandial sleepiness is common in untreated **diabetes**. A malaise, with reduced mental activity, and gradual sleepiness possibly leading to a state of obnubilation or coma, may signify subacute, **hypoglycemia** which is either organic (Langerhans' adenoma, hypophysial tumour) or functional (prediabetic state, gastrectomy).

Hepatic encephalopathy is initially accompanied by abnormal sleepiness which may regress under treatment or alternatively, develop into a state of obnubilation or coma. Patients with **renal insufficiency** subjected to chronic haemodialysis sometimes report excessive sleepiness [65, 66, 93].

Hypothyroidism is accompanied by fatigue, and mental and physical slowing down. If sleepiness is present it should prompt the search for an associated sleep apnoea syndrome [63, 73, 85]. It regresses in parallel with apnoeas under the influence of substitutive opotherapy.

Acromegalia may also be accompanied by sleep apnoea syndrome [68].

Finally, **Prader-Willi syndrome** characterised by hypotonia, mental retardation, obesity, hypogonadism and height-weight retardation, is usually accompanied by excessive daytime sleepiness [25, 40]. Episodes evoking cataplexy are sometimes reported.

CONCLUSION

In comparison with other hypersomnias, hypersomnias associated with medical or surgical disorders stand apart. Indeed hypersomnia in these cases is not the predominant symptom or at least trivial compared with the other symptoms of the causal disease and its prognosis. However these conditions are of definite interest in trying to disentangle the various possible origins of hypersomnia.

REFERENCES

1. ALDRICH M.S., NAYLOR M.W. – Narcolepsy associated with lesions of the diencephalon. *Neurology* 39, 1505-1508, 1989.
2. AMERICAN PSYCHIATRIC ASSOCIATION – *Diagnostic and statistical manual of mental disorders* (4th edn). American Psychiatric Association, Washington D.C., 1994.
3. ANDERSON M., SALMON M.V. – Symptomatic cataplexy. *J. Neurol. Neurosurg. Psychiatry* 40, 186-191,1977.
4. AUTRET A., LUCAS B., CHATEL M., DEGIOVANNI E. - Cataplexy, sleep disorganization and apnea in a probable striatonigral degeneration. *In: Sleep 1986*, W.P.Koella, F. Obal, H. Schulz, P. Visser (eds). Gustav Fischer Verlag, Stuttgart, 402-404, 1988.
5. AUTRET A., LUCAS B., HENRY-LEBRAS F., DE TOFFOL B. - Symptomatic narcolepsies. *Sleep*, 17, S21-S24, 1994.
6. BALK R.A., HIMLLER F.C., LUCAS E.A., SCRIMA L., WILSON F.J., WOOTEN V. - Sleep apnea and the Arnold-Chiari malformation. *Am. Rev. Respir. Dis.* 132, 929-930, 1985.
7. BASSETTI C., ALDRICH M., CHERVIN R., QUINT D. - Sleep apnea in patients with transient ischemic attack and stroke: a prospective study of 59 patients. *Neurology* 47,1167-1173, 1996.
8. BASSETTI C., MATHIS J., GUGGER M., LÖBLAD K.O., HESS C.W. - Hypersomnia following paramedian thalamic stroke: a report of 12 patients. *Ann. Neurol.* 39,471-480, 1990.
9. BENTIVOGLIO M., GRASSI-ZUCCONI G., OLSSON T., KRISTENSSON K. - Trypanosoma brucei and the nervous system. *Trends Neurosci.* 17, 325-329, 1994.
10. BERG O., HANLEY J. - Narcolepsy in two cases of multiple sclerosis. *Acta Neurol. Scand.* 39, 252-257, 1963.
11. BERT J., COLLOMB H., FRESSY J., GASTAUT H. - Etude électro-encéphalographique du sommeil nocturne au cours de la Trypanosomiase humaine africaine. *In: Etudes électro-encéphalographiques*, Masson, Paris. 334-352, 1965.
12. BILLIARD M., CARLANDER B., LE GALLOU A., CADILHAC J. - Narcolepsy associated with neurofibromatosis (Von Recklinghausen's Disease). *Sleep Res.* 19: 315, 1990.
13. BILLIARD M., DOLENC L., ALDAZ C., ONDZE B., BESSET A. - Hypersomnia associated with mood disorders: a new perspective. *Journal of Psychosomatic Research* 38 (Suppl.1), 41-47, 1994.
14. BLIWISE D.L. - Dementia. *In:* Kryger M.H., Roth T., Dement W.C. (eds). *Principles and Practice of Sleep Medicine*, ed. 2. WB Saunders, Philadelphia. 790-800, 1994.
15. BLIWISE D.L., WATTS R.L., WATTS N., RYE D.B., IRBE D., HUGUES M. - Disruptive nocturnal behavior in Parkinson's disease and Alzheimer's disease. *J. Geriatr. Psychiatry Neurol.* 8, 107-110, 1995.
16. BLIWISE D.L, YESAVAGE J.A., TINKLENBERG J.R. - Sundowning and rate of decline in mental function in Alzheimer's disease. *Dementia* 3, 335-341, 1992.

17. BONDUELLE M., DEGOS C.F. - Symptomatic narcolepsies: a critical study. *In:* Guilleminault C., Dement W.C., Passouant P. (eds) *Narcolepsy*, Spectrum Publication, New-York, 313-332,1976.
18. BUGUET A., BERT J., TAPIE P., TABARAUD F., DOUA F., LONSDORFER J., BOGUI P., DUMAS M. - Sleep-wake cycle in human African Trypanosomiasis. *J. Clin. Neurophysiol.* 10,190-196, 1993.
19. BUGUET A., GATI R., SEVRE J.P., DELEVOUX M., BOGUI P., LONSDORFER J. - 24 hour polysomnographic evaluation in a patient with sleeping sickness. *Electroenceph. Clin. Neurophysiol.* 72, 471-478, 1989.
20. CASTAIGNE P., ESCOUROLLE R. - Etude topographique des lésions anatomiques dans les hypersomnies. *Rev. Neurol.* (Paris) 116, 547-584, 1967.
21. CASTAIGNE P., LHERMITTE F., BUGE A., ESCOUROLLE R., HAUW J.J., LYON-CAEN O. - Paramedian thalamic and midbrain infarcts: clinical and neuropathological study. *Ann. Neurol.* 10, 127-148, 1981.
22. CHOKROVERTY S. - Sleep, breathing and neurological disorders. *In:* Chokroverty (ed): *Sleep Disorders Medicine: Basic Science, Technical consideration and clinical aspects.* M.A. Butterworth-Heinemann, Boston, 295, 1994.
23. CIRIGNOTTA F., MONDINI S., ZUCCONI M., BARRUT-CORTES E., STURANI C., SCHIAVINA M., COCCAGNA G., LUGARESI E. - Sleep-related breathing impairment in myotonic dystrophy. *J. Neurol.* 235, 80-85, 1987.
24. CLAGHORN J.L., MATHEW R.J., WEINMAN M.L., HRUSKA N. - Daytime sleepiness in depression. *J. Clin. Psychiatry* 42, 342-343, 1981.
25. CLARKE D.J., WATERS J., CORBETT J.A. - Adults with Prader-Willi syndrome: abnormalities of sleep and behaviour. *J. R. Soc. Med.* 82, 21-24, 1989.
26. CLAVELOU P., TOURNILHAC M., VIDAL C., GEORGET A.M., PICARD L., MERIENNE L. - Narcolepsy associated with arteriovenous malformation of the diencephalon. *Sleep* 18, 202-205, 1995.
27. COCCAGNA G., MANTOVANI M., PARCHI C., MIRONI F., LUGARESI E. - Alveolar hypoventilation and hypersomnia in myotonic dystrophy. *J. Neurol. Neurosurg. Psychiat.* 38, 977-984, 1975.
28. CRAVIOTO H., SILBERMAN J., FEIGIN I. - A clinical and pathologic study of akinetic mutism. *Neurology* 10: 10-21, 1960.
29. CULEBRAS A., FELDMAN R.G., MERK F.B. - Cytoplasmatic inclusion bodies within neurons of the thalamus in myotonic dystrophy. A light and electron microscope study. *J. Neurol. Sci.* 19, 319-329, 1973.
30. DARKO D.F., Mc CUTCHAN J.A., KRIPKE D.F., GILLIN J.C., GOLSHAN S. - Fatigue, sleep disturbance, disability and indices of progression of HIV infection. *Am. J. Psychiatry* 149, 514-520, 1992.
31. DETRE T., HIMMELHOCH J., SWARTZBURG M., ANDERSON C.M., BYCK R., KUPFER D.J. - Hypersomnia and manic-depressive disease. *Am. J. Psychiatry* 128, 1303-1305, 1972.
32. DOHERTY M.J., SPENCE D.P., YOUNG C., CALVERLEY P.M. - Obstructive sleep apnoea with Arnold-Chiari malformation. *Thorax* 50, 690-691, 1995.
33. DOLENC L., BESSET A., BILLIARD M. - Hypersomnia in association with idiopathic hypersomnia and normal controls. *Pflugers Arch.* 431(6 Suppl.2), R 303-304, 1996.
34. DUMAS M., BOA F.Y. - Human African Trypanosomiasis. *In: Handbook of clinical Neurology,* A.A.Haris (ed) , vol 8 (52): Microbial disease, Elsevier, Amsterdam, 339-344, 1988.
35. EKBOM K. - Familial multiple sclerosis associated with narcolepsy. *Arch. Neurol.* 15, 377-384, 1966.
36. ELY E.W., Mc CALL W.V., HAPONIK E.F. - Multifactorial obstructive sleep apnea in a patient with Chiari malformation. *J. Neurol. Sci.* 126, 232-236, 1994.
37. FACON E., STERIADE M., WERTHEIM W. - Hypersomnie prolongée engendrée par des lésions bilatérales du système activateur medial. Le syndrome thrombotique de la bifurcation du tronc basilaire. *Rev. Neurol.* (Paris) 98, 117-133, 1958.
38. FRUCHT S., ROGERS J.D., REENE P.E., GORDON M.F., FAHN S. - Falling asleep at the wheel: motor vehicle mishaps in persons taking pramipexole and ropinirole. *Neurology* 52, 1908-1910, 1999.
39. GOOD J.L., BARRY E., FISHMAN P.S. - Post-traumatic narcolepsy: the complete syndrome with tissue typing. *J. Neurosurg.* 71, 765-767, 1989.
40. GREENSWAG L.R. - Adults with Prader-Willi syndrome . A survey of 232 cases. *Dev. Med. Child Neurol.* 29, 145-152, 1987.
41. GUILLEMINAULT C., CUMMISKEY J., MULTA J., LYNNE-DAVIS P. - Respiratory and hemodynamic study during wakefulness and sleep in myotonic dystrophy. *Sleep* 1,19-31, 1978
42. GUILLEMINAULT C., FAULL K.F., MILES L., VAN den HOED J. - Post-traumatic excessive daytime sleepiness: a review of 20 patients. *Neurology* 33, 1584-1589, 1983.
43. GUILLEMINAULT C., MONDINI S. - Infectious mononucleosis and excessive daytime sleepiness. A long term follow-up study. *Arch. Intern. Med.* 146, 1333-1335, 1986.
44. HANSOTIA P., FRENS D. - Hypersomnia associated with alveolar hypoventilation in myotonic dystrophy. *Neurology* 31: 1336-1337, 1981.
45. HAWKINS D.R., TAUB J.M., VAN DE CASTLE R.L. - Extended sleep (hypersomnia) in young depressed patients. *Am. J. Psychiatry* 142, 905-910, 1985.
46. HAYAISHI O. - Prostaglandin D2 and sleep. *In: Sleep and Sleep Disorders: from molecule to behavior,* Hayaishi O., Inoué S. (eds) Academic Press / Harcourt Brace Japan, Tokyo, 3-10, 1997.
47. HEYCK H., HESS R. - Zur Narkolepsie Frage, Klinik und Electroenzephalogramm. *Fortsch. Neurol. Psychiatr.* 12, 531-579, 1954.
48. HOWARD R.S., LEES A.J. - Encephalitis lethargica. A report of four recent cases. *Brain* 110, 19-33, 1987.
49. KOWATCH R.A., PARMELEE D.X., MORIN C.M. - Narcolepsy in a child following removal of a craniopharyngioma. *Sleep Res.* 18: 250, 1989.

50. KUPFER D.J., HIMMELHOCH J.M., SWARTZBURG M., ANDERSON C., BYCK R., DETRE T.P. - Hypersomnia in manic-depressive disease. *Dis. Nerv. System* 33, 720-724, 1972.

51. LANKFORD D.A., WELLMAN J.J., O'HARA C. - Post-traumatic narcolepsy in mild to moderate closed head injury. *Sleep* 17, S25-S28, 1994.

52. LEYGONIE-GOLDENBERG F., PERRIER M., DUIZABO Ph., BOUCHAREINE A., HARF A., BARBIZET J., DEGOS J.D. - Troubles de la vigilance, du sommeil et de la fonction respiratoire dans la maladie de Steinert. *Rev. Neurol.* (Paris) 133, 255-270, 1977.

53. LITTMAN P., ROSENSTOCK J., GALE G., KRISCH R.E., MEADOWS A., SATHER H., COCCIA P., DECAMARGO B. - The somnolence syndrome in leukemic children following reduced daily dose fractions of cranial radiation. *Int. J. Radiat. Oncol. Biol. Phys.* 10: 1851-1853, 1984.

54. MACCARIO M., RUGGLES K.H., MERIWETHER M.W. - Post-traumatic narcolepsy. *Military Med.* 152, 370-371, 1987.

55. McNICHOLAS W.T., RUTHERFORD R., GROSSMAN R., MOLDOFSKY H., ZAMEL N., PHILLIPSON E.A. - Abnormal respiratory pattern generation during sleep in patients with autonomic dysfunction. *Am. Rev. Resp. Dis.* 128, 429-433, 1983.

56. MANNI R., ZUCCA C., MARTINETTI M., OTTOLINI A., LANZI G., TARTARA A. - Hypersomnia in dystrophia myotonica: a neurophysiological and immunogenetic study. *Arch. Neurol. Scand.* 84, 498-502, 1991.

57. MICHAELIS R., HOFMANN E. - Zur Phänomenologie und Atiopathogenese der Hypersomnie bei endogen-phasischen Depressionen. *In: The Nature of Sleep*, Jovanovic U.J. (ed) Gustav Fischer Verlag, Stuttgart. 190-193, 1973.

58. NOFZINGER E.A., THASE M.E., REYNOLDS C.F. 3D, HIMMELHOCH J.M., MALLINGER A., HOUCK P., KUPFER D.J. - Hypersomnia in bipolar depression: a comparison with narcolepsy using the multiple sleep latency test. *Am. J. Psychiatry* 148, 1177-1181, 1991.

59. NORDGREEN R.E., MARKESBERY W.R., FUKUDA K., REEVES A.G. - Seven cases of cerebro-medullospinal disconnection: the " locked-in "syndrome. *Neurology* 21, 1139-1148, 1971.

60. OLANOW C.W., SHAPIRA A.H.V., ROTH T. - Falling asleep at the wheel: Motor vehicle mishaps in people taking pramipexole and ropinirole. *Neurology* 54, 274, 2000.

61. ONO S., KURISAKI H., SAKUMA A., NAGAO K. - Myotonic dystrophy with alveolar hypoventilation and hypersomnia: a clinico-pathological study. *J. Neurol. Sci.* 114, 68-75, 1993.

62. ONOFRJ M., CURATOLA L., FERRACCI F., FULGENTE T. - Narcolepsy associated with primary temporal lobe β-cells lymphoma in a HLA DR2 negative subject. *J. Neurol. Neurosurg. Psychiatry* 55, 852-853, 1992.

63. ORR W.C., MALES J.L., IMES N.K. - Myxedema and obstructive sleep apnea. *Am. J. Med.* 70, 1061-1066, 1981.

64. PARK Y.D., RADTKE R.A. – Hypersomnolence in myotonic dystrophy: demonstration of sleep onset REM sleep. *J. Neurol. Neurosurg. Psychiatry* 58, 512, 1995.

65. PARKER KP., BLIWISE D.L. - Timing of hemodialysis affects daytime sleepiness. *Sleep Res.* 25: 435, 1996.

66. PARKER K.P., BLIWISE D.L. - Excessive daytime sleepiness in chronic hemodialysis and sleep apnea patients. *Sleep Res.* 26, 580, 1997.

67. PENTREATH V.W. - African sleeping sickness and the nervous system. *In: Sleep and Sleep Disorders: from molecule to behavior,* Hayaishi O., Inoué S. (eds) Academic Press / Harcourt Brace Japan, Tokyo, 11-27, 1997.

68. PERKS W.H., HORROCKS P.M., COOPER R.A., BRADBURY S., ALLEN A., BALDOCK N., PROWSE K., Van't HOFF W. - Sleep apnea in acromegaly. *Br. Med. J.* 280, 894-897, 1980..

69. PHEMISTER J.C., SMALL J.M. - Hypersomnia in dystrophia myotonica. *J. Neurol. Neurosurg. Psychiat.* 24, 173-175, 1961.

70. POIRIER G., MONTPLAISIR J., DUMONT M., DUQUETTE P., DECARY F., PLEINES J., LAMOUREUX G. - Clinical and sleep laboratory study of narcoleptic symptoms in multiple sclerosis. *Neurology* 37, 693-695, 1987.

71. PRITCHARD P.B. III, DREIFUSS F.E., SKINNER R.L., PICKETT J.B., BIGGS P.J. - Symptomatic narcolepsy. *Neurology* 37, 693-695, 1983.

72. RABEC C., LAURENT G., BAUDOUIN N., MERATI M., MASSIN F., FOUCHER P., BRONDEL L., REYBET-DEGAT O. - Central sleep apnoea in Arnold-Chiari malformation: evidence of pathophysiological heterogeneity. *Eur. Respir. J.* 12, 1482-1485, 1998.

73. RAJAGOPAL K.R., ABBRECHT P.H., DERDERIAN S.S., PICKETT C., HOFELDT F., TELLIS C.J., ZWILLICH C.W. - Obstructive sleep apnea in hypothyroidism. *Ann. Intern. Med.* 101, 491-494, 1984.

74. RIVERA V.M., MEYER J.S., HATA T., HISHIKAWA Y., IMAI A. - Narcolepsy following cerebral hypoxic ischemia. *Ann. Neurol.* 19, 505-508, 1986.

75. ROBLES A., ALDREY J.M., RODRIGUEZ-FERNANDEZ R.M., SUAREZ C., CORREDERA E., LEIRA R., NOYA M. - Paramedian bithalamic infarct syndrome: report of five new cases. *Rev. Neurol.* (Paris) 23, 276-284, 1995.

76. ROSENTHAL N.E., SACK D.A., GILLIN J.C., LEWY A.J., GOODWIN F.K., DAVENPORT Y., MUELLER P.S., NEWSOME D.A., WEHR T.A. - Seasonal affective disorder: a description of the syndrome and preliminary findings with light therapy. *Arch. Gen. Psychiatry* 41, 72-80, 1984.

77. RYE D.B., BLIWISE D.L., DIHENIA B., GURECKI P. – First track: daytime sleepiness in Parkinson's disease. *J. Sleep Res.* 9 (1), 63-69, 2000.

78. SCHLUTER B., AGUIGAH G., ANDLER W. - Hypersomnia in multiple sclerosis. *Klin. Pediatr.* 208: 103-105, 1996.

79. SCHOTT B., MICHEL D., MOURET J., RENAUD B., QUENIN P., TOMMASI M. - Monoamines et régulation de la vigilance II. Syndromes lésionnels du système nerveux central. *In: " les Médiateurs*

Chimiques " rapports présentés à la XXIXème Réunion Neurologique Internationale, Paris. Masson, 157-171, 1972.

80. SCHRADER H., GOTLIBSEN O.B. - Multiple sclerosis and narcolepsy-cataplexy in a monozygotic twin. *Neurology* 30, 105-108, 1980.
81. SCHWARTZ B.A., ESCANDE C. - Sleeping sickness: sleep study of a case. *Electroencephalogr. Clin. Neurophysiol.* 29: 83-87, 1970.
82. SCHWARTZ W.J., STAKES J.W., HOBSON J.A. - Transient cataplexy after removal of a craniopharyngioma. *Neurology* 34: 1372-1375, 1984.
83. SHIIHARA T., SHIMIZU Y., MITSUI T., SAITOH E., SATO S. - Isolated sleep apnea due to Chiari type I malformation and syringomyelia. *Pediatr. Neurol.* 13, 266-267, 1995.
84. SHIMIZU A., HIYAMA H., YAGASAKI A., TALASHASHI H., FUJIKI A., YOSHIDA I. - Sleep of depressed patients with hypersomnia: a 24 hour polygraphic study. *Waking and Sleeping* 3, 335-359, 1979.
85. SKATRUD J., IBER C., EWART R., THOMAS G., RASMUSSEN H., SCHULTZE B. - Disordered breathing during sleep in hypothyroidism. *Am. Rev. Resp. Dis.* 124, 325-329, 1981.
86. SOUTHMAYD S.E., CAIRNS J., DELVA N.J., LETEMENDIA F.J.J., BRUNET D.G. - Awake, perchance asleep? *Br. J. Psychiatry* 148, 748-749, 1986.
87. STAHL S.M., LAYZER R.B., AMINOFF M.J., TOWNSEND J.T., FELDON S. - Continuous cataplexy in a patient with a midbrain tumor: the limp man syndrome. *Neurology* 30,1115-1118, 1980.
88. TANBERG E., LARSEN J.P., KARLSEN K. – Excessive daytime somnolence and sleep benefit in Parkinson's disease: a community-based study. *Mov. Disord,* 14 (6), 922-927, 1999.
89. TELERMAN-TOPPET N. - Les accidents de narcose dans la dystrophie myotonique. *Acta Neurol. Belg.* 70, 551-560, 1970.
90. THIEBAUT F., ROHMER F., KURTZ D. - Les hypersomnies symptomatiques continues. *Rev. Neurol.* (Paris) 116, 491-546, 1967.
91. VERMERSCH P., CAPARROS-LEFEBVRE D. - Encéphalites d'origine virale. *Encycl. Med. Chir.* Neurologie, Elsevier, Paris. 17-050-A-10, 12, 1997.
92. VON ECONOMO C. - Grippe, Encephalitis und Encephalitis lethargica. *Klinisch Wochens* (Wien) 32, 15, 393-396, 1919.
93. WALKER S., FINE A., KRYGER M.H. - Sleep complaints are common in a dialysis unit. *Am. J. Kidney Dis.* 26, 751-756, 1995.
94. YOUNGER D.S., PEDLEY T.A., THORPY M.J. - Multiple sclerosis and narcolepsy: possible similar genetic susceptibility. *Neurology* 41, 447-448, 1991.

Chapter 35

Hypersomnia in children

M. J. Challamel

Unité de Sommeil de l'Enfant, Explorations Neurologiques, Centre Hospitalier Lyon-Sud, Pierre-Bénite, Lyon, France

A diagnosis of hypersomnia is rarely applied to young children for a variety of reasons.

- A degree of uncertainty remains as to the normal duration of sleep in relation to age, varying widely from one individual to another.

- Excessive daytime sleepiness is expressed in young children by paradoxical or misleading signs.

- Sleep needs are often overestimated by parents: prolonged sleep in children does not cause parents much concern, and diagnosis is often sought at the start of schooling, resulting from marked difficulties in morning arousal and irresistible sleep episodes which disrupt school activities.

The two main causes of sleep disorder only will be dealt with: obstructive sleep apnoea syndrome (OSAS) and narcolepsy.

FREQUENCY

No systematic epidemiological study has been carried out to indicate the frequency of hypersomnia in children, but two inquiries suggest that excessive daytime sleepiness is relatively common. A Finnish study by Saarenpää-Heikkilä [39], involving 547 children aged 7 to 17, showed that 20% presented excessive daytime sleepiness; another study conducted by Patois [37] of 11,923 boys and 12,870 girls aged 15 to 20, revealed that 77% of boys and 88% of girls complained of excessive sleepiness, particularly in the morning. The majority of these cases of abnormal daytime sleepiness are probably secondary, as with adults, to a chronic lack of sleep since 74% of the children in Patois' study complained of lack of sleep.

AETIOLOGIES OF HYPERSOMNIA IN CHILDREN

Here again data is lacking; the most comprehensive study is that of the Stanford team covering 197 children aged 2 to 11 and 153 children aged 11 to 16, scored for excessive daytime sleepiness (table 35.1) [17]. In both age groups, obstructive sleep apnoea syndrome is, as with adults, the primary cause of excessive daytime sleepiness, representing 47% and 45% respectively, of cases for children aged 2 to 11 and aged 11 to 16.

For children under aged 11: narcolepsy is rare (5 cases), whereas hypersomnia associated with a neurological problem, with or without mental retardation and/or epilepsy, represents 89 cases out of 197. After age 11, narcolepsy becomes the second cause of excessive daytime sleepiness, as in the case of adults.

In a private statistical study of 50 children recorded for hypersomnia (excluding obstructive syndromes and insufficient sleep) (table 35.2) 25 patients were diagnosed with narcolepsy, 9 of which were associated with neurological damage. In 2 children the only recognised cause was periodic leg movements, and in 3 a very likely diagnosis of idiopathic hypersomnia. The aetiology of 13 patients in this study was still unknown.

Table 35.1. Stanford study [17] aetiology of hypersomnia in children

Children aged 2 to 11	**197**
Neurological illnesses with or without mental retardation or epilepsy	89
Obstructive sleep apnoea syndromes	92
Narcolepsy	5
Sleep disorder associated with medical treatment	4
Sleep disorder associated with a medical illness	5
Sleep disorder of unknown origin	2
Children aged 11 to 16	**153**
Neurological illnesses with or without mental retardation or epilepsy	14
Obstructive sleep apnoea syndromes	69
Narcolepsy	38
Idiopathic hypersomnia	11
Sleep disorder associated with medical treatment	6
Sleep disorder associated with a medical illness	5
Sleep delayed phase syndrome	3
Kleine-Levin syndrome	4
Sleep disorder of unknown origin	3

Comments

The author's view is that all the causes of hypersomnia in adults also exist in children and need to be investigated. Certain relatively simple diagnostic aetiologies in adults, such as hypersomnia secondary to chronic lack of sleep or delayed or advances sleep phase syndromes, are far more difficult to diagnose in children, as the child and his family may be completely unaware of their existence.

Table 35.2. Private data on 50 cases of hypersomnia in children

- Idiopathic narcolepsy: 16
- Symptomatic narcolepsy: 9
- Periodic leg movements: 2
- Circadian rhythm sleep disorders: 2
- Isolated cataplexy: 1
- Neurological aetiologies: 7
- Unrecognised aetiologies: 13

In prepubescent children, certain neurological disorders (table 35.3) should be investigated, as at this age neurological causes and OSAS are the most common aetiologies of sustained hypersomnia [17].

In adolescents, chronic lack of sleep, hypersomnia symptomatic of school phobia and above all depressive syndrome, must be considered; sedative consumption in the form of drugs or alcohol should not be ruled out either.

Finally, in all cases, narcolepsy, idiopathic hypersomnia, Kleine-Levin syndrome or, in young girls, menstruation-related hypersomnia, should all be considered, as these syndromes may appear from the start of the second decade and even occasionally during the first ten years of life [2].

DIAGNOSING HYPERSOMNIA IN CHILDREN

Hypersomnia is to be suspected where:

- *Time asleep or at least spent in bed, is prolonged* by 2 or 3 hours compared with the average time for the age (*see chapter2*).

- *Regular napping persists or recommences*, in children of over 6 years of age.

- *Irresistible episodes of daytime sleep occur*, brought on by monotonous activity: schoolwork, television, short car journeys, meals. When excessive sleep is recognised in children it is often severe; parents fairly often report onset of sleep while the child is talking, eating, playing or even walking.

- *Excessive daytime sleepiness occurs*. This symptom is often difficult to recognise in young children, and may be replaced by abnormal hyperactivity associated with sudden naps at inappropriate times or places or in unlikely sleeping positions. This symptom sometimes goes unrecognised, even by adolescents; they complain of tiredness, disturbed vision in the form of diplopia or blurred vision; tiredness and focusing problems are associated with abnormally frequent yawning. Finally, this abnormal sleepiness is often expressed as behavioural or emotional disturbance: abnormal aggression, in particular, toward peers, impulsiveness, inability to tolerate frustration, pathological timidity. Sleepiness is almost always associated with reduced academic performance.

- Additional symptoms may occur: abnormally agitated sleep, nocturnal terrors and/or unaccustomed episodes of sleepwalking, frequent talking in sleep; difficult morning arousal, sometimes with sleep drunkenness.

Table 35.3. OSAS in children

Nocturnal symptoms
snoring
breathing through the mouth
apnoeas with noisy respiration resumption
thoraco-abdominal asynchronicity, signs of retraction
abnormal sleeping position: seated, kneeling…
hypersudation
secondary enuresis
sleep disorders with abnormal agitation, nightmares, confusion on arousal, sleep talking, difficult arousal

Daytime symptoms		
Young child:	-	aggression, hyperactivity, pathological timidity
	-	frequent rhino-pharyngitis and otitis
	-	breath holding spells
Older child:	-	morning headaches
	-	abnormal sleepiness
	-	diminished academic performance

OBSTRUCTIVE SLEEP APNOEA IN CHILDREN (OSAS)

The symptom of excessive daytime sleepiness is far less common in child than in adult OSAS, and is often even absent. Nevertheless, investigating abnormal sleepiness in a child must include looking for signs of OSAS. It is not known exactly how common OSAS is in children; it was estimated by Benediktsdottir [1] at between 1.6 and 3.4% of a child population aged 6 months to 6 years.

Diagnostic criteria for OSAS in children

The diagnosis of OSAS is based on nocturnal and diurnal symptoms (table 35.3) and on the results of clinical examination. Complementary investigations (polygraphic, ENT, cardio-vascular) can be used to assess the severity, determine aetiology and indicate subsequent action.

Nocturnal symptoms. - Continuous loud snoring, which can be heard from one room to another, is pathological; in children it may occur only in the second part of the night and is found in 76 to 100% of cases [3, 13, 19, 20, 29]; it may exclude, be replaced by or linked to a stridor, wheezing or coughing [20].

Daytime symptoms. - Daytime sleepiness is reported in literature in 14% to 84% of cases [3, 13, 19, 20, 29]. This sleepiness is not always detected in the youngest children, in whom it is expressed as behavioural and/or emotional disturbances. Anamnesis should, finally, be oriented to frequent upper airway infections, and recurrent ear infections.

The symptoms of older children are closer to those of adults. They complain of morning headaches, abnormal sleepiness; secondary enuresis or difficulties at school should also be looked into [19, 20].

Can a diagnosis of OSAS be established clinically?

It would appear that the diagnosis of obstructive sleep apnoea syndrome can be clinically detected in a certain number of cases and even its severity may be assessed clinically [4].

According to Vaara *et al.* [43] the symptomatic association between snoring, open-mouthed breathing, apnoeas and hypersudation is linked with severe nocturnal breathing difficulties in 1 case out of 4. On the other hand, Carroll *et al.* [7] insist on the fact that it is impossible to differentiate clinically between children who present with snoring alone and those who present with snoring associated with an obstructive sleep apnoea syndrome.

Clinical examination

This should concentrate on the following:

- Adenoid-tonsil hypertrophia: this is the main cause of OSAS in children, being present in 3 out of 4 children [3, 13, 20, 29].

- Anomalous height-weight development: lagging height-weight development is present in 27 to 56% of cases [3, 13, 20, 29], this is more common than obesity: 45% of children in Leach's study [29] had weight and height below the 10th percentile, whereas only 12% were obese. This anomaly sometimes shows as only a slight dip in the height-weight curve, which requires close observation.

- Thorax malformation: funnel-shaped thorax, abnormally wide base of thorax.

- a nasal or oropharyngeal abnormality: nasal obstruction, malformation of the wall, macroglossia, abnormal palate.

- Maxillofacial malformation. In children presenting OSAS, cranio-facial dysmorphia is common: 32% of the 50 children in Guilleminault's study [20] had a maxillofacial abnormality. These abnormalities are sometimes part of the symptomatology of a neuro-muscular illness and/or a weight disease (thesaurismosis) which should be investigated.

Numerous genetic and chromosomal abnormalities are accompanied by obstructive sleep apnoea syndrome. This possibility should be considered if the child shows the slightest breathing difficulties during sleep (table 35.4).

Table 35.4. Diseases linked to OSAS in children

Neuromuscular illnesses
-      cerebral palsy
-      poliomyelitis, muscular dystrophy
Craniofacial anomalies
-      syndromes: Crouzon, Pierre Robin, Apert, Treacher-Collin, Goldenhar, Francheschetti
-      acromegalia, achondroplasia, Arnold Chiari
-      micro- and retrognathia
Thesaurismosis
-      Cushing, myxoedema, Hurler, Hunter
Chromosomal abnormalities
-      Trisomia 21 and Trisomia 18

Laboratory tests

Polygraphic sleep recording

This can be used to confirm the diagnosis of total or partial (upper airway resistance) OSAS, to assess its severity and to analyse the mechanism of respiratory difficulties.

The severity of the syndrome is estimated on the index for obstructive apnoeas and/or hypopneas. This assessment is sometimes difficult as few reliable norms exist for children [31]; an index of >1/hour apnoea is very likely to be pathological [6] but some children present considerable breathing difficulties below this figure.

Other children presenting nocturnal breathing difficulties may not suffer from obstructive sleep apnoeas. For these children the possibility of an upper airway resistance syndrome should be investigated [23], objectified by the presence of paradoxical inspiratory thoracic breathing, of recruitment at the level of the accessory respiratory muscles (genioglossus, sternocleidomastoid and abdominal muscles accompanied by signs of retraction: susternal and xiphoidian funnel, intercostal strain. [14].

- The severity of the OSAS is assessed by studying the effects of disrupted respiration on gas exchange [16] using transcutaneous electrodes to measure oxygen saturation, expired CO_2 or partial pressures of CO_2 and O_2. Child OSAS is not always accompanied by gas exchange alterations.

- Finally, the severity of OSAS is assessed by evaluating the effects of disrupted respiration on sleep organisation: the increase of stage I, decrease in stages III - IV and REM sleep, but primarily, by examining arousals. This is because very often, sleep fragmentation is only revealed in children by an increase in arousals (increase in K complexes, increase in EEG waking reactions with the appearance of alpha or slow waves).

Should polygraphic sleep recordings be obligatory?

This test would not appear to be absolutely necessary if the aetiology of an OSAS is obvious, consisting of an isolated case of enlarged tonsils and adenoids, but should be envisaged in cases of:
- enlarged tonsils linked to maxillofacial malformation;
- isolated enlarged tonsils and adenoids combined with another ailment likely to increase surgical morbidity;
- an aetiology other than enlarged tonsils and adenoids ;
- a child under 3 years of age.

Should recording take place at night?

We do not believe greatly in polygraphic sleep recordings during napping for children over 2 years of age, because abnormal respiratory pauses and breathing difficulties may only appear during the second part of the night. For children under 2 years of age, if the recording is made during a nap, the latter must comprise a sufficiently long REM sleep time [30] for the detection of breathing difficulties to be reliable; these naps must not be encouraged by sleep deprivation or induced by sedatives [13].

Other investigations

Neuroimaging - this is indispensable where a maxillofacial anomaly is suspected: cephalometry, an MRI or CT scan in axial or coronal sections giving an accurate assessment of these anomalies.

Cardio-vascular tests - these tests are necessary as cardio-vascular repercussions have been described in 15 to 54% of cases [3, 13, 19, 20, 29], but OSAS which are secondary to amygdaloid hypertrophia are rarely accompanied by cardiac features. Long ECG recording would reveal any problems in rhythm [20]. Echocardiography could reveal right ventricle hypertrophia, or left ventricle hypertrophia in much rarer cases where there is very high blood pressure, uncommon in children.

Treatment

Adenoido-amygdalectomy - Amygdalectomy is widely indicated for children presenting OSAS or an upper airway resistance syndrome. This operation should be considered if there is a marked enlargement of amygdala and adenoids. It is not only advisable in cases of amygdala hypertrophia alone; it is also advised for children with maxillofacial anomalies: non-hypertrophic amygdala in a narrow oropharynx may induce or aggravate an obstructive syndrome. An early amygdalectomy for a child with an obstructive syndrome may have a favourable effect on the anatomy and the development of the oropharynx.

Other surgical interventions - Reparatory surgery for maxillofacial anomalies, uvulopalato-pharyngoplasty.

Indications must be carefully considered in the case of children and surgical intervention should only be carried out after scrupulous ENT and maxillofacial evaluation when growth has terminated.

Continuous positive airway pressure (CPAP) - This has replaced tracheostomy and may sometimes be used while waiting for surgical intervention; it should be considered more frequently, despite certain difficulties of use in children [21].

Medical treatment. - Analeptic treatment is rarely indicated for children, oxygenotherapy is not advisable [5]; it is nonetheless important for an obese child to lose weight.

NARCOLEPSY

For children as for adults, narcolepsy is a syndrome which, when complete, associates episodes of irresistible daytime sleep, cataplexy, sleep paralysis and hypnagogic hallucinations. Excessive daytime sleepiness and episodes of sleep which occur in inappropriate circumstances must be considered as possible signs of narcolepsy; association with cataplexy would clinically confirm the diagnosis.

Frequency

The frequency of narcolepsy in children has not been evaluated. Narcolepsy is rarely diagnosed before puberty: of 400 narcoleptics diagnosed at the Mayo clinic, only 4% were under age 15 [48]. A discrepancy exists, in fact, between the number of children actually diagnosed and the number of adults who trace their first symptoms back to childhood. Indeed, an analysis of the 3 studies grouping together 235 adult patients for whom the onset of symptoms was studied by age bracket, reveals that 16.2% of these patients traced the onset of symptoms to before the age of 10, and 39% before the age of 15 [9]. According to Billiard and Guilleminault, the diagnosis would have been made 10 to 15 years after the first symptoms appeared.

Published studies on children remain rare. Of these, 77 cases of "idiopathic" narcolepsy were selected [10, 12, 17, 24, 28, 33, 34, 38, 40, 41, 46, 47, 48] and 20 cases of "symptomatic" narcolepsy. Twelve of the latter are associated with a type C Nieman Pick disease [25], 2 others with Turner's syndrome [16] and one with cranio-pharyngioma [42].

Some of these observations are debateable: indeed if all symptomatic narcolepsies are associated with irresistible sleep episodes and cataleptic attacks, the same cannot be said of idiopathic cases for which many observations are clinically incomplete or poorly inventoried, polygraphically.

The age of onset of the illness is not always precise; in our study [9] the average age is 9 years 9 months, (SEM: 3 years 9 months) for idiopathic narcolepsy; 6 years (SEM: 4 years 2 months) for symptomatic narcolepsy.

Clinical features

All the classic symptoms described in the adult can be found in children, the only difference being that children are rarely aware of their disorder.
- Irresistible sleep episodes are constant, longer than in adults, reaching up to 2-3 hours and are generally non restorative.
- Cataleptic attacks are reported in 80.5% of idiopathic cases [9]; this symptom is thus as common as in adult narcolepsy, suggesting that its presence facilitates early diagnosis. As in the case for adults, attacks tend to be brought on by laughter and emotion. They may be reduced to a simple speech block.
- Hypnagogic hallucinations and sleep paralysis have been noted, respectively, in 39% and 29% of children presenting idiopathic narcolepsy. These symptoms may be present in normal children and thus have less diagnostic value than for adults.
- Excessive daytime sleepiness is often marked. It may alternate with abnormal hyperactivity. It may take the form of behavioural problems, neurological features like ptosis, diplopia, slurred speech or automatism. It is accompanied by learning and social integration problems which sometimes disrupt education.

- Accessory symptoms. Parasomnia is frequent but not particularly specific in children. Polygraphic recordings almost always show marked dysomnia, even if the child and his family are usually unaware of this. Arousal difficulties probably represent one of the earliest signs of the disorder; sleep drunkenness appears to be more common than in adults. Finally the beginning of obesity is linked with the onset of the illness, in half the cases.

"Symptomatic" narcolepsy (table 35.5)

In adults, "symptomatic" narcolepsy, or narcolepsy associated with a neurological illness, is rare. In children under 10 years of age it is perhaps more common than "idiopathic" narcolepsy [17]. In a personal study of 25 children presenting narcolepsy, 9 had a neurological disease. In 3 cases, narcolepsy was associated with a type C Niemann-Pick illness, 3 were associated with an unclassified neurological syndrome and one had a tumour in the floor of the third ventricle, 1 had Steinert's disease, and 1 had Willi-Prader syndrome. In all these children, the two major symptoms – irresistible sleep episodes and cataleptic attacks – were present. The latter were frequent and were always recorded by polysomnography (fig. 35.1). Nevsimalova [35] has also insisted on the presence of symptomatic narcolepsy in young children.

Table 35.5. Neurological causes of excessive daytime sleepiness

-	Niemann-Pick type C disease
-	Myotonic dystrophy
-	Prader-Willi syndrome
-	Kearns-Sayre syndrome
-	Multiple sclerosis
-	Autosomal dominant cerebellar ataxia
-	Deafness and narcolepsy syndrome [32]
-	3rd ventricle, pineal and hypothalamic tumours
-	Hydrocephaly
-	Encephalitis (Lyme, Guillain-Barré, mononucleosis…)
-	Right temporal and thalamic strokes

Diagnosing narcolepsy syndrome

While recent studies insist on the association of several of the tetrad features present at the onset of the illness in children [16, 22, 28, 38, 45, 47], earlier studies [33, 48] and adult retrospective studies indicate that narcoleptic syndromes in children tend to be incomplete. Cataleptic attacks may appear very late, but may also precede excessive daytime sleepiness [9, 38, 44].

Diagnosis is thus difficult:

- in the absence of cataleptic attacks;
- when excessive daytime drowsiness is expressed by neurological features, with behavioural or emotional disorders: 2 children in the study by Vecchierini [43], several children in the study by Dahl [12] were treated for psychological disorders;
- when narcolepsy is associated with a neurological illness or mental retardation: cataleptic attacks are taken to be atonic or gelastic seizures and automatic behaviour, for complex partial seizures.

The most common diagnoses made at the onset of the illness were hypothyroidism, myasthenia, epilepsy or encephalitis.

To sum up, symptomatology differs for children. Irresistible sleep episodes are longer, sleep drunkenness is more common, dyssomnia is unnoticed and sleepiness sometimes alternates with abnormal hyperactivity and behavioural problems. Vision and speech disorders are common.

Laboratory tests

The diagnosis of narcolepsy may be aided by the use of sleep diaries and/or actigraphy, polygraphic recordings of nocturnal and diurnal sleep, the multiple sleep latency test (MSLT) and HLA typing.

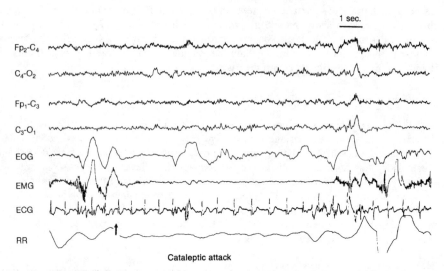

Figure 35.1. Cataleptic attack recorded on an ambulant monitor in a 12 year old child presenting type C Niemann-Pick disease. EEG on the first four derivations. EOG = elecro-oculogram, EMG = chin-electromyogram, ECG: electrocardiogram, RR = thoracic respiratory rate by impedance.

Sleep diary and actigraphy

A sleep schedule, recorded over at least 15 days, will provide an estimate of how much time is spent in bed, the times and frequency of naps or of irresistible sleep episodes and those of cataplexy. False hypersomnia linked to delayed or advanced sleep phase syndrome is thereby eliminated. We also use actigraphy to obtain an objective assessment of the presence of dyssomnia, often unrecognised by the child and his family (fig. 35.2).

Multiple sleep latency test (MSLT)

This is easy to carry out on normal children aged 4-5 onwards. Normal values differ between children and adults. Adolescents between 12 and 20 (Tanner stages 3 to 5) present physiological hypersomnia. At this age mean sleep latency of under 5 minutes is not always pathological [8]. Conversely, in preadolescents, a mean sleep latency of less than 10 or even 15 minutes is probably abnormal [18, 36]. Palm [36], having studied 18 normal preadolescents, considers sleep latency to be abnormal if it lasts less than 10 minutes for 2 or more tests. He suggests prolonging tests for children to 30 minutes rather than 20, and has developed a measure: the Daily Average Sleep Latency DAST, which calculates the tendency to fall asleep. This is based on the number of tests during which sleep latency occurs, divided by time of wakefulness i.e.; the sum of sleep latencies observed: if latencies are of 30, 19, 14 and 22 minutes, The DAST will be 3/85 = 3.5%. The first test was not counted as it was interrupted before the child fell asleep. Normal values would be 1.3 for the median, with a 95% confidence interval between 0 and 2.1%.

Daytime and overnight polygraphic sleep recordings

- Continuous daytime recordings in the form of ambulant monitoring are used instead of MSLT, for the youngest children or those with mental deficiencies. They can provide a recording of cataleptic attacks (fig. 35.1).
- Overnight recordings eliminate other causes of abnormal sleepiness: sleep apnoea syndrome, periodic limb movement disorder or epilepsy.

However it is important to note that, in children as in adults, narcolepsy may be accompanied by abnormal respiratory pauses and periodic limb movements. For Young [47] the latter are more frequent in children than in adults.

Polygraphic criteria for narcolepsy in children

"Idiopathic narcolepsy"

The most recent studies on children [9, 28, 27, 26, 33, 38, 44, 47] insist on the fact that all the polygraphic criteria used to diagnose narcoleptic syndrome in adults are found in children.

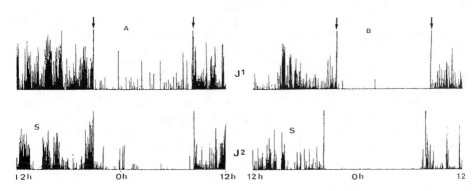

Figure 35.2. Actigraphy over 2 successive days (J1 and J2) for two children A and B. S = siesta (nap). The arrows mark the nocturnal portion of recordings. A = a 6 year old boy presenting idiopathic narcolepsy; note daytime hyperactivity associated with siestas during the second day, the existence of marked dyssomnia during the first day. B = a 10 year old girl presenting idiopathic hypersomnia; note the existence of long periods of daytime hypoactivity and of very stable sleep overnight, with few body movements.

MSLT generally scores more than 2 sleep onset REM periods and a considerable drop in average sleep latency, which is generally much shorter than 5 minutes, even for preadolescents. During the night, the latency to the first REM sleep phase is shortened [26] and the quantity of REM sleep maintained despite a dyssomnia that may at times be important. These results conflict with remarks made by Guilleminault [18] in relation to 25 teenage narcoleptic patients, 4 of whom had a narcolepsy-cataplexy association, as he notes that only one of his patients experienced several sleep onset REM periods on the MSLT and no dyssomnia. In children, the absence of sleep onset REM periods and of dyssomnia does not disqualify a diagnosis of narcolepsy. Polysomnography and MSLT should thus be repeated within a space of 12 to 24 months [27], particularly in the absence of cataplexy.

"Symptomatic" narcolepsy

Our 6 patients had one or several 24 to 36 hour polygraphic recordings. Significant differences were noted between idiopathic and symptomatic narcolepsy with, in the latter case particularly, far greater dyssomnia, diminished REM sleep time and no shortened REM latency [9]. These differences should provide an indication for diagnosis between idiopathic and symptomatic narcolepsy, when this is not clinically evident.

HLA typing

In our view, it is most important to carry out HLA typing for all hypersomniac children (excluding OSAS) so that diagnosis for children presenting hypersomnia with no cataleptic attack, can be oriented toward narcolepsy syndrome or another form of hypersomnia, since it is known that the former would, in principle be associated with HLA DQB1*0602 in Caucasian children. The absence of HLA DQB1*0602 could indicate "symptomatic" narcolepsy in the young child.

Neuroimaging

In view of the frequency of symptomatic hypersomnia in children, a CT scan or preferably an MRI would appear to be indispensable.

Treatment

There is no specific treatment for narcolepsy in children as compared with that for adults. This relies on taking psychosocial measures: psychological treatment, academic assistance. Narcolepsy must be explained to parents and teachers in particular, emphasising the importance of allowing for 20 to 30 min. naps at midday and at the end of the afternoon.

It is vital, in our view, to treat the illness as early as possible, avoiding the development of academic problems which, once they arise appear to be fairly refractory to treatment [9, 44].

The most common medicines used for children are: methylphenidate, chlorimipramine and modafinil. In children, as in adults, chlorimipramine is remarkably effective in treating cataleptic attacks, sometimes even at very weak doses. All these treatments should thus be initially prescribed at the weakest possible dose. Amphetamine derivatives should be avoided [41].

Symptomatic narcolepsy must also be treated. Indeed we have noted that narcolepsy unquestionably aggravates the neurological symptoms of the illness in question [9].

Psychosocial measures have an important role to play e.g. psychological support, academic support: parents and teachers need to be made aware of the importance of arranging naps at midday and toward the end of the afternoon.

OTHER HYPERSOMNIAS

Recurrent hypersomnia

Kleine-Levin syndrome and menstruation-related hypersomnia are rare conditions occurring in adolescence.

Idiopathic hypersomnia

This is suspected in adults who experience long, non restorative naps, automatic behaviour and sleep drunkenness. These symptoms are not uncommon in children presenting narcolepsy. Diagnosis must therefore be carried out with care and may not be possible before several years' course of illness, given that cataleptic seizures may appear late and that the polygraphic criteria for narcolepsy may be lacking in the early stages of the illness.

REFERENCES

1. BENEDIKTSDOTTIR B., GISLASON T. – Sleep related breathing disturbances among children 6 months to 6 years old. *10th Congress of the Eur. Sleep Res. Society.* Strasbourg, 427, 1990.
2. BILLIARD M. – The Kleine-Levin Syndrome, *In: Sleep 1980,* W.P. Koella (ed). Karger, Basel, 124-127, 1981.
3. BROUILLETTE R.T., FERNBACK S.K., HUNT C.E. – Obstructive sleep apnea in infants and children. *J. Pediatr.* 105, 10-14, 1984.
4. BROUILLETTE R.T., HANSON D., DAVID R., KLEMLA L., SZATKOWSKI A., FERNBACK S., HUNT C.E.A. – A diagnosis approach to children with suspected obstructive sleep apnea. *J. Pediatr.* 105, 10-14, 1984.
5. BROUILLETTE R.T., WATERS K.- Oxygen therapy for pediatric obstructive sleep apnea syndrome : How safe ? How effective? Editorial. *Am.J.Resp. Crit. Care Med.* 153, 1-2, 1996.
6. CARROL J.L., LOUGHLIN G.M. – Diagnostic criteria for obstructive sleep apnea in children. *Ped. Pulmonol.* 14, 71-74, 1992.
7. CARROLL J.L., McCOLLEY S.A., MARCUS C.L., CURTIS S., LOUGHLIN G.M. – Inability of clinical history to distinguish primary snoring from obstructive sleep apnea syndrome in children. *Chest,* 108, 610-618, 1995.
8. CARSKADON M.A., HARVEY K., DUKE P., ANDERS T.F., LITT I.F., DEMENT W.C. – Pubertal change in daytime sleepiness. *Sleep,* 2, 4, 453-460, 1980.

9. CHALLAMEL M.J., MAZZOLA M.E., NEVSIMALOVA S., CANNARD C., LOUIS J., REVOL M. - Narcolepsy in children. *Sleep,* 17, S17-S20, 1994.

10. CHISHOLM R.C., BROOK C.J., HARRISON G.F., LYON L., ZUKAITIS D. - Prepubescent narcolepsy in a six year old girl. *Sleep Res.,* 1, 15, 113, 1985.

11. DAHL R.E., HOLTTUM J., TRUBNICK L. - A clinical picture of child and adolescent narcolepsy. *J. Am. Acad. Child Adolesc. Psychiatry.,* 6, 834, 1994.

12. DUCHOWNY M.S. - Narcolepsy cataplexy and gelastic atonic seizure. *Neurology* 35, 775-776, 1985.

13. GAULTIER C., BOBIN S., PRAUD J.P., DELAPERCHE F. - Syndrome d'apnées obstructives du nourrisson et de l'enfant. *In :* Gaultier C. (ed.), *Pathologie respiratoire du sommeil du nourrisson et de l'enfant.* Vigot, Paris, 69-78, 1989.

14. GAULTIER C. - Clinical and therapeutic aspects of obstructive sleep apnea syndrome in infants and children. *Sleep,* 15, S36-38, 1992.

15. GAULTIER C. - Obstructive sleep apnoea syndrome in infants and children: facts and unsettled issues. *Thorax,* 50, 1204-1210, 1995.

16. GEORGE C.F.P., SINGH S.M. - Juvenile onset narcolepsy in an individual with Turner syndrome. A case report. *Sleep,* 14,3,267-269, 1991.

17. GUILLEMINAULT C. - Disorders of excessive daytime sleepiness. *Sleep and its Disorders in Children.* C. Guilleminault (ed), Raven Press, New York, 177-179, 1987.

18. GUILLEMINAULT C. - Narcolepsy and its differential diagnosis. *Sleep and its Disorders in Children.* C. Guilleminault (ed), Raven Press, New York, 181-194, 1987.

19. GUILLEMINAULT C. ELDRIDGE F., SIMMONS F., DEMENT W.C. - Sleep apnea in 8 children. *Pediatrics,*58, 28-31, 1976.

20. GUILLEMINAULT C., KOROBKIN R., WINKLE R. - A review of 50 children with OSAS. *Lung,* 159, 275-287, 1981.

21. GUILLEMINAULT C., NINO-MURCIA G., HELDT G., BALDWIN R., HUTCHINSON D. - Alternative treatment to tracheostomy in obstructive sleep apnea syndrome : Nasal Continuous Positive Airway Pressure in young children. *Pediatrics, 78, 797-802, 1986.*

22. GUILLEMINAULT C., PELAYO R. - Narcolepsy in prepubertal children. *Ann. Neuro.* 43, 135-142, 1998.

23. GUILLEMINAULT C., WINKLE R., KOROBKIN R., SIMMONDS B. - Children and nocturnal snoring: evaluation of the effects of sleep related respiratory resistive load and daytime functioning. *Eur. J. Pediatr.,* 139, 165-171, 1982

23. JACOME D.E., RISKO M. - Pseudo-cataplexy: gelastic atonic seizure, *Neurology,* 34, 1381-1383, 1984.

24. KANDT R.S., EMERSON R.G., SINGER H.S., VALLE D.L., MOSER H.W. - Cataplexy in variant forms of Neimann Pick Disease. *Ann. Neurol.,* 12, 284-288, 1982.

25. KOTAGAL S., GOULDING P.M. - Characteristics of nocturnal sleep in children with narcolepsy. *Sleep Res.,* 24, 265, 1995.

26. KOTAGAL S., GOULDING P.M. - The laboratory assessment of daytime sleepiness in childhood. *J. Clin. Neurophysiol.,* 13, 208-218, 1996.

27. KOTAGAL S., HARTSE K.M., WALS J.K.- Characteristics of narcolepsy in preteen-aged children, *Pediatrics,* 85, 205-209, 1990.

28. LEACH J., OLSON J., HERMANN J., MANNING S.- Polysomnographic and clinical findings in children with obstructive sleep apnea. *Arch. of Otolaryngology Head and Neck Surgery,* 118, 741-744, 1992.

29. MARCUS C.L., KEENS T.G., WARD S.L. - Comparison of nap and overnight polysomnography in children. *Pediatr. Pulmonol.,* 13, 16-21, 1992.

30. MARCUS C.L., OMLIN K.J., BASINSKI D.J., BAILEY S.L., RACHEL A.B., KEENS T.G., WARD S.L.D. - Normal polysomnographic values for children and adolescents. *Am. Rev. Respir. Dis.,* 146, 1235-1239, 1992.

31. MELBERG A., HETTA J., DAHL N., NENNESMO I., BENGTSSON M., WILBOM R., GRANT C., GUSTAVSON K.H., LUNDBERG P.O. - Autosomal dominant cerebellar ataxia deafness and narcolepsy. *J. of the Neurol. Sciences.,* 134, 119-129, 1995.

32. NAVELET Y., ANDERS T., GUILLEMINAULT C. - Narcolepsy in children. *In: Narcolepsy,* Guilleminault C., Dement W.C., Passouant P. (eds), Plenum, New York, 1976.

33. NAVELET Y., BENOÎT O. - Somnolence diurne chez l'enfant et l'adolescent, *Neurophysiol. Clin.,* 21, 223-224, 1991.

34. NEVSIMALOVA S., ROTH B., TAUBEROVA A. - Perinatal risk factors as a cause of excessive sleepiness in early childhood. *In: Sleep 1990,* Horne J. (ed), Pontenagel Press, Bochum, 155-157, 1990.

35. PALM L., PERSSON E., ELMQVIST D., BLENNOW G. - *Sleep and wakefulness in normal preadolescent children.* Thesis, Lund, 1991.

36. PATOIS E., VALATX J.L., ALPEROVITCH A. - Prevalence des troubles du sommeil et de la vigilance chez les lycéens de l'académie de Lyon. *Rev. Epidémiol. Santé Publ.,* 41, 380-388, 1993.

37. REIMAO R., LEMMI H. - Narcoleptic children and adolescent sleep structure and multiple sleep latency test. *Sleep Res.,*20A, 382, 1991.

38. SAARENPAA-HEIKKILA O., KOIVIKKO M. - A questionnaire study of sleep habits and sleep disturbances among school children. *10th Congress of the Eur. Sleep Res. Society,* Strasbourg, May 20-25, 235, 1990.

39. SALFIELD D.J. - Narcolepsy in a child under 7 years old. *Arch. Dis. Child.,* 34, 538-539, 1959.

40. SAUVAGE D. - Tics apparus chez un enfant de 8 ans au cours du traitement d'une hypersomnie (syndrome de Gélineau) par un psychostimulant amphétaminique. *Neuropsychiatre de l'enfance et de l'adolescence,* 4, 176-177, 1983.

41. SCHWARTZ W.J., STAKES J.W., HOBSON J.A. - Transient cataplexy after removal of a craniopharyngioma. *Neurology,* 34, 1372-1375, 1984.

42. VAARA P., PARTINEN M., SEPPALA P., MICHELSSON K., PIHL S. – Snoring and sleep apnea among 5 years-old children in Finland. *J. Sleep Res.,* 1, Supplement 1, 476, 1992.
43. VECCHIERINI-BLINEAU M.F., TIBERGE M., CALVET U., JALIN C., DELEERSNYDER H. - Narcolepsie chez l'enfant : à propos de dix cas. *Neurophysiol. Clin.,* 22, 338, 1992.
44. WISE M.S., - Childhood Narcolepsy. *Neurology,* suppl 1: S37-S42, 1998.
45. WITTIG R., ZORICK F., ROEHRS T., SICKELESTEEL J., ROTH T. – Narcolepsy in a pediatric population. *Amer. J. Dis. Child.,* 102, 725-727, 1983.
46. YOUNG D., ZORICK F., COTTIG R., ROEHRS T., ROTH T. - Narcolepsy in a pediatric population. *Amer. J. Dis. Child.,* 142, 210-214, 1988.
47. YOSS R., DALY D. – Narcolepsy in children. *Pediatrics,* 1025-1033, 1960.

Chapter 36

Circadian rhythm sleep disorders
Introduction

M. Billiard
Service de Neurologie B, Hôpital Gui de Chauliac, Montpellier, France

The grouping of circadian rhythm sleep disorders dates back to the Diagnostic Classification of Sleep Disorders in 1979 [4], based on a shared cardinal feature: all these disorders represent one form or another of initial misalignment between the patient's sleep and wake behaviours and that which is desired or regarded as the societal norm. In this Classification the circadian rhythm sleep disorders were divided into two groups, transient and persistent disorders. Nowadays, in the International Classification of Sleep Disorders [3], this subdivision has not been retained. However some of these disorders are only of an extrinsic type whereas the others depend on intrinsic and extrinsic factors (table 36.1).

Table 36.1. Circadian rhythm sleep disorders

Disorders of an extrinsic type	Shift Work Sleep Disorder
	Time Zone Change (Jet Lag) Syndrome
Disorders with extrinsic and intrinsic factors	Delayed sleep phase syndrome
	Advanced sleep phase syndrome
	Irregular sleep-wake pattern
	Non-24-hour sleep-wake disorder

The first category includes two conditions, shift work sleep disorder and time zone change (jet lag) syndrome.

Examples of the former have been with us since ancient times. According to Roman Law delivery men had to work at night to restrict daytime traffic. Work on board ship has always been characterised by four hour watches. Bakers work during the night to provide fresh bread for their customers in the morning. Security guards work in relays over a 24 hour period. But shift work really began to expand with Edison's invention of electric light at the end of the 19th century, together with the development of increasingly expensive machinery needing to operate 24 hours a day to ensure rapid cost-effectiveness. Today more than a quarter of the working population is engaged in shift work. Yet a large number of those who work outside normal working hours complain of poor sleep and/or falling asleep at work [2]. This is due to the fact that the sleep and work times of these subjects are at variance with the external cycles of light/darkness, and those of social/consumer activities, named "Zeigebers" (from the German "time givers"). The consequences are not inconsiderable, as evidenced by the major catastrophes that have occurred in industry and transportation (Bhopal, Chernobyl, Exxon Valdez oil tanker) linked to the operators' sleepiness, given that the accidents happened at night. Nor must we overlook the reduction in professional performance levels and the conjugal or social difficulties experienced by those with alternate work schedules. Treatment for these disorders is based on reorganising work and rest hours and on therapies of which the two principle ones, melatonin and light therapy, are still at the experimental stage

Time zone change (jet lag) syndrome is naturally more recent, dating from the development of intercontinental air travel. This syndrome is designated by a series of disorders, difficulties of sleep onset or early morning awakening, headaches, a lowering of physical and cognitive performance, gastro-intestinal disorders, mood disorders, that affect intercontinental flight staff and their passengers on arrival at destination. Here again, these disorders are extremely common, considering that the number of passengers in transit through the main 22 airports worldwide in the year 2000 amounted to over a billion [1]. The factors involved include those common to all flights at high altitude, sleep deprivation before and during the flight, and the conflict between the individual's internal clock and the synchronizers at the place of destination. As for shift work, treatment for these disorders is still far from satisfactory. It consists of advice to flight staff and passengers essentially in regard to their behaviour before, during and after the flight, to avoid excessive sleep deprivation and manipulating the speed of resynchronisation. The use of melatonin is recommended for eastbound flights and a hypnotic or wakefulness-promoting drug may help in adapting to local conditions during the first days after arrival at the destination.

The second category of circadian rhythm sleep disorders comes within the same framework of misalignment between the patient's sleep and wake behaviours and that which is regarded as the societal form. There are however several points of difference.

In comparison with the disorders previously referred to, the latter are far less common or even, for some of them, quite exceptional.

These disorders may depend on factors within the subject's control, as in the case of the adolescent whose bedtime gets later and later as a result of watching television or listening to music, the student who studies at night, the artist whose rhythm is dictated by inspiration; or by intrinsic factors, the reduced sensitivity of the circadian system to light, a genetic predisposition, blindness, degenerative brain dysfunction, extreme night or morning typology, or an association of both.

They frequently occur in a particular psychological context, a taste for night life, a rejection of school, or a psychiatric context, schizoid personality disorder, or avoidant personality disorder.

The different disorders are misleading, as a rule: delayed sleep phase syndrome is taken to be a disorder of sleep initiation if the subject is not questioned as to the quality of his sleep (normal) and the time of his spontaneous morning awakening (late); and advance sleep phase syndrome is taken to be a disorder of sleep maintenance if no attempt is made to determine whether there is abnormal sleepiness in the early evening. Irregular sleep-wake pattern is taken to be a severe form of insomnia if the subject's daily schedule is not examined closely, while non-24-hour sleep-wake syndrome will fail to be recognised unless the daily shift of sleep-wake times in relation to conventional sleep-wake times is detected.

Another feature of this category of disorders is that consultation is not necessarily sought, and if a disorder is diagnosed, it is not always followed by treatment. Indeed quite a number of subjects affected by delayed sleep phase syndrome are not particularly bothered by it and do not really want to return to conventional times of sleep and wakefulness. The same is true for irregular sleep-wake pattern, whether this is a choice of lifestyle or part of a degenerative brain dysfunction.

Laboratory tests are called for to confirm the disorder and decide on treatment. These comprise at the least a morningness/eveningness questionnaire, a sleep-wake diary, an actigraphic recording of rest/activity over several weeks, a temperature curve for a 24 hour period, a personality test, and at the most a 24-hour polysomnography in an ad lib situation, a dim light melatonin onset (DLMO) , a 24-hour secretory pattern of growth hormone and cortisol, and a neuroimaging of the suprasellar region.

Treatment is based on chronotherapy, light therapy or the administration of melatonin, a melatonin agonist or vitamin B12. As a general rule, treating these disorders remains delicate, especially for subjects presenting a degenerative brain dysfunction or an associated psychiatric disorder, and no long term study is yet available.

REFERENCES

1. ACI – Airports Council International – P.O. Box 16, 1215 Geneva 15- Airport, Switzerland
2. AKERSTEDT T. – Sleepiness as a consequence of shift work. *Sleep,* 11, 17-34, 1988
3. AMERICAN SLEEP DISORDERS ASSOCIATION. ICSD – *International classification of sleep disorders, revised:* Diagnostic and coding manual. American Sleep Disorders Association, 1977
4. ASSOCIATION OF SLEEP DISORDERS CENTERS - *Diagnostic Classification of Sleep and Arousal Disorders,* First Edition, prepared by the Sleep Disorders Classification Committee, H.P. Roffwarg, Chairman, *Sleep* 2, 1-137, 1979

Chapter 37

A decision tree approach to the diagnosis of a circadian rhythm sleep disorder

M. Billiard
Service de Neurologie B, Hôpital Gui de Chauliac, Montpellier, France

INTRODUCTION

The differential diagnosis of circadian rhythm sleep disorder may be immediate or prove more delicate, depending on the case, i.e., immediately apparent from the clinical interview in the case of shift work sleep disorder or time zone change (jet lag) syndrome, but more delicate in the case of a disorder due to the circadian sleep rhythm pathologically escaping the influence of environmental synchronisers (endogenous circadian rhythm sleep disorder with an exogenous component), implying that the disorder must be evoked and then confirmed by the relevant additional tests.

DIAGNOSTIC CONSIDERATIONS OF THE DECISION TREE

To evoke a circadian rhythm sleep disorder with an endogenous component laboratory tests are required, the minimum being tests carried out in consultation i.e., morningness-eveningness questionnaire, handicap scale (some subjects affected by these syndromes experience no discomfort), psychological evaluation (a large number of patients suffer from a psychiatric illness) and at home (sleep diary, actigraphy, to ascertain the hours of sleep and wakefulness). But for added assurance, it is often advisable to supplement these initial tests with examinations carried out in a specialised sleep laboratory, including 24 hour polysomnography coupled with 24 hour temperature recording in an ad lib situation, dim light melatonin onset (DLMO) and 24-hour secretory pattern of melatonin and cortisol.

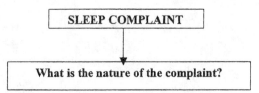

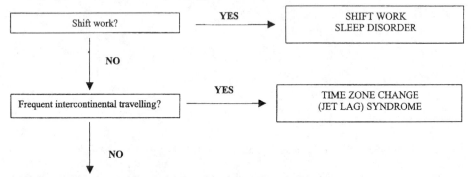

1. Is the complaint of insomnia and/ or hypersomnia related to:

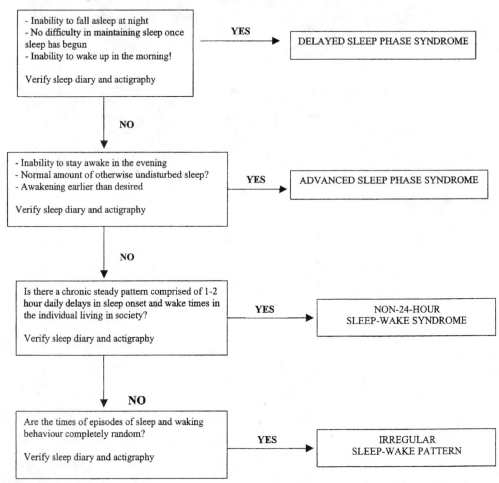

Chapter 38

Shift work sleep disorder

D. Leger
Centre du Sommeil et Consultation de Pathologie Professionelle, Hotel-Dieu de Paris, Paris, France

SHIFT WORK, DEFINITIONS, EPIDEMIOLOGY

Shift work applies to work in successive alternating time shifts, and may be organised in one of a number of different ways.

It developed out of the need to maintain activities related to human or material safety, over a 24 hour period. The police and the army, health care services, and transport (maritime, initially) have long been part of this tradition. Occupations like those related to food, *i.e.* bakers, wholesalers, also had to prepare goods at night to ensure maximum freshness for daytime consumption. From the 19th century onwards, shift work spread to industrial activities in which the interruption of processes each night, would seriously disrupt production. The simplest example is that of high furnaces which require continuous stoking. But shift work really began to be organised in the 20th century, in the bid to derive maximum potential from industrial equipment which quickly became obsolete, within a context of fierce international competition.

New IT and "communications" jobs also demand "continuous" work or take up long periods of time. Elsewhere, night work carried out by women, which up to 1990 was restricted to the health sector or child care, has spread to professions of all types in keeping with European regulations on job equality between men and women. Work "teams" nevertheless are chiefly composed of men, and operate mainly in big firms (500 employees and over) in the industrial sector. More than one wage earner in four works in a shift system. Work as two teams is the most frequent, involving nearly two thirds of shift workers, while work in three teams (or three 8-hour shifts) involves roughly 23% of shift workers.

Definitions and descriptions of shift work

Shift work is defined by criteria such as continuity, type of rotation and shift alternation.

Continuity is a function of the activity.

Certain types of shift work cannot be interrupted, such as health care or security services. Others can be suspended for the weekend or during vacations: these are industrial activities which do not rely on "continuous stoking", but where shift work is justified by the wish to maximise equipment potential.

Continuous shift work. This is conducted without interruption and relies on three active shifts and one resting shift. A time overlap must be planned between each shift to allow for the transmission of information from one team to the next. The length of the shift usually fluctuates between 8 hours 10 minutes and 8 hours 20 minutes, depending on the complexity of the task and transmission. These rhythms may be referred to as 3X8, 4X8, 5X8.

Semi-continuous shift work comprises three teams with an interruption at the weekend and often during the holidays (3X8). No additional team is needed therefore, to cover the resting shift. Some workers alternate their rest periods, because of the recuperation time imposed by the 35 hour working week.

Discontinuous shift work. This applies to certain industrial tasks or services which require an extended period of work, but which do not provide a public service or need to maintain operation for 24 hours. Many work schedules can be individually adapted: 2X8, 2X10, 2X8 + 5, 2X10 + 4 etc. The implementation of the 35 hour working week in France has encouraged the development of this type of rotation in firms.

The type of rotation is defined by three factors:

- rhythm, relating to the number of days spent in the same shift. The rhythm is said to be short when the shift only lasts from 1 to 3 days. A rhythm is long if the same shift lasts more than 5 days.
- orientation *i.e.* whether clockwise or anticlockwise.
- rotation cycle describing the alternation of shifts *e.g.* morning, afternoon, evening, or morning, evening, evening, morning etc.

Fixed or alternating teams

- the teams are fixed when the worker always works the same shift. This type of organisation is common in hospitals, for instance, where the person is always on duty at the same time: daytime, evening, night shift.
- teams are said to alternate when workers regularly change shifts.

The duration of each shift

The standard length of a shift is roughly 8 hours. Recent negotiations regarding the 35 hour week in France have sometimes led to shifts lasting 10 or even 12 hours.

Factors of adjustment, pathophysiological aspects

Shift work or night work is not physiological. Most physiological rhythms are designed to function in modes of daytime/wakefulness, night time/sleep. Man is nevertheless able to adjust to different schedules, but not without difficulties in the short- or long-term. Three sets of factors contribute to adjustment (or to maladjustment) to shift or night work: chronobiological factors, "sleep" factors, and domestic factors. T. Monk has clearly demonstrated the way in which these three interlink. A disturbance to one set of factors may result in an imbalance and maladjustment [34].

Chronobiological factors

When an individual is exposed to a phase shift of 6 hours, for instance, circadian rhythms do not readjust to the new timetable at the same rate. The sleep-wake rhythm is usually the fastest, taking two to three days to adapt, and temperature rhythm is the slowest (average adjustment in 8 days). Hormone rhythms may sometimes take three weeks to adjust. This results in the desynchronisation of circadian rhythms, which no doubt explains the discomfort felt by individuals exposed to a phase shift.

Shift workers are regularly exposed to this type of shift in timetable. In cases of three 8 hour shifts, for example, they submit to an 8 hour change in timetable every week, to which is added the change at the weekend, when the worker usually tends to comply with daytime hours.

The days immediately following the shift alteration are the most unpleasant. If a worker goes from an afternoon to a night time shift, for example, the period from 2 to 5 am will be particularly difficult. Indeed, the worker will need to carry out an important task during a period of minimal temperature, with a maximum risk of low vigilance. The risk of error or accident is considerable. Alternatively, it will be more difficult for him to get to sleep the following morning, simply because his temperature is too high to obtain really refreshing sleep.

Moreover, readjusting to circadian rhythms after a change in shift work routine is made all the more difficult by the lack of synchroniser or "Zeitgeber" action. Social rhythms and bright light are the two most powerful synchronisers of circadian rhythms. When the shift worker changes to a night routine, these two synchronisers are no longer present. Darkness and the lack of social stimulation (outside work) counteract adjustment to the new timetable. Conversely, readjusting to the daytime routine is facilitated by the action of these synchronisers.

A number of chronobiological characteristics may serve to predict potential adjustment to time shifts and thus to changes in work schedule:
- a clockwise shift in rotation appears to be better tolerated than one in the opposite direction [6, 21].
- being a "night" person rather than a "morning" person [44]. This tendency is sometimes pronounced and probably involves a genetic etiological component,
- being of a young age [39].

We are still unable, however, to obtain indicators predicting a subject's adaptability to phase shifts. According to Quera-Salva *et al.* [44], the way the internal melatonin rhythm adjusts may nonetheless provide a possible indication of shift work adaptability (observed among a group of night nurses).

The "sleep" factor

Sleeping well is an essential tolerance factor in relation to shift work. But, for various reasons, it appears that in the shift worker population, sleep is often disturbed both in quantity and quality.

This is firstly for chronobiological reasons: in fact as explained earlier, it is not possible to sleep in the same way at any time of the day or night. Deep NREM sleep occurs mainly at the beginning of the night in a subject with a diurnal schedule. When a change occurs to this rhythm, the sleep deficit or debt will result in NREM sleep, but in reduced quality and quantity in relation to the norm during the first two or three days of adjustment. REM sleep cannot occur at any time during the 24 hour cycle. The night worker may obtain a little REM sleep during the morning after a night's work, but not during the sleep period which ensues, particularly in the afternoon at the end of his sleep period, for the first days of adjustment, at least [2].

Secondly, for environmental reasons. In fact, it is very difficult to isolate oneself sufficiently to sleep during the daytime. Subjects wishing to sleep during the day are disturbed by the light, the noise, temperatures in summer and by the social routine.

And thirdly for social and family reasons. The role of partner or of mother or father, involves sharing time with the family or with friends, which often interferes with opportunities to recuperate sleep.

The result is often a "chronic sleep debt", in shift workers, who have an average of one to two hours less sleep per day than "daytime" subjects [22].

Sleep problems in shift workers include all types of sleep disturbance: sleep onset disorders, sleep maintenance insomnia with the presence of nocturnal awakenings and difficulty getting back to sleep, and insomnia due to premature awakening. These disorders can arise early or late, sometimes after years of working a shift schedule. It is therefore important to be especially attentive, as they may be the warning signs of maladjustment to shift work.

Domestic factors

Tolerance to shift work also requires a favourable domestic environment. Shift work is called "work at asocial hours" in Denmark [7] and all the inquiries reveal this disruption of family and social relations in regard to alternate hours.

The domestic problems linked to shift work particularly affect young women with small children, who have the greatest difficulty recuperating normal sleep duration and who often have sleepless nights. This was particularly well illustrated for part time night nurses [15].

But the family life of fathers is also affected. Lenzing and Nacgreiner [30] show for example, that the children of shift workers have fewer friends and play less easily than the children of day workers.

The three types of factors we have just considered jointly affect a subject's tolerance to shift work. If one of these is disrupted, the worker, even though well adjusted for several years, sometimes, may quickly become maladjusted to the work and cease to tolerate shift work. Vigilance is needed on the part of the clinician or the work doctor to try to anticipate maladjustment by detecting signs of intolerance at an early stage. Sleep disorders appear to be one of the best preliminary signs of de-adaptation to working shift hours.

Sleep disorders related to shift work

Sleep is a good indicator of work tolerance even for those who work in the daytime. Leigh showed that insomnia was the most significant predictive factor for absenteeism (among the 37 independent variables studied) [23]. Jacquinet-Salord also found insomnia to be a major predictive factor in the employee's poor perception of work [20].

In the context of shift work, these sleep disorders are often evoked by workers, and principally include:

Shortened sleep time in relation to that of daytime workers is a standard complaint in studies on shift work [16, 43, 45]. These studies are based either on the simple (but not always reliable) use of sleep diaries, or on more objective parameters (actigraphy, polysomnography [16, 45]. Shortened sleep time each day gradually creates a "sleep debt". This debt may be compensated by some, at least partially, during days off. For others it represents a net loss of one or two hours per 24 hours, compared with the daytime worker. Recuperating this debt is an important factor in adjustment. The young are often better able to recuperate their sleep debt at the weekend, than are older workers.

Sleep quality is perturbed both for environmental and for chronobiological reasons. The main complaints evoked by shift workers are those of difficulties getting to sleep and the presence of frequent awakenings, followed by the complaint of non-refreshing sleep with tiredness on getting up.

Other clinical complaints from shift workers

Fatigue

Chronic fatigue even at the moment of getting up is reported by a large number of night workers (from 36 to 68% depending on the group) [36]. In a review of the consequences of shift work, Akerstedt [3] reported that the fatigue of shift workers has often been confused with sleepiness. The prevalence of sleepiness during shift work is impressive. Akerstedt [3] reported having found the complaint of fatigue or sleepiness at work among 80 to 90% of workers studied in different occupations (police, steelworkers and meteorologists). In another study involving 1000 locomotive drivers, Akerstedt *et al.* [4] show that 11% admitted drowsiness during night trips. 59% admitted having experienced drowsiness at least once. An objective study (with MEDILOG sleep recording) carried out on locomotive drivers demonstrated the reality of this sleepiness, with several drivers falling asleep while driving [2].

Somnolence can have an effect on work and the safety of workers [26]. In a keynote study Bjerner *et al.* [7] showed the effect of night work on the risk of errors in a gas works firm. On a more alarming note, in terms of safety, Hildebrandt *et al.* [19] also found that the reactions of locomotive drivers to safety signals were less accurate at night than in the daytime. The time of work considerably affects the risk of accidents, particularly among drivers of heavy vehicles [18]. We have also demonstrated an increase in accident risk during chronobiological periods of lowered vigilance, based on automobile accident statistics in the USA, in 1988 [24]. The accident risk is a function of the number of hours worked and the circadian period.

Little attention has been focused on accident risk linked to sleepiness in the normal work setting. The reports issued after industrial catastrophes such as Chernobyl or Three Mile Island nevertheless suggest that these accidents were linked to a human error related to work times [5].

On the whole, chronic fatigue and sleepiness are a habitual consequence of shift work. This is partly related to the influence of chronobiological factors but is mainly influenced by sleep debt. The consequences in terms of accidents are a measure of their seriousness.

Digestive disorders

These represent a habitual complaint among shift or night workers. The most common disorders are those of dyspepsy, flatulence and intestinal transit. A number of recent studies have shown a higher prevalence of helicobacter pylori infection in shift workers [50]. Indeed, shift work hours make it more difficult to organise regular canteen meals and the traditional "snack" is not always well balanced.

These disorders are mainly due to the lack of nutritional balance in these meals. The chronobiological factor probably plays a part also. Poor food assimilation may stem from the lack of synchronisation between the sleep-wake rhythm and the other biological rhythms. This would also account for the weight gain often found in these workers, particularly among night nurses.

Other disorders

Over-consumption of coffee and tobacco is also common among shift workers. This may explain the increase in cardio-vascular morbidity often evoked among shift workers. However, to our knowledge, no case control study respecting the covariables (tobacco, coffee, obesity) has demonstrated a higher incidence of cardio-vascular disorders solely related to shift work.

Obesity is in fact higher among shift workers [41].

The increased risk of miscarriage and hypofertility has been evoked among night nurses. But prospective studies have found no indication of a higher risk of miscarriage linked to shift work [13]. Fenster in a prospective study of 5144 pregnant women in California, found no higher risk of spontaneous abortion in relation to shift work [13].

Degenerative desynchronisation-maladaptation syndrome in the older shift worker

The shift worker who returns to a day job after years of alternating work schedules, either because of retirement, or a later change in work shifts, may present desynchronisation-maladaptation syndrome. Sleep disorders are the first to appear with the inability to adapt to a normal daytime routine. This leads to daytime fatigue and sleepiness. The sleep recordings we have conducted with some of these patients [11] reveal destructured sleep (with the disappearance of a cyclic pattern), poor in deep NREM sleep and presenting numerous short arousals. This period of de-adaptation is reported by many retired shift workers. It may nevertheless improve after several months.

Organisation of shift work

The way in which shift work is organised varies considerably from one firm to another. It is often the result of tight wage negotiations between bosses and unions and acquired advantages related to the socio-demographic characteristics of workers and financial realities. It is important to bear in mind the contradiction between physiological regulations and the interests expressed by the employees. It is not uncommon to find two or more different rotations in the same firm.

Even the best organisation needs to adopt fairly standard patterns because of the dual necessities of:
- on one hand, ensuring the maintenance of installations 24 hours a day and often 7 days a week,
- and on the other, respecting regulations in regard to maximum working hours per week (48 hours according to the orientation laws for the 35 hour week).

It is thus important to avoid being too theoretical when faced with the task of advising on work organisation. Any organisation of shift work must fully respect the chronobiological, "sleep" and domestic factors already referred to. These ultimately relate to the type of work and population concerned. A pragmatic approach will consider the added benefits which compensate for work discomfort and which would probably be challenged if the organisation were changed. Lowden and Akerstedt give the example of the positive influence, on workers in a supermarket chain, of choosing their own work schedules, in terms of work satisfaction and quality of life [33].

In the early 80s, European chronobiology experts tended to favour a system of fast alternating shifts. This pattern had the advantage of entraining fewer chronobiological perturbations, as the subject remained essentially orientated to the daytime rhythm. But this type of rhythm is somewhat difficult to maintain in terms of sleep and is hard to adopt from the social point of view.

North American chronobiologists for their part tended to advise fixed shifts. These have the advantage of exerting a less traumatising effect on sleep. On the other hand, there are drawbacks in terms of social integration and chronobiology (indeed, subjects reorientate to daytime patterns at the weekend and are thus out of phase after each period of rest).

The shift should not exceed 10 hours in duration, except for work which requires low physical or mental input: surveillance or intermittent responsibility with long periods with no responsibility. Several studies, reviewed by Rosa [40] have shown that an increase in the duration of a shift may be accompanied by a significant decrease in performance and a greater risk of accidents [17, 35]. This is particularly marked at night. It is also why it is so important to analyse the following factors before extending any work hours: hierarchical level, physical or mental input, type of rotation, physical environment, accident risk, breaks, time the shift starts, domestic and social responsibilities. The directives issued by the European Union recommend, for instance, that night shifts should not exceed 8 hours of consecutive work [12].

Shift work is thus never ideal. Many experts today think that the least harmful solution consists in associating alternating daytime schedules of long periods of at least a week, with fixed night work reserved for volunteers (with the possibility of changing shifts on request). Night work would naturally be associated with compensation in the form of financial rewards or additional days off, compared to the other shifts.

In the absence of ideal shift work, improving working conditions as advised by the Committee for Hygiene and Safety and for Working Conditions, as well as by work doctors, may help to improve tolerance to shift work. In fact, these bodies rarely include night workers. The under-representation of this category of workers is generally detrimental to the respect for regulations. The most useful of these are the following:

Improved meals. The ideal would be to serve a hot meal at night, as is the case in the daytime. Failing this, there must be easy access to well-stocked distributors, containing snacks and hot or cold drinks. It has been shown that a carbohydrate-protein ratio of about 3 is optimal to maintain a good mood and good performance during shift work [37].

Light must be bright. Light is essential to maintaining alertness. It can also help biological rhythms to adjust (see part 6). This type of lighting may sometimes be difficult for night workers to accept, as they tend to shun bright lights, and employers consider it expensive. Its importance needs to be explained.

Arrangements for breaks must be as flexible as possible, but allow for a proper break from work rather than being systematically placed at the end of the shift to "gain" work time. In some monotonous working conditions (surveillance or testing) a longer break is recommended with the possibility of taking a good nap.

Noise. Noisy working conditions disturb alertness and is an accident risk factor. It is thus important to advise action against noise during shift work.

Transportation. This is a social complement to night work, and the employer must ensure the accessibility of his firm. Facilities for individual transportation (creating car parks, for example) will also help to improve the length of sleep and thus adjustment.

Child care. This is no longer simply a question of improving working conditions but involves taking responsibility for shift work which has an effect on the quality of life of the shift worker and thus on his tolerance for this work.

Therapy and surveillance

Is it necessary to propose therapy for shift workers when signs of intolerance develop? Surely the best attitude is to remove the risk by offering a return to normal rhythms on a daytime shift. This is the ideal solution, but one which is often hard to apply today, partly because many firms do not have a collective agreement providing for the automatic reclassification of shift workers who are unable to adjust – even for short periods of readjustment *i.e.* after sick leave or maternity leave. The other reason is that most shift workers do not have insurance arrangements allowing for compensation a drop in salary linked to changing to a daytime shift.

Because of this, shift workers often refuse to consider themselves as ill adjusted to the work. However, most of them present clinical signs of intolerance at some stage in their lives, which could be countered by therapeutic means for some time at least.

Hypnotic drugs

When sleep disorders are related to the disturbance of one of the three shift work adjustment factors, treatment may be envisaged with hypnotic drugs such as Zolpidem or Zopiclone for a maximum period of 4 weeks.

This period should provide an opportunity to assess the worker's sleeping habits using a sleep diary in which the patient records his sleep periods each day.

Melatonin

Melatonin has been tried with success in therapeutic protocols in subjects with shift work schedules.

Melatonin is a hormone which is both secreted and released by the pineal or epiphysial gland. Its release is eminently circadian as the nocturnal rate is 10 to 20 times higher than the diurnal. Light stimulation entrains the inhibition of melatonin secretion through the intermediary of the retino-hypothalamic pathways. Conversely, suppressing light stimulation will release epiphysial melatonin.

It was only in 1980 that Alfred Lewy *et al.* [32] demonstrated the effect of exposure to bright light (over 3000 lux) on the suppression of melatonin secretion. Parallel to this, Rosenthal *et al.* [42] demonstrated the effectiveness of bright light in the treatment of seasonal depression. Evidence was then found of the effect of bright light on resynchronising circadian rhythms in shift work, jet lag, and phase shifts (phase advance and delay).

Is melatonin really an external synchroniser enabling the modulation of circadian rhythms or just an indicator offering an easy measurement of rhythm changes, as described by Lewy *et al.* [31]?

Recently, Quera-Salva *et al.* [38] conducted a study of 40 nurses, 20 of whom worked at night, showing that a fast adjustment of internal melatonin rhythm (measured by the urine dosage of 6-sulfatoxymelatonin) could act as an indicator of better tolerance to night work. The faster the melatonin adjusted, the better the workers' tolerance of rhythm changes.

Various experiments have used melatonin directly, to assess its effect on resynchronising circadian rhythms in subject whose rhythms were artificially shifted due to jet lag or shift work [14]. These studies show that administering melatonin enables faster adjustment to rhythm changes.

It is thus indicated, theoretically, for incidents of this type, its role being to resynchronise the biological clock. To do so, it should be administered two to three hours before bedtime.

Despite this, no treatment based on melatonin has received authorisation for use in the French market at the time of writing. In addition to its effect on rhythms the potential indications for melatonin appear to be multiple but poorly understood: antioxidant effect, antidepressant, contraceptive... The *Agence du Médicament* in France and drug control authorities in most European countries currently prohibit the importation of melatonin from the USA. Its use is reserved for a future date once the tests have been completed.

The role of light in adjusting to and preventing disorders related to shift work

The alternations between light and darkness exert a marked effect on our circadian rhythms. Experiments of temporal isolation in underground caves have demonstrated the existence in man of a distinct endogenous rhythmicity which evolves with a periodicity of roughly 24 hours, known as circadian (circa-diem = around a day) [46, 47]. A lack of exposure to light will desynchronise the sleep-wake rhythm. This rhythm is regulated by an internal clock situated in the suprachiasmatic para-ventricular nuclei of the hypothalamus. The clock is affected by external factors which may vary its periodicity. These stimulant factors are referred to as "zeitgebers" (from the German: time givers) or synchronisers [49]. Synchronisers have only a limited effect on entraining endogenous rhyhms, however.

The influence of light as a synchroniser was discussed at a very early stage, if only because of the logical relationship between wakefulness and daytime and sleep and night. To determine the influence of light, experiments were practised for a long time using weak light, but with little success [49] before the discovery was made recently of the decisive effects of high intensity bright light.

Phase shift experiments using normal light

Phase shift experiments alter light rhythms to simulate the shift work situation with a shift of 6 to 8 hours. Thus situations related to day and night are used (maintaining autonomous lighting), as they are closer to reality. It is possible for example, after a reference period of several days, to simulate an anticlockwise shift of 6 hours, and several days, to simulate a new 6 hour shift in a clockwise direction to regain the initial position. Results show that temperature rhythm adapts to the synchroniser rhythm, even though this is slower than for the sleep-wake rhythm. Adjustment is more difficult in the anticlockwise than in the clockwise direction.

Six studies with identical protocols have formed the basis for the following remarks [49]:

1) Resynchronisation is faster for a clockwise shift (1.2 hour/day) than for an anticlockwise shift (0.9 hour/day). Subjects show greater fatigue after an anticlockwise shift than after a clockwise shift.

2) Resynchronisation varies in duration with the amplitude of temperature rhythms. If the rhythm is of weak amplitude, resynchronisation is short. Likewise, there is a relationship between temperature rhythm amplitude and the proportion of sleep during the day: short sleepers adapt more quickly than long sleepers.

3) Fatigue is more marked if the original phase of temperature rhythm decrease occurs early. "Night" people suffer less than "morning" people. The last two remarks provide an easy prediction of tolerance to a shift in hours among individuals and may be of practical interest. The shorter the sleep and the more the sleeper tends to be a "night" person, the better his adjustment to phase shifts.

However the above studies have demonstrated that behavioural signals in a constant light environment (whether lit or in darkness) constituted a powerful synchroniser whose entrainment interval was no lower than that of absolute day-night (in ordinary light).

Effects of artificial bright light (> 2500 lux)

However a contradiction exists between the apparently marginal effect of light on human circadian rhythms, and the fact that the same light represents one of the most powerful stimuli in animal circadian studies.

New perspectives were opened when Lewy [32] observed that only light of high intensity (over 2500 lux) could entrain the suppression of melatonin secretion in man, whereas in the animal, suppression can be obtained with light of much lower intensity. Light acts through the optic pathways sending projections to the suprachiasmatic nuclei. The latter act on the epiphysis through the hypothalamo-epiphysial pathways.

But all the preceding studies were conducted using light of a much lower intensity than that required to suppress melatonin. Thus the interest of repeating these studies using bright light of above threshold intensity, involving heavy technical equipment.

Czeisler *et al.* [9, 10] demonstrated the efficacy of bright light in experimental shift work situations. Resynchronisation occurs significantly faster than in placebo conditions.

Light in the workplace

Article R.232-6-2 of the Work Code, amended by the decree of August 2, 1983 (O. J. of August 5) recommends the minimal values for lighting in the workplace:

- in windowless premises used for permanent work: 200 lux
- in work places, changing rooms and sanitation facilities: 200 lux
- in external areas where work is permanently conducted: 40 lux.

The memo issued on the 11th of April 1984 defined the minimal level of lighting for certain activities, light being derived from lighting focused on the work area, in addition to general lighting:

- general mechanics, typing, office work: 200 lux
- small instruments, drawing office, mechanography: 300 lux
- fine mechanics, engraving, colour comparison, complex drawings, clothing industry: 400 lux
- precision mechanics, fine electronics, various testing: 600 lux
- highly complex tasks in industry and laboratories: 800 lux.

The use of daylight is nevertheless recognised by legislation, particularly by decree n° 83-722 of the 2nd of August 1983, fixing "the obligations of site overseers in regard to plans for construction operations in the interests of hygiene and work safety". This decree stipulates in articles R.235-2 and R.235-3 that "unless there is incompatibility with the nature of the activity, justified by the site overseer, the design of new working premises must include the use of natural light and views to the exterior".

Despite recent regulations, many workers find themselves in working conditions which include little or no natural light.

This is either because they are engaged in activities which are exempted by the decree, as the activity is incompatible with natural light: this is the case for instance of miners, underground transport employees, sewage workers, photographers, radiology operators …

Or else, as is usually the case, their workplace was built before the decree or is located in restricted urban conditions to the extent that every area, including windowless premises, is exploited.

Quite apart from the disorders linked to the lack of rapid resynchronisation in shift workers, dim lighting may give rise to other disorders. Eye complaints and psychological or behavioural functional signs have been found in studies conducted in this type of light environment. Woodson *et al.* [48] reported for example that subjects working in artificial light complained more often of fatigue and irritability, than did subjects working in natural light.

Their complete inability to perceive light gives rise to severe problems of alertness in the blind [28]. For practical purposes, we studied 21 subjects working as radiology operators in Parisian hospitals and 15 control subjects in the same professional category but who worked in daylight, both in summer and in winter [25, 27]. The study looked at sleep-wake rhythms and made an assessment of fatigue and the coupling of vigilance-sleepiness using a validated subjective scale. We found no differences between the groups in terms of sleep. By contrast, subjects exposed to daylight felt more alert, more satisfied by their day and less often sleepy than subjects working under electric light.

Bright light and shift work: practical aspects

Light, as we have seen, is one of the most powerful synchronisers of circadian rhythms. It must be used (and avoided) carefully to favour a better adjustment to shift work rhythms [29].

The use of high intensity light is still at the experimental stage in the work setting. Nevertheless it is certainly advisable to use it more systematically in the years to come (without obligation), recommending for example:

- to the worker taking up a night shift: to expose himself to light as much as possible in the evening and to use maximum lighting at night until 2 or 3 am. In the morning, on the contrary, he should avoid exposure to daylight before going to bed (wearing sunglasses if possible, on the way home in the summertime, and blocking the light out of the bedroom).
- To the worker taking up a morning shift: to expose himself to as much light as possible in the morning and to avoid it at night after 5 pm.

Arranging rest periods: naps. Surveillance of shift workers

Naps

Because the shift worker's sleep deteriorates, compared to that of the daytime worker, (on average one or two hours at least), he should be advised to take naps. The positive effect of these naps has been demonstrated by several authors [2, 8]. They help to reduce the sleep debt and the anxiety generated by sleep disorders affecting a single episode. They also provide better adaptation to social rhythms. Naps may vary in duration. To avoid be awoken from deep sleep and taking a long time to regain alertness (due to what is referred to as sleep inertia) two types of nap can be advised.
-   Short naps lasting less than 20 minutes, the subject rests but does not go into deep sleep (the alarm must be set at 20 minutes).
-   Naps lasting at least a cycle (1 hour 15 minutes to 1 hour 45 minutes) during which the subject goes into deep sleep. This does not necessarily require setting the alarm.

The timing of naps should be arranged with the subject according to the shift work hours, travelling time and the usual family and social activities. It is best to set a time which is not too rigid and involves simply resting with or without sleeping, in comfortable clothing, without complete darkness.

Surveillance

The law in France calls for special medical checkups carried out by the work doctor, in the case of employees engaged in shift work. In our view this ruling must be adhered to in order to detect maladjustment to shift work. Checkups should ideally be carried out twice a year. In the course of checkup, the doctor will consider the following aspects:

Factors of adjustment to shift work

This should on no account be considered as a criterion for selecting employees for shift work and individuals should maintain their flexibility in choice of schedule [1]. But certain factors are known to make adjustment more difficult:
-   the age of 50 years or over
-   presence of heavy domestic work
-   the "morning" rather than the "evening" type
-   previous history of sleep disorders.

Recent apparition of sleep disorders

The extent of these can be determined by use of a sleep diary for three weeks.

Work or traffic accidents

The interrogation will assess the times of occurrence and any relationship to lowered vigilance.
Surveillance is also intended as a reminder of the rules of prevention and advice on hygiene and diet.

CONCLUSION

In addition to standard shift work, many working situations today involve time constraints affecting the hours of sleep and rest. The transformation of modes of communication, through internet and cell phones, also has repercussions on the time commitments of many categories of worker. Many firms are reconsidering these issues in the light of the 35 hour working week in France, stimulating a closer look at the concept of "working hours". This often reveals the existence of expandable availability for work, making increasing incursions into leisure time and family life.

Sleep deprivation resulting from professional commitments may have dramatic consequences, on the individual scale as well as on a collective scale (catastrophes such as Three Mile Island or Bhopal) [3]. It is the responsibility of every doctor to be able to detect situations of this kind.

REFERENCES

1. ASHKENAZI I E., REINBERG A.-E., MOTOHASHI Y. - Interindividual differences in the flexibility of human temporal organization: pertinence to jet-lag and shiftwork. *Chronobiol. Int.* 14,99-113, 1997.
2. AKERSTEDT T., TORSVALL L., FROBERG J.-E. - A questionnaire study of sleep-wake disturbances and irregular schedules. *Sleep Res.* 12, 358, 1983.
3. AKERSTEDT T. - Adjustment of physiological circadian rhythms and the sleep-wake cycle to shiftwork, in: *Hours of work-Temporal factors in work scheduling*, Folkard S., Monk T.H. (eds), John Wiley and Sons, New York, 119-210, 1985.
4. AKERSTEDT T. - Sleepiness as a consequence of shift work. *Sleep* 11,17-34, 1989.
5. A.P.S.S. (Association of Professional Sleep Societies) Committee on catastrophes - Sleep an public policy. Consensus report. *Sleep* 11,23-26, 1988.
6. ASCHOFF J., HOFFMANN K., POHL H., WEVER R. - Re-entrainment of circadian rhythms after phase-shifts of the Zeitgeber, *Chronobiologica* 2, 23-78, 1975.
7. BJERNER B., HOLM A., SWENSSON A. - Diurnal variation of mental performance. A study of three shift workers, *Br. J. Ind. Med.* 12, 103-110, 1955.
8. BONNET MH., ARAND DL. - Consolidated and distributed nap schedules and performance. *J. Sleep Res.* 1997;4:71-77.
9. CZEISLER C.-A., JOHNSON M.-P., DUFFY J.-F., BROWN E.-N., RONDA J.-M., KRONAUER R.-E. - Exposure to bright light and darkness to treat physiologic maladaptation to night work. *N. Eng. J. Med.* 322, 1253-1259, 1990.
10. CZEISLER C.-A., SHANAHAN T.-L., KLERMAN E.-B., MARTENS H., BROTMAN D.-J., EMENS J.-S., KLEIN T., RIZZO J.-F. - Suppression of melatonin secretion in some blind patients by exposure to bright light. *N. Eng. J. Med.* 332, 6-11, 1995.
11. DEVIENNE A., LEGER D., HABERT C., CASSUTO D., CHAHBENDERIAN M., VICTOR V., DOMONT A., PROTEAU J. - Aspects dégénératifs du travail de nuit. *Arch. Mal. Prof.* 54, 660-662, 1993.
12. EUROPEAN UNION. - 93/104/EU. Directive of the council of the EU concerning certain aspects of working hours, 1993.
13. FENSTER L., HUBBARD A.-E., WINDHAM G.-C., WALLER K.-O., SWAN S.-H. - A prospective study of work-related physical exertion and spontaneous abortion, *Epidemiology* 8, 66-74, 1997.
14. FOLKARD S., MONK T.-H., LOBBAN M.C. - Short and long-term adjustment of circadian rhythms in "permanent" night nurses. *Ergonomics* 21, 785-799, 1978.
15. FOLKARD S., ARENDT J., CLARK M. - Can melatonin improve shift workers' tolerance to the night shift? Some preliminary findings. *Chronobiol. Int.* 10, 315-320, 1995.
16. FORET J., BENOIT O. - Sleep recordings of shiftworkers adhering to a three to four day rotation, *Chronobiologia* suppl 1, 45-56, 1978.
17. HAENECKE K., TIEDEMANN S., GRZECH-SUKALO H., NACHREINER F. - Accident risk as a function of hours of work and time of the day as determined from accident data and exposure models for the German working population. *Scand. J. Work Environ. Health* 24, 43-48, 1998.
18. HAMELIN P. - Les conditions temporelles de travail des conducteurs routiers et la sécurité routière. *Trav. Hum.* 44 ,5-21, 1981.
19. HILDEBRANDT G., ROHMERT W., RUTENFRANZ J. - 12 and 24 hour rythms in error frequency of locomotive divers and influence of tiredness, *Int. J. Chronobiol.* 2, 175-180, 1974.
20. JACQUINET-SALORD M.-C.; LANG T., FOURIAUD C., NICOULET I., BINGHAM A. - Sleeping tablet consumption, self reported quality of sleep, and working conditions. *J. Epidemiol. Community Health* 47, 64-68, 1993.
21. KLEIN K.-E., WEGMANN H.-M., HUNT B.-I. - Desynchronization as a function of body temperature and performance circadian rhythm as a result of outgoing and homecoming transmeridian flights. *Aerospace Med.* 43, 119-132, 1972.
22. KNAUTH J., LANDAU K., DROGE C., SCHWITTECK M., WIDYNSKI M., RUTENFRANZ J. - Duration of sleep depending on the type of shift work, *Int. Arch. Occup. Environ. Health* 46,167-177, 1980.
23. LEIGH P. - Employee and job attributes and predictors of absenteeism in a national sample of workers: The importance of health and dangerous working conditions, *Soc. Sci. Med.* 33, 127-137, 1991.
24. LEGER D., PROTEAU J. - Sommeil, lumière et travail, *Arch. Mal. Prof.* 52, 513-515, 1991.
25. LEGER D., DAURAT A., BENOIT O., PROTEAU J. - Seasonal evolution of sleep and alertness of radiologist as working without exposition to bright outdoor light. *Sleep Res.* 22, 247, 1993.
26. LEGER D., de LA GICLAIS B., HAMMO S., DEHAYE J., VOISIN C., RAIX A., PROTEAU J. - Accidents mortels liés à la somnolence chez des chauffeurs routiers: de la réglementation à l'application, *Arch. Mal. Prof.* 55, 291-292, 1993.
27. LEGER D. - The cost of sleep-related accidents: a report for the national commission on sleep disorders research, *Sleep* 17, 84-93, 1994.
28. LEGER D., DOMONT A., BENOIT O. - Evolution du sommeil et de la vigilance d'opérateurs radiologiques travaillant sans exposition à la lumière du jour, *Arch. Mal. Prof.* 56, 324-326, 1995.

29. LEGER D., GUILLEMINAULT C., DEFRANCE R., DOMONT A., PAILLARD M. - Blindness and sleep patterns, *Lancet* 348, 830-831, 1996.
30. LENZING K., NACHREINER F. - Effects of fathers' shift work on children-results of an interview study with children of school age, in *Shiftwork in the 21st century*, Hornberger S, Knauth P, Costa G, Folkard S. (eds), Peter Lang, Frankfurt, p 399-404, 2000.
31. LEWY A.-J., WEHR T.-A., GOODWIN F.-K., NEWSOME D.-A., MARKEY S.-P. - Light suppresses melatonin secretion in humans. *Science* 210, 1267-1269, 1980.
32. LEWY A.-J., SACK R. - The dim light melatonin onset as a marker for circadian phase position. *Chronobiol. Int.* 6, 93-102, 1989.
33. LOWDEN A., AKERSTEDT T. - Self selected work hours-work satisfaction, health and social life, in *Shiftwork in the 21st century*, Hornberger S, Knauth P, Costa G, Folkard S. (eds), Peter Lang, Frankfurt, p 345-350, 2000.
34. MONK T.H. - Shift work, in *Principles and practice of sleep medicine*, Kryger M., Roth T., Dement W. C., W.B. Saunders, Philadelphia, p.332-337, 1989.
35. NACHREINER F., AKKERMANN S., HAENECKE K. - Fatal accident risk as a function of hours into work, in *Shifwork in the 21st century*, Hornberger S, Knauth P, Costa G, Folkard S. (eds) Peter Lang, Frankfurt, p 19-24, 2000.
36. PARTINEN M., KAPRIO J., KOSKENVU O.M., PUTKONEN P. - The genetic and sleep. *Sleep* 1983, 6, 186-192.
37. PAZ A., BERRY EM. - Effects of meal composition on alertness and performance of hospital night-shift workers. Do mood and performance have different determinants. *Ann. Nutr. Metab.* 41, 291-297, 1997.
38. QUERA-SALVA M.-A., DEFRANCE R., CLAUSTRAT B., DE LATTRE J., GUILLEMINAULT C. - Rapid shift in sleep time and acrophase of melatonin secretion in short shift work schedule. *Sleep* 19,539-543, 1996.
39. REINBERG A., ANDLAUER P., de PRINS J. - Desynchronization of the oral temperature circadian rhythm and intolerance to shift work. *Nature* 308, 272-274, 1984.
40. ROSA RR. - Extended worshifts and excessive fatigue. *J. Sleep Res.* 4, 51-56, 1997.
41. ROSMOND R., LAPIDUS L., BJORNTORP P. - The influence of occupational and social factors on obesity and body fat distribution in middle-aged men. *Int. J. Obes. Relat. Metab. Disord.* 20,599-607, 1996.
42. ROSENTHAL N.-E., SACK D.-A., JAMES S.P. - Seasonal affective disorders and phototherapy, in Wurtman RJ. Saum MJ. Potts Jr JT. (eds), *The medical and biological effects of light* The New-York Academy of Sciences, New York, p.260-269, 1985.
43. TEPAS D.-L., CARVALHAIS A.B. - Sleep patterns of shiftworkers, *Occupational Health* 5, 199-208, 1990.
44. TORSWALL L., AKERSTEDT T. - A diurnal type scale: construction, consistency and validation in shift work, *Scand. J. Work Environ. Health* 6, 283-290, 1980.
45. TORSWALL L., AKERSTEDT T., GILLANDER K., KNUTSSON A. - Sleep on the night shift: 24-hour EEG monitoring of spontaneous sleep/wake behavior. *Psychophysiology* 26, 352-358, 1989.
46. WEVER R.A. - *The circadian system of man*. Springer Verlag, Berlin, 276p, 1979.
47. WEVER R.A. - Characteristics of circadian rhythms in human functions. *J. Neural Trans.* suppl. 21, 323-373, 1986.
48. WOODSON W.E., CONOVER D.W. - L'ambiance de travail, éclairage. *Guide d'Ergonomie*. Les éditions d'Organisation, Paris, 1978, 2-176-2-209.
49. WURTMAN R.-J., SAUM M.J., POTTS Jr. J.-T. - *The medical and biological effects of light*. The New-York Academy of Sciences, New York, 407p, 1985.
50. ZOBER A., SCHILLING D., OTT M.G. - Helicobacter pylori infection: prevalence and clinical relevance in a large company, *J. Occup. Environ. Med.* 40, 586-594,1998.

Chapter 39

Time zone change (jet lag) syndrome

M. Tiberge
Service d'Explorations Fonctionnelles du Système Nerveux, Hôpital de Rangueil, Toulouse, France

INTRODUCTION

One merely has to consider the number of flights crossing the Atlantic, the Pacific, the USA or Russia, to see that the number of travellers taking long-distance flights for professional or recreational reasons is growing every year. Most, if not all of these travellers and flight crews are affected by jet lag syndrome, caused by rapidly crossing time zones. The symptomatology is varied and influenced by a series of factors, and is sufficiently unpleasant to warrant the search for preventive and curative counter measures.

CLINICAL FEATURES

The clinical features of jet lag syndrome are inconsistent, multiple and variable in their presentation, duration and intensity. The severity and duration of these symptoms depend on the number of time zones crossed, the direction of the trip (eastbound or westbound), departure and arrival times as well as individual factors (typology, age).

These clinical features mainly consist of sleep disorders characterised by a reduction in sleep duration, sleep fragmentation and as a corollary, excessive daytime sleepiness. These sleep disorders become increasingly marked with the number of time zones crossed. They are usually more intense if the flight is eastbound.

Other clinical features may be encountered: mood disorders such as irritability, increased levels of anxiety and the appearance of depressive phenomena [13, 31, 37, 43], a drop in cognitive [19, 33] and sports [24, 53, 54] performance. This last point is important to consider when the subject is a top athlete due to participate in a competition within hours of arriving at his/her destination.

Other aspecific clinical features may be seen such as gastro-intestinal signs [30, 43, 51] involving diarrhoea or constipation, which must not however be confused with disorders linked to changes in diet or infectious factors. Finally, certain clinical features may be accentuated or revealed by time differences i.e. affective symptoms [22, 23, 50], particularly depression, or complications linked to the imbalance of chronic illnesses such as diabetes [48].

FACTORS WHICH DETERMINE JET LAG SYNDROME

Jet lag symptoms appear to be due to a number of factors:

- *Tiredness* linked to travelling and poor environmental conditions (high altitude flight, reduced oxygen, flight anxiety, uncomfortable seating...). While these adverse environmental conditions obviously have an effect on sleep they do not fully account for maladjustment syndrome. Indeed, in the case of long north-south flights, the same environment-related disturbances are present [20] but do not give rise to the group of features associated with maladjustment syndrome; hence the search for other causes.

- *A sudden change of external synchronisers*: in the normal subject, the various biological rhythms result from a mechanism controlled by one (or several) internal oscillator(s), and the action of natural external synchronisers (alternation between light and darkness, outside temperature...)

and social synchronisers (professional or family constraints...). These synchronisers act in entraining the internal biological clock at a 24 hour rhythm.

The internal clock can thus adjust itself each day as a function of environmental stimuli. The speed with which the different rhythms adjust will vary depending on the parameters studied. Although sleep resynchronises fairly quickly, other circadian rhythms, particular hormone rhythms, take much longer [1, 15, 18, 36]. For cortisol, for example, an hour of resynchronisation is required per day for time shifts of over six hours.

But whatever the case, adaptation is limited and asymmetrical [16, 24, 44].

It is easier in fact to readjust the sleep-wake cycle for a long distance journey in a westbound direction, than for one which is eastbound. Flight direction will determine the direction of phase shift. Westbound flights lengthen the day, requiring phase delay i.e. the traveller's circadian rhythm period needs to extend beyond 24 hours to keep in phase with external synchronisers. Eastbound flights, on the contrary, shorten the day, provoking a phase advance. Even though readjustment is the same after east or westbound flights, where the number of time zones crossed is identical, the speed of readjustment is different. Readjustment is asymmetrical. After a westbound flight crossing 8 time zones, 5.1 days are needed to reach 95% of the total resynchronisation of psychomotor performance rhythm, whereas 6.5 days are needed after an eastbound flight crossing the same number of time zones [26]. Taking an average of the readjustment effects reported in seven flight studies [4, 25, 26], the mean readjustment period for all the studied variables is 92 minutes per day after a westbound flight compared with only 57 minutes per day for an eastbound flight. Readjustment is faster during the first 24 hours then diminishes exponentially.

The phenomenon of the different readjustment times observed between westbound and eastbound flights is referred to as the "asymmetrical effect". This refers to an inherent tendency for the biological clock to extend its period to over 24 hours after a westbound flight, and to resist synchronising its period to under 24 hours after an eastbound flight. Resynchronisation generally occurs more quickly with phase shifts which are similar to synchroniser shifts; in such cases the biological clock follows the direction of the flight. This is not always the case with eastbound flights. The internal clock does not always take account of phase shortening and paradoxically, a phase delay is observed after crossing only 8 to 9 time zones. Certain travellers' circadian rhythms adjust by lengthening their period by 15 or 16 hours to attain the time of their destination. The result is that a week after their arrival, some of their circadian rhythms are advanced (hormones) and others are delayed (such as core temperature) [17, 26]. This resynchronisation process is called partition resynchronisation and is observed in severe cases of jet lag syndrome following eastbound flights. However individuals vary greatly in their ability to adapt to new time schedules. Several individual factors account for this: the power of synchronisers, the stability and amplitude of circadian rhythms, typology of subject (morning or evening person), behavioural traits (extroverts apparently adapting more quickly than introverts [30]), age, motivation, sleep habits.

One important factor contributing to jet lag maladjustment syndromes is lack of sleep. This lack of sleep results, again, from environmental disturbances during the flight, as well as the subjects' inability to present satisfactory sleep phases in the course of post flight sleep. Indeed, during a westbound flight the subject is in phase delay, with the possibility of falling asleep quickly followed by premature awakening which occurs with the increase in core temperature (fig. 39.1).

Contrary to this, eastbound flights induce a phase advance syndrome, with difficulty getting to sleep at the moment when core temperature is at a maximum.

These sleep alterations thus induce chronic sleep deprivation, which is responsible for the majority of clinical features observed, particularly that of excessive daytime sleepiness. Clinical sleep alterations correlate with those observed in polygraphic sleep recordings. NREM sleep increases and REM sleep diminishes after an eastbound flight [42, 44]. After a westbound flight, there is only an increase in REM sleep [43]. Sleep efficiency is reduced, whatever the flight direction [12, 35, 52] and total sleep duration will always be reduced, with an increase in intra-sleep wakefulness.

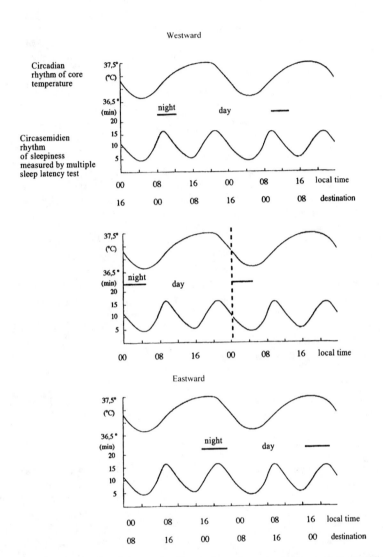

Figure 39.1. Night time situation in relation to circadian rhythms of core temperature and circasemidian rhythm of sleepiness, in a normal situation and after westward and eastward flights. *In the centre*: In normal conditions the subject falls asleep when his core temperature decreases, his level of sleepiness increases and he is in phase with external synchronisers, particularly with the approach of darkness (———). Wakefulness occurs when body temperature increases with the approach of daybreak. *Top*: During a westward flight, the subject's core temperature and sleepiness are in desynchronisation with the external synchronisers: there is a phase delay. The desired onset of sleep occurs at around the core temperature minimum, favouring the appearance of REM sleep. Sleep is short lasting because core temperature rises provoking the subject's arousal. A certain degree of sleepiness persists on awakening. *Bottom*: During an eastward flight, the subject is required to sleep at a time when temperature is rising, possibly creating difficulties getting to sleep. Sleep contains little REM sleep. Temperature decreases when the subject is required to get up, creating marked sleepiness on rising.

Other factors intervene in this maladjustment syndrome:

The number of time zones crossed: clinical symptomatology may appear after crossing five time zones in an eastward direction and intensifies with the number of time zones crossed.

Individual factors: these are important, particularly that of the age of the subject. The elderly present a greater risk of difficulties adjusting to long distance flights, particularly in the case of eastbound flights [32]. This phenomenon may be explained by the reduction in circadian flexibility related to age [19], an increase in rhythm amplitudes [40, 41] (greater amplitudes being provoked by milder phase shifts), by a period of recuperation which lengthens with age [1] and finally by the increased frequency of sleep disorders in the elderly [19], particularly with the occurrence of fragmentation and intra-sleep wakefulness.

THERAPEUTIC ASPECTS

Drug treatment

At present, no known substance is capable of resynchronising jet lag of over 5 hours. Only theophylline (in the rat) [14] and triazolam (in the hamster) [50] have proved their capacity to resynchronise moderate time shifts but neither substance is capable of resetting the biological clock in man. Nevertheless, the use of hypnotics such as the benzodiazepines [39,] zopiclone, [10] or zolpidem [47], produces satisfactory periods of sleep duration after a long-distance flight. These products may also increase sleep quantity and reduce the effects of sleep deprivation, but may also, by the same action, delay physiological resynchronisation. They must therefore only be prescribed exceptionally. They may be indicated to improve sleep during a flight. In this context, short half-life products appear to be the best.

Other families have been studied: slow release caffeine has proved to be effective in resynchronising biological rhythms, by maintaining a good level of daytime wakefulness with no harmful effects [28, 38, 46]. A molecule which induces wakefulness (modafinil) may prove to be an interesting approach, but the product most often studied is melatonin. This is an endogenous hormone synthesised by the pineal gland, presenting a secretion rhythm which depends on the alternation between light and darkness. Maximum secretion occurs in the middle of the night. Random double-blind trials using low doses have shown melatonin to have a mild hypnotic effect, with capacities for resynchronisation [8, 37] which could be used to treat the effects of jet lag [3].

Preventing jet lag syndrome after an eastbound flight requires presynchronisation with melatonin which must be administered in late afternoon on the day of the departure or several days beforehand, and for the 4 following days at bedtime, i.e. at 10 to 11 pm, local time [2, 7].

Conversely, when the journey is from east to west, improvement is less marked with melatonin and treatment does not usually commence until after arrival [37].

Studies found a 60% reduction in the effects of jet lag measured by subjective tests in the case of eastbound flights and 40% for westbound flights.

Nevertheless, the routine use of melatonin raises certain difficulties: the long term side effects are unknown, particularly with regard to reproduction [44]: cases of primary hypogonadism or amenorrhoea have been reported, associated with high serum levels of melatonin. Pharmacological doses of melatonin also increase nocturnal serum levels of prolactin. Lastly, effects have been reported on temperature rhythm.

A difficulty arises in prescribing melatonin in terms of when it should be taken in relation to temperature rhythm. Administering melatonin before the minimum thermal point will actually induce an advanced phase, whereas administering it after the minimum thermal point will induce a delayed phase. Even if variations in temperature rhythm are fully understood in normal conditions, they are much more difficult to determine when the same individual experiences a time zone change, as the core temperature rhythm also changes in an attempt to resynchronise. Thus paradoxical effects may be seen in cases of marked changes in time zone. The same applies to the light effect.

The second difficulty raised by the routine use of melatonin is the lack of control in regard to production sources [27]. Wide variations exist in product quality: some pills contain no trace even of melatonin and there are no indications as yet of production quality assurance, particularly with regard to animal sources. It would be unwise to propose this treatment before more intensive scientific research has been done on the possible human side effects.

Phototherapy

Light is a physical agent which plays an important role in synchronising sleep-wake circadian rhythm and in treating certain pathologies involving seasonal affective disorders (seasonal depression).

It was first proposed in the 1980s on the basis that human circadian rhythms respond to bright light stimulation [21, 29]. The technique proposed, notably by Czeisler [9], involves relatively long exposure of about 5 hours to very bright light, particularly at the minimum thermal point and just afterwards. This technique shows signs of promise although the practical modalities of phototherapy, particularly in terms of the criteria to be applied (light intensity, length of exposure, number of days of treatment), need to be better defined. Chesson *et al.* recently issued recommendations [5] for the use of phototherapy in treating sleep disorders including jet lag. They confirm the variability in the intensities of light applied and the length of exposure. They also warn against the difficulties of applying this technique in the case of marked time zone changes, the moment of exposure to light being particularly hard to determine (see below). Furthermore, they draw attention to the many biases in the studies leading to these recommendations.

While the beneficial effects of prescribing melatonin and phototherapy appear to be genuine, a number of studies are still needed to refine their efficacy.. Some authors [2] suggest linking the two.

Sleep hygiene

Several solutions have been put forward for treating disorders caused by jet lag.
1) Some authors [11, 33] have recommended gradually modifying sleep times before the journey. They propose altering normal bedtime by an hour each day, going to bed and getting up an hour earlier if the trip is in an eastbound direction, or delaying it for a westbound flight. This preventive strategy, made difficult by the time needed to implement it, usually creates a situation which is incompatible with the subject's social life. It can even, in extreme cases, lead to symptoms like those of jet lag, before the journey starts. Nevertheless, this may be considered as a partial solution in reducing the adverse effects of crossing several time zones, as in the case of top athletes required to perform in competition shortly after arriving at their destination.
2) *Reinforcing social synchronisers*: this is the main factor to which subjects must adapt. On arrival at the destination, the subject should immediately adapt to the social synchronisers of the country. Naps should be avoided, especially at inappropriate times in relation to the sleep-wake circadian rhythm.

Nevertheless, in certain very specific cases (short trip for flight crews) subjects should be advised to maintain the schedules of their country of origin.

CONCLUSION

The symptomatology of desynchronisation syndrome due to jet lag is varied, aspecific and dominated by a clinical picture of sleep deprivation. Physiopathogeny is highly dependent on chronobiological disturbances. Treatment varies but drugs should only be prescribed exceptionally, while treatments such as melatonin require further evaluation and testing against harmful effects.

REFERENCES

1. AKERSTEDT T., TORSVALL L. – Shift Work: Shift dependent well-being and individual differences. *Ergonomics*, *24*, 265-273, 1981.
2. ARENDT J., ALDHOUS M., MARKS V. – Alleviation of jet lag by melatonin: preliminary results of controlled double-blind trial. *Br. Med. J.*, *292*, 1170, 1986.
3. ARENDT J., SKENE D.J., MIDDLETON B., LOCKLEY S.W., DEACON S. – Efficacy of melatonin treatment in jet lag, shift work and blindness. *J. Biol. Rhythms,* 12, 604-607, 1997.
4. ASCHOFF J., HOFFMAN K., POHL H., WEVER R. – Re-entrainment of circadian rhythms after phase-shifts to the Zeitgeber. *Chronobiologia*, *2*, 23-78, 1975.

5. CHESSON A.L., LITTNER M., DAVILLA D., ANDERSON W., GRIGG-DAMBERGER M., HARTSE K., JOHNSON S., WISE M. – Practice parameters for the use of light therapy in the treatment of sleep disorders. *Sleep*, 22, 641-660, 1999.
6. CHO K., ANNACEUR A., COLE J.C., SUH C.K. – Chronic jet lag produces cognitive deficits, *J. Neurosci. 20*, RC66, 2000.
7. CLAUSTRAT B., BRUN J., DAVID M., SASSOLAS G., CHAZOT G. – Melatonin and jet lag: confirmatory result using a simplified protocol. *Biol. Psychiatry*, 32, 705-711, 1992.
8. CLAUSTRAT B., GEOFFRIAU M., BRUN J., CHAZOT G. – Melatonin in humans: a biochemical marker of ₁he circadian clock and an endogenous synchroniser. *Neurophysiol. Clin.*, 25, 351-359, 1995.
9. CZEISLER C.A., KRONAUER R.E., ALLAN J.S., DUFFY J.F., JEWETT M.E., BROWN E.N., RONDA J.M. – Bright light induction of strong (type O) resetting of the human circadian pacemaker. *Science*, 244, 1328-1333, 1989.
10. DAURAT A., BENOIT O., BUGUET A. – Effects of zopiclone on the rest/activity rhythm after a westward flight across five time zones. *Psychopharmacology* (Berlin), 149, 241-245, 2000.
11. DAVIS J.O. – Strategy for managing athletes jet lag. *Sport Psychologist*, 2, 154-160, 1988.
12. DEMENT W.C., SEIDEL W.F., COHEN S.A., BLIWISE N.G., CARSKADON M.A. – Sleep and wakefulness in aircrew before and after transoceanic flights. *Aviat. Space Environ. Med.*, 57, B17-B28, 1986.
13. DESIR D., VAN CAUTER E., FANG V.S., MARTINO E., JADOT C. – Effects of jet lag on hormonal patterns. I. Procedures, variations in total plasma proteins and disruption of adrenocorticotropin-cortisol periodicity. *J. Clin. Endocrinol. Metab.*, 52, 628-641, 1981.
14. EHRET C.F., POTTER V.R., DOBRA K.V. – Chronotopic action of theophylline and of pentobarbital as circadian Zeitgebers in the rat. *Science*, 188, 1212-1215, 1975.
15. FEVRE-MONTANGE M., VAN CAUTER E., REFETOFF S., DESIR D., TOURNIAIRE J. – Effects of jet lag on hormonal patterns. II. Adaptation of melatonin circadian periodicity. *J. Clin. Endocrinol. Metab.*, 52, 642-649, 1981.
16. GANDER P.H., MYHRE G., GRAEBER R.C. – *Crew factors in flight operations. I. Effects of 9-hour time zone changes on fatigue and the circadian rhythms of sleep/wake and core temperature*. NASA Technical Memorandum 88197. Moffett Field, CA, NASA Ames Research Center, 1985.
17. GANDER P.H., MYHRE G., GRAEBER R.C., ANDER-SEN H.T., LAUBER J.K. – Adjustment of sleep and the circadian temperature rhythm after flights across nine time zones. *Aviat. Space Environ. Med.*, 60, 733-743, 1989.
18. GOLSTEIN J., VAN CAUTER E., DESIR D., NOEL P., SPIRE J.P. – Effects of jet lag on hormonal patterns. IV. Time shifts increase growth hormone release. *J. Clin. Endocrinol. Metab.*, 56, 433-440, 1983.
19. GRAEBER R.C. – Alterations in performance following rapid transmeridian flight. *In:* Brown F.M., Graeber R.C. (eds), *Rhythmic aspects of behavior*. Lawrence Erlbaum Associates: Hillsdale, N.J., 173-212, 1982.
20. HAUTY G.T., ADAMS T. – Phase shifts of the human circadian system and performance deficits during the periods of transition: III. North-South flight. *Aerospace Med.*, 37, 1257-1262, 1966.
21. HONMA K., HONMA S. – Circadian rhythm: its appearance and disappearance in association with a bright light pulse. *Experientia*, 44, 981-983, 1988.
22. JAUHAR P., WELLER M.P.I. – Psychiatric morbidity and time zone changes: a study of patients from Heathrow Airport. *Br. J. Psychiatry*, 140, 231-235, 1982.
23. KATZ G., DURST R., ZISLIN Y., BAREL Y., KNOBLER H.Y. – Psychiatric aspects of jet lag : Review and hypothesis. *Medical hypotheses*. 56, 20-23, 2001.
24. KLEIN K.E., WEGMANN H.M., HUNT B.I. – Desynchronization of body temperature and performance circadian rhythms as a result of outgoing and homegoing transmeridian flights. *Aerospace Med.*, 43, 119-132, 1972.
25. KLEIN K.E., WEGMANN H.M. – The resynchronization of human circadian rhythms after transmeridian flights as a result of flight direction and mode of activity. *In:* Sceving L.E., *et al.* (eds.), *Chronobiology*, Igaku-Shoin, Tokyo, 564-570, 1974.
26. KLEIN K.E., WEGMANN H.M. – *Significance of circadian rhythms in aerospace operations*. Nato Agardograph Number 247. Neuilly-sur-Seine, France, Nato Agard 1980.
27. KRYGER M.H. – Controversies in sleep medicine: Melatonin. *Sleep*, 20, 10, 898, 1997.
28. LAGARDE D., BATEJAT D., SICARD B., TROCHERIE S., CHASSARD D., ENSLEN M., CHAUFFARD F. – Slow release caffeine : a new response to the effects of a limited sleep deprivation. *Sleep*, 23, 651-661, 2000.
29. LEVY A.J., WEHR T.A., GOODWIN F.K., NEW-SOME D.A., MARKEY S.P. – Light suppresses melatonin secretion in humans. *Science*, 210, 1267-1269, 1980.
30. MINORS D.S., WATERHOUSE J.M. – Circadian rhythms and their application to occupational health and medecine. *Rev. Environ. Health*, VII, 1-64, 1987.
31. MOLINE M.L., POLLAK C.P., WAGNER D.R., ZEN-DELL S.M., MONK T., GRAEBER R.C., LESTER L.S., SALTER C.A., HIRSCH E.– Effects of age on the ability to sleep following an acute phase advance. *Sleep Res.*, 19, 400, 1990.
32. MONK T.H., BUYSSE D., CARRIER J., KUPFER D. – Inducing jet lag in older people : directional asymmetry. *J. Sleep Research* 9, 101-116, 2000.
33. MONK T.H., MOLINE M.L., GRAEBER R.C. – Inducing jet lag in the laboratory: patterns of adjustment to an acute shift in routine. *Aviat. Space Environ. Med.*, 59, 703-710, 1988.
34. MONK T.H. – Coping with the stress of jet lag. *Work Stress*, 1, 163-166, 1987.
35. NICHOLSON A.N., PASCOE P.A., SPENCER M.B., STONE B.M., GREEN R.L. – Nocturnal sleep and daytime alertness of aircrew after transmeridian flights. *Aviat. Space Environ. Med.*, 57, B43-B52, 1986.

36. NOWAK R. – The salivary melatonin diurnal rhythm may be abolished after transmeridian flights in an eastward but not a westward direction. *Med. J. Australia, 149*, 340-341, 1988.

37. PETRIE K., CONAGLEN J.V., THOMPSON L., CHAMBERLAIN K. – Effect of melatonin on jet lag after long haul flights. *Br. Med. J., 298*, 705-707, 1989.

38. PIERARD C., BEAUMONT M., ENSLEN M., CHAUFFARD F., TAN D.X., REITER R.J., FONTAN A., FRENCH J., COSTE O., LAGARDE D. – Resynchronization of hormonal rhythms after an eastbound flight in humans: effects of slow-release caffeine and melatonin. *Eur. J. of Applied Physiol.* 85, 144-150, 2001.

39. REILLY T., ATKINSON G.F., BUDGETT R. – Effects of low-dose temazepam on physiological variables and performance tests following a westerly flight across five time zones . *Int. J. of Sports Med.* 22, 166-174, 2001.

40. REINBERG A., ANDLAUER P., GUILLET P., NICOLAI A. – Oral temperature, circadian rhythm amplitude, aging and tolerance to shift-work. *Ergonomics, 23*, 55-64, 1980.

41. REINBERG A., VIEUW N., GHATA J., CHAUMONT A.J., LAPORTE A. – Is a rhythm amplitude related to the ability to phase-shift circadian rhythms of shift-workers? *J. Physiol.*, 74, 405-409, 1978.

42. SAMEL A., WEGMANN H.M., VEJVODA M. – Jet lag and sleepiness in aircrew. *J. Sleep Res.*, 4 (suppl. 2), 30-36, 1995.

43. SASAKI M., ENDO S., NAKAGAWA S., KITAHARA T., MORI A. – A chronobiological study on the relation between time zone changes and sleep. *Jikeikai Med. J.*, 32, 83-100, 1985.

44. SASAKI T., TSUZUKI S. – Directional asymmetry of phase shift following transmeridian flight. *Sangyo Ika Daigaku Zasshi, 7* (suppl.), 113-121, 1985.

45. SASAKI M., KUROSAKI Y., MORI A., ENDO S. – Patterns of sleep-wakefulness before and after transmeridian flight in commercial airline pilots. *Aviat. Space Environ. Med.*, 57, B29-B42, 1986.

46 SICARD B., PERAULT M., ENSLEN M. – The effects of 600 mg of slow release caffeine on mood and alertness. *Aviat. Space Environ. Med.* 67, 859-862, 1996.

47. SUHNER A., SCHLAGENHAUF P., HÖFER I., JOHNSON R., TSCHOPP A., STEFFEN R. – Effectiveness and tolerability of melatonin and zolpidem for the alleviation of jet lag. *Aviat. Space Environ. Med.* 72, 638-646, 2001.

48. TASHIMA C.K., FILLHART M., CUNANAN A. – Jet lag ketoacidosis. *JAMA*, 227, 328, 1974.

49. TEC L. – Depression and jet lag. *Am. J. Psychiatry, 138*, 6, 1981.

50. TUREK F.W., LOSEE-OLSON S. – A benzodiazepine used in the treatment of insomnia phase shifts the mammalian circadian clock. *Nature, 321*, 167-168, 1986.

51. VENER K.J., SZABO S., MOORE J.G. – The effect of shift work on gastrointestinal (GI) function: a review. *Chronobiologia, 16*, 421-439, 1989.

52. WEGMANN H.M., GUNDEL A., NAUMANN M., SAMEL A., SCHWARTZ E., VEJVODA M. – Sleep, sleepiness and circadian rhythmicity in aircrews operating on transatlantic routes. *Aviat. Space Environ. Med.*, 57, B53-B64, 1986.

53. WINGET C.M., DEROSHIA C W., HOLLEY D.C. – Circadian rhythms and athletic performance. *Med. Sci.Sports Exerc.*, 17, 498-516, 1985.

54. WRIGHT J.E., VOGEL J.A., SAMPSON J.B., KNAPIK J.J., PATTON J.F. – Effects of travel zones (jet lag) on exercise capacity and performance. *Aviat. Space Environ. Med.*, 54, 132-137, 1983.

55. YOUNGSTEDT S.D., O'CONNOR P.J. – The influence of air travel on athletic performance. *Sports Med.*, 28, 197-207, 1999.

56. ZHDANOVA I., LYNCH H.J., WURTMAN – Melatonin: A sleep-promoting hormone. *Sleep, 20*, 10, 899-907, 1997.

Chapter 40

Circadian rhythm sleep disorders related to an abnormal escape of the sleep-wake cycle

D. Boivin and J. Santo
Centre for Study and Treatment of Circadian Rhythms, Douglas Hospital Research Centre, Department of Psychiatry, McGill University, Verdun, Quebec, Canada

DELAYED SLEEP PHASE SYNDROME

Historical Background

In 1981 Weitzman and colleagues gave the first description of a new disorder, the delayed sleep phase syndrome (DSPS), after an extensive evaluation of a subgroup of insomniac patients at the Montefiore Hospital Medical Center [115]. They reported delayed sleep patterns in 30 out of 450 insomniac patients (7%) studied between 1976-1979. These patients were younger (33 ±14 y.o.) than the general population of insomniacs (46 ±15 y.o.) and were described as having extreme evening chronotypes. The sleep problems were often present for several years and frequently began during childhood. In 7 of these patients, the disturbed sleep pattern was reported to have started prior to the age of 10, though occurrence at other ages is also possible. In Weitzman's study, no sex differences were reported although most subsequent accounts have stressed a male predominance of about 10:1. Since then, over 75 cases have been described [3, 29, 90].

Epidemiology

DSPS is one of the most common circadian rhythm sleep disorders [19, 25, 77, 79] . It is generally estimated that 7-10% of insomniac patients might suffer from DSPS and that this number might even be underestimated [24, 114]. Indeed, sleep onset insomnia is one of the most frequent types of insomnia and is more prevalent among younger patients with the peak prevalence among high school and college students [10, 44, 69, 92]. An epidemiological study in the San Francisco Bay area revealed that sleep-onset insomnia had an earlier age of onset (37 y.o.) as opposed to early morning awakening (40.2 y.o.). The most prevalent pattern of sleep disruption in younger patients (18-30 y.o.) was sleep-onset insomnia and the least prevalent was early morning awakening [11]. The incidence of DSPS is estimated to be ≥ 7% in adolescents compared to less than 0.7% in the general population [90, 111]. Several epidemiological studies have been conducted in populations of students using sleep questionnaires. Students with DSPS went to bed more than 2 hours later on weekends and suffered from sleep-onset insomnia and insufficient sleep on week nights [53]. Interestingly, a sleep questionnaire administered to 277 students revealed that poor sleepers generally felt less tired in the evening than good sleepers [50] suggesting a chronobiological basis. The Standford Sleep Inventory, administered to 639 adolescents revealed chronic and severe sleep disturbances in 12.6% of the sample population [88]. Difficulty falling asleep was observed in 74.8% of those reporting chronic sleep difficulties and in 60% of those reporting occasional difficulties. A higher proportion of adolescents suffering from chronic sleep problems enjoyed staying up at night (79.5%) compared to healthy sleepers (59.9%) [88].

Positive diagnosis

According to the International Classification of Sleep Disorders [104]. DSPS is characterised by an inability to fall asleep or to awaken spontaneously at the desired times and a phase delay in the main sleep episode. Sleep quality, sleep stage distribution, and sleep duration are normal when patients are not forced to maintain a strict schedule and instead allowed to sleep at their desired times. However, sleep latency is frequently longer than 30 minutes even if patients go to bed at times of their choosing [110]. When sleep is planned earlier (e.g. 11 pm), a significant increase in sleep latency and wake time during the first part of the night is observed [2]. These patients are entrained to a 24-hour day and go to bed at about the same clocktime every night. However, patients are extreme night owls such that sleep onset and wake times are intractably later than desired [24]. Bedtimes are frequently observed around 3 – 6 am with waketimes around 10 am – 6 pm if the patients are not disturbed. In DSPS, the average sleep duration on weekdays varies from 2 to 5 hours per night with a tendency to recuperate on the week-ends by sleeping 9-18 hours per night. Excessive sleepiness is generally experienced in the morning following awakening [93] such that severe difficulties awakening at socially acceptable times are reported. As a result, a significant proportion of DSPS patients report disrupted work or social functioning [3, 90]. A diagnosis of DSPS should be confirmed by a sleep/wake log and ideally wrist actigraphy monitoring for at least one month [93]. Polysomnographic recording of 1 or 2 nights should be planned at the patient's preferred sleep time [113]. The disturbances of their sleep/wake cycle should not be explained by other psychiatric or medical conditions. Some patients have delayed sleep patterns but are able to advance their bedtime earlier due to social pressure. These patients should rather be assigned a diagnosis of motivated sleep phase delay rather than DSPS [99].

Pathophysiology

In humans, it is possible to induce sleep-onset insomnia by scheduling subjects to live on shorter than 24-hour days, presumably because bedtime would coincide with the so-called evening wake maintenance zone [40, 101]. Morris suggested that DSPS could result from delayed core body temperature rhythms that would then align the wake maintenance zone with bedtime [69]. Indeed, delayed rhythms of plasma melatonin, urinary 6-sulfatoxy melatonin, and core body temperature have been reported in DSPS [80, 82, 83] but one cannot determine if these are a consequence rather than a cause of delayed sleep times. Recent studies suggest that individual differences in the increase of the circadian period might lead to a different phase relationship between the endogenous circadian pacemaker and the sleep/wake cycle [34]. Based on core body temperature and melatonin onset, a circadian period slightly longer than normally reported in adults was found in adolescents [17]. Despite these results of an increased period, it is improbable that this sole difference could account for the daily difficulties in falling asleep. Another possibility is that the duration or strength of the evening wake maintenance zone is enhanced in patients with DSPS or that patients have stronger wake mechanisms (perhaps a faulty homeostatic drive for sleep) that pushes them to delay sleep onset to a time of increased sleep propensity, namely later at night. Indeed, there is compelling evidence to suggest that DSPS patients go to bed on the rising limb of their core body temperature cycle close to the circadian temperature nadir [82] (fig. 40.1). The phase angle between the time of awakening and the temperature nadir was found to be significantly larger in DSPS patients (3.78 hours) compared to controls (2.86 hours) [83] (fig. 40.2). This indicates that DSPS patients tend to go to sleep at later circadian phases, a situation that results in a shielding of the phase advance portion of the phase response curve to light and thus a worsening of their tendency to delay sleep times. This could also explain why DSPS patients are at risk to develop a non-24-hour sleep/wake syndrome.

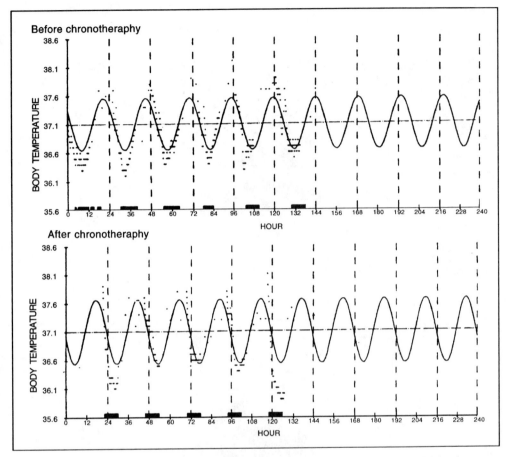

Figure 40.1. Distribution of core temperature in a patient with DSPS across a 10-day period with best-fitted cosine curve by least squares method before (above) and after (below) chronotherapy. Dots stand for rectal temperature data while bars refer to sleep times. The temporal relationship between the temperature cycle and sleep onset is more appropriate after chronotherapy.

Differences among individuals might exist with respect to their sensitivity to light as a circadian synchroniser [115]. In DSPS, it was proposed that the phase advance portion of the phase-response curve to light is weaker than normal but that some phase advance still occurs in order to stabilise the sleep/wake cycle to the 24-hour environment [23, 24, 115]. Alternatively, one could also hypothesise an increased sensitivity to evening light. This could delay the start of melatonin secretion and the so-called opening of the sleep gate until later at night. In support of this hypothesis, the percentage of melatonin suppression by a bright light stimulus of 1,000 lux administered 2 hours prior to the melatonin peak has been reported to be greater in 15 DSPS patients than in 15 controls [4].

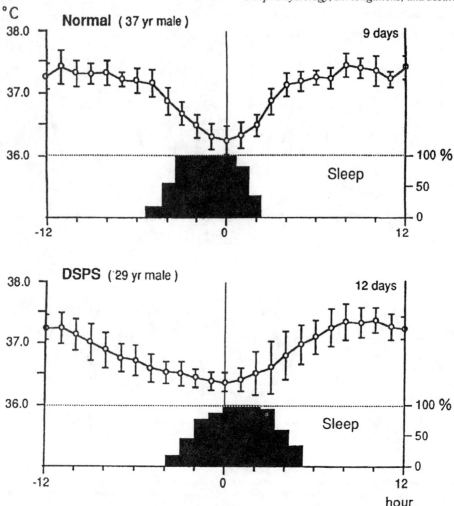

Figure 40.2 Records of temperature and sleep phase across a 24-hour period in a control subject (above) and a patient with DSPS (below) using habitual sleep-wake schedules. Temperature values were averaged for each 60-minute period relative to the temperature minima for a 9 day period for the control subject and 12 days for the patient with DSPS. Standard error is reported using the error bars. Sleep phase histograms were also averaged using the occurrences of sleep episodes within each 60-minute period. It is worth noting that in the control subject the majority of sleep occurs before the temperature minimum meanwhile in the patients with DSPS, it follows the minimum.

Several case studies have been published reporting of patients who developed DSPS after traumatic brain injuries or whiplash [70, 86, 89]. It has been hypothesised that suprachiasmic nucleus (SCN) lesions or lesions of the cervical projections from the SCN to the pineal gland could be responsible for the deficits. However, so far, no group has been able to relate this condition to any specific lesion.

Whether patients with DSPS have a higher incidence of personality disorders that could contribute to their delayed sleep pattern remains unclear. So far, no specific profile of personality disorders has been identified [25, 110, 114, 115]. About 50% of DSPS patients have some psychopathology [3, 115] and depression is the most frequent psychiatric condition associated with DSPS [77, 90]. Conversely, there is a high incidence of DSPS in depressed patients and there is a good possibility that antidepressants such as MAOI and SSRIs might contribute to the appearance of DSPS by promoting alertness and delaying sleep patterns [90]. The analyses of 63 hospitalised adolescents in a psychiatric ward revealed that 10 patients (15.9%) had DSPS [28]. It has been suggested that sleep deprivation associated with DSPS could lead to depression or that late sleeping could aggravate or precipitate depression [90]. This possibility cannot be excluded even though sleep deprivation often exerts an antidepressant effect in depressed populations. The general interpretation is that sleep-onset insomnia is not dependent on psychological problems but psychological problems might rather be a consequence of sleep disturbances [92]. Indeed, an improvement in social and psychological functioning was reported in many DSPS patients after chronotherapy [115]. This was at least the case of a 24 y.o. man in the navy who had been court-martialed for absenteeism due to DSPS [30]. The psychiatric evaluation was reported to be normal.

Treatment

Hypnotic medications are usually ineffective in entraining the sleep/wake cycle of DSPS patients and the use of drugs or alcohol may even worsen the problem of morning drowsiness [115]. Weitzman and colleagues reported the first successful cases of chronotherapy [24, 115]. This was achieved by scheduling the patients to live on several consecutive 27-hour days until the desired bedtime was achieved (fig. 40.3). This particular duration was chosen in order to exceed the natural tendency of the endogenous circadian system to delay while staying at the upper limit of its range of entrainment. This resulted in a disappearance of sleep disturbances and the withdrawal of the hypnotic medication for several consecutive weeks and months. The bedtime and waketime advanced 1-4 hours earlier [24] and alertness levels significantly improved in the morning. The patients fell asleep faster and slept better on the new socially acceptable schedule. In comparison, acute phase advances of the sleep schedule result in increased sleep disturbances on the second night despite substantial sleep deprivation [24]. Chronotherapy using 27-hour days was able to successfully advance the core body temperature rhythm from 4 am to 10 pm in 4 DSPS patients [82]. Prior to the intervention, the sleep onset occurred 2.7 hours prior to the temperature minimum. Following the intervention, it occurred at a more conventional phase, namely 5.3 hours before the temperature minimum. However, the delayed sleep pattern tended to relapse and several patients have found it necessary to repeat chronotherapy at 6- to 12-month intervals. After the report of DSPS patients who developed a non-24-hour sleep-wake syndrome with delayed chronotherapy, it was argued that small advances of the sleep schedule with bright light exposure would be safer [81]. However, circadian rhythm disorders are typically difficult to treat, and others believe it is easier to utilise the natural tendency of endogenous circadian rhythms to delay, than to fight against its current. Variants of the initial chronotherapy have been reported and comprise a combination of sleep deprivation and smaller advances (e.g. 15-minutes) of the sleep/wake cycle. These approaches were rarely effective by themselves and were combined with other circadian synchronisers such as morning exposure to bright light, sunlight or evening administration of exogenous melatonin [3, 90, 111]. For example, a 29 y.o. woman with DSPS from the age of 16 with delayed sleep onset until 4 am had made unsuccessful attempts to advance her sleep times. She was treated using a combination of light restriction in the evening for 2 hours and a 2-hour exposure to 2,500 lux upon awakening. Her bedtimes were slowly and progressively advanced by 15 minutes each day and naps were forbidden. After 3 weeks of treatment, she was sleeping from 11 pm to 7 am and exposing herself to 1 hour of bright light daily [116]. However, it seems difficult to keep the therapeutic effect of chronotherapy for a long time, even when other therapeutic approaches are added [44]. Overall, the failure rate for treatment of DSPS (52%) is greater than that for narcolepsy (10%) or for mixed insomnia (36%) [90].

In 1990, Rosenthal and colleagues [93] investigated the therapeutic effects of bright light in a crossover study of 20 DSPS patients. The active condition consisted of a 2-hour exposure to 2,500 lux full spectrum light from 6 am – 9 pm and the use of dark goggles in the evening from 4 pm until dusk [93]. After dusk, the lighting was restricted to 1 or 2 bedside lamps. In the control condition, patients were exposing themselves to 300 lux full spectrum bright light from 6am – 9 pm and used clear goggles in the evening. The patients rated the treatment condition as superior. Morning alertness levels improved during the 2nd week of treatment and earlier sleep times were noticed under the active condition. A significant phase advance of the diurnal rhythm of core body temperature was observed in the treatment condition. Other cases of successful treatment with morning bright light or sunlight exposure have subsequently been reported [27, 47, 90].

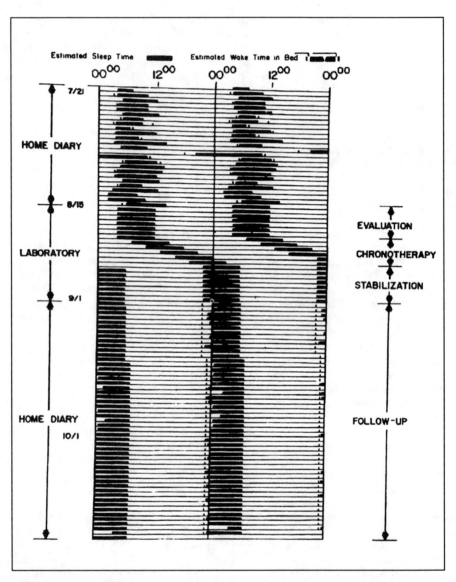

Figure 40.3 Sleep period distribution of a 34 year old woman with DSPS before chronotherapy and after. Black bars indicate sleep periods.

An animal model of DSPS can be produced by keeping rats in constant darkness for several months and returning them to a light/dark cycle of 12h/12h [7]. Among these rats, the onset of activity often lags behind the onset of darkness by 3-4 hours. Exogenous administration of a melatonin agonist, S20098, just prior to darkness can eliminate this negative phase angle. Successful treatment of DSPS was also reported in humans using 5 mg of exogenous melatonin in the evening for several consecutive weeks [28, 29, 70, 79]. The medication significantly advanced sleep onset by 82 minutes and waketime by 117 minutes [29]. Several studies indicate that the timing of melatonin administration is crucial and should be planned several hours prior to bedtime for maximal efficiency [94]. However across studies no consistent changes in sleep structure were reported and both melatonin and S20098 failed to permanently change the phase angle between circadian markers and the sleep schedule. These results suggest that melatonin may exert a direct hypnotic effect on the sleep/wake cycle rather than phase shifting the endogenous circadian system to an earlier time. Similar results have been reported with triazolam, a short acting benzodiazepine, which has been shown to successfully advance the sleep schedule of DSPS patients [46].

In the last 10 years, several patients with DSPS successfully treated with vitamin B12 have been published [72, 74]. It has been hypothesised that B12 administration could phase advance and improve the human circadian system's sensitivity to light [43]. However, animal studies do not clearly support this hypothesis [36] and a recent multicentre double blind study with 55 DSPS patients suggest that prior reported success might be largely explained by a placebo effect [77]. Unfortunately, the sleep/wake schedule was not properly controlled in this last study and partial improvement has been reported in another double blind study [103]. Further experiments are thus needed to clarify this issue.

ADVANCED SLEEP PHASE SYNDROME

Historical background and epidemiology

Advanced sleep phase syndrome (ASPS) was originally reported in older individuals and is extremely rare in younger patients unless depression is present. In 1986, Moldofsky and colleagues reported the case of a 62 y.o. man who had been suffering from early sleep-onset and early morning awakenings for 8 years [67]. The patient had also been treated for hypothyroidism and depression 4 years before. No depressive symptoms were noted on the SCL-90 questionnaire although an in-depth psychological evaluation was not performed and the initial polysomnographic recording revealed an abbreviated REM sleep latency. The patient also exhibited a sleep apnea index of 17/hour and his mean sleep latency was pathological during the multiple sleep latency test (MSLT) with one sleep onset REM period. Chronotherapy was used as an attempt to delay his bedtimes. He was scheduled to live on 21-hour days for 2 weeks and stabilised for one week on a 11 pm – 7 am sleep schedule. He was able to maintain this schedule for 5 months without early morning awakenings or daytime somnolence. However, sleep stage transition did not improve after chronotherapy. This case has thus confounding diagnoses. At the same time, Czeisler and colleagues reported the case of a 66 y.o. woman with early sleep times and an early endogenous temperature minimum [20]. They were able to phase delay the endogenous component of her core body temperature cycle without changing her sleep schedule.

Later on, Singer and Lewy, described the case of a 38 y.o. woman who struggled with evening sleepiness and early morning awakening [100]. If the patient went to bed past midnight, she would still be awake at 4 – 5 am. They treated her with bright light exposure of 2,500 lux from 8 to 10 pm for 2 weeks then from 9 to 11 pm for 2 additional weeks. The timing of her main sleep episode was free but naps were forbidden. Her sleep episodes delayed from 10:52 pm – 3:28 am to 11:11 pm – 5:01 am and her dim light melatonin onset delayed by 50 minutes from 7:50 pm to 8:40 pm. In 1993, Billiard and colleagues published the first case of a 15 y.o. caucasian girl with early evening sleepiness, early bedtime (6 – 8 pm) and early morning awakening (4 – 6 am) [9]. The disturbed sleep schedule was present since childhood and similar symptoms were reported in her mother and maternal grandfather. The diagnosis was confirmed by 3 weeks of actigraphy recording. A polysomnographic recording revealed a sleep onset at 8:09 pm, an abbreviated REM latency of 6 minutes, and an early morning awakening at 1:39 am resulting in reduced total sleep time of 4 hours 49 minutes. Some recuperation was noted on the second night. She was successfully treated using chronotherapy with a 3-hour advance of her bedtime each day. The success was short lived as she relapsed after 2 weeks.

Positive diagnosis

ASPS is often described as been the old-age equivalent of DSPS but with sleep times that are intractably earlier than desired. ASPS is characterised by evening sleepiness, sleep onset around 6 – 9 pm and early morning awakening around 1 – 3 am [111]. Sleep recording performed at the patient's desired sleep time is normal. However, when the patient attempts to sleep at later times, evening sleepiness, early morning awakening and reduced total sleep time are reported. The diagnosis should be confirmed by wrist actigraphy for several weeks at home and a laboratory investigation comprised of 2 polysomnographic recordings at the patients' best time. A MSLT could also be useful to confirm the diagnosis (ICSD, 2001). Intrinsic sleep disorders such as sleep apnea/hypopnea should be ruled out and no pathological sleepiness should be present during the MSLT. Sleep structure and efficiency should be normal if the patient sleeps at their desired time. A major depressive illness should be suspected if reduced REM sleep latency is present [111].

Pathophysiology

Complaints about disturbed sleep and insomnia appear more frequently in the elderly than in any other age group and its prevalence is estimated to be between 15 and 50% [13, 32, 59]. Older subjects tend to go to bed at an earlier time, get up early, and sleep less at night than younger subjects [14, 21, 91] . These changes appear progressively in life as they start to be observable in middle-aged individuals [16, 68]. We might hypothesise that ASPS is an exaggeration of the normal tendency to advance sleep time with aging or we might hypothesise that ASPS patients are extreme and pathological cases of morning larks. However, more studies are needed to clarify the phase angle between circadian markers and the sleep/wake cycle in patients with ASPS. Changes in circadian propensity with age cannot be summarised by simply considering the elderly as morning-type subjects. We know from the studies by Duffy and colleagues [33] that the differences observed between old and young subjects in the phase relationship between awakening and the temperature nadir is different from that observed between young morning-type and evening-type subjects. The authors reported that older subjects tend to wake up earlier than younger subjects but that they would also wake up at an earlier circadian phase than young morning-type subjects would. They would thus expose themselves to light earlier in the circadian cycle at a phase when it would further reinforce advances of the circadian system. This phase relationship is more comparable to that of young evening-type subjects who would also wake up early in their circadian cycle, enhancing any phase advancing effects of light exposure. This phase relationship in young evening-type subjects could possibly compensate for a weaker tendency to phase advance with light exposure. In young morning-type subjects, the temperature minimum is observed earlier within the sleep episode, which implies that a larger delay portion of the phase-response curve (PRC) is exposed to evening light. This phase relationship in young morning-type subjects could possibly compensate for weaker phase delaying mechanisms. It has been suggested that a reduced circadian period and/or partial defect in phase delaying mechanisms could be involved in the pathophysiology of ASPS [111]. Interestingly, a significant correlation between circadian period and the chronotype was observed, with young morning- and evening-type subjects having the shortest and longest periods, respectively [34]. A familial form of ASPS has also recently been associated with a defect in the phosphorylation of the human homologue of the *period* gene in Drosophila (hPer2) [105]. However, not all patients are equally affected and familial ASPS is thought to be a heterogeneous condition.

Treatment

Historically, chronotherapy in patients with ASPS has been designed to work in the opposite direction than for patients with DSPS and to exploit the presumed greater phase advancing capacity of these patients. Typically, patients are scheduled to live on 21-hour days until they go backwards around the clock and reestablish socially acceptable bedtimes and waketimes. Small, progressive delay shifts in bedtimes by ~15 minutes/day may also be attempted [8]. As for DSPS, patients should rigorously comply with the new schedule as the risk of relapse seems high [9]. Any deviation from the establish routine (e.g. by going to bed too early one night) would then shield the delay portion of the phase response curve to light and reinforce the natural tendency of the patient to advance the timing of his sleep episodes. This would lead to early morning awakenings and light exposure on the most sensitive part of the advance portion of the phase response curve to light. Since the light/dark cycle is the most powerful synchroniser of human circadian rhythms [12, 22], therapeutic manipulations of the sleep/wake cycle should be reinforced by judicious exposure to light and darkness. Bright light in the evening will promote delay shifts and be an effective

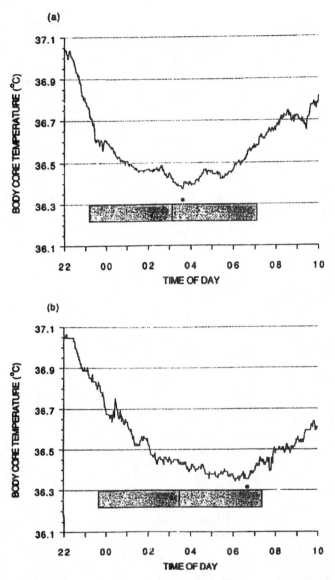

Figure 40.4 Distribution of temperature across a 12-hour period during baseline (a) and treatment (b) in patients with ASPS. The shaded box represents the time spent in bed with the vertical line signifying the midpoint of the sleep period. The closed circles stand for the temperature minima.

treatment of ASPS. Indeed, Campbell and colleagues [14] have successfully treated 16 ASPS patients with this approach. A 2-hour session of exposure to 4,000 lux prior to bedtime for 10 days produced a significant phase delay shift of their core body temperature rhythm (-3.13 hours). The bright light exposure tended to delay the onset of sleep and waketime by 29 and 18 minutes, respectively. The temperature minimum regained a more appropriate position within the sleep episode after the bright light treatment (fig. 40.4). As a result, patients treated with bright light saw their sleep efficiency increase from 77.5% to 90.1%, with fewer awakenings especially in the second third of the night. REM sleep latency also increased. Similar results have been reported in 5 women and 10 men aged 32-77 years complaining of early morning awakening [54]. Core body temperature and urinary 6-sulfatoxymelatonin was assessed prior to and after bright light treatment using 26-hour constant routine procedures. In the treatment condition, patients were exposed to 2,500 lux from 8 – 12 pm. They remained in ordinary room light of 150 lux in the control condition. Bright light exposure phase delayed the core body temperature rhythm by 2-4 hours and the urinary 6-sulfatoxymelatonin rhythm by 1-2 hours. Sleep onset remained unchanged but patients woke up 1 hour later.

Exogenous melatonin administered in the morning was also proposed based on the PRC to melatonin [57]. However, there is a lack of strong data supporting phase delaying properties of S20098 or melatonin such that their potential use in ASPS is uncertain [7].

NON-24 HOUR SLEEP-WAKE (OR HYPERNYCHTHEMERAL) SYNDROME

Historical Background

In 1970, Eliott and colleagues [37] described the first case of a man living on 26-hour days. Unfortunately, the authors did not clarify whether the patient was blind, suffered from a psychiatric disorder, or made voluntary attempts to entrain to the 24-hour day. So far, most cases have been described in either blind patients [39, 55, 62, 71, 75] or in patients with schizoid and introvert personality disorders [35, 38, 41] . For instance, Miles and colleagues [62] described the case of a psychologically healthy 28 y.o. blind man, actively working, with severe and cyclic sleep-wake disturbances (fig. 40.5). Circadian rhythms of core body temperature, alertness, performance, cortisol, and urinary electrolytes excretion were free running with a period longer than 24 hours. The patient was suffering from insomnia and excessive daytime sleepiness every 20 weeks or so. When he slept at his desired times, he spontaneously adopted a sleep-wake cycle of 24.9 hours and was asymptomatic. Attempts to entrain his sleep-wake cycle with hypnotics and stimulants were unsuccessful. A 10-day trial on a 24-hour schedule was attempted when his core body temperature arrived in phase with his sleep-wake cycle. Sleep recordings were normal during the free-running section but sleep significantly deteriorated during the attempts to synchronise to the 24-hour day. In this case, sleep onset and wake after sleep onset increased whereas total sleep time, stage 4 sleep, and REM sleep decreased. Growth hormone secretion remained associated with the occurrence of sleep whereas cortisol remained more tightly coupled with the endogenous circadian system and free-ran regardless of the imposed sleep-wake cycle.

Other cases of non-24-hour sleep-wake syndrome have been described. For instance, Kokkoris and colleagues [52] described the case of a 34 y.o. man who maintained a sleep-wake cycle of about 24 hours 20 minutes for an 8-year period. Core body temperature was recorded over 3.5 months and the patient filled out a sleep-wake log for 6 months. He reported better sleep and alertness when his sleep episodes coincided with the low points of his core body temperature rhythm. The patient also had a schizoid personality that might have contributed to his lack of entrainment to the 24-hour day. In 1980, Weber and colleagues [112] described the case study of two sighted college students who maintained a longer-than-24-hour sleep-wake cycle for nearly 4 years. John, a psychology student kept a sleep-wake diary from the age of 24 to 28 years in order to analyse the environmental determinants of his own behavior. Mary, apparently his friend, was a research assistant in psychology with a flexible work schedule. She kept a diary for 1 year primarily due to John's influence. There is no clear indication that these subjects suffered from their unusual schedule, which seemed to depend on a voluntary decision. The subjects were sleeping on average 7.6 hours per day. When faced with social demands, John attempted to adjust to a 24-hour rhythm by keeping regular waketimes but he needed several alarm clocks to wake up in the morning for his classes. The analysis of his sleep-wake log revealed some sections, especially during the summer vacation period, during which he maintained a regular 24-hour schedule, although at a delayed time. His disturbance is thus more compatible with a DSPS than with a hypernychthemeral syndrome. Similar observations can be made of Mary. In 1986, Wollman and Lavie [117] reported the case of a 26 y.o. man who for over 4 years had an inability to wake up and a greater than 24-hour sleep-wake cycle. The patient was described as extremely introvert with neurotic, narcissistic and borderline personality traits. The neurological exam was normal. A sleep diary kept for 4 years revealed an average day length of 27.4 hours. However, the analyses included long disrupted days and substantial sections where 24-hour rhythmicity was observable. There was a bimodal

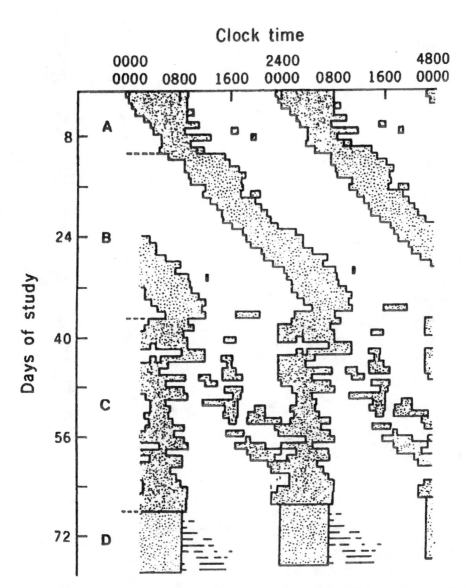

Figure 40.5 Sleep/wake pattern of a 28 y.o. blind man. The shaded area denotes sleep. (A) In home environment with the use of an alarm clock. (B) Ad-lib sleep/wake, work, and meal schedule, without time isolation. (C) Return to unusual home, work, and drug environment. (D) Entrainment attempt, with strict schedule of nocturnal sleep, meals and activity. The small trailing lines in (D) show episodes during the entrainment attempt in which the Stanford Sleepiness Scale was greater than 3.

distribution of sleep onset and a clustering around the "main sleep gate" (2 – 4 am) and the "secondary sleep gate" (2 – 6 am). Almost no sleep episodes were initiated around the "forbidden zone for sleep" namely, around 11 pm (fig. 40.6). Other cases of patients with introvert personality traits have been reported and responded to antidepressant therapy such as amitriptyline 50 mg taken at bed time [35].

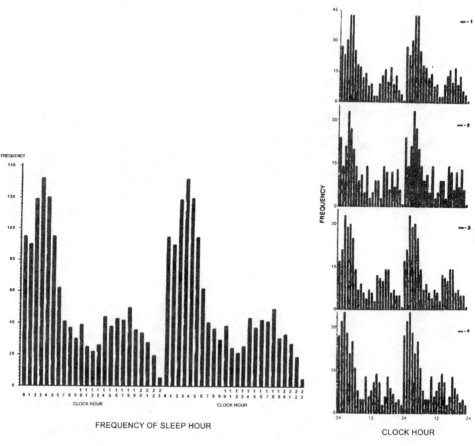

FREQUENCY OF SLEEP HOUR

CLOCK HOUR

Figure 40.6 Left: Frequency distribution of clock hours (double plot) during which a 26 y.o. male patient with a hypernychthemeral syndrome initiated a sleep period, across a 4-year period. Right: Frequency distribution of clock hours during which the patient initiated a sleep period, separated yearly for the 4-year period.

In 1983, Kamgar-Parsi and colleagues [48] described the case study of a young man who developed a hypernychthemeral sleep-wake syndrome after a stressful chess tournament associated with migraines and sleep disruption. Following the tournament, the patient was unable to get up at 7:30 am and missed all his morning classes of the semester. Before this episode, he was regularly sleeping from 11:30 pm to 7:30 am. Delayed chronotherapy successfully reentrained him to a socially acceptable schedule but he tended to drift back to his delayed position. His actigraphic data revealed a longer than 24-hour sleep-wake cycle when he was free to do so, but overall he was able to maintain a regular 24-hour schedule although at late times. Again, the clinical picture is more suggestive of a DSPS than a hypernychthemeral syndrome. It was found that his TSH was elevated with normal T3 and T4 concentrations. Because of the high incidence of B12 deficiency in patients with thyroid disease, he decided to self-medicate himself with B12 supplements (up to 0.2 mg/day taken at waketime). No B12 dosage was done prior to treatment. The patient reported an improvement in sleep quality and claimed that B12 supplements helped him to maintain a regular 24-hour sleep-wake cycle. A tendency to relapse to a 25-hour rhythm was reported when the B12 was stopped. It is possible that the chronotherapy used for his DSPS could have worsened his sleep-wake cycle disturbances and led him to live on 25-hour "days".

A few cases have been reported following traumatic brain injury or in brain damaged children [78, 84]. For instance, Okawa and colleagues [76] reported the case of a severely brain-damaged 12 y.o. boy with an apparent non-24-hour sleep-wake syndrome. His sleep disturbances appeared to cycle every 10-15 days with no obvious periodic changes in his cortisol rhythm. Sleep structure was substantially disrupted and no sleep spindles or K-complexes were observed on serial polysomnographic recordings. However, this case is unclear since the analysis of his sleep-wake cycle shows a rather consistent 24-hour rhythm with some periods of desynchronisation.

Epidemiology

Non-24-hour sleep-wake syndrome is extremely uncommon in sighted subjects living under normal conditions. Lack of entrainment is more frequently observed in blind patients or in astronauts and sub-mariners living under artificial light-dark cycles [49]. It can also be associated with psychiatric conditions such as schizoid or avoidant personality disorders. The incidence of sleep-wake disturbances in blind patients is apparently 5 times that observed in healthy subjects and about 76% of blind patients complain of some sleep-wake cycle disturbances [63, 111]. It is estimated that between 17 to 50% of blind patients have free-running circadian rhythms [56, 95].

Positive diagnosis

Patients with non-24-hour sleep-wake syndrome often report irregular sleep-wake cycles and periodic insomnia mixed with daytime sleepiness. The condition may go unrecognised for several years as apparent periods of successful treatment with large doses of hypnotics can occur in a recurrent fashion [104]. In the disturbed periods, the patient suffers from severe sleep-onset insomnia and difficulty awakening in the morning. Sleep onset tends to be delayed by 1-2 hours from one day to the other. The sleep-wake cycle disturbances need to have been present for at least 6 weeks and should not be better explained by another intrinsic sleep disorder. It can be intrinsic if abnormal circadian mechanisms or entrainment are suspected or extrinsic if socially or environmentally induced. A sleep-wake log and wrist actigraphy monitoring for several consecutive weeks (or even months) is suggested to confirm the diagnosis. This will reveal longer than 24-hour rest-activity cycles that may be interrupted by periods of relative coordination to the 24-hour day. Periods of long days with 24-40 hours without sleep followed by 14-24 hours of uninterrupted sleep may also occur [108, 110]. Relative coordination and a long sleep-wake cycle have been initially observed in healthy individuals kept for months in time free environments and undergoing a phenomenon known as spontaneous desynchronisation. Serial polysomnographic recordings are required to rule out intrinsic sleep disorders and to confirm the cyclical nature of the disturbance. If these recordings are planned at the same clock time, they will reveal a pattern of cyclic sleep disruption. Sighted individuals should also undergo a neurological evaluation with neuroimaging of the suprasellar region [60, 110].

Pathophysiology

It has been proposed that in sighted individuals, the hypernychthemeral syndrome might be a more severe form of DSPS [115]. Indeed, some cases of DSPS have converted to a non-24-hour sleep-wake syndrome after chronotherapy on a 27-hour day [48, 81]. Animal studies suggest that the hypernychthemeral syndrome can occur as a physiological aftereffect of lengthening the rest-activity cycle [87]. This could increase the endogenous circadian period until the 24-hour days falls outside of the new range of entrainment. It is interesting to note that several of the DSPS patients stated that they found themselves going to bed later from one night to the other [115]. A non-24-hour sleep-wake syndrome could thus be seen as a late complication of a sleep disturbance in which the patient deliberately decides to go to bed later to increase his sleepiness and ability to fall asleep [48]. In DSPS, a 1-2 hour increase of the waking episode might induce a tendency to delay sleep onset simulating a hypernychthemeral pattern for short periods of time. The difference between the two syndromes may also depend on the ability to phase advance, with greater impairments seen in the hypernychthemeral syndrome. Moreover, the maintenance of a greater-than-24-hour day often worsens the tendency to delay since the patient is exposed to light and darkness at times that favour further phase delays. The weakening of social time cues has been implicated [117] and this could play a role in some patients with personality disorders, if they attempt to minimise social interactions. Studies have also been carried out in people living in Antarctica. When free to choose their own sleep times, humans can present free-running melatonin, cortisol and sleep rhythms [49]. However, they are able to entrain to a 24-hour schedule if they keep regular bedtimes and waketimes [42]. In blind patients, the free-running rhythms of core body temperature, subjective alertness, urinary electrolytes, cortisol, and melatonin [5, 51, 62, 71, 95] are presumed to result from an inability to respond to the light-dark cycle as a circadian synchroniser. No substantial reduction of nocturnal melatonin secretion was observed with exposure to light of 500 or 1,000 lux in a sighted 41 y.o. man with hypernychthemeral syndrome [61]. These results suggest that a reduced sensitivity of the human circadian system to light could contribute to this syndrome.

Treatment

In children, non-24-hour sleep-wake schedules can be successfully treated by non-pharmacological approaches. For instance, Ferber and Boyle [38] reported the case of an 11 month old girl whose parents let her set her sleep-wake cycle on a 25-hour schedule. Her sleep schedule normalised with regular feeding and parental interactions. In adults, the treatment of non 24-hour sleep-wake syndrome is much more difficult and once entrainment is achieved, relapses are frequent. Very large doses of hypnotics and stimulants have been tried but the results alternate between efficacy and progressive loss of efficacy [110]. Exogenous melatonin 0.5-6 mg administered per os in the evening has been successfully used to entrain the non-24-hour sleep-wake cycle of a retarded boy [84], and the free-running rhythms of a sighted man [61] and several blind patients [5, 55, 85, 96, 97] . In one blind man, exogenous melatonin was able to stabilise the sleep-wake cycle but cortisol and temperature rhythms were still free-running [39]. These results suggest that the therapeutic efficacy of exogenous melatonin could result from an action on the rest-activity cycle rather than a direct chronobiotic effect on the endogenous circadian system. Eastman and colleagues [35] published the case of a man who had been free-running since high school. He was described as introverted and resistant to any psychological evaluation. His sleep schedule was temporarily stabilised on amitriptyline 50 mg taken at bed time for about 3 months. Exposure to bright light for 2-3 hours upon awakening also appeared to stop his free-running pattern although a small drift of a few minutes per day was reported. It is Kamgar-Parsi and colleagues [48] who reported in 1983 the first successful trial of a non-24h sleep-wake syndrome with B12 supplement. Since then other successful treatment cases have been reported [72, 74], although it still remains unclear whether this success is due to a placebo effect or to a physiological action of unknown mechanisms.

IRREGULAR SLEEP-WAKE PATTERN

Historical background

Irregular sleep-wake pattern (ISWP) is more likely found in nursing homes and demented populations [73, 107]. For instance, a progressive and increasing disruption of the sleep-wake cycle was reported in patients with senile dementia of Alzheimer type [109] (figure 40.7). In these patients, progressive disruption of SWS, REM sleep, sleep efficiency, increased nightmares, and daytime napping is observed. In ISWP, total sleep time per 24-hour day is about normal and consists of 3-4 bouts of sleep lasting 4 hours or less each day. Disrupted circadian rhythms of body temperature, melatonin, or cortisol have been reported in demented patients or in severely brain-damaged patients. A few cases of ISWP were described in cognitively intact patients with several years of prolonged bedrest [66, 98, 107, 111] but results vary greatly between studies. Irregular sleep patterns were also reported in patients with severe congenital, developmental or degenerative brain dysfunction [110].

Epidemiology

The epidemiology of ISWP or its sex prevalence is unknown [104]. It is most frequently described in patients with neurological conditions such as diffuse brain dysfunction, head injury, hypothalamic lesions or developmental disabilities [60].

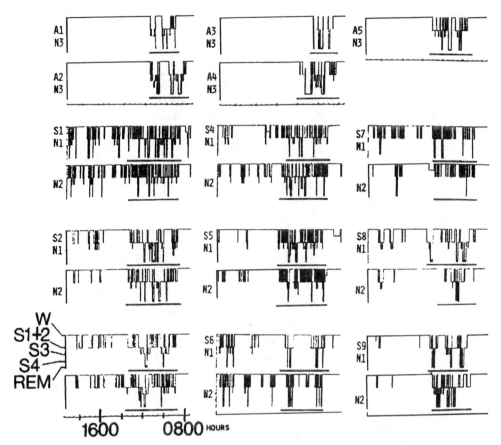

Figure 40.7 Distribution of sleep-waking pattern in severe stage Alzheimer's disease (AD) patients and normal aged controls. Nine AD patients (S1-S9) in the last two nights of three consecutive 24-hour monitoring periods along with typical data from five controls (A1-A5) for the three consecutive nights. Time spent in bed is represented by a horizontal bar under each plot.

Positive diagnosis

A diagnosis of ISWP is made when there are more than three sleep episodes per 24-hour day. The condition must be present for at least 3 consecutive months and is associated with chronic insomnia and daytime sleepiness. The sleep-wake cycle is irregular and disorganised, although the total amount of sleep per day is often normal. Extended polysomnographic recordings (e.g. for 72 hours), in the presence of conventional environmental cues, show irregular periods of sleep with bouts of 2-3 hours [2, 78]. The overt rhythmicity of circadian markers such as that of core body temperature and melatonin secretion can be disrupted by the disorganised sleep-wake pattern [106].

Pathophysiology

It is possible to create an animal model of ISWP by increasing the photoperiod by about 30 minutes each week [6, 31]. The rest-activity cycle becomes unstable and disrupted when the light/dark ratio reaches 19 hours/5hours. In demented patients, a reduction of light exposure rather than an extension of the photoperiod was hypothesised to contribute to the appearance of ISWP. In these patients, it was proposed that a substantial reduction of light exposure in nursing homes or reduced retinal sensitivity to light with age could lead to disrupted sleep patterns and nocturnal wandering [6, 15]. Degenerative brain lesions affecting the SCN or brain areas involved in sleep-wake regulation could also play a role in the appearance of ISWP in demented patients [18, 102]. In cognitively intact patients, daytime napping and irregular sleep patterns appear principally as a disorder of sleep hygiene [92].

Treatment

ISWP is typically very difficult to treat [110]. Non-pharmacological approaches should aim at increasing the regularity of environmental synchronisers. These consist of regular activities such as meals, outdoor activities and social interactions. Sleep restrictions to nighttime hours is encouraged to help consolidate the main sleep episode [73, 111,]. Exposure to bright light, either in the morning [65] or in the evening [98] can reduce nocturnal agitation and improve nighttime sleep in demented patients. Therapeutic improvement was also reported with exogenous melatonin 2.5-10 mg at bedtime in children resistant to hypnotics [45]. Vitamin B12 was tried successfully for ISWP, although a placebo effect is plausible [77].

REFERENCES

1. ALLEN R., ROSENTHAL N., JOSEPH-VANDERPOLL J.R., NADEAU J., KELLY K., SCHULZ P., SOUETRE E. - Delayed sleep phase syndrome: polysomnographic characteristics. *Sleep Research*, 18, 133-133, 1989.
2. ALLEN S.R., SEILER W.O., STAHELIN H.B., SPIEGEL R. - Seventy-two hour polygraphic and behavioral recordings of wakefulness and sleep in a hospital geriatric unit: Comparison between demented and non demented patients. *Sleep*, 10(2), 143-159, 1987.
3. ALVAREZ B., DAHLITZ M.J., VIGNAU J., PARKES J.D. - The delayed sleep phase syndrome: clinical and investigative findings in 14 subjects. *Journal of Neurology, Neurosurgery & Psychiatry*, 55, 665-670, 1992.
4. AOKI H., OZEKI Y., YAMADA N. - Hypersensitivity of melatonin suppression in response to light in patients with delayed sleep phase syndrome. *Chronobiology International*, 18(2), 263-271, 2001.
5. ARENDT J., ALDHOUS M., WRIGHT J. - Synchronisation of a disturbed sleep-wake cycle in a blind man by melatonin treatment. *Lancet*, 2(1), 772-773, 1988.
6. ARMSTRONG S.M. - Melatonin as a chronobiotic for circadian insomnia. Clinical observations and animal models. *Advances in Experimental Medicine & Biology*, 283-297, 1999.
7. ARMSTRONG S.M., McNULTY O.M., GUARDIOLA-LEMAITRE B., REDMAN J.R. - Successful use of S20098 and melatonin in an animal model of delayed sleep-phase syndrome (DSPS). *Pharmacology, biochemistry and behavior*, 46, 45-49, 1993.
8. BAKER S.K., PHYLLIS C.Z. - Circadian disorders of the sleep-wake cycle. In: M. H. Kryger, T. Roth, & W. C. Dement (Eds.), *Principles and practice of sleep medicine*. W.B. Saunders Company, Philadelphia, Pennsylvania, U.S.A., 606-614, 2001.
9. BILLIARD M., VERGÉ M., ALDAZ C., CARLANDER B., TOUCHON J., BESSET A. - A case of advanced-sleep phase syndrome. *Sleep Research*, 22, 109, 1993.
10. BILLIARD M., VERGÉ M., TOUCHON J., CARLANDER B., BESSET A. - Delayed sleep phase syndrome: subjective and objective data chronotherapy and follow-up. *Sleep Research*, 22, 172-172, 1993.
11. BIXLER E.O., KALES A., SOLDATOS C.R., KALES J.D., HEALEY S. - Prevalence of sleep disorders in the Los Angeles metropolitan area. *American Journal of Psychiatry*, 136(10), 1257-1262, 1979.
12. BOIVIN D.B., DUFFY J.F., KRONAUER R.E., CZEISLER C.A. - Dose-response relationships for resetting of human circadian clock by light. *Nature*, 379(6565), 540-542, 1996.
13. BUYSSE D.J., REYNOLDS III, C.F., MONK T.H., HOCH C.C., YEAGER A.L., KUPFER D.J. - Quantification of subjective sleep quality in healthy elderly men and women using the Pittsburgh sleep quality index (PSQI). *Sleep*, 14(4), 331-338, 1991.
14. CAMPBELL S.S., DAWSON D., ANDERSON M.W. - Alleviation of sleep maintenance insomnia with timed exposure to bright light. *Journal of the American Geriatric Society*, 41, 829-836, 1993.
15. CAMPBELL S.S., KRIPKE D.F., GILLIN J.C., HRUBOVCAK J.C. - Exposure to light in healthy elderly subjects and Alzheimer's patients. *Physiology & Behavior*, 42(2), 141-144, 1988.
16. CARRIER J., MONK T.H., BUYSSE D.J., KUPFER D.J. - Sleep and Morningness-Eveningness in the "middle" years of life (20y-59y). *Journal of Sleep Research*, 6(4), 230-237, 1997.
17. CARSKADON M.A., LABYAK S.E., ACEBO C., SEIFER R. - Intrinsic circadian period of adolescent humans measured in conditions of forced desynchrony. *Neuroscience Letters*, 260(2), 129-132, 1999.
18. COHEN R.A., ELLIOTT ALBERS H. - Disruption of human circadian and cognitive regulation following a discrete hypothalamic lesion: A case study. *Neurology*, 41, 726-729, 1991.
19. COLEMAN R.M. - Diagnosis, treatment, and follow-up of about 8,000 sleep/wake disorder patients. In: C. Guilleminault & E. Lugaresi (Eds.), *Sleep/Wake Disorders: Natural History, Epidemiology and Long Term Evolution*. Raven Press, New York: 87-97, 1983.
20. CZEISLER C.A., ALLAN J.S., STROGATZ S.H., RONDA J.M., SANCHEZ R., RIOS C.D., FREITAG W.O., RICHARSON G.S., KRONAUER R.E. - Bright light resets the human circadian pacemaker independent of the timing of the sleep-wake cycle. *Science*, 233, 667-671,1986.

21. CZEISLER C.A., DUMONT M., DUFFY J.F., STEINBERG J.D., RICHARSON G.S., BROWN E.N., SANCHEZ R., RIOS C.D., RONDA J.M. - Association of sleep-wake habits in older people with changes in output of circadian pacemaker. *Lancet*, 340, 933-936, 1992.

22. CZEISLER C.A., KRONAUER R.E., ALLAN J.S., DUFFY J.F., JEWETT M.E., BROWN E.N., RONDA J.M. - Bright light induction of strong (Type 0) resetting of the human circadian pacemaker. *Science*, 244, 1328-1333, 1989.

23. CZEISLER C.A., MOORE-EDE M.C., COLEMAN R.M. - Resetting circadian clocks: Applications to sleep disorders medicine and occupational health. In: C. Guilleminault & E. Lugaresi (Eds.), *Sleep/Wake Disorders: Natural History, Epidemiology, and Long- Term Evolution.* Raven Press, New York: 243-260, 1983.

24. CZEISLER C.A., RICHARSON G.S. - Detection and assessment of insomnia. *Clinical Therapeutics*, 13, 663-679, 1991.

25. DAGAN Y., SELA H., OMER H., HALLIS D., DAR R. - High prevalence of personality disorders among circadian rhythm sleep disorders (CRSD) patients. *Journal of Psychosomatic Research*, 41(4), 357-363, 1996.

26. DAGAN Y., STEIN D., STEINBOCK M., YOVEL I., HALLIS D. - Frequency of delayed sleep phase syndrome among hospitalized adolescent psychiatric patients. *Journal of Psychosomatic Research*, 45(1), 15-20, 1998.

27. DAGAN Y., TZISCHINSKY O., LAVIE P. - Sunlight treatment for delayed sleep-phase syndrome: Case report. *Sleep Research*, 20, 451, 1991.

28. DAGAN Y., YOVEL I., HALLIS D., EISENSTEIN M., RAICHIK I. - Evaluating the role of melatonin in the long-term treatment of delayed sleep phase syndrome (DSPS). *Chronobiology International*, 15(2), 181-190, 1998.

29. DAHLITZ M., ALVAREZ B., VIGNAU J., ENGLISH J., ARENDT J., PARKES J.D. - Delayed sleep phase syndrome response to melatonin. *Lancet*, 337, 1121-1124, 1991.

30. DEBECK T.W. - Delayed sleep phase syndrome - criminal offense in the military? *Military Medicine*, 155(1:14), 14-15, 1990.

31. DELAGRANGE P., GUARDIOLA-LEMAITRE B. - Melatonin, its receptors, and relationships with biological rhythm disorders. *Clinical Neuropharmacology*, 20(6), 482-510, 1997.

32. DEMENT W.C., MILES L.E., CARSKADON M.A. - "White paper" on sleep and aging. *Journal of the American Geriatric Society*, 30, 25-50, 1982.

33. DUFFY J.F., DIJK D.J., HALL E.F., CZEISLER C.A. - Relationship of endogenous circadian melatonin and temperature rhythms to self-reported preference for morning or evening activity in young and older people. *Journal of Investigative Medecine*, 47(3), 141-150, 1999.

34. DUFFY J.F., RIMMER D.W., SILVA E.J., CZEISLER C.A. - Correlation of intrinsic circadian period with morningness-eveningness in young men. *Sleep*, 22(Supplement), S92-S92, 1999.

35. EASTMAN C.I., ANAGNOPOULOS C.A., CARTWRIGHT R.D. - Can bright light entrain a free-runner? *Sleep Research*, 17, 372-372, 1988.

36. EBIHARA S., MANO N., KURONO N., KOMURO G., YOSHIMURA T. - Vitamin B12 affects non-photic entrainment of circadian locomotor activity rhythms in mice. *Brain Research*, 727(1-2), 31-39, 1996.

37. ELIOTT A.L., MILLS J.N., WATERHOUSE J.M. - A man with too long a day. *Proceedings of the Physiological Society*, 30-31, 1970.

38. FERBER R., BOYLE M.P. - Persistence of a free-running sleep-wake rhythm in a one year old girl. *Sleep Research*, 12, 364, 1983.

39. FOLKARD S., ARENDT J., ALDHOUS M., KENNETT H. - Melatonin stabilises sleep onset time in a blind man without entrainment of cortisol or temperature rhythms. *Neuroscience Letters*, 113, 193-198, 1990.

40. FOOKSON J.E., KRONAUER R.E., WEITZMAN E.D., MONK T.H., MOLINE M.L., HOEY E. - Induction of insomnia on a non-24 hour sleep-wake schedule. *Sleep Research*, 13, 220, 1984.

41. GIEDKE H., ENGELMANN W., REINHARD P. - Free-running circadian rest-activity cycle in a normal environment. A case study. *Sleep Research*, 12, 365-365, 1983.

42. GRIFFITHS P.A., FOLKARD S., BOJKOWSKI C., ENGLISH J., ARENDT J. - Persistant 24-h variations of urinary 6-hydroxy melatonin sulphate and cortisol in Antarctica. *Experientia*, 42, 430-432, 1986.

43. HONMA K., KOHSAKA M., FUKUDA N., MORITA N., HONMA S. - Effects of vitamin B12 on plasma melatonin rhythm in humans: increased light sensitivity phase-advances the circadian clock? *Experientia*, 48, 716-720, 1992.

44. ITO A., ANDO K., HAYAKAWA T., IWATA T., KAYUKAWA Y., OHTA T., KASAHARA Y. - Long-Term Course of Adult Patients with Delayed Sleep Phase Syndrome. *The Japanese Journal of Psychiatry and Neurology*, 47(3), 563-567, 1993.

45. JAN J.E., ESPEZEL H., APPLETON R.E. - The treatment of sleep disorders with melatonin. *Developmental Medicine & Child Neurology*, 36, 97-107, 1994.

46. JOSEPH-VANDERPOOL J.R., KELLY K.G., SCHULZ P.M., ALLEN R., SOUETRE E., ROSENTHAL N.E. - Delayed sleep phase syndrome revisited: Preliminary effects of light and triazolam. *Sleep Research*, 17, 381-381, 1988.

47. JOSEPH-VANDERPOOL J.R., ROSENTHAL N.E., LEVENDOSKY A.A., JOHNSTON S.H., ALLEN R., KELLY K.A., SOUETRE E., SCHULZ P.M., STARZ B.K. - Phase-shifting effects of bright morning light as treatment for delayed sleep phase syndrome. *Sleep Research*, 18, 422-422, 1989.

48. KAMGAR-PARSI B., WEHR T.A., GILLIN J.C. - Successful treatment of human non-24-hour sleep-wake syndrome. *Sleep*, 6(3), 257-264, 1983.

49. KENNAWAY D.J., VAN DORP C.F. - Free-running rhythms of melatonin, cortisol, electrolytes, and sleep in humans in Antarctica. *American Journal of Physiology*, 260, R1137-R1144, 1991.

50. KIRMIL-GRAY K., EAGLESTON J.R., GIBSON E., THORESEN C.E. - Sleep disturbance in adolescents: sleep quality, sleep habits, beliefs about sleep, and daytime functioning. *Journal of Youth and Adolescence*, 13(5), 375-384, 1984.

51. KLEIN T., MARTENS H., DIJK D.J., KRONAUER R.E., SEELY E.W., CZEISLER C.A. - Circadian sleep regulation in the absence of light perception: chronic non-24-hour circadian rhythm sleep disorder in a blind man with a regular 24-hour sleep-wake schedule. *Sleep*, 16(4), 333-343, 1993.

52. KOKKORIS C.P., WEITZMAN E.D., POLLAK C.P., SPIELMAN A.J., CZEISLER C.A., BRADLOW H. - Long-term ambulatory temperature monitoring in a subject with a hypernychthemeral sleep-wake cycle disturbance. *Sleep*, 1, 177-190, 1978.

53. LACK L.C. - Delayed sleep and sleep loss in university students. *Journal of American College Health*, 35, 105-110, 1986.

54. LACK L., WRIGHT H. - The effect of evening bright light in delaying the circadian rhythms and lengthening the sleep of early morning awakening insomniacs. *Sleep*, 16(5), 436-443, 1993.

55. LAPIERRE O., DUMONT M. - Melatonin treatment of a non-24-hour sleep-wake cycle in a blind retarded child. *Biological Psychiatry*, 38(2), 119-122, 1995.

56. LEGER D., GUILLEMINAULT C., DEFRANCE R., DOMONT A., PAILLARD - Blindness and sleep patterns. *Lancet*, 348, 830-831, 1996.

57. LEWY A.J., AHMED S., JACKSON J.M.L., SACK R.L. - Melatonin shifts human circadian rhythms according to a phase- response curve. *Chronobiology International*, 9(5), 380-392, 1992.

58. LEWY A.J., BAUER V.K., AHMED S., THOMAS K.H., CUTLER N.L., SINGER C.M., MOFFIT M.T., SACK R.L. - The human phase response curve (PRC) to melatonin is about 12 hours out of phase with the PRC to light. *Chronobiology International*, 15(1), 71-83, 1998.

59. LORRAIN D., BOIVIN D.B. - Troubles du sommeil. In: M. Arcand & R. Hébert (eds.), *Précis Pratique de Gériatrie*. (pp. 239-258). Edisem Inc , Sherbrooke: 239-258, 1997.

60. MAHOWALD M.W., ETTINGER M.G. - Circadian Rhythm Disorders. In S. Chokroverty (Ed.), *Sleep Disorders Medicine: Basic Science, Technical Considerations, and Clinical Aspects*. (pp. 619-634). Butterworth Heinemann, Boston: 1999.

61. McARTHUR A.J., LEWY A.J., SACK R.L. - Non-24-hour sleep-wake syndrome in a sighted man: circadian rhythm studies and efficacy of melatonin treatment. *Sleep*, 19(7), 544-553, 1996.

62. MILES L.E., RAYNAL D.M., WILSON M.A. - Blind man living in normal society has circadian rhythms of 24.9 hours. *Science*, 198, 421-423, 1977a.

63. MILES L.E., WILSON M.A. - High incidence of cyclic sleep/wake disorders in the blind. *Sleep Research*, 6, 192-192, 1977b.

64. MINORS D.S., WATERHOUSE J.M., WIRZ-JUSTICE A. - A human phase-response curve to light. *Neuroscience Letters*, 133, 36-40, 1991.

65. MISHIMA K., OKAWA M., HISHIKAWA Y., HOZUMI S., HORI H., TAKAHASHI K. - Morning bright light therapy for sleep and behaviour disorders in elderly patients with dementia. *Acta Psychiatrica Scandinavica*, 89, 1-7, 1994.

66. MISHIMA K., TOZAWA T., SATOH K., MATSUMOTO Y., HISHIKAWA Y., OKAWA M. - Melatonin secretion rhythm disorders in patients with senile dementia of Alzheimer's type with disturbed sleep-waking. *Biological Psychiatry*, 45(4), 417-421, 1999.

67. MOLDOFSKY H., MUSICI S., PHILLIPSON E.A. - Treatment of a case of advanced sleep phase syndrome by phase advance chronotherapy. *Sleep*, 9(1), 61-65, 1986.

68. MONK T.H., BUYSSE D.J., REYNOLDS III, C.F., KUPFER D.J., HOUCK P.R. - Circadian temperature rhythms of older people. *Experimental Gerontology*, 30(5), 455-474, 1995.

69. MORRIS M., LACK L., DAWSON D. - Sleep-onset insomniacs have delayed temperature rhythms. *Sleep*, 13(1), 1-14, 1990.

70. NAGTEGAAL J.E., KERKHOF G.A., SMITS M.G., SWART A.C.W., VAN DER MEER Y.G. - Traumatic brain injury-associated delayed sleep phase syndrome. *Functional Neurology*, 12(6), 345-348, 1997.

71. NAKAGAWA H., SACK R.L., LEWY A.J. - Sleep propensity free-runs with the temperature, melatonin and cortisol rhythms in a totally blind person. *Sleep*, 15(4), 330-336, 1992.

72. OHTA T., ANDO K., IWATA T., OZAKI N., KAYUKAWA Y., TERASHIMA M., OKADA T., KASAHARA Y. - Treatment of persistent sleep-wake schedule disorders in adolescents with methylcobalamin (vitamin B12). *Sleep*, 14(5), 414-418, 1991.

73. OKAWA M., MISHIMA K., HISHIKAWA Y., HOZUMI S., HORI H., TAKAHASHI K. - Circadian rhythm disorders in sleep-waking and body temperature in elderly patients with dementia and their treatment. *Sleep*, 14(6), 478-485, 1991.

74. OKAWA M., MISHIMA K., NANAMI T., SHIMIZU T., IIJIMA S., HISHIKAWA Y., TAKAHASHI K. - Vitamin B12 treatment for sleep-wake rhythm disorders. *Sleep*, 13(1), 15-23, 1990.

75. OKAWA M., NANAMI T., WADA T., SHIMIZU T., HISHIKAWA Y., SASAKI H., NAGAMINE H., TAKAHASHI K. - Four congenitally blind children with circadian sleep-wake rhythm disorder. *Sleep*, 10(2), 101-110, 1987.

76. OKAWA M., SASAKI H., NAKAJIMA S., TAKAHASHI K. - Altered sleep rhythm: a patient with a 10 to 15 day cycle. *Journal of Neurology*, 266, 63-71, 1981.

77. OKAWA M., TAKAHASHI K., EGASHIRA K.,; FURUTA H., HIGASHITANI Y., HIGUCHI T., ICHIKAWA H., ICHIMARU Y., INOUE Y., ISHIZUKA Y et al. - Vitamin B12 treatment for delayed sleep

phase syndrome: A multi-center double-blind study. *Psychiatry and Clinical Neurosciences,* 51, 275-279, 1997.

78. OKAWA M., TAKAHASHI K., SASAKI H. - Disturbance of circadian rhythms in severely brain-damaged patients correlated with CT findings. *Journal of Neurology,* 233, 274-282, 1986.

79. OLDANI A., FERINI-STRAMBI L., ZUCCONI M., STANKOV B., FRASCHINI F., SMIRNE S. - Melatonin and delayed sleep phase syndrome: ambulatory polygraphic evaluation. *Neuroreport,* 6, 132-134, 1994.

80. OREN D.A., TURNER E.H., WEHR T.A. - Abnormal circadian rhythms of plasma melatonin and body temperature in the delayed sleep phase syndrome. *Journal of Neurology, Neurosurgery and Psychiatry,* 58, 379-385, 1995.

81. OREN D.A., WEHR T.A. - Hypernychthemeral syndrome after chronotherapy for delayed sleep phase syndrome. *New England Journal of Medecine,* 327, 1762, 2001

82. OZAKI N., IWATA T., ITOH A., KOGAWA S., OHTA T., OKADA T., KASAHARA Y. - Body temperature monitoring in subjects with delayed sleep phase syndrome. *Neuropsychobiology,* 20, 174-177, 1988.

83. OZAKI S., UCHIYAMA M., SHIRAKAWA S., OKAWA M. - Prolonged interval from body temperature nadir to sleep offset in patients with delayed sleep phase syndrome. *Sleep,* 19(1), 36-40, 1996.

84. PALM L., BLENNOW G., WETTERBERG L. - Correction of non-24-hour sleep/wake cycle by melatonin in a blind retarded boy. *Annals of Neurolology,* 29, 336-339, 1991.

85. PALM L., BLENNOW G., WETTERBERG L. - Long term melatonin treatment in blind children and young adults with circadian sleep-wake disturbances. *Developmental.Medicine & Child Neurology,* 39(5), 319-325, 1997.

86. PATTEN S.B., LAUDERDALE W.M. - Delayed sleep phase disorder after traumatic brain injury. *Journal of the American Academy of Child and Adolescent Psychiatry,* 31(1), 100-102, 1992.

87. PITTENDRIGH C.S., DAAN S. - A functional analysis of circadian pacemakers in nocturnal rodents I. the stability and lability of spontaneous frequency. *Journal of Comparative Physiology* A, 106, 223-252, 1976.

88. PRICE V.A., COATES T.J., THORESEN C.E., GRINSTEAD O.A. - Prevalence and correlates of poor sleep among adolescents. *American Journal of Diseases of Children,* 132, 583-586, 1978.

89. QUINTO C., GELLIDO C., CHOKROVERTY S., MASDEU J. - Posttraumatic delayed sleep phase syndrome. *Neurology,* 54(1), 250-252, 2000.

90. REGENSTEIN Q.R., MONK T.H. - Delayed sleep phase syndrome: A review of its clinical aspects. *American Journal of Psychiatry,* 2, 602-608, 1995.

91. REYNER A., HOME J.A. - Gender- and aging-related differences in sleep determined by home-recorded sleep logs and actimetry from 400 adults. *Sleep,* 18(2), 127-134, 1995.

92. RICHARDSON G.S., MALIN H.V. - Circadian rhythm sleep disorders: pathophysiology and treatment. *Journal of Clinical Neurophysiology,* 13(1), 17-31, 1996.

93. ROSENTHAL N.E., JOSEPH-VANDERPOOL J.R., LEVENDOSKY A.A., JOHNSTON S.H., ALLEN R., KELLY K.A., SOUETRE E., SCHULTZ P.M., STARZ K.E. - Phase-shifting effects of bright morning light as treatment for delayed sleep phase syndrome. *Sleep,* 13(4), 354-361, 1990.

94. SACK R.L., LEWY A.J. - Melatonin as a chronobiotic: treatment of circadian desynchrony in night workers and the blind. *Journal of Biological Rhythms,* 12(6), 595-603, 1997.

95. SACK R.L., LEWY A.J., BLOOD M.L., KEITH L.D., NAKAGAWA H. - Circadian rhythm abnormalities in totally blind people: incidence and clinical significance. *Journal of Clinical Endocrinology and Metabolism,* 75, 127-134, 1992.

96. SACK R.L., LEWY A.J., BLOOD M.L., STEVENSON J., KEITH L.D. - Melatonin administration to blind people: Phase advances and entrainment. *Journal of Biological Rhythms,* 6(3), 249-261, 1991.

97. SARRAFZADEH A., WIRZ-JUSTICE A., ARENDT J., ENGLISH J. - Melatonin stabilises sleep onset in a blind man. In: J. Horne (Ed.), *Sleep 90.* (pp. 51-54). Pontenagel Press, Bochum: 51-54, 1990.

98. SATLIN A., VOLICER L., ROSS V., HERZ L., CAMPBELL S. - Bright light treatment of behavioral and sleep disturbances in patients with Alzheimer's disease. *American Journal of Psychiatry,* 149, 1028-1032, 1992.

99. SCHRADER H., BOVIM G., SAND T. - Depression in the delayed sleep phase syndrome. *American Journal of Psychiatry,* 153(9), 1238-1238.

100. SINGER C.M., LEWY A.J. - Case report: Use of the dim light melatonin onset in the treatment of ASPS with bright light. *Sleep Research,* 18,445, 1989.

101. STROGATZ S.H., KRONAUER R.E., CZEISLER C.A. - Circadian pacemaker interferes with sleep onset at specific times each day: Role in insomnia. *American Journal of Physiology,* 253, R172-R178, 1987.

102. SWAAB D.F., FLIERS E., PARTIMAN T.S. - The suprachiasmatic nucleus of the human brain in relation to sex, age and senile dementia. *Brain Research,* 342, 37-44, 1985.

103. TAKAHASHI K., OKAWA M., MARSUMOTO M., MISHIMA K., YAMADERA H., SAAKI M., ISHIZUKA Y., YAMADA K., HIGUCHI T., OKAMOTO N., et al. - Double-blind test on the efficacy of methylcobalamin on sleep-wake rhythm disorders. *Psychiatry and Clinical Neurosciences,* 53, 211-213, 1999.

104. *The International Classification of Sleep Disorders Revised Diagnostic and Coding Manual* (1997 Revised ed.). Lawrence, Kansas, Allen Press Inc, 2001.

105. TOH K.L., JONES C.R., HE Y., EIDE E.J. HINZ W.A., VIRSHUP D.M., PTACEK L.J., FU Y.M. - An hPer2 phosphorylation site mutation in familial advanced sleep phase syndrome. *Science,* 291, 1040-1043, 2001.

106. TOUITOU Y., SULON J., BOGDAN A., TOUITOU C., REINBERG A., BECK H., SODOYEZ J.C., DEMEY-PONSART E., VAN CAUWENBERGE H. - Adrenal circadian system in young and elderly human subjects: a comparative study. *Journal of Endocrinology*, 93, 201-210, 1982.
107. UCHIDA K., OKAMOTO N., OHARA K., MORITA Y. - Daily rhythm of serum melatonin in patients with dementia of the degenerate type. *Brain Research*, 717, 154-159, 1996.
108. UCHIYAMA M., OKAWA M., OZAKI S., SHIRAKAWA S., TAKAHASHI K. - Delayed phase jumps of sleep onset in a patient with non-24-hour sleep-wake syndrome. *Sleep*, 19(8), 637-640, 1996.
109. VITIELLO M.V., PRINZ P.N. - Sleep/wake patterns and sleep disorders in Alzheimer's disease. In: M. J. Thorpy (Ed.), *Handbook of Sleep Disorders*. Marcel Dekker, New York: 703-718, 1990.
110. WAGNER D.R. - Circadian rhythm sleep disorders. In: M. J. Thorpy (ed.). *Handbook of Sleep Disorders*. Marcel Dekker, New York: 493-527, 1990.
111. WAGNER D.R. - Disorders of the circadian sleep-wake cycle. *Neurologic clinics*, 14(3), 651-670, 1996.
112. WEBER A.L., CARY M.S., CONNOR N., KEYES P. - Human non-24-hour sleep -wake cycles in an everyday environment. *Sleep*, 2(3), 347-354, 1980.
113. WEITZMAN E.D. - Sleep and its disorders. *Annual Review of Neuroscience,* 4, 381-417, 1981.
114. WEITZMAN E.D., CZEISLER C., COLEMAN R., DEMENT W., RICHARDSON G., POLLAK C.P. - Delayed sleep phase syndrome: A biological rhythms sleep disorder. *Sleep Research*, 8, 221, 1979.
115. WEITZMAN E.D., CZEISLER C.A., COLEMAN R.M., SPIELMAN A.J., ZIMMERMAN J.C., DEMENT W.C. - Delayed sleep phase syndrome: a chronobiologic disorder associated with sleep onset insomnia. *Archives of General Psychiatry*, 38, 737-746, 1981.
116. WEYERBROCK A., TIMMER J., HOHAGEN F., BERGER M., BAUER J. - Effects of light and chronotherapy on human circadian rhythms in delayed sleep phase syndrome: cytokines, cortisol, growth hormone, and the sleep-wake cycle. *Biological Psychiatry*, 40(8), 794-797, 1996.
117. WOLLMAN M., LAVIE P. - Hypernychthemeral sleep-wake cycle: Some hidden regularities. *Sleep*, 9(2), 324-334, 1986.

Chapter 41

Parasomnias

M.F. Vecchierini
Service d'Explorations Fonctionnelles, Hôpital Bichat, Paris, France

DEFINITION

Parasomnias are a group of clinical disorders observed during sleep. They often involve undesirable motor disorders. These fairly heterogeneous physical disorders are not due to alterations in the processes responsible for sleep and awake states, but occur in subjects who are predisposed, at the moment of sleep-wake transition phases or during the different stages of sleep. They have little effect, in general, on the quality of wakefulness.

INTERNATIONAL CLASSIFICATION OF SLEEP DISORDERS

This classification subdivides the parasomnia group into 4 subsets according to when they occur in the course of the night.

Thus *arousal disorders* correspond to parasomnias which occur during NREM sleep:
- confusional disorders
- sleepwalking
- sleep terrors.

These parasomnias are common and their name emphasises their pathophysiological mechanism realising a partial arousal still referred to as dissociated arousal reactions.

The *sleep-wake transition disorders* are:
- rhythmic movement disorder
- sleep starts
- sleep talking
- nocturnal leg cramps.

These disorders mainly occur in the transition from wakefulness to sleep, even if rhythmic movement disorder is also observed in awake states and sleep talking can occur not only on arousal but also during NREM sleep or during REM sleep. It thus seems justifiable to deal with sleep talking in the subchapter covering "other parasomnias". Finally, certain authors include hypnagogic hallucinations in this subset of parasomnias. The classification system mentions them as a possible sign of narcolepsy. What is known is that these often strong perceptual experiences, occur in the transition from wakefulness to sleep and may be accompanied not only by sensorial but also by tactile and kinaesthetic phenomena, and are found to occur occasionally in 15 to 45% of subjects in the general population.

There are six *parasomnias usually associated with REM sleep*:
- nightmares
- sleep paralysis
- impaired sleep-related penile erections
- sleep-related painful erections
- REM sleep sinus arrest
- REM sleep behaviour disorder.

Sleep paralysis tends to occur during the transition from sleep to wakefulness but all the parasomnias in this subset have a consistent link with REM sleep.

This chapter will deal neither with impaired sleep-related penile erections nor with sleep-related painful erections. REM sleep sinus arrest will only be referred to briefly since this manifestation is reported in the more general chapter on dysautonomia.

The *other parasomnias* include a group of highly heterogeneous manifestations assembled in an arbitrary and no doubt temporary fashion.
- sleep bruxism
- sleep enuresis
   sleep-related abnormal swallowing syndrome
- nocturnal paroxysmal dystonia
- sudden unexplained nocturnal death syndrome
- primary snoring
- infant sleep apnoea
- congenital central hypoventilation syndrome
- sudden infant death syndrome
- benign neonatal sleep myoclonus.

The rarer phenomena are thus grouped together either because they can occur during several stages of sleep, precluding them from the parasomnia subsets previously referred to, or because they are still ill-defined, their sole link being the fact that they occur during the night and cannot be included in other classification sections. However since 1990 our understanding of certain entities has improved and these will probably cease to be included in the subset; nocturnal paroxysmal dystonia, increasingly considered as a particular form of focalised epilepsy, being a case in point.

Several of these pathological entities are dealt with in other chapters of this book: enuresis, snoring, apnoeas and sudden infant death syndrome. Nocturnal paroxysmal dystonia is studied in the chapter on abnormal sleep postures and movements; congenital central hypoventilation syndrome is studied in the more general chapter on central alveolar hypoventilation.

Consequently, three very different pathologies will be described here: sleep bruxism, sleep-related abnormal swallowing syndrome and benign neonatal sleep myoclonus to which will be added a description of sleep talking.

Hence a hypnogram can be used to illustrate the preferential occurrence of the main parasomnias (fig. 41.1).

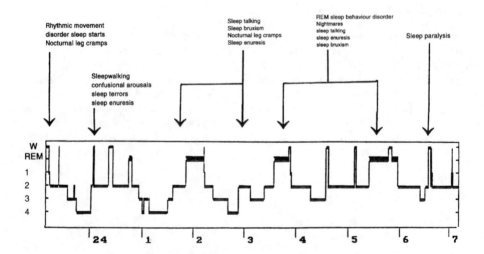

Figure 41.1. Sleep histogram: stages at which the main parasomnias occur.

Some of these parasomnias are particularly related to the age of the subject. Very broadly speaking it can be said that parasomnias involving disorders of arousal, notably sleepwalking and night terrors, tend to be childhood parasomnias, whereas parasomnias which are usually associated with REM sleep tend to affect adult and elderly subjects. This age factor is important for clinical diagnosis of parasomnia and for the investigations which follow.

Hence, a clinical diagnosis of parasomnia and its type is mainly oriented by three clinical factors:

1) the subject's age,
2) the stage during the night when the disorder occurs,
3) an accurate description of the manifestation.

In function of these criteria, a simple classification can be proposed (table 41.1) of the most common parasomnias. A diagnosis can fairly often be made on the basis of the interrogation; nevertheless in certain difficult cases or for certain parasomnias, laboratory tests are indispensable. These points will be clarified later.

For a given parasomnia, the *intensity of symptomatology and frequency of occurrence* are the determining elements for treatment. In fact, many parasomnias, notably those which occur in childhood, only take on a pathological character if they are particularly intense (constituting a danger for the subject himself or for those around him) or very frequent. The anxiety generated in the entourage is also important to take into consideration.

Table 41.1. Parasomnias

	Sleep-wake transition	NREM sleep stages 3-4	2nd part of night REM sleep	Any time of night
Child	Rhythmic movement disorder	Confusional arousals Sleepwalking Sleep terrors		Sleep enuresis Neonatal sleep myoclonus
Adult	Sleep starts		Sleep paralysis REM SLEEP behaviour disorder	Abnormal swallowing syndrome
Any age	Sleep talking		Nightmares	Sleep bruxism

THE DIFFERENT PARASOMNIAS

The parasomnias or groups of parasomnias will be presented in turn according to their clinical aspects and epidemiological data, contributing factors, laboratory tests, differential diagnosis, pathophysiological mechanisms and approaches to treatment.

AROUSAL DISORDERS

The first to be presented are the "arousal disorders", previously referred to as parasomnias in deep NREM sleep. They consist of three parasomnias which begin during a phase of deep NREM sleep and are considered to be due to an arousal disorder. In addition, confusional arousals, episodes of sleepwalking and night terrors will be seen to share common symptoms: confused mental state, automatic behaviour, absence of reactivity to external stimuli and amnesia of the episode.

Confusional arousals

Clinical features

Marked confusional episodes can occur at the point of arousal from deep NREM sleep or just afterwards. They may also be referred to as "sleep drunkenness" or excessive sleep inertia.

They affect children or more rarely adults of both sexes, who, on arousal from deep NREM sleep, remain confused, with a slowing of the thought and speech processes, and varying degrees of spatiotemporal disorientation. The child is restless, whimpers and sometimes tries to get out of bed. He pushes away anyone trying to comfort him. He appears to be awake but is not conscious and has no later recollection of the incident. Memory disorder is retrograde and anterograde; the subject has difficulty perceiving the outside world. Subjects are only partially awake, have difficulty responding to questions and commands, may display strange behaviour and perform inappropriate acts.

These episodes begin gradually, usually occurring during the first part of the night, but may be repeated during the same night, or occur on arousal, especially after an early afternoon nap. They are favoured by forced awakening.

They last from several minutes to an hour and are sometimes very long in children.

Prevalence

Before the age of five years, this disorder is extremely common, and is likely to occur in all children.

After this age, it is less frequent but remains benign, disappearing gradually.

In adults, the prevalence of this parasomnia is unknown. Confusional arousals are more disquieting than in children, requiring a search for contributing factors or an associated pathology. The evolution of the disorder will thus depend on the nature and treatment of the associated factors.

Both sexes are equally affected irrespective of age.

Clinical variants

Familial forms exist of idiopathic confusional arousal. There are no known genetic factors. This parasomnia is particularly common in patients with idiopathic hypersomnia in its polysymptomatic form [4]. It may be a symptom of symptomatic hypersomnias and is also found in patients with narcolepsy or sleep apnoea syndrome.

In adults, the use of tranquillisers (chlorpromazine), tricyclic antidepressants, a sudden discontinuation of hypnotics (particularly benzodiazepines), alcohol or an association of these factors will contribute to the occurrence of confusional arousals.

Finally, stressful situations can provoke confusional arousals. These vary greatly in severity, from an unpleasant awakening with several moments of confusion to prolonged forms, involving the display of unusual behaviour (e.g. the subject who puts his coffee cup away in the chest of drawers) which may also be dangerous, incidents being reported of subjects injuring themselves or others. These "dangerous" forms [8] are more common where confusional arousals are associated with another parasomnia in deep NREM sleep, i.e. sleepwalking or night terrors. These will be referred to under a later heading.

Sleepwalking

Clinical features

This affects children (70% of whom are boys) aged between 4 and 15 years (with maximum frequency between 7 and 12) who, in the three hours after the onset of sleep, get out of bed, walk around the bedroom or even the apartment or house and may also go outside. They walk with their eyes open, with a vacant expression, clumsily perform habitual, unusual or inappropriate acts (such as the little boy who urinates in his chest of drawers) which can be dangerous (such as the child who steps out of the bedroom window). The child may spontaneously return to bed or lie down anywhere and continue to sleep. The child can be gently led back to bed without him waking up. Hitting an obstacle may wake him up but it is usually very difficult to get the child to wake up; when there are attempts to do so, the child remains confused or may express violent behaviour: he struggles, tries to escape and may even try to attack the person holding him. If he does wake up, he has no recollection of dreaming. Nor does he usually have any recollection of the episode the following day. Sleepwalking attacks can occur in adults who remember the incident more often than is the case with children, partially recalling a mental activity inciting them to move [54]. It is hence an episode of sleepwalking which is unconscious, non stereotyped and which usually only occurs once a night (twice on rare occasions) but which may be repeated on subsequent nights.

The episode lasts from five to thirty minutes; its frequency is equally variable, ranging from several nights a week to less than once a month, sometimes triggered by contributing factors.

A diagnosis of parasomnia of this type will depend on a description of the disorder, the time it occurs, age and sex as well as on previous family history. Indeed, the incidence of sleepwalking increases from 45%, if one parent was a sleepwalker, to 60 or even 80% if both parents were sleepwalkers [35, 50].

Prevalence

There is a high prevalence of sleepwalking in the general population, estimated at 1 to 6%. This parasomnia occurs occasionally in 15 to 40% of children, depending on the statistics [16, 55]; and frequently (several episodes a month) in 2.5% of children. Sleepwalking usually ceases spontaneously around the age of 12 or during adolescence at the latest. It rarely persists into adulthood. Its prevalence in adults is approximately 2.5%, and only 0.4% in frequent forms.The aetiology of sleepwalking is not fully understood; it includes maturational and sometimes psychopathological genetic factors, the latter being frequent mainly in adults.

Cohort studies of twins have shown six times greater incidence of sleepwalking in homozygotic as compared with heterozygotic twins, and in the Finnish study [48] girls were slightly more affected than boys (6.9 *vs* 5.7% for occasional forms, 2.8 *vs* 2% for frequent forms). An association is found with HLA DQB1*05 in 58% of cases of sleepwalking [69].

Between 85 and 89% of adult sleepwalkers have a previous history of sleepwalking in childhood; only 0.6% of subjects fail to report this. Studies have shown the importance of genetic factors in sleepwalking. The environmental factors contributing to episodes of sleepwalking are partially understood and will be referred to later, as these factors are common to the three arousal disorder parasomnias. There is no psychopathology associated with classic sleepwalking in children. Subjects have a normal personality profile, occasionally showing a slightly introverted character, with higher than average sensitivity and less emotional control. In adults, on the contrary, marked hysterical traits are often revealed, which are not expressed in day to day behaviour [19], aggressive impulses are poorly mastered, with low tolerance to frustration and a lack of imaginative creativity.

Clinical variants

These are numerous. Depending on the frequency of the episodes, the forms may be mild (less than one episode a month), moderately severe (more than one episode a month and less than one a week) with no discomfort to the subject, or severe, with episodes occurring several times a week, or even nightly or during which the subject injures himself [50]. These severe forms are included under "risky sleepwalking" described by de Villard *et al.* [15, 21].

This risky form of sleepwalking involves episodes which are frequent (2 to 3 per week) long (over ten minutes) and accompanied by dangerous activities. This form carries a far greater risk of accidents or of falling from a window as in classic Elpenor syndrome. The episodes often persist into adulthood and are encouraged by stress and anxiety.

The same authors describe "night terror sleepwalking" associating night terrors with sleepwalking; the episode starts with a night terror which ceases when the subject gets up, or may continue throughout the sleepwalking episode. These attacks are violent and may recur two to five times a night, which is unusual in the classic form. They are as frequent in girls as in boys and often occur before the age of 6 or after the age of 10 in children with obsessive, highly anxious traits. This form of sleepwalking probably corresponds to a night terror associated with the reflex to escape [15, 35]. These forms tend to be prolonged and persist beyond puberty into adulthood. Such severe forms of sleepwalking may involve the subject injuring himself or even falling out of a window, the risks of which must not be underestimated [65] at any age. In adults, there are reports of violent behaviour inflicted on others, such as attempts to strangle and acts of homicide, creating the medico-legal problems focused on in several articles in recent years [9, 11, 13, 38, 81, 90, 115].

This aggressive, violent behaviour does not only occur during sleepwalking episodes or confusional arousals and will be reviewed later in the context of REM sleep behaviour disorders. As to the issue of the legal responsibility of the author of the crime, verdicts vary according to countries, judges and individual cases.

Sleepwalking is often associated with other parasomnias, particularly sleep talking, which as we will see, is very common in children and is easily triggered by calling the subject by name or by talking to him. Its association with night terrors is well known; these two parasomnias no doubt share common hereditary factors [35].

Eating disorders may accompany episodes of sleepwalking or confusional arousal in adults, particularly in women [102].

Finally, Schenck et al. [98] have reported 33 observations of parasomnia overlap disorder, i.e. an association between arousal disorder parasomnia and REM sleep behaviour disorder. This parasomnia association is idiopathic in 2/3 of the cases and the psychopathological mechanisms involved will be discussed with reference to REM sleep-related parasomnias.

Sleepwalking is also commonly associated with migraines: 66% of children who sleepwalk also suffer from migraines: adult migraine sufferers are more likely to have a previous history of sleepwalking than those suffering from other forms of headache [35, 37].

Similarly, sleepwalking is classically associated with Gilles de la Tourette's syndrome.

Sleep apnoea syndrome is known to contribute to confusional arousals and episodes of sleepwalking; cases of night terrors or sleepwalking have been reported after continuous positive airway pressure [79, 90].

Night terrors

Clinical features

The typical subject is a child of under 12 years, usually a boy aged 3 to 6, who, one to three hours after sleep onset i.e. in deep NREM sleep, will suddenly scream or cry and is found sitting up in bed, eyes wide open and staring, with dilated pupils, sometimes speaking incoherently or gesticulating wildly. The child is erythrosic, with high tachycardia and sudation, and occasional breathing difficulties. He has all the symptoms of intense fear. It is extremely difficult to wake him up and any attempts to calm or console the child will often aggravate the symptoms, even leading to aggressive reactions and an escape reflex. This escape reflex is naturally observed in the older child, who will get up, jump out of bed, run out of the room and may injure himself (falling down the stairs, or hitting an obstruction…).

In the case of adults, men are twice as likely to be affected as women. Only one night terror episode usually occurs a night, although some adults have been known to experience several episodes in one night [18].

Subjects usually have no recollection of the episode, or if they do wake up during the parasomnia, will only vaguely recall cries and simple mental images. Crisp [18] reported adults describing vivid memories of terror. The episode lasts from one to twenty minutes after which the subject falls asleep. The intensity of the attack appears to be proportional to the duration of the preceding stage 4 [35]. This parasomnia is thus particular in terms of the intensity of its accompanying neurovegetative features; micturition is often observed in addition to vocalisation.

Prevalence

The figures on the prevalence of night terrors in children vary considerably in the literature, from 1 to 17.3% [59]. In the adult, the prevalence of this type of parasomnia is 1 to 2.2% [85]. Night terrors begin in adults between the ages of 20 and 30 and are often associated with episodes of sleepwalking [54].

A previous family history of arousal disorder parasomnia is found in 96% of cases, indicating the importance of genetic factors, as in sleepwalking. Several members of the same family may present night terrors but family incidence does not decrease in function of how far removed the relationship is, unlike the cases reported in sleepwalking. Heredity appears to be multifactorial. The expression of the disorder is influenced by the surroundings: stress and daytime anxiety will contribute to the onset of night terror episodes.

When the onset of this parasomnia is late, after the age of 15, a psychological trigger factor is found in half the cases.

No particular psychological profile is found in children, whereas in the adult, the MMPI and other batteries of tests show psychological disorders, usually with obsessive traits, occasional depressive traits, repressed aggression which is often directed inward [18, 19] and a particularly high anxiety score [18, 72]. These affective disorders may be associated with drug addiction.

Clinical variants

As in the case of sleepwalking, these depend on the frequency of attacks. Severe forms involve night terror episodes occurring almost nightly, which may be harmful to the subject himself or to others.

The forms of terror-sleepwalking have already been described, including the reflex to escape and violent forms as potential sources of accident.

Night terrors may be associated with the same parasomnias and pathologies as sleepwalking. Enuresis is particularly common, during an episode of night terror.

In its classic form, night terrors tend to disappear spontaneously at 12 to 15 years of age; they may reappear between the ages of 20 and 30. More rarely, they persist into adulthood or appear for the first time at 20 to 30 years of age.

Contributing factors

These three deep NREM sleep-related parasomnias may occur in the absence of any identifiable trigger factor; they are nonetheless encouraged by all the circumstances provoking a deepening and/or increase of deep NREM sleep as well as by any circumstances which fraction or disturb sleep, particularly by increasing intero- and/or exteroceptive stimulation.

Thus the onset of these parasomnias will be encouraged not only by disruptions to the sleep-wake rhythm and sleep deprivation, but also by intense physical exercise late in the evening, or by fever. Certain medication such as valproic acid, thioridazine, tricyclics and antiarhythmics, prescribed to the elderly, [54], as well as the abuse of alcohol, will have the same effect.

A distended bladder from excessive drinking in the evening, external stimuli (noise, calling the sleeper's name, a conversation with the subject, moving or trying to lift the child...) and any pathology which fragments sleep, is likely to trigger episodes of the parasomnia. Finally it is important to underline the negative effect of stress and anxiety in the daytime or during a previous period in life. A group of adults who had experienced traumatic events in the past (Vietnam war veterans) was thus identified as presenting this type of parasomnia [54].

Some observations have been reported of night terrors and sleepwalking linked to episodes of menstruation [104].

Laboratory tests

1) When is it important to record an episode of parasomnia?

This is not necessary when the diagnosis is obvious i.e. in typical episodic attacks in young children.

On the other hand, it is useful to record an episode of parasomnia when attacks are frequent, atypical, or severe and when the parasomnia is likely to be associated with another pathology, or when onset occurs in adulthood.

2) Which method should be used?

A complete polysomnographic recording in hospital, including a video recording has the advantage of covering all the desired parameters, giving accurate information on the subject's behaviour and screening any associated pathologies. The difficulty is in capturing an episode of parasomnia in only one or two nights. In an attempt to raise the chances of recording an episode in hospital, some advocate applying tactics to induce the episodes (sleep deprivation the night before, excessive drinking in the evening, calling or moving the subject in stage 4 of NREM sleep).

Other authors prefer a Holter's recording at home, allowing the subject to remain in familiar surroundings, making repeated recordings easier.

A simple video recording can also be carried out at home. In difficult cases, polysomnography can be used in association with a Holter's recording to look for other pathologies if the parasomnia episode was not previously recorded in hospital.

3) What information do these recordings supply?

Recording the parasomnia episode shows that it starts in NREM sleep stage 4, often at the end of the 1st or 2nd phase of deep NREM sleep, i.e. just before the first phase of REM sleep or at the point when one would expect to find the first phase of REM sleep.

The parasomnia episode is accompanied by bursts of hypersynchronous slow waves, or sometimes with a series of fairly monomorphous slow waves. In adults, lighter sleep is accompanied by theta rhythms or even the reappearance of a sequence of alpha activity, followed by a muscular and artefact overload.

The duration of the bursts of hypersynchronous slow waves at the start of the episode correlates with the severity of the parasomnia. Sometimes a few peaks are recorded, especially in certain forms of night terrors, with no sign of epileptic manifestations.

During these night terrors, an abrupt acceleration is also recorded in cardiac and respiratory rhythms as well as a drop in skin resistance. These parasomnias are thus clearly linked to NREM sleep stage 4 and do not constitute the acting out of a dream.

Apart from the parasomnia episodes, sleep remains correctly structured; there are inconsistent reports of alterations in deep NREM sleep levels; likewise a high number of partial or incomplete REM sleep episodes have been described. In any case, sleep microstructure is modified. Bursts of hypersynchronous slow waves interrupting NREM sleep, with no clinical manifestation, are more numerous than in the normal subject. These are purely electrical arousal reactions to slow waves [7]. Deep NREM sleep is thus more fragmented than in controls and a study of the cyclic alternating pattern confirms this sleep instability [134].

4) The other laboratory tests to be carried out mainly consist of a personality test [85] in adults' parasomnias, or in the case of frequent episodes in children, and a Holter cardiac recording to search for any cardiac arrhythmias during sleep, which may be associated with these parasomnias, particularly if initial onset occurs in adulthood.

Differential diagnosis

An accurate clinical picture and polysomnographic recording will eliminate most diagnoses.

Confusional arousals and night terrors must not be confused with *hypnagogic hallucinations* which occur at the onset of sleep, may be frightening but are usually complained of by the subject. Cases of sleepwalking associated with hallucinations have however been described [53].

Night terrors are often mistaken for *nightmares* by parents. The section dealing with nightmares will examine the main differences between the two parasomnias (time of onset; arousal after the nightmare accompanied by dream recall).

One of the diagnoses sometimes given is that of *partial complex sleep epilepsy,* particularly if the subject grinds his teeth or if activity is stereotyped. These critical episodes have been variously termed in the literature as paroxysmal arousal, night wandering episodes or nocturnal paroxysmal dystonia [37, 82, 88]. Critical episodes usually recur several times a night, at any time, start very abruptly with typical stereotyped wandering for each patient, and unintelligible muttering. During the critical episode, the patient does not react to any appeal for attention and contact is only resumed at the end of the episode. Finally, epileptics rarely have a previous history of sleepwalking, although cases do exist [112]. Diagnosis is easy if critical daytime manifestations exist. Day and night electroencephalographic recordings will reveal spike discharges of notably temporal localisation in at least 2/3 of cases where there is epilepsy. The slow and sharp elements observed in partial arousals must not be confused with paroxysmal discharges [14]. However authentic spikes have been found on the EEG of subjects with parasomnias [121], and epileptic seizures can be facilitated by the neurovegetative phenomena of night terrors or preceded by a sleepwalking episode. Despite the existence of these borderline or associated forms [120], a distinction must be made between these arousal disorder parasomnias, which are not epileptic manifestations, and partial complex seizures occurring at night.

In elderly subjects, these parasomnias must be distinguished from a *confusional state during the night*, whose frequency increases with age and medication but which often occurs after a free interval between arousal and the start of confused wanderings. These nocturnal confusions in elderly subjects correspond to several aetiologies [12].

As regards *ŘEM sleep behaviour disorder*, this often occurs during the second part of the night, in elderly men, who act out their dreams which are often terrifying. A diagnosis of certain forms of sleepwalking in adults, if the subject recalls mental activity, is also possible. Whatever the case, a recording of the episode of parasomnia will clarify the diagnosis.

On very rare occasions, the *hysterical manifestations* or *dissociative disorders* [108] of young adult subjects will imitate episodes of sleepwalking. In such cases, the attack of hysteria is longer and adapted behaviour persists. One case of false parasomnia of psychological origin was reported in a child [80]. Recording the episode will settle any doubts.

Pathophysiology

These parasomnias appear to exist on a clinical continuum, extending from the least severe, as in typical sleepwalking, to a major expression as in the reflex to escape sleepwalking-terror [30]. They correspond to similar pathophysiological mechanisms, indicated by their identical clinical characteristics [76]. These parasomnias involve a partial, dissociated arousal: partial arousal from deep NREM sleep, notably at the end of the stage, before the onset of REM sleep, at a moment when sleep is most unstable; partial arousal is witnessed by the electroencephalographic picture and there is amnesia of the episode [77].

This partial arousal reaction may have a purely electroencephalographic expression, but may also take on the aspect of a dissociated arousal, accompanied by motor activation which is elementary or complex, as in sleepwalking, neurovegetative, as in nocturnal terrors, or again, motor activation associated with neurovegetative activation as in terror sleepwalking. It is thus a question of psychobehavioural dissociation; this cortical "sub-arousal" state is accompanied by motor and behavioural arousal bringing the different neuronal and neuromediator networks into play.

This explains why stimulation or pathologies which fraction sleep favour their onset by provoking incomplete arousal. These parasomnias occur, in fact, at a particular point in the arousal process (between deep NREM sleep and arousal), which is why they are now classified as arousal disorders.

Treatment

Treatment will depend on the age of the subject, the frequency and severity of attacks as well as their impact on the family and social context.

In young children, if the episodes of parasomnia remain infrequent, no medical treatment is prescribed. It is necessary, however, to explain the nature of the disorder to the parents, indicating its transient nature, reassuring them while pointing out any potential dangers. They must thus be helped in devising measures of protection (removal of dangerous objects, sometimes placing the mattress on the floor, covering the windows with curtains, placing locks on doors and windows...). Equally, intense physical exercise should be avoided as well as the over-consumption of liquids in the evening. It is important to explain the futility of constraining or waking the child during an episode of parasomnia [24]. Some authors [32, 118] advise treatment in the form of programmed arousals at the times when episodes of parasomnia habitually occur.

If parasomnia episodes are very frequent, accompanied by dangerous acts which disrupt family life or impair the normal socialisation of the child (refusal to sleep at a friend's house or go away to summer camp, on the part of the child or his parents), it may be necessary to prescribe medical treatment in the form of a benzodiazepine [107]. Among the benzodiazepines, diazepam (2 to 10 mg in the evening), alprazolam as well as clonazepam (0.25 to 1 mg/d) or clobazam (5 to 10 mg in the evening) are prescribed discontinuously, for periods when episodes are to be avoided. These drugs act by suppressing stage 4 of NREM sleep [35, 92].

In the case of adults, apart from benzodiazepines, certain authors have obtained good results using treatment based on paroxetine, a serotonin reuptake inhibitor [70]. Depending on the results of the personality test, psychological support is often arranged, either in the form of psychotherapy, behaviour therapy or hypnosis therapy - very popular in the USA - with good short and medium-term results [49, 54]. Hypnosis has also been used in children [113] or in severe cases [56]. In the elderly, it is important to look for a state of depression or any medication liable to favour the onset of parasomnia episodes. If treatment needs to be prescribed, benzodiazepines, which may favour mental confusion, must be prescribed with caution at moderate doses, tybamate can be recommended for subjects who fail to respond to benzodiazepines [54].

SLEEP-WAKE TRANSITION DISORDERS

Rhythmic movement disorder

Clinical features

These consist of repetitive, stereotyped movements affecting the muscles of the trunk and neck.

The term *jactatio capitis nocturna* introduced by Zappert in 1905 [117] illustrates the fact that the head is often involved in this motor activity: either as a back and forth movement repeatedly hitting the forehead or occipital region at the back of the head (head banging) or side to side (head rolling). Head banging is the most severe form as the head can be thrust violently, not only against the pillow, but also against the bars of the bed or the wall. It may also consist of rocking the trunk back and forth (body rocking) with the child in a sitting position, or of the whole body if the child is on its knees or standing.

The different types of movement can be observed in the same child but at some point the same movement will be repeated in rhythmic episodes of several minutes, not usually exceeding 15 minutes, at 0.5 to 2 second intervals. Prolonged forms have however been described [40, 71]. The movements are sometimes accompanied by sounds.

These rhythmic movements usually begin before the onset of sleep and persist into light NREM sleep or after nocturnal arousals. Some may also occur during the daytime just before sleep onset as well as in calm, inactive wakeful periods (car rides or listening to music...). During these rhythmic movement episodes, the child appears calm, enjoys the activity, does not wish to be disturbed but remains sensitive to external stimuli. Thus his mother or a third party may get him to stop the activity; if woken up he rapidly regains consciousness with no prolonged confusion.

Prevalence

This parasomnia starts early in life, always before 18 months and usually before the age of one year, i.e. between six and ten months. Body rocking is equally common in both sexes [59] while head banging tends to occur more often in boys (2 to 3 times more common). Found occasionally in 20 to 66% of children aged nine months, depending on the study, it is only observed in 50% of them at 18 months and 8% at 4 years [117].

The figures vary depending on the definition of "occasional", and according to the type of rhythmic movement. Hence body rocking occurs at an earlier age (6 months on average) than head banging or head rolling (9 months on average). The different forms of rhythmic movement may be isolated, successive or associated in the course of evolution of a given child.

Clinical variants

Familial forms have been described whose frequency can be as high as 20% of cases in certain series; observations of head banging have also been reported in twins.

This parasomnia varies in intensity, ranging from several crude forms to the most severe in terms of frequency, duration of episodes, and complications. Indeed repeated shocks to the head, particularly from head banging, may lead to skull fractures, external or subdural haematoma or even retinal lesions in the form of petechia [129]. A case has been described of repeated tongue biting [123]. Even in such severe forms, the child never complains or cries [109, 127].

These sleep onset rhythmic movements occur in the majority of cases in young children with perfectly normal psychomotor development. Contributing factors are occasionally found, either organic (cranial traumatism [23], otitis [10]), or psychological: a change of routine or an anxiogenic situation. However, cases associated with mental deficiency, autism or other psychiatric disorders, or with blindness have been known for some time. These children usually have serious psychological disorders with difficulties in developing relationships. In forms such as these, repetitive movements often predominate during wakefulness or are virtually continuous; they lead more often to severe lesions, than in the case of healthy children, and furthermore persist into adolescence or even adulthood rather than diminishing and finally disappearing at around 4 to 5 years of age. Whatever the case, this parasomnia is often poorly tolerated by the family, being both a source of annoyance because of the noise it can create and a source of anxiety for the parents, who often mistakenly interpret it as "abnormal" behaviour or a sign of the "lack of affection".

Laboratory tests

The clinical history, a description of the disorder, the age of onset and time of occurrence usually suffice to establish a diagnosis precluding the need for laboratory tests, notably that of polysomnographic recording coupled with video [26]. The recordings which have been made confirm the occurrence of these stereotyped movements before and at the onset of sleep and during NREM sleep stage 2. These episodes do not provoke arousal. Very rare cases have been reported of rhythmic movements in deep NREM sleep and in REM sleep [91]. Heart and respiration rates are stable during episodes of parasomnia. Sleep remains well organised during the night. No paroxysmal anomalies or rhythmic peak discharges are recorded, either during or outside the episodes of this parasomnia.

Only very severe forms and those which persist beyond early childhood require searching for an aetiology and arranging for psychological testing.

No real differential diagnosis exists for this parasomnia. It is not normally associated with other parasomnias such as bruxism or other rhythmic movements like thumb sucking or periodic leg movements.

Pathophysiology

Several pathophysiological hypotheses have been advanced to explain rhythmic movements in the healthy child. No sociological factor has been found; children may come from wealthy or disadvantaged socio-economic backgrounds. Some authors have suggested that there is an attempt to relive the sensations experienced *in utero*, or when rocked in the parent's arms, particularly to encourage sleep.

These ideas have not been demonstrated, but are not inconsistent with the hypothesis that this parasomnia satisfies an innate need for rhythmic movements generated by the central nervous system [92], is expressed at a particular age, when the child shifts from a crawling to an upright position.

These rhythmic movements are thought to call upon the vestibular system, creating sensations which are a source of intense pleasure, likely to calm the anxiety of going to sleep, the rocking thus encouraging sleep onset. Hence the interaction between these repetitive movements, the vestibular system, and the development of motor and sleep control form areas of research. More exclusively psychological hypotheses have also been advanced: the search for erotic pleasure, search for fusion with the mother, compensation for a lack of stimulation or movement... with little supporting evidence. When these manifestations persist into adolescence, with no sign of neuropsychiatric disorder, they are considered as acquired manifestations, with no apparent objective.

Treatment

1) In infants and young children with normal development, no treatment is usually required.

At the same time it is important to appreciate the parents' anxiety, and to explain the harmless nature of this behaviour, the fact that it usually improves rapidly, and the absence of any neurological or psychoaffective pathology.

Rhythmic movements, even those which are strong and prolonged are only very rarely accompanied by serious lesions, although they may lead to head bumps and occasional bleeding. They can also cause noise and may damage the bed.

Whatever the case, a few simple measures can be introduced e.g. reducing the time spent in bed before the onset of sleep and after awakening, padding the bed and wall, sometimes placing the mattress directly on the floor if the child is only subject to body rocking, fixing the bed to the floor. Some authors also advise reinforcing rhythmic activity during the daytime and rhythmic sounds at bedtime to try to reduce rhythmic movements (the metronome is recommended but may make as much noise as the movements themselves). In any event, one should try to convince the parents to be patient, that there is nothing serious and that the disorder will disappear spontaneously [30].

2) Violent rhythmic movements in older children may relate to high anxiety and problems of affect. An attempt should be made to find the source of the psychological factors maintaining this parasomnia and to instigate supportive attitudes (reassurance, talking about any conflict or tension...) likely to help the child; psychological support for the child and his family may be necessary.

Medical treatment has only been tried by some authors in severe cases in older children or adults: sedative benzodiazepines (triazolam, diazepam or oxazepam) or tricyclic antidepressants (imipramine) at moderate doses for short periods lasting no more than two weeks.

3) For the handicapped child, when rhythmic movements are marked or lead to severe lesions, behavioural therapy has been used with a certain degree of success.

Sleep starts or predormita myoclonus

Clinical features

These starts consist of sudden short contractions of the lower limbs, sometimes extending to the arms and head, occurring at the onset of sleep.

There may be only one start, affecting part of the body or the whole body in an asymmetrical fashion. Sometimes several starts occur in succession. The phenomena are occasionally observed by the partner and are seldom recalled by the subject. It is only in the case of massive or repeated starts waking the subject, that he becomes aware of the phenomenon; these may also wake the partner. When the subject recalls a start, he sometimes describes a visual, auditory or sensory hallucination occurring at the same time. These starts occur at the moment of transition between wakefulness and sleep. They may be spontaneous or encouraged by stimulation: they are found at all ages, but only adults worry or complain about them. Men and women are equally affected. It is a very common phenomenon, present in 60 to 70% of people. No genetic study exists and no familial form has been described.

The phenomena normally go undetected, the starts are described by the partner, when the interview is systematically carried out. They are rarely strong enough to wake the subject or recur frequently enough to stop the subject getting to sleep – which is itself a cause of anxiety contributing to sleep onset insomnia. In a few cases, myoclonia are strong enough to result in kicks to the partner. They occur in patients devoid of any neurological or psychiatric pathology. They may nonetheless be affected by the over consumption of stimulants, particularly coffee, by intense physical exercise and situations of stress or anxiety.

Laboratory tests

Diagnosis is usually made purely on the basis of the interview and requires no laboratory tests. Polysomnography is hence only practised in extreme cases and above all where diagnosis is unclear. This examination includes a surface recording of several muscles, particularly of the lower limbs, showing the occurrence of bursts of high amplitude muscle activity during stage 1 of sleep, lasting 75 to 250 milliseconds, which may be isolated or follow in rapid succession, associated with a vertex peak on the EEG. After the start, a micro-arousal may occur accompanied by a brief moment of tachycardia, or even complete arousal. No starts are recorded in the other stages of sleep; the EEG is normal and sleep once it sets in, is unaltered.

Differential diagnosis

This examination may serve to eliminate a *hyperplexia syndrome* in the young child during which starts are triggered by stimulation in all states of sleep and wakefulness [119].

Benign neonatal myoclonia consist of myoclonia of the extremities and face, but which occur in sequences during NREM sleep.

Finally, regardless of age, *epileptic myoclonia* are recorded in wakefulness and NREM sleep and are not favoured by the transition from wakefulness to sleep; these are usually accompanied by paroxysmal discharges on the EEG. *Restless leg syndrome* is clinically different: an unpleasant sensation in the legs requiring the subject to move his legs or even to get up and walk around. The *periodic leg movements* which they often accompany recur during long passages of NREM sleep, particularly in stage 2.

Pathophysiology

These starts are nothing more than intensified physiological phenomena occurring at sleep onset. Sleep structures actively inhibit the centres responsible for wakefulness, creating an unstable state before sleep is fully installed. The instability of the control mechanisms of sleep and motoricity explain the symptoms observed during this period.

Treatment

Most cases require no treatment i.e. in hidden or moderate forms. It is only when sleep starts directly or indirectly lead to sleep onset insomnia, that measures relating to general hygiene should be considered (avoiding the over consumption of stimulants or alcohol, intense physical activity in the evening, taking measures against excessive stress...), and in the most severe cases, benzodiazepine treatment with a weak dose of clonazepam. The dose can be very gradually increased at bedtime and stopped as soon as the effective dose is found.

Nocturnal leg cramps

Clinical features

These are painful sensations of stiffness or tension usually affecting the calf or sometimes the foot. Cramps may last from several seconds to several minutes. They may occur with the onset of sleep but often appear during sleep, causing arousal. Cramps vary considerably in severity, ranging from short cramps occurring occasionally at night, to cramps which last longer, recurring two or three times nightly and several nights a week. These nocturnal cramps bear no relation to cramps which occur during the daytime. They have a harmful effect on sleep quality when they lead to micro-arousals or repeated arousals.

Prevalence

They often begin in adulthood but can also appear in elderly subjects. They are more frequent in women and are favoured by pregnancy; they are also more frequent in the aged.

Few epidemiological studies exist but nocturnal cramps are thought to occur in 16% of healthy adults, sometimes brought on by physical exercise. The incidence is raised in elderly subjects [50]. Sporadic cases are the most common; a few familial forms may exist [51]. There is no known mode of transition.

Evolution is somewhat capricious, tending to be chronic with moments of exacerbation and periods of remission, for several years.

Laboratory tests

These are unnecessary in the majority of cases.

Polysomnography may record one or several cramps expressed as ample bursts of muscular activity, which are more or less prolonged, on the gastrocnemius, followed by varying periods of arousal. It may sometimes show periodic leg movements associated with cramps. No sleep alteration occurs outside these arousals.

The search for a metabolic or endocrine disorder often proves deceptive, and blood, Ca, D vitamin tests etc. are usually normal.

It may sometimes be useful to test venous circulation.

Pathophysiology

This has not been defined but the circumstances contributing to its occurrence are known: pregnancy, the use of oral contraceptives, hydroelectrolytic or endocrine disorders, as well as violent muscular exercise or conversely, reduced motoricity due to neuromuscular, extrapyramidal or other disorders.

Certain authors support the notion that cramps are linked to calcium metabolism anomalies.

Treatment

This consists of localised measures: massage, movements, heating tetanized muscles.

When cramps are very severe, some advise benzodiazepine (clonazepam); others quinine benzoate (hexaquine) or quinine sulfate, which may provide relief. [5, 17].

Finally, medication based on trace elements, calcium, and magnesium [33] are often tried, with very variable results.

Obviously any contributing cause found must be treated in its own right (diabetic balance, correcting an endocrine disorder).

PARASOMNIAS USUALLY ASSOCIATED WITH REM SLEEP

Nightmares

Clinical features

The nightmare is a dream which is usually long, complex and frightening, accompanied by increasing anxiety which finally leads to the subject waking up in spite of being in REM sleep. The subject quickly becomes lucid after waking up, with clear and accurate recall of his anguished dream. He is frightened by the experience and may take several minutes to recover. The dreams vary in content, but are always complex, detailed, oppressive, with vivid scenes rich in auditory sensations during which violent and dangerous events occur to the subject, giving him the feeling he is in vital danger. Sometimes the dreamer is the central actor or victim of criminal acts.

More rarely, the nightmare is experienced simply as a fear of imminent death with no representation, or where the dreamer experiences changes of identity, with the body breaking up or changing. Whatever the contents of the nightmare, it is always a disquieting source of terror and oppression.

Nightmares are not usually accompanied by any substantial neurovegetative disorder, or sleep talking; the subject may cry out or utter sounds.

Nightmares often end with the subject waking up and being unable to get back to sleep due to anxiety; they may lead to insomnia with its well known daytime consequences.

Occurring as they do in REM sleep, nightmares are situated in the second part of the night.

Prevalence

Nightmares occur at all ages.

It is difficult to define their frequency in very young children who are not yet able to speak; 5% of children in this age range are thought to have nightmares.

Between the ages of 3 and 6, occasional nightmares are very common, found in 10 to 50% of children depending on estimations [42].

They persist in children aged 7 to 10. Some authors have found a predominance in girls for this age range (47 to 57%) compared with boys (33 to 37%), while others have found no differences between sexes. Secondary school pupils commonly report occasional nightmares and 5% of them have frequent nightmares, occurring at least once a week [131]. Nightmares diminish in adulthood; only 50% of students who previously suffered from nightmares continue to have them in adulthood. Nevertheless some adolescents go on to have nightmares throughout life. In adults, 50% present occasional nightmares, whereas only 1 to 5% have them frequently. It is in these cases of frequent nightmares that 50% of sufferers report nightmares starting before the age of 10 and family incidence [35]. Women report nightmares more often than men. Cases in which nightmares occur more than once a month or week, involve 2/3 women to 1/3 men. The question arises as to whether this relates to a biological difference or a psychological one, men possibly being more reticent about reporting nightmares.

Clinical variants

There have been numerous descriptions of clinical forms of nightmare; we will focus particularly on posttraumatic nightmares [1] i.e. those which are subsequent to highly traumatic events, threatening the person's life and often causing the death of others, some of whom were close.

Studies have been carried out on posttraumatic nightmares, using different methodologies, among survivors of the holocaust, ex soldiers, Vietnam war veterans and children who were subject to kidnapping. Here, the dream content is directly related to the causal events. Nightmares repeat the traumatic events in a stereotyped fashion, provoking a feeling of distress. They may be isolated or part of a group of disorders defining the posttraumatic stress condition [125]. They persist for a minimum of a month after the accident, often for a lot longer, only fading away very gradually. They may disappear for a time and then return; in fact as they occur spontaneously after the event, they are brought on by any random combination of events evoking the trauma [78]. The more fragile or poorly integrated the subject before the traumatic event or the more difficulty he had in expressing emotions or relating to others, the longer the nightmares tend to persist. In some cases, nightmares disappear quickly, the subjects then dreaming very little as if they had cancelled out their dreams [35, 66]. The contents of the dream gradually evolve; recall of the trauma becoming interwoven with mental associations of other aspects of life and normal dream processes. This interweaving process and the substitution of the contents of the nightmare by other mental and dream activities is also found in children [114]. It is gradually followed by a process of symbolisation, whereby the nightmare slowly sheds its clear links with the traumatic event.

These processes can help to free the person from nightmares; if nightmares fail to carry out this reparative work they may become chronic. In a state of posttraumatic stress, nightmares can be associated with a high number of periodic leg movements, which also occur in REM sleep [93].

Basic nightmares are unlike posttraumatic nightmares in the sense that they do not integrate elements of day to day reality and are not triggered by a traumatic event. They are nevertheless increased by stress. When they are very frequent some authors refer to "nightmare syndrome". Personality tests on these subjects show no neurotic profile, but do indicate a high degree of fragility and vulnerability similar to that of borderline personalities; these it will be recalled, are cases indicating more psychological difficulties, and nightmares experienced by other members of the family.

Nightmares can be the symptom of a psychiatric pathology. It is important to be particularly wary in the case of adolescents who regularly experience nightmares, accompanied by insomnia. These features may herald a psychotic event occurring several weeks later.

Lastly, alcohol and several medications [84] such as anti-Parkinson treatment [116] or antihypertensors (beta blockers) can induce nightmares or increase their frequency.

Depending on their intensity, and the tolerance shown by the subject, treatment should either be continued or replaced by a substitute medication.

Laboratory tests

A personality test is useful in the case of an adolescent presenting an unusual case of nightmares, which he is unable to cope with. It is also of use in some cases of posttraumatic nightmares as an aid to treatment.

Polysomnography is generally of no help in diagnosis.

When it is carried out, it often shows manifestations of sympathetic activation during a nightmare [84], before arousal during a phase of REM sleep which is often long and rich in eye movements [125]. There is a parallel increase in the variability of heart and respiration rhythms with no accompanying major neurovegetative phenomenon. Motor control disorders in REM sleep are sometimes noted (periodic movements, persistence of muscular activity).

Differential diagnosis

Nightmares need to be differentiated from night terrors. The younger the child, the more often this diagnosis is made. Parents often consult thinking their child is suffering from nightmares, which in certain cases are night terrors [30].

A probing interview should attempt to look for semiological nuances differentiating between the two phenomena.

The parents are usually awoken after the nightmare when the child begins to cry and is terrified; this fear continues even when the child is awake. Conversely, during the night terror, the child is agitated, screams, talks and shouts; afterwards he becomes calm.

Nightmares are not accompanied by significant neurovegetative signs, unlike night terrors during which the child breathes very fast, his heart beats quickly and he is covered in sweat.

When the parents get to the child's bedroom after a nightmare, he is conscious and can, if he is old enough, recount the terrifying dream either straight away or the following morning. In the case of night terrors, the child is unaware of the parents' presence, may push them away or start to shout and become more agitated if they try to hold or constrain him. When the child is fully awake, he has no recollection of the episode or of the dream.

Finally, after a nightmare, the upset child may take a long time to get back to sleep, whereas after a night terror he will fall asleep quickly.

The distinction is not easy in some cases, but in general, the interview suffices to establish the diagnosis and it is only exceptional cases which require polysomnography with video or a Holter recording at home.

In adults or the elderly, nightmares must be distinguished from REM sleep-related behaviour disorders in which motor activity may occur which can be violent and is not followed by quick, lucid awakening as in nightmares; the subject appears distant and less frightened.

Some authors insist on the possibility of a continuum between these two parasomnias. A REM sleep-related behaviour disorder is often preceded by a period in which dreams change in content and intensity, become abundant and frightening and the REM sleep-related behaviour disorder is, as will be seen later, simply the acting out of a dream which is often violent.

Pathophysiology

Certain neurotransmitters are probably involved in generating nightmares. Thus any increase in the metabolism of dopamine or an L-Dopa treatment encourages nightmares, which are diminished by dopamine receptor antagonists [35]. Any central cholinergic hyperactivity would contribute to the occurrence of nightmares [42]. Views differ on the noradrenergic mechanisms involved, some authors asserting that reduced central noradrenergic activity is a source of nightmares [42] while others [78] postulate, according to Crocq's hypothesis, that posttraumatic nightmares are due to noradrenergic hyperactivation, influenced by the autonomous nervous system linked by the locus coeruleus which sends projections up to the limbic system toward the cortex, allowing traumatic memories to intrude during both wakefulness and sleep.

Nightmares have been subject to numerous psychological [43] and psychoanalytical interpretations; their common pathogenic mechanism is thought to be a psychic trauma, easily identified in the case of posttraumatic nightmares; more difficult to discern in the case of basic nightmares as these probably go back to childhood. The same infantile conflicts also come into play in forming the personality, possibly explaining the psychological characteristics of subjects who present frequent nightmares. In the child, a nightmare also reflects anxiety producing the emotional conflicts which are a normal part of development: fear of separation, fear of losing the parents' love, the struggle against aggressive and sexual impulses, fear of death, stress of puberty... depending on the physical and psychoaffective stage of the child's development. The contents of the nightmare also evolve with age (simple dream content reproducing a frightening event from the previous day; followed by more symbolic dreams featuring wild animals or historical characters; and finally more complex dreams, substitution phenomena etc). The child begins to mark the difference between dreams and reality from the age of 3 years but the emotional content of the nightmare remains very vivid for him [30].

Whatever the age and whatever the type of nightmare, pleasure is no longer at the centre of the psychological process. The dream situation is unbearable for the subject; waking up saves him from threatening dangers, putting an end to the nightmare and to sleep [36].

Treatment

If nightmares fall within a pathological context (psychiatric or otherwise) the causal illness must be treated.

Nightmares favoured by certain medications may require adjusting the dose or switching to a substitute, depending on their intensity and the demands of the patient.

In the case of occasional nightmares, treatment is not normally prescribed.

In the child, certain attitudes can be advised such as avoiding violent scenes (TV or books) and conflict before bedtime; reassuring him and giving him affection, helping him to get back to sleep; explaining that the dream is separate from reality. In the daytime, the child should be allowed to express his feelings and communication should be established.

Behaviour therapy is only envisaged for recidivist or chronic forms, whatever the age of the subject (using drawings with young children, and desensitising techniques, relaxation, hypnosis, guided imagery or lucid dreams in the case of adults or psychotherapy [84, 126]).

Certain authors suggest using medical treatment, either with sulpiride or thioridazine, which are dopamine receptor antagonists [35] or by cyproheptadine which inhibits serotonin [39, 106].

However, treatment of this kind (psychotherapy or pharmacological) is only resorted to in a limited number of cases.

Sleep paralysis

Clinical features

These consist of episodes during which the subject is incapable of making voluntary movements, even though he is conscious. The subject is quite unable to move his limbs, trunk and head although eye and respiratory movements continue to function. The patient is awake or semi-awake and clearly recalls his vain efforts to move. The contrast between maintained consciousness and a paralysed body provokes a feeling of anxiety and fear. Paralysis lasts from several seconds to several minutes and stops spontaneously or in response to external stimulation. It may be accompanied by hallucination. Sleep paralysis can occur at the onset of sleep or on returning to sleep during the night (hypnagogic form) or during wakefulness (hypnopompic form). It may start at any age but tends to be seen in the adolescent or middle aged adult. It is equally common in men and women, in its isolated forms. However in 20% of cases there is a previous family history [20]. These familial forms particularly affect women, with an X-related mode of transmission. Sleep paralysis may form part of the narcoleptic tetrade, although it may be associated with excessive daytime sleepiness with no sign of narcolepsy [20].

Prevalence

The prevalence of this disorder is very high in occasional forms, 40 to 50% of normal subjects experiencing at least one episode of sleep paralysis during their lives. Differences exist according to ethnic group [20] and depending on questionnaires [34]. Chronic forms affect only 3 to 6% of subjects. 20 to 40% of narcoleptics experience sleep paralysis at some stage of their illness. Sleep paralysis occurs randomly, with no rhythmicity or periodicity. It may be brought on by irregularities in sleep-wake rhythm or sleep deprivation as well as by psychological stress and overwork. It may cause anxiety but present no complications and once the episode has passed the subject returns to his normal state.

Laboratory tests

Polysomnographically, the multiple sleep latency test and HLA typing are only useful in confirming or challenging a diagnosis of narcolepsy.

If by chance an episode of sleep paralysis is recorded, it is seen to occur during direct entry to REM sleep or during a REM sleep-wake transition. It is characterised by the abolition of muscle tone albeit with abundant twitches, wakeful EEG, and abundant movements of the eyelids and eyes. The EEG may sometimes show stage 1 sleepiness.

The abolished H reflex during sleep paralysis testifies to the monosynaptic inexcitability of the motor neurones of the anterior horn, as is also observed in cataplexies.

Differential diagnosis

Diagnosis usually presents no problem. The important thing is to distinguish isolated sleep paralysis from that which occurs as part of the semiology of narcolepsy. The problem may arise in certain cases, of a manifestation of hysteria.

Pathophysiology

Considering the time at which sleep paralysis occurs, it corresponds to the perception of all muscle tone being abolished (sleep onset or the persistence of REM sleep muscular atonia) although the patient is still conscious. Thus the mental content of wakefulness is associated with paralysis of the body [5]. It does not appear to be due to the effect of lesion; it may be a neurochemical dysfunction.

Treatment

In its isolated forms, in cases of irregular sleep schedules, treatment consists of adopting stricter sleep hygiene.

In its chronic forms, particularly those which are familial and in cases where sleep paralysis is associated with narcolepsy, treatment by tricyclic antidepressants or serotonin reuptake inhibitors [57] (as in attacks of cataplexy) may be useful.

REM sleep behaviour disorder

Clinical features

REM sleep behaviour disorder is defined as the intermittent disappearance of muscle atonia characteristic of REM sleep, and the onset of dream activity [50]. This disorder, first described in the cat by Jouvet and Delorme [52] after experimental lesions of the pontine tegmentum, was later isolated and identified in man, by Schenck *et al.* in the years 1985-1987 [99]. This pathological entity was finally included in the international classification of sleep disorders in 1990.

Behaviour disorders may be moderate, i.e. speech, shouts or laughter sometimes accompanied by grabbing movements of the hand, hitting or kicking; they may be more intense and elaborate: the subject sitting up in bed, getting up, crawling around the bed or walking around the room and even inflicting violence on nearby objects, other people or himself. These accompany dreams which are often frightening, described as nightmares [74, 75]. Hence it is the acting out of nightmarish dream activity. This violent behaviour may lead to injuries to the partner or the subject himself: ecchymosis, fractures, laceration, subdural haematosis [25, 27]. The disorder generally occurs in the second half of the night, during REM sleep.

Prevalence

The disorder affects older subjects, at a mean age of 60, 87.5% of these being men. Women are thus more rarely affected (12.5% of cases) but at a younger age (M = 45 = 22.5 years) [103]. Cases have also been recorded of young people [111]; motor activity appears to be less intense and violent, in these cases.

The interview reveals, in 25% of cases, prodromes which often go back a long way, such as sleep talking, screaming, limb movements or starts and body jerks.

The sign which heralds the parasomnia by several weeks is the apparition of dreams which are very different in tone to the usual dreams; they become violent, aggressive and anxiety provoking. It is these dreams which are later acted out.

Conversely, it is rare to find a previous history of sleepwalking in patients or night terrors in childhood [101].

With a picture of this kind, careful clinical examination is called for to look for any associated neurological or psychiatric disorders.

Laboratory tests

Polysomnography will provide a more accurate picture of the behaviour disorder, confirming whether it occurs during a phase of REM sleep; it will look outside the episode for any irregular increases in muscle tone activity of the chin during REM sleep and verify the frequent increase of limb muscle phasic activity (80% of cases).

Polysomnography reveals the absence of significant vegetative disorder during the episode of parasomnia (notably the absence of tachycardia). No paroxysmal activity or critical phenomenon of an epileptic nature is ever recorded.

Other tests may be programmed according to the clinical context: a brain MRI, a neuropsychological and psychiatric check up to detect any memory disorders, deterioration, state of depression or other mental pathology and to evaluate a state of posttraumatic stress.

Clinical variants

Chronic forms. Approximately 290 observations have been reported in literature throughout the world; many cases have undoubtedly been detected but unpublished and some observations are no doubt incorrectly classified. The parasomnia is thus relatively uncommon but no doubt underestimated.

1) Among these chronic forms, roughly 50% are idiopathic; aging in men is a contributing factor.

2) Chronic forms are associated with pathologies interfering with the parasomnia in over half the cases [86].

a. These chiefly consist of the degenerative diseases of the nervous system in 29% of cases. Parkinson's disease is found in about 13% of elderly patients with REM sleep-related behaviour disorder, Parkinson's sometimes appearing several years after the onset of the parasomnia [100]. Likewise, a high percentage of patients affected by Parkinson's disease, with no complaints of sleeping disorder, have motor behaviour disorders at an infraclinical or clinical stage. Other neurodegenerative diseases can be associated with this parasomnia i.e. multiple system atrophy, spinocerebellar degeneration [110].

Dementia, whether or not it is associated with Parkinson's disease is also found in a considerable number of cases, calling for the need to explore the cognitive functions of elderly subjects.

b. The other neurological pathologies are far more rare, whether vascular (cerebrovascular accident, subarachnoid haemorrhage), tumorous, demyelinating (multiple sclerosis) or again, postanoxic.

c. Narcolepsy is the second illness, in order of frequency, to be associated with this parasomnia. We are familiar with the motor anomalies appearing in the polysomnography of narcoleptic patients, with no associated parasomnia [77]. Roughly 14% of subjects affected by REM sleep behaviour disorders are also narcoleptic. In these patients, who are often young, narcolepsy is full blown, the cardinal symptoms are present and parasomnia occurs at the onset of the clinical features of narcolepsy [103].

d. This parasomnia is rarely associated with a psychiatric illness; conversely, the triggering of a REM sleep behaviour disorder by intense stress has been described - the link referred to earlier between nightmares and posttraumatic stress. This initially acute form may become chronic [93].

The acute forms of this parasomnia have also been described and are usually of toxic and/or metabolic origin.

Observations have thus been reported with biperiden intoxication, after massive alcohol intoxication associated with meprobamate, after suspending amphetamines or cocaine, during or after suspending treatment with fluoxetine, with tricyclic antidepressants or with monoamine oxydase inhibitors, i.e. in circumstances where there is often a rebound of REM sleep [75]. Anticholinesterasics have been known to trigger this parasomnia, in a few rare cases.

REM sleep related parasomnia can also be associated, albeit rarely, with other parasomnias with motor disorders occurring in NREM sleep. Thus sleepwalking, night terrors and behaviour disorders were recently shown to be possibly associated in the same subjects [98]. The onset of these disorders occurs at roughly the age of 15 years, i.e. later than the usual age of onset of arousal disorder parasomnias, while also affecting young subjects. The syndrome is idiopathic in 66% of cases; in the other cases, it occurs in subjects with various neurological disorders or with sleep-related cardiac disorders (auricular fibrillation…) or psychiatric disorders (posttraumatic stress, major depression) or after stopping or resuming various forms of intoxication (alcohol, cocaine, amphetamines).

Differential diagnosis

This arises with certain forms of sleep-related epilepsy. Such a diagnosis requires an in-depth interview followed by a polysomnographic recording of the behaviour disorder.

A state of posttraumatic stress must be recognised, as this may only cause nightmares or provoke acted out nightmares, as a continuum of expression.

Pathophysiology

In animals, the landmark work of Jouvet and Delorme [52] followed by the work of the Jouvet's team [95, 96] showed that bilateral, symmetrical lesions to the pontine tegmentum in the cat, provoked a loss of muscle atonia during REM sleep and exploratory type motor behaviour with occasional attacking behaviour. This hallucinatory behaviour of the REM sleep-related dream type, is not altered by external stimulation, never includes food related or sexual behaviour and is thus very close to the expression of the parasomnia in man. The work of Morrison's team [44, 83, 96] defined the different types of dream motor behaviour depending on the site and extent of the lesion.

A lesion of the locus coeruleus and sometimes of the adjacent part of the medial reticular formation leads to an irregular loss of muscle atonia (25 to 100% of the time) with jerks and sometimes violent movements of the head and the proximate part of the limbs and trunk, although the animal cannot hold its head up or express co-ordinated behaviour.

When the lesion is more ventral, on the locus coeruleus and ventro-medial reticular formation, the animal expresses exploratory and orientation behaviour; the head is raised, the eyes fixed and it turns its head to look around.

Episodic forms of behaviour i.e. violent, attacking, hunting, stalking imaginary prey or an object, or attacking the animal's own tail (self-mutilation), are observed several days after a rostro-ventral lesion of the mesencephalon or after a lesion to the bundle of the central amygdaloid nucleus. The lateral hypothalamus through the intermediary of the ventral tegmental area of Tsai, is the organisation site for attacking motor behaviour, the more rostral structures modulating this behaviour.

An even more ventral lesion, of the fibres leading to the superior colliculus, will trigger locomotive behaviour and wandering in cats, as though the animal were awake but with no response to external stimuli. REM sleep muscle atonia is also known to result from the active inhibition of motoneurones through bundles descending from the locus coeruleus and perilocus coeruleus , but suprapontine structures are known to influence these centres and participate in controlling REM sleep-related muscle tone. What these animal experiments do show is that the loss of REM sleep-related muscle atonia, whatever the mechanism, is a necessary condition but one which is insufficient to explain the occurrence of motor behaviour. Indeed, the loss of muscle atonia must be accompanied by considerable muscle phasic activity, with the expression of more or less pre-wired motor behaviour in the mesencephalic region. But we know that the centres responsible for motor inhibition, on the one hand, and those responsible for motor programming on the other, are both located in the pons and bring into play the different neuromediators [60]. This explains why the same anatomical lesion can be responsible for disorders to both systems or why a lesion can individually affect one system or the other.

Finally, experiments on the decerebrate cat have shown that lesions, dysfunctioning or excessive use of N-methyl-D aspartate receptors at the site of the ventral mesopontine junction or cellular group A8, behind the red nucleus, will result in phasic movements, rhythmic limb movements and locomotor movements, both in wakefulness and sleep [61]. These anatomical zones contain dopaminergic neurones which project onto the caudal nucleus and the putamen and glutaminergic neurones which project onto the pontine reticular formation and the macrocellular medullary nucleus [62] whose role was seen in motor control. Hence a dysfunction (disrupted balance of influx to NMDA receptors), hypoactivity or a lesion to these anatomical zones may explain the occurrence of these motor disorders (periodic movements or abnormal motor behaviour) during sleep (NREM and REM) and help to explain the efficacy of certain types of treatment (L-Dopa in treating periodic limb movements, for example). The pedunculopontine region, with its afferents and efferents thus play a major role in motor control [94].

In man, certain neurological illnesses, whether degenerative, tumorous or traumatic may be accompanied by lesions in the centres or pathways which are directly or indirectly involved in REM sleep-related motor control explaining the origin of the disorder. Thus Parkinson's disease which often seems to be associated with this parasomnia, is accompanied by the degeneration of the locus niger. Moreover, autopsies carried out on two patients who had presented a chronic form of REM sleep behaviour disorder (one case with Lewy bodies [124] and another with a variant of Alzheimer's disease with Lewy bodies) showed either extensive depigmentation or the loss of aminergic neurones in the locus coeruleus. In this case, it is as though the noradrenergic neurones in the locus coeruleus were unable to exert their inhibiting influence on the pontopeduncular and laterodorsal tegmental nuclei, allowing the expression of REM sleep motor behaviour. In any case, too few autopsies have been carried out to date. Conversely, many patients are affected by chronic idiopathic forms of REM sleep behaviour disorder but show no anatomoradiological indication of lesions to the pons, locus coeruleus or pericoeruleus.

This motor disorder may thus be secondary to a neurophysiological and/or neuroaminergic dysregulation rupturing the balance of inhibitory mechanisms (muscle atonia) and excitation (motor expression, simple movements or complex behaviour) exerted at the level of the brain stem, controlled by underlying neural structures. Reduced dopamine fixation on the striatum has been shown in SPECT, in subjects affected by this parasomnia in its idiopathic form [28]. And finally, these changes in cerebral synaptic functioning occur in the course of aging. The functional dysregulation of neurone populations situated at the level of the cerebral cortex and not only in the brain stem may account for the occurrence of this parasomnia, associated with mental activity in the form of nightmares, after the model suggested by Hobson *et al.* [45, 46].

Treatment

Pharmacological treatment:

Clonazepam is administered at bedtime, at an average dose of 0.5 to 1 mg/day (0.25 to 4 mg). The treatment is effective in 90% of patients (completely effective in 78% and partially so in 12% of subjects). Treatment starts with a dose of 0.25 or 0.5 mg and increased after several days in stages of 1.25 mg until the effective dose is reached i.e. suppression of the motor disorder and the accompanying anxious dreams. This treatment usually acts quickly from the first night or couple of nights [75, 106]. A partial escape may occasionally occur with the recurrence of some motor phenomena. When treatment is stopped, the symptomatology always reappears. Clonazepam can continue to be effective over long periods of up to 10 years. It has no unpleasant side effects, tolerance or addiction [107]. This benzodiazepine acts on the system by generating locomotor activity and reducing excessive phasic motor activity, possibly due to a serotoninergic property, but has no action on the loss of REM sleep muscle atonia [63].

Where clonazepam is ineffective or poorly tolerated, alprazolam (0.25 to 0.50 mg/day), triazolam (0.125 to 0.25 mg/day) have proved to be effective in certain cases. Tricyclic antidepressants, imipramine or desipramine, can paradoxically give good results by partially suppressing REM sleep. Other products have been tried with less consistent results: clonidine, carbamazepine [4] which suppress violent and aggressive motor behaviour, L-Dopa and L tryptophane [106]. Finally melatonin has recently been tried with success, on a limited number of subjects [58].

Adapting the environment to reduce the risks of injury is also useful: eliminating dangerous objects, protecting the windows, placing cushions around the bed, and the mattress on the floor etc.

Causal or associated pathologies must of course be treated in their own right, wherever possible.

Hence this noisy, sometimes dangerous pathology is often the easiest and most effectively treated, both in the long and short term, once it has been diagnosed.

OTHER PARASOMNIAS

These parasomnias group together a number of heterogeneous disorders which may occur at any stage of sleep and share no identical or close patholophysiological mechanisms [50].

Sleep talking

Clinical features

Sleep talking consists in fact of talking or uttering words and sounds during sleep, without the subject being aware of this or recalling it later.

Sleep talking can thus be expressed in the form of uttering murmurs or unintelligible sounds, but also by uttering words, phrases and long conversations which are entirely comprehensible. Very often spontaneous, sleep talking can be induced or encouraged by someone talking to the sleeper.

Sleep talking episodes are generally short, lasting a few seconds, but may be longer, frequent and repeated every night. They then become a cause for complaint from the partner or others living under the same roof.

Episodes of sleep talking can occur both in NREM and REM sleep. The contents of the words are often linked to recent events and the voice often monotonous in NREM sleep, whereas in REM sleep they are more complex and accompanied by dreaming [4] and expression in the voice [80]. Some subjects sleep talk at fixed times but this may vary. The contents may be banal in tone or full of feeling, anger or hostility.

Prevalence

Sleep talking may occur at any age. It is very widespread among children, adolescents and young adults; observed in boys as much as in girls. In adults, it is more common in men than in women.

Minor expressions of sleep talking are very common but it occurs more rarely on a daily basis or in a manner which is discomforting to others.

There are family tendencies in sleep talking. There is no known mode of transmission. Sleep talking can be completely isolated; it is sometimes triggered or favoured by an episode of fever, by problems, daily stress and conflicts as well as by other sleep disorders (other parasomnias: night terror, REM sleep-related behaviour disorder or sleep apnoea syndrome). It may be associated with a psychopathological disorder, particularly in adults.

Its evolution is highly variable. Usually benign, it disappears spontaneously after a few days or months, only reappearing in short episodes.

Although sometimes isolated, it can occur in long episodes every night, often at the same time, waking the partner. It is subjects who are affected by these serious forms who tend to consult.

Laboratory tests

Polysomnography is chiefly carried out to look for another sleep pathology associated with sleep talking.

In such cases, the sleep recording will confirm that these episodes tend to occur in NREM sleep (65% of these in light NREM sleep, 25% in deep NREM sleep) and roughly 10% in REM sleep [92].

Pathophysiology

This is not known.

Treatment

Treatment is not usually necessary for this essentially benign disorder. In severe forms particularly if there is a notion of emotional stress, treatment with moderate doses of benzodiazepine for several weeks, may improve the disorder.

Sleep bruxism

Clinical features

This refers to a grinding of the teeth, secondary to a stereotyped jaw movement linked either to phasic (rhythmic) or tonic (sustained) contractions of the masseter and other temporal and pterygoid muscles, during sleep.

The disorder may also occur during wakefulness, corresponding to a different pathological entity consisting of a quasi voluntary activity, producing no noise.

The subject consults because of the grinding sounds he makes which are often disagreeable and annoying for the partner who is woken up by them. The sounds are often dry, grinding, violent, unpleasant, occurring in episodes of 5 to 10 seconds or more, and recurring dozens or even hundreds of times a night.

The episodes do not waken the subject who is unaware of the disorder and is usually informed of it by someone close to him or by his dentist. This parasomnia wears away the teeth, causes parodontological lesions which may be severe with inflammation and retraction of the gums, alterations to dental alveoli and temporomandibular articulation disorder. When severe or long standing, it can lead to pain in the muscles of the cheeks or face, even headaches, principally on morning arousal.

Bruxism can occur at any time of night since it may be present at any stage of sleep, chiefly in stage 2 of NREM sleep but also in REM sleep [67].

Prevalence

It can occur at any age and is very common in children and young adults. In the child, onset may occur very early, as soon as the incisors appear. In the adult, onset usually occurs between the ages of 10 and 20. It affects girls and women as much as boys and men [59] but is more often reported by women [67].

Up to 85% of the population may experience teeth grinding episodes, to a varied extent, in the course of their lifetime [50]. Bruxism occurs in 5.1 to 21% of normal adults [27, 67]; it may be debilitating for 5% of subjects.

Clinical variants

These are extremely variable, ranging from mild, transient forms to those which are severe and chronic. Bruxism can develop in the healthy child or adult, but is often encountered in subjects suffering from mental retardation (up to 58%) [37] or psychotic states.

In the child [59] and in the adult it may be brought on by anxiogenic situations or stress although not invariably [87]. It is increased by the excessive consumption of alcohol, drugs, (cocaine, amphetamines), by the intake of antidopaminergic medications such as haloperidol and certain serotonin reuptake inhibitors like fluoxetine or sertraline [67].

Several familial forms and one genetic component [1] have been reported.

Bruxism may be associated with other parasomnias. In the child it is most commonly associated with sleep talking [59, 128]. In the adult, restless leg syndrome and periodic limb movements are found in roughly 10% of subjects with bruxism [67, 68]. An association has also been found with cramps, upper airway resistance syndrome and sleep apnoea syndrome.

Laboratory tests

1) A dental check up is indispensable to look for any contributing factors and to determine the impact of bruxism on the teeth and surrounding soft tissue.

2) Polysomnography with audio-visual recording is rarely carried out to confirm a diagnosis which is often clinically evident. It is nevertheless useful in certain complex, associated forms or to quantify the disorder. It would then include a recording of the electromyographic activity of the masseter and temporal muscles, using attached electrodes. It may also be instructive to record other muscles. A diagnosis of bruxism is sometimes made by chance based on the EMG of the chin muscle. This shows the rhythmic intensification of basic muscular phasic activity; these muscular bursts are repeated at a rhythm of roughly 1/sec, in sequences of several seconds recurring dozens of times during the night. The activity is more frequent in stage 2 of NREM sleep, but has been described in REM sleep [67] and can be registered during any stage of sleep. These episodes tend to get longer as sleep deepens [3].

In severe forms, these episodes are accompanied by a lightening of sleep with a stage shift [74% of cases] and a micro-arousal (28% of cases). Bader *et al.* [3] noted that brief reappearances of alpha rhythm may precede these passages of bruxism. There is accompanying tachycardia persisting for about 10 seconds after the episode of bruxism.

As it may occur one night and disappear the next, the problem lies in picking it up in the laboratory, which may require several nights' recording.

3) In certain cases, a psychological check up is useful (MMPI, projection tests) in the search for personality traits of the obsessive type. Some authors [3, 31] have noted a tendency to anxiety in children [59] and repressed aggression in adults, but others have failed to corroborate this [41]. Obvious mental pathologies will already have been diagnosed, in general.

Pathophysiology

Local anatomical factors contribute to the occurrence of bruxism: malformations of the jaw, dental occlusion disorders.

Neurophysiological mechanisms: the loss during sleep of the inhibiting control exerted on certain neural centres and pathways can give rise to rhythmic movements. On the basis of animal experiments, the nigro-striatal dopaminergic pathways have been shown to be involved in generating bruxism [67]. Moreover, high doses of L-Dopa administered to man can produce episodes of bruxism. Similarly, certain psychoses or encephalopathies associated with dopaminergic hyperactivity may be accompanied by bruxism.

Psychological factors: periods of stress or increased anxiety can trigger or aggravate sleep-related bruxism, especially in sensitive subjects or at the onset of a psychotic episode.

It may be the expression of an endogenous physiological arousal mechanism, as suggested by polysomnographic findings [2].

Treatment

This parasomnia seldom calls for treatment. Only 10% of patients affected by bruxism require treatment [37].

Depending on the case:

Treatment may be *preventive*: correcting minor malformations of articulation may be advised, although this will not necessarily stop the occurrence of bruxism. Local treatment may be used to counteract the excessive dryness of the mouth which favours bruxism.

Treatment is often *symptomatic*. To prevent teeth grinding, causing an unpleasant sound and the risk of wearing down the teeth, the patient can be advised to wear a soft or hard device placed over the teeth at bedtime (reviewed in [2]). This procedure often proves very successful, at any age. The results of the other therapeutic procedures are often disappointing [47]. In severe forms, medication treatment based on L. Dopa, levocarbidopa and levobenserazide, have improved bruxism. Finally propranolol (beta adrenergic receptor) may be effective in idiopathic and iatrogenic cases of bruxism but this medication is known to have a detrimental effect on the quality of sleep [67].

It is sometimes necessary to treat *anxiety* either through the behavioural approach, relaxation or biofeedback [6], the results of which remain debateable [67, 122]; or through psychotherapy, depending on the results of the psychological check up; or in rarer cases, by anxiolytics, prescribed for short periods in situations of stress.

Stomatological and psychological treatment can be combined [133].

Finally, in very severe cases causing pain and masseter hypertrophy, the use of botulinic toxin (BXTA) has been advised.

This is a disorder whose pathophysiology is multifactorial and whose treatment remains controversial [29].

Benign neonatal sleep onset myoclonus

Clinical features

These refer to short, involuntary myoclonic jerks, triggered by asynchronous muscular contractions of the trunk and extremities.

These rare manifestations are observed in the newborn, regardless of sex, usually between fifteen and thirty five days old, in good health, born at full term after a normal pregnancy and birth, with no previous perinatal history or family history of epilepsy. The movements always occur during sleep, whether in the daytime or at night.

The baby's psychomotor development is normal. Both clinical and para-clinical tests, notably the transfontanelle CT scan or even the brain scan are normal. These myoclonic jerks have no familial character, although two cases have been reported in twins. The myoclonia may become more pronounced during the first three or four weeks, on average and then gradually diminish. There is positive spontaneous evolution, with myoclonia diminishing in number and intensity and finally disappearing after a period ranging from two weeks to three months.

Laboratory tests

Polysomnographic recording is almost always carried out in practice, during rest periods in the daytime or at night and involves recording the various muscles (biceps, deltoids, the rear calf, in particular and sometimes the quadriceps) using surface electrodes. It shows myoclonia to be generalised on the whole, but affecting the upper limbs more than the lower limbs and the distal more than the proximal extremities. These myoclonic jerks, usually massive and bilateral, are rhythmical, and can be intense causing the whole body to jump. They may be segmentary, involving only one segment of the limb or half of the body, while the face is hardly ever affected.

Authors differ as to when myoclonia occur, some asserting that they only occur at the onset of sleep and during NREM sleep [22] whatever the EEG tracing, alternating or slow continuous, and others that they occur in all stages of sleep [27], disappearing when the stage shifts and on arousal. Myoclonia do not occur with every onset of sleep or light NREM sleep phase. They last between 40 and 300 milliseconds, at a rhythm of 4 to 5 Hz, in one second sequences recurring irregularly in a series of twenty to thirty minutes in total. No EEG anomalies are recorded either during the myoclonia or between them.

Differential diagnosis

Are these epileptic manifestations?

The problem which most often arises is that of the *neonatal convulsive fit* within a context of benign idiopathic, as opposed to familial epilepsy [89]. Clinically, these manifestations often resemble partial clonic convulsions in form, passing sometimes from one half of the body to the other, with or without apnoea, but rarely with a pattern of bilateral, synchronous jerks. Electrically, they are accompanied by critical epileptic discharges. The interictal trace is non specific.

Other diagnoses include the beginning of *myoclonic encephalopathy*, and more rarely, a neonatal subintrant crisis or pyridoxine deficiency. These problems are fairly easy to resolve thanks to both clinical and electroencephalographic data.

Are these non epileptic paroxysmal manifestations of a different nature, from which the following may be distinguished?

a) *Benign myoclonus of early infancy* [73] which consists of a short contraction of the entire body, occurring during wakefulness, as opposed to during sleep, during the first months of life rather than during the neonatal period.

b) *Hyperplexia or a pathological jerking reaction*, with prolonged, generalised muscle contraction or with massive myoclonus in response to a stimulus, occurs both during wakefulness and sleep. These manifestations may lead to apnoea with bracycardia and cyanosis [119]. They are triggered by an external stimulus. This is a genetic disease of dominant autosomic transmission, quite different to benign sleep myoclonus.

c) As for fragmentary myoclonus, this chiefly affects adult males, and consists of light, asymmetrical, asynchronous contractions mainly of the hands and face. The EMG shows muscular discharges of 75 to 150 milliseconds, with very little repetition, during light NREM sleep, which may persist during REM sleep. It is often accompanied by excessive daytime sleepiness.

d) Finally, sleep onset myoclonus is different, occurring mainly in light NREM sleep. This consists of short, bilateral synchronous, isolated movements, with little repetition and which predominate in the lower limbs. They are very widespread throughout the population.

Pathophysiology

Myoclonus is of the same significance as clonus, resulting from a facilitating influence issuing from the brain stem on segmental reflex activity, this mechanism being favoured by NREM sleep. Other authors have put forward the hypothesis that a transitory disorder in serotoninergic balance could be responsible for the myoclonus. It remains to be seen whether this entity should be classed as a parasomnia.

Sleep-related abnormal swallowing syndrome

Clinical features

This disorder listed among the parasomnias, is rarely encountered; it consists of difficulty swallowing saliva during sleep, provoking aspiration with coughing and suffocation, followed by a micro-arousal and/or arousal.

The patient complains of a sensation of blocked respiration and suffocation, and sometimes of insomnia. The episodes stop on arousal.

This parasomnia chiefly affects elderly subjects in whom it can lead to respiratory infections, through aspiration and saliva passing into the respiratory tract.

The disorder is isolated and spontaneous and sometimes favoured by the administration of sedative treatment having a depressive effect on the central nervous system.

Laboratory tests

Sleep polygraphy is used to check for sleep apnoeas, whether obstructive or central. In certain cases the same complaint may be accompanied by apnoeas.

There is no sleep apnoea with this parasomnia, but an objective recording of brief, recurrent arousals with a very low level of NREM sleep.

An ENT examination while the disorder is occurring shows an accumulation of saliva in the hypopharynx and the passage of saliva into the trachea. Tests are also carried out to determine whether the airways are free and the vocal cords functioning properly.

Differential diagnosis

During SAS, the subject is usually unaware of having breathing difficulties and complains of hypersomnia.

A gastro-oesophageal reflux may show similar symptomatology during sleep, but is typically accompanied by a burning sensation or heartburn and epigastric pain.

Finally, night terror attacks may be accompanied by the sensation of breathing difficulties and suffocation but are preceded by a cry, frightened or panic-stricken behaviour, marking the difference between these two parasomnias.

Impaired swallowing disorder is presumed to be rare but no epidemiological study has been done. Its spontaneous evolution is largely unknown. The aetiology and pathophysiology are essentially unknown. It may be a disorder of psychological origin.

REFERENCES

1. ABE K., SHIMAKAWA M. – Genetic and developmental aspects of sleep talking and teethgrinding. *Acta Paedopsychiatrica*, 33, 339-344, 1966.
2. ATTANASIO R. – An overview of bruxism and its management. *Dent. Clin. North Am.* 41, 2, 229-241, 1997
3. BADER G., KAMPE T., TAGDOE T., KARLSSON S., BLOMQVIST M. – Desriptive physiological data on a sleep bruxism population. *Sleep.* 20, 982-990, 1997.
4. BAMFORD C.R. – Carbamazepine in REM sleep behavior disorder. *Sleep*, 16, 33-34, 1993.
5. BILLIARD M. – Les parasomnies. *In: Sommeil et Eveil*, Espaces 34, Montpellier, 193-215, 1997.
6. BIONDI M., PICARDI A. – Temporomandibular joint pain dysfunction syndrome and bruxism: etiopathogenesis and treatment from a psychosomatic integrative viewpoint. *Psychoter. Psychosom.*, 59, 84-98, 1993.
7. BLATT I., PELED R., GADOTH N., LAVIE P. – The value of sleep recording in evaluating somnambulism in young adults. *Electroencephalog. Clin. Neurophysiol.* 78, 407-412, 1991.
8. BONKOLO A. – Impulsive acts and confusional states during incomplete arousal from sleep: criminological and forensic implications. *Psychiatr. Q. ,* 48, 400-409, 1974.
9. BORNSTEIN S., GUEGUEN B., HACHE E. – Elpenor's syndrome or somnambulistic murder? *Ann. Med. Psychol.* (Paris), 154, 195-201, 1996.
10. BRAMBLE D. – Two cases of severe head-banging parasomnias in peripubertal males resulting from otitis media in toddlerhood. *Child: Care, Health & Development*, 21, 247-253, 1995.
11. BROUGHTON R., BILLINGS R., CARTWRIGHT R., DOUCETTE D., EDMEADS J., EDWARDH M., ERVIN F., ORCHARD B., HILL R., TURRELL G. – Homicidal somnambulism: a case report. *Sleep*, 17, 253-264, 1994.
12. BROUGHTON R., SHAPIRO C.M. – Parasomnias in sleep solutions. In: C.M. Shapiro (ed) *Manual – Communication Publication*. Quebec, 169-188, 1995.
13. BROUGHTON R.J., SHIMIZU T. – Sleep-related violence: a medical and forensic challenge. *Sleep*, 18, 727-730, 1995.
14. CHALLAMEL M.J. – Les parasomnies de l'enfant. *B.V.S.*, 9, 8-13, 1992
15. CHALLAMEL M.J. – Les parasomnies. *Revue du Praticien*, 46, 2448-2451, 1996.
16. CLORE E.R., HIBEL J. – The parasomnias of childhood, *J. Pediat. Health Care*, 7, 12-16, 1993.

17. CONNOLLY P.S., SHIRLEY E.A., WASSON J.H., NIERENBERG D.W. – Treatment of nocturnal leg cramps. A crossover trial of quinine vs vitamin E. *Arch. Intern. Med.,* 152, 1877-1880, 1992.
18. CRISP A.H. – The sleepwalking night terrors syndrome in adults. *Postgrad. Med.,* 72, 599-604, 1996.
19. CRISP A.H., MATTHEWS B.M., OAKEY M., CRUCHFIELD M. – Sleepwalking, night terrors, and consciousness. *BMJ,* 300, 360-362, 1990.
20. DAHLITZ M., PARKES J.D. – Sleep paralysis. *Lancet,* 341, 406-407, 1993.
21. DE VILLARD R., BASTUJI H., GARDE P., MAUGUIERE F., DALERY J., MAILLET J., REVOL O. – Les terreurs nocturnes, le somnambulisme. *Revue du Praticien.* 39, 10-14, 1989.
22. DI CARPUA M., FUSCO L., RICCI S., VIGEVANO F., – Benign neonatal sleep myoclonus: clinical features and video-polygraphic recordings. *Mov. Disord.,* 8, 191-194, 1993.
23. DRAKE M.E. –Jactatio nocturna after head injury. *Neurology,* 36, 867-868, 1986.
24. DRIVER H.S., SHAPIRO C.M.- ABC of sleep disorders. Parasomnias. *BMJ,* 306, 921-924, 1993.
25. DYKEN M.E:, LIN-DYKEN D.C., SEABA P., YAMADA T. – Violent sleep-related behavior leading to subdural hemorrhage. *Arch. Neurol.,* 52, 318-321, 1995.
26. DYKEN M.E., LIN-DYKEN D.C., YAMADA T. – Diagnosing rhythmic movement disorder with video-polysomnography. *Pediatr. Neurol.,* 16, 37-41, 1997.
27. DYKEN M.E., RODNITZKY R.L. – Periodic, aperiodic, and rhythmic motor disorders of sleep. *Neurology,* 42, 68-74, 1992.
28. EISENSEHR I., LINKE R., NOACHTARS, SCHWARZ J., GILDEHAUS F.J., TATSCH K. – Reduced striatal dopamine transporters in idiopathic rapid eye movement sleep behaviour disorder. Comparison with Parkinson's disease and controls. *Brain,* 123, 1155-1160, 2000.
29. FAULKNER K.D. – Bruxism: a review of the literature. Part II. *Australian Dental Journal,* 35, 355-361, 1990.
30. FERBER R. – *Protéger le sommeil de votre enfant.* ESF, Paris, 237 p., 1990
31. FISCHER W.F., O'TOOLE E.T. – Personality characteristics of chronic bruxers. *Behavioral Medicine,* 19, 82-86, 1993.
32. FRANK N.C., SPIRITO A., STARK L., OWENS-STIVELY J. – The use of scheduled awakenings to eliminate childhood sleepwalking. *J. Pediatr. Psychol.,* 22, 345-353, 1997.
33. FRUSSO R., ZARATE M., AUGUSTOVSKI F., RUBINSTEIN A. - Magnesium for the treatment of nocturnal leg cramps: a crossover randomized trial. *J. Fam. Pract.* 48, 11, 868-871, 1999.
34. FUKUDA K. – One explanatory basis for the discrepancy of reported prevalences of sleep paralysis among healthy respondents. *Percept. Mot. Skills,* 77, 803-807, 1993.
35. GAILLARD J.M. – Les parasomnies. *Le sommeil, ses mécanismes , ses troubles,* Doin, Paris, 233-252, 1990.
36. GARMA L. – Approches critiques de la clinique du rêve et du sommeil. *Confrontations Psychiatriques,* 38, 115-140, 1997.
37. GUILLEMINAULT C. – Disorders of arousal in children. Somnambulism and night terrors. In: C. Guilleminault (ed), *Sleep and its Disorders in Children,* Raven Press, New York, 243-252, 1987.
38. GUILLEMINAULT C., MOSCOVITCH A., LEGER D. – Forensic sleep medicine: nocturnal wandering and violence. *Sleep,* 18, 740-748, 1995.
39. GUPTA S., AUSTIN R., CALI L.A., BATHARA V. – Nightmares treated with cyproheptadine. *J. Am. Acad. Child Adolesc. Psychiatry* 37, 6, 570-571, 1998.
40. HAMEURY L., GARREAU B. – Troubles du sommeil du nourrisson et du jeune enfant (0-3 ans). *Neuro-Psychiatr. Enf.* 38, 1-6, 1990.
41. HARNESS D.M., PELTIER B. – Comparison of MMPI scores with self-report of sleep disturbance and bruxism in the facial pain population. *Cranio,* 10, 70-74, 1992.
42. HARTMANN E. – Nightmares and other dreams. *In:* M.H. Kryger, T. Roth, W.C. Dement (eds), *Principles and Practice of Sleep Medicine,* 2nd edition, W.B. Saunders, Philadelphia, 36, 407-410, 1994.
43. HARTMANN E., Nightmare after trauma as paradigm for all dreams: A new approach to the nature and functions of dreaming. *Psychiatry,* 61, 223-238, 1998.
44. HENDERICKS J.C., MORRISON A.R., MANN G.L. – Different behaviours during paradoxical sleep without atonia depend on pontine lesion site. *Brain Res.,* 239, 81-105, 1982.
45. HOBSON J.A., LYDIC R., BAGHDOYAN M.A. – Evolving concepts of sleep cycle generation: from brain centers to neuronal population. *Behav. Brain. Sci.,* 9, 371-448, 1986.
46. HOBSON J.A., McCARLEY R.W. – The brain as a dream state generator: an activation-synthesis hypothesis of the dream process. *Am. J. Psychiat.,* 134, 1335-1348, 1977.
47. HOLMGREN K., SHEIKHOLESAM A., RIISE C. – Effect of full-arch maxillary occlusal splint on parafunctional activity during sleep in patients with nocturnal bruxism and signs and symptoms of craniomandibular disorders. *Journal of Prosthetic Dentistry,* 69, 293-297, 1993.
48. HUBLIN C., KAPRIO J., PARTINEN M., HEIKKILA K., KOSKENVUO M. – Prevalence and genetics of sleepwalking: a population-based twin study. *Neurology,* 48, 177-181, 1997.
49. HURWITZ T.D., MAHOWALD M.W., SCHENCK C.H., SCHLUTER J.L., BUNDLE S.R. – A retrospective outcome study and review of hypnosis as treatment of adults with sleepwalking and sleep terror. *J. Nerv. Ment. Dis.,* 179, 228-233, 1991.
50. ICSD- *International Classification of Sleep Disorders. Diagnostic and Coding Manual.* Diagnostic Classification Steering Committee, Thorpy M.J., Chairman. Rochester, Minnesota: American Sleep Disorders Association, 1990.
51. JACOBSEN J.H., ROSENBERG R.S., HUTTENLOCHER P.R., SPIRE J.P. – Familial nocturnal cramping. *Sleep,* 9, 54-60, 1986.

52. JOUVET M., DELORME J.F. – Locus coeruleus et sommeil paradoxal. *C.R. Soc. Biol.,* Paris, 159, 895-899, 1965.
53. KAVEY N.B., WHYTE J. – Somnambulism associated with hallucinations . *Psychosomatics,* 34, 86-90, 1993.
54. KEEFAUVER S.P., GUILLEMINAULT C. – Sleep terrors and sleepwalking . *In:* M.H.Kryger, T. Roth, W.C. Dement (eds). *Principles and Practice of Sleep medicine.* 2nd edition, W.B. Saunders, Philadelphia, 56, 567-573, 1994.
55. KLACKENBERG G. – Incidence of parasomnia in children in a general population. *In:* C. Guilleminault (ed), *Sleep and its Disorders in Children,* Raven Press, New York, 99-113, 1987.
56. KOHEN D.P., MAHOWALD M.W., ROSEN G.M. – Sleep-terror disorder in children: the role of self-hypnosis in management. *Am. J. Clin. Hypn.,* 34, 233-244, 1992.
57. KORAN L.M., RAGHAVAN S. – Fluoxetine for isolated sleep paralysis. *Psychosomatics,* 34, 184-187, 1993.
58. KUNTZ D., BES F. – Melatonin as therapy in REM sleep behavior disorder patients: an open labelled pilot study on the possible influence of melatonin on REM sleep regulation. *Mov. Disord.* 14, 3, 507-511, 1999.
59. LABERGE L., TREMBLAY R.E., VITARO F., MONTPLAISIR J. Development of parasomnias from childhood to early adolescence. *Pediatrics,* 106, 1, 67-74, 2000.
60. LAI Y.Y., SIEGEL J.M. – Muscle tone suppression and stepping produced by stimulation of midbrain and rostral pontine reticular formation. *J. Neurosci.,* 10, 2727-2734, 1990.
61. LAI Y.Y., SIEGEL J.M. – Brainstem-mediated locomotion and myoclonic jerks. I. Neural substances. *Brain Res.* 745, 257-264, 1997.
62. LAI Y.Y., SIEGEL J.M. – Brainstem-mediated locomotion and myoclonic jerks. II Pharmacological effects. *Brain Res.* 745, 265-270, 1997b.
63. LAPIERRE O., MONTPLAISIR J. – Polysomnographic features of REM sleep behavior disorder: development of a scoring method. *Neurology,* 42, 1371-1374, 1992.
64. LARSON C.P., PLESS I.B., MIETTINEN O. – Preschool behavior disorders: their prevalence in relation to determinants. *J. Pediatr.* 113, 278-285, 1988.
65. LAUERMA H. – Fear of suicide during sleepwalking. *Psychiatry,* 59, 206-211, 1996.
66. LAVIE P., HEFEZ A., HALPERIN G., ENOCH D. – Longterm effects of traumatic war related events on sleep. *Am. J. Psychiat.* 136, 175-178, 1979.
67. LAVIGNE G.J., MANZINI C. - Bruxism. *In* M.H. Kryger, T. Roth, W.C. Dement (eds). *Principles and Practice of Sleep Medicine,* 3rd edition, W.B. Saunders, Philadelphia, 67, 773-785, 2000.
68. LAVIGNE G.J., MONTPLAISIR J.Y. – Restless legs syndrome and sleep bruxism: prevalence and association among Canadians. *Sleep,* 17, 739-743, 1994.
69. LECENDREUX M., MAYER G., BASSETTI C., NEIDHART E., CHAPPUIS R., TAFTI M. HLA class II association in sleepwalking. *Sleep,* Supp. 2, 23, A13, 2000.
70. LILLYWHITE A.R., WILSON S.J., NUTT D.J. – Successful treatment of night terrors and somnambulism with paroxetine. *Br. J. Psychiat.,* 164, 551-554, 1994.
71. LINDSAY S.J., SALKOVSKIS P.M., STOLL K. – Rhythmical body movement in sleep: a brief review and treatment study. *Behav. Res. Ther.,* 20, 523-526, 1982.
72. LLORENTE M.D., CURRIER M.B., NORMAN S.E., MELLMAN T.A. – Night terrors in adults: phenomenology and relationship to psychopathology. *J. Clin. Psychiat.,* 53, 392-394, 1992.
73. LOMBROSO C.T., FEJERMAN N. – Benign myoclonus of early infancy. *Ann. Neurol..* I, 138-143, 1977.
74. MAHOWALD M.W., SCHENCK C.H. - REM sleep behavior disorder. *In* M J. Thorpy (ed), *Handbook of Sleep Disorder.* Marcel Dekker, New York, 25, 567-593, 1990.
75. MAHOWALD M.W., SCHENCK C.H. – REM sleep behavior disorder. *In:* M.H. Kryger, T. Roth, W.C. Dement (eds) *Principles and Practice of Sleep Medicine,* 2nd edition, W.B. Saunders, Philadelphia. 57, 574-588, 1994.
76. MAHOWALD M.W., SCHENCK C.H. – NREM sleep parasomnias. *Neurol. Clin.* 14, 675-696, 1996.
77. MAYER G., MEIER-EVERT K. – Motor discontrol in sleep of narcoleptic patients (a lifelong development). *J. Sleep Res.* 2, 143-148, 1993.
78. MENNY J.C., SAUTERAND A., BOURGEOIS M. – Les rêves en psychologie. Relation avec la pensée diurne. *Confrontations Psychiatriques,* 38, 307-336, 1997.
79. MILLMAN R.P., KIPP G.J., CARSKADON M.A. – Sleepwalking precipitated by treatment of sleep apnea with nasal CPAP. *Chest,* 99, 750-751, 1991.
80. MOLAIE M., DEUTSCH G.K. –Psychogenic events presenting as parasomnias. *Sleep,* 20, 402-405, 1997.
81. MOLDOFSKY H., GILBERT R., LUE F.A., MCLEAN A.W. – Sleep-related violence. *Sleep,* 18, 731-739, 1995.
82. MONTAGNA P. – Nocturnal paroxysmal dystonia and nocturnal wandering. *Neurology,* 42, 61-67, 1992.
83. MORRISSON A.R. – Mechanism underlying oneiric behaviour released in REM sleep by pontine lesions in cats. *J. Sleep Res.* 2, 4-7, 1993.
84. NIELSEN T.A., ZADRA A. - Dreaming disorders *In* M.H.Kryger, T. Roth, W.C. Dement (eds), *Principles and Practice of Sleep Medicine.* 3rd edition, W.B. Saunders, Philadelphia 66, 753-772, 2000.
85. OHAYON M.M., GUILLEMINAULT C., PRIEST R.G. – Night terrors, sleep-walking and confusional arousals in the general population: their frequency and relationship to other sleep and mental disorders. *J. Clin. Psychiatry* 60, 4, 268-276, 1999.
86. OLSON E.J., BOEVE B.F., SILBER M.M. – Rapid eye movement sleep behaviour disorder: demographic, clinical and laboratory findings in 93 cases. *Brain,* 123, 331-339, 2000.
87. PIERCE C.J., CHRISMAN K., BENNETT M.E., CLOSE J.M. – Stress, anticipatory stress, and psychologic measures related to sleep bruxism. *Journal of Orofacial Pain,* 9, 51-56, 1995.

88. PLAZZI G., TINUPER P., MONTAGNA P., PROVINI F., LUGARESI E. – Epileptic nocturnal wanderings. *Sleep*, 18, 749-756, 1995.

89. PLOUIN P. – Les convulsions néonatales idiopathiques bénignes, familiales ou non. In: J. Roger, M. Bureau, Ch. Dravet, F.E. Dreifuss, A. Perret, P. Wolf (eds), *Les syndromes épileptiques de l'enfant et de l'adolescent*, 2nd edition, John Libbey, London, Paris, Rome, 3-12, 1992.

90. PRESSMAN M.R., MEYER T.J., KENDRICK-MOHAMED J., FIGUEROA W.G., GREENSPON L.W., PETERSON D.D. – Night terrors in an adult precipitated by sleep apnea. *Sleep*, 18, 773-775, 1995.

91. REGESTEIN Q.R., HARTMANN E., REICH P. – A head movement disorder occurring in dreaming sleep. *J. Nerv. Ment. Dis.* 164, 432-436, 1977.

92. REIMAO R. – les parasomnies. In: M. Billiard (ed), *Le sommeil normal et pathologique*, Masson , Paris 346-357, 1994.

93. ROSS R.J., BALL W.A., DINGES D.F., KRIBBS N.B., MORISSON A.R., SILVER S.M., MULVANEY F.D. – Motor dysfunction during sleep in posttraumatic stress disorder. *Sleep*, 17, 723-732, 1994.

94. RYE D., - Contributions of the pedonculopontine region to normal and altered REM sleep. *Sleep*, 20, 9, 757-788, 1997.

95. SAKAI K., SASTRE J.P., SLAVERT D., TOURET M., TOHYAMA M., JOUVET M. – Tegmentoreticular projection with special reference to the muscular atonia during paradoxical sleep in the cat. *Brain Res.*, 176, 233-253, 1979.

96. SANFORD L.D., MORRISON A.R., MANN G.L., HARRIS J.S., YOO L., ROSS R.J. – Sleep patterning and behaviour in cats with pontine lesions creating REM without atonia. *J. Sleep Res.*, 3, 233- 240, 1994.

97. SASTRE J.P., JOUVET M. – Le comportement onirique du chat. *Physiol.Behav.* 22, 233-240, 1994.

98. SCHENCK C.H., BOYD J.L., MAHOWALD M.W. – A parasomnia overlap disorder involving sleepwalking, sleep terrors and REM sleep behaviour disorder in 33 polysomnographically-confirmed cases. *Sleep*, 11, 972-981, 1997.

99. SCHENCK C.H., BUNDLIE S.R., ETTINGER M.G., MAHOWALD M.W. – Chronic behavioral disorders of human REM sleep: a new category of parasomnia. *Sleep*, 9, 293-308, 1986.

100. SCHENCK C.H., BUNDLIE S.R., MAHOWALD M.W. – Delayed emergence of a parkinsonian disorder in 38% of 29 older men initially diagnosed with idiopathic rapid eye movement sleep behaviour disorder. *Neurology*, 46, 388-393, 1996.

101. SCHENCK C.H., HURWITZ T.D., MAHOWALD M.W. – REM sleep behaviour disorder: an update on a series of 96 patients and a review of the world literature. *J. Sleep Res.* 2, 224-231, 1993.

102. SCHENCK C.H., HURWITZ T.D., O'CONNOR K.A., MAHOWALD M.W. – Additional categories of sleep-related eating disorders and the current status of treatment. *Sleep*, 16, 457-466, 1993.

103. SCHENCK C.H., MAHOWALD M.W. – Motor dyscontrol in narcolepsy: rapid-eye-movement (REM) sleep without atonia and REM sleep behaviour disorder. *Ann. Neurol.* 32, 3-10, 1992.

104. SCHENCK C.H., MAHOWALD M.W. – Two cases of premenstrual sleep terrors and injurious sleep-walking. *Journal of Psychosomatics Obstetrics and Gynaecology*. 16,79-84, 1995.

105. SCHENCK C.H., MAHOWALD M.W. – A polysomnographically documented case of adult somnambulim with long-distance automobile driving and frequent nocturnal violence: parasomnia with continuing danger as a noninsane automatism? *Sleep* 18, 765-772, 1995.

106. SCHENCK C.H., MAHOWALD M.W. – REM sleep parasomnias. *Neurologic Clinics*, 14, 697-720, 1996.

107. SCHENCK C.H., MAHOWALD M.W. – Long-term, nightly benzodiazepine treatment of injurious parasomnias and other disorders of disrupted nocturnal sleep in 170 adults. *Am. J. Med.*, 100, 333-337 , 1996.

108. SCHENCK C.H., MILNER D.M., HURWITZ T.D., BUNDLIE S.R., MAHOWALD M.W. - Dissociative disorders presenting as somnambulism: polysomnographic, video and clinical documentation. *Dissociation*, II, 4, 194-204, 1989.

109. SELBST S.M., BAKER M.D., SHAMES M. – Bunk bed injuries. *Am. J. Dis. Child.*, 144, 721-723, 1990.

110. SEPTIEN L., DIDI ROY R., MARIN A., GIROUD M. – Troubles du comportement du sommeil paradoxal et atrophie olivo-ponto-cérébelleuse: un cas. *Neurophysiologie Clinique*, 22, 459-464, 1992.

111. SHELDON S.H., GARAY A., JACOBSEN J.H. – REM sleep motor disorder in children. *Sleep Research*, 23, 173, 1994.

112. SILVESTRI R., DE DOMENICO P., MENTO G., LAGANA A., DI PERRI R. – Epileptic seizures in subjects previously affected by disorders of arousal. *Neurophysiol. Clin.* 25, 19-27, 1995.

113. SURGAMAN L.I. – Hypnosis teaching children self-regulation. *Pediatrics in Review*, 17, 1, 5-11, 1996.

114. TERR L.C. – Nightmares in children. In: C. Guilleminault (ed), *Sleep and its Disorders in Children*, Raven Press, New York, 231-242, 1987.

115. THOMAS T.N. – Sleepwalking disorder and mens rea: a review and case report. *J. Forensic Sci.*, 42, 17-24, 1997.

116. THOMSON D.F., PIERCE D.R.- Drug-induced nightmares. *Ann. Pharmacother.* 33, 1, 93-98, 1999.

117. THORPY M.J. – Rhythmic movement disorder. *In:* M. Thorpy (ed), *Handbook of Sleep Disorders*, Marcel Dekker, New York, Basel, 27, 609-629, 1990.

118. TOBIN J.D. – Treatment of somnambulism with anticipatory awakening. *J. Pediatr.*, 122, 426-427, 1993.

119. TOHIER C., ROZE J.C., DAVID A., VECCHIERINI M.F., RENAUD P., MOUZARD A. – Hyperplexia or stiff baby syndrome (editorial). *Arch. Dis. Child.*, 66, 460-461, 1991.

120. TOUCHON J., BESSET A., BILLIARD M., CADILHAC J. – Parasomnie et epilepsie: problème de diagnostic. *Neurophysiol. Clin. (*abstract), 18, 478, 1988.

121. TOUCHON J., PAVY A., DE LUSTRAC G., BESSET A., BILLIARD M. – Parasomnia and epileptic activity. *In:* W.P. Koella, H. Ruther, H. Schulz (eds), *Sleep 84,* Gustav Fisher Verlag, Stuttgart, New York, 425-427, 1985.

122. TRACY K. – Awareness/relaxation training and transcutaneous electrical neural stimulation in the treatment of bruxism. *J. Oral Rehabil.,* 26, 280-287, 1999.
123. TUXHORN I., HOPPE M. – Parasomnia with rhythmic movements manifesting as nocturnal tongue biting. *Neuropediatrics,* 24, 167-168, 1993.
124. UCHIYAMA M., ISSE K., TANAKA K., YOKOTA N., HAMAMOTO M., AIDA S., ITO Y., YOSHIMURA M., OKAWA M.- Incidental Lewy body disease in a patient with REM sleep behavior disorder. *Neurology,* 45, 709-712, 1995.
125. UHDE T.W. – The anxiety disorders. *In:* M.H. Kryger, T. Roth, W.C. Dement. (eds), *Principles and Practice of Sleep Medicine,* 2nd edition, W.B. Saunders, Philadelphia, 871-898, 1994.
126. VECCHIERINI-BLINEAU M.F. – Le traitement des parasomnies. *Rev. Neurol.* 2001 (in press).
127. VINSON R.P., GELINAS-SORREL D.F. – Head banging in young children. *Am. Fam. Physician,* 43, 1625-1628, 1991.
128. WEIDEMAN C.L., BUSH D.L., YAN-GO F.L., CLARK G.T., GORNBEIN J.A. – The incidence of parasomnias in child bruxers versus non bruxers. *Pediatric Dentistry,* 18, 456-460, 1996.
129. WHYTE J., KAVEY N.B., GIDRO-FRANK S. – A self destructive variant of jactatio capitis nocturna. *J. Nerv. Ment. Dis.* 179, 49-50, 1991.
130. WING Y.K., LEE S.T., CHEN C.N. – Sleep paralysis in Chinese: ghost oppression phenomenon in Hong Kong. *Sleep,* 17, 609-613, 1994.
131. WOOD J., BOOTZIN R. – The prevalence of nightmares and their independance from anxiety. *J. Abnor. Psychol.* 99, 64-68, 1990.
132. WRIGHT E.F. – Using soft splints in your dental practice. *Gen. Dent.* Sept-Oct, 506-512, 1999.
133. YUSTIN D., NEFF P., RIEGER M.R., HURST T. – Characterization of 86 bruxing patients with long-term study of their management with occlusal devices and other forms of therapy. *J. Orofacial Pain,* 7, 54-60, 1993.
134. ZUCCONI M., OLDANI A., FERINI-STRAMBI L., SMIRNE S. – Arousal fluctuations in non-rapid eye movement parasomnias: the role of cyclic alternating pattern as a measure of sleep instability. *J. Clin. Neurophysiol.,* 12, 147-154,1995.

Chapter 42

Enuresis

M. Averous
Service d'Urologie, Hôpital Lapeyronie, Montpellier, France

It may be as old as mankind, but enuresis is still a topical issue. Aristotle may already have recorded the fact that children lose urine during sleep, in the era of robots and the demystification of all our taboos, man's offspring still loses urine during the night... to the great dismay of his/her parents and those who are supposed to offer a solution, and notably doctors. We will focus our attention on bedwetting during the night ... and sometimes during napping. The term "primary enuresis" refers to cases where the disorder develops without the slightest stage of continence. Secondary enuresis, on the other hand, is referred to when there is a period of several months of being "dry". In this case, it is important to look for an affective or social factor which might have triggered the situation.

Fifteen percent of children are enuretic after the age of 5 years and each year 10% of these will spontaneously cure themselves, up until puberty. The army recognises 1% of enuretics among its conscripts.

In fact, it is more appropriate to speak in terms of enuresis in its several childhood forms, rather than simple enuresis. Enuresis in the case of a little 7 year old boy, differs from that of the 10 year old girl who still wets the bed, presenting slight urine leakage during the day, or that of the adolescent who starts urinating in bed in response to family problems. The common denominator is the symptom of enuresis to which several physiopathological components are linked.

We will present our understanding of the enuresis syndrome [6], which is subject to a number more or less inter-related factors. Recognising the various factors by interrogating the child and his/her family often proves a better approach to the situation, thus resulting in better therapeutic treatment. Indeed, enuresis must no longer be considered as an inevitability which will eventually disappear of its own accord, nor can we hide behind the major, insurmountable alibi which consists of saying "It's psychological" and which is often synonymous with the powerlessness or abdication of treatment.

THE STAGES TO BECOMING CONTINENT

The study of the control of miction and bowel movements has been greatly assisted by the use of urodynamic explorations. Three stages have been isolated [1, 3, 4, 9, 10].

The infantile, automatic or reflex stage

The bladder of the newborn and infant in the first year of life is a purely reflex organ in which receptors sensitive to parietal distension induce a contraction of the detrusor at very low volume from the medullar miction centre (S2 S3 S4). The infant does not appear to perceive the sensation of a full bladder; he can neither initiate nor stop micturating. The voluntary central and cortical modulating influences are totally absent at this age. The urodynamic picture of this infantile bladder is characterised by a cystometric trace of hyperactivity in which each contraction reaching the critical threshold induces miction in a context of perfect bladder-sphincter synergy, with silent mictional electromyography (fig. 42.1).

The immature bladder

The first stage in gaining control of miction usually occurs between the ages of one and two years and relies on becoming aware of the bladder filling and thus of the urge to urinate, and on the child's ability to realise that he can avoid a urine leak if, at the moment he gets the urge, he voluntarily contracts the pelvic floor, and thereby his striated sphincter, setting up a detrusor inhibition reflex. This acquisition gradually leads to the child increasing his bladder capacity to become continent and clean in the daytime. This physiological phase of acquisition involves *imperiosity* at peaks of hyperpressure, *pollakiuria* if the child decides to urinate or cannot resist the urge, and sometimes a *leak of urine* if bladder pressure is greater than the sphincter muscle's capacity of resistance (particularly in situations where sphincter vigilance is low: play, laughter...). But at night when sleep eliminates the capacity to voluntarily control the perineal floor, bladder hyperactivity will cause the bladder to empty: this is sleep enuresis. It explains why continence is first acquired during the daytime even if leakage persists... longer. The urodynamic picture of the "immature bladder" (fig. 42.2) is characterised by a tracing of bladder hyperactivity in which not every detrusor contraction necessarily leads to miction, to the extent that the child is able to reinforce the activity of his perineal floor. But once the command is given to urinate, miction continues to occur in perfect bladder-sphincter synergy [2, 7, 12].

The adult bladder

In time, the inhibiting effect on the bladder-sphincter apparatus, originating from the high level centre, leading to the increase in bladder volume and a reduction of bladder hyperactivity. Continence is ensured as much by central inhibition as by the voluntary action on the striated sphincter of the urethra. Miction then always occurs in complete bladder-sphincter synergy with a cystometric tracing devoid of any hyperactivity accidents (fig. 42.3).

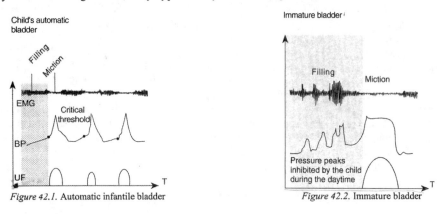

Figure 42.1. Automatic infantile bladder   Figure 42.2. Immature bladder

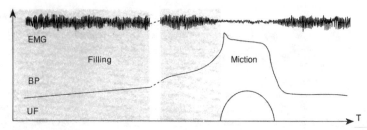

Figure 42.3. Adult bladder. EMG: Electromyographic activity of striated sphincter, BP: Bladder pressure, UF: Urine flow, T: Time.

THE PRINCIPLE FACTORS IN ENURESIS

The genetic factor

Heredity is a determining factor. Indeed, the disorder affects 77% of children where both parents were enuretic. The proportion drops to 44% if only one parent has a past history of enuresis and 15% in the case of neither parent [15, 19]. A gene has been indicated for enuresis on the short arm of chromosome 13 [13].

"Psychological" factors and family context

These are always present, often becoming evident from the first contact with the patient. It is important to avoid relying on an exclusively psychological explanation. Other factors may be associated.

- Trigger factors, which are often emotional, must be recognised as being responsible for certain secondary cases of enuresis: anxiety related to school, separation, overwork, mourning or family conflict.

- In the same way, there are often consistent secondary psychological consequences of enuresis. The feelings of shame, rejection and emotional withdrawal are well known in enuretics, as is the often radical effect of encouragement after the first positive results.

The bladder factor: bladder immaturity

The bladder of the newborn is automatic, hyperactive, uninhibited and responsible for reflex as opposed to controlled miction [4, 9]. The adult bladder or that of the continent child is perfectly controlled by the inhibiting intervention of the high level centres.

Between these two states, there is an intermediary state through which the child passes more or less happily, situated between the ages of two and...puberty. This is the phase of the physiologically immature bladder. When it goes on too long, the child is prone to a number of problems, responding to the immature bladder syndrome, some of the consequences of which may become truly pathological, such as lower urinary tract infection in young girls or certain vesicorenal refluxes [2, 4, 17]. Urodynamic data, particularly those of cystomanometry, have taught us that the common denominator is bladder hyperactivity i.e. a pressure increase which may exceed 100 cm of H_2O in the daytime and 300cm of H_2O at night, whereas the normal bladder is full at a pressure of roughly 10 to 30 cm of H_2O.

Basing our study on several thousand recordings [2, 4, 14], we have been able to establish a very close correlation between urodynamic exploration and the elementary clinical interrogation of the child and his family.

- **The typical form** of immature bladder syndrome associates sleep enuresis with daytime disorders: imperiosity, pollakiuria, minor urine leakage responsible for "wet pants".

- **The urine retention form** is less common [2, 4] After too much urine retention, the child may develop striated sphincter hypertonia, and become retentionist. The bladder increases its capacity, only emptying once or twice every 24 hours, sometimes incompletely, leaving a residue which is a source of many lower urinary tract infections, especially in girls.

We have dwelt heavily on the daytime signs and their physiopathological significance, in order to assist in searching, recognising and integrating these within the framework of certain sleep enuresis.

The sleep factor

Sleep enuresis occurs at night, and in keeping with its name, many parents complain of their children sleeping so soundly that nothing will waken them, not even their urine leakage.

Taking into account the many studies reported in literature, notably those of Gastaut and Broughton [8], Mikkelsen and Rappoport [18], we made a study of the relationship between enuresis, sleep and bladder immaturity, based on two homogeneous groups of sleep enuretic children aged 7 to 17 years [5].

- Group I comprised 16 children affected by immature bladder disorder, with clinical features in the daytime.

- Group II comprised 14 children having no clinical features in the daytime.

The exploration of these subjects included daytime cystomanometry and sleep polysomnography as well as continuous bladder pressure measurement. An alarm was installed outside the room to determine the exact moment of leakage, in relation to sleep stages and cycles and to evaluate the bladder's manometric context.

This study has shown the following results:

1. Neither group showed anomalies of sleep architecture and organisation. Nevertheless, a globally irritative aspect was noted in EEG activity in 29 out of 30 cases. But it remains hard to discern whether this aspect is pathological.

2. In the group of children affected by clinical bladder immaturity, it came as no surprise to find a hyperactive bladder, during the daytime.

It was thus noted that all these children had increased bladder hyperactivity at night, sometimes to a considerable extent, with three children having values of over 250 cm of H_2O.

Hence during the sleep phase, there appears to be a lowering of nocturnal inhibition of overall bladder-sphincter behaviour which, the child controls better in the daytime, probably because of a better control of the detrusor inhibition reflex which does not work when the child is asleep. The phenomenon may recur several times a night resulting in several enuretic episodes (fig. 42.4).

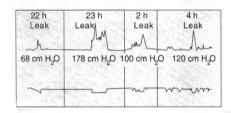

Figure 42.4. Four episodes of hyperactivity during the nigh in an immature bladder. This child presents four episodes of nocturnal leakage in the course of a night.

3. Conversely we were very surprised to note manifestations of nocturnal bladder hyperactivity in 12 of the 14 children in group II (with no daytime signs of immaturity, with apparently isolated enuresis). These are lower in amplitude than those noted for children in the first group, rarely exceeding 60 cm of H_2O, but are virtually consistent. During the daytime these children have bladder manometric tracings devoid of any hyperactivity. This behaviour probably testifies to residual bladder immaturity, well compensated for in the daytime, but still free at night, due to lack of inhibition.

All this suggests that a substantial number of children have residual global immaturity, which is more or less active, both of the bladder-sphincter and perhaps in terms of sleep. The gradual acquisition of maturation would explain why enuresis disappears spontaneously each year in 10% of the children affected.

4. In all cases, the enuretic episode included:

a) a phase of initiation in which bladder pressure increased,

b) a phase of leakage per se,

The initiation phase often starts during the first part of the night, sometimes quickly, a few minutes after the onset of sleep, with a sudden or gradual increase in bladder pressure occurring eight times out of ten during NREM sleep (stages 3 and 4).

During this period of NREM sleep, the child is clearly very vulnerable. This would moreover account for leakage occurring during daytime naps, in certain children who quickly sink into deep NREM sleep.

Uninhibited bladder hyperpressure occurring during deep NREM sleep, sets off the control reflex which results in the child having an awakening reaction. This successively mounts the stages of sleep as he emerges to the wakefulness state. There is a veritable race between increasing bladder pressure and the emergence from sleep, allowing the inhibitory reflexes to become more effective and adequate to protect the child from leakage.

Two situations may arise.

1. If the child quickly reaches a state of lighter sleep, he can:
- either effectively inhibit his bladder hyperactivity and continue sleeping with no leakage;
- or wake up to urinate, becoming pollakiuric rather than enuretic.

2. If the child does not manage to lighten his sleep enough, inhibition remains inadequate and the leakage phase follows, provoked by uncontrolled bladder hyperpressure (which, as we have seen, may reach 300 cm of H_2O.) As for the moment when leakage occurs, this varies, occurring at any point during the emergence from sleep, thus explaining its polymorphic nature when described in literature, whereas the initiation of this leakage occurs during a stage of deep NREM sleep.

The sleep factor comprises three important notions

1. The depth of sleep itself.
2. Bladder hyperpressure which was seen to be linked to manifest or latent bladder immaturity and which can itself be aggravated by deep sleep.
3. The threshold of awakening, which represents a fundamental given on which the setting up of the inhibition reflex appears to depend. This is unfortunately difficult to ascertain and varies according to the multiple factors.

Psychological factors may affect this awakening threshold. It is common to find situations in which daily circumstances disrupt sleep and alter the threshold of awakening, and as a result, bladder-sphincter control (physical fatigue, tiredness at school, family problems etc.).

The hormonal factor

The question arose as to whether the sleep enuretic secretes urine because of overproduction during the night, thus exceeding the bladder's functional capacity. This has led some authors to refer to an impairment of nocturnal DHA secretion in certain enuretics [11, 16].

The hormonal factor responsible for nocturnal polyuria is perhaps decisive in certain isolated sleep enuretics. This may well be an aggravating factor but one which fails to account for leakage which occurs very early, sometimes only minutes after the onset of sleep.

Other factors

a) Urethral instability [15].

This refers to the sudden drop in urethral tone which relies on the tonic action of the striated sphincter muscle, and which, with no significant increase in bladder pressure, can result in urine leakage. This results from α stimulant treatment.

b) Encopresis [15].

Bowel movement control usually precedes the acquisition of miction control. Most encopretics are enuretic.

c) Constipation. This is very common in enuresis and must always be treated in parallel.
Intestinal parasitosis. Lamblia and oxyuriasis.

d) Certain abnormal psychomotor behaviour.

e) and no doubt many others…

TREATING THE ENURETIC CHILD

Bedwetting is very common and is not an illness. But the enuretic symptom, i.e. the involuntary loss of urine is one of the clinical manifestations of urological and neurological pathology.

So, it is indispensable to run a basic clinical test to make sure there is no organicity? This includes an interrogatory search for mictional disorders associated in particular, with dysuria, permanent loss of urine and a history of urinary infection. There must be a full uro-genital examination with observation of miction. The neuro-urological test must eliminate any neuropathic

550 Sleep: Physiology, Investigations, and Medicine

bladder condition: checking for an abolition of the bone tendon reflexes of the lower limbs, cavus foot, cutaneous anomalies along the median line in relation to the lumbosacral column (indentation, pilosity, spina bifida) which might evoke lesion of the medullary cone. This overall aim of examination is to formally eliminate any uropathy due to malformation or any neurological lesion involving enuresis. If doubts persist, one must not hesitate to supplement clinical investigation with specialised diagnostic tests (echography, IVU, cystography, urodynamic explorations, or even MRI).

Non specific therapeutic support

Treatment is extremely varied. But all forms of treatment rely on establishing a relationship of confidence between the therapist and the child [16]. The child must be made aware and encouraged to co-operate through:
- trying to explain the syndrome of enuresis, which should be placed within the context of the person requesting help and the family circle.
- the child keeping a daily diary, noting down enuretic episodes, daytime disorders if any exist, as well as any improvements. Parents may be advised to give rewards from the first stages of progress.
- limiting the intake of drinks after 6 pm, and advising the child to empty his bladder, just before going to sleep.
- banishing diapers even if they are seen to be useful.
- getting the child up at night, which is sometimes suggested for those whose leakage occurs at "fixed times". But this method has been criticised as altering the quality of sleep and disrupting the family atmosphere.
- acupuncture, auriculotherapy and homeopathy are sometimes used with encouraging results, but no sufficiently accurate study has been published to date.

Specific therapeutic support

These forms are, by definition, directed at the various dominant factors, which have been identified in the enuretic.

Psychological factors: Above all, it should be recalled that setting up any treatment for an enuretic is in itself a certain form of psychotherapy. However, when the psychological component appears to be serious and (or) overriding, one must not hesitate to refer the child to a psychotherapist.

The immature bladder factor relies on two possible forms of treatment: pharmacological treatment, and mictional training or retraining.

Pharmacological treatment

Bladder hyperactivity linked to detrusor contraction depends on acetylcholine. It can thus be reduced by prescribing anticholinergic drugs.
The "relaxing" effect of anticholinergics on the detrusor will result in diminishing the amplitude of uninhibited bladder contractions, reducing imperiosities and leaks during the day (the child increases his bladder capacity) and at night, aiding the bladder to shift to the stage of sleep without urine leakage.
These anticholinergics are essentially represented by oxybutynine (*Ditropan*).
- The usual doses prescribed range from 5 mg to 15 mg/day, regularly spread over a 24 hour period to "calm" the bladder and increase its functional capacity.
- Treatment varies in duration and is carried out in sequences of several months, with several therapeutic windows to review the situation.
Some hyperactive bladders do however resist the anticholinergics. The failure is due to:
- the product's very short half life, only protecting the child at the beginning of the night. Despite the impression of failure, the treatment should be continued, all the more so if there has been an

improvement in mictional disorders during the day. It is reasonable in the cases of enuresis at the end of the night, to associate desmopressine treatment which, by reducing nocturnal dioresis, supplements the protection ensured at the start of the night by the anticholinergics;

- considerable intravesical pressure (over 300 cm of H_2O) which the anticholinergics are not sufficient to reduce.

However it is important to avoid prescribing anticholinergics to a child with a marked tendency for retention. They would only serve to increase this with the risk of mictional disorders and urinary infection.

Mictional training and retraining

The simple, common sense approach is referred to as simple mictional training:
For the pollakiural forms, the subject is reasonably asked to resist the urge as much as possible, and to space out mictions, leading to an increase in bladder capacity.
For retentive forms, the subject is asked to increase the frequency of miction and to keep a regular diary of miction.

Retraining per se relies on biofeedback processes which help the child to become aware of detrusor contractions and of the tonic action of the perineal floor, notably of his urethral sphincter.

Retraining of this kind relies largely on the detrusor inhibition reflex, whereby any reinforcement of sphincter tone, whether voluntary or induced by electrostimulation, will lead to a reduction in detrusor activity. Also, by becoming aware of his striated sphincter the child is better able to "manage" his bladder in the daytime, thus improving nocturnal instability.

The sleep factor

This factor is far more sensitive. It appears that the child's sleep need only be lightened or the awakening threshold affected, for bladder stimulation to enable the child either to wake up or to trigger a detrusor inhibition reflex in time [9, 20]. But altering a child's sleep is a serious matter, potentially
Exposing him to disturbances which involve unreasonable risk.

Pharmacological treatment

This relies on the tricyclic antidepressants i.e. imipramine. Their efficacy is indisputable, but may include a number of escapes during and at the end of treatment. The mode of action involved however, is disputable. While there is no apparent doubt about their central effect on sleep, their efficacy is certainly enhanced by their peripheral anticholinergic properties [16].

However, this medication is dangerous and can even be fatal, after an overdose. Needless to say there is resistance to prescribing medication of this sort for a functional symptom which will eventually be cured anyway, in 99% of cases!

Alarms

If the child agrees, a conditioning method "pee pee stop" aims at modifying bladder-sphincter behaviour during sleep. When a leak of urine sets off the alarm, the child wakes up, thus activating the detrusor inhibitor reflex, gets up, finishes miction, changes his sheets and goes back to bed, having reset the alarm. Little by little, conditioning leads to better control. The method has a considerable rate of success: up to 66% for MacKendry, referred to by Robert [19].

The hormonal factor

This factor is responsible for nocturnal polyuria due to impaired DHA secretion and responds to desmopressine treatment administered nasally at evening bedtime at doses of 20 to 40µgv. Desmopressine reduces the amount of urine produced during the night and adds potential to anticholinergic action, resulting in the increased functional capacity of the bladder.

CONCLUSION

1) Treating enuretic children involves a dual approach [6], consisting of:
The formal elimination of any organic cause.

The clinical examination must be very thorough, including a neurological evaluation of the lower limbs and the perineum. It is always helpful to be present during a miction, if possible.

Cases do not normally call for diagnostic procedures, other than to search for a urinary or proteinuric infection, i.e. a simple vesicorenal CT scan. On the other hand, if an abnormal symptom does appear, notably a urinary infection (with the well-known chance of this occurring in the immature bladder context) other investigations, e.g. imaging, may be necessary.

2) Interrogation to assess the enuretic's specific situation, as well as his relationships with the socio-familial environment, through accurate, directive, almost detective-like questioning, while maintaining a kindly, attentive attitude. It is vital to establish confidence and to propose a veritable contract whereby the child feels the therapist is an ally who can understand and help him find a cure. Searching for the various factors will uncover the overriding ones. It thus becomes easier to suggest a course of treatment, associating one or a number of the therapeutic modalities we have listed. On the other hand, the question arises as to whether it is worth wasting time with a child who neither accepts a minimum of treatment nor wishes to be cured.

Finally, it must never be forgotten that time is also an ally and that every year, 10% of enuretic children recover spontaneously.

REFERENCES

1.   ARCHIMBAUD J.P. – Les dysfonctionnements vésico-sphintériens neurologiques. *Rapport du 68ème Congrès de l'Association Française d'Urologie* Paris, 1974.
2.   AVEROUS M. – Le syndrome d'immaturité vésicale. A propos de 1097 observations. *J. Urol.* (Paris), 91, 257-267, 1985.
3.   AVEROUS M. – Enurésies de l'enfant. *Encycl. Med. Chir., Paris.* Néphrologie-Urologie-Pédiatrie, 18207 E10 et 4085 C10, 8-1992.
4.   AVEROUS M. - *Guide pratique de l'énurésie*, SDI, Paris, 1992.
5.   AVEROUS M., ROBERT M., BILLIARD M., GUITER J., GRASSET D. - Le contrôle mictionnel au cours du sommeil. *Rev. Prat.* (Paris), 41, 2282-2287, 1991.
6.   AVEROUS M., LOPEZ C. – Bilan de 20 ans de réflexion sur l'énurésie de l'enfant. *Progrès en Urologie*, 7, 476-483, 1997.
7.   BLAIVAS J.G., LABIB K.L., BAUERS B., RETIK A.B. – Changing concepts in the urodynamic evaluation of children. *J. Urol.*, 117, 778-781, 1977.
8.   BROUGHTON R.J., GASTAUT H. – Etude polygraphique de l'énurésie nocturne. *Rev. Neurol.*, 16, 246-247, 1963.
9.   BUZELIN J.M. – *Urodynamique. Bas appareil urinaire*. Masson, Paris, 1984.
10. BUZELIN J.M., LE COGNIC C., ETIENNE Ph. – *Apport de l'urodynamique dans la pathologie fonctionnelle du bas appareil urinaire de l'enfant*. Congrès de la Société Internationale d'Urologie, Paris, 14-19 juin 1979.
11. COCHAT P. – L'énurésie et les troubles mictionnels communs de l'enfance. *Pédiatrie*, 44, 523-530, 1989.
12. Von GARRELTS B. – Micturition in the normal male. *Acta Chir. Scand.*, 114, 197-210, 1957.
13. Von GONTARD A., EIBERG H., HOLLMANN E., RITTIG S., LEHMKUHL G. - Molecular genetics of nocturnal enuresis and genetic heterogeneity. *Acta Paediatr.* 87, 487-488, 1998.
14. GRASSET D. – *La cysto-sphinctérométrie: exploration dynamique de l'appareil vésico-sphinctérien,* Masson, Paris, 1961.
15. KOFF S.A. - *Enuresis*, In: Campbell's Urology, 6th edition. W.B. Saunders, Philadelphia, 1621-1633, 1992.
16. LENOIR G., TURBERG-ROMAIN C. – Les traitements de l'énurésie. *Entretiens de Bichat*. Expansion Scientifique Française, Paris, 151-158, 1986.
17. LYON R.P., SMITH D.R. – Distal urethral stenosis. *J. Urol.*, 89, 414, 1963.
18. MIKKELSEN E.J., RAPOPORT J.L. – Enuresis: psychopathology, sleep stage and drug response. *Urol. Clin. North Am.*, 7, 361-377, 1980.
19. ROBERT M. – *Sommeil et énurésie: Intérêt de l'étude simultanée du sommeil et du comportement manométrique de la vessie. A propos de 20 cas*. Thesis, Montpellier, 1989
20. TURNER-WARWICK R. – Observations on the function and dysfunction of the sphincter and detrusor mechanisms. *Urol. Clin. North Am.*, 6, n°1, 1979.

PART 4

MEDICAL DISORDERS ASSOCIATED WITH
SLEEP OR WORSENED DURING SLEEP

Chapter 43

Snoring

L. Crampette
Service ORL et Chirurgie Cervico-faciale, Hôpital Gui de Chauliac, Montpellier, France

This chapter chiefly concerns simple snoring (SS). Nevertheless other forms of snoring will be evoked in relation to diagnosis, *i.e.* snoring related to obstructive sleep apnoea hypopnea syndrome (OSAHS) or upper airway resistance syndrome (UARS) as it is essential to distinguish between simple snoring and snoring related to breathing abnormalities, before starting treatment.

In the section devoted to treatment, only SS treatment will be envisaged, as surgical treatment for OSAHS or UARS are dealt with elsewhere. (see chapter 30).

HISTORY

From the time of antiquity until recent times snoring was evoked as a sign of deep, restorative sleep. Its possible link with pathological sleep has only been known for a matter of years, since the discovery of sleep apnoea syndrome. Interest has focused more recently on the existence of non apnoea respiratory abnormalities (hypopnoeas, periodic breathing) in some snorers, or an increase in upper airway resistance. The term rhonchopathy is currently used to refer to simple snoring (SS), sleep apnoea syndrome (OSAHS) and a third entity known as upper airway resistance syndrome (UARS) [18, 31].

In terms of treatment, the first attempts at surgery can be attributed to Ikematsu in 1952 [21]. It should however be stressed that noise was the motive prompting Ikematsu to practise a partial resection of the mucous membrane of the velum and pharynx. From 1981 onwards, several American teams began proposing surgical treatment for SAS. These notably include Fujita *et al.* in Detroit [16], who "invented" the term uvulopalatopharyngoplasty (UPPP), Hernandez in Miami [19] and Blair-Simmons *et al.* in Stanford [3]. In 1983, the Montpellier school began practising this form of surgery [11, 15], which was later extended to other parts of France, through the work of Claude-Henri Chouard *et al.* [7, 8] after the world congress on ENT in Miami, in 1986.

EPIDEMIOLOGY

Snoring is an extremely common symptom. In 1980, Lugaresi *et al.* [24] carried out a survey of 5,713 inhabitants of the San Marino Republic and found that snoring was a common occurrence in 19% of the subjects interviewed, affecting 24% of men and 14% of women. Gislason *et al.* [17] obtained similar percentages, after studying a population of 4,064 inhabitants in Uppsala aged 39 to 69, distinguishing between regular snoring (incidence of 15.5%) and occasional snoring (incidence of 29.6%). Some authors have come up with different percentages, no doubt due to a bias in recruitment or epidemiological inquiry criteria; hence, for example, Kwan *et al.* [23] found 57% of snorers among 1,381 Canadian workers between the ages of 20 and 65. 51% of these snorers caused annoyance to their sleeping partners and 7%, to persons sleeping in the adjacent room.

It is a well-known fact that snoring increases with age. In a series reported by Lugaresi *et al.* [24], 60% of men and 40% of women aged between 60 and 65, were regular snorers. However, snoring also affects the young: in a population of young men (58,162 military draftees in selection centers in Tarascon and Vincennes), Billiard *et al.* [2] found 13.6% of snorers.

Many other factors influence snoring: overweight, alcohol, hypnotic drugs, fatigue, tobacco and any inflammation of the upper airways.

What proportion of snorers are affected by OSAHS or UARS? There is no precise answer to this question but it can be estimated in view of the prevalence of OSAHS (1 to 3% of the general population), that one every fifteen to twenty snorers is apnoeic. Data is lacking, however, on the prevalence of UARS among the snoring population.

PATHOPHYSIOLOGY

Origins of snoring

Snoring is generated by the inspiratory vibration of the soft parts of the oropharyngeal walls. For a pharyngeal element to vibrate, its compliance and the speed of the inspiratory air in contact with it must be sufficient [12, 13, 14]. A structure's compliance depends on its morphology and trophicity; with the same trophicity, a large soft palate will vibrate more easily, as will the fatty infiltration of an organ of the same size, or conversely, its thinness or hypotonia will also facilitate vibration. The speed of the inspiratory air is conditioned by Poiseuille's law, whereby at a constant rate, the speed of a fluid is inversely proportional to the section of its vector.

Pathophysiology shows that the velopharyngeal sector is where most snoring originates, because the soft palate is such a compliant organ and because anatomical narrowing, often observed at this point leads to an acceleration in the rate of inspiratory air. But other structures may also be affected by the preceding criteria: the base of the tongue or the vestibule of the larynx may also begin to vibrate. If no sound can be emitted by the nose, nasal dyspermeability is a contributing factor of rhonchopathy by accelerating the nasal airflow [9], enlarging the inspiratory pharyngeal depression, or compelling oral respiration [26, 27]; breathing through the mouth is a predisposing factor of snoring as it leads to the backward movement of the base of the tongue.

Repercussions of snoring

Local repercussions: to what extent does snoring modifiy the oropharyngeal structures? The enlargement of the uvula, for example, is as much the result as the cause of snoring. Moreover, the inspiratory effort during snoring may favour a gastro-oesophageal reflux which, in turn, may cause pharyngeal inflammatory lesions which again, facilitate snoring.

General repercussions: in 1980 [24] systemic hypertension was discovered to be more frequent in snorers (independent of apnoeas) than in control groups.

INTERVIEW

The aim of the interview is above all to distinguish between simple snoring, and apnoea-related or non apnoea-related breathing abnormalities. In practise, the advisability of a sleep recording is decided on the basis of the interview, thus indicating the importance of carrying out a detailed and thorough investigation. It should be based on a search for nocturnal and diurnal features, and for any associated pathological history, which should be included in the indications for treatment [12, 14].

Nocturnal signs

Mode of sleep onset

Very rapid sleep onset and the inability to stay up at night are suspicious indications of pathological sleep.

Nocturnal sleep

a) Snoring: how long has it been going on? How intense is it? Is it noticed by the partner only or by those in adjacent rooms? Is it intermittent or constant throughout the night? Occasional or every night? Is it affected by changes in body posture?

b) Is sleep calm or disturbed? Are there any breathing pauses? If so, an attempt must be made to determine their frequency, duration, mode of inspiratory uptake (a loud noise, start…), although those close to the subject are often vague about this. Other nocturnal manifestations may be reported such as repetitive leg movements.

c) Nocturnal arousals are important to analyse. Frequent arousals or nocturnal polyuria with no urological cause should be considered as possible signs of OSAHS.

Morning arousal

Asthenia on arousal, sometimes associated with headaches, the feeling of not having had a restful night's sleep are symptoms which strongly indicate nocturnal breathing abnormalities.

Daytime activity

The subject should be questioned about his professional activity, any abnormal episodes of fatigue during the day, particularly daytime sleepiness. Although these are non significant when occasional, after meals or in the evening in front of the television, they become so when they occur during a period of relative inactivity, seated in a waiting room, for example, or at a desk and of course, when driving a vehicle. Epworth's questionnaire provides an evaluation of daytime sleepiness. It is also important to obtain a general idea of the subject's physical, intellectual and sexual activity.

Previous history and habits

The height, weight, body mass index (weight/height^2) must be recorded.

Respiratory, cardio-vascular or neurological disorders must be carefully searched for.

A note must be made of alcohol and tobacco consumption, eating habits and any prescribed treatment (particularly tranquillisers and hypnotics which encourage snoring and sleep apnoeas).

At the outcome of the interview

Broadly speaking, three situations are encountered:

No serious feature is present. The patient falls into the SS category. Sleep recording is not required. A private study of 100 consecutive snorers, who were all recorded, showed that a full interrogation which perfectly coincided with SS, was not later refuted by polysomnography [12].

Likelihood of OSAHS. Polysomnographic recording is required, partly to confirm the diagnosis, as 20% of clinical diagnoses are estimated to be unconfirmed by polysomnographic recording, and partly to define the characteristics.

The findings of the interrogation remain vague or discordant, as in 25% of cases in the study referred to above, [12]. Screening processes are particularly valuable in cases of this kind.

ENT EXAMINATION

The severity of snoring and its respiratory repercussions cannot be ascertained on the basis of the degree of ENT abnormality, on an individual scale. Subjects whose pharynx is extremely narrow have been known to be affected by SS, while others, whose pharynx is normal may be affected by

OSAHS (2%, in our experience) [15]. Thus, the ENT examination, however important it may be, can in no way replace the interrogation or sleep exploration.

The ENT examination is nonetheless indispensable from the pathophysiological aspect (the anatomical structure responsible for snoring) and will influence treatment.

CLINICAL EXAMINATION

Prior to the ENT examination, the general morphotype should be recorded. The patient may either be:

- clearly overweight, with a relatively wide face, small mouth and short neck,

- or, more rarely, of a non-evocative morphotype: a slim or athletic subject, sportsperson, young woman.

The ENT examination will focus first on the upper oropharynx, commonly known as the velo pharynx, which is where most snoring occurs, and then on the lower part of the oropharynx also referred to as the retrobasilingual pharynx, and finally, on the rhinopharyngeal and laryngeal sectors [11, 14, 15].

Velopharynx

Separate examinations should be made of the lateral, tonsil, median and velar regions (fig. 43.1).

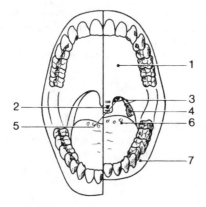

Figure 43.1. Pharyngeal morphotype of snorer (on the left, the control pharynx; on the right the snorer's pharynx)

      1. Abnormally long soft palate
      2. Enlarged uvula
      3. Soft palate pillars thick and abnormally close to the median line
      4. Tonsil enlargement
      5. Membranous folds at the junction of the uvula and pillars abnormally developed
      6. Macroglossia
      7. Retrognathia

The tonsil region: tonsil size must be determined, as well as the size of the superior intravelar poles, and that of the lower pole, which is often neglected, the appearance and position of the posterior pillars. Close pillars correspond to a narrow pharynx.

The soft palate: the length of the palate i.e. the distance between the posterior edge of the palate and the root of the uvula, which can be examined with the use of a tongue depressor. This examination should be carried out with the subject breathing freely, with a relaxed palate, then with phonation ("Ah"); in addition to the size, it is important to determine the trophicity of the soft palate, which is often thick and infiltrated with fatty tissue or oedema, or sometimes, on the contrary, thin and flaccid. Finally, the position of the soft palate in the buccopharyngeal cavity should be noted; some soft palates are of subnormal size, but in a retro-position, which is conducive to snoring.

The uvula is also very often abnormal, enlarged and may touch the epiglottis. It may be thick or thin and flaccid.

The membrane folds at the junction between the uvula and the pillars may be abnormally developed, with a "webbed uvula" commonly found in snorers.

Retrobasilingual oropharynx

This is more difficult to examine, requiring an indirect view, using a pharyngeal mirror, or a fiberscope as will be discussed in a later section.

The elements to be looked for include:
- Enlargement of the inferior pole of the palatine tonsils
- Enlargement of the lingual tonsils
- Enlargement of the muscle mass at the base of the tongue
- Retropositioning of the base of the tongue.

An assessment of dental articulation may point to tongue retropositioning due to retrognathia.

Larynx

The larynx is systematically checked as part of the examination of the base of the tongue. Nevertheless, even if redundant supraglottic mucosa in the vestibule of larynx, or a deflexion of the epiglottis are identified as causing a vibratory noise, snoring rarely originates in the larynx.

Nasal cavities – Cavum

The nasal examination will look for septum abnormalities, dysfunctioning of the nasal valve or a nasal-sinus inflammatory pathology.

In the case of septal abnormalities, it is important to stress that these are extremely common and many are asymptomatic or only inhibit the flow in one nasal cavity with no repercussions on the overall permeability of the nose. Consequently, the only septal abnormalities to be taken into consideration are those which are bilateral and correlated to a symptom of nasal obstruction. If doubt persists, a study of nasal respiratory resistance by rhinomanometry may be useful.

Nasal valve dysfunctions are often poorly understood. They lead to the inspiratory collapse of the nostrils. This diagnosis calls for an inspection under forced inspiration. In rhinomanometry, a normalisation of nasal respiratory resistance after opening the nostrils would indicate a pathology of this kind.

An enlarged inferior turbinate may result from surplus turbinate bones or a particular state of mucous membrane vasodilation (a veritable cavity plexus) in the nasal structure. Patients often describe a shifting nasal obstruction which becomes bilateral in decubitus. They often overuse vasoconstrictors.

Chronic sinusitis lies at the origin of a number of symptoms: nasal obstruction, rhinorrhea, olfactory disorders, headaches. However the clinical table may be virtually devoid of symptoms. This calls for the routine examination of the middle nasal meatus using rigid optics or a fiberscope: the appearance of suppuration or inflammation in the course of examination would then warrant referring the patient for a sinus scan to confirm diagnosis.

FIBEROPTIC INVESTIGATION

A fiberscope introduced through the nasal cavities and down the throat complete the investigation of the upper airways. Its main interest lies in evaluating the posterior pharyngeal wall in its natural position, without the tongue protraction required for mirror examination. The dynamic manoeuvres are then carried out: the patient is asked to speak, to say "K" to better evaluate velar excess, he is asked to snore to ascertain which pharyngeal structure vibrates [14]. The best known dynamic test is the Müller's manoeuvre which attempts to reproduce an apnoeaic pharyngeal collapse. This consists of asking the patient, once the fiberoptic nasalpharyngoscope has been introduced, to practise forced inhalation, with the nose pinched by the examiner, and the mouth closed. The glottis must stay open, thereby forming a relative pharyngeal depression. The fiberscope enables the examiner to observe the extent of collapse (nil, partial, total) its antero-posterior or transversal character, both at the level of the velopharynx and of the retrobasilingual oropharynx. The predictive value of Müller's manoeuvre is far from absolute, owing to false negatives (absence of collapse because of inept manoeuvre, particularly the closing of the glottis) and false positives (formation of a large pharyngeal depression, which is bigger than that formed in sleep).

RADIOGRAPHY

This is of no use for patients with SS. Indeed, any corrective information would only apply to the velar pharyngeal sector, which is carefully checked in the course of clinical examination. On the other hand, pharyngeal imagery is clearly of value in patients with OSAHS or UARS in providing an inventory of the retrobasilingual oropharynx. This is dealt with in detail in chapter 30.

LABORATORY TESTS DURING SLEEP

The tests are presented in chapters 9 and 10. A number of remarks should be added however, in regard to the choice of test. The standard reference in matters of sleep-related breathing disorders is polysomnographic recording, comprising an evaluation of sleep, breathing, electrocardiogram and leg movements. This is the most accurate way of testing for apnoeas and hypopnoeas, their type and repercussions for oxygen saturation and heart rate. However it does not measure the intensity of snoring, or non periodic sleep-related breathing abnormalities such as upper airway resistance syndrome. The advantage with systems like Mesam IV or Polymesam, is that they test snoring, monitored by a tracheal sound sensor *i.e.* duration, distribution throughout the night, and intensity. Oesophageal pressure recording is essential for investigating UARS.

CLINICAL VARIANTS

In the elderly

Snoring is frequent in subjects aged 60 and over. While it is far more frequent in young men than in young women, snoring occurs almost equally in both sexes in the elderly subject. It is more difficult to detect a sleep pathology through interrogation in the case of the elderly owing to the associated pathologies, reduced sleep, various medical treatments etc).

In children

Snoring is rare in young children and is liable to indicate a breathing difficulty which needs clarifying. Occasional snoring, especially during episodes of rhinopharyngitis (due to nasal obstruction, or enlarged adenoids) is no cause for concern. However, permanent snoring must be accurately tested as it is often accompanied by pathological sleep. The search should centre around:
    - nocturnal features: snoring, respiratory pauses, restless sleep, night sweating;
    - diurnal features: character disorders, difficulties at school.
Brouillette *et al.* [5] suggested clinical scoring for the different symptoms likely to lead to a diagnosis of sleep apnoea syndrome. The latter may lead to height-weight retardation and right ventricle insufficiency. The seriousness of the clinical picture contrasts with the simplicity and efficacy of treatment, which is generally limited to removing the tonsils and adenoids. More rarely, a malformation syndrome of the base of the skull, upper maxillary or of the mandible, are found to be responsible for OSAHS.

TREATMENT OF SIMPLE SNORING

Measures of hygiene and diet

Patients should first be advised to lose weight, avoid alcohol (notably in the evening) and discontinue hypnotic medication. It is sometimes necessary to suggest a change of lifestyle for patients who are overworked and/or chronically deprived of sleep. However these contributing factors are not always present ... nor is the patient always inclined to follow the advice.

Postural treatment

Developed by Pieyre [30], this consists of using conditioning to obtain sleep in a ventral or lateral position. If snoring is related to position, this method can be effective. Limits to postural treatment include obesity, arthritis, or the failure of one or both members of the couple to comply with treatment.

Prosthetic treatment

Two categories of anti-snoring device exist:
-those which deteriorate sleep and should be prohibited: systems triggered by a vibration sensor, stimulating the snorer with a flashing light, noise, electric charge etc.;
-those which do not alter sleep. These include silicon nasal appliances, which effectively treat nasal valve insufficiency [28]. Some systems are designed to obtain an anti-snoring position without using Pieyre's conditioning method, e.g. a simple rubber ball sewn into the back of the pyjamas or Ikematsu's pillow which gradually modifies the position of the head when a snoring sound is registered. Dental devices maintaining the mandible in an anterior position have been tried in patients with simple snoring of hypopharynx origin. These prosthetic devices may be tried but there is no guarantee of their success. They can also be uncomfortable which is why they may be rejected or abandoned by the patient.

Treatment for nasal obstruction

Nasal obstruction is treated in function of the causes:
- septoplasty
- reduction of the inferior turbinates, by cauterisation or applying radiofrequency, conducted in the course of a simple consultation, or using surgery (partial inferior turbinectomy)
- surgical treatment of a nasal valve abnormality
- medico-surgical treatment of chronic rhinosinusitis.

Uvulopalatopharyngoplasty (UPPP)

UPPP is now well codified. It should not be seen as simply the partial amputation of the soft palate but rather as a multiple aim operation: the soft palate may require shortening, thinning, or advancing, or the pharynx may need enlarging [3, 8, 11, 13, 14, 16, 22, 25, 29].. The technique and post operative period are marked by fairly intense pain comparable to that of acute tonsillitis (see chapter 30).

Laser treatment of the palate

The CO_2 laser is a treatment option for simple snoring [34, 35]. The laser is applied along two vertical lines from one side of the uvula, vaporising the anterior, the submucous tissue and part of the azygos muscles of the uvula (fig. 43.2). Some authors also vaporise the posterior mucous membrane, thus creating two complete "trenches" in the para-uvular region. Cicatrisation then forms a continuation of the mucous membrane and constitutes sub-mucous membrane fibrosis [10] which "rigidifies" the soft palate. Thus laser treatment for chronic rhonchopathy does not alter the dimensions of the soft palate but renders it less compliant. Laser treatment should not be applied to the lateral regions of the pharynx, due to the secondary risk of stenosis.

Laser treatment is carried out under local anaesthetic in outpatient conditions. However, it has the drawback of requiring several sessions at intervals of three to four weeks, and which, like UPPP can cause intense pain for several days.

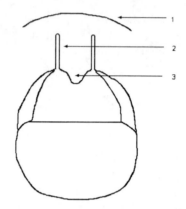

Figure 43.2. Laser application to the para-uvular region.
1.   Boundary between the hard palate and the soft palate
2.   Para-uvular "trenches"
3.   Uvula

Radiofrequency treatment of the palate

This is the most recent treatment available to patients with a simple snoring complaint; the first series being published in 1998 [32].

The effects of "radiosurgery" are produced by a mild current with low thermal tissue level (40° to 90°C). When radiofrequency is applied it sets up ionic agitation in the water molecules responsible for the tissue lesions in the volatilisation, section and above all coagulation of tissue proteins. The cicatrisation phase generates fibrosis [10] intended to reduce the compliance of the palate. The initial expectation was that there would be a reduction in velar volume [32], but it has now been established that radiofrequency chiefly acts by "rigidifying "the soft palate [20].

Radiofrequencies are emitted by an electrode which is inserted into the thickness of the soft palate. Most authors use three impacts per session (fig.43.3) [33]
-    one is sagittal and median, the electrode being inserted close to the posterior nasal spine and directed toward the base of the uvula, to fibrose the uvula azygos muscle region
-    the other two are lateral, the electrode being inserted close to the posterior nasal spine and directed outwards, to fibrose the pharyngo-staphylion muscle region

Some radiofrequency generators produce low frequencies (465 Hz) with controlled thermal and energy discharge (Somnoplasty-Somnus Inc, Sunnyvale, CA, USA); the radiofrequency session is conducted at fixed temperature but for variable duration, cutting off once the required energy, generally 1500 to 2100 joules, has been discharged. Other generators discharge higher frequencies, of over 1 MHz, without thermal control, since high frequencies reduce the risk of thermal increase; the energy discharged is estimated in proportion to the time of exposure (Surgitron-Ellman, New York, USA).

The application of velar radiofrequencies is an outpatient measure which can be conducted after a simple local anaesthetic. As the mucous membranes are left largely intact, there is virtually no pain following treatment. Patients nevertheless experience slight discomfort. A number of incidents have been described:

- ulceration of the mucous membrane, linked to placing the electrode in too superficial a position; the incident causes pain but cicatrisation is normal.
- oedema of the uvula, linked to placing the electrode in too distal a position in patients with a long uvula; the incident causes dyspnoea requiring corticoid treatment.

Several sessions are usually required at intervals of six to eight weeks [33]. It is possible to apply radiofrequencies to the inferior turbinates [33]. Moreover, although the results are insufficient when the uvula is too long, uvulectomy can be practised, by means of the same device [33].

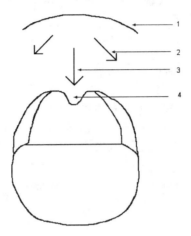

Figure 43.3. Radiofrequency application to the soft palate.
1. boundary between the hard palate and the soft palate
2. lateral impacts
3. median impact
4. uvula

Indications and results

Measures involving hygiene and diet, or postural and prosthetic treatments do not always prove effective. If these fail in the case of a simple snorer, who is adversely affected socially or conjugally, and is well informed of the after effects, laser or radiofrequency treatment or UPPP are warranted, possibly reinforced by nasal surgery. Broadly speaking, if the soft palate is hypertrophic and the pharynx is narrow, UPPP is the most appropriate treatment. If the palate is abnormally small, there are contraindications for surgery, and laser or radiofrequency treatment is indicated. The latter is in the process of taking over from laser treatment due to the absence of pain in the case of radiofrequency treatment.

UPPP is an effective short to medium-term option in cases of simple snoring. Katsantonis *et al.* [22], have reported good results in 86.48% of cases. The successes include roughly half the patients for whom sleep has become noiseless, and half whose snoring has been replaced by non vibratory, heavy respiration, causing little annoyance.

Short and medium term results after laser treatment are comparable to those of UPPP for some [35], and inferior for others: 54% for Séquert *et al.* [34] who obtained 78% with UPPP.

The short and medium term results of velar radiofrequency are encouraging with 75% [1] to 85% [4, 33] of good results.

It is the long term rather than the immediate effects which pose the greatest problems in the case of all the therapeutic methods relating to simple snoring. Snoring tends to reappear with time, often after weight gain. This finding is well established in the case of UPPP and CO2 laser treatment whose efficacy after 3 years (50%) is lower than after 1 year [6]. This trend is also appearing in the case of radiofrequency: 13 months after radiofrequency, half the patients reported a recurrence of snoring or an increase in snoring which had initially diminished [4]. Data is not yet sufficient to ascertain whether it is feasible to propose a series of laser or radiofrequency treatments several years after the initial treatment.

REFERENCES

1. ATTAL P., POPOT B., LE PAJOLEC C., ALFANDARRY D., MARUANI M., AGEEL M., ESCOURROU P., BOBIN S. – Evaluation à court terme d'une nouvelle méthode de traitement du ronflement simple: l'énergie radiofréquentielle. *Ann. Oto-Laryngol. Chir. Cervicofac.* 117, 259-265, 2000.
2. BILLIARD M., ALPEROVITCH A., PERROT C., JAMMES A. – Excessive daytime somnolence in young men: prevalence and contributing factors. *Sleep,* 10, 297-305, 1987.
3. BLAIR SIMMONS F., GUILLEMINAULT C., SILVESTRI R. – Snoring and some obstructive sleep apnea syndrome can be cured by oropharyngeal surgery. *Arch. Otolaryngol. Head Neck Surg.* 109, 503-507, 1983.
4. BLUMEN M.B., DAHAN S., DE DIEULEVEULT T., WAGNER I., CHABOLLE F. – Le traitement du ronflement par radiofréquence avec contrôle thermique (somnoplastie). Tolérance et efficacité à court terme. *Ann. Oto-Laryngol. Chir. Cervicofac.* 117, 349-358, 2000.
5. BROUILLETTE R., HANSON D., DAVID R., KLEMKA L., SZATKOWSKI A., FERNBACH S., HUNT C. – A diagnostic approach to suspected obstructive sleep apnea in children. *J. Pediatr.,* 105, 10-14, 1984.
6. CHABOLLE F., DE DIEULEVEULT T., CABANES J., SEQUERT C., DAHAN S., DRWESKI P., ENGALENC D. – Résultats à long terme de la pharyngotomie chirurgicale classique (uvulopalatopharyngoplastie) versus laser CO2 en ambulatoire dans le traitement de la ronchopathie simple. *Ann. Oto-Laryngol. Chir. Cervicofac.* 115, 196-201, 1998.
7. CHOUARD C.H., MEYER B., CHABOLLE F., FLEURY B. – Résultats cliniques du traitement chirurgical dans 1222 cas de rhonchopathie chronique. *Ann. Oto-Laryngol. Chir. Cervicofac.* 107, 154-158, 1990.
8. CHOUARD C.H., MEYER B., CHABOLLE F., de LARMINAT J.M., VERICEL R., LACCOURREYE O. – Le traitement chirurgical du ronflement. Principe et technique. *Ann. Oto-Laryngol. Chir. Cervicofac.* 103, 329-333, 1986.
9. COLEMAN R.F., SCHECHTER G.L. – A basic model to study acoustic evaluation of airway obstruction. *Arch. Otolaryngol. Head Neck Surg.* 117, 1144-1149, 1991.
10. COUREY M.S., FOMIN D., SMITH T., HUANG S., SANDERS D., REINISCH L. - Histologic and physiologic effects of electrocautery, CO2 laser, and radiofrequency injury in the porcine soft palate. *Laryngoscope,* 109, 1316-1319, 1999.
11. CRAMPETTE L. – *Syndrome d'apnées au cours du sommeil (SAS) de l'adulte.* Thèse Médecine, Montpellier, 1986.
12. CRAMPETTE L., DEJEAN Y., LEFEBVRE P., DJEMEL L. – La consultation ORL pour ronflement . Intérêt et limites. *Cahiers d'ORL,* 5, 349-356, 1988.
13. DEJEAN Y., CHOUARD C.H. – La ronchopathie chronique. Ronflement et Syndrome d'Apnée du Sommeil. *Rapport de la Société Française d'ORL.* Arnette, Paris, 348p. 1993.
14. DEJEAN Y., CRAMPETTE L. – Insuffisance respiratoire d'origine pharyngée (ronchopathie et syndrome d'apnées au cours du sommeil). *Encycl. Med. Chir. Oto-Rhino-Laryngologie.* Editions techniques, Paris, 20, 621, A10, 1992, 10p.
15. DEJEAN Y., CRAMPETTE L., BILLIARD M., GROSS F. – Intérêt de l'examen oto-rhino-laryngologique dans le syndrome d'apnées au cours du sommeil. Traitement chirurgical et indications. *Cahiers d'ORL,* 8, 571-584, 1985.
16. FUJITA S., CONWAY W., ZORICK F., ROTH T. – Surgical correction of anatomic abnormalities in obstructive sleep apnea syndrome: uvulopalatopharyngoplasty. *Otolaryngol. Head Neck Surg.,* 89, 923-934, 1981.
17. GISLASON T., ABERG H., TAUBE A. – Snoring and systemic hypertension: an epidemiological study. *Acta Med. Scand.,* 222, 415-421, 1987.
18. GUILLEMINAULT C., STOOHS R., CLERK A., CETEL M., MAISTROS P. – A cause of excessive daytime sleepiness. The upper airway resistance syndrome. *Chest,* 104, 781-787, 1993.
19. HERNANDEZ S.F. – Palatopharyngoplasty for the obstructive sleep apnea syndrome. Technique and preliminary report of results in ten patients. *Am. J. Otolaryngol.* 3, 229-234, 1982.
20. HUKINS C.A., MITCHELL I.C., HILLMAN D.R. – Radiofrequency volume tissue reduction of the soft palate in simple snoring. *Arch. Otolaryngol. Head Neck Surg.* 126, 602-606, 2000.
21. IKEMATSU T.- Study of snoring. Fourth report:therapy. *J. Jpn. Otorhinolaryngol.,* 64, 434-435, 1964.
22. KATSANTONIS G.P., FRIEDMAN W.H., ROSENBLUM B.N., WALSH J.K. – The surgical treatment of snoring: a patient perspective. *Laryngoscope,* 100, 138-140, 1990.
23. KWAN S., FLEETHAM J.A., ENARSON D.A., CHAN YEUNG N. – Snoring, obesity, smoking and systemic hypertension in a working population in British Columbia. *Am. Rev. Respir. Dis.* A 380, 1991.
24. LUGARESI E., CIRIGNOTTA F., COCCAGNA G., PIANA C. – Some epidemiological data on snoring and cardiocirculatory disturbances. *Sleep,* 3, 221-224, 1980.
25. MONDAIN M., DEJEAN Y., CRAMPETTE L. – Pharyngoplastie et fonction vélaire dans la chirurgie du ronflement. *J. Fr. ORL.* 43, 429-431, 1994.
26. NIINIMA V., COLE P., SHEPHARD R.J. – The switching point from nasal to oronasal breathing. *Respir. Physiol.,* 42, 61-71, 1980.
27. OLSEN K.D., KERN E.B. – Snoring. *Mayo Clin. Proc.* 65, 1095-1105, 1990.
28. PETRUSON B. – Increased nasal breathing decreases snoring and improves oxygen saturation during sleep apnoea. *Rhinology,* 32, 87-89, 1994.

29. PICHE J., GAGNON N.B. – Snoring: physiopathology, surgical treatment and a modified uvulo-palato-pharyngoplasty. *J. Otolaryngol.,* 18, 36-43, 1989.

30. PIEYRE J.M. – Le traitement chirurgical du ronflement. *Med. et Hyg.,* 41, 3752-3754, 1983.

31. POLO O., BRISSAUD L., FRAGA J., DEJEAN Y., BILLIARD M. – Partial upper airway obstruction in sleep after uvulopalatopharyngoplasty. *Arch. Otolaryngol. Head Neck Surg.* 115, 1350-1354, 1989.

32. POWELL N.B., RILEY R.W., TROELL R.J., BLUMEN M.B., GUILLEMINAULT C. - Radiofrequency volumetric tissue reduction of the palate in subjects with sleep-disordered breathing. *Chest,* 113, 1163-1174, 1998.

33. SCHMITT E., PETELLE B., MEYER B. – Radiofréquence vélaire et vélo-turbinale pour rhonchopathie ; à propos de 256 cas. *Rev. Soc. Fr. ORL,* 67, 7-11, 2001

34. SEQUERT C., CARLES P., KAMAMI P.Y., GIRSCHIG H., FLEURY B., CHABOLLE F. – Traitement de la ronchopathie simple. Pharyngoplastie chirurgicale versus pharyngotomie au laser CO2. *Ann. Oto-Laryngol. Chir. Cervicofac.* 109, 317-322, 1992.

35. WALKER R., GRIGG-DAMBERGER M., GOPALSAMI C., TOTTEN M. – Laser-assisted uvulopalatopharyngoplasty for snoring and obstructive sleep apnea: results in 170 patients. *Laryngoscope,* 105, 938-943, 1995.

Chapter `44

Nocturnal hypoxemia in chronic obstructive pulmonary disease

E. Weitzenblum*, A. Chaouat*, C. Charpentier*, R. Kessler*, and J. Kreiger**
* Service de Pneumologie, Hôpital de Hautepierre. ** Service des Explorations Fonctionnelles du Système Nerveux et de Pathologie du Sommeil, Strasbourg, France

The notion that sleep can accentuate hypoxemia in patients with chronic obstructive pulmonary disease (COPD) has been understood for almost 40 years [44]. The earliest nocturnal polygraphic studies of patients go back to 1975-1976 [32, 33] and include intermittent blood gas tension measurements during sleep. But it was only with the advent of reliable transcutanous oximeters in 1976, that the respiratory "events" occurring during sleep could be clearly defined. After the landmark study of Flick and Block in 1977, [27] several studies [5, 7, 15, 17, 35, 42, 47] went on to show that COPD patients experienced a worsening of hypoxemia, particularly during REM sleep. The initial studies focused almost entirely on patients with severe cases who were clearly hypoxemic during the daytime. It was only later [25, 34, 45] that data appeared for nocturnal hypoxemia in COPD patients with little or no hypoxemia during the day ($PaO_2 > 60$ mmHg).

In this short review, we will attempt to recall the characteristics of nocturnal hypoxemia in COPD patients, its mechanisms and consequences, and to consider the treatment options currently available.

QUALITY OF SLEEP IN COPD PATIENTS

Most authors are struck by the poor quality of sleep in these patients: increased latency to sleep onset; reduced total sleep time; increased duration of light NREM sleep (stages 1-2) with a concomitant reduction in deep NREM sleep (stages 3 - 4), to the point of disappearing altogether in some cases; reduction of REM sleep; increased intra-sleep arousals. Sleep architecture is thus altered compared with that of healthy subjects, but it is hard to determine how much this is due to causal factors, how much to the age of the patient, to the influence of certain treatments or the conditions of recording.

Some authors maintain that sleep quality is poorer in patients suffering from emphysema ("PP" or pink puffer) than in those who are bronchitic ("BB" or blue-bloated) [5]. These results have not been confirmed by later studies [7, 13] which make no distinction between "PP", "BB" and healthy subjects. Sleep fragmentation, in particular, was found to be no more marked in hypoxic and hypercapnic subjects, or episodes of intense, sleep-related oxygen desaturation [1]. It is not yet clear whether the poor quality of sleep experienced by advanced COPD patients is directly linked to nocturnal hypoxemia.

OXYGEN DESATURATION DURING SLEEP IN COPD PATIENTS

Episodes of oxygen desaturation are frequent in patients with advanced COPD. They are usually defined by a fall in $SaO_2 > 4\%$ in relation to its baseline level in calm respiration, immediately preceding the hypoxemic episode [45]. Episodes of oxygen desaturation are characterised by number, duration, severity (minimum SaO_2 reached). The most representative parameters of nocturnal desaturation are mean nocturnal SaO_2, reflecting the mean desaturation level during the night, and the time (or % of time) spent in sleep below a given saturation threshold (90%, 80% etc.). Most software used for processing the data supplied by pulse oximeters provides easy access to this data.

The longest and most severe episodes of desaturation occur during REM sleep, lasting from several minutes up to 15 or even 30 minutes. Desaturation is not in fact specific to REM sleep, and may occur during the unstable period of sleep onset and during light NREM sleep (stages 1-2), but it must be stressed that these episodes of desaturation are not as intense as those recorded in REM sleep and that they do not exceed several minutes, sometimes lasting no more than one minute.

An example of this is presented in figure 44.1 which shows a continuous recording of SaO_2 in a patient with advanced COPD and severely disrupted arterial blood gas tension during wakefulness. The most severe periods of desaturation are seen to occur during REM sleep in which SaO_2 falls below 60%.

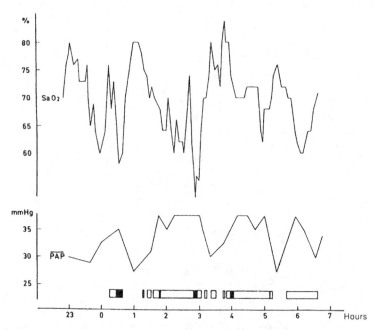

Figure 44.1. Evolution of oxygen saturation (SaO_2) measured by transcutaneous oximetry (upper part of figure) and pulmonary arterial pressure (PAP) (lower part of figure) in a patient presenting advanced COPD with marked hypoxemia during the day. The sleep stages are shown at the bottom of the figure; the black rectangles correspond to REM sleep. Note that the patient experiences episodes of marked oxygen desaturation. The most severe episodes occur during REM sleep; these are accompanied by a rise in arterial pulmonary pressure, with a "mirror effect": the lower the oxygen saturation, the higher the pulmonary arterial pressure.

Episodes of oxygen desaturation are mo.e often seen in the bronchitic or *blue and bloated* type of patient than in the emphysematous or *pink puffer* type [5, 7, 13]. The severity of nocturnal hypoxemia is in fact closely aligned to the level of daytime hypoxemia: subjects who are the most hypoxemic during the day are those whose oxygen desaturation is the most marked during sleep; desaturation peaks reaching 50% of SaO_2 and even more have been observed in patients with severe COPD whose daytime SaO_2 is well below 90% [5, 7](see fig. 44.1).

Healthy subjects also experience a physiological drop in PaO_2 during sleep [37], particularly in REM sleep. This is generally less than 10 mmHg and the effect on SaO_2 is thus minimal, being on the horizontal portion of the haemoglobin dissociation curve. By contrast, a fall of 10 mmHg in PaO_2 may have more serious repercussions for a COPD patient, whose PaO_2 is 50 mmHg during wakefulness, appearing on the steep portion of the dissociation curve. The same drop in PaO_2, particularly during REM sleep, would thus result in oxygen desaturation, which varies greatly depending on whether the subject is healthy, emphysematous (little hypoxia) or chronic bronchitic (marked hypoxia). Flenley's group [7, 19] put forward the hypothesis that the fall in PaO_2 during REM sleep was roughly equal for healthy subjects, "bronchitic" type COPD patients and "enphysematous" type COPD patients. The difference appeared to be determined by the position of PaO_2 on the haemoglobin dissociation curve. The theory is tempting but is doubtless an oversimplification. It fails to explain, for instance, why some patients with a low incidence of hypoxia during the daytime (PaO_2 > 60 mmHg) show marked desaturation at night, while others who are comparable in every respect, do not desaturate at night [34]. Daytime PaO_2 is clearly the best predictive indication of nocturnal SaO_2 [45], but significant correlation in a series of patients has nevertheless failed to accurately predict nocturnal desaturation in individual cases.

In the case of patients who present little or no daytime hypoxia (PaO_2 > 60 mmHg) it was suggested that nocturnal desaturation be defined as $\geq$ 30% of the recording time showing SaO_2 > 90% [34]. The definition proposed by Fletcher *et al.* [25], $\geq$ 5 min. with SaO_2 > 90% and a desaturation peak of at least 85%, does not appear to be marked enough.

What can be said about the incidence of marked sleep-related hypoxemia in COPD patients with only slight hypoxia in the daytime? Fletcher *et al.* [25] observed that 37/135 (27%) of patients whose daytime PaO_2 was > 60 mmHg, experienced significant nocturnal desaturation. Levi-Valensi *et al.* [34] investigated a smaller series (n = 40) but with a more homogeneous group of COPD patients (daytime PaO_2 from 60 to 70 mmHg). Significant nocturnal desaturation occurred in 18/40 (45%) of patients. It has thus been confirmed that nocturnal desaturation tends to be frequent in COPD patients with little hypoxia during the daytime, but it is important to emphasise that desaturation is usually moderate in these cases (average nocturnal Sa O_2 approximately 88 – 89% [25, 34, 45]).

Little data is available on how daytime $PaCO_2$ evolves during the night in COPD patients. Continuous $PaCO_2$ measurement with transcutaneous electrodes is not common practise, due to the technical and metrological problems it poses. The measurement is fairly unreliable, at least for adults, with a long reaction time, which is noticeably slower with oximetry. A small catheter can be fixed to the radial artery, several samples being taken during sleep [32, 33], but this ceases to be a continuous measurement, and sample taking is liable to disturb sleep. The results, (arterial blood and transcutaneous PCO_2) show that the rise in PCO_2 during REM sleep is generally less marked than the fall in SaO_2 [36]. $PaCO_2$ may not even rise during certain episodes of desaturation [46]. Variations in $PaCO_2$ will depend in fact on the desaturation mechanism: marked in cases of alveolar hypoventilation but absent or minimal in ventilation-perfusion mismatching.

SLEEP-RELATED MECHANISMS OF HYPOXEMIA

Two main mechanisms explain the accentuation of sleep hypoxemia: alveolar hypoventilation and the occurrence, or worsening, of ventilation-perfusion mismatching. The two mechanisms may also be associated. Apnoeas are not a feature, and COPD nocturnal hypoxemia is not a sleep apnoea syndrome [7], although this does not prevent the two conditions from coexisting (see below).

As alveolar hypoventilation is observed during certain stages of sleep in healthy subjects [4, 37], it is hardly surprising to find in respiratory insufficient COPD patients. It is evoked by a rise in $PaCO_2$ (or transcutaneous PCO_2) or a decrease in minute ventilation [6, 21, 31]. The "hypopnoeas" frequently observed in these patients during sleep, are no doubt synonymous with alveolar hypoventilation: these are in fact episodes characterised by decreased ventilation (often poorly quantified unless a pneumotachograph is used) associated with increased hypoxemia and a drop in respiratory effort. These are unlikely to consist of obstructive hypopnoeas, in COPD patients, contrary to cases of sleep apnoea syndrome.

The role of alveolar hypoventilation appears to predominate in determining hypoxemia, especially in REM sleep [6, 18, 28]. Alveolar hypoventilation is explained by a reduction in "central command" linked to the diminished sensitivity to hypoxia and hypercapnia, by increased upper airway resistance, and by an abolition of the intercostal muscles and accessory respiratory muscles, during phasic REM sleep [28].

The other nocturnal desaturation mechanism is ventilation-perfusion mismatching. The presence or accentuation of ventilation-perfusion mismatching is suggested by the discrepancy sometimes observed between a marked fall in SaO_2 and a slight (or even non existent) rise in $PaCO_2$ [46] and by an increase in the arterial-alveolar difference in PO_2 and venous intake [21, 31]. It is in fact very difficult to perform detailed studies of gas transfer during sleep, and the data available is correspondingly limited, referring to a limited series of patients [21, 31]. The importance of ventilation-perfusion mismatching, especially in REM sleep, may be accounted for by reduced mucociliary clearance, resulting in the accumulation of bronchial secretions; and above all, by decreased functional residual capacity, already favoured by the supine position, becoming more marked in REM sleep [30]. This decrease leads to the closing of the small airways in the lower lung area, which themselves generate a shunting effect.

Sleep hypoxemia in COPD patients is thus linked to a variable combination of alveolar hypoventilation and additional ventilation-perfusion mismatching. Alveolar hypoactivity predominates in REM sleep.

CONSEQUENCES OF SLEEP-RELATED HYPOXEMIA

Haemodynamic pulmonary repercussions

Acute alveolar hypoxemia, which characterises severe episodes of nocturnal desaturation, typically causes vasoconstriction of the lung and a rise in mean pulmonary arterial pressure (PAP). Episodes of nocturnal hypoxemia, especially when marked and protracted, can lead to pulmonary hypertensive jolts. Few attempts have been made to study pulmonary haemodynamics, due to the invasive nature of the former, which has an adverse effect on sleep quality. These studies [3, 11, 22, 46] consistently show that episodes of hypoxemia, particularly in REM sleep, tend to be accompanied by PAP increases, which can be substantial.

Hence in the series of twelve patients with severe chronic bronchitis explored by Coccagna and Lugaresi [11], the mean PAP of the group rose from 37 mmHg in the wakeful state to 55 mmHg in REM sleep, although it is important to stress that pulmonary hypertension was already high in wakefulness. Boysen *et al.* [3], Fletcher and Levin [22], reported a marked but less severe rise in PAP during episodes of nocturnal desaturation, chiefly in REM sleep [22]. We have noted discrepancies of up to 25 mmHg between baseline PAP in wakefulness and the maximal peak of pulmonary hypertension, usually present in REM sleep [46]. Patients tend to show a fair degree of consistency between the fall in SaO_2 and the rise in PAP, producing a "mirror effect" [46]. While most COPD patients are considered as "prone" to nocturnal hypoxemia (fig. 44.2), it is important to add that there are the "non prone" and the "weakly prone". The rise in PAP is chiefly due to that of pulmonary vascular resistance, but an increase in heart rate or an association of both mechanisms has also been reported in some patients [22].

The rare studies on pulmonary haemodynamics during sleep were carried out on patients with very advanced COPD, who usually presented daytime pulmonary hypertension. The question remains as to whether these results can be applied to subjects with minor hypoxemia outside sleep and whose wakeful PAP is still normal. It is difficult to answer this question because of the lack of adequate studies. One tempting hypothesis [2, 18] describes daytime pulmonary hypertension as starting with episodes of nocturnal desaturation: these are then thought to give rise to transient increases in PAP, with pulmonary hypertension eventually becoming established even though nocturnal hypoxemia remains low or non existent. This hypothesis [2, 18] has not yet been tested but COPD patients are known to show a strong correlation between mean nocturnal SaO_2 and daytime PAP [38]. The studies by Fletcher *et al.* [24] and Levi-Valensi *et al.* [34] have shown that among patients with low daytime hypoxemia ($PaO_2 > 60$ mmHg), those who desaturate significantly during sleep have a higher risk of daytime pulmonary hypertension. However a more recent European multicentric study [9], on a wider patient group failed to confirm the initial studies: PAP was shown to be identical for patients with and without nocturnal desaturation. Hence there is no clear evidence to date that isolated nocturnal hypoxemia, in the absence of marked daytime hypoxemia, is a determining factor for permanent (daytime) pulmonary hypertension.

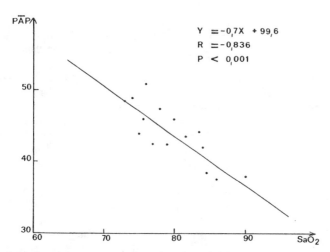

Figure 44.2. Correlation between oxygen saturation values (measured by transcutaneous oximeter) and those for pulmonary arterial pressure (PAP) during sleep, in a patient with severe COPD. Note the strong correlation. This patient may be considered as being "prone" to nocturnal hypoxemia. The lower the SaO_2, the higher the PAP. A discrepancy can also be observed of almost 20 mmHg between the lowest PAP and the highest PAP during sleep.

Cardiac dysrhythmias

Tachycardia and other cardiac dysrhythmias have been observed in COPD patients during sleep [39, 43]. Supraventricular and ventricular extra systoles appear to be particularly frequent between 3 and 5 am [27]. The study of a group of 42 severely affected COPD patients showed that ventricular extra systoles were present during sleep in 60% of cases, and were more frequent in cases of severe nocturnal desaturation ($SaO_2 > 80$ %) [39]. Certain unexplained deaths during sleep may be due to severe dysrhythmias, which in turn are favoured by severe hypoxemia, although the frequency and extent of these remain controversial.

Repercussions on coronary circulation and myocardium

For Shepard *et al.* [40], episodes of nocturnal hypoxemia lead to a strong increase in myocardium oxygen consumption. In roughly 30% of their patients, the maximal myocardial blood flow during sleep was greater than the flow observed in the course of considerable effort during the daytime. The myocardium thus experienced "hypoxemic stress" during sleep which may account for the higher incidence of nocturnal mortality in COPD patients. In fact it is not known whether nocturnal hypoxemia is a direct cause of mortality in COPD and the hypothesis of Shepard *et al.* [40] has not been sustained due to the lack of studies in this domain.

Polycythemia

Experimental, intermittent hypoxemia, is known to lead to polycythemia in animals after several weeks, but it is not known whether nocturnal hypoxemia in COPD patients has a similar effect. Two studies [25, 34] found identical haemoglobin levels for nocturnal desaturator and nondesaturator COPD patients. A further study [16] has shown that daytime hypoxemia had to be present to provoke an increase in the production of erythropoietin in COPD patients. Isolated nocturnal hypoxemia does not appear to be sufficient to induce polycythemia which would require the presence of daytime hypoxemia.

COPD ASSOCIATED WITH OBSTRUCTIVE SLEEP APNOEA HYPOPNEA SYNDROME

COPD is a common condition. Obstructive sleep apnoea hypopnea syndrome (OSAHS) is also common, affecting roughly 5% of middle aged adult males [48]. It is thus not exceptional to encounter an association between these two conditions, with no causal link. Flenley [19] suggested calling this association "overlap syndrome". In a study carried out [8] on 265 OSAHS patients, an associated COPD was found in 30 cases (11%). Overlap syndrome is thus far from exceptional.

In COPD patients, the OSAHS is to be suspected in the presence of clinical features e.g. sleepiness in a subject who snores, breathing pauses reported by the partner, but also when the respiratory insufficiency picture, whether or not this is complicated by polycythemia and cardiac repercussions, appears to be disproportionate in relation to ventilatory deficiency which remains moderate. This diagnosis should also be evoked where marked nocturnal hypoxemia persists with oxygen therapy.

Patients affected by overlap syndrome have an increased risk of daytime respiratory insufficiency and pulmonary hypertension compared with "simple" OSAHS patients [8]. Their nocturnal hypoxemia is more marked [19]. In severe cases, treatment should combine oxygen therapy with continuous positive pressure by nasal mask. This illustrates the importance of detecting cases of COPD-OSAHS association as early as possible and one should not hesitate to call for a polysomnographic examination - even if this test is not normally indicated for COPD, where hypoxemia does not increase significantly during sleep and can thus be checked by simple oximetric recording. Polysomnography, which is onerous and expensive, should be reserved for cases strongly suggesting an association of COPD and an obstructive sleep apnoea hyponea syndrome.

TREATMENT OF NOCTURNAL HYPOXEMIA IN COPD

One of the prime elements in the treatment of advanced COPD at the respiratory insufficiency stage, is to correct hypoxemia, but nocturnal hypoxemia is rarely treated on its own and sleep is usually completely covered by standard oxygen therapy. While the indications for oxygen therapy are well codified, the same cannot be said of isolated, nocturnal oxygen therapy, and there is still controversy as to the importance of treating COPD patients who are not severely hypoxemic during the day, but who experience marked desaturation during sleep at night.

Medication for nocturnal hypoxemia

No specific treatment exists for sleep-related hypoxemia in COPD, but some substances (analeptics or others) designed to improve daytime hypoxemia will also have a favourable effect on nocturnal hypoxemia, because of the more advantageous position of PaO_2 on the O_2 dissociation curve. The improvement of sleep-related hypoxemia thus simply results from the improvement of daytime hypoxemia.

Progesterone, a respiratory analeptic, will improve the $PaCO_2$ and nocturnal SaO_2 of some COPD patients [41] although these results have not been confirmed by a controlled study, which would be more satisfying from a methodological standpoint [14]. Almitrine will significantly improve daytime hypoxemia in stable COPD patients so it is hardly surprising that it also has a positive effect on episodes of nocturnal desaturation, as shown by Connaughton *et al.* [12] who observed a marked improvement in nocturnal hypoxemia with no deterioration of sleep quality.

Standard long-term oxygen therapy

This is required for COPD patients with marked, persistent hypoxemia (daytime PaO_2 < 55-60 mmHg). It is precisely these patients who present the severest nocturnal hypoxemia. To be effective, long-term oxygen therapy should be applied for at least 16/24 hours and if possible, 18/24 hours. This must include the sleep period.

The usual rate of oxygen is 1.5-3 litres/min. But are these rates sufficient to correct episodes of desaturation which may be severe, during sleep? In the study by Calverley *et al.* [5], a rate of 2 litres/min. was seen to be effective, with mean transcutaneous SaO_2 rising from 53 ± 29 to $90 \pm 9\%$. Episodes of desaturation persist, but are few in number and less severe. The efficacy of this standard oxygen therapy at a normal rate has been confirmed by other studies [17, 22, 29]. Hence it is not normally necessary to adopt a specific rate for nocturnal oxygen therapy. However, it is useful to use nocturnal oximetry to check that the selected rate of oxygen ensures SaO_2 >90% during sleep. If this is not the case, the rate must be increased in stages of 0.5 litre/min. until the required level is reached.

There may be fears of a progressive rise in $PaCO_2$ during sleep under oxygen, as the hypoxic stimulus is suppressed by hypoxemia correction, and of a diminished ventilatory response to CO_2 during sleep. In fact, the study by Goldstein *et al.* [29] has clearly demonstrated that this risk is minimal, at least in the case of stable COPD patients. The increase in trancutaneous $PaCO_2$ is usually moderate (<6 mmHg), occurring early in the night and self-stabilising. Patients of the overlap type may show a greater rise in $PaCO_2$ [29], underlining the importance of detecting the association with obstructive apnoeas.

Nocturnal oxygen therapy has a favourable effect on pulmonary hypertensive jolts [22] and on cardiac rhythm disorders [43]. Oxygen therapy is also said to improve the quality of sleep [5], although some authors have reported otherwise, observing an identical number of EEG arousals in ambient air and under oxygen [17].

Oxygen therapy during sleep only

A number of COPD patients whose condition does not justify standard oxygen therapy, as their daytime PaO_2 is > 60 mmHg, do suffer severe hypoxemia at night. One study [34] found that 18/40 COPD patients, with PaO_2 of the order of 60-65 mmHg, spent over 30% of their sleep time with SaO_2 < 90%. Oxygenation might be envisaged for such patients during sleep (8-10 hours/24 hours).

The harmful effects of isolated nocturnal hypoxemia have not been demonstrated convincingly. Some studies have indicated that nocturnal hypoxemia increases the risk of permanent pulmonary arterial hypertension [24, 34] but a more recent multifactorial study aimed at a greater number of patients failed to confirm these findings [9]. Fletcher *et al.* [20] observed that, among COPD patients with little or no daytime hypoxemia (PaO_2 > 60 mmHg), those with significant nocturnal hypoxemia had a shorter life expectancy. This in fact refers to a retrospective study, which presents a number of methodological insufficiencies, and prospective prognostic studies are essential before any firm conclusions can be reached. The same authors [24] studied the long-term effects (3 years) of nocturnal oxygen therapy administered to patients with desaturation compared with a control group of patients with desaturation (9 and 7 respectively). No difference was reported between the two groups in terms of survival [24].

Table 44.1. Evolution of gasometric and pulmonary haemodynamic data in nocturnal desaturation COPD patients treated and non treated with oxygen therapy during sleep [10].

	Patient group treated with nocturnal oxygen therapy (n = 24)		Control group (n = 22)		Statistical significance
	T0	T2	T0	T2	
PaO_2 (mmHg)	63.0 ± 3.3	62.2 ± 7.4	63.1 ± 2.8	64.5 ± 5.7	NS
$PaCO_2$ (mmHg)	45.0 ± 5.6	46.3 ± 5.9	44.3 ± 4.2	44.9 ± 5.6	NS
SaO_2 mean nocturnal %	87.9 ± 2.7	87.9 ± 4.2	88.6 ± 2.0	89.3 ± 2.9	NS
$TsaO_2$ < 90%	62.5 ± 25.3	57.9 ± 31.9	64.7 ± 24.8	51.2 ± 36.2	NS
PAP (mmHg)	18.3 ± 4.7	19.5 ± 5.3	19.8 ± 5.6	20.5 ± 6.5	NS
PAP exercise (mmHg)	35.2 ± 7.2	38.3 ± 10.3	36.2 ± 11.7	37.1 ± 11.3	NS

T0 corresponds to the start of the study, T2 to the monitoring test carried out 2 years later.

$TsaO_2$ < 90%: percentage of recording time spent at oxygen saturation < 90%, PAP: mean pulmonary arterial pressure.

Note that there is no significant difference between the evolution of the principle variables from T0 to T2 in either group. Thus nocturnal oxygen therapy does not appear to have a beneficial effect on pulmonary haemodynamics.

A more recent European multicentric study [10], based on a greater number of patients failed to confirm these results: 76 nocturnal desaturation COPD patients were monitored; 41 received nocturnal oxygen therapy and 35 acted as controls. Life expectancy was comparable for both groups but statistical analysis was limited by the number of patients included and the number of deaths. 46 patients (24 in the treated group and 22 in the control group) had a right cardiac catheterisation monitoring after 2 years: MPP evolution was identical for both groups (table 44.1).

It appears then, that nocturnal oxygen therapy has no positive effect on either the prognosis or evolution of pulmonary haemodynamics in COPD patients with low daytime hypoxemia and with nocturnal desaturation. No current evidence exists therefore to justify setting up oxygen therapy during sleep only for these patients, although this treatment might be considered in particular cases (e.g. overlap syndrome).

REFERENCES

1.   ARAND D.L., McGINTY D.J., LITTNER M.R. – Respiratory patterns associated with hemoglobin desaturation during sleep in chronic obstructive pulmonary disease. *Chest,* 80, 183-190, 1981.

2.   BLOCK A.J., BOYSEN P.G., WYNNE J.W. – The origin of cor pulmonale. A hypothesis (editorial). *Chest,* 75, 109-110, 1979.

3.   BOYSEN P.G., BLOCK A.J., WYNNE J.W., HUNT L.A., FLICK M.R. – Nocturnal pulmonary hypertension in patients with chronic obstructive pulmonary disease. *Chest,* 76, 536-542, 1979.

4.   BULOW K. – Respiration and wakefulness in man. *Acta Physiol. Scand.,* 59, 1-110, 1963.

5.   CALVERLY P.M.A., BREZINOVA V., DOUGLAS N.J., CATTERALL J.R., FLENLEY D.C. – The effect of oxygenation on sleep quality in chronic bronchitis and emphysema. *Am. Rev. Respir. Dis.,* 126, 206-210, 1982.

6.   CATTERALL J.R., CALVERLY P.M.A., McNEE W., WARREN P.M., SHAPIRO C.M., DOUGLAS N.J., FLENLEY D.C. – Mechanism of transient nocturnal hypoxemia in hypoxic chronic bronchitis and emphysema. *J. Appl. Physiol.,* 59, 1698-1703, 1985.

7.   CATTERALL J.R., CALVERLEY P.M.A., SHAPIRO C.M., BREZINOVA V., BRASH H.M., FLENLEY D.C. – Transient hypoxemia during sleep in chronic obstructive pulmonary disease is not a sleep apnea syndrome. *Am. Rev. Respir. Dis.,* 128, 24-29, 1983.

8.   CHAOUAT A., WEITZENBLUM E., KRIEGER J., IFOUNDZA Th., OSWALD M., KESSLER R. – Association of chronic obstructive pulmonary disease and sleep apnea syndrome. *Am. J. Respir. Crit. Care Med.,* 151, 82-86, 1995.

9.   CHAOUAT A., WEITZENBLUM E., KESSLER R., CHARPENTIER C., EHRHART M., LEVI-VALENSI P., ZIELINSKI J., DELAUNOIS L., CORNUDELLA R., MONTINHO DOS SANTOS J. – Sleep-related O2 desaturation and daytime pulmonary haemodynamics in COPD patients with mild hypoxaemia. *Eur. Respir J.,* 10, 1730-1735, 1997.

10. CHAOUAT A., WEITZENBLUM E., KESSLER R., CHARPENTIER C., EHRHART M., SCHOTT R., LEVI-VALENSI P., ZIELINSKI J., DELAUNOIS L., CORNUDELLA R., MONTINHO DOS SANTOS J. -   A randomized trial of nocturnal oxygen therapy in chronic obstructive pulmonary disease patients. *Eur. Respir. J.,* 14, 1002-1008, 1999.

11. COCCAGNA G., LUGARESI E. – Arterial blood gases and pulmonary and systemic arterial pressure during sleep in chronic obstructive pulmonary disease. *Sleep,* 1, 117-124, 1978.

12. CONNAUGHTON J.J., DOUGLAS N.J., MORGAN A.D., SHAPIRO C.M., CRITCHLEY J.A., PAULY N., FLENLEY D.C. – Almitrine improves oxygenation when both awake and asleep in patients with hypoxia and carbon dioxide retention caused by chronic bronchitis and emphysema. *Am. Rev. Respir. Dis.,* 132, 206-210, 1985.

13. DE MARCO F.J., WYNNE J.W., BLOCK A.J., BOYSEN P.G., TASSAN V.C. – Oxygen desaturation during sleep as a determinant of the « blue and bloated » syndrome. *Chest,* 79, 621-625, 1981.

14. DOLLY F.R., BLOCK A.J. – Medroxyprogesterone acetate and COPD. Effect on breathing and oxygenation in sleeping and awake patients. *Chest,* 83, 469-472, 1983.

15. DOUGLAS N.J., CALVERLEY P.M.A., LEGGETT R.J.E., BRASH H.M., FLENLEY D.C., BREZINOVA V. – Transient hypoxaemia during sleep in chronic bronchitis and emphysema. *Lancet,* 1, 1-4, 1979.

16. FITZPATRICK M.F., MACKAY T., WHYTE D.K., ALLEN M., TAM R.C., DORE C.J., HENLEY M., COTES P.M., DOUGLAS N.J. – Nocturnal desaturation and serum erythropoietin: a study in patients with chronic obstructive pulmonary disease and in normal subjects. *Clin. Sci.,* 84, 319-324, 1993.

17. FLEETHAM J., WEST P., MEZON B., CONWAY W., ROTH T., KRYGER M.H. – Sleep, arousals and oxygen desaturation in chronic obstructive pulmonary disease. The effect of oxygen therapy. *Am. Rev. Respir. Dis.,* 126, 429-433, 1982.

18. FLENLEY D.C. – Clinical hypoxia: causes , consequences, and correction. *Lancet,* 1, 542-546, 1978

19. FLENLEY D.C. – Sleep in chronic obstructive lung disease. *Clin. Chest Med.,* 6, 651-661, 1985.

20. FLETCHER E.C., DONNER C., MIDGREN B., ZIELINSKI J., LEVI-VALENSI P., BRAGHIROLI A., RIZA Z., MILLER C. – Survival in COPD patients with a daytime PaO2 > 60 mmHg with and without nocturnal oxyhemoglobin desaturation. *Chest,* 101, 649-655, 1992.

21. FLETCHER E.C., GRAY B.A., LEVIN D.C. – Non apneic mechanism of arterial oxygen desaturation during rapid eye-movement sleep. *J. Appl. Physiol.,* 54, 632-639, 1983.

22. FLETCHER E.C., LEVIN D.C. – Cardiopulmonary hemodynamics during sleep in subjects with chronic obstructive pulmonary disease: the effect of short and long-term O2. *Chest*, 85, 6-14, 1984.

23. FLETCHER E.C., LUCKETT R.A., GOODNIGHT-WHITE S., MILLER C.C., QIAN W., COSTARANGOS-GALARZA C. – A double-blind trial of nocturnal supplement oxygen for sleep desaturation in patients with chronic obstructive pulmonary disease and a daytime PaO2 above 60 mmHg. *Am. Rev. Respir. Dis.*, 145, 1070-1076, 1992.

24. FLETCHER E.C., LUCKETT R.A., MILLER T., COSTA-RANGOS C., KUTKA N., FLETCHER J.G. – Pulmonary vascular hemodynamics in chronic lung disease patients with and without oxyhemoglobin desaturation during sleep. *Chest*, 95, 757-764, 1989.

25. FLETCHER E.C., MILLER J., DIVINE G.W., FLETCHER J.G., MILLER T. – Nocturnal oxyhemoglobine desaturation in COPD patients with arterial oxygen tensions above 60 torr. *Chest*, 92, 604-608, 1987.

26. FLICK M.R., BLOCK A.J. – Continuous *in vivo* monitoring of arterial oxygenation in chronic obstructive lung disease. *Ann. Intern. Med.* 86, 725-730, 1977.

27. FLICK M.R., BLOCK A.J. – Nocturnal *vs* diurnal cardiac arrhythmias in patients with chronic obstructive pulmonary disease. *Chest*, 75, 8-11, 1979.

28. GEORGE C.F., WEST P., KRYGER M.H. – Oxygenation and breathing pattern during phasic and tonic REM in patients with chronic obstructive pulmonary disease. *Sleep*, 10, 234-243, 1987.

29. GOLDSTEIN R.S., RAMCHARAN V., BOWES G., McNICHOLAS W.T., BRADLEY D., PHILLIPSON E.A. – Effect of supplemental nocturnal oxygen on gas exchange in patients with severe obstructive lung disease. *N. Engl. J. Med.*, 310, 425-429, 1984.

30. HUDGEL D.W., DEVADATTA P. – Decrease in functional residual capacity during sleep in normal humans. *J. Appl. Physiol.*, 57, 1319-1322, 1984.

31. HUDGEL D.W., MARTIN R.J., CAPEHART M., HOHNSON B., HILL P., - Contribution of hypoventilation to sleep oxygen desaturation in chronic obstructive pulmonary disease. *J. Appl Physiol.*, 55, 669-677, 1983..

32. KOO K.W., SAX D.S., SNIDER G.L. – Arterial blood gases and pH during sleep in chronic obstructive pulmonary disease. *Am. J. Med.*, 58, 663-670, 1975.

33. LEITCH A.G., CLANCY L.J., LEGGETT R.J.E., TWEED-DALE P., DAWSON P., EVANS J.J. – Arterial blood gas tensions, hydrogen ion, and electrencephalogram during sleep in patients with chronic ventilatory failure. *Thorax*, 31, 730-735, 1976.

34. LEVI-VALENSI P., WEITZENBLUM E., RIDA Z., AUBRY P., BRAGHIROLI A., DONNER C., APPRILL M., ZIELINSKI J., WURTEMBERGER G. – Sleep-related oxygen desaturation and daytime pulmonary haemodynamics in COPD patients. *Eur. Respir. J.*, 5, 301-307, 1992.

35. LITTNER M.R., McGINTY D.J., ARAND D.L. – Determinants of oxygen desaturation in the course of ventilation during sleep in chronic obstructive pulmonary disease. *Am. Rev. Respir. Dis.*, 122, 849-857, 1980.

36. MIDGREN B., ALRIKKALA P., RYDING E., ELMQVIST D. – Transcutaneous CO2 monitoring and disordered breathing during sleep. *Eur. J. Respir. Dis.*, 65, 621-628, 1984.

37. ROBIN E.D., WHALEY R.D., CRUMP C.H., TRAVIS D.M. – Alveolar gas tensions, pulmonary ventilation and blood pH during physiologic sleep in normal subjects. *J. Clin. Invest.*, 37, 981-989, 1958.

38. SAUTEGEAU A., HANNHART B., BEGIN P., POLU J.M., SCHRIJEN F. – Pression artérielle pulmonaire basale diurne et niveau nocturne d'oxygénation chez les bronchitiques chroniques. *Bull. Europ. Physiopathol. Resp.*, 20, 541-545, 1984.

39. SHEPARD J.W.Jr., GARRISON M.W., GRITHER D.A., EVANS R., SCHEITZER P.K. – Relationship of ventricular ectopy to nocturnal oxygen desaturation in patients with chronic obstructive pulmonary disease. *Am. J. Med.*, 78, 28-34, 1985.

40. SHEPARD J.W. Jr., SCHWEITZER P.K., KELLER C.A., CHUN D.S., DOLAN G.F. – Myocardial stress. Exercise versus sleep in patients with COPD. *Chest*, 86, 366-374, 1984.

41. SKATRUD J.B., DEMPSEY J.A., IBER C., BEERSSENBRUGGE A. – Correction of CO2 retention during sleep in patients with chronic obstructive pulmonary disease. *Am. Rev. Respir. Dis.*, 124, 260-268, 1981.

42. STRADLING J.R., LANE D.J. – Nocturnal hypoxaemia in chronic obstructive pulmonary disease. *Clin. Sci.*, 64, 213-222, 1983.

43. TIRLAPUR V.G., MIR M.A. – Nocturnal hypoxemia and associated electrocardiographic changes in patients with chronic obstructive airways disease. *N. Engl. J. Med.*, 306, 125-130, 1982.

44. TRASK C.H., CREE E.M. – Oxymeter studies on patients with chronic obstructive emphysema, awake and during sleep. *N. Engl. J. Med.*, 266, 639-642, 1962.

45. VOS P.J.E., FOLGERING Th.M., VAN HERWAARDEN C.L.A. – Predictors for nocturnal hypoxaemia (mean SaO2 < 90%) in normoxic and midly hypoxic patients with COPD. *Eur. Respir. J.* 8, 74-77, 1995.

46. WEITZENBLUM E., MUZET A., EHRHART J., SAUTEGEAU A., WEBER L. – Variations nocturnes des gaz du sang et de la pression artérielle pulmonaire chez les bronchitiques chroniques insuffisants respiratoires. *Nouv. Presse Méd.*, 11, 1119-1122, 1982.

47. WYNNE J.W., BLOCK A.J., HEMENWAY J., HUNT L.A., FLICK M.R. – Disordered breathing and oxygen desaturation during sleep in patients with chronic obstructive lung disease. *Am. J. Med.* 66, 573-579, 1979.

48. YOUNG T., PALTA M., DEMSEY J., SKATRUD J., WEBER S., BADR S. – The occurrence of sleep-disordered breathing among middle-aged adults. *N. Engl. J. Med.*, 328, 1230-1235, 1993.

Chapter 45

Central alveolar hypoventilation syndrome

E. Weitzenblum

Service de Pneumologie, Hôpital de Hautepierre, Strasbourg, France

Alveolar hypoventilation of central origin and primary in appearance was described over 30 years ago: it is characterised by considerable daytime hypoxemia-hypercapnia, but which is often well-tolerated, and by the abolition of the ventilatory response to a hypercapnic stimulus. Central chemoreceptor dysfunctioning, demonstrated by an abnormal response to the CO_2 stimulus, is the most characteristic trait of this syndrome [2], but peripheral chemoreceptor deficiency, responsible for the ventilatory response to hypoxia, has also been observed in some cases. The syndromes of central alveolar hypoventilation and central apnoeas are not absolutely synonymous, even if the former can elicit the latter: central apnoeas have been observed in the absence of any daytime hypoventilation, and central hypoventilation syndrome is not necessarily accompanied by central apnoeas during sleep. Nor is central alveolar hypoventilation synonymous with obesity-hypoventilation syndrome (the current term for Pickwick's syndrome) even though these conditions have several points in common. Central apnoea syndrome and obesity-hypoventilation syndrome are described in other chapters of the present volume.

Severinghaus and Mitchell [3] used the term "Ondine's curse" in 1962, in reference to Giraudoux (Ondine, act three, scene six), to describe a picture of central alveolar hypoventilation, which is iatrogenic, observed in three patients after chordotomy. In fact, various neurological conditions such as acute bulbar poliomyelitis, can lead to central hypoventilation but this is usually an idiopathic condition, sometimes, but not always, with a neurological history (encephalitis, cranial traumatism). But a detailed macroscopic, histological study of the central nervous system has failed to show any significant abnormalities, in most cases

There is an infant and childhood form which is often congenital, and an adult form which is probably acquired. It is rare in infants and only slightly more common in adults. The term "Ondine's curse" is most appropriate in the case of alveolar hypoventilation in infants, as blood gases can be normal during the daytime but progressively worsen during sleep, due to the disappearance of the behavioural (cerebral) control of breathing, while automatic (metabolic) control is seriously disrupted. Sleep is characterised by deep hypoventilation (major hypoxemia-hypercapnia) with occasional central apnoeas. Hypercapnia is easily corrected by assisted ventilation, but also by crying; it is aggravated by inhaling oxygen. The former observation reported by Mellins in 1970 [2] referred to an infant who was monitored and treated up to the age of 14 months, the age at which the infant died as a result of major right cardiac insufficiency.

The adult form is slightly more common as Mellins [2] collated 30 cases, published from 1970 onwards, even though the *princeps* description only dates from 1957. Roughly a hundred cases have been published to date and the number of observed cases is certainly higher. Diagnosis is often made between the ages of 20 and 60. There is a clear male predominance (80%). Hypoventilation may be discovered by chance during investigations for various reasons (simple bronchitis, preoperative check up etc.) which involve a measurement of arterial blood gases: one is often surprised to discover hypoxemia-hypercapnia ($PaO_2 < 70$, $PaCO_2 > 45$) in subjects with no previous respiratory history, who do not complain of dyspnoea during effort and whose pulmonary auscultation is generally normal (there is sometimes a diminished vesicular murmur or crepitation at the base). There is no severe obesity and so idiopathic central alveolar hypoventilation must be distinguished from obesity-hypoventilation syndrome, even if these have a number of points in common. Spirographic exploration shows no marked abnormalities and notably, no obstructive ventilatory deficiency.

In other, probably more common cases, alveolar hypoventilation is discovered when complications develop in the syndrome, such as polycythemia, attacks (sometimes iterative) of right cardiac failure, acute respiratory insufficiency with occasional respiratory encephalopathy. The most frequent symptoms include headaches, insomnia, daytime sleepiness, night snoring (non specific), and tiredness. It must be stressed that clinical tolerance of hypoxemia-hypercapnia can remain very high for several years.

Voluntary hyperventilation normalises blood gases, at least in the early stages of the illness. Moreover patients vary widely in arterial blood gas levels, even during wakefulness. The dissociation between blood gas disruption (which may be intense) and the absence of any notable ventilatory deficiency is characteristic. CO transfer is normal. Pulmonary arterial hypertension is virtually the norm, and is linked to the effects of prolonged alveolar hypoxia on pulmonary circulation; this must be examined for and quantified by non invasive means (Doppler echocardiography) or by right cardiac catheterisation.

Table 45.1. Main results of supplementary explorations in a patient presenting a typical picture of central alveolar hypoventilation

Mr. F... Robert, 63 years old.
History of severe cranial traumatism at the age of 50. Hypoxemia-hypercapnia and polycythemia (Hb = 19g %), known for 10 years. No known respiratory history. Non smoker. No dyspnoea. Weight = 82 kg. Height = 1m 69 cm.

Pulmonary function tests: vital capacity = 3,025 ml (81% of theoretical), residual volume. = 1750 (91%) total lung capacity = 4,730 (84%), forced expiratory volume = 2,375 (94%) forced expiratory volume/vital capacity = 78% (111%). Peak expiratory rate: 9 l/s; at 50% vital cap.: 5.2 l/s.
Total airway resistance: 1.7 cm H_2O (normal).
Maximal static inspiratory pressure: 110 cm H_2O (normal).
CO transfer (inspiratory apnoea): normal.

Arterial blood gases: PaO_2= 48 mmHg, $PaCO_2$= 50mmHg, pH = 7.39.
During exercise (40 watts): PaO_2 = 41 mmHg, $PaCO_2$= 60 mmHg, pH = 7.33.

Ventilatory response to CO_2 stimulus: $\Delta VE/\Delta PaCO_2$ = 0.23 l/mm/mmHg (normal: 1.5-5).
ΔP 0.1/$\Delta PaCO_2$ = 0.037 cm H_2O/mmHg (normal: 0.2-1.6).

Right cardiac catheterisation:
At rest: mean pulmonary arterial pressure (PAP) = 22 mmHg, capillary pulmonary pressure = 6.5 mmHg.
Effort (40 watts): PAP = 38 mmHg.

Polysomnography: accentuated hypoventilation, especially in REM sleep. Some central apnoeas. No obstructive apnoeas.

Note the substantial ventilatory deficiency, maintenance of good ventilatory mechanics and collapse of the response to CO_2, whether expressed in terms of a variation in ventilation (ΔVE) or variation in P 0.1 (ΔP 0.1). Absence of obstructive apnoeas during sleep.

Diagnosis always relies on the collapse, sometimes even the total abolition, of the ventilatory response to CO_2. In fact, it is not easy to interpret a diminished ventilatory response to CO_2 as this may denote the effect rather than simply the cause of hypercapnia. Hypersensitivity to CO_2 is of high diagnostic value when it is marked (variation of minute ventilation/variation of PaO_2 < 0.5 l/mmHg) and when this is observed in a patient with no marked ventilatory deficiency. The ventilatory response to a hypoxic stimulus (more rarely studied) is diminished in roughly two thirds of cases. Administering oxygen will lead to a major drop in ventilation and a worsening of hypercapnia, due to the combination of suppressing the hypoxic stimulus and hyposensitivity to the hypercapnic stimulus.

Polysomnographic explorations must be carried out systematically in the case of these patients. Because the syndrome is rare, little data is available. What is most commonly observed is the presence of alveolar hypoventilation during all stages of sleep [1], usually more marked than in the wakefulness state, and sometimes more severe during REM sleep. Central apnoeas do not generally occur [1]. They are observed in some patients, particularly during light NREM sleep (stages 1-2);

these may be prolonged. In the case of one patient, who we examined on several occasions, the results of the polysomnography were not reproducible, and depending on the recording, showed isolated alveolar ventilation, hyperventilation associated with central apnoeas and sometimes an association between hypoventilation, central apnoeas and obstructive apnoeas. Polysomnography is also useful in assessing the effectiveness of the various types of treatment.

The most logical treatment for central alveolar hypoventilation is the use of respiratory stimulants, but few substances have proved effective. Progesterone can, on rare occasions correct hypoxemia-hypercapnia. Doxapram has non-negligible toxicity (convulsive effects). Almitrine, which stimulates peripheral chemoreceptors, has little effect when taken orally over a long period. Assisted ventilation, either non invasive through nasal mask, or invasive by endotracheal probe, is used only in severe forms and during certain episodes of acute aggravation; continuous nasal positive pressure has provided good results in cases which include obstructive sleep apnoeas [4]. It is not indicated for the usual forms of alveolar hypoventilation. Electrostimulation of the phrenic nerves appears to be effective but experiments refer to a limited number of patients only; it may favour the onset of obstructive apnoeas. Lastly, long term oxygen therapy can be prescribed provided that clinical tolerance and $PaCO_2$ are closely monitored; this has been used on one of our patients for over ten years now, with very good results.

REFERENCES

1. COCCAGNA G., CIRIGNOTTA F., ZUCCONI M., GERARDI R., MEDORI R., LUGARESI E. – A polygraphic study of one case of primary alveolar hypoventilation (Ondine's curse). *Bull. Eur. Physio. Path. Resp.*, 20, 157-161, 1984.
2. MELLINS R.B., BALFOUR H.H., TURINO G.M., WINTERS R.W. – Failure of automatic control of ventilation (Ondine's curse). Report of an infant born with this syndrome and review of the literature. *Medicine*, 49, 487-504, 1970.
3. SEVERINGHAUS J.W., MITCHELL R.A. – Ondine's curse. Failure of respiratory center automaticity while awake. *Clin. Res.*, 10, 122, 1962.
4. SULLIVAN C.E., BERTHON-JONES M., ISSA F.G. – Remission of severe obesity-hypoventilation syndrome after short-term treatment during sleep with nasal continuous positive airway pressure. *Am. Rev. Respir. Dis.*, 128, 177-181, 1983.

More detailed references may be found in: CHAOUAT A., WEITZENBLUM E., KESSLER R., SCHOTT R. - Syndrome d'hypoventilation alvéolaire centrale et syndrome d'apnées centrales. *In Encyc. Med. Chir.* Editions Scientifiques et Médicales, Elsevier, SAS, Paris., 6-040-K-10, 5p, 2000.

Chapter 46

Sleep breathing abnormalities in neuromuscular diseases

M.A. Quera Salva, G. Mroue, Ph. Gajdos, J. C Raphael, and F. Lofaso
Service de Réanimation Médicale et Explorations Fonctionnelles, Hôpital Raymond Poincaré, Garches, France,

In neuromuscular diseases, respiratory muscle paralysis, and the thoracic deformities which often accompany these diseases, may lead to a respiratory deficiency resulting in alveolar hypoventilation [8, 16, 21, 23, 46]. During sleep, these breathing abnormalities may worsen and apnoeas or hypopnoeas may appear. Thus patients with moderate ventilatory function restrictions may present episodes of severe haemoglobin desaturation, during sleep. These patients also complain of nocturnal arousals with a sensation of asphyxia, as well as daytime sleepiness. Nocturnal mechanical ventilation will correct nocturnal alveolar hypoventilation, improve sleep quality and daytime haematosis [16, 19, 20, 26, 36, 38, 48]. It is thus important to detect sleep breathing abnormalities by polysomnographic recording in subjects presenting symptoms suggesting sleep breathing abnormalities [1].

PATHOPHYSIOLOGY

During sleep onset and NREM sleep, airway resistance increases with the increased activity of the upper airway dilatory muscles [30, 33, 40]. On the other hand, in NREM sleep, intercostal muscle activity increases and trans-diaphragmatic pressure is greater than during wakefulness [45]. In normal subjects, the balance is maintained between the suction forces of the inspiratory muscles (accessory inspiratory muscles, particularly the diaphragm) and those of the dilatory muscles of the upper airways (genioglossus, cricoarytenoid and geniohyoid).

In cases where the oropharyngeal muscles are affected by certain neuromuscular pathologies, and in the frequent cases of macroglossia in patients affected by Duchenne's dystrophy, upper airway resistance is increased and may provoke obstructive apnoeas.

REM sleep is characterised by postural muscle atonia [35]. This atonia affects the intercostal muscles and the accessory respiratory muscles [11, 55]. Thus during REM sleep, ventilation is chiefly maintained by the diaphragm. When these neuromuscular diseases cause diaphragmatic paralysis, apnoeas or non obstructive hypopnoeas may be provoked during REM sleep. These apnoeas are often referred to as central. As their main mechanism is generally linked to a diaphragmatic deficiency, the term "diaphragmatic" apnoeas is more appropriate.

SLEEP BREATHING ABNORMALITIES IN DIFFERENT PATHOLOGIES

Myasthenia

This is characterised by muscular weakness linked to a block in neuromuscular transmission. The muscular deficiency may affect the respiratory muscles, particularly the diaphragm, causing respiratory insufficiency [44]. Myasthenics may present apnoeas or hypnopoeas during sleep, resulting in considerable desaturation. These episodes chiefly occur during REM sleep [49, 52]. We used polysomnography to study twenty patients with generalised myasthenia, whose condition was stable at the time of study [49]. Eleven patients had a distinctly pathological rate of apnoeas and hypopnoeas per hour of REM sleep (mean 45) and four of these had more than ten apnoeas and hypopnoeas per hour of sleep (all sleep stages included). With the exception of one patient these

581

episodes were of the central type, probably due to the diaphragm being affected. All these patients presented symptoms suggesting a sleep apnoea syndrome. The occurrence of apnoeas, hypopnoeas and nocturnal desaturation correlated to the extent of total pulmonary volume reduction.

Acid maltase deficiency

This often leads to the early development of diaphragmatic paralysis [53]. Sleep abnormalities with nocturnal awakenings accompanied by a sensation of asphyxia may be signs of the disease [29]. Apnoeas predominate or appear exclusively in REM sleep and are of the central (or diaphragmatic) type.

Myotonic dystrophy

Myotonic dystrophy (MD) is a progressive multisystem disorder with an estimated prevalence of 2.5 to 5.5 per 100,000. The abnormal gene for this autosomal dominant disorder has been localised to chromosome 19. It has been known for some time that sleep disorders occur with Steinert's myotonic dystrophy. These patients present alveolar hypoventilation due to inspiratory muscle deficiency or myotonia [9, 10, 24] and often complain of daytime sleepiness. Polysomnographic studies show sleep apnoeas of both the obstructive and non obstructive (central and diaphragmatic) types [15, 16, 27, 39]. Non obstructive apnoeas are explained by the fact that the diaphragm is affected. Obstructive apnoeas may occur in relation to myotonia but with the mandibular alterations frequently observed in these patients [29]. Kilburn *et al.* [37] suggest that daytime sleepiness is linked to alveolar hypoventilation. Coccagna *et al.* [16] and Leygonia-Goldenberg *et al.* [39] suggest that the syndrome occurs because the central nervous system is affected, as several patients present direct sleep onset REM periods and diminished hypercapnia sensitivity. On the other hand, in the series presented by Guilleminault *et al.* [27], daytime sleepiness was related directly to sleep fragmentation due to apnoeic episodes and for these authors, diminished sensitivity to hypercapnia is secondary rather than primary.

MD is a disease involving abnormalities in many organs, including the blood vessels, smooth muscle and uterus. These patients present neuropsychological deficits such as mental impairment and personality changes. Hypersomnia and cognitive impairment may be secondary to sleep apnoea, but daytime sleepiness is common in patients with myotonic dystrophy, even in patients without sleep apnoea and or fragmented sleep. In fact brain abnormalities such as enlarged ventricules on computer tomographic scans [3], white-matter lesions on magnetic resonance images [25, 32] and decreased cerebral blood flow on positron emission tomography [14] have been described in these patients, demonstrating significant brain involvement in MD. Furthermore, the fact that some of these patients present a disordered respiratory rhythm during wakefulness and light sleep, but no deep NREM sleep [13, 56] indicates a central participation which is insufficiently corrected by the chemical control mechanism. Finally, Damian *et al.* [18] have demonstrated the effectiveness of modafinil from 200 to 400 mg/day for excessive daytime sleepiness in an open study in 11 MD patients.

Duchenne's muscular dystrophy

This starts in childhood and follows a stereotyped pattern of development. The respiratory muscles, and the diaphragm in particular, are affected relatively late. If scoliosis is present this may further affect the respiratory function. Sleep breathing abnormalities have been reported [7, 41, 50, 54] often during REM sleep and may result in serious desaturation. These apnoeas are of the central or diaphragmatic type, apparently due to the diaphragm being affected. Obstructive apnoeas exist in relation to the macroglossia and the pharyngeal muscles are also affected.

Limb-girdle dystrophy

This leads to the fairly early onset of diaphragmatic paralysis with hypoventilation. Of the five patients we studied with limb-girdle dystrophy, two presented diaphragmatic apnoeas predominating in REM sleep.

Patients with severe kiphoscoliosis

Patients may present non obstructive apnoeas and nocturnal hypoventilation particularly during REM sleep [42, 43, 51]. The drop in SaO_2 during sleep correlates with the daytime SaO_2 level and the extent of pulmonary volume restriction. The fact that episodes of nocturnal desaturation, with or without associated apnoeas, often occur during REM sleep is perhaps due to thoracic deformity and accessory respiratory muscle atonia, which are responsible for a paradoxical thoracic movement during this stage of sleep.

RESULTS

Patients affected by a neuromuscular pathology often present sleep-related disorders (apnoeas, hypopnoeas and severe desaturation). This directly results in sleep fragmentation, daytime sleepiness and may have an adverse effect on daytime alveolar hypoventilation. The extent and duration of nocturnal desaturation may contribute to the occurrence of pulmonary arterial hypertension and right cardiac insufficiency. The precise frequency of these ventilatory disturbances during sleep is difficult to ascertain, and is likely to depend on the stage of the disease. In a series of forty-five patients suffering from the various neuromuscular diseases discussed here (table 46.1 and 46.2), fifteen had more than ten apnoeas or hypopnoeas per hour of sleep, leading to a substantial drop in SaO_2 chiefly in REM sleep. The extent of these abnormalities correlates with reductions in pulmonary volume and maximal expiratory pressure as well as with daytime hypoxia and. hypercapnia (table 46.3). Bye *et al.* [12] found a similar link between the severity of the restrictive syndrome and the occurrence of apnoeas and nocturnal desaturation. It is interesting to note that these authors show a correlation between desaturation during REM sleep and the extent to which vital capacity is reduced by shifting from a seated to a supine position, thus underlining the major role likely to be played by diaphragmatic paralysis in generating these abnormalities.

Table 46.1. Results of functional respiratory and polysomnographic explorations in 45 patients affected by neuromuscular diseases

Variable	Patients (N = 45) Mean ± SD
Age	39 ± 16
BMI	21 ± 5
VC (% norm)	56 ± 28
TPC (% norm)	70 ± 20
FRC (% norm)	85 ± 24
Pimax (% norm)	42 ± 25
Pemax (% norm)	41 ± 28
PaO₂ mmH	80 ± 9
PaCO₂ mmHg	41 ± 6
RDI	12 ± 18
REMRDI	33 ± 39
TB 90%	38 ± 39
TB 5 % BL	24 ± 27

BMI: weight in kg x 10,000/height in cm²
% norm: percentage of normal values
VC: vital capacity
TPC: total pulmonary capacity
FRC: functional residual capacity
Pimax and Pemax: maximal inspiratory and expiratory statistical pressures
RDI: number of apnoeas and hypopnoeas per hour of sleep
REMRDI: number of apnoeas and hypopnoeas per hour of REM sleep
TB 90%: time spent with SaO_2 below 90%
B 5 % BL: time spent with diminished $SaO_2 \geq 5$ in relation to baseline SaO_2

Table 46.2. Neuromuscular pathology, frequency and type of nocturnal ventilatory disorder

Pathology	Number of patients with RDI > 10/total number of patients	Type of nocturnal ventilatory disorder (number of patients)
Myasthenia	4/20	Diaphragmatic (4)
Duchenne	4/6	Obstructive (3)
Limb-girdle dystrophy	2/5	Diaphragmatic (1)
Steinert	2/3	Diaphragmatic (2)
Idiopathic scoliosis	1/3	Obstructive (1)
Postpoliomyelitic scoliosis	0/2	Diaphragmatic (1)
Spinal amyotrophia	0/2	Diaphragmatic (1)
Dystrophy from maltase deficiency	1/2	Diaphragmatic (1)
Post-traumatic tetraplegia	1/2	Obstructive (2)

In terms of practice, it is essential to establish the extent of these sleep breathing abnormalities. The inquiry systematically look for features evocative of sleep apnoea syndrome. The presence of these signs and/or an important restrictive syndrome, would call for nocturnal oximetry or even polysomnography.

Table 46.3. Spearman's correlations between nocturnal ventilatory parameters and the results of functional respiratory explorations for the group of 45 neuromuscular disease patients

	RDI r (p)	REMRDI r (p)	TB 5% BL r (p)	TB 90% r (p)
Age	.13 (.38)	-.11 (.48)	.19 (.20)	.25 (.09)
BMI	-.06 (.69)	-.17 (.26)	.08 (.62)	.09 (.52)
TLC	-.43 (**.003**)	-.53 (**.0002**)	-.54 (**.0001**)	-.42 (**.003**)
VC	-.32 (**.02**)	-.52 (**.0002**)	-.36 (**.01**)	-.35 (**.01**)
FRC	-.30 (**.04**)	-.40 (**.006**)	-.43 (**.003**)	-.30 (.04)
Pemax	-.43 (**.002**)	-.43 (**.003**)	-.37 (**.01**)	-.37 (**.01**)
Pimax	-.23 (.13)	-.28 (.06)	-.22 (.14)	-.29 (.06)
PaO_2	-.57 (**.0001**)	-.34 (**.01**)	-.63 (**.0001**)	-.63 (**.0001**)
$PaCO_2$	.30 (.04)	.42 (**.003**)	.52 (**.0003**)	.59 (**.0001**)

r: Spearman's correlation factor
BMI: weight in kg x 10,000/height in cm
TLC: total lung capacity
VC: vital capacity
FRC: functional residual capacity
Pimax and Pemax: maximal inspiratory and expiratory statistical pressures
RDI, REMRDI, TB 5% BL, TB 90% as in table 46.1.

Evidence of an important nocturnal ventilatory disorder is an element which may indicate treatment by intermittent positive pressure ventilation (IPPV) to correct nocturnal desaturation, and improve both the quality of sleep and daytime haematosis [6].

TREATMENT OF SLEEP BREATHING ABNORMALITIES

Patients without a moderate ventilatory restriction and predominantly obstructive and mixed apnoeas can be treated as a first step with continuous positive airway pressure through a nasal mask. Guilleminault *et al.* [128] also reported the efficacy in treating a group of neuromuscular patients with bilevel positive airway pressure by nasal mask. However the authors studied 19 patients who were seen in the sleep clinic for symptoms of sleepiness or nocturnal sleep disturbances, which impaired work and socio-familial activities. Their patients had a moderate ventilatory restriction which was not associated with daytime hypoventilation. One patient could not tolerate the nasal mask from the start. Four patients had to be switched to volume cycled nocturnal IPPV during the first two years, 10 patients needed adjustment of the initially determined pressures. Six patients needed low flow oxygen fed into their masks, three of whom subsequently switched to IPPV.

Nocturnal IPPV applied by nasal mask or through tracheostomy is recommended in patients with neuromuscular or chest-wall diseases presenting a severe ventilatory restriction and daytime hypercapnia. Nocturnal IPPV is easy to implement and in stable patients with a slowly progressive disease, nocturnal IPPV effects may be prolonged after a very long period of ventilation [2, 6, 22].

This long term nocturnal ventilation improves chronic hypoventilation during daytime spontaneous ventilation [4, 6, 19, 31] and produces improved respiratory drive both asleep and awake and improved arousal responses to abnormal blood gases [2, 47]. The improvement in ventilatory response to CO2 probably relates to the central chemoreceptors adapting to the reduction of hypercapnia overnight [2, 34] since IPPV does not affect respiratory muscle strength [6, 26, 31] or lung compliance [2, 6].

The clinical indicators [17] for non-invasive positive pressure ventilation in restrictive thoracic diseases include symptoms (such as fatigue, dyspnea, morning headache etc.) and one of the following physiological criteria: 1. PaCO2 >/= 45 mm Hg; 2. nocturnal oxymetry demonstrating oxygen saturation </= 88% for 5 consecutive minutes; and 3. for progressive neuromuscular disease, maximal inspiratory pressures < 60 cm/H2O or FVC < 50%.

Despite the implementation of nocturnal IPPV some patients may continue to present symptoms of fatigue or daytime sleepiness or inadequate oxygenation in nocturnal oxymetry. In these cases, sleep studies may be helpful to disclose the origins of the defective ventilation such as oro-nasal leaks (see figure 46.1) or an asynchrony between the patient and the ventilator (see figure 46.2).

Furthermore when the hypercapnia is not well controlled by IPPV, a tracheotomy should be considered.

Finally, in chronic and rapidly progressive neuromuscular diseases, such as Duchenne muscular dystrophy, mechanical support of minute ventilation is usually associated with deterioration to the point of complete ventilator dependency for survival.

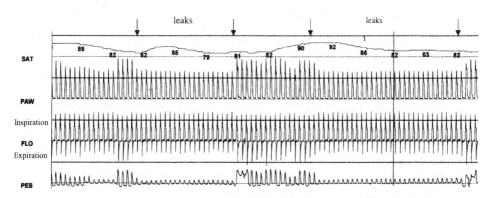

Figure 46.1. Night-time hypopnoeas due to mask leaks in a neuromuscular patient treated with IPPV. PES: oesophageal pressure; FLO: airway flow; PAW: airway pressure. Leaks are characterised by a decrease of airway pressure (PAW) produced by the ventilator and a decrease of expiratory flow.

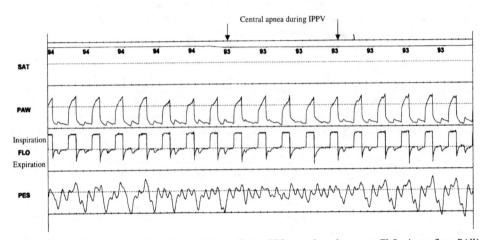

Figure 46.2. Asynchrony between the patient and the ventilator. PES: oesophageal pressure; FLO: airway flow; PAW: airway pressure. Interestingly expiratory effort occurred during mechanical expiration. This induced an asynchrony of expiration (biphasic expiration) which disappeared during a central apnoea.

REFERENCES

1. AMERICAN THORACIC SOCIETY – Medical section of the American lung association. Indications and standards for cardiopulmonary sleep studies. *Am. Rev. Respir. Dis.,* 139, 559-568, 1989
2. ANNANE D., QUERA-SALVA M.A., LOFASO F., VERCKEN J.B., LESIEUR O., FROMAGEOT C., CLAIR B., GAJDOS P., RAPHAEL J.C. – Mechanisms underlying effects of nocturnal ventilation on daytime blood gases in neuromuscular diseases. *Eur. Respir. J.* 1999, 13, 157-162.
3. AVRAHAMI E., KATZ A., BORNSTEIN N., KORCZYN A.D. – Computed tomographic findings of brain and skull in myotonic dystrophy. *J. Neurol. Neurosurg. Psychiatry,* 50, 435-438, 1987.
4. BACH J.R., ALBA A., BOHATIUK G., SAPORITO L., LEE M. – Mouth intermittent positive pressure ventilation in the management of post-polio respiratory insufficiency. *Chest,* 91, 859-864, 1987.
5. BACH J.R., ALBA A., MOSHER R., DELAUBIER A. – Intermittent positive pressure ventilation via nasal access in the management of respiratory insufficiency. *Chest,* 92, 168-170, 1987.
6. BARBE F., QUERA-SALVA M.A., DE LATTRE J., GADJOS Ph, AGUSTI A.G.N. – Long term effects of nasal intermittent positive-pressure ventilation on pulmonary function and sleep architecture in patients with neuromuscular diseases. *Chest,* 110, 1179-1183, 1996.
7. BARBE F., QUERA-SALVA M.A., McCANN C., GAJDOS Ph., RAPHAEL J.C., de LATTRE J., AGUSTI A.G.N. – Sleep-related respiratory disturbances in patients with Duchenne muscular dystrophy. *Eur. Respir. J.,* 7, 1403-1408, 1994.
8. BAYDUR A. – Respiratory muscle strength and control of ventilation in patients with neuromuscular disease. *Chest,* 99, 330-338, 1991.
9. BEGIN R., BUREAU A., LUPIEN L., BERNIER J.P., LEMIEUX B. – Pathognesis of respiratory insufficiency in myotonic dystrophy. *Am. Rev. Respir. Dis.,* 125, 312-318, 1982.
10. BENAIM S., WORSTER-DROUGHT C. – Dystrophia myotonica: with myotonia of the diaphragm causing pulmonary hypoventilation with anoxaemia and secondary polycythaemia. *Med. Illustrated,* 8, 221-226, 1954.
11. BERGER R. – Tonus of the extrinsic laryngeal muscles during sleep and dreaming. *Science,* 134, 840, 1961.
12. BYE P.T.B., ELLIS E.R., ISSA F.G., DONNELLY P.M., SULLIVAN C.E. – Respiratory failure and sleep in neuromuscular disease. *Thorax,* 45, 241-247, 1990
13. CALABRESE P., GRYSPEERT N., AURIANT I., FROMAGEOT C., RAPHAEL J.C., LOFASO F., BENCHETRIT G. - Postural breathing pattern changes in patients with myotonic dystrophy. *Respir. Physiol.* 122, 1-13, 2000.
14. CHANG L., ANDERSON T., MIGNECO O.A., BOONE K., MEHRINGER C.M., VILLANUEVA-MEYER J., BERMAN N., MENA I. – Cerebral abnormalities in myotonic dystrophy. Cerebral blood flow, magnetic resonance imaging, and neuropsychological tests. *Arch. Neurol.* 50, 917-923, 1993.
15. CIRIGNOTTA F., MONDINI S., ZUCCONI M., BARROT-CORTES E., STURANI C., SCHIAVINA M., COCCAGNA G., LUGARESI E. – Sleep-related breathing impairment in myotonic dystrophy. *J. Neurol. ,* 235, 80-85, 1987.
16. COCCAGNA G., MANTOVANI M., PARCHI C., MIRONI F., LUGARESI E. – Alveolar hypoventilation and hypersomnia in myotonic dystrophy. *J. Neurol. Neurosurg. Psychiatry,* 38, 977-984, 1975.
17. CONSENSUS CONFERENCE – Clinical indications for non-invasive positive pressure ventilation in chronic respiratory failure due to restrictive lung diseases, COPD, and nocturnal hypoventilation - A consensus conference report. *Chest,* 116, 521-534, 1999.

18. DAMIAN M.S., GERLACH A., SCHMIDT F., LEHMANN E., REICHMANN H. – Modafinil for excessive daytime sleepiness in myotrophic dystrophy. *Neurology,* 56, 794-796, 2001.
19. ELLIS E.R., BYE P.T.B., BRUDERER J.W., SULLIVAN C.E. – Treatment of respiratory failure during sleep in patients with neuromuscular disease. Positive-pressure ventilation through a nose mask. *Am. Rev. Respir. Dis.,* 135, 148-152, 1987.
20. ELLIS E.R., GRUNDSTEIN R.R., SHU CHAN M.D., BYE P.T.B., SULLIVAN C.E. – Noninvasive ventilatory support during sleep improves respiratory failure in kyphoscoliosis. *Chest,* 94, 811-815, 1988.
21. ESTENNE M., HEILPORN A., DELHEZ L., YERNAULT J.C., de TROYER A. – Chest wall stiffness in patients with chronic respiratory muscle weakness. *Am. Rev. Respir. Dis.,* 115, 389-395, 1977.
22. GARAY S.M., TURINO G.M., GOLDRING R.M. – Sustained reversal of chronic hypercapnia in patients with alveolar hypoventilation syndromes. Long-term maintenance with non-invasive nocturnal mechanical ventilation. *Am. J. Med.* , 70, 269-274, 1981.
23. GIBSON G.J., PRIDE N.B., NEWSON DAVIS J., LOH L.C. – Pulmonary mechanics in patients with respiratory muscle weakness. *Am. Rev. Respir. Dis.,* 115, 389-395, 1977.
24. GILLAM P.M.S., HEAF P.J.D., KAUFMAN L., LUCAS B.G.B. – Respiration in dystrophia myotonica. *Thorax,* 19, 112-120, 1964.
25. GLANZ R.H., WRIGHT R.B., HUCKMAN M.S., GARRON D.C., SIEGEL I.M. – Central nervous system magnetic resonance imaging findings in myotonic dystrophy. *Arch. Neurol.* 45, 36-37, 1988.
26. GOLDSTEIN R.S., de ROSIE J.A., AVENDANO M.A., DOLMAGE T.E. – Influence of non-invasive positive pressure ventilation on inspiratory muscles. *Chest,* 99, 408-415, 1991.
27. GUILLEMINAULT C., CUMMISKEY J., MOTTA J., LYNNE-DAVIES P. – Respiratory and hemodynamic study during wakefulness and sleep in myotonic dystrophy. *Sleep,* 1, 19-31, 1978.
28. GUILLEMINAULT C., PHILIP P., ROBINSON A. – Sleep and neuromuscular disease: bilevel positive airway pressure by nasal mask as a treatment for sleep disordered breathing in patients with neuromuscular disease. *J. Neurol. Neurosurg. Psychiatry,* 65, 225-232, 1998.
29. GUILLEMINAULT C., STOOHS R., QUERA-SALVA M.A. – Sleep related obstructive and nonobstructive apneas and neurologic disorders. *Neurol.,* 42(suppl. 6), 53-60, 1992.
30. HENKE K., DEMPSEY J., BADR M., KOWITZ J., SKATRUD J. – Effect of sleep induced increases in upper airway resistance on respiratory muscle activity. *J. Appl. Physiol.,* 70, 158-168, 1991.
31. HOEPPNER V.H., COCKCROFT D.W., DOSMAN J.A., COTTON D.J. – Night-time ventilation improves respiratory failure in secondary kyphoscoliosis. *Am. Rev. Respir. Dis.,* 129, 240-243, 1984.
32. HUBER S.J., KISSEL J.T., SHUTTLEWORTH E.C., CHAKERES D.W., CLAPP L.E., BROGAN M.A. – Magnetic resonance imaging and clinical correlates of intellectual impairment in myotonic dystrophy. *Arch. Neurol.* 46, 536-540, 1989.
33. HUDGEL D., MARTIN R., JOHNSON B., HILL P. – Mechanics of the respiratory system and breathing pattern during sleep in normal humans. *J. Appl. Physiol.,* 56, 133-137, 1984.
34. HUKINS C.A., HILLMAN D.R. – Daytime predictors of sleep hypoventilation in Duchenne muscular dystrophy. *Am. J. Respir. Crit. Care Med.* 161, 166-170, 2000.
35. JOUVET M. – Recherches sur les structures nerveuses et les mécanismes responsables des différentes phases du sommeil physiologique. *Arch. Ital. Biol.,* 100, 125-206, 1962.
36. KERBY G.R., MAYER L.S., PINGLETON S.K. – Nocturnal positive pressure ventilation by nasal mask. *Am. Rev. Respir. Dis.,* 92, 738-760, 1987.
37. KILBURN K., EAGAN J., HEYMAN A. – Cardiopulmonary insufficiency associated with myotonic dystrophy. *Am. J. Med.,* 26, 929-935, 1959.
38. LEGER P., MADELON J., JENNEQUIN J., GERARD M., ROBERT D. – Non-invasive home IPPV via nasal mask in nocturnal ventilator dependent patients with musculoskeletal disorders. *Chest,* 92, 109S, 1987.
39. LEYGONIE-GOLDENBERG F., PERRIER M., DUIZABO P.H., BOUCHAREINE A., HARF A., BARBIZET J., DEGOS J.D. – Troubles de la vigilance, du sommeil et de la fonction respiratoire dans la maladie de Steinert. *Rev. Neurol.,* 33, 255-270, 1977.
40. LOPES J., TABACHNICK E., MULLER N., LEVISON H., BRYAN A. – Total airway resistance and respiratory muscle activity during sleep. *J. Appl. Physiol.,* 54, 773-777, 1983.
41. MANNI R., ZUCCA C., GALIMBERTI C.A., OTTOLINI A., CERVERI I., BRUSCHI C., ZOIA M.C., LANZI G., TARTARA A. – Nocturnal sleep and oxygen balance in Duchenne muscular distrophy. *Eur. Arch. Psy. Clin. Neurosci.,* 240, 255-257, 1991.
42. MEZON B., WEST P., ISRAELS J., KRIEGER J. – Sleep breathing abnormalities in kiphoscoliosis. *Am. Rev. Resp. Dis.,* 122, 617-621, 1980.
43. MIDGREN B., PETERSON K., HANSSON L., ERIKSSON L., AIRIKKALA P., ELMQVIST D. – Nocturnal hypoxaemia in severe scoliosis. *Br. J. Dis. Chest,* 82, 226-236, 1988.
44. MIER-JEDRZEJOWICZ A., BROPHY C., GREEN M. – Respiratory muscle function in myasthenia gravis. *Am. Rev. Respir. Dis.,* 138, 867-873, 1988.
45. NAIFEH K.Y., KAMIYA J. – The nature of respiratory changes associated with sleep onset. *Sleep,* 4, 49-59, 1981.
46. NEWSON DAVIS J., GOLDMAN M., LOH L., CASSON M. – Diaphragm function and alveolar hypoventilation. *Q.J. Med.,* 45, 87-100, 1976.
47. PIPER A.J., SULLIVAN C.E. – Effects of long-term nocturnal ventilation on spontaneous breathing during sleep in neuromuscular chest wall disorders. *Eur. Respir. J.* 9, 1515-1522, 1996.
48. QUERA SALVA M.A., BARBE F., McCANN C., BOUVET F., de LATTRE J., GAJDOS Ph. – Long term nocturnal intermittent positive pressure ventilation through a nose mask in neuromuscular patients. *Am. Rev. Respir. Dis.,* 145, A865, 1992.

49. QUERA-SALVA M.A., GUILLEMINAULT C., CHEVRET S., TROCHE G., FROMAGEOT C., CROWE McCANN C., STOOHS R., de LATTRE J., RAPHAEL J.C., GAJDOS P. – Breathing disorders during sleep in myasthenia gravis. *Ann. Neurol.,* 31, 86-92, 1992.
50. REDDING G., OKAMOTO G., GUTHRIE R., ROLLEVSON D., HILSTERN J. – Sleep patterns in non ambulatory boys with Duchenne muscular distrophy. *Arch. Psy. Med. Rehabil.,* 66, 818-825, 1985.
51. SAWICKA E., BRANTHWAITE M. – Respiration during sleep in kyphoscoliosis. *Thorax,* 42, 801-808, 1987.
52. SHIOZAWA E., BRANTHWAITE M. – Sleep apnea in well-controlled myasthenia gravis. *Sleep Res.* 12, 84, 1987.
53. SIVACK E., SALANGA V., WILBOURN A., MITSUMOTO H., GOLISH J. – Adult onset maltase deficiency presenting as diaphragmatic paralysis. *Ann. Neurol.,* 9, 613-615, 1981.
54. SMITH P.E.M., CALVERLEY P.M.A., EDWARDS R.H.T. – Hypoxemia during sleep in Duchenne muscular dystrophy. *Am. Rev. Respir. Dis.,* 137, 884-888, 1988.
55. TUSIEWICZ K., MOLDOFSKI H., BRYAN A., BRYAN M. – Mechanics of the rib cage and diaphragm during sleep. *J. Appl. Physiol: Respirat. Environ. Exercise Physiol.,* 43, 500-602, 1977.
56. VEALE D., COOPER B.G., GILMARTIN J.J., WALLS T.J., GRIFFITH C.J., GIBSON G.J. – Breathing pattern awake and asleep in patients with myotonic dystrophy. *Eur. Respir. J.,* 8, 815-818, 1995.

Chapter 47

Cardiocirculatory disorders and sleep

G. Coccagna and C. Scaglione
Instituto di Clinica Neurologica, Universita di Bologna, Bologna, Italia

Since the advent of polysomnography, cardiocirculatory changes during sleep in normal subjects have been extensively monitored by recording a wide range of physiological parameters. These recordings demonstrated not only that the influence of the autonomic nervous system varied widely in sleep compared with wakefulness, but that it also behaved differently in NREM sleep and REM sleep. There is general agreement that vagal tone prevails in each sleep stage with respect to wakefulness, but how this occurs remains controversial. According to some [27], sympathetic tone diminishes in NREM sleep and rises again in REM sleep. On the basis of experiments in the cat, others have postulated an increase in vagal tone mainly in NREM sleep [2]. The phasic changes in neurovegetative function during REM sleep in man are thought to be due to both sudden inhibitions in vagal activity and abrupt increases in sympathetic activity. Since most pathological events involving the cardiocirculatory apparatus are influenced by the autonomic nervous system, it would be logical to assume that sleep also plays a prominent role.

The technical and ethical problems encountered in the polysomnographic investigation of patients with cardiocirculatory diseases have persistently hindered research and led to divergent and even contradictory findings. Imagine, for instance, the problems entailed in recording a patient in a coronary care unit or an angina sufferer left treatment free for the duration of the study.

After a short survey of the cardiocirculatory changes occurring during sleep under physiological conditions, this chapter will summarise the key findings on the relations between cardiocirculatory disorders and sleep.

HAEMODYNAMIC CHANGES DURING SLEEP IN NORMAL SUBJECTS

Under physiological conditions blood pressure drops rapidly in the early hours of sleep, reaching a nadir about two and a half hours after falling asleep. The maximum fall in pressure values is 20-23% compared with wakefulness [7, 75]. Pressure varies in different sleep stages. There is a progressive, significant decrease during the successive phases of NREM sleep, while values are similar to stage 2 sleep during REM sleep. Systolic pressure falls more than diastolic so that differential pressure gradually decreases during NREM sleep [3, 4, 7]. According to other studies the difference in blood pressure values in different sleep stages was not significant [40, 71].

Periodic phasic pressure oscillations may occur regularly in NREM sleep stages 1-2 with oscillation periods lasting from 20 to 30 seconds; these periodic changes are part of a wider sphere of concomitant fluctuations which affect different autonomic and somatic functions [36]. In REM sleep, abrupt irregular pressure increases often coincide with irregular breathing and, more seldom, clusters of rapid eye movements. Heart rate during NREM sleep drops 5% to 10% compared with wakefulness [27, 75]. Unlike blood pressure, heart rate does not reach minimum values in the early hours of sleep but continues to diminish throughout the night [65, 75]. During REM sleep heart rate is faster than in NREM sleep but tends to vary widely.

To what extent the changes in heart rate and/or arteriolar tone contribute to pressure variations during sleep is still unknown. Some have reported a fall in total peripheral resistance without changes in cardiac output [4], while others found a decreased cardiac output in all sleep stages without significant changes in total peripheral resistance [27]. It is likely that the mechanisms underlying phasic cardiocirculatory changes in REM sleep are highly complex, since polygraphic evidence has demonstrated that heart rate variations, sudden pressure increases and the onset of peripheral vasoconstrictions can all occur independently [7, 75].

ARTERIAL HYPERTENSION

In hypertensive patients blood pressure and heart rate decrease in a similar fashion to normal subjects [4, 28, 45]. Likewise, cardiac output and total peripheral resistance behave identically both in normal and hypertensive subjects [4, 28].

Some report no percentage difference in blood pressure fall during sleep compared with wakefulness in normal and hypertensive subjects [4, 28, 34] whereas others find that arterial pressure falls far less in hypertensive than in normotensive subjects [3, 62].

In hypertensive patients, Williams and Cartwright [81] noted a marked variability in blood pressure values even during NREM sleep, mimicking the trend in REM sleep observed in normal subjects.

It was recently reported that during sleep, between 17% and 40% of hypertensive subjects do not present the normal drop in pressure values agreed to be around 10% with respect to wakefulness levels. These individuals were called "non-dippers", to differentiate them from the "dippers" who undergo normal falls in pressure. In addition, "non-dippers" present significantly greater hypertrophy of the left ventricle [31, 78] and are also significantly more at risk for stroke [56].

As recently emphasised by Pickering [58], the pathophysiological mechanism underlying the failure to lower nocturnal pressure levels remains unsettled. Different biochemical and humoral explanations have been postulated, but none can be said to apply to all cases. Twenty years ago, through polysomnographic recordings, we identified snoring as a definite cause of persistently elevated pressure values during sleep [37]. These findings obtained in a limited number of patients were followed by an epidemiological survey which disclosed the link we had anticipated between snoring and arterial hypertension [38]. These results were subsequently confirmed by others [18,53,70]. Researchers at other laboratories performed polysomnography in patients with essential hypertension and found that many more of them presented nocturnal apnoeas than did a control group of normotensive subjects [15,24,32,82]. We do not know to what extent snoring and nocturnal obstructive apnoeas occur among hypertensive "non-dippers". However, the finding that habitual heavy snoring affects 24% of men and 14% of women in the general population should be something of an alarm bell for those concerned with cardiocirculatory problems, particularly essential hypertension.

A recent study investigated a group of patients with essential hypertension classified by cardiologists as "non-dippers" on the basis of 24-hour ambulatory blood pressure monitoring. These patients subsequently underwent polysomnographic recording including concomitant monitoring of respiratory and cardiocirculatory parameters. Similar recordings were made in a group of hypertensive patients classified as dippers. Of the 15 "non-dippers", two failed to present any breathing disorder during sleep while another two proved to be "dippers" at polysomnographic recording; the remainder presented snoring and/or apnoeas during sleep [60].

An epidemiological survey of 7901 workers [22] demonstrated a significant correlation between poor sleep quality and arterial hypertension. This does not necessarily mean that a cause-effect relationship exists between the two conditions, since external factors (e.g. many antihypertensive drugs) may interfere with sleep architecture. Surprisingly, this study failed to establish a positive correlation between hypertension and snoring.

ANGINA PECTORIS

Episodes of angina during sleep can occur in patients with classic or effort-induced angina (detected at ECG by a characteristic S-T segment depression) and in patients with resting angina or Prinzmetal's variant angina (with S-T segment elevation at ECG) or a mixed form. This fact has always aroused interest on the part of clinicians since it appeared to contradict the concept of sleep being a condition of absolute rest and reduced energy demand by the organism. However, as early as 1923 MacWilliam [41] pinpointed possible relationship between cardiocirculatory events and what he called "periods of disturbed sleep" characterised by dreams and nightmares that may be accompanied by striking neurovegetative changes. His simple clinical observation heralded the discovery of REM sleep some thirty years later.

The idea that the emotional content of dreams could trigger an attack of angina seemed to be confirmed by the study of Nowlin *et al.* [54] involving four patients with coronary disease and a history of nocturnal angina attacks, monitored in twelve polysomnographic recordings. In all, thirty nine episodes of angina were recorded accompanied by typical ECG changes, thirty two of which were associated with REM sleep. In two patients probability analysis demonstrated that the correlation between REM sleep and angina was significantly higher than a chance association. Although patients were not questioned immediately after awakening from a nocturnal angina attack, the following morning they reported having had dreams involving violent physical activity or strong emotions like fear, anger or frustration. Another patient with variant angina studied by King *et al.* [29] presented several episodes of S-T segment elevation during polysomnographic recordings, some of which were accompanied by pain during REM sleep. However, the strict correlation between angina and REM sleep has not been confirmed, or confirmed only in part by other authors. Snyder [74], for example, noted the correlation in one patient, whereas angina attacks were confined to the early hours of the night and unrelated to REM sleep in another case. Other studies failed to establish any relationship between angina attacks and different sleep stages [6, 39, 66].

Murao *et al.* [51] observed 58 ischaemic episodes with S-T segment elevation or depression in fourteen polysomnographic recordings from twelve patients. Most attacks occurred between 4 and 6 a.m., 21 episodes were in wakefulness and of the 37 attacks during sleep, 24 arose in REM sleep or in close proximity to REM periods. The fewest episodes were recorded in deep NREM sleep, but in one patient attacks occurred only during NREM sleep and in three cases only during wakefulness. During 45 nocturnal polygraphic recordings Maggini *et al.* [42] found 30 ischaemic episodes characterised by S-T segment elevation or depression, 42% of which were accompanied by pain. Most attacks occurred between midnight and 4 a.m. with a peak between 3 and 4 a.m. However, there was no significant difference between the likelihood of an ischaemic episode arising in wakefulness and the different sleep stages, although the attacks tended to be more frequent in wakefulness, light NREM and REM sleep. In patients with variant angina, Water *et al.* [80] found a circadian trend in the onset of ischaemic episodes, with a greater incidence between midnight and 8 am.

On the other hand, other authors [48, 64] found a radically different circadian rhythm at the onset of coronary ischaemic episodes with or without pain: peak incidence occurred in the morning on awakening while the lowest incidence was noted in the last part of the night. The patients studied by Rocco *et al.* [64] had classic angina while those of Mulcahy *et al.* [48] were a mixed population.

In conclusion, although angina attacks tend to appear in REM sleep in certain patients, this is certainly not the rule; nor does the type of angina favour the onset of an attack in one sleep stage or another. What does appear to be present, particularly in variant angina, is a circadian rhythm of attack frequency, with peak incidence in the last part of the night and the early hours of the morning.

The few studies which have investigated sleep structure in angina patients show that sleep is disturbed in a number of ways. In Maggini's patients [42], monitored in a coronary care unit, total sleep time was reduced due to an increase in the number of nocturnal awakenings and early morning awakening; stages 1 and 2 were increased to the detriment of stages 3 and 4. During the nights with ischaemic attacks sleep was slightly but not significantly more disturbed than on attack-free nights. Similar results have been reported by Karakan *et al.* [25]. On the other hand, alongside an increase in the time spent awake, Katayama *et al.* [26] found an increased stage 4 to the detriment of other NREM sleep stages, with normal percentages of REM sleep.

The pathophysiological correlates of angina attacks during sleep have yet to be established. Some authors have postulated an increased oxygen demand by the myocardium, given evidence of increased heart rate during the onset of S-T segment depression in sleep [61]. Increased oxygen consumption was also suggested by Mohr *et al.* [46] who found a marked decrease in angina episodes during sleep in one patient by keeping the bed in a reverse Trendelenburg position (feet down). They assumed that in this position diminished venous return and hence reduced filling of the left ventricle, cut down the myocardial oxygen demand. However, others failed to detect any significant change in haemodynamic parameters (heart rate, arterial pressure, left ventricular end diastolic pressure) before the onset of the ischaemic episode [13, 42, 51]. We can therefore speculate that the angina episodes occur without changes in the major determinants of myocardial oxygen consumption, due to a primary reduction in myocardial perfusion. This may be caused by a humoral factor which would explain why the myocardial ischaemic threshold follows a circadian rhythm, being lower in the early morning [84] or in the second half of the night [14]. Nocturnal changes in cardiac perfusion could therefore be linked to circadian variations in circulating platelet aggregates due to changes in the fibrinolytic activity of the blood or changes in coronary arterial tone and coronary vasoconstrictive response which are both enhanced in the early hours of the morning [80, 84]. In turn, these changes in coronary artery reactivity would be tied to circadian variations in the concentration of calcium and hydrogen ions [84].

MYOCARDIAL INFARCTION

The relations between sleep and myocardial infarction can be established indirectly by epidemiological surveys. Muller et al. [50] investigated 2999 patients and their results confirmed earlier findings on smaller cohorts that the peak incidence of infarction is between 6 and 12 a.m. Time of infarction onset was fixed as the time when pain first appeared and confirmed in a large percentage of cases by serial determinations of plasma creatine kinase MB (isomer originating from myocardium) activity which notoriously increases 4 hours after infarction. From these investigations, it is apparent that peak onset of infarction occurs later in comparison with the peak of angina attacks which mainly arise during the second half of the night. This discrepancy may be explained by the fact that myocardial infarction is probably the final event in a series of coronary changes already underway during the night [72]. Resumption of physical and mental activity, together with changes in many physiological variables (body temperature, plasma catecholamine level, heart rate and systemic arterial pressure, etc.) on awakening could be triggering factors.

The mental stress which may be associated with work has been implicated as a possible factor triggering myocardial ischemia [10,69]. As regards the circadian changes in blood coagulability and platelet aggregation already referred to in treating angina pectoris, it must be stressed that these activities reach a maximum peak in the morning on awakening [11, 55, 68].

The importance of platelet aggregates in the onset of myocardial infarction in patients with unstable angina is also demonstrated by the protective effect of aspirin [33,63]. Ridker et al. [63] offered a convincing explanation of the importance of the rise in platelet aggregation in the early hours of the morning at the onset of myocardial infarction. A group of physicians enrolled in a study lasting 60 months during which they took aspirin while another group received placebo. In addition to a lower incidence of infarction in absolute terms in the treated group, there was also a more selective reduction in the morning peak between 4 and 10 a.m. Spectral power analysis of heart rate variability recently disclosed that the vagal and sympathetic components of circadian regulation of cardiac activity by the autonomic nervous system is impaired in patients with coronary disease [23]. Moreover, power spectral analysis of hear rate variability by Vanoli et al. [77] showed that in patients who have recently suffered from myocardial infarction, the normal activation of the spectrum's vagal component represented by high frequencies is missing during NREM sleep so that the sympathetic component characterised by low frequencies prevails. According to the authors, this major imbalance in the LF/HF ratio towards the sympathetic component could be implicated in the pathogenesis of sudden death at night in these patients. Given the close pathophysiological links between myocardial infarction and sudden cardiac death, some studies have investigated a possible temporal relation between the onset of the two events. Although it is difficult to establish the exact time of sudden cardiac death with respect to myocardial infarction, as witnesses are often lacking, widescale epidemiological surveys have shown that its circadian curve conforms exactly to that of myocardial infarction [49,83].

Last but not least, the circadian incidence of another ischaemic event is of interest even though it is not directly linked to our topic. We refer to acute ischaemic stroke. The frequency of this event follows a circadian rhythm identical to that of coronary ischaemic attacks. In fact, there is a consensus in the literature that the peak incidence occurs in the morning on awakening between 6 a.m. and noon [1, 16, 30, 43, 44].

Broughton and Baron [6] studied the evolution of sleep architecture in 12 patients with myocardial infarction followed for up to 13 consecutive days, first in the Intensive Care Unit (ICU) and then in the ward. As expected, sleep was very disturbed on the nights spent in the ICU, with a wide interindividual variability. Sleep was characterised by reduced efficiency; increased latency, more wakefulness, awakenings and stage 1, decreased REM sleep both in percentage and in duration of individual REM episodes. On the other hand, increased stages 3-4 were recorded from the second night onwards. Similar results were reported by Karakan *et al.* [25] but there was a total lack of stages 3-4 in their patients. In Broughton and Baron's patients nocturnal sleep gradually improved starting from the second day in the ICU, along with a progressive reduction in daytime sleep periods (estimated only visually) which initially could last up to 8-10 hours.

Poor sleep quality seemed to be due to the stress of infarction more than the environmental disturbance of the ICU. This is supported by the sleep improvement which started early during the patients' stay in the ICU and by the fact that there were no significant differences between sleep on the last ICU night and the first night spent in the ward.

CARDIAC ARRYTHMIAS

During sleep young, apparently healthy subjects may present different types of arrythmia, particularly bradyarrhythmias (sinus arrest, I or II degree atrio-ventricular block, severe sinus bradycardia [4, 48]).

In exceptional circumstances, arrythmias may be particularly severe as found by Guilleminault *et al.* [19] who noted sinus arrests lasting up to 6 seconds invariably occurring in REM sleep; or Hein *et al.*'s report [21] in which episodes of ventricular fibrillation, sometimes accompanied by syncope, occurred on arousal from sleep by auditory stimuli. The peak incidence of first and second degree A-V blocks has been found to coincide with both REM [52] and NREM sleep [57].

Literature data on the behaviour of hyperkinetic arrhythmias during sleep are virtually confined to ventricular arrhythmias and results are contradictory. Most studies found a reduction in premature ventricular contractions (PVCs), even greater than that obtained with antiarrythmic drugs [35]; in the rare cases monitored polysomnographically, there were no significant differences in the degree of reduction between REM and NREM sleep [17, 35, 59]. Some investigations have also demonstrated that the curtailment in PVCs was correlated to changes in heart rate rather than sleep per se [17, 59]. Gillis *et al.* [17] also identified two groups of arrhythmic patients: a heart rate-dependent group presenting a linear increase in PVCs with increasing heart rate and a heart rate-independent group which lacked this correlation. Only patients in the first group had a reduction in PVCs during sleep. However, these results cannot explain why the frequency of ventricular extrasystoles during REM sleep when heart rate varies widely, does not change with respect to NREM sleep. Others have demonstrated that sleep, particularly REM sleep, facilitates ventricular arrhythmias [67]; while some failed to find significant differences between wakefulness and sleep [47, 73]. These discrepancies could in part be due to varying patient severity, the different places in which patients were monitored and hence the emotional influence of the surrounding environment, administration of antiarrhythmic drugs or even differing cardiac conditions (e.g. only Gillis' heart rate-dependent patients had structural heart diseases [17]. Moreover, the reduction in PVCs during sleep contradicts the finding that these arrhythmias are favoured by slow heart rates [20] but supports evidence, in other studies, that an increase in heart rate may lead to an increased incidence of ventricular arrhythmias [12, 79].

As regards hyperkinetic atrial arhythmias, a study on a patient population with paroxysmal atrial fibrillation (PAF) without evident heart disease, failed to detect any statistically significant differences in the number of FAP episodes per time unit, between wake and sleep or between different sleep stages (unpublished findings). In addition, the LF/HF (low frequency/high frequency) ratio (sympathovagal balance) of the power spectra in the two minutes preceding the start of PAF episodes tended to be greater than that observed in the same sleep stages some time away from the start (basal values). Moreover, separate analysis of the behaviour of the two power spectra (before the start of PAF episodes) showed that the LF spectra values were constantly higher than basal levels, whereas HF values varied, being higher, lower or unchanged.

These results fail to demonstrate the influence of different sleep stages on PAF and the shift towards the sympathetic limb of the sympathovagal balance in the short period before FAP episodes. This evidence contradicts the hypothesis that FAP episodes arising during sleep are usually heralded by vagal activation [9], and rather supports the idea of ongoing sympathetic activation before the episode, both in wakefulness and in each sleep stage.

MULTIPLE SYSTEM ATROPHY

In this syndrome, degeneration of the structures regulating neurovegetative functions accounts for the disorders affecting the cardiocirculatory system (fixed pulse orthostatic hypotension), sphincters (urinary and faecal incontinence), and skin (anhydrosis).

Unlike the trend in normal subjects, systemic arterial pressure rises during sleep, particularly in REM sleep and sudden phasic increases are encountered in all sleep stages [8]. Given the autonomic failure present in this syndrome, these pressure increases are probably caused by biochemical factors such as the rennin-angiotensin system whose activity could be enhanced during sleep as an over-response to physiological hypnogenic hypotension. Shy-Drager patients not only present an exaggerated susceptibility to angiotensin infusion, but also a marked hypertensive response with elevated plasma renin activity on resuming the horizontal position after an episode of orthostatic hypotension. Typically, when Shy-Drager syndrome is associated with obstructive sleep apnoea syndrome, systemic arterial pressure oscillations during apnoeas and when breathing is resumed are minimal [8].

All polygraphic studies have shown that sleep structure in Shy-Drager syndrome is highly impaired. Sleep is shortened by increased sleep latency and numerous nocturnal awakenings; only a few deep NREM sleep stages can be detected and above all, there is a marked reduction in REM sleep.

Likewise in other plurisystemic degenerative diseases of the central nervous system, also in Shy-Drager syndrome, the particular type of parasomnia known as REM behaviour disorder is often present.

REFERENCES

1. ARGENTINO C., TONI D., RASURA M., VIOLI F., SACCHETTI M.L., ALLEGRETTA A., BALSAMO F., FIESCHI C. – Circadian variation in the frequency of ischaemic stroke. *Stroke*, *21*, 387-389, 1990.

2. BAUST W., BOHNERT B. – The regulation of heart rate during sleep. *Exp. Brain Res.*, *7*, 169-180, 1969.

3. BEVAN A.T., HONOUR A.J., STOTT F.F. – Direct arterial pressure recording in unrestricted man. *Clin. Sci.*, *36*, 329-344, 1969.

4. BRISTOW J.D., HONOUR A.J., PICKERING T.G., SLEIGHT P. – Cardiovascular and respiratory changes during sleep in normal and hypertensive subjects. *Cardiovasc. Res.*, *3*, 476-485, 1969.

5. BRODSKY M., WU D., DENES P., KANAKIS C., ROSEN K.M. – Arrhythmias documented by 24 hour continuous electrocardiographic monitoring in 50 male medical students without apparent heart disease. *Am. J. Cardiol.*, *39*, 390-395, 1977.

6. BROUGHTON R., BARON R. – Sleep patterns in the intensive care unit and on the ward after acute myocardial infarction. *Electroenceph. Clin. Neurophysiol.*, *45*, 348-360, 1978.

7. COCCAGNA G., MANTOVANI M., BRIGNANI F., MAN-ZINI A., LUGARESI E. – Arterial pressure changes during spontaneous sleep in man. *Electroenceph. Clin. Neurophysiol.*, *31*, 277-281, 1971.

8. COCCAGNA G., MARTINELLI P., ZUCCONI M., CIRIGNOTTA F., AMBROSETTO G. – Sleep related respiratory and haemodynamic changes in Shy-Drager syndrome: case report. *J. Neurol.*, *232*, 310-313, 1985.

9. COUMEL P. – Role of the autonomic nervous system in paroxysmal atrial fibrillation. *In:* Touboul P. Waldo A.L. (eds.), *Atrial arrhythmias. Current concepts and managment.* Mosby-Year Book, St Louis, 248-261, 1990.

10. DEANFIELD J.E., SHEA M., KENSETT M., HORLOCK P., WILSON R.A., DE LANDSHEERE C.M., SELWYN A.P. – Silent myocardial ischema due to mental stress. *Lancet*, *2*, 1001-1005, 1984.

11. DECOUSUS H.A., CROZE M., LEVI F.A., JAUBERT J.G., PERPOINT B.M., DE BONADONA J.F., REINBERG A., QUENEAU P.M. – Circadian changes in anticoagulant effect of heparin infused at a constant rate. *Br. Med. J.*, *290*, 341-344, 1985.

12. EPSTEIN S.E., REDWOOD D.R., SMITH E.R. – Atropine and acute myocardial infarction. *Circulation*, *45*, 1273-1278, 1972.

13. FIGUERAS J., SINGH B.N., GANZ W., CHARUZI Y., SWAN H.J.C. – Mechanism of rest and nocturnal angina: observations during continuous hemodynamic and electrocardiographic monitoring. *Circulation*, *59*, 955-968, 1979.

14. FIGUERAS J., CINCA J., BALDA F., MOYA A., RIUS J. – Resting angina with fixed coronary artery stenosis: nocturnal decline in ischaemic threshold. *Circulation, 74*, 1248-1254, 1986.

15. FLETCHER E.C., DE BEHUKE R.D., LOVOI M.S., GORIN A. – Undiagnosed sleep apnea in patients with essential hypertension. *Ann. Int. Med., 103*, 190-195, 1985.

16. GALLERANI M., MANFREDINI R., RICCI L., COCU-RULLO A., GOLDONI C., BIGONI M., FERSINI C. – Chronobiological aspects of acute cerebrovascular diseases. *Acta Neurol. Scand., 87*, 482-487, 1993.

17. GILLIS A.M., GUILLEMINAULT C., PARTINEN M., CONNOLLY S.J., WINKLE R.A. – The diurnal variability of ventricular premature depolarizations: influence of heart rate, sleep, and wakefulness. *Sleep, 12*, 391-399, 1989.

18. GISLASON T., ABERG H., TAUBE A. – Snoring and systemic hypertension: an epidemiological study. *Acta Med. Scand., 222*, 415-421, 1987.

19. GUILLEMINAULT C., POOL P., MOTTA J., GILLIS A.M. – Sinus arrest during REM sleep in young adults. *N. Engl. J. Med., 311*, 1006-1010, 1984.

20. HAN J., DE TRAGLIA J., MILLET D., MOE G.K. – Incidence of ectopic beats as a function of basic rate in the ventricle. *Am. Heart J., 72*, 632-639, 1966.

21. HEIN J., WELLENS J., VERMEULEN A., DURRER D. – Ventricular fibrillation occurring on arousal from sleep by auditory stimuli. *Circulation, 46*, 661-665, 1972.

22. HOUYEZ F., DEGOULET P., CITTEE J., FOURIAND C., JACQUINET-SALORD M.C., LANG T., AIMÉ F. – Sommeil et hypertension arterielle (une étude épidémiologique chez 7 901 salariés). *Arch. Mal. Coeur, 83*, 1085-1088, 1990.

23. HUIKURI H.V., NIEMELA M.J., OJALA S., RANTALA A., IKAHEIMO M.J., AIRAKSINEN K.E. – Circadian rhythms of frequency domain measures of heart rate variability in healty subjects and patients with coronary artery (Effects of arousal and upright posture). *Circulation, 90*, 121-126, 1994.

24. KALES A., BIXLER E.O., CADIEUX R.J., SCHNECK D.W., SHAW L.C., LOKE T.V., VELA BUENO A., SOLDATOS C.R. – Sleep apnoea in a hypertension population. *Lancet, 2*, 1005-1008, 1984.

25. KARACAN I., WILLIAMS R.L., TAYLOR W.J. – Sleep characteristics of patients with angina pectoris. *Psychosomatics, 10*, 280-284, 1969.

26. KATAYAMA S., HARUMI K., SHIMOMURA K., MURAO S. – Nocturnal attacks of angina pectoris in relation to sleep stages and sleep characteristics: a polygraphic study. *Electroenceph. Clin. Neurophysiol., 34*, 716, 1973.

27. KHATRI I.M., FREIS E.D. – Hemodynamic changes during sleep. *J. Appl. Physiol., 22*, 867-873, 1967.

28. KHATRI I.M., FREIS E.D. – Hemodynamic changes during sleep in hypertensive patients. *Circulation, 39*, 785-790, 1969.

29. KING M.J., ZIR L.M., KALTMAN A.J., FOX A.C. – Variant angina associated with angiographically demonstrated coronary artery spasm and REM sleep. *Am. J. Med. Sci., 265*, 419-422, 1973.

30. KUBOTA K., SAKURAI T., TAMURA J., SHIRAKURA T. – Is the circadian change in hematocrit and blood viscosity a factor triggering cerebral and myocardial infarction? *Stroke, 18*, 812-813, 1987.

31. LANG R.M., BOROW K.M., WEINERT L., BEDNARZ J., NEUMAN A., NELSON K., DAVID D., MURPHY M.B. – Circadian blood pressure patterns in hypertensive patients: importance of nocturnal blood pressure. *Circulation, 78* (suppl. II), 517, 1988 (Abstract).

32. LAVIE P., BEN-YOSEF R., RUBIN A.E. – Prevalence of sleep apnea syndrome among patients with essential hypertension. *AM. Heart J., 108*, 373-376, 1984.

33. LEWIS H.D. Jr., DAVIS J.W., ARCHIBALD D.G., STEINKE W.E., SMITHERMAN T.C., DOHERTY J.E., SCHNAPER H.W., LE WINTER M.M., LINARES E., POUGET J.M., SATHAR-VAL S.C., CHESLER E., DE MOTS H. – Protective effect of aspirin against acute myocardial infarction and death in men with unstable angina. *N. Engl. J. Med., 309*, 396-403, 1983.

34. LITTLER W.A., HONOUR A.J., CARTER R.D., SLEIGHT P. – Sleep and blood pressure. *Brit. Med. J., 3*, 346-348, 1975.

35. LOWN B., TYKOCINSKI M., GARFEIN A., BROOKS P. – Sleep and ventricular premature beats. *Circulation, 48*, 691-701, 1973.

36. LUGARESI E., COCCAGNA G., MANTOVANI M., LEBRUN R. – Some periodic phenomena arising during drowsiness and sleep in man. *Electroenceph. Clin. Neurophysiol., 32*, 701-705, 1972.

37. LUGARESI E., COCCAGNA G., FARNETI P., MANTO-VANI M., CIRIGNOTTA F. – Snoring. *Electroenceph. Clin. Neurophysiol., 39*, 59-64, 1975.

38. LUGARESI E., CIRIGNOTTA F., COCCAGNA G., PIANA C. – Some epidemiological data on snoring and cardiocirculatory disturbances. *Sleep, 3*, 221-224, 1980.

39. LUNGER M., SHAPIRO A. – Continuous electrocardiographic monitoring in nocturnal angina. *Am. J. Cardiol., 13*, 119, 1964.

40. LYON A., TIBERGE M., ELEFTERION A. – Influence du sommeil sur la pression artérielle. *Arch. Mal. Coeur, 83*, 1075-1079, 1990.

41. McWILLIAM J.A. – Blood pressure and heart action in sleep and dreams: their relation to haemorrhages, angina and sudden death. *Brit. Med. J., 2*, 1196-1200, 1923.

42. MAGGINI C., GUAZZELLI M., MAURI M., CHIERCHIA S., CASSANO G.B., MASERI A. – Relation of transient myocardial ischemia to the sleep pattern in patients with «primary» angina. *In: Primary and secondary angina pectoris*, A. Maseri, G.A. Klassen, M. Lesch (eds.), Grune and Stratton Inc., New York, San Francisco, London, 157-167, 1978.

43. MARLER J.R., PRICE T.R., CLARK G.L., MULLER J.E., ROBERTSON T., MOHR J.P., HIER D.B., WOLF P.A., KAPLAN L.R., FOULKES M.A. – Morning increase in onset of ischaemic stroke. *Stroke, 20,* 473-476, 1989.

44. MARSH E.E.III, BILLER J., ADAMS H.P., MARLER J.R., HULBERT J.R., LOVE B.B., GORDON D.L. – Circadian variation in onset of acute ischaemic stroke. *Arch. Neurol., 47,* 1178-1180, 1990.

45. MILLAR-CRAIG M.W., BISHOP C.N., RAFTERY E.B. – Circadian variation of blood pressure. *Lancet, 1,* 795-797, 1978.

46. MOHR R., SMOLINSKY A., GOOR D.A. – Treatment of nocturnal angina with 10° reverse Trendenlemburg bed position. *Lancet, i,* 1325-1327, 1982.

47. MONTI J.M., FOLLE L.E., PELUFFO C., ARTUCIO R., ORTIZ A., SEVRINI O., DIGHIERO J. – The incidence of premature contractions in coronary patients during the sleep-awake cycle. *Cardiology, 60,* 257-264, 1975.

48. MULCAHY D., KEEGAN J., CUNNINGHAM D., QUYYUMI A., CREAN P., PARK A., WRIGHT C., FOX K. – Circadian variation of total ischaemic burden and its alteration with anti-anginal agents. *Lancet, 2,* 755-759, 1988.

49. MULLER J.E., LUDMER P.L., WILLICH S.N., TOFLER G.H., AYLMER G., KLANGOS I., STONE P.H. – Circadian variation in the frequency of sudden cardiac death. *Circulation, 75,* 131-138, 1987.

50. MULLER J.E., STONE P.H., TURI Z.G., RUTHERFORD J.D., CZEISLER C.A., PARKER C. *et al.* – Circadian variation in the frequency of onset of acute myocardial infarction. *N. Engl. J. Med., 313,* 1315-1322, 1985.

51. MURAO S., HARUMI K., KATAYAMA S., MASHIMA S., SHIMOMURA K., MURAYAMA M., MATMO H., YAMA-MOTO H., KATO R., CHEN C. – All-night polygraphic studies of nocturnal angina pectoris. *Jap. Heart J., 13,* 295-306, 1972.

52. NEVINS D.B. – First and second degree A-V heart block with rapid eye movement sleep. *Ann. Int. Med., 76,* 981-983, 1972.

53. NORTON P.G., DUNN E.V. – Snoring as a risk factor for disease: an epidemiological survey. *Brit. Med. J., 291,* 630-632, 1985.

54. NOWLIN J.B., TROYER W.G., COLLINS W.S., SILVER-MAN G., NICHOLS C.R., MCINTOSH D., ESTES E.H., BOGDONOFF M.D. – The association of nocturnal angina pectoris with dreaming. *Ann. Intern. Med., 63,* 1040-1046, 1965.

55. NUBILE G., D'ALONZO L., CONSOLI A., MONTZOU-RIDIS G., SENSI S. – Variazione circadiana dell'aggregazione piastrinica indotta da adrenalina e collagene nell'uomo. *Boll. Soc. It. Biol. Sper., 62,* 947-950, 1982.

56. O'BRIEN E., SHERIDAN J., O'MALLEY K. – Dippers and non-dippers. *Lancet, 2,* 397, 1988.

57. OTSUKA K., ICHIMARN Y., YANAGA T. – Studies of arrhythmias by 24-hour polygraphic recordings: relationship between atrioventricular block and sleep states. *Am. Heart J., 105,* 934-940, 1983.

58. PICKERING T.G. – Clinical significance of diurnal blood pressure variations. *Circulation, 81,* 700-702, 1990.

59. PICKERING T.G., JOHNSTON J., HONOUR A.J. – Comparison of the effects of sleep, exercise and autonomic drugs on ventricular extrasystoles, using ambulatory monitoring of electroencephalogram. *Am. J. Med., 65,* 575-583, 1978.

60. PROVINI F., PORTALUPPI F., PLAZZI G., CORTELLI P., MONTAGNA P., BERTOZZI N., MANFREDINI R., VER-GNANI L., FERSINI C., LUGARESI E. – Sleep breathing disorders in non-dippers with «essential» hypertension. (Abstract 503). APSS, 10th Annual Meeting, May 28, June 2, 1966, Washington D.C. *Abstract Book, 255,* 1966.

61. QUYYUMI A.A., WRIGHT C.A., MOCKUS L.J., FOX K.M. – Mechanisms of nocturnal angina pectoris: importance of increased myocardial oxygen demand in patients with severe coronary artery disease. *Lancet, 1,* 1207-1209, 1984.

62. RICHARDSON D.W., HONOUR A.J., FENTON G.W., STOTT F.H., PICKERING G.W. – Variation in arterial pressure throughout the day and night. *Clin. Sci., 26,* 445-460, 1964.

63. RIDKER P.M., MANSON J.E., BURING J.E., MULLER J.E., HENNEKENS C.H. – Circadian variation of acute myocardial infarction and the effect of low-dose aspirin in a randomized trial of physicians. *Circulation, 82,* 897-902, 1990.

64. ROCCO M.B., BARRY J.B., CAMPBELL S., NABEL E., COOK E.F., GOLDMAN L., SELWYN A.P. – Circadian variation of transient myocardial ischemia in patients with coronary artery disease. *Circulation, 75,* 395-400, 1987.

65. ROHMER F., SCHAFF G., COLLARD M., KURTZ D. – La motilité spontanée, la fréquence cardiaque et la fréquence respiratoire au cours du sommeil chez l'homme normal. *In: Le sommeil de nuit normal et pathologique.* Masson, Paris, 192-207, 1965.

66. ROSENBLATT G., ZWILLING G., HARTMANN E. – Electrocardiographic changes during sleep in patients with cardiac abnormalities. *Psychophysiol., 6,* 233, 1969.

67. ROSENBLATT G., HARTMANN E., ZWILLING G.R. – Cardiac irritability during sleep and dreaming. *J. Psychosom. Res., 17,* 129-134, 1973.

68. ROSING D.R., BRAKMAN P., REDWOOD D.R., GOLDSTEIN R.E., BEISER G.D., ASTRUP T., EPSTEIN S.E. – Blood fibrinolytic activity in man: diurnal variation and the response to varying intensities of exercise. *Circ. Res., 27,* 171-184, 1970.

69. ROZANSKI A., BAIREY C.N., KRAUTZ D.S., FRIEDMAN J., RESSER K.J., MORELL M., HILTON-CHALFEN S., HESTRIN L., BIETENDORF J., BERMAN D.S. – Mental stress and the induction of silent myocardial ischemia in patients with coronary artery disease. *N. Engl. J. Med., 318,* 1005-1012, 1988.

70. SCHMIDT-NOWARA W.W., COULTAS D.B., WIGGINS C., SKIPPER B.E., SAMET J.M. – Snoring in a Hispanic – American population. Risk factors and association with hypertension and other morbidity. *Arch. Int. Med., 150,* 597-601, 1990.

71. SCHNEIDER-HELMERT D., SCHENKER J. – Die normale arterielle Hypotension im Schlaf. *Schweiz. med. Wschr.*, *110*, 563-570, 1980.

72. SCHWAB L.E.M. – Circadian variation in myocardial infarction (Letter to the Editor). *N. Engl. J. Med.*, *314*, 1188, 1986.

73. SMITH R., JOHNSON L., ROTHFELD D., ZIR L., THARP B. – Sleep and cardiac arrhythmias. *Arch. Intern. Med.*, *130*, 751-753, 1972.

74. SNYDER F. – Autonomic nervous system manifestations during sleep and dreaming. *In: Sleep and altered states of consciousness*. The Williams and Wilkins Company, Baltimore, 469-487, 1967.

75. SNYDER F., HOBSON J.A., MORRISON D.F., GOLD-FRANK F. – Changes in respiration, heart rate, and systolic blood pressure in human sleep. *J. Appl. Physiol.*, *19*, 417-422, 1964.

76. SOBOTKA P.A., MAYER J.H., BAUERNFEIND R.A., KANAKIS C., ROSEN K.M. – Arrhythmias documented by 24-hour continuous 'ambulatory electrocardiographic monitoring in young women without apparent heart disease. *Am. Heart J.*, *101*, 753-759, 1981.

77. VANOLI E., ADAMSON P.B., BA-LIN, PINNA G.D., LAZZARA R., ORR W.C. – Heart rate variability during specific sleep stages (A comparison of heathy subjects with patients after myocardial infarction). *Circulation*, *91*, 1918-1922, 1995.

78. VERDECCHIA P., SCHILLACI G., GUERRIERI M., GATTESCHI C., BENEMIO G., BOLDRINI F., PORCELLATI C. – Circadian blood pressure changes and left ventricular hypertrophy in essential hypertension. *Circulation, 81*, 528-536, 1990.

79. VERRIER R.L., KIRBY D.A. – Sleep and cardiac arrhythmias. *In: The sudden infant death syndrome: cardiac and respiratory mechanisms and interventions*. Annals of N.Y. Academy of Sciences, *533*, 238-251, 1988.

80. WATERS D.D., MILLER D.D., BOUCHARD A., BOSCH X., THEROUX P. – Circadian variation in variant angina. *Am. J. Cardiol.*, *54*, 61-64, 1984.

81. WILLIAMS D.H., CARTWRIGHT R.D. – Blood pressure changes during EEG-monitored sleep (A comparative study of hypertensive and normotensive negro women). *Arch. Gen. Psychiat.*, *20*, 307-314, 1969.

82. WILLIAMS A.J., HOUSTON D., FINBERG S., LAM C., KINNEY J.L., SANTIAGO S. – Sleep apnea syndrome and essential hypertension. *Am. J. Cardiol.*, *55*, 1019-1022, 1985.

83. WILLICH S.N., LEVY D., ROCCO M.B., TOFLER G.H., STONE P.H., MULLER J.E. – Circadian variation in the incidence of sudden cardiac death in the Framingham Heart Study Population. *Am. J. Cardiol.*, *60*, 801-806, 1987.

84. YASUE H., OMOTE S., TAKIZAWA A., NAGAO M., MIWA K., TANAKA S. – Circadian variation of exercise capacity in patients with Prinzmetal's variant angina: role of exercise-induced coronary arterial spasm. *Circulation*, 59, 938-948, 1979.

Chapter 48

Restless legs syndrome in wakefulness and periodic leg movements in sleep

J. Montplaisir*, A. Nicolas+, and O. Lapierre*
* Centre d'Etude du Sommeil, Département de Psychiatrie, Hôpital du Sacré-Cœur, Montréal, Québec, Canada.
+Unité de sommeil, Unité Clinique de Psychiatrie Biologique du Vinatier, Bron, France

RESTLESS LEGS SYNDROME IN WAKEFULNESS

We owe it to Thomas Willis for the first description of restless legs syndrome (RLS) in wakefulness. Written in Latin, this study dates from 1672 and is one of the oldest observations in the history of sleep medicine [68]. But it was only in 1945 that Eckbom identified these disorders as a distinct condition [15].

The prevalence of the illness has been estimated at about 5% [16] of the general population, but we consider this estimate to be too low. In an epidemiological study published in 1994 in Canada, 15% of the interviewed subjects reported having frequent or very frequent episodes of restless legs (RL) in wakefulness [26].

Clinical features

Four clinical features are considered essential in diagnosing this disorder [64]:
- the irresistible urge to move, associated with paresthesias. This is usually described as a crawling or gnawing sensation or as tension;
- motor restlessness: repetitive movements which may or may not be associated with paresthesias;
- symptoms must be more severe at rest and at least partially relieved by movement especially by walking;
- symptoms usually or only appearing in the evening, particularly at bedtime.

In 50% of cases, restlessness is also noted in the lower limbs. Moreover, in 42% of cases there is symptomatological lateralisation, either left or right [37]. The effect is also brought on by tiredness. Patients usually report difficulties falling asleep, and 85% report waking up during the night with sensorimotor symptoms of RLS. Laboratory studies have confirmed these clinical findings. Indeed patients who report difficulties falling asleep present high latencies to sleep, while those who report the presence of symptomatology during the night have reduced sleep effectiveness [37].

Course

In a study of 133 patients, diagnosed in our centre, the mean age of onset of symptoms was 27 years. In most cases, symptoms had worsened with age. Many patients report periods of spontaneous remission which may last several weeks or even months. Various factors can affect the course of the illness, and thus RLS may worsen during pregnancy, particularly during the last three months [18].

PERIODIC LEG MOVEMENTS IN SLEEP

History and epidemiology

A laboratory study of 131 patients with RLS revealed the presence of periodic leg movements in sleep (PLMs) in 80% of cases. In this study, the diagnostic threshold was placed at 5 periodic movements per hour of sleep [37]. The association between the two phenomena increases considerably with age.

PLMs also represent a separate entity and may appear in isolation with no clinical sign of RLS in wakefulness. PLMs used to be listed as nocturnal myoclonia or Symond's disease. This term was dropped due to the number of different types of nocturnal myoclonia, added to which PLMs are not true myoclonia. PLMs last longer and their main characteristic is precisely that of periodicity.

The prevalence of PLMs increases substantially with age [5, 7]. It is estimated that 5% of normal subjects of between 30 and 50 years of age have a pathological index of PLMs with polygraphic recording, while this percentage may increase to 29% in subjects of over 50 years of age and 44% for those over 65 years.

Clinical features

PLMs take the form of an extension of the big toe and dorsiflexion of the foot, occasionally with flexion of the knee and hip. Movements usually last a few seconds, occurring periodically roughly every 30 seconds and are more numerous during the first half of the night. These movements often cause arousals, although some patients present a large number of PLMs with no arousals or substantial modification of the general organisation of sleep. Periodic movements also occur in the upper limbs [12]. The phenomenon may appear in normal subjects but it is often associated with a wide variety of sleep disorders such as insomnia, idiopathic hypersomnia, narcolepsy, sleep apnoea syndrome or REM sleep behaviour disorder. It is worth noting that PLMs are no more frequent in cases of insomnia or hypersomnia, but rather in pathologies linked to a dopaminergic deficit. Thus a diagnosis of PLMs may not be satisfactory in excessive daytime sleepiness and the possibility of idiopathic hypersomnia should be considered [41, 45].

DIAGNOSIS OF RLS AND PLMs

A diagnosis of RLS and PLMs should be considered for any patient consulting for a sleep disorder.

RLS

A diagnosis of RLS is now entirely based on clinical history [13, 34]. The examination will thus attempt to reveal the principle manifestations of the disorder. The two symptoms which appear to be the most specific are sensorimotor manifestations which tend to occur at rest and which are relieved by various leg movements, particularly walking. These two features are distinct from the sensorimotor symptoms associated for example with arterial insufficiency which are aggravated by walking, or those appearing in venous insufficiency in which symptoms are aggravated by remaining in an upright position and relieved by lying down with the legs extended. Relief by walking also distinguishes RLS from akathisia. The latter usually appears in patients treated with neuroleptics; it is characterised by an irresistible urge to move but symptoms are not alleviated by walking.

Clinical examination also enables identification of one or another of the medical conditions associated with RLS (table 48.1, class A). Three associations are usually described:
-   renal failure (15 to 40% of patients under haemodialysis).
-   Iron-deficiency or megaloblastic anaemia. Iron deficiency in fact appears to be the main agent responsible, as shown in a study by O'Keefe *et al.* [48] in which iron supplements improved RLS in patients with low ferritin. The authors suggest that iron acts by re-

establishing dopaminergic transmission. In fact the D2 receptor uses iron as a cofactor and hyposideraemia leads to its hypofunctioning.

- Damaged peripheral nerves. A recent study carried out on 144 patients presenting a clinical picture of polyneuropathy, showed 5.2% of them as presenting the clinical symptoms of RLS, a prevalence which is close to that observed in the general population [50].

RLS symptoms have been found in a large number of other illnesses (table 48.1, class B), but these associations have not been subjected to systematic study and in the light of the high prevalence of RLS, the specificity of these observations remains uncertain [15, 34].

Table 48.1. Conditions associated with RLS and PLMs

Class A
- renal failure
- anaemia
- damaged marrow and peripheral nerves
Class B
- rheumatoid arthritis
- primary amyloidosis
- diabetes
- chronic obstructive pulmonary disease
- leukaemia
- Parkinson's disease
- Huntington's chorea
- Amyotrophic lateral sclerosis

RLS is a familial illness. Several recent studies have indeed shown that over 60% of those suffering from RLS had at least one direct relative who suffered from the illness [29, 32, 65]. The family tree of these families strongly indicates an autosomal dominant transmission [58].

The diagnosis of RLS is clinical today. Several tests have however been developed over the past years, one of which is becoming recognised as a diagnostic tool. This is the Suggested Immobilisation Test (SIT). This consists of asking the patient to sit motionless in bed, with the torso at a 45° angle to the outstretched legs. Patients must keep their eyes open and move as little as possible. During the test, the leg movements are recorded using electrodes placed on the anterior tibialis. The sensory events are indicated by the patient by means of a remote control which marks the tracing. EEG and other physiological signals are recorded to monitor the patient's state of alertness. The test lasts 60 minutes and must be carried out in the evening before bedtime (9 or 10 pm) to coincide with the period with the most symptomatology [9]. In severe cases, movements last longer and continuous muscular activity is occasionally observed, to the extent that the patient is no longer able to score the individual movements. This test can be repeated to monitor the development of symptoms or the efficacy of treatment. Refining the test and recording additional groups of muscles should lead to improving its sensitivity and power of discrimination. Moreover, recent results obtained from patients with RLS and recorded for two consecutive nights show that 81% of these have a PLM index of more than 11 per hour of sleep for at least one of the two nights [38].

In recent years efforts have been made to develop scales to facilitate the diagnosis and evaluation of the severity of RLS. Some studies use informal scales [57] or visual analogical evaluations of discomfort [30]. John Hopkins Hospital recently proposed an interesting severity scale [4] and the international RLS study group is in the process of developing and refining a tool to standardise diagnosis and the evaluation of symptomatology.

PLMs

It is hard to make a clinical diagnosis of PLMs. In the absence of RLS, patients may report tiredness or heavy legs on morning arousal. It is sometimes the subject's partner who complains of repeated kicking during the night. It is thus very helpful for the partner to be present at the interview. But polysomnographic recording is still necessary to gain a definite diagnosis of this condition.

In addition to the polysomnographic variables required for sleep staging, the polysomnographic study will include a recording of the left and right anterior tibialis muscles Certain cases may call for an additional recording of other muscle groups of the upper and lower limbs.

The movement is characterised by a sustained contraction or a polyclonic burst [28]. These muscle contractions are often associated with polygraphic signs of arousal. The most common method of quantifying PLMs, is that of Coleman. According to this method, the only movements scored are those which last from 0.5 to 5 seconds and occur in a series of at least four consecutive movements, separated by intervals of 4 to 90 seconds. A movement index of more than five per hour of sleep is considered pathological. PLMs are especially abundant during stages of NREM sleep, decreasing strongly in REM sleep. During deep NREM sleep, the number decreases and the mean duration of intervals between movements increases slightly. The characteristics of periodic movements thus vary according to the sleep stage as well as the age of the patients [46].

Numerous arousals or stage shifts are seen in certain patients, with a reduction in the proportion of deep NREM sleep stages to the benefit of lighter stages (NREM sleep stages 1 and 2). Physiological reactions (notably cardiac) secondary to the movements reflect the extent of sleep disturbance and the effort of adaptation to which the organism is subjected [53]. We recently observed the presence of other PLMs-related rhythmic motor events. Thus rhythmic masseter muscle movements have been observed in certain patients. These contractions occur at the rate of roughly one per second and last roughly five seconds.

This rhythmic activity usually occurs after the leg movements. The observation is of specific interest in studying the pathophysiology of PLMs.

Other types of motor activity affect the legs during sleep. These are easily distinguished in polysomnography by their temporal distribution and duration. No other movement shows the periodicity characteristic of PLMs. Moreover, the other motor events are distributed in a particular way. Sleep starts, for example, occur only during the transition from wakefulness to sleep. On the other hand fragmentary myoclonus is more abundant in REM sleep and does not exceed 200 ms in duration. In the case of PLMs, maximum activity is observed at about 2 am, with a corresponding circadian component that is not strictly linked to the presence of a specific sleep stage [44, 59].

PATHOPHYSIOLOGY

Several mechanisms have been evoked to account for this illness. Damage to bone marrow, peripheral nerves or the muscle itself, has been observed in several patients but, in most cases, the studies of evoked brainstem potentials and nerve conduction and biopsies of muscle structures have proved normal [42]. However a recent study suggests the possibility of the axonal atrophy of the sural nerve in this condition [25]. Other authors have noted an abnormal delayed reflex in shutting the eyelids, after electrical or mechanical stimulation [66], suggesting central damage, which is probably pontic. This delayed reflex was not found in all the studies [10] but the analyses carried out by Briellman *et al.* show that the particularity of PLMs subjects tends to occur in the form of alterations to the habituation mechanisms. The authors link this motor control modification with a dysfunctionning of the pallido-striatal dopaminergic system [8]. Videoscopic studies have revealed the similarities between PLMs and the Babinski sign [54], although it is true that an extension of the big toe is observed in half the normal subjects in response to stimulation of the sole of the foot during NREM sleep. There is no consensus today in regard to the presence or absence of cortical activity accompanying or preceding leg movements [23, 56]. And these studies are still unable to define the nerve structures implicated in the pathophysiology of RLS or PLMs.

Even if the neurophysiological mechanisms are still unknown, the periodic nature of diurnal and nocturnal leg movements does suggest the presence of a rhythmic generator in the central nervous system. Studies carried out on cats and humans, either asleep or in a coma, have helped to reveal the regular variations in blood pressure, heartbeat, respiration rate, intraventricular pressure and EEG during the course of sleep. The mean variation period is comparable to that of PLMs, varying from 20 to 60 seconds. Moreover, we have noted that in RLS patients with PLMs, periodic arousals persist during the night after the suppression of movements by pharmacological treatment [36]. These results suggest that the periodic arousals may not be secondary to movements but, on the contrary, directly involved in the mechanisms responsible for their occurrence [36]. It is as though the brain periodically receives excitatory influxes. In the normal subject, these would be inhibited during sleep so as not to provoke arousals, appearing only as transitory electrical events like K complexes. An increase in the intensity of these influxes or a deficiency in the inhibitory mechanisms could thus be responsible for the arousals and the periodic movements such as PLMs or bruxism. Moreover the persistence of periodic movements in patients having undergone traumatic spinal cord section, suggests the presence of a generator located in the spinal cord, which may cease to be controlled by the cerebral centres and function according to its own rhythm [14,43].

From a neurochemical point of view, there is substantial evidence to suggest that dopamine is implicated in the pathophysiology of RLS and PLMs [28]. First of all, L-dopa, which predominantly increases concentrations of dopamine, and bromocriptine, a D2 receptor agonist, both inhibit RLS and PLMs [33, 61]. Furthermore, symptoms are worsened by administering pimozide, a D2 receptor antagonist [1]. These results suggest that RLS and PLMs are caused by a reduced dopamine transmission. This hypothesis would account for the high frequency of PLMs in Parkinson's disease and in narcolepsy, a disease in which we believe there is a reduction in dopaminergic transmission. A recent study using PET functional imagery showed that in 13 patients presenting RLS, the link to D2 receptors was significantly reduced in the striatum [60]. So far, the measurement of this deficit remains simple and quantifiable, which is unfortunate considering its valuable potential as a tool for diagnosis and evaluation.

In terms of imaging, a recent positron emission tomography (PET) study using deoxyglucose showed the cerebral metabolism to be normal in RLS patients [63]. Moreover, Staedt *et al.* have used single photon emission computerized tomography SPECT to demonstrate a reduction in the dopamine density of the D2 receptor subtype in the striata of patients suffering from PLMs [55]. Finally in the case of RLS, a magnetic resonance imagery (MRI) study showed no anatomical lesions at cerebral level [11]. The same authors later carried out a functional MRI study which revealed thalamic and cerebellar activation concomitant with the sensory symptoms of patients suffering from RLS. During PLMS, activation was observed in the red nucleus and pons. These results suggest hyperactivity in the cerebellum, red nucleus and pons in patients with RLS and PLMs. These formations may exert a disinhibiting effect on the spinal passages and provoke an effect close to that observed in startle reactions [11].

TREATMENT

Most medications recognised as being effective against RLS [35] have a therapeutic effect on PLMs. This observation supports the hypothesis of a close association between the mechanisms responsible for both clinical manifestations. Three families of medication have been subject to more systematic study and are commonly prescribed.

Benzodiazepines

Several benzodiazepines, particularly clonazepam, have been used in treating RLS with PLMs [35]. The recommended dose at the onset of treatment is 0.5 mg, with a gradual increase to a maximum of approximately 2 mg.

Over 50% of patients report an initial therapeutic effect when treated with clonazepam although several develop tolerance after several months. It is not unusual however to observe a pharmacological effect persisting over several weeks after withdrawing treatment. Moreover, we have never observed a major rebound effect of RLS or PLMs on withdrawal. In patients treated with clonazepam, polysomnography shows shorter latency to sleep and a reduction in the number of nocturnal arousals. Numerous studies [24, 35] have confirmed the effectiveness of this treatment.

Roughly 25% of patients report adverse effects, particularly morning sleepiness during the first weeks of treatment. Other patients report erection problems or a drop in libido. It must also be borne in mind that benzodiazepines depress the nervous system and may worsen sleep apnoeas.

Dopaminergic agents

L-dopa administered with a decarboxylase inhibitor (carbidopa or benserazide) is undoubtedly the most effective treatment for RLS and PLMs [1, 9, 33]. One study showed the persistence of therapeutic effects after two years of L-dopa treatment [61].

Whatever the case, weak doses should be prescribed at the start of treatment. We suggest administering L-dopa at a dose of 100 mg, half a tablet at bedtime to be repeated once in the middle of the night. The maximum dose should not exceed 600 mg per day. It would be logical to consider using L-dopa for long term action, but the efficacy of this medication has not yet been clearly demonstrated.

It is important to emphasise certain difficulties which are encountered by certain patients during long-term treatment with L-dopa [35]. Two adverse effects warrant special attention. First, the administration of a single dose of L-dopa at bedtime is accompanied by a substantial rebound of PLMs in the second half of the night i.e. at the point when the substance ceases to have an effect [33]. When a second dose is administered in the middle of the night, roughly a third of patients report RLS during the day, even though these symptoms had previously only been present in the evening or at night [2, 20]. This RLS rebound effect during wakefulness rapidly ceases when treatment is withdrawn. The rebound effect is the main reason why L-dopa is not automatically prescribed as first line treatment for all patients. Another adverse effect, found in roughly 10% of patients, is insomnia due to the stimulating effect of L-dopa, which can appear even in patients reporting the complete disappearance of sensorimotor symptoms.

Clinicians often question the possibility of delayed dyskenesias resulting from prolonged use of L-dopa. Contrary to observations made among Parkinsonian patients, we have never noted dyskenesias in RLS patients treated with L-dopa for over five years. There are several ways of accounting for this phenomenon. Firstly, the dose is generally weaker in RLS than in Parkinson's. Moreover, delayed dyskenesias in Parkinson's probably result from denervation hypersensitivity, and is thus closely associated with the cellular loss of dopaminergic neurons in the nigrostriatal system, which has never been demonstrated in patients with RLS or PLMs.

Other dopaminergic agents have a considerable therapeutic effect. Bromocriptine, a D2 receptor agonist, used at a dose of 7.5 mg at bedtime, has therapeutic effects similar to those of L-dopa [62]. Therapeutic effects have also been observed with selegiline, a monoamine B oxydase inhibitor. These effects have not however been verified in a controlled study. Finally pergolide, a D1/D2 non selective dopaminergic agent, ten times more powerful than bromocriptine, appeared to offer satisfactory results in an open study of six resistant cases of RLS [27].

More recently, more specific dopaminergic agents, notably D3 receptors, have demonstrated remarkable sustained efficacy in patients with RLS and PLMs. These consist of ropinirole and pramipexole. Our team has experimented with the second product in particular, which leads to a dramatic reduction in the nocturnal periodic movements index and clearly improves patients' comfort [39, 40, 67]. In regard to ropinirole, studies are encouraging, with good tolerance and minimal adverse effects (0.25 to 4 mg at bedtime) [17, 49, 52]. This product has been accused of causing sudden sleep episodes in patients presenting RLS or PLMs [51], probably due to these patients taking a dose which is far too weak and to the intrinsic harmful effect of Parkinson's on vigilance.

Opiates

The therapeutic effects of opiates were noted in the original study by Ekbom [15]. Later studies confirmed the therapeutic effects of morphine, methadone and other opiates in the treatment of RLS [22]. On the other hand, the therapeutic effect of these substances has not yet been clearly demonstrated for PLMs. The rapid development of tolerance and dependence on these substances considerably limits their clinical usefulness.

Other pharmacological treatment

Table 48.2 summarises the list of medication currently used in treating RLS and PLMs. Carbamazepine for example, is used successfully for a large number of patients with RLS [69].

However this substance does not appear to significantly reduce PLMs. Baclofene, a GABAergic agonist, slightly increases the number of PLMs but reduces the intensity of the movements and the associated arousals [19]. Contradictory results have been obtained with clonidine [6, 21]. Iron, vitamin B12 or folic acid can have therapeutic effects but these substances have not been quantitatively evaluated in non anaemic patients presenting RL or PLMS. Finally, GABApentine, a new anticonvulsant agent, was initially used in patients with RLS with success [29] but this does not seem to have been confirmed by later structured studies [3].

Table 48.2. Pharmacological treatment for RLS and PLMS

- Benzodiazepines, particularly clonazepam (*Rivotril®*)
- L-dopa (*Sinemet®* or *Sinemet CR®*)
- Dopaminergic agonists: pergolide (*Permax®*), ropinirole (*Requip®*) and pramipexole (*Mirapex®*)
- Opiates
- Carbamazepine (*Tegretol®*)
- Gabapentine (*Neurontin®*)
- Baclofen (*Lioresal®*)
- Clonidine (*Catapresson®*)
- Iron, vitamin B12, folic acid

Choice of medication

Considerable difficulties arise in the long-term treatment of severe cases of RLS, despite the marked immediate therapeutic effects of many medications. While clonazepam was the standard initial treatment until recent years, the effectiveness of dopaminergic agents like ropinirole or pramipexole is tending to favour their use as first line treatment. Prolonged studies are clearly still lacking to evaluate the persistency of the effects of these products and any long term side effects. Pergolide, used in doses increasing to 0.5mg (in association with domperidone, a peripheral dopamine antagonist) has proved highly effective in severe cases or those presenting a daytime RL rebound effect. As administering L-dopa in the evening may result in symptoms appearing during the daytime, it is mainly used in treating severe cases resistant to dopaminergic agonists and benzodiazepines. The opiates are very effective in some patients but prolonged administration of these substances causes problems which considerably reduce their clinical indication. Finally the antiepileptics such as carbamazepine or gabapentine may offer an alternative in cases which resist or escape treatment with the products previously referred to.

REFERENCES

1. AKPINAR S. – Restless legs syndrome treatment with dopaminergic drugs. *Clinical Neuropharmacology, 10,* 69-79, 1987.
2. ALLEN R.P., EARLEY C.J. - Augmentation of the restless legs syndrome with carbidopa/levodopa. *Sleep, 19,* 205-213, 1996.
3. ALLEN R.P., EARLEY C.J. - An open label clinical trial with structured subjective reports and objective leg activity measures comparing gabapentin with alternative treatment in the restless legs syndrome. *Sleep Research, 25,* 184, 1996.
4. ALLEN R.P., EARLEY C.J. – Validation of the Johns Hopkins restless legs severity scale. *Sleep Medicine, 2(3),* 239-242, 2001.
5. ANCOLI-ISRAËL S., KRIPKE D.F., MASON W., KAPLAN O.J. – Sleep apnea and periodic movements in sleep in an aging population. *Journal of Gerontology, 40,* 419-425, 1985.
6. BAMFORD C.R., SANDYK R. – Failure of clonidine to ameliorate the symptoms of restless legs syndrome [letter]. *Sleep, 10,* 398-399, 1987.
7. BIXLER E.O., KALES A., VELA-BUENO A., JACOBY J.A., SCARONE S., SOLDATOS C.R. - Nocturnal myoclonus and nocturnal myoclonic activity in a normal population. *Research Communications in chemical pathology and pharmacology, 36,* 129-140, 1982.
8. BRIELLMANN R.S., ROSLER K.M., HESS C.W. – Blink reflex excitability is abnormal in patients with periodic leg movements in sleep. *Movement Disorders, 11,* 710-714, 1996.
9. BRODEUR C., MONTPLAISIR J., GODBOUT R. – Treatment of restless legs syndrome and periodic movements during sleep with L-dopa: a double-blind, controlled study. *Neurology, 35,* 1845-1848, 1988.
10. BUCHER S.F., TRENKWALDER C., OERTEL W.H. – Reflex studies and MRI in the restless legs syndrome. *Acta Neurologica Scandinavica, 94,* 145-150, 1996.
11. BUCHER S.F., SEELAS K.C., OERTEL W.H., REISER M., TRENKWALDER C. – Cerebral generators involved in the pathogenesis of the restless legs syndrome. *Annals of Neurology, 41,* 639-645, 1997.

12. CHABLI A, MICHAUD M, MONTPLAISIR J. – Periodic arm mouvements in patient with restless legs syndrome. *European Neurology, 44(3),* 133-138, 2000.

13. COCCAGNA. G – Restless legs syndrome/periodic movements in sleep. *In:* M.J. Thorpy (ed), *Handbook of sleep disorders,* Dekker, New York, 457-478, 1990.

14. DICKELL MJ, RENFROW SD, MOORE PT. – Rapid eye movement sleep leg movements in patients with spinal cord injury. *Sleep, 17,* 733-738, 1994.

15. EKBOM K.A. – Restless legs syndrome. *Neurology, 10,* 868-873, 1960.

16. EKBOM K.A. – Restless legs. *Acta Medica Scandinavica, suppl. 158,* 1-123, 1945.

17. ESTIVILL E., DE LA FUENTE V. – The efficacy of ropinirole in the treatment of chronic insomnia secondary to restless legs syndrome: polysomnographic data. *Revista Neurologica, 29(9),* 805-807, 1999.

18. GOODMAN J.D., BRODIE C., AYIDA G.A. – Restless leg syndrome in pregnancy. *British Medical Journal, 297,* 1101, 1988.

19. GUILLEMINAULT C., FLAGG W. – Effect of baclofen on sleep-related periodic leg movements. *Annals of Neurology, 15,* 234-239, 1984.

20. GUILLEMINAULT C., CETEL M, PHILIP P. – Dopaminergic treatment of restless legs and rebound phenomenon. *Neurology 43:* 445, 1993.

21. HANDWERKER J.V., PALMER R.F. – Clonidine in the treatment of "Restless leg" syndrome. *New England Journal of Medicine, 313,* 1228-1229, 1985.

22. HENING W.A., WALTERS A., KAVEY N., GIDRO-FRANK S., CÔTÉ L., FAHN S. – Dyskinesias while awake and periodic movements in sleep in restless legs syndrome: treatment with opioids. *Neurology, 36,* 1363-1366, 1986.

23. HENING W.A., CHOKROVERTY S., ROLLERI M., WALTERS A. – The cortical premovement potentials in RLS jerks. *Sleep Research, 20,* 255, 1991.

24. HENING W.A. - Treatment of the restless legs syndrome: current practice of sleep experts. *Neurology, 45 (suppl 4),* 445S, 1995.

25. IANNACCONE S., ZUCCONI M., MARCHETTINI P. – Evidence of peripheral axonal neuropathy in primary restless legs syndrome. *Movement Disorders, 10,* 2-9, 1995.

26. LAVIGNE G.J., MONTPLAISIR J.Y. – Restless legs syndrome and sleep bruxism: prevalence and association among Canadians. *Sleep, 17,* 739-743, 1994.

27. LIN S.-C., KAPLAN J., BURGER C., FREDRICKSON P. – The effect of pergolide on the treatment of resistant restless legs syndrome. *Sleep Research, 24,* 277, 1995.

28. LUGARESI E., CIRIGNOTTA F, COCCAGNA G., MONTAGNA P. – Nocturnal myoclonus and restless legs syndrome. *In:* S. Fahn *et al.* (eds), *Advances in Neurology, vol.43, Myoclonus,* Raven Press, New York, 295-307, 1986.

29. MELLICK G.A., MELLICK L.B. – Successful treatment of restless legs syndrome with gabapentin (Neurontin). *Sleep Research, 24,* 290, 1995.

30. MICHAUD M., LAVIGNE G., PAQUET J., MONTPLAISIR J. – Discomfort scale during the suggested immobilization test: a new method for diagnosing RLS. *Sleep, 22 (supp 1),* 67-68, 1999.

31. MICHAUD M., CHABLI A., LAVIGNE G., MONTPLAISIR J. – Arm restlessness in patients with restless legs syndrome. *Movement Disorders, 15,* 289-293, 1999.

32. MONTPLAISIR J., GODBOUT R., BOGHEN M.D., DECHAMPLAIN J., YOUNG S.N., LAPIERRE G. – Familial restless legs with periodic movements in sleep: Electrophysiological, biochemical, and pharmacological study. *Neurology, 35,* 130-134, 1985.

33. MONTPLAISIR J., GODBOUT R., POIRIER G., BÉDARD M.A. – Restless legs syndrome and periodic movements in sleep: physiopathology and treatment with L-dopa. *Clinical Neuropharmacology, 9,* 456-463, 1986.

34. MONTPLAISIR J., GODBOUT R., PELLETIER G., WARNES H. – Restless legs syndrome and periodic limb movements during sleep. *In:* M.H. Kryger, T. Roth, W.C. Dement (eds), *Principles and practice of sleep medicine,* Second Edition, Saunders, New York, 589-597, 1992.

35. MONTPLAISIR J., LAPIERRE O., WARNES H., PELLETIER G. – The treatment of the restless leg syndrome with or without periodic leg movements in sleep. *Sleep, 15,* 391-395, 1992.

36. MONTPLAISIR J., BOUCHER S., GOSSELIN A, POIRIER G., LAVIGNE G. – Persistence of repetitive EEG arousals (K-alpha complexes) in RLS patients treated with L-Dopa. *Sleep, 19,* 196-199, 1996.

37. MONTPLAISIR J., BOUCHER S., POIRIER G., LAVIGNE G., LAPIERRE O., LESPÉRANCE P. – Clinical, polysomnographic, and genetic characteristics of restless legs syndrome: A study of 133 patients diagnosed with new standard criteria. *Movement disorders, 11,* 61-65, 1997.

38. MONTPLAISIR J, BOUCHER S, NICOLAS A, LESPERANCE P, GOSSELIN A, ROMPRÉ P, LAVIGNE G. - Immobilization tests and periodic leg movements in sleep for the diagnosis of restless leg syndrome. *Movement Disorders, 13,* 324-359, 1998.

39. MONTPLAISIR J, NICOLAS A, DENESLE R, GOMEZ-MANCILLA B. - Pramipexole alleviates sensory and motor symptoms in restless legs syndrome. *Neurology, 51(1),* 311-312, 1998.

40. MONTPLAISIR J., NICOLAS A, DENESLE R, GOMEZ-MANCILLA B. - Restless legs syndrome improved by pramipexole: A double-blind randomized trial. *Neurology, 52(5),* 938-943, 1999.

41. MONTPLAISIR J., MICHAUD M., DENESLE R., GOSSELIN A. – Periodic leg movements are not more prevalent in insomnia or hypersomnia but are specifically associated with sleep disorders involving a dopaminergic impairment. *Sleep Medicine, 1(2),* 163-167, 2000.

42. MOSKO S.S., NUDLEMAN K.L. – Somatosensory and brainstem auditory evoked responses in sleep-related periodic leg movements. *Sleep, 9,* 399-404, 1986.

43. NICOLAS A, GAUDREAU H, PETIT D, MONTPLAISIR J. - Peripheral or central control of periodic leg movements: a case study of complete spinal cord transection. *Sleep Research, 26,* 442, 1997.
44. NICOLAS A, LAVIGNE G, MONTPLAISIR J. - Temporal distribution of periodic leg movements during nocturnal sleep in RLS patients. *Sleep, 20 (suppl 1),* 69, 1998.
45. NICOLAS A, LESPERANCE P, MONTPLAISIR J. - Is excessive daytime sleepiness with periodic leg movements during sleep a specific diagnostic category? *European Neurology, 40;* 22-26, 1998.
46. NICOLAS A, LAVIGNE G, MONTPLAISIR J. - The influence of sex, age and sleep/wake state on characteristics of periodic leg movements in RLS patients. *Clinical Neurophysiology, 110(7),* 1168-1174, 1999.
47. OHANNA N., PELED R., RUBIN A.H.E., ZOMER J., LAVIE P. - Periodic leg movements in sleep: effect of clonazepam treatment. *Neurology, 35,* 408-411, 1985
48. O'KEEFE S.T., GAVIN K., LAVAN J.N. – Iron status and restless legs syndrome in the elderly. *Age and Ageing, 23,* 200-203, 1994.
49. ONDO W. - Ropinirole for restless legs syndrome. *Movement Disorders, 14(5),* 890-892, 1999.
50. RUTKOVE S.B., MATHESON J.K., LOGIGIAN E.L. – Restless legs syndrome in patients with polyneuropathy. *Muscle and Nerve, 19,* 670-672, 1996.
51. SALETU B., GRUBER G., SALETU M., BRANDSTATTER N., HAUER C., PRAUSE W., RITTER K., SALETU-ZYHLARZ G. - Sleep laboratory studies in restless legs syndrome patients as compared with normals and acute effects of ropinirole. 1. Findings on objective and subjective sleep and awakening quality. *Neuropsychobiology, 41(4),* 181-189, 2000.
52. SALETU M., ANDERER P., SALETU B., HAUER C., MANDL M., OBERNDORFER S., ZOGHLAMI A., SALETU-ZYHLARZ G. - Sleep laboratory studies in restless legs syndrome patients as compared with normals and acute effects of ropinirole. 2. Findings on periodic leg movements, arousals and respiratory variables. *Neuropsychobiology, 41(4),* 190-199, 2000.
53. SFORZA E, NICOLAS A, LAVIGNE G, GOSSELIN A, PETIT D, MONTPLAISIR J. - EEG and cardiac activation during leg movements in sleep: support for a hierarchy of arousal responses. *Neurology, 52(4),* 786-791, 1999.
54. SMITH R.C. – Relationship of periodic movements in sleep (nocturnal myoclonus) and the Babinski sign. *Sleep, 8,* 239-243, 1985.
55. STAEDT J., STOPPE G., KOGLER A., RIEMANN H., HAJAK G., MUNZ D.L., EMRICH D., RUTHER F. – Single photon emission tomography (SPET) imaging of dopamine D_2 receptors in the course of dopamine replacement therapy in patients with nocturnal myoclonus syndrome (NMS). *Journal of Neural Transmission (GenSect), 99,* 187-193, 1995.
56. TRENKWALDER C., BUCHER S.F., OERTEL W.H. - Bereitschafts potential in idiopathic and symptomatic restless legs syndrome. *Electroencephalography and Clinical Neurophysiology, 89,* 95-103, 1993.
57. TRENKWALDER C., STIASNY K., POLLMACHER T. – L-Dopa therapy of uremic and idiopathic restless legs syndrome: a double blind crossover trial. *Sleep, 18,* 681-688, 1995.
58. TRENKWALDER C., SEIDEL V.C., GASSER T., OERTEL W.H. – Clinical symptoms and possible anticipation in a large kindred of familial restless legs syndrome. *Movement disorders, 11,* 389-394, 1996.
59. TRENKWALDER C., HENING W.A., WALTERS A.S., CAMPBELL S.S., RAMAN K., CHOKROVERTY S. – Circadian rhythm of periodic limb movements and sensory symptoms of restless legs syndrome. *Movement Disorders, 14(1),* 102-110, 1999.
60. TURJANSKI N., LEES A.J., BROOKS D.J. – Striatal dopaminergic function in restless legs syndrome:[18]F-dopa and ^{11}C-raclopride PET studies. *Neurology, 25,* 932-937, 1999.
61. VON SCHEELE C., KEMPI V. – Long-term effect of dopaminergic drugs in restless legs. A 2-year follow-up. *Archives of Neurology, 47,* 1223-1224, 1990.
62. WALTERS A.S., HENING W.A., CHOKROVERTY S., GIDRO-FRANCK S. – A double blind randomized crossover trial of bromocriptine and placebo in restless leg syndrome. *Annals of Neurology, 24,* 455-458, 1988.
63. WALTERS A.S., TRENKWALDER C., HENING W.A. - Fluorodeoxyglucose PET scanning in 5 patients with restless legs syndrome. *Sleep Research, 24,* 365, 1995.
64. WALTERS A.S. - Toward a better definition of the restless legs syndrome. *Movement disorders, 10,* 634-642, 1995.
65. WALTERS A.S., HICKEY K., MALTZMAN J., VERRICO T., JOSEPH D., HENING W., WILSON V., CHOKROVERTY S. – A questionnaire study of 138 patients with restless legs syndrome: The 'Night-Walkers' survey. *Neurology, 46,* 92-95, 1996.
66. WECHSLER L.R., STAKES J.W., SHAHANI B.T., BUSIS N.A. - Periodic leg movements of sleep (nocturnal myoclonus): an electrophysiological study. *Annals of Neurology, 19,* 168-173, 1986.
67. WEIMERSKIRCH P.R., ERNST M.E. - Newer dopamine agonists in the treatment of restless legs syndrome. *Annual Pharmacotherapy, 35(5),* 627-630, 2001.
68. WILLIS T. De Animae Brutorum. Wells and Scott, London, 1672, p339.
69. ZUCCONI M., COCCAGNA G., PETRONELLI R,. GERARDI R., MONDINI S., CIRIGNOTTA F. – Nocturnal myoclonus in restless legs syndrome effect of carbamazepine treatment. *Functional Neurology, 4,* 263-271, 1989.

Chapter 49

Abnormal postures and movements during sleep

E. Hirsch, B. Maton, F. Sellal, and C. Marescaux
Unité d'Explorations Fonctionnelles des Epilepsies, Clinique Neurologique, Hôpital Civil, Strasbourg, France

Paroxysmal movements of many types occur during sleep. These correspond either to sleep epileptic seizures or to parasomnias (sleepwalking, night terrors, REM sleep-related behaviour disorders etc.). These parasomnias, defined by strict clinical and paraclinical criteria, are described elsewhere in this book and so will not be dealt with here. The present chapter will focus on nocturnal motor phenomena referred to in the literature as hypnogenic paroxysmal dystonia, nocturnal paroxysmal dystonia, paroxysmal arousals, paroxysmal nightmares, episodic wanderings etc. Their pathophysiology has long been subject to debate. The 1997 classification of sleep disorders defined these complex abnormal movements as parasomnias [1]; most authors do not consider them as forming a separate entity, but as a clinical form of partial epileptic seizures [9, 19 20, 26, 31, 35].

HISTORY

In 1981, Lugaresi and Cirignotta [14] reported observations of five patients with brief episodes of abnormal "dystonic" movements, occurring intermittently during NREM sleep. During the seizure, the EEG tracing showed no abnormalities suggesting an epileptic seizure. The title of the publication "Hypnogenic paroxysmal dystonia, epileptic seizures or a new syndrome?" reflected the nosological questions which would still be asked 10 years later. Later, the Bologna school proposed the term "nocturnal paroxysmal dystonia" (NPD) with short-lasting seizures [15]. Numerous publications have provided accurate accounts of the clinical symptoms of this illness and discussed its pathophysiology [3, 5, 8, 16, 19, 20, 31, 35].

EPIDEMIOLOGY

It is difficult to determine the incidence and true prevalence of NPD, as a number of corresponding cases have been reported under different names [17, 21, 24, 25, 36]. Roughly sixty cases were reported between 1981 and 1992, 20 of these by the Bologna team. At the Clinical Epileptology Unit in Strasbourg between 1985 and 1992, we observed 40 patients presenting symptoms which were compatible with a diagnosis of NPD. These 40 cases represent about 10% of the patients referred to us for investigations for nocturnal paroxysmal phenomena.

CLINICAL AND POLYGRAPHIC FEATURES

The clinical picture is very uniform for NPD, as described by the Bologna school [5, 14, 15, 16, 20, 21, 32]. Seizures start at any age in patients with no particular family or personal history. Clinical, neuropsychological and neuroradiological tests are normal. Both sexes are equally affected.

The attacks last between 15 seconds and 2 minutes. They are frequent, usually occur nightly, and often several dozen times a night. Seizures occur in the middle of sleep. The patient suddenly opens his eyes, looking around with anxiety or astonishment. After several seconds or in just a fraction of a second, the dystonic postures and movements of the head, trunk and limbs and/or choreic or ballic movements of the upper and lower limbs appear. The motor phenomena are sometimes accompanied by vocalisation (cries, blurred speech, murmurs, sobs etc.). The patient appears to be awake, but is not usually capable of reacting to external stimuli. The movements stop suddenly but reappear after a short period of normal sleep. There is no confusion after the seizure. Patients vaguely recall the seizure if questioned about it afterwards. They will often describe a sensation of fright, but never report dreaming.

NPD semiology varies from one patient to another, but patients each have stereotyped episodes, which do not change during the night or from one night to another. Most patients have quick movements (flexion-extension of fingers, abduction of the arms, rotation or extension of the head, touching the urogenital zone, etc.) for 1 to 5 seconds. The episodes often appear to coincide with the initial phase of complete seizures [16, 28].

A sizeable proportion of these patients also present diurnal seizures. They may be identical to nocturnal seizures, or limited to focal sensory motor symptoms. Finally, over a third of patients suffer from generalised tonic-clonic epileptic convulsions during sleep.

From a polysomnographic point of view, the seizures usually arise during NREM sleep, during stages 2, 3 or 4. More rarely, they begin during an REM sleep transition phase. An arousal is initially noted on the EEG, with vegetative alterations (tachycardia and tachypnoea, or alternatively, bradycardia and central apnoea). During the seizure itself, tracings usually only show artefacts. No post seizure depression occurs. Interictal EEG tracings of wakefulness or sleep are generally normal.

The consistently normal EEG tracings probably reflect a recruitment bias. Indeed, for the Italian authors, NPDs are never accompanied by EEG abnormalities, by definition. For this reason, patients presenting identical motor symptoms associated with disturbed EEG tracings, are not considered as having NPD. Many teams disagree with this view, however (see pathophysiology and nosological situation).

COURSE OF ILLNESS

The age of onset of seizures ranges from 3 to 50 years. Without treatment, the seizures persist in the same way throughout the patient's life, and chronic cases lasting over 30 years have been reported. Anti-epileptic treatment, particularly carbamazepine, reduces or completely suppresses seizures in nearly 70% of patients.

CLINICAL VARIANTS

Two close syndromes have been described in the literature.

Paroxysmal arousals

Montagna *et al.* [21] observed six cases of repeated nocturnal paroxysmal arousals occurring in NREM sleep. These extremely short episodes start with a cry and gestures of fear or astonishment. These are associated with dystonic postures, choreic movements and trembling. After a few seconds the patient goes back to sleep. Polysomnographic tracings show tachycardia, ample respiration or tachypnoea, and the EEG records an arousal followed by electromyographic artefacts. These nocturnal phenomena are not clearly recalled, but patients do complain of poor sleep and recall a number of fearful arousals. Peled and Lavie [25] reported fourteen patients presenting very similar symptomatology, associated with daytime sleepiness. In this series, some patients presented interictal EEG abnormalities leading to a suspected diagnosis of epilepsy. "Paroxysmal arousals" are improved by anti-epileptics, particularly carbamazepine. Hence, apart from the duration of episodes, considerable similarities exist between "paroxysmal arousals" and NPD.

Episodic nocturnal wanderings (or paroxysmal nightmares)

Pedley and Guilleminault [24] reported six observations under this heading, with sudden arousals followed by a cry, unintelligible speech, elaborate gesticulation and motor activity of unaccustomed violence. Patients appear frightened, oblivious of their surroundings and may be dangerous to themselves or others; they have no recall of these episodes. The same symptomatology was observed in fifteen patients by Maselli *et al.* [17] and by Boller *et al.* [4]. During the seizures, polysomnographic tracing shows only artefacts, but during interictal phases, epileptiform discharges are sometimes observed. These spectacular phenomena respond favourably to anti-epiletics, particularly carbamazepine.

Isolated cases have been described with clinical variants. Some patients present atypical symptoms: seizures involving only vocalisation [32], seizures induced as a reflex to sensory stimulation [13], seizures manifesting as daytime sleepiness [18]. Lee *et al.* [12] have reported a familial form of NPD. In this family, NPD was of autosomal dominant transmission. Carbamazepine had little effect. A few cases of family history have also been reported in patients suffering from paroxysmal arousals [17, 18].

DIFFERENTIAL DIAGNOSIS

NPDs are easily distinguishable from generalised sleep epileptic seizures, but less so for certain partial epileptic seizures. In fact, this distinction no longer seems relevant, since most authors now recognise the epileptic nature of NPD (see pathophysiology and nosological situation).

Certain parasomnias, sleepwalking, night terrors and REM sleep behaviour disorders may pose diagnostic problems. As NPDs occur several times per day, one or two overnight polysomnographic recordings with video control are usually sufficient to record an episode and to correct the diagnosis. Similarly, findings from the interrogation and polysomnographic recordings provide a clear distinction between a diagnosis of NPD and that of sleep starts, periodic leg movements during sleep, or restless leg syndrome during wakefulness.

Nocturnal paroxysmal dystonias with long-lasting seizures may pose a difficult problem for differential diagnosis. Long lasting dystonias are extremely rare. Two isolated cases have been reported to date by Lugaresi's team [5, 16]. This syndrome, which it is no doubt too soon to identify clearly, is characterised by abnormal movements lasting 2 to 60 minutes which are repeated several times a night for years; no abnormal daytime movements occur. All the additional tests, particularly polysomnography, are normal; these long-lasting nocturnal paroxysmal dystonias do not respond to anti-epileptic medication. They may constitute the early phase of certain movement pathologies. Indeed, one patient developed typical Huntingdon's chorea, after a period of twenty years.

PATHOPHYSIOLOGY AND NOSOLOGICAL SITUATION

Numerous pathophysiological hypotheses have been put forward. NPDs have been variously thought to be hysterical manifestations [11, 21], parasomnias [1, 14, 21], or abnormal movements close to certain forms of paroxysmal dystonia observed during the daytime [14, 21].

The tendency for the personality of patients to be normal, the stereotyped nature of episodes, the fact they occur only during sleep and the absence of secondary benefits rules out a neurotic pathology.

A diagnosis of parasomnia also appears unlikely. Classic parasomnias last longer than NPDs. They occur less frequently, usually once a night, with long intervals between attacks; they tend to occur during NREM sleep stages 3-4. For a considerable number of NPD patients, the existence of diurnal episodes eliminates this hypothesis.

NPDs are unlike abnormal movements of extrapyramidal origin. In fact the latter diminish or disappear during sleep, which allows only pathological motor activities of epileptic origin to persist. NPDs do however present some analogies to diurnal paroxysmal dystonias. These paroxysmal choreo-athetoses, whether or not induced by movement, may be of an epileptic nature [5, 31].

Most authors, including those in the Bologna team, now agree that NPDs with short seizures correspond to partial epileptic fits. Four arguments tend to support this hypothesis:
1. A large proportion of patients have a personal history of generalised convulsive nocturnal epileptic seizures.

2. EEG epileptic abnormalities have been recorded for a sizeable percentage of subjects, after repeated interictal scalp recordings [3, 9, 31] or using sphenoid electrodes [8] or deep electrodes implanted in stereotaxic conditions [28, 31].
3. NPDs respond to anti-epileptic medication.
4. Semiological similarities exist between NPD motor manifestations and those of certain focal epileptic seizures. Thus internal frontal region epilepsy is characterised by short seizures associating vocalisation (cries, laughter, swearing), coordinated motor activity of the upper limbs (hand clapping, rubbing the hands) or the lower limbs (flexing the thighs, pedalling), changes to axial posture and tonic spasms at the extremities. Patients continue to react to external stimulation and are only slightly if at all preoccupied after short seizures, which occur in bursts chiefly at night, are usually electrically silent, and respond relatively well to treatment with carbamazepine [22, 34, 36, 37]. Partial temporal seizures, particularly those which are hippocampal, may also be accompanied by unilateral dystonic movements [2, 10].

Several subsequent publications have conclusively confirmed the epileptic nature of NPD. Tinuper *et al.* [35] recorded two patients presenting typical NPD in whom a normal attack was followed by tonic-clonic convulsions. This secondary generalisation occurred at the same time as an EEG charge at the level of the scalp. Mejerkord *et al.* [19] compared the video recordings of nine patients presenting seizures compatible with a diagnosis of NPD, with no EEG abnormalities and eight patients with partial sleep epilepsy. There were no semiological differences between the two groups (age of onset, men/women ratio, motor symptoms etc.). The findings collected in Strasbourg since 1985 point to similar conclusions [9, 31]. Among the thirty patients in whom abnormal nocturnal paroxysmal movements were identified, between 1985 and 1990, we distinguished three groups in according to EEG seizure patterns. Group I (9 patients) had no history of EEG abnormalities during a nocturnal episode. Group II (11 patients) showed marked EEG abnormalities evocative of partial seizures during one or several nocturnal episodes (repetitive, focal spikes or spike-waves). Group III (10 patients) sometimes presented doubtful EEG alterations (transitory focalised flattening, postictal slowing) which were insufficient in themselves to confirm the epileptic nature of the motor phenomena. Patients in groups I and II were compared for personal history of epilepsy, clinical semiology of nocturnal episodes, neuroradiological tests and data from multiple interictal EEG recordings. The results are summarised in tables 49.1, 49.2 and 49.3. No difference is found between the two groups of patients as regards sex, age of onset, response to treatment, history of epilepsy, the characteristics of abnormal nocturnal paroxysmal movements, or the frequency and duration of seizures. EEG recordings were carried out 10 to 15 times in wakefulness and sleep, for each patient. Seventy seven percent of the patients in group I presented abnormalities in interictal EEG, at one time or another, confirming a diagnosis of epilepsy. Five patients (two from group I and three from group II) showed a secondary generalisation on recordings. The EEG video findings showed that tonic-clonic convulsions were preceded by an episode of abnormal paroxysmal movements similar to the usual seizures of these patients. Seven patients also had PET scans, using deoxyglucose. In four of these cases, the examination showed frontal or temporal zones of hypometabolism which were very similar to those of patients who clearly presented partial epilepsy. The results were identical for two patients who had never shown characteristic EEG abnormalities, and two patients whose EEG clearly demonstrated epilepsy.

Table 49.1. General characteristics of patients presenting NPD without (group I) and with (group II) ictal EEG abnormalities

Characteristics	Groups	
	I	II
Number of patients	9	11
Men	5	4
Women	4	7
Age of onset	15.3 ± 2.4	13.6 ± 5.7
History of partial epilepsy	1/9	1/11
History of tonic-clonic generalisation	6/9	5/11

Table 49.2. Comparative semiology of movements observed during nocturnal episodes in patients presenting NPD without (group I) and with (group II) ictal EEG abnormalities

Symptoms of nocturnal seizure	Groups	
	I (n = 9)	II (n = 11)
Altered axial posture	6	9
Tonic and/or dystonic movements	8	8
Grimaces, eyes open	8	9
Automatisms	5	6
Pelvic movements, pedalling	4	7
Purposeless agitation	4	5
Choreic, ballic-like movements	1	2
Vocalisation	7	8
Secondary generalisation recorded on video-EEG	2	3

Table 49.3. EEG type for patients presenting NDP without (group I) and with (group II) ictal EEG abnormalities

EEG abnormalities	Groups	
	I	II
Ictal EEG abnormalities	0/9	11/11
Interictal EEG abnormalities	7/9	7/11
Number of abnormal interictal tracings	26/102	35/179

Most authors recognise seizures as originating in the deep frontal lobe. This would account for the clinical semiology and rarity of EEG abnormalities. Nevertheless EEG findings and SPECT data suggest that seizures may sometimes start in the temporal or parietal regions, even in the parieto-occipital regions, only gaining the frontal structures secondarily. The only way to verify the different hypotheses is by multiplying stereo EEG recordings.

The epileptic nature of hypnogenic paroxysmal dystonia is further corroborated by genetic findings. Indeed a familial form of nocturnal frontal epilepsy, by autosomic dominant transmission began to be reported in 1965 [23, 29, 30]. The symptoms observed in these forms are identical to those described in sporadic cases. If these forms are characterised by phenotypic homogeneity, this is not the case in genotypic terms: familial genetic studies have indeed demonstrated different mutations on variable loci of chromosomes 20, 4 and 1. The first, discovered by Steinlein *et al.* [33] in an Australian family, showed a mutation in the gene coding for the nicotinic acetylcholine receptor alpha 4 subunit (heterodimer formed of 2 alpha 4 subunits and 2 beta 2 subunits). The latest familial studies have demonstrated, in Italian [6, 7] and Scottish [26] families, a new mutation on the gene coding for the beta 2 subunit of the same nicotinic receptor. There may be a common pathophysiological mechanism underlying this genotype heterogeneity, affecting cholinergic transmission. Some sporadic cases may also be genetically determined [27].

TREATMENT

Without treatment, NPD will persist throughout life. Seizures are not affected by behaviour therapy or psychotherapy. On the other hand, NPD are sensitive to anti-epileptics. Carbamazepine appears to be the most effective treatment. This is the first line treatment prescribed by all the teams. It must be introduced gradually. Weak doses (5 to 10 mg/kg) are sometimes effective. In other cases NPD is only affected by strong doses (>20 mg/kg) with high serum levels (> 10 mg/L). The response to treatment is usually good in subjects treated early who have had no previous anti-epileptic treatment, and 50 to 75% of patients are controlled by well monitored monotherapy. The efficacy of treatment does not wane with time, but seizures recur on its withdrawal.

When carbamazepine has no effect, a second monotherapy or bitherapy can be envisaged. Second line anti-epileptics have not been clearly defined in the literature. In our unit we use lamotrigine, topiramate, phenytoin or phenobarbital. The pharmaco-resistant forms correspond to 25-30% of cases. The persistence of seizures with social and personal repercussions may be an argument for surgical treatment. Very few surgical operations have been performed and they often show disappointing results.

CONCLUSION

NPDs are involuntary nocturnal movements characterised by the association of dystonic postures and movements, tonic movements in the upper and lower limbs and the trunk, automatisms, affective poses and vocalisation. In some cases EEG recordings show abnormalities evocative of epilepsy, in others they appear normal. However the high degree of homogeneity in terms of clinical and paraclinical symptoms as well as the effects of treatment, appear to obviate the value of forming subgroups according to EEG findings. Data available in the literature suggest that these preponderantly nocturnal seizures are related to partial epilepsy rather than to a movement pathology or a parasomnia. This epilepsy is sometimes pharmaco-resistant. A better understanding of the mechanisms by which the sleep-wake cycle controls NPD should help in developing new treatment strategies. A surgical approach may be worth considering in certain cases. There is substantial semiological and EEG evidence to suggest that these seizures involve the internal frontal regions at an early or a later stage. Only stereo EEG recordings will provide an understanding of the true topography of the epileptogenous zone enabling us to define the cortical and subcortical regions involved in the genesis and propagation of seizures.

Finally, behind the clinical homogeneity, genetic studies have demonstrated genotypic heterogeneity with mutations affecting variable loci, in familial forms or sporadic cases. It appears however, that the loci so far identified concern the genes coding for the subunits which form the nicotinic cholinergic receptor, suggesting that there is a common pathophysiological mechanism.

REFERENCES

1. AMERICAN SLEEP DISORDERS ASSOCIATION. ICSD – *International classification of sleep disorders, revised: Diagnostic and coding manual.* American Sleep disorders Association, 1997.
2. BENNETT D.A., RISTANOVIC R.K., MORRELL F., GOETZ C.G. – Dystonic posturing in temporal lobe seizures. *Neurology,* 39, 1270-1272, 1989.
3. BESSET A., BILLIARD M. – Nocturnal paroxysmal dystonia. In: *Sleep 1984,* W.P. Koella, E. Rüther, H. Schulz (eds), Gustave Fischer Verlag, Stuttgart, 383-385, 1985.
4. BOLLER F., WRIGHT D.G., CAVALIERI R., MITSUMOTO H. – Paroxysmal « nightmares ». Sequel of a stroke responsive to diphenylhydantoin. *Neurology,* 25, 1026-1028, 1975.
5. CIRIGNOTTA F., LUGARESI E., MONTAGNA P. – Nocturnal paroxysmal dystonia. In: *Principles and practice of sleep medicine.* M.H. Kryger, T. Roth, W.C. Dement (eds), Saunders Company, New York, 410-412, 1989.
6. DE FUSCO M., BECHETTI A., PATRIGNANI A., ANNESI G., GAMBARDELLA A., QUATTRONE A., BALLABIO A.,WANKE E., CASARI G. – The nicotinic receptor beta 2 subunit is mutant in frontal nocturnal lobe epilepsy. *Nat. Genet.,* 26, 275-276, 2000.
7. GAMBARDELLA A., ANNESI G., DE FUSCO M., AGUGLIA U., PASQUA A.A., SPADAFORA P., OLIVERI R.L., VALENTINO P., ZAPPIA M., BALLABIO A., CASARI G., QUATTRONE A. - A new locus for autosomal dominant nocturnal frontal lobe epilepsy maps to chromosome 1. *Neurology,* 55, 1567-1471, 2000.
8. GODBOUT R., MONTPLAISIR J., ROULEAU I. – Hypnogenic paroxysmal dystonia: epilepsy or sleep disorder? A case report. *Clin. Electrencephalogr.,* 16, 136-142, 1985.
9. HIRSCH E. – Abnormal paroxysmal postures and movements during sleep: partial epilepsy or paroxysmal hypnogenic dystonia ? *In: Sleep 1990,* J. Horne (ed.), Pontenagel Press, Bochum, 471-473, 1900.
10. KOTAGAL P., LÜDERS H., MORRIS H.H., DINNER D.S., WYLLI E., GODOY J., ROTHNE R.D. – Dystonic posturing in complex partial seizures of temporal lobe onset: a new lateralizing sign. *Neurology,* 39, 196-201, 1989.
11. KOVACEVIC-RISTANOVIC R., GOLBIN A., CARTWRIGHT R. – Nocturnal conversion disorder and nocturnal paroxysmal dystonia. Similarities and treatments. *Sleep Res.,* 17, 204, 1988.
12. LEE B.I., LESSER R.P., PIPPENGER C.E., MORRIS H.H., LÜDERS H., DINNER D.S., CORRIE W.S., MURPHY W.F. – Familial paroxysmal hypnogenic dystonia. *Mov. Disord.,* 3, 290-294, 1988.
13. LEHKUNIEC E., MICHELI F., DE ARBELAIZ R., TORRES M., PARADISO G. – Concurrent hypnogenic and reflex paroxysmal dystonia. *Mov. Disord.,* 3, 290-294, 1988.
14. LUGARESI E., CIRIGNOTTA F. – Hypnogenic paroxysmal dystonia: epileptic seizures or a new syndrome ? *Sleep,* 4, 129-138, 1981
15. LUGARESI E., CIRIGNOTTA F., MONTAGNA P. – Nocturnal paroxysmal dystonia. . *J. Neurol. Neurosurg. Psychiatry,* 49, 375-380, 1986.
16. LUGARESI E., CIRIGNOTTA F., MONTAGNA P. – Nocturnal paroxysmal dystonia. *Epilepsy Res.,* suppl. 2, 137-140, 1991.
17. MASELLI R.A., ROSENBERG R.S., SPIRE J.P. – Episodic nocturnal wandering in nonepileptic young patients. *Sleep,* 11, 156-161, 1988.

18. McCARIO M., LUSTMAN L.I. – Paroxysmal nocturnal dystonia presenting as excessive daytime somnolence. *Arch. Neurol.,* 47, 291-294, 1990.

19. MEIERKORD H., FISH D.R., SMITH S.J.M., SCOTT C.A., SHORU S.D., MARSDE C.D.- Is nocturnal paroxysmal dystonia a form of frontal lobe epilepsy? *Mov. Disord.,* 7, 38-42, 1992.

20. MONTAGNA P. – Nocturnal paroxysmal dystonia and nocturnal wandering. *Neurology,* 42, suppl. 6, 61-67, 1992.

21. MONTAGNA P., SFORZA E., TINUPER P., CIRIGNOTTA F ;, LUGARESI E. – Paroxysmal arousals during sleep. *Neurology,* 40, 1063-1066, 1990.

22. NIEDERMEYER E., WALKER A.E. – Mesio-frontal epilepsy. *Electroencephalogr. Clin. Neurophysiol.,* 31, 103-106, 1971.

23. OLDANI A., ZUCCONI M., FERINI-STRAMBI L., BIZZOZERO D., SMIRNE S. – Autosomal dominant nocturnal frontal lobe epilepsy ; electroclinical picture. *Epilepsia,* 37, 964-976, 1996.

24. PEDLEY T.A., GUILLEMINAULT C. – Episodic nocturnal wandering responsive to anticonvulsant drug therapy. *Ann. Neurol.,* 2, 30-35, 1977.

25. PELED R., LAVIE P. – Paroxysmal awakenings from sleep associated with excessive daytime somnolence. A form of nocturnal epilepsy. *Neurology,* 36, 95-98, 1986.

26. PHILLIPS H.A., FAVRE I., KIRKPATRICK M.,ZUBERI S.M., GOUDIE D., HERON S.E., SCHEFFER I.E., SUTHERLAND G.R., BERKOVIC S.F., BERTRAND D., MULLEY J.C. – CHRNB2 is the second acetylcholine receptor subunit associated with autosomal dominant nocturnal frontal lobe epilepsy. *Am. J. Hum. Genet.* 68, 225-231, 2000.

27. PHILLIPS H.A., MARINI C., SCHEFFER I.E., SUTHERLAND G.R., MULLEY J.C., BERKOVIC S.F. - A de novo mutation in sporadic nocturnal frontal lobe epilepsy. *Ann. Neurol.* 48, 264-267, 2000.

28. RAJNA P., KUNDRA O., HALASZ P. – Vigilance level-dependant tonic seizures-epilepsy or sleep disorder ? A case report. *Epilepsia,* 24, 725-733, 1983.

29. SCHEFFER L.E., BHATIA K.P., LOPES-CENDES I., FISH D.R., MARSDEN C.D., ANDERMANN F., ANDERMANN E., DESBIENS R., CENDES F., MANSON J.I. – Autosomal frontal epilepsy misdiagnosed as sleep disorder. *Lancet,* 343, 515-517, 1994.

30. SCHEFFER L.E., BHATIA K.P., LOPES-CENDES I., FISH D.R., MARSDEN C.D., ANDERMANN E., ANDERMANN F., DESBIENS R., KEENE D., CENDES F. – Autosomal dominant nocturnal frontal lobe epilepsy. A distinctive clinical disorder. *Brain,* 118, 61-73, 1995.

31. SELLAL F., HIRSCH E., MAQUET P., SALMON E., FRANCK G., COLLARD M., KURTZ D., MARESCAUD C. – Postures et mouvements anormaux paroxystiques au cours du sommeil: dystonie paroxystique hypnogénique ou épilepsie partielle? *Rev. Neurol.* (Paris), 147, 121-128, 1991.

32. SILVESTRI R., DE DOMENICO P., CASELLA C., MENTO G., DI PERRI R. – Nocturnal paroxysmal dystonia with atypical short-lasting attacks. *Neurophysiol. Clin.* 20, 217-219, 1990.

33. STEINLEIN O.K., MULLEY J.C., PROPPING P., WALLACE R.H., PHILLIPS H.A., SUTHERLAND G.R., SCHEFFER I.E., BERKOVIC S.F. – A missense mutation in the neuronal nicotinic acetylcholine receptor alpha 4 subunit is associated with autosomal dominant nocturnal frontal lobe epilepsy. *Nat. Genet.* 11, 201-203, 1995.

34. THARP B.R. – Orbital frontal seizures. An unique electroencephalographic and clinical syndrome. *Epilepsia,* 13, 627-642, 1972.

35. TINUPER P., CERULLO A., CIRIGNOTTA F., CORTELLI P., LUGARESI E., MONTAGNA P. – Nocturnal paroxysmal dystonia with short-lasting attacks: three cases with evidence for an epileptic frontal lobe origin of seizures. *Epilepsia,* 31, 549-556, 1990.

36. WADA J.A. – Nocturnal recurrence of brief, intensely affective vocal and facial expression with powerful bimanual, bipodal, axial and pelvic activity with rapid recovery as manifestations of mesial frontal lobe seizure. *Epilepsia,* 29, 209, 1988.

37. WATERMAN K., PURVES S.J., KOSAKA B., STRAUSS E., WADA J. – An epileptic syndrome caused by mesial frontal lobe seizure foci. *Neurology,* 37, 577-582, 1987.

Chapter 50

Night epilepsies

M. Baldy-Moulinier
Service d'Explorations Neurologiques et Epileptologie, Hôpital Gui de Chauliac, Montpellier, France

The sudden, unexpected nature of seizures is a disturbing aspect of all epilepsies. But from the time of Antiquity, it has been noted that the seizures of some epileptics occur at more favourable times in the sleep-wake cycle, sometimes exclusively so, particularly during nocturnal sleep. This observation suggests that sleep mechanisms may influence the activation of epileptogenesis, raising the question as to whether sleep epilepsies can be considered as an entity. Night epilepsies are included in the many studies devoted to the inter-relationship between sleep and epilepsy, whether the studies are clinical [22, 24, 28, 39, 55], or electroclinical and polysomnographic [21, 23, 30, 51, 54]. The multiple, complex data now available on this subject has been the focus of several general reviews [5, 16, 43]. The pathophysiological analysis presented in this chapter will refer to the electroclinical parameters of states of wakefulness, the nosological reference being the Commission on Classification and Terminology of the League against Epilepsy [16, 17, 60].

SLEEP-RELATED EPILEPTIC PHENOMENA

Sleep has an effect on all forms of epilepsy by modifying the frequency and expression of seizures, and modulating interictal electroencephalographic (EEG) discharges. Each type of epilepsy has its own modality of expression.

Generalised idiopathic epilepsy

Generalised seizures

Tonic-clonic seizures (Grand Mal) are observed during sleep with two frequency peaks: one occurring two hours after the onset of sleep, the other at the end of the night, between 5 and 6 am. These occur during NREM sleep only, most often during stages 2 and 3, and are usually associated with a short awakening or stage shift. The seizure momentarily interrupts the course of sleep, which later resumes with little or no REM sleep. The subject suspects having had a seizure, on awakening the following morning, due to the signs of urination, tongue biting or muscular pain.

Clonic seizures rarely occur in sleep and are always associated with arousals during NREM sleep.

Myoclonic seizures may also occur during sleep, either spontaneously at sleep onset or after a provoked arousal [44, 67]. These myoclonic seizures should be distinguished from periodic limb movements, which are a purely segmentary, motor phenomenon affecting the lower limbs, occurring intermittently during the night and occasionally provoking insomnia.

The short absences in *absence epilepsy* in childhood or Petit Mal, are hard to identify during sleep. In some cases seizure events are detected such as blinking of the eyelids, at the start of sleep.

Interictal EEG discharges

The paroxysmal activities: interictal, spike, spike-wave, polyspike-wave both bilateral and synchronous, which characterise generalised idiopathic epilepsies increase during NREM sleep progressing from stage 1 to stages 3-4. Conversely, they diminish or disappear during REM sleep [21, 50]. NREM sleep activation of interictal discharges is particularly marked at the beginning of the night. The morphology of paroxysmal abnormalities is transformed, interfering with the spindles, K complexes and slow waves. The inverse effects of NREM and REM sleep on synchronous bilateral activity can be seen in experimental models of generalised epilepsy [30]. The setting up of synchronisation systems [53] and thalamocortical mechanisms in generating sleep spindles and bilateral spike-waves may explain the effects of NREM sleep and the interference of bioelectrical signals [35, 37].

Cryptogenic or secondary generalised epilepsy

West's syndrome

This is characterised by infantile spasms with disorganised interictal activity corresponding to hypsarrythmia. Spasms recur more often during NREM sleep, parallel to an amplification in the paroxysmal elements of hypsarrhythmia. Inversely, amplitude is seen to decrease in REM sleep, with background activity sometimes appearing as pseudonormal [35, 51]. Signs of localisation may sometimes become apparent during REM sleep [33].

Lennox-Gestaut syndrome

Epileptic seizures during wakefulness are polymorphic, associating tonic seizures, atonic seizures and atypical absences.

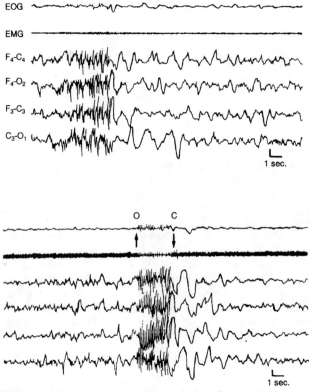

Figure 50.1. Bursts of diffuse rapid spikes during NREM sleep in Lennox-Gastaut syndrome.

Characteristically, tonic seizures are strongly activated in this syndrome during sleep, showing diffuse bursts of rapid spikes [7, 9] (fig. 50.1). This may appear as status epilepticus due to the succession of seizures with reduced clinical, or merely electrical, expression. Activation occurs during NREM sleep; it may persist to a lesser extent in REM sleep, particularly in severe cases.

Symptomatic generalised epilepsies

Seizures take various forms and are not specifically nocturnal or diurnal. Interictal discharges are distinguishable from those in idiopathic generalised epilepsies by their bilateral, asymmetrical character, with a less rhythmic, more irregular pattern, and the fact that their modulation is less affected by sleep. No defined etiological expression exists. However, attention should be drawn to the particularity of the Unverricht-Lundborg type of progressive myoclonic epilepsy, including both the symptoms of Baltic myoclonus and of Mediterranean myoclonus, an autosomal recessive genetic disease associating myoclonic jerks, truncal ataxia and progressive intellectual deterioration. In this form of epilepsy, generalised interictal discharges of polyspike-waves are not activated by NREM sleep, rapid polyspikes affecting the central regions and vertex appear during REM sleep [61]. The discovery of mutations in cystatin B gene coding for an antiproteinase, localised on chromosome 21 [38] should help to determine the mechanism of myoclonic jerks, the way in which NREM sleep is not conducive to epileptic discharges, and the increase in focal discharges during REM sleep.

Partial idiopathic epilepsies

Partial idiopathic epilepsy with midtemporal spikes

Benign childhood epilepsy is the form in which ictal and interictal manifestations are the most sleep-sensitive.

Characteristic seizures, which are orofacial, possibly with tonic-clonic generalisation, occur only during sleep, in 70 to 80% of cases [13, 31, 40]. Sleep seizures tend to be generalised whereas seizures occurring during wakefulness, or waking seizures, are almost always partial.

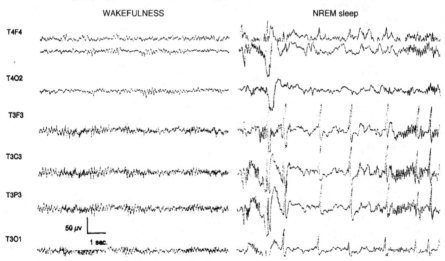

Figure 50.2. Bursts of midtemporal spikes during NREM sleep in benign infantile epilepsy

The interictal discharges of negative or bipolar spikes in the midtemporal area, rhythmical in aspect and essential for diagnosis, may only appear during sleep [10, 20] (fig. 50.2). In 30% of benign epilepsies, EEG abnormalities are only seen during sleep [29]. Discharges consistently progress from stage 1 to stages 3-4 of NREM sleep, with frequent generalisation and possible electrical status epilepticus. During REM sleep, the discharges wane, but may persist in certain cases. This form of epilepsy disappears at puberty. Prognosis can be established early, on the basis of the criteria for nocturnal seizure and sleep-sensitive EEG abnormalities. The most significant benign feature is the production of bipoles, negative temporal spikes with simultaneous positive frontal spikes [29].

Benign infantile epilepsy with occipital paroxysms

This form of epilepsy is characterised by seizures which usually occur at night, with visual semiology followed by a hemiconvulsion and sometimes generalisation. EEG tests reveal interictal discharges of spike-waves affecting the temporo-occipital regions, whose particularity lies in only occurring when the eyes are closed [25]. The activation of occipital discharges during NREM sleep appears to be an element which distinguishes between idiopathic and symptomatic epilepsy [26].

Symptomatic partial epilepsies

Complex partial epilepsies (CPE)

Sleep or a diminution of the level of wakefulness is considered to be conducive to partial seizures. The conducive effect of sleep varies according to the origin of the seizures. The nocturnal onset of complex partial seizures was initially proposed as a localising criterion in favour of temporal lobe epilepsy. In fact, in this type of epilepsy, it is rare to find nocturnal seizures only (10%), as compared to specifically diurnal seizures (60%) or both diurnal and nocturnal (30%) [12].

A comparative study using EEG video monitoring was carried out on two cohorts of patients affected with epilepsy of frontal or midtemporal origin, validated by in-depth preoperative investigations and confirmed by postoperative results. The study demonstrated the significant differences between these two topographical types [18]. Frontal lobe epilepsy is characterised by the occurrence of seizures during sleep which are not preceded by awakening, often with recurring seizures (clusters) during the same night, with sleep architecture remaining relatively intact. Midtemporal structure epilepsy differs from frontal lobe epilepsy in the rarity of seizures during sleep, the awakening reaction which precedes the electro-clinical manifestations of the seizure by a few seconds, and the frequent secondary tonic-clonic generalisation when seizures occur during sleep. Sleep architecture is also altered, with numerous awakenings and sleep stage shifts [19]. In function of such differential characteristics, frontal lobe epilepsy may be considered as sleep epilepsy and midtemporal epilepsy as waking epilepsy.

Nonetheless sleep seizures of frontal and midtemporal origin do share certain aspects, i.e. the fact that they essentially occur during NREM sleep, and rarely during REM sleep [36, 51] (fig. 50.3). The possibility of midtemporal seizures specifically activated during REM sleep, as observed in certain experimental models [1], does not apply in man, even if some patients experience recurrent seizures during each phase of REM sleep [53].

Interictal discharge in epilepsies with complex partial seizures

These are distinctly activated by sleep. This phenomenon, discovered by Gibbs and Gibbs [27] from simple EEG recordings, has been confirmed by several authors on the basis of nocturnal polygraphic recordings with both surface and deep electrodes. During NREM sleep, discharges increase and tend to become diffuse or bilateral. During REM sleep, discharges persist or weaken as focalisation is optimised [8, 41, 46, 62]. Activation by NREM sleep and focalisation during REM sleep are elements in favour of prolonged monitoring as part of the explorations carried out prior to surgery for drug resistant epileptics with complex partial seizure (fig. 50.4). The information provided by polysomnography is particularly useful in determining the lateralisation of the primary focus in the case of secondary bilateral synchrony frequent in frontal lobe and posterior temporal lobe epilepsies. Polysomnography helps to determine primary focus localisation in epileptics whose surface EEG shows several foci. Discharges relating to a primary focal lesion are subjected to less fluctuation during sleep than those which relate to a projected focus or to primary focus diffusion [62].

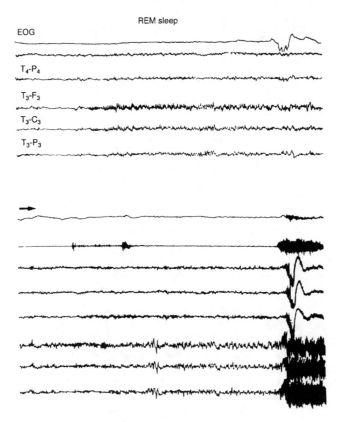

Figure 50.3. Complex partial seizure during REM sleep in partial temporal lobe epilepsy.

Particular sleep-related epileptic syndromes

Continuous spike-wave syndrome in NREM sleep, initially defined as "encephalopathy with infraclinical epileptic status in NREM sleep" [56, 66, 67], is observed in children between 4 and 10 years old. It is characterised by continuous spike-wave activity with the onset of sleep, persisting throughout the phases of NREM sleep, occupying over 85% of this type of sleep, and disappearing with REM sleep. This activity is associated with the destructuring of the intellectual functions concomitant with the appearance of continuous spike-waves, and behaviour disorders with occasional nocturnal seizures, generally with spontaneous remission before the age of 15. The disappearance of continuous spike-waves during the second decade is accompanied by an improvement in behaviour disorders and cognitive skills. Neuropsychological sequelae are often severe. Benzodiazepines prescribed as first line treatment have a transitory effect. Corticosteroid treatment proposed as second line treatment may lead to an improvement [11, 15]. This syndrome, defined as a separate entity, is similar to Landau-Kleffner's syndrome or acquired aphasia in which the same type of EEG epileptic abnormality appears in NREM sleep, with localisation sometimes limited to the dominant hemisphere or the left temporal region. Seizures do not occur systematically but epileptic activity is thought to be implicated in the development of aphasia. Sodium valproate, benzodiazepines and early, prolonged corticosteroid treatment may have an effect on epileptic activity and aphasia. The pathophysiology of continuous spike-wave syndrome and Landau-Kleffner's syndrome remains unclear. There is no consistent relationship between EEG abnormalities and cognitive disorders. Electrophysiological, haemodynamic and metabolic studies suggest the emergence of a pathological process affecting the associative areas of the cerebral cortex during maturation [40].

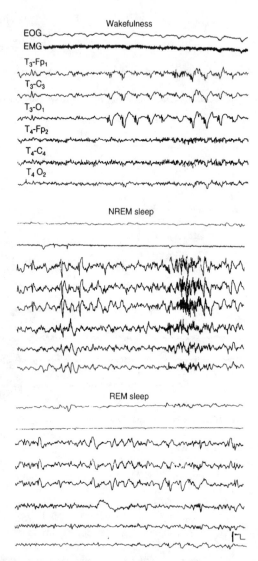

Figure 50.4. Modulation of focal discharges (midtemporal and right posterior) during sleep and wakefulness (W: wakefulness, NREM: non REM sleep, REM: REM sleep), in temporal lobe epilepsy.

SLEEP EPILEPSIES

The fact that some epileptics only experience seizures during the night, has led some authors to refer to sleep epilepsies. Delmas-Marsalet [23] was the first to attempt to define the clinical and electrical characteristics of this type of epilepsy. For Janz [34], sleep epilepsies correspond to epilepsies with complex partial seizures. They are distinct from awakening epilepsies, absence epilepsies or Petit Mal and myoclonic epilepsies, or from diffuse epilepsies whose seizures are indifferently distributed throughout the night or day and which are of organic aetiology.

The findings presented in the previous chapter, showed that no epilepsy is specifically linked to sleep, apart from continuous spike-wave syndrome in NREM sleep. Sleep-sensitivity is greater in some epilepsies, chiefly benign infantile epilepsy with midtemporal spikes, Lennox-Gastaut's syndrome, certain generalised idiopathic epilepsies with tonic-clonic seizures and partial symptomatic epilepsies of the temporal lobe and frontal lobe. The exclusive character of seizures during the night does not correlate with any particular electroclinical form and does not particularly relate to the aetiology and evolution of the epilepsy. Next to idiopathic and benign forms, such as epilepsy with midtemporal spikes, sleep-sensitivity is found in severe syndromes such as West's syndrome, Lennox-Gastaut's syndrome and encephalopathy with continuous spike-waves in NREM sleep.

In certain epilepsies, such as benign infantile epilepsy, generalised idiopathic epilepsy, or certain partial symptomatic epilepsies, the occurence of exclusively nocturnal seizures is usually associated with normal wakefulness EEG activity, but this is not an absolute criterion. In the case of generalised idiopathic epilepsies with Grand Mal seizures, most patients react to anti-epileptic treatment. But, in some patients the seizures are resistant to drugs, and become more frequent, sometimes occurring daily and several times during the night, but with no disorders during the day [16]. The variability in the expression, aetiology, and evolution of epilepsies of an exclusively nocturnal character, has precluded sleep epilepsies from being classified as such in the recent international classification of epilepsies. Some original indications obtained by a genetic study of certain forms of focal familial epilepsy may help to explain the nocturnal exclusivity of seizures in certain epileptics. A mutation of the gene in the alpha-4 subunit of the nicotinic acethylcholine receptor has been identified in cases of familial frontal lobe epilepsy, of which one of the main characteristics is the fact that seizures are exclusively nocturnal [63].

Some important observations have been made in regard to Autosomal Dominant Nocturnal Frontal Lobe Epilepsy: ADNFLE. This syndrome described by Scheffer *et al.* (1995) was subjected to advanced neurophysiological and genetic studies. The expression of seizures is polymorphic, with a spectrum ranging from simple awakenings accompanied by dystonic or dyskinetic movements to complex behavioural manifestations [Provini *et al.* 1999]. The infra-familial and inter-familial diversity of the electroclinical characteristics of ADNFLE seizures contrasts with intra-individual stereotyping. Clinical heterogeneity is associated with genetic heterogeneity, with the identification of several loci (20q13.2, 15q24) and several types of mutation [Steinlein *et al.*, 1995; Philip *et al.*, 1998; Hirose *et al.*, 1999].

A dysfunctioning of the acetylcholine neuron receptor sub units may account for the ADFNLE phenotype.

AWAKENING EPILEPSIES

Unlike sleep epilepsies, awakening epilepsies have relatively consistent electroclinical features. These are generalised idiopathic epilepsies: absence seizures in children, or pyknolepsy, juvenile myoclonic epilepsy and epilepsy with Grand Mal on awakening. Only the latter form is included as such in the present classification of epilepsies. The fundamental feature of this form of epilepsy, as in the two others, is the occurrence of seizures and/or epileptic discharges shortly after morning awakening. Another consistent feature is sensitivity to sleep deprivation, voluntary or otherwise, both in terms of quantity and quality, particularly when induced by alcohol [14]. This is related to sleep rather than times of seizure, in juvenile myoclonic epilepsies, as evidenced by the presence of polyspike bursts on the multiple provoked arousal test [67].

For Niedermeyer [47, 48], awakening epilepsies derive from an arousal process triggered during sleep, or "dyshormia", which normally underscores the production of K complexes. The closeness of awakening epilepsies to sleep epilepsies further illustrates the complexity of interactions between sleep and epilepsy. In addition to exerting an activating or provoking effect on seizures, sleep is also thought to have a protective effect in certain conditions.

NON-EPILEPTIC NOCTURNAL SEIZURES WHICH MAY BE ASSOCIATED WITH EPILEPSY

This category includes the parasomnias, notably episodes of sleepwalking and night terrors, as well as nocturnal paroxysmal dystonias [39] (fig. 50.5) and migraine attacks. These nocturnal disorders occur during sleep with no paroxysmal EEG manifestation, but may be associated with true epileptic seizures [39, 48]. In the case of nocturnal paroxysmal dystonias, the efficacy of carbamazepine, as well as certain EEG and isotopic indications, suggest an epilepsy-related mechanism, or even epilepsy itself, possibly originating in the frontal lobe.

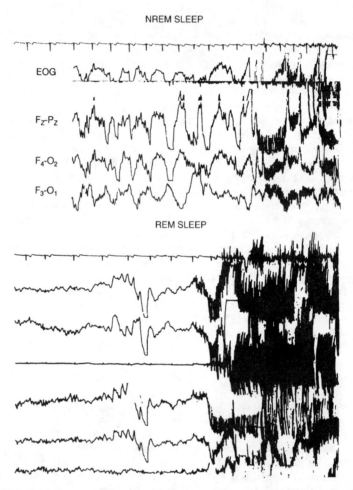

Figure 50.5. Paroxysmal dystonia during nocturnal sleep. Absence of epileptic discharges during the episode of dystonia in NREM and REM sleep.

THE SLEEP OF EPILEPTICS

The occurrence of nocturnal seizures alters sleep architecture [4]. The total duration of REM sleep is reduced after a generalised tonic-clonic seizure. There is a substantial reduction in REM sleep in encephalopathies of the Lennox-Gastaut syndrome type, in relation to the number of tonic seizures. The alterations observed after a simple or a complex partial seizure are minimal, often limited to a short period of sleep, followed by the normal course of sleep. REM sleep may be reduced in partial symptomatic epilepsies, where multiple seizures are observed [3]. Single partial seizures cause no alteration to the course of sleep, even if they occur at the start of the night with sleep onset.

The sleep of epileptics outside nocturnal seizures

Sleep architecture is altered in most epilepsies outside the seizures themselves. Some alterations are general: sleep cycle instability, repeated arousals during sleep, frequent sleep stage shifts. The causes are multiple: underlying cerebral lesions, anti-epileptic treatment, the subject's anxiety, the epileptic process. A distinct relationship between awakening and epileptic discharges was identified in drug resistant temporal lobe epileptics explored using stereo-EEG [45]. Sleep instability may favour the occurrence of seizures, thereby helping to sustain epilepsy. A reduction in the sleep efficiency index due to repeated nocturnal arousals is one of the most consistent factors of daytime sleepiness complained of by certain epileptics, irrespective of treatment [2].

A relationship has been demonstrated between the reduction of NREM sleep spindles and the frequency of Grand Mal seizures [59], a phenomenon which indicates the involvement of thalamocortical circuits in the epileptogenesis of idiopathic epilepsies. Sleep spindles are associated with a long period of thalamocortical neuron inhibition and thus a reduction in the input of external signals. The reduced abundance of spindles observed in some epilepsies may explain the frequent shifts in stage and the poor quality of sleep [59].

In severe encephalopathies such as Lennox-Gastaut's syndrome, sleep architecture can be extremely disturbed to the point of no longer differentiating between sleep and wakefulness and the different cycles of sleep.

The degree of disorganisation depends on the evolution of the epileptic, neurological and mental syndrome. Polysomnographic findings can be used as a guide in setting up and monitoring corticosteroid treatment and in obtaining a prognostic value.

More specific alterations occur in absence epilepsies, with irregularities in the cyclic alternating pattern (CAP) [6].

REFERENCES

1. ARIAS L.P., PASSOUANT P. – Etude des décharges limbiques au cours du sommeil rapide. *C.R. Soc. Biol.* (Paris), 165, 1345-1389, 1971
2. BALDY-MOULINIER M. – Sleep architecture and childhood absence epilepsy. In: Benign localized and generalized epilepsies of early childhood (*Epilepsy Res., Suppl.6*), R.Degen, F.E. Dreifus (eds), Elsevier Science Publ., Amsterdam, 1992.
3. BALDY-MOULINIER M. – Temporal lobe epilepsy and sleep organization. In: *Sleep and Epilepsy,* M.B. Sterman, M.N. Shouse, P. Passouant p. (eds.), Academic press, New-York, 349-359, 1992.
4. BALDY-MOULINIER M. – Inter-relationships between sleep and epilepsy. In: *Recent advance in epilepsy,* T.A. Pedley, B.S. Meldrum (eds), Livingstone, Amsterdam, 37-56, 1986.
5. BALDY-MOULINIER M., BELAIDI M. – Influence du traitement antiépileptique sur les fonctions cognitives et la vigilance. *Epilepsies,* 4, 89-97, 1992.
6. BALDY-MOULINIER, TOUCHON J., BESSET A., BILLIARD M., CADILHAC J., PASSOUANT P. – Sleep architecture and epileptic seizures. In: *Epilepsy, Sleep and Sleep deprivation.* R. Degen, C. Niedermeyer (eds). Elsevier, Amsterdam, 109-118, 1984.
7. BALDY-MOULINIER M., TOUCHON J., BILLIARD M., CARRIÈRE A., BESSET A. – Nocturnal sleep study in the Lennox-Gastaut Syndrome. In: *The Lennox-Gastaut Syndrome.* E. Niedermeyer, R. Degen (eds), Alan R. Liss, New-York, 243-260, 1988.
8. BANCAUD J., TALAIRACH J., BORDAS-FERRER M., AUBER J.L., MARCHAND H. – Les accès épileptiques au cours du sommeil de nuit (étude stéréoélectroencéphalographique). In: *Le Sommeil de nuit normal et pathologique,* H. Fishgold (ed.), Masson, Paris, 255-274, 1965.
9. BEAUMANOIR A. – The Lennox-Gastaut Syndrome: a personal study. *EEG Clin. Neurophysiol.,* Suppl. 35, 85-89, 1982.
10. BEAUMANOIR A., BASSLIS T., VARGFIS G., ANSARI K. – Benign epilepsy of childhood with rolandic spikes. *Epilepsia,* 15, 301-315, 1974.
11. BILLARD C. – Paroxysmes intercritiques et déficits cognitifs chez l'enfant. Quelles indications à les traiter ? *ANAE.* Hors Série: 23-27, 1996
12. BILLIARD M. – Epilepsies and the sleep-wake cycle. In: *Sleep and Epilepsy,* M.B. Sterman. M.N., Shouse, P. Passouant (eds), Academic Press, New York, 481-494, 1982.
13. BLOM S., HEIJBEL J., BERFORW P.G. – Benign epilepsy of children with centro-temporal EEG foci: prevalence and follow-up study of 40 patients. *Epilepsia,* 13, 609-619, 1972.

14. BROUGHTON R.J. – Sleep and sleep deprivation studies in epilepsy. In:*Clinical Neurophysiology of Epilepsy, EEG handbook* (revised series, Vol. 4), J.A. Wada, R.J. Ellingson (eds.), Elsevier Sci. Publ., Amsterdam, 89-119, 1990.

15. BULTEAU C., JAMBAQUE I., KIEFFER V., DULAC O. – Aspects neuropsychologiques du syndrome de pointes ondes continues pendant le sommeil lent. *ANAE*, 41, 5-9, 1997

16. COMMISSION ON CLASSIFICATION AND TERMINOLOGY OF THE INTERNATIONAL LEAGUE AGAINST EPILEPSY. – Proposal for revised clinical and electroencephalographic classification of epileptic seizures. *Epilepsia*, 22, 489-501, 1981.

17. COMMISSION ON CLASSIFICATION AND TERMINOLOGY OF THE INTERNATIONAL LEAGUE AGAINST EPILEPSY. – Proposal for revised classification of epilepsies and epileptic seizures. *Epilepsia*, 30, 389-399, 1989.

18. CRESPEL A., BALDY-MOULINIER M., COUBES P. – Comparison of relationship between sleep and epilepsy in frontal and temporal lobe epilepsies: practical and physiopathological considerations. *Epilepsia*, 39, 150-157, 1998.

19. CRESPEL A., COUBES Ph., BALDY-MOULINIER M. – Sleep influence on seizures and epilepsy effects on sleep in partial frontal and temporal lobe epilepsies. *Clin. Neurophysiol.* 111, Suppl. 2, S54-S59, 2000.

20. DALLA BERNARDINA B., TASSINARI C.A. – Nocturnal EEG of a patient with « benign epilepsy of childhood with epilepsy spikes ». *Epilepsia*, 16, 497-501, 1975.

21. DELANGE M., CASTAN P., CADILHAC J., PASSOUANT P. – Etude du sommeil de nuit au cours d'épilepsies centrencéphaliques et temporales. *Rev. Neurol.*, 106, 106-113, 1962.

22. DELASIAUVE – *Traité de l'épilepsie*. Baillère, Paris, 1854.

23. DELMAS-MARSALET P., FAVRE J. – Etude électroencéphalographique des épilepsies de structure morphéique. *Rev. Neurol.* 82, 546-550, 1950.

24. FERE C. – *Les épilepsies et les épileptiques*. Alcan, Paris, 1890.

25. GASTAUT H. – L'épilepsie bénigne de l'enfant à pointes ondes occipitales. *Rev. EEG Neurophysiol. Clin.*, 12, 179-201, 1982.

26. GASTAUT H. – Epilepsie bénigne de l'enfant avec paroxysmes occipitaux. In: *Les syndromes épileptiques de l'enfant et de l'adolescent*. J. Roger, M. Bureau, Ch. Dravet, F.E. Dreyfuss, A. Perret, P. Wolf (eds), John Libbey, London, 101-217, 1992.

27. GIBBS E., GIBBS F.A.- Diagnostic and localizing value of electroencephalographic studies in sleep. *Res. Publ. Ass. Nerv. Ment. Dis.*, 26, 366-376, 1947.

28. GOWERS W. – *Epilepsy and other convulsive diseases*. Churchill, London, 1881.

29. GREGORY D.L., WONG P.K.H. – Clinical relevance of a dipole field in rolandic spikes. *Epilepsia*, 33, 36-44, 1992.

30. GUBERMAN A., GLOOR P. – Cholinergic drug studies of generalized penicillin epilepsy in the cat. *Brain Res.*, 78, 203-222, 1974.

31. HEIJBEL J. – Benign partial epilepsies in childhood. In: *Paediatric epilepsy*. M. Sillanpää, S.I. Johannessen, G. Blenno, M. Dam (eds.), Wrightson Biomedical, Petersfield, 111-122, 1990.

32. HIROSE S., IWATA H., AKIYOSHI H., KOBAYASHI K., ITO M., WADA K., KANEKO S., MITSUDOME A. - A novel mutation of CHRNA4 responsible for autosomal dominant nocturnal frontal lobe epilepsy. *Neurology*, 53, 1749-1753, 1999.

33. HRACHOVY R.A., FROST J.D.Jr., KELLAWAY P. – Sleep characteristics in infantile spasms. *Neurology*, 31, 688-694, 1981.

34. JANZ D. – The Grand-Mal epilepsies and the sleep-waking cycle. *Epilepsia*, 3, 69-109, 1962.

35. KELLAWAY P. – Sleep and epilepsy. *Epilepsia*, 26 (suppl. 1), S15-S30, 1985.

36. KIKUCHI S. – An electroencephalographic study of nocturnal sleep in temporal lobe epilepsy. *Folia Psychiatr. Neurol. Jap.*, 23, 59-81, 1969.

37. KOSTOPOLOUS G., GLOOR P. – A mechanism for spike-wave discharge in feline penicillin epilepsy and its relationship to spindle generation. In: *Sleep and Epilepsy*, M.B. Sterman, M.N. Shouse, P. Passouant (eds), Academic Press, New York, 11-27, 1982.

38. LALIOTI M., SCOTT R.S., BURESI C., ROSSIER C., BOTTANI A., MORRIS M.A., MALAFOSSE A., ANTONORAKIS S.E. – Dodecanner repeat expression in cystatin B gene in progressive myoclonus epilepsy (EPMI), *Nature*, 386, 847-851, 1997.

39. LANGDON-DOWN M., BRAIN W.R. – Time of day in relation to convulsions in epilepsy. *Lancet*, (ii), 1028-1032, 1929.

40. LERMAN P., KIVITY S. – The benign focal epilepsies of childhood. In: *Recent advances in epilepsy*. T.A. Pedley, B.S. Meldrum (eds.), Livingstone, Edinburgh, 137-156, 1986.

41. LIEB J.P., JOSEPH J.P., ENGEL J.JR., ALKER J., CRANDALL P.H. – Sleep state and seizure foci related to depth spike activity in patients with temporal lobe epilepsy. *Electroenceph. Clin. Neurophysiol.*, 49, 538-557, 1980.

42. LUGARESI E., CIRIGNOTTA F. – Nocturnal paroxysmal dystonia: epileptic seizure or a new syndrome ? *Sleep*, 4, 129-138, 1981.

43. MAQUET P., HIRSCH E., METZ-LUTZ M., MOTTE J., DIVE D., MARESCAUX C., FRANCK G. – Regional cerebral glucose metabolism in children with deterioration of one or more cognitive functions and continuous spike and wave discharges during sleep. *Brain*, 118, 1497-1530, 1995.

44. MEIER-EWERT K.H., BROUGHTON R. – Photomyoclonic response of epileptic and non-epilepstic subjects during wakefulness, sleep and arousal. *Electroenceph. Clin. Neurophysiol.* 23, 142-151, 1967.

45. MONTPLAISIR J., LAVERDIÈRE M., SAINT-HILAIRE J.M. – Sleep and epilepsy. In: *Long term monitoring in epilepsy* (EEG, suppl. 37), J. Gotman, J.R. Ives, P. Gloor (eds.), Elsevier Sci. Publ., Amsterdam, 215-239, 1985.

46. MONTPLAISIR J., SAINT HILAIRE J.M., LAVERDIÈRE M., WALSH J., BOUVIER G. – Contribution of all-night poloygraphic recording to the localization of primary epileptic foci. In: *Advances in epileptology*, R. Canger, F. Angeleri, J.K. Penzi (eds.), Raven Press, New York, 135-138, 1980.

47. NIEDERMEYER E. – Sleep electroencephalogram in Petit-Mal. *Arch. Neurol.*, 12, 625-630, 1965.

48. NIEDERMEYER E. – Awakening epilepsy revisited 30 years later. In: *Epilepsy, sleep and sleep deprivation*, R. Degen, E. Niedermeyer (eds.), Elsevier, Amsterdam, 85-96, 1984.

49. OLDANI A., ZUCCONI M., CASTRONOVO C., FERINI-STRAMBI L. – Nocturnal frontal lobe epilepsy misdiagnosed as sleep apnea syndrome. *Acta Neurol. Scand.* 98, 67-71, 1998.

50. PASSOUANT P. – Epilepsie temporale et sommeil. *Rev. Roum. Neurol.*, 4, 151-163, 1967.

51. PASSOUANT P., BESSET A., CARRIERE A., BILLIARD M.- Night sleep and generalized epilepsies. In: *Sleep, 1974*, Karger, Basel, 185-196, 1975.

52. PASSOUANT P., BILLIARD M., PAQUET J. – Terreurs nocturnes et crises épileptiques chez l'enfant. *Revue Int. Pediatr.*, 31, 1-10, 1972.

53. PASSOUANT P., CADILHAC J. – Décharges épileptiques et sommeil. *Mod. Prob. Pharmaco. Psychiat.*, 4, 87-104, 1970.

54. PASSOUANT P., LATOUR H., CADILHAC J. – L'épilepsie morphéique. *Ann. Med. Psychol.*, 109, 526-540, 1951.

55. PATRY F.L. – The relation of time of day, sleep and other factors to the incidence of epileptic seizures. *Am. J. Psychiat.* 87, 789-813, 1931.

56. PATRY F.L., LYAGOUBI S., TASSINARI C.A. – Subclinical « electrical status epilepticus » induced by sleep in children. *Arch. Neurol.*, 24, 242-252, 1971.

57. PHILLIPS H.A., SCHEFFER I.E., CROSSLAND K.M., BHATIA K.P., FISH D.R., MARSDEN C.D., HOWELL S.J., STEPHENSON J.B., TOLMIE J., PLAZZI G. *et al.* – Genetic heterogeneity and evidence for a second locus at 15q24. *Am. J. Hum. Genet.* 63, 1108-1116, 1998

58. POMPEIANO O. – Sleep mechanisms. In: *Basic mechanisms of the epilepsies*, A.R. Ward, A. Pope (eds), Little Brown and Co., Boston, 453-473, 1969.

59. PROVINI F., PLAZZI G., TINUPER P., VANDI S., LUGARESI E., MONTAGNA P. – Nocturnal frontal lobe epilepsy. A clinical polygraphic overview of 100 consecutive cases. *Brain* 122, 1017-1031, 1999.

60. ROGER J., GENTON P., BUREAU M. – La classification internationale des épilepsies et des syndromes épileptiques adoptée au Congrès de New Delhi (Octobre 1989): Commentaires et traduction. *Epilepsies*, 2, 183-197, 1990.

61. ROGER J., GENTON P., BUREAU M., DRAVET C. - Les épilepsies myocloniques progressives de l'enfant et de l'adolescent. In: *Les syndromes épileptiques de l'enfant et de l'adolescent*. J. Roger, M. Bureau, Ch. Dravet, F.E. Dreifuss, A. Perret, P. Wolf (eds.), John Libbey, London, 381-400, 1992.

62. ROSSI G., COLLICCHIO G., POLA P., ROSELLI R. – Sleep and epileptic activity. In: *Epilepsy, sleep and sleep deprivation*, R. Degen, E. Niedermeyer (eds.), Elsevier, Amsterdam, 35-46, 1984.

63. SCHEFFER J.E., BHATIA K.P., LOPES-CENDES I., FISCH D.R., MARSDEN C.D., ANDERMANN F. – Autosomal dominant nocturnal frontal lobe epilepsy. A distinctive clinical disorder. *Brain*, 118, 61-73, 1995.

64. STEINLEIN O., MULLEY J., PROPPING T., WALLACE R., PHILIPS H., SUTHERLAND G., SCHEFFER I., BERKIVIC S.E. – A missense mutation in the neuronal nicotinic acetylcholine receptor α4 subunits is associated with autosomal dominant nocturnal frontal lobe epilepsy. *Nature Genet.*, 11, 201-203, 1995.

65. STERMAN M.B. – Power spectral analysis of EEG characteristics during sleep in epileptics. *Epilepsia*, 22, 95-106, 1981.

66. TASSINARI C.A., BUREAU M., DRAVET C., DALLA BERNARDINA B., ROGER J. – Epilepsie avec pointes-ondes continues pendant le sommeil lent antérieurement décrite sous le nom d'ESES (épilepsies avec état de mal électroencéphalographique pendant le sommeil lent). In: *Les syndromes épileptiques de l'enfant et de l'adolescent*. J. Roger, M. Bureau, Ch. Dravet, M. Dreifuss, A. Peret, P. Wolf (eds.), John Libbey, London, 245-256, 1992.

67. TOUCHON J. – Effect of awakening in primary generalized myoconic epilepsy. In: *Sleep and epilepsy*, M.B. Sterman, M.N. Shouse, P. Passouant (eds.), Academic Press, New York, 239-248, 1982.

Chapter 51

Sleep-related headaches

B. Carlander
Service de Neurologie B, Hôpital Gui de Chauliac, Montpellier, France

This chapter deals not only with headaches and cluster headaches which occur during sleep, but also with those which are temporally linked to sleep, such as morning headaches. In terms of standard clinical reports, very few polysomnographic observations are available, making it difficult to deduce definite pathophysiological conclusions. The relative lack of data no doubt reflects the fact that practical indications for overnight recording are rare in cases of headache, even those which tend to occur at night.

HEADACHES IN OBSTRUCTIVE SLEEP APNOEA-HYPOPNOEA SYNDROME

Morning headaches form part of the classical clinical picture for this illness [19]; but are fairly inconsistent [35], insufficiently correlated to the severity of the illness [23], and above all, limited in specificity: in practice, they are far more common in other sleep disorders, particularly in non-apnoeic respiratory pathology [1, 27]. When they are very marked, however, blood pressure should be examined on morning awakening. The suggested explanations include an increase in blood pressure often observed during the night, hypercapnic bursts (hence vasodilation), hypoxia, as in high altitude headaches, neck movements occurring at the end of apnoeas and finally, sleep deprivation (see below). The question is whether the headaches remit with treatment for obstructive sleep apnoea/hypopnoea syndrome. Dexter [12] answers in the affirmative in the case of uvulopalatopharyngoplasty, for which the failure rate is nevertheless well known. As for nasal continuous positive airway pressure (CPAP), studies seldom focus on this particular aspect. Several authors have suggested that there is an improvement [21, 29]; conversely, the Stanford team reported an incidence of 16% of headaches, at non specified times, in a survey of 144 patients under continuous positive airway pressure for an average of a year [26].

MIGRAINE AND SLEEP

The usual clinical situation is that of remission of migraine attacks by voluntary sleep recovery [4], which is made easier when photophobia and phonophobia lead to the patient isolating himself in a dark room. The pain disappears more completely if the patient actually manages to sleep [36]. Sleepiness also forms part of the picture of hypothalamic dysfunctioning, which tends to herald the migraine attack (in parallel with a general malaise, yawning, a sudden desire for sweet foods etc.). The most developed explanation as to why the attack is improved by sleep, is that of a vegetative failure at hypothalamic level, even if the pathogeny of migraine remains highly mysterious [6].

In other cases, sleep appears to be temporally linked to the onset of the migraine attack, either by the patient waking up with a headache in the middle of the night, or at the time the patient normally gets up in the morning. Several decades ago, Gans [17] noted a relationship between the onset of migraine attacks at night and how deeply the patient slept, and that by lightening sleep (through light tactile stimulation) it was possible to reduce the severity and frequency of attacks. The first polysomnographic study [15] later demonstrated the correlations to sleep stages, particularly with REM sleep: most arousals accompanied by headaches occurred during a period of REM sleep, and the others during the subsequent minutes. It must however be noted that only a small number of patients and migraine attacks were studied (2 and 8 respectively) and that two other subjects studied by Cirignotta *et al.* [9] woke up with headaches during NREM sleep. However, after studying napping in five subjects, Dexter [11] found that episodes of REM sleep clearly predominated when the nap was accompanied by a headache on awakening (8 over 9 for a total of 17 episodes of the recorded naps of five subjects). This migraine-REM sleep coincidence may be related to the role of noradrenergic projections from the locus coeruleus in the physiology of this type of sleep; considering the effect on blood flow of stimulating these neurones, i.e. a reduction in the intracerebral compartment and an increase in the extracerebral compartment, these alterations are similar to those observed in classic migraine [18].

The second study by Dexter [11], carried out on four patients over five consecutive nights, supports Gans's earlier hypothesis i.e. that arousals with headaches were associated with the deepest periods of sleep with a predominance of NREM sleep stages 3 and 4 and REM sleep. By reducing the depth of sleep using dextroamphetamine and imipramine, an "excellent" long-term control of migraine attacks was obtained.

More recently, the advantages of better sleep hygiene have been demonstrated in a population of children and adolescents suffering from migraine [7].

Finally, a relationship has been shown between migraine and arousal disorders, with a marked prevalence of histories of night terrors, sleepwalking and enuresis in migraine sufferers [2, 13].

CLUSTER HEADACHE AND SLEEP

Pain characteristically occurs at a fixed time in over half the patients, usually at night: either while relaxing in the evening or during sleep, and particularly during the first phase of REM sleep, as has been demonstrated by three studies [14, 22, 24]. This association is not systematic, especially when subjects suffering from the clinical variant paradoxically known as "chronic cluster headache" are included in the study. The same authors reported oxygen desaturation or apnoeas, leading to the hypothesis that hypoxia is instrumental in triggering cluster headaches, all the more so as inhalation of a strong flow of oxygen through the nostril which is homolateral to the pain has been shown to have an abortive effect on the cluster headache episode. Epidemiologically, the association has been shown between sleep apnoea syndrome and cluster headache [8], with the contingent possibility of improving headaches using continuous positive pressure [37].

As in the case of nocturnal migraine, the anxious anticipation of a headache can lead to true insomnia [31].

CHRONIC PAROXYSMAL HEMICRANIA

This is a rare form of headache, isolated by Sjaastad and Dale [34], distinguished from cluster headache by shorter lasting pain, female predominance, more frequent attacks and a tendency to be sensitive to indomethacine. "Sympathetic" features are present in this form, (rhinorrhea, redness of the face, weeping eyes). The nocturnal tendency of these attacks is even more marked, particularly in relation to REM sleep [20] which has led to use of the term "REM-sleep locked headache" (fig 51.1). Moreover, in the only subject studied, over four nights, this type of sleep was particularly fragmented; the vegetative instability of REM sleep ("autonomic storm") possibly provides a favourable terrain for the expression of this type of headache, considering the accompanying features referred to above.

A related syndrome has been described in the elderly, known as "hypnic headache syndrome" [25, 28]. The patient is woken at the same time by a diffuse headache, which is often pulsating but with no dysautonomic features. Lithium is remarkably effective in these cases, which, like cluster headaches, may be akin to the suppressive action of this ion on REM sleep [3]; its side effects have nevertheless led to the search for alternatives such as caffeine, verapamil, flunarizine and above all indomethacine [16].

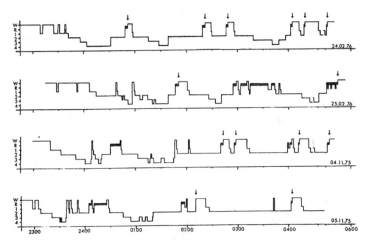

Figure 51.1. Hypnogramme of a subject affected by chronic paroxysmal haemicrania (after Kayed *et al.*, 1978, with permission), showing the concomitance between painful attacks and episodes of REM sleep.

HEADACHES FROM SLEEP DEPRIVATION OR EXCESS

The organic consequences of chronic sleep deprivation are remarkably discrete, although a retro-ocular stinging sensation is often reported, as well as trembling or nystagmus [32]. The frequency of headaches due to sleep deprivation may have been underestimated, according to Blau [5], who reported a figure of 38.8% among 327 medical students. A more detailed interrogation of 25 subjects eliminated the habitual role of alcohol (hangover) or of any associated migraine. The satisfactory effect obtained with usual analgesics is used as an argument against including this type of headache in the group referred to as tension headaches.

It is also possible to include in this picture a symptomatic complex known as fibromyalgia or fibrositis syndrome (see chapter 60). This involves an alteration in sleep continuity and architecture represented by relative sleep deprivation, with diurnal symptoms such as headaches as a possible consequence.

Less frequent no doubt, but also typical, are migraines triggered by excessive sleep, described by Blau [6]. A hypothesis advanced by this author refers to hypoglycaemia, or another metabolic deficiency provoked by prolonged sleep.

EXPLODING HEAD SYNDROME

This refers to a borderline clinical situation in relation to headaches: the patient describes an abrupt awakening with violent flashes and intense noise, causing anxiety but no real pain. However polysomnographic study [30] has shown that the attacks only occur during wakefulness, in fact, and in conditions of stress, thus seeming to indicate a predominantly psychogenic involvement.

NOCTURNAL POSTEPILEPTIC HEADACHES

It is not rare for headaches to occur in the aftermath of a seizure [33]. In sleep epilepsy, the patient may be woken by a postictal headache, which may be the only evidence of the seizure [10].

INTERCRANIAL HYPERTENSION HEADACHE

Whether this is caused by a tumour, a cerebral oedema, hydrocephalus, or so-called benign intracranial hypertension (wrongly so as this may lead to visual prognosis) the headache tends to occur spontaneously on morning awakening, or to cause early awakening, and is typically accompanied by projectile vomiting. This is by no means systematic, however, and of low specificity as it is also encountered with conditions such as depression, the morning after an episode of acute alcohol intoxication or hypertension. Finally, no direct relationship appears to exist between sleep and this form of headache, but rather, a coincidence in timing linked to the cortisol circadian rhythm.

REFERENCES

1. ALDRICH M.S., CHAUNCEY J.B. – Are morning headaches part of the obstructive sleep apnea syndrome? *Arch. Intern. Med.,* 150, 1265-1267.
2. BARABAS G., FERRARI M., MATTHEWS W. – Childhood migraine and somnambulism. *Neurology,* 33, 948-949, 1983.
3. BILLIARD M. – Lithium carbonate: effects on sleep patterns of normal and depressed subjects and its use in sleep-wake pathology. *Pharmacopsychiatry,* 20, 195-196, 1987.
4. BLAU J.N. - Resolution of migraine attacks: sleep and the recovery phase. *J. Neurol. Neurosurg. Psychiatry,* 45, 223-226, 1982.
5. BLAU J.N. - Sleep deprivation headache. *Cephalalgia,* 10, 157-160, 1990.
6. BLAU J.N. – Migraine: theories of pathogenesis. *Lancet,* 339, 1202-1207, 1992.
7. BRUNI O., GALLI F., GUIDETTI V. - Sleep hygiene and migraine in children and adolescents. *Cephalagia,* 19 suppl 25, 57-59, 1999.
8. CHERVIN R.D., ZALLEK S.N., LIN X., HALL J.M., SHARMA N., HEDGER K.M. – Sleep disordered breathing in patients with cluster headache. *Neurology,* 54, 2302- 2306, 2000.
9. CIRIGNOTTA F., COCCAGNA G., SACQUENGNA G., SFORZA E., LAMONTANARA M., ZUCCONI P., CORTELLI E., LUGARESI E. – Nocturnal headache: systemic arterial pressure and heart rate during sleep. *Cephalalgia,* suppl.1, 54-57, 1983.
10. CULEBRAS A. – Neuroanatomic and neurologic correlates of sleep disturbance. *Neurology,* 42 (suppl 6), 19-27, 1992.
11. DEXTER J.D. – The relationship between stage III + IV + REM-sleep and arousals with migraine. *Headache,* 19, 364-369, 1979.
12. DEXTER J.D. – Headache as a presenting complaint of sleep apnea syndrome. *Headache,* 24, 171, 1984.
13. DEXTER J.D. – The relationship between disorders of arousal from sleep and migraine. *Headache,* 26, 322, 1986.
14. DEXTER J.D., RILEY T.R. – Studies in nocturnal migraine. *Headache,* 15, 51-62, 1975.
15. DEXTER J.D., WEITZMAN E.D. – The relationship of nocturnal headaches to sleep stage patterns. *Neurology,* 20, 513-518, 1970.
16. DODICK D.W., JONES J.M., CAPOBIANCO D.J. - Hypnic headache: another indomethacin-responsive headache syndrome? *Headache,* 40, 830-835, 2000.
17. GANS M. – Treating migraine by sleep rationing. *J. Nerv. Ment. Dis., 113, 405-429, 1951.*
18. GOADSBY P.J., LAMBERT G.A., LANCE J.W. - Differential effects on the internal and external carotid circulation of the monkey evoked by locus coeruleus stimulation. *Brain Res.* 249, 247-254, 1982.
19. GUILLEMINAULT C., VAN DEN HOED J., MITLER M.M. – Clinical overview of the sleep apnea syndromes. *In*: C. Guilleminault, W.C. Dement (eds) *Sleep apnea syndrome,* Alan R. Liss, New-York, 1-21, 1978.
20. KAYED K., GODTLIBSEN O., SJAASTAD O. - Chronic paroxysmal hemicrania IV: « REM-sleep locked » nocturnal attacks. *Sleep,* 1, 91-95, 1978.
21. KRIEGER J. – les syndromes d'apnées du sommeil. *Encycl. Méd. Chir.* Neurologie, Paris,17025 C10, 9, 1-4, 1988.
22. KUDROW L., McGINTY D.J., PHILLIPS E., STEVENSON M. – Sleep apnea and cluster headache. *Cephalalgia,* 4, 33-38, 1984.
23. LOH N.K., DINNER D.S., FOLDVARY N., SKOBIERANDA F., YEW W.W. – Do patients with obstructive sleep apnea wake up with headaches? *Arch. Intern. Med.,* 159, 1765-1768, 1999.
24. MATHEW N.T., GLAZE D., FROST J. – Sleep apnea and other sleep abnormalities in primary headache disorders. *In*: F.C. Rose (ed) *Proceedings of the 50th International Migraine Symposium,* Karger, Basel, 40, 1985.
25. NEWMAN L.C., LIPTON R.B., SOLOMON S. – The hypnic headache syndrome: a benign headache disorder of the elderly. *Neurology,* 40, 1904-1905, 1990.
26. NINO-MURCIA G., CROWE McCANN C., BLIWISE D.L., GUILLEMINAULT, DEMENT WC. – Compliance and side effects in sleep apnea patients treated with nasal continuous positive airway pressure. *West J. Med.,* 150, 165-169, 1989.
27. POCETA J.S., DALESSIO D.J. – Identification and treatment of sleep apnea in patients with chronic headache. *Headache,* 35, 586-589, 1995.
28. RASKIN N.H. – The hypnic headache syndrome. *Headache,* 28, 534-536, 1988.
29. RODENSTEIN D., AUBERT-TULKENS G. – Traitement instrumental du syndrome d'apnées du sommeil. *Rev. Mal. Resp.,* 7, 459-465, 1990.
30. SACHS C., SVANBORG E. – The exploding head syndrome: polysomnographic recordings and therapeutic suggestions. *Sleep,* 14, 263-266, 1991.
31. SAHOTA P.K., DEXTER J.D. – Transient recurrent situational insomnia associated with cluster headache. *Sleep,* 16, 255-257, 1993.

32. SASSIN J.F. – Neurological findings following short-term sleep deprivation. *Arch. Neurol.,* 22, 54-56, 1970.
33. SCHON F., BLAU J.N. – Post-epileptic headache and migraine. *J. Neurol. Neurosurg. Psychiatry.* 50, 1148-1152, 1987.
34. SJAASTAD O., DALE I. – Evidence for a (?) new headache entity. *Headache,* 14, 105-108, 1974.
35. ULFBERG J., CARTER N., TALBACK M., EDLING C. - Headache, snoring and sleep apnoea. *J. Neurol.,* 243, 621-625, 1996.
36. WILKINSON M., WILLIAMS K., LEYTON M. – Observations on the treatment of an acute attack of migraine. *Res. Clin. Stud. Headache.* Basel, Karger. 6, 141-146, 1978.
37. ZALLEK S.N., CHERVIN R.D. - Improvement in cluster headache after treatment for obstructive sleep apnea. *Sleep Med., 1, 135-138, 2000.*

Chapter 52

Fatal familial insomnia

E. Lugaresi and P. Montagna
Instituto di Clinica Neurologica, Universita di Bologna, Bologna, Italia

Fatal familial insomnia (FFI) is a rapidly progressing familial prion disease, clinically characterised by insomnia which resists treatment, dysautonomia, motor signs and, anatomicopathologically, by the selective degeneration of the ventral anterior (VA) and dorsomedial (DM) thalamic nuclei. The disease is transmitted in an autosomal dominant fashion [8].Until now, more than 30 kindreds are known to be affected with FFI, making it one of the most prevalent hereditary prion diseases. More recently, sporadic cases with clinico-pathological features quite similar to FFI but in the absence of mutations in the PRNP gene have also been reported (so-called sporadic FI).

CLINICAL FEATURES

The average age at the start of the disease in the 14 cases confirmed by histo-pathological tests, is 51 years (extremes: 36 – 62 years) [13]. Progression is gradual, leading to death within an average of 18 months (extremes: 8 – 72 months). The cardinal signs are insomnia, dysautonomia and motor disorders.

Insomnia is present from the outset, developing to the complete or virtually complete disappearance of sleep. Dreamlike activity and automatic behaviour appear, associated with increasingly prolonged episodes of stupor. An irreversible coma lasting several days or weeks precedes the death of the patient.

Dysautonomic disorders consist of urinary difficulties, impotence in men, watering of the eyes, salivation and perspiration, high body temperature, elevated heart rate and blood pressure. Breathing becomes particularly laborious in the final phase of the disease, with tachypnoea, paradoxical breathing, frequent apnoeas and snoring.

Motor manifestations, dysarthria and ataxic gait, are the earliest neurological symptoms. Speech, which is slow and poorly articulated, becomes unintelligible. The gait becomes increasingly ataxic progressing to complete abasia-astasia with retropulsion. Patients later have segmental or diffuse myoclonus, both spontaneous and evoked. Diplopia and spasmodic eye movements occur transiently in the later stages of the disease. Generalised tonic-clonic seizures and attacks of dystonia may develop late in certain subjects. Two main clinical courses are apparent, one short lasting, ending in death in 8 – 11(mean 9.1) months, and a long lasting one 11 – 72 (mean 30.8) months. The latter group is characterised by less prominent sleep and autonomic abnormalities, and by early ataxia and somamotor manifestations, and epileptic seizures [13].

NEUROPHYSIOLOGY

Background electroencephalographic activity becomes progressively slower and less reactive [15]. In some cases, periodic or pseudo-periodic activities at 1 – 2 Hz, synchronous with myoclonus, can be observed in the terminal stage. Visual, brainstem and background somatosensory-evoked potentials are normal. The sleep time gradually decreases progressing to the complete or virtually complete absence of electroencephalographic sleep patterns (fig. 52. 1).

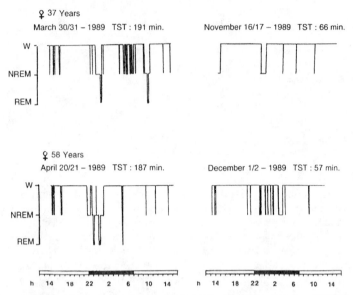

Figure 52.1. 24 hour sleep wake histogram of two subjects affected by FFI. Total sleep time (TST) is severely and progressively reduced, and cyclic sleep architecture is lost.

Sleep spindles and delta activity disappear, replaced by weak amplitude 4 Hz theta activity.

The transition from wakefulness to NREM sleep and vice versa is immediate, associated with rapid heart-circulation changes

REM sleep episodes are short, associated with motor activities and semipurposeful gestures mimicking the content of a dream. Sleep (or coma) pharmacologically induced by barbiturates and benzodiazepines is associated with flattened EEG tracings, devoid of the rapid or slow activities typically produced by these agents (fig. 52.2).

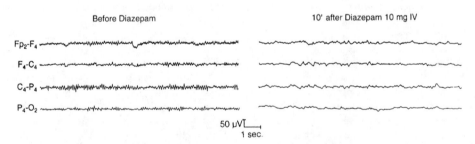

Figure 52.2. Samples of an electroencephalogram in a patient affected by FFI before and 10 minutes after diazepam *i.v.* diazepam leads to the slowing down and flattening of EEG but not to an EEG activity typical of sleep.

AUTONOMIC FUNCTIONS

Body temperature, heart and breathing rates and systemic arterial pressure are persistently elevated, with enhanced baseline and increased sympathetic activities (fig. 52.3) [2]. Catecholamine concentrations (adrenaline and norepinephrine) are persistently high.

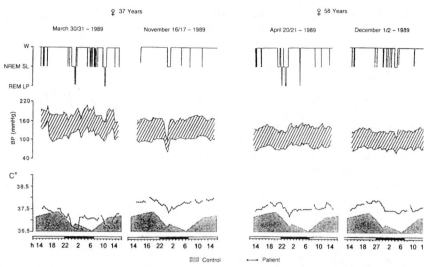

Figure 52.3. 24 hour sleep-wake recordings (see fig. 52.1) of the blood pressure (BP) and core temperature (C°) of two patients with FFI. Blood pressure and temperature are persistently elevated with no physiological circadian fluctuations. Sudden drops in blood pressure and temperature occur during the transition from wakefulness to NREM sleep (second part of the figure).

ENDOCRINE FUNCTIONS

The hypothalamus-pituitary axis is normal. Hourly measurements of plasmatic values nevertheless show diminished sleep-related (GH, PRL) circadian fluctuations and sleep-unrelated hormones (cortisol, ACTH).

Moreover, cortisol concentration was constantly above the maximum normal level [7]. Melatonin secretion shows a gradual decrease in circadian amplitude up to complete rhythm loss.

NEUROPSYCHOLOGICAL DATA

The disease is characterised by the gradual impairment of vigilance, attention and visual and motor performance, associated with selective memory loss, in the absence of a decline of global intellectual faculties [5].

PET SCAN

All the patients we studied (7 cases) showed a striking diminution of the thalamic metabolism (particularly in the anterior portion) associated or not with hypometabolism in the cerebral cortex (cingular cortex). Additional glucose hypometabolism was found in the frontal and temporal cortices and basal nuclei, especially in patients with a long-lasting course [3].

NEUROPATHOLOGICAL ASPECTS [8, 9]

The main pathological lesion consisted of severe degeneration of the DM and VA thalamic nuclei with neuron loss and reactive gliosis. Other thalamic nuclei (centro-median, pulvinar etc.) were less persistently and less severely affected.

The extra-thalamic regions comprised:
a) atrophy-hypertrophy of the lower olivae;
b) formation of "torpedoes" (enlargement of the Purkinje cell axons);
c) minimal gliosis in the deep layers of the cerebral cortex and surface white substance in the frontal and temporal lobes. The hypothalamus and brainstem were relatively spared.

Slight spongiosis was found in the cerebral cortex especially in cases with a long lasting disease course. Apoptotic neurons have been found in the brains of FFI patients especially in the areas with neuronal loss [4].

GENETICS AND MOLECULAR BIOLOGY

Fragments of prion protein resistant to K proteinase are present in FFI (so-called PrPres type 2). They differ in size from those in the sporadic form of Creuzfeldt-Jakob disease. In all cases of FFI, a point mutation occurred at codon 178 of the prion protein gene with substitution of aspartic acid by asparagine. Polymorphism at codon 129 (methionine or valine) of the same gene as the prion protein explains the phenotype diversity of the two diseases; FFI has methionine 129 and asparagine 178 alleles, whereas the familial form of the Creuzfeldt-Jakob disease has valine 129 and asparagine 178 alleles [6]. In addition, patients with a short lasting disease course are characterised by a methionine polymorphism at codon 129 of the non mutated allele, whereas long lasting cases display a valine polymorphism [13]. FFI has been transmitted to transgenic mice, which precisely replicate the size and the profile lesion of the human disease [14]. Similar results have been obtained with brain homogenates from sporadic FI brains [11].

PATHOPHYSIOLOGICAL CONSIDERATIONS

In FFI, not only the sleep-wake cycle, but all the circadian functions (autonomic and hormonal) are affected. Thus it might be deduced that a lesion, principally or exclusively limited to the thalamic nuclei (VA and DM) is responsible for the fragmentation, diminution and even the disappearance of sleep, of the gradual activation of the autonomous nervous system involving only the sympathetic component and an alteration to vegetative and endocrine circadian fluctuations. In 1972, Villablanca *et al.* [16, 17] insisted on the crucial role played by the cerebral cortex and the thalamus in sleep-wake architecture: indeed severe and persistent insomnia develops in diencephalic cats (decorticated) and in athalamic cats.

The role of the DM thalamic nucleus (but not that of the VA nucleus) in sleep physiology was confirmed by Marini *et al.* [10], who demonstrated that the local application of ibotenic acid leads to sustained insomnia in the laboratory animal. More recently, bicuculline administration into the MD thalamic nucleus of rats elicited increased arterial pressure and heart rates [1] consistent with FFI. These data indicate that several nervous structures, located in the cerebral cortex, the thalamus, hypothalamus and brainstem take part in sleep-wake physiology. The different nervous structures have a caudorostral organisation within the phylogenetic hierarchy, as was evoked by Jackson for numerous aspects of cerebral functions. The primary and more archaic structures co-ordinating the rest-activity cycles can be placed at the level of the lowest reticular formation in the brainstem (medullar pontine). The other structures responsible for sleep and wakefulness were later placed above the preceding ones, in the mesodiencephalic region. In the subsequent stages of evolution (in mammals and man) the increasing complexity of instinctive behaviour and the gradual development of the telencephalon led to a third, higher level located in the cortical regions which controls the lower levels of integration.

The clinical complexity of FFI is consistent with the view that the VA and DM nuclei represent the visceral or limbic part of the thalamus. Their alteration, separating the limbic and the paralimbic cortex of the hypothalamus, probably elicits a permanent functional imbalance resulting in the loss of sleep, adrenal and adrenal cortical activation and the disappearance of vegetative and endocrine circadian fluctuations observed in FFI. By affecting other neuronal circuits involved in controlling the motor and cognitive functions, alterations of the VA and DM thalamic nuclei equally produce an effect on selective memory, and cause cognitive disturbance and motor disorders.

Chronic sleep deprivation could moreover contribute to the other disorders and ultimately to the death of the patient, even if, as yet, there is no conclusive evidence of this.

REFERENCES

1. BENARROCH E.E., STOTZ-POTTER E.H. - Dysautonomia in fatal familial insomnia as an indicator of the potential role of the thalamus in autonomic control. *Brain Pathol.* 8, 527-530, 1998.
2. CORTELLI P., PARCHI P., CONTIN M., PIERANGELI G., AVONI P., TINUPER P., MONTAGNA P., BARUZZI A., GAMBETTI P., LUGARESI E. - Cardiovascular dysautonomia in fatal familial insomnia. *Clinical Autonomic Research*, I, 15-21, 1991.
3. CORTELLI P., PERANI D., PARCHI P., GRASSI F., MONTAGNA P., DE MARTIN M., CASTELLANI R., TINUPER P., GAMBETTI P., LUGARESI E., FAZIO F. - Cerebral metabolism in fatal familial insomnia: relation to duration, neuropathology and distribution of protease-resistant prion protein. *Neurology* 49, 126-133, 1997.
4. DORANDEU A., WINGERTSMANN L., CHRETIEN F., DELISLE M.B., VITAL C., PARCHI P., MONTAGNA P., LUGARESI E., IRONSIDE J.W., BUDKA H., GAMBETTI P., GRAY F. - Neuronal apoptosis in fatal familial insomnia. *Brain Pathol.* 8, 531-537, 1998.
5. GALLASSI R., MORREALE A., MONTAGNA P., GAMBETTI P., LUGARESI E. – Fatal familial insomnia: Behavioral and cognitive features. *Neurology* 46, 835-839, 1996.
6. GOLDFARB L.V., PETERSEN R.B., TABATON M., BROWN P., LE BLANC A.C., MONTAGNA P., CORTELLI P., JULIEN J., VITAL C., PENDELBURY W.W., HALTIA M., WILLS P.R., HAUW J.J., McKEEVER P.E., MONARI L., SCHRANK B., SWERGOLD G.D., AUTILIO-GAMBETTI L., GAJDUSEK D.C., LUGARESI E., GAMBETTI P. – Fatal familial insomnia and familial Creutzfeldt-Jakob disease: disease phenotype determined by a DNA polymorphism. *Science*, 258, 806-808, 1992.
7. LUGARESI A., BARUZZI A., CACCIARI E., CORTELLI P., MEDORI R., MONTAGNA P., TINUPER P., ZUCCONI M., ROITER I., LUGARESI E. – Lack of vegetative and endocrine circadian rhythms in fatal familial thalamic degeneration. *Clinical Endocrinology*, 126, 573-580, 1987.
8. LUGARESI E., MEDORI R., MONTAGNA P., BARUZZI A., CORTELLI P., LUGARESI A., TINUPER P., ZUCCONI M., GAMBETTI P. – Fatal familial insomnia and dysautonomia with selective degeneration of thalamic nuclei. *New Eng. J. Med.*, 315, 997-1003, 1986.
9. MANETTO V., MEDORI R., CORTELLI P., MONTAGNA P., TINUPER P., BARUZZI A., RANCUREL G., HAUW J.J., VANDERHAEGEN J.J., MAILLEUX P., BUGIANI O., TAGLIAVINI F., BOURAS C., RIZZUTO N., LUGARESI E., GAMBETTI P. – Fatal familial insomnia: clinical and pathological study of five new cases. *Neurology,* 42, 312-319, 1992.
10. MARINI G., IMERI L., MANCIA M. - Changes in sleep waking cycle induced by lesions of medialis dorsalis thalamic nuclei in the cat. *Neurosci. Lett.*, 85, 223-227, 1988.
11. MASTRIANNI J.A., NIXON R., LAYZER R., TELLING G.C., HAN D., DE ARMOND S.J., PRUSINER S.B. – Prion protein conformation in a patient with sporadic fatal insomnia. *New Eng. J. Med.* 340, 1630-1638, 1999.
12. MEDORI R., TRITSCHLER H.J., LE BLANC A., VILLARE F., MANETTO V., YING CHEN H., XUE R., LEAL S., MONTAGNA P., CORTELLI P. - Fatal familial insomnia: a prion disease with a mutation at codon 178 of the prion protein gene. *New Eng. J. Med.* 326, 444-487, 1992.
13. MONTAGNA P., CORTELLI P., AVONI P., TINUPER P., PLAZZI G., GALLASSI R., PORTALUPPI F., JULIEN J., VITAL C., DELISLE M.B., GAMBETTI P., LUGARESI E. – Clinical features of fatal familial insomnia: phenotype variability in relation to a polymorphism at codon 129 of the prion protein gene. *Brain Pathol.* 8, 515-520, 1998.
14. TELLING.C., PARCHI P., DE ARMOND S.J., CORTELLI P., MONTAGNA P., GABIZON R., MASTRIANNI J., LUGARESI E., GAMBETTI P., PRUSINER S.B. – Evidence for the conformation of the pathologic isoform of the prion protein enciphering and propagating prion diversity. *Science* 274, 2079-2082, 1996.
15. TINUPER P., MONTAGNA P., MEDORI R., CORTELLI P., ZUCCONI M., BARUZZI A, LUGARESI E. – The thalamus participates in the regulation of the sleep-waking cycle. A clinico-pathological study in fatal familial thalamic degeneration. *Electroencephalogr. Clin. Neurophysiol.* 73, 117-123, 1989.
16. VILLABLANCA J., MARCUS R. – Sleep-wakefulness, EEG and behavioural studies of chronic cats without neocortex and striatum: the « diencephalic » cat. *Arch. Ital. Biol.* 110, 348-382, 1972.
17. VILLABLANCA J., SALINAS-ZEBALLOS M.E. – Sleep-wakefulness EEG and behavioural studies of chronic cats without the thalamus: the « athalamic » cat. *Arch. Ital. Biol.* 110, 383-411, 1972.

Chapter 53

Sleep and the gastrointestinal tract

P. Ducrotté
Service de Gastroentérologie, Hôpital Charles Nicolle, Rouen, France

INTRODUCTION

In gastro-intestinal (GI) practice, the onset of gastrointestinal nocturnal symptoms or their predominance during the night is often reported by patients. These nocturnal symptoms may be heartburn, epigastric or abdominal pain, diarrhoea or episodes of faecal incontinence. Patients report either that symptoms interfere with sleep quality or, in some instances, that disturbed sleep worsens their symptoms.

Despite these clinical observations, the GI system has been largely ignored in the study of sleep physiology, most likely because of the inability to study gut function with non invasive techniques during sleep. Thus, the study of GI functioning during sleep is not a part of the usual practice of gastroenterologists while, conversely, the description of major alterations of respiratory function during sleep has led to the understanding of the relationship between respiration and sleep becoming a key point in the training of pneumonologists.

Nevertheless, knowledge about the relationship between GI function and sleep has somewhat improved in recent years. The purpose of this chapter is to review the available data about this relationship and to highlight how this better understanding could influence the practical management of the patients.

SLEEP AND OESOPHAGEAL FUNCTION

Without any doubt, the influence of sleep on gastro-oesophageal reflux (GOR) is the topic where our knowledge is the most important.

GOR is a physiological phenomenon that occurs mainly after the ingestion of a meal [4]. GOR becomes a disease (GORD) when it promotes recurrent and troublesome symptoms and/or oesophagitis.

Combined pH and manometric studies have demonstrated that most physiological GOR episodes follow transient relaxations of the lower oesophageal sphincter (LOS) that are spontaneous relaxations, not triggered by swallowing and unrelated to a sequence of oesophageal peristalsis [38]. After a GOR episode, both oesophageal peristalsis and saliva clear the acid refluxate from the oesophagus.

Symptoms and/or oesophageal mucosal lesions are usually associated with an abnormally high oesophageal acid exposure. A first cause of excessive oesophageal acidification is an impaired anti-reflux barrier, mainly represented by LOS [4]. In GORD, frequent transient LOS relaxations are the most prevalent mechanism, accounting for more than 70 % of reflux episodes in patients [4]. Sleep consequences on LOS function are limited. A gradual rise in LOS pressure has been observed during sleep [38]. Transient LOS relaxations are the most common mechanism of nocturnal reflux [4] but occur relatively infrequently during the nocturnal period and have been shown to be most commonly associated with a transient arousal from sleep [4]. However, some reflux events in the absence of an arousal response have also been identified.

In some patients, particularly in the elderly, the pathophysiological mechanism of GORD is a permanent hypotensive LOS [8]. Ulcerated oesophagitis is more often observed in this subgroup of patients [29]. In this form of GORD, reflux episodes occur postprandially but also when the patient is in a supine position. Therefore the sleeping period may be a critical period to consider, mainly when GORD is complicated by erosive or ulcerated mucosal oesophageal lesions.

An abnormally high oesophageal acid exposure can also be related to a delayed oesophageal acid clearance time, that is defined as the time during which the oesophageal mucosa remained acidified to a pH < 4 after a GOR episode. A normal oesophageal clearance during the nocturnal period is a key factor preventing oesophagitis. Indeed, a delayed oesophageal acid clearance time is an important factor in promoting erosive or ulcerated oesophageal lesions [29] and several studies have shown an abnormally high acid mucosal contact time in patients with oesophagitis, mainly in the supine position or during the sleeping interval. For instance, Orr *et al.* have observed that prolonged acid clearance during sleep and a higher number of GOR lasting more than 5 minutes in the supine position were two parameters to differentiate patients with oesophagitis from those without oesophageal mucosal damage [31].

The two major potential causes of prolonged oesophageal acid clearance are impaired oesophageal emptying due to peristaltic dysfunction and/or a non reducing hiatal hernia, and diminished salivary neutralising capacity. Nocturnal oesophageal motor activity is dependant on sleep stage. In the lower third of the oesophagus, the frequency of primary oesophageal contractions during sleep stages 2 to 4 is thirtyfold lower than during arousal period and almost fiftyfold lower during REM sleep [3]. This could explain why oesophageal acid clearance time is correlated with the percentage of arousal periods during the night. The higher this percentage, the shorter the oesophageal acid clearance time.

The marked effects of sleep on salivary function is a second important explanation for the effects of sleep on acid clearance time. Helm *et al.* have demonstrated the importance of saliva for acid clearance [14]. During sleep, salivation and swallowing frequency are significantly reduced, leading to markedly prolonged acid clearance time [30] (fig. 53.1).

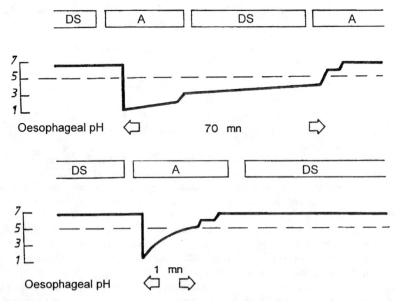

Figure 53.1. Impact of sleep on duration of a GOR episode. Above: when this episode occurs mainly during a period of deep sleep (DS), its duration is long. Below: At variance, when its onset coincides with a period of arousal (A), GOR episode is short because of an effective oesophageal acid clearance (salivation, swallowing, oesophageal peristalsis).

The most common clinical symptoms of GORD are heartburn and acid regurgitations while some patients complain of extraoesophageal manifestations, mainly asthma and ENT symptoms, such as posterior laryngitis. These extraoesophageal symptoms are related, at least partly, to GOR episodes occurring during sleep. Proximal oesophageal acid exposure is significantly increased in patients with laryngeal symptoms with nocturnal proximal oesophageal acidification present in over half of these patients but in none of the control patients [17]. Orr and colleagues have shown that sleep is a significant risk factor for acid migration to the proximal oesophagus for even minute volumes. For instance, 40 % of 1 ml perfusions during sleep migrated to the proximal oesophageal sensor compared with < 1 % during waking [34].

The arguments for a relationship between nocturnal GOR during sleep and asthma are less convincing. Proposed mechanisms of reflux induced-asthma are either aspiration of the gastric content into the lung with consequent bronchoconstriction or activation of a vagal reflex from the oesophagus to the lung causing bronchoconstriction [12, 40]. The proximal migration of refluxed gastric contents certainly increases the risk of pulmonary aspiration, *a fortiori* when oesophageal peristalsis is impaired. The risk of micro-aspiration is increased when proximal migration of the gastric content occurs during NREM sleep stages 3 and 4 or during REM sleep. Indeed, during these nocturnal periods, upper oesophageal sphincter pressure is at its lowest [18]. The vagal reflex from the oesophagus to the lung causing bronchoconstriction could be enhanced during the night. In children, one study showed that oesophageal acid infusion caused more airways response at 4 am than at 12 pm and asthmatic children with nocturnal asthma symptoms have a higher reflux score with a positive correlation between reflux score and nighttime-associated wheezing. Similar data were never obtained in adults.

Therefore, in patients with oesophagitis and those with aerodigestive manifestations of GORD, the control of GOR during sleep is one of the main objectives of treatment. This is sometimes difficult to obtain because of the inability of the proton pump inhibitor to block completely gastric acid secretion due to a nocturnal breakthrough (see below). The sleep position could also be a parameter affecting the number and duration of nocturnal GOR episodes. A recent study has found that GOR are more frequent in the supine position when right lateral decubitus was associated with a greater percentage of time with pH < 4 and longer oesophageal acid clearance compared to the left, supine and prone [21]. These results suggest that the left lateral position could be the recommended position for patients with nocturnal GORD, particularly in case of low LOS pressure.

Even if sleep is associated with a more prolonged acid clearance time compared to waking, secondary to the natural consequences of a depressed level of consciousness, protective mechanisms exist to limit the duration of GOR during the night. Responsiveness to oesophageal acid mucosal contact has been shown to be enhanced under the circumstances of depressed consciousness, which would present a much greater risk of pulmonary aspiration, while the communication between oesophageal receptors and the central nervous system (CNS) seems enhanced during sleep by comparison with waking. An arousal from sleep following the infusion of acid is an important parameter in producing prompt acid clearance. During sleep, the infusion of acid produces a significantly greater percentage of arousal responses than does water [33]. Moreover, both volume and acidity of the refluxate affect arousal latency and acid clearance time, the largest volume and the more acid reflux episodes being associated with the shortest arousal latency and the shortest clearance time [31]. These data suggest that drugs, i.e. hypnotics, able to lengthen arousal latency after reflux episodes, could have deleterious consequences on the consequences of nocturnal GOR episodes and may promote oesophagitis or the onset of extraoesophageal manifestations of GORD.

SLEEP AND GASTRIC FUNCTIONS

Our knowledge of gastric motility and emptying during sleep is very limited. The only study focused on gastric motility was an indirect study based on the recording of gastric electrical activity (electrogastrogram). This study revealed that normal 2-4 cycles per minute (cpm) electrical gastric activity is reduced during sleep while gastric dysrhythmia is more frequent [6]. How these electrical changes affect gastric motor activity during this period remains to be elucidated. In healthy controls, Goo *et al* have reported a delayed solid-phase emptying in the evening (8 pm) compared to the morning (8 am) [10]. This slow solid-phase emptying could be an additional factor promoting the onset of GOR episodes during the night, mainly after a late evening meal [32].

By contrast, nycthemeral variations of gastric pH are well documented. Gastric pH monitorings have shown that the profile of median pH values in healthy volunteers is characterised by periods of postprandial alkalinisation and a highly acid profile during the fasting nocturnal period (24 pm – 4 am), with a decline in acidity in the early morning hours [37] (fig. 53.2). To counteract this high acidity, the secretion of trefoil peptide with cytoprotective effects on the gastric mucosa, such as TFF2, is the highest during sleep when it is lowest in the early evening, when neither pepsin or protein in the gastric juice vary significantly over 24 hours [39]. Some relationship exists between sleep stages, acid secretion and gastric pH. Compared to the waking state, sleep was found to be associated with significantly lower levels of acid secretion. In an early study, Stacher *et al.* have reported that, even if there were no significant differences during NREM sleep stages 1 to 4 and REM sleep, acid secretion decreased with deeper stages of sleep. During REM phases, only small amounts of acid were produced. Arousal or periods of waking in the course of the night, as well as morning waking, were associated with an increase in acid output [41]. More recently, in healthy controls, monitoring of intragastric pH showed a more acidic gastric juice with lower pH values during wakefulness than during NREM sleep stages 1-4 and REM sleep [43]. Similar results to those of healthy controls were obtained in patients with gastric ulcer when no obvious changes of pH values with sleep stages were reported in duodenal ulcer patients [43], who have a higher acid level (lower pH) in both the nocturnal and the postprandial periods [37].This dysfunction may play an important role in the occurrence of duodenal ulcer.

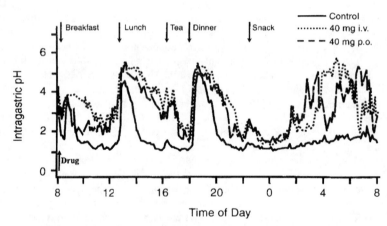

Figure 53.2. Nycthemeral variations of gastric pH values after placebo (dark line) or an antiH2 inhibitor given either intravenously (dotted line) or orally (dashed line). The figure shows that, without any antisecretory drug, gastric pH is around 1.5 during a large part of the nocturnal period.

From a practical point of view, for the treatment of both GORD and peptic ulcer, it is important to remember that proton pump inhibitors given twice daily do not suppress gastric acid secretion sufficiently overnight to prevent recovery of gastric acidity to potentially harmful pH levels. Nocturnal acid breakthrough, defined as intragastric pH < 4 for more than one hour overnight, occurs in about two-thirds of the patient with GOR disease and normal volunteers treated by omeprazole [36]. This decrease in intragastric pH occurs approximately between 1am and 2am [36] when oesophageal acid exposure is common and potentially injurious for the oesophageal mucosa because of delayed oesophageal clearance during the night (see above). Helicobacter pylori eradication increases this nocturnal acid breakthrough [42]. This nocturnal acidity can be controlled more effectively with a nighttime dose of histamine-2 receptor antagonists than with a third dose of omeprazole [36].

SLEEP AND THE SMALL BOWEL

Several studies have documented the effects of sleep on small bowel motility.

When recordings are performed for a sufficiently long period of time, two main motor patterns are always present in healthy subjects. These are the migrating motor complex (MMC) that occur during fasting and the postprandial pattern of motility that follows the intake of a nutrient meal [16, 19].

The MMC consists of three phases, of which phase III is the most distinctive, with a succession of regular contractions that migrates aborally from the upper gut to the ileum. Phase III is usually succeeded by quiescence or phase I and then irregular contractile activity of phase II recurs before the next phase III completes the MMC cycle. Phase III recurs at average intervals of 90 to 110 minutes. Food intake interrupts the MMC and induces irregular contractile activity in the small bowel soon after the meal. Postprandial state is defined as the elapsed time between the start of the meal intake and the return of Phase III [16].

In healthy volunteers, small bowel motility shows nycthemeral variations. Overall motor activity decreases during deep NREM sleep while it increases during REM sleep. Sleep affects both MMC characteristics and duration of postprandial state. Almost half of the MMC phases are recorded during the night and the interval between phases III is shorter during the night than during daytime (19). Moreover, during the nocturnal period, fewer phases III start from the oesophagus [19] when their migration velocity along the small bowel is significantly reduced [22] (fig. 53.3). The final difference is that Phase I is markedly increased during reduced CNS arousal, when Phase II activity is significantly reduced. By acute reversal of the sleep cycle, Kumar *et al.* showed that it is wakefulness rather than the diurnal phase that is responsible for this modulation [24]. The 90 minute periodicity of REM sleep stage, very similar to that of MMC, led to attempts to correlate these two major biorhythms. Most studies, found no clear relationship between these two biorhythms which seem independent [24]. Postprandial activity is diminished during sleep, whereas the consumption of a late meal, 15 minutes before going to bed restores the phase II activity usually absent during sleep, abolishes the nocturnal reduction of MMC cycle length but maintains the circadian variation in the propagation velocity of the MMC cycle [23]. These results suggest that small bowel postprandial motor activity is modulated both by CNS arousal and by nutrient stimulation of receptors in the small bowel. Whether these changes in motor patterns, when sleep occurs very early after a late meal, are accompanied by equivalent changes in exocrine secretory responses and absorption kinetics, remain to be demonstrated.

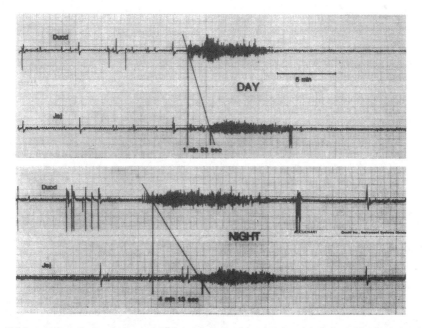

*Figure 53.3.*Propagation velocity of a human MMC during the day (above) and at night (below). The time taken for each Phase III to travel between the two sensors is about threefold longer during the night (from reference 22).

The consequences of the modulation of small bowel motility by sleep remain unclear. As phase III of the MMC is considered as the "housekeeper" of the gut, one may speculate that the withdrawal of meal residues from the intestinal lumen by the regular occurrence of phase III during the night could be a protective mechanism against bacterial overgrowth [16]. This may be of particular importance in subjects allowed to eat *ad libitum* during the day, a condition during which phase III may be absent altogether during the diurnal period [16]. Altered nocturnal motility could also be a substratum for diurnal abdominal symptoms. Mertz *et al* have shown that non-ulcer dyspepsia patients with severe symptoms had a considerably decreased number of MMC during the nocturnal period, associated with a reduced phase I and increased phase II. These findings were specific of severe non-ulcer dyspepsia patients when patients with non-ulcer dyspepsia plus irritable bowel syndrome (IBS) were no different from normal subjects [25]. The lack of disturbed nocturnal motility in irritable bowel syndrome confirmed the results of 72-h ambulant recordings of duodenojejunal contractility showing that motility disturbances in patients with IBS by comparison with healthy controls are confined to the waking state [20]. In a prospective study, Goldsmith and Levin observed a significant correlation between morning symptoms of IBS symptoms and the quality of the prior night's sleep but, in this study, intestinal motility was not monitored [9].

SLEEP AND THE LARGE BOWEL

The functions of the colon include facilitation of water, electrolyte and short chain fatty acid absorption; storage of excreta by delaying aboral flow; effective voluntary evacuation when convenient and acceptable.

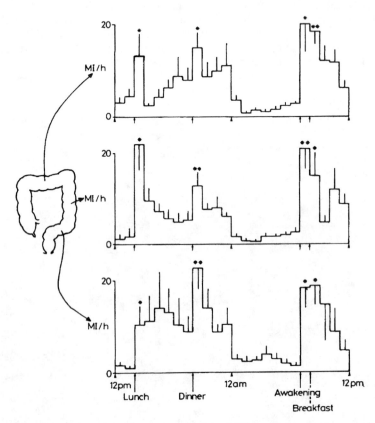

Figure 53.4. Nycthemeral variations of hourly colonic motor index (MI/h) recorded in the transverse, descending and sigmoid colon. During the night, the motor index is the lowest with a quasi absence of motor activity (from reference 26).

During the night, effective colonic storage is essential to avoid rectal filling with stools and possible episodes of faecal incontinence. The importance of the colon for a comfortable sleep is highlighted by the regular onset of nocturnal stools, sometimes urgent, and of episodes of faecal incontinence in patients operated for total colectomy despite the persistence of the rectum or the realisation of an ileal reservoir.

CONCLUSIONS

Our knowledge of gut function during the night has improved. Even if it is sometimes difficult to determine whether nocturnal changes in gut physiology are related to sleep stages rather than to circadian rhythms, these changes must not be overlooked in attempts to understand symptoms or to adapt treatment in some patients. For instance, the effective control of GORD disease during the night appears crucial in patients with oesophagitis or extraoesophageal symptoms. Further studies are needed to understand the underlying mechanisms of these changes but also to determine how treatments that affect sleep quality can interfere with gut physiology.

REFERENCES

1. BASSOTTI G., GABURRI M. - Manometric investigation of high-amplitude propagated contractile activity of the human colon. *Am. J. Physiol., (Gastrointest. Liver Physiol.)* 255, G660-G664, 1988.
2. BASSOTTI G., IANTORNO G., FIORELLA S., BUSTOS-FERNANDEZ L., BILDER C.R. - Colonic motility in man: features in normal subjects and in patients with chronic idiopathic constipation. *Am. J. Gastroenterol.,* 94, 1760-1770, 1999.
3. CASTIGLIONE F., EMDE C., ARMSTRONG D., SCHNEIDER C., BAUERFEIND P., STACHER G., BLUM A.L. - Nocturnal oesophageal motor activity is dependent on sleep stage. *Gut,* 34, 1653-1659, 1993.
4. DENT J., HOLLOWAY R.H., TOOULI J., DODDS W.J. - Mechanisms of lower oesophageal sphincter incompetence in patients with symptomatic gastrooesophageal reflux. *Gut,* 29, 1020-1028, 1988.
5. ELSENBRUCH S., HARNISH M.J., ORR W.C. - Subjective and objective sleep quality in irritable bowel syndrome. *Am. J. Gastroenterol.,* 94: 2447-2452, 1999.
6. ELSENBRUCH S., ORR W.C., HARNISH M.J., CHEN J.D. - Disruption of normal gastric myoelectrical functioning by sleep. *Sleep,* 15, 453-458, 1999.
7. FURUKAWA Y., COOK I.J., PANAGOPOULOS V., MCEVOY R.D., SHARP D.J., SIMULA M. - Relationship between sleep patterns and human colonic motor patterns. *Gastroenterology,* 107, 1548-1550, 1994.
8. GALMICHE J.P., JANSSENS J. - The pathophysiology of gastrooesophageal reflux disease: an overview. *Scand. J. Gastroenterol.,* suppl (211), 7-18, 1995.
9. GOLDSMITH G., LEVIN J.S. - Effect of sleep quality on symptoms of irritable bowel syndrome. *Dig. Dis. Sci.,* 38, 1809-1814, 1993.
10. GOO R.H., MOORE J.G., GREENBERG E., ALAZRAKI N.P. - Circadian variation in gastric emptying of meals in humans. *Gastroenterology,* 93, 515-518, 1987.
11. GUÉDON C., DUCROTTÉ P., ANTOINE J.M., DENIS P., COLIN R., LEREBOURS E. - Does chronic supplementation of the diet with dietary fibre extracted from pea or carrot affect colonic motility in man? *Br. J. Nutr.,* 76, 51-61, 1996.
12. HARDING S.M. - Nocturnal asthma: role of nocturnal gastroesophageal reflux. *Chronobiol. Int.* 16, 641-662, 1999.
13. HEBDEN J.M., GILCHRIST P.J., BLACKSHAW E., FRIER M.E., PERKINS A.C., WILSON C.G., SPILLER R.C. - Night-time quiescence and morning activation in the human colon: effect on transit of dispersed and large single unit formulations. *Eur. J. Gastroenterol. Hepatol.,* 11, 1379-1385, 1999.
14. HELM J.F., DODDS W.J., PELC L.R., PALMER D.W., HOGAN W.J., TEETER B.C. - Effects of esophageal emptying and saliva on clearance of acid from the esophagus. *N. Engl. J. Med.* 310, 284-288, 1984.
15. HERVE S., LEROI A.M., MATHIEX-FORTUNET H., GARNIER P., KAROUI S., MENARD J.F., DUCROTTÉ P., DENIS P. - Effects of polyethylene glycol 4000 on 24-h manometric recordings of left colonic motor activity. *Eur. J. Gastroenterol. Hepatol.* 13, 647-654, 2001.
16. HUSEBYE E. - The patterns of small bowel motility: physiology and implications in organic disease and functional disorders. *Neurogastroenterol. Mot.,* 11, 141-161, 1999.
17. JACOB P., KAHRILAS P.J., HERZON G. - Proximal esophageal pH-metry in patients with 'reflux laryngitis'. *Gastroenterology,* 100, 305-310, 1991.
18. KAHRILAS P.J., DODDS W.J., DENT J., HAEBERLE B., HOGAN W.J., ARNDORFER R.C. - Effect of sleep, spontaneous gastrooesophageal reflux, and a meal on upper esophageal sphincter pressure in normal human volunteers. *Gastroenterology,* 92, 466-471, 1987.
19. KELLOW J.E., BORODY T.J., PHILLIPS S.F., TUCKER R.L., HADDAD A.C. - Human interdigestive motility: variations in patterns from esophagus to colon. *Gastroenterology,* 91: 386-395, 1986.
20. KELLOW J.E., GILL R.C., WINGATE D.L. - Prolonged ambulant recordings of small bowel motility demonstrate abnormalities in the irritable bowel syndrome. *Gastroenterology* 98, 1208-1218, 1990.

21. KHOURY R.M., CAMACHO-LOBATO L., KATZ P.O., MOHIUDDIN M.A., CASTELL D.O. - Influence of spontaneous sleep positions on nightime recumbent reflux in patients with gastroesophageal reflux disease. *Am. J. Gastroenterol.* 94, 2069-2073, 1999.
22. KUMAR D., WINGATE D., RUCKEBUSH Y. - Circadian variation in the propagation velocity of the migrating motor complex. *Gastroenterology*, 91, 926-930, 1986.
23. KUMAR D., SOFFER E.E., WINGATE D.L., BRITTO J., DAS-GUPTA A., MRIDHA K. - Modulation of the duration of human postprandial motor activity by sleep. *Am. J. Physiol.*, (*Gastrointest. Liver Physiol.*), 256, G851-G855, 1989.
24. KUMAR D., IDZIKOWSKI C., WINGATE D.L., SOFFER E.E., THOMPSON P., SIDERFIN C. - Relationship between enteric migrating motor complex and the sleep cycle *Am. J. Physiol.*, (*Gastrointest. Liver Physiol.*) 259, G983-G990, 1990.
25. MERTZ D.D., FEFER L., SYTNIK B., RAEEN H., NIAZI N., KODNER A., MAYER E.A. - Sleep and duodenal motor activity in patients with severe non-ulcer dyspepsia. *Gut*, 35, 916-925, 1994.
26. NARDUCCI F., BASSOTTI G., GABURRI M., MORELLI A. - Twenty four hour manometric recording of colonic motor activity in healthy man. *Gut*, 28, 17-25, 1987.
27. ORKIN B.A., HANSON R.B., KELLY K.A., PHILLIPS S.F., DENT J. - Human anal motility while fasting, after feeding and during sleep. *Gastroenterology*, 100; 1016-1023, 1991.
28. ORKIN B.A., SOPER N.J., KELLY K.A., DENT J. - Influence of sleep on anal sphincteric pressure in health and ileal pouch-anal anastomosis. *Dis. Colon Rectum*, 35, 137-144, 1992.
29. ORLANDO R.C. - Reflux esophagitis: overview. *Scand. J. Gastroenterol.*, suppl 210, 36-37, 1995.
30. ORR W.C., JOHNSON L.F., ROBINSON M.G. - The effect of sleep on swallowing, esophageal peristalsis, and acid clearance. *Gastroenterology*, 86, 814-819, 1984.
31. ORR W.C., ROBINSON M.G., JOHNSON L.F. - The effect of esophageal acid volume on arousals from sleep and acid clearance. *Chest*, 99, 351-354, 1991.
32. ORR W.C., HARNISH M.J. - Sleep-related gastrooesophageal reflux: provocation with a late evening meal and treatment with acid suppression. *Aliment. Pharmacol. Ther.*, 10, 1033-1038, 1998.
33. ORR W.C., JOHNSON L.F. - Responses to different levels of esophageal acidification during waking and sleep. *Dig. Dis. Sci.*, 43, 241-245, 1998.
34. ORR W.C., ELSENBRUCH S., HARNISH M.J., JOHNSON L.F. - Proximal migration of esophageal acid perfusions during waking and sleep. *Am. J. Gastroenterol.*, 95, 37-42, 2000.
35. ROARTY T.P., SURRATT P.M., HELLMANN P., MCCALLUM R.W. - Colonic motor activity in women during sleep. *Sleep*, 21, 285-288, 1998.
36. ROBINSON M. - Drugs, bugs, and esophageal pH profiles. *Yale J. Biol. Med.*, 72, 169-172, 1999.
37. SAVARINO V., MELA G.S., SCALABRINI P., SUMBERAZ A., FERA G., CELLE G. - 24-hour study of intragastric acidity in duodenal ulcer patients and normal subjects using continuous intraluminal pH-metry. *Dig. Dis. Sci.*, 33, 1077-1080, 1988.
38. SCHOEMAN M.N., TIPPET M.D., AKKERMANS L.M., DENT J., HOLLOWAY R.H. - Mechanisms of gastroesophageal reflux in ambulant healthy human subjects. *Gastroenterology*, 108, 83-91, 1995.
39. SEMPLE J.L., NEWTON J.L., WESTLEY B.R., MAY F.E. - Dramatic diurnal variations in the concentration of the human trefoil peptide TFF2 in the gastric juice. *Gut*, 48, 648-655, 2001.
40. SONTAG S., O'CONNELL S., KHANDEWAL S., MILLER T., NEMCHAUSKY B., SCHNELL T.G., SERLOVSKY R. - Effect of positions, eating, and bronchodilatators on gastroesophageal reflux in asthmatics. *Dig. Dis. Sci.*, 35, 849-856, 1990.
41. STACHER G., PRESSLICH B., STARKER H. - Gastric acid secretion and sleep stages during natural night sleep. *Gastroenterology*, 68 , 1449-55, 1975.
42. VAN HERWAARDEN M.A., SAMSON M., SMOUT A.J.P.M. - Helicobacter pylori eradication increases nocturnal acid breakthrough. *Aliment. Pharmacol. Ther.*,14 , 961-962, 2000.
43. WATANABE M., NAKAZAWA S., YOSHINO J., YAMAO K., INUI K., YAMACHIKA H., KANEMAKI N., WAKABAYASHI T., FUJIMOTO M., WATARAI K. - A study of the relationship between nocturnal intragastric pH and sleep stages of peptic ulcer. *Nippon Shokakibyo Gakkai Zasshi*, 92 , 1241-1249, 1995.

Chapter 54

Sleep and sudden infant death syndrome

A. Kahn, J. Groswasser, M. Sottiaux, E. Rebuffat, and P. Franco
Hôpital Universitaire des Enfants Reine Fabiola, Bruxelles, Belgique

This chapter deals with the relationship between sleep and sudden infant death. We will not refer to associated conditions, such as the characteristics of deceased infants' siblings, or situations referred to as "unexplained serious illnesses".

DEFINITIONS

Sudden infant death (SID) is defined as the sudden death of an infant, which is unaccounted for by previous history and which remains unexplained despite the examinations and enquiries carried out after the death. The aetiological diagnosis is carried out by a process of elimination, based on the child's previous history, clinical data and the findings of a complete autopsy, involving microscopic study, and testing for infection and metabolic disorders. This examination is crucial for diagnosis, enabling medical or surgical causes to be identified in over 15% of cases of unexpected death [48]. A medical enquiry carried out at the place of death may provide useful information regarding the circumstances of death as, for example in cases of accidental strangling or suffocation.

EPIDEMIOLOGICAL CHARACTERISTICS

The incidence of death has been substantially reduced since the launching of prevention campaigns at the beginning of the 1990s. In our regions, sudden death now affects approximately 0.7 infants per 1000 live births.

These unexplained deaths represent almost 20 to 30% of deaths occurring in the postneonatal period. The victims are aged between 2 and 6 months in over 80% of cases. Less than 1% of deaths occur after the first year of life. Boys are 1.7 times more affected than girls. The younger siblings of infants who died from sudden infant death suffer 3 times the risk of becoming victims themselves. The deaths are often linked to birth weight of less than 1,900 g or to a mother of under 19 years of age. These factors partly reflect the influence of an underprivileged socio-economic background [20, 51]. In 70% of cases, death is preceded by a benign respiratory or digestive infection. Death almost always occurs during the night or at a time when the child is supposed to be sleeping. Deaths occur at the end of the night, for the most part. The infant had sometimes been seen awake or in good health, or had cried in an unaccustomed way just before the death was discovered [20, 51]. The "risk factors" in the infant's environment have been defined, and will be discussed later.

MECHANISMS RESPONSIBLE FOR SUDDEN DEATH

Our understanding of the mechanisms involved in sudden death is still partial, and these deaths represent failures in our capacities of diagnosis and prevention. While there is no specific illness, or "sudden death syndrome", the causes of these deaths can nevertheless be grouped into three general categories. The first of these is illness. In much the same way as a viral or bacterial infection which is harmless to adults or older children can lead to a sudden illness in an infant, an apparently harmless infection may help to further destabilise an infant whose cardio-respiratory control is still unstable.

The second group covers the mechanisms of maturation. These involve the maturation of the systems which control the vital functions, such as breathing, cardiac activity, neurovegetative reactions or the digestive processes. These control mechanisms are immature in some infants, exposing them more than others to cardio-respiratory abnormalities during sleep. The abnormalities found in this group remain undetected unless tests are carried out during the infant's sleep.

The third group relates to the environment. Unfavourable conditions in the infants' environment are experienced as a form of stress which he finds difficult to cope with. Some examples will be given of this.

SLEEP AND SUDDEN INFANT DEATH

As most deaths occur at night, many research centres have focused on studying the sleep of infants. Some of the children whose sleep had been studied, later succumbed to sudden death and their sleep tracings have been studied in depth.

Sleep characteristics

The sleep of the infants who died showed few differences in comparison with that of normal children [21]. Subsequent victims nevertheless manifest fewer body movements [21, 39], less frequent and shorter arousals, particularly at the end of the night [21, 39]. Subsequent victims spend more time in NREM sleep stages 2 and 3 throughout the night and in active sleep (REM sleep) at the end of the night [39]. The clinical significance of this sleep profile is not yet understood. Subsequent victims may have a higher threshold of arousal than normal children and thus have less tendency to awaken spontaneously. This observation was confirmed by a study of twins, where one of the infants was a victim of sudden death [22].

Some subsequent victims of sudden death present particular signs during sleep. Heavy perspiration is reported in 25% of infants [4, 22] with 10% showing episodes of pallor [51]. Such signs are only two to three times less common in healthy children.

Sleep respiration

Anatomo-pathological tests have shown signs of tissue hypoxia in nearly 40% of sudden death victims, in the form of an excess of brown adrenal fat, hepatic haematopoiesis [33] or zones of cerebral leukomalacia [48]. These signs of hypoxia probably result from recurrent episodes of hypoxemia [49], attributed to the apnoeas observed during the sleep of some infants [16, 17, 28, 46]. These observations have led to the hypothesis that children die during sleep from the effects of prolonged respiratory arrest [16, 46].

The clinical importance of these first observations has not however been confirmed by later research, several studies showing that subsequent victims of sudden death show no more central apnoeas than do normal children [21, 33, 44].

Other work has demonstrated that some infants who later die, show upper airway blockages during sleep, termed as "obstructive apnoeas" [17, 21, 29, 37]. These are manifested as an interruption of the passage of air through the nose and mouth despite persistent thoracic and abdominal movement. Almost 40% of subsequent victims show more than 0.6 respiratory obstruction per hour of recording, a figure which is significantly higher than that observed in controls [21]. These obstructive apnoeas chiefly occur during agitated sleep of the REM type [17, 21] affecting boys more often than girls among sudden death victims [21]. They last a minimum of 3 seconds and are accompanied by cardiac slowing and arterial desaturation [17, 21].

The presence of respiratory obstructions during sleep recordings does not account for sudden death. It does however indicate a number of particular respiratory features which are rarely observed in the sleep of healthy children [17, 21]. Obstructive apnoeas appear to result from disorders affecting the control of the upper airways. [17]. The occurrence of obstructive apnoeas is favoured by the presence of obstacles in the upper airways, such as an autonomic abnormality [18], tongue blocking, maxillary recession [49], a respiratory infection [2] or mechanical blocking of the nasal passages [38, 47]. The development of obstructive apnoeas is also favoured by epilepsy, hyperthyroidism, overweight or unstable neurovegetative control [24]. Respiratory obstructions also result from factors such as several hours of sleep deprivation [5], or administering sedative and cough-relieving drugs, such as the phenothiazines [25] or type H-1 antihistamines.

The autopsy of infants which are victims of sudden death reveals evidence of repeated obstruction of the upper airways, such as a thickening of the vocal chords [42] or the tongue [43].

Even though studies appear to indicate the presence of abnormalities in the control of upper airways in certain infants who are subsequent victims of sudden death, they nevertheless refute the hypothesis that all sudden deaths are attributable to respiratory disorders during sleep.

Cardiac rhythm during sleep

A lengthening of the Q-T space [40] has been found in some infants who later died during sleep, as well as in their parents. This anomaly has not however been confirmed by other studies [28, 44]. Other cardiac rhythm anomalies have been described in subsequent sudden death victims, such as reduced instantaneous cardiac rhythm variability [39], bradycardia or episodes of tachycardia [28, 44]. These observations may reveal the existence of anomalies in autonomic control of cardiac rhythm in certain sudden death victims [9, 19]. In some of these, spectral analysis of their cardiac rhythm at the end of the night had shown a preponderance of orthosympathetic control not observed in normal cases.

Other factors observed during sleep

Control of digestive movements

Some infants who succumb to sudden death had previously presented signs of poor co-ordination of the oro-oesophageal junction, or of digestion, such as problems of deglutition [42], abnormal tiredness during feeding [22, 35], or postprandial regurgitation [21]. These disorders were associated with bradycardia and apnoeas. During sleep, the lack of co-ordination in digestive transit would favour an oesophageal acid reflux. This mainly occurs during agitated (REM) sleep, in a dorsal position, and when there is increased intra-abdominal pressure [52]. If severe, the reflux will result in the acid product being inspired by the upper airways. This accident does not appear to be a common cause of death, however [51]. The study of sleeping infants has not shown any direct link between acid reflux and the manifestation of cardiac or respiratory abnormalities [21].

Role of sleeping position

Infants who succumb to sudden death, are found sleeping on their stomachs in an abnormally high number of cases [1, 7, 8, 30, 39]. Statistical studies have shown the risk of death to be 3 to 9 times greater if the infant sleeps on his stomach. In countries where information campaigns have reduced the frequency of infants sleeping in the ventral position, there has been a considerable reduction in infant mortality. The reasons why body position contributes to sudden death are still unknown. Laboratory studies show that in a ventral position the infant has a risk of suffocating in the bedding, breathing in too much carbon dioxide or being less able to control his temperature [36]. When an infant lies face down, he is also less receptive to outside stimulation. He sleeps more deeply and is less easily awakened [26]. It is not known whether the characteristics linked to position contribute to death.

Role of ambient temperature during sleep

Studies have shown a correlation between high temperature in the room where the infant sleeps and the incidence of accidents of sudden death [45]. The temperature may only exceed the infant's ideal temperature by a few degrees. Thermal excess is favoured by central heating in the room, the use of duvet quilts or too many covers, or clothing that is too warm [14]. A hyperthermal situation also occurs when the child's face is covered by the bedding during sleep. It is not known how hyperthermia favours sudden death, but an increase in the infant's rectal temperature has been shown to lead to the deregulation of respiratory control during sleep [15] and to raise arousal thresholds.

Role of tobacco during and after pregnancy

Epidemiological studies have shown that tobacco consumption during pregnancy increases the risk of sudden death [31-34]. The effect is directly linked to the number of cigarettes smoked by the pregnant mother and is reinforced by the father's tobacco consumption. The mechanisms explaining this relationship between tobacco consumption during the prenatal period and death during the first months of life are still unknown. Tobacco consumption may lead to intra-uterine hypoxia, as evidenced by the risks of premature birth or low birth weight. Tobacco consumption also causes alterations to the respiratory control of the newborn and young infant, shown by respiratory blockages during sleep [23]. It also favours higher arousal thresholds during the sleep of these children [11]. Passive smoking during pregnancy also alters the autonomic control of the heart, in favour of the predominance of orthosympathetic controls [12].

Role of sedatives

Medication which has a sedative effect favours the occurrence of sudden death. This relationship has been demonstrated in the case of phenothiazines, substances used to relieve coughs and restlessness in infants. The drug can be directly administered to the child orally or rectally. It can also be transmitted through breast milk, if the nursing mother is taking this type of medication. Laboratory studies have shown that these substances cause heavier sleep, reduce the capacity for arousal and can lead to respiratory blockages [25]. Similar observations have been made in relation to other medications with sedative effects, such as antihistamines.

Arousal thresholds

Sudden deaths may result from a raised threshold of arousal, whereby the infant fails to awaken during a cardio-respiratory incident. Tests of hypoxia and hypercapnia nevertheless give contradictory results [3, 53]. The threshold levels measured by auditory stimulation prove to be abnormally high in the different situations regarded as increasing the risk of sudden death. This applies to cases of infant exposure to maternal tobacco consumption during pregnancy, sleep deprivation, sleeping in the ventral position, sleeping in an atmosphere which is too hot or receiving sedatives. A rise in arousal threshold may thus lead to the occurrence of sudden death. The notion of an association between "sleep and sudden death" is probably being replaced by the association between "sleep, arousal and sudden death".

Other lines of research

Infants who died at home have been recorded on monitors with data memory. Their tracings taken before death show prolonged episodes of hypoxia and bradycardia, although the causes of these alterations were unknown. To clarify the mechanisms responsible for unexpected deaths, research is being pursued in areas as varied as modes of autonomic system control, the molecular effects of infection, the genetics of respiratory anomalies or metabolic disorders [50]. These may all influence the relationships between "sleep, arousals and sudden death", which are still inadequately understood.

PRACTICAL CONSEQUENCES OF THE SLEEP-AROUSAL-SUDDEN DEATH ASSOCIATION

For several years now, many countries have been issuing recommendations aimed at reducing the risk of accidents during sleep. It is recommended not to place infants on their stomachs for sleep, unless there are particular medical indications to the contrary, such as the presence of upper airway abnormalities.

Parents are also advised to stop smoking before and after the birth of their child, to opt for breast feeding and to avoid covering the child too much or allowing him to sleep in a room which is too hot. It is also recommended that young infants be regularly checked while sleeping.

Infants are placed far less often on their stomachs, in countries like France or Belgium, where the lateral or dorsal position is considered preferable. A 50 to 60% reduction has since been witnessed in the incidence of infant mortality. There has been no parallel increase in the incidence of inhalation, due to gastro-oesophageal reflux.

The association between a problem arising during sleep and the incidence of sudden death calls for three types of prevention. The first concerns general recommendations given to families to avoid accidents occurring during sleep. The second involves recognising the signs which families should report to the doctor in order to detect any abnormalities arising during sleep. And the third relates to additional tests to be carried out on infants who are symptomatic, to confirm the presence of any disorders.

SAFETY DURING SLEEP

The advice aims at ensuring the well-being and safety of infants. Parents are advised to comply with the following rules:

Basic recommendations

Sleeping position

The child should be laid on his back, avoiding the ventral position unless medical advice is to the contrary.

Temperature control

Check that the infant is neither too hot nor too cold. The young child is more sensitive to variations in temperature, than the adult.

- At home, the temperature of the child's room should not exceed 20° C if he is under 8 weeks old and 18°C if he is over 8 weeks. The temperature should be checked in any room in which the child sleeps.

- Before the age of one year, the child should not be too covered up during sleep. It is sufficient to cover him with a sheet and blanket, leaving the face uncovered. Avoid the use of quilts. The blanket may be placed in such a way that only part of the bed is covered, avoiding the risk of the child rolling under the covers. A light sleeping bag may also be used instead of a blanket.

-The child should be dressed according to the temperature of the room as opposed to the outdoor temperature.

Bedding

The child should sleep on a firm mattress, with no gaps between the mattress and the bed frame. Do not use a pillow. Avoid any risk of strangling or suffocation by removing objects such as cords around the neck, cords in the bed, plastic sheeting or other objects liable to cover the child's face. If the child sleeps in a cot, make sure that the space between the bars does not exceed 8 cm. Avoid situations which restrict the child from moving his arms and legs freely. The child's bed must be stable. Avoid letting the infant sleep in a carry cot.

Tobacco

It is strongly advised to avoid smoking, both during and after pregnancy. Do not smoke in the room in which the child sleeps.

Medication

Administering sedatives to an infant must be avoided (such as certain cough syrups or suppositories). No other medication should be given to the infant or nursing mother without a doctor's approval that the substance is not harmful.

General Recommendations

As a general rule, it is important to respect the infant's daily rhythm and to establish a regular schedule, to avoid depriving him of sleep.

More specific points to be aware of include the following:
- when the infant falls asleep after crying, check to see that he is alright;
- the room in which the infant sleeps should be well ventilated;
- avoid allowing animals into the infant's room
- in summer, ensure that the infant drinks regularly to avoid dehydration;
- respect the infant's sleep and feeding schedules;
- breast feed the infant if possible;
- ensure that the infant has regular check ups with the doctor.

WARNING SIGNS DURING SLEEP

It is important to draw the parents' attention to a number of signs which should prompt them to call the doctor. Certain signs herald a cardiac or respiratory disorder, others the presence of an undiagnosed illness. It is thus important for parents to report these signs. Parent should seek advice if, during sleep the infant
- turns blue or white,
- does not appear to be breathing,
- perspires to the extent that his clothing is wet,
- groans during sleep or wakefulness,
- snores or makes breathing noises other than when he has a cold.

FURTHER TESTS

If a child presents a clinical disorder or suspicious signs during sleep, tests must be carried out. These will depend on the child's previous history, on clinical observation and include the search for:
- infection (blood test, urine test, cerebro-spinal fluid),
- oesophageal reflux (pH monitoring, manometry, oesophageal opacification),
- neurovegetative instability (cardiac Holter test, scan),
- respiratory abnormalities during sleep (polysomnography).

CONCLUSIONS

No "sudden death syndrome" exists as such. The sudden deaths of infants result from a combination of illnesses or circumstances which can lead to the sudden death of a young child. These unexpected deaths testify to the failure of our present means of diagnosis and prevention. Epidemiological and clinical studies suggest that several unfavourable factors combine to bring about a fatal accident in certain infants. Pejorative influences arise during foetal life or in the neonatal period which may render the infant more fragile. During the weeks following birth, an already delicate balance may be seriously disturbed by additional stress. Sleep is a particularly vulnerable time during which the vital controls are more prone to attack; a serious accident occurs at a moment when the child is not under surveillance.

Diverse factors in the prenatal and postnatal environment affect the infant's cardio-respiratory or neurological state. These may exert a harmful effect on an infant who is particularly fragile, due to retarded maturation or a passing illness. Unexpected death during sleep may result from a combination of circumstances. Many of these factors can be changed, provided their roles are identified and the parents adequately informed [6, 27].

A clearer understanding of normal child development and the risks to which the child is exposed, will no doubt lead to reducing the number of unexpected and unexplained deaths. The French *Centres de Référence sur la Mort Subite du Nourrisson*, and the European Society for the Study and Prevention of Infant Death are engaged in work on research and prevention. These societies bring together researchers from different scientific disciplines and national horizons.

Thanks to their collaboration, studies are being carried out on a scale designed to add significantly to our understanding of the causes of sudden death and to develop the right strategies for effective prevention.

REFERENCES

1. AAP – Task Force on Infant Positioning and SIDS. Positioning and SIDS. *Pediatrics*, 89, 1120-1126, 1992
2. ABREU E., SILVA F.A., McFAYDEN U.M., WILLIAMS A., SIMPSON H. – Sleep apnoea during upper respiratory infection and metabolic alkalosis in infancy. *Arch. Dis. Child* 61, 1056-1062, 1986.
3. ARIAGNO R., NAGEL L., GUILLEMINAULT C. – Waking and ventilatory responses during sleep in infants near-miss for sudden infant death syndrome. *Sleep*, 3, 351-359, 1980.
4. BEAL S.M. – Some epidemiological factors about sudden infant death syndrome (SIDS) in South Australia. In: *Sudden Infant Death Syndrome*. J.T. Tildon, L.M. Roeder, A. Steinschneider (eds). Academic Press, New York, 15-28, 1983.
5. CANET E., GAULTIER C., D'ALLEST A.M., DEHAN M. – Effects of sleep deprivation on respiratory events during sleep in healthy infants. *J. Appl. Physiol.* 66, 1158-1163, 1989.
6. DALTVEIT A.K., IRGENS L.M., OYEN N., SKJAERVEN R., MARKESTADT T., ALM B., WENNERGREN G., NORVENIUS G., HELWEG-LARSEN K. – Sociodemographic risk factors for sudden infant death syndrome: associations with other risk factors. *Acta Paediatr.*, 77, 284-290, 1998.
7. DWYER T., PONSONBY A.L., GIBBONS L.E., NEWMAN N.M. – Prone sleeping position and SIDS: evidence from recent case-control and cohort studies in Tasmania. *J. Paediatr. Child Health*, 27, 340-343, 1991.
8. ENGELBERTS A.C., DEJONGE G.A., KOSTENSE P.J. - An analysis of trends in the incidence of sudden infant death syndrome in the Netherlands, 1969-1989. *J. Pediatr. Child Health*, 27, 329-333, 1991.
9. FRANCO P., GROSWASSER J., SOTTIAUX M., BROADFIELD E., KAHN A. – Decreased cardiac responses to auditory stimulation during prone sleep. *Pediatrics*, 97, 174-178, 1996.
10. FRANCO P., SZLIWOWSKI H., DRAMAIX M., KAHN A. – Polysomnographic study of the autonomic nervous system in potential victims of sudden infant death syndrome. *J. Clin. Autonom. Res.*, 8, 243-249, 1998.
11. FRANCO P., GROSSWASSER J., HASSID S., LANQUART J.P., SCAILLET S., KAHN A. – Prenatal exposure to cigarettes is associated with a decreased arousal in infants. *J. Pediatr.*, 135, 34-38, 1999.
12. FRANCO P., CHABANSKI S., SZLIWOWSKI H., DRAMAIX M., KAHN A. – Influence of maternal smoking on autonomic nervous system in healthy infants. *Pediatr. Res.*, 47, 215-220, 2000.
13. FRANCO P., SZLIWOWSKI H., DRAMAIX M., KAHN A. – Influence of ambient temperature on sleep characteristics and autonomic nervous system in healthy infants. *Sleep*, 23, 401- 407, 2000.
14. GILBERT R., RUDD P., BERRY P.J., FLEMING P.J., HALL E., WHITE D.G., OREFFO V.O.C., JAMES P., EVANS J.A. – Combined effects of infection and heavy wrapping on the risk of sudden unexpected infant death. *Arch. Dis. Child*, 67, 171-177, 1992.
15. GOZAL D., COLIN A.A., DASKALOVIC Y.I., JAFFE M. – Environmental overheating as a cause of transient respiratory chemoreceptor dysfunction in an infant. *Pediatrics*, 82, 738-740, 1988.
16. GUILLEMINAULT C., PERAITA R., SOUQUET M., DEMENT W.C. – Apneas during sleep in infants: possible relationship with sudden infant death syndrome. *Science*, 190, 677-679, 1975.
17. GUILLEMINAULT C., ARIAGNO R.L., FORNO L.S., NAGEL L., BALDWIN R., OWEN M. – Obstructive sleep apneas and near miss for SIDS: I. Report of an infant with sudden death. *Pediatrics*, 63, 837-843, 1979.
18. GUILLEMINAULT C., HELAT G., POWELL N., RILEY R.O. – Small airway in near-miss sudden infant death syndrome infants and their families. *Lancet*, 1, 402-407, 1987.
19. GUNTHEROTH W.G., SPIERS P.S. – Prolongation of the QT interval and the sudden infant death syndrome. *Pediatrics*, 103, 813, 1999.
20. HOFFMAN H.J., DAMUS K., HILLMAN L., KRONGARD E. – Risk factors for SIDS. Result of the National Institute of Child Health and Human Development SIDS Cooperative Epidemiological study. In: *The Sudden infant Death Syndrome*. P.J. Schwartz, D.P. Southall, M. Valdes-Dapena (eds), Annals of the New York Academy of Sciences, 533, 13-30, 1988.
21. KAHN A., GROSSWASSER J., REBUFFAT E., SOTTIAUX M., BLUM D., FOERSTER M., FRANCO P., BOCHNER A., ALEXANDER M., BACHY A., RICHARD P., VERGHOTE M., LE POLAIN D., WAYENBERG J.L. – Sleep and cardiorespiratory characteristics of infant victims of sudden death: a prospective case-control study. *Sleep*, 15, 287-292, 1992.
22. KAHN A., BLUM D., MULLER M.F., MONTAUX L., BOCHNER A., MONOD N., PLOUIN P., SAMSON-DOLLFUS D., DELAGREE E.H. – Sudden infant death syndrome in a twin: a comparison of sibling histories. *Pediatrics*, 93, 778-783, 1994.

23. KAHN A., GROSSWASSER J., SOTTIAUX M., KELMANSON I., REBUFFAT E., FRANCO P., DRAMAIX M., WAYENBERG J.L. – Prenatal exposure to cigarettes in infants with obstructive sleep apneas. *Pediatrics,* 93, 778-783, 1994.

24. KAHN A., REBUFFAT E., SOTTIAUX M., MULLER M.F., BOCHNER A., GROSWASSER J. – Prevention of airway obstructions during sleep in infants with breath-holding spells by means of oral belladonna. A prospective double-blind cross-over evaluation. *Sleep,* 14, 432-438, 1991.

25. KAHN A., HASAERTS D., BLUM D. – Phenothiazine-induced sleep apnea in normal infants. *Pediatrics,* 75, 844-847, 1985.

26. KAHN A., GROSSWASSER J., SOTTIAUX M., REBUFFAT E., FRANCO P., DRAMAIX M. – Prone or supine position and sleep characteristics in infants. *Pediatrics* 91, 1112-1115, 1993.

27. KAHN A., BAUCHE P., GROSSWASSER J., DRAMAIX M., SCAILLET S. – Working group of the Groupe Belge de Pédiatres Francophones. *Eur. J. Pediatr.* 160, 505-508, 2001.

28. KELLY D.H., GOLUB H., CARLEY D., SHANNON D.C. - Pneumograms in infants who subsequently died of sudden infant death syndrome. *J. Pediatr.* 109, 249-254, 1986.

29. KELLY D.H., SHANNON D.C. – Episodic complete airway obstruction in infants. *Pediatrics,* 67, 823-827, 1981

30. MITCHELL E.A., SCRAGG R., STEWART A.W., BECROFT D.M.O., TAYLOR B.J., FORD R.P.K., HASSALL I.B., BARRY D.M.J., ALLEN E.M., ROBERTS A.P. – Results from the first year of the New Zealand cot death study. *N.Z.Med. J.* 104, 71-76, 1991.

31. MITCHELL E.A., FORD H.J., STEWART A.W. – Snoring and the sudden infant death syndrome. *Pediatrics,* 91, 893-896, 1993.

32. MALLOY M.H., HOFFMAN H.J., PETERSON D.R. – Sudden infant death syndrome and maternal smoking. *Am. J. Public Health,* 82, 1380-1382, 1992.

33. MONOD N., PLOUIN P., STERNBERG B., PEIRANO P., PAJOT N., FLORES R., LINNETT S., KASTLER B., SCAVONE C., GUIDASCI S. – Are polygraphic and cardiopneumographic respiratory patterns useful tools for predicting the risk for sudden infant death syndrome? *Biol. Neonate,* 50, 147-153, 1986.

34. NAYE R.L. – Brain-stem and adrenal abnormalities in the sudden infant death syndrome. *Am. J. Clin. Path.,* 66, 526-530, 1976.

35. NAYE R.L., LADIS B., DRAGE J.S. – Sudden infant death syndrome: a prospective study. *Am. J. Dis. Child,* 130, 1207-1210, 1976

36. NELSON E.A.S., TAYLOR B.J., WEATHERALL I.L. – Sleeping position and infant bedding may predispose to hyperthermia and the sudden infant death syndrome. *Lancet,* i, 199-201, 1989.

37. ROBERTS J.L., MATHEW O.P., THACH B.T. – Upper airway obstruction and the sudden infant death syndrome. *Pediatr. Res.* 15, 729, 1981.

38. SCHAFFER A.T., LEMKE R., ALTHOFF H. – Airway resistance of the posterior nasal pathways in sudden infant death victims. *Eur. J. Pediatr.,* 150, 595-598, 1991.

39. SCHECHTMAN V.L., RAETZ S.L., HARPER R.K., HOFFMAN H.J., SOUTHALL D.P. – Dynamic analysis of cardiac R-R intervals in normal infants and in infants who subsequently succumbed to the sudden infant death syndrome. *Pediatr. Res.* 31, 606-612, 1992.

40. SCHWARTZ P.J. – Cardiac sympathetic innervation and the sudden infant death syndrome: a possible pathological link. *Am. J. Med.* 60, 167-172, 1976.

41. SENECAL J., ROUSSEY M., DEFAWE G., DELAHAYE M., PIQUEMAL B. – Prone position and unexpected sudden infant death. *Arch. Fr. Pediatr.* 44, 131-136, 1987.

42. SHATZ A., HISS J., ARENSBURG B. – Basement membrane thickening of the vocal cords in sudden infant death syndrome. *Laryngoscope,* 101, 484-486, 1991.

43. SLEBERT J.R., HAAS J.E. – Enlargement of the tongue in sudden infant death syndrome. *Pediatr. Pathol.,* 11, 813-826, 1991.

44. SOUTHALL D.P., RICHARDS J.M., STEBBENS V., WILSON A.J., TAYLOR V., ALEXANDER J.R. – Cardiorespiratory function in 16 full-term infants with sudden infant death syndrome. *Pediatrics,* 78, 787-796, 1986.

45. STANTON A.N. – Sudden infant death. Overheating and cot death. *Lancet,* 2, 1199-1201, 1984.

46. STEINSCHNEIDER A. – Prolonged apnea and the sudden infant death syndrome: clinical and laboratory observations. *Pediatrics,* 50, 654-664, 1972.

47. SWIFT P.G., EMERY J.L. – Clinical observation on response to nasal occlusion in infancy. *Arch. Dis. Child,* 48, 947-951, 1973.

48. TAKASHIMA S., ARMSTRONG D., BECKER L.E., BRYAN C. – Cerebral white matter lesions in sudden infant death syndrome. *Pediatrics,* 62, 155-159, 1978.

49. TONKIN S. – Sudden infant death syndrome: hypothesis of causation. *Pediatrics,* 55, 650-660, 1975.

50. TREACEY E.P., LAMBERT D.M., BARNES R. – Short-chain hydroxyacyl-coenzyme. A dehydrogenase deficiency presenting as unexpected infant death: a family study. *J. Pediatrics,* 137, 257-259, 2000.

51. VALDES-DAPENA M. – Sudden Infant Death syndrome: a review of the medical literature 1974-1979. *Pediatrics,* 66, 597-614, 1980.

52. VANDENPLAS Y., SACRE-SMITS L. – Continuous 24-hour esophageal pH monitoring in 285 asymptomatic infants 0-15 months old. *J. Pediatr. Gastroenterol. Nutr.* 6, 220-224, 1987.

53. WARD S.L.D., BAUTISTA D.B., KEENS T.G. – Hypoxic arousal response in normal infants. *Pediatrics,* 89, 860-864, 1992.

Chapter 55

Sleep related painful erections

U. Calvet

Laboratoire de Neurophysiologie Clinique, Clinique St. Jean Languedoc, Toulouse, France

Painful nocturnal erection is characterised by internal penile pain which only occurs during sleep-related erections [4, 9]. This is a rare disorder, often poorly understood and lacking in documentation. It specifically affects men: only one woman has been reported suffering from severe painful clitoral tumescence, and this was during wakefulness [1]. The most informed work on neurophysiological and neurochemical erectile mechanisms [5, 8] has shown the importance of intercavernous smooth muscle tonicity in controlling penile haemodynamics in the flaccid and erect states, but has failed to elucidate the origin of the syndrome. A hypothesis has been put forward of altered autonomic nervous control [2].

The rarity of published observations is no reflection of the incidence of this problem, which affects 1% of the population consulting for sexual disorders.

CLINICAL FEATURES

Essential or major features

- Painful nocturnal erection is characterised by internal penile pain, chiefly occurring during erections linked to REM sleep [9].

Repeated arousals occur during sleep, due to the unpleasant tension or even the unbearable pain of the erect penis.

The intensity of the pain and duration of arousals increase during the second half of the night. The erections last several minutes after awakening, but they may persist for several hours despite various attempts to make them disappear: walking about, cold showers, exercise bike etc.

The repeated arousals lead to sleep fragmentation, or even true insomnia, with the onset of fatigue, tension and irritability during the day.

- The erections are never painful during sexual relations or masturbation.

Associated features

Anxiety is consistently found - either focused on the genital organs and expressed as refractory testicular or pelvic pain, which has already been the motive of previous consultations, due to fear of a local degenerative disorder;

- or in a more diffuse, general form, with the usual list of somatic features which are more or less developed, colopathy, gastralgia, precordialgia and sleep disorders which, because of their refractory nature in the context of these painful erections, will have prompted the patient to follow psychotherapy or psychiatric treatment, well before the diagnosis is evoked.

Conjugal misunderstanding, may be at the forefront, as a trigger or reactive factor of the symptom, with an unsatisfactory sex life contrasting with the intensity of the reported nocturnal erections.

Local anatomical state is always normal, with no nodules, fibrosis or modification of penile elasticity.

LABORATORY TESTS

Plethysmography coupled with polysomnography

Diagnosis can be confirmed by full polysomnography and a recording of nocturnal erections using mercury gauge plethysmography, and a test of rigidity carried out over two nights (fig. 55.1a and b).

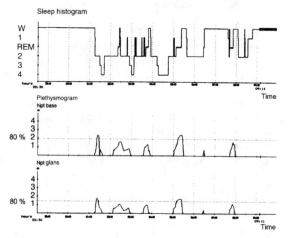

Figure 55.1a. Night 1. Long latency to sleep, five arousals in REM sleep. Six erections, one of which during NREM sleep, all abnormal: the subject wakes up before maximum tumescence. Npt (nocturnal penile tumescence), base (gauge placed around the penile base), glans (gauge placed behind the glans, at the coronal sulcus). The ordinates in the lower diagrams indicate circumference change.

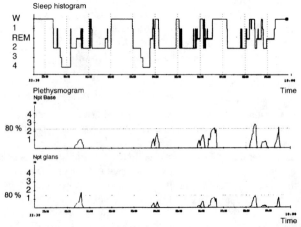

Figure 55.1b. Night 2. Short latency to sleep onset, increase of REM sleep phases. Fragmentation with six arousals in the last six periods of REM sleep. Erections always unstable with the subject waking up before maximum erection.

The erections are observed during each period of REM sleep and sometimes even during NREM sleep.

Each period of REM sleep is associated with an arousal reaction. Arousals usually occur during the period of tumescence, rarely during the peak plateau. After the arousal, the erection disappears or, exceptionally, becomes full.

The plethysmogramme is abnormal with unstable, fluctuating erections. A reduction is seen in the total time of peak plateau, with a reduction in the maximum number of erections.

Sleep is modified and fragmented, with an increased number of intermittent arousals, an increase in time spent awake between periods of sleep. The sleep efficiency index is reduced. Arousals lead to a reduction in the time spent in REM sleep while REM latency is normal [3].

Further tests

These are of no use (penile scan, vascular check-up, endocrine tests, cavernography).

COURSE OF ILLNESS

The onset of painful nocturnal erections occurs very early in the life of the couple, a few years after marriage, but can also be seen in single subjects: happy marriages in some cases and disharmonious relationships in others.

The onset of the symptom will lead to multiple consultations, sometimes even explorations, yielding neither diagnosis nor effective treatment.

If the relations change between the couple, intermittent erectile incapacity may occur. There may be spontaneous remissions lasting several months or years, often favoured by changes in the subject's emotional and sexual life. The symptom reappears as a reaction to trigger factors, which are also emotional.

The lapse of nearly 20 years between the onset of the symptom and the moment the diagnosis is made, confirms not only that the disorder is rare, but also that it is poorly understood.

CLINICAL VARIANT

Painful nocturnal erections are sometimes associated with prolonged erections during wakefulness. These are spontaneous erections, acting as a reflex to sensory stimulation (penis rubbing against underwear, for example) or vibratory stimulation (vibration of a moving car, truck or tractor).

Erections last for several hours and may sometimes last more than 24 hours.

These diurnal erections, when they are recurrent, are the more preoccupying. They spontaneously remit, giving way to painful nocturnal manifestations only.

DIFFERENTIAL DIAGNOSIS

Fibrosis of the cavernous body or Peyronie's disease

This is observed in older subjects. The pain is different. It involves an internal tension which occurs during sleep erections, but also and chiefly, during sexual relations. When the pain appears, a local anatomical modification is always noted, retraction, acquired bending, reduction in the volume of the erect penis, reduced elasticity, or sometimes even a palpable nodule.

The erections rapidly become unstable, short lasting, lead to sexual failure and complete impotence.

Nocturnal plethysmography shows unstable erections, with a shorter peak plateau and chiefly, altered base and distal erection compliance with no associated sleep disorder. A weak intracavernous injection of a vaso-active product, such as prostaglandin E1 will quickly confirm a diagnosis of local fibrosis.

PATHOPHYSIOLOGY

Painful nocturnal erection and the neurophysiological and neurochemical mechanisms of erection

Better understanding of the mechanisms of erection [5,8] may help to determine whether the disorder is local or central.

At local level

Permanent orthosympathetic hypertonia in the restful state will account for detumescence or flaccidity. During tumescence there is a reduction in orthosympathetic tonicity and parasympathetic cholinergic activation.

Tumescence is favoured by secretions at endothelial level of prostaglandins E1, nitrogen monoxide and calcitonine gene related peptide (CGRP).

Rigidity depends on a number of factors: venous and pericavernous compression, linked to intracavernous filling and the lengthening of the penis which diminishes the light of the venous plexus, and finally, perineal compression due to bulbocavernous and ischiocavernous muscle activity [11].

Ferini-Strambi *et al.* [2] suggested the hypothesis of a disorder of the autonomic nervous system during sleep, after observing 10 patients affected by painful nocturnal erections with reduced cardiac vagal tonicity during sleep. The hypothesis of an endothelial disorder has also been raised, as the intracavernous prostaglandin E1 injection [12] provoked long lasting erections, which were often painful, with internal tension similar to that reported by patients. A final possible explanation of the prolonged and painful nature of these erections is that of a disorder of bulbocavernous and ischiocavernous muscle control, with hyperactivation of the perineal chamber. Further study is warranted using a strain gauge [7] to provide a clear indication of peak pressure levels.

At central level

Certain dopaminergic or agonist substances may have a stimulating effect on libido, favouring prolonged diurnal erections but never painful nocturnal erections. Prolonged diurnal erections and priapisms have been observed with certain antidepressants or neuroleptics [15]. In women, the only observation of painful clitoral tumescence was secondary to taking a dopaminergic agonist, after stopping breast feeding [1].

Painful nocturnal erection and REM sleep

In men, 85 to 90% of REM phases are associated with erections. If painful nocturnal erection were dependant on REM sleep, suppressing the latter with drugs would cause nocturnal erections to disappear. Tricyclic antidepressants modify REM sleep and the erections associated to it, but there is a rapid escape phenomenon [10]. Brofaremine can suppress REM sleep without modifying sleep erections [13]. Some traumatic lesions of the central nervous system are accompanied by the disappearance of REM sleep, with no modification to erections [6]. These different observations suggest that the REM sleep-nocturnal erection relationship is not as close as it was thought to be.

Painful nocturnal erection, anxiety, pathology of the couple

Karacan reports a notion of anxiety, in his observation [4]. Other authors [9, 14] also refer to this. All our patients showed anxiety and the couples experienced difficulties.

TREATMENT

- The various types of medication proposed always prove ineffective.
  Anxiolytics (diazepam, clobazam) have been tried without success.

Antidepressants, clomapramine, amitriptyline, have no effect. Clozapine may be effective [14] but this still lacks confirmation.

Carbamazepine is also ineffective.

ENT vaso-constrictors (norephedrine) have no effect.

Beta blockers are effective but for short periods, with escape and secondary effects of impotence [9].

- Treatment must take the key elements into account i.e. anxiety and the nature of the relationship within the couple. Any treatment for anxiety, whether behavioural or chemical, will only be effective if the symptom is made to seem less dramatic and the patient is reassured.

Treating the couple helps to resolve some of the tension which clearly seems to generate and sustain the symptom.

Aggressive or surgical treatment should be avoided as this is dangerous and can lead to impotence.

CONCLUSION

Painful nocturnal erection is a somatic disorder in which anxiety is expressed through a symptom directly linked to a psychological problem in the subject's sex life, relating it to a conversion disorder.

Instead of reducing anxiety, it increases it, but with the advantage that it leads the subject to seek help for the complaint.

This illustrates the importance of decoding the symptom during the first consultation to avoid setting up heavy and intrusive additional tests: the only useful additional test in the aim of authenticating the syndrome and reducing the emotional connotation, is sleep plethysmography, but even this is not absolutely essential for diagnosis.

The medical treatment so far available has proved ineffective. A better understanding of the mechanisms of erection, both at local and central levels, should help in developing treatment for these patients.

REFERENCES

1. BLIN O., SCHWERTSCHLAG U.S., SERRATRICE G. – Painful clitoral tumescence during bromocriptine therapy. *Lancet*, 337, 1231-1232, 1991.
2. FERINI-STRAMBI L., MONTORSI F., ZUCCONI M., OLDANI A., SMIRNE S., RIGATTI P. – Cardiac autonomic nervous activity in sleep-related painful erections. *Sleep*, 19, 136-138, 1996.
3. FERINI-STRAMBI L., OLDANI A., ZUCCONI M., CASTRONOVO V., MONTORSI F., RIGATTI P., SMIRNE S. – Sleep-related painful erections: clinical and polysomnographic features. *J. Sleep Res.* 5, 195-197, 1996.
4. KARACAN I. – Painful nocturnal penile erections. *Sleep*, 19, 136-138, 1996.
5. KARACAN I., ASLAN C., HIRSHKOWITZ M. – Erectile mechanisms in man. *Science*, 220, 1080-1082, 1983.
6. LAVIE P. – Penile erections in a patient with near total absence of REM: a follow-up study. *Sleep, 13, 276-278, 1990.*
7. LAVOISIER P., PROULX J., COURTOIS F., DE CARUFEL F., DURAND L.G. – Relation between perineal muscle contractions, penile tumescence and penile rigidity during nocturnal erections. *J. Urol.*, 139, 176-179, 1988.
8. LAVOISIER P., ALOUI R., IWAZ J., KOKKIDIS M.J. – Considération sur la physiologie de l'érection pénienne. *Prog. Urol.*, 2, 119-127, 1992.
9. MATTHEWS B.J., CRUTCHFIELD M.B. – Painful nocturnal penile erections associated with rapid eye movement sleep. *Sleep*, 10, 184-187, 1987.
10. MITCHELL J.E., POPKIN M.K. – Antidepressant drug therapy and sexual dysfunction in men. A review. *J. Clin. Psychopharmacol.*, 3, 76-79, 1983.
11. SCHMIDT M.H., SCHMIDT H.S. – The ischiocavernous and bulbospongious muscles in mammalian penile rigidity. *Sleep*, 16, 171-183, 1993.
12. STACKL W., HASUN R., MARBERGER M. – Intracavernous injection of prostaglandin E1 in impotent men. *J. Urol.*, 140, 66-68, 1988.
13. STEIGER A., HOLSBOER F., BENKERT O. – Effects of brofaremine (CGP 11 305A), a short-acting, reversible, and selective inhibitor of MAO-A on sleep, nocturnal penile tumescence and nocturnal hormone secretion in three healthy volunteers. *Psychopharmacology*, (Berlin), 92, 110-114, 1987.
14. STEIGER A., BENKERT O. – Examination and treatment of sleep related painful erections. A case report. *Arch. of Sex Behav.*, 18, 263-267, 1989.

15. VELEK M., STANFORD G.K., MARCO J. – Priapism associated with concurrent use of thioridazine and metoclopramide. *Am. J. Psychiatry,* 144, 827-828, 1987.

PART 5

SLEEP AS A SPECIAL CIRCUMSTANCE IN INVESTIGATING SOME MEDICAL DISORDERS

Chapter 56

Mood disorders and sleep

M. Kerkhofs

Laboratoire de Sommeil, Centre Hospitalo-Universitaire de Charleroi, Hôpital A. Vésale, Montigny-le-Tilleul, Belgique

Disturbed sleep has long been recognised as one of the classic symptoms of mood disorder. Moreover, a high proportion of patients presenting sleep complaints are in fact suffering from an affective disorder [21]. Furthermore, sleep appears to sustain close links with mood disorders [54]. Indeed, if sleep is suppressed or reduced by total or partial deprivation, it proves to have antidepressant properties [26].

This chapter will begin by describing the sleep alterations related to mood disorders, with discussion of their specificities. This will be followed by the indications for polysomnographic examinations in affective disorders, and the interpretation of findings. The use of sleep recordings in predicting the response to antidepressant treatment will also be discussed, as will the interest of sleep studies as a tool for exploring the hypotheses and pathophysiological models relating to mood disorder.

SLEEP ALTERATIONS IN MOOD DISORDERS

A number of sleep alterations are observed in patients presenting mood disorders [13, 20, 28, 44, 45]. According to Reynolds and Kupfer [62] roughly 90% of depressed patients present sleep disturbances which can be objectified by polysomnographic recording (fig. 56.1). These disturbances are of three types: sleep continuity disturbances, sleep architecture disturbances and an alteration in the organisation of REM sleep (table 56.1). Sleep continuity disturbance is characterised as increased latency to sleep, increased number of arousals and early morning

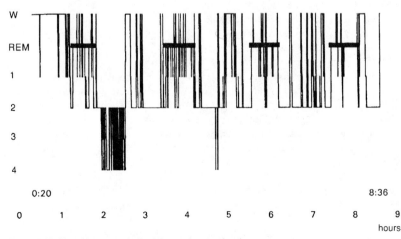

Figure 56.1. Sleep histogram obtained from a depressed patient.

awakening, leading to a reduction in total sleep time. A minority of patients (10-15%) present sleep of normal or prolonged duration, in association with a lack of energy and psychomotor slowing [14, 32]. In terms of sleep architecture, a reduction in NREM sleep (stages 3 and 4) has been demonstrated, particularly during the first period of NREM sleep. Quantified analyses of sleep EEG and of delta activity in particular, have served to confirm and clearly define these anomalies [3, 4]. Alterations are reported in the organisation of REM sleep, including its early onset (reduced REM sleep latency), a greater quantity of REM sleep at the beginning of the night, and increased eye movement activity. These alterations increase with age [27, 69] as well as with the severity of the depressive state [11, 38].

Various abnormalities are reported, depending on the type of depression. More specifically, depressives of the endogenous type are distinguishable from the reactive type in terms of their reduced REM sleep latency [18, 38, 59]. Moreover, according to Giles *et al.* [22] the reduction in REM sleep latency is linked to endogenous symptoms such as insomnia at the end of the night, the loss of pleasure, the absence of mood reactivity and the loss of appetite. However, contrary to Kupfer's original description [44, 45], polysomnographic recording data shows no distinction between primary and secondary depressives [70]. Psychotic depression relates, for its part, to a particular subset of endogenous depression, characterised by a weak response to tricyclic treatment and substantial psychomotor slowing. In terms of sleep, psychotic depression is distinguished from non-psychotic depression by a reduction in the quantity of REM sleep and eye movement activity, with nevertheless a higher frequency of very short REM sleep latencies (less than 20 min) even after testing for the effects of age, the severity of depressive state and restlessness [71].

Table 56.1. Sleep alterations in mood disorders

Sleep continuity
Sleep time ↘ : ↗ latency to sleep, ↗ nocturnal arousals, early morning awakening.
Sleep time = or ↗ : in 10-15% of patients (young, bipolar)
Sleep architecture
↘ NREM sleep (stages 3 + 4) especially during the first cycle of sleep.
REM sleep
↘ latency to the first phase of REM sleep,
↗ REM sleep at the beginning of the night with prolonged first phase of REM sleep
↗ eye movement activity

SPECIFICITY OF ABNORMALITIES

Sleep abnormalities have also been reported in other psychiatric disorders (table 56.2). Sleep continuity disturbance as well as a reduction in deep NREM sleep have been described in anxiety disorders [61, 64], obsessive syndrome [35], schizophrenia [11, 18, 23, 33, 39], alcoholism [30] and dementia [63]. Reduced REM sleep latency has been observed in patients presenting with obsessive syndrome [35] and, for certain authors, in schizophrenic patients [76] although others reported no abnormalities [29, 39]. In patients presenting a syndrome of generalised anxiety, sleep continuity disturbance and reduced NREM sleep have been observed [61, 64] with no abnormalities in REM sleep latency (fig. 56.2). Sleep continuity disturbances and reduced NREM sleep have also been reported in patients affected by panic disorder [16, 73]. However considerable discrepancies have been observed in regard to REM sleep latency: reduced REM sleep latency observed in certain cases has been linked to the presence of a personal and/or family history of depression [16]. Elsewhere, the most detailed study of the first cycle of sleep comparing depressed patients with controls and patients suffering from panic attacks, reported depressed patients as having less NREM sleep, more arousals as well as a shorter first sleep cycle, while REM sleep latency was reduced in both groups of patients [50].

As a conclusion, REM sleep abnormalities are not clearly specific to mood disorders. However it would appear that the simultaneous presence of reduced REM sleep latency and a prolonged first phase of REM sleep are more characteristic of mood disorders [62]. According to Benca *et al.* [6], no sleep variable appears to be completely specific to a given psychiatric disorder, even if affective disorders are recognised as focusing a maximum number of important abnormalities [6].

Table 56.2. Specificity of sleep alterations observed in mood disorders: presence of these alterations in other psychiatric disorders.

	↘ Sleep time	↘ NREM sleep	↘ Latency to 1st phase of REM sleep
Anxiety disorders	+	-	-
Panic disorders	+	-?	-?
Obsessive syndromes	+	+	+
Schizophrenia	+	+?	?
Alcoholism	+	+	?
Dementia	+	+	-
Eating disorders	+?	-	-

+ present
- absent
? controversial

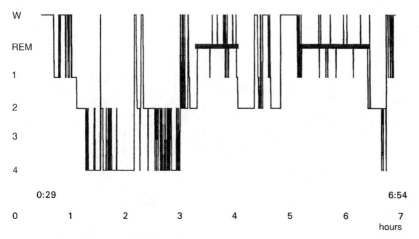

Figure 56.2. Sleep histogram obtained in a patient presenting a generalised anxiety disorder.

POLYSOMNOGRAPHIC EXAMINATIONS - INDICATIONS, REALISATION AND INTERPRETATION FOR PATIENTS PRESENTING MOOD DISORDERS

A polysomnographic examination may be indicated for a patient presenting either a mood disorder with doubtful diagnosis or resistant and/or recurrent depression. Furthermore the classic indication for polysomnographic examination for suspected apnoeas and/or periodic leg movements remain applicable. Although sleep apnoeas syndromes are no more frequent in patients affected by mood disorders, it is important to exclude this pathology in patients with suspected depression, as depressive characteristics (lack of energy, fatigue, sadness, low spirits, irritability) are sometimes the chief complaints of the apnoeic patient, and may wrongly orientate the diagnosis toward one of a depressive state.

For polysomnographic examinations to be correctly interpreted, they must be carried out in strict conditions of medical withdrawal. The recommended duration of suspension of treatment is two weeks [62]. This refers to all psychotropes (antidepressants, neuroleptics, hypnotics, anxiolytics) in view of their well known effect on sleep architecture [12].

It is also crucial to check that there are no somatic pathologies before sleep recording, as these may alter sleep in one way or another.

It is usually advisable to carry out recordings over two consecutive nights (at a minimum) to objectify any variability caused by rebound effects secondary to sleep destructuration: a night of almost total insomnia may be followed by a night of better sleep quality but one which includes a rebound of REM sleep at the beginning of the night expressed by reduced REM sleep latency (fig.56.3). Note that Ansseau *et al.* [2] advise extending the recordings over three nights to obtain better sensitivity.

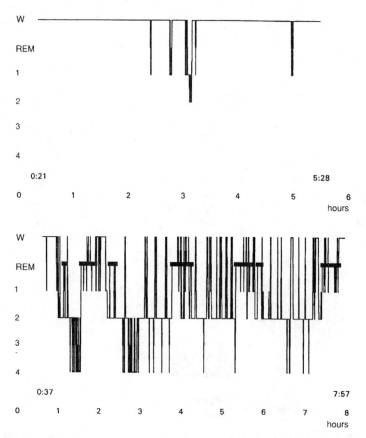

Figure 56.3. Illustration of recording variability from one night to another: sleep histograms of two consecutive nights recorded for the same depressed patient.

It is also important to exclude daytime sleepiness, which is a relatively frequent phenomenon [67, 41], in view of its potential effect on sleep structure the following night [1, 20, 38].

Polysomnographic data must be interpreted within the context of the clinical findings. To do so, the tools of diagnosis and evaluation must be valid and reliable. Moreover, it is essential to take the age of the patient into account and refer to the norms pertaining to this. For REM sleep latency, Ansseau *et al.* [2] suggested following the "90 minute rule" proposed by Kupfer *et al.* [47], by which a significant reduction of REM sleep latency should only be recorded if age plus REM sleep latency add up to sum of under 90.

The sensitivity of this abnormality is limited and a certain proportion of patients affected by mood disorders show no REM sleep latency abnormalities.

Moreover, it is important to take into account any alterations in REM sleep latency or other sleep variables both during the episode and when the depressive state is in remission. Indeed, sleep disturbances seem to be more pronounced at the beginning of the episode [15]. Also, where there are no changes in clinical state, sleep variables remain stable in patients treated with placebo [13]. Several studies [24, 65] have shown that sleep abnormalities, particularly reduced REM sleep latency, persist during remission. These sleep abnormalities may thus be a characteristic marker of depression. Consequently, reduced REM sleep latency may be the sign of a subject's vulnerability to affective disorder, as opposed to indicating the symptomatic episode itself [62]. However one cannot dismiss the notion of reduced REM sleep latency as a consequence of a depressive state which will take longer to return to normal than the other symptoms of the depression. Finally, when this abnormality is encountered in non-affected subjects (past or present) it may indicate a potential risk of developing the illness [24, 62]. This question will only be elucidated by longitudinal studies including sleep evaluations during, after and before an episode of depression [43, 51].

POLYSOMNOGRAPHIC RECORDINGS AND ANTI-DEPRESSANT TREATMENT

Polysomnographic recordings have provided an accurate picture of the effects of antidepressant treatment on sleep continuity and architecture. These produce a rapid reduction of REM sleep and lead to prolonging REM sleep latency, and reducing eye movement activity [12, 22, 42, 60]. Furthermore, polysomnographic recordings carried out before and at the start of treatment may help to indicate the likely therapeutic response. Indeed, Kupfer *et al.* [48] in a study of 82 depressed patients, reported that reduced REM sleep and prolonged REM sleep latency during the first two nights of treatment allowed them to predict the response to amitryptiline treatment. Gillin *et al.* [31] also observed that a reduction of REM sleep at the onset of treatment provided a good indication of likely clinical response. Similarly, Hochli *et al.* [34] showed a significant correlation between the early reduction of REM sleep induced by clomipramine treatment and potential clinical response. Rush *et al.* [65] showed that reduced REM sleep latency (latency under 65 min) before treatment predicted a positive response to tricyclic antidepressant treatment. According to Mendlewicz *et al.* [58] depressed patients who respond to treatment by amitriptyline present less stage 4 in the polysomnographic recordings made before treatment.

Sleep disturbances are also a means of predicting the response or lack of response to psychotherapy. According to Thase *et al.* [72] patients presenting a profile of disturbed sleep respond less well to psychotherapy.

PATHOPHYSIOLOGICAL MODELS

Polysomnographic recordings carried out on patients presenting mood disorders have yielded hypotheses or pathophysiological models of depression and its accompanying sleep or biological rhythm disturbances. Several models have been proposed, to this effect, the most important of which are set out as follows.

The cholinergic-aminergic imbalance hypothesis

Clinical and pharmacological observations have suggested that in depression, cholinergic activity may be increased in relation to noradrenergic activity [36]. On the other hand, several studies suggest that REM sleep can be triggered by cholinergic mechanisms, localised in the pontine reticular formation and inhibited by aminergic mechanisms [29, 56]. It is thus possible that in mood disorders an increase in cholinergic activity lies at the origin of reduced REM sleep latency and perhaps of reduced sleep time [55, 68]. Sitaram *et al.* [63] and Gillin *et al.* [28] infused arecoline during the second period of NREM sleep and observed that the second episode of REM sleep began earlier in depressed patients than in normal patients. The difference persisted when the experience was carried out during a period of remission, indicating that hypersensitivity persists in remission. Sleep manipulation by administering muscarinic agonist RS-86 yielded similar results [7].

Chronobiological models

According to Wehr *et al.* [75] disturbances in the temporal organisation of REM sleep in depression results from a phase-advance in the circadian oscillator which regulates the rhythms of REM sleep, temperature and cortisol plasma level. Phase-advance of these rhythms is thought to be present in some patients [52, 53], although this is not systematic. Moreover it appears that advancing the hour of bedtime may have an antidepressant effect; as has been cited as evidence in favour of phase-advance [75]. However, Elsenga and Van Den Hoofdakker [17], failed to confirm the antidepressant effects of phase-advance. It is important to emphasise that sleep itself is likely to influence temperature and possibly cortisol rhythms. Avery *et al.* [5] reported a reduction in the amplitude of circadian temperature rhythm with a higher mean temperature nadir. For Schulz and

Lund [66] the alterations of REM sleep in depression may be explained by a reduction in wakeful circadian rhythm amplitude. This possibility was also evoked after observing the occurrence of episodes of daytime sleepiness in depressed patients [41].

The S-deficiency hypothesis

The model proposed by Borbély [8] suggests that sleep is dependent upon two processes: a homeostatic, sleep-inducing process (termed process "S"), and a circadian process ("C"). According to Borbély and Wirz Justice [9], the S process is deficient in patients presenting mood disorders. This model suggests that, during wakefulness, the S process in subjects with depression does not reach the level observed in healthy subjects. This deficiency is shown by the reduction in sleep time and NREM sleep as well as that of slow wave activity (particularly during the first phase of NREM sleep) [8]. It may be that reduced NREM sleep at the beginning of the night allows for the earlier onset of REM sleep. Moreover, this hypothesis is also supported by the beneficial effects of total or partial sleep deprivation on depressive symptomatology. In fact process S is increased by prolonging wakefulness.

CONCLUSIONS

Polysomnographic recordings carried out on patients suffering from mood disorders have allowed us to identify a group of sleep alterations. However, the specificity of these abnormalities is not clear in view of observations of similar abnormalities in other psychiatric disorders. This warrants more in depth study of sleep alterations during the first sleep cycle. Quantified EEG analyses are a promising development in specifying these anomalies. Longitudinal studies are indispensable for a better understanding of the abnormalities observed, both in the light of the pathophysiological hypotheses developed and in terms of the potential predictive value of sleep disturbances observed in mood disorders.

REFERENCES

1. ABER W., WEBB W.B. – Effects of a limited nap on night sleep in older subjects. *Psychology and Aging,* 1, 4, 300-302, 1986.
2. ANSSEAU M., VON FRENCKELL R., FRANCK G., REYNOLDS III C.F., KUPFER D.J. – Sommeil et dépression: vers une standardisation de l'utilisation de la latence du sommeil paradoxal en tant que marqueur biologique de la dépression majeure. *Rev. E.E.G. Neurophysiol. Clin.,* 17, 411-424, 1987.
3. ARMITAGE R., CALHOUN J.S., RUSH A.J., ROFFWARG H.P. – Comparison of the delta EEG in the first and second non-REM periods in depressed adults and normal controls. *Psychiatry Res.,* 41, 65-72, 1992.
4. ARMITAGE R., HOFFMANN R., TRIVEDI M., RUSH A.J. – Slow-wave activity in NREM sleep: sex and age effects in depressed outpatients and healthy controls.*Psychiatry Res.,* 95, 201-213, 2000.
5. AVERY D., WILDSCHODTZ G., RAFAELSEN O. – REM latency and temperature in affective disorder before and after treatment. *Biol. Psychiatry,* 17, 4, 463-470, 1982.
6. BENCA R.M., OBERMEYER W.H., THISTED R.A., GILLIN J.CH. – Sleep and psychiatric disorders. A metaanalysis. *Arch. Gen. Psychiatry,* 49, 651-668, 1992.
7. BERGER M., RIEMANN D., HOCHLI D., SPIEGEL R. – The cholinergic rapid eye movement sleep induction test with RS-86. *Arch. Gen. Psychiatry,* 46, 421-428, 1989.
8. BORBELY A.A. – A two process model of sleep regulation. *Hum. Neurobiol.,* 1, 195-204, 1982.
9. BORBELY A.A., WIRZ-JUSTICE A. – Sleep, sleep deprivation and depression. *Hum. Neurobiol.,* 1, 205-210, 1982.
10. BORBELY A.A. – The S-Deficiency hypothesis of depression and the two-process model of sleep regulation. *Pharmacopsychiat.,* 20, 23-29, 1987.
11. CALDWELL D.F., DOMINO E.F. – Electroencephalographic and eye-movement pattern during sleep in chronic schizophrenic patients. *Electroenceph. Clin. Neurophysiol.* 22, 414-420, 1967.
12. CHEN C.N. – Sleep, depression and antidepressants. *Brit. J. Psychiat.* 135, 385-402, 1979.
13. COBLE P.A., FOSTER F.G., KUPFER D.J. – Electroencephalographic sleep diagnosis of primary depression. *Arch. Gen. Psychiatry,* 33, 1124-1127, 1976.
14. DETRE T., HIMMELHOCH J., SWARTZBURG M., ANDERSON C.M., BYCK R., KUPFER D.J. – Hypersomnia and manic-depressive disease. *Amer. J. Psychiat.,* 128, 10, 1303-1305, 1972.
15. DEW M.A., REYNOLDS Ch.F., BUYSSE D.J., HOUCK P.R., HOCH C.C., MONK T.H., KUPFER D.J. – Electroencephalographic sleep profile during depression. *Arch. Gen. Psychiatry,* 53, 148-156 , 1996.
16. DUBE S., JONES D.A., BELL J., DAVIES A., ROSS E., SITARAM N. – Interface of panic and depression: Clinical and sleep EEG correlates. *Psychiatry Res.,* 19, 119-133, 1986.

17. ELSENGA S., VAN DEN HOOFDAKKER R.H. – Clinical effects of several sleep-wake manipulations on endogenous depression. *Sleep Res.* 12, 326, 1983.

18. FEINBERG I., BRAUN M., KORESKO R.L., GOTTLIEB F. – Stage 4 sleep in schizophrenia. *Arch. Gen. Psychiat.,* 21, 262-266, 1969.

19. FEINBERG M., GILLIN J.C., CARROLL B.J., GREDEN J.F., ZIS A.P. – EEG studies of sleep in the diagnosis of depression. *Biol. Psychiatry,* 17, 305-316, 1982.

20. FEINBERG I., MARCH J.D., FLOYD T.C., JIMISON R., BOSSOM-DEMITRACK L., KATZ P.H. – Homeostatic changes during post-nap sleep maintain baseline levels of delta EEG. *Electroenceph. Clin. Neurophysiol.* 61, 134-137, 1985.

21. FORD D.E., KAMEROW D.B. – Epidemiologic study of sleep disturbances and psychiatric disorders: an opportunity for prevention? *JAMA,* 262, 1479-1484, 1989.

22. GAILLARD J.M. – Sommeil et psychotropes. *J. Pharmacol.,* 15, 4, 389-399, 1984.

23. GANGULI R., REYNOLDS III Ch.F., KUPFER D.J. – Electroencephalographic sleep in young never-medicated schizophrenics. *Arch. Gen. Psychiatry,* 44, 36-44, 1987.

24. GILES D.E., JARRETT R.B., ROFFWARG H.P., RUSH A.J. – Reduced rapid eye movement latency: a predictor for recurrence in depression. *Neuropsychopharmacology,* 1, 1, 33-39, 1987.

25. GILES D.E., ROFFWARG H.P., SCHLESSER M.A., RUSH A.J. – Which endogenous depressive symptoms relate to REM latency? *Biol. Psychiatry,* 21, 473-482, 1986.

26. GILLIN J.Ch. - The sleep therapies of depression. *Prog. Neuro-Psychopharmacol. Bol. Psychiat.,* 7, 351-364, 1983.

27. GILLIN J. Ch., DUNCAN W.C., MURPHY D.L., POST R.M., WEHR T.A., GOODWIN F.K., WYATT R.J., BUNNEY Jr.W.E. – Age related changes in sleep in depressed and normal subjects. *Psychiatry Res.* 4, 73-78, 1981.

28. GILLIN J.Ch., DUNCAN W.C., PETTITGREW K.D., FRANKEL B.L., SNYDER F. – Successful separation of depressed, normal and insomniac subjects by EEG sleep data. *Arch. Gen. Psychiatry,* 36, 85-90, 1979.

29. GILLIN J.Ch., SITARAM N., DUNCAN W.C. – Muscarinic supersensitivity: A possible model for the sleep disturbance of primary depression? *Psychiatry Res.,* 1, 17-22, 1978.

30. GILLIN J.Ch., SMITH T.L., IRWIN M., KRIPKE D.F., SCHUCKIT M.- EEG sleep studies in « pure » primary alcoholism during subacute withdrawal: relationships to normal controls, age and other clinical variables. *Biol. Psychiatry,* 27, 477-488, 1990.

31. GILLIN J.Ch.WYATT R.J., FRAM D., SNYDER F. – The relationship between changes in REM sleep and clinical improvement in depressed patients treated with amitriptyline. *Psychoparmacology,* 59, 267-272, 1978.

32. HAWKINS D.R., TAUB J.M., VAN DE CASTLE R.L. – Extended sleep (hypersomnia) in young depressed patients. *Am. J. Psychiatry,* 142, 905-910, 1985.

33. HIATT J.F., FLOYD T.C., KATZ P.H., FEINBERG I. – Further evidence of abnormal non-rapid-eye-movement sleep in Schizophrenia. *Arch. Gen. Psychiatry,* 42, 797-802, 1985.

34. HOCHLI D., RIEMANN D., ZULLEY J., BERGER M. – Initial REM suppression by clomipramine: a prognostic tool for treatment response in patients with a major depressive disorder. *Biol. Psychiatry,* 21, 1217-1220, 1986.

35. INSEL Th.R., GILLIN J.Ch., MOORE A., MENDELSON W.B., LOEWENSTEIN R.J., MURPHY D.L. – The sleep of patients with obsessive-compulsive disorder. *Arch. Gen. Psychiatry,* 39, 1372-1377, 1982.

36. JANOWSKY D.L., EL-YOUSEF M.K., DAVIS J.M. – A cholinergic-adrenergic hypothesis of mania and depression. *The Lancet, ii,* 632-635,1972.

37. KARACAN I., FINLEY W., WILLIAMS R.L., HURSCH C.J. – Changes in stage 1 – REM and stage 4 sleep during naps. *Biol. Psychiatry,* 2, 391-399, 1970.

38. KARACAN I., WILLIAMS R.L., FINLEY W.W., HURSCH C.J. – The effects of naps on nocturnal sleep: Influence on the need for stage-1 REM and stage 4 sleep. *Biol. Psychiatry,* 2, 391-399, 1970.

39. KEMPENAERS C., KERKHOFS M., LINKOWSKI P., MENDLEWICZ J. – Sleep EEG variables in young schizophrenic and depressive patients. *Biol. Psychiatry,* 24, 828-833, 1988.

40. KERKHOFS M., KEMPENAERS C., LINKOWSKI P., DE MAERTELAER V., MENDLEWICZ J. – Multivariate study of sleep EEG in depression. *Acta Psychiatr. Scand.,* 77, 463-468, 1988.

41. KERKHOFS M., LINKOWSKI P., LUCAS F., MENDLEWICZ J. – 24-hour patterns of sleep in depression. *Sleep,* 14, 6, 501-506, 1991.

42. KERKHOFS M., MENDLEWICZ J. – The effects of antidepressant treatment on sleep disturbances in depression. *Eur. Psychiatry,* 10, 391-396, 1995.

43. KRIEG J.C., LAUER C.J., SCHREIBER W., MODELL S., HOLSBOER F. – Neuroendocrine, polysomnographic and psychometric observations in healthy subjects at high familial risk for affective disorders: the current state of the « Munich vulnerability study ». *J. of Affective Dis.,* 62, 33-37, 2001.

44. KUPFER D.J. – REM latency: A psychobiologic marker of primary depressive disease. *Biol. Psychiatry,* 11, 159-174, 1976.

45. KUPFER D.J., FOSTER F.G. – *EEG sleep and depression. Sleep disorders diagnosis and treatment.* R.L. Williams, I. Karacan (eds), John Wiley, New York, 1978.

46. KUPFER D.J., FOSTER F.G., COBLE P., McPARTLAND R.J., ULRICH R.F. – The application of EEG sleep for the differential diagnosis of affective disorders. *Am. J.Psychiatry,* 135, 69-74, 1978.

47. KUPFER D.J., REYNOLDS III C.F., ULRICH R.F., SHAW D.J., COBLE P.A. – EEG sleep, depression and aging. *Neurobiol. Aging.,* 3, 351-360, 1982.

48. KUPFER D.J., SPIKER D.G., COBLE P.A., NEIL J.F., ULRICH R.F., SHAW D.H. – Depression, EEG sleep and clinical response. *Compr. Psychiatry,* 21, 3, 212-220, 1980.

49. KUPFER D.J., SPIKER D.G., COBLE P.A., NEIL J.F., ULRICH R.F., SHAW D.H. – Sleep and treatment prediction in endogenous depression. *Am. J. Psychiatry,* 138, 429-434, 1981.

50. LAUER J. Ch., KRIEG J.Ch., GARCIA-BORREGUERO D., OZDAGLAR A., HOLSBOER F. – Panic disorder and major depression: a comparative electroencephalographic sleep study. *Psychiatry Res.,* 44, 41-54, 1992.

51. LAUER J.Ch., SCHREIBER W., HOLDSBOER F., KRIEG J.Ch. – In quest of identifying vulnerability markers for psychiatric disorders by all-night polysomnography. *Arch. Gen. Psychiatry,* 52, 145-153, 1995.

52. LINKOWSKI P., MENDELWICZ J., KERKHOFS M., LECLERCQ R., GOLDSTEIN J., BRASSEUR M., COPINSCHI G., VAN CAUTER E. – 24-hour profiles of adrenocorticotropin, cortisol and growth hormone in major depressive illness: effect of antidepressant treatment. *J. Clin. Endocrinol. Metab.,* 65, 141-151, 1987.

53. LINKOWSKI P., MENDLEWICZ J., LECLERCQ R., BRASSEUR M., HUBAIN P., GOLDSTEIN J., COPINSCHI G., VAN CAUTER E. – The 24 hour profile of adrenocorticotropin and cortisol in major depressive illness. *J. Clin. Endocrinol. Metab.,* 61, 429-438, 1987.

54. LUSTBERG L., REYNOLDS Ch.F.III – Depression and insomnia: questions of cause and effect. *Sleep Medicine Rev.,* 4, 253-262, 2000.

55. McCARLEY R.W. – Sleep and depression: common neurobiological control mechanisms. *Am. J. Psychiatry,* 139, 565-570, 1982.

56. McCARLEY R.W., HOBSON J.A. – Sleep-cycle oscillation: reciprocal discharge by two brainstem neuronal groups. *Science,* 189, 55-60, 1975.

57. McCARLEY R.W., MASSAQUOI S.G. – A limit cycle mathematical model of the REM sleep oscillator system. *Amer. J. Physiol.,* 251, R1011-R1029, 1986.

58. MENDELWICZ J., KEMPENAERS C., DE MAERTELAER V. – Sleep EEG and amitriptyline treatment in depressed inpatients. *Biological Psychiatry,* 30, 691-702, 1991.

59. MENDELWICZ J., KERKHOFS M. – Sleep EEG in depressive illness: A world health organization collaborative study. *Brit. J. Psychiatry,* 159, 505-509, 1991.

60. OBERNDORFER S., SALETU-ZYHLARZ G., SALETU B. – Effects of selective serotonin inhibitors on objective and subjective sleep quality. *Neuropsychobiology,* 42, 69-81, 2001.

61. PAPADIMITRIOU G.N., KERKHOFS M., KEMPENAERS C., MENDELWICZ J. – EEG sleep studies in patients with generalized anxiety disorder. *Psychiatry Res.,* 26, 183-190, 1988.

62. REYNOLDS III Ch.F., KUPFER D.J. - Sleep research in affective illness: state of the art circa 1987. *Sleep,* 10, 3, 199-215, 1987.

63. REYNOLDS III Ch.F., KUPFER D.J., TASKA L.S., HOCH C.C., SPIKER D.G., SEWICH D.E., ZIMMER B., MARIN R.S., NELSON J.P., MARTIN D., MORYCZ R. – EEG sleep in elderly depressed, demented and healthy subjects. *Biol. Psychiatry,* 20, 431-442, 1985.

64. REYNOLDS III Ch.F., SHAW D.H., NEWTON T.F., COBLE P.A., KUPFER D.J. – EEG sleep in outpatients with generalized anxiety: a preliminary comparison with depressed outpatients. *Psychiatry Res.* 8, 81-89, 1983.

65. RUSH A.J., GILES D.E., JARRETT R.B., PELDMAN-KOFFLER F., DEBUS J.R., WEISSENBURGER J., ORSULAK P.J., ROFFWARG H.P. – Reduced REM latency predicts response to tricyclic medication in depressed outpatients. *Biol. Psychiatry,* 26, 61-72, 1989.

66. SCHULZ H., LUND R. – On the origin of early REM episodes in the sleep of depressed patients: A comparison of three hypothesis. *Psychiatry Res.* 16, 65-77, 1985.

67. SHIMIZU A., HIYAMA H., YAGASAKI A., TAKAHASHI H., FUYIKI A., YOSHIDA I. – Sleep of depressed patients with hypersomnia: A 24-h polygraphic study. *Waking and Sleeping,* 3, 335-339, 1979.

68. SITARAM N., NURNBERGER J.L., GERSHON E.S., GILLIN J.C. – Cholinergic regulation of mood and REM sleep: a potential model and marker for vulnerability to depression. *Am. J. Psychiatry,* 139, 571-576, 1982.

69. THASE M.E., HIMMELHOCH J.M., MALLINGER A.G., JARRETT D.B., KUPFER D.J. – Sleep EEG and DST findings in anergic bipolar depression. *Am. J. Psychiatry,* 146, 3, 329-333, 1986.

70. THASE M.E., KUPFER D.J., SPIKER D.G. – Electroencephalographic sleep in secondary depression: A revisit. *Biol. Psychiatry,* 19, 6, 805-810, 1984.

71. THASE M.E., KUPFER D.J., ULRICH R.F. – Electroencephalographic sleep in psychotic depression. A valid subtype? *Arch. Gen. Psychiatry,* 43, 886-893, 1986.

72. THASE M.E., SIMONS A.D., REYNOLDS Ch.F. III – Abnormal electroencephalographic sleep profile in major depression. *Arch. Gen. Psychiatry,* 53, 99-108, 1996.

73. UHDE Th.W., ROY-BYRNE P., GILLIN J. Ch., MENDELSON W.B., BOULENGER J.Ph., VITTONE B.J., POST R.M. – The sleep of patients with panic disorder: A preliminary report. *Psychiatry Res.* 12, 251-259, 1984.

74. ULRICH R.F., SHAW D.H., KUPFER D.J. – Effects of aging on EEG sleep in depression. *Sleep,* 3, 31-40, 1980.

75. WEHR T.A., WIRZ-JUSTICE A., GOODWIN F.K., DUNCAN W., GILLIN J.C. – Phase advance of the sleep-wake cycle as an antidepressant. *Science,* 206, 710-713, 1979.

76. ZARCONE V.P., BENSON K.L., BERGER P.A. – Abnormal rapid eye movement latencies in schizophrenia. *Arch. Gen. Psychiatry,* 44, 45-48, 1987.

Chapter 57

Sleep and lesions in the central nervous system

A. Autret

Clinique Neurologique, Hôpital Bretonneau, Tours, France

The clinical and polygraphic phenomenology of the different sleep-wake states and their alternation, clearly rely on the anatomical and functional integrity of the neuronal network. Hence the pathologies affecting these structures may alter sleep. Conversely NREM or REM sleep, through their accompanying electrophysiological and biochemical alterations, can alter or even create the semiology of certain neurological conditions.

In practice, the study of sleep in neurological patients can take different forms. The first relates to anatomical functions, and consists of looking for sleep-wake abnormalities in human subjects with lesions similar to those experimentally created in animals, in order to validate in man the theories developed in animals. The second approach is to search for sleep-wake abnormalities within a given context of illness, in order to better understand their pathophysiology. The third approach is different, consisting of observing the alterations of an illness or symptomatology during the different sleep-wake states and interpreting these in neurochemical and neurophysiological terms.

One must proceed with great caution in establishing any specific relationship between sleep-wake disturbances and a given pathology. Indeed, sleep-wake states are highly sensitive to external factors, leading to wide intra- and inter-individual variability [54]. Outside factors play a major role in pathological conditions: discomfort created by the symptoms, associated pathology, anxiety, additional treatment etc. These factors of confusion are difficult to control. In practice, it is necessary to use the closest possible group of references to the studied pathology. Finally, trying to establish a correlation between the morphological abnormalities revealed either by imagery or anatomical examination and electrophysiological data is a perilous exercise. Indeed, certifying the presence of abnormalities does not automatically imply dysfunction, just as morphological normality does not preclude dysfunction.

In this chapter, we will analyse the sleep studies carried out in the different contexts of neurological pathology and attempt to assess the value of including polysomnographic recording in the examination of central nervous system lesions.

VASCULAR PATHOLOGY

Acute phase

During the acute phase of a cerebrovascular accident, or stroke, the presence of a sleep-wake disorder is of clinical interest. Hypovigilance is part of a picture of cerebral involvement, whether cingulated, central or temporal. The most common explanation given is the bilateral dysfunction of the arousal systems due to a causal lesion and the phenomena of distal and controlateral compression [128]. But hypovigilance may also be due to a haemorrhagic or ischaemic process interrupting the arousal system at the midbrain reticular formation. It is thus part of a basilar artery syndrome. However this is usually a transitory disorder, like that observed in animals after lesion in this region [81], no doubt due to the rapidity with which the neurochemical replacement systems operate.

An early reduction of stage 2 signifies a poor prognosis [70]. In sylvian artery malacia, the reported increase in NREM and decrease in REM has been noted to be more significant in right lesions than in left lesions, suggesting that the right hemisphere is involved in regulating REM sleep [86]. REM sleep is reduced in relation to the degree of malacia [61] and is transitory [70, 71].

Sleep breathing disorders are observed with a high prevalence just after a stroke [41, 19, 62]. An obstructive sleep apnoea hypopoea syndrome (OSAHS) is the most frequent anomaly and is associated with a high mortality at 4 years [41] and a poorer functional outcome [62]. It is not yet known if the OSAHS precedes or is the consequence of the stroke. The persistence of a high level of obstructive apnoeas (OA) three months after a stroke with a decreasing level of central apnoea favours the idea that OA are more likely to be associated with the cause of the stroke, and central apnoea with its consequence [125].

Anatomo-electrophysiological correlations in the sequela phase

The role of the cerebral cortex has been studied on the basis of several recordings of a case of laminar necrosis of the entire cerebral cortex [15]. In this case, the disappearance of NREM sleep and spindles and the disturbance of ultradian cyclic organisation indicate the importance of the cortex in their formation.

Insomnia has been reported after bilateral [147] or left [58] thalamic ischaemia. These two observations support the notion of Villablanca and Marcus [165] for whom the reduction of total sleep time in athalamic or diencephalic cats is caused by the disappearance of rostral hypnogenic influences. It should be noted that these lesions probably affected the subthalamic region interrupting the connections between the anterior and posterior hypothalamus which play a major role in inhibiting arousal activity [141].

Following a bilateral paramedian thalamic infarct, several studies mention the occurrence of hypersomnia on clinical grounds [44, 59, 30] or after actigraphic measurement [20]. These patients present behavioural abnormalities: they are apragmatic, aboulic, remain immobile or supine for long periods, in a sleeping posture, with eyes closed [68]. Considering that this state is associated with a NREM sleep stage 1 pattern, Bassetti kept the denomination of hypersomnia [20]. Guilleminault however classified this as a waking state because the patients only exhibited a slow rhythm and were able to react quickly to slight stimulation; he called this "de-arousal" [68], a concept close to subwakefulness used earlier by Schott [147]. During this abnormal waking, a reduction in frontal, cingular and anterior temporal metabolism has been observed using positron emission tomography [160].

There is a reduction in the percentage of spindles and NREM sleep stages 3 and 4, while that of REM sleep seems to be little affected [20]. In the complete form of this syndrome, associated with sleep and waking anomalies there is vertical gaze paralysis, memory disturbance, mood changes and bulimia [59]. In one case, bromocriptine induced an increase in motor activity with so-called "instruction dependent behaviour" [32], and another by modafinil, a drug thought to act through a noradrenergic mechanism [17].

Thus following bilateral thalamo-subthalamic ischaemia, two functional defects appear to coexist, each of variable intensity. The first concerns the waking system and explains the more or less profound decrease in the waking quality termed as de-arousal in pseudo-hypersomnia, or stage 1 hypersomnia. The second affects the thalamic networks, producing spindles and slow oscillations of sleep [158]. It is tempting to attribute the behavioural disturbances to reduced cortical solicitations secondary to the defect in the arousal systems, some of which are dopaminergic, and others noradrenergic, as suggested by the therapeutic effects of bromocriptine [32] and modiodal [17].

Pontine lesions lead to substantial sleep alterations, with reduced NREM and REM sleep [6, 46, 36, 101]. Their association with lateral gaze paresis which is also observed in other aetiologies, lead to the notion that this is a topographical syndrome of the brain stem [16]. These reports confirm the role of the paramedial reticular formation in man, in generating REM sleep and lateral eye movements. It is somewhat intriguing to note stage 1 and 2 hypersomnia occurring after a right tegmental pontine haematoma affecting the medial part of the pontis caudalis and oralis reticular nuclei [11]. The same applies in the publication submitted by Kuschida *et al.* [88] observing in the course of a protuberant haematoma, a homolateral slowing of the electroencephalogram during REM sleep in the form of sawtooth delta sharp waves: this observation led to the notion of homolateral pathways responsible for cortical REM sleep activation.

Finally, Askenasy and Goldhammer [12] have reported the occurrence of sleep apnoea syndrome after a bulbar stroke.

Mention should also be made of the startling observation of narcoleptic episodes related to ischaemic accidents in the vertebrobasilar area which disappeared after treatment with aspirin [113] and that of an authentic case of narcolepsy after a cardiac arrest [136].

Effects of sleep on cerebrovascular pathology

The time variability in cerebrovascular strokes is not a matter of chance: there is less risk in the evening and an increased risk in the morning, particularly of cerebral ischaemia, between 8 and 10 am [102, 169] and during the 30 minutes after morning awakening [124]. Snoring, regardless of age and weight, is a factor of cerebrovascular strokes [111, 123, 156]. The latter may be related to the marked prevalence of both sleep apnoea syndrome and high blood pressure in this population. Sleep apnoea syndrome is known to increase mortality, no doubt due to the vascular risk cofactors. Apnoeas have not been confirmed as heightening the risk of cerebral ischaemic strokes, and the mechanism through which they act has not been documented (ischaemia of hypoxemic origin during prolonged apnoeas or cruoric embolism induced by rhythm disorders) [93].

An interesting study by Nakamura *et al.* [109] showed the presence of a nocturnal drop in blood pressure favouring silent strokes, in patients presenting cerebral ischaemic lesions.

DEGENERATIVE DISORDERS

Amyotrophic lateral sclerosis

Abnormal sleep architecture has been described since 1975 [104], with a poor sleep syndrome, lack of atonia in REM sleep. Bulbar forms of the disease seem prone to significant breathing disorders [84]. When a diaphragmatic dysfunction is present, there is a significant decrease in REM sleep percentages, with a poorer prognosis [10]. Patients with sleep breathing disorders seem to respond favourably to positive airway pressure treatment [38].

Parkinson's disease

Those with Parkinson's disease specifically suffer from poor sleep. This is fragmented and agitated. It is accompanied by intense dreaming activity. These alterations increase during the course of the disease and are more often encountered in subjects presenting levodopa dyskenesia, on-off phenomena and anticholinergic treatment [110]. The dream activity of patients treated with L-dopa is particularly rich [145]: this presents in the form of vivid dreams, night terrors, nightmares and above all, REM sleep-related behaviour disorders. These may herald the onset of Parkinson's disease by several years [146].

Polysomnographic recordings objectify the reduction in total sleep time, NREM sleep stages 3 and 4 and REM sleep, with an increase in intermittent arousals. There are also qualitative alterations in sleep pattern, frequent blinking during arousal, reduced spindles in stage 2, possible rapid eye movements occurring in stages 3 and 4 and frequent bursts of alpha rhythms and absence of atonia in REM sleep [106]. The same author, having noted that patients presenting reduced REM sleep and the absence of atonia at this stage were not the same as those who blinked, suggested that there were two subgroups of Parkinson's sufferers, the first with more severely affected locus coeruleus and the second, as in the reserpine cat, with reduced catecholamines at the site of the facial nerve nuclei.

The marked beneficial effect of subthalamic nuclei stimulation is an argument strongly in favour of the poor sleep of these patients being due more to motor impairment than to dopaminergic deficit [8].

With DOPA therapy one witnesses a reduction in sleep fragmentation parallel to that seen in motor hyperactivity [52, 24]. Some authors have also reported an increase in the density of rapid eye movements which has been attributed to the increased dream activity seen with this treatment [14]. Several works pointed out the occurrence of daytime sleep episodes either in treated [50, 143] or in non treated patients [139]. On the other hand, Fenelon *et al.* [47] stressed the frequency of visual hallucinations in these patients. It seems to be hazardous to privilege a specific explanation, be it lack of sleep, an iatrogenic effect or a direct consequence of the lesion. However, the hypothesis of a narcoleptic-like dysfunction has been advanced, considering the co-occurrence of hallucinations and sleep onset REM episodes in the multiple sleep latency test in some patients [9].

Sleep quality does not appear to affect the type of daytime motor performance variation [45]. Parkinsonian trembling disappears during sleep [14]. It is replaced by more diffused rhythmic infraclinical contractions whose amplitude reduces between stage 1 and stage 4, finally being cancelled out in REM sleep. Furthermore, during sleep, Parkinson's sufferers experience anarchic muscular spasms which can be objectified either with surface electrodes [14] or by means of an accelerometer [52]. The movements are a function of the level of arousal. These authors showed that movement diminished in frequency depending on the level of arousal, in the following order: arousal, awakening, shift from stage 2 to stage 1, stage 1, stage 2 and REM sleep. They are absent in stages 3 and 4. They are far more likely to occur where there has been a prior arousal, this likelihood diminishing with bursts of slow waves exceeding 75 mV.

Under L-dopa, there is a reduction in sleep fragmentation parallel to that of motor hyperactivity [19, 24]. There may also be an increase in the density of rapid eye movements, which has been linked to the high level of dream activity with this treatment [91].

Hence the disappearance of agonist and antagonist contractions during sleep can be interpreted either as a reduction of the activity of a central system oscillator which generates trembling, or by the regression of peripheral reciprocal inhibitory phenomena [168]. The increase in anarchic motor phenomena during sleep may be interpreted as a reduction of motor activity inhibitory phenomena during sleep.

Multiple system atrophy

A high frequency of principally central apnoeas was initially reported [7, 31, 67]. Later, the frequency of laryngeal stridors was noted with more or less complete paralysis of the vocal chords, possibly accounting for sudden nocturnal death [76, 108, 140], the complications of which can be surgically prevented.

Striatonigral degeneration can substantially affect sleep. We published a clinical case involving narcoleptic episodes, cataplexy, sleep onset REM episodes and a marked reduction in NREM sleep [16, 127]. Perret *et al.* [127] reported a case of major insomnia ; this was associated with a slight increase in hydroxyindolacetic acid (5 HIAA) in the spinal fluid after injecting probenecide, suggesting a serotonin synthesis deficiency. Pathological examination showed a marked atrophy of the putamen, locus niger and, to a lesser degree, the locus coeruleus; no neuronal depopulation was seen at the site of the raphe nuclei. Hence there was no obvious anatomical explanation for the major insomnia presented by this patient.

Olivopontocerebellar atrophy sleep has been widely studied [1, 89, 112]. A syndrome of poor sleep is observed, notably in terms of the extent to which REM sleep and eye movement density are reduced. Recurrent REM sleep-related behaviour disorders include one case associated with nocturnal laryngeal stridor [85, 133, 150, 153].

Dystonia

In cranial dystonia, poor sleep is correlated with the number of spasms and severity of the illness [152, 154]. The spasms diminish in frequency and duration in the following order: stage 1, stage 2, REM sleep and stages 3 and 4. During sleep, they can occur at any stage outside EEG arousal. Thus this illness presents a non-specific reduction of abnormal movements during sleep, probably linked to diminished arousal mechanisms.

Sleep has been described in dystonia musculorum deformans, showing an increase of spindles in stage 2, which is particularly marked in the severe forms [79]. The same year this author published a case in which intense spindle activity had disappeared after unilateral thalamotomy [79]. These facts are interesting as they reveal encephalic disturbances in this illness with no pathophysiological substratum. However the observations did not stand up to the statistical analysis carried out by Fish *et al.* [51] on 14 primary dystonias, 10 secondary dystonias and 10 controls, which showed no spindle predominance in any of the subgroups.

In this study, movements diminished in the following order: arousal, awakening, lighter sleep, stage 1 and stage 2, with an increased probability of occurring just after a polysomnographic arousal and a diminished probability after a spindle.

Hence no specific relationship appears to exist between sleep and dystonia musculorum deformans.

Subcortical dementia

Progressive supranuclear palsy

Sleep has been widely studied in progressive supranuclear palsy [65, 89, 90, 94, 126]. The authors report a syndrome of poor sleep. This may be preceded by an initial increase in NREM sleep stages 3 and 4. Multiple abnormalities have also been reported in sleep microstructure: theta activity at 6 Hertz in stage 1, absence of spindles, bursts of alpha and eye movements in NREM sleep, the absence of atonia during REM sleep. Electromyographic activity is intense during wakefulness. Rapid eye movements in wakefulness and REM sleep have a very specific appearance due to the microbursts against a background of slow waves. At the end of evolution the wakefulness tracing is composed of 6 Hertz activity saturated with slow rhythms. REM sleep-related behaviour disorders were first reported in these patients in 1980 [153].

For Perret and Jouvet [114], the reduction of REM sleep is accounted for by the fact that the locus coeruleus is initially affected, and the increase in stages 3 and 4 by the fact that an inhibiting control is removed; later, insomnia may be explained by the rostral raphe being affected.

Huntington's chorea

Patients affected by this disease suffer from poor sleep: an increase in sleep latency, reduced sleep efficiency, frequent nocturnal arousals and reduced duration of deep NREM sleep, which correlate with caudal nuclei atrophy. There is also an increase in spindles [155, 166].

Dementia with Lewy bodies

Two observations of REM sleep-related behaviour disorders have been reported [163, 164]. Sleep shows no other abnormalities. The pathological study of the former case revealed Lewy bodies in the locus coeruleus, interpreted by the authors as the cause of a tonic inhibition defect during REM sleep.

Alzheimer's disease

Substantial work has been devoted to sleep in Alzheimer's disease. The alterations increase simply as a result of ageing, i.e. a deficiency in sleep continuity and a reduction of NREM sleep stages 3 and 4 and REM sleep [131, 166]. REM sleep phases are shorter. There is a marked slowing of EEG rhythm during arousal and REM sleep; for Montplaisir *et al.* [105], these anomalies are linked to the particularly marked cholinergic dysfunction in this disease. It is also worth noting the alteration in sleep microstructure: reduced K complexes, vertex sharp waves and spindles [135].

Sleep alterations form part of an overall disturbance of biological rhythms as evidenced by the collapse of thermal variation amplitudes [118].

The percentage of wakefulness, stages 3 and 4 and REM sleep marks the distinction between subjects with moderate dementia and controls of the same age. Sleep abnormalities also distinguish dementia from depression [134] characterised by reduced REM sleep latency and an increased density of REM sleep eye movements. Dementia, contrary to depression, is not accompanied by true recuperation after sleep deprivation, resulting in a lower percentage of REM sleep and eye movements [134]. Alternatively, sleep disturbances in Alzheimer's seem less extensive than in multi-infarct dementia of the same degree of severity [2].

TUMOURS

Work has tended to focus on narcolepsies symptomatic of a cerebral tumour. Granuloma will be dealt with at the same time as tumours, as the consequences for sleep may be similar.

Narcolepsy symptomatic of a cerebral tumour

The first observation is attributed to André-Thomas in 1923 [5]. In their review, Bonduelle and Degos [26], before the age of polysomnographic recordings, referred to 7 cases of tumour at the base of the skull (3rd ventricle, hypothalamus, pituitary type) and 2 frontal tumours. However lethargic states and narcoleptic episodes were poorly differentiated at the time, the two terms apparently being used indifferently. From 1979 onward, published observations included polysomnographic recordings. The semiology is more richly described, and the term symptomatic narcolepsy was used to refer only to patients who presented irresistible sleep episodes or possibly cataplexies, hypnagogic hallucinations and/or sleep paralysis, and sleep onset REM periods whether nocturnal, or during a multiple sleep latency test. A dozen tumours or expanding processes were thus published when narcolepsy was first defined. These essentially involve the region of the 3rd ventricle and the diencephalon. Their histology varies considerably: craniopharyngioma [3, 149], glioma [4, 157], two cases of chromophobic adenoma [132], colloid cyst [132], sarcoidal granuloma [132, 138] or toxoplasmic granuloma [49], a haemangioblastoma [162] and a fourth ventricular subependymoma [99], an arteriovenous aneurysm [34]. A case of temporal lymphoma [119] was also reported.

From a semiological point of view, it can be noted that two cases presented subintrant cataplexies [157]. HLA DR subtypes were studied in 4 of the 12 cases previously cited; in 2 cases the HLA DR2 subtypes were absent [35, 119]. These are of considerable value, as they tend to prove that encephalic lesions are themselves capable of creating symptoms of narcolepsy aside from any genetic predetermination.

Recurrent hypersomnia of tumoural origin

This section includes a case of recurrent hypersomnia associated with a midbrain craniopharyngioma [97] and another due to sarcoidosis [3]. The first patient, unlike the second, carried the HLA DR2 subtype.

Sleep-related breathing disorders of tumoural origin

This chiefly concerns alveolar hypoventilation: ganglioma [121], acoustic neurinoma [94], ganglioneuroblastoma [40]. Ito et al. [77], reported two cases of glioma chiefly affecting the medulla associated with central origin sleep apnoea. But it is also interesting to note the strange observation by Jaeckle et al. [78] of neurogenous hyperventilation due to an anaplastic astrocytoma of the brainstem which was reduced by opiates.

Finally it is worth noting the observation by Barros-Ferreira et al. [21] of a brainstem tumour where REM sleep was abnormal due to the absence of tonic abolition, sleep talking and facial movements, possibly related to a syndrome of REM sleep-related behaviour disorder, and that of Etzioni et al. [43] of severe insomnia in a child with a pineal tumour, corrected by melatonin.

SLEEP AND HEAD TRAUMATISM

Prognostic value of initial polysomnography

During post-traumatic coma, the reappearance of polysomnographic sleep patterns, observed by Chatrian et al. [33] is a sign of favourable prognosis [23]. A parallel exists [137] between cognitive and REM sleep recuperation. The latter is probably no more than a distant witness of the recuperation of encephalic functions. Head traumatism patients later present marked sleep disorders essentially in the form of an increased number of nocturnal awakenings [130].

Correlations between post-traumatic lesions and sleep

The distant lesional sequelae of head traumatism may induce sleep disorders of pathophysiological interest. Thus Guilleminault *et al.* [66] reported a considerable reduction of NREM and REM sleep, which was reversed after administering 5 HTP, following a posterior fossa traumatism. An absence of REM sleep after pontine lesion was also reported by Lavie *et al.* [92]. There is also the startling observation by Bricolo [28], of total insomnia lasting 5 days after a stereotaxic lesion which had initially affected the medioventral portion of the ventrolateral nucleus, the intermediary ventral nucleus and the rostral part of the posterior ventral nucleus, and later, on the other side, the intermediary lateral and ventral nuclei, medial centre and rostral part of the posterior ventral nucleus as well as the subthalamic region.

Finally, sleep-related breathing disorders have been reported after medullary trauma, anterolateral cordotomies or Arnold-Chiari malformation [18], and the strange observation of a narcolepsy episode after atlantoaxial dislocation [72].

In contrast to these cases in which lesions are relatively focalised, persistent post-traumatic vegetative states have no characteristic polygraphic pattern; Billiard *et al.* [25] noted frequent disturbances of the sleep-wake cycles as well as the absence of any correlation between the functional state and sleep pattern normality.

Posttraumatic hypersomnia and narcolepsy

Post-traumatic narcolepsies are not exceptional even though few publications have been devoted to this subject [55, 62, 68, 100]. They raise a medico-legal problem which has no simple solution. Theoretical evidence in favour of a causal link focuses on the chronology of events, the association between the brainstem and lesions, and the absence of DR2 DQl haplotype [100]. Next to these cases, is the surprising publication describing non narcoleptic post-traumatic cataplexy in a non HLA DR2 DQl subject [151].

Post-traumatic hypersomnias are more frequent but require closer analysis. In a series of 20 cases of post-traumatic sleepiness [68], polysomnography showed 8 obstructive sleep apnoea syndromes , 1 narcolepsy and 9 cases of daytime sleepiness.

SLEEP AND MULTIPLE SCLEROSIS

An association has been reported between narcolepsy and multiple sclerosis in a limited number of cases. The first of these are the three cases of narcolepsy in familial multiple sclerosis, an exceptional form of the disease [42]. Eight cases have been reported in the usual sporadic form [22, 73, 129, 147, 170]. They raise the question as to whether there is a fortuitous coexistence of the two ailments or whether a causal relationship exists between them. One patient was HLA DR2 [169]. Conversely, the observation by Schrader *et al.* [148] is remarkable inasmuch as the multiple sclerosis-narcolepsy relationship occurred in one monozygotic twin who did not carry the HLA DR2 DQl haplotype.

Two cases of sleep apnoea syndrome also occurred [56] and a single case of hypersomnia which signalled the presence of multiple sclerosis [146].

SLEEP AND INFECTIOUS DISORDERS

HIV infection

Several publications have dealt with the sleep of HIV seropositive patients. Kubicki *et al.* [87] reported poor sleep in non selected patients with increased arousals in stage 1 and reduced NREM sleep, REM sleep and particularly, of spindles, attributed to nerve destruction. Weigand *et al.* [167], in a controlled study of fourteen seropositive patients with no opportunist infections, noted poor sleep with low density of spindles in two cases, and reduced REM sleep latency correlated to the extent of depression. These results have been confirmed by Ferini-Strambi *et al.* [48], who noted a reduction in NREM sleep associated with increasing periods of alternating cyclic patterns. In contrast to this, Norman *et al.* [114, 115, 116, 117] found an increase and predominance of NREM sleep in the second part of the night, associated with disturbances in the cyclic organisation of sleep, bearing no relation to underlying psychopathological disorders. These divergences are not without significance as specific sleep abnormalities suggest a possible infraclinical encephalic effect. This possible alteration of sleep has been related to tumour necrosis factor (TNF) α in seropositive patients, which is a hypnogenic factor in animals (secretion which may be induced by the viral envelope) [37, 120].

Prion diseases

Thalamic degeneration, known of since Stern's description [159], is characterised by a bedridden and demented state, associated with abnormal motor activity (abnormal movements, fasciculations) with subacute fatal evolution in under a year. Familial forms have been described by Little *et al.* [96], and by Julien *et al.* [83]. But Lugaresi *et al.* [98] are attributed with having established the link between these fatal familial forms and the coexistence of a particularly serious pathological insomnia, referred to as fatal familial insomnia. Certain cases of Morvan's fibrillar chorea may possibly come within this framework [53].

Clinically, the onset of the illness occurs between 30 and 60 years, the course of the illness lasting under 3 years. The first characteristic is insomnia. Polysomnographic recordings show the absence of spindles and delta activity and the occurrence of short periods of hallucinatory dreaming associated with electroencephalographic desynchronisation and bursts of phasic muscular activity. This coincides with intense vegetative activity, including sweating, tachycardia, hyperthermia, high blood pressure and motor hyperactivity associating diverse abnormal movements, ataxia, dysarthria, myoclonia and fasciculation. Lesions are restricted to the ventral anterior (VA) and dorso-medial (DM) thalamic nuclei, associating neuronal depopulation, gliosis and spongiosis [98]. This disease was linked to a prion disease after the discovery of an abnormal protein-prion resistant to protein kinase due to a mutation at codon 178 of the prion protein gene [103].

This observation, relating to a prion disease, is similar to the very early disturbances of sleep-wake states observed in Creutzfeldt-Jakob disease [64, 89]. This was subject to a detailed analysis by Terzano *et al.* [161].

Trypanosomiasis

Trypanosomiasis leads at the initial stage of meningo-encephalitis, to a disturbance of circadian sleep-wake alternation, without hypersomnia [29]. At a later stage, polysomnographic recordings are saturated with slow waves and the various stages of sleep are indistinguishable from one another [29]. Parallel to the alteration of sleep-wake patterns, circadian rhythm disorders develop (prolactine and cortisol secretion, and plasmatic renine activity) [27].

Other infections

Sleep apnoeas have been variably observed in medullary forms of poliomyelitis [74, 128, 142]. Listeria monocytogenes brain stem encephalitis has been reported in association with Ondine's curse [80].

POLYSOMNOGRAPHIC RECORDINGS AS A MEANS OF STUDYING ENCEPHALIC DISTURBANCES

At the end of this analytical review of the polysomnographic disturbances linked to the different neurological illnesses, we question the extent to which polysomnographic recordings provide an estimate of the anatomical and functional state of the encephalon.

Polysomnographic recording to define lesion topography

A ponto-medullary infarct may take the form of apnoeas and respiratory rhythm disorders [12, 39]. We have already seen the intense quantitative and qualitative disturbance to sleep caused by pontine lesions: reduced NREM and REM sleep, disappearance of lateral eye movements [16, 122]. But reduced NREM and REM sleep are witnessed in all types of poor sleep; hence it is only a permanent reduction in stable conditions which is of any significance. REM sleep-related behaviour disorders often associated with the absence of atonia in REM sleep, can be induced by bilateral lesions of the tegmental reticular bundle [35]. Lesions in this region are also responsible for symptomatic narcolepsies and cataplexies (cf. section on tumours and multiple sclerosis).

According to the work of Jouvet [82] a medial peduncular or posterior hypothalamic lesion normally leads to the reduced duration of arousal, corresponding to an a-arousal syndrome [147] and hypersomnia [20], pseudohypersomnias [69] of thalamoperforated infarcts. Although this is certainly a familiar picture, it is one which is transitory, as in the case of animals. Other systems which are still to be discovered, ensure the recuperation of the arousal functions.

An anterior hypothalamic lesion should, according to experimental data [82], lead to insomnia. Cases of this type of lesion-related insomnia have never been published. However insomnias from thalamo-subthalamic lesions described by Bricolo [28], Schott *et al.* [147] and Lugaresi [98] which interrupt the subthalamic rostral caudal connections, no doubt represent the clinical equivalents of this experimental model.

As regards the thalamus, if we set aside the possibility of thalamic insomnia, we could expect to find an alteration of spindles, but insufficient work has been done on this to date.

Can polysomnographic recordings testify to neurochemical dysfunction?

Is it possible to apply the model set up by Hobson *et al.* [75] to man, whereby an aminergic system inhibits REM sleep, this inhibition being a permissive factor, while a cholinergic system facilitates triggering? This notion is partly reinforced by the human sleep alterations induced by cholinergic [60] or catecholaminergic [57] agonists.

Several attempts have been made to confirm this in neuropathology. Hence, on the basis that Parkinson's sufferers present a reduction of REM sleep and an absence of tonicity during this stage, Mouret [106] suggests that there is a subgroup of Parkinson's patients suffering from dopaminergic deficiency. An analogous observation was made for a subgroup of endogenous depression which was thus considered to be dopamine-dependent. Reduced dopaminergic activity is expressed by qualitative polysomnographic abnormalities (blepharospasm, presence of alpha rhythm and absence of abolition of muscle tone during REM sleep episodes) [107].

In the same way, reduced cholinergic activity has been evoked to account for the prolonged REM sleep latency observed in Alzheimer's disease, by inverted analogy with the effects of injecting atropine or eserine.

Deficient serotoninergic functioning may account for certain reductions in total sleep time, due to lesions in the serotoninergic raphe nuclei. This hypothesis is confirmed by the fact that sleep time is restored to an acceptable level by administering serotonin precursor 5-HTP, as observed by Guilleminault *et al.* [66], Fischer-Perroudon *et al.* [53] or Perret *et al.* [127]. This hypothesis is further supported by the increase in serotonin catabolite 5-HIAA in the spinal fluid after administering probenecide. Increased wakefulness (i.e. insomnia) evokes the possibility of noradrenergic hyperfunctioning, as suggested by the alterations induced by amphetamines. Conversely, it is the hyperfunctioning of this system which is considered to be at the origin of a-arousals secondary to medial midbrain lesions, providing insight into the anatomical and functional integrity of the multiple systems which control the occurrence and morphology of arousal, NREM and REM sleep. The problem with this method lies in the wide inter- and intra-individual variability of sleep parameters, compromising the interpretation of results for a given patient. While these results are of interest for clinical research, their application is only justified in exceptional cases of diagnosis and treatment.

REFERENCES

1.   ADELMAN S., DINNER D., GOREN H., LITTLE J., NICKERSON P. – Obstructive sleep apnea in association with posterior fossa neurologic disease. *Arch. Neurol.,* 41, 509-510, 1984.
2.   AHARON-PERETZ J., MASIAH A., PILLAR T., EPSTEIN R., TZISCHINSKY O., LAVIE P. – Sleep-wake cycles in multi-infarct dementia and dementia of the Alzheimer type. *Neurology,* 41, 1616-1619, 1991.
3.   ALDRICH M.S., NAYLOR M.W. – Narcolepsy associated with lesions of the diencephalon. *Neurology,* 39, 1505-1508, 1989.
4.   ANDERSON M., SALMON M.V. – Symptomatic cataplexy. *J. Neurol. Neurosurg. Psychiatry,* 40, 186-191, 1977.
5.   ANDRE-THOMAS A., JUMENTIE J., CHAUSSEBLANCHE – Léthargie intermittente traduisant l'existence d'une tumeur du 3ème ventricule. *Rev. Neurol.,* 40, 67-73, 1923.
6.   APPENZELLER C., FISCHER A.P. – Disturbances of rapid eye movement sleep in patients with lesions of the nervous system. *Electroencephalogr. Clin. Neurophysiol.,* 25, 29-32, 1968.
7.   APPS M., SCHEAFF P., INGRAM D., KENNARD C., EMPEY D.W. – Respiration and sleep in Parkinson's disease. *J. Neurol. Neurosurg. Psychiatry,* 48, 1240-1245, 1985.
8.   ARNULF I., BEJJANI B.P., GARMA L., BONNET A.M., HOUETO J.L., DAMIER P., DERENNE J.P., AGID Y. - Improvement of sleep architecture in Parkinson's disease with subthalamic nucleus stimulation. *Neurology,* 55, 1732-1734, 2000.
9.   ARNULF I., BONNET A.M., DAMIER P., BEJJANI B.P., SEILHEAN D., DERENNE J.P., AGID Y. – Hallucinations, REM sleep, and Parkinson's disease. A medical hypothesis. *Neurology,* 55, 281-288, 2000.
10.   ARNULF I., SIMILOWSKI T., SALACHAS F., GARMA L., MEHIRI S., ATTALI V., BEHIN-BELLHESEN V., MEININGER V., DERENNE J.P. – Sleep disorders and diaphragmatic function in patients with amyotrophic lateral sclerosis. *Am. J. Respir. Crit. Care Med.,* 161, 849-856, 2000.
11.   ARPA J., RODRIGUEZ-ALBARINO A., IZAL E., SARRIA J., LARA M., BARREIRO P. – Hypersomnia after tegmental pontine hematoma: Case report. *Neurologia,* 10, 140-144, 1995.
12.   ASERINSKI J.J., GOLDHAMMER I. – Sleep apnea as a feature of a bulbar stroke. *Stroke,* 19, 637-639, 1988.
13.   ASKENASY J., YAHR M.- Reversal of sleep disturbance in Parkinson's disease by antiparkinsonian therapy: a preliminary study. *Neurology,* 35, 527-532, 1985.
14.   ASKENASY J., YAHR M. – Parkinsonian tremor loses its alternating aspect during non-REM sleep and is inhibited by REM sleep. *J. Neurol. Neurosurg. Psychiatry,* 53, 749-753, 1990.
15.   AUTRET A., CARRIER H., THOMMASI M., JOUVET M., SCHOTT B. – Etude physiopathologique et neuropathologique d'un syndrome de décortication cérébrale. *Rev. Neurol.,* 131, 491-504, 1975.
16.   AUTRET A., LAFFONT F., DE TOFFOL B., CATHALA H.P. – A syndrome of REM and non-REM sleep reduction and lateral gaze paresis after medial tegmental pontine stroke: CT scans and anatomical correlations in four patients. *Arch. Neurol.,* 45, 1236-1242, 1988.
17.   AUTRET A., LUCAS B., MONOON, K., HOMMET C., CORCIA P., SAUDEAU, D., DE TOFFOL B. – Sleep and brain lesions: a review of the literature and additional new cases. *Neurophysiol. Clin.,* 31, 356-375, 2001.
18.   BALK R.A., HILLER F.C., LUCAS E.A., SCRIMA L., WILSON F.J., WOOTEN V. – Sleep apnea and the Arnold-Chiari malformation. *Am. Rev. Resp. Dis.,* 132, 929-930, 1985.
19.   BASSETTI C., ALDRICH M.S., QUINT D. - Sleep-disordered breathing in patients with acute supra- and infratentorial strokes. A prospective study of 39 patients. *Stroke,* 28, 1765-1771, 1997.
20.   BASSETTI C., MATHIS J., GUGGER M., LOVBLAD K.O., HESS C.W. – Hypersomnia following paramedian thalamic stroke: a report of 12 patients. *Ann. Neurology,* 39, 471-480, 1996.
21.   DE BARROS-FERREIRA M., CHODKIEWICZ J.P., LAIRY G., SALZARULO P. – Disorganized relations of tonic and phasic events of REM sleep in a case of brain-stem tumor. *Electroencephalogr. Clin. Neurophysiol.,* 38, 203-207, 1975.
22.   BERG O., HANLEY J. – Narcolepsy in two cases of multiple sclerosis. *Acta Neurol. Scand.,* 39, 252-257, 1963.
23.   BERGAMASCO B., BERGAMINI L., DORIGUZZI T. - Clinical value of the sleep electroencephalographic patterns in posttraumatic coma. *Acta Neurol. Scand.,* 44, 495-511, 1968.
24.   BERGONZI P., CHIURULLA L., GAMBI D., MENNUNI G., PINTO F. – L-Dopa plus dopa-decarboxylase inhibitor: sleep organization in Parkinson's syndrome before and after treatment. *Acta Neurol. Belg.,* 75, 5-10, 1975.
25.   BILLIARD M., NEGRE C., BESSET A., BALDY-MOULINIER M., ROQUEFEUIL B., PASSOUANT P. – Organisation du sommeil chez les sujets atteints d'inconscience postraumatique chronique. *Rev. EEG Neurophysiol. Clin.,* 9, 171-178, 1979.
26.   BONDUELLE M., DEGOS C. – Symptomatic narcolepsies: a critical study. *In: Narcolepsy.* C. Guilleminault, W. Dement, P. Passouant (eds). Spectrum. New York. 313-322, 1976.
27.   BRANDENBERGER G., BUGUET A., SPIEGEL K., STANGHELLINI A., MOUANGA G., BOGUI P., MONTMAYEUR A., DUMAS M. – Maintenance of the relation between the pulsed secretion of hormones and the internal sleep structure in human African trypanosomiasis. *Bull. Soc. Pathol. Exot.,* 87, 383-389, 1994.
28.   BRICOLO A. – Insomnia after bilateral stereotactic thalamotomy in man. *J. Neurol. Neurosurg. Psychiatry,* 30, 154-158, 1967.
29.   BUGUET A., BERT J., TAPIE P., TABARAUD F., DOUA F., LONSDORFER J., BOGUI P., DUMAS M. – Sleep-wake cycle in human African trypanosomiasis . *J. Clin. Neurophysiol.* 10, 190-196, 1993.
30.   CASTAIGNE P., ESCOUROLLE R. – Etude topographique des lésions anatomiques dans les hypersomnies. *Rev. Neurol.* Paris, 116, 547-584, 1967.

31. CASTAIGNE P., LAPLANE D., AUTRET A., BOUSSER M.G., GRAY F., BARON J.C. – Syndrome de Shy et Drager avec troubles du rhythme respiratoire et de la vigilance: à propos d'un cas anatomo-clinique. *Rev. Neurol.* Paris. 133, 455-466, 1977.
32. CASTMAN-BERREVOETS C.E., HARSKAMP F.V. - Compulsive pre-sleep behavior and apathy due to bilateral thalamic stroke: response to bromocriptine. *Neurology*, 38, 647-649, 1988.
33. CHATRIAN G.E., WHITE L.E., DALY D. – Electroencephalographic patterns resembling those of sleep in certain comatose states after injuries to the head. *Electroencephalogr. Clin. Neurophysiol.* 15, 272-280, 1963.
34. CLAVELOU P., TOURNILHAC M., VIDAL C., GEORGET A.M., PICARD L., MERIENNE L. – Narcolepsy associated with arteriovenous malformation of the diencephalon. *Sleep,* 18, 202-205, 1995.
35. CULEBRAS A. – Neuroanatomic and neurologic correlates of sleep disturbances. *Neurology,* 42, 19-27, 1992.
36. CUMMINGS J.L., GREENBERG R. – Sleep patterns in the « locked » in syndrome. *Electroencephalogr. Clin. Neurophysiol.,* 43, 270-271, 1977.
37. DARKO D.F., MILLER J.C., GALLEN C., WHITE J., KOZIOL J., BROWN S.J., HAYDUK R., ATKINSON J.H., ASSMUS J., MUNNEL D.T., *et al.* – Sleep electroencephalogram delta-frequency amplitude, night plasma levels of tumor necrosis factor alpha, and human immunodeficiency virus infection. *Proc. Natl. Acad. Sci. USA,* 92, 12080-12084, 1995.
38. DAVID W.S., BUNDLIE S.R., MAHDAVI Z. - Polysomnographic studies in amyotrophic lateral sclerosis. *J. Neurol. Sci.* 152, 29-35, 1997.
39. DEVEREUX M., KEANE J., DAVIS R. – Automatic respiratory failure associated with infarction of the medulla. *Arch. Neurol.,* 29, 46-52, 1973.
40. DIEZ GARCIA R., CARILLO A., BARTOLOME M., CASANOVA A., PRIETO M. – Central hypoventilation syndrome associated with ganglioneuroblastoma. *Eur. J. Pediatr. Surg.* 5, 292-294, 1995.
41. DYKEN M.E., SOMERS V.K., YAMADA T., REN Z.Y., ZIMMERMAN M.B. -Investigating the relationships between stroke and obstructive sleep apnea. *Stroke,* 27, 401-407, 1996.
42. EKBOM K. – Familial multiple sclerosis associated with narcolepsy. *Arch. Neurol.,* 15, 337-344, 1966.
43. ETZIONI A., LUBOSHITZKY R., TIOSANO D., BEN HARUSH M., GOLDSHER D., LAVIE P. – Melatonin replacement corrects sleep disturbances in a child with pineal tumor. *Neurology,* 46, 261-263, 1966.
44. FACON E., STERIADE M., WERTHEIM N. – Hypersomnie prolongée engendrée par des lesions bilatérales du système activateur médial. Le syndrome thrombotique de la bifurcation du tronc basilaire. *Rev. Neurol.* Paris, 98, 117-133.
45. FACTOR S.A., McALARNEY T., SANCHEZ-RAMOS J.R., WEINER W.J. – Sleep disorders and sleep effect in Parkinson's disease. *Mov. Disord.,* 5, 280-285, 1990.
46. FELDMAN M.H. – Physiological observations in a chronic case of « locked-in » syndrome. *Neurology,* 21, 459-478, 1971.
47. FENELON G., MAHIEUX F., HUON R., ZIEGLER M. - Hallucinations in Parkinson's disease. Prevalence, phenomenology and risk factors. *Brain,* 123, 733-745, 2000.
48. FERINI-STRAMBI L., OLDANI A., TIRLONI G., ZUCCONI M., CASTAGNA A., LAZZARIN A., SMIRNE S. – Slow wave sleep and cycling alternating pattern (CAP) in HIV-infected asymptomatic men. *Sleep,* 18, 6, 446-450, 1995.
49. FERNANDEZ J.M., SADABA F., VILLAVERDE F.J., ALVARO L.C., CORTINA C. – Cataplexy associated with midbrain lesion. *Neurology,* 45, 393-395, 1995.
50. FERREIRA J.J., GALITZKY M., MONTASTRUC J.L., RASCOL O. - Sleep attacks and Parkinson's disease treatment. *Lancet,* 355, 1333-1334, 2000.
51. FISH D.R., ALLEN P.J., SAWYERS D., BLACKIE J.D., MARSDEN C.D.- Sleep spindles in torsion dystonia. *Arch. Neurol.,* 47, 216-218, 1990.
52. FISH D.R., SAWYERS D., ALLEN P.J., BLACKIE J.D., LEES A.J., MARSDEN C.D. – The effect of sleep on the dyskinetic movements of Parkinson's disease, Gilles de la Tourette syndrome, Huntington's disease, and torsion dystonia. *Arch. Neurol.* 48, 210-214, 1991.
53. FISHER-PERROUDON C., MOURET J., JOUVET M. – Sur un cas d'agrypnie (4 mois sans sommeil) au cours d'une maladie de Morvan. Effet favorable du 5-hydroxytryptophane. *Electroencephalogr. Clin. Neurophysiol.,* 36, 1-18, 1974.
54. FORET J. – Variations spontanées et expérimentales. *In: Le Sommeil humain.* O. Benoît, J. Foret (eds), Masson, Paris, 61-75, 1992.
55. FRANCISCO G.E., IVANHOE C.B. – Successful treatment of posttraumatic narcolepsy with methylphenidate: a case report. *Am. J. Phys. Med. Rehabil., 75, 63-65, 1996.*
56. FUNAKAWA I., HARA K., YASUDA T., TERAO A. – Intractable hiccups and sleep apnea syndrome in multiple sclerosis: report of two cases. *Acta Neurol. Scand.* 88, 401-405, 1993.
57. GAILLARD J.M., KAFI S. – Involvement of pre-and post-synaptic receptors in catecholaminergic control of paradoxical sleep in man. *Europ. J. Clin. Pharmacol.,* 15, 83-89, 1979.
58. GARREL S., FAU R., PERRET J., CHATELAIN R. – Troubles du sommeil dans deux syndromes vasculaires du tronc cérébral dont l'un anatomo-clinique. *Rev. Neurol.* 115, 575-584, 1966.
59. GENTILINI M., DE RENZI E., CRISI G. – Bilateral paramedian thalamic artery infarcts: report of eight cases. *J. Neurol. Neurosurg. Psychiatry,50,* 900-909, 1987.
60. GILLIN J.C., SITARAM N., ; MENDELSON W.B., WYATT R.J. – Physostigmine alters onset but not duration of REM sleep in man. *Psychopharmacology,* 58, 111-114, 1978.
61. GIUBELEI F., IANNILLI M., VITALE A., PIERALLINI A., SACCHETTI M.L., ANTONINI G., FIESCHI C. – Sleep patterns in acute ischaemic stroke. *Acta Neurol. Scand.,* 86, 567-571, 1992.
62. GOOD J.L., BARRY E., FISHMAN P.S. – Posttraumatic narcolepsy: the complete syndrome with tissue typing. *J. Neurosurg.,* 71, 765-767, 1989.

63. GOOD D.C., HENKLE J.Q., GELBER D., WELSH J., VERHULST S. – Sleep-disordered breathing and poor functional outcome after stroke. *Stroke,* 27, 252-259, 1996.

64. GOTO K., UMEZAKI H., SUETSUGU M. – Electroencephalographic and clinicopathological studies on Creutzfeldt-Jakob syndrome. *J. Neurol. Neurosurg. Psychiatry.,* 39, 931-940, 1976.

65. GROSS R., SPEHLMANN R., DANIELS J.C. – Sleep disturbances in progressive supranuclear palsy. *Electroencephalogr. Clin. Neurophysiol.,* 45, 16-25, 1978.

66. GUILLEMINAULT C., CATALA H.P., CASTAIGNE P. – Effects of 5 hydroxytryptophane on sleep of a patient with a brain stem lesion. *Electroencephalogr. Clin. Neurophysiol.,* 34, 177-184, 1973.

67. GUILLEMINAULT C., TILKIAN A., LEHRMAN K., FORNO L., DEMENT W.C. – Sleep apnoea syndrome: states of sleep and autonomic dysfunction. *J. Neurol. Neurosurg. Psychiatry,* 40, 718-725, 1977.

68. GUILLEMINAULT C., FAULL K.F., MILES L., VAN DEN HOED J. – Posttraumatic excessive daytime sleepiness: a review of 20 patients. *Neurology.,*33, 1584-1598, 1983.

69. GUILLEMINAULT C., QUERA-SALVA M.A., GOLDBERG M.P. – Pseudo-hypersomnia and pre-sleep behaviour with bilateral paramedian thalamic lesions. *Brain,* 116, 1549-1563, 1993.

70. HACHINSKI V., MAMELAK M., NORRIS J.W. – Sleep morphology and prognosis in acute cerebrovascular lesions. *In: Cerebral Vascular Disease,* Meyer J.S., Lechner H., Reivich M. (eds), Excerpta Medica, Amsterdam, 69-71, 1977.

71. HACHINSKI V., MAMELAK M., NORRIS J.W. – Clinical recovery and sleep architecture degradation. *Can. J. Neurol.,* 17, 332-335, 1990.

72. HALL C.W., DANOFF D. – Sleep attacks-apparent relationship to atlantoaxial dislocation. *Arch. Neurol.,* 32, 57-58, 1975.

73. HEYK., HESS R. – Zur Narkolepsiefrage: Klinik und Electroenzephalogramm. *Fortschr. Neurol. Psychiatr.,* 12, 531-579, 1954.

74. HILL R., ROBBINS A., MESSING R., AROA N. – Sleep apnea after poliomyelitis. *Ann. Rev. Respir. Dis.,* 127, 129-131, 1983.

75. HOBSON J.A., McCARLEY R.W., WYZINSKI OP.W. – Sleep cycle oscillation reciprocal discharge by two brain stem neuronal groups. *Science,* 189, 55-58, 1975.

76. ISOZAKI E., HAYASHI M., HAYASHIDA T., TANABE H., HIRAI S. – Vocal cord abductor in multiple system atrophy-paradoxical movement of vocal cords during sleep. *Rinsho Shinkeigaku,* 36, 533-539, 1996.

77. ITO K., MIROFUSCHI T., MIZUNO M., SEMBA T. – Pediatric brain stem gliomas with the predominant symptom of sleep apnea. *Int. J. Pediatr. Otorhinolaryngol.* 37, 53-64, 1996.

78. JAECKLE K.A., DIGRE K.B., JONES C.R., BAILEY P.L., McMAHILL P.C.- Central neurogenic hyperventilation: Pharmacologic intervention with morphine sulfate and correlative analysis of respiratory , sleep, and ocular motor dysfunction. *Neurology,* 40, 1715-1720, 1990.

79. JANKEL W.R., NIEDERMEYER E., GRAF M., KALSHER M. – Case report: polysomnographic effects of thalamotomy for torsion dystonia. *Neurosurgery,* 14, 495-498, 1984.

80. JENSEN T.H., HANSEN P.B., BRODERSEN P. – Ondine's curse in listeria monocytogenes brain stem encephalitis. *Acta Neurol. Scand.,* 77, 505-506, 1988.

81. JONES B., HARPER S., HALARIS A. – Effect of locus coeruleus lesion upon cerebral mono-amine content, sleep-wakefulness states and the responses to amphetamine in the cat. *Brain Res.,* 24, 473-496, 1977.

82. JOUVET M. – Mécanismes des états de sommeil. *In: Physiologie du Sommeil.* O. Benoît (ed). Masson, Paris. 1-18, 1984.

83. JULIEN J., VITAL C., DELEPLANQUE B., LAGUENY A., FERRER W. – Atrophie thalamique subaiguë familiale. Troubles mnésiques et insomnie totale. *Rev. Neurol.* 146, 173-178, 1990.

84. KIMURA K., TACHIBANA N., KIMURA J., SHIBASAKI H . Sleep-disordered breathing at an early stage of amyotrophic lateral sclerosis. *J. of the Neurological Sciences.* 164, 37-43, 1999.

85. KNEISLEY L.W., REDERICH G.L.- Nocturnal stridor in olivopontocerebellar atrophy. *Sleep,* 13, 362-368, 1990.

86. KORNER E., FLOOH E., REINHART B., WOLF R., OTT E., KRENN W., LECHNER H. – Sleep alterations in ischaemic stroke. *Eur. Neurol.,* 25, 104-110, 1986.

87. KUBICKI S., HENKES H., TERSTEGGE K., RUF B. – AIDS related sleep disturbances. A preliminary report. *In: HIV and nervous system.* Kubicki S., Henkes H., Bienzle U. and Pohle H.D. (eds). Gustav Fischer, Stuttgart, New York, 97-105, 1988.

88. KUSCHIDA C.A., RYE D.B., NUMMY D., MILTON J.G., SPIRE J.P., RECHTSCHAFFEN A. – Cortical asymetry of REM sleep EEG following unilateral pontine hemorrhage. *Neurology,* 41, 598-601, 1991.

89. LAFFONT F., AUTRET A., MINZ M., BEILLEVAIRE T., GILBERT A., CATHALA H.P. – Polygraphic study of nocturnal sleep in three degenerative diseases: ALS, olivo-ponto-cerebellar atrophy and progressive supranuclear palsy. *Waking and Sleeping.* 3, 17-29, 1979.

90. LAFFONT F., LEGER J.M., PENICAUD A., MINZ M., CHAINE P., BERTRAND P., CATHALA HP. – Etude des anomalies du sommeil et des potentiels évoqués (PEV-PEA-PES) dans la paralysie supranucléaire progressive (PSP), *Electroencephalogr. Clin. Neurophysiol.* 18, 255-269, 1988.

91. LAVIE P., BENTAL E., GOSHEN H., SCHARF B. – REM ocular activity in Parkinsonian patients chronically treated with levodopa. *J. Neurol. Transm.* 47, 61-67, 1980.

92. LAVIE P., PRATT H., SCHARF B., PELED R., BROWN J. - Localized pontine lesion: near total absence of REM sleep. *Neurology,* 34, 118-120, 1984.

93. LAVIE P., HERER P., PELED R., BERGER I., YOFF N., ZOMER J., RUBIN A.H.E. – Mortality in sleep apnea patients: a multivariate analysis of risk factors. *Sleep,* 18, 149-157, 1995.

94. LEE D.K., WAHL G.W., SWINBURNE A.J., FEDULLO A.J.- Recurrent acoustic neuroma presenting as central alveolar hypoventilation. *Chest,* 105, 949-950, 1994.

95. LEYGONIE F., THOMAS J., DEGOS J.D., BOUCHAREINE A., BARBIZET J. – Troubles du sommeil dans la maladie de Steele-Richardson. Etude polygraphique de 3 cas. *Rev. Neurol.* 125-136, 1976.

96. LITTLE B.W., BROWN P.W., RODGERS-JOHNSON R., PERL D.P., GADSUSK D.C. – Familial myoclonic dementia masquerading as Creutzfeld-Jakob disease. *Ann. Neuro.* 20, 231-239, 1986.

97. LUCAS B., LAFFONT F., VELUT S., JAN M., DEGIOVANNI E., AUTRET A. – Hypersomnie périodique due à un cavernome mésencéphalique. *Electroencephalogr. Clin. Neurophysiol.* 19, 349, 1989.

98. LUGARESI E. – The thalamus and insomnia. *Neurology,* 42, 28-33, 1992.

99. MA T.K., ANG L.C., MAMELAK M., KISH S.J., YOUNG B., LEWIS A.J. – *Can. J. Neurol. Sci.* 23, 59-62, 1996.

100. MAEDA M., TAMAOKA A., HAYASHI A., MIZUSAWA H., SHOJI S. - A case of HLA-DR2, DQw1 negative posttraumatic narcolepsy. *Rinsho Shinkeigaku,* 35, 811-813, 1995.

101. MARKAND O.N., DYKEN M.L. - Sleep abnormalities in patients with brain stem lesions. *Neurology,* 26, 769-776, 1976.

102. MARLER J.R., PRICE T.R., CLARK G.L., MULLER J.E., ROBERTSON T., MOHR J.P., HIER D.B., WOLF P.A., CAPLAN L.R., FOULKES M.A. – Morning increase in onset of ischaemic stroke. *Stroke,* 20, 473-476, 1989.

103. MEDORI R., TRITSCHLER H.J., LE BLANC A., VILLARE F., MANETTO V., YING CHEN H., XUE R., LEAL S., MONTAGNA P., CORTELLI P. – Fatal familial insomnia, a prion disease with a mutation at codon 178 of the prion protein gene. *New Engl. J. Med.,* 13, 444-449, 1992.

104. MINZ M., AUTRET A., LAFFONT F., BEILLEVAIRE T., CATHALA H.P., CASTAIGNE P. - A study on sleep in amyotrophic lateral sclerosis. *Biomedicine,* 30, 40-46, 1978.

105. MONTPLAISIR J., PETIT D., LORRAIN D., GAUTHIER S., NIELSEN T. – Sleep in Alzheimer's disease: Further considerations on the role of brainstem and forebrain cholinergic populations in sleep-wake mechanisms. *Sleep,* 18, 145-148, 1995.

106. MOURET J. – Differences in sleep in patients with Parkinson's disease. *Electroencephalogr. Clin. Neurophysiol.* 38, 653-657, 1975.

107. MOURET J., LEMOINE P., MINUIT M.P. – Marqueurs polygraphiques , cliniques et thérapeutiques des dépressions dopamino-dépendantes (DDD). *C.R. Acad. Sci.* (Paris), 305, 301-306, 1987.

108. MUNSCHAUER F.E., LOH L., BANNISTER R., NEWSOM-DAVIS J. - Abnormal respiration and sudden death during sleep in multiple system atrophy with autonomic failure. *Neurology,* 40, 677-679, 1990.

109. NAKAMURA K., OITA J., YAMAGUCHI T.- Nocturnal blood pressure dip in stroke survivors: a pilot study. *Stroke,* 26, 1373-1378, 1995.

110. NAUSIEDA P.A., WEINER W.J., KAPLAN L.R., WEBER S., KLAWANS H.L.- Sleep disruption in the course of chronic levodopa therapy: an early feature of the levodopa psychosis. *Clin. Neuropharmacol.,* 5, 183-194, 1982.

111. NEAU J.P., MEURICE J.C., PAQUEREAU J., CHAVAGNAT J.J., INGRAND P., GIL R. - Habitual snoring as a risk factor for brain infarction. *Acta Neurol. Scand.* 92, 63-68, 1995.

112. NEIL J.F., HOLZER B.C., SPIKER D.G., COBLE P., KUPFER D.J. – EEG sleep alteration in olivopontocerebellar degeneration. *Neurology,* 30, 660-662, 1980.

113. NIEDERMEYER E., COYLE P.K., PREZIOZI T.S. – Hypersomnia with sudden sleep attacks, « symptomatic narcolepsy » on the basis of vertebrobasilar artery insufficiency. A case report. *Waking and Sleeping,* 3, 361-364, 1979.

114. NORMAN S.E., CHEDIAK A.D., FREEMAN C., KIEL M., MENDEZ A., DUNCAN R., SIMONEAU J., NOLAN B. – Sleep disturbances in men with asymptomatic human immunodeficiency (HIV) infection. *Sleep,* 15, 150-155, 1992.

115. NORMAN S.E., CHEDIAK A., KIEL M., GAZEROGLU H., MENDEZ A. – HIV infection and sleep: follow-up studies. *Sleep Res.* 19, 339, 1990.

116. NORMAN S.E., DEMIROZU M.C., CHEDIAK A.D. – Distorted sleep architecture (NREM/REM sleep cycles) in HIV healthy infected men. *Sleep Res.,* 19, 340, 1990.

117. NORMAN S.E., RESNICK L., COHN M.A., DUARA R., HERBST J., BERGER J.R., - Sleep disturbances in HIV-seropositive patients (letter). *JAMA,* 260 (7), 922, 1988.

118. OKAWA M., MISHIMA K., HISHIKAWA Y., HOZUMIS S., HORI H., TAKAHASHI K. – Circadian rhythm disorders in sleep-waking and body temperature in elderly patients with dementia and their treatment. *Sleep,* 14, 478-485, 1991.

119. ONOFRJ M., CURATOLA L., FERRACCI F., FULGENTE T. - Narcolepsy associated with primary temporal lobe B-cells lymphoma in a HLA DR2 negative subject. *J. Neurol., Neurosurg. and Psychiatry,* 55, 852-853, 1992.

120. OPP M.R., RADY P.L., HUGHES T.K., CADET P., TYRING S.K., SMITH E.M. - Human immunodeficiency virus enveloppe glycoprotein 120 alters sleep and induces cytokine mRNA expression in rats. *Am. J. Physiol.,* 270, 963-970, 1996.

121. OSANAI S., IIDA Y., NOMURA T., TAKAHASHI F., TSUJI S., FUJIUCHI S., AKIBA Y., NAKANO H., YAHARA O., KIKUCHI K. – A case of unilateral brain-stem tumor and impaired ventilatory response. *Nippon Kyobu Shikkan Gakkai Zasshi,* 32, 10, 990-995, 1994.

122. OSORIO I., DAROFF R.B. – Absence of REM and altered NREM sleep in patients with spinocerebellar degeneration and slow saccades. *Ann. Neurol.,* 7, 277-280, 1980.

123. PALOMAKI H., PARTINEN M., ERKINJUNTTI T., KASTE M. - Snoring, sleep apnea syndrome, and stroke. *Neurology,* 42, 75-82, 1992.

124. PALOMAKI H. - Snoring and the risk of ischaemic brain infarction. *Stroke,* 22, 1021-1025, 1991.

125. PARRA O., ARBOIX A., BECHICH S., GARCIA-EAROLES L., MONTSERRAT J.M., ANTONI LOPEZ J., BALLESTER E., GUERRA J.M., SOPENA J.J. – Time course of sleep-related breathing disorders in first-ever stroke or transient ischaemic attack. *Am. J. Respir. Crit. Care Med.,* 161, 375-380, 2000.
126. PERRET J.L., JOUVET M. – Etude du sommeil dans la paralysie supranucléaire progressive. *Electroencephalogr. Clin. Neurophysiol.* 49, 323-329, 1980.
127. PERRET J.L., TAPISSIER J., JOUVET M.- Insomnie et mémoire. A propos d'une observation de dégénérescence striatonigrique. *Electroencephalogr. Clin. Neurophysiol.* 47, 499-502, 1979.
128. PLUM F., POSNER J.- *Diagnostic de la stupeur et des comas.* Masson. Paris, 261 p., 1973.
129. POIRIER G., MONTPLAISIR J., DUMONT M., DUQUETTE P., DECARY F., PLEINES J., LAMOUREUX G. – Clinical and sleep laboratory study of narcoleptic symptoms in multiple sclerosis. *Neurology,* 37, 693-695, 1987.
130. PRIGATANO G.P., STAHL M.L., ORR W.C., ZEINER H.K., - Sleep and dreaming disturbances in closed head injury patients. *J. Neurol. Neurosurg. Psychiatry,* 45, 78-80, 1982.
131. PRINZ P., POCETA J.S., VITIELLO M. – Sleep in the dementing disorders. *In:* Boller and Graftman J. (eds), *Book of Neuropsychology,* Elsevier Science Pub., Amsterdam, 4, 335-347, 1990.
132. PRITCHARD P.B., DREIFUSS F.E., SKINNER R.L., PICKETT J.B., BIGGS P.J. – Symptomatic narcolepsy (abstract). *Neurology,* 33, 329, 1983.
133. QUERA SALVA M.A., GILLEMINAULT C. – Olivopontocerebellar degeneration, abnormal sleep, and REM sleep without atonia. *Neurology,* 36, 576-577, 1986.
134. REYNOLDS C., KUPFER D., HOUCK P., HOCH C., STACK J., BERMAN S., ZIMMER B. – Reliable discrimination of elderly depressed and demented patients by electroencephalographic sleep data. *Arch. Gen. Psychiatry,* 18, 258-264, 1988.
135. REYNOLDS C., SPIKER D., HANIN I., KUPFER D. – Electroencephalographic sleep, aging and psychopathology: new data and state of the art. *Biol. Psychiatry,* 18, 139-155, 1983.
136. RIVERA V.M., MEYER J.S., HATA T., HISHIKAWA Y., IMAI A. – Narcolepsy following cerebral ischaemia. *Ann. Neurol.* 19, 505-508, 1986.
137. RON S., ALGOM D., HARY D., COHEN M. – Time-related changes in the distribution of sleep stages in brain injured patients. *Electroencephalogr. Clin. Neurophysiol.* 48, 432-441, 1980.
138. RUBINSTEIN I., GRAY T.A., MOLDOFSKY H., HOFFSTEIN V. – Neurosarcoidosis associated with hypersomnolence treated with corticosteroids and brain irradiation. *Chest,* 94, 205-206, 1988.
139. RYE D.B., BLIWISE D.L., DIHENIA B., GURECKI H. – Daytime sleepiness in Parkinson's disease. *J. Sleep Res.,* 9, 63-69, 2000.
140. SADAOKA T., KAKITSUBA N., FUJIWARA Y., KANAI R., TAKAHASHI H. – Sleep-related breathing disorders in patients with multiple system atrophy and vocal fold palsy. *Sleep,* 19, 479-484, 1996.
141. SAKAI K. - Mécanismes cholinergiques: veille et sommeil. *In: Etats de veille et de sommeil.* P. Meyer, J.L. Elghozi, A. Quera-Salva (eds). Masson. Paris. 13-24, 1990.
142. SARNOFF S., WITTENBERGER J., APPELDT J. – Hypoventilation syndrome in poliomyelitis . *JAMA,* 174, 30-34, 1951.
143. SCHAPIRA A.H. – Sleep attacks (sleep episodes) with pergolide. *Lancet,* 355, 1332-1333, 2000.
144. SCHARF B., MOSKOVITZ C., LUPTON M.D., KLAWANS H.L. – Dream phenomena induced by chronic levodopa therapy. *J. Neural Transm.,* 43, 143-151, 1978.
145. SCHENCK C.H., BUNDIE S.R., MAHOWALD M.W. – Delayed emergence of a parkinson disorder in 38% of 29 older men initially diagnosed with idiopathic rapid eye movement sleep behavior disorder. *Neurology,* 46, 388-392, 1996.
146. SCHLÜTER B., AGUIGAH G., ANDLER W. – Hypersomnia in multiple sclerosis. *Klin. Pediatr.,* 208, 103-105, 1996.
147. SCHOTT B., MICHEL D., MOURET J., RENAUD B., QUENIN P., TOMMASI M. – Monoamines et régulation de la vigilance. II – Syndromes lésionels du système nerveux central. *Rev. Neurol.,* 127, 157-171, 1972.
148. SCHRADER H., GOTLIBSEN O.B., SKOMEDAL G.N. – Multiple sclerosis and narcolepsy/cataplexy in a monozygotic twin. *Neurology,* 30, 105-108, 1980.
149. SCHWARTZ W.J., STAKES J.W., HOBSON J.A. – Transient cataplexy after removal of a craniopharyngioma. *Neurology,* 34, 1372-1375, 1984.
150. SEPTIEN L., DIDI-ROY R., MARIN A., GIROUD M. – REM-sleep behavior disorder and olivo-ponto-cerebellar atrophy: A case report. *Neurophysiol. Clin.* 22, 459-464, 1992.
151. SERVAN J., MARCHAND F., GARMA L., RANCUREL G. – A case on non-narcoleptic posttraumatic cataplexy (letter), *Neurophysiol. Clin.* 26, 115-116, 1996.
152. SFORZA E., ZUCCONI M., PIETRONELLI R., LUGARESI E., CIRIGNOTTA F. – REM sleep behavior disorders. *Eur. Neurol. ,* 28, 295-300, 1988.
153. SHIMIZU T., SUGITA Y., IIJIMA Y., TESHIMA Y., HISHIKAWA Y. – Sleep study in patients with spinocerebellar degeneration and related diseases. *In: Sleep 1980,* W.P. Koella (ed), Karger, Basel, 435-437, 1981.
154. SILVESTRI R., DE DOMENICO P., DI ROSA A.E., BRAMANTI P., SERRA S., DI PERRI R. – The effect of nocturnal physiological sleep on various movement disorders. *Mov. Disord.,* 5, 8-14, 1990.
155. SILVESTRI R., RAFFAELE M., DE DOMENICO P., TISANO A., MENTO G., CASELLA C., TRIPOLI M.C., SERRA S., DI PERRI R. – Sleep features in Tourette's syndrome, neuroacanthocytosis and Huntington's chorea. *Neurophysiol. Clin.* 25, 66-77, 1995.
156. SPRIGGS D.A., FRENCH J.M., MURDY J.M., CURLESS R.H., BATES D., JAMES O.F.W. – Snoring increases the risk of stroke and adversely affects prognosis. *Q.J. MED.,* 83, 303, 555-562, 1992.

157. STAHL S.M., LAYZER R., AMINOFF M., TOWNSEND J., FELDON S. – Continuous cataplexy in a patient with a midbrain tumor: the limp man syndrome. *Neurology,* 30, 1115-1118, 1980.
158. STERIADE M., AMZICA F. – Slow sleep oscillations, rhythmic k-complexes, and their paroxysmal developments. *J. Sleep Res.* 7, suppl. 1, 30-35, 1998.
159. STERN K. – Severe dementia associated with bilateral symmetrical degeneration of the thalamus . *Brain,* 62, 157-171, 1939.
160. TERAO Y., SAKURAI Y., SAKUTA M., ISHII K., SUGISHITA M. - FDG-PET in an amnestic and hypersomnic patient with bilateral paramedian thalamic infarction. *Clin. Neurol.,* 33, 951-956, 1993.
161. TERZANO M.G., PARRINO L., PIETRINI V., MANCIA D., SPAGGIARI M.C., ROSSI G., TAGLIAVINI F. – Precocious loss of physiological sleep in a case of Creutzfeldt Jakob disease: a serial polygraphic study. *Sleep,* 18, 10, 849-858, 1995.
162. TRIDON P., MONTAUT J., PICARD L., WEBER M., ANDRE J.M. – Syndrome de Gelineau et hémangioblastome kystique du cervelet. *Rev. Neurol.* 118, 186-189, 1969.
163. TURNER R.S., CHERVIN R.D., FREY K.A., MINOSHIMA S., KUHL D.E. – Probable diffuse Lewy body disease presenting as REM sleep behavior disorder. *Neurology,* 49, 2, 523-527, 1997.
164. UCHIYAMA M., ISSE K., TANAKA K, YOKOTA N., HAMAMOTO M., AIDA S., ITO Y., YOSHIMUR M., OKAWA M. – Incidental Lewy body disease in a patient with REM sleep behavior disorder. *Neurology,* 45, 709-712, 1995.
165. VILLABLANCA J., MARCUS R. – Sleep wakefulness EEG and behavioral studies of chronic cats without neocortex and striatum: the diencephalic cat. *Arch. Ital. Biol.,* 11a, 348-382, 1972.
166. VITIELLO M.V., POCETA J.S., PRINZ P.N. – Sleep in Alzheimer's disease and other dementing disorders. *Can. J. Psychology,* 45, 221-239, 1991.
167. WIEGANG M., MOLLER A.A., SCHREIBER W., KRIEG J.C., FUCHS D., WACHTER H., HOLSBOER F. – Nocturnal sleep EEG in patients with HIV infection. *Eur. Arch. Psychiatry Clin. Neurosci. ,* 240, 153-158, 1991.
168. WINCHMANN T., DELONG M. – Physiopathology of parkinsonian motor abnormalities. *In:* H. Narabayashi, T. Nagatsu, Y. Mizuno (eds). *Advances in Neurology.* Raven Press, New York, 60, 53-61, 1993.
169. WROE S.J., SANDERCOCK P., BAMFORD J., DENNIS M., SLATTERY J., WARLOW C. – Diurnal variation in incidence of stroke: Oxfordshire community stroke project. *Br. Med. J.,* 304, 6820, 155-157, 1992.
170. YOUNGER D.S., PEDLEY T.A., THORPY M.J.- Multiple sclerosis and narcolepsy: possible similar genetic susceptibility. *Neurology,* 41, 447-448, 1991.

Chapter 58

Sleep as a tool for investigating epilepsies

M. Baldy-Moulinier
Service d'Explorations Neurologiques et Epileptologie, Hôpital Gui de Chauliac, Montpellier, France

Epileptogenesis, whatever the form of epilepsy, is modulated by the states of sleep and wakefulness. The role of sleep in activating interictal discharges, known of since the beginnings of electroencephalography, has led to the use of sleep recordings to explore epilepsy. The modalities of use are varied, ranging from spontaneous sleep onset, during routine electroencephalographic (EEG) tests, to polysomnography, including the techniques of inducing sleep pharmacologically or as a result of sleep deprivation. It is widely indicated for diagnosis, prognosis, pathophysiology, and the determination and localisation of an epileptogenous focus.

METHODOLOGY

The many methods used can be distinguished in terms of how sleep is obtained and the duration and techniques of recording.

Routine EEG recording

Sleep onset is often observed during routine EEG testing. The test conditions *i.e.* supine position, dark, soundproofed room, the relaxing effect of regular hyperventilation are all factors which favour the onset of sleep. If spontaneous sleep onset does occur, the recording should naturally be prolonged, with no untimely interruptions.

Recording naps

This involves finding a suitable time in the sleep-wake cycle to obtain the spontaneous onset of sleep, taking account of fluctuations in wakefulness, particularly between 12 noon and 2 pm. This type of test is particularly useful for infants and young children, to avoid the muscular artefacts during wakefulness and when wakeful activity is normal.

Pharmacologically induced sleep

Sleep can be induced by sedatives administered orally or rectally (promethazine, [11], alimemazine 0.5-1 mg/kg [16]). Benzodiazepines should be avoided because of the rapid, ample rhythms induced by these drugs.

Polysomnography

Polysomnographic recordings of nocturnal sleep, possibly coupled with video recording, cover all the sleep recording indications necessary for exploring epilepsies [6]. Polysomnography (PSG) with video recording is essential for certain forms of epilepsy, particularly in frontal lobe epilepsy, in which seizures are generally linked to sleep and sometimes will only occur during nocturnal sleep.

In these cases of epilepsy, PSG is necessary to establish differential diagnosis with non-epileptic nocturnal paroxysmal manifestations, particularly the different forms of parasomnia. PSG is also useful in suspected sleep apnoea syndrome [23]. These usually refer to obstructive apnoeas which can aggravate epilepsy [24, 37] and add to excessive daytime sleepiness which is already facilitated by certain antiepileptic medications. PSG is usually carried out in a sleep disorders centre. It should be systematically considered in units specialising in the exploration of epilepsy. Likewise, a video recording coupled with PSG should become routine procedure in Sleep Disorders Centres to facilitate diagnosis of epileptic seizures which are often confused with manifestations of parasomnia, due to the absence of surface EEG in mesial frontal epilepsies. As well as the indications specific to epilepsy, there may be the relative indication of obstructive sleep apnoea-hypopnoea syndrome, which may complicate epilepsy and worsen its effects [37]. PSG is generally carried out in a sleep laboratory. Patients must be selected for this test as it is both cumbersome and costly.

Ambulatory recording

This is a cassette recording, which is particularly indicated for subjects who present nocturnal seizures, the epileptic nature of which is uncertain, and where problems arise each time they are admitted to a sleep laboratory. This type of test is particularly valuable for children, provided certain technical precautions are taken and ensuring that the distinction between epileptiform abnormalities and artefacts is determined on the basis of wide experience [13].

Recording after sleep deprivation

Sleep deprivation may be total (24 hours without sleep), particularly in the case of adults, or partial, between 2 and 6 hours, depending on the age of the child [11, 21]. The use of sleep deprivation to activate epileptogenicity through sleep deprivation itself, is advocated by some [15] and rejected by others [35].

Post-deprivation sleep, the duration of the recording after sleep deprivation, and the type of epilepsy are all factors to be taken into consideration [11, 20]. Even though sleep deprivation does not appear to have any real advantage over nocturnal sleep recording [5], this form of testing is widely used for practical purposes, particularly in paediatrics. Certain indications should be mentioned *i.e.* juvenile myoclonic epilepsy [31] and epilepsy with generalised tonic clonic seizures on awakening. The protocols for presurgical investigation for pharmaco-resistant partial epilepsies often include sleep deprivation in association with withdrawal from antiepileptic treatment.

EEG video monitoring

EEG video monitoring consists of continuously recording the patient's behaviour during periods of sleep and wakefulness, in correlation with EEG activity. The indications for this can be considered in the light of two aspects [3]:

- Non surgical indications for diagnosis. These essentially concern children, in the case of infants, to differentiate between epileptic and non epileptic paroxysmal events; in school age children, to determine the type of epilepsy, evaluate the effect of interictal epileptic discharges on the level of consciousness and cognitive performance. Sleep recordings are particularly useful in distinguishing between epileptic and parasomniac manifestations.

- Indications for diagnosis and surgical orientation. These apply to any form of epilepsy which is difficult to diagnose and/or refractory. In the latter case, the EEG video serves to authenticate pharmacoresistance and localise the epileptogenous zone. Recordings during sleep and wakefulness are particularly useful in establishing a diagnosis of pseudo-seizures or psychogenetic seizures.

INDICATIONS

Diagnosis of epilepsy

Defining the epileptic nature of seizures

Epileptic seizures are easily confused with non epileptic seizures of cardiac, vagal or metabolic origin, or of psychogenic incidence. The greatest difficulties arise with nocturnal epileptic seizures. These must be distinguished from sleep-related paroxysmal phenomena, episodes of parasomnia (night terrors, sleepwalking, enuresis), nocturnal paroxysmal dystonias, periodic leg movements, sleep apnoeas and migraine attacks. A diagnosis of epileptic seizures must be based on the findings from the interview, reports of previous history, an analysis of the circumstances, a description of the seizures, usually from the patient's family, but occasionally from the patient himself who will evoke indirect signs of generalised seizures (tongue biting, urination, muscular pain), precursory signs (aura) or certain aspects of the seizure in the case of complex partial seizures. The elements compiled are often insufficient to establish a diagnosis, or may be misleading. A sudden loss of consciousness, axial hypertonia, clonic contractions, urination or even tongue biting do not necessarily imply a generalised epileptic seizure. These elements must be compared against interictal EEG data; but EEG tests may appear normal or conversely, they may indicate epileptiform abnormalities in non epileptic patients, particularly in the case of children [7, 14]. Hesitations may thus often arise [38] and can only be resolved by EEG/video recording of the seizures and interictal activity during sleep and wakefulness.

Defining the type of epileptic seizure

A formal diagnosis of epilepsy may be made for certain patients; nevertheless the type of seizure needs to be defined in whatever circumstances, particularly when the seizures only occur at night and when they prove resistant to antiepileptic medication. The existence of epileptic seizures associated with non epileptic seizures, as in the case of night terrors, sleepwalking, migraine, sleep apnoea syndrome and paroxysmal dystonia all warrant polygraphic sleep recordings. Polysomnographic data shows the continuum which exists between sleep-related paroxysmal arousal phenomena and nocturnal epileptic seizures [25].

Recording interictal EEG discharges in subjects with suspected epilepsy but normal wakeful EEG

Overall, routine EEG tests in the wakeful state show generalised or localised epileptiform abnormalities (spike discharges, spike waves or polyspike waves, vertex sharp slow waves), occurring regularly in 35% of epileptic patients and occasionally in 50%. For 15%, no paroxysmal abnormality is seen [1]. There is a 50% chance of obtaining interictal EEG abnormalities during the first test [27%]. Although the percentage rises with repeated tests, reaching as much as 84% in the third recording and 92% in the fourth, the abnormalities appear to occur all the more frequently when sleep onset occurs during routine testing. For some epilepsies, waking EEG is frequently normal, the ictal and interictal abnormalities sometimes only occurring during sleep. This is often the case in "benign" juvenile epilepsy with midtemporal spikes, in certain symptomatic partial epilepsies and in febrile convulsions. Children are now more often treated after their first seizure, in response to observations which tend to indicate that epilepsy is worsened by repeated seizures [4]. The risk of recurrence estimated at 27% after a first seizure, rises to 60% after a second seizure. This underlines the need for the early diagnosis of epilepsy and the importance of recording EEG signs which have already proven their value in predicting recidivism [34].

The diagnostic indications for PSG particularly apply to:

-"Benign" juvenile epilepsies with midtemporal spikes, due to the fact that in 75% of cases the seizures only occur at night and in 30% of cases, the midtemporal spikes only appear during sleep [17].

- Lennox-Gastaut syndrome to show up:

1. Interictal activity of slow spike waves (2-2.5 Hz) characteristic of this syndrome which may only appear at night.

2. Generalised tonic seizures and/or bursts of rapid polyspikes also characteristic of the syndrome which may present an aspect of subintrant seizures during NREM sleep.

- West's syndrome: in the search for hypsarrhythmia an interictal EEG characteristic of the syndrome, which may only be visible during sleep, with muscular spasms - critical manifestations characterising this epileptic syndrome.

-A syndrome of continuous spike waves in NREM sleep: in this case diagnosis is directly related to the findings of overnight EEG recording. Spike wave discharges, usually generalised, occupy almost 85% of NREM sleep. This continuous epileptic activity is associated with mental retardation in children of 4 to 10 years, with rare partial seizures [27, 31]. Similar activity is seen in Landau-Kleffner's syndrome (fig. 58.1) with deep aphasia occurring at the same time [18]. Polysomnographic findings are also useful in identifying the many forms of epileptic encephalopathies in children and adolescents. Unverricht-Lundborg's disease combining Baltic myoclonus and Mediterranean myoclonus is unusual in that, unlike most epilepsies, the epileptic discharges regress during NREM sleep and increase in REM sleep with the appearance of bursts of multiple spikes in motor sensory areas [32].

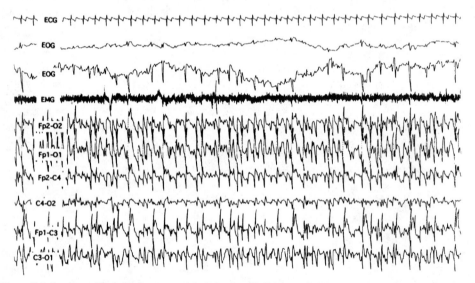

Figure 58.1. Continuous bilateral spike wave activity in Landau-Kleffner's syndrome.

Prognosis for epilepsy

The prognostic indications rely for the most part on a study of the organisation of nocturnal sleep. The absence or reduction of EEG grapho-elements and polysomnographic characteristics such as spindles, regular slow waves in stages 3 and 4, the duration of NREM sleep, the frequency and amplitude of eye movements during REM sleep phases, and the number and duration of REM sleep phases are all of obvious prognostic value. Alterations result from lesions or from the cerebral dysfunction responsible for the epilepsy, particularly in epileptic encephalopathies, as well as from the epilepsy itself. Other alterations such as sleep instability with frequent stage shifts or recurrent arousals during the various stages, reflect the influence of epilepsy itself on sleep (fig. 58.2). The alterations in sleep architecture linked to epilepsy are seen in all forms of epilepsy but are particularly marked in pharmaco-resistant temporal lobe epilepsies [2]. The failure of the different stages to modulate epileptic discharges may be considered as a criterion of possible severity and thus of a poor prognosis.

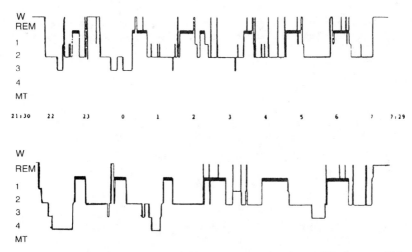

Figure 58.2. Unstable organisation of nocturnal sleep with numerous arousals and stage shifts, in temporal lobe epilepsy. Upper sleep histogram: before treatment. Lower sleep histogram: partial correction of instability after setting up treatment with carbamazepine.

Localising the focus or epileptogenous zone

EEG video monitoring combined with PSG is commonly used in the context of the presurgical investigation of pharmaco-resistant epilepsies. Depending on the case, investigation is carried out using intra-cranial electrodes (subdural grids, deep electrodes implanted by stereotaxy, stereo-EEG). The interictal EEG abnormalities will indicate the irritative zone. The seizures must be recorded, paying particular attention to the first signs at the start of the seizure, in order to determine the epileptogenous zone, the zone from which the seizure emanates and which will require exeresis or disconnection to suppress the seizures.

In temporal lobe epilepsy, epileptic seizures during sleep are rare or occur in relation to an awakening. Sleep recording findings are nevertheless important to determine the laterality of the irritative zone in regard to localising the epileptic discharges. The most accurate indications are provided by interictal EEG activity during REM sleep [36]. Several criteria have been selected to localise the primary focus: the persistence of discharges during the three stages of wakefulness, NREM and REM sleep [28], activation of discharges in REM sleep [22], and optimised localisation during REM sleep [26] (fig. 58.3).

PSG is indispensable in frontal lobe epilepsies due to the specifically conducive effect of sleep on seizures [9], the frequent absence of EEG abnormalities; or the presence of secondary bilateral synchrony [6]. It should be recalled that it was deep electrode recording which elucidated the epileptic nature of nocturnal paroxysmal dystonia.

Pathophysiological study

The mechanisms which come into play in generalised epilepsy are unlike those of partial epilepsies, as evidenced by the different ways in which the two classes of epilepsy alter interictal discharges during sleep. The mechanisms of epileptogenesis can be understood by considering the anatomical and neurochemical systems which control the states of wakefulness and sleep. Several reviews have been made of the findings on the mechanisms generating sleep and their implication in sleep disorders [10, 30]. The modulation of epileptic phenomena by sleep depends on the fluctuating effects of the monoaminergic and cholinergic systems on neuronal excitability and postsynaptic propagation. In partial epilepsies, the increase in discharges during NREM sleep has been related to a gradual reduction in noradrenergic tonic influence, resulting in disinhibition. In generalised epilepsies, the disinhibition of noradrenergic tonicity is accompanied by a thalamocortical excitatory process of phasic nature, underlying the production of sleep spindles [19]. Current thinking in relation to generalised absence epilepsy, suggests the presence of an oscillatory mechanism in the thalamocortical circuitry, abnormally heightened GABAergic inhibition [12] and particular neurones with voltage-dependent calcium canals and low threshold [8].

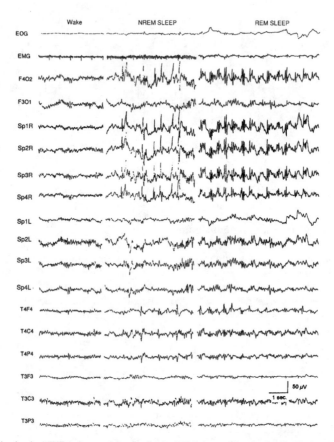

Figure 58.3. Activation by NREM sleep and focalisation by REM sleep of interictal spike discharges in a case of right temporal lobe epilepsy (Sp: sphenoidal electrodes, R: right, L: left).

REFERENCES

1. AIMONE-MARSAN S., ZIVIN L. – Factors related to the occurrence of typical paroxysmal abnormalities in the EEG records of epileptic patients. *Epilepsia*, 11, 361-381, 1970.
2. BALDY-MOULINIER M. – Temporal lobe epilepsy and sleep organization. In: *Sleep and epilepsy.* B. Sterman, M.N. Shouse, P. Passouant (eds), Academic Press, New York, 347-359, 1982
3. BALDY-MOULINIER M. – *Epilepsies en questions.* J. Libbey Eurotext, Paris, 250, 1997.
4. BALDY-MOULINIER M., LERNER-NATOLI M., RONDOUIN G. – La répétition des crises aggrave-t-elle l'épilepsie ? *Epilepsies,* 213, 233-243, 1991.
5. BILLIARD M., ECHENNE B., BESSET A., TOUCHON J., BALDY-MOULINIER M., PASSOUANT P. – Intérêt de l'enregistrement polygraphique du sommeil de nuit chez l'enfant suspect de crises épileptiques lorsque les EEG de routine et après privation de sommeil demeurent normaux. *Rev. EEG Neurophysiol.*, 11, 450-456, 1981.
6. BROUGHTON R.J. – Sleep and sleep deprivation studies in epilepsy. In: *Clinical neurophysiology of epilepsy. EEG Handbook* (revised series, Vol. 4), J.A. Wada, R.J. Ellingson (eds.), Elsevier Science Publ., Amsterdam, 89-119, 1990.
7. CAVAZUTTI G.B., CAPELLA L., NALIN A. – Longitudinal study of epileptiform EEG patterns in normal children. *Epilepsia*, 21, 43-55, 1980.

8. COULTER D.A., ZHANG Y.F. – Thalamocortical generation in vitro: physiological mechanisms, pharmacological control and relevance to generalized absence epilepsy. In: Malafosse *et al.* (eds), *Idiopathic generalized epilepsies: clinical, experimental and genetic aspects*. J. Libbey. London, 123-131, 1994.

9. CRESPEL A., BALDY-MOULINIER M., COUBES P. – The relationship between sleep and epilepsy in frontal and temporal lobe epilepsies: practical and physiopathological considerations. *Epilepsia*, 39, 150-157, 1998.

10. CULEBRAS A. – Neuroanatomic and neurologic correlates of sleep disturbances. *Neurology*, 42 (suppl. 6), 19-27, 1992.

11. DEGEN R., DEGEN H.E. – Sleep and sleep deprivation in epileptology. In: *Epilepsy, sleep and sleep deprivation* (Epil. Res. Suppl.2), R. Degen, E.A. Rodin (eds). Elsevier Science Publ. Amsterdam, 235-260, 1991.

12. DEPAULIS A., DERANSART C., VERGNES M., MARESCAUX C. – Mécanismes GABAergiques dans les épilepsies généraliséees: la dimension neuroanatomique. *Rev. Neurol.*, 153, 15, 8-13, 1997.

13. EBERSOLE J.S. – Outpatient monitoring: ambulatory cassette recording. In: *Clinical neurophysiology of epilepsy. EEG Handbook* (revised series. Vol. 4), J.A. Wada, R.J. Ellingson (eds.), Elsevier Science Publ., Amsterdam, 155-184, 1990.

14. EEG- OLOFSSON O., PETERSEN I., SELDEN U. – The development of the electroencephalogram in normal children from the age of 1 through 15 years: paroxysmal activity. *Neuropaediatrie*, 2, 375-404, 1971.

15. ELLINGSON R.J., WILKEN K., BENNET D.R. – Efficacy of sleep deprivation as an activation procedure in epilepsy patients. *J. Clin. Neurophysiol.*, 1, 83-101, 1984.

16. GASTAUT H., GOMEZ-ALMANZAR M., TAURY M. – The enforced nap: a simple effective method of inducing sleep activation in epileptics. In: *Epilepsy and sleep and sleep deprivation* (Epil. Res. Suppl. 2), R. Degen, E.A. Rodin (eds.), Elsevier Science Publ., Amsterdam, 31-36, 1991.

17. HEUBEL J. – Benign partial epilepsies in childhood. In: *Paediatric epilepsy*, M. Sillanpää, S.I. Johannessen, G. Blennow, M. Daur (eds), Wrightson Biomed. Publ., Petersfield, 111-122, 1990.

18. HIRSCH E., MARESCAUX C., NAQUET P. – Landau-Kleffner syndrom: a clinical and EEG study of five cases. *Epilepsia*, 31, 756-767, 1990.

19. KELLAWAY P. – Sleep and epilepsy. *Epilepsia*, 26 (suppl.1), S15-S30, 1985.

20. KLINGER D., TRÄGNER H., DEISENHAMMER E. – The nature of the influence of sleep deprivation on the EEG. In: *Epilepsy, sleep and sleep deprivation* (Epil. Res. Suppl.2), R. Degen, E.A. Rodin (eds.). Elsevier Science Publ., Amsterdam, 231-234, 1991.

21. KUBICKI D., SCHEULER W., WITTENBECHER H. – Short-term sleep EEG recordings after partial sleep deprivation as a routine procedure in order to uncover epileptic phenomena: an evaluation of 719 EEG recordings. In: *Epilepsy, sleep and sleep deprivation* (Epil. Res. Suppl. 2), R Degen, E.A. Rodin (eds.), Elsevier Science Publ., Amsterdam, 217-230, 1991.

22. LIEB J.P., JOSEPH J.P., ENGEL J.Jr., WALKER J., CRANDALL P.R. – Sleep state and seizure foci related to depth spike activity in patients with temporal lobe epilepsy. *Electroenceph. Clin. Neurophysiol.*, 49, 538-557, 1980.

23. MALOW B.A., FROMES G.A., ALDRICH M.S. - Usefulness of polysomnography in epilepsy patients. *Neurology*, 48, 1389-1394, 1997.

24. MALOW B.A., LEVY K., MATUREN K., BOWES R. – Obstructive sleep apnea is common in medically refractory epilepsy patients. *Neurology*, 55, 1002-1007, 2000.

25. MONTAGNA P. – Nocturnal paroxysmal dystonia and nocturnal wandering. *Neurology*, 42 (suppl. 6), 61-67, 1992.

26. MONTPLAISIR J., LAVERDIÈRE M., SAINT HILAIRE J.M. – In: *Long-term monitoring in epilepsy* (EEG suppl. 37), J. Gotman, J.R. Ives, P. Gloor (eds.), Elsevier, Amsterdam, 215-239, 1985.

27. PATRY G., LYAGOUBI S., TASSINARI C.A. – Sub-clinical electrical status epilepticus induced by sleep in children. A clinical and electroencephalographic study of six cases. *Arch. Neurol.*, 24, 242-252, 1971.

28. ROSSI G.F., COLICCHIO G., POLA P. – Interictal epileptic activity during sleep: a stereo-EEG study in patients with partial epilepsy. *Electroenceph. Clin. Neurophysiol.*, 106, 58-97, 1984.

29. SALINSKY M., KANTER R., DASHEIFF R.M. – Effectiveness of multiple EEGs in supporting the diagnosis of epilepsy: an operational curve. *Epilepsia*, 28, 331-334, 1987.

30. STERIADE M. – Basic mechanism of sleep generation. *Neurology*, 42 (suppl. 6), 9-18, 1992.

31. TASSINARI C.A., BUREAU M., DRAVET C., DALLA BERNARDINA B., ROGER J. – Epilepsie avec pointes-ondes continues pendant le sommeil lent-antérieurement décrite sous le nom d'ESES (épilepsie avec état de mal électroencéphalographique pendant le sommeil lent). In: *Les syndromes épileptiques de l'enfant et de l'adolescent*. J. Roger, C. Dravet, M. Bureau, F.E. Dreifuss, P. Wolf (eds.) John Libbey Eurotext, Paris, 198-209, 1984.

32. TASSINARI C.A., MICHELUCCI R., DANIELE O., DRAVET C., BUREAU M., DALLA BERNARDINA B., RUBBOLI G., VOLDI L., VIGEVANO F., ROGER J. – Sleep polygraphic findings in epileptic encephalopathies from infancy and adolescence. In: *Epilepsy, sleep and sleep deprivation* (Epil. Res. Suppl. 2), R. Degen, E.A. Rodin (eds.) Elsevier Science Publ., Amsterdam, 141-151, 1991.

33. TOUCHON J. – Effects of awakening on epileptic activity in primary generalized myoclonic epilepsy. In: *Sleep and epilepsy*, M.B. Sterman , M.N. Shouse, P. Passouant (eds.). Academic Press, New York, 239-248, 1982.

34. VAN DONSELAAR C.A., SCHIMSHEIMER R.J., GEERTS A.T., DECLERCK A.C. – Value of the EEG in adult patients with untreated idiopathic first seizures. *Arch. Neurol.* 49, 231-237, 1992.

35. VELDUIZEN R., BINNIE C.D., BEINTEMA D.J. – The effect of sleep deprivation on the EEG in epilepsy. *Electroenceph. Clin. Neurophysiol.*, 55, 505-512, 1983.

36. WEISER H.G. – Temporal lobe epilepsy, sleep and arousal stereo-EEG findings. In: *Epilepsy, Sleep and sleep deprivation* (Epil. Res. Suppl. 2), R. Degen, E.R. Rodin (eds.). Elsevier Science Publ., Amsterdam, 97-119, 1991.

37. WYLER A.R., EYMULLER E.A. – Epilepsy complicated by sleep apnea. *Ann. Neurol.*, 9, 403-404, 1981.

38. ZIVIN L., AJMONE-MARSAN C. – Incidence and prognostic significance of « epileptiform » activity in the EEG of non epileptic subjects. *Brain*, 91, 751-758, 1968.

Chapter 59

Dysautonomias

L. Ferini-Strambi and S. Smirne
Centro del Sonno, Ospedale San Raffaele, Miilano, Italia

INTRODUCTION

When a patient refers symptoms mainly suggesting a sleep disorder, such as excessive daytime sleepiness, all-night polysomnography in the sleep lab is fundamental for differential diagnosis. However there are several diseases usually investigated only during wakefulness, but in which a sleep disorder or the nocturnal symptoms may be more important than the diurnal symptoms and may markedly contribute to the severity and evolution of the disease. In these cases the polysomnographic evaluation may contribute to better understanding the disease and act as an aid to therapeutic strategy. An example of this condition is the identification of the cardiac and respiratory disturbances that may only be present during sleep in some stages of diseases such as amyotrophic lateral sclerosis, myasthenia gravis, myotonic muscular dystrophy, Duchenne muscular dystrophy.

It is well known that sleep and the autonomic nervous system (ANS) are closely linked, both anatomically and physiologically. Most of the autonomic alterations occurring during normal sleep involve the cardiovascular and respiratory systems [16]. As extensively reported in chapter 5, nocturnal autonomic alterations are important both in quantitative and qualitative terms. In non-REM sleep, blood pressure, heart rate and respiratory chemosensitivity are reduced. REM sleep is associated with phasic irregularities of heart rate, blood pressure and respiration [5]. In several diseases the delicate balance between the ANS and the neuronal system controlling sleep and wakefulness may be altered.

CENTRAL AND PERIPHERAL DYSAUTONOMIAS

The ANS is involved in the vital regulatory functions. The organ systems (cardiorespiratory, gastrointestinal, genitourinary) regulated by the ANS are largely independent of volitional control, though they are subject to volitional and emotional inputs. The ANS also plays a crucial part in maintaining internal homeostasis by regulating different activities such as blood pressure and body temperature. These multiple responsibilities are accomplished by close integration of the excitatory and inhibitory impulses carried by axons of the parasympathetic (PS) and sympathetic (S) nervous systems: the two divisions of the ANS, which are craniosacral and thoracolumbar in location, are closely integrated and regulated by the central autonomic network (CAN). The CAN has both direct and indirect reciprocal connections with PS and S cranial and spinal cord outflow, as well as with more rostral neuronal aggregates located in the hypothalamus, amygdala and forebrain [3]. The efferent autonomic pathway of PS and S divisions consists of two types of neurons, preganglionic and postganglionic.

ANS disorders may be divided into "primary" and "secondary" autonomic failures [2]. The primary (degenerative) failures include pure autonomic failure (PAF) and multiple-system atrophy (MSA). Autonomic dysfunction in PAF and MSA result from the loss of ganglionic and of preganglionic neurons, respectively. The following three principle forms of motor disturbance occur in MSA but are entirely absent in PAF: 1) striatonigral degeneration (predominance of rigidity without much tremor; poor response to levodopa; 2) olivopontocerebellar atrophy (predominant disturbance of gait with truncal ataxia); 3) pyramidal lesion.

Secondary autonomic failures include vascular, traumatic, inflammatory, metabolic, neoplastic processes that involve the anatomic territories of the central ANS structures; in these cases a hypofunction of the ANS is usually observed, but ANS overactivity is not uncommon. Hypertension and other manifestations of the hyperadrenergic state may be observed in patients with decortication second to head injuries, increased intracranial pressure with brainstem compression or brainstem ischemia. Table 59.1. shows the main causes of central dysautonomias.

Table 59.1 Dysautonomias affecting the central nervous system

Progressive dysautonomias: - pure autonomic failure - multiple-system atrophy
Parkinson's disease Wernicke's encephalopathy Fatal familial insomnia Brainstem disorders: - tumour - stroke - syringobulbia - multiple sclerosis Spinal cord lesions

The ANS is affected in many peripheral neuropathies, although the clinical manifestations may be mild. The peripheral nerve comprises somatic motor, somatic sensory and unmyelinated (somatic and sympathetic) fibres. The composition and type of unmyelinated fibres may vary depending on the nerve involved. Acute and chronic autonomic neuropathies may be distinguished. There is a spectrum of acute autonomic neuropathies. At one extreme is acute panautonomic neuropathy characterised by widespread and severe PS and S failure. Guillain-Barré syndrome is at the other end of the spectrum, where the brunt of the disorder falls on the somatic nervous system. Table 59.2 shows a classification of autonomic neuropathies.

Table 59.2. Classification of the main acute and chronic autonomic neuropathies

Acute autonomic neuropathies	**Chronic autonomic neuropathies**
Pandysautonomia (panautonomic neuropathy) Acute paraneoplastic neuropathy Guillain-Barré syndrome Botulism Porphyria Drug-induced neuropathies (cis-platinum, vincristine, amiodarone) Toxic neuropathies (heavy metals, acrylamide)	Distal sympathetic neuropathies Pure cholinergic neuropathies (Lambert-Eaton syndrome, Adie's syndrome) Combined sympathetic and parasympathetic failure: - amyloid neuropathy* - diabetic neuropathy* - chronic paraneoplastic neuropathy - hereditary motor and sensory neuropathy - Friedreich's ataxia - connective tissue diseases (rheumatoid arthritis, systemic lupus erythematosus) - infections (leprosy, AIDS) * *autonomic dysfunction is often clinically important*

The specific aims of the clinical evaluation of ANS are mainly to: 1) recognise the presence and distribution of an autonomic dysfunction; 2) recognise patterns of autonomic failure that can be related to specific syndromes.

The aims of laboratory evaluation are to detect autonomic failure, quantify its severity, identify the type (cardiovagal, adrenergic, sudomotor) and distribution of deficits, and determine the site of the autonomic lesion. Although many of the tests performed during wakefulness have been in relatively common use for a long time, some of these do not appear to have sufficient sensitivity or specificity to warrant their routine use as tests of autonomic function [17]. On the contrary, it may be useful to evaluate ANS during sleep to clarify certain pathological conditions.

ROUTINE TESTS OF AUTONOMIC FUNCTION DURING WAKEFULNESS

A variety of simple tests have been proposed for assessing different autonomic functions. Routine tests do not necessarily indicate that they are well validated, but they are considered sufficiently useful to warrant being included as part of any complete ANS laboratory [17]. The routine tests include:

1) Orthostatic blood pressure (BP) and heart rate (HR) response to tilt
2) HR response to deep breathing
3) Valsalva manoeuver
4) Beat-to-beat BP to the Valsalva manoeuvre, tilt and deep breathing
5) Sustained handgrip
6) Quantitative sudomotor axon-reflex test
7) Salivation test
8) Plasma catecholamine levels

The most widely used tests measure variations in HR during short periods of wakefulness either at rest or after various stimuli. However, most tests have their limitations: a) the results may be influenced by the emotional state of the patient, and they show large inter- and intra-individual variations [10]; b) in some tests a high degree of cooperation from the patient is required; c) some tests may put certain patients at risk, such as the handgrip test for patients with cerebrovascular or cardiac disorders, and the Valsalva manoeuvre for patients with diabetic retinopathy.

EVALUATION OF AUTONOMIC FUNCTION DURING SLEEP

Sleep acts as a suitable model for studying the neural modulation of ANS under natural and repeatable conditions. The emotional state of the subject has very little influence on ANS activity during sleep, and ANS function during sleep may be studied in all patients, including the demented, unlike some of the traditional tests during wakefulness. It has been determined that the decrease in HR observed when the subject changes from wakefulness to non-REM sleep reflects increased parasympathetic influence on HR, while increased HR and HR variability during REM sleep derives from combined S and PS system influences [21]. Normal HR variability is currently regarded as a reliable index of ANS integrity [22].

An increase in the blood flow to the corpora cavernosa is essential for normal penile erection. This is related to the relaxation of the corporeal and arteriolar smooth muscle under the control of ANS. Penile erection is primarily controlled by PS fibres but S erector pathways are known to exist within the hypogastric nerves [4]. Sexual dysfunction may clearly result from ANS disorders, but psychological factors may be important in many patients. Since nocturnal erections normally occur during REM sleep, sleep may provide an objective assessment of erectile capability.

Moreover, there are other aspects which justify ANS evaluation. In some diseases the autonomic dysfunction may appear or worsen during sleep, and it has been recognised as a possibly severe, or even lethal complication. Sudden infant death syndrome is an example, as well as the nocturnal sudden death in some patients with Duchenne's muscular dystrophy.

Impaired sleep-related penile tumescence

Since the 1940s evidence has shown that men experience several penile erection cycles during sleep. In 1955, Aserinsky and Kleitman [1] demonstrated a strong association between these cycles of nocturnal penile erections and REM sleep. In 1970 Karacan [15] proposed the evaluation of nocturnal penile erections to discriminate between organic and psychogenic sexual impotence. Deficient nocturnal penile tumescence (NPT) indicates an organic dysfunction, but NPT itself will not reveal the aetiology. However, ANS dysfunction in both central and peripheral components has been recognised as one of the main causes of organic impotence.

Cardiac autonomic dysfunctions

Since the 1980s we have systematically studied HR variability in sleep in normal subjects and in patients, to gain a better understanding of the ANS function and to develop procedures for clinical application. Different methodologies have been applied in our studies:
  a) 24-hour ECG monitoring
  b) the evaluation of tonic (vagal activity) and phasic (sympathetic activity) HR modifications in relation to spontaneous body movements (BM) during non-REM and REM sleep
  c) the evaluation of HR variability by power spectral analysis, by measuring the high- (HF, Vagal activity) and low-frequency (LF sympathetic activity) components and the LF/HF ratio (sympatho-vagal balance) in the different stages of sleep.

Methodology a)

Ambulatory 24-hour ECG monitoring is widely used for the diagnosis of cardiac disease, but it has rarely been applied to evaluate ANS activity. To assess the possible use of this method in screening autonomic dysfunction, we first evaluated HR at different ages (from 15 – 19 to 65 – 69 years) by considering the following parameters: mode, mean and standard deviation, distribution by 50 msec of R-wave to R-wave (R-R) intervals during wakefulness and sleep [20]. During wakefulness, R-R interval mode became longer with advancing age, i.e. HR mode became lower; the mean and distribution of R-R intervals were quite similar in the different ages. During sleep, the R-R interval mode progressively increased after the age of 35-39, but the distribution of R-R intervals showed that the balance between PS and S activities changes with advancing age: PS activity becomes stronger, but during sleep loses some of its power to modulate bradycardia.

In diabetic patients with autonomic neuropathy, the R-R interval mode during sleep was longer than during wakefulness, as it was in controls. The distribution pattern of R-R intervals showed that HR variability was decreased in diabetic patients compared to controls during sleep more than during wakefulness.

Methodology b)

During a 20-s period of quiet wakefulness the mean R-R interval before sleep onset was measured. During sleep, the shortest R-R interval in the 20-s period after the onset of BM, and the longest and mean R-R intervals between 30 and 10 s before BM were recorded. The following indices were calculated during sleep: 1) the ratio of mean R-R interval before BM: mean R-R interval during wakefulness (sleep/wakefulness ratio, Rs/w); 2) the ratio of longest R-R interval before BM to the shortest R-R interval after BM (body movement ratio, Rb/m).
Rs/w was considered an index of tonic HR decrease, induced by sleep (predominantly vagal activity). Rb/m was considered an index of phasic HR increase, induced by BM (mainly sympathetic activity). Using this methodology, patients with Alzheimer's disease (AD), Parkinson's disease (PD), multiple sclerosis (MS) and REM sleep behaviour disorder (RBD) were studied.

In AD [14] we found that more than one-third of patients had defective cardiac ANS control, which was predominantly sympathetic in origin (Rb/m values were significantly lower in AD patients than in controls).

In our PD untreated patients [6] a predominantly parasympathetic dysfunction has been observed; this result from patients in early stages of PD suggests that neuronal damage to the vagal dorsal nucleus occurs almost at the same time as that in the substantia nigra.

In MS patients with normal data in the conventional autonomic tests during wakefulness, significantly reduced Rs/W values have been found [7], suggesting that ANS evaluation during sleep may show impairment earlier than the traditional tests during wakefulness.

This finding was confirmed in another study we performed in RBD [8]. A high proportion (64%) of RBD patients had abnormal results in the conventional tests during wakefulness. During sleep these patients showed a reduction in Rs/w and Rb/m values: the reduced nocturnal HR variability was correlated with autonomic impairment during wakefulness. However, patients with normal ANS function during wakefulness were also found to have reduced nocturnal HR variability.

Methodology c)

The HR variability signal was processed using an autoregressive algorithm. The spectral calculations were performed on all the successive 300-s segments of ECG recordings.

In normal subjects a minimal value of the LF component (as well as LF/HF ratio) was found during NREM sleep stages 3 and 4, while LF value (as well as LF/HF ratio) in REM sleep was significantly higher than in other sleep stages. Maximal values of the HF component were characteristic of NREM sleep stages 3 and 4, while minimal values were observed during REM sleep.

Using spectral analysis we studied patients affected by narcolepsy and patients with panic disorder. In narcoleptics [11] a normal HR variability during sleep was found; however, a higher LF/HF ratio and a reduced HF value compared with controls were observed during wakefulness before sleep. This suggests that in healthy subjects, but not in narcoleptic patients, an initial increase in PS activity may be observed in the wakefulness state immediately preceding sleep.

Also patients with panic disorder [9] showed normal HR variability during sleep; an increased LF and a decreased HF were observed during wakefulness before sleep; this high S tone during the wakefulness state is probably the result of cognitive activity and might explain the association of cardiovascular illness and panic disorder.

It is well known that arousal events during sleep are accompanied by marked increases in muscle S activity, HR and blood pressure. Studies concerning the aggregation of the arousal-related phasic events within NREM sleep have led to the identification of a double modality of arousal control [23]: the cyclic alternating pattern (CAP), which corresponds to a prolonged oscillation of the arousal level between two reciprocal functional states termed phase A (greater arousal) and phase B (lesser arousal); the complementary condition non-CAP, was closely related to a degree of stability in sleep depth.

We evaluated the effects of CAP and non-CAP conditions during NREM sleep on HR variability in healthy young subjects [12]. A significant difference between CAP and non-CAP conditions was found in LF component (increased in CAP) and HF component (decreased in CAP). LF/HF ratio was increased in CAP. These data suggest that the studies on nocturnal ANS function should take account of microstructural sleep changes, other than the conventional polysomnographic parameters.

The importance of the evaluation of ANS activity in relation to EEG arousal induced by periodic limb movements (PLMs) or obstructive sleep apnoea hypopnoea (OSAH) has been reported in several studies. It is well known that PLMs are associated with cardiac acceleration; however, a significant rise in HR may be observed in relation to PLMs even in the absence of traditional EEG arousal [24]. This finding suggests the presence of "autonomic" arousals.

Acute apnoea is associated with numerous autonomic responses including the elevation of BP, tachycardia/bradycardia, high S output [13]. The alternating cyclic pattern of sleep and arousal typifying OSAH suggests that other physiological aspects related to apnoeic events and to interval phase arousals, might be monitored to provide an index of sleep disordered breathing. Some authors have shown that auditory-induced arousal from NREM sleep resulted in bursts of S activity associated with increased HR, BP and decreased cardiac output [18]. Auditory stimuli, which failed to induce EEG frequency changes were nonetheless still associated with haemodynamic changes consistent with increased peripheral resistance and transient tachycardia (again, the suggestion of "autonomic arousals"). Other authors recently reported a novel approach to the determination of sleep apnoea based on measuring peripheral circulatory responses in OSAH patients [19]. The apparatus is a finger plethysmograph coupled to a constant volume, variable pressure, pneumatic system. The obtained results suggest that this approach may have important implications for the screening of OSAH, but also provide valuable insight into the cardiovascular consequence of OSAH, as well as of other nocturnal phenomena that usually cause sleep fragmentation.

The evaluation of the relationship between microstructural aspects of sleep and ANS function could change the diagnostic and therapeutic approaches to some sleep disorders, in the near future.

REFERENCES

1. ASERINSKY E., KLEITMAN N.A. - A motility cycle in sleeping infants as manifested by ocular and gross bodily activity. *J. Appl. Physiol.* 8, 11-18, 1955

2. BANNISTER R. - *Autonomi failure: a textbook of the autonomic nervous system.* Oxford Medical Publications, Oxford, 1988.

3. BARRON K.D., CHOKROVERTY S. - Anatomy of the autonomic nervous system: brain and brainstem. In: Low P.A. (ed) *Clinical autonomic disorders,* Little, Brown and Company, Boston, 3-15, 1993.

4. BETTS C.D., FOWLER C.J. – Investigation and treatment of bladder and sexual dysfunction in diseases affecting autonomic nervous system. In: Bannister R., Mathias C.J. (eds), *Autonomic failure,* IIIrd Edition, Oxford Medical Publications, Oxford, 462-478, 1993.

5. CHOKROVERTY S. - Sleep apnea and autonomic failure. In: Low P.A. (ed) « *Clinical autonomic disorders »,* Little, Brown and Company, Boston, 589-603, 1993.

6. FERINI-STRAMBI L., FRANCESCHI M., PINTO P., ZUCCONI M., SMIRNE S. – Respiration and heart rate variability during sleep in untreated Parkinson patients. *Gerontology,* 38, 92-98, 1992.

7. FERINI-STRAMBI L., ROVARIS M., OLDANI A., MARTINELLI V., FILIPPI M., SMIRNE S., ZUCCONI M., COMI G. – Cardiac autonomic function during sleep and wakefulness in multiple sclerosis. *J. Neurol.* 242, 639-643, 1995.

8. FERINI-STRAMBI L., OLDANI A., ZUCCONI M., SMIRNE S. – Cardiac autonomic activity during wakefulness and sleep in REM sleep behavior disorder. *Sleep* 19, 367-369, 1996.

9. FERINI-STRAMBI L., SPERA A., OLDANI A., BATTAGLIA M. – Cardiac autonomic regulation during sleep in panic disorder. *J. Neurol. Neurosurg. Psychiatry* 61, 421-422, 1996.

10. FERINI-STRAMBI L., SMIRNE S. – Cardiac autonomic function during sleep in several neuropsychiatric disorders. *J. Neurol.* 244 (Suppl 1), S29-S36, 1997.

11. FERINI-STRAMBI L., SPERA A., OLDANI A., ZUCCONI M., BIANCHI A., CERUTTI S., SMIRNE S. – Autonomic function in narcolepsy: power spectrum analysis of heart rate variability. *J. Neurol.* 244, 252-255, 1997.

12. FERINI-STRAMBI L., BIANCHI A., ZUCCONI M., OLDANI A., CASTRONOVO V., SMIRNE S. – The impact of cyclic alternating pattern on heart rate variability during sleep in healthy young adults. *Clin. Neurophysiol.* 111, 99-101, 2000.

13. FLECHTER E.C. – Cardiovascular consequences of obstructive sleep apnea: experimental hypoxia and sympathetic activity. *Sleep* 23 (Suppl. 4), S127-S131, 2000.

14. FRANCESCHI M., FERINI-STRAMBI L., MINICUCCI F., SFERRAZZA-PAPA A., SMIRNE S. – Signs of cardiac autonomic dysfunction during sleep in patients with Alzheimer's disease. *Gerontology* 32, 327-334, 1986.

15. KARACAN I. – The developmental aspects and the effect of certain clinical conditions upon penile erection during sleep. *Excerpta Medica International Congress Series* 4, 27-34, 1970.

16. LOEWY A.D., SPYER K.M. – *Central regulation of antonomic functions,* Oxford University Press, Oxford, 1990.

17. LOW P.A. – Laboratory evaluation of autonomic failure. In: Low P.A. (ed), *Clinical autonomic disorders,* Little, Brown and Company, Boston, 169-195, 1993.

18. MORGAN B.J., CRABTREE D.C., PULEO D.S., BADR M.S., TROIBER F., SKATRUD J.B. – Neurocirculatory consequences of abrupt change in sleep stage in humans. *J. Appl. Physiol.* 80, 1627-1636, 1996.

19. SCHNALL R.P., SHLITNER A., SHEFFY J., KEDAR R., LAVIE P. – Periodic, profound peripheral vaso-constriction. A new marker of obstructive sleep apnea. *Sleep* 22, 939-946, 1999.

20. SMIRNE S., MONTANARI C., PINTO P., FERINI-STRAMBI L. – Cardiac autonomic dysfunction during sleep in diabetic patients. In: Smirne S., Franchesci M., Ferini-Strambi L. (eds) *Sleep in medical and neuropsychiatric disorders..* Masson, Milan, 3-12, 1988.

21. SOMERS V.K., PHIL D., DYKEN M., ALLYN L.M., ABBOUND F.M. – Sympathetic-nerve activity during sleep in normal subjects. *N. Engl. J. Med.* 328, 303-307, 1993.

22. STEIN P.K., BOSNER M.S., KLEIGER R.E., CONGER B.M. – Heart rate variability: a measure of cardiac autonomic tone. *Am. Heart J.* 127, 1376-1381, 1994.

23. TERZANO M.G., PARRINO L., SPAGGIARI M.C. – The cyclic alternating pattern sequences in the dynamic organization of sleep. *Electroenceph. Clin. Neurophysiol.* 699, 437-444, 1988.

24. WINKELMAN J.W. – The evoked heart rate response to periodic leg movements of sleep. *Sleep* 22, 575-580, 1999.

Chapter 60

Fibromyalgia and chronic fatigue syndrome: the role of sleep disturbances

H. Moldofsky

Sleep Disorders Clinic of the Centre for Sleep and Chronobiology Ltd., Toronto, Ontario, Canada

HISTORY

Fibromyalgia is a descriptive diagnosis attributed to patients who have generalised muscular pain, multiple areas of tenderness, fatigue and unrefreshing sleep. The term was coined to replace the etiological term, "fibrositis". In 1904 William Gower suggested this term because he speculated that the cause of the pain and tenderness of "lumbago", "muscular rheumatism", or "brachial myalgia" was inflammation of fibrous or connective tissue. However, subsequent studies did not confirm the presence of inflamed tissues so that the diagnosis fell into disrepute, having become a wastebasket diagnosis for poorly understood nonarticular rheumatic complaints. In 1975 Moldofsky *et al.* [47] first described a specific syndrome of generalised pain, tenderness in specific anatomical regions and fatigue that accompanied unrefreshing sleep and disorder sleep physiology. Subsequently, the criteria were refined as the result of a multicentre study of patients with the disorder, which permitted differentiation from patients with known rheumatic or articular disease. These 1990 criteria by Wolfe *et al* [68] have become the standard diagnostic American College of Rheumatology criteria for the diagnosis of the fibromyalgia disorder, descriptive term that replaced "fibrositis" or "fibromyositis".

"Chronic Fatigue Syndrome"(CFS) is also a descriptive label that served to describe patients who complained of persistent fatigue. This term came into existence to describe patients who became chronically ill with fatigue but did not have a defined medical or psychiatric disease. The disabling fatigue is accompanied by joint and muscle pains, headaches, poor concentration, impaired short-term memory, disturbed sleep, recurrent subjective feverish feelings and sore throat. Similar features had been previously described in the 1930's in patients with what became known as "Myalgic Encephalomyelitis". Originally, such people were thought to have become ill as the result of a viral infection, hence the term post-viral fatigue or post infectious neuromyaesthenia. Epidemics of the ailment were given various labels according to the setting in which they occurred, e.g., "Akureyi Disease", "Royal Free Disease". Subsequently in the 1980's other labels that were used included "Epstein Barr Syndrome" or "Chronic Mononucleosis". Such labels were used because of the belief that persistence of the Epstein Barr viral antigens that had been identified in patients with Infectious Mononucleosis were the cause of the chronic illness. When no viral agent could be identified to be specific for the ailment, the United States Centre for Disease Control (CDC) convened a consensus conference where the descriptive label was recommended and criteria approved in 1988 [31]. Subsequently, the CDC criteria were revised so that currently the CDC 1994 Fukuda *et al.* criteria are employed internationally [22].

It soon became evident that there was considerable overlap of the clinical features of CFS and Fibromyalgia [24]. Indeed patients with similar symptoms were first described at the end of the nineteenth century and were subsumed under the umbrella diagnosis of neurasthenia. All these various diagnostic terms share similar features of variable diffuse musculoskeletal pain, fatigue, and unrefreshing sleep. Commonly such patients experience psychological distress with difficulties in concentration, forgetfulness and depression [13].

EPIDEMIOLOGY

Fibromyalgia is a common occurrence. After osteoarthritis, it is the second most common rheumatic ailment. Between 1-2% of the population are affected, with 80% being women who are in the prime of their lives. Chronic fatigue occurs in about 24% of the population [22]. According to the current criteria for CFS, between 0.4% to 1.5% of people are affected in US and Japan [33, 35].

CLINICAL FEATURES

Fibromyalgia is a syndrome that is characterised by chronic diffuse musculoskeletal pain and multiple areas of tenderness in specific anatomic regions (see table 60.1).

Table 60.1

The American College of Rheumatology 1990 Criteria for the Classification of Fibromyalgia [68]
1. A history of widespread musculoskeletal pain for at least three months.   2. Tenderness is found in at least 11 of the 18 anatomical sites with the application of 4 kg. pressure by palpation at the following bilateral anatomical regions: • Occiput: at the suboccipital muscle insertions.   • Low cervical: at the anterior aspects of the intertransverse spaces at C5-C7.   • Trapezius: at the midpoint of the upper border.   • Supraspinatus: at origins, above the scapula spine near the medial border.   • Second rib: at he second costochondral junctions, just lateral to the junctions on upper surfaces.   • Lateral epicondyle: 2 cm distal to the epicondyles.   • Gluteal: in upper outer quadrants of buttocks in anterior fold of muscle.   • Greater trochanter: posterior to the trochanteric prominence.   • Knee: at the medial fat pad proximal to the joint line.

In addition to the diffuse pain and tenderness, there are complaints of chronic fatigue, unrefreshing sleep, cognitive difficulties, and psychological distress. These symptoms may appear for no apparent reason, or may follow any of the following events: a viral illness, an emotionally distressing event, a work-related soft tissue injury, or a whiplash neck injury following an automobile accident.

Often patients describe other pain problems. These include such disorders as migraine or chronic headache, irritable bowel syndrome, atypical chest pain, irritable bladder syndrome, and temporomandibular joint pain. While at times the pain might focus upon a particular region of the body, usually the pain is generalised or without any specific boundaries in segments of the body. The pain varies not only in distribution but also in quality and severity. The diffuse pain is aggravated by unaccustomed physical exertion or exercise, emotional distress, and prolonged stationary posture. Symptoms are often worse with the menses or emerge during the menopause. Climatic conditions such as cold, damp, humid weather, and autumn or winter months are especially troublesome. The symptoms may improve in warm, dry weather, and during the spring or summer season. Temporary relief is often achieved with applications of heat (heating pad, hot shower, a warm bath), massage, acupuncture, stretching, and mild exercise. Fibromyalgia may occur in the context of other potentially disabling conditions, e.g., rheumatic (e.g., rheumatoid arthritis) or connective tissue disease (e.g., systemic lupus erythematosis), HIV, or Lyme disease.

Daytime energy is reduced. The person may fatigue easily with minimal physical exertion and usually require a prolonged time to recover with rest. Strenuous chores become difficult. Physical unfitness ensues. There may be daytime variation in symptoms so that the person may experience reduced pain and fatigue, often during mid-day between 10 a.m. and 3 p.m.

Cognitive functioning is impaired with reduced concentration, word lapses, and forgetfulness. There are difficulties in performing multiple tasks resulting in a reduction in the speed of performance [13]. Work performance declines and disability may be a morbid outcome of the disorder. Often patients are frustrated with their limitations and are irritable. Between 20 and 50% of patients suffer from depression. Some describe anxiety and phobic symptoms.

They may complain of environmental sensitivities. They may describe symptoms suggestive of autonomic disturbances, e.g. unsteadiness, faintness with sudden shift in posture, and heart palpitations.

Fibromyalgia patients use more outpatient health services and medications than aged-matched patients who seek health care and do not have such chronic ailments [67]. Furthermore, there is considerable concern with regard to the social and economic burden for such patients. Insurance claims for disability contribute to the patient's emotional distress. Because of the general impairment to effectively functioning at home and at work, such disability claims result in considerable economic costs to society and to the insurance industry.

The symptoms of fibromyalgia share many of the features of CFS. Indeed there is considerable overlap between these diagnostic groupings [24, 66]. Whereas pain is the focus of interest in the former, fatigue is of concern in the latter diagnosis. To meet the CDC 1994 criteria, CFS patients must have the features described in table 60.2.

Table 60.2

Centre for Disease Control 1994 Criteria for Chronic Fatigue Syndrome [22]

A. Fatigue: Severe, unexplained fatigue that is not relieved by rest, which can cause disability and which has an identifiable onset (i.e., not lifelong fatigue). It must be persistent or relapsing fatigue that lasts for at least six or more consecutive months.
 	
B. Four or more of the following symptoms:
 	
    • impaired memory or concentration problems
    • tender cervical or axillary lymph nodes
    • sore throat
    • muscle pain
    • multi-joint pain
    • new onset headaches
    • unrefreshing sleep
    • post-exertional malaise

DIAGNOSTIC PROCEDURES

Diagnosis depends upon a careful clinical history and physical examination. The history must include the onset and clinical course of symptoms, i.e., pain, fatigue and sleep difficulties (e.g. post infection, trauma, stress etc.), duration, severity, and modulating influences (climate, physical exertion and psychological stress, menses). The following frequent co-morbid conditions should be reviewed: irritable bowel, irritable bladder syndrome, temporomandibular joint disorder, migraine or persistent headache, as well as orthostatic symptoms. Tender point examination is essential to determine if the patient fulfils criteria for fibromyalgia. Neuropsychiatric assessment is required to determine whether there is a brain or peripheral neuromotor disease that could explain the fatigue and weakness, or a primary psychiatric illness, e.g., major depression, behavioural difficulties and psychological disturbance. Such psychosocial concerns as current stressors, personality, interpersonal and economic problems need to be assessed in order to work out a suitable plan of management.

Specific enquiry about the quality of sleep is essential (i.e., presence of light and unrefreshing sleep). The sleep-wake habits, quantity of sleep and use of caffeine, alcohol, and drugs can be assessed with the aid of a sleep diary maintained for at least one week. There should be a specific enquiry about the presence of snoring and interruptions of breathing (sleep apnoea), dysaesthesia and uncontrollable leg movements in the evening and restlessness or kicking during sleep, that might suggest restless legs syndrome/periodic involuntary limb movements disorder, and teeth grinding or bruxism. Significant or irresistible daytime sleepiness should be differentiated from physical and mental exhaustion. Routine lab tests will help to exclude an underlying chronic infectious, metabolic, endocrine, rheumatic, connective tissue or oncological disease. The routine use of expensive tests, e.g. imaging studies, is not required. However, fibromyalgia and CFS patients may be accompanied by concomitant painful and fatiguing conditions, e.g., systemic lupus erythematosis (SLE), rheumatoid arthritis and osteoarthritis that require careful assessment.

Procedures for assessment of sleep disturbances

Where the diagnosis is uncertain or a primary sleep disorder is suspect, overnight polysomnography provides objective evidence for unrefreshing sleep and a rationale for treatment of the underlying sleep disorder. Not only is this laboratory technique helpful for clarifying the diagnosis, it is also useful for properly diagnosing a suspected sleep disorder for formulating a proper rationale of management. These sleep disorders include: the tonic and phasic alpha electroencephalographic (EEG) sleep disorder, periodic K-alpha disorder, restless legs and sleep-related periodic limb movement, and sleep apnoea. The multiple sleep latency test [30] can help to differentiate physical and mental fatigue from excessive daytime sleepiness that results in impaired concentration and deficiency in alertness. The test is especially important in some patients with features of fibromyalgia or CFS who have narcolepsy [3, 15].

COURSE OF ILLNESS

Often the course varies in severity so that subjectively patients describe seasonal variation in pain and fatigue. Although objectively there is no yearly variation in measures of pain in fibromyalgia, such patients complain of troublesome symptoms during damp or humid weather or with the cold weather of the winter season, i.e. during January and February in northern climates [28]. Long-term outcome studies indicate that the pain, fatigue, sleep difficulties and psychological distress of fibromyalgia patients remain unchanged over seven years [70]. Similarly the long-term prognosis for CFS is unfavourable with less than 10% of patients returning to pre-morbid levels of functioning [34].

DIFFERENTIAL DIAGNOSIS

Clinical history, physical examination and standard laboratory tests will exclude chronic pain and fatiguing illnesses that are features of medical and psychiatric disorders.

As indicated above, fibromyalgia and CFS frequently co-exist. The presence of a second clinical disorder, e.g., osteoarthritis, SLE, does not exclude the diagnosis of fibromyalgia.

Fibromyalgia and CFS should be differed from somatoform chronic pain disorder where psychological stresses and depressive symptoms are prominent features in the aetiology or course of illness. Such patients do not have the prominent alpha EEG sleep disorder or any sleep disorder that would result in unrefreshing sleep [44].

Fatigue that occurs as the result of sleep restriction can be determined by history and a sleep diary. Chronic fatigue and morning headache may be the result of a primary sleep disorder, which may be suspected by clinical history and can be confirmed by overnight polysomnography, e.g., loud snoring, interruptions to breathing during sleep or restless legs.

Disorders of excessive daytime sleepiness, e.g., narcolepsy or idiopathic hypersomnolence may be differentiated from CFS and fibromyalgia by history and sleep lab tests. In narcolepsy the MSLT shows an abbreviated onset to sleep over the course of the day and abnormal intrusion of rapid eye movement (REM) sleep on at least 2 of 4 or 5 daytime nap opportunities.

PATHOPHYSIOLOGY

Human experiments on the inter-relationships of sleep and musculoskeletal pain

The type of pain stimuli during sleep in human subjects affects the features in the sleep EEG and the stages of sleep [22]. For example, muscle stimuli when applied during sleep cause a decrease in delta (0.5-3.5 Hz) and sigma (12-14 Hz) and increases in alpha 1 (8-10 Hz) and beta (14.5-25 Hz) brain wave frequencies. During joint pain stimulation the delta, theta (3.5-8 Hz) and alpha 1EEG frequency bands in sleep are decreased. The higher EEG frequencies [alpha 2 (10-12 Hz), sigma and beta bands] are increased. Cutaneous stimuli do not affect the background EEG. Sleepiness does not modulate experimental joint pain.

In the mid 1970's Moldofsky *et al* [47,48] showed that the disruption of stage 4 NREM in normal sedentary people by noise stimuli resulted in complaints of unrefreshing sleep, variable aching, and fatigue. Furthermore, they showed increased sensitivity to the application of a pressure gauge to specific anatomical regions that had been identified in fibromyalgia patients. Other researchers have replicated many of these seminal findings. Older *et al* [51] confirmed that noise-induced disruption of deep NREM sleep was followed by generalised aching and fatigue in healthy subjects, but in their modified protocol did not demonstrate changes in tenderness. However, Lentz *et al.* [38] confirmed the induction of increased tenderness, diffuse myalgia and fatigue over three nights of noise- induced disruption of deep NREM sleep in normal middle-aged women.

PATHOGENESIS

No specific cause of fibromyalgia or CFS is known. None of the routine laboratory procedures show any specific abnormality. Various etiological hypotheses have been proposed for fibromyalgia and CFS. These include the possibilities of any one or a combination of the following factors: a genetic predisposition, infectious agents, neurotransmitter, neuroendocrine, neuroimmune and autonomic disturbances, and psychological stress factors. Although no specific causative factor or biological agent is evident, the fatigue and bodily hypersensitivity are related to disturbance in central nervous system (CNS) functions. In particular, the myalgia and tender points in specific anatomic regions and fatigue are related to the unrefreshing sleep. The poor quality of sleep is related to number of tender points in patients with fibromyalgia, but not to psychological factors [71]. A theoretical mechanism for the induction of pain and fatigue symptoms is that disturbance in the 24-hour pattern of sleep/wakefulness results in metabolic changes that adversely affect the harmonic integration of the various metabolic functions of the body. That is, there is evidence in fibromyalgia and CFS patients for perturbations in circadian sleep/wake-related autonomic, neuroendocrine, and immune functions of the body and alterations in neurotransmitter functions that affect Substance P, catecholamine, serotonin and neuroendocrine metabolism [35, 53].

The role of unrefreshing sleep in the pathogenesis of fibromyalgia and CFS

More than 90% of patients with fibromyalgia and CFS describe disturbed sleep. The sleep is often perceived to be light and unrefreshing irrespective of its duration. Some patients may be aware of restlessness with kicking and involuntary leg movements or sleep-related breathing disorder such loud snoring and interruptions to breathing. On the rare occasion that sleep is restful, there is substantial improvement in daytime symptoms.

Unrefreshing sleep and the alpha EEG sleep disorder

Sleep laboratory studies show disordered EEG sleep physiology, which is the basis of the unrefreshing sleep experience. Most people with fibromyalgia have the alpha (7.5-11 Hz) EEG disorder during non-rapid eye movement (NREM) sleep. Normally the alpha frequency occurs in the EEG during quiet wakefulness. The moment sleep begins the alpha frequency disappears to be replaced by slower frequencies patterns that characterise specific stages of NREM sleep. The alpha pattern reappears with arousals from sleep, by external noise or awakening. In 1975, Moldofsky *et al*. described this alpha EEG NREM sleep anomaly in patients with fibrositis (fibromyalgia) that persisted during stages 2, and/or 3 and 4 NREM sleep. They proposed that this EEG sleep disorder is related to the poor quality of sleep that is perceived to be light and unrefreshing, which is associated with the diffuse myalgia, numerous localised areas of tenderness in specific anatomic areas and mood symptoms [47, 48].

They and subsequently other researchers showed that disruption of stage 4 NREM sleep induces unrefreshing sleep, fatigue, and muscular pain in normal healthy people [37, 50]. This supposed arousal disturbance in NREM sleep, or alpha-delta sleep, pain, and fatigue symptoms that were artificially induced in healthy people are similar to the sleep disturbance and symptoms that are observed in patients with fibromyalgia and CFS [66]. The alpha-delta EEG sleep disorder is seen to reflect a vigilant state during sleep and results in the daytime symptoms of nonrestorative sleep [4]. That is, this sleep physiological disturbance and the coincident perception of light, unrefreshing sleep are not only accompanied by a daytime hyperalgesic state, but also the pervasive fatigue and cognitive impairment that are observed in patients with fibromyalgia and CFS [65]. However, the symptoms were not induced by the noise-induced disruption of stage 4 sleep in a small group of physically fit long distance runners [48]. This observation suggests that physical fitness plays a significant role in the pathogenesis and management of fibromyalgia. Indeed, cardiovascular fitness treatment programmes reduce the pain and weakness in such patients.

A number of investigators have reported on the computerised analyses of the alpha EEG Non REM sleep disorder in patients with fibromyalgia [7, 20, 56, 57]. In particular, the frequency analyses of the sleep EEG by Roizenblatt *et al*. [57] demonstrate three varieties of alpha EEG sleep: phasic (50% of patients vs. 7 % normals), tonic (20 % of patients vs. 9% of normals), and low alpha in 30% of patients vs. 84% of normals. Those with the phasic pattern of the alpha intrusion in deep NREM sleep or alpha-delta sleep are more likely to have increased tenderness after awakening from the overnight sleep, more subjective pain, poor sleep efficiency and less deep NREM sleep than the other groups. Furthermore, morning stiffness, diffuse pain and discomfort after awakening commonly occur in fibromyalgia patients with phasic alpha sleep. The data suggest that the phasic alpha sleep pattern is associated with a longer duration of pain symptoms, the perception of poor sleep and morning pain. The finding of the alpha EEG sleep anomaly in children and their mothers suggests the possibility of a familial or genetic influence in the pathogenesis of the disorder [56].

Although the alpha EEG sleep may be found in non-complaining people, and is not specific for patients with fibromyalgia and CFS, this sleep anomaly may be a sensitive indicator for the nonrestorative sleep and daytime symptoms [45]. Furthermore, as with CFS, and in patients with temporomandibular joint disorder, patients with fibromyalgia have cognitive impairment that relate to the chronic disturbance in sleep [14]. Indeed, such patients are more likely than normal subjects to meet lifetime symptom and diagnostic criteria for CFS, fibromyalgia, irritable bowel syndrome, multiple chemical sensitivities, and headache [1]. Overall the disturbances in sleep physiology contribute to the poor quality of sleep and the unrefreshing sleep features of morning aching, stiffness, fatigue and /or sleepiness [4,7, 20, 44, 47, 48, 56, 57].

Periodic, involuntary, arousal disturbances during sleep

In addition to the alpha EEG sleep patterns, there are some patients with fibromyalgia who have fragmented sleep as a result of sleep-related periodic, involuntary, arousal disturbances that occur over the course of the night. These periodic sleep-related disturbances include: periodic involuntary limb movements (PLMS), sleep-related periodic K-alpha or frequent cyclic alternating EEG sleep pattern [39], and sometimes sleep apnea [2, 37].

1. Periodic Involuntary Limb Movements (PLMS) and Restless Legs Syndrome

One of these periodic arousal disturbances in sleep is the periodic involuntary limb movements (PLMS) disorder. PLMS and Restless Legs Syndrome are reported to occur in patients with fibromyalgia [55, 64]. This primary sleep disorder, which involves movements of the limbs, especially of the lower limbs, occurs at approximately 20 to 40 second intervals during sleep. The PLMS may extend into the waking daytime, and manifest as restless legs syndrome where the person may be aware of an unpleasant sensation in the lower limbs that oblige them to move, massage or stretch the legs in order to achieve temporarily relief. This discomfort in the legs usually becomes troublesome in the evening and interferes with falling asleep, but becomes quiescent in the latter part of the night when sleep is less disturbed. If severe, restless legs syndrome results in considerable daytime discomfort in the legs and sometimes the upper limbs so that the person is unable to sit still for any length of time [49].

2. Periodic K Alpha of the Cyclic Alternating Pattern

Periodic K alpha, which is a variety of the cyclic alternating pattern (CAP), is another variety of periodic arousal disturbance in the EEG sleep that is not accompanied by involuntary limb movements. This periodic disturbance in the sleep EEG is associated with poor quality of sleep in patients with fibromyalgia/CFS and chronic insomnia [39]. The typical stage 2 NREM sleep K-complex is not followed by a sleep spindle (approximately 15 Hz) as occurs normally in quiet sleep, but it is followed, immediately, by an activation of the EEG comprising a burst of alpha activity lasting less than 5 seconds. Subsequently, there is a quiescent periodic in stage 2 NREM sleep. Then in about 30 seconds the cycle is repeated, hence the term CAP. Their frequency is often, more than 10/hour of sleep [39].

3. Sleep Apnea

Sometimes, sleep may be disrupted by another periodic arousal disturbance involving respiration, where there are interruptions to the breathing, or sleep apnea. Whereas fibromyalgia is uncommon in sleep apnea patients, a majority of whom are male [2, 16, 37, 42, 62], sleepy fibromyalgia female patients are more likely to have a greater number of sleep-related disturbances in breathing with reductions in arterial oxygen saturation than a comparative group of control subjects [2]. This sleepy subgroup of fibromyalgia patients has more periodic breathing and greater impairment in the transfer factor for carbon monoxide from the lungs. These patients also have more tender points, about twice as many arousals per hour of sleep and lower sleep efficiency than those patients who do not report sleepiness [61].

Circadian Autonomic Disturbance

Electrocardiographic analyses show increased overnight sympathetic activity in patients where normally sympathetic activity declines during sleep [12, 40]. This observation is consistent with the notion of a chronobiological disorder, which is characterised as an arousal disturbance during the sleep of patients with fibromyalgia that affects not only the pain and fatigue during the waking period [45, 53], but also contributes to various autonomic symptoms including faintness, unsteadiness, palpitations, paraesthesias and blurring of vision.

TREATMENT

Theoretically, treatment that would improve the alpha-delta or phasic alpha sleep disorder, should improve the symptoms of fibromyalgia and CFS. However, to date no specific treatment is known to reduce this specific EEG sleep disorder and have long lasting remedial benefits on subjective sleep quality, pain and fatigue symptoms. Therefore, non-specific remedial measures are used, which empirically are found to show some benefit. These treatments involve pharmacological and psychological methods that aim to improve sleep quality, reduce pain and improve energy.

Pharmacological

Specific Sleep Treatments

Drugs for Sleep

Overall the sleep hygiene and non-specific methods provide a suitable background regimen that is helpful for regularising and facilitating sleep. Medications that aim to improve sleep appear to help in some patients but have not been demonstrated to provide lasting benefit on pain. Those pharmacological treatments that have been well studied include tricyclic antidepressant agents, which facilitate CNS serotonin metabolism. Such drugs as cyclobenzaprine and amitriptyline have continuing favourable effects on sleep up to 2 or 3 months and 5 months respectively [5, 9, 60]. However, in one prospective, long-term, double-blind study where placebo showed a dominant effect, cyclobenzaprine and amitriptyline lost their efficacy in reducing pain symptoms after one month [9]. Furthermore, neither amitriptyline nor cyclobenzaprine reduce the alpha EEG sleep disorder [10, 55]. The results of employing selective serotonin uptake inhibitors, e.g., fluoxetine, show no specific benefit [69], possibly because these drugs tend to be disruptive to sleep. However, fibromyalgia symptoms may be improved when amitriptyline, which is sedative, is combined with fluoxetine [25].

It would be expected that agents that increase deep NREM sleep and reduce the alpha EEG sleep disorder would benefit both the quality of sleep, pain and fatigue symptoms. These symptoms are not directly helped by standard sedatives/hypnotic benzodiazepines [54], but a combination of alprazolam with an analgesic, e.g., ibuprofen, may be helpful [58]. Nonbenzodiazepine hypnotic drugs such as zopiclone [19, 26] and zolpidem [46] improve subjective sleep and daytime tiredness. However, these drugs do not modify alpha EEG sleep nor do they benefit pain symptoms. Other sedatives that are available over the counter such as antihistamines, e.g., diphenhydramine and herbal agents, e.g. valerian, have not been systematically assessed. L-Tryptophan 5 Gms at bedtime facilitates sleep, but there is no effect on alpha EEG sleep, pain and mood symptoms in patients with fibromyalgia [44]. However, 5-hydroxytryptophan, 100 mg, a direct precursor of brain serotonin, tends to improve pain and sleep quality [11], but its effect on sleep physiology of fibromyalgia patients is unknown. Phenothiazines have both sedative and analgesic properties. Chlorpromazine 100 mg decreases pain and improves sleep physiology by reducing the alpha EEG sleep and increasing deep NREM sleep, but the potential for untoward effects over long term use does not make this a desirable drug [44]. Anticonvulsant medications, e.g., gabapentin, pregabalin and carbamazepine also have sedative and antinociceptive effects [65] but have not been systematically assessed in terms of sleep physiology and symptoms of fibromyalgia and CFS. Finally, a preliminary uncontrolled study of gammahydroxybutyrate proved beneficial in reducing the alpha EEG sleep disorder and the pain of patients with fibromyalgia [61].

Sleep-related neuroendocrine methods of treatment

It would be expected that when a patient establishes a regular schedule of favourable sleep-wake habits, the coincident neuroendocrine and neuroimmune rhythms should be facilitated. Melatonin, which has mild hypnotic and chronostatic effects in humans, is not abnormally secreted at night in fibromyalgia patients nor does its administration benefit their symptoms [53]. Cortisone drugs, e.g. prednisone, have no beneficial effect for the pain and fatigue. Furthermore, the use of morning bright light treatment, which tends to modify the timing of the nocturnal secretion of melatonin, does not improve sleep, pain or mood symptoms [52]. On the other hand, nocturnal growth hormone, which is reduced in fibromyalgia [35], does improve the fibromyalgia symptoms [6]. High costs and the current need for injection do not make growth hormone a suitable substance for the treatment of the disorder.

Management of primary sleep disorders

For those patients that are found to have specific primary sleep disorders, e.g., restless legs and PLMS or sleep apnoea, specific remedial measures that have been demonstrated to provide relief for these conditions should be considered. Unfortunately, as yet there are no systematic studies of their benefits on sleep pathologies for the pain and fatigue symptoms in patients with fibromyalgia.

(a) Treatment of PLMS and Restless Legs Syndrome

While there are no published drug studies for control of PLMS and/or restless legs syndrome for patients with fibromyalgia and CFS, such medications may be helpful for the fragmented and disturbed sleep and daytime symptoms. Serum ferritin should be assessed before initiating treatment because low levels serve to unmask the restlessness. Identification and treatment of the underlying anemia may benefit the restlessness and disturbed sleep [29]. Pramipexole 0.25 mg given early in the evening, around 8pm controls the restless legs and sleep disturbance so that the patient is more rested, alert and less fatigued during the day [50]. This dopamine agonist functions better than L-DOPA/carbidopa, which often loses its effect after 3 or 4 hours and may lead to rebound restlessness and sleep disturbances [29]. Usually a dose range of pramipexole from 0.25 to 0.75 mg is adequate. Clonazepam, a benzodiazepine with anticonvulsant properties, reduces the involuntary limb movements, facilitates sleep and reduces nocturnal awakenings. An initial dose of 0.5 mg is used at 8pm, but the dose may need to be adjusted over time according to the individual sensitivity and tolerance. If too much is given, or if this drug is taken late in the evening, there is likelihood of an unpleasant sedative sensation during the next morning. Other drugs that have been reported to be helpful include gabapentin [29], and various long-lasting dopamine agonists i.e., pergolide [17], ropinirole [59], and talipexole [32]. Patient sensitivity to such drugs is variable so that unpleasant side-effects might prevent their use. Opiates, e.g. oxycontin or codeine before bedtime, which sometimes are used if the other drugs are not tolerated, are helpful in not only dulling nocturnal pain, but also in suppressing restless legs [29]. Their use is compromised by their potential addictive properties and unpleasant side-effects.

(b) Treatment of sleep apnoea

Where the patient is found to have sleep apnoea, 3 possible options, medical, surgical, and dental, need to be reviewed to achieve optimum benefit. Once again, while treatments with these techniques are helpful in selected circumstances, their benefit for improving the quality of sleep, fatigue and pain remains to be determined. Nasal continuous positive air pressure delivered by a nasal mask (nasal CPAP) is one of the most commonly used treatment for sleep apnoea. Where there is any anatomical pathology that interferes with the upper airway, e.g. enlarged tonsils, adenoids, enlarged tongue or uvula, appropriate surgery may provide remedial benefit. However, given the sensitivity to pain and postoperative discomfort as well as the possibility for recurrence of the disorder ENT surgeons may prefer to be cautious about intervention. The use of a dental appliance employed during the night that keeps the jaw advanced is helpful for control of snoring and possibly sleep apnoea. Caution is required about the use of any sedatives (e.g., alcohol, or potent hypnotics) and narcotic analgesics that can reduce upper airway muscle tone, blunt the respiratory drive, and depress arousals during sleep with resulting aggravation of apnoeas and arterial oxygen desaturations. However, a preliminary study reveals that zaleplon, which has a short elimination half-life, does not have an adverse effect on sleep apnoea [23].

Behavioural

Non-specific treatments for sleep

Sleep Hygiene

Based upon the principle that there is a deregulation of the sleep and rhythms of the body, management requires regularising both the patient's behavioural and physiological functions. The psychological methods for improving circadian sleep-wake behaviour are central to any cognitive behavioural approach in the management of fibromyalgia. Sleep hygiene enables regulation of the circadian rhythm of sleep-wakefulness where the pattern is disorganised because of faulty sleep habits. Two of the most common faulty habits are variable bedtimes and inadequate nocturnal sleep time that result in disorganisation of circadian rhythms and sleep deprivation with sleepiness and fatigue. Efforts should be directed to stabilise the regulatory functions of sleep by going to bed and awakening at suitable specific times to assure adequate duration of sleep. While sleep needs vary from person to person, the average adult requires approximately 7.5 to 8.5 hours of sleep. The establishment of a regular daily routine not only includes meeting the sleep requirements of the body, but also requires a regular daily schedule of eating nutritious meals and engaging in a suitable exercise programme. A gentle graded aerobic fitness routine is helpful during that part of the day when there is least fatigue and pain, i.e. commonly between 10am and 3pm. Vigorous physical activities should not occur before bedtime because their stimulating effect interferes with falling asleep.

Any disturbances in the sleep environment should be reduced if possible. Behavioural management, e.g. counselling or psychotherapy may be needed to reduce psychological distress where anxiety or depression interferes with sleep. Often patients are sensitive to the effects of caffeine and alcohol. If consumed, these substances should be discouraged because they may interfere with sleep. Other behavioural methods that reduce psychological distress, such as hypnosis and biofeedback treatments, are reported to be helpful in modulating symptoms [27].

Physical methods, which include massage and acupuncture, provide temporary improvement [18]. An appropriate neck support pillow may be helpful to reduce cervical nerve entrapment where there is evidence for cervical injury and referred pain distribution in from C5-C6 and C6-C7 [63]. Based on the postulate that cardiovascular fitness training should improve sleep and symptoms of fibromyalgia [48], a daily gentle graded aerobic fitness program that is tailored to individual needs and circumstances reduces pain threshold measures [43] and improves symptoms [8, 40].

REFERENCES

1. AARON L.A., BURKE M.M., BUCHWALD D. - Overlapping conditions among patients with chronic fatigue syndrome, fibromyalgia, and temporomandibular disorder. *Arch. Intern. Med.*, 24, 221-227, 2000.
2. ALVAREZ LARIO B., ALONSO VALDIVIELSO J.L., ALERE LOPEZ J., MARTEL SOTERES C., VIEJO BANUELOS J.L., MARANON CABELLO A. - Fibromyalgia syndrome: overnight falls in arterial oxygen saturation. *Am. J. Med.*, 101,54-60,1997.
3. AMBROGETTI A., OLSON L.G. - Consideration of narcolepsy in the differential diagnosis of chronic fatigue syndrome. *Medical Journal of Australia.* 160, 426-9, 1994.
4. ANCH A.M., LUE F.A., MAC LEAN A.W., MOLDOFSKY H. - Sleep physiology and psychological aspects of the fibrositis (fibromyalgia) syndrome. *Can. J. Psychol.* 45, 178-84, 1991.
5. BENNETT R.M., GATTER R.A., CAMPBELL S.M., ANDREWS R.P., CLARK S.R., SCAROLA J.A. - A comparison of cyclobenzaprine and placebo in the management of fibrositis. A double-blind controlled study. *Arth. Rheum.* 31, 1535-1542, 1988.
6. BENNETT R.M., CLARK S.R., WALCZYK J. - A randomized, double-blind, placebo-controlled study of growth hormone in the treatment of fibromyalgia. *Am. J. Med.* 104, 227-231, 1998.
7. BRANCO J., ATALAIA A., PAIVA T. - Sleep cycles and alpha-delta sleep in fibromyalgia syndrome. *J. Rheumatol.* 21, 1113-1117,1994.
8. BURCKHARDT C.S., MANNERKORPI K., HEDENBERG L., BJELLE A. - A randomized, controlled clinical trial of education and physical training for women with fibromyalgia. *J. Rheumatol.* 21, 714-720, 1994.
9. CARETTE S., BELL M.V., REYNOLDS W.J., HARAOUI B., McGAIN G.A., BYKERK V.P., EDWORTHY S.M., BARON M., KOEHLER B.E., FAM A.G., *et al.* - Comparison of amitriptyline, cyclobenzaprine and placebo in the treatment of fibromyalgia: A randomized double blind clinical trial *Arth. Rheum.* 37, 32-40, 1994.
10. CARETTE S., OAKSON G., GUIMONT C., STERIADE M. - Sleep electroencephalography and the clinical response to amitriptyline in patients with fibromyalgia. *Arth. Rheum.* 38, 1211-1217, 1995.
11. CARUSO I., SARZI PUTTINI P., CAZZOLA M., AZZOLINI V. - Double-blind study of 5-hydroxytryptophan versus placebo in the treatment of primary fibromyalgia syndrome. M., *Journal of International Medical Research.* 18, 201-9, 1990.
12. COHEN H., NEUMAN L., SHORE M., AMIR CASSUTO Y., BUSKILA D. D. - Autonomic dysfunction in patients with fibromyalgia: application of power spectral analysis of heart rate variability. *Semin. Arth. Rheum.*, 29, 217-27, 2000.
13. CÔTE K.A., MOLDOFSKY H. - Sleep, daytime symptoms, and cognitive performance in patients with fibromyalgia. *J. Rheumatol.*. 24,2014-2023, 1997.
14. DAO T.T., REYNOLDS W.J., TENEBAUM H.C. - Comorbidity between myofascial pain of the masticatory muscles and fibromyalgia. *Journal of Orofacial Pain.* 11, 232-241, 1997.
15. DISDIER P., GENTON P., BOLLA G., VERROT D., CHRISTIDES C., HARLE J.R., WEILLER P.J. - Clinical screening for narcolepsy/cataplexy in patients with fibromyalgia. *Clinical Rheumatology.* 13, 132-4, 1994.
16. DONALD F., ESDAILE J.M., KIMOFF J.R. - Musculoskeletal complaints and fibromyalgia in patients attending a respiratory sleep disorders clinic. *J. Rheumatol.* 23, 1612-16, 1996.
17. EARLEY C.J., YAFFEE J.B., ALLEN R.P. - Randomized, double-blind, placebo-controlled trial of pergolide in restless legs syndrome. *Neurology* 51, 1599-1602, 1998.
18. DELUZE C., BOSIA L., ZIRBS A., CHANTRAINE A., VISCHER T.L. - Electroacupuncture in fibromyalgia: results of a controlled trial. *BMJ* 305, 1249-1252, 1992.
19. DREWES A.M., ANDREASEN A., JENNUM P., NIELSEN K.D. - Zopiclone in the treatment of sleep abnormalities in fibromyalgia. *Scandinavian J. Rheumatol.,* 20, 288-293, 1991.
20. DREWES A.M., NIELSEN K.D., TAAGHOLT S.J., BJERREGARD K., SVENDSEN L., GADE J. - Sleep intensity in fibromyalgia: focus on the microstructure of the sleep process. *Br. J. Rheumatol.*, 34,629-635,1995.
21. DREWES A.M., NIELSEN K.D., ARENDT-NIELSEN L., BIRKET-SMITH L., HANSEN L.M. - The effect of cutaneous and deep pain on the electroencephalogram during sleep: an experimental study. *Sleep* 20, 632-640, 1997.

22. FUKUDA K., STRAUSS S.E., HICKIE I., SHARPE M.C., DOBBINS J.G., KOMAROFF A. - The chronic fatigue syndrome: a comprehensive approach to its definition and study. *Ann. Int. Med.* 121, 953-959, 1994.

23. GEORGE C.F. - Perspectives on the management of insomnia in patients with chronic respiratory disorders. *Sleep.* 23 Suppl 1, S31-35, 2000.

24. GOLDENGERG D.L.- Fibromyalgia and its relation to chronic fatigue syndrome, viral illness and immune abnormalities. *J. Rheumatol.,* 16(Suppl.19), 91-93, 1989.

25. GOLDENBERG D.L., MAYSKIY M., MOSSEY C., RUTHAZER R., SCHMID C. - A randomized, double-blind crossover trial of fluoxetine and amitriptyline in the treatment of fibromyalgia. *Arthr. Rheum..* 39,1852-1859, 1996.

26. GRONBLAD M., NYKANEN J., KONTINNEN Y., JARVINEN E., HELVE T. - Effect of zopiclone on sleep quality, morning stiffness, widespread tenderness and pain and general discomfort in primary fibromyalgia patients. A double-blind randomized trial. *Clinical Rheumatology*, 12, 186-191, 1993.

27. HAANEN H.C.M., HOENDERDOS H.T.W., VAN ROMUNDE L.K.J., HOP W.C., MALLEE C., TERWIEL J.P., HEKSTER G.B. - Controlled trial of hypnotherapy in the treatment of refractory fibromyalgia. *J. Rheumatol.* 18,72-75, 1991.

28. HAWLEY D.J., WOLFE F., LUE F.A., MOLDOFSKY H. - Seasonal symptom severity in patients with rheumatic diseases: A study of 1424 patients. *J. Rheumatol.* 28, 1900-1909, 2001.

29. HENING W., ALLEN R., EARLEY C., KUSHIDA C., PICCHIETTI D., SILBER M. - The treatment of restless legs syndrome and periodic limb movement disorder. An American Academy of Sleep Medicine Review. *Sleep.* 22, 970-999, 1999.

30. HODDES E., ZARCONE V., SMYTHE H., PHILLIPS R., DEMENT W.C. - Quantification of sleepiness: a new approach. *Psychophysiology.* 10, 431-436, 1973.

31. HOLMES G.P., KAPLAN J.E., GANTZ N.M., KOMAROFF A.L., SCHONBERGER L.B., STRAUSS S.E., JONES J.F., DUBOIS R.E., CUNNINGHAM-RUNDLES C., PAHWA S. *et al.* - Chronic fatigue syndrome: a working case definition. *Ann. Intern. Med.* 108, 387-389, 1988

32. INOUE Y., MOTANI H., NANBA K., KAWAHARA R. - Treatment of periodic leg movement disorder and restless leg syndrome with talipexole. *Psychiatry Clin. Neurosci.* 53, 283-285, 1999.

33. JASON L.A., RICHMAN J.A., RADEMAKER A.W., JORDAN K.M., PLIOPLYS A.V., TAYLOR R.R., McCREADY W., HUANG C.F., PLIOPYS S. - A community-based study of chronic fatigue syndrome. *Arch. Intern. Med.* 159,2129-37, 1999.

34. JOYCE J., HOTOPF M., WESSELY S. - The prognosis of chronic fatigue and chronic fatigue syndrome: a systematic review. *QJM.* 90, 223-33, 1997.

35. KAWAKAMI N., IWATA N., FUJIHARA S., KITAMURA T. - Prevalence of chronic fatigue syndrome in a community population in Japan. *Tohoku Journal of Experimental Medicine.* 186, 33-41, 1998.

36. LANDIS C.A., LENTZ M.J., ROTHERMEL J., RIFFLE S.C., CHAPMAN D., BUCHWALD D., SHAVER J.L. - Decreased nocturnal levels of prolactin and growth hormone in women with fibromyalgia. *J. Clin. Endocrinol. & Metabol.* 86, 1672-8, 2001.

37. LARIO B.A., TERAN J., ALONSO J.L. - Lack of association between fibromyalgia and sleep apnoea syndrome. *Ann. Rheumat. Dis.* 51, 108-11, 1992.

38. LENTZ M.J., LANDIS C.A., ROTHERMEL J. SHAVER J.L. - Effects of selective slow wave sleep disruption on musculoskeletal pain and fatigue in middle aged women. *J. Rheumatol.* 26,1586-1592, 1999.

39. MAC FARLANE J.G., SHAHAL B., MOLDOFSKY H. - Periodic K-alpha sleep EEG activity and periodic leg movements during sleep: comparisons of clinical features and sleep parameters. *Sleep* 19, 200-204, 1996.

40. MARTIN L., NUTTING A., MAC INTOSH B.R., EDWORTHY S.M., BUTTERWICK D., COOK J. - An exercise program in the treatment of fibromyalgia. *J. Rheumatol.* 23, 1050-1053, 1996.

41. MARTINEZ-LAVIN M., HERMOSILLO A.G., ROSAS M., SOTO M.E. - Circadian studies of autonomic nervous balance in patients with fibromyalgia: a heart rate variability analysis. *Arth. Rheum.* 41, 1966-71, 1998.

42. MAY K.P., WEST S.G., BAKER M.R., EVERETT D.W. - Sleep apnea in male patients with fibromyalgia syndrome. *Am. J. Med.* 94, 505-508, 1993.

43. McCAIN G.A., BELL D.A., MAI F.M., HALLIDAY P.D. - A controlled study of the effects of a supervised cardiovascular fitness training program on the manifestations of primary fibromyalgia. *Arth. Rheum.*, 31, 1135-1141, 1988.

44. MOLDOFSKY H., LUE F.A. - The relationship of alpha and delta EEG frequencies to pain and mood in 'fibrositis' patients treated with chlorpromazine and L-tryptophan. *Electroencephalogr. Clin. Neurophysiol.* 50, 71-80, 1980.

45. MOLDOFSKY H. - Sleep and Pain. *Sleep Medicine Reviews,* 2001 (in press).

46. MOLDOFSKY H., LUE F.A., MOUSLY C., ROTH-SCHECHTER B., REYNOLDS W.J. - The effect of zolpidem in patients with fibromyalgia: a dose ranging, double blind, placebo controlled, modified crossover study. *J. Rheumatol.* 23, 529-533, 1996.

47. MOLDOFSKY H., SCARISBRICK P., ENGLAND R., SMYTHE H. - Musculoskeletal symptoms and NonREM sleep disturbance in patients with fibrositis syndrome and healthy subjects. *Psychosom. Med.* 37, 341-351,1975.

48. MOLDOFSKY H., SCARISBRICK P. - Induction of neurasthenic musculoskeletal pain syndrome by selective sleep stage deprivation. *Psychosom. Med.* 38, 35-44,1976.

49. MONTPLAISIR J., BOUCHER S., POIRIER G., LAVIGNE G., LAPIERRE O., LESPERANCE P. - Clinical, polysomnographic, and genetic characteristics of restless legs syndrome: a study of 133 patients diagnosed with new standard criteria. *Mov. Disord.* 12, 61-65, 1997.

50. MONTPLAISIR J., NICOLAS A., DENESLE R., GOMEZ-MANCILLA B. - Restless legs syndrome improved by pramipexole: a double-blind randomized trial. *Neurology* 52, 938-943, 1999.

51. OLDER S.A., BATTAFARANO D.F., DANNING C.L., WARD J.A., GRADY E.P., DERMAN S., RUSSEL I.J. - The effects of delta wave sleep interruption on pain thresholds and fibromyalgia-like symptoms in healthy subjects; correlations with insulin-like growth factor I. *J. Rheumatol.* 25,1180-1186, 1998.

52. PEARL S.J., LUE F., MAC LEAN A.W., HESLEGRAVE R.J., REYNOLDS W.J., MOLDOFSKY H. - The effects of bright light treatment on the symptoms of fibromyalgia. *J Rheumatol*, 23,896-902, 1996.

53. PILLEMER S.R., BRADLEY L.A., CROFFORD L.J., MOLDOFSKY H., CHROUSOS G.P. - The neuroscience and endocrinology of fibromyalgia. *Arth. Rheum.* 40, 1928-39, 1997.

54. QUIJADA-CARRERA J., VALENZUELA-CASTANO A., POVEDANO-GOMEZ J., FERNANDEZ-RODRIGUEZ A., HERNANZ-MEDIANO W., GUTIERREZ-RUBIO A., de la IGLESIA-SALGADO J.L., GARCIA-LOPEZ A. - Comparison of tenoxicam and bromazepan in the treatment of fibromyalgia: a randomized, double-blind, placebo-controlled trial. *Pain.* 65, 221-5, 1996.

55. REYNOLDS W.J., MOLDOFSKY H., SASKIN P., LUE F.A. - The effects of cyclobenzaprine on sleep physiology and symptoms in patients with fibromyalgia. *J. Rheumatol.* 18,452-454, 1991.

56. ROIZENBLATT S., TUFIK S., GOLDENBERG J., PINTO L.R., HILARIO M.D., FELDMAN D. - Juvenile fibromyalgia. Clinical and polysomnographic aspects. *J. Rheumatol.* 24, 579–585,1997.

57. ROIZENBLATT S., MOLDOFSKY H., BENEDITO-SILVA A.A., TUFIK S. - Alpha sleep characteristics in fibromyalgia. *Arth. Rheum.* 44, 222-30, 2001.

58. RUSSELL I.J., FLECHTER E.M., MICHALEK J.E., McBROOM P.C., HESTER G.G. - Treatment of primary fibrositis/fibromyalgia syndrome with ibuprofen and alprazolam: A double-blind, placebo-controlled study. *Arth. Rheum.* 34,552-560, 1991.

59. SALETU M., ANDERER P., SALETU B., HAUER C., MANDL M., OBERNDORFER S., ZOGHLAMI A., SALETU-ZYHLARZ G. - Sleep laboratory studies in restless legs syndrome patients as compared with normals and acute effects of ropinirole. 2.Findings on periodic leg movements, arousals and respiratory variables. *Neuropsychobiology.* 41,190-9, 2000.

60. SANTANDREA S., MONTRONE F., SARZI-PUTTINI P., BOCCASSINI L., CARUSO I. - A double-blind crossover study of two cyclobenzaprine regimens in primary fibromyalgia syndrome. *J. Intern. Med. Res.* 21,74-80,1993.

61. SCHARF M.B., HAUCK M., STOVER R., McDANNOLD M., BERKOWITZ D. - Effect of gamma-hydroxybutyrate on pain, fatigue, and the alpha sleep anomaly in patients with fibromyalgia. Preliminary report. *J. Rheumatol.* 25, 1986-1990, 1998.

62. SERGI M., RIZZI M., BRAGHIROLI A., PUTTINI P.S., GRECO M., CAZZOLA M., ANDREOLI A. - Periodic breathing during sleep in patients affected by fibromyalgia syndrome. *Eur. Respir. J.* 14, 203-208, 1999.

63. SMYTHE H.A. - The C6-7 syndrome--clinical features and treatment response. *J. Rheumatol.* 21, 1520-1526, 1994.

64. TAYAG-KIER C.E., KEENAN G.F., SCALZI L.V., SCHULTZ B., ELIOTT J., ZHAO R.H., ARENS R. - Sleep and periodic limb movement in sleep in juvenile fibromyalgia. *Pediatrics.* 106,E70, 2000.

65. TREMONT-LUKATS I.W., MEGEFF C., BACKONJA M.M. - Anticonvulsants for neuropathic pain syndromes: mechanisms of action and place in therapy. *Drugs.* 60, 1029-52, 2000.

66. WHELTON C.L., SALIT I., MOLDOFSHY H. - Sleep, Epstein-Barr virus infection, musculoskeletal pain, and depressive symptoms in chronic fatigue syndrome. *J. Rheumatol.,* 19, 939-943, 1992.

67. WHITE K.P., SPEECHLEY M., HARTH M., OSTBYE T. - The London Fibromyalgia Epidemiology Study:direct health care costs of fibromyalgia syndrome in London, Canada. *J. Rheumatol.* 26, 885-9, 1999.

68. WOLFE F., SMYTHE H.A., YUNUS M.B., BENNETT R.M., BOMBARDIER C., GOLDENBERG D.L., TUGWELL P., CAMPBELL S.M., ABELES M., CLARK P. *et al.* - The American College of Rheumatology 1990 criteria for the classification of fibromyalgia: report of the multicenter criteria committee. *Arth. Rheum.* 33,160-172, 1990.

69. WOLFE F., CATHEY M.A., HAWLEY D.J. - A double-blind placebo controlled trial of fluoxetine in fibromyalgia. *Scandinavian J. Rheumat.* 23, 255-259, 1994.

70. WOLFE F., ANDERSON J., HARKNESS D., BENNETT R.M., CARO X.J., GOLDENBERG D.L., RUSSEL I.J., YUNUS M.B. - Health status and disease severity in fibromyalgia: results of a six-center longitudinal study. *Arth. Rheum.* 40, 1571-9, 1997.

71. YUNUS M.B., AHLES T.A., ALDAG J.C., MASI A.T. - The relationship of the clinical features with psychological status in primary fibromyalgia. *Arth. Rheum.* 34,15-21, 1991.

72. YUNUS M.B., ALDAG J.C. - Restless legs syndrome and leg cramps in fibromyalgia syndrome: a controlled study. *BMJ.* 312 (7042), 1339, 1996.

ANNEXES

Questionnaires and Scales

The questionnaires and scales presented in the following pages are diagnostic aids.

1. General questionnaire on sleep
- Morning questionnaire
Questionnaire given in the morning on awakening to assess sleep continuity and quality.

2. Questionnaires used in the evaluation of insomnia
- Sleep impairment index
Questionnaire used to assess the severity of insomnia.
- Beliefs and Attitudes about Sleep Scale
Questionnaire on the perpetuating factors of insomnia.
- The Pittsburg Sleep Quantity Index (PSQI)
This index assesses the quality of sleep and its disorders during the preceding month. It was specially designed for subjects affected by psychiatric illnesses.
- The Leeds Sleep Evaluation Questionnaire (SEQ)
The aim is to make a quantitative and qualitative assessment of sleep, after the administration of medication acting on sleep.
- The Beck Depression Inventory
An inventory for measuring depression.

3. Scales used in the evaluation of hypersomnia
- The Stanford Sleepiness Scale (SSS)
This was designed to evaluate sleepiness after a night of total sleep deprivation in a normal subject. Its reliability in assessing sleepiness in excessive daytime sleepiness disorders is only relative.
- The Karolinska Sleepiness Scale (KSS)
This provides an appreciation of the degree of subjective sleepiness, at several points during the day.
- The Epworth Sleepiness Scale (ESS)
Self questionnaire assessing the habitual level of daytime sleepiness.

4. Questionnaire on circadian rhythm sleep disorders
- The Horne and Östberg Questionnaire
Self questionnaire to determine circadian typology.

1. General questionnaire on sleep

MORNING QUESTIONNAIRE

First Name   Age
Surname   Date

Instructions: Fill out the different questions of the questionnaire below

1. How long did it take you to fall asleep last night after the lights were turned out? ___hr ____ min

2. How does this compare with the length of time it usually takes you to fall asleep at home? (circle one)

> much longer than usual
> longer than usual .
> same as usual
> shorter than usual
> much shorter than usual

3. How long do you feel you slept last night? _____ hr ____ min

4. How does this compare with the length of time you usually sleep at home? (circle one)

> much longer than usual
> longer than usual
> same as usual
> shorter than usual
> much shorter than usual

5. How many times do you remember waking up last night? _____ times

6. How do you feel right now?

> very tired and sleepy
> awake but not alert
> rested
> alert and wide awake

7. Do you have any physical complaints this morning?
    Describe:

8. Rate the quality of your sleep last night by circling one number in each of the five categories listed below:

My sleep last night was:

	Very	Very	
a. Deep	1 2 3 4 5 6 7	Light	
b. Short	1 2 3 4 5 6 7	Long	
c. Interrupted	1 2 3 4 5 6 7	Uninterrupted	
d. Dreamless	1 2 3 4 5 6 7	Many dreams	
e. Restless	1 2 3 4 5 6 7	Restful	

9. Do you remember any dreams from last night?   yes   n o
If yes, please describe them in detail:

10. What awakened you this morning?
    noise
    discomfort
    technician
    spontaneous
    other _____

11. In general, how would you say your sleep last night compared with your usual sleep at home?

    much worse than usual
    worse than usual
    same as usual
    better than usual
    much better than usual

12. Please add any additional comments or information:

2. Questionnaires used in the evaluation of insomnia

SLEEP IMPAIRMENT INDEX

First Name   Age
Surname   Date

Instructions: Circle the most relevant number in each question of the questionnaire below

1. Please rate the current severity of your insomnia problem(s):

	none	mild	moderate	severe	very
Difficulty falling asleep	1	2	3	4	5
Difficulty staying asleep	1	2	3	4	5
Problem waking up too early	1	2	3	4	5

2. How satisfied are you with your current sleep pattern?

Very satisfied		Moderately satisfied		Very dissatisfied
1	2	3	4	5

3. To what extent do you consider your sleep problem to INTERFERE with your daily functioning (e.g. daytime fatigue, ability to function at work/daily chores, concentration, memory, mood, etc.)?

Not at all	A little	Somewhat	Much	Very much
1	2	3	4	5

4. How NOTICEABLE to others do you think your sleep problem is in terms of impairing the quality of your life?

Not at all	A little	Somewhat	Much	Very much
1	2	3	4	5

5. How CONCERNED are you about your current sleep problem?

Not at all	A little	Somewhat	Much	Very much
1	2	3	4	5

6. To what extent do you believe the following factors are contributing to your sleep problem?

	None		Some		Much
Cognitive disturbance (racing thoughts at night):	1	2	3	4	5
Somatic disturbances (muscular tension; pain):	1	2	3	4	5
Bad sleeping habits:	1	2	3	4	5
Natural aging process:	1	2	3	4	5

7. After a poor night's sleep, which of the following problems do you experience on the next day? Circle all those that apply.

a. Daytime fatigue: tired, exhausted, washed out, sleepy.
b. Difficulty functioning: performance impairment at work/daily chores, difficulty concentrating, memory problems.
c. Mood problems: irritable, tense, nervous, groggy, depressed, anxious, grouchy, hostile, angry, confused.
d. Physical symptoms: muscle aches/pain, light-headed, headache, nausea, heartburn, muscle tension.
e. None.

Score:

A cut-off score of 15 is judged to reflect an insomnia complaint which is below clinical threshold (minimal or no sleep difficulties, minimal impairment, and no or little distress).

References:
Taken from: MORIN C.M. - Sleep impairment Index. In: *Insomnia: Psychological Assessment and Management.* 199-200, 1993, with permission of Guilford Publications, Inc.
BASTIEN C.H., VALLIÈRES A., MORIN C.M. – Validation of the insomnia severity index as an outcome measure for insomnia research, *Sleep Medicine* 2, 297-307, 2001.

BELIEFS AND ATTITUDES ABOUT SLEEP SCALE

First Name   Age
Surname   Date

Instructions: Several statements reflecting people's beliefs and attitudes about sleep are listed below. Please indicate to what extent you personally agree or disagree with each statement. There is no right or wrong answer. For each statement, place a mark (/) along the line wherever your personal rating falls. Try to use the whole scale, rather than placing your marks at one end of the line.

A. I need 8 hours of sleep to feel refreshed and function well during the day.
Strongly disagree _____ Strongly agree

B. When I don't get a proper amount of sleep on a given night, I need to catch up on the next day by napping or on the second night by sleeping longer.
Strongly disagree _____ Strongly agree

C. Because I am getting older, I need less sleep.
Strongly disagree _____ Strongly agree

D. I am worried that if I go for one or two nights without sleep, I may have a nervous breakdown.
Strongly disagree _____ Strongly agree

E. I am concerned that chronic insomnia may have serious consequences for my physical health.
Strongly disagree _____ Strongly agree

F. By spending more time in bed, I usually get more sleep and feel better the next day.
Strongly disagree _____ Strongly agree

G. When I have trouble getting to sleep, I should stay in bed and try harder.
Strongly disagree _____ Strongly agree

H. I am worried that I may lose control over my abilities to sleep.
Strongly disagree _____ Strongly agree

I. Because I am getting older, I should go to bed earlier in the evening.
Strongly disagree _____ Strongly agree

J. After a poor night's sleep, I know that it will interfere with my daily activities on the next day.
Strongly disagree _____ Strongly agree

K. In order to be alert and functioning well during the day, I am better off taking a sleeping pill rather than having a poor night's sleep.
Strongly disagree _____ Strongly agree

L. When I feel irritable, depressed, or anxious during the day, it is mostly because I did not sleep well the night before.
Strongly disagree _____ Strongly agree

M. Because my bed partner falls asleep as soon as his or her head hits the pillow and stays asleep through the night, I should be able to do so too.
Strongly disagree _____ Strongly agree

N. I feel that insomnia is basically the result of aging, and that there isn't much that can be done about the problem.
Strongly disagree _____ Strongly agree

O. I am sometimes afraid of dying in my sleep.
Strongly disagree _____ Strongly agree

P. When I have a good night's sleep, I know that I will have to pay for it on the following night.
Strongly disagree _____ Strongly agree

Q. When I sleep poorly on one night, I know it will disturb my sleep schedule for the whole week.
Strongly disagree _____ Strongly agree

R. Without an adequate night's sleep, I can hardly function the next day.
Strongly disagree _____ Strongly agree

S. I can't ever predict whether I'll have a good or poor night's sleep.
Strongly disagree _____ Strongly agree

T. I have little ability to manage the negative consequences of disturbed sleep.
Strongly disagree _____ Strongly agree

U. When I feel tired, have no energy, or just seem not to function well during the day, it is generally because I did not sleep well the night before.
Strongly disagree _____ Strongly agree

V. I get overwhelmed by my thoughts at night and often feel I have no control over my racing mind.
Strongly disagree _____ Strongly agree

W. I feel I can still lead a satisfactory life despite sleep difficulties.
Strongly disagree _____ Strongly agree

X. I believe insomnia is essentially the result of a chemical imbalance.
Strongly disagree _____ Strongly agree

Y. I feel insomnia is ruining my ability to enjoy life and prevents me from doing what I want.
Strongly disagree _____ Strongly agree

Z. I avoid or cancel obligations (social, family, occupational) after a poor night's sleep.
Strongly disagree _____ Strongly agree

AA. A "nightcap" before bedtime is a good solution to sleeplessness.
Strongly disagree _____ Strongly agree

BB. Medication is probably the only solution to sleeplessness.
Strongly disagree _____ Strongly agree

CC. My sleep is getting worse all the time, and I don't believe anyone can help.
Strongly disagree _____ Strongly agree

DD. It usually shows in my physical appearance when I haven't slept well.
Strongly disagree _____ Strongly agree

This questionnaire is designed at identifying dysfunctional sleep-related cognitions.

Reference:

Taken from: MORIN C.M. – Beliefs and Attitudes about Sleep Scale. In: *Insomnia: Psychological Assessment and Management.* 201-204, 1993, with permission of Guilford Publications, Inc.

THE PITTSBURGH SLEEP QUALITY INDEX (PSQI)

First Name   Age
Surname   Date

Instructions: The following questions relate to your usual sleep habits during the past month only. Your answers should indicate the most accurate reply for the majority of days and nights in the past month. Please answer all questions

1. During the past month, when have you usually gone to bed at night?
   USUAL BED TIME _____
2. During the past month, how long (in minutes) has it usually take you to fall asleep each night?
   NUMBER OF MINUTES _____
3. During the past month, when have you usually gotten up in the morning?
   USUAL GETTING UP TIME _____
4. During the past month, how many hours of *actual sleep* did you get at night? (This may be different than the number of hours you spend in bed.)
   HOURS OF SLEEP PER NIGHT _____

For each of the remaining questions, check the one best response. Please answer *all* questions.

5. During the past month, how often have you trouble sleeping because you...

(a) Cannot get to sleep within 30 minutes

| Not during the past month _____ | Less than once a week _____ | Once or twice a week _____ | Three or more times a week _____ |

(b) Wake up in the middle of the night or early morning

| Not during the past month _____ | Less than once a week _____ | Once or twice a week _____ | Three or more times a week _____ |

(c) Have to get up to use the bathroom

| Not during the past month _____ | Less than once a week _____ | Once or twice a week _____ | Three or more times a week _____ |

(d) Cannot breathe comfortably

| Not during the past month _____ | Less than once a week _____ | Once or twice a week _____ | Three or more times a week _____ |

(e) Cough or snore loudly

| Not during the past month _____ | Less than once a week _____ | Once or twice a week _____ | Three or more times a week _____ |

(f) Feel too cold

| Not during the past month _____ | Less than once a week _____ | Once or twice a week _____ | Three or more times a week _____ |

(g) Feel too hot

| Not during the past month _____ | Less than once a week _____ | Once or twice a week _____ | Three or more times a week _____ |

(h) Had bad dreams

| Not during the past month _____ | Less than once a week _____ | Once or twice a week _____ | Three or more times a week _____ |

(i) Have pain

| Not during the past month _____ | Less than once a week _____ | Once or twice a week _____ | Three or more times a week _____ |

(j) Other reason(s), please describe _____

_____

How often during the past month have you had trouble sleeping because of this?

| Not during the past month _____ | Less than once a week _____ | Once or twice a week _____ | Three or more times a week _____ |

6. During the past month, how would you rate your sleep quality overall?
   Very good   _____
   Fairly good   _____
   Fairly bad   _____
   Very bad   _____

7. During the past month, how often have you taken medicine (prescribed or "over the counter") to help you sleep?

| Not during the past month _____ | Less than once a week _____ | Once or twice a week _____ | Three or more times a week _____ |

8. During the past month, how often have you had trouble staying awake while driving, eating meals, or engaging in social activity?

| Not during the past month _____ | Less than once a week _____ | Once or twice a week _____ | Three or more times a week _____ |

9. During the past month, how much of a problem has it been for you to keep up enough enthusiasm to get things done?
   No problem at all   _____
   Only a very slight problem _____
   Somewhat of a problem   _____
   A very big problem   _____

10. Do you have a bed partner or roommate?
    No bed partner or roommate   _____
    Partner / roommate in other room   _____
    Partner in same room, but not same bed _____
    Partner in same bed   _____

If you have a roommate or bed partner, ask him / her how often in the past month you have had...

a. Loud snoring

| Not during the past month _____ | Less than once a week _____ | Once or twice a week _____ | Three or more times a week _____ |

b. Long pauses between breaths while asleep

| Not during the past month _____ | Less than once a week _____ | Once or twice a week _____ | Three or more times a week _____ |

c. Legs twitching or jerking while you sleep

| Not during the past month _____ | Less than once a week _____ | Once or twice a week _____ | Three or more times a week _____ |

d. Episodes of disorientation or confusion during sleep

Not during the	Less than	Once or	Three or more
past month _____	once a week _____	twice a week _____	times a week _____

e. Other restlessness while you sleep: please describe _____

_____

Not during the	Less than	Once or	Three or more
past month _____	once a week _____	twice a week _____	times a week _____

Scoring instructions: The Pittsburgh Sleep Quality Index (PSQI) contains 19 self-rated questions and 5 questions rated by the bedpartner or roommate (if one is available). Only self-rated questions are included in the scoring. The 19 self-rated items are combined to form seven "components" scores, each of which has a range of 0-3 points. In all cases, a score of "0" indicates no difficulty, while a score of "3" indicates severe difficulty. The seven component scores are then added to yield one "global" score, with a range of 0-21 points, "0" indicating no difficulty and "21" indicating severe difficulties in all areas.
    Scoring proceeds as follows:

Component 1: Subjective sleep quality
  Examine question #6, and assign scores as follows:

Response	Component 1 score
"Very good"	0
"Fairly good"	1
"Fairly bad"	2
"Very bad"	3

                                               Component 1 score: _____

Component 2: Sleep latency
1. Examine question #2, and assign scores as follows:

Response	Score
< or = 15 minutes	0
16 - 30 minutes	1
31 – 60 minutes	2
> 60 minutes	3

                               Question #2 score: _____

2. Examine question #5a, and assign scores as follows:

Response	Score
Not during the past month	0
Less than once a week	1
Once or twice a week	2
Three or more times a week	3

                               Question #5a score: _____

3. Add #2 score and #5a score

                               Sum of #2 and #5a: _____

4. Assign component 2 score as follows:

Sum of #2 and #5a	Component 2 score
0	0
1-2	1
3-4	2
5-6	3

                               Component 2 score: _____

Component 3: Sleep duration
   Examine question #4, and assign scores as follows:

Response	Component 3 score
> 7 hours	0
6-7 hours	1
5-6 hours	2
< 5 hours	3

   Component 3 score: _____

Component 4: Habitual sleep efficiency
(1) Write the number of hours slept (question # 4) here: _____
(2) Calculate the number of hours spent in bed:
   Getting up time (question #3): _____
   Bedtime (question #1): _____
   Number of hours spent in bed: _____
(3) Calculate habitual sleep efficiency as follows:
   (Number of hours slept / Number of hours spent in bed) x 100 = Habitual sleep efficiency (%)
   (_____ / _____) x 100 = _____%
(4) Assign component 4 score as follows:

Habitual sleep efficiency %	Component 4 score
> 85%	0
75-84%	1
65-74%	2
< 65%	3

   Component 4 score: _____

Component 5: Sleep disturbances
(1) Examine questions # 5b-5j, and assign scores for each question as follows:

Response	Score
Not during the past month	0
Less than once a week	1
Once or twice a week	2
Three or more times a week	3

#5b score	_____
c score	_____
d score	_____
e score	_____
f score	_____
g score	_____
h score	_____
i score	_____
j score	_____

(2) Add the scores for questions # 5b-5j:

   Sum of # 5b-5j: _____

(3) Assign component 5 scores as follows:

Sum of # 5b-5j	Component 5 score
0	0
1 – 9	1
10 -18	2
19 - 27	3

   Component 5 score: _____

Component 6: Use of sleeping medication
   Examine question #7 and assign scores as follows:

Response	*Component 6 score*
Not during the past month	*0*
Less than once a week	*1*
Once or twice a week	*2*
Three or more times a week	*3*

Component 6 score: _____

Component 7: Daytime dysfunction
(1) Examine question #8, and assign scores as follows:

Response	*Score*
Never	*0*
Once or twice	*1*
Once or twice each week	*2*
Three or more times each week	*3*

Question #8 score: _____

(2) Examine question #9, and assign scores as follows:

Response	*Score*
No problem at all	*0*
Only a very slight problem	*1*
Somewhat of a problem	*2*
A very big problem	*3*

Question # 9 score: _____

(3) Add the scores for question #8 and #9:

Sum of #8 and #9: _____

(4) Assign component 7 score as follows:

Sum of #8 and #9	*Component 7 score*
0	*0*
1-2	*1*
3-4	*2*
5-6	*3*

Component 7 score: _____

Global PSQI Score
   Add the seven component scores together:

Global PSQI Score: _____

Reference:
Taken from: BUYSSE D.J., REYNOLDS III C.F., MONK T.H., BERMAN S.R., KUPFER D.J. The Pittsburgh sleep quality index: a new instrument for psychiatric practice and research. *Psychiatry Research*, 28, 193-213, 1989 – *with permission from Elsevier Science.*

THE LEEDS SLEEP EVALUATION QUESTIONNAIRE (SEQ)

First Name   Age
Surname   Date

Instructions: A 10-cm line separates the two halves of each question. The questionnaire instructions are: "Each question is answered by placing a vertical mark on the answer line. If no change was experienced then place your mark in the middle of the line. If a change was experienced then the position of your mark will indicate the nature and extent of the change, i.e. large changes near the ends of the line, small changes near the middle. As an example, a mark placed like this would indicate a limited change

How would you compare getting to sleep using the medication with getting to sleep normally, i.e. without medications?

1) Harder than usual /
   easier than usual

2) Slower than usual /
   quicker than usual

3) Felt less drowsy than usual /
   felt more drowsy than usual

How would you compare the quality of sleep using the medication with non-medicated (your usual) sleep?

4) More restless than usual /
   more restful than usual

5) More periods of wakefulness
   than usual / fewer periods of
   wakefulness than usual

How did your awakening after medication compare with your usual pattern of awakening?

6) More difficult than usual /
   easier than usual

7) Took longer than usual /
   took shorter than usual

How did you feel on waking?

8) Tired / alert

How do you feel now?

9) Tired / alert

How was your sense of balance and coordination upon getting up?

10) More clumsy than usual /
    less clumsy than usual

Score:

This questionnaire contains 10 questions referring to 4 different items: sleep onset (1,2,3) sleep quality (4,5), morning awakening (6,7) and state after awakening (9,10).

The score for each item is the sum of the distances separating the mark (\) from the central grey box for each question referring to the item. The distances are negative to the left of the box, positive to the right of the box.

References:

Taken from: PARROTT A.C., HINDMARCH I. The Leeds Sleep Evaluation Questionnaire in Psychopharmacological Investigations – a Review. *Psychopharmacology,* II, 173-179, 1980 – *with permission of Springer-Verlag GmbH & Co. KG.*

THE BECK DEPRESSION INVENTORY

First Name   Age
Surname   Date

Instructions: Circle the most relevant number in each question of the questionnaire below

A (Mood)

    0   I do not feel sad
    1   I feel blue or sad
    2a I am blue or sad all the time and I can't snap out of it.
    2b I am so sad or unhappy that it is very painful
    3   I am so sad or unhappy that I can't stand it

B (Pessimism)

    0   I am not particularly pessimistic or discouraged about the future
    1a I feel discouraged about the future
    2a I feel I have nothing to look forward to
    2b I feel that I won't ever get over my troubles
    3   I feel that the future is hopeless and that things cannot improve.

C (Sense of Failure)

    0 I do not feel like a failure
    1 I feel I have failed more than the average person
    2a I feel I have accomplished very little that is worthwhile or that means anything
    2b As I look back on my life all I can see is a lot of failures
    3 I feel I am a complete failure as a person (parent, husband, wife)

D (Lack of Satisfaction)

    0 I am not particularly dissatisfied
    1a I feel bored most of the time
    1b I don't enjoy things the way I used to
    2 I don't get satisfaction out of anything any more
    3 I am dissatisfied with everything

E (Guilty Feeling)

    0 I don't feel particularly guilty
    1 I feel bad or unworthy a good part of the time
    2a I feel quite guilty
    2b I feel bad or unworthy practically all the time now
    3 I feel as though I am very bad or worthless

F (Sense of Punishment)

    0 I don't feel I am being punished
    1 I have a feeling that something bad may happen to me
    2 I feel I am being punished or will be punished
    3a I feel I deserve to be punished
    3b I want to be punished

G (Self Hate)

0   I don't feel disappointed in myself
1a I am disappointed in myself
1b I don't like myself
2   I am disgusted with myself
3   I hate myself

H (Self Accusation)

0   I don't feel I am any worse than anybody else
1   I am very critical of myself for my weakness or mistakes
2a I blame myself for everything that goes wrong
2b I feel I have many bad faults

I (Self-punitive Wishes)

0   I don't have any thoughts of harming myself
1   I have thoughts of harming myself but I would not carry them out.
2a I feel I would be better off dead
2b I have definite plans about committing suicide
2c I feel my family would be better off if I were dead
3   I would kill myself if I could

J (Crying Spells)

0   I don't cry any more than usual
1   I cry more now than I used to
2   I cry all the time now. I can't stop it.
3   I used to be able to cry but now I can't cry at all even though I want to

K (Irritability)

0   I am no more irritated now than I ever am
1   I get annoyed or irritated more easily than I used to
2   I feel irritated all the time
3   I don't get irritated at all at the things that used to irritate me

L (Social Withdrawal)

0   I have not lost interest in other people
1   I am less interested in other people now than I used to be
2   I have lost most of my interest in other people and have little feeling for them
3   I have lost all my interest in other people and don't care about them at all

M (Indecisiveness)

0   I make decisions about as well as ever
1   I am less sure of myself now and try to put off making decisions
2   I can't make decisions any more without help
3   I can't make any decisions at all any more

N (Body Image)

0   I don't feel I look any worse than I used to
1   I am worried that I am looking old or unattractive
2   I feel that there are permanent changes in my appearance and they make me look unattractive
3   I feel that I am ugly or repulsive looking

O (Work Inhibition)

0   I can work about as well as before
1a It takes extra effort to get started at doing something
1b I don't work as well as I used to
2   I have to push myself very hard to do anything
3   I can't do any work at all

P (Sleep Disturbance)

0   I can sleep as well as usual
1   I wake up more tired in the morning than I used to
2   I wake up 1-2 hours earlier than usual and find it hard to get back to sleep
3   I wake up earlier every day and can't get more than 5 hours sleep

Q (Fatigability)

0   I don't get any more tired than usual
1   I get tired more easily than I used to
2   I get tired from doing anything
3   I get too tired to do anything

R (Loss of Appetite)

0   My appetite is no worse than usual
1   My appetite is not as good as it used to be
2   My appetite is much worse now
3   I have no appetite at all any more

S (Weight Loss)

0   I haven't lost much weight, if any, lately
1   I have lost more than 5 pounds
2   I have lost more than 10 pounds
3   I have lost more than 15 pounds

T (Somatic Preoccupation)

0   I am no more concerned about my health than usual
1   I am concerned about aches and pains *or* upset stomach *or* constipation *or* other
    unpleasant feelings in my body
2   I am so concerned with how I feel that it's hard to think of much else
3   I am completely absorbed in what I feel

U (Loss of Libido)

0   I have not noticed any recent change in my interest in sex
1   I am less interested in sex than I used to be
2   I am much less interested in sex now
3   I have lost interest in sex completely

Score

Depth of depression	Mean	S.D.
0: none	10.9	8.1
1: mild	18.7	10.2
2: moderate	25.4	9.6
3: severe	30.0	10.6

Reference:

Taken from: BECK A.T., WARD C.H., MENDELSON M., MOCK J., ERBAUGH J. An Inventory for Measuring Depression. *Arch. Gen. Psychiat.* 4, 561-571, 1961 – *with permission of the American Medical Association.*

3. Scales used in the evaluation of hypersomnia

THE STANFORD SLEEPINESS SCALE (SSS)

First Name   Age
Surname   Date

How many hours did you sleep last night? _____
This morning, do you feel your amount of sleep was (check one):
        more than sufficient   _____
        sufficient   _____
        insufficient   _____

Instructions: Choose the statement below which best describes your state of sleepiness. Then place the number of that statement in the appropriate box. Write down under "comments" if you have taken medication, indulged in exercise, or have done anything else that you feel may have influenced your sleepiness. If you have forgotten to record your sleepiness for a particular time period, then leave that time blank.

Rate your sleepiness for each of the time periods indicated. Please record your sleepiness during 15 minute intervals, and note that times are presented in 24 hour notation, e.g. 5:00 A.M. is 0500, and 3:00 P.M. is 1500.

    Code: *1 Feeling active and vital; alert; wide awake.*
           2 Functioning at a high level; but not at peak; able to concentrate
           3 Relaxed; awake; not at full alertness; responsive
           4 A little foggy, not at peak; let down
           5 Fogginess, beginning to lose interest in remaining awake; slowed down
           6 Sleepiness; prefer to be lying down; fighting sleep, woozy
           7 Almost in reverie, sleep onset soon; lost struggle to remain awake
           X Asleep (If you are sleeping during any of the time periods, then write "X" for these periods)

	0-15	15-30	30-45	45-60	Comments
Midnight 0000					
0100					
0200					
0300					
0400					
0500					
0600					
0700					
0800					
0900					
1000					
1100					
Noon 1200					
1300					
1400					
1500					
1600					
1700					
1800					
1900					
2000					
2100					
2200					
2300					

Score: The scores obtained for each 15 minute period are transferred to a diagram with scores in ordinate and time in abscissa. The mean score for one day is the sum of scores divided by the number of scores.

Reference:

HODDES E., ZARCONE V., SMYTHE H., PHILLIPS R., DEMENT W.C. Quantification of sleepiness: A new approach. *Psychophysiology, 10,* 431-436, 1973.

THE KAROLINSKA SLEEPINESS SCALE (KSS)

First Name   Age
Surname   Date

*Instructions: Choose the statement below which best describes your state of sleepiness.
The steps in between have a scale value but no verbal label.*

Code:

☐ extremely alert   1

                                         2

☐ alert   3

                                         4

☐ neither alert nor sleepy   5

                                         6

☐ sleepy but no difficulty remaining awake   7

                                         8

☐ extremely sleepy, fighting sleep   9

Score: The score is the mean of the sum of scores obtained.

Reference:
AKERSTED T., GILBERG M. Subjective and objective sleepiness in the active individual. *Int. J. Neuroscience, 52,* 29-37, 1990.

THE EPWORTH SLEEPINESS SCALE (ESS)

First Name   Age
Surname   Date

Instructions: How likely are you to doze off or fall asleep in the following situations, in contrast to feeling just tired? This refers to your usual way of life in recent times. Even if you have not done some of these things recently try to work out how they would have affected you. Use the following scale to choose the most appropriate number for each situation:

0 = would *never* doze
1 = *slight* chance of dozing
2 = *moderate* chance of dozing
3 = *high* chance of dozing

Situation	**Chance of dozing**
Sitting and reading	_____
Watching TV	_____
Sitting, inactive in a public place (e.g. a theatre or a meeting)	_____
As a passenger in a car for an hour without a break	_____
Lying down to rest in the afternoon when circumstances permit	_____
Sitting and talking to someone	_____
Sitting quietly after a lunch without alcohol	_____
In a car, while stopped for a few minutes in the traffic	_____

Thank you for your cooperation

Score: A score of over 10 is generally taken to indicate excessive daytime sleepiness.

Reference:
JOHNS M.W. – A new method for measuring daytime sleepiness: the Epworth Sleepiness Scale. *Sleep, 14,* 540-545, 1991.

4. Questionnaire used in the evaluation of circadian rhythm sleep disorders

THE HORNE AND ÖSTBERG QUESTIONNAIRE

First Name   Age
Surname   Date

Instructions:
1.Please read each question very carefully before answering
2. Answer ALL questions
3.Answer questions in numerical order
4. Each question should be answered independently of others. DO NOT go back and check your answers
5. All questions have a selection of answers. For each question place a cross alongside ONE answer only. Some questions have a scale instead of a selection of answers. Place a cross at the appropriate point along the scale
6. Please answer each question as honestly as possible. Both your answers and the results will be kept, in strict confidence.
7. Please feel free to make any comments in the section provided below each question.

The questionnaire, with scores for each choice

1.Considering only your own "feeling best" rhythm, at what time would you get up if you were entirely free to plan your day?

05   06   07   08   09   10   11   12

2. Considering only your own "feeling best" rhythm, at what time would you go to bed if you were entirely free to plan your evening?

20   21   22   23   00   01   02   03

3. If there is a specific time at which you have to get up in the morning, to what extent are you dependent on being woken up by an alarm clock ?

- Not at all dependent   ☐ 4
- Slightly dependent   ☐ 3
- Fairly dependent   ☐ 2
- Very independent   ☐ 1

4. Assuming adequate environmental conditions, how easy do you find getting up in the mornings?

- Not at all easy   ☐ 1
- Not very easy   ☐ 2
- Fairly easy   ☐ 3
- Very easy   ☐ 4

5. How alert do you feel during the first half hour after having woken in the mornings?

      - Not at all alert   ☐ 1
      - Slightly alert   ☐ 2
      - Fairly alert   ☐ 3
      - Very alert   ☐ 4

6. How is your appetite during the first half-hour after having woken in the mornings?

      - Very poor   ☐ 1
      - Fairly poor   ☐ 2
      - Fairly good   ☐ 3
      - Very good   ☐ 4

7. During the first half-hour after having woken in the morning, how tired do you feel?

      - Very tired   ☐ 1
      - Fairly tired   ☐ 2
      - Fairly refreshed   ☐ 3
      - Very refreshed   ☐ 4

8. When you have no commitments the next day, at what time do you go to bed compared to your usual bedtime?

      - Seldom or never later   ☐ 4
      - Less than one hour later   ☐ 3
      - 1 – 2 hours later   ☐ 2
      - More than two hours later   ☐ 1

9. You have decided to engage in some physical exercise. A friend suggests that you do this one hour twice a week and the best time for him is between 7.0 – 8.0 am. Bearing in mind nothing else but your own "feeling best" rhythm how do you think you would perform?

      - Would be on good form   ☐ 4
      - Would be on reasonable form   ☐ 3
      - Would find it difficult   ☐ 2
      - Would find it very difficult   ☐ 1

10. At what time in the evening do you feel tired and as a result in need of sleep?

8PM   9   10   11   12AM   1   2   3

◄— 5 —►◄— 4 —►◄— 3 —►◄— 2 —►◄— 1 —►

11. You wish to be at your peak performance for a test which you know is going to be mentally exhausting and lasting for two hours. You are entirely free to plan your day and considering only your own "feeling best" rhythm which ONE of the four testing times would you choose?

      - 8:00 – 10:00 AM   ☐ 6
      - 11:00 AM – 1:00 PM   ☐ 4
      - 3:00 – 5:00 PM   ☐ 2
      - 7:00 – 9:00 PM   ☐ 0

12. If you went to bed at 11:00 PM at what level of tiredness would you be?

      - Not at all tired   ☐ 0
      - A little tired   ☐ 2
      - Fairly tired   ☐ 3
      - Very tired   ☐ 5

13. For some reason you have gone to bed several hours later than usual, but there is no need to get up at any particular time the next morning. Which ONE of the following events are you most likely to experience?

      - Will wake up at usual time and will NOT fall asleep   ☐ 4
      - Will wake up at usual time and will doze thereafter   ☐ 3
      - Will wake up at usual time but will fall asleep again   ☐ 2
      - Will NOT wake up until later than usual   ☐ 1

14. One night you have to remain awake between 4:00 – 6:00 AM in order to carry out a night watch. You have no commitments the next day. Which ONE of the following alternatives will suit you best?

      - Would NOT go to bed until watch was over   ☐ 1
      - Would take a nap before and sleep after   ☐ 2
      - Would take a good sleep before and nap after   ☐ 3
      - Would take ALL sleep before watch   ☐ 4

15. You have to do two hours of hard physical work. You are entirely free to plan your day and considering only your own "feeling best" rhythm which ONE of the following times would you choose?

      - 8:00 – 10:00 AM   ☐ 4
      - 11:00 – 1:00 AM   ☐ 3
      - 3:00 – 5:00 PM   ☐ 2
      - 7:00 – 9:00 PM   ☐ 1

16. You have decided to engage in hard physical exercise. A friend suggests that you do this for one hour twice a week and the best time for him is between 10:00 – 11:00 PM. Bearing in mind nothing else but your own "feeling best" rhythm how well do you think you would perform?

      - Would be on good form   ☐ 1
      - Would be on reasonable form   ☐ 2
      - Would find it difficult   ☐ 3
      - Would find it very difficult   ☐ 4

17. Suppose that you can choose your own work hours. Assume that you worked a FIVE hour day (including breaks) and that your job was interesting and paid by results. Which FIVE CONSECUTIVE HOURS would you select?

☐ ☐
12 1 2 3 4 5 6 7 8 9 10 11 12 1 2 3 4 5 6 7 8 9 10 11 12
MIDNIGHT   NOON   MIDNIGHT

18. At what time of the day do you think that you reach your "feeling best" peak?

☐ ☐
12 1 2 3 4 5 6 7 8 9 10 11 12 1 2 3 4 5 6 7 8 9 10 11 12
MIDNIGHT   NOON   MIDNIGHT

19. One hears about "morning" and "evening" types of people. Which ONE of these types do you consider yourself to be?

- Definitely a "morning" type	☐ 6
- Rather more a "morning" than an "evening" type	☐ 4
- Rather more an "evening" than a "morning" type	☐ 2
- Definitely an "evening" type	☐ 0

Score:
The score is the sum of the figures marked beside or above the boxes.

Score	Typology
70 – 86	definitely morning type
59 - 69	moderately morning type
42 – 58	neither type
31 – 41	moderately evening type
6 – 30	definitely evening type

Reference:
Taken from: HORNE J.A., ÖSTBERG O. – A self-assessment questionnaire to determine morningness-eveningness in human circadian rhythms. *Int. J. Chronobiol., 4,* 97-110, 1976 – *with permission from Taylor & Francis Ltd.*

Glossary

Absorption: process by which a drug goes from the site of application to the plasma. The rate of absorption determines onset of action.

Active sleep: term used in the phylogenetic and ontogenic literature for the state of sleep that is considered to be equivalent to REM sleep.

Allele: alternative forms of a gene. There may be many alleles for a given gene, but each person possesses only 2 alleles for each gene, receiving one of each pair of alleles from each parent. A person with a pair of similar alleles is a homozygote; one with a dissimilar pair is a heterozygote.

Alpha-delta sleep: sleep in which alpha activity occurs during NREM sleep stages 3 and 4

Apnoea-hypopnoea index: the number of apnoeic episodes (obstructive, central, and mixed) plus hypopnoeas per hour of sleep as determined by all-night polysomnography. Synonymous with: respiratory disturbance index.

Arousal: an abrupt change from a deeper stage of non-REM sleep to a lighter stage, or from REM sleep to wakefulness, with the possibility of awakening as the final outcome.

Balistocardiography: the balistocardiographic wave is caused by the recoil action of the human body in response to the movement of the blood from the ventricle towards the aorta, the pulmonary artery and the other main arteries. While the electrocardiogram reflects electrical stimulation, the balistocardiogram reflects the mechanical response of the heart to this stimulation.

Basic rest activity cycle (BRAC): the reliable alternation of REM and NREM sleep every 70 to 110 minutes throughout the night prompted Kleitman (1963) to suggest that this cycle was a sleep-independent ultradian (90-min) rhythm that might be present in many aspects of human physiology and function throughout the 24-h day. He referred to it as the basic rest activity cycle (BRAC).

Biological rhythm: almost all physiological functions, from the most elementary to the most complex, are subject to a rhythm *i.e.* a regular, cyclic variation comparable to a sinusoidal function. A biological rhythm is characterised by its period, an interval of time which regularly separates the identical states of the function, its amplitude or half the total measured variability between the acrophase (peak or zenith of the variation), and the batyphase (trough or nadir of the variation), its phase or value of the variable at a given moment, in hours and in minutes or degrees, in relation to a phase of reference, and its mesor or the mean adjusted level of the rhythm for a set period.

Candidate gene strategy: an approach used in the study of genetic linkage. It relies on the pathophysiology of the disease concerned. It consists of using probes to recognise the genes which may be implicated in the pathophysiology of the disease, and in studying whether these genes cosegregate with the disease.

Central apnoea: cessation of airflow at the nostrils and mouth lasting at least 10 seconds, associated with a cessation of all respiratory movements.

Cheyne-Stokes breathing syndrome: a cyclical fluctuation in breathing with periods of central apnoeas or hypopnoeas alternating with periods of hyperpnoeas in a gradual waxing and waning fashion. It occurs in patients with cardiac dysfunction usually in association with severe congestive heart failure or neurological disease/dysfunction; usually cerebrovascular. Cheyne-Stokes breathing is present during sleep, and in more severe cases may also be observed during wakefulness.

Chip technology: this technology involves hybridising DNA labelled with a fluorescent nucleotide to oligonucleotides that have been synthesised directly onto glass.

Chronobiotic: an agent used to treat a specific subtype of insomnia and directly correct the underlying problem (in contrast, a hypnotic improves insomnia regardless of aetiology).

Circadian process C: modality of sleep regulation controlled by the circadian system, independent of sleep-wake alternation.

Circadian rhythm: a biological rhythm with a period ranging from 21 to 27 hours.

Circasemidian rhythm: a biphasic sleep-wake behavioural pattern characteristic of humans and reflecting an endogenous two per day rhythm of sleep facilitation. The two sleep zones (nocturnal sleep and mid-afternoon nap time) have very unequal durations.

Clinophilie: literally "love of bed". This applies to persons who spend a large part of their time in bed, out of preference rather than the need for sleep.

Cloning: in vitro recombination of a DNA fragment in a vector capable of self-replication.

Congenic strain: a congenic strain is created when a piece of genomic material of interest is brought from one inbred strain into another inbred genetic background. Congenic strains are produced by repeated backcrossing to an inbred strain and selection for a phenotype determined by a single gene, followed by brother-sister mating of animals homozygous for the introduced gene.

Constant routine: a chronobiological test of the endogenous pacemaker that involves a 36-hour baseline monitoring period followed by a 40- hour waking period of monitoring with the individual on a constant routine of food intake, positioning, activity and light exposure.

Contingent negative variation: slow potential whose amplitude varies with attention and wakefulness. Contingent negative variation amplitude is linked to wakefulness on a bell curve.

Co-sleeping: in relation to the child, sleeping in his parents' bed.

Crossing over: the exchange of corresponding segments between maternal and paternal homologous chromosomes, occurring when maternal and paternal homologous chromosomes are paired during the prophase of the first meiotic division.

Cross-tolerance and cross-dependence: the ability of one drug to suppress the manifestations of dependence produced by another drug and to maintain the physical dependence (e.g., between alcohol and benzodiazepines).

Cyclic alternating pattern (CAP): periodic EEG activity composed of the alternation of 2 EEG patterns (phase A and phase B), each lasting more than 2 seconds and generally less than 60 seconds. According to the specific stage, phase A consists of intermittent alpha rhythms in S1, sequences of two or more K-complexes in S2, runs of K-complexes or reactive slow waves showing an amplitude at least one-third above the EEG background in S3 and S4. The phase B patterns correspond to the periodic replacement of the background EEG activities, peculiar to the single NREM stages.

Deletion: loss of a segment of DNA when the continuity of the molecule is re-established.

Derivation: a pair of electrodes from which a difference of potential is recorded.

Diallelic analysis: study of several inbred strains and of all the possible reciprocal crossings between these strains.

Distribution: process by which a drug migrates from the blood to the organs and tissues.

Dominant gene: a gene that expresses its effect even when it is present on only one chromosome.

Elimination: process by which a drug is eliminated from the body either by biotransformation (in the liver) or excretion (through the kidneys).

Elimination half-life: time required for the plasma drug concentration to decrease by 50%.

Entrainment: synchronisation of a biological rhythm by a stimulus such as an environmental time cue.

Epistasis: effects of interaction between genes situated on different loci.

Epoch: a measure of duration of the sleep-recording, typically 20-30 seconds in duration.

« Executive » neurons: neuronal systems commanding a type of sleep (NREM or REM).

Exon: a segment of a gene represented in the mature RNA messenger and which codes for a portion of the structure of a protein.

First night effect: the effect of the environment and polysomnographic recording apparatus on the quality of the subject's sleep during the first night of recording.

Forced desynchrony studies: in these studies sleep and wakefulness are scheduled to alternate at a period distinctly different from 24 hours.

Free-running: a chronobiological term that refers to the natural endogenous period of a rhythm when Zeitgebers are removed.

Gene: set of DNA sequences involved in producing a polypeptidic chain.

Genetic linkage: cosegregation of two or several genes due to their proximity on a chromosome. The degree of linkage is measured by the genetic distance expressed in centimorgans.

Genetic marker: a genotype or phenotype trait enabling a locus to be found.

Genetic probe: DNA sequence, cloned or obtained synthetically, identical to a genomic DNA sequence, coding or not, with which it hybridises in a stable manner.

Genotype: the full set of genes carried by an individual.

Haplotype: assortment of alleles between different loci, which are in proximity on the same chromosome.

Hertz: a unit of frequency; preferred to the synonymous expression: cycles per second.

Heterozygote: an individual possessing differing alleles at a given locus on a pair of homologous chromosomes.

Hjorth parameters: normalised sleep descriptors (NSDs) used in sleep EEG processing for data reducing and/or automatic sleep scoring.

HLA antigen: membrane protein implied in antigenic presentation and in triggering immune responses.

Homeostatic process (S): modality of sleep regulation dependent on sleep, increasing during wakefulness and decreasing during sleep.

Homozygote: an individual possessing a pair of identical alleles at a given locus on a pair of homologous chromosomes.

Hot spot: region of genomic DNA where the frequency of mutations or recombination events is abnormally high.

Hybridisation: the process whereby two strands of genomic DNA, separated chemically or by heat (the DNA is said to be denatured), will pair to their complementary sequences, which can be either their original partner or any other DNA with the same complementary sequence.

Hypopnoea: an event characterised by a clear decrease (> 50%) from baseline in the amplitude of a valid measure of breathing during sleep, lasting 10 seconds or longer, associated with either an oxygen desaturation of > 3 sec or an arousal.

Hypsarhythmia: uninterrupted succession of slow waves and spikes of variable duration, associated in all proportions, showing no precise relations of phase and spread across the scalp in a non-synchronous manner. This type of EEG activity is encountered in West's syndrome.

Immediate early genes (IEG): genes that encode regulatory proteins that control the transcriptional response of cells to environmental stimuli. The best-studied member of this class is c-*fos*.

Infradian rhythm: biological rhythm with a period ranging from 28 hours to 6 days.

Intron: a segment of a gene that is initially transcribed but then spliced out of the messenger RNA: It is an intervening segment of DNA between 2 exons.

K complex: a sharp negative EEG wave followed by a high-voltage slow wave. The complex duration is at least 0.5 second, and may be accompanied by a sleep spindle. K complexes occur spontaneously during non-REM sleep, and begin and define stage 2 sleep. They are thought to be evoked responses to internal stimuli. They can also be elicited during sleep by external (particularly auditory) stimuli.

Knock-out mice: mice whose genetic heritage has been altered by deleting the gene of interest.

Linkage disequilibrium: this occurs when two alleles corresponding to two distinct loci on the same chromosome are more commonly associated in the studied population than they would be at random.

Linkage study: search for the linkage between a genetic marker and a disease. It confirms whether a genetic region is implicated, and calculate the recombination fraction and the lod score.

Locus: a position on a chromosome: this general term can refer to a gene or a segment of DNA with no known function.

Lod score: method of genetic study developed in the framework of monogenic diseases, with the dual function of testing the genetic linkage between two loci (that of the pathological gene and that of the marker) and estimating the genetic distance between these two loci. A lod score maxima value of less than − 2 signifies genetic independence, whereas a value of over + 3 signifies genetic linkage: in this case the pathological gene is localised in the same region as the marker.

Masking effect: result of the effect of an external stimulus on a biological parameter, which does not relate to its rhythmic variable. For example, physical activity, at any hour of the day, will induce an increase in temperature, testifying to the activity of the circadian clock.

Microsleep: brief intrusion of EEG indications of sleep.

Microstructure of sleep: besides macrostructure or sleep architecture as states and stages, emphasis is placed today on the importance of the microstructure of sleep corresponding to graphoelements of short duration, whether normal (K complexes, spindles etc.) or abnormal (K alpha complexes, alpha delta rhythm, microarousals etc.).

Mini and microsatellites: genomic DNA sequences from tens to hundreds of base pairs (mini) or 2 to 4 base pairs (micro) repeated at regular intervals along the genome.

Mixed apnoea: cessation of airflow at the nostrils and mouth lasting at least 10 seconds, with both a central component (at the beginning) and an obstructive component (at the end).

Monogenetic: caused by a single gene or a single pair of genes found at a specific locus.

Montage: the particular arrangement by which a number of derivations are displayed simultaneously in a polysomnogram.

mRNA differential display (*m*RNA-DD): A polymerase chain reaction (PCR)- based technique that has been shown to be at least as sensitive as substractive hybridisation in detecting rare m-RNA.

Mutagenesis: technique which consists of inducing (chemical) mutations in the genome as a whole, followed by screening to identify the individual carrying a mutation on a gene responsible for the studied characteristic.

Mutation: abrupt, transmissible change in the genomic DNA sequence.

Nap: sleep episode of short duration, intentional or unintentional, and occurring at some point during a habitual period of wakefulness.

Non rapid eye movement sleep (NREM sleep): a period of sleep without rapid eye movements.

Obstructive apnoea: cessation of airflow at the nostrils and mouth lasting at least 10 seconds, secondary to upper airway obstruction.

Overlap syndrome: association of chronic obstructive bronchopneumopathy and sleep apnoea syndrome.

Penetrance: the frequency of phenotypic expression of a dominant gene or a homozygous recessive gene.

« Permissive » neurons: neuronal systems allowing a certain type of sleep (NREM or REM) to become established.

Phasic event (-activity): brain, muscle, or autonomic events of a brief and episodic nature occurring in sleep; characteristic of REM sleep such as eye movements, or muscle twitches; the duration is usually from milliseconds to 1-2 sec.

Phase response curve: graphic representation of the value of a phase discrepancy of a rhythm induced by a stimulus, in function of the time of application of the stimulus in relation to the phase of the rhythm.

Phenocopy: an individual with all the hallmarks of a particular genetic disorder but with no hereditary cause apparent in his pedigree.

Phenotype: the sum of all observable features of an individual (including his anatomic, physiologic, biochemical, and psychological make up, and his disease reactions, potential or actual). The phenotype is the result of interaction between the genotype and the environment.

Photoperiod: the period of light in a light-dark cycle.

PLM index: The number of periodic leg movements per hour of total sleep time as determined by all-night polysomnography.

Polymorphism: DNA sequence variations found on the genome. Most polymorphisms are phenotypically mute. There are two main types of restriction polymorphisms (alternative versions of the size of DNA fragments obtained with a given restriction enzyme) and a variable number of repetitions of short DNA sequences (microsatellite repetition).

Ponto-geniculo-occipital (PGO) spikes: These spikes are generated in the pons, propagate rostrally through pathways in the vicinity of the brachium conjonctivum and project through the laterate geniculate and other thalamic nuclei to the cortex. PGO spikes are one of several phasic events of REM sleep.

Positional cloning: set of molecular techniques providing the chromosomic identification of a gene.

Prion protein: a glycoprotein anchored to cell membranes and expressed in most cell types. Since mutations in this protein lead to severe neurodegeneration and death in humans and animals, it is possible that the loss of its normal function contributes to the development of the pathology. Little is known about its normal function.

Quantitative trait locus (QTL) analysis: statistical method allowing the detection of minor and/or modifier genes, as well as the major genes, influencing different quantitative traits (allelic variation responsible for the increase or decrease of a trait).

Quiet sleep: a term used to describe non-REM sleep in infants and animals when specific non-REM sleep stages 1-4 cannot be determined.

REMs: rapid eye movements.

Rebound insomnia: a transient insomnia caused by the abrupt discontinuation of short and intermediate-acting benzodiazepine hypnotics.

Recessive: a trait is recessive if it is expressed only in individuals who are homozygous for the gene concerned.

Rechtschaffen and Kales (1968): editors of a manual of sleep analysis published under the patronage of the National Institute of Health in the USA.

REM density: a function that expresses the frequency of eye movements per unit of time during REM sleep.

REM efficiency (REM efficiency index): ratio of total REM sleep time to total REM sleep episodes duration.

REM-off neurons: neurons which cease their activity for the duration of REM sleep.

REM-on neurons: neurons selectively active for the duration of REM sleep.

REM sleep episode: set of REM sleep sequences uninterrupted by a sequence of wakefulness or of NREM sleep > or = 15 minutes.

Respiratory effort-related arousal (RERA) event: a sequence of breaths characterised by increasing respiratory effort leading to an arousal from sleep, but which does not meet criteria for an apnoea or hypopnoea.

Restriction endonucleases: endonuclease of bacterial origin specifically splitting the two strands of DNA at defined sequences (4 to 8 base pairs) for each enzyme.

Sawtooth waves: a form of theta rhythm that occurs during REM sleep and is characterised by a notched waveform. Occurs in bursts lasting up to 10 seconds.

Segregation: the separation of the two alleles of a pair of allelic genes during meiosis, so that they pass to different gametes.

Segregation study: study of the distribution of cases in pedigrees. They enable the mode of transmission of a disease to be deduced, as well as the number of genes involved, penetrance and frequency.

Siesta: a sleep episode of variable duration, occurring in the morning and afternoon in children under 2 years of age, and in the early afternoon after the age of 2 years.

Simple backcross segregation analysis: analysis of the distribution of F2 or backcross according to parental types (mendelian distribution analysis: in the case of a gene for example, 50% of one of the parental types and 50% of the heterozygote types should be found in a backcross).

Sleep cycle: a cycle comprises a phase of non-REM sleep followed by a phase of REM sleep. Adult sleep is comprised of 4 to 5 sleep cycles.

Sleep efficiency (sleep efficiency index): ratio of total sleep duration/time in bed with the light off (including latency to sleep onset and time awake before switching the light on).

Sleepiness: a subjective and objective state of lowered physiologic alertness, and a greater inclination to sleep or to doze.

Sleep onset REM period: any sleep onset in which REM sleep appears within a period of 0 to 15 minutes after the onset of sleep.

Sleep spindle: spindle-shaped bursts of 11.5 to 15.0 Hz waves lasting $0.5 - 1.5$ second. Generally diffuse but of highest voltage over the central regions of the head. The amplitude is generally less than $50\mu V$ in the adult. One of the identifying EEG features of NREM stage 2 sleep; may persist into NREM stages 3 and 4; not generally seen in REM sleep.

Slow wave activity, SWA: mean power density in the 0.75 to 4.5 Hz band.

Southern blot: a technique for transferring DNA fragments separated by gel electrophoresis to nitrocellulose paper for molecule hybridisation to labelled probes.

Spectral analysis: the EEG signal may be considered as a continuous succession of amplitude variations in time. Fast Fourier transform and spectral analysis break down this signal (whose amplitudes are a function of time) into a spectrum whose amplitudes are a function of frequency.

Substractive hybridisation: technology aimed at enriching differentially expressed mRNA sequences by subtracting one sequence population from another.

TaqMan method: a novel fluorescence-based method for quantification of gene expression and gene copy number variations as gene amplification and locus haploinsufficiencies.

Ten-twenty: international system for placing EEG electrodes diagonally across the scalp, corresponding to intervals of 10% or 20% of the distance from nasion to inion.

Tolerance: following repeated exposure to a drug, a given dose of the drug produces a decreasing effect, or, conversely higher doses are needed to obtain the effects observed with the original dose.

Tonic activity: brain, muscle, or autonomic events of a sustained nature occurring in sleep; characteristic of REM sleep such as muscle atonia, sleep-related erection.

Tracé alternant: EEG pattern of sleeping newborns, characterised by bursts of slow waves, at times intermixed with sharp waves, and intervening periods of relative quiescence with extreme low-amplitude activity.

Transduction: synthesis of a polypeptide chain from an RNA messenger.

Transcription: RNA synthesis from a DNA matrix.

Transgenic mice: mice whose genetic heritage has been altered by adding supplementary copies of a given gene.

Transitional sleep: state of sleep of the newborn, intermediate to active and quiet sleep.

Translocation: a change in location of genetic material, either for a chromosome or from one chromosome to another.

Ultradian rhythm: biological rhythm whose period ranges from 30 minutes to 20 hours.

Vertex sharp transient: sharp negative potential maximal at the vertex, occurring spontaneously during sleep or in response to sensory stimulus during sleep or wakefulness. Amplitude varies but rarely exceeds 250 μV.

Vigilance: sustained capacity to detect changes in the environment and respond appropriately (performance-orientated concept).

Withdrawal symptoms: new physical and psychological manifestations following abrupt cessation of a dependence-producing drug.

Wrist actigraphy: technological method allowing for 24-hour (or more) recording of movement. The interest of the method is that during sleep there is little movement, whereas during wakefulness movement increases.

Zeitgeber: German term for an environmental time cue that usually helps entrainment to the 24-hour day, such as sunlight, noise, social interaction, alarm clock.

Index

Abdominal straps, respiratory measurements, 132–133

Abnormal postures and movements: see Nocturnal paroxysmal dystonia

Aciclovir, insomnias, medication-linked, 253

Acid maltase deficiency, sleep breathing abnormalities, 582

Acromegaly, hypersomnias, 453

ACTH: see Adrenocorticotropic hormone (ACTH)

Active sleep (AS), 14

Adenoido-amygdalectomy, obstructive sleep apnoea in children (OSAS), 461

Adenosine, wakefulness, neurobiology, 35

Adenotonsillectomy, 365–366

Adjustment insomnia, transient/short term insomnia, 203

Adolescents: see also Childhood: hypersomnias, Kleine-Levin syndrome, 466; insomnias, aetiologies, 287–288; normal sleep in, 24–25

Adrafinil, insomnias, medication-linked, 249

Adrenocorticotropic hormone (ACTH): circadian rhythmicity, 52; stress, aging effects, 308

Advanced sleep phase syndrome, 499–502: diagnosis, 499–500; epidemiology, 499; genetics, 100; historical perspective, 499; melatonin, 75; pathophysiology, 500; treatment, 500–502

Adverse effects: aging, pharmacologic agents, 317–318; benzodiazepines, 260; zaleplon, 264; zolpidem, 263; zopiclone, 261

Aetiology: hypersomnias, 334; childhood, 457–458; obstructive sleep apnoea-hypopnoea syndrome and upper airway resistance syndrome, 366; insomnias: childhood insomnias, 285–288; primary insomnia, 211–212

African trypanosomiasis (sleeping sickness): infection-related hypersomnias, 452; sleep in, 680

Aging, 297–332; aging effects, 305–310; caffeine, 308–309; phase angle misalignment, 305–307; sleep pathology and comorbid disease, 309–310; stress, 307–308; circadian rhythm sleep disorders: irregular sleep-wake pattern, 506–507; shift work sleep disorder, 477; demographics, 297; descriptive studies, 297–298; gender differences, 311–315; hypersomnias: medication and alcohol dependent sleepiness, 352; obstructive sleep apnoea-hypopnoea syndrome and upper airway

Aetiology (cont.): resistance syndrome, 365; insomnia treatments, 315–321; melatonin, 320–321; pharmacologic, 315–318; phototherapy, 319–320; medical disorders: headache, 630; snoring, 560; parasomnias, REM sleep behavior disorder, 531; two-process model, 298–304; circadian process (process C), 302–304; homeostatic sleep process (process S), 298–302

AIDS: see HIV/AIDS infection

Alcohol: see also Medication and alcohol dependent sleepiness; hypersomnias, medication and alcohol dependent sleepiness, 354–355; insomnias, medication-linked, 253; parasomnias: bruxism, 536; nightmares, 527

Almitrine, chronic obstructive pulmonary disease, 572

Alpha attenuation test, sleepiness assessment, 178

Alpha-methyl dopa, hypersomnias, medication and alcohol dependent sleepiness, 354

Altitude insomnia, 203

Alveolar hypoventilation syndrome, 577–579

Alzheimer's disease: see also Dementia; circadian rhythm sleep disorders, irregular sleep-wake pattern, 507; neurological hypersomnias, 449; sleep in, 677–678

Ambient temperature, sudden infant death syndrome, 651–652, 653; see also Thermoregulation

Ambulatory sleep recording, 139–149; see also Automatic sleep analysis system; Polysomnography: development of, 141–142; epilepsy investigation, 690; four to eight channel systems, 143–146; future developments, 147; indications for, 140–141; one to three channel systems, 142–143; role of, 139–140; ten or more channel systems, 146–147

American Academy of Sleep Medicine (AASM), 7

American Sleep Disorders Association (ASDA), 7, 140, 141

Amineptine, insomnias, 272

Amino acids, insomnias, 275–276

Amisulpride, insomnias, medication-linked, 251

Amitriptyline, insomnias, 272, 273

Amoxapine, insomnias, 271

Amphetamines: hypersomnias, narcolepsy, 419–420; insomnias, medication-linked, 249; parasomnias, bruxism, 536

Amyotrophic lateral sclerosis, 675, 697

Analgesics, hypersomnias, medication and alcohol dependent sleepiness, 353

Angina pectoris, cardiocirculatory disorders, 590–592

Anterior pituitary hormones, sleep architecture, 52

Antibiotics, insomnias, medication-linked, 252–253

Anticholinergics: enuresis, 550–551; hypersomnias, medication and alcohol dependent sleepiness, 354

Anticonvulsants, insomnias, medication-linked, 251

Antidepressants: see also Tricyclic antidepressants; benzodiazepines interactions with, 274; insomnias, 269–274; hypnotic effect, 271; listing of drugs, 270; medication-linked, 249–250; NREM sleep, 272; prescribing of, 273–274; REM sleep, 271–272; medication and alcohol dependent sleepiness, 351–352; mood disorders, 669; parasomnias: REM sleep behaviour disorder, 534; rhythmic movement disorder, 524; sleep paralysis, 530

Antiepileptic drugs: hypersomnias, medication and alcohol dependent sleepiness, 353; nocturnal paroxysmal dystonia, 613

Antifungal agents, insomnias, medication-linked, 253

Antihistamines: hypersomnias, medication and alcohol dependent sleepiness, 353; insomnias, 274–275

Anti-infectious drugs, insomnias, medication-linked, 252–253

Anti-inflammatory drugs, insomnias, medication-linked, 252

Antimigraine drugs, hypersomnias, medication and alcohol dependent sleepiness, 353

Anti-Parkinsonian drugs: hypersomnias, medication and alcohol dependent sleepiness, 353; insomnias, medication-linked, 251; parasomnias, nightmares, 527

Anti-rheumatic drugs, insomnias, medication-linked, 252

Antituberculosis drugs, insomnias, medication-linked, 252

Antiviral drugs, insomnias, medication-linked, 253

Anxiety disorders: see also Psychiatric disorders; generalised anxiety disorder, insomnias, 234–235; obsessive-compulsive disorder, insomnias, 237–238; panic disorder, insomnias, 236–237; post-traumatic stress disorder, insomnias, 238–239; social/specific phobias, insomnias, 236

Anxiolytics: insomnias, medication-linked, 251; medication and alcohol dependent sleepiness, 349–350; parasomnias, bruxism, 537

Apnoeas: see Obstructive sleep apnoea-hypopnea syndrome and upper airway resistance syndrome; Obstructive sleep apnoea in children (OSAS); Sleep apnoeas

Arboviruses, infection-related hypersomnias, 451

Arginine-L-aspartate, insomnias, 276

Arnold-Chiari malformation, neurological hypersomnias, 449

Arousal disorders, 515–522: confusional arousals, 515–516; fibromyalgia and chronic fatigue syndrome, 708–709; night terrors, 518–522: clinical features, 518; clinical variants, 519; diagnosis, 520–521; epidemiology, 518–519; factors in, 519; laboratory tests, 519–520; pathophysiology, 521; treatment, 521–522; sleepwalking, 516–518

Arousal threshold, sudden infant death syndrome, 652

Arterial hypertension, cardiocirculatory disorders, 590

Association studies, narcolepsy, 412–414

Asthma, insomnia, 222–223

Automatic sleep analysis system, 159–167; see also Ambulatory sleep recording; Polysomnography: autonomic variables, 165; characteristics of, 159–160; data processing, 166; electroencephalogram (EEG), 160–161; electromyogram (EMG), 162; electrooculogram (EOG), 161–162; fine analysis, 164; muscular and respiratory activity, 165; sleep stage, 163–164; software classification, 163

Autonomic nervous system: see Dysautonomias

Autosomal dominant nocturnal frontal lobe epilepsy, 623

Awakening epilepsies, 623

Baclofen: hypersomnias, medication and alcohol dependent sleepiness, 353; periodic leg movements, 605; restless legs syndrome, 605

Barbiturates: fatal familial insomnia, 636; insomnias, 275

Base command, sleep-wake architecture, neurobiology, 37–39

Beck Depression Inventory, 732–735

Bedding, sudden infant death syndrome, 653

Bedwetting: see Enuresis

Behavioural assessment, sleepiness, 170–171

Beliefs and Attitudes about Sleep Scale, 722–724

Benign infantile epilepsy, 620, 691

Benign neonatal sleep onset myoclonus, 537–538

Benzodiazepine(s): antidepressant interactions with, 274; epilepsy, 621; fatal familial insomnia, 636; hypersomnias, medication and alcohol dependent sleepiness, 349, 352, 353; insomnias, 257–261; medication-linked, 251; parasomnias: night terrors, 521–522; nocturnal leg cramps, 526; rhythmic movement disorder, 524; sleep talking, 535; periodic leg movements, 603; restless legs syndrome, 603

Beta blockers: insomnias, medication-linked, 252; parasomnias, nightmares, 527

Bladder control, 545–546; see also Enuresis

Bladder pressure, polysomnography, 137

Brain: see Neurobiology

Brain tumor: see also Central nervous system lesions; neurological hypersomnias, 447–448; recurrent symptomatic hypersomnias, 442–443

Bronchial asthma, insomnia, 222–223

Bruxism, 535–537; clinical features, 535; clinical variants, 536; epidemiology, 535–536; genetics, 100; laboratory tests, 536; pathophysiology, 536–537; treatment, 537

Caffeine: hypersomnias, insufficient sleep syndrome, 344–345; insomnias: aging effects, 308–309; medication-linked, 249; shift work sleep disorder, 477

Candidate gene studies, normal sleep (animal studies), 85

Carbon dioxide measurement, polysomnography, 136

Cardiac arrhythmias: cardiocirculatory disorders, 593–594; insomnia, 222

Cardiac autonomic dysfunctions, sleep evaluation, 700–701

Cardiac dysrhythmias, chronic obstructive pulmonary disease, hypoxemia, 571

Cardiocirculatory disorders, 589–597: angina pectoris, 590–592; arterial hypertension, 590; cardiac arrhythmias, 593–594; haemodynamic changes during normal sleep, 589–590; multiple system atrophy, 594; myocardial infarction, 592–593; overview, 589

Cardio-vascular function, 48–50; arterial pressure, 49; cardiac output and peripheral circulation, 50; circadian

Cardio-vascular function (*cont.*): rhythm sleep disorders, shift work sleep disorder, 477; heart rate, 49; obstructive sleep apnoea in children (OSAS), 461; rhythmicity, 50

Cardio-vascular medications: hypersomnias, medication and alcohol dependent sleepiness, 353–354; insomnias, medication-linked, 251–252

Carpipramine, insomnias, medication-linked, 251

Cataplexy: genetics, 101; hypersomnias, narcolepsy, 405, 411, 421–422

Cavernous body, fibrosis of, painful erections, 659

Cavum, snoring, 559

Central alveolar hypoventilation syndrome, 577–579

Central nervous system, recurrent symptomatic hypersomnias, 442–443

Central nervous system lesions, 673–687; *see also* Brain tumor; Head trauma: degenerative disorders, 675–678; head trauma, 678–679; infectious disorders, 679–680; multiple sclerosis, 679; overview, 673; polysomnography, 680–681; tumours, 678; vascular pathology, 673–675

Central sleep apnoea syndromes, obstructive sleep apnoea-hypopnea syndrome and upper airway resistance syndrome, 366–368; *see also* Obstructive sleep apnoea-hypopnea syndrome and upper airway resistance syndrome

Cephalometric radiograph, obstructive sleep apnoea-hypopnea syndrome and upper airway resistance syndrome, 393

Cerebral evoked potentials, sleepiness assessment, vigilance, 177

Cerebral metabolism, physiological functions, 56

Cerebral tumour, sleep in, 678

Channel setting: electrocardiogram (ECG), 132; electroencephalogram (EEG), 128–129; electromyogram (EMG), 130–131; electrooculogram (EOG), 130

Childhood: enuresis, 545–546; epilepsy, 619–620; medical disorders, snoring, 560; parasomnias: benign neonatal sleep onset myoclonus, 537–538; bruxism, 535, 536; nightmares, 526–527, 528; night terrors, 518–519; sleepwalking, 516–517; parasomnias, rhythmic movement disorder, 522–524

Childhood hypersomnias, 457–468; aetiology, 457–458; diagnosis, 458–459; epidemiology, 457; idiopathic hypersomnia, 466; Kleine-Levin syndrome, 466; narcolepsy, 462–466; clinical features, 462–463; diagnosis, 463; epidemiology, 462; laboratory tests, 463; symptomatic, 463; treatment, 466; obstructive sleep apnoea-hypopnea syndrome and upper airway resistance syndrome, 365–366; obstructive sleep apnoea in children (OSAS), 459–462; clinical features, 460; diagnosis, 459–460; laboratory tests, 460–461; treatment, 461–462

Childhood insomnias, 283–295; *see also* Normal sleep (in children): aetiologies, 285–288; infants and young children, 285–287; schoolchildren and adolescents, 287–288; clinical investigation, 288; definition, 283; epidemiology, 284; environmental factors, 288–289; pharmacology, 291–292; psycho-emotional factors, 290–291; rituals and separation anxiety, 289–290; sleep hygiene and prevention, 288; sleep-wake cycle, 289; treatment, 288–292

Chloroquine, insomnias, medication-linked, 252

Cholinergic-aminergic imbalance hypothesis, mood disorders, 669

Chromosomal abnormalities, sleep disorders, genetics, 102

Chronic fatigue syndrome: *see* Fibromyalgia and chronic fatigue syndrome

Chronic obstructive pulmonary disease (COPD), 567–575; hypoxemia, 569–571; consequences of, 570–571; sleep-related mechanisms of, 569–570; insomnia, 223; obstructive sleep apnoea hypopnea syndrome, 572; overview, 567; oxygen desaturation, 567–569; sleep quality in, 567; treatment, 572–574

Chronic paroxysmal hemicrania, headache, 630–631

Chronotherapy, circadian rhythm sleep disorders: advanced sleep phase syndrome, 500; delayed sleep phase syndrome, 497

Circadian rhythms: aging, 297–332; *see also* Aging; disorders of, genetics, 100–101; endocrine function, 52–53; genetics: animal studies, 86–90; human studies, 95–96; melatonin, 71–82; *see also* Melatonin; mood disorders, 670; normal sleep, in children, 25; two-process model; aging, 298–304; sleep regulation, 62–67

Circadian rhythm sleep disorders, 469–512; advanced sleep phase syndrome, 499–502; diagnosis, 499–500; epidemiology, 499; historical perspective, 499; pathophysiology, 500; treatment, 500–502; delayed sleep phase syndrome, 493–499; diagnosis, 494; epidemiology, 493; historical perspective, 493; pathophysiology, 494–497; treatment, 497–499; differential diagnosis, decision tree approach, 471–472; fibromyalgia and chronic fatigue syndrome, 709; irregular sleep-wake pattern, 506–508: diagnosis, 507; epidemiology, 506; historical perspective, 506; pathophysiology, 507–508; treatment, 508; jet lag syndrome, 485–491: clinical features, 485; factors in, 485–488; pharmacology, 488; phototherapy, 489; sleep hygiene, 489; non-24 hour sleep-wake (hypernychthemeral) syndrome, 502–506: diagnosis, 505; epidemiology, 505; historical perspective, 502–505; pathophysiology, 505–506; treatment, 506; overview, 469–470; questionnaires and scales, Horne and Östberg Questionnaire, 740–743; shift work sleep disorder, 473–484: clinical features, 476–477; definitions, 473–474; epidemiology, 473; factors in, 474–476; light role, 479–481; melatonin, 479; napping, 482; pharmacology, 479; therapy and surveillance, 478–479; work organisation, 477–478

Clinical trials: hypersomnias, medication and alcohol dependent sleepiness, 348; insomnias, medication-linked, 247–248

Clomipramine: hypersomnias, cataplexy, 421; insomnias, 272

Clonazepam: parasomnias, REM sleep behaviour disorder, 534; periodic leg movements, 605; restless legs syndrome, 605

Clonidine: hypersomnias, medication and alcohol dependent sleepiness, 353–354; insomnias, medication-linked, 251; parasomnias, REM sleep behaviour disorder, 534; periodic leg movements, 605; restless legs syndrome, 605

Cocaine, parasomnias, bruxism, 536

Cognition, obstructive sleep apnoea-hypopnea syndrome and upper airway resistance syndrome, 377

Cognitive-behavioral therapies, primary insomnia, 214–218

Cognitive therapy, primary insomnia, 216

Comorbid disease, aging effects, 309–310

Complex partial epilepsies, 620–621

Computed tomography (CT): see also Radiology: narcolepsy, childhood, 466; obstructive sleep apnoea-hypopnea syndrome and upper airway resistance syndrome, 393–394; pharyngeal, obstructive sleep apnoea-hypopnea syndrome and upper airway resistance syndrome, 392

Congestive heart failure, insomnia, 221–222

Contiguity, rule of, visual sleep analysis, sleep stage scoring, 152

Continuous arterial pressure, polysomnography, 137

Continuous positive airway pressure (CPAP): obstructive sleep apnoea-hypopnea syndrome and upper airway resistance syndrome, 378–379, 381; obstructive sleep apnoea in children (OSAS), 462

Coronary circulation, chronic obstructive pulmonary disease, hypoxemia, 571

Cortisol: circadian rhythmicity, 52; Kleine-Levin syndrome, 441; stress, aging effects, 307–308

Creutzfeldt-Jakob disease, 99

Cyclic alternating pattern (CAP), polysomnography, 7–8, 24

Cyclopyrrolone derivatives (zopiclone), insomnias, 261–262

Daytime sleepiness: see also Sleepiness assessment: narcolepsy, 405, 411, 419; obstructive sleep apnoea-hypopnea syndrome and upper airway resistance syndrome, 368–369

Decision tree approach: circadian rhythm sleep disorders, differential diagnosis, 471–472; hypersomnias, differential diagnosis, 337–339; insomnias, differential diagnosis, 191–199

Degenerative desynchronisation-maladaptation syndrome, circadian rhythm sleep disorders, shift work sleep disorder, 477

Delayed sleep phase syndrome, 493–499: diagnosis, 494; epidemiology, 493; genetics, 100–101; historical perspective, 493; melatonin, 75; pathophysiology, 494–497; treatment, 497–499

Dementia: see also Alzheimer's disease; circadian rhythm sleep disorders, irregular sleep-wake pattern, 506–508; neurological hypersomnias, 449; parasomnias, REM sleep behavior disorder, 531; sleep in, 677–678

Diabetes: hypersomnias, 453; insomnias, 224

Diachronic hypothesis, hypnogenic substances, neurobiology, 40–42

Diagnosis and differential diagnosis: circadian rhythm sleep disorders: advanced sleep phase syndrome, 499–500; decision tree approach, 471–472; delayed sleep phase syndrome, 494; irregular sleep-wake pattern, 507; non-24 hour sleep-wake (hypernychthermeral) syndrome, 505; epilepsy, 691–692; hypersomnias: childhood, 458–459; obstructive sleep apnoea in children (OSAS), 459–460; childhood narcolepsy, 463–466; decision tree approach, 337–339; idiopathic hypersomnia, 430–431, 432; insufficient sleep syndrome, 343–344; narcolepsy, 405–407, 409; insomnias: decision tree approach, 191–199; primary

Diagnosis and differential diagnosis (cont.): insomnia, 212–213; medical disorders: fibromyalgia and chronic fatigue syndrome, 705–706; nocturnal paroxysmal dystonia, 611; painful erections, 659; periodic leg movements, 601–602; restless legs syndrome, 600–601; snoring, 556–560; parasomnias: benign neonatal sleep onset myoclonus, 538; nightmares, 528; night terrors, 520–521; REM sleep behaviour disorder, 532; sleep paralysis, 530; sleep-related abnormal swallowing syndrome, 539; sleep starts (predormital myoclonus), 524–525

Differential diagnosis: see Diagnosis and differential diagnosis

Digestive disorders, shift work sleep disorder, 477

Digestive function: sleep, 53–54; sudden infant death syndrome, 651

Dihydralazine, insomnias, medication-linked, 251

Diphenylhydantoine, insomnias, medication-linked, 251

Dissociated REM sleep, genetics, 101

Dopaminergic agents: periodic leg movements, 604; restless legs syndrome, 604

Dopaminergic system, wakefulness, neurobiology, 34

Dreams, 113–123; content control, 118–119; effects of, 118; historical perspective, 113; normative data, 117–118; physiological correlation of, 115; psychopathologies, 119–120; recall and collection methods, 115–116; REM and NREM sleep, 113–115; sources of, 116–117

Duchenne's muscular dystrophy, 582, 697

Dysautonomias, 697–702: central and peripheral, 697–698; sleep evaluation, 699–701; wakefulness evaluation, 699

Eating disorders: insomnias, 241; sleepwalking and, 518

Echocardiography, polysomnography, 137

EEG: see Electroencephalogram (EEG)

Electrical stimulation, obstructive sleep apnoea-hypopnea syndrome and upper airway resistance syndrome, 380

Electrocardiogram (ECG): ambulatory sleep recording systems, 142; polysomnography, 6, 131–132

Electroencephalogram (EEG), 127–129: aging, 301–302; ambulatory sleep recording systems, 141; automatic sleep analysis system, 160–161; channel setting, 128–129; derivations, 128; electrode placement, 127; epilepsy investigation, 689, 690; fibromyalgia and chronic fatigue syndrome, 707–708; generalised idiopathic epilepsy, 617–618; genetics, animal studies, 90–91; Kleine-Levin syndrome, 439; nocturnal paroxysmal dystonia, 612–613; normal sleep, 4–5; normal sleep (in children), 11–12; one to three years, 21–22; six to twelve years, 23–24; three to six years, 23; polysomnography, 6–7; sleepiness assessment, 177; subject position calibration, 131

Electromyogram (EMG), 130–131; automatic sleep analysis system, 162; channel setting, 130–131; derivations, 130; electrode placement, 130; restless legs syndrome; ambulatory sleep recording systems, 146; polysomnography, 137; subject position calibration, 131

Electrooculogram (EOG), 129–130; automatic sleep analysis system, 161–162; channel setting, 130; derivations, 129–130; electrode placement, 129; subject position calibration, 131

Endocrine function, 50–53; enuresis, 549, 551; fatal familial insomnia, 637; Kleine-Levin syndrome, 441;

Endocrine function (*cont.*): obstructive sleep apnoea-
    hypopnea syndrome and upper airway resistance
    syndrome, 375–376, 377; sleep architecture, 51–52
Endocrine–metabolic disorders: hypersomnias, 453;
    insomnias, 224; obstructive sleep apnoea-hypopnea
    syndrome and upper airway resistance syndrome, 364–
    365
ENT examination, snoring, 557–558
Enuresis, 545–552: *see also* Parasomnias: factors in, 547–
    549; overview, 545; stages of control, 545–546;
    treatment, 549–551
Environmental factors: circadian rhythm sleep disorders,
    shift work sleep disorder, 475–476; hypersomnias,
    narcolepsy, 415; insomnias: childhood insomnias, 288–
    289; transient/short term insomnia, 202–203; sudden
    infant death syndrome, ambient temperature, 651–652
Epidemic encephalitis, infection-related hypersomnias, 451
Epidemiology: arousal disorders, night terrors, 518–519;
    circadian rhythm sleep disorders: advanced sleep phase
    syndrome, 499; delayed sleep phase syndrome, 493;
    irregular sleep-wake pattern, 506; non-24 hour sleep-
    wake (hypernychthermeral) syndrome, 505; shift work
    sleep disorder, 473; enuresis, 545; hypersomnias, 334:
    childhood, 457, 462; idiopathic, 429–430; insufficient
    sleep syndrome, 342; medication and alcohol
    dependent sleepiness, 348; narcolepsy, 403–404;
    obstructive sleep apnoea-hypopnea syndrome and upper
    airway resistance syndrome, 357; recurrent, Kleine-
    Levin syndrome, 437–438; insomnias: aging,
    pharmacologic agents, 315–317; childhood, 284;
    medication-linked, 248; primary insomnia, 207, 210–
    211; psychiatric disorders and, 227; transient/short term
    insomnia, 201–202; medical disorders: fibromyalgia
    and chronic fatigue syndrome, 704; headache, 630;
    nocturnal paroxysmal dystonia, 609; periodic leg
    movements, 600; snoring, 555–556; sudden infant
    death syndrome, 649; parasomnias: arousal disorders;
    confusional arousals, 516; sleepwalking, 517; bruxism,
    535–536; nightmares, 526–527; nocturnal leg cramps,
    525; REM sleep behavior disorder, 531; rhythmic
    movement disorder, 522; sleep paralysis, 530; sleep
    talking, 534–535
Epilepsy, 617–627; awakening epilepsies, 623; benign
    infantile epilepsy, 620; diagnosis, 691–692; generalised
    idiopathic epilepsy, 617–618; investigations of, 689–
    696; localisation, 693; methodology, 689–690;
    nocturnal paroxysmal dystonia, 611–614; nocturnal
    postepileptic headache, 631; non-epileptic seizures
    associated with, 624; parasomnias, benign neonatal
    sleep onset myoclonus, 538; partial idiopathic epilepsy
    with midtemporal spikes, 619–620; pathophysiology,
    693–694; prognosis, 692–693; Lennox-Gastaut
    syndrome, 618–619; sleep epilepsies, 622–623; sleep of
    epileptics, 624–625; sleep-related syndromes, 621–622;
    symptomatic generalized, 619; symptomatic partial,
    620–621; West's syndrome, 618
Epinephrine system, wakefulness, neurobiology, 33
Epiphysial activity, melatonin plasma profile and, 71–72
Episodic nocturnal wandering, 610–611; *see also* Nocturnal
    paroxysmal dystonia
Epoch, visual sleep analysis, sleep stage scoring, 152
Epstein-Barr virus, infection-related hypersomnias, 451

Epworth Sleepiness Scale, 739
Erections; *see also* Painful erections: dysautonomias, 699;
    REM sleep, 54
Evoked potentials, sleepiness assessment, vigilance, 177
Evoked responses measurement, sleepiness assessment,
    vigilance, 172–173
Executive REM sleep, sleep-wake architecture,
    neurobiology, 38
Exploding head syndrome, headache, 631
Eye movements, sleepiness assessment, vigilance, 176

Fatal familial insomnia, 635–639; autonomic functions,
    636–637; clinical features, 635; endocrine functions,
    637; genetics, 98–99, 638; neuropathological aspects,
    637–638; neurophysiology, 635–636; neuropsychology,
    6371 pathophysiology, 638; PET scan, 637
Fatigue, shift work sleep disorder, 476
Fiberoptic investigation, snoring, 559–560
Fibromyalgia and chronic fatigue syndrome, 703–714:
    clinical features, 704–705; course of illness, 706;
    diagnosis, 705–706; epidemiology, 704; headache, 631;
    historical perspective, 703; pathogenesis, 707–709;
    pathophysiology, 707; treatment, 709–712
Fibrosis, of cavernous body, painful erections, 659
Finger tapping test, sleepiness assessment, vigilance, 172–
    173
Flow limitation, respiratory measurements, quantitative,
    135–136
Fluoxetine: hypersomnias, cataplexy, 421; insomnias, 271,
    272, 274
Fluvoxamine: hypersomnias, cataplexy, 421; insomnias, 271,
    272
Folic acid: periodic leg movements, 605; restless legs
    movement, 605
Follicle-stimulating hormone (FSH), Kleine-Levin
    syndrome, 441
Frequency: *see* Epidemiology

GABAergic neurons, NREM sleep, neurobiology, 36
Gain: electroencephalogram (EEG), 128; electromyogram
    (EMG), 130
Gamma-hydroxybutyrate, hypersomnias, cataplexy, 421, 422
Gastrointestinal diseases, insomnias, 223–224
Gastrointestinal tract, 641–648; gastric functions, 643–644;
    large bowel, 646–647; oesophageal function, 641–643;
    overview, 641; small bowel, 644–646
Gastro-oesophageal reflux disease (GORD), 223–224, 641–
    643, 644
Gayet-Wernicke's encephalopathy, neurological
    hypersomnias, 449
Generalised idiopathic epilepsy, 617–618
Genetics, 83–111; bruxism, 536; enuresis, 547; fatal familial
    insomnia, 638; normal sleep (animal studies), 83–94;
    candidate gene studies, 85; circadian rhythms, 86–90;
    EEG activity, 90–91; factor determination, 83; gene
    expression alterations, 93–94; gene localization
    methods, 84; genetically modified mice studies, 85;
    homeostatic process, 93; linkage analysis studies, 84–
    85; mutagenesis studies, 85–86; nycthemer, 91–93;
    quantitative trait loci (QTL) analysis, 86; normal sleep
    (human studies), 94–96; circadian rhythms, 95–96;
    factor determination, 94–95; gene localization methods,

Genetics (*cont.*): 95; overview, 83; sleep, 3; sleep disorders, 96–102; chromosomal abnormalities, 102; circadian rhythm disorders, 100–101; dissociated REM sleep, 101; hypersomnias, 96–98; idiopathic hypersomnia, 433–434; insomnias, 98–99; narcolepsy, 412–415, 418–419; parasomnias, 99–100; periodic limb movements and restless legs syndrome, 101

Genioglossus advancement procedure, 396

Gilles de la Tourette's syndrome, sleepwalking and, 518

Global motor activity, sleepiness assessment, vigilance, 176

Gold compounds, insomnias, medication-linked, 252

Griseofulvine, insomnias, medication-linked, 253

Growth hormone (GH): Kleine-Levin syndrome, 441; obstructive sleep apnoea-hypopnea syndrome and upper airway resistance syndrome, 375, 377; sleep architecture, 51

Haemodynamic pulmonary complications, chronic obstructive pulmonary disease, hypoxemia, 570–571

Haemodynamics, obstructive sleep apnoea-hypopnea syndrome and upper airway resistance syndrome, 374–377

Hallucination: *see* Hypnagogic hallucinations

H$_1$ antihistamines, hypersomnias, medication and alcohol dependent sleepiness, 353

H$_2$ antihistamines, hypersomnias, medication and alcohol dependent sleepiness, 354

Headache, 629–633; chronic paroxysmal hemicrania, 630–631; cluster headache, 630; exploding head syndrome, 631; intracranial hypertension headache, 631–632; migraine, 629–630; nocturnal postepileptic headache, 631; obstructive sleep apnoea-hypopnea syndrome and upper airway resistance syndrome, 629; sleep deprivation, 631

Head trauma: neurological hypersomnias, 450; polysomnography, 678–679; sleep in, 679

Heart, chronic obstructive pulmonary disease, hypoxemia, 571; *see also* Cardiac dysrhythmias

Heart rate: automatic sleep analysis system, 165; cardiovascular function, 49; normal sleep, 3; normal sleep (in children), 11–12; sudden infant death syndrome, 651

Histaminergic system, wakefulness, neurobiology, 34

HIV/AIDS infection: drug-linked insomnias, 253; infection-related hypersomnias, 451; sleep in, 679–680

HLA testing: adult narcolepsy, 407, 412–413; childhood narcolepsy, 463, 465–466

Homeostatic process: genetics, animal studies, 93; normal aging, two-process model, 298–302

Hormones: *see* Endocrine function; Endocrine–metabolic disorders; specific hormones

Horne and Östberg Questionnaire, 740–743

Human immunodeficiency virus (HIV): *see* HIV/AIDS infection

Huntington's chorea, sleep in, 677

Hydrocephalus, normal pressure, neurological hypersomnias, 449

Hypernychthemeral syndrome: *see* Non-24 hour sleep-wake (hypernychthermeral) syndrome

Hypersomnias: aetiology, 334, 457–458; childhood, 457–468: aetiology, 457–458; diagnosis, 458–459; epidemiology, 457; idiopathic hypersomnia, 466; Kleine-Levin syndrome, 466; narcolepsy, 462–466:

Hypersomnias (*cont.*): clinical features, 462–463; diagnosis, 466; epidemiology, 462; symptomatic, 463; treatment, 466; obstructive sleep apnoea in children (OSAS), 459–462: clinical features, 460; diagnosis, 459–460; laboratory tests, 460–461; treatment, 461–462; differential diagnosis, decision tree approach, 337–339; endocrine–metabolic disorders, 453; epidemiology, 334; genetics, 96–98; head trauma, 679; idiopathic hypersomnia, 429–435; clinical features, 430; course of illness, 431; diagnosis, 430–431; differential diagnosis, 432; epidemiology, 429–430; historical perspective, 429; pathogenesis, 433–434; pathophysiology, 433; treatment, 434; infection-related, 451–452: African trypanosomiasis (sleeping sickness), 452; encephalitis, 451; Epstein-Barr virus, 451; human immunodeficiency virus (HIV), 451; insufficient sleep syndrome, 341–346: clinical features, 342; clinical variants, 343; course of illness, 343; differential diagnosis, 343–344; epidemiology, 342; laboratory tests, 342–343; overview, 341; pathophysiology, 344; treatment, 344–345; medication and alcohol dependent sleepiness, 347–356: alcohol, 354–355; analgesics, 353; anticholinergics, 354; antiepileptic drugs, 353; antihistamines, 353, 354; antimigraine drugs, 353; anti-Parkinson drugs, 353; available data, 348; cardiovascular medications, 353–354; data sources, 347–348; mood regulators, 352; myorelaxants, 353; overview, 347; progesterone and progestins, 354; psychotropic drugs, 349–352; narcolepsy, 403–428: clinical features, 405; clinical variants, 408–409; course of illness, 408; diagnosis, 405–407; differential diagnosis, 409; epidemiology, 403–404; historical perspective, 403; pathogenesis: animal studies, 415–418; human studies, 412–415; hypocretin, 418; mode of transmission, 418–419; pathophysiology, 409–411; treatment, 419–422; neurological, 447–450: brain tumor, 447–448; degenerative disease, 449; head injury, 450; stroke, 448–449; obstructive sleep apnoea-hypopnea syndrome and upper airway resistance syndrome, 357–401: central sleep apnoea syndromes, 366–368; clinical features, 358–359; clinical variants, 364–366; course of illness, 363–364; differential diagnosis, 368–369; epidemiology, 357; historical perspective, 357; laboratory tests, 359–363; pathophysiology, 369–377; treatment (nonsurgical), 378–381; treatment (surgical), 391–401; overview, 333–336; psychiatric, 450–451: depression, 450; seasonal affective disorder (SAD), 450–451; questionnaires and scales, 736–739; Epworth Sleepiness Scale, 739; Karolinska Sleepiness Scale, 738; Stanford Sleepiness Scale, 736–737; recurrent (Kleine-Levin syndrome), 437–445: brain tumour, 678; clinical features, 438–439; clinical variants, 440; epidemiology, 437–438; historical perspective, 437; laboratory tests, 439–440; pathogenesis, 441–442; pathophysiology, 440–441; treatment, 442; recurrent symptomatic, 442–443; organic origin, 442–443; psychiatric origin, 443

Hypertension, cardiocirculatory disorders, 590; *see also* Haemodynamics

Hypnagogic hallucinations: genetics, 101; narcolepsy, 405

Hypnogenic substances, sleep-wake architecture, 40–42

Hypnotic drugs: *see also* Pharmacology; specific drugs:

Hypnotic drugs (*cont.*): benzodiazepines, 257–261; childhood, 291–292; long-term use, 264; medication and alcohol dependent sleepiness, 349–350; medication-linked insomnias, 251

Hypocretin, narcolepsy, 416–418

Hypocretine neurons, wakefulness, neurobiology, 34

Hypoglycemia, hypersomnias, 453

Hypopnoeas, respiratory measurements, 135–136; *see also* Obstructive sleep apnoea-hypopnea syndrome and upper airway resistance syndrome

Hypothalamic-pituitary-adrenal (HPA) axis, stress, aging effects, 307–308

Hypothalamus: NREM sleep, neurobiology, 35–36; REM sleep, neurobiology, 40; thermoregulation, 55–56

Hypothyroidism, hypersomnias, 453

Hypoxemia; *see also* Chronic obstructive pulmonary disease (COPD); chronic obstructive pulmonary disease, 569–571; obstructive sleep apnoea-hypopnea syndrome and upper airway resistance syndrome, 374

Idiopathic hypersomnia, 429–435; childhood, 466; clinical features, 430; course of illness, 431; diagnosis, 430–431; differential diagnosis, 432; epidemiology, 429–430; historical perspective, 429; pathogenesis, 433–434; pathophysiology, 433; treatment, 434

Idiopathic infiltrative diseases of lung, insomnia, 223

IMAOs, medication and alcohol dependent sleepiness, 351

Imidazole, insomnias, medication-linked, 253

Imidazopyridine derivatives (zolpidem), insomnias, 262–263

Imipramine: insomnias, 271, 272; medication-linked, 249; parasomnias: REM sleep behaviour disorder, 534; rhythmic movement disorder, 524

Impedence plethysmography, respiratory measurements, 134–135

Incidence: *see* Epidemiology

Inductance plethysmography, respiratory measurements, quantitative, 135

Infection-related hypersomnias, 451–452; African trypanosomiasis (sleeping sickness), 452; encephalitis, 451; Epstein-Barr virus, 451; human immunodeficiency virus (HIV), 451

Infectious disorders, 679–680; HIV infection, 679–680; prion diseases, 680; trypanosomiasis, 680

Infiltrative diseases of lung, idiopathic, insomnia, 223

Insomnias: childhood, 283–295: aetiologies, 285–288; clinical investigation, 288; definition, 283; epidemiology, 284; treatment, 288–292; differential diagnosis, decision tree approach, 191–199; genetics, 98–99; medical disorders associated with, 221–226: cardiovascular disorders, 221–222; chronic obstructive lung diseases, 222–223; endocrine–metabolic disorders, 224; gastrointestinal diseases, 223–224; renal disorders, 224; rheumatic disorders, 224; medication-linked, 247–255: alcohol, 253; amphetamines, 249; anticonvulsants, 251; antidepressants, 249–250; anti-infectious drugs, 252–253; anti-inflammatory drugs, 252; anti-Parkinsonian drugs, 251; anti-rheumatic drugs, 252; caffeine, 249; cardio-vascular medications, 251–252; hypnotics and anxiolytics, 251; information sources, 247–248; neuroleptic disinhibitors, 251; psychostimulants, 249; retinoids, 253; thyroid hormonotherapy, 253; overview of, 187–189;

Insomnias (*cont.*): pharmacology, 257–281: amino acids, 275–276; antidepressants, 269–274; antihistamines, 274–275; barbiturates, 275; benzodiazepines, 257–261; chloral hydrate, 275; melatonin, 276–277; neuroleptics, 275; phytotherapy, 277; ritanserin, 277; S-adenosyl-homocytein (SAH), 277; sleep-inducing peptide (DSIP), 276; tiagabine, 277; zaleplon, 263–264; zolpidem, 262–263; zopiclone, 261–262; primary insomnia, 207–220: aetiology, 211–212; clinical presentation, 208–209; course and prognosis, 211; epidemiology, 210–211; evaluation and diagnosis, 212–213; overview, 207; subtypes of, 209–210; treatment, 213–218; psychiatric disorders associated with, 227–245: anxiety disorders: generalised anxiety disorder, 234–235; obsessive-compulsive disorder, 237–238; panic disorder, 236–237; post-traumatic stress disorder, 238–239; social/specific phobias, 236; eating disorders, 241; epidemiology and classification, 227; mood disorders: major depressive disorder, 227–232; manic episode, 232; treatments, 232–234; personality disorders, 241–242; schizophrenia, 239–240; questionnaires and scales, 720–735: Beck Depression Inventory, 732–735; Beliefs and Attitudes about Sleep Scale, 722–724; Leeds Sleep Evaluation Questionnaire, 730–731; Pittsburgh Sleep Quality Index, 725–729; Sleep Impairment Index, 720–721; transient/short term insomnia, 201–205: adjustment, 203; altitude, 203; environmental factors, 202–203; epidemiology, 201–202; physical stress, 204; rebound insomnia, 204, 260–261; sleep hygiene, 202

Insufficient sleep syndrome, 341–346: clinical features, 342; clinical variants, 343; course of illness, 343; differential diagnosis, 343–344; epidemiology, 342; laboratory tests, 342–343; overview, 341; pathophysiology, 344; treatment, 344–345

Intracranial hypertension headache, headache, 631–632

Intracranial pressure, obstructive sleep apnoea-hypopnea syndrome and upper airway resistance syndrome, 375

Intrathoracic pressure, obstructive sleep apnoea-hypopnea syndrome and upper airway resistance syndrome, 375

Iron: periodic leg movements, 605; restless legs movement, 605

Irregular sleep-wake pattern, 506–508: diagnosis, 507; epidemiology, 506; historical perspective, 506; pathophysiology, 507–508; treatment, 508

Isocarboxazide, insomnias, 272

Jet lag syndrome, 485–491: clinical features, 485; factors in, 485–488; melatonin, 76; pharmacology, 488; phototherapy, 489; sleep hygiene, 489

Karolinska Sleepiness Scale, 738

Kiphoscoliosis, sleep breathing abnormalities, 583

Kleine-Levin syndrome: childhood, 466; clinical features, 438–439; clinical variants, 440; epidemiology, 437–438; genetics, 98; historical perspective, 437; laboratory tests, 439–440; pathogenesis, 441–442; pathophysiology, 440–441; treatment, 442

Lamotrigine, nocturnal paroxysmal dystonia, 613

Landau-Kleffner's syndrome, 621

Larynx, snoring, 559

L-Dopa, parasomnias, REM sleep behaviour disorder, 534

Leeds Sleep Evaluation Questionnaire, 730–731

Lennox-Gastaut syndrome, epilepsy, 618–619, 623, 625, 691–692

Lesions: *see* Central nervous system lesions

Letter cancellation tasks, sleepiness assessment, vigilance, 173

Lewy body disease, sleep in, 677

Light effects: circadian rhythm sleep disorders, shift work sleep disorder, 479–481; melatonin secretion, 73–74

Limb-girdle dystrophy, sleep breathing abnormalities, 583

Limb movements, periodic: *see* Periodic limb/leg movement disorder

Lingual reduction surgery, 398

Listeria monocytogenes, 680

Lithium: headache, 630; hypersomnias, medication and alcohol dependent sleepiness, 352; insomnias, 272

Locus coeruleus, wakefulness, neurobiology, 33

Low-pass filters: electroencephalogram (EEG), 128; electromyogram (EMG), 131

L-tryptophane, insomnias, 275–276

Luteinising hormone (LH): circadian rhythmicity, 53; Kleine-Levin syndrome, 441

Magnetic resonance imaging (MRI): *see also* Radiology; narcolepsy, childhood, 466; obstructive sleep apnoea-hypopnea syndrome and upper airway resistance syndrome, 393–394; parasomnias, REM sleep behavior disorder, 531; restless legs syndrome, 603

Maintenance of wakefulness test (MWT), sleepiness assessment, 180

Mandibular advancement appliances, obstructive sleep apnoea-hypopnea syndrome and upper airway resistance syndrome, 379, 397

Mandibular osteotomy, obstructive sleep apnoea-hypopnea syndrome and upper airway resistance syndrome, 396

Manic episode: described, 232; treatment, 234

Maxillomandibular advancement appliances, obstructive sleep apnoea-hypopnea syndrome and upper airway resistance syndrome, 397–398

Mazindol, hypersomnias, cataplexy, 421

Medical disorders, insomnia: aging effects, comorbid disease, 309–310; cardiovascular disorders: arterial hypertension, 590; cardiac arrhythmias, 222, 593–594; congestive heart disease, 221; ischemic heart disease, 222, 590–593; chronic fatigue syndrome, 703–714; endocrine–metabolic disorders, 224; fibromylagia, 703–714; gastrointestinal disordres: gastroesophageal reflux diseases, 223–224, 641–643; peptic ulcer disease, 224, 643–644; small and large bowels, 644–646; neurological disorders: cerebrovascular disorders, 673–675; degenerative disor4ders: Alzheimer's disease, 677–678; amyotrophic lateral sclerosis, 675, 697; dementia with Lewy bodies, 677; Huntington's chorea, 677; multiple system atrophy, 594, 676; Parkinson's disease, 675–676; progressive supranuclear palsy, 677; dystonia, 676–677; epilepsy, 624–625; head traumatism, 678–679; headache, 629–633; infectious disorders: fatal familial insomnia, 635–639, 680; HIV infection, 679–680; prion diseases, 680; neuromuscular diseases, 581–588; overview, 221; painful erections, 657–662; pulmonary disorders: bronchial asthma, 222–

Medical disorders, insomnia (*cont.*): 223; chronic obstructive pulmonary diseases, 223, 567; idiopathic infiltrative diseases of the lungs, 223; renal disorders: chronic renal failure, 224; rheumatic disorders, 224

Medication and alcohol dependent sleepiness, 347–356; *see also* Alcohol; Pharmacology: alcohol, 354–355; analgesics, 353; anticholinergics, 354; antiepileptic drugs, 353; antihistamines, 353; antimigraine drugs, 353; anti-Parkinson drugs, 353; available data, 348; cardio-vascular medications, 353–354; data sources, 347–348; mood regulators, 352; myorelaxants, 353; overview, 347; progesterone and progestins, 354; psychotropic drugs, 349–352: antidepressants, 351–352; anxiolytics and hypnotics, 349–350; neuroleptics, 350–351

Medication-linked insomnias: *see also* Pharmacology: alcohol, 253; amphetamines, 249; anticonvulsants, 251; antidepressants, 249–250; anti-infectious drugs, 252–253; anti-inflammatory drugs, 252; anti-Parkinsonian drugs, 251; anti-rheumatic drugs, 252; caffeine, 249; cardio-vascular medications, 251–252; hypnotics and anxiolytics, 251; information sources, 247–248; neuroleptic disinhibitors, 251; psychostimulants, 249; retinoids, 253; thyroid hormonotherapy, 253

Medulla, wakefulness, neurobiology, 33

Melatonin, 71–82; aging, treatments, 320–321; circadian rhythmicity, 53; circadian rhythm sleep disorders: advanced sleep phase syndrome, 502; delayed sleep phase syndrome, 498–499; irregular sleep-wake pattern, 508; shift work sleep disorder, 479; endogenous synchroniser, 74–75; insomnias, 276–277; light effects, 73–74; pharmacology of, 75–79; plasma profile and epiphysial activity, 71–72; plasma rhythm, 74; secretion regulation system, 72–73

Menopause: obstructive sleep apnoea-hypopnea syndrome and upper airway resistance syndrome, 366; sleep effects, 314–315

Menstruation, Kleine-Levin syndrome, 440, 466

Mental activity: *see* Dreams; Psychiatric disorders

Metabolism: *see* Endocrine–metabolic disorders

Methyldopa, insomnias, medication-linked, 251

Methylxanthine, insomnias, medication-linked, 252

Methysergide, hypersomnias, medication and alcohol dependent sleepiness, 353

Meynert's nucleus: NREM sleep, neurobiology, 36; wakefulness, neurobiology, 32–33

Mianserin: insomnias, 272; medication and alcohol dependent sleepiness, 352

Miction: *see* Enuresis

Migraine headache, 629–630

Migrating motor complex (MMC), 644–646

Mirtazapine, medication and alcohol dependent sleepiness, 352

Moclobemide: insomnias, 272; medication and alcohol dependent sleepiness, 351

Modafinil: idiopathic hypersomnia, 434; insomnias, medication-linked, 249; narcolepsy, 419–420, 466; recurrent hypersomnias, 442

Monoamine assays, Kleine-Levin syndrome, 440

Monoamine-oxydase inhibitors (MAOIs): insomnias, 272; medication-linked insomnias, 249

Mood disorders, 228–234, 665–672; antidepressant treatment, 669; major depressive disorder, insomnias,

Mood disorders (*cont.*): 228–232; manic episode, insomnias, 232; pathophysiology, 669–670; polysomnography, 667–669; sleep alterations, 665–666; specific types of disorders, 666–667; treatments, 232–234

Morning Questionnaire, 718–719

Morphine, hypersomnias, medication and alcohol dependent sleepiness, 353

Movement time, visual sleep analysis, 157

Multiple sclerosis: neurological hypersomnias, 449; sleep in, 679

Multiple sleep latency test (MSLT): narcolepsy, childhood, 463, 464; obstructive sleep apnoea-hypopnea syndrome and upper airway resistance syndrome, 363; sleepiness assessment, 178–180

Multiple system atrophy: cardiocirculatory disorders, 594; neurological hypersomnias, 449; sleep in, 676

Muscular activity, automatic sleep analysis system, 165

Mutagenesis studies, normal sleep (animal studies), 85–86

Myasthenia, 581–582, 697

Myocardial infarction, cardiocirculatory disorders, 592–593

Myoclonus, benign neonatal sleep onset myoclonus, 537–538

Myorelaxants, hypersomnias, medication and alcohol dependent sleepiness, 353

Myotonic dystrophy, 582, 697

Napping: circadian rhythm sleep disorders, shift work sleep disorder, 482; hypersomnias: insufficient sleep syndrome, 344; narcolepsy, 421; normal sleep, in children, 22

Narcolepsy, 403–428; cerebral tumour, 678; childhood, 462–466: clinical features, 462–463; diagnosis, 463; epidemiology, 462; laboratory tests, 463; symptomatic, 463; treatment, 466; clinical features, 405; course of illness, 408; diagnosis, 405–407; differential diagnosis, 409; epidemiology, 403–404; genetics, 96–97; head trauma, 679; historical perspective, 403; idiopathic hypersomnia and, 429, 430; parasomnias, REM sleep behavior disorder, 532; pathogenesis, 412–419: animal studies, 415–418; human studies, 412–415; hypocretin, 418; mode of transmission, 418–419; pathophysiology, 409–411; treatment, 419–422

Nasal cavities, snoring, 559

Nasal obstruction: obstructive sleep apnoea-hypopnea syndrome and upper airway resistance syndrome, 398; snoring, treatment, 561

Nefazodone, insomnias, 272

Neonatal period, 12–16; full-term newborn, 13–15: polysomnography, 13–14; sleep-wake architecture, 14–15; premature newborn: polysomnography, 15–16; sleep-wake architecture, 16

Neurobiology, 31–43; hypnogenic substances, 40–42; NREM sleep, 35–36; REM sleep, 37–40; theories, 31–32; wakefulness, 32–35

Neurohumoral aspect, sleep-wake architecture, 39

Neuroleptic disinhibitors, insomnias, medication-linked, 251

Neuroleptics: insomnias, 275; medication and alcohol dependent sleepiness, 350–351

Neurological hypersomnias, 447–450: brain tumor, 447–448; childhood narcolepsy, 463; degenerative disease, 449; head injury, 450; stroke, 448–449

Neuromuscular disease: *see also* Sleep breathing

Neuromuscular disease (*cont.*): abnormalities: neurological hypersomnias, 449–450; obstructive sleep apnoea-hypopnea syndrome and upper airway resistance syndrome, 365; sleep breathing abnormalities in, 581–588

Neuropsychology, primary insomnia, 209

Niaprazine, insomnias, childhood, 291–292, 292

Nightmares, 120, 526–529: clinical features, 526; clinical variants, 527; differential diagnosis, 528; epidemiology, 526–527; laboratory tests, 528; pathophysiology, 528–529; REM sleep behavior disorder, 531; treatment, 529

Night terrors, 518–522: clinical features, 518; clinical variants, 519; diagnosis, 520–521; dreams, 120; epidemiology, 518–519; factors in, 519; genetics, 99; laboratory tests, 519–520; pathophysiology, 521; treatment, 521–522

Night workers, melatonin, 77; *see also* Shift work sleep disorder

Nigro-striatal system, wakefulness, neurobiology, 34

Nocturnal erections: *see* Painful erections

Nocturnal hypoxemia: *see* Chronic obstructive pulmonary disease (COPD)

Nocturnal leg cramps, 525–526

Nocturnal paroxysmal dystonia, 609–615: clinical and polygraphic features, 609–610; clinical variants, 610–611; course of illness, 610; differential diagnosis, 611; epidemiology, 609; historical perspective, 609; pathophysiology and nosology, 611–613; treatment, 613

Non-24 hour sleep-wake (hypernychthemeral) syndrome, 502–506: diagnosis, 505; epidemiology, 505; historical perspective, 502–505; pathophysiology, 505–506; treatment, 506

Non-rapid eye movement (NREM) sleep: aging, 298–299, 305, 307, 309, 311–313; antidepressant effects, insomnias, 272; cardio-vascular function, 49–50; cerebral metabolism, 56; circadian rhythm sleep disorders, shift work sleep disorder, 475; digestive function, 53–54; dreams, 113–115; endocrine function, 51–52; enuresis, 548–549; epilepsy, 692; fibromyalgia and chronic fatigue syndrome, 707–708; idiopathic hypersomnia, 433; narcolepsy, 409–410; neurobiology, 31, 35–36; obstructive sleep apnoea-hypopnea syndrome and upper airway resistance syndrome, 361–362; parasomnias: night terrors, 519, 520; sleep talking, 535; sleep regulation, 61; sleep studies, 3–5; thermoregulation, 54–56; ventilatory function, 47; visual sleep analysis, 154–156

Non steroid anti-inflammatory drugs, insomnias, medication-linked, 252

Norepinephrine system, wakefulness, neurobiology, 33

Normal aging: *see* Aging

Normal pressure hydrocephalus, neurological hypersomnias, 449

Normal sleep (in adults), 3–9; definition, 3; genetics, animal studies, 83–94; *see also* Genetics: haemodynamic changes during, 589–590; polysomnography, 6–8: cyclic alternating pattern (CAP), 7–8; scoring of, 7; technique of, 6–7; studies of, 4–6

Normal sleep (in children), 11–30; *see also* Childhood insomnias: infancy to one year, 16–21: polysomnography, 16–19; sleep-wake architecture, 19–21; neonatal period, 12–16: full-term newborn, 13–15;

Normal sleep (in children) (*cont.*): polysomnography, 13–14; sleep-wake architecture, 14–15; premature newborn, 15–16: polysomnography, 15–16; sleep-wake architecture, 16; one to three years, 21–23: EEG characteristics, 21–22; sleep-wake architecture, 22–23; six to twelve years, 23–24: EEG characteristics, 23–24; sleep-wake architecture, 24; studies of, 11–12; three to six years, 23: twelve to fifteen years, 24–25: polysomnography, 24; sleep-wake architecture, 24–25

NREM sleep: *see* Non-rapid eye movement (NREM) sleep

Nycthemer, genetics, animal studies, 91–93

Obsessive-compulsive disorder, insomnias, 237–238

Obstructive sleep apnoea-hypopnea syndrome and upper airway resistance syndrome, 357–401; *see also* Chronic obstructive pulmonary disease (COPD); Obstructive sleep apnoea in children (OSAS); Sleep apnoeas; Snoring: aetiology, 366; central sleep apnoea syndromes, 366–368; clinical features, 367; mechanisms, 367–368; treatment, 368; chronic obstructive pulmonary disease, 572; clinical features, 358–359; clinical variants, 364–366; course of illness, 363–364; differential diagnosis, 368–369; epidemiology, 357; headache, 629; historical perspective, 357; laboratory tests, 359–363; pathophysiology, 369–377; treatment (nonsurgical), 378–381; treatment (surgical), 391–401: basis for, 392–394; indications and outcomes, 398–400; methods, 394–398; principles, 391–392

Obstructive sleep apnoea in children (OSAS), 459–462; *see also* Obstructive sleep apnoea-hypopnea syndrome and upper airway resistance syndrome; Sleep apnoeas: clinical features, 460; laboratory tests, 460–461; treatment (surgical), 461–462

Oesophageal function, gastrointestinal tract, 641–643

Ondine's disease, 680

Opiates: periodic leg movements, 604; restless legs syndrome, 604

Orexine neurons, wakefulness, neurobiology, 34

Oro-nasal sensors, respiratory measurements, qualitative, 133–134

Outcomes: obstructive sleep apnoea-hypopnea syndrome and upper airway resistance syndrome, 398–400; primary insomnia, 217

Oxford sleep resistance test, sleepiness assessment, vigilance, 173

Oxygen therapy: hypersomnias, obstructive sleep apnoea-hypopnea syndrome and upper airway resistance syndrome, 380–381; medical disorders, chronic obstructive pulmonary disease, 572–574

Oxymetry (SaO$_2$ measurement), polysomnography, 136

Painful erections, 657–662; clinical features, 657–658; clinical variant, 659; course of illness, 659; differential diagnosis, 659; laboratory tests, 658–659; pathophysiology, 660; treatment, 660–661

Panic disorder, 236–237, 274

Paramedian infarcts, neurological hypersomnias, 448

Parasomnias, 513–543; *see also* Enuresis: arousal disorders, 515–522; confusional arousals, 515–516; night terrors, 518–522; sleepwalking, 516–518; benign neonatal sleep onset myoclonus, 537–538; classification of, 513–515;

Parasomnias (*cont.*): definition, 513; differential diagnosis, 197; genetics, 99–100; REM sleep, 526–534; nightmares, 526–529; REM sleep behavior disorder, 530–534; sleep paralysis, 529–530; sleep bruxism, 535–537; clinical features, 535; clinical variants, 536; epidemiology, 535–536; genetics, 100; laboratory tests, 536; pathophysiology, 536–537; treatment, 537; sleep-related abnormal swallowing syndrome, 538–539; sleep talking, 534–535; sleep-wake transition disorders, 522–526: nocturnal leg cramps, 525–526; rhythmic movement disorder, 522–524; sleep starts (predormital myoclonus), 524–525

Parkinson's disease: anti-Parkinsonian drugs: hypersomnias, medication and alcohol dependent sleepiness, 353; insomnias, medication-linked, 251; parasomnias, nightmares, 527; neurological hypersomnias, 449; parasomnias, REM sleep behavior disorder, 531; sleep in, 675–676

Paroxetine: hypersomnias, cataplexy, 421; insomnias, 271, 272, 274

Partial idiopathic epilepsy with midtemporal spikes, 619–620

Pediatrics: *see* Normal sleep (in children)

Penile plethysmogram, polysomnography, 137; *see also* Erections; Painful erections

Peptic ulcer, 224, 644

Perdue pegboard test, sleepiness assessment, 173, 174

Pergolide: periodic leg movements, 605; restless legs movement, 605

Periodic limb/leg movement disorder: *see also* Restless legs syndrome; ambulatory sleep recording systems, 146; automatic sleep analysis system, 165; clinical features, 600; diagnosis, 601–602; epidemiology, 600; fibromyalgia and chronic fatigue syndrome, 708–709, 710–711; genetics, 101; insomnias, differential diagnosis, 197; pathophysiology, 602–603; polysomnography, 137; treatment, 603–605

Permissive REM sleep, sleep-wake architecture, neurobiology, 38–39

Peyronie's disease, painful erections, 659

Pharmacology: *see also* Medication-linked insomnias; *specific drugs*: aging, 315–318; circadian rhythm sleep disorders: delayed sleep phase syndrome, 497; irregular sleep-wake pattern, 508; jet lag syndrome, 488; shift work sleep disorder, 479; enuresis, 550–551; epilepsy, 621; epilepsy investigation, 689; fatal familial insomnia, 636; genetic linkage analysis, normal sleep (animal studies), 85; genetics, nycthemer, animal studies, 91–93; headache, 630; hypersomnias: idiopathic, 434; narcolepsy, 410–411, 419–422; childhood, 466; obstructive sleep apnoea-hypopnea syndrome and upper airway resistance syndrome, 380; insomnias, 257–281: amino acids, 275–276; antidepressants, 269–274; antihistamines, 274–275; barbiturates, 275; benzodiazepines, 257–261; childhood, 291–292; chloral hydrate, 275; eating disorders, 241; generalised anxiety disorder, 235; melatonin, 276–277; mood disorders, 232–234; neuroleptics, 275; obsessive-compulsive disorder, 237–238; panic disorder, 236–237; phytotherapy, 277; post-traumatic stress disorder, 239; primary insomnia, cognitive-behavioral therapies combined with, 216–217;

Pharmacology (*cont.*): ritanserin, 277; S-adenosyl-
homocytein (SAH), 277; schizophrenia, 240; sleep-
inducing peptide (DSIP), 276; social/specific phobias,
236; tiagabine, 277; transient/short term insomnia,
rebound insomnia, 204, 260–261; zaleplon, 263–264;
zolpidem, 262–263; zopiclone, 261–262; medical
disorders: chronic obstructive pulmonary disease, 572;
fibromyalgia and chronic fatigue syndrome, 709–711;
painful erections, 660–661; melatonin, 75–79; mood
disorders, 669; nocturnal paroxysmal dystonia, 613;
obstructive sleep apnoea in children (OSAS), 462;
parasomnias: bruxism, 536, 537; nightmares, 527, 529;
night terrors, 521–522; nocturnal leg cramps, 526;
REM sleep behaviour disorder, 533–534; rhythmic
movement disorder, 524; sleep paralysis, 530; sleep
starts (predormital myoclonus), 525; sleep talking, 535;
periodic leg movements, 603–605; restless legs
syndrome, 603–605; sudden infant death syndrome,
652, 654
Phenelzine, medication and alcohol dependent sleepiness, 351
Phenobarbital: insomnias, medication-linked, 251; nocturnal
paroxysmal dystonia, 613
Phenytoin, nocturnal paroxysmal dystonia, 613
Phototherapy: aging, 319–320; circadian rhythm sleep
disorders: advanced sleep phase syndrome, 500–502;
delayed sleep phase syndrome, 497–498; jet lag
syndrome, 489
Physical stress, transient/short term insomnia, 204; *see also*
Stress
Physiological functions, 45–60; cardio-vascular, 48–50:
arterial pressure, 49; cardiac output and peripheral
circulation, 50; heart rate, 49; rhythmicity, 50; cerebral
metabolism, 56; digestive, 53–54; endocrine, 50–53;
circadian rhythmicity, 52–53; sleep architecture, 51–52;
sleep-wake architecture, 51; sexual, 54;
thermoregulation, 54–56; ventilatory, 45–48: NREM
sleep, 47; REM sleep, 47–48; sleep onset, 46–47;
wakefulness, 46
Phytotherapy, insomnias, 277
Pickwickian syndrome: *see* Obstructive sleep apnoea-
hypopnea syndrome and upper airway resistance
syndrome
Pittsburgh Sleep Quality Index, 725–729
Pituitary hormones, anterior, sleep architecture, 52
Plethysmography: impedence, respiratory measurements,
quantitative, 134–135; inductance, respiratory
measurements, quantitative, 135; painful erections,
658–659
Pleural pressure, respiratory measurements, quantitative, 135
Pneumotachography, respiratory measurements, quantitative,
135
Poliomyelitis, 680
Polycythemia, chronic obstructive pulmonary disease,
hypoxemia, 571
Polysomnography, 6–8, 127–138; *see also* Ambulatory sleep
recording; Automatic sleep analysis system: cyclic
alternating pattern (CAP), 7–8; electrocardiogram
(ECG), 131–132; electroencephalogram (EEG), 127–
129: channel setting, 128–129; derivations, 128;
electrode placement, 127; electromyogram (EMG),
130–131: channel setting, 130–131; derivations, 130;
electrode placement, 130; electrooculogram (EOG),

Polysomnography (*cont.*): 129–130: channel setting, 130;
derivations, 129–130; electrode placement, 129;
encephalic disturbances and, 680–681; epilepsy
investigation, 689–690; head trauma, 678–679;
idiopathic hypersomnia, 430–431; indications for, 140;
Kleine-Levin syndrome, 439–440; medical disorders:
painful erections, 658–659; mood disorders, 667–669;
narcolepsy, 406–407; childhood, 464–465; nocturnal
paroxysmal dystonia, 609–610; normal sleep (in
children): full-term newborn, 13–14; infancy to one
year, 16–19; premature newborn, 15–16; twelve to
fifteen years, 24; obstructive sleep apnoea-hypopnea
syndrome and upper airway resistance syndrome, 359–
363, 369; obstructive sleep apnoea in children (OSAS),
460–461; parasomnias: benign neonatal sleep onset
myoclonus, 537–538; bruxism, 536; night terrors, 519–
520; nocturnal leg cramps, 526; REM sleep behavior
disorder, 531; rhythmic movement disorder, 523; sleep
paralysis, 530; sleep-related abnormal swallowing
syndrome, 539; sleep starts (predormita myoclonus),
524–525; sleep talking, 535; primary insomnia, 208–
209; pulse transit time (PTT), 136; respiratory
measurements, 132–136: qualitative, 132–134;
quantitative, 134–136; scoring of, 7; snoring, 560;
subject position calibration, 131; technique of, 6–7
Ponto-geniculo-occipital (P.G.O.) spikes, sleep studies, 4–5
Positron emission tomography (PET): *see also* Radiology:
fatal familial insomnia, 637; restless legs syndrome,
603
Posterior hypothalamus, wakefulness, neurobiology, 34
Post-traumatic stress disorder, insomnias, 238–239
Postural treatment, snoring, 561
Prader-Willi syndrome, hypersomnias, 453
Pramipexole: periodic leg movements, 605; restless legs
syndrome, 605
Predormital myoclonus (sleep starts), 524–525
Pregnancy, obstructive sleep apnoea-hypopnea syndrome
and upper airway resistance syndrome, 366
Premature newborn, normal sleep, 15–16
Primary insomnia, 207–220; aetiology, 211–212; clinical
presentation, 208–209; course and prognosis, 211;
epidemiology, 210–211; evaluation and diagnosis, 212–
213; overview, 207; subtypes of, 209–210; treatment,
213–218
Prion diseases, sleep in, 680
Progesterone: hypersomnias, medication and alcohol
dependent sleepiness, 354; medical disorders, chronic
obstructive pulmonary disease, 572
Progressive supranuclear palsy, sleep in, 677
Prolactin: Kleine-Levin syndrome, 441; sleep-wake cycle, 51
Promethazine, insomnias, 275
Prosthetic treatment: obstructive sleep apnoea-hypopnea
syndrome and upper airway resistance syndrome, 379,
397–398; snoring, 561
Protriptyline: hypersomnias, cataplexy, 421; insomnias, 273
Psychiatric disorders: anxiety disorders: generalised anxiety
disorder, insomnias, 234–235; obsessive-compulsive
disorder, insomnias, 237–238; panic disorder,
insomnias, 236–237; post-traumatic stress disorder,
insomnias, 238–239; social/specific phobias, insomnias,
236; dreams, 119–120; eating disorders, insomnias, 241;
hypersomnias, 450–451: depression, 450; seasonal

Protriptyline (*cont.*): affective disorder (SAD), 450–451;
    mood disorders, 228–234, 665–672: antidepressant
    treatment, 669; major depressive disorder, insomnia
    and, 227–232; manic episode, insomnias, 232;
    pathophysiology, 669–670; polysomnography, 667–
    669; sleep alterations, 665–666; specific types of
    disorders, 666–667; treatments, insomnias, 232–234;
    painful erections, 657, 660; personality disorders,
    insomnias, 241–242; recurrent symptomatic
    hypersomnias, 443; schizophrenia, insomnias, 239–240
Psychomotor measures, vigilance, sleepiness assessment,
    171–176
Psychopathologies, dreams, 119–120
Psychoses, dreams, 119–120
Psychotropic drugs, medication and alcohol dependent
    sleepiness, 349–352
Pulse transit time (PTT), polysomnography, 136
Pupillometry, sleepiness assessment, vigilance, 176–177
Pyrazolopyrimidine derivatives (zaleplon), insomnias, 263–
    264

Qualitative respiratory measurements, 132–134
Quantitative respiratory measurements, 134–136
Quantitative trait loci (QTL) analysis, normal sleep (animal
    studies), 86
Questionnaires and scales, 717–743; circadian rhythm sleep
    disorders: Horne and Östberg Questionnaire, 740–743;
    hypersomnias, 736–739: Epworth Sleepiness Scale,
    739; Karolinska Sleepiness Scale, 738; Stanford
    Sleepiness Scale, 736–737; insomnias, 720–735: Beck
    Depression Inventory, 732–735; Beliefs and Attitudes
    about Sleep Scale, 722–724; Leeds Sleep Evaluation
    Questionnaire, 730–731; Pittsburgh Sleep Quality
    Index, 725–729; Sleep Impairment Index, 720–721;
    Morning Questionnaire, 718–719
Quinine, parasomnias, nocturnal leg cramps, 526
Quinupramine, insomnias, 272

Radiology: fatal familial insomnia, 637; Kleine-Levin
    syndrome, 439; narcolepsy, childhood, 466; obstructive
    sleep apnoea-hypopnea syndrome and upper airway
    resistance syndrome, 392, 393–394; obstructive sleep
    apnoea in children (OSAS), 461; parasomnias, REM
    sleep behavior disorder, 531; restless legs syndrome,
    603; snoring, 560
Raphe system, wakefulness, neurobiology, 34
Rapid eye movement (REM) sleep: cardio-vascular function,
    49–50; cerebral metabolism, 56; circadian rhythm sleep
    disorders, shift work sleep disorder, 475; digestive
    function, 53; dreams, 113–115; endocrine function, 51–
    52; idiopathic hypersomnia, 433; insomnias,
    antidepressant effects, 271–272; narcolepsy, 409–410,
    411, 421–422; neurobiology, 31, 37–40; obstructive
    sleep apnoea-hypopnea syndrome and upper airway
    resistance syndrome, 362; painful erections, 660;
    parasomnias, 526–534: nightmares, 526–529; REM
    sleep behavior disorder, 530–534; sleep paralysis, 529–
    530; sexual function, 54; sleep regulation, 61; sleep
    studies, 3–5, 11; thermoregulation, 54–56; ventilatory
    function, 47–48; visual sleep analysis, 156
Rapid eye movement (REM) sleep behaviour disorder, 530–
    534: clinical features, 530–531; clinical variants, 531–

Rapid eye movement (REM) sleep behaviour disorder
    (*cont.*): 532; differential diagnosis, 532; epidemiology,
    531; genetics, 99; laboratory tests, 531;
    pathophysiology, 532–533; treatment, 533–534
Reaction time tests, sleepiness assessment, vigilance, 171–
    172
Rebound insomnia: benzodiazepines, 260–261; transient/
    short term insomnia, 204
Recurrent hypersomnias (Kleine-Levin syndrome), 437–445;
    brain tumour, 678; clinical features, 438–439; clinical
    variants, 440; epidemiology, 437–438; historical
    perspective, 437; laboratory tests, 439–440;
    pathogenesis, 441–442; pathophysiology, 440–441;
    treatment, 442
Recurrent symptomatic hypersomnias, 442–443; organic
    origin, 442–443; psychiatric origin, 443
Relaxation-based interventions, primary insomnia, 215–216
REM sleep: *see* Rapid eye movement (REM) sleep
Renal disorders: hypersomnias, 453; insomnias, 224
Renin-angiotensin-aldosterone system, sleep architecture,
    51–52
Respiration: *see* Sleep breathing abnormalities; Ventilatory
    function
Respiratory disorders; *see also* Sleep breathing
    abnormalities: obstructive sleep apnoea-hypopnea
    syndrome and upper airway resistance syndrome, 365,
    377; sudden infant death syndrome, 650–651
Respiratory measurements, 132–136; *see also* Ventilatory
    function: ambulatory sleep recording systems, 143–146;
    automatic sleep analysis system, 165; obstructive sleep
    apnoea-hypopnea syndrome and upper airway
    resistance syndrome, 359–363; polysomnography:
    qualitative, 132–134; quantitative, 134–136
Restless legs syndrome: ambulatory sleep recording
    systems, 146; automatic sleep analysis system, 165;
    diagnosis, 600–601; fibromyalgia and chronic fatigue
    syndrome, 708–709, 710–711; genetics, 100, 101;
    insomnias, differential diagnosis, 197;
    polysomnography, periodic limb movements, 137;
    treatment, 603–605; in wakefulness, 599
Reticular formation, wakefulness, neurobiology, 32
Reticular nucleus, NREM sleep, neurobiology, 36
Retinoids, insomnias, medication-linked, 253
Retrobasilingual surgery, obstructive sleep apnoea-hypopnea
    syndrome and upper airway resistance syndrome, 396–398
Rheumatic disorders, insomnias, 224
Rhythmicity, cardio-vascular function, 50
Rhythmic movement disorder, 522–524: clinical features,
    522; clinical variants, 522–523; epidemiology, 522;
    laboratory tests, 523; pathophysiology, 523; treatment,
    523–524
Ropinirole: periodic leg movements, 605; restless legs
    syndrome, 605
Rule of contiguity, visual sleep analysis, sleep stage scoring,
    152
Rule of unity, visual sleep analysis, sleep stage scoring, 152

S-adenosyl-homocytein (SAH), insomnias, 277
Salbutamol, insomnias, medication-linked, 252
Scales: *see* Questionnaires and scales
Schizophrenia, insomnias, 239–240
Scopolamine, hypersomnias, 354

Seasonal affective disorder (SAD), psychiatric hypersomnias, 450–451

Sedatives, sudden infant death syndrome, 652, 654

Seizures: *see also* Epilepsy: generalised, 617; nocturnal paroxysmal dystonia, 611–614

Selective norepinephrine reuptake blockers, insomnias, medication-linked, 250

Selegiline, hypersomnias, cataplexy, 421

Separation anxiety, childhood insomnias, 289–290

Serotoninergic system: NREM sleep, neurobiology, 36; wakefulness, neurobiology, 34–35

Serotonin reuptake blockers: hypersomnias, cataplexy, 421; insomnias, medication-linked, 250; medication and alcohol dependent sleepiness, 351–352; parasomnias, sleep paralysis, 530

Sertraline, hypersomnias, cataplexy, 421

Sexual differences: *see* Gender differences

Sexual function, 54; *see also* Erections; Painful erections

Shift workers, melatonin, 77

Shift work sleep disorder, 473–484; circadian rhythm sleep disorders, light role, 479–481; clinical features, 476–477; definitions, 473–474; epidemiology, 473; factors in, 474–476; melatonin, 479; napping, 482; pharmacology, 479; therapy, 478–479; work organisation, 477–478

Short term insomnia: *see* Transient/short term insomnia

Shy-Drager syndrome, 594

Signal detection tests, sleepiness assessment, vigilance, 172

Simple snoring: *see* Snoring

Sleep, physiological functions during, 45–60; *see also* Normal sleep (in adults); Normal sleep (in children); Physiological functions

Sleep apnoeas: *see also* Obstructive sleep apnoea-hypopnea syndrome and upper airway resistance syndrome; Obstructive sleep apnoea in children (OSAS): antidepressants, insomnias, 273; fibromyalgia and chronic fatigue syndrome, 709, 711; genetics, 97–98; insomnias, differential diagnosis, 197; sleepwalking and, 518

Sleep-arousal-sudden death association, sudden infant death syndrome, 652–653

Sleep breathing abnormalities, 581–588; acid maltase deficiency, 582; Duchenne's muscular dystrophy, 582; kiphoscoliosis, 583; limb-girdle dystrophy, 583; myasthenia, 581–582; myotonic dystrophy, 582; pathophysiology, 581; results, 583–584; treatment, 584–586

Sleep bruxism: *see* Bruxism

Sleep deprivation, headache, 631

Sleep disorders: *see* Circadian rhythm sleep disorders; Genetics; Hypersomnias; Insomnias; specific disorders

Sleep epilepsies, 622–623; *see also* Epilepsy

Sleep hygiene: childhood insomnias, 288; circadian rhythm sleep disorders, jet lag syndrome, 489; fibromyalgia and chronic fatigue syndrome, 711–712; insomnias, childhood, 288; parasomnias, sleep paralysis, 530; primary insomnia, 214; transient/short term insomnia, 202

Sleep Impairment Index, 720–721

Sleep-inducing peptide (DSIP), insomnias, 276

Sleepiness assessment, 169–184: overview, 169; polygraphic measurement, 177–180: alpha attenuation test, 178; continuous recording, 177–178; electroencephalogram

Sleepiness assessment (*cont.*): (EEG), 177; maintenance of wakefulness test (MWT), 180; multiple sleep latency test (MSLT), 178–180; sleepiness index, 180; subjective, 169–171: behavioural, 170–171; scales: visual analogue scales (VAS), 169–170; vigilance, 171–177; evoked potentials, 177; eye movements, 176; global motor activity, 176; limitations of, 176; performance tests, 172–175; pupillometry, 176–177: signal detection tests, 172; spontaneous or evoked responses measurement, 172–173

Sleeping position, sudden infant death syndrome, 651, 653

Sleeping sickness (African trypanosomiasis): infection-related hypersomnias, 452; sleep in, 680

Sleep onset, ventilatory function, 46–47

Sleep paralysis: genetics, 101; parasomnias, 529–530

Sleep regulation, 61–70; *see also* Sleep-wake cycle; overview, 61; process interactions, 63–67; two-process model, 62–63

Sleep-related abnormal swallowing syndrome, 538–539

Sleep restriction therapy, primary insomnia, 215

Sleep starts (predormital myoclonus), 524–525

Sleep talking: bruxism and, 536; described, 534–535; sleepwalking and, 518

Sleep-wake cycle; *see also* Circadian rhythms; Circadian rhythm sleep disorders: aging, 297–332; *see also* Aging: childhood insomnias, 289; endocrine function, 51; melatonin, 71–82; *see also* Melatonin: mood disorders, 670; neurobiology, 31–43: hypnogenic substances, 40–42; NREM sleep, 35–36; REM sleep, 37–40; theories, 31–32; wakefulness, 32–35; normal sleep (in children): full-term newborn, 14–15; infancy to one year, 19–21; one to three years, 22–23; premature newborn, 16; six to twelve years, 24; three to six years, 23; twelve to fifteen years, 24–25

Sleep-wake transition disorders, 522–526; nocturnal leg cramps, 525–526; rhythmic movement disorder, 522–524; clinical features, 522; clinical variants, 522–523; epidemiology, 522; laboratory tests, 523; pathophysiology, 523; treatment, 523–524; sleep starts (predormita myoclonus), 524–525

Sleepwalking: genetics, 99; parasomnias, arousal disorders, 516–518

Snoring, 555–565; clinical examination, 558–559; clinical variants, 560; ENT examination, 557–558; epidemiology, 555–556; fiberoptic investigation, 559–560; historical perspective, 555; interview, 556–557; laboratory tests, 560; obstructive sleep apnoea-hypopnea syndrome and upper airway resistance syndrome, 368; pathophysiology, 556; radiography, 560; treatment, 561–564

Spontaneous responses measurement, sleepiness assessment, vigilance, 172–173

Stanford Sleepiness Scale, 736–737

Steinert's myotonic dystrophy, genetics, 98

Stimulus control therapy, primary insomnia, 215

Stress: aging effects, 307–308; parasomnias, nightmares, 527, 528–529; physical, transient/short term insomnia, 204

Stroke, neurological hypersomnias, 448–449; *see also* Central nervous system lesions

Stroop color-word test, sleepiness assessment, vigilance, 173, 175

Subcortical dementia, sleep in, 677–678

Sudden infant death syndrome, 649–656: ambient temperature, 651–652; arousal threshold, 652; cardiac rhythm, 651; definitions, 649; digestive movements, 651; epidemiology, 649; mechanisms in, 649–650; safety recommendations, 653–654; sedatives, 652; sleep-arousal-sudden death association, 652–653; sleep characteristics, 650; sleeping position, 651; sleep respiration, 650–651; tests for, 654; tobacco use, 652; warning signs, 654

Surgical management: obstructive sleep apnoea-hypopnea syndrome and upper airway resistance syndrome, 391–401; obstructive sleep apnoea in children (OSAS), 461–462; snoring, 561–563

Surveillance, circadian rhythm sleep disorders, shift work sleep disorder, 482

Swallowing, sleep-related abnormal swallowing syndrome, 538–539

Symbol cancellation tasks, sleepiness assessment, vigilance, 173

Symptomatic generalized epilepsy, 619

Symptomatic partial epilepsies, 620–621

Tegmental infarct, neurological hypersomnias, 449

Temperature, sudden infant death syndrome, 651–652, 653; see also Thermoregulation

Testosterone: circadian rhythmicity, 53; Kleine-Levin syndrome, 441

Theophylline, insomnias, medication-linked, 252

Thoracic straps, respiratory measurements, qualitative, 132–133

Thyroid hormonotherapy, insomnias, medication-linked, 253

Thyrotropin (TSH): circadian rhythmicity, 52; Kleine-Levin syndrome, 441; sleep-wake cycle, 51

Tiagabine, insomnias, 277

Time constant: electroencephalogram (EEG), 128; electromyogram (EMG), 130

Time zone change syndrome: see Jet lag syndrome

Tobacco: shift work sleep disorder, 477; sudden infant death syndrome, 652, 653

Tonsillectomy, obstructive sleep apnoea-hypopnea syndrome and upper airway resistance syndrome, 365–366

Topiramate, nocturnal paroxysmal dystonia, 613

Tracheal sounds, respiratory measurements, qualitative, 134

Tracheostomy, obstructive sleep apnoea-hypopnea syndrome and upper airway resistance syndrome, 394

Transient/short term insomnia, 201–205; adjustment insomnia, 203; altitude, 203; environmental factors, 202–203; epidemiology, 201–202; physical stress, 204; rebound insomnia, 204, 260–261; sleep hygiene, 202

Trauma: head injury: neurological hypersomnias, 450; sleep in, 679; parasomnias: nightmares, 528–529; REM sleep behavior disorder, 530–531; rhythmic movement disorder, 522–523; polysomnography, 678–679

Trazodone: insomnias, 271, 272; medication and alcohol dependent sleepiness, 352

Triazolam, parasomnias, REM sleep behaviour disorder, 534

Tricyclic antidepressants: see also Antidepressants; hypersomnias, cataplexy, 421; medication and alcohol dependent sleepiness, 351; parasomnias: REM sleep behaviour disorder, 534; rhythmic movement disorder, 524; sleep paralysis, 530

Trimipramine, insomnias, 271, 272, 273

Trypanosomiasis (African sleeping sickness): infection-related hypersomnias, 452; sleep in, 680

Tryptophane, parasomnias, REM sleep behaviour disorder, 534

Tumours (brain): neurological hypersomnias, 447–448; recurrent symptomatic hypersomnias, 442–443; sleep in, 678

Two-process model: aging, 298–304; mood disorders, 670; sleep regulation, 62–67

Ultradian rhythms, normal sleep, in children, 25

Unity, rule of, visual sleep analysis, sleep stage scoring, 152

Upper airway resistance, respiratory measurements, quantitative, 135–136

Uvulopalatopharyngoplasty (UPPP), snoring, 561–563

Vascular pathology, central nervous system lesions, 673–675

Velopharyngeal surgery, 394–396

Velopharynx, snoring, 558–559

Venlafaxine: hypersomnias, cataplexy, 421; insomnias, 272

Ventilatory function, 45–48; see also Respiratory measurements; Sleep breathing abnormalities: brain tumor, 678; NREM sleep, 47; obstructive sleep apnoea-hypopnea syndrome and upper airway resistance syndrome, 369–377; REM sleep, 47–48; sleep onset, 46–47; wakefulness, 46

Vigilance, insufficient sleep syndrome, 344–345

Vigilance assessment, 171–177: aging, 300–301; evoked potentials, 177; eye movements, 176; global motor activity, 176; limitations of, 176; performance tests, 172–175; pupillometry, 176–177; reaction time tests, 171–172; signal detection tests, 172; spontaneous or evoked responses measurement, 172–173

Viloxazine: hypersomnias, cataplexy, 421; insomnias, medication-linked, 250

Viral infection, infection-related hypersomnias, 451; see also Infectious disorders

Visual analogue scales (VAS), sleepiness assessment, 169–170

Visual sleep analysis, 151–158; diagnosis, 153–157; movement time, 157; NREM sleep, 154–156; REM sleep, 156; wakefulness, 153; overview, 151; sleep stage scoring, 151–152

Vitamin B_{12}: circadian rhythm sleep disorders: delayed sleep phase syndrome, 499; irregular sleep-wake pattern, 508; periodic leg movements, 605; restless legs syndrome, 605

Wakefulness: dysautonomias, 699; sleep-wake architecture, neurobiology, 32–35; ventilatory function, 46; visual sleep analysis, 153

Weight loss, hypersomnias, 380

West's syndrome, epilepsy, 618, 623, 692

Wisconsin card sorting task, sleepiness assessment, vigilance, 173

Workplace: see Shift work sleep disorder

Zaleplon, 263–264

Zolpidem, 262–263

Zopiclone, 261–262